Progress in Clean Energy, Volume 1

Ibrahim Dincer • C. Ozgur Colpan
Onder Kizilkan • M. Akif Ezan
Editors

Progress in Clean Energy, Volume 1

Analysis and Modeling

Springer

Editors
Ibrahim Dincer
Department of Mechanical Engineering
University of Ontario
Oshawa, ON, Canada

C. Ozgur Colpan
Department of Mechanical Engineering
Dokuz Eylul University
Izmir, Turkey

Onder Kizilkan
Department of Energy Systems Engineering
Suleyman Demirel University
Isparta, Turkey

M. Akif Ezan
Department of Mechanical Engineering
Dokuz Eylul University
Izmir, Turkey

ISBN 978-3-319-16708-4 ISBN 978-3-319-16709-1 (eBook)
DOI 10.1007/978-3-319-16709-1

Library of Congress Control Number: 2015943852

Springer Cham Heidelberg New York Dordrecht London

Printed on acid-free paper

Springer International Publishing AG Switzerland is part of Springer Science+Business Media (www.springer.com)

Preface

During the past few decades, we have been faced various critical challenges in various dimensions, ranging from energy to environment and from economy to sustainability. In all these dimensions, energy plays the most critical roles since the way that we produce, transfer, transport, convert, and use energy affects all other dimensions significantly. There is a strong need for clean energy solutions to overcome environmental, resource, efficiency, cost, energy security, and sustainability issues.

The first volume of this edited book entitled "Progress in Clean Energy" focuses on analysis and modeling of clean energy systems and applications and contains 67 uniquely selected papers out of the conference papers presented in the 13th International Conference on Clean Energy (ICCE-2014) which was held in Istanbul, Turkey from June 8 to 12, 2014. This distinctive event was not only organized to bring all the researchers, scientists, policy makers, and engineers conducting research on the clean energy field together but also to honor Dr. Veziroglu's 90th birthday in Istanbul, which is also his birthplace. This conference provided a forum for the exchange of latest technical information, the dissemination of the high-quality research results on the issues, the presentation of the new developments in the area of clean energy, and the debate and shaping of future directions and priorities for better environment, sustainable development, and energy security. The recent research findings in the clean energy topics including solar, wind, hydropower, nuclear, and hydrogen energy technologies, fuel cells, biomass and biofuels, clean fossil fuels, carbon sequestration and carbon tax, energy storage and energy conservation, environmental impact and remediation, and sustainable development and energy management were presented and discussed in this conference.

This volume covers a number of major topics focusing on design, analysis, modeling, optimization, assessment, evaluation, and improvement. Some significant subjects covered include sustainability modeling, thermodynamic analysis through energy and exergy methods, forecasting approaches, numerical studies, kinetic studies, fabrication and characterization of new energy materials, advanced optimization, performance evaluation, system integration, and multigeneration.

We hope that the first volume of this edited book will provide a unique source of clean energy systems and applications with a prime focus on analysis, modeling, and optimization. We sincerely appreciate the help and assistance provided by various individuals and conference organizing committee members who deserve a clear acknowledgement.

Oshawa, ON, Canada — Ibrahim Dincer
Izmir, Turkey — C. Ozgur Colpan
Isparta, Turkey — Onder Kizilkan
Izmir, Turkey — M. Akif Ezan

Contents

Chapter 1
Sustainability Assessment of Hybrid Community Energy Systems

Kevork Hacatoglu, Ibrahim Dincer, and Marc A. Rosen

Abstract The goal to achieve a sustainable society that will endure over the long term is generally regarded as a positive evolutionary course. One of the challenges with this goal is developing a quantitative assessment of the sustainability of a system. Despite the different measures available in the literature, a standard and universally accepted index for assessing sustainability does not yet exist. Here, we develop a novel integrated sustainability index (ISI) for energy systems that considers critical multidimensional sustainability criteria. The originality of this new index is that it incorporates fundamental thermodynamic, economic, and environmental constraints to combine indicators from multiple dimensions into a single-score evaluation of sustainability. The index is therefore unique because it can assess sustainability relative to an ideal reference state instead of being limited to ranking systems via relative assessments. The ISI is applied to a stand-alone solar-PV-battery system designed to meet the needs of a small community in Southern Ontario. The ISI of the system ranges from 0.52 to 0.66, where one is considered to be a sustainable system. The weighting factors associated with critical economic and global environmental criteria have the greatest effect on the ISI. This index is expected to prove useful as a high-level, multi-criteria decision analysis tool for understanding and fostering sustainable energy systems, alone or in concert with other approaches.

Keywords Energy system • Integrated sustainability index • Life-cycle assessment • Sustainability • Solar-PV-battery system

Nomenclature

a	Azimuth angle, °
A	Dimensional sustainability indicator
B	Nondimensional sustainability indicator
Cn	Clearness number

K. Hacatoglu (✉) • I. Dincer • M.A. Rosen
University of Ontario Institute of Technology, 2000 Simcoe Street,
North Oshawa, Canada, L1H 7K4
e-mail: kevork.hacatoglu@uoit.ca; ibrahim.dincer@uoit.ca; marc.rosen@uoit.ca

I. Dincer et al. (eds.), *Progress in Clean Energy, Volume 1*,
DOI 10.1007/978-3-319-16709-1_1

k_C	Local extinction coefficient
$\dot{Q}$	Heat rate, kW
W	Weighting factor

Greek Letters

β	Collector tilt angle, °
θ	Elevation angle, °
ϕ	Incidence angle, °

Subscripts

Col	Collector
ETR	Extraterrestrial radiation
i	Sub-indicator
j	Category indicator
m	Number of sub-indicators
n	Number of category indicators
T	Target

Abbreviations

ADP	Abiotic depletion potential
AF	Affordability
APP	Air pollution potential
CFC	Chlorofluorocarbon
CV	Commercial viability
EF	Economic factor
EnER	Energy efficiency ratio
EP	Eutrophication potential
ER	Efficiency ratio
ExER	Exergy efficiency ratio
FAETP	Freshwater aquatic ecotoxicity potential
GEIP	Global environmental impact potential
GWP	Global warming potential
IPCC	Intergovernmental panel on climate change
ISI	Integrated sustainability index
MAETP	Marine aquatic ecotoxicity potential
PM	Particulate matter
SF	Size factor
SODP	Stratospheric ozone depletion potential
WPP	Water pollution potential

Chemical Compounds

CO	Carbon monoxide
CO_2	Carbon dioxide
N_2O	Nitrous oxide
NO_2	Nitrogen dioxide
O_3	Ozone
Pb	Lead
SO_2	Sulphur dioxide

1.1 Introduction

The struggle to achieve a sustainable society is not unique to the modern age. Sustainability has been a goal since the earliest human civilizations. Ever since the Neolithic Revolution approximately 10,000 years ago, when human beings transitioned from mobile hunter-gatherers to agriculture and settlements, the sustainability of the local lifestyle has been essential to avoid societal collapse [1].

One of the most ambitious goals of a society is to achieve sustainability. Making sustainability operational as opposed to a grand but ambiguous idea is a challenge. For example, various definitions of sustainability exist, none of which applies to all circumstances. Even after selecting a definition, there is no universal method of measuring sustainability, which makes it extremely difficult to track progress towards sustainability.

Although there are numerous methods of assessing the sustainability of energy systems, a standard and universally accepted approach does not exist. Some studies comment on the sustainability of an energy system from a thermodynamic [2–6] or environmental [7–9] perspective. More comprehensive approaches that consider different aspects of sustainability but rank indicators without normalization with respect to sustainability target values are better suited to relative assessments of energy systems [10, 11]. Other studies develop quantitative sustainability assessment tools that address technical, economic, social, and environmental criteria [12–15]. However, sustainability indicators are not normalized with respect to a reference state that represents limits on, for example, emissions of pollutants.

The original contribution of this research is the development of a novel sustainability assessment index for energy systems that considers several critical multidimensional sustainability criteria such as exergy efficiency, affordability, land area, greenhouse gas (GHG) emissions, stratospheric ozone depletion, air pollution, and water pollution. This new integrated sustainability index (ISI) incorporates fundamental thermodynamic, economic, and environmental constraints to combine indicators from multiple dimensions into an overall composite index. The index is therefore unique because it can assess the sustainability of a system relative to an ideal reference state. Other approaches are limited to relative assessments between systems that are useful for ranking purposes but provide little insight with respect to overall sustainability.

1.2 Assessment Methodology

Sustainability is a multidimensional concept that links the economy, society, and environment. A sustainability analyst must combine results from disparate fields to make an assessment on the sustainability of a system. The field of multi-criteria decision analysis is therefore very applicable to sustainability assessment.

1.2.1 *Normalization*

Normalization transforms a sustainability indicator into a nondimensional value between zero and one through a comparison to a sustainable reference level. Normalization is a necessary precursor to weighting and aggregation procedures in compensatory multi-criteria decision analysis.

Nondimensional sustainability sub-indicators ($B_{i,j}$) between zero (undesired) and one (desired) are derived by comparing actual sustainability sub-indicators ($A_{i,j}$) to smaller target values ($A_{i,j,T}$). In special cases where an actual sub-indicator is less than its target value, the nondimensional sub-indicator is adjusted to one. A nondimensional sub-indicator is therefore calculated by

$$B_{i,j} = \begin{cases} \dfrac{A_{i,j,T}}{A_{i,j}}, & A_{i,j} > A_{i,j,T} \\ 1, & A_{i,j} \leq A_{i,j,T} \end{cases} \tag{1.1}$$

where $A_{i,j}$ represents sub-indicator i for category j and $A_{i,j,T}$ is the associated target value with the same units as $A_{i,j}$.

The selection of $A_{i,j,T}$ is of critical importance to the sustainability assessment. For a given sub-indicator, $A_{i,j,T}$ represents the threshold beyond which a system may exhibit negative impacts from an economic, social, or environmental perspective. The selection of $A_{i,j,T}$ often depends on local characteristics but may also be of a global nature. For example, a sub-indicator related to freshwater consumption is strongly linked to the local context and its sustainable target value will vary across geographic regions. On the other hand, climate change has a global impact and international carbon budgets have been proposed to limit the rise in global average surface temperatures below certain levels. A sustainable target value for GHG emissions should therefore exhibit less variability although the method of allocating a carbon budget has a strong effect on local target values.

1.2.2 *Weighting*

Determining the weighting factors of indicators is an essential but often controversial step in a sustainability assessment. Many sustainability assessments circumvent the pitfalls associated with weighting by assuming equivalent weighting factors for all indicators or by not assigning weights at all, which is analogous to the equivalency assumption [16].

There are many different approaches to determine weighting factors, all of which have advantages and drawbacks. One popular approach is the panel method, where a panel of experts and stakeholders are asked to weight the importance of various criteria. A typical approach in life-cycle impact assessment for deriving weighting factors is the distance-to-target method [17], where weights are derived based on political or other targets. Monetary valuation methods are another class of weighting approaches. The relative importance of criteria in monetary valuation methods is assessed depending on market prices, willingness to pay, or avoidance costs [18]. However, there are well-known challenges associated with ecosystem valuation.

The exact meaning of a weight depends on whether a compensatory or non-compensatory aggregation method is being implemented [16]. Weights used with compensatory approaches represent trade-off factors or substitution rates. They describe the capacity for trade-offs between indicators. Weights used with non-compensatory approaches represent importance coefficients that describe the relative importance of an indicator in comparison to others.

Eliciting the relative importance of an indicator or its trade-off with respect to other indicators is a subjective, inexact science. One possibility is to use the relative importance of an indicator to then determine trade-offs with respect to other indicators. Relative importance coefficients can then be utilized in non-compensatory sustainability assessments while trade-off factors can be utilized in compensatory approaches.

The suggested approach for evaluating the relative importance of sustainability indicators is based on time, space, and receptor criteria. The time and physical scales of an indicator are important characteristics of sustainability, which are related to intergenerational and intragenerational equity, respectively. The receptor criterion is related to the extent of the indicator's impact on human or ecosystem receptors.

The importance of a sustainability indicator is evaluated on a scale of 1–5 ("very unimportant" to "very important") with respect to each of the aforementioned criteria. This type of rating scale is similar to Likert responding formats, which are commonly used in questionnaires across many different fields [19]. These Likert-type rating scales provide the most benefit when respondents are presented with 5–7 response categories [20]. Questionnaires with less than five categories do not provide much range to differentiate between responses, whereas more than seven is confusing to respondents. In addition, response formats that feature an odd number of response categories (equal number of "positive" and "negative" responses plus a "neutral" middle option) provide the most balance [20]. The uncertainty associated with evaluating the importance of sustainability sub-indicators led to the decision to adopt a five-point rating scale, which is easier to manage and still within the recommended range.

Table 1.1 Summary of the different archetypes for scoring and evaluating weighting factors

Archetype	Time	Space	Receptor
Individualist	Short	Local	Humans
Egalitarian	Long	Global	Ecosystems
Hierarchist	Medium	Regional	Both

The score assigned by a sustainability analyst to an indicator with respect to time, space, and receptor criteria is a function of the perspective of the decision-maker. A summary of the different archetypes and how they relate to the weighting factor criteria is presented in Table 1.1.

The individualist perspective is self-seeking and uninterested in inter- and intragenerational equity. Consequently, the evaluation of indicators is of a short-term and local perspective. The individualist view of nature is as resilient and plentiful, which translates into concerns regarding human as opposed to ecosystem receptors.

The egalitarian perspective is concerned with inter- and intragenerational equity and thus exhibits a long-term and global perspective. The egalitarian view of nature as fragile leads to concerns with respect to ecosystem receptors.

The hierarchist perspective is more moderate, believing that nature and natural resources can be managed within certain limits. It leads to a balanced approach to decision-making, predicated on negotiation and compromise.

The process of developing importance coefficients and trade-off factors is described in [21].

1.2.3 Aggregation

The aggregation of sustainability indicators into a single-value composite index is attractive to decision and policymakers because it is simple and can be easily communicated to stakeholders and the general public.

Compensatory aggregation procedures combine sub-indicators into a composite index but allow for the possibility of offsetting a disadvantage on some indicators by a sufficiently large advantage on other indicators [16]. Nondimensional sub-indicators can be aggregated into a category indicator (B_j) using a linear aggregation procedure. Consequently,

$$B_j = \sum_{i=1}^{m} \left(B_{i,j} \times W_{i,j}\right) \tag{1.2}$$

$$\sum_{i=1}^{m} W_{i,j} = 1,$$

where m represents the number of sub-indicators in a category and $W_{i,j}$ represents the weight associated with sub-indicator i in category j.

Overall sustainability is represented by the ISI, which is determined by aggregating category indicators and their respective weights using a linear aggregation procedure. This leads to

$$\mathrm{ISI} = \sum_{j=1}^{n} (B_j \times W_j) \tag{1.3}$$

$$\sum_{j=1}^{n} W_j = 1,$$

where n represents the number of categories in an assessment and W_j represents the weight associated with category j.

1.2.4 Components of the Integrated Sustainability Index

The scope of the assessment is limited to several key indicators and sub-indicators as shown in Fig. 1.1. The ISI is composed of six category indicators, each of which consists of at least two sub-indicators. Each of the category indicators and sub-indicators are described in [21].

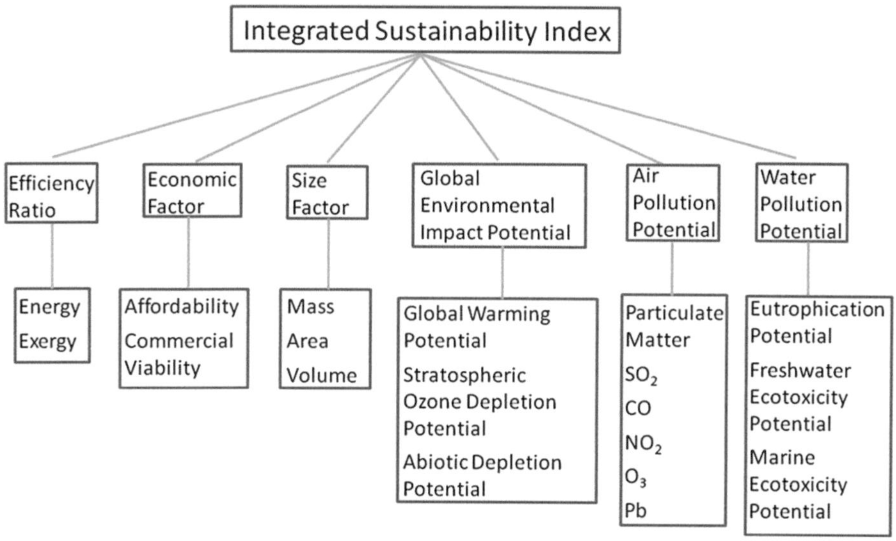

Fig. 1.1 Components of the integrated sustainability index

1.3 Case Study

A stand-alone, solar-based photovoltaic (PV) system integrated with lead-acid battery storage is proposed to meet the heat, cooling, and electricity needs of a 50-household community in Ontario.

1.3.1 System Description

The system consists of solar PV panels that collect solar radiation and convert photons to electricity, a lead-acid battery for storage, and a heat pump for heating and cooling needs. The integrated system is presented in Fig. 1.2. A solar-PV-battery system is entirely free of fossil fuels during operation, but the absence of a fossil-based backup system means that a very large storage system is required to reliably meet the energy needs of a community. When the power delivered by the PV panels is greater than the load, the battery enters a charging mode. When there is unmet demand, the battery discharges. The area of the PV panels is selected such that the year-end net charge of the battery is positive. Similarly, the size of the battery needs to be sufficient to ensure a reliable supply of energy for the community.

1.3.2 Analysis

A thermodynamic model of the system developed through the Engineering Equation Solver (EES) was run for 365 days with input variables that define the electricity (Fig. 1.3) and heat (Fig. 1.4) demand over 1 year for a typical household in Ontario [22].

Fig. 1.2 General layout of a solar-PV-battery system with an air-source heat pump

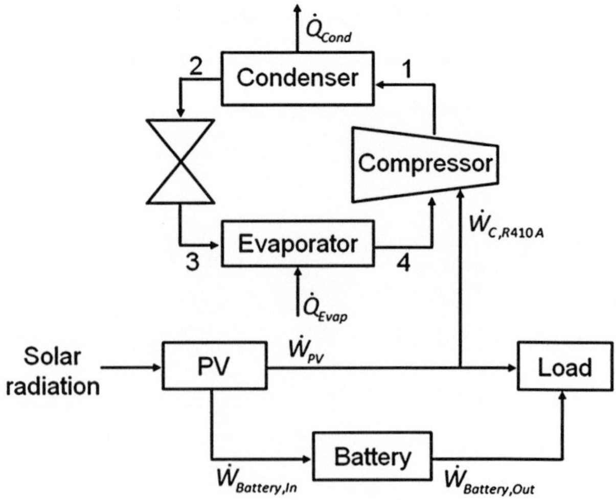

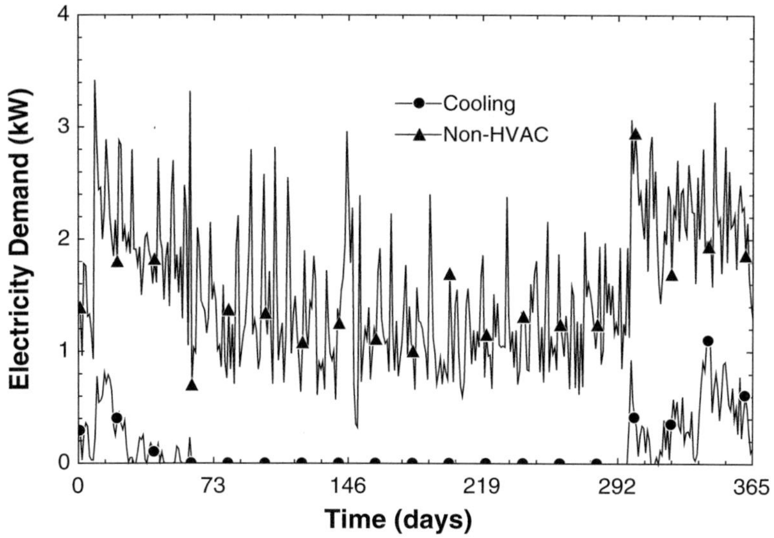

Fig. 1.3 Cooling and non-HVAC electricity demand over 1 year for a typical household in Ontario (day "1" corresponds to August 1, 2009)

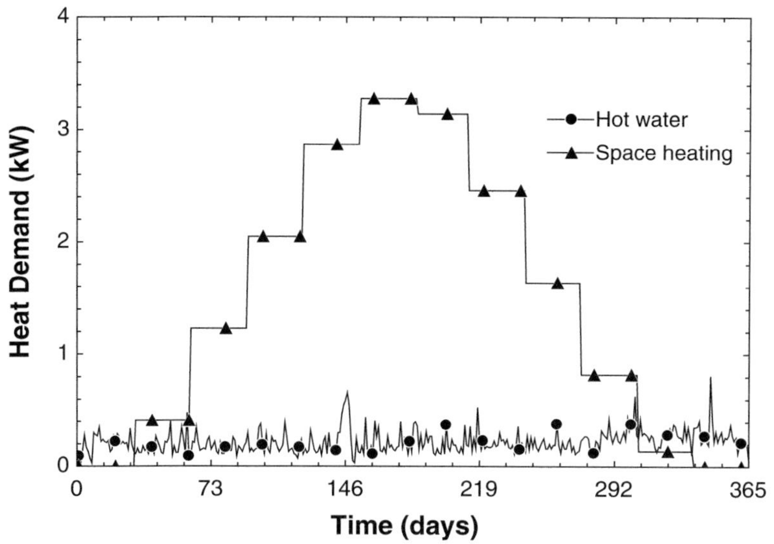

Fig. 1.4 Daily average domestic hot water and space heating demand over 1 year for a typical household in Ontario (day "1" corresponds to August 1, 2009)

The following assumptions were made to model the solar-PV-battery system:

- Energy efficiency of a PV cell is constant.
- Charging and discharging efficiencies of the lead-acid battery are constant.
- Fugitive emissions of refrigerant (R-410A) are negligible.

The solar irradiance available to the city of Toronto can be approximated by a semiempirical model [23]. The solar irradiance ($\dot{Q}_{\text{Solar}}$) on a tilted plane is a function of the extraterrestrial solar irradiance ($\dot{Q}_{\text{ETR}}$), clearness number (Cn), local extinction coefficient (k_c), elevation angle (θ), and incidence angle (ϕ). Thus,

$$\dot{Q}_{\text{Solar}} = \dot{Q}_{\text{ETR}} Cn \exp(-k_c \sin\theta)\cos\phi. \tag{1.4}$$

The incidence angle is a function of the elevation angle, solar azimuth angle (a_{Solar}), collector azimuth angle (a_{Col}), and collector tilt angle (β). Consequently,

$$\cos\phi = \cos\theta\cos(a_{\text{Solar}} - a_{\text{Col}})\sin\beta + \sin\theta\cos\beta. \tag{1.5}$$

Elevation and solar azimuth angles can be determined from the day of the year, latitude, and longitude of the location. The parameters used to calculate solar irradiance can be found in Tables 1.2 and 1.3.

Combining Eqs. (1.4) and (1.5) with the data in Tables 1.2 and 1.3 yields the daily average solar irradiance for Toronto, Canada, over the course of 1 year (Fig. 1.5).

Table 1.2 Monthly parameters required to model solar irradiance for Toronto, Canada

Month	k_c	Cn	β (°)	a_{Col} (°)
January	0.142	0.85	70	0
February	0.144	0.85	60	0
March	0.156	0.85	50	0
April	0.180	0.85	30	0
May	0.196	0.85	20	0
June	0.205	0.85	20	0
July	0.207	0.85	20	0
August	0.201	0.85	30	0
September	0.177	0.85	40	0
October	0.160	0.85	60	0
November	0.149	0.85	70	0
December	0.142	0.85	70	0

Table 1.3 Parameters required to model solar irradiance for Toronto, Canada

Parameter	Value	Reference
Extraterrestrial solar irradiance	1,353 W m^{-2}	[23]
Latitude (Toronto)	43.7°N	[24]
Longitude (Toronto)	79.4°W	[24]
Solar noon	720 min	[23]
Time zone (eastern)	75°W	[23]

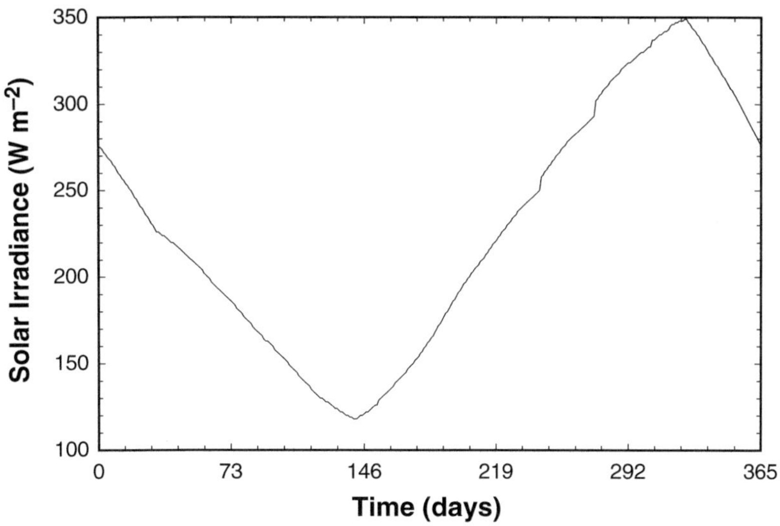

Fig. 1.5 Daily average solar irradiance in Toronto, Canada, over 1 year (day "1" corresponds to August 1, 2009)

1.4 Results and Discussion

The behaviour of the lead-acid battery over the course of 1 year for a panel area of 3,500 m^2 is illustrated in Fig. 1.6. The battery is continuously discharged after day 80 as the solar irradiance declines throughout the winter months. The battery is then continuously charged after day 220, when the solar irradiance begins to increase in intensity. The net increase in the charge of the battery after 1 year is 0.8 MWh. The capacity of the battery to ensure a reliable supply of energy to the community needs to be at least 130 MWh.

Each component of the solar-PV-battery system is associated with a certain amount of exergy destruction. The annual exergy destruction for each subsystem over 365 days is presented in Fig. 1.7.

The largest share of exergy destruction is attributed to the solar PV subsystem (21 TJ per year) due to the low electric conversion efficiency of a PV panel. Since the total exergy destruction is 22 TJ per year, the solar PV subsystem is responsible for 95 % of total exergy destruction. The energy and exergy efficiencies of the system are 24 % and 12 %, respectively.

A thermodynamic analysis is a precursor to sustainability assessment. Thermodynamic, cost, and life-cycle emission factors are combined with weighting factors for three different perspectives to yield the sustainability assessment results presented in Tables 1.4, 1.5, and 1.6.

The ISI for the solar-PV-battery system ranges from 0.52 to 0.66, where the primary determinant of the score depends on the perspective. The EF category, specifically the AF sub-indicator, is the largest contributor to the score for the

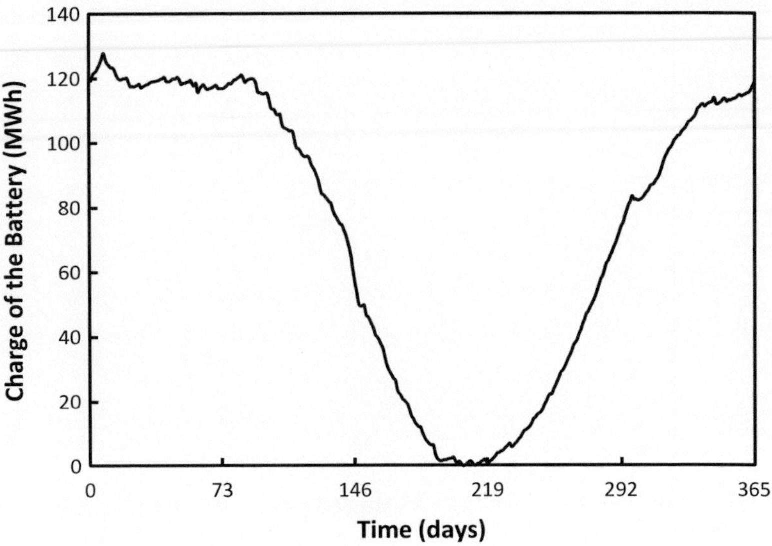

Fig. 1.6 Variation in the charge of the battery over the course of 1 year for the solar-PV-battery system (day "1" corresponds to August 1, 2009)

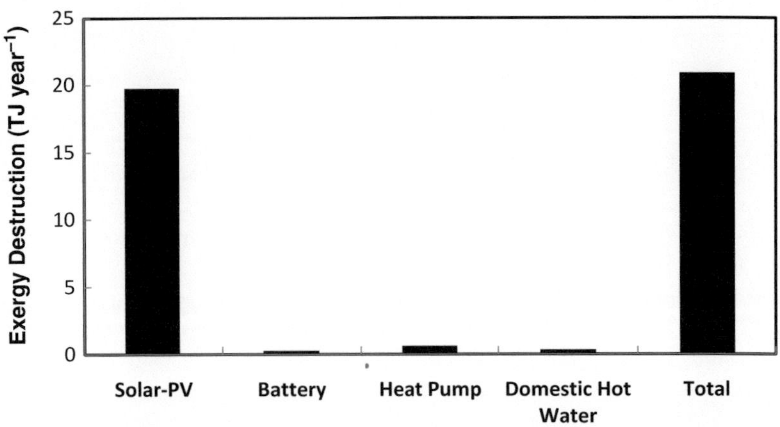

Fig. 1.7 Annual exergy destruction of subsystems in the solar-PV-battery system over a 1-year period

individualist perspective. The expected annual cost of a stand-alone solar-PV-battery system to a household is approximately $64,000, whereas the median after-tax income of a household in Ontario is $69,300 [25], of which no more than 10 % should be allocated to energy needs [26].

Sustainability sub-indicators in Tables 1.4, 1.5, and 1.6 with a $B_{i,j}$ value equal to one have no negative effect on ISI. Sub-indicators with a $B_{i,j}$ value less than one will have a negative effect on ISI, which is a function of the actual $B_{i,j}$ value and its

Table 1.4 Sustainability assessment results for the solar-PV-battery system from the individualist perspective

Category	Sub-indicator	$B_{i,j}$	$W_{i,j}$	B_j	W_j	$B_j \times W_j$
ER	EnER	0.63	0.00	0.00	0.07	0.000
	ExER	0.84	1.00	0.84		0.056
EF	AF	0.11	0.50	0.05	0.42	0.023
	CV	1.00	0.50	0.50		0.211
SF	Mass	0.00	0.00	0.00	0.04	0.000
	Area	1.00	1.00	1.00		0.042
	Volume	0.00	0.00	0.00		0.000
GEIP	GWP	0.11	0.24	0.03	0.11	0.003
	SODP	0.07	0.69	0.05		0.005
	ADP	1.00	0.07	0.07		0.007
APP	$PM_{2.5}$	1.00	0.21	0.21	0.30	0.062
	PM_{10}	1.00	0.21	0.21		0.062
	SO_2	1.00	0.04	0.04		0.013
	CO	1.00	0.11	0.11		0.033
	NO_2	1.00	0.11	0.11		0.033
	O_3	1.00	0.11	0.11		0.033
	Pb	1.00	0.21	0.21		0.061
WPP	EP	0.74	0.08	0.06	0.07	0.004
	FAETP	0.30	0.78	0.23		0.015
	MAETP	0.03	0.14	0.00		0.000
ISI						0.66

weight. The range of the negative effect of each sustainability sub-indicator on ISI is graphed in Fig. 1.8.

The sub-indicators that can have the greatest potential reduction in the ISI of the system are AF and GWP. The effects of the weighting factors associated with these sub-indicators as well as SODP are therefore studied in Figs. 1.9 and 1.10.

The ISI from the individualist perspective (Fig. 1.9) is highly sensitive to the weighting factor attached to the AF sub-indicator. The ISI decreases from 0.86 to 0.51 as the weighting factor associated with the AF sub-indicator varies from zero to one. As a comparison, the actual individualist ISI for the system is 0.66 (Table 1.4). The effect of varying the GWP and SODP weighting factors on the individualist ISI is much less significant.

The ISI from the egalitarian perspective (Fig. 1.10) is much more sensitive to changes in the weighting factors associated with the GWP and SODP sub-indicators. The egalitarian ISI decreases from 0.74 to 0.43 as the weighting factor associated with the GWP sub-indicator varies from zero to one. Similarly, the egalitarian ISI decreases from 0.62 to 0.29 when the weighting factor associated with the SODP sub-indicator varies from zero to one. As a comparison, the actual egalitarian ISI for the system is 0.52 (Table 1.5). Unlike with the individualist ISI, the effect of varying the AF weighting factor on the egalitarian ISI is insignificant.

Table 1.5 Sustainability assessment results for the solar-PV-battery system from the egalitarian perspective

Category	Sub-indicator	$B_{i,j}$	$W_{i,j}$	B_j	W_j	$B_j \times W_j$
ER	EnER	0.63	0.00	0.00	0.16	0.000
	ExER	0.84	1.00	0.84		0.134
EF	AF	0.11	0.67	0.07	0.04	0.003
	CV	1.00	0.33	0.33		0.015
SF	Mass	0.00	0.00	0.00	0.25	0.000
	Area	1.00	1.00	1.00		0.247
	Volume	0.00	0.00	0.00		0.000
GEIP	GWP	0.11	0.65	0.07	0.37	0.025
	SODP	0.07	0.25	0.02		0.007
	ADP	1.00	0.11	0.11		0.040
APP	$PM_{2.5}$	1.00	0.04	0.04	0.03	0.001
	PM_{10}	1.00	0.04	0.04		0.001
	SO_2	1.00	0.15	0.15		0.004
	CO	1.00	0.04	0.04		0.001
	NO_2	1.00	0.15	0.15		0.004
	O_3	1.00	0.04	0.04		0.001
	Pb	1.00	0.52	0.52		0.013
WPP	EP	0.74	0.08	0.06	0.15	0.009
	FAETP	0.30	0.19	0.06		0.009
	MAETP	0.03	0.72	0.02		0.003
ISI						0.52

Figures 1.8, 1.9, and 1.10 demonstrate the importance of GWP and SODP to the sustainability of the system. These sub-indicators are studied in more detail in Figs. 1.11 and 1.12.

The dashed horizontal line in Fig. 1.11 represents actual annual per capita emissions from the solar-PV-battery system, while the solid line represents allowable emissions under the Intergovernmental Panel on Climate Change (IPCC) scenario RCP2.6 [27], where the global carbon budget is equally distributed amongst the population. The intersection of the two functions is the maximum allowable global population to stay within the international RCP2.6 carbon budget. For example, the allowable emissions curve intersects the horizontal solar-PV-battery system line at a population of approximately 800 million people. If everyone in the world had an energy-demand profile as a typical Ontario household and met that demand through a solar-PV-battery system, the planet would be constrained to 800 million people to stay within the confines of the RCP2.6 carbon budget. The intersection point can be shifted to the right by increasing the carbon budget or reducing per capita GHG emissions.

Table 1.6 Sustainability assessment results for the solar-PV-battery system from the hierarchist perspective

Category	Sub-indicator	$B_{i,j}$	$W_{i,j}$	B_j	W_j	$B_j \times W_j$
ER	EnER	0.63	0.00	0.00	0.15	0.000
	ExER	0.84	1.00	0.84		0.129
EF	AF	0.11	0.67	0.07	0.17	0.012
	CV	1.00	0.33	0.33		0.057
SF	Mass	0.00	0.00	0.00	0.15	0.000
	Area	1.00	1.00	1.00		0.153
	Volume	0.00	0.00	0.00		0.000
GEIP	GWP	0.11	0.57	0.06	0.31	0.019
	SODP	0.07	0.33	0.02		0.007
	ADP	1.00	0.10	0.10		0.030
APP	$PM_{2.5}$	1.00	0.10	0.10	0.08	0.008
	PM_{10}	1.00	0.10	0.10		0.008
	SO_2	1.00	0.10	0.10		0.008
	CO	1.00	0.05	0.05		0.004
	NO_2	1.00	0.17	0.17		0.013
	O_3	1.00	0.05	0.05		0.004
	Pb	1.00	0.44	0.44		0.036
WPP	EP	0.74	0.08	0.06	0.13	0.008
	FAETP	0.30	0.58	0.17		0.022
	MAETP	0.03	0.34	0.01		0.001
ISI						0.52

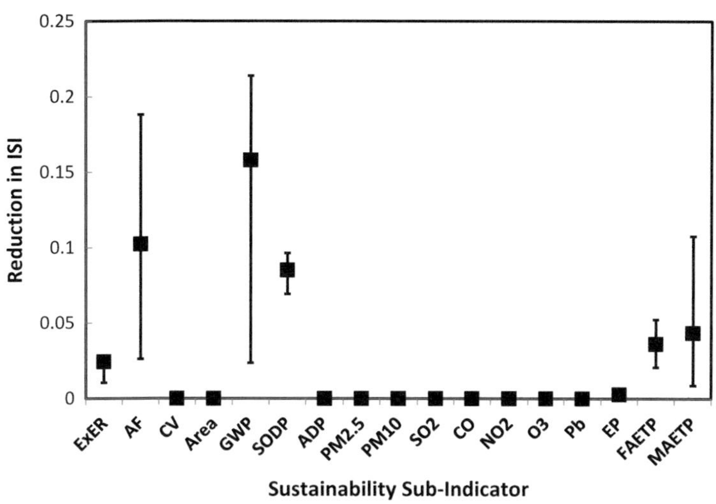

Fig. 1.8 Reduction in ISI for each sustainability sub-indicator for the solar-PV-battery system

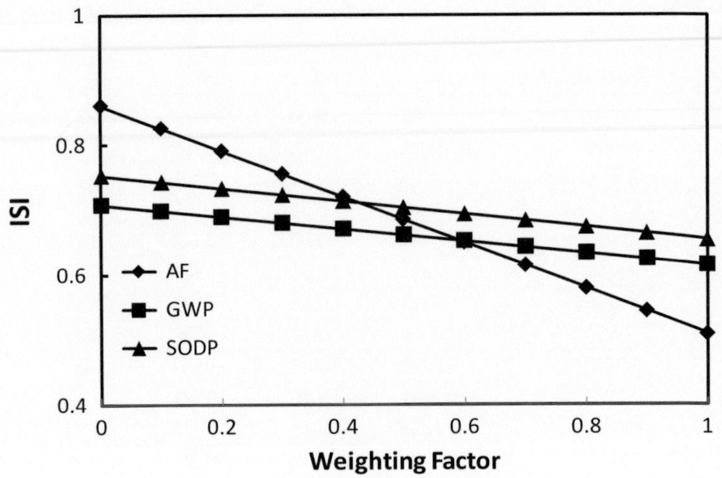

Fig. 1.9 Variation of the individualist ISI with respect to weighting factor for the solar-PV-battery system

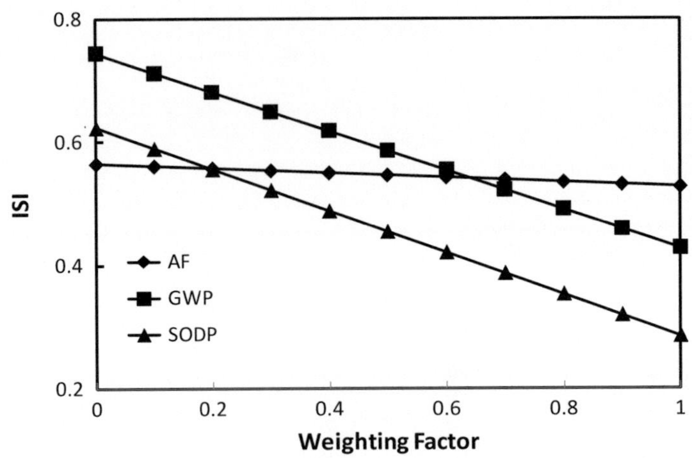

Fig. 1.10 Variation of the egalitarian ISI with respect to weighting factor for the solar-PV-battery system

The graph in Fig. 1.12 estimates the allowable amount of ozone-depleting substance emissions per capita per year as a function of the percent loss in stratospheric ozone over the time scale for considering sustainability. For example, a 2 % loss in ozone over 50 years is equivalent to 0.04 % per year. Actual annual per capita emissions from the solar-PV-battery system are approximately 0.0009 kg CFC-11, most of which is due to the effect of nitrous oxide (N_2O) [28]. This corresponds to approximately 26 % stratospheric ozone depletion over 50 years, which is much higher than the 2 % target value. Although the actual effect on the

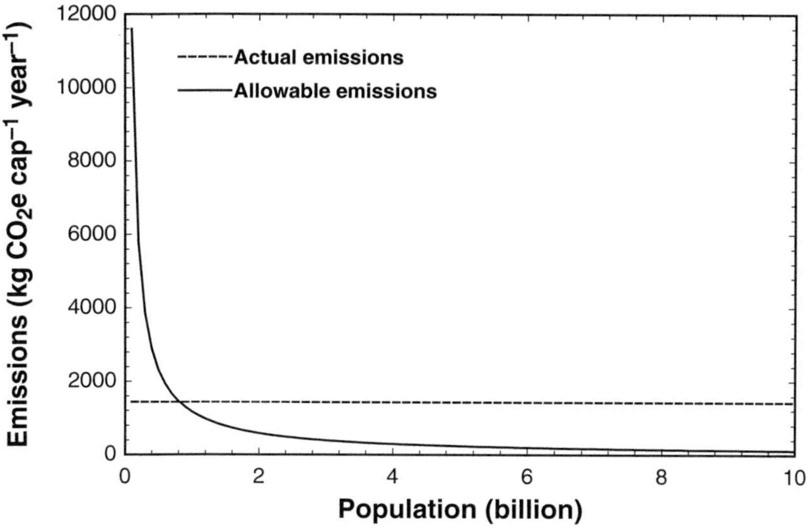

Fig. 1.11 Actual and allowable annual per capita GHG emissions for the solar-PV-battery system based on the lower limit of the representative concentration pathway (RCP2.6) carbon budget

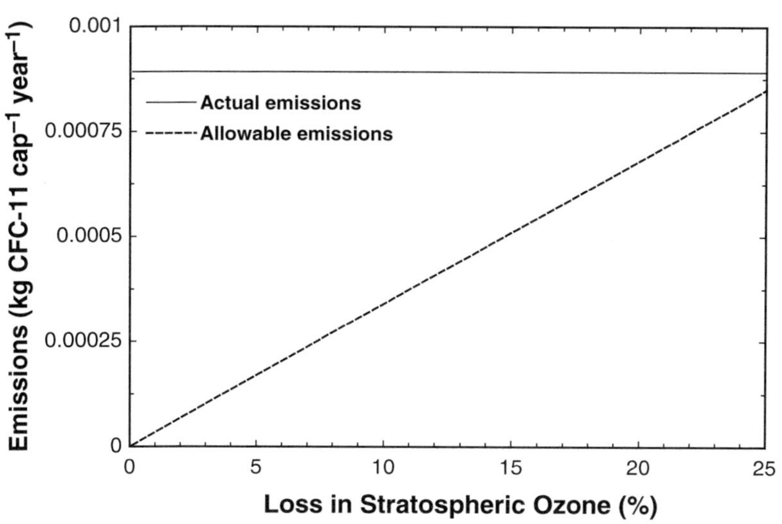

Fig. 1.12 Actual and allowable annual per capita ozone-depleting substance emissions (including N_2O) with respect to the percent loss in stratospheric ozone over the time scale for considering sustainability

ozone layer is hard to determine due to the complexities associated with the photochemical reactions [29], this is a useful approximation of potential impact.

The photovoltaic efficiency and battery charging efficiency are important parameters for the solar-PV-battery system. The impacts of these parameters on ISI are illustrated in Figs. 1.13 and 1.14. At higher efficiencies, both of these parameters have a modest but positive effect on ISI. However, there may potentially be higher costs associated with more advanced batteries, which could have a negative effect on ISI.

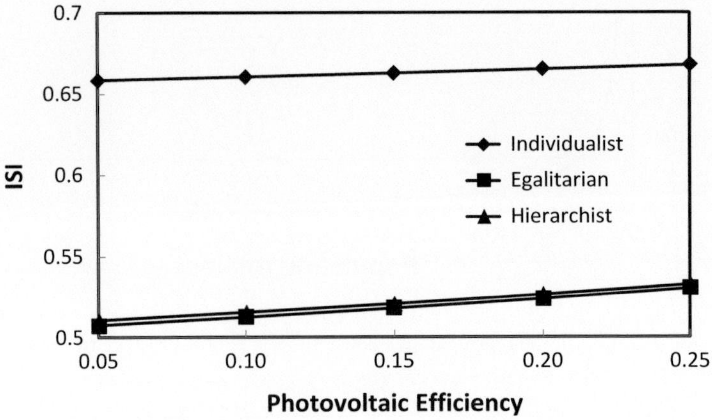

Fig. 1.13 Variation of ISI with respect to photovoltaic efficiency for the solar-PV-battery system

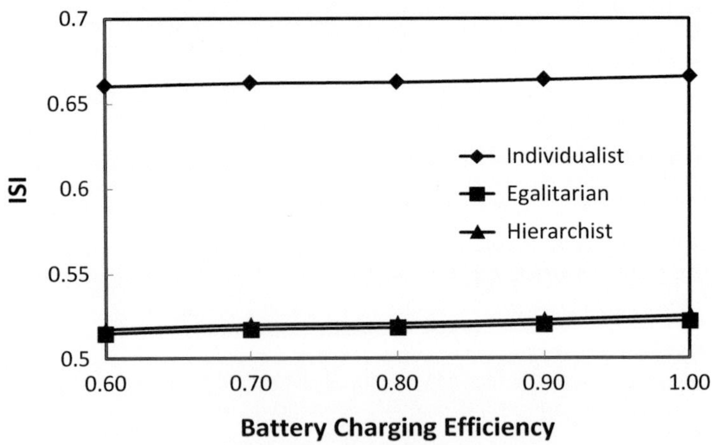

Fig. 1.14 Variation of ISI with respect to battery charging efficiency for the solar-PV-battery system

1.5 Conclusions

This study introduces a new approach for assessing the sustainability of energy systems based on normalizing, weighting, and aggregating several different criteria. The index is applied to a stand-alone solar-PV-battery system, which is estimated to require a 130 MWh lead-acid battery to ensure a reliable supply of energy to the community. The results of the analysis demonstrate that a solar-PV-battery system has an ISI that ranges from 0.52 to 0.66. The ISI of the system is strongly affected by the weighting factors associated with the AF, GWP, and SODP sub-indicators. Improvements in the electric conversion efficiency of PV cells and lead-acid batteries can modestly increase the ISI of the system. An assessment of life-cycle GHG emissions reveals that widespread adoption of a solar-PV-battery system at Southern Ontario energy demand levels can only meet the needs of 800 million people while staying within a stringent global carbon budget that limits the rise in average surface temperature to 2 °C by the year 2100.

Acknowledgement The authors gratefully acknowledge the support provided by the Natural Sciences and Engineering Research Council of Canada.

References

1. Tainter JA (1988) The collapse of complex societies. Cambridge University Press, Cambridge
2. Dewulf H, Van Langenhove H, Mulder J, van den Berg MMD, van der Kooi HJ, de Swaan Arons J (2000) Illustrations towards quantifying the sustainability of technology. Green Chem 2:108–114
3. Ferrari S, Genoud S, Lesourd J (2001) Thermodynamics and economics: towards exergy-based indicators of sustainable development. Swiss J Econ Stat 137:319–336
4. Midilli A, Dincer I (2009) Development of some exergetic parameters for PEM fuel cells for measuring environmental impact and sustainability. Int J Hydrogen Energy 34:3858–3872
5. Rosen MA, Dincer I, Kanoglu M (2008) Role of exergy in increasing efficiency and sustainability and reducing environmental impact. Energy Policy 36:128–137
6. Sciubba E, Zullo F (2011) Is sustainability a thermodynamic concept? Int J Exergy 8:68–85
7. Dincer I, Zamfirescu C (2012) Potential options to greenize energy systems. Energy 46:5–15
8. Rosen MA (2009) Energy sustainability: a pragmatic approach and illustrations. Sustainability 1:55–80
9. Zvolinschi A, Kjelstrup S, Bolland O, van der Kooi HJ (2007) Exergy sustainability indicators as a tool in industrial ecology. J Ind Ecol 11:85–98
10. Evans A, Strezov V, Evans TJ (2009) Assessment of sustainability indicators for renewable energy technologies. Renew Sustain Energy Rev 13:1082–1088
11. Gnanapragasam NV, Reddy BV, Rosen MA (2010) A methodology for assessing the sustainability of hydrogen production from solid fuels. Sustainability 2:1472–1491
12. Afgan NH, Carvalho MG, Hovanov NV (2000) Energy system assessment with sustainability indicators. Energy Policy 28:603–612
13. Afgan NH, Carvalho MG (2002) Multi-criteria assessment of new and renewable energy power plants. Energy 27:739–755
14. Afgan NH (2010) Sustainability paradigm: intelligent energy system. Sustainability 2:3812–3830

15. Frangopoulos CA, Keramioti DE (2010) Multi-criteria evaluation of energy systems with sustainability considerations. Entropy 12:1006–1020
16. Rowley HV, Peters GM, Lundie S, Moore SJ (2012) Aggregating sustainability indicators: beyond the weighted sum. J Environ Manage 111:24–33
17. Guinée JB (2002) Handbook on life cycle assessment: Operation guide to the ISO standards. Kluwer Academic, New York
18. Ahlroth S, Nilsson M, Finnveden G, Hjelm O, Hochschorner E (2011) Weighting and valuation in selected environmental systems analysis tools—suggestions for further developments. J Cleaner Prod 19:145–156
19. Carifio J, Perla RJ (2007) Ten common misunderstandings, misconceptions, persistent myths and urban legends about Likert scales and Likert response formats and their antidotes. J Soc Sci 3:106–116
20. Neuman WL (2010) Social research methods: qualitative and quantitative approaches. Pearson PLC, London
21. Hacatoglu K (2014) A systems approach to assessing the sustainability of hybrid community energy systems. Unpublished doctoral dissertation, University of Ontario Institute of Technology (UOIT), Oshawa
22. Saldanha N, Beausoleil-Morrison I (2012) Measured end-use electric load profiles for 12 Canadian houses at high temporal resolution. Energy Build 49:519–530
23. Kreith F, Kreider JF (2011) Principles of sustainable energy. CRC, Boca Raton
24. Weather Network (2014) Statistics: Toronto, Ontario, Canada. http://past.theweathernetwork.com/statistics/CL6158350. Accessed 10 Sept 2014
25. Statistics Canada (2012) Median after-tax income, by economic family type, 2010 constant dollars, annual (CANSIM Table 202-0605). Ottawa
26. Fankhauser S, Tepic S (2007) Can poor consumers pay for energy and water? An affordability analysis for transition countries. Energy Policy 35:1038–1049
27. IPCC (2013) Summary for policymakers. In: Stocker TF, Qin D, Plattner GK, Tignor M, Allen SK et al (eds) Climate change 2013: The physical science basis. Contribution of working group I to the fifth assessment report of the intergovernmental panel on climate change. Cambridge University Press, Cambridge
28. Ravishankara AR, Daniel JS, Portmann RW (2009) Nitrous oxide (N_2O): the dominant ozone-depleting substance emitted in the 21st century. Science 326:123–125
29. Goedkoop M, Spriensma R (2000) The Eco-Indicator 99: a damage oriented method for life cycle impact assessment. PRe Consultants, Amersfoort

Chapter 2
Thermodynamic Modeling of an Integrated Energy System for Poly-generation Design

Yunus Emre Yuksel and Murat Ozturk

Abstract Integrated energy production systems based on the renewable or fossil energy sources for poly-generation applications are inevitable in the near future for both environmental and sustainability concerns. Increasing the overall efficiency by combining system decreases the energy consumption and increases the system outputs such as electricity, heat, hot water, cooling, hydrogen, oxygen, and ext. Considering the global energy demands, the increase of efficiency of poly-generation systems will decrease emissions and therefore helps to protect the environment. Moreover, not only decreasing the emissions but also reducing the energy consumption is very important to achieve more sustainable systems, and it is again possible with integrated systems. In this chapter, thermodynamic assessment formulations and energy and exergy efficiency of a new poly-generation design which consists of biomass gasification, solid oxide fuel cell (SOFC), organic Rankine cycle (ORC), and double-effect absorption cooling and heating systems are given and analyzed in detail through energy, exergy, and sustainability approaches.

Keywords Thermodynamic analysis • Energy analysis • Exergy analysis • Integrated system • Poly-generation • Rankine cycle • Gasification

Nomenclature

E	Energy, kJ
$\dot{E}$	Energy rate, kW
ex	Specific exergy, kJ/kg
$\dot{E}x$	Exergy rate, kW
G	Gibbs free energy, kJ
h	Specific enthalpy, kJ/kg

Y.E. Yuksel (✉)
Afyon Kocatepe University, Afyonkarahisar, Turkey
e-mail: yeyuksel@aku.edu.tr

M. Ozturk
Suleyman Demirel University, Isparta, Turkey

I. Dincer et al. (eds.), *Progress in Clean Energy, Volume 1*,
DOI 10.1007/978-3-319-16709-1_2

$\dot{m}$ Mass flow rate, kg/s
n Mole number
Q Heat, kJ
$\dot{Q}$ Heat rate, kW
s Entropy, kJ/kg
T Temperature, K
x Mole fraction
W Work, kJ
$\dot{W}$ Power, kW

Greek Letters

Δ Change in variable
ξ Chemical exergy coefficient
η Energy efficiency
ψ Exergy efficiency

Subscripts

A Ash
a Ambient
c Coal
ch Chemical
D Destruction
f Fuel
H Hydrogen
in Inlet
ke Kinetic energy
O Oxygen
out Outlet
pe Potential
ph Physical
S Sulfur
W Water

Acronyms

HHV Higher heating value
LHV Lower heating value

2.1 Introduction

Energy supply is one of the significant subjects for human economy and social development. Increased energy production and consumption rate especially in manufacturing and standards of living in the growing population causes some global problems, such as greenhouse gas effects, global warming and climate change, melting ice, acid precipitation, and stratospheric ozone depletion. Nowadays, most of the world's energy is produced and consumed in ways that it cannot be sustained if energy technologies remain unchanged and rates of consumption increased substantially. Thermal energy available in sustainable energy source should be used for different aims, directly in the process plant or converted into mechanical work to generate electrical power.

Methods of converting sustainable thermal energy into useful mechanical work represent a very important issue regarding the achievement of developing a clean, nonpolluting, non-exhaustible energy supply system for future generations. Changes are required in energy systems, partly through the adoption of advanced energy technologies and systems where advantageous, to address serious environmental concerns. Thus, there is a direct link between energy efficiencies and renewable energies on one hand and sustainable development on the other hand. In this manner, integrated energy systems for poly-generations are very significant for effective energy production and environmental friendly ecological energy. Cogeneration systems produce both the power and heat energy, but integrated systems offer more useful outputs, such as power, heat, fresh water, cooling, hydrogen, chemicals, drying, or fuel. Poly-generation energy systems have many advantages such as higher system performance, decreased thermal losses and wastes, decreased operating cost, decreased greenhouse gas and pollution emissions, better use of sources, shorter transmission lines, fewer distribution units, multiple generation options, increased reliability, and less grid failure [1]. The development in performance is often the most important parameter in implementing a poly-generation energy system. Further evaluations before choosing poly-generation systems, such as evaluations of initial capital and operating costs, are needed to ensure performance and economic poly-generation system construction and efficiency [2].

Thermodynamic assessment of the energy production system should be given for system efficiency, cost, and also environmental effects. Exergy analysis according to the second law of thermodynamics is a tool for understanding and improving efficiency and is used throughout this investigation in addition to energy analysis. Rosen and Scott [3] have performed energy and exergy analysis of a system for methanol facility production from natural gas sources. The system involved production of synthesis gases through steam methane reforming process, compression of the product gases, methanol synthesis, and distillation of the methanol. They have found the energy and exergy efficiency for the methanol production system as 39 % and 41 %, respectively.

Bilgen and Kaygusuz [4] have given the credit method to the energy and exergy analysis, which is the effective thermodynamic method for using the conservation of mass and energy principle together with the second laws of thermodynamic for the design and analysis of integrated system. This particular study has presented the completed methodologies and mainly focused on exergy analysis that helped to identify the process of estimating the physical, chemical exergy, and exergy destruction.

According to the exergy analysis, two basic indicators as chemical and physical exergies are needed to be determined, before processing to the overall exergy efficiency. Kamate and Gangavati [5] have described the methodology in determining fuel chemical exergy, and the limitation was made when the mass ratio of oxygen to carbon (O/C) is in the range from 0.667 to 2.67.

The exergy analysis which has been investigated by Gao et al. [6] on a coal-based poly-generation system for power and chemical production shows significant improvement in energy savings when compared to individual systems. The results of the analysis indicate that the combination of a power system with a chemical process achieved a 3.9 % increase in energy savings. Also, it is indicated that the synthesis on the basis of thermal energy cascade utilization is the main contribution to the performance benefit of the poly-generation system. A key criterion of a poly-generation system is the capacity ratio of the chemical process to the power system. This ratio strongly affects the matching of the two sides involved in the poly-generation system. Moreover, besides the thermal energy integration, the cascade utilization of the chemical exergy is likely to be a key issue in further studies of poly-generation systems.

Carvalho et al. [7] designed a tri-generation system to be installed in a hospital to minimize total annual cost and CO_2 emissions. Among those systems, the most cost-effective and the most environmentally benign configurations were then compared to each other and one configuration was selected that meets the optimum criteria. Based on the size of the hospital and its geographical location, they have assumed the annual electricity consumption, the cooling demand, and the heat requirements. They have ended up with several configurations; in each case, an inversely proportional relationship was observed with emission reduction and cost-effectiveness.

Khaliq et al. [8] have studied an energy and exergy analyses of an integrated electric production and refrigeration cycle, as well as a parametric study of the impacts of exhaust gases inlet temperature, pinch point and gas composition on energy and exergy efficiency, electricity to cold production ratio, and exergy destruction rate for the components of system.

Ahmadi et al. [9] have studied a new integrated tri-generation energy system consisting of a gas turbine, a double-pressure heat recovery steam generator, and a single effect absorption chiller and an organic Rankine cycle. The authors have also performed a parametric study to see the variation of exergy efficiency, cooling and heating load, and cost of environmental impact. The results of this study demonstrated that system performance is notably affected by the compressor pressure ratio, the gas turbine inlet temperature, and the gas turbine isentropic efficiency.

Fossil energy sources are still the most used source for power production, particularly coal, which has the highest portion among these sources. Coal is the most carbon-intensive fossil fuel source in the world as well as the most common primary energy source used in power plants. It also emits significant amount of greenhouse gas emissions based on common power production systems.

For these reason, increasing performance and applying low-carbon emission systems can decrease the associated environmental impacts considerably. Nowadays, gasification technology is one of the most promising methods to make use of coal with higher performance and lower environmental impacts. Solar energy technologies are one of the emerging energy production systems based on its availability, low cost, and environmental impact during the operation, especially for off-grid utilization.

Ozturk and Dincer [10] conducted a thermodynamic analysis of a solar-based multi-generation system with hydrogen production. The solar-based multi-generation considered for this analysis consists of four main subsystems: a Rankine cycle, an organic Rankine cycle, an absorption cooling and heating, and a hydrogen production and utilization. The exergy efficiency and exergy destruction rate for the subsystems and the overall system show that the parabolic dish collectors have the highest exergy destruction rate among constituent parts of the solar-based multi-generation system.

Ozturk and Dincer [11] have studied a multi-generation energy production system supported by solar energy and coal gasification. Also, parametric studies have given by the change in design parameters, such as an ambient temperature, a compressor pressure ratio, a nitrogen supply ratio for the combustion chamber, and a gas turbine inlet temperature.

The main purposes of this chapter are to investigate a thermodynamic analysis of the poly-generation system supported by a solar concentrating collector and coal gasification system, with a gasification process, a concentrating collector, a double-effect absorption cooling system, a Rankine cycle, a methanol production plant, and a water-gas shift reactor, and to decrease environmental impacts and cost. In order to understand the relation among various parameters affecting the performance of an integrated energy system for poly-generation, parametric studies are presented. The other purposes of this chapter should be given as follows:

- To develop an advanced Engineering Equation Solver (EES) software code and carry out parametric studies for system components
- To calculate the exergy content for each stream of the process including the chemical exergy for the gasification-based plant
- To determine the exergy destruction rate and exergy efficiency of each system component
- To perform a complete parametric study and the performance assessment of the system

2.2 System Analysis

The schematic diagram of the integrated energy system for poly-generation is shown in Fig. 2.1. This integrated system consists of five main subsystem as (1) concentrating collector, (2) coal gasification, (3) Rankine cycle, (4) double-effect absorption cooling system, and (5) methanol production. The significant role of the solar concentrating concentrator is to minimize the area of the solar receiver with respect to the aperture area. These methods are utilized to reduce heat losses through radiation with respect to the useful heat flux, increase the solar collector efficiency, and reach the higher temperatures.

Gasification technologies convert solid fuel such as coal and biomass into product gas that can be used in different technologies as an input. The product gas is mainly composed of CO, CO_2, CH_4, and H_2 and can be combusted to generate heat and work. It can also be used as a feedstock for the production of synthesis gas, liquid fuel, and chemicals. The lower heating value (LHV) of the synthesis gas is calculated by using the gas composition data. The outputs of the gasifier also include particulates, tar, ammonia, and hydrogen sulfide.

The lithium bromide-water absorption system is used instead of a conventional refrigeration system to utilize surplus heat in the system. In this chapter, it is assumed that the required energy for the double-effect absorption cooling and heating subsystem can be provided by the heat recovered from the solar thermal collector subsystem. As it can be seen in Fig. 2.1, it contains the most significant components of the double-effect absorption system: high- and low-pressure generator, high- and low-temperature heat exchanger, solution and refrigerant pump, absorber, condenser, and evaporator. Also, it involves three pressure levels: high, medium, and low. The high-pressure generator functions at high pressure and high temperature, whereas the low-pressure generator and condenser operate at medium pressure, and the evaporator and absorber work at low pressures. It should be noted

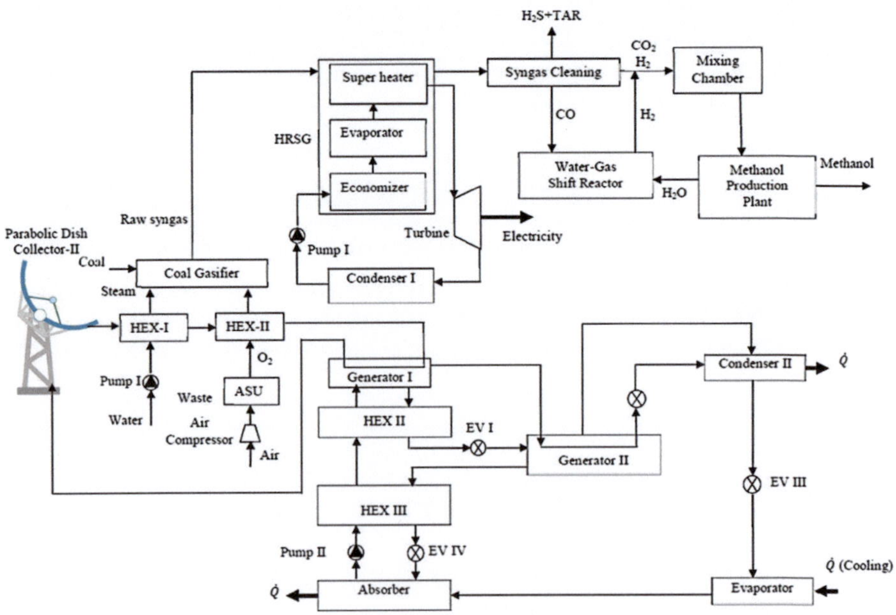

Fig. 2.1 Schematic diagram of the integrated energy system for poly-generation

that the integrated system is modeled according to the optimum operating parameters for the double-effect absorption subsystem.

The significant purpose of the integrated system is to minimize waste energy and improve the sustainability of useful products generation. The heat taken from a high temperature process is converted into mechanical work through a heat engine, which has a maximum efficiency equal to the Carnot cycle efficiency. Heat engines must reject heat, which can be made useful by integrating with another system to yield a higher efficiency. Heat exchangers are devices that are used to transfer thermal energy between media without mixing the fluid themselves. The media may be fluid to fluid, fluid to solid, and solid to solid at different temperatures and in thermal contact directly or through another conductive material. Depending on the process or application, heat exchangers are designed to achieve an efficient transfer of heat.

2.3 Assumptions

The following simplified assumptions are made in the present work:

- All the components operate at steady-state conditions and the heat losses from piping are also ignored.
- All the gases are ideal gases.
- Ambient air is considered as 79 % nitrogen and 21 % oxygen on the volume basis.
- Ambient temperature and pressure of the reference environment are chosen as 25 °C and pressure 1.013 bar, respectively.
- All the components of the steam cycle have adiabatic boundaries. In addition, pressure drops, kinetic energy, and the change in potential energy of system components are negligible.
- Real steam turbines have isentropic efficiencies in the range of 60–90 %, while pumps have isentropic efficiencies in the range of 75–100 %. It is assumed that isentropic efficiency of each turbine is 85 %, while each pump has an isentropic efficiency of 88 %.
- Mechanical efficiency of turbine is in the range of 95–99 %. Mechanical efficiency of each steam turbine is assumed to be 99 % in the present analysis.
- Generator efficiency is in the range of 98–99 %. In the present work, the generator is assumed to be 98 % efficient.
- The proton exchange membrane (PEM) electrolyzer operates at 30 bar with 65 % efficiency.
- The hydrogen generated by the electrolyzer is stored at 25 bar on a seasonal storage basis.
- The size of the storage tank is determined based on the seasonal need for hydrogen.

2.4 Fuel Characteristics

Lignite sample has been chosen for the system analysis. The lower heating value (LHV) of coal is calculated from the following formula proposed by Ghamarian and Cambel [12]:

$$\mathrm{LHV}_c = 427.0382n_C + 90.88110n_H - 207.46424n_O + 297.0116n_S \tag{2.1}$$

where subscripts c and n represent coal and number of moles of respective constituents, C, H, S, and O, which are the carbon, hydrogen, sulfur, and oxygen contents of coal sample in weight %, respectively. For the given coal sample, the higher heating value should be calculated as follows:

$$\mathrm{HHV} = \mathrm{LHV} + 21.978n_H \tag{2.2}$$

Proximate and ultimate analyses (wt%) of lignite, which is used as coal sample in this chapter, are given in Tables 2.1 and 2.2, respectively. Lower and higher heating values of the lignite are calculated as 19,070 kJ/kg and 20,070 kJ/kg, respectively, from Eqs. (2.1) and (2.2).

2.5 Thermodynamic Analysis

General thermodynamic assessments involving the mass, energy and exergy balance equations, and energy and exergy efficiencies are presented to investigate for the integrated system improvement potentials. In the most general viewpoint, a balance equation for a given quantity in a process should be written as follows:

$$\text{Input} + \text{Generation} - \text{Output} - \text{Consumption} = \text{Accumulation} \tag{2.3}$$

Equation (2.3) is known as the quantity balance for the process and used for determination of accumulated quantity. In order to investigate the poly-generation system, the principle of the mass, energy, and exergy conservation and also energy

Table 2.1 Proximate analysis (wt%) of lignite utilized in the analysis [13]

Ash	Fixed carbon	Volatile matter	Moisture
8.1	35.0	44.5	12.4

Table 2.2 Ultimate analysis (wt%) of lignite utilized in the analysis [13]

Ultimate analysis (received basis)					Ultimate analysis (dry ash-free basis)				
C	O	H	N	S	C	O	H	N	S
51.0	23.8	4.1	0.4	0.16	64.2	29.9	5.2	0.5	0.2

and exergy efficiency is given in this chapter. These approaches are applied to each component of the system and whole system.

2.5.1 Mass Balance Analysis

The mass balance analysis is a basic principle in analyzing any thermodynamic system. In the steady-state conditions, the rate form of the mass balance for the steady-state condition can be given as follows:

$$\sum \dot{m}_{\text{in}} = \sum \dot{m}_{\text{out}} \tag{2.4}$$

2.5.2 Energy Balance Analysis

The energy balance equation is applied to the description of a wide variety of the given process. The energy balance of a considered process is always conserved within the system according to the first law of thermodynamics. The sum of all energy types is constant in an isolated system. The rate form of the energy balance equation for the steady-state condition with negligible kinetic and potential energy effects can be given below as the total energy input equal to the total energy output, with all energy types, including heat, power, and material flows, as follows:

$$\sum_i \dot{E}_i + \dot{Q}_i = \sum_e \dot{E}_e + \dot{W} \tag{2.5}$$

where $\dot{Q}$ and $\dot{W}$ are the heat and work transfer rate, respectively. Neglecting potential and kinetic energy, the above equation can be written as follows:

$$\dot{Q} + \sum \dot{m}_i h_i = \dot{W}_{\text{net}} + \sum \dot{m}_e h_e \tag{2.6}$$

where h is the specific enthalpy.

2.5.3 Exergy Balance Analysis

Nowadays, exergy analysis of the energy production system is considered as an important tool in the engineering system designs. In any system, when energy is converted into another less useful type, the rest of the useful part of the energy cannot be recovered again, a portion that is not conserved as total energy of the process. According to this procedure, the amount of useful work can be given in the exergy (also called available energy or availability) terms for a better understanding of the energy conversion. The properties of a reference environment have to be given in the exergy analysis. This is generally done by given the temperature, pressure, and chemical proportion of the reference environment [14]. When a system is in equilibrium with reference environment, its exergy content is zero. In this study, real environment cases are accepted as the reference condition or environment properties.

Exergy analysis based on the second law of the thermodynamics can support to create the strategies and managements for more economical and effective use of energy sources and is utilized to study integrated systems. The sources of the irreversibility and proposed approaches to increase the whole efficiency of the given system should be considered by using exergy analysis. Exergy content of matter is generally divided into four parts which are physical exergy (ex_{ph}), chemical exergy (ex_{ch}), kinetic exergy (ex_k), and potential exergy (ex_p). In this study, kinetic and potential exergy are negligible, as the elevation difference is low and speeds in the process are small. In general, physical exergy is represented by the maximum effective work available as a process interacts with the environment. Chemical exergy is connected with the departure of the chemical configuration of a matter from its chemical equilibrium and also is considered necessary in systems including combustion and whole chemical variations [15]. Using the first and second laws of thermodynamics, balance equation for exergy analysis can be written as follows:

$$\sum_i \dot{m}_{\text{in}} ex_{\text{in}} + \dot{E}x_Q = \sum_e \dot{m}_{\text{out}} ex_{\text{out}} + \dot{E}x_W + \dot{E}x_D \tag{2.7}$$

$\dot{E}x_Q$ and $\dot{E}x_W$ are the heat and work exergy flow rates through the boundary at temperature T_j at location j, respectively.

$$\dot{E}x_Q = \left(1 - \frac{T_o}{T_i}\right)\dot{Q}_i \tag{2.8}$$

where T_i is the temperature in the ith given state.

$$\dot{E}x_W = \dot{W} \tag{2.9}$$

ex is the specific exergy of the process inlet and outlet flow and $\dot{E}x_D$ is the exergy destruction rate. The specific exergy can be given as follows:

$$ex = ex_{ke} + ex_{pe} + ex_{ph} + ex_{ch} \tag{2.10}$$

Any substance of which the temperature, pressure, or composition is different from the thermodynamic equilibrium with the surroundings (thermal, mechanical, and chemical) has the possibility to produce a change. As given before, since the variations of the kinetic and potential exergy are accepted negligible in this study, both physical and chemical exergies are defined for the system components. The physical exergy or general flow exergy of the *i*th flow is given as

$$ex_{ph,i} = (h_i - h_o) - T_o(s_i - s_o) \tag{2.11}$$

where h is the specific enthalpy and s is the specific entropy, respectively. The chemical exergy of an ideal gas mixture can be determined by

$$ex_{ch} = \sum x_i ex^o_{ch,i} + RT_o \sum x_i \ln(x_i) + \Delta G_e \tag{2.12}$$

where $ex^o_{ch,i}$ is the standard chemical exergy of a pure chemical combination and x_i is the mole fraction of the *i*th component, R is the gas constant, and ΔG_e is the excess free Gibbs energy [16]. In this study, free Gibbs energy is negligible, as the amount difference for free Gibbs energy is small at low pressure and at a given gas mixture. The specific chemical exergies of the usual fuels are defined using the corresponding ratio of simplified chemical exergy as follows:

$$\dot{E}x^f_{ch} = \xi_i \text{LHV}_f \tag{2.13}$$

where the factor ξ_i gives the division of the chemical exergy to the LHV (lower heating value) of the *i*th fuel components. The factor ξ_i of the most usual gaseous fuels with C_xH_y can be calculated as [17]

$$\xi_i = 1.22 + 0.0169\frac{y}{x} - \frac{0.0698}{x} \tag{2.14}$$

In the literature [18], the numerical correlation for ξ of the biomass-based fuels is given as follows:

$$\xi = \frac{1.044 + 0.016\frac{X_{H_2}}{X_C} - 0.34493\frac{X_{O_2}}{X_C}\left(1 + 0.053\frac{X_{H_2}}{X_C}\right)}{1 - 0.4124\frac{X_{O_2}}{X_C}} \tag{2.15}$$

where X_i is the mass fraction of the *i*th element. The specific chemical exergy of a dry assumption coal involving sulfur, assuming that it occurs as a free element, is given as [19]

$$\dot{E}x^{\text{coal}}_{ch} = (\text{LHV}_{\text{coal}} + h_W)\xi_{\text{coal}} + \left(\dot{E}x^S_{ch} - h_S\right)X_S + \dot{E}x^A_{ch}X_A + \dot{E}x^W_{ch}X_W \tag{2.16}$$

where W, S, and A subscripts refer to water, sulfur, and ash, respectively, h_i is the specific enthalpy of the *i*th chemical components, $\dot{E}x^i_{ch}$ is the standard chemical exergy of the *i*th chemical components, and ξ_{coal} is the chemical exergy coefficient of the selected coal type.

Table 2.3 The values of standard molar exergy of the substance [26]

Substances	Standard enthalpy (MJ/mol)	Standard exergy (MJ/mol)
O_2(g)	0	3.97
N_2(g)	0	0.72
CO_2(g)	−393.52	19.87
H_2O(g)	−241.82	9.5
H_2O(l)	−285.83	0.9
SO_2(g)	−297.10	313.40
NO(g)	90.59	88.90
NO_2(g)	33.72	55.60

$$\xi_{\text{coal}} = \frac{(\text{LHV}_{\text{coal}} + 2442X_W)\xi_{\text{dry}} + 941X_{S_2}}{\text{LHV}_{\text{coal}}} \tag{2.17}$$

where ξ_{dry} is the chemical exergy coefficient in dry assumption [18]. Equation (2.17) is valid only for a mass ratio $(x_{O_2}/x_C) \leq 0.667$ [11]. Many researchers [20–23] have proposed methods to calculate the chemical exergy of elements, organic, and inorganic substances. However, the chemical values used in this chapter have been taken from Szargut [24]. The values given in this model are based on standard temperature as 25 °C. These values were corrected for the reference temperature by the following equation [25]:

$$ex_i^{ch} = \frac{T_o}{T^s} ex_{ch}^i - \overline{h}_f^o \frac{T^s - T_o}{T^s} \tag{2.18}$$

where ex_i^{ch}, $\overline{h}_f^o$, T^s, and T_o are standard molar chemical exergy, enthalpy of formation, standard temperature, and reference temperature, respectively. The values of standard molar exergy of the substance used in this chapter are given in Table 2.3.

2.5.4 *Thermodynamic Efficiencies*

A thermodynamic efficiency can be based only on the first law of thermodynamics (energy efficiency) or consider both laws of thermodynamics (exergy efficiency). In order to evaluate the performance of the poly-generation system, energy and exergy analyses are commonly used to establish the whole system. Energy and exergy efficiency of the components of the system and whole system should be given for detailed thermodynamic analysis.

2.5.4.1 Energy Efficiency

The energy efficiency (η) of the process can be defined as the ratio of useful energy produced by the process to the total energy input. The useful produced energy represents the desired results produced by the system components. The energy efficiency for the single production option can be given as follows:

$$\eta = \frac{\text{useful energy output rate with products}}{\text{total energy input rate}} = \frac{\dot{E}_{\text{out, useful}}}{\sum \dot{E}_{\text{in, total}}} \tag{2.19}$$

2.5.4.2 Exergy Efficiency

In order to evaluate the performance of the poly-generation system from the exergy analysis point of view, it is essential to determine both the product and the fuel for the components of the system and whole system. The exergy efficiency (ψ) of the process should be defined as the divided exergy output rate ($\dot{E}x_{\text{out, useful}}$) that is created by the considered system to the overall exergy inlet rate $\left(\sum \dot{E}x_{\text{in, total}}\right)$ that crosses the boundaries of the system, as follows:

$$\psi = \frac{\text{total useful exergy output rate with products}}{\text{total exergy input rate}} = \frac{\sum \dot{E}x_{\text{out, total}}}{\sum \dot{E}x_{\text{in, total}}} \tag{2.20}$$

The exergy efficiency for the process should also be given in terms of exergy destruction rate as follows:

$$\psi = \frac{\sum \dot{E}x_{\text{out}}}{\sum \dot{E}x_{\text{in}}} = \frac{\sum \dot{E}x_{\text{in}} - \dot{E}x_D}{\sum \dot{E}x_{\text{in}}} = 1 - \frac{\dot{E}x_D}{\sum \dot{E}x_{\text{in}}} \tag{2.21}$$

Using the derived equations, a model is generated by using EES. With the help of this model, the system performance and emissions are investigated by varying a series of input conditions.

2.6 Results and Discussion

The investigation of exergy destruction for each component is given in this analysis, besides the assessment of system components efficiencies. The thermal energy is lost from the components during the process that has been approved in many literatures. In the definition of exergy destruction, the generation of entropy always destroys exergy, because of that the useful work is reduced and cannot be recovered, and sometimes exergy destruction can refer to irreversibility such as chemical reaction, expansion work, compression work, and heat transfer through the temperature difference that always involves entropy generation. In order to evaluate the overall cogeneration performance, the study of exergy destruction can help to identify the defect within the cogeneration plant, and the performance can be improved by the redesign or modification of the existing cogeneration plant from considering reduction of heat loses in the further improvement and development.

The results of the energy and exergy analysis, including the energy and exergy efficiency, and exergy destruction rate for the system components are reported. It is shown that the inlet and outlet exergy flows of the coal gasification plant are mainly attributed to the energy and exergy inlet with the coal sample and process air. In addition, the exergy loss is determined to be due to emission and internal consumption associated with chemical reaction, especially those related to gasification. Note that inlet exergy values are evaluated for coal on an LHV basis.

Also, the results in energy and exergy efficiencies variation by changing operating conditions are investigated in depth using the thermodynamic laws. Besides the energy and exergy efficiencies examinations, the thermodynamic analysis is also going to examine the influences of the work output and process heat from operating conditions though the parametric studies. Also, the exergy destruction rate is another important matter to be emphasized because the study of irreversibility can help to identify where the work or energy is lost during the operation. Various factors identified in the design section can influence both the energy and exergy efficiencies of the integrated system.

The effect of varying ambient temperature from 10 to 30 °C on the exergy destruction rate and exergy efficiencies for the concentrating solar collector, coal gasification system, Rankine cycle system, double-effect absorption cooling system, methanol production system, and whole system is illustrated in Figs. 2.2, 2.3,

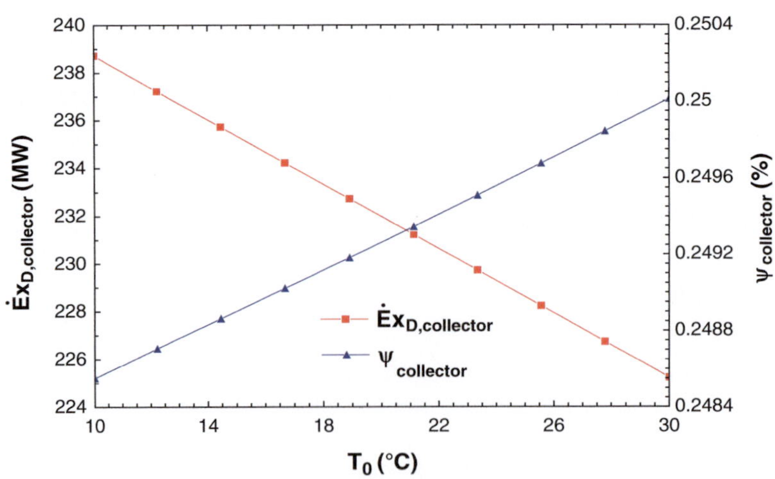

Fig. 2.2 Variations with ambient temperature of the exergy destruction rate and exergy efficiency for the concentrating collector subsystem

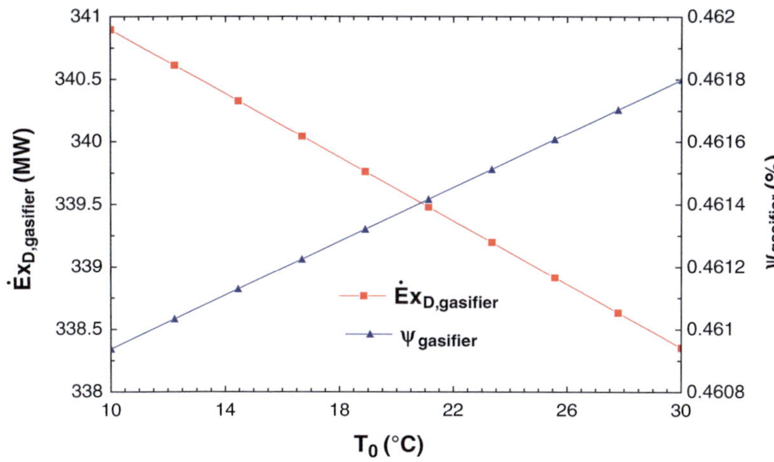

Fig. 2.3 Variations with ambient temperature of the exergy destruction rate and exergy efficiency for the coal gasification subsystem

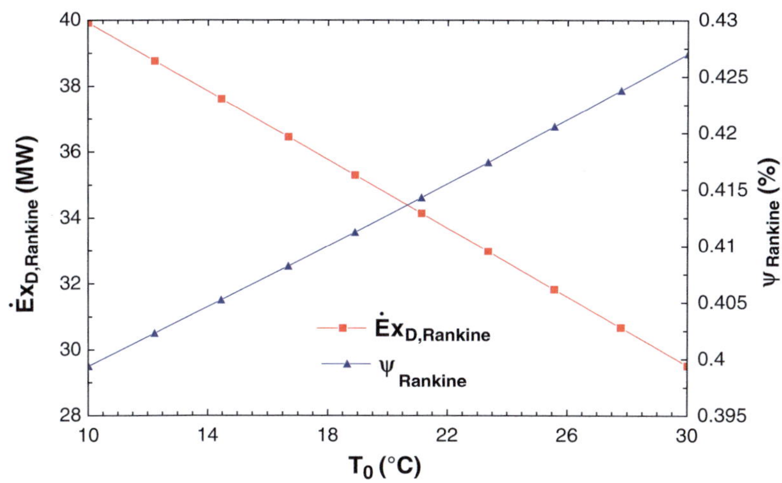

Fig. 2.4 Variations with ambient temperature of the exergy destruction rate and exergy efficiency for the Rankine cycle subsystem

2.4, 2.5, 2.6, and 2.7, respectively. Figure 2.2 shows that the exergy destruction rate of the concentrating solar collector subsystem decreases with increasing ambient temperature, but its exergy efficiency rate increases. Figures 2.3, 2.4, and 2.6 show that varying the ambient temperature in the coal gasification system, Rankine cycle system and methanol production system studies, respectively, have the same effect on the exergy destruction rate and exergy efficiencies. The exergy destruction rate of these subsystems decreases with increasing ambient temperature, while their

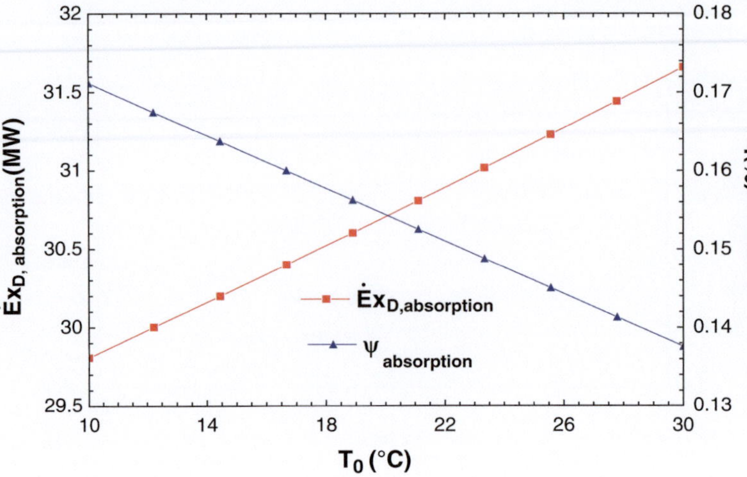

Fig. 2.5 Variations with ambient temperature of the exergy destruction rate and exergy efficiency for the double-effect absorption cooling subsystem

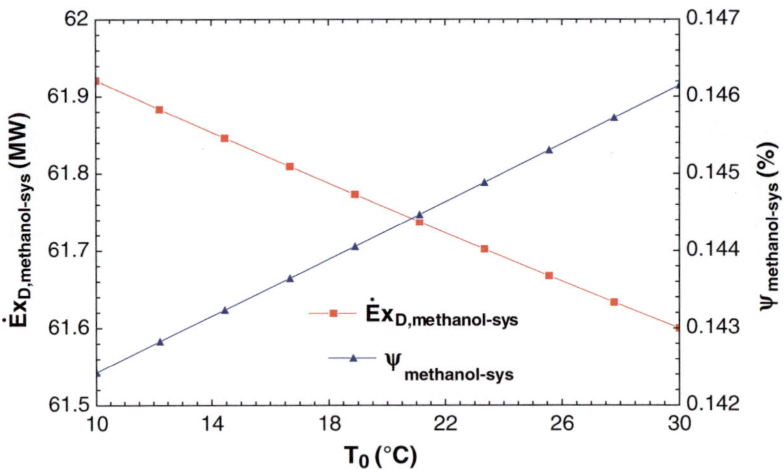

Fig. 2.6 Variations with ambient temperature of the exergy destruction rate and exergy efficiency for the methanol production subsystem

exergy efficiencies increase. The variations of exergy destruction rate and exergy efficiencies of these subsystems remain almost linear depending in the ambient air. In contrast, as seen in Figs. 2.5 and 2.7, the exergy destruction rate in the double-effect absorption cooling system and whole system increases with the increasing ambient temperature, but their exergy efficiencies decrease.

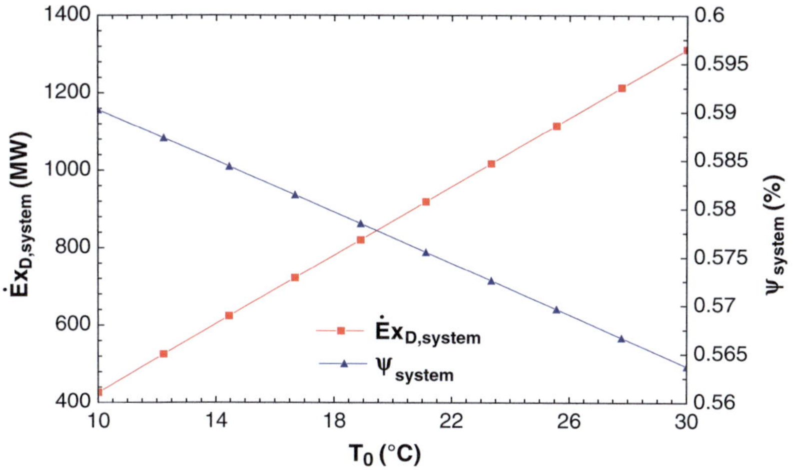

Fig. 2.7 Variations with ambient temperature of the exergy destruction rate and exergy efficiency for the whole system

2.7 Conclusions

Poly-generation systems can make significant supports for energy and other useful outlet products due to their potential for high performance as well as low operating costs and greenhouse gas and pollution emissions. Environmental problems such as fossil fuel depletion and climate change amplify the advantages and significance of performance poly-generation energy systems. An important reserve of low-quality lignite in many countries such as Turkey should be used with clean coal technology for sustainable development. Turkish lignite sources have high ash, volatile matter, and sulfur content, which draw a non-feasible picture for conventional combustion techniques. In this chapter, energy and exergy analysis is carried out for the coal- and solar-based poly-generation system to evaluate the system performance and exergy destruction rates for system components. This integrated coal and solar energy-based system is developed to accommodate different poly-generation options. In addition to that, the energy and exergy analyses are performed and applied to the poly-generation system components to calculate the rates of energy and exergy flows, the efficiencies, and the unit exergy destruction rate of flows in the components. The main conclusions drawn from the present study are as follows:

- The capacities and performances of the integrated system depend on solar weather data and on the design parameters of the system components.
- The coal gasification system is the source of 18 % of the total exergy destruction and 14 % of the annual exergy destruction in the concentrating collector system.
- The overall energy and exergy efficiencies of the integrated system are calculated as 67 % and 58 %, respectively.

- Integrated energy system for poly-generation is a promising technology for electricity, hydrogen, chemicals, hot water production, and volume heating and cooling application for rural areas. They are undergoing research and experimentation, and further research is needed to improve understanding of integrated systems.

The poly-generation design model can also be used to help decision makers in selecting the optimum size and/or efficiencies of the components when designing integrated fossil and renewable power generation systems.

References

1. Dincer I, Zamfirescu C (2012) Renewable-energy-based multigeneration systems. Int J Energy Res 36(15):1403–1415
2. Ahmadi P, Rosen MA, Dincer I (2012) Multi-objective exergy-based optimization of a polygeneration energy system using an evolutionary algorithm. Energy 46(1):21–31
3. Rosen MA, Scott DS (1988) Energy and exergy analyses of a production process for methanol from natural gas. Int J Hydro Energy 13(10):617–623
4. Bilgen S, Kaygusuz K (2008) Second law (Exergy) analysis of cogeneration system. Energy Sources, Part (A) 30(13):267–1280
5. Kamate SC, Gangavati PB (2009) Cogeneration in sugar industries technology options and performance parameters—a review. Cogeneration Distributed Generation J 24(4):6–33
6. Gao L, Jin H, Liu Z, Zheng D (2004) Exergy analysis of coal-based poly-generation system for power and chemical production. Energy 29(12–15):2359–2371
7. Carvalho M, Lozano MA, Serra LM (2012) Multicriteria synthesis of trigeneration systems considering economic and environmental aspects. Appl Energy 91(1):245–254
8. Khaliq A, Kumar R, Dincer I (2009) Performance analysis of an industrial waste heat-based trigeneration system. Int J Energy Res 33(8):737–744
9. Ahmadi P, Rosen MA, Dincer I (2011) Greenhouse gas emission and exergo-environmental analyses of a trigeneration energy system. Int J Greenhouse Gas Control 5(6):1540–1549
10. Ozturk M, Dincer I (2013) Thermodynamic analysis of a solar-based multi-generation system with hydrogen production. Appl Therm Eng 51(1–2):1235–1244
11. Ozturk M, Dincer I (2013) Thermodynamic assessment of an integrated solar power tower and coal gasification system for multi-generation purposes. Energy Conversion Manage 76:1061–1072
12. Ghamarian A, Cambel AB (1982) Biomass exergy analysis of Illinois No. 6 coal. Energy 7(6):483–488
13. Vassilev SV, Vassilev CG (2009) A new approach for the combined chemical and mineral classification of the inorganic matter in coal 1, chemical and mineral classification systems. Fuel 88(3):235–245
14. Dincer I, Rosen MA (2013) Exergy: energy, environment, and sustainable development, 2nd edn. Elsevier, Oxford
15. Bejan A, Tsatsaronis G, Moran M (1996) Thermal design and optimization. Wiley, New York
16. Kotas TJ (1985) The exergy method of thermal plant analysis, 1st edn. Buttworth Scientific, London
17. Mansouri TM, Ahmadi P, Kaviri AG, Jaafar MNM (2012) Exergetic and economic evaluation of the effect of HRSG configurations on the performance of combined cycle power plants. Energy Conversion Manage 58:47–58

18. Szargut J, Styrylska T (1964) Approximate evaluation of the exergy of fuels. Brennst Waerme Kraft 16(12):589–596 (In German)
19. Szargut J (2005) Exergy method: technical and ecological applications. WIT, Southampton
20. Kameyama H, Yoshinda K, Yamauchi S, Fueki K (1982) Evaluation of reference exergies for the elements. Appl Energy 11(1):69–83
21. Lozano MA, Valero A (1982) Evaluation of reference exergies for the elements. Appl Energy 11(1):69–83
22. Gallo WLR, Milanez LF (1990) Choice of a reference state for exergetic analysis. Energy 15(2):113–121
23. Rivero R, Garfias M (2006) Standard chemical exergy of elements updated. Energy 31(15):3310–3326
24. Szargut J (1990) Chemical exergies of the elements. Appl Energy 32(4):269–286
25. Kotas TJ (1980) Exergy concepts for thermal plant: first of two papers on exergy techniques in thermal plant analysis. Int J Heat Fluid Flow 2(3):105–114
26. Shoaib M (2011) Energy and exergy analyses of biomass Co-firing based pulverized coal power generation. A Master Thesis, University of Ontario Institute of Technology

Chapter 3
Assessment of Maisotsenko Combustion Turbine Cycle with Compressor Inlet Cooler

Hakan Caliskan, Ibrahim Dincer, and Arif Hepbasli

Abstract In this study, a Maisotsenko combustion turbine cycle (MCTC) with compressor inlet cooling system is proposed and studied through energy, exergy and exergoeconomic analysis methods. The present system consists of a Maisotsenko air cooler, a compressor, a turbine, a generator, a combustor, and a compressed air saturator. The results show that an exergy efficiency of 58.27 % is higher than the corresponding energy efficiency of 51.55 % for the MCTC system, due to the fact that the exergy content of the fuel fed into the combustion chamber is lower than its energy content. Also, the maximum exergy destruction rates occur in the compressor and turbine with the values of 166.964 kW and 150.864 kW, respectively. Furthermore, the exergoeconomic results indicate that the highest exergetic cost factor defined as the destruction in the component per cost is determined to be 0.013148 kW/$ for the turbine, while the Maisotsenko cycle air cooler has a minimum rate of 0.000006 kW/$. The better optimization of this component may be considered. It is concluded that Maisotsenko cycle systems can be effectively integrated to turbine cycle systems. Also, energy, exergy and exergoeconomic analyses give more useful information together about assessing the MCTC system and minimizing the thermodynamic inefficiencies.

Keywords Energy • Exergy • Exergoeconomics • Maisotsenko combustion turbine cycle • Maisotsenko compressor inlet cooler • Thermodynamics

H. Caliskan (✉)
Faculty of Engineering, Department of Mechanical Engineering,
Usak University, Usak 64200, Turkey
e-mail: hakan.caliskan@usak.edu.tr

I. Dincer
Faculty of Engineering and Applied Science, University of Ontario Institute of Technology,
2000 Simcoe Street North, Oshawa, ON, Canada L1H 7K4

A. Hepbasli
Faculty of Engineering, Department of Energy Systems Engineering,
Yaşar University, Izmir 35100, Turkey

I. Dincer et al. (eds.), *Progress in Clean Energy, Volume 1*,
DOI 10.1007/978-3-319-16709-1_3

3.1 Introduction

According to the International Energy Agency (IEA), energy demand will increase steadily from 2000 to 2030 with an increase of 1.7 % per year, and electricity demand will grow by 2.4 % per year. So advanced technologies for power generation are necessary [1]. In recent years, humid air turbine (HAT) cycle, in which the humid air is used to generate work and/or electricity, has been used as an advanced technology for power production. It is environmentally friendly and its working fluid is not harmful to the atmosphere [2]. But generally, the efficiency of a HAT is lower than other turbines, because it cannot work at high pressures, and as a result, it generates less power. In HATs, increasing humidity requires heating the compressed air at high temperatures. So more fuel consumption and a separate boiler are needed to do this operation. As a result, a high-cost additional component is necessary for a HAT [3].

Practical limitations have prevented the HAT cycle from being commercialized. So there is a new developed technology named the Maisotsenko cycle (M-cycle) that can be a solution for limitations of the HAT cycle [2]. M-Cycle systems are commonly known with air conditioning systems, and this cycle includes the wet and dry sides of a plate similar to indirect evaporative coolers, but with a much different airflow creating a new thermodynamic cycle resulting in product temperatures which approach the dew point temperature of the air [4]. The enthalpy difference of the air at the dew point temperature and the air saturated at a higher temperature to reject the heat from the product is used in this M-cycle. Also, it allows the product fluid to be cooled into the dew point temperature of the incoming air ideally. The air is then precooled before passing into the heat rejection stream where the water is evaporated [5–8].

The HAT cycle and the M-cycle-integrated Maisotsenko combustion turbine cycle (MCTC) have some similar specification/works. They may increase power output while maintaining the optimum system efficiency by supplying the moist air to the turbine without an additional compressor work. They can also recover the turbine exhaust heat by putting it back into the cycle and lower emissions by providing the combustor with moist air. But the MCTC has additional advantages by further increasing the moisture in the compressed air stream, which increases power and efficiency. In addition, the M-cycle uses souped-up heat transfer process and minimizes the size of the heat transfer surface, the pressure losses, and the capital cost. Also, the M-cycle air cooler is used to increase the compressor efficiency by supplying the compressor with the cool air that has not had humidity added to it. Thus, this cooler air can be compressed easily using less compressor work.

Thermodynamic systems can be understood better with advanced analysis methods. Energy analysis is based on the first law of thermodynamics, but there are some limitations, such as not taking into account the environment properties, the degradation of the energy quality, and the process irreversibility. On the other hand, exergy analysis is based on both of the first and second laws of thermodynamics, and it characterizes the work potential of a system. Exergy is generally defined as a maximum work that can be obtained from the system, when its state is brought to the dead state (also known as environmental condition) [9]. Also, there is exergoeconomic analysis methodology that is a combination of thermodynamics and cost accounting disciplines. Normally, energy-based unit costs are used for the cost accounting. But exergy cost accounting is essential in evaluating the system input and output [7].

In the open literature, there have been numerous studies on the M-cycle for cooling applications and two studies about the MCTC. In this regard, Wicker [2] theoretically explained the possible combination of the M-cycle with a turbine. In this regard, MCTC with a Maisotsenko compressor inlet cooler was designed and studied theoretically. Alsharif et al. [3] presented energy and exergy analyses of the M-cycle, and the theoretical information of Wicker [2] was used with additional assumptions for analyzing purposes.

In this study, energy, exergy and exergoeconomic analysis methods are applied to an empirically designed MCTC with a compressor inlet cooler. So this study differs from the previously conducted ones as follows: (1) the new data (assumptions) are considered for the MCTC system, (2) entropy generation rates are determined, (3) exergy destruction ratios of the system components are calculated, and (4) exergoeconomic analysis is applied to the system for evaluation.

3.2 System Description

The system (the MCTC with Maisotsenko compressor inlet cooler) consists of a Maisotsenko air cooler, a compressor, a turbine, a generator, a combustion chamber (combustor), and a Maisotsenko compressed air saturator. The schematic layout of the system is illustrated in Fig. 3.1.

The MCTC takes the dry, hot, and compressed air and cools it toward its dew point temperature without adding the humidity. Then the cool air extracts the heat from the exhaust gases of the turbine, bringing them to a lower temperature at which even more heat may be extracted. Two streams occur from the cooled air. One of these streams cools itself by being passed in the counterflow to the air being cooled and by adding the moisture to it through an indirect evaporative heat and mass transfer process. The other stream is then passed in the counterflow to the turbine exhaust gas stream while evaporating the water into the air again in an indirect evaporative heat and mass exchanger. The two high-humidity air streams are then recombined and sent to the combustor for heating before entering the turbine. The humid air, however, has a higher specific heat and takes more fuel to increase the temperature. However, the additional heat in the turbine exhaust gases allows additional water to be evaporated and more heat to be recovered [2].

The process can be explained as follows: the air enters the M-cycle air cooler at 25 °C temperature and 101.325 kPa pressure. Hence, it exchanges its heat with working air and the air is cooled down to a compressor inlet temperature of 8 °C, while the pressure is 101.325 kPa, the air is fully dry (without moisture) in this condition. In the compressor, the pressure and temperature of the air are increased

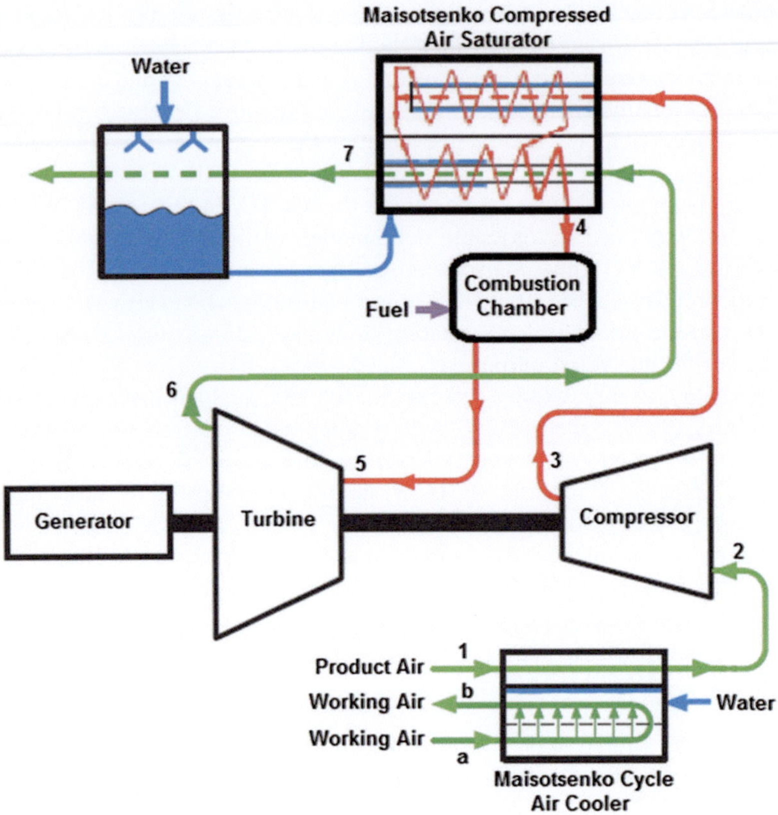

Fig. 3.1 Maisotsenko combustion turbine cycle with Maisotsenko compressor inlet cooler

to 1,200 kPa and 462.1 °C, respectively. The air then reaches the Maisotsenko compressed air saturator and exchanges its heat with the secondary air coming from the turbine and the water from the water reservoir. After the Maisotsenko compressed air saturator, the air enters the combustion chamber at 600 °C temperature and 1,200 kPa pressure. In the combustion chamber, the temperature of the air is raised to 1,127 °C using the energy of the fuel. Thus, the air enters the turbine at 1,127 °C temperature and 1,200 kPa pressure to obtain the power and leaves the turbine to enter the Maisotsenko compressed air saturator (as a secondary air) at 674.1 °C temperature and 101.325 kPa pressure. After the Maisotsenko compressed air saturator, the air exits at 487.61 °C temperature and 101.325 kPa pressure. Hence, this air heats the water in the reservoir.

It is assumed that the isentropic efficiency of the compressor is 70 %, the isentropic efficiency of the turbine is 71.1 %, the combustion gases are air, the combustion chamber temperature is equal to the turbine inlet temperature, and there is no energy loss in the combustion chamber. The assumed data of the system can be seen in Table 3.1.

Table 3.1 Data considered for the system

State	No	m (kg/s)	T (°C)	T (K)	P (kPa)
Dead state—air	0	–	25	298.15	101.325
Air	1	2	25	298.15	101.325
Air	2	2	8	281.15	101.325
Air	3	2	462.1	735.25	1,200
Air	4	2	650	923.15	1,200
Air	5	2	1,127	1,400.15	1,200
Air	6	2	674.1	947.25	101.325
Air	7	2	487.6	760.75	101.325
Air	a	3.01	3.5	276.65	101.325
Air	b	3	12	285.15	101.325
Dead state—water	0	–	25	298.15	101.325
Water	Water	0.01	25	298.15	101.325

3.3 Analyses

The general flow energy rate ($\dot{E}n_i$) can be written as follows:

$$\dot{E}n_i = \dot{m}_i \ h_i \tag{3.1}$$

where "$\dot{m}_i$" and "h_i"are the mass flow rate and the enthalpy of the related component in the given condition, respectively.

The general flow exergy rate ($\dot{E}x_i$) can be determined by [10]

$$\begin{aligned} \dot{E}x_i &= \dot{m}_i \ ex_i = \dot{m}_i[(h_i - h_0) - T_0(s_i - s_0)] \\ &= \dot{m}_i \left[c_p(T_i - T_0) - c_p T_0\left(\ln\frac{T_i}{T_0}\right) + T_0 R\left(\ln\frac{P_i}{P_0}\right)\right] \end{aligned} \tag{3.2}$$

where "ex_i" is the specific exergy, "h_i" is the enthalpy in the given condition, "h_0" is the enthalpy in dead-state condition, "s_i" is the entropy in the given condition, "s_0" is the entropy in dead-state condition, "T_0" is the dead-state (reference) temperature, "T_i" is the temperature in the given condition, "R" is the universal gas constant, "P_i" is the pressure in the given condition, "P_0" is the dead-state pressure, and "c_p" is the specific heat.

3.3.1 *Maisotsenko Cycle Air Cooler*

The mass balance of the M-cycle air cooler is written as [3]

$$\dot{m}_1 = \dot{m}_2 \tag{3.3}$$

$$\dot{m}_a = \dot{m}_b + \dot{m}_{\text{water}} \tag{3.4}$$

The energy balance of the M-cycle air cooler is given by

$$\dot{E}n_1 + \dot{E}n_a = \dot{E}n_2 + \dot{E}n_b + \dot{E}n_{\text{water}} + \dot{E}n_{\text{loss},M\text{-cycle}} \tag{3.5}$$

where "$\dot{E}n_{\text{loss},M\text{-cycle}}$" is the energy loss rate of the M-cycle air cooler.

The exergy balance of the M-cycle air cooler becomes

$$\dot{E}x_1 + \dot{E}x_a = \dot{E}x_2 + \dot{E}x_b + \dot{E}x_{\text{water}} + \dot{E}x_{\text{loss},M\text{-cycle}} + \dot{E}x_{\text{dest},M\text{-cycle}} \tag{3.6}$$

where "$\dot{E}x_{\text{loss},M\text{-cycle}}$" and "$\dot{E}x_{\text{dest},M\text{-cycle}}$" are the exergy loss and destruction rates of the M-cycle air cooler, respectively.

$$\dot{E}x_{\text{loss},M\text{-cycle}} = \dot{E}n_{\text{loss},M\text{-cycle}}\left[1 - \frac{T_0}{T_{M\text{-cycle}}}\right] \tag{3.7}$$

where "T_0" is the dead-state temperature and "$T_{M\text{-cycle}}$" is the M-cycle operation temperature (25 °C).

3.3.2 *Compressor*

The mass balance of the compressor is given by

$$\dot{m}_2 = \dot{m}_3 \tag{3.8}$$

The energy balance of the compressor is written as follows:

$$\dot{E}n_2 + \dot{W}_{\text{comp}} = \dot{E}n_3 \tag{3.9}$$

where "$\dot{W}_{\text{comp}}$" is the necessary electrical energy rate of the compressor.

$$\frac{P_{r,3}}{P_{r,2}} = \frac{P_3}{P_2} \tag{3.10}$$

where "P_r" is the relative pressure and "P" is the pressure.

$$\eta_{\text{comp}} = \frac{h_{3s} - h_2}{h_{3a} - h_2} \tag{3.11}$$

where "η_{comp}" is the isentropic efficiency of the compressor. Also, "h_{3a}" is the actual enthalpy and "h_{3s}" is the isentropic enthalpy of the air at the compressor outlet.

The exergy balance of the compressor is given as

$$\dot{E}x_2 + \dot{W}_{comp} = \dot{E}x_3 + \dot{E}x_{dest,comp} \tag{3.12}$$

where "$\dot{E}x_{dest,comp}$" is the exergy destruction rate of the compressor.

3.3.3 *Combustion Chamber (Combustor)*

The mass balance of the combustion chamber (air) is expressed as

$$\dot{m}_4 = \dot{m}_5 \tag{3.13}$$

The energy balance of the combustion chamber is given by

$$\dot{E}n_4 + \dot{E}n_{comb} = \dot{E}n_5 \tag{3.14}$$

where "$\dot{E}n_{comb}$" is the fuel energy input rate of the combustion chamber.
The exergy balance of the combustion chamber is written as follows:

$$\dot{E}x_4 + \dot{E}x_{comb} = \dot{E}x_5 + \dot{E}x_{dest,comb} \tag{3.15}$$

where "$\dot{E}x_{comb}$" and "$\dot{E}x_{dest,comb}$" are the fuel exergy input rate and the exergy destruction rate of the combustion chamber.

$$\dot{E}x_{comb} = \dot{E}n_{comb}\left[1 - \frac{T_0}{T_{comb}}\right] \tag{3.16}$$

where "T_{comb}" is the combustion room temperature (as 1,127 °C).

3.3.4 *Turbine*

The mass balance of the turbine is as follows:

$$\dot{m}_5 = \dot{m}_6 \tag{3.17}$$

The energy balance of the turbine is given by

$$\dot{E}n_5 = \dot{E}n_6 + \dot{W}_{turb} \tag{3.18}$$

where "$\dot{W}_{turb}$" is the electrical energy output rate of the turbine.

$$\frac{P_{r,6}}{P_{r,5}} = \frac{P_6}{P_5} \tag{3.19}$$

where "P_r" is the relative pressure and "P" is the pressure.

$$\eta_{turb} = \frac{h_5 - h_{6a}}{h_5 - h_{6s}} \tag{3.20}$$

where "η_{turb}" is the isentropic efficiency of the turbine. Also, "h_{6a}" is the actual enthalpy and "h_{6s}" is the isentropic enthalpy of the air at the turbine outlet.

The exergy balance of the turbine becomes

$$\dot{E}x_5 = \dot{E}x_6 + \dot{W}_{turb} + \dot{E}x_{dest,\,turb} \tag{3.21}$$

where "$\dot{E}x_{dest,turb}$" is the exergy destruction rate of the turbine.

3.3.5 Maisotsenko Compressed Air Saturator

The mass balance of the Maisotsenko compressed air saturator is expressed as

$$\dot{m}_3 = \dot{m}_4 \tag{3.22}$$

$$\dot{m}_6 = \dot{m}_7 \tag{3.23}$$

The energy balance of the Maisotsenko compressed air saturator is given by

$$\dot{E}n_3 + \dot{E}n_6 + \dot{E}n_{water} = \dot{E}n_4 + \dot{E}n_7 + \dot{E}n_{loss,\,sat} \tag{3.24}$$

where "$\dot{E}n_{loss,sat}$" is the energy loss rate of the Maisotsenko compressed air saturator.

The exergy balance of the Maisotsenko compressed air saturator is written as follows:

$$\dot{E}x_3 + \dot{E}x_6 + \dot{E}x_{water} = \dot{E}x_4 + \dot{E}x_7 + \dot{E}x_{loss,\,sat} + \dot{E}x_{dest,\,sat} \tag{3.25}$$

where "$\dot{E}x_{loss,sat}$" and "$\dot{E}x_{dest,sat}$" are the exergy loss and destruction rates of the Maisotsenko compressed air saturator.

$$\dot{E}x_{loss,\,sat} = \dot{E}n_{loss,\,sat}\left[1 - \frac{T_0}{T_{sat}}\right] \tag{3.26}$$

where "T_{sat}" is the Maisotsenko compressed air saturator temperature.

3.3.6 Overall Efficiency

The energy efficiency of the overall system (η) is defined as

$$\eta = \frac{\dot{W}_{\text{turb}}}{\dot{W}_{\text{comp}} + \dot{E}n_{\text{comb}}} \tag{3.27}$$

The exergy efficiency of the overall system (Ψ) is calculated from the following equation:

$$\Psi = \frac{\dot{W}_{\text{turb}}}{\dot{W}_{\text{comp}} + \dot{E}x_{\text{comb}}} \tag{3.28}$$

3.3.7 Entropy Generation

The entropy generation rate ($\dot{S}_i$) is the ratio of the exergy destruction rate to the reference state temperature:

$$\dot{S}_i = \frac{\dot{E}x_{\text{dest},\,i}}{T_0} \tag{3.29}$$

3.3.8 Exergy Destruction Ratio

The exergy destruction ratio of the system component ($\dot{E}x_{\text{dest},R,i}$) is found to be

$$\dot{E}x_{\text{dest},R,i} = \frac{\dot{E}x_{\text{dest},\,i}}{\dot{E}x_{\text{dest, tot}}} \tag{3.30}$$

where "$\dot{E}x_{\text{dest},i}$" is the exergy destruction rate of the related component of the system and "$\dot{E}x_{\text{dest,tot}}$" is the total exergy destruction of the system.

3.3.9 Exergoeconomic Analysis

The exergoeconomic analysis is applied by exergetic cost factor (R_i) as follows:

$$R_i = \frac{\dot{E}x_{\text{dest},\,i}}{Z_i} \tag{3.31}$$

where "Z_i" is the capital cost of the component.

The capital cost of the compressor (Z_{comp}) is defined as [11, 12]:

$$Z_{\text{comp}} = \frac{(71.1)\,m_2\left(\frac{P_3}{P_2}\right)\ln\left(\frac{P_3}{P_2}\right)}{(0.9) - \eta_{\text{comp}}} \quad (3.32)$$

where "η_{comp}" is isentropic efficiency of the compressor (70 %).

The capital cost of the combustion chamber (Z_{comb}) is obtained from

$$Z_{\text{comb}} = \frac{(46.08)\,m_4\,[1 + \exp((0.018)T_4 - 26.4)]}{\left|(0.995) - \left(\frac{P_5}{P_4}\right)\right|} \quad (3.33)$$

The capital cost of the turbine (Z_{turb}) is determined using the following equation:

$$Z_{\text{turb}} = \frac{(479.34)\,m_5\ln\left(\frac{P_5}{P_6}\right)[1 + \exp((0.036)T_5 - 54.4)]}{(0.92) - \eta_{\text{turb}}} \quad (3.34)$$

where "η_{turb}" is isentropic efficiency of the turbine (taken as 71.1 %).

3.4 Results and Discussion

The energy, exergy, and exergoeconomic analyses are applied to the MCTC with a Maisotsenko compressor inlet cooler system. The energy and exergy analysis results of each point in the system are tabulated in Table 3.2. The maximum energetic and exergetic rates are found before the turbine as 3,030 kW and 1,862.14 kW, respectively. Because the temperature of the gas is maximum after the combustion chamber, and the difference between this temperature and the reference environment temperature is maximum. On the other hand, the minimum energetic and exergetic rates of the system are determined for the water works in the M-cycle systems to be 1.048 kW and 0 kW, respectively. The temperature of the

Table 3.2 Some results of energy and exergy analyses

No	Energy rates (kW)	Exergy rates (kW)
1	597.2	0
2	563	0.98
3	1,502.78	773.79
4	1,918	1,039.35
5	3,030	1,862.14
6	1,972.33	653.60
7	1,558.15	384.33
a	833.47	1.99
b	856.50	0.95
Water	1.048	0

water is assumed to be equal to the reference environment. So the exergetic rate is zero and the energetic rate is minimum among the results. Also, the energy and exergy efficiencies of the overall system are shown in Fig. 3.2. As seen, the energy and exergy efficiencies are 51.55 % and 58.27 %, respectively. The exergetic efficiency is found higher than the corresponding energetic efficiency. Because the exergy rate of the fuel given to the combustion chamber (875.21 kW) is lower than its corresponding energy rate (1,112 kW), this is inversely proportional to the efficiency. In addition, the consumed energy rate for the compressor is 939.78 kW, while the produced energy rate by the turbine is 1,057.67 kW.

The exergy destruction and entropy generation results are given in Table 3.3. As seen from the table, the maximum exergy destruction and entropy generation rates are found to be 166.964 kW and 0.56 kW/K for the compressor, while the minimum corresponding rates are determined as 0.058 kW and 0.000196 kW/K for the Maisotsenko cycle air cooler used before the compressor, respectively. The compressor and turbine are the most enforced components in the system. So it is predictable that the most exergy destruction occurs in these components. In addition, the entropy generation is directly proportional to the exergy destruction, and the results verify

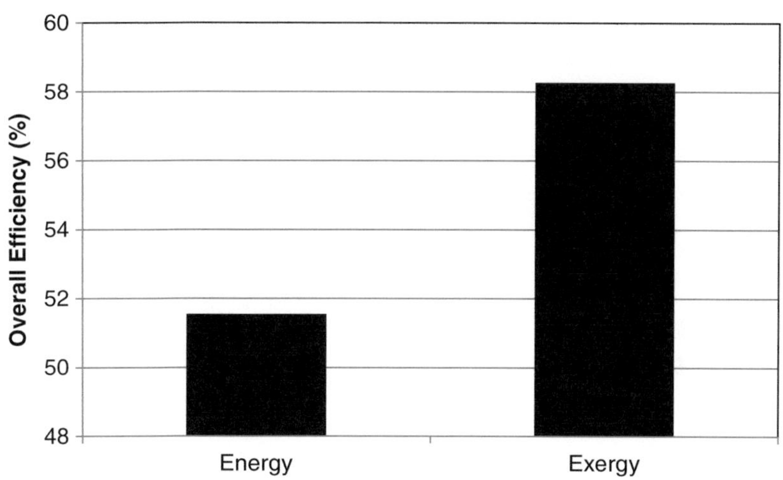

Fig. 3.2 Energy and exergy efficiencies of the overall system

Table 3.3 Exergy destruction and entropy generation rate results

Component	Exergy destruction rate (kW)	Entropy generation rate (kW/K)
Maisotsenko cycle air cooler	0.058	0.000196
Compressor	166.964	0.560
Combustion chamber	52.415	0.176
Turbine	150.864	0.506
Maisotsenko compressed air saturator	3.722	0.0125

it. After the compressor, the second highest exergy destruction and entropy generation rates are obtained for the turbine as 150.864 kW and 0.506 kW/K, respectively.

On the other hand, the exergy destruction ratios of the system components are shown in Fig. 3.3. The exergy destruction ratio of the Maisotsenko cycle air cooler, the compressor, the combustion chamber, the turbine, and the Maisotsenko compressed air saturator are found to be 0.00016, 0.44640, 0.14014, 0.40335, and 0.00995, respectively. Among the system components, the compressor has the maximum ratio, while the Maisotsenko cycle air cooler has the minimum ratio. The reason for this is the magnitude of the exergy destruction rate explained above.

The capital costs of the system components are calculated and given in Fig. 3.4. The capital costs of the Maisotsenko cycle air cooler, the compressor, the

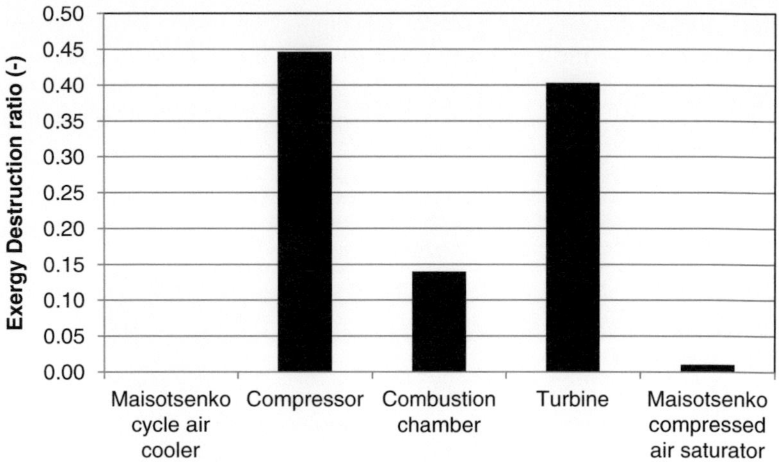

Fig. 3.3 Exergy destruction ratios of the system components

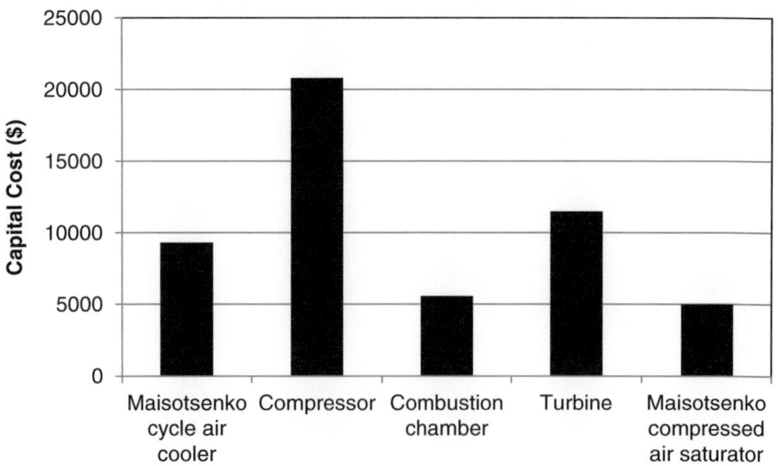

Fig. 3.4 Capital costs of the system components

combustion chamber, the turbine, and the Maisotsenko compressed air saturator are determined as $9,316, $20,813.14, $5,566.62, $11,474.37, and $5,000, respectively. As seen, the compressor and the turbine, which are the most important parts, are the most costly components in the system.

Finally, the exergoeconomic analysis results are illustrated in Table 3.4. The exergoeconomic analysis is connected with the cost of the components and their exergetic results. So the assessment is done for defining all of the necessary parameters correctly. The highest exergetic cost factor is found to be 0.013148 kW/$ for the turbine, while the Maisotsenko cycle air cooler has a minimum rate of 0.000006 kW/$. The variation of the exergetic cost factor by components can be seen in Fig. 3.5.

Table 3.4 Exergoeconomic analysis results

Component	Z_i ($)	R_i (kW/$)
Maisotsenko cycle air cooler	9,316	0.000006
Compressor	20,813.14	0.008022
Combustion chamber	5,566.621	0.009416
Turbine	11,474.37	0.013148
Maisotsenko compressed air saturator	5,000	0.000744

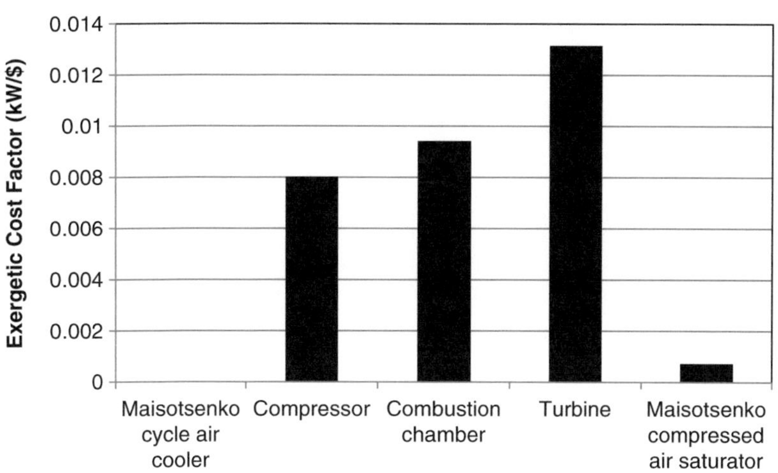

Fig. 3.5 Exergoeconomic results

3.5 Conclusions

In the present study, the MCTC with a compressor inlet cooler system is analyzed through the energy, exergy, and exergoeconomic analyses. According to the results, the following conclusions can be summarized:

(a) The exergy efficiency (58.27 %) is higher than the corresponding energy efficiency (51.55 %) for the MCTC system because the exergy of the fuel used in the combustion chamber is lower than its energetic rate. So when the MCTC systems are analyzed using exergetic approach, their efficiencies can be more reliable due to taking into account the reference environment conditions.
(b) The maximum exergy destruction rates occur in the compressor (166.964 kW) and the turbine (150.864 kW). The compressor supplies the necessary working fluid and increases its temperature and pressure. On the other hand, the turbine produces power. So these two components are the main parts and the most imposed ones. Parallel to this situation, their exergy destruction ratios are the highest ones in the system.
(c) The turbine has the highest exergetic cost factor (0.013148 kW/$) that explains the destruction in the component per cost. So the better optimization of this equipment may be considered.

As a final remark, it can be remarked that Maisotsenko cycle systems can be effectively integrated to turbine cycle systems. Also, energy, exergy and exergoeconomic analyses give useful information about assessing the MCTC system and minimizing the thermodynamics inefficiencies.

References

1. Mohapatra AK (2014) Sanjay, “Analysis of parameters affecting the performance of gas turbines and combined cycle plants with vapor absorption inlet air cooling”. Int J Energy Res 38:223–240
2. Wicker K (2003) Life below the wet bulb: the Maisotsenko cycle. Turbine Technol November/December (Power): 1–7, www.powermag.platts.com
3. Alsharif A, Gadalla M, Dincer I (2011) Energy and exergy analyses of Maisotsenko cycle. ASME 2011 5th international conference on energy sustainability & 9th fuel cell science, August 7–10, 2011, Washington, DC, ES2011-54753, pp. 1–7
4. Gillan L (2008) Maisotsenko cycle for cooling processes. Int J Energy Clean Environ 9 (1–3):47–64
5. Caliskan H, Dincer I, Hepbasli A (2011) Exergetic and sustainability performance comparison of novel and conventional air cooling systems for building applications. Energy Buildings 43:1461–1472
6. Caliskan H, Hepbasli A, Dincer I, Maisotsenko V (2011) Thermodynamic performance assessment of a novel air cooling cycle: Maisotsenko cycle. Int J Refrigeration 34(4):980–990
7. Caliskan H, Dincer I, Hepbasli A (2012) Exergoeconomic, enviroeconomic and sustainability analyses of a novel air cooler. Energy Buildings 55:747–756

8. Caliskan H, Dincer I, Hepbasli A (2012) A comparative study on energetic, exergetic and environmental performance assessments of novel M-Cycle based air coolers for buildings. Energy Conversion Manage 56:69–79
9. Regulagadda P, Dincer I, Naterer GF (2010) Exergy analysis of a thermal power plant with measured boiler and turbine losses. Appl Thermal Eng 30:970–976
10. Ghaebi H, Amidpour M, Karimkashi S, Rezayan O (2011) Energy, exergy and thermoeconomic analysis of a combined cooling, heating and power (CCHP) system with gas turbine prime mover. Int J Energy Res 35:697–709
11. Memon AG, Harijan K, Uqaili MA, Memon RA (2013) Thermo-environmental and economic analysis of simple and regenerative gas turbine cycles with regression modeling and optimization. Energy Conversion Manage 76:852–864
12. Shirazi A, Najafi B, Aminyavari M, Rinaldi F, Taylor RA (2014) Thermal-economic-environmental analysis and multi-objective optimization of an ice thermal energy storage system for gas turbine cycle inlet air cooling. Energy 69:212–226

Chapter 4
Modeling of Fluidized Bed Gasification of Rice Straw in Egypt

Rami Salah El-Emam and Ibrahim Dincer

Abstract Useful energy can be extracted from agricultural waste and residues instead of the uncontrolled burning that causes severe harm to the environment. Rice straw can be combusted in a fluidized bed producing useful heat with better control of emissions. One of the most promising alternatives is converting biomass into a more energy-dense fuel with a higher heating value through gasification. This study presents a one-dimensional model of atmospheric bubbling fluidized bed and investigates the effect of different operating parameters on the temperature profile and gas species concentrations through the fluidized bed in the gasification process through concentration and energy balance equations. The model is validated with data and measurements from the literature. The temperature profiles show a peak temperature value occurring early in the freeboard zone. Both the value and position of the peak occurrence are affected by the operation and fluidization condition.

Keywords Biomass • Gasification • Fluidized bed

4.1 Introduction

Since 1999, The Nile Delta and Cairo in Egypt have been suffering from thick layers of smog, forming black clouds, caused by burning of rice straw after rice harvest season and persist for days or weeks. In Egypt, rice is second in priority cereal crop after wheat and occupies almost one quarter of the cultivated area during summer, engaging about 10 % of Egyptian population in rice farming [1]. The black clouds appear once a year in between September and October, bringing the pollution levels

R.S. El-Emam (✉)
Clean Energy Research Laboratory (CERL), University of Ontario Institute of Technology, 2000 Simcoe Street North, Oshawa, ON, Canada L1H 7K4

Faculty of Engineering, Mansoura University, Mansoura, Egypt
e-mail: rami.elemam@uoit.ca

I. Dincer
Clean Energy Research Laboratory (CERL), University of Ontario Institute of Technology, 2000 Simcoe Street North, Oshawa, ON, Canada L1H 7K4
e-mail: Ibrahim.Dincer@uoit.ca

I. Dincer et al. (eds.), *Progress in Clean Energy, Volume 1*,
DOI 10.1007/978-3-319-16709-1_4

Fig. 4.1 Burning of rice straw in Egypt [1]

up to 10 times the limits set by the World Health Organization [2]. It is reported by Marey [3] that the black clouds lurk low to the ground, concentrating in the lower 500 m of the atmosphere. This gives the chance for the pollution to be trapped by the temperature inversion at the cities' atmosphere. Annually, more than half of the 3–5 million tons of the produced rice straw residues tend to be burned by Egyptian farmers; see Fig. 4.1. This gives them a clear cultivated land within 1 or 2 weeks after harvesting for sowing the following crop.

Considering the combustion or gasification of rice straw for energy useful production is one of the most promising alternatives that are to be integrated to solve the black cloud problem. It also gives an energy solution for the depleted fossil fuels. Rice straw needs pretreatment process to be ready for energy conversion process. Pelletizing in compact pellets is considered the most appropriate form of utilizing rice straw as a fuel. The produced pellets have higher energy density and results in a reduction in transportation and storage cost. It can also be mixed with other fuels to improve the pellet characteristics.

Gasification of biomass works on producing a syngas with higher fuel heating value and better hydrogen-to-carbon ratio. This results in less greenhouse gas emissions when the fuel is used. Also, removal of sulfur is a great benefit of this process where limestone is to be added to capture it, preventing the formation of sulfur oxides. Nitrogen oxides are minimized as the temperature is relatively low compared with other combustion processes. Fluidized bed gasifiers are widely used for converting biomass into energy which also has a great potential to be integrated with other energy conversion systems for more useful products and enhancing the overall performance of the system [4]. The interest in this study is to investigate the gasification of rice straw in fluidized bed. The temperature profile and gas concentrations through the gasifier are investigated.

Table 4.1 Ultimate analysis of the biomass pellets considered in the current study

	Rice straw (%)	Bitumen (%)	Straw-bitumen (%) (2:1)
Carbon	38.2	85.5	60
Hydrogen	5.5	9.4	7.3
Oxygen	41.1	0.1	22.5
Nitrogen	0.5	0.6	0.5
Sulfur	0.2	3.5	1.7
Ash	14	0.9	7.9
LHV (kJ/kg)	14,100	40,600	25,800

Fig. 4.2 Rice straw pellets and straw-bitumen pellets [5, 6]

The fuel pellets considered in this study are prepared of rice straw and bitumen with two-to-one ratio. The analysis of rice straw, bitumen, and the produced pellets are shown in Table 4.1. Shaaban [5] illustrated the formation of straw-bitumen pellets process. Rice straw is to be chopped in a couple of centimeter parts and mixed with bitumen at 120 °C. The mixture is then pressed under about 150 bar after being cooled to ambient temperature. Figure 4.2 shows rice straw pellets and rice straw-bitumen pellets made in cylinders of 12 mm height and 10 mm diameter.

4.2 Modeling and Analyses

In this section, one-dimensional model of a fluidized bed is introduced and formulated. The model is designed to analyze the temperature and gas concentration through the fluidized bed (FB) at different operating conditions. The model considers combustion (FBC) and gasification (FBG) of biomass fuels. In the presented work, fluidized bed gasification is considered. The fluidized bed gasifier is modeled in three different zones as stated by Basu [7]: dense bed, splashing zone, and the freeboard zone. The species concentrations and the axial temperature distribution through the bed height are investigated by applying the species concentrations and energy balance equations over the three different zones of the fluidized bed. The model considers the bubble and emulsion phases, which is known as the two-phase theory of fluidization.

For the dense bed zone, where volatilization and drying occur, bubbling fluidized-type zone is considered. The axial variation of the species concentration through the bed height can be described by the following differential equations for the bubble and emulsion phases, respectively [8–11]:

$$u_b d\bar{x}_{b,i} - \left[K_{be}\left(\bar{x}_{e,i} - \bar{x}_{b,i}\right) + \sum R_{b,i} + \frac{1}{A_{bed}\,.\,\delta}\left(\frac{dC_i}{dz}\right)\right]dz = 0$$

$$u_e d\,\bar{x}_{e,i} - \left[K_{be}.\,\delta.\,(\bar{x}_{b,i} - \bar{x}_{e,i}) + \varepsilon_{mf}(1-\delta).\sum R_{e,i}\right]dz = 0$$

where dx/dz expresses the rate of change in the concentration of the different species through the bed height, dC/dz is the local release rate of the species in kmol/ms, R is the reaction rate, and $\bar{x}$ is the species concentrations expressed in kmol/m^3. The other parameters and coefficients in the species mass balance equations are illustrated in the hydrodynamics equations as described in Table 4.2. The main parameters in this table are shown as a function of the bed diameter, D, the fluidization velocity, U_f, and the minimum fluidization velocity, U_{mf}.

The splashing zone is located over the dense bed, and better mixing occurs as a result of the bubble bursting [16]. For the splashing zone, the species balance equation can be written as follows where the reaction rates are considered as for

Table 4.2 Hydrodynamics equations considered for the bubbling fluidized bed model

Name	Formula	Ref.
Bubble diameter	$d_B = d_l - (d_l - d_o)\exp\left(\frac{-0.3\ z}{D}\right)$	[12]
Limiting bubble diameter	$d_l = 0.24\left[(U_f - U_{mf}).D^2\right]^{0.4}$	[12]
Initial bubble diameter, m	$d_o = 0.794\left[\frac{(U_f - U_{mf})\cdot D^2}{N}\right]^{0.4}$	[12]
Bubble velocity, m/s	$U_B = U_f - U_{mf} + 0.227\sqrt{d_B}$	[13]
Bed fraction in bubbles	$\delta_B = \frac{U_f - U_{mf}}{U_B} = 1 - \frac{U_{mf}}{U_e}$	[13]
Superficial gas velocity through emulsion, m/s	$U_e = (U_f + 2U_{mf})/3$	[11]
Bubble-cloud transfer coefficient, 1/s	$K_{B-C} = \left(\frac{4.5\ U_{mf}}{d_B}\right) + \left(\frac{10.35\ \sqrt{D}}{{d_B}^{1.25}}\right)$	[10]
Cloud-emulsion transfer coefficient, 1/s	$K_{C-E} = 3.226\sqrt{\frac{\mathrm{Dif.}\ \varepsilon_{mf}}{{d_B}^{2.5}}}$	[10]
Bubble-emulsion transfer Coefficient, 1/s	$K_{B-E} = \frac{K_{C-E}.K_{B-C}}{K_{C-E} + K_{B-C}}$	[10]
Void fraction at minimum fluidization	$\varepsilon_{mf} = 0.478/A^{0.018}$	[14]
Archimedes number	$A = \frac{g.\rho_g(\rho_p - \rho_g)d_B{}^3}{\mu_g{}^2}$	[15]

the emulsion phase without considering the items related to the coarse char based on the assumption that this zone is coarse char free [8]:

$$u_{sp,g} d\overline{x}_{sp,i} = \left(\sum R_{sp,i}\right) dz$$

The following equation is used to determine the temperature profile. This is applied to the splashing zone where the dense bed zone is considered to have a constant temperature equal to the bed temperature:

$$\frac{dT_{g,sp}}{dz} = \frac{\left[A_{sp}\left(\sum R_{sp,i}\, HR_i\right) - Q_s - Q_w\right]}{N_g.c_{p,g}}$$

where HR is the heat of reactions that occur through the gasifier. The other terms in the previous equation are illustrated as follows: Q_s is the energy exchange with the ejected sand.

$$Q_s = M_s\, .A_{sp}.c_{p,s}\left(\frac{dT_s}{dz}\right)$$

and Q_w expresses the loss through the walls of combustion area as

$$Q_w = L_{sp}\, .h_{ov}\left(T_{sp} - T_{am}\right)$$

where the peripheral of the splashing zone cross section, L_{sp}, and the overall heat transfer coefficient, h_{ov}, are given in Table 4.3. Both Q_s and Q_w are calculated per unit length of the gasifier.

The sand temperature gradient in the calculation of the energy exchange with the ejected sand can be calculated as follows:

$$\frac{dT_s}{dz} = 6\, .h_{s,sp}.\frac{T_{g,sp} - T_s}{d_s\, .\, \rho_s.\, c_{p,s}}$$

where the sand specification and properties are shown in Table 4.3.

The following exponentially decay function expresses the mass flux of ejected sand particles:

$$M_s = M_{s,o}\exp\left[-C_{ej}.(z - z_{eb})\right]$$

where M_s is in kg/m^{2}s and is the function of the initial mass flux at the bed surface which is calculated as follows [11]:

$$M_{s,o} = \beta\rho_s(U_f - U_{mf})(1 - \varepsilon_{mf})$$

Table 4.3 Parameters and constants for the formulas at the basic case

Parameter	Value
Bed height above distributer, m	$z = 3$
Bed diameter, m	$D = 0.3$
Number of bed orifices	$N_{or} = 800$
Extended bed height, m	$z_{eb} = 0.6$
Specific heat of sand, kJ/kg·K	$c_{p,s} = 0.83$ [7]
Diameter of sand particles, mm	$d_s = 0.45$ [17]
Density of sand particles, kg/m^3	$\rho_s = 2,650$ [17]
Average diameter of char particle, μm	$d_{ch} = 500$
Density of char, kg/m^3	$\rho_{ch} = 470$
Peripheral of splashing zone cross section, m	$L_{sp} = 0.3$
Parameter in sand initial mass flux	$\beta = 0.1$ [17]
Parameter in sand mass flux	$C_{ej} = 18$ [8]
Diameter of fine char particles, μm	$d_{ch,fine} = 100$ [18]
Minimum fluidization velocity, m/s	$U_{mf} = 0.175$ [17]
Air equivalence ratio coefficient	0.20–0.30
Steam to biomass coefficient	0.7
Gas diffusivity, m^2/s	$\mathrm{Dif}_g = 1.88 \times 10^{-4}$ [7]
Splitting factor for oxygen reacting to CO	$\varepsilon_{CO} = 0.3$ [19]
Splitting factor for C reacting to tar	$\varepsilon_{tar} = 5 \times 10^{-3}$ [19]
Partial combustion splitting coefficient	$\alpha_1 = 1.3$ [19]
Heterogeneous water-gas shift splitting coefficient	$\alpha_2 = 1.2$ [19]

The upper limit of the splashing zone is the bed identified by setting the decay of flux of the ejected sand particles to 0.001 [8]. The mass and energy equations of the splashing zone are also applied to the freeboard zone.

The mass and energy balance equations are applied on the proposed gasification system over the three described zones considering the composition of the gasification product to be as follows: CO, CO_2, H_2, CH_4, H_2O, and O_2.

4.2.1 Devolatilization

During the devolatilization process, biomass fuel particles tend to concentrate near the bed surface [25]. Volatiles can be assumed to be released in the bubble phase as described by Okasha et al. [8]. The volatiles are considered as CH_y [20] and can be calculated as tabulated in Table 4.4. The volatiles release distribution as a function of bed height can be given in kmol/ms as

$$\frac{dC_i}{dz} = 4\,\overline{C}_i \frac{z^3}{{z_{eb}}^4}$$

Table 4.4 Devolatilization and char consumption equations

Name	Formula	Ref.
Devolatilization		
Molar releasing rate, kmol/s	$\overline{C}_{O_2} = \frac{\dot{m}_f \cdot X_O}{MW_{O_2}}$	
Released hydrocarbon mole fraction	$y = \frac{X_H MW_C}{(X_C - X_{FC}) MW_H}$	[20]
Rate of hydrocarbon primary reaction to CO (step 1), kmol/m^3·s	$R_{CH_y} = -k_{CH_y} \left(\overline{x}_{CH_y}\right)^{0.7} \left(\overline{x}_{O_2}\right)^{0.8}$	[20]
	where $k_{CH_y} = 4.74 \times 10^{11} \exp\left(\frac{-2415}{T}\right)$	
Reaction rate of CO (step 2), kmol/m^3·s	$R_{CO} = -k_{CO}\, \overline{x}_{CO} \sqrt{\overline{x}_{O_2} \overline{x}_{H_2O}}$	[21]
	where $k_{CO} = 1.3 \times 10^{11} \vartheta \exp\left(\frac{-15,088}{T}\right)$	
Char combustion		
Combustion rate of char particles, kg/s	$R_{ch} = -72 m_{ch} \,.\, \overline{x}_{O_2} / \left[\rho_{ch} . d_{ch} \cdot \left(\frac{1}{K_{ch}} + \frac{1}{K_m}\right)\right]$	[22]
Char kinetic coefficient, m/s	$K_{ch} = 1.045\, T_{ch} \exp\left(\frac{-70,400}{R_\mu T_{ch}}\right)$	
Mass transfer coefficient, m/s	$K_m = 0.03$	
Char uniform temperature, K	$T_{ch} = 0.84\, T_{bed} A^{0.05}$	[22]
Carbon fines generation rate	$R_{ch,fine} = \alpha_{fine} R_{ch}$	[23]
	$\alpha_{fine} = \left(\frac{1 - V_{ash} - \theta_{cr}}{\theta_{cr} - \theta_0}\right)$	[24]

where z_{eb} is the extended bed height and $\overline{C}_i$ is the molar releasing rate of moisture, oxygen, and volatile, which is expressed as follows:

$$\overline{C}_i = \frac{\dot{m}_f \cdot X_i}{MW_i}$$

Here, the volatile combustion is considered to occur in two steps, where it reacts to form carbon monoxide and water vapor and then carbon monoxide oxidize to carbon dioxide in the second step [20, 21]. The rate of hydrocarbon conversion and the reaction rate of carbon monoxide for these two steps are shown in Table 4.4.

The volatiles can be given according a stoichiometric reaction of a biomass fuel of the formula $C_{\alpha_c} H_{\alpha_H} O_{\alpha_O} S_{\alpha_S} N_{\alpha_N}$ as follows:

$$\begin{aligned} \text{Biomass} \rightarrow & A_1 \cdot CO + A_2 \cdot CO_2 + A_3 \cdot CH_4 + A_4 \cdot H_2 + A_5 \cdot H_2S + A_6 \cdot O_2 \\ & + A_7 \cdot N_2 + A_8 \cdot \text{tar} \end{aligned}$$

where the mole faction can be written as follows:

$$\begin{aligned} A_1 &= \varepsilon_{CO} . \alpha_O \\ A_2 &= \alpha_O (1 - \varepsilon_{CO})/2 \\ A_3 &= \alpha_c (1 - A_2 - \varepsilon_{CO} + \varepsilon_{tar}) \end{aligned}$$

$$A_4 = \frac{\alpha_H}{2} - \alpha_S - 2A_3$$

$$A_5 = \alpha_H$$

$$A_6 = \alpha_O(1 - A_2 - \varepsilon_{CO})$$

$$A_7 = \varepsilon_{tar}\alpha_S$$

These equations are built considering some assumptions and considerations. It is assumed that partial combustion producing CO or complete combustion to CO_2 may occur. This is controlled by the splitting factor ε_{CO}. The rest of the carbon volatiles are released in the form of methane and additional tar which are controlled by the splitting factor ε_{tar} which is experimentally measured. All sulfur and nitrogen will be released as H_2S and N_2 considering that no formation of sulfur dioxide or ammonia in the products.

4.2.2 Gasification and Char Consumption

After devolatilization, char gets involved into different heterogeneous gasification reactions. The char consumption occurs in the gasifier and is happening through the reaction with oxygen, water vapor, or carbon dioxide. The shrinking particle model is used to describe the char oxidation in this model. This is because the ash layer gets stripped off from the particles as a result of the mechanical stress when colliding with the sand particles [9]. The oxidation of char particles is described by the gas-solid reaction introduced by Leckner et al. [22]. They defined the mechanical factor, α_1, which has a minimum value of one when a complete oxidation occurs and a maximum value of two when the reaction results in a complete production of carbon monoxide. The char consumption while reacting with water vapor is known as the heterogeneous water-gas shift reaction which can be shown, introducing the splitting factor α_2 given between 1 and 2, in Table 4.5. At the same time, with the three char consumption reactions, the oxidation of methane, carbon monoxide, and hydrogen occur. The water-gas shift reaction occurs when the available oxygen is consumed where carbon dioxide and hydrogen are produced through an equilibrium reaction of carbon monoxide and water vapor, depending on the reaction temperature. Methane steam reforming and dry reforming reactions are also considered. Table 4.5 shows the gasification reactions considered in the present model.

Table 4.4 lists the combustion rates of char and other char consumption parameters. It also shows the empirical formula used to calculate the uniform temperature of char. During the char burning process, it yields fines which are considered as a function of the char combustion rate as shown in Table 4.4. The parameter α_{fine} depends on fuel particle properties. The given relationship to calculate α_{fine} is the function of the volume fraction of ash in the unconverted char particle, V_{ash}, particles' initial porosity, θ_0, and porosity at the percolation threshold, θ_{cr}.

Table 4.5 Gasification reactions in the model

Reaction	Reaction rate [Ref.]
$\alpha_1 C + O_2 \rightarrow 2(\alpha_1 - 1)CO + (2 - \alpha_1)CO_2$ (combustion)	[27]
$C + \alpha_2 H_2O \rightarrow (2 - \alpha_2)CO + (\alpha_2 - 1)CO_2 + \alpha_2 H_2$ (steam gasification)	[19]
$C + CO_2 \rightarrow 2CO$ (Boudouard reaction)	[10]
$CO + 1/2\ O_2 \leftrightarrow CO_2$ (carbon monoxide oxidation)	[28]
$H_2 + 1/2\ O_2 \leftrightarrow H_2O$ (hydrogen oxidation)	[10]
$CH_4 + 1/2\ O_2 \leftrightarrow CO + 2H_2$ (methane oxidation)	[19]
$CO + H_2O \leftrightarrow CO_2 + H_2$ (water-gas shift reaction)	[10]
$CH_4 + H_2O \leftrightarrow CO + 3H_2$ (methane steam reforming)	[29]
$CH_4 + CO_2 \leftrightarrow 2CO + 2H_2$ (CO_2 reforming)	[19]

4.3 Results and Discussion

The presented model facilitates the investigation of axial temperature and gas species concentrations distribution through an atmospheric bubbling fluidized bed gasifier. It also gives the ability to perform parametric studies to study the influence of different operating parameters, such as bed temperature, fluidization velocity, steam to biomass ratio, air equivalence ratio, and bed material, on the gasification process.

The model is validated in two different aspects as shown in Fig. 4.3 and Table 4.6. The results of the temperature profile, considering the combustion of rice straw pellets at 0.5 m/s of fluidization velocity, a bed height of 3 m above distribution, and a static bed height of 30 cm, are validated with data measurements and show good agreement as shown in Fig. 4.3. The bed temperature is 850 °C, and excess air of 0.3 is considered in this case. Also, gasification gas species concentrations are validated with the results reported in the literature. The results listed in Table 4.6 give the validation of the fluidized bed gasification model with respect to the gas species concentrations at different gasification temperature values. The results show good agreement with both cases of comparison.

The effects of gasifier operating parameters on the axial temperature distribution of the fluidized bed gasifier are investigated. The results presented in Figs. 4.4, 4.5, and 4.6 show the axial temperature profile at values of operating parameters of the fluidized bed. The bed height in these figures is considered from above bed zone height, where isothermal operation is considered for this zone. Bed temperature is taken as the base case condition of 800 °C for Figs. 4.4 and 4.5. For Fig. 4.6, the temperature is varied when all other operating parameters are kept at base case condition. The results show that the temperature profile starts with an increasing trend through 15 % of the gasifier height where the freeboard zone exists. Temperature then starts to decrease as a result of the endothermic reactions occurring through the bed height and the heat interaction with the surrounding environment. The overheat that occurs in all the represented curves demonstrates the effect of oxidation reaction, and the position of this zone appears to be affected by the

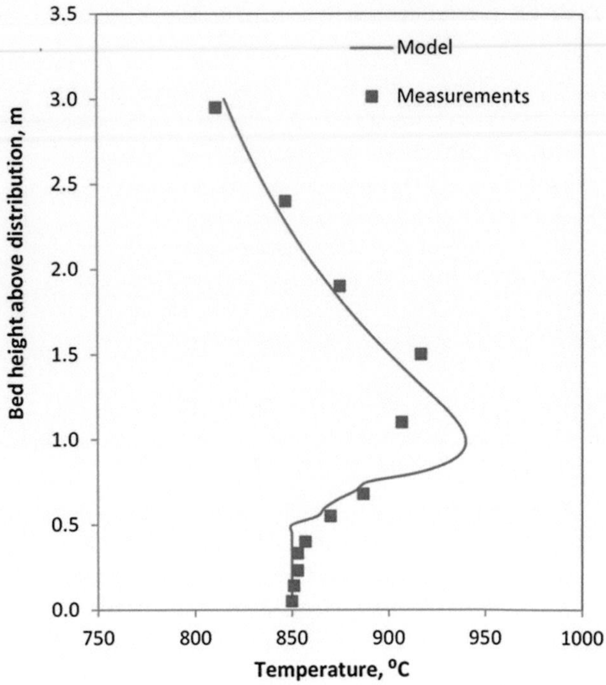

Fig. 4.3 Validation of axial temperature profile with experimental measurements reported by El-Emam et al. [26]

Table 4.6 Fluidized bed model validation with Kaushal et al. [14]

Composition of dry gas (%)		Gasification temperature				Error (RMS) (%)
		700 °C	750 °C	800 °C	850 °C	
H_2	Model	41.7	43.9	45.9	48.2	1.6
	Ref.	38.7	44.2	46.6	47.2	
CO	Model	27.2	29.5	30.4	31.9	2.6
	Ref.	25.1	26.4	27.9	29.4	
CO_2	Model	19.6	16.3	14.7	12.2	2.2
	Ref.	16.2	14.1	12.8	12.1	
CH_4	Model	6.9	6.2	5.8	5.0	1.0
	Ref.	8.4	6.2	4.9	3.9	

operating conditions. Figure 4.4 demonstrates the effect of varying fluidization velocity on the axial temperature profile through the gasifier height. Three values of fluidization velocity are considered as shown in the figure. At higher fluidization velocity, better mixing occurs which is expected to result in a more uniform temperature through the splashing and freeboard zones. Overheat temperatures of 22.29, 42.5, and 68 °C over bed temperature occur when varying the velocity of fluidization values as 0.75, 1, and 1.25 m/s, respectively. This results in higher output temperature as well for the gases leaving the fluidized bed.

Fig. 4.4 Temperature profile at different fluidization velocity values

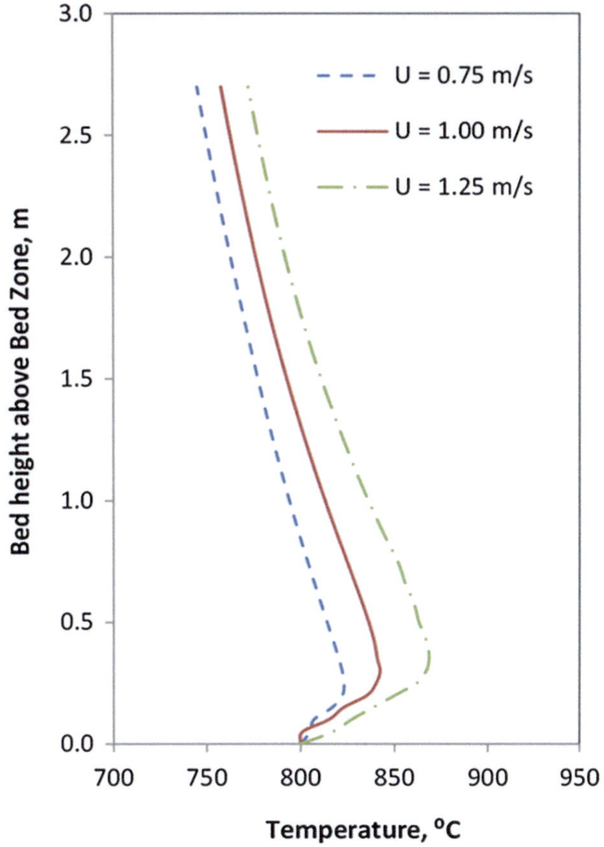

Figure 4.5 demonstrates the influence of air equivalence ratio on the gases' temperature profile through the fluidized bed gasifier. Fluidization velocity is kept constant at 1 m/s and the temperature of the bed at 800 °C. Increasing air equivalence ratio causes the maximum temperature, representing the peak of oxidation period, to shift toward higher temperature value. This is caused by the excess oxygen flowing through the gasifier. However, the overheating values are not so big compared with the results represented in Fig. 4.4 of the effect of fluidization velocity.

The results shown in Fig. 4.6 represent the effect of changing bed temperature on the profile of axial temperature through the fluidized bed gasifier. Three temperature values are tested; 750, 800, and 850 °C, where 800 °C is the bed temperature at the base case. Other operating conditions are kept constant at 0.25 for equivalence ratio and 1 m/s for fluidization velocity. It appears that at lower bed

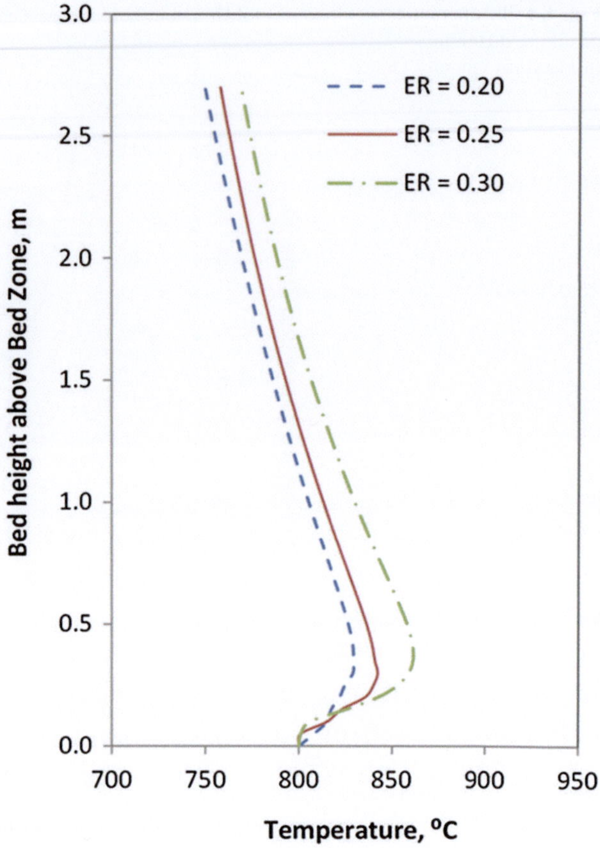

Fig. 4.5 Temperature profile at different values of air equivalence ratio

temperature, the temperature levels are relatively high through the freeboard zone where some oxidation still occurs to some extent compared with higher bed temperature where oxidation happens right after the splashing zone and then the temperature tends to decrease with the effect of the endothermic reactions after that.

Figure 4.7 shows the gas species concentrations through the fluidized bed height represented in molar concentration percentage. With the consumption of oxygen, it is noticed, as shown in the temperature profile figures, that an increase of temperature occurs with solid fuel oxidation reaction near the distributor zone. This is reflected on the composition of the gas at this zone where carbon dioxide has a peak of concentration and oxygen is consumed near the distributor. The produced gas compositions for the base case condition are found to be as follows: 18.9, 11.7, 13.53, and 4.6 % for carbon dioxide, carbon monoxide, hydrogen, and methane, respectively.

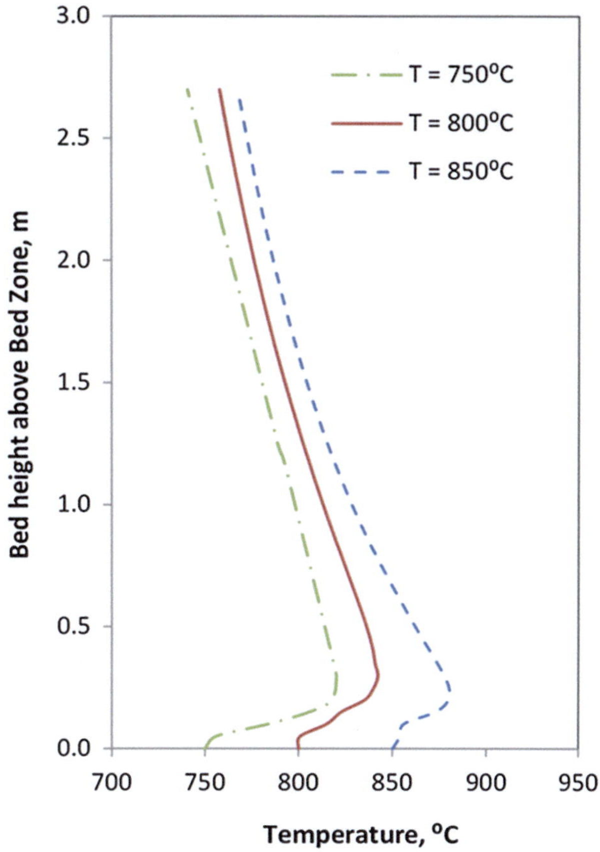

Fig. 4.6 Temperature profile at different bed temperature values

4.4 Conclusions

In this study, a model of atmospheric bubbling fluidized bed is introduced with considering combustion and gasification of biomass-driven fuel. The pellets of rice straw and bitumen are considered as the fuel. The results obtained from this work show the effects of fluidization velocity and air equivalence ratio on the fluidized bed axial temperature for the gasification process. Also, different gasifier bed temperature values are taken and studied. The model is validated with temperature measurements and other results from the literature. The results show that increasing the fluidization velocity shifts the oxidation to higher temperature causing a higher temperature through the gasifier height. The values of overheat are 22.29, 42.5, and 68 °C with fluidization velocity of 0.75, 1, and 1.25 m/s, respectively. The equivalence ratio has the same influence on the temperature profile, however, with lower level of overheating compared with the effect of fluidization velocity. The values of

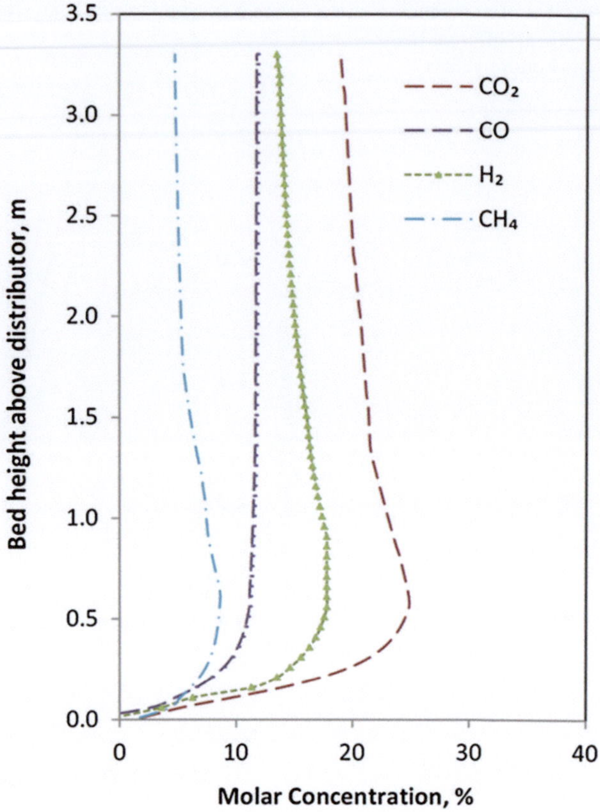

Fig. 4.7 Gases concentration profile through the fluidized bed height

the peak temperature for the equivalence ratio of air at 20 and 30 % are 29 and 60 °C, respectively. The gasification gas compositions through the gasifier are obtained, and it is affected by the reaction rates of oxidation and gasification reactions through its height.

References

1. FAO (2008) Conservation agriculture in Egypt. In: Rice straw management and conservation of environment
2. Leitzell K (2011) A black cloud over Cairo. EOSDIS 2011, cited 2014; https://earthdata.nasa.gov/featured-stories/featured-research/black-cloud-over-cairo
3. Marey HS et al (2010) Study of the formation of the "black cloud" and its dynamics over Cairo, Egypt, using MODIS and MISR sensors. J Geophys Res 115(D21)
4. El-Emam RS, Dincer I, El-Emam SH (2014) Efficiency and environmental assessment of biomass gasification for hydrogen and power production. In: Global conference on global warming, Beijing
5. Shaaban W (2010) Fluidized bed combustion of rice straw and Bitumen pellets. MSc thesis, Mansoura University, Egypt

6. Okasha FM, El-Emam SH, Zaatar G (2005) Fluidized bed combustion of an agriculture waste, case study: combustion of rice straw. Mansoura Eng J 30(2):M1–M12
7. Basu P (2006) Combustion and gasification in fluidized bed. CRC, Boca Raton
8. Okasha F (2007) Modeling combustion of straw–bitumen pellets in a fluidized bed. Fuel Process Technol 88(3):281–293
9. Nemtsov DA, Zabaniotou A (2008) Mathematical modelling and simulation approaches of agricultural residues air gasification in a bubbling fluidized bed reactor. Chem Eng J 143 (1–3):10–31
10. Radmanesh R, Chaouki J, Guy C (2006) Biomass gasification in a bubbling fluidized bed reactor: experiments and modeling. AIChE J 52(12):4258–4272
11. El-Emam RS, Dincer I (2015) Thermal modeling and efficiency assessment of an integrated biomass gasification and solid oxide fuel cell system. Int J Hydrog Energy. http://dx.doi.org/10.1016/j.ijhydene.2015.02.061
12. Kunii D, Levenspiel O (1991) Fluidization engineering. Butterworth Heinemann, Boston
13. Davidson JF, Harrison O (1963) Fluidized particles. Cambridge University Press, Cambridge
14. Kaushal P, Abedi J, Mahinpey N (2010) A comprehensive mathematical model for biomass gasification in a bubbling fluidized bed reactor. Fuel 89(12):3650–3661
15. Gómez-Barea A, Leckner B (2010) Modeling of biomass gasification in fluidized bed. Progr Energy Combustion Sci 36(4):444–509
16. Pemberton ST, Davidson JF (1984) Turbulence in the freeboard of a gas-fluidised bed: the significance of ghost bubbles. Chem Eng Sci 39(5):829–840
17. Milioli FE, Foster PJ (1995) Entrainment and elutriation modelling in bubbling fluidized beds. Powder Technol 83(3):233–244
18. Chirone R et al (1999) Fluidized bed combustion of high-volatile solid fuels: an assessment of char attrition and volatile matter segregation. In: The 15th international conference on FBC, ASME
19. Petersen I, Werther J (2005) Experimental investigation and modeling of gasification of sewage sludge in the circulating fluidized bed. Chem Eng Process 44(7):717–736
20. Philippek C et al (1997) NOx formation and reduction during combustion of wet sewage sludge in the circulating fluidized bed—measurement and simulation. In: 14th international conference on FBC
21. Howard JB, Williams GC, Fine DH (1973) Kinetics of carbon monoxide oxidation in postflame gases. International Symposium on Combustion 14(1):975–986
22. Leckner B, Palchonok GI, Andersson BA (1992) Representation of heat and mass transfer of active particles. In: The IEA-FBC, mathematical modelling meeting. Turku, Finland
23. Salatino P, Scala F, Chirone R (1998) Fluidized-bed combustion of a biomass char: the influence of carbon attrition and fines postcombustion on fixed carbon conversion. In: 27th symposium on combustion. The Combustion Institute
24. Scala F, Chirone R (2006) Combustion and attrition of biomass chars in a fluidized bed. Energy Fuels 20:91–102
25. Fiorentino M, Marzocchella A, Salatino P (1997) Segregation of fuel particles and volatile matter during devolatilization in a fluidized bed reactor—I. Model development. Chem Eng Sci 52(12):1893–1908
26. El-Emam RS, Okasha FM, El-Emam SH (2014b) Clean combustion of low quality fuel in fluidized bed combustor. In: Dincer I, Midilli A, Kucuk H (eds) Progress in sustainable energy technologies: creating sustainable development, vol II. Springer, Geneva, pp 531–546
27. Luecke K, Hartge E, Werther J (2004) A 3D model of combustion in large-scale circulating fluidized bed boilers. Int J Chem Reactor Eng 2(1)
28. Bryden KM, Ragland K (1996) Numerical modeling of deep fixed bed combustor. Energy Fuel 10:269–275
29. Biba V et al (1978) Mathematical modeling for the gasification of coal under pressure. Ind Eng Chem Process Des Dev 17:92–98

Chapter 5
Thermodynamic Evaluation of an Integrated System with Concentrating Collector

Umran Cevrimli, Yunus Emre Yuksel, and Murat Ozturk

Abstract In this paper, thermodynamic analysis of an integrated system with parabolic collector which produces a number of outputs, such as heating, cooling, hot water, and electricity, is investigated. This integrated system consists of four main subsystems: concentrating collector, energy storage, Rankine cycle, and double-effect absorption cooling and heating. The renewable energy-based system is operated in two modes, which are solar mode and storage mode. Exergy destruction ratios and rates, power or heat transfer rates, and energy and exergy efficiencies of the system components and whole system are carried out. From the results, energy and exergy efficiencies for solar mode are found as 51.32 and 46.75 %, whereas for storage mode these efficiencies are calculated as 47.44 % and 45.43 %, respectively. Additionally, parametric studies, including the thermodynamic performance of the system and its components, are conducted by the change in some design parameters, as variation of the ambient temperature changes from 0 to 30 °C.

Keywords Thermodynamic analysis • Energy analysis • Exergy analysis • Exergy efficiency • Integrated system • Multi-generation

Nomenclature

E	Energy, kJ
$\dot{E}$	Energy rate, kW
ex	Specific exergy, kJ/kg
$\dot{E}x$	Exergy rate, kW
h	Specific enthalpy, kJ/kg
$\dot{m}$	Mass flow rate, kg/s
P	Pressure, kPa

U. Cevrimli • M. Ozturk
Suleyman Demirel University, Isparta, Turkey

Y.E. Yuksel (✉)
Afyon Kocatepe University, Afyonkarahisar, Turkey
e-mail: yeyuksel@aku.edu.tr

I. Dincer et al. (eds.), *Progress in Clean Energy, Volume 1*,
DOI 10.1007/978-3-319-16709-1_5

$\dot{Q}$ Heat rate, kW
s Specific entropy, kJ/kgK
T Temperature, K
$\dot{W}$ Work transfer rate, kW

Greek Letters

η Energy efficiency
ψ Exergy efficiency

Superscripts

D Destruction
in Inlet
o Reference state
out Outlet

Acronyms

HEX Heat exchanger
PDC Parabolic dish collector
PTSC Parabolic trough solar collector

5.1 Introduction

As fast increase in energy depletion is nearly related to an increase in a population, people face several environmental problems such as decreasing energy resources, global climate change, and greenhouse emission. The integrated systems for combined generation of useful outputs are spreading around the world fast, since they offer diverse advantages, such as higher efficiency and low operating cost. Energy application choices are restricted by thermodynamic principles. Comprehension of exergy application methods can help identifying the efficiency and losses of an energy production systems. An integrated system for energy production refers to a system more than three different purposes, including electricity, cooling, heating, hot water, fresh water, hydrogen, oxygen and air, chemical with the same sources of the input energy.

Analyses of the integrated systems process including trigeneration and polygeneration have increased throughout the past decade so as to reduce energy consumption and also succeed more sustainable and economic energy production system [1]. There are lots of advantages of integrated energy manufacture systems, such as minimized energy and exergy losses, reduced greenhouse gas effects, decreased material waste, reduced fuel use and material consumption relying on increased exergetic efficiency, and decreased harmful gas emissions compared to other systems for electricity generation, heating, cooling, drying, and air-conditioning [2–4]. Energy generation and utilization choices are controlled by the thermodynamic laws. For a process design, evaluation of the exergy destruction and efficiency analysis can help identifying and figuring out the highly efficient energy generation systems.

Solar energy is a reliable energy resource. It is a renewable energy resource with no gas emissions. Solar radiation can be used directly to produce electricity utilizing photovoltaic (PV) collectors or to get thermal energy as well as operating electrical power by heat engine. There are a few solar thermal systems that can be used to manufacture electrical power via thermal power plants, such as solar towers, parabolic dish collector (PDC), and parabolic trough solar collector (PTSC). Al-Sulaiman et al. [5] have used exergy modeling in order to evaluate the exergetic performance of a novel trigeneration system utilizing a PTSC. Solar PDCs concentrate solar radiation to a single focal point by means of one or more parabolic dishes. PDCs have concentration ratios in the range of 600–2,000. As a result, they are rather efficient at thermal energy absorption and power conversion processes [6]. Thus, PDC is selected to be the prime mover for the integrated system in this study.

Absorption chillers can make use of renewable energy resources, such as solar, geothermal, and biomass, in order to generate cooling. The other benefits of absorption refrigeration systems are that they do not cause ozone layer retrenchment, they make use of natural refrigerants possibly having less CO_2 emissions, and they are independent of electric grids. The most common commercially suitable absorption refrigeration systems are single- and double-effect systems. In double-effect absorption refrigeration systems, a secondary fluid (absorbent) is used to circulate and to absorb the primary fluid (refrigerant). The success of the absorption relies on the selection of an appropriate combination of absorbent and refrigerant [7]. The most widespread absorbent and refrigerant combinations in absorption refrigeration systems have been LiBr–H_2O and ammonia–water. The LiBr–H_2O pair is the most suitable for air-conditioning and chilling applications. Ammonia–water is utilized for cooling and low-temperature freezing applications.

Farshi et al. [8] have presented a computational model for crystallization phenomena in double effect for three absorption refrigeration systems and showed the range of operating conditions without crystallization risks in the parallel configuration. Reverse parallel configurations are broader than those of the series flow system. The most improved effort is being directed to a single-effect absorption system though some research has been published on double-effect absorption systems. Furthermore, other types of multi-effect (triple, quadruple) systems have been analyzed. For a single-effect system, COP is not as high as the number of effect systems (in other words, COP is not doubled when a single-effect system turns into a double effect). In addition, a higher number of effects cause system complexity. Double-effect systems have received important care and are suitable commercially.

Absorption cooling systems have been conducted utilizing energy analysis. However, it is increasingly accepted that exergy analysis supplies more meaningful information when assessing the performance of conversion systems, as it supplies meaningful efficiencies and sources of thermodynamic inefficiencies in these systems. Farshi et al. [9] evaluated exergoeconomic analyses in which they compared flow double effect and combined ejector-double-effect system in order to investigate the effect of various parameters on investment cost of the complete systems. As a result, the combined cycle works more economically as compared to the double-effect system.

Several researchers have studied the utilization of integrated systems in energy generation to improve the thermodynamic and environmental performance. Ozturk and Dincer [10] have researched on the integrated systems gaining rising interest in the last few decades so as to reduce energy consumption and accomplish more sustainable and energy production. Buck and Friedmann [11] have evaluated experimentally the efficiency of a trigeneration system based on a turbine and assessed by a solar tower cycle. They have investigated the benefits of using single- and double-effect absorption systems. The authors indicated that using the double-effect chiller had more advantages since it gave better thermal performance and lower operating cost compared to the single-effect absorption system. The specific objectives of this paper are given as follows:

- To investigate exergy efficiencies and destructions of subsystems and whole system for solar mode
- To evaluate the impact of ambient temperature on the exergy efficiencies and respective exergy destructions of a subsystem and whole system for solar mode
- To develop a comprehensive thermodynamic assessment of a multi-generation system based on a concentrating solar collector, Rankine cycle for electricity production, and double-effect absorption system for heating and cooling application
- To investigate the effects of environmental condition on the efficiency assessment of a subsystem and whole system for solar mode

5.2 System Description

The brief explanation of the process components and starting conditions is given to evaluate for the thermodynamic assessment for solar-based multi-generation purposes as power, cooling, heating, and hot water generation. Solar-based multi-generation system can be split into three main subsystems: (1) concentrating collector, (2) Rankine cycle, and (3) double-effect absorption cooling and heating.

The schematic of the solar-based multi-generation energy system is shown in Fig. 5.1. Thermal energy of solar radiation is first collected and concentrated by

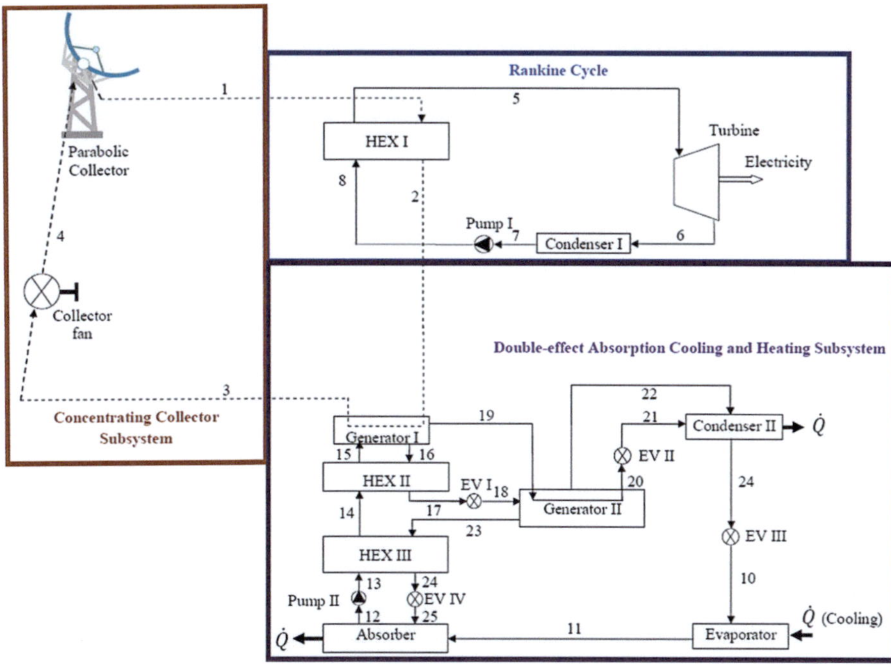

Fig. 5.1 Schematic diagram of a solar-based multi-generation system

utilizing a PDC in order to boil water. Parabolic dish concentrating collector cycle is used to supply high-temperature water vapor (600 °C) and shown in Fig. 5.1a. Solar water cycle considered in this paper supplies the heat input to the integrated system. As it can be seen from Fig. 5.1a, water leaves the collector fan at point 4, and then it is heated by a PDC. At point 1 the water vapor leaves the collector and goes through the heat exchanger (HEX-I) to be heated up. At point 2 this hot water leaves HEX-I with temperature around 200 °C, and then it directly goes through the generator to separate water from the solution of LiBr–H_2O.

As it can be seen in Fig. 5.1b, the Rankine cycle considered here has one turbine and a feedwater heater unit. Heat input to the cycle comes from the PDC. Steam leaves the HEX-I with temperature around 420 °C at point 5, and then it directly goes through a Rankine turbine to be expanded and generate power. The temperature and pressure of steam drop during this process to the values at state 6, where steam enters the condenser. At this state, steam is generally a saturated liquid–vapor mixture with a high quality. Steam is condensed at constant pressure in the condenser and leaves the condenser as saturated liquid. Water enters the pump at state 7 as saturated liquid and is compressed isentropically to the operating pressure of the boiler. As the water temperature goes up during this isentropic compression, the specific volume of water increases negligibly. The output of the Rankine cycle is the power; in this integrated system, a portion of the electric output is used to run the system device.

The lithium bromide–water absorption system is used instead of a conventional refrigeration system to utilize surplus heat in the system. In this paper, it is assumed that the required energy for the double-effect absorption cooling and heating subsystem can be provided by the heat recovered from the Rankine cycle. Figure 5.1c shows a schematic diagram of the flow double-effect lithium bromide–water absorption refrigeration system. As it can be seen in this figure, it contains the most significant components of the double-effect absorption system: high- and low-pressure generator, high- and low-temperature heat exchanger, solution and refrigerant pump, absorber, condenser, and evaporator. Also, it involves three pressure levels: high, medium, and low. The high-pressure generator functions at high pressure and high temperature, whereas the low-pressure generator and condenser operate at medium pressure, and the evaporator and absorber work at low pressures. It should be noted that the integrated system is modeled according to the optimum operating parameters for the double-effect absorption subsystem.

5.3 Thermodynamic Analyses

By using the thermodynamic analysis approach, both the development of renewable and conventional energy conversion systems and the development potential of its operations should be investigated at any category such as systems, processes, process units, or components. In this section, the general thermodynamic approaches including the mass, energy, and exergy balance equations are given to investigate for the analysis and improvement potential of the solar-based integrated system. In the most general principle, a balance equation for a quantity in the given process should be written as follows:

$$\text{Input} + \text{Generation} - \text{Output} - \text{Consumption} = \text{Accumulation} \tag{5.1}$$

This equation is considered as the quantity balance for a process. The accumulated quantity in the process is equal to the difference between the input and output quantity by the system boundary plus the generated and consumed quantity within the system boundary.

In the steady-state condition, the accumulation terms in Eq. (5.1) are equal to zero, because all properties in the process are unchanging with time [12]. Therefore, the mass balance equation for the steady-state conditions can be written in the rate form as given:

$$\sum \dot{m}_{\text{in}} = \sum \dot{m}_{\text{out}} \tag{5.2}$$

where $\dot{m}$ is the mass flow rate and subscripts in and out indicate the inlet and outlet of the matter, respectively. The energy balance equation is applied to the description of a wide variety of investigative processes. The energy balance equation for the steady-state condition with negligible potential and kinetic energy terms can be given in the rate form as follows:

$$\dot{Q} + \sum \dot{m}_{\text{in}} h_{\text{in}} = \dot{W}_{\text{net}} + \sum \dot{m}_{\text{out}} h_{\text{out}} \tag{5.3}$$

where $\dot{Q}$ and $\dot{W}$ are the heat and work transfer rate, respectively, and h is the specific enthalpy. In any process, when energy is transferred into another less useful quality, the rest of the useful quality of the energy cannot be recovered again, a portion that is not conserved as total energy of the process. In agreement with this approach, the sum of useful work can be defined in the exergy terms according to the second law of thermodynamics for a better understanding of the energy transfer. The characteristics of a reference environment of the given process have to be defined in the exergy modeling. This necessity is usually done by using the temperature, pressure, and chemical composition of the reference environment. In this paper, actual environment conditions are accepted as the reference environment properties. The exergy balance equation for the steady-state condition can be given in the rate form as follows:

$$\sum_i \dot{m}_{\text{in}} ex_{\text{in}} + \dot{E}x_Q = \sum_e \dot{m}_{\text{out}} ex_{\text{out}} + \dot{E}x_W + \dot{E}x_D \tag{5.4}$$

where $\dot{E}x_Q$ and $\dot{E}x_W$ are the heat and work exergy flow rates. They are defined as follows, respectively:

$$\dot{E}x_Q = \left(1 - \frac{T_o}{T_i}\right)\dot{Q}_i \tag{5.5}$$

where T_i is the temperature in the ith given state.

$$\dot{E}x_W = \dot{W} \tag{5.6}$$

where $\dot{E}x_D$ in Eq. (5.4) is the exergy destruction rate, and it can be expressed in the rate form as

$$\dot{E}x_D = T_o \dot{S}_{\text{gen}} \tag{5.7}$$

where ex in Eq. (5.4) is the specific exergy of the process inlet and outlet flow. The exergy content of a stream of matter can be expressed by three categories, such as thermal exergy content (resulting from the temperature difference between the substance and environment), mechanical exergy content (resulting from the pressure difference between the substance and environment), and chemical exergy content (resulting from the composition difference between the substrate and environment). According to these exergy contents, the determination of the specific exergy can be written as given:

Table 5.1 Equations for exergy destruction rate for the components of the solar-based integrated system

System components	Exergy destruction rate equations
Parabolic dish collector	$\dot{E}x_{D,\text{PDC}} = \dot{E}x_4 - \dot{E}x_1 + \dot{E}x^Q_{\text{Solar}}$
Collector fan	$\dot{E}x_{D,\text{CF}} = \dot{E}x_3 - \dot{E}x_4 + \dot{W}_{\text{CF}}$
HEX-I	$\dot{E}x_{D,\text{HEX-V}} = \dot{E}x_1 + \dot{E}x_8 - \dot{E}x_2 - \dot{E}x_5$
Turbine	$\dot{E}x_{D,\text{turbine}} = \dot{E}x_5 - \dot{E}x_6 - \dot{W}_T$
Condenser-I	$\dot{E}x_{D,\text{Con}-I} = \dot{E}x_6 - \dot{E}x_7 - \dot{E}x^Q_{\text{Con}-I}$
Pump-I	$\dot{E}x_{D,\text{pump-I}} = \dot{E}x_7 - \dot{E}x_8 + \dot{W}_{P-I}$
Generator-I	$\dot{E}x_{D,\text{Gen-I}} = \dot{E}x_8 + \dot{E}x_{15} - \dot{E}x_5 - \dot{E}x_{16} - \dot{E}x_{19}$
Expansion valve-I	$\dot{E}x_{D,\text{EV-I}} = \dot{E}x_{17} - \dot{E}x_{18}$
Generator-II	$\dot{E}x_{D,\text{Gen-II}} = -\dot{E}x_{18} - \dot{E}x_{19} + \dot{E}x_{20} + \dot{E}x_{22} + \dot{E}x_{23}$
Condenser-II	$\dot{E}x_{D,\text{Con-II}} = \dot{E}x_{21} - \dot{E}x_{22} - \dot{E}x_9 - \dot{E}x^Q_{\text{Con-II}}$
Evaporator	$\dot{E}x_{D,\text{Eva}} = \dot{E}x_{10} - \dot{E}x_{11} + \dot{E}x^Q_{\text{Eva}}$
Absorber	$\dot{E}x_{D,\text{Ab}} = \dot{E}x_{11} + \dot{E}x_{25} - \dot{E}x_{12} - \dot{E}x^Q_{\text{Ab}}$

$$ex = ex_{ke} + ex_{pe} + ex_{ph} + ex_{ch} \tag{5.8}$$

When the temperature, pressure, or composition of a substance is different from the thermodynamic equilibrium with the environment, this situation can produce a useful change. Since the variations of the mechanical and chemical exergy are accepted negligible in this paper, the physical or flow exergy can be given as follows:

$$ex_{ph} = (h - h_o) - T_o(s - s_o) \tag{5.9}$$

where h is the specific enthalpy and s is the specific entropy in the process. According to the given above procedure, the exergy destruction rate of the whole system and its components are given in Table 5.1 and used in parametric studies for the investigation of the improvement potential of the integrated system.

5.3.1 Energy and Exergy Efficiency

The thermodynamic efficiencies can be given by the first law of thermodynamics (energy efficiency) and by both laws of thermodynamics (exergy efficiency). In order to investigate the operating performance of the solar-based integrated system, energy and exergy analyses are commonly used to the detailed analysis of the whole system and its components. The performance analysis of the integrated system has great importance due to the limited supply of solar energy from the photons to the

Rankine and absorption cooling cycle as well as the overall impact on the system operating performance.

The energy efficiency of the given process can be described as the ratio of useful energy output from the system boundary to the total energy input to the system boundary. The useful output energy gives the desired results produced by the whole system and its components. In the general manner, the energy efficiency can be given as follows:

$$\eta = \frac{\text{useful energy output rate}}{\text{total energy input rate}} = \frac{\dot{E}_{\text{out, useful}}}{\sum \dot{E}_{\text{in, total}}} \tag{5.10}$$

In order to investigate the performance of the solar-based integrated system by using the exergy analysis, it is essential to determine both the product and the fuel for the whole system and its components. The exergy efficiency of the process can be defined as the division of exergy output rate by the overall exergy inlet rate. According to this definition, the exergy efficiency of the process can be given as follows:

$$\psi = \frac{\text{total useful exergy output rate with products}}{\text{total exergy input rate}} = \frac{\sum \dot{E}x_{\text{out, total}}}{\sum \dot{E}x_{\text{in, total}}} \tag{5.11}$$

Another concept of the exergy efficiency for the process should also be given in terms of exergy destruction rate as follows:

$$\psi = \frac{\sum \dot{E}x_{\text{out}}}{\sum \dot{E}x_{\text{in}}} = \frac{\sum \dot{E}x_{\text{in}} - \dot{E}x_D}{\sum \dot{E}x_{\text{in}}} = 1 - \frac{\dot{E}x_D}{\sum \dot{E}x_{\text{in}}} \tag{5.12}$$

According to the exergy efficiency description or Eq. (5.11), the exergy efficiencies of the system components are shown in Table 5.2.

5.4 Results and Discussion

The values of mass flow rate (kg/s), pressure (kPa), temperature (°C), enthalpy (kJ/kg), entropy (kJ/kgC), energy rate (kW), specific exergy (kJ/kg), and exergy rate (kW) for the components of the solar-based integrated system for the solar mode are given in Table 5.3. Reference temperature is taken to be 25 °C and the pressure is taken as 100 kPa. The thermodynamic properties of the working fluids are calculated via utilizing the Engineering Equation Solver (EES) software [13].

The exergy analysis is applied to the whole system and its components by using the first and second laws of thermodynamics because exergy analysis generates more outcomes than energy analysis in designing and assessing the processes.

Table 5.2 Equations for exergy efficiency for the components of the solar-based integrated system

System components	Exergy efficiency equations
Parabolic dish collector	$\Psi_{\text{PDC}} = \frac{\dot{E}x_1 - \dot{E}x_4}{\dot{E}x^Q_{\text{Solar}}}$
Collector fan	$\Psi_{\text{CF}} = \frac{\dot{E}x_4 - \dot{E}x_3}{\dot{W}_{\text{CF}}}$
HEX-I	$\Psi_{\text{HEX-V}} = \frac{\dot{E}x_5 - \dot{E}x_8}{\dot{E}x_1 - \dot{E}x_2}$
Turbine	$\Psi_{\text{Turbine}} = \frac{\dot{W}_T}{\dot{E}x_5 - \dot{E}x_6}$
Condenser-I	$\Psi_{\text{Con-I}} = \frac{\dot{E}x^Q_{\text{Con-I}}}{\dot{E}x_6 - \dot{E}x_7}$
Pump-I	$\Psi_{\text{Pump-I}} = \frac{\dot{E}x_8 - \dot{E}x_7}{\dot{W}_{\text{Pump-I}}}$
Generator-I	$\Psi_{\text{Gen-I}} = \frac{\dot{E}x_{16} + \dot{E}x_{19} - \dot{E}x_{15}}{\dot{E}x_2 - \dot{E}x_3}$
Expansion valve-I	$\Psi_{\text{EV-I}} = \frac{\dot{E}x_{18}}{\dot{E}x_{17}}$
Generator-II	$\Psi_{\text{Gen-II}} = \frac{\dot{E}x_{19} - \dot{E}x_{20}}{\dot{E}x_{22} + \dot{E}x_{23} - \dot{E}x_{18}}$
Condenser-II	$\Psi_{\text{Con-II}} = \frac{\dot{E}x^Q_{\text{Con-II}}}{\dot{E}x_{21} + \dot{E}x_{22} - \dot{E}x_9}$
Evaporator	$\Psi_{\text{Eva}} = \frac{\dot{E}x^Q_{Col}}{\dot{E}x_{11} - \dot{E}x_{10}}$
Absorber	$\Psi_{\text{Ab}} = \frac{\dot{E}x^Q_{Ab}}{\dot{E}x_{11} + \dot{E}x_{25} - \dot{E}x_{12}}$

The exergy analysis results based on the values for exergy destruction ratio (%), exergy efficiency (%), exergy destruction rates (kW), and the heat transfer or power rate of the integrated system for the solar mode, corresponding to the data in Table 5.3, are given in Table 5.4. Exergy destruction rate demonstrates the reduction in energy availability; but it cannot be used to investigate the energy and exergy utilization efficiency of the integrated system. The exergy efficiencies of the system components are more useful for determining exergy losses. As seen in Table 5.4, the exergy destruction rates and exergy loss ratios of the PDC are higher than the other components although the exergy efficiency of the collector is 80.73 %. According to exergy loss ratios and exergy destruction rates, it is significant to supply development applications on this collector type. Therefore, minimization of exergy destruction rate in the PDC is vital to flourish the performance of the integrated system, which in turn diminishes the cost and emissions and thus assists in lessening the associated environmental impact.

Table 5.4 demonstrates that expansion valve-I and expansion valve-IV have high exergy efficiency values of 99.34 % and 99.77 %, respectively. The expansion valves are a highly effective device because the refrigerant pressure is diminished without any energy exit from the system. In addition, the exergy efficiency of generator-II is less than that of the other components in the solar-based integrated system, suggesting that it should likely be worthwhile to focus development efforts on this component. In addition to that, the exergy analysis results demonstrate that the single-effect absorption system does not exhibit significant exergy destruction, because this system does not directly use the mainly produced energy but instead utilize the steam exhaust by the Rankine cycle.

Table 5.3 Thermodynamic data for flows in the solar mode of the solar-based integrated system

State no.	Fluid	Mass flow rate $\dot{m}$ (kg/s)	Pressure, P (kPa)	Temperature T (°C)	Enthalpy h (kJ/kg)	Entropy s (kJ/kgC)	x LiBr %	Energy rate $\dot{E}$ (kW)	Specific exergy ex (kJ/kg)	Exergy rate $\dot{E}_x$ (kW)
0	H_2O		100	25	104.8	0.3669	–	–	–	–
1	H_2O	14	4,500	600	3,670	7.311	–	49,916	20,932	1,495
2	H_2O	14	4,500	199.1	849.4	2.317	–	10,422	733.8	507.4
3	H_2O	14	4,500	257.5	2,251	4.989	–	30,050	10,756	768.3
4	H_2O	14	4,500	257.5	2,295	5.071	–	30,661	11,024	787.4
5	H_2O	4.385	14,200	420	3,071	6.04	–	13,005	5,588	1,274
6	H_2O	4.385	7	39.01	1,994	6.424	–	8,284	365.7	83.39
7	H_2O	4.385	7	38	159.1	0.5455	–	238.1	4.642	1.059
8	H_2O	4.385	14,200	39	175.9	0.5535	–	311.5	67.61	15.42
9	H_2O	0.127	2.547	35	146.6	0.5051	–	5.308	0.07445	0.5862
10	H_2O	0.127	1.069	4	131.4	0.4746	–	16,69	0.7025	5.532
11	H_2O	0.127	1.069	4	2,508	9.051	–	305.2	23.58	185.6
12	H_2O/LiBr	1.737	2.547	35	87.63	0.2081	0.5588	64	48.03	27.65
13	H_2O/LiBr	1.737	5.616	35.2	88.03	0.2094	0.5588	64.7	48.05	27.66
14	H_2O/LiBr	1.737	5.616	62.5	143.2	0.381	0.5588	160.4	54.94	31.63
15	H_2O/LiBr	1.737	5.616	107	235.9	0.6394	0.5587	321.6	82.23	47.34
16	H_2O/LiBr	1.671	5.616	130	288.6	0.7365	0.5806	397.3	118.7	71.04
17	H_2O/LiBr	1.671	5.616	83.09	192.5	0.4844	0.5806	236.8	83.75	50.12
18	H_2O/LiBr	1.671	2.547	82.09	190.5	0.4788	0.5806	233.4	83.2	49.79
19	H_2O	0.066	5.616	130	2,745	8.862	–	174.2	7.064	107
20	H_2O	0.066	5.616	82.45	345.2	1.104	–	15.86	1.354	20.52
21	H_2O	0.066	2.547	77	322.3	1.04	–	14.36	1.12	16.96
22	H_2O	0.061	2.547	51.79	2,597	8.819	–	152	7.478	122.6
23	H_2O/LiBr	1.61	2.547	80.92	197.7	0.4557	0.6029	236.5	102.8	63.84
24	H_2O/LiBr	1.61	2.547	48.67	136	0.273	0.6029	137.2	91.24	56.67
25	H_2O/LiBr	1.61	1.069	47.67	134.1	0.2671	0.6029	134.1	91.02	56.54

Table 5.4 Thermodynamic analysis results for solar mode of the solar-based integrated system devices

Devices	Exergy destruction rate (kW)	Exergy destruction ratio (%)	Exergy efficiency (%)	Power or heat transfer rate (kW)
Parabolic dish collector	2,042	51.30	80.73	19,255
HEX-I	473.8	11.90	92.1	12,694
Collector fan	342.9	8.61	43.81	610.3
Pump-I	10.46	0.26	85.76	73.43
Condenser-I	357.8	8.98	72.22	8,046
Turbine	501.1	12.58	90.41	4,721
Generator-I	33.85	0.85	56.27	253.5
Generator-II	21.36	0.53	21.09	158.4
HEX-II	7.676	0.19	78.05	161.1
HEX-III	4.658	0.11	59.68	95.75
Pump-II	12.08	0.30	60.12	55.36
Condenser-II	8.523	0.21	72.68	161.1
Expansion valve-I	0.553	0.013	99.34	3.366
Expansion valve-II	0.2346	0.005	82.68	1.509
Expansion valve-III	0.6281	0.015	84.77	1.932
Expansion valve-IV	0.2144	0.005	99.77	3.039
Evaporator	27.44	0.68	45.46	301.9
Absorber	8.264	0.20	22.64	49.57

As another part of the thermodynamic assessment analysis, some variable major design parameters are examined. The effects on the exergy efficiency and exergy destruction of the integrated system components and the whole system variation of exergy destruction rate and exergy efficiency by changing the reference temperature from 10 to 30 °C for the solar mode subsystems and the whole system are shown in Figs. 5.2, 5.3, 5.4, and 5.5. The exergy destruction ratio and exergy efficiency of the integrated system and subsystems are directly proportional to the environmental temperature variations.

Exergy destruction rates and exergy efficiencies of the PDC and Rankine subsystem are investigated by using Tables 5.1, 5.2, and 5.3, and results are given in Figs. 5.2 and 5.3, respectively. The variation of these exergetic variables as exergy efficiency and exergy destruction rate for subsystems remains nearly linear relying on the ambient air. As seen in Fig. 5.5, the integrated whole system has similar results between the exergy efficiency and exergy destruction rate based on the variable reference temperature. In contrast, as seen in Fig. 5.4, exergy destruction rate in the absorption refrigeration system increases with the increasing reference temperature. Thus, exergy efficiency dwindles with the increasing reference temperature, thanks to the increasing temperature difference between the refrigeration system and ambient air.

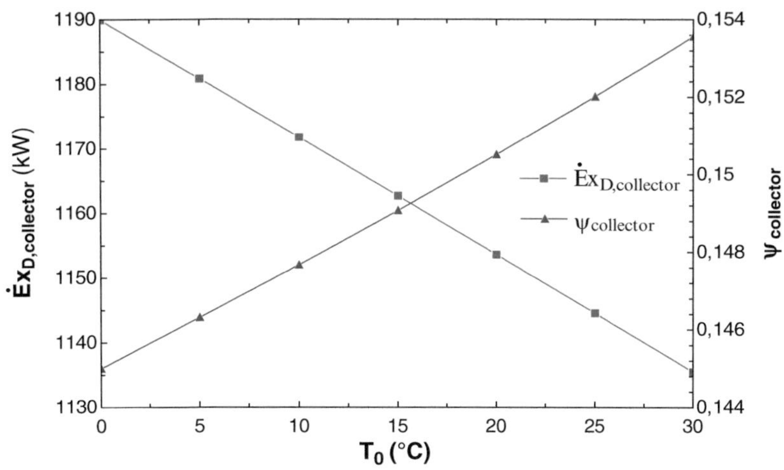

Fig. 5.2 Exergy destruction rates and exergy efficiencies of the parabolic dish collector system for solar mode depending on the reference temperature changes

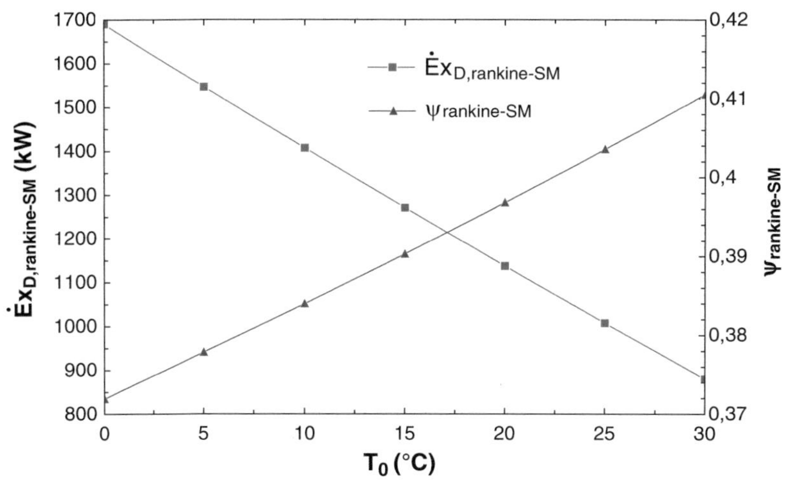

Fig. 5.3 Exergy destruction rates and exergy efficiencies of the Rankine system for solar mode depending on the reference temperature changes

It can be seen in Fig. 5.4 that the integrated system has higher exergy efficiency than the collector, Rankine, and double-effect absorption cycles. This investigation results from the large quantity of energy lost to the environment from the solar collector and Rankine cycle. The exergy efficiency of the integrated system is slightly higher than that of the Rankine cycle because the exergy associated with the cooling and heating load by the double-effect absorption system is small. Also, the main reason is utilizing the waste energy from the Rankine cycle to produce cooling and heating by using the double-effect absorption cycle. It is clear that the

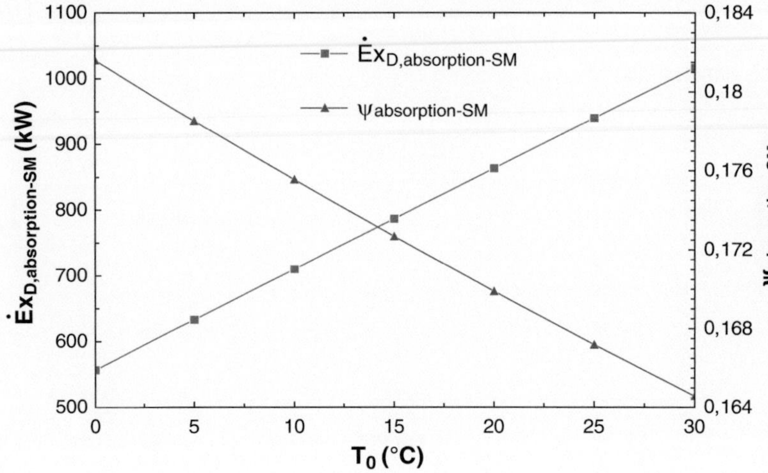

Fig. 5.4 Exergy destruction rates and exergy efficiencies of the absorption system for solar mode depending on the reference temperature changes

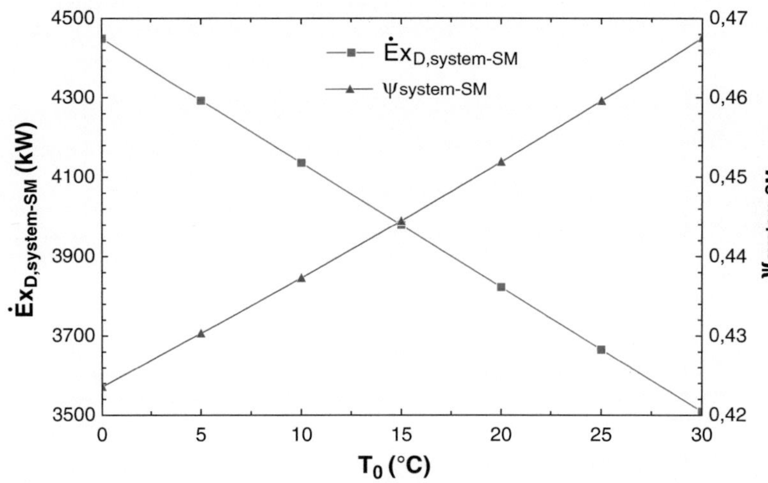

Fig. 5.5 Exergy destruction rates and exergy efficiencies of the whole system for solar mode depending on the reference temperature changes

environmental impacts of the integrated system are less than the conventional systems because cooling and heating applications are produced without using additional fuel.

The effects of the reference temperature on the exergy destruction rate and exergy efficiency of the PDC are shown in Fig. 5.2. This figure indicates that the exergy efficiency of the solar collector is lower than the other subsystems. As given in the literature [14], the exergy efficiency of the solar collectors is considerably less than the energy efficiency. This difference is associated with the high temperature difference between the solar and working temperatures in the solar systems.

On the other hand, the exergy efficiency of the double-effect absorption system decreases from around 18 % at 0 °C to 16 % at 30 °C. In this case, the exergy destruction rate of the absorption system decreases nearly by two times between reference temperature differences.

As seen in Fig. 5.5, the increase in the exergy efficiency of the integrated system is not observed to be a significant change with increasing reference temperature. The maximum integrated system exergy efficiency is calculated as nearly 47 % at a reference temperature of 30 °C.

5.5 Conclusions

In the present study, a thermodynamic analysis of the multi-generation energy system is investigated so as to better determine the true magnitude of losses and the true efficiencies by determining the exergy efficiency and exergy destruction rate in each system component. The parametric studies are given to analyze the effect of the reference temperature on the exergy efficiency and exergy destruction rate of the subsystem in general. A thermodynamic analysis (both first and second laws) showed that integrating systems to obtain multiple outputs significantly increases the overall efficiency. Moreover, the following concluding remarks can be drawn from the thermodynamic analyses:

- Energy efficiencies of the Rankine cycle, dish collector, absorption subsystems, and whole system for solar mode are calculated as 40.19 %, 28.87 %, 30.73 %, and 51.32 %, respectively.
- Exergy efficiencies of the Rankine cycle, dish collector, absorption subsystems, and whole system for solar mode are found as 43.05 %, 15.36 %, 16.46 %, and 46.75 %, respectively.
- With regard to exergy analysis results, the highest exergy destruction occurs in the PDCs as 51.30 %.
- The expansion valve demonstrates the lowest exergy destruction rates in the multi-generation system because the absorption cooling and heating components create relatively miniscule exergy destruction in the proposed system.

References

1. Ozturk M, Dincer I (2013) Thermodynamic assessment of an integrated solar power tower and coal gasification system for multi-generation purposes. Energy Conversion Manage 76:1061–1072
2. Dincer I, Zamfirescu C (2012) Renewable-energy-based multigeneration systems. Special issue on developments in nuclear and renewable energy. Int J Energy Res 36(15):1403–1415

3. Ahmadi P, Rosen MA, Dicer I (2011) Greenhouse gas emission and exergo-environmental analyses of trigeneration energy system. Int J Greenhouse Gas Control 5(6):1540–1549
4. Malico I, Carvalhinho AP, Tenreiro J (2009) Design of a trigeneration system using a high-temperature fuel cell. Int J Energy Res 33(2):144–151
5. Al-Sulaiman FA, Dincer I, Hamdullahpur F (2011) Exergy modeling of a new solar driven trigeneration system. Solar Energy 85(9):2228–2243
6. Al-Sulaiman FA, Hamdullahpur F, Dincer I (2011) Trigeneration: A comprehensive review based on prime movers. Int J Energy Res 35(3):233–258
7. Minciuc E, Corre OL, Athanasovici V, Tazerout M, Bitir I (2003) Thermodynamic analysis of trigeneration with absorption chilling machine. Appl Thermal Eng 23(5):1391–1404
8. Farshi LG, Mahmoudi SM, Rosen MA (2011) Analysis of crystallization risk in double effect absorption refrigeration systems. Appl Thermal Eng 31(10):1712–1717
9. Farshi LG, Mahmoudi SM, Rosen MA (2013) Exergoeconomic comparison of double effect and combined ejector-double effect absorption refrigeration systems. Appl Energy 103:700–711
10. Ozturk M, Dincer I (2013) Thermodynamic analysis of a solar-based multi-generation system with hydrogen production. Appl Thermal Eng 51(1–2):1235–1244
11. Buck R, Friedmann S (2007) Solar-assisted small solar tower trigeneration systems. J Solar Energy Eng 129(4):349–354
12. Dincer I, Rosen MA (2013) Exergy: energy, environment, and sustainable development, 2nd edn. Elsevier, Oxford
13. Klein SA (2007) Engineering Equation Solver (EES). Academic commercial, F-Chart Software, www.fChart.com
14. Petela R (2005) Exergy analysis of the solar cylindrical parabolic cooker. Solar Energy 79(3):221–233

Chapter 6
Design and Optimization of an Integrated System to Recover Energy from a Gas Pressure Reduction Station

Shoaib Khanmohammadi, Pouria Ahmadi, Kazem Atashkari, and Ramin Kouhi Kamali

Abstract This chapter deals with thermodynamic modeling, parametric analysis, and optimization of an integrated system to recover energy from pressure reduction station in city gate station (CGS). This chapter aims to fully cover the thermodynamic modeling of an integrated system consisting of a turbo expander, an organic Rankine cycle (ORC) and a proton exchange membrane (PEM) electrolyzer to produce and store hydrogen. The pressure of natural gas in transmission pipeline in Iran gas system is high which sometimes go beyond 7 MPa. This pressure needs to be reduced near the cities pipeline pressure to 1.7 MPa. This pressure reduction results in ample potential to recover energy to generate electricity. In the proposed integrated system in this chapter, a comprehensive parametric analysis including the effect of main parameters such as natural gas preheat temperature, the natural gas pressure inlet to turbo expander, the heater mass fuel flow rate, and high temperature of ORC on the system performance is investigated.

The results show that although the natural gas inlet pressure has a trivial effect on exergy efficiency of the ORC cycle, PEM electrolyzer, and turbo expander, it has a significant effect on turbo expander power output and hydrogen production rate. Besides, results indicate that increase in preheat temperature of natural gas from 130 to 165 °C has the favorable effect on the turbo expander power output and exergy destruction rate of ORC components. Also, it is concluded that an increase in preheat temperature leads to a decrease in hydrogen production from 15.9 to 14.8 kg/day due to decrease in ORC output electricity. In order to determine the

S. Khanmohammadi (✉) • K. Atashkari • R.K. Kamali
Department of Mechanical Engineering, Faculty of Engineering, University of Guilan, Rasht, Iran
e-mail: khshoaib@webmail.guilan.ac.ir

P. Ahmadi
Fuel Cell Research Lab, School of Mechatronic System Engineering, Simon Fraser University, Vancouver, BC, Canada
e-mail: pahmadi@sfu.ca

I. Dincer et al. (eds.), *Progress in Clean Energy, Volume 1*,
DOI 10.1007/978-3-319-16709-1_6

optimum value of design parameters, an optimization method is applied. The genetic algorithm optimization results show there are acceptable values for five design parameters which guarantee the optimum performance of the novel proposed integrated system.

Keywords Pressure reduction • City gate station (CGS) • Turbo expander • Hydrogen production • PEM electrolyzer

Nomenclature

a	Anode
c	Cathode
E	Activation energy (kJ/mole)
$\dot{E}x$	Exergy flow rate (W)
h	Specific enthalpy (kJ/kg)
LHV	Lower heating value (kJ/kg)
$\dot{m}$	Mass flow rate, kg s^{-1}
ORC	Organic Rankine Cycle
P	Pressure (bar)
PEM	Polymer exchange membrane
$\dot{W}$	Work, (W)
ΔP	Pressure drop, (kPa)

Greek Letter

λ	Number of mole of fuel to number of mole of air (−)
η	Efficiency (−)

Subscripts

a	Actual
Cond	Condenser
Eva	Evaporator
ex	Exergy
f	Fuel
is	Isentropic
TE	Turbo expander
Tur	Turbine

6.1 Introduction

Worldwide energy crisis and increasing the energy demand obligate scientist and engineers for seeking ways to recover energy from useless source to generate electricity. One of the potential sources to recover energy to generate electriicty is to utalize wase energy from pressure reduction station in natural gas transmission. The pressure of natural gas in transmission pipeline is high varying from 5 to7 MPa depending on the location and limitations. In order to utilize the natural gas pressure, this gas pressure should drop to 1.7 MPa in a city gate stations (CGS). This source of energy can meet the demand of electrical load of residential buildings or it can be used to produce other energy products such as fresh water and hydrogen production for later usage. In such systems, usually a turbo expander is used to recover energy from high-pressure natural gas stream.

Since the flow rate of natural gas may vary widely, designing a suitable combined system can assist us to provide desired constant output electricity. Some researchers have demonstrated the energy recovery from pressure reduction in CGS. The power generation from pressure reduction in the natural gas supply chain in Bangladesh is investigated by Mohammad Mahbubur Rahman [1].

Taheri Seresht et al. [2] studied energy recovery process using expansion turbine as well as achieve pressure reduction in natural gas transmission pipeline. Their results indicated that energy loss in the natural gas CGS is about 38.443 GWhr per year. In addition, their analysis showed that energy recovery using a turbo expander is 96 % with payback period of 2 years. In gas pressure reduction stations, the aim of using turbo expanders to achieve pressure reduction leads to a drastic temperature drop in natural gas; therefore the need for preheating the natural gas is vital to dispense the natural gas to the end-use stations. Preheating is typically applied using a gas-fired boiler. Howard [3] proposed a hybrid system includes a molten carbon fuel cell (MCFC) combined with a turbine in order to preheat the natural gas and provide additional low-emission electrical power.

Kostowski Wojeciech [4] studied the possibility of electricity generation from pressure reduction in a conventional natural gas transportation system. A thermodynamic modeling and exergy analysis is used to understand potential of energy recovery. Moreover, an economic analysis carried out to investigate the impact of the design parameter on the feasibility of turbo expander installation.

Farzaneh et al. [5] considered cooling and electricity generation from pressure reduction in CGS in Iran. Various scenarios considered to assess cooling production from throttle process and electricity production using a turbine. Kostowski and Uson [6] proposed and evaluated an innovative system for exergy recovery from natural gas expansion based on the integration of an organic Rakine cycle and an internal combustion engine. Their results indicate that proposed system achieved the favorable exergy efficiency up to 52.6 %. Furthermore, they show that outstanding performance ratio (0.69–0.77) relating to power generated to locally combusted fuel in the system can be attained.

He and Ju [7] reported exergy analysis and optimization of natural gas liquefaction process utilizing gas pipeline pressure energy. A novel liquefaction process was designed and the effect of key parameters on performance assessment of the system was investigated. Also their results showed that the process has a low unit energy consumption of 0.03975 kWh/N m^3.

In this chapter, an integrated ORC, PEM, and turbo expander system is proposed to recover energy from natural gas reduction station. Exergy analysis is conducted as a potential tool to magnify the magnitude and location of the losses in the system. A main problem in energy recovery from reduction pressure station is variation of pressure and mass flow rate of natural gas which leads to the variation of power output. A proposed system can solve this problem by supplying a base load and using surplus electricity to produce hydrogen and store it using. This hydrogen can be later used in a fuel cell to generate both electricity and heat for different applications.

6.2 System Description

Figure 6.1 illustrates a CGS which reduces natural gas pressure via throttle process. After passing through a dry gas filter the natural gas is heated up in a fired heater in order to increase its temperature. It should be mentioned that there are three lines in which two of them are in service with 50 % capacity and the other is standby.

Also Fig. 6.2 shows the integrated system includes a turbo expander, an ORC, and a PEM electrolyzer to produce hydrogen. The gas fired heater increases the temperature of high pressure natural gas stream. Then, it expands in a turbo expander with isentropic efficiency of 86 % to achieve a desired pressure and produce some work. The waste heat from gas-fired heater is utilized in ORC to produce more work. The organic fluid used in organic Rankine cycle is R123 which has the critical temperature 457 K.

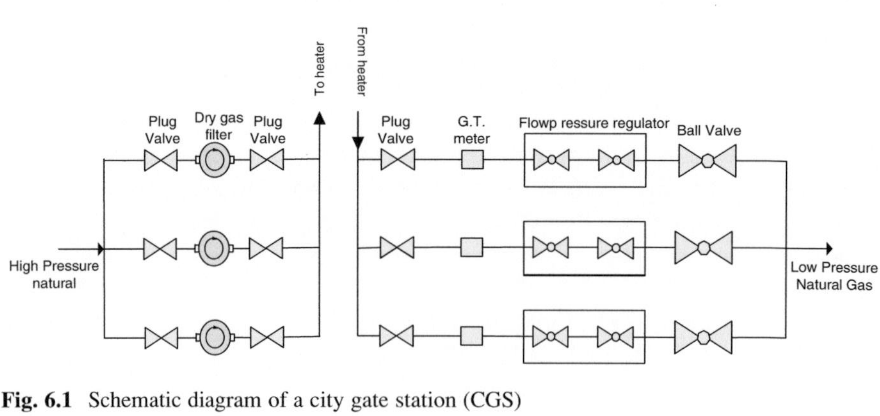

Fig. 6.1 Schematic diagram of a city gate station (CGS)

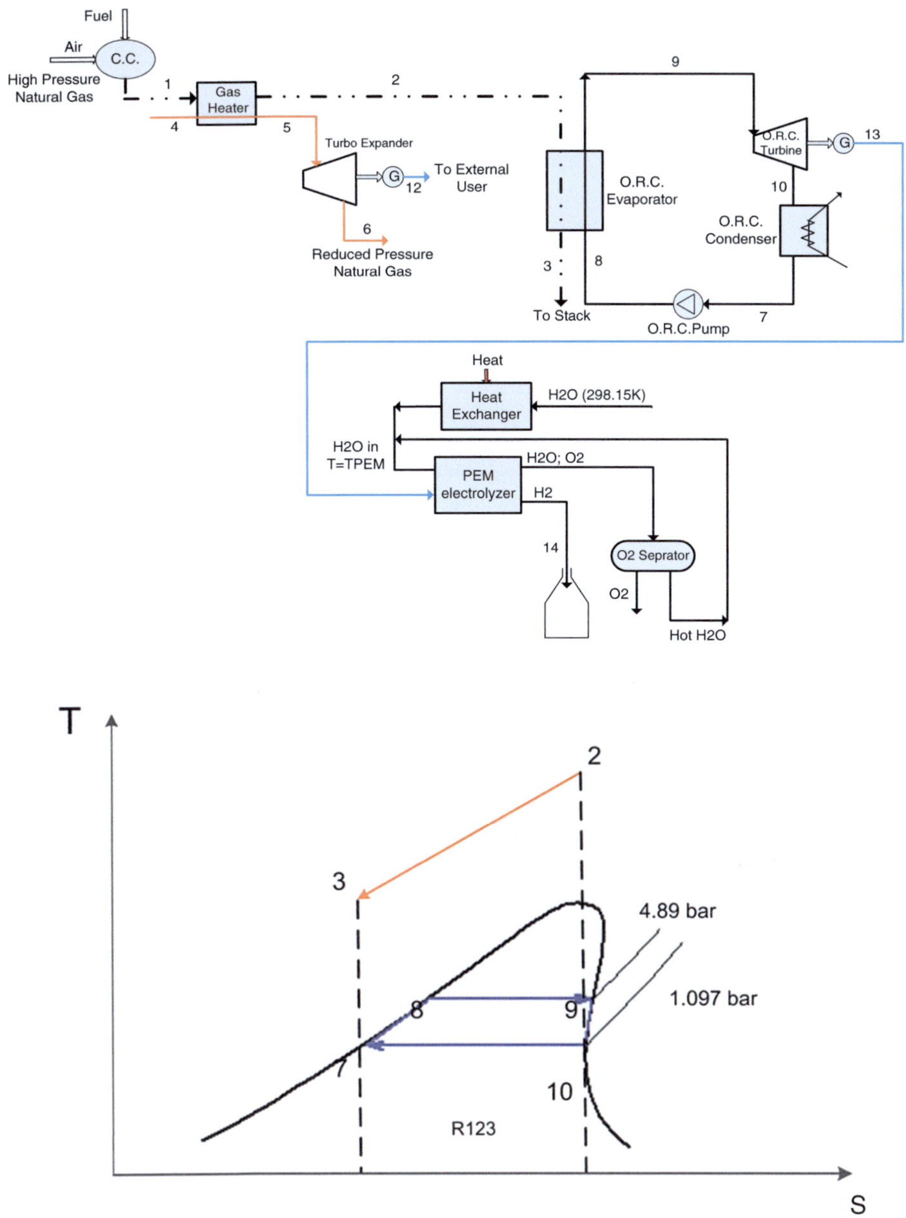

Fig. 6.2 Schematic of the integrated system to produce electricity and hydrogen from pressure reduction process

In order to model the system, a computer code using Engineering Equation Solver (EES) is developed to calculate the thermo-physical properties of each state of the system as well as desired outputs such as output power, exergy efficiency, and hydrogen production rate. The input parameters for the simulation are listed in Table 6.1.

Table 6.1 Input parameter to model an integrated system [8, 9]

Parameter	Unit	Value
Organic turbine isentropic efficiency	%	0.83
Turbo expander isentropic efficiency	%	0.85
Organic pump isentropic efficiency	%	0.80
Heater effectiveness	–	0.85
Combustion chamber heat loss	%	2
Generator mechanical efficiency	%	98
Reference environmental temperature	K	293
Reference environmental pressure	Bar	1.013
Stack temperature	°C	130
Natural gas preheat temperature	°C	135
Natural gas pressure inlet	Bar	20–69
Natural gas temperature inlet	°C	40
Natural gas temperature outlet	°C	130
Fuel mass flow rate	kg/s	0.3
Organic working fluid	–	R123
TPEM	K	352
Activation energy for anode (Eac,a)	kJ/mol	76
Activation energy for cathode (Eac,c)	kJ/mol	18
J_a^{ref}	$\left(\frac{A}{m^2}\right)$	1.76E6
J_c^{ref}	$\left(\frac{A}{m^2}\right)$	4.6E6
P_{O_2}	Bar	1.0
P_{H_2}	Bar	1.0

6.3 Thermoeconomic Modeling

The thermodynamic modeling of the integrated system considered here (Fig. 6.2) can be divided into three subsystems: Turbo expander and gas-fired heater, organic Rankine cycle, and proton exchange membrane (PEM) electrolyzer. The relevant thermodynamic relation includes enthalpy of flows, flows exergy, exergy destruction rate, exergy efficiency, and energy and exergy balance and governing equations of the main subsystems of hybrid system shown in Fig. 6.2 are described in the following sections.

6.3.1 Combustion Chamber

As shown in Fig. 6.2, air and fuel are burnt in combustion chamber. The outlet properties of the combustion chamber are a function of air mass flow rate, lower heating value (LHV) of fuel, and combustion efficiency and are related as follows:

$$\dot{m}_a h_a + \dot{m}_f \mathrm{LHV} = \dot{m}_1 h_1 + (1 - \eta_{cc}) \dot{m}_f \mathrm{LHV} \tag{6.1}$$

The chemical reaction of combustion process in general form can be written as follows:

$$\begin{aligned}\lambda C_{x1}H_{y1} &+ [x_{O_2}O_2 + x_{N_2}N_2 + x_{H_2}H_2O + x_{CO_2}CO_2] \\ &\rightarrow y_{CO_2}CO_2 + y_{N_2}N_2 + y_{O_2}O_2 + y_{H_2O}H_2O + y_{NO}NO + y_{CO}CO]\end{aligned} \tag{6.2}$$

where:

$$y_{CO_2} = (\lambda x_1 + x_{CO_2} - y_{CO}) \tag{6.3}$$

$$y_{N_2} = (x_{N_2} - y_{NO}) \tag{6.4}$$

$$y_{H_2O} = \left(x_{H_2O} + \frac{\lambda \times y_1}{2}\right) \tag{6.5}$$

$$y_{O_2} = \left(x_{O_2} - \lambda \times x_1 - \frac{\lambda \times y_1}{4} - \frac{y_{CO}}{2} - \frac{y_{NO}}{2}\right) \tag{6.6}$$

$$\lambda = \frac{n_{fuel}}{n_{air}} \tag{6.7}$$

6.3.2 *Turbo Expander and Gas-Fired Heater*

6.3.2.1 Gas-Fired Heater

An energy balance for gas-fired heater can be written as follows:

$$\dot{m}_5 h_5 - \dot{m}_4 h_4 = \dot{m}_1 h_1 - \dot{m}_2 h_2 \tag{6.8}$$

6.3.2.2 Turbo Expander

The turbo expander isentropic efficiency is assumed to be 86 %. The net power output can be found as

$$\dot{W}_{TE,is} = \dot{m}_5(h_5 - h_{6s}) \tag{6.9}$$

$$h_{6a} = h_{5a} + \dot{W}_{TE,is}/\eta_{TE} \tag{6.10}$$

$$\dot{W}_{TE,a} = \dot{m}_5(h_5 - h_{6a}) \tag{6.11}$$

6.3.3 *Organic Rankine Cycle*

The hot flow gases leaving the gas-fired heater give their energy to the organic fluid in the evaporator in the organic Rankine cycle to produce more electricity. Energy balances and governing equation for various components of ORC are described here.

6.3.3.1 Organic Pump

The ORC pump work can be expressed using an energy balance equation for a control volume around the ORC pump as follows:

$$\dot{W}_{\mathrm{ORC,pump}} = \dot{m}_7(h_8 - h_7) \tag{6.12}$$

6.3.3.2 Organic Evaporator

It is assumed that hot flue gases give their energy to organic fluid with effectiveness of 85 %. The outlet temperature of organic fluid can be expressed as follows:

$$\mathrm{Eff}_{\mathrm{Eva}} = \frac{\dot{m}_9(h_{10} - h_9)}{\dot{m}_2(h_2 - h_3)} \tag{6.13}$$

6.3.3.3 Organic Condenser

An energy balance gives outlet properties of organic fluid as follows:

$$\dot{m}_{11}h_{11} = \dot{Q}_{\mathrm{cond.}} - \dot{m}_7 h_7 \tag{6.14}$$

6.3.3.4 Organic Turbine

The energy balance for organic turbine shown in Fig. 6.2 and the isentropic efficiency equation are written as follows:

$$\dot{m}_{11}h_{11} = \dot{m}_{10}h_{10} - \dot{W}_{\mathrm{ORC,Tur}} \tag{6.15}$$

$$\eta_{\mathrm{is,Turb}} = \frac{\dot{W}_{\mathrm{is,Turb}}}{\dot{W}_{\mathrm{a,Turb}}} \tag{6.16}$$

6.3.4 PEM Electrolyzer

The PEM electrolyzer duty in the system is H_2 production unit which is illustrated on the bottom side of Fig. 6.2. As it is seen, the electricity and heat are both supplied to the electrolyzer to drive the electrochemical reactions. The liquid water enters a heat exchanger that heats it to the PEM electrolyzer temperature before it enters the electrolyzer. The H_2 produced dissipates heat to the environment and stores in the storage tank for later use. The oxygen gas produced at the anode is separated from the water and oxygen mixture and then cooled to the reference environment temperature. The remaining water is returned to the water supply stream for the next hydrogen production cycle.

Thermochemical modeling and energy and exergy analyses carried out here. The total energy needed by the electrolyzer can obtain as

$$\Delta H = \Delta G + T\Delta S \tag{6.17}$$

where ΔG is Gibb's free energy and TΔS represents the thermal energy requirement. The mass flow rate of hydrogen is determined by [10]

$$\dot{N}_{H_2,Out} = \frac{J}{2F} = \dot{N}_{H_2,reacted} \tag{6.18}$$

Here, J is the current density and F is the Faraday constant. The PEM electrolyzer voltage can be expressed as

$$V = V_\circ + V_{act,a} + V_{act,c} + V_{ohm} \tag{6.19}$$

where $V_\circ$ is the reversible potential and it can be obtained with the Nernst equation as follows:

$$V_\circ = 1.229 - 8.5 \times 10^{-4}(T_{PEM} - 298) \tag{6.20}$$

Here, $V_{act,a} V_{act,c}$ and V_{ohm} are the activation overpotential of the anode, the activation overpotential of the cathode, and the ohmic overpotential of the electrolyte, respectively. The local ionic conductivity $\sigma(x)$ of the PEM is expressed as [9]

$$\sigma_{PEM}[\lambda(x)] = [0.5139\lambda(x) - 0.326]\exp\left[1,268\left(\frac{1}{303} - \frac{1}{T}\right)\right] \tag{6.21}$$

where x is the distance into the membrane measured from the cathode membrane interface and $\lambda(x)$ is the water content at a location x in the membrane. The value of $\lambda(x)$ can be calculated in terms of the water content at the membrane electrode edges:

$$\lambda(x) = \frac{\lambda_a - \lambda_c}{D} x + \lambda_c \tag{6.22}$$

Here, D is the membrane thickness, and λ_a and λ_c are the water contents at the anode membrane and the cathode membrane interfaces, respectively. The overall ohmic resistance and ohmic overpotential can be expressed as [9]

$$R_{PEM} = \int_0^D \frac{dx}{\sigma_{PEM}[\lambda(x)]} \tag{6.23}$$

$$V_{ohm,PEM} = JR_{PEM} \tag{6.24}$$

The activation overpotential, V_{act}, caused by a deviation of net current from its equilibrium, and also an electron transfer reaction, must be differentiated from the concentration of the oxidized and reduced species [9, 10]. Then,

$$V_{\mathrm{act,i}} = \frac{RT}{F}\sinh^{-1}\left(\frac{J}{2J_{0,\mathrm{i}}}\right) \quad i = a, c \tag{6.25}$$

Here, J_0 is the exchange current density, which is an important parameter in calculating the activation overpotential. The exchange current density for electrolysis can be expressed as [9]

$$J_{0,\mathrm{i}} = J_{\mathrm{i}}^{\mathrm{ref}}\exp\left(-\frac{E_{\mathrm{act,i}}}{RT}\right) \quad i = a, c \tag{6.26}$$

where J_{ref} i is the pre-exponential factor and $E_{\mathrm{act,i}}$ is the activation energy for the anode and cathode. More details about PEM electrolysis modeling can be found elsewhere [11].

6.4 Performance Assessment

In order to enhance understanding of energy system, a performance assessment is needed. In this regard the variation of some of the major design parameters on the system performance is discussed and analyzed. This helps to increase the knowledge of the designers to predict the system when a change is inserted according to any causes. In order to conduct performance assessment, we need to consider those outputs where the variation of design parameters has significant effects. In this chapter we consider the exergy efficiency, total exergy destruction, and output power of the integrated system as our criteria. Exergy efficiency is a real measure of how well the system works and differs from the ideal system with the maximum efficiency. Exergy efficiency is a good indicator for energy systems and assists us to come up us ideas and improvements for the system to reduce the losses. Exergy efficiency for a system is defined as the exergy of desired output divided by the exergy of the input of the system. Exergy efficiency and hydrogen production rate for this integrated system for natural gas reduction station are expressible as

$$\psi_{\mathrm{system}} = \frac{\dot{E}x_{\mathrm{H_2}} + \dot{W}_{\mathrm{TE}}}{\dot{E}x_{\mathrm{Fuel}}} \tag{6.27}$$

$$\dot{m}_{\mathrm{H_2}} = 2\dot{N}_{H_2} \tag{6.28}$$

For more details about the calculation of exergy in different points of system, the study carried out by [12–17] can be considered.

6.5 Results and Discussion

6.5.1 Exergy Analysis Results

In the following sections the main goals are overall thermodynamic appraisal of the system and conduct a comprehensive parametric analysis to determine the effect of main effective parameters on the system performance. The major output parameters of simulation are listed in Table 6.1. Result indicates that exergy efficiency of system is 38.8 % at the design points. Also, it is seen that the total power output is 5.56 MW.

The results of thermodynamic analysis are presented here. Figure 6.3 shows the exergy destruction rate of each component. It is determined that combustion chamber, evaporator, and condenser have the highest exergy destruction rate among the system devices. The highest exergy destruction rate in combustion

Table 6.2 Output values from modeling of integrated system

Parameter	Unit	Value
Exergetic efficiency of system	%	38.8
Exergetic efficiency of ORC	%	11.25
Exergetic efficiency of PEM	%	53.92
Exergetic efficiency of turbo expander	%	90.7
Net power output of turbo expander	MW	3.73
Net power output of organic Rankine cycle	MW	1.83
Total exergy destruction rate	kW	7,075
Hydrogen production rate	k/day	15.89

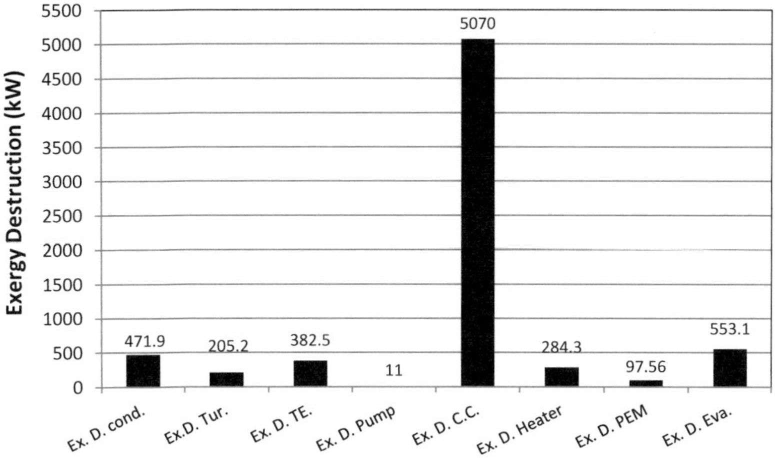

Fig. 6.3 Exergy destruction rate of system component

chamber is due to temperature difference between the burner and working fluid which causes the large entropy generation. The exergy destruction rate in evaporator and condenser has the second and third rank after combustion chamber. The main reason is due to the phase change and temperature difference between the hot and cold streams which causes high entropy generation rate which subsequently results in a high exergy destruction.

6.5.2 Parametric Analysis

6.5.2.1 Natural Gas Pressure Effects

Figure 6.4 shows the effect of natural gas inlet pressure on the exergetic efficiency of each subsystems and the integrated system. As it is seen an increase in natural gas pressure has more effect on the integrated system exergetic efficiency.

To better understand the effect of natural gas pressure inlet on exergetic efficiency, the variation of power output and hydrogen production rate is illustrated in Fig. 6.5.

As shown in this figure an increase in natural gas pressure leads to a positive effect on turbo expander power output which increases the power output from 461.8 to 3.731 kW. Since, there is a decrease from 1,890 to 1,836 kW in organic Rankine cycle power output, a decrement in hydrogen production rate can be seen.

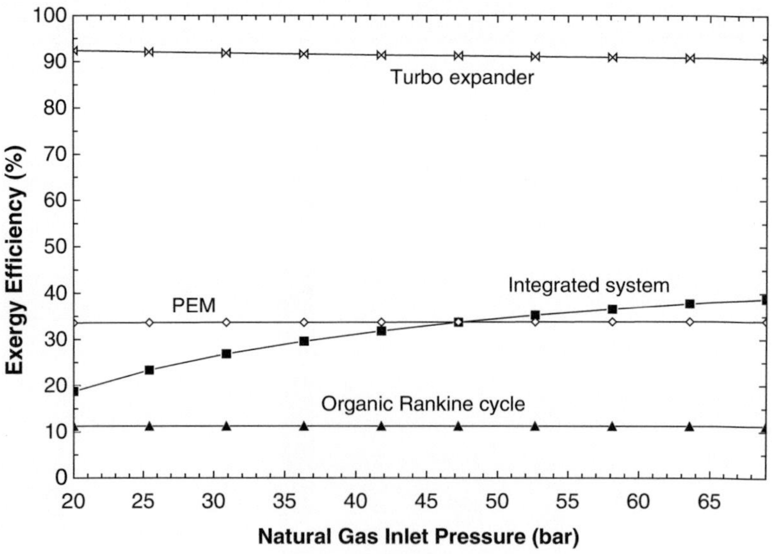

Fig. 6.4 The effect of natural gas pressure on exergetic efficiency

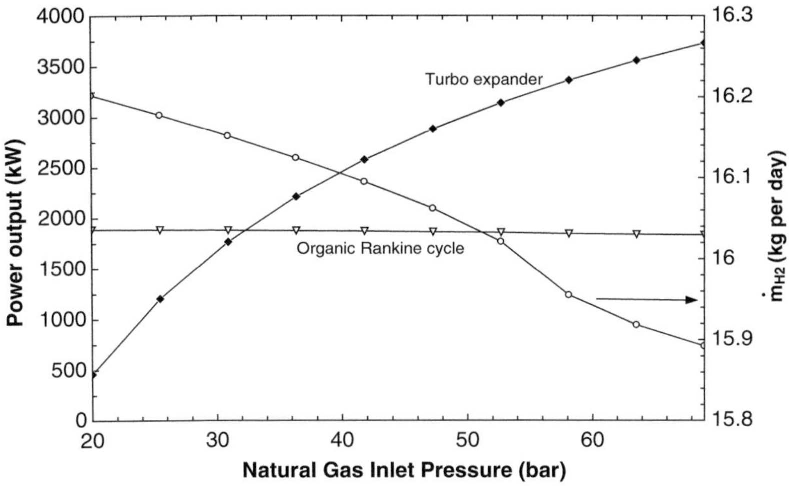

Fig. 6.5 Variation of power output and hydrogen production rate with natural gas pressure

6.5.2.2 Natural Gas Preheat Temperature Effects

One of the important design parameters which has a substantial effect on integrated system performance is natural gas preheat temperature. In the CGS natural gas is preheated to an appropriate temperature to avoid low temperature in CGS outlet. The low temperature has harmful effects such as formation of hydrates and ice which has bad drawbacks on the performance of system. Figure 6.6 shows the variation of exergy destruction rate of components with natural gas preheat temperature. This figure indicate that increase the natural gas preheat temperature has different effect on evaporator, condenser, and heater exergy destruction rate while others have insignificant change.

Although natural gas preheat temperature leads to an increase in turbo expander power output due to an increase in inlet enthalpy (h_5), it leads to an increase in natural gas heater exergy destruction rate. As shown in Fig. 6.7, an increase in natural gas preheat temperature, the organic Rankine cycle power output reduces from 1,836 to 1,651 kW which eventually decreases the hydrogen production rate from 15.89 to 14.79 kg/day.

6.5.2.3 Fuel Mass Flow Rate

The fuel mass flow rate is one of the important parameters in the integrated system. At constant preheat temperature, increase in fuel mass flow rate leads to an increase in ORC power output and hydrogen production rate. It is due to the fact that fuel mass flow rate play an important role in the control hydrogen production rate and ORC power output to meet the electrical demand (Fig. 6.8).

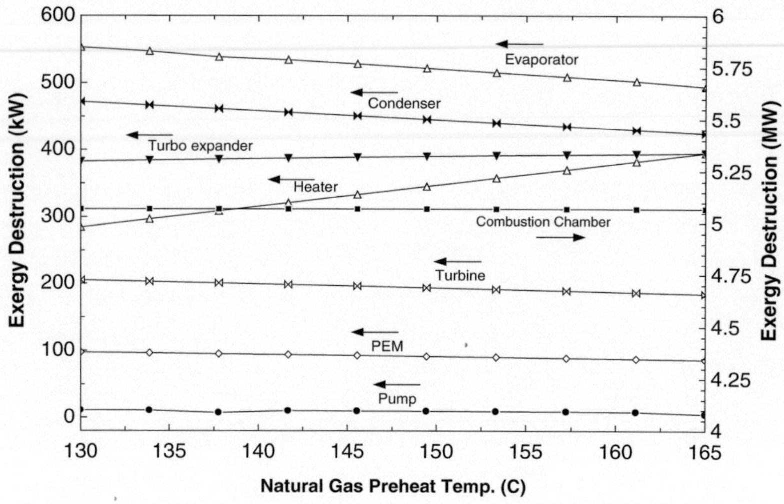

Fig. 6.6 Variation of exergy destruction rate with natural gas preheat temperature

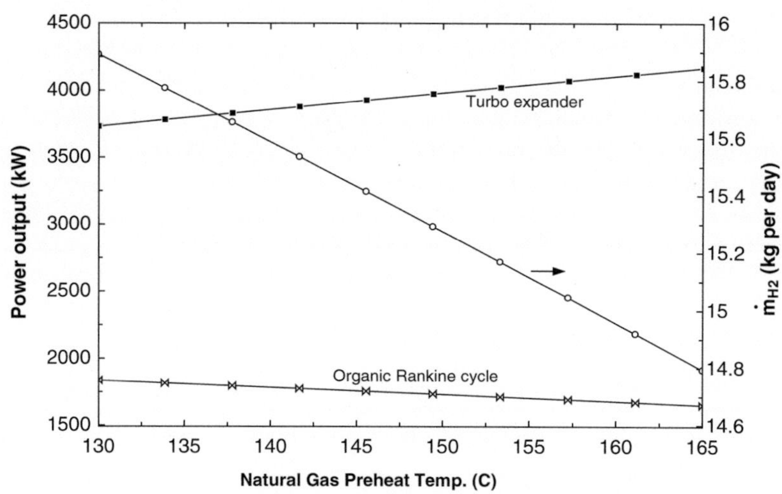

Fig. 6.7 Variation of power output and hydrogen production rate with natural gas preheat temperature

Despite the fact that an increase in fuel mass flow rate has a positive effect on power output and hydrogen production, it imposes more exergy destruction to the system (Fig. 6.9). Results show that with increased fuel mass flow rate to 0.45 kg/s, the exergy destruction rate of combustion chamber can be more than 7.5 MW.

It should be mentioned that the CGS can be available in various regions with different climate and ambient temperature. To provide environmental insight, the effect of ambient temperature on integrated system is investigated here. As it can be

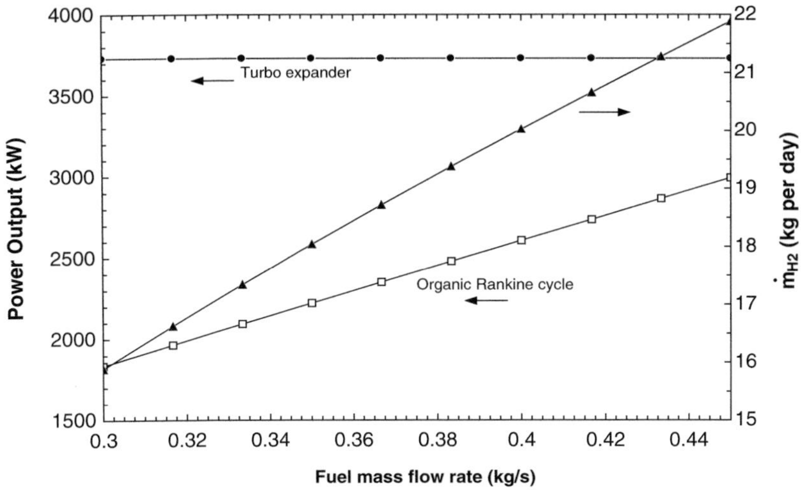

Fig. 6.8 Variation of power output and hydrogen production rate with the fuel mass flow rate

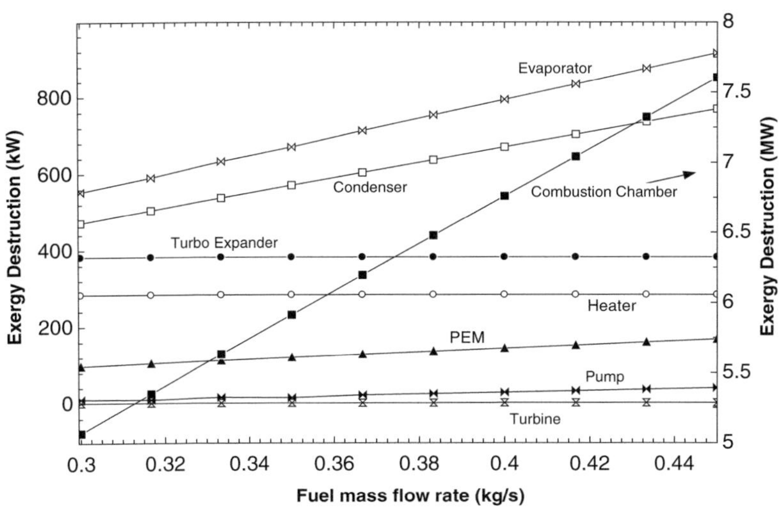

Fig. 6.9 Variation of exergy destruction rate with the fuel mass flow rate

seen the exergy destruction rate of each subsystem decreases with a rise in ambient temperature. An increase in $T_\circ$ has major effects on the exergy destruction rate for the turbo expander, heater, and combustion chamber while for polymer exchange membrane electrolyzer is almost constant (Fig. 6.10).

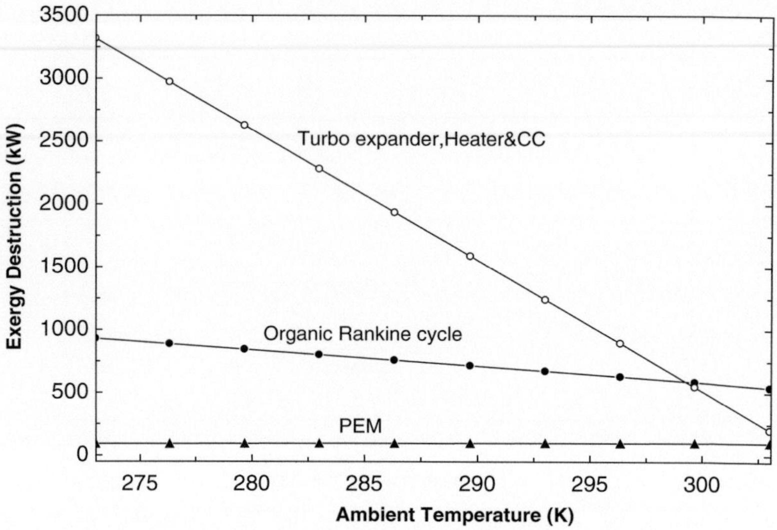

Fig. 6.10 Variation of exergy destruction rate with variation ambient temperature

6.5.2.4 Organic Rankine Cycle High Temperature

The high temperature of organic Rankine cycle is another major parameter in ORC because it affects the exergy input to the turbine. Although an increase in high temperature of ORC decreases mass flow rate of working fluid in a constant heat feed to evaporator, higher inlet enthalpy (h_{10}) leads to increase power output of ORC. Increase the power output of ORC has a positive effect on hydrogen production rate of PEM electrolyzer which can improve the integrated system performance (Fig. 6.11). Moreover, high-temperature variation has a different effect on PEM electrolyzer and turbo expander exergy efficiency as is shown in the Fig. 6.11.

6.5.3 Genetic Algorithm Optimization

Genetic algorithm is one of the efficient methods based on evolutionary algorithm using natural selection. More details are given [18]. As the parametric study showed, the parameter variations have complicated effects on the integrated system performance. To find the optimized parameters which ensure maximum performance of the integrated system, a genetic algorithm based optimization is applied. The decision variables and constraints are presented in Table 6.3. The constraints are selected based on the allowable actual data variation range.

The objective function in this problem is exergy efficiency of the integrated system. The optimization results for independent variables are presented in the Table 6.4.

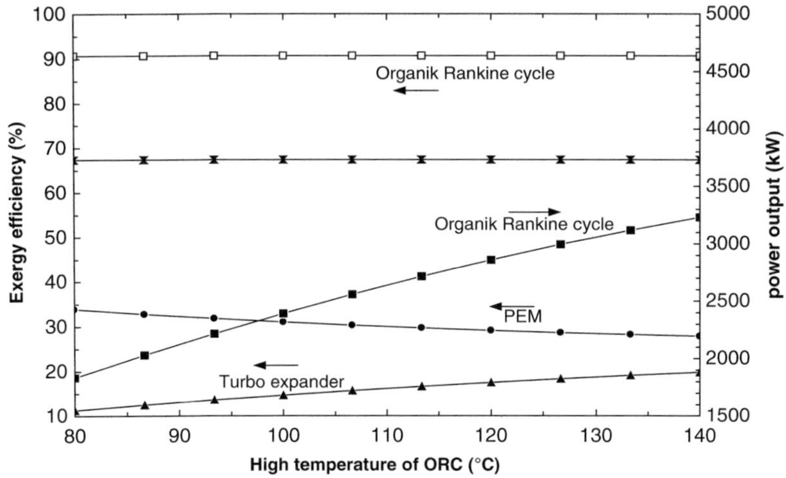

Fig. 6.11 Variation of exergy efficiency and power output with variation ORC high temperature

Table 6.3 Decision variables and constraints

Decision variables	Unit	Constrains
Natural gas preheat temperature	°C	$130 < T_5 < 165$
Fuel mass flow rate	kg/s	$0.25 < \dot{m}_f < 0.4$
High temperature of ORC	°C	$80 < T_{10} < 140$
Natural gas inlet pressure	Bar	$17.5 < P_4 < 69$
Stack temperature	°C	$130 < T_3 < 160$

Table 6.4 Optimized value using GA method to independent variables

	T_5(°C)	$\dot{m}_f$ (kg/s)	P_4 (bar)	T_{10} (°C)	T_3(°C)
Design value	130	0.3	69	80	130
Optimized value	155.8	0.309	66.31	139	131.2

It is concluding that maximum performance of the proposed system for hydrogen production rate and electricity generation can be achieved by optimized value. As Table 6.5 shows, in the optimized state, integrated exergy efficiency of system has an appropriate increment of about 9.8 %. Also, increase ORC power output causes a positive effect on hydrogen production rate.

6.6 Conclusions

In this study, the thermodynamic modeling and genetic algorithm-based optimization for practical integrated energy system provide more useful information. The parametric studies reveal that variation of design parameters has different effects on

Table 6.5 Optimized value using genetic algorithm method

	Exergy efficiency (%)			ORCPower output (kW)	Turbo expander power output (kW)	Hydrogen production rate (kg/day)
	Integrated system	ORC	PEM			
Design values	38.8	11.3	53.9	1,836	3,731	16.21
Optimized values	48.6	19.78	28.26	3,106	4,008	22.39
Variation (%)	25.25	75	−47.6	69.2	7.42	38.12

system performance. In order to determine the best values of design parameter which ensure the optimum performance of system, a genetic algorithm optimization method is applied. Optimization results show that in optimum decision variables, the exergetic efficiency of proposed integrated system can be enhanced from 38.8 to 48.6 %. Other concluding remark result can be listed as follows:

- The natural gas pressure inlet to turbo expander has a positive effect on turbo expander power output while decreases hydrogen production rate.
- Although increase fuel mass flow rate of combustion chamber increases power output, it imposes more exergy destruction to the proposed integrated system.
- Preheating of natural gas to higher temperature has two effects on the proposed integrated system. However, natural gas preheat temperature increases turbo expander power output, and it can diminish hydrogen production rate severely from 15.9 to 14.8 kg/day.
- Increasing the high-temperature of ORC to a higher level lower than critical temperature of organic fluid has a favorable impact on integrated system.
- The optimum values of five design parameters are natural gas preheat temperature of 155.8 °C, fuel mass flow rate of 0.309 kg/s, natural gas inlet pressure of 66.31 bar, high temperature of ORC of 139 °C, and stack temperature of 131.2 °C.

The thermodynamic analysis and optimization of proposed integrated system can provide a useful base to design a more efficient and sustainable energy system to recover energy from pressure reduction in gas network system.

References

1. Rahman MM (2010) Power generation from pressure reduction in the natural gas supply chain in Bangladesh. J Mech Eng 41(2):89–95
2. Seresht RT, Ja H (2010) Retrofit of Tehran City gate station (CGS No. 2) by using turboexpander

3. Howard CR (2009) Hybrid turboexpander and fuel cell system for power recovery at natural gas pressure pressure reduction stations. Department of Mechanical and Materials Engineering, Queen's University, Kingston
4. Kostowski W (2010) The possibility of energy generation within the conventional natural gas transport system. Strojarstvo: časopis za teoriju i praksu u strojarstvu 52(4):429–440
5. Farzaneh-Gord M, Sadi M (2008) Enhancing energy output in Iran's natural gas pressure drop stations by cogeneration. J Energy Institute 81(4):191–196
6. Kostowski WJ, Usón S (2013) Comparative evaluation of a natural gas expansion plant integrated with an IC engine and an organic Rankine cycle. Energy Conversion Manage 75:509–516
7. He T, Ju Y (2013) Design and optimization of natural gas liquefaction process by utilizing gas pipeline pressure energy. Appl Thermal Eng 57(1):1–6
8. Ahmadi P, Dincer I, Rosen MA (2012) Energy and exergy analyses of hydrogen production via solar-boosted ocean thermal energy conversion and PEM electrolysis. Int J Hydrogen Energy 38:1795–1805
9. Ni M, Leung MK, Leung DY (2008) Energy and exergy analysis of hydrogen production by a proton exchange membrane (PEM) electrolyzer plant. Energy Conversion Manage 49 (10):2748–2756
10. Khanmohammadi S, Ahmadi P, Mirzei D (2014) Thermodynamic modeling and optimization of a novel integrated system to recover energy from a gas pressure reduction station. In: Proceedings of the 13th international conference of clean energy, Istanbul, Turkey, pp 2357–2369
11. Nieminen J, Dincer I, Naterer G (2010) Comparative performance analysis of PEM and solid oxide steam electrolysers. Int J Hydrogen Energy 35(20):10842–10850
12. Khanmohammadi S, Azimian AR (2013) Exergoeconomic evaluation of a two-pressure level fired combined-cycle power plant. J Energy Eng doi:10.1061/(ASCE)EY.1943-7897.0000152
13. Khanmohammadi S, Azimian AR, Khanmohammadi S (2013) Exergy and exergo–economic evaluation of Isfahan steam power plant. Int J Exergy 12(2):249–272
14. Ahmadi P, Dincer I, Rosen MA (2011) Exergy, exergoeconomic and environmental analyses and evolutionary algorithm based multi-objective optimization of combined cycle power plants. Energy 36(10):5886–5898
15. Ameri M, Ahmadi P, Khanmohammadi S (2008) Exergy analysis of a 420 MW combined cycle power plant. Int J Energy Res 32(2):175–183
16. Meigounpoory MR, Ahmadi P, Ghaffarizadeh AR, Khanmohammadi S (2008) Optimization of combined cycle power plant using sequential quadratic programming. In: ASME heat transfer summer conference collocated with the fluids engineering, energy sustainability, and 3rd energy nanotechnology conferences
17. Ahmadi P, Dincer I, Rosen MA (2013) Performance assessment and optimization of a novel integrated multi-generation system for residential buildings. Energy Build 67:568–578
18. Ghaffarizadeh A, Eftekhari M, Esmailizadeh AK, Flann NS (2013) Quantitative trait loci mapping problem: an extinction-based multi-objective evolutionary algorithm approach. Algorithms 6(3):546–564

Chapter 7
Performance Improvement of Adsorption Cooling System by Heat Recovery Operation

W. Chekirou, N. Boukheit, and A. Karaali

Abstract A detailed thermodynamic analysis of simple and regenerative cycles of adsorption refrigeration is presented. Two functions of the incoming and outgoing energy for the regenerative cycle using two isothermal adsorbers have been calculated in order to obtain the heated adsorber temperature at the end of heat recovery. Results are presented in terms of performances for the pair activated carbon AC-35 as adsorbent and methanol as adsorbate. These results demonstrated that the performance coefficient of double-bed adsorption refrigeration cycle increases with respect to the single-bed configuration. Several main factors affecting the performances of cycles are discussed according to the results of computer simulations.

Keywords Adsorption system • Heat regenerative • Performance • Thermodynamics

Nomenclature

COP_s, COP_d	Performance coefficient in single bed and double bed, respectively
Cp_{ml}, Cp_{mg}	Specific heat of the adsorbate in liquid and vapour state, respectively, J/kg k
Cp_a, Cp_g	Specific heat of the adsorbent and the metal of the adsorber, respectively, J/kg k
L	Latent heat of evaporation, KJ/kg
m, m_a, m_g	Adsorbed mass, mass of the adsorbent and metallic mass of the adsorber, respectively, kg
$m_{\max}$, $m_{\min}$	Adsorption mass at adsorbed and desorbed state, respectively, kg/kg
T_a, T_g	Adsorption and regenerating temperature, respectively, °C
T_{c1}, T_{c2}	Limit temperature of desorption and adsorption, respectively, °C
T_r	Heated adsorber temperature at the end of heat recovery, °C

W. Chekirou (✉) • N. Boukheit • A. Karaali
Laboratoire de thermodynamique et traitement de surface de matériaux, Université Constantine1, Route Ain El Bey, Constantine 25000, Algeria
e-mail: chekirouw@yahoo.fr

I. Dincer et al. (eds.), *Progress in Clean Energy, Volume 1*,
DOI 10.1007/978-3-319-16709-1_7

T_e, T_c	Evaporation and condensation temperature, respectively, °C
Q_f, Q_c	Cooling power and total heat necessary for heating the adsorber, respectively, KJ/kg
Q_r	Heat recovered, KJ/kg
q_{st}	Isosteric heat of adsorption, KJ/kg
r	Heat recovery ratio
ΔT_r	Two-adsorber temperature difference at the end of heat recovery, °C

7.1 Introduction

Adsorption refrigeration is a technology of environmental-friendly refrigeration, which can be driven by solar energy or waste heat. In recent years, adsorption refrigeration has already been widely studied. With the development of its technologies and theories [1–4], the application of adsorption refrigeration is getting extensive; research indicates that adsorption is friendlier for the environment than HFC vapour compression in several cases, such as waste heat-powered adsorption cycle [5], natural gas-fired air conditioning [6], reversible heat pump [7], cold stores for transportation [1], etc.

However, the basic cycle of adsorption refrigerating machines presents two main drawbacks: the production of cold is intermittent, and the efficiency is low. To attain higher efficiencies and to achieve continuous production of cold, it is necessary to use advanced cycles. Several kinds of advanced cycles have been proposed and tested. Two main technologies have been developed: regenerative processes with uniform temperature adsorbers and regenerative processes with temperature fronts (or thermal waves).

The basic idea is to use the heat discarded by one adsorber under cooling to preheat another adsorber under heating. In so doing, the COP is enhanced and the production of cold is continuous.

In this paper, we are interested to the uniform temperature adsorbers. A numerical analysis was carried out, studying the influence of the main parameters on both regeneration and performance coefficients of the machine. A detailed thermodynamic and parametric analysis of a double adsorptive cycle is given. The basic fundamentals of the adsorption process are discussed, where the Dubinin-Astakhov equation is used to describe the isotherm of adsorption.

7.2 System Description

Adsorption refrigeration systems have been initially proposed of a single adsorbent bed alternately connected to a condenser and evaporator. Theoretically, the corresponding cycle consists of two isosters and two isobars, as illustrated in the Clapeyron diagram (Fig. 7.1). The process starts at point **a**, where the adsorbent is at a low temperature T_a and at low pressure P_e (evaporation pressure). While the adsorbent is heated, the temperature and the pressure increase along the isoster which the mass of the adsorbate in the adsorbent remains constant at m_{max}. The adsorber is still isolated until the pressure reaches the condenser pressure at point **b**. At this time, the adsorber is connected with the condenser and the progressive heating of the adsorbent from point **b** to **c** causes a desorption of methanol and its vapour is condensed in the condenser and collected in a receiver. When the adsorbent reached its maximum temperature value T_g (regenerating temperature) and the adsorbed mass decreases to its minimum value m_{min} (point c), the adsorbent starts cooling along the isoster at a constant mass m_{min} to point **d**. During this isosteric cooling phase, the adsorbent pressure decreases until it reaches the evaporator pressure P_e. After that, the adsorber is connected to the evaporator, and both adsorption and evaporation occur while the adsorbent is cooled from point **d** to **a**. In this phase, the adsorbed mass increases up to its maximum m_{max} at point **a**, and the adsorbent is cooled until the adsorption temperature T_a. During this phase also, the cold is produced.

The discontinuity of the useful effect makes the system with single adsorber commercially unsuitable. Later, two adsorbent bed machines have been proposed [8, 9], in order to operate regenerative cycles. By these, both the results of a higher overall efficiency and a continuous useful energy production can be obtained. The idealized thermodynamic cycle of a double-bed adsorption machine is plotted in Fig. 7.2. The adsorbent beds operate the same cycle but in counter phase, so that the heat recovery is obtained by transferring the heat from one bed to the other, until a fixed difference of temperature ΔT_r between beds is reached.

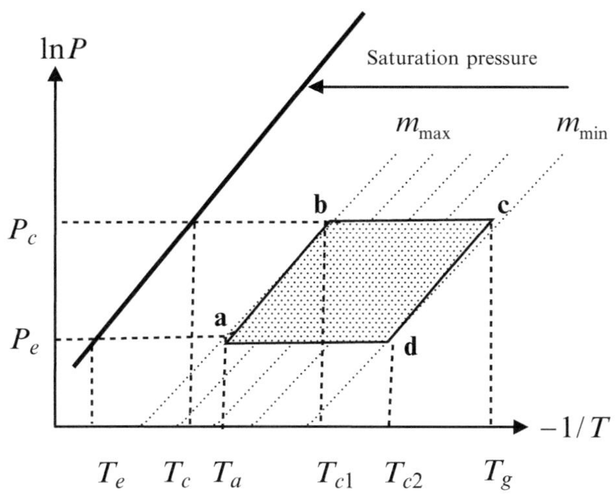

Fig. 7.1 Clapeyron diagram of a single-bed adsorption cycle

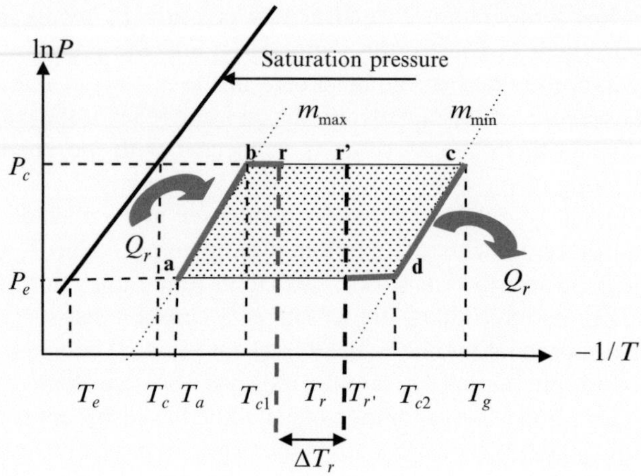

Fig. 7.2 Clapeyron diagram of a regenerative adsorption cycle

7.3 Thermodynamic Model

In the following, the developed model is described with reference to a single-bed adsorption system (Fig. 7.1), and the main difference introduced in modelling of a two-bed system is also presented.

In order to calculate the coefficient of performance of the machine, all the thermal contributions must be calculated in detail; the most important equations used in the model are described below.

The heat that must be supplied to the adsorber for its heating is:

$$Q_c = Q_{ab} + Q_{bc} \tag{7.1}$$

where Q_{ab} is the heat that must be supplied to the adsorber for its isosteric heating:

$$Q_{ab} = \int_{T_a}^{T_{c1}} \left[Cp_a + m_g Cp_g + m_{\max} Cp_{ml}(T) \right] m_a \, \mathrm{d}T \tag{7.2}$$

and Q_{bc} is the heat needed for the desorption phase:

$$Q_{bc} = \int_{T_{c1}}^{T_g} \left[Cp_a + m_g Cp_g + m(T) Cp_{ml}(T) + q_{st} \frac{\partial m}{\partial T} \right] m_a \, \mathrm{d}T. \tag{7.3}$$

During the isosteric cooling phase, only the sensible heat is withdrawn from the bed:

$$Q_{cd} = \int_{T_{c2}}^{T_g} \left[Cp_a + m_g Cp_g + m_{\min} Cp_{ml}(T) \right] m_a \, \mathrm{d}T. \tag{7.4}$$

While during the adsorption phase, the energy released is equal to the heat of adsorption, plus the sensible heat obtained from cooling of adsorbent, adsorber and adsorbate and from critical adsorption temperature to adsorption temperature, minus the energy needed to heat up the vapour from evaporation to adsorption temperature:

$$Q_{da} = \int_{T_a}^{T_{c2}} \left[Cp_a + m_g Cp_g + m(T) Cp_{ml}(T) + q_{st} \frac{\partial m}{\partial T} - (T - T_e) Cp_{mg}(T) \frac{\partial m}{\partial T} \right] m_a \, \mathrm{d}T. \tag{7.5}$$

The energy that must be supplied to the evaporator Q_f is calculated as the latent heat of evaporation of the cycled adsorbate, minus the sensible heat of the adsorbate that is entering the evaporator at condensation temperature:

$$Q_f = m_a (m_{\max} - m_{\min}) \left[L(T_e) - \int_{T_e}^{T_c} Cp_{ml}(T) \, \mathrm{d}T \right]. \tag{7.6}$$

The adsorbed mass is obtained from the state equation of the bivariant solid-vapour equilibrium using the Dubinin-Astakhov model.

On the basis of the previous equations, the coefficient of performance of single adsorbent bed can be calculated as the ratio of useful effect produced and energy supplied to the machine:

$$\mathrm{COP}_s = \frac{Q_f}{Q_c} = \frac{Q_f}{Q_{ab} + Q_{bc}}. \tag{7.7}$$

In the regenerative cycle, the above parameters have been calculated between $m_{\max}$ and a generic isosteric line $m_r \geq m_{\min}$; the function Q_{ar} is calculated in order to obtain the energy incoming to the system when its temperature increases from T_a to a generic value T_r. The thermodynamic cycle of the heat regenerative adsorption cycle can be expressed by Fig. 7.2. The function Q_{ar} can be represented by the following expression, for $T_{c1} \prec T_r \prec T_{c2}$:

$$Q_{ar} = Q_{ab} + \int_{T_{c1}}^{T_r} \left[Cp_a + m_g Cp_g + m(T) Cp_{ml}(T) + q_{st} \frac{\partial m}{\partial T} \right] m_a \, \mathrm{d}T. \tag{7.8}$$

In the same way, the function $Q_{cr'}$ has been calculated. It represents the energy out coming from the system when its temperature decreases from T_g to a generic value $T_{r'}$ where $T_{r'} = T_r + \Delta T_r$; ΔT_r is the two-adsorber temperature difference at the end of heat recovery. The temperature at which Q_{ar} is equal to the absolute value of $Q_{cr'}$ is the final temperature of the regenerative phase T_r, and the corresponding value of the function Q_{ar} or $Q_{cr'}$ is the regenerative energy Q_r. In this case, the performance of the machine can be calculated by the following formulae:

$$\mathrm{COP}_d = \frac{\mathrm{COP}_s}{1 - r} \tag{7.9}$$

where r is the coefficient of regeneration (or the heat recovery ratio); it can be calculated by:

$$r = \frac{Q_r}{Q_{ab} + Q_{bc}}. \tag{7.10}$$

7.4 Results

We admit the following data: the pair activated carbon AC-35/methanol is an adsorptive pair, this last has proved to be the best pair among those studied [10]; copper is a material of construction of the adsorber; $m_g = 5\,\mathrm{kg}$ its mass; $m_a = 1$ kg is an adsorbent mass; two-adsorber temperature difference at the end of heat recovery $\Delta T_r = 2\,°\mathrm{C}$; adsorption temperature $T_a = 25\,°\mathrm{C}$; condensation temperature $T_c = 30\,°\mathrm{C}$; evaporation temperature $T_e = 0\,°\mathrm{C}$; and regenerating temperature $T_g = 105\,°\mathrm{C}$. The performances under this conditions are predicted as $\mathrm{COP}_d = 0.682$, $\mathrm{COP}_s = 0.483$ and $r = 0.291$ ($r = 29.1\,\%$). The influence of the regenerating temperature on performance coefficients, heat recovery ratio and temperature at the end of heat recovery is shown in Figs. 7.3, 7.4 and 7.5, respectively.

From Fig. 7.3, with the increase of regenerating temperature, COP_d increases all along, while COP_s decreases from regenerating temperature $T_g = 105\,°\mathrm{C}$. This behaviour can be justified by the fact that after certain regenerating temperature, the energy of heating only serves to increase the activated carbon temperature, the adsorber metal part temperature and the methanol temperature; nevertheless, the desorbed mass of methanol $m_{\min}$ becomes more and more weak. Where, the heat provided to the adsorber Q_c increases more than the quantity of cold produced at evaporator level Q_f. It is clear that, in the case of two adsorbers, the performance of the system increases in an appreciable manner compared to the case of one adsorber.

The regenerating temperature is a design variable that must be optimized. Generally, it is stipulated to obtain a large amount of cycled masses at lower levels of T_g. In the case of the adsorption of methanol in activated carbon AC-35, this temperature is limited by 150 °C, because the methanol would decompose, the process of adsorption is blocked and the adsorption power of activated carbon decreases sharply beyond the abovementioned temperature.

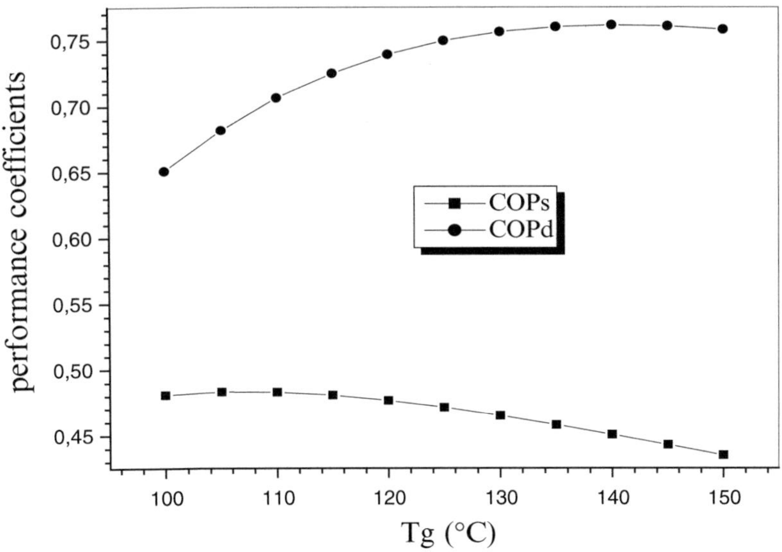

Fig. 7.3 Influence of regenerating temperature T_g on performance coefficients COP_s and COP_d

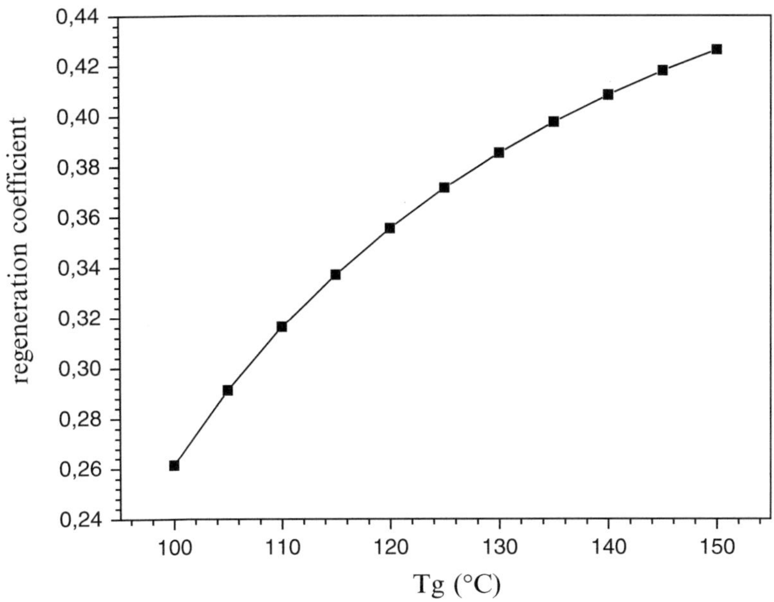

Fig. 7.4 Influence of regenerating temperature T_g on heat recovery ratio r

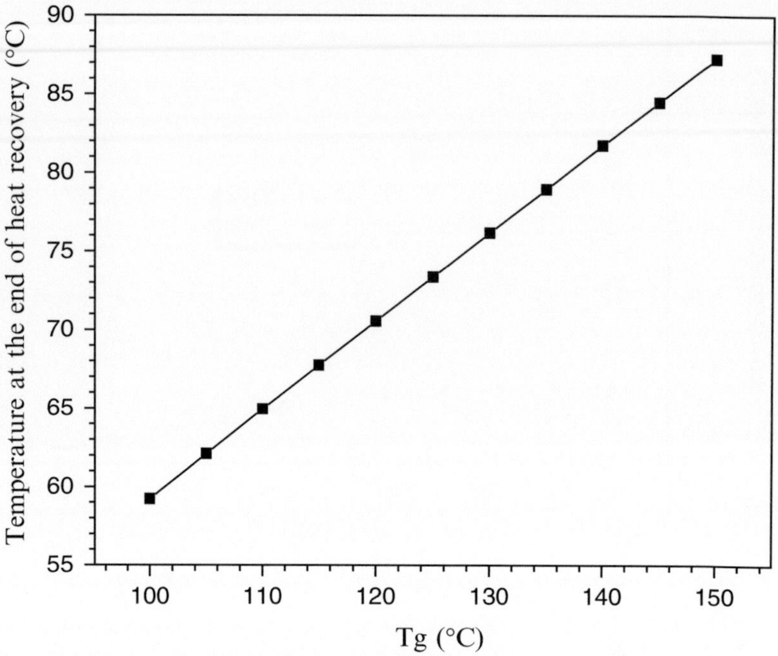

Fig. 7.5 Influence of regenerating temperature T_g on temperature at the end of heat recovery T_r

The heat recovery ratio r and the temperature at the end of heat recovery T_r increase with the increase of T_g (Figs. 7.4 and 7.5).

The influence of the adsorber mass on the performance coefficients of system COP_s and COP_d has been evaluated, by varying the mass m_g.

From Fig. 7.6, if the adsorber metal mass increases, COP_s and COP_d reduce. The reason is that the bigger the heat capacity of the non-adsorbent materials is, the more the heat will be consumed for inert materials. Thus, a part of heating power becomes energy loss due to the switch between two adsorbers. This kind of energy loss leads to the decrease of performance coefficients of system COP_s and COP_d.

On the contrary, the heat recovery ratio r and the temperature at the end of heat recovery T_r are on the increase (Figs. 7.7 and 7.8).

7.5 Conclusions

A detailed thermodynamic model has been developed for the analysis of solid adsorption refrigeration system (single bed and two beds), using the pair activated carbon AC-35/methanol as an adsorptive pair. This model was usually expressed in algebraic or relatively simple equations based on the representation of Dubinin-Astakhov.

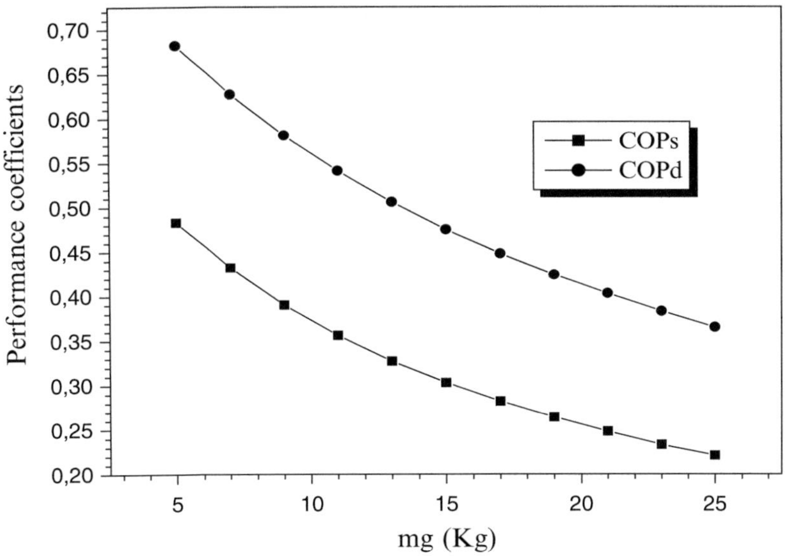

Fig. 7.6 Influence of adsorber mass m_g on performance coefficients COP_s and COP_d

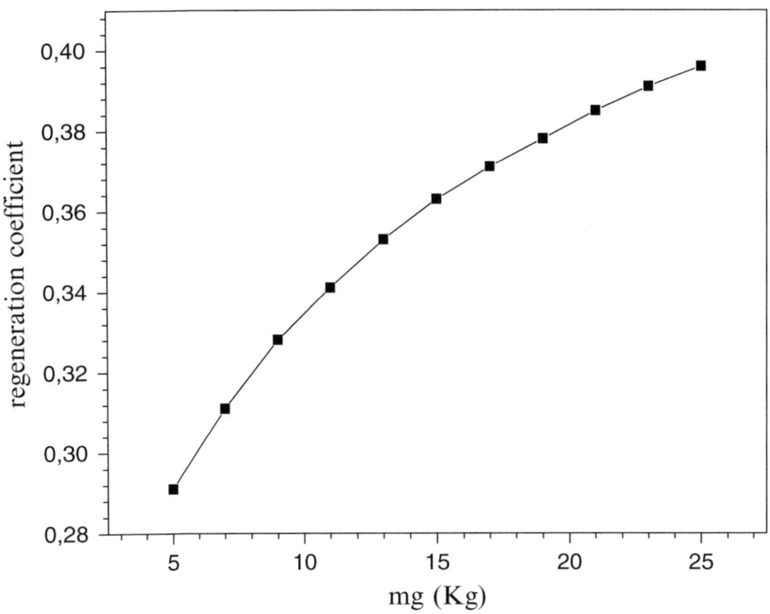

Fig. 7.7 Influence of adsorber mass m_g on heat recovery ratio r

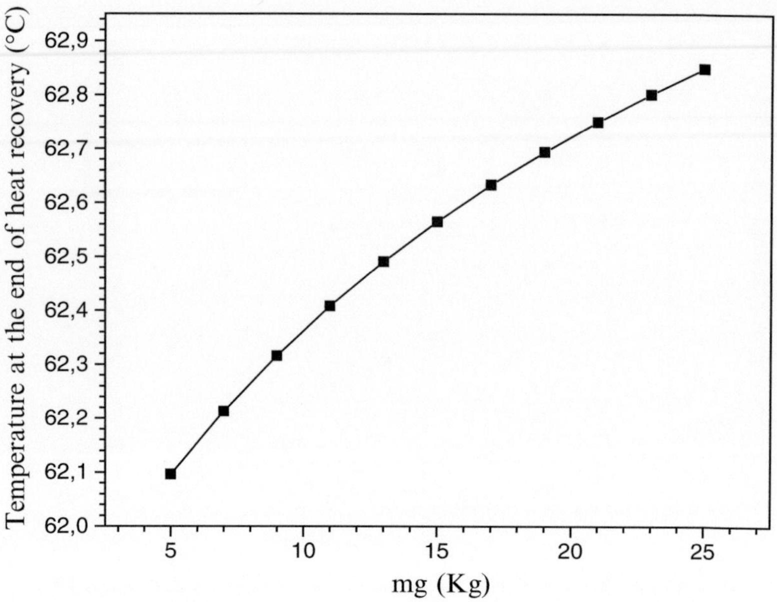

Fig. 7.8 Influence of adsorber mass m_g on temperature at the end of heat recovery T_r

The developed model is described with reference to a single-bed adsorption system, and the main differences introduced in modelling of a two-bed system are also presented.

By the computer program based on this model, it is possible to calculate the operative conditions and coefficient of performance and heat regenerative ratio.

References

1. Meunier F (1999) Adsorption heat pump technology: possibilities and limits. In: Proceedings of the international sorption heat pump conference, Munich, p. 25
2. Cacciola G, Restuccia G (1994) Progress on adsorption heat pump. Heat Recovery Syst CHP 4:409–420
3. Chua HT et al (1999) Modeling the performance of two-bed silica gel–water adsorption chillers. Int J Refrigeration 22:194–204
4. Critoph RE (1999) Forced convection adsorption cycle with packed bed heat regeneration. Int J Refrigeration 22(1):38–46
5. Meunier F (1999) Adsorption heat pump technology: present and future. In: 20th international congress of refrigeration, IIR/IIF, Sydney, p 522
6. Critoph RE (1996) Gas fired air conditioning using a carbon–ammonia convective thermal wave cycle. In: Proceedings of absorption, Montreal, p 353
7. Miles DJ, Shelton SV (1996) Design and testing of a solid-sorption heat-pump system. Appl Therm Eng 16(5):389–394

8. Douss N, Meunier F (1989) Experimental study of cascading adsorption cycles. Chem Eng Sci 44:225–235
9. Miles DJ, Shelton SV (1991) Coupled heat transfer and thermodynamic adsorption heat pump analysis AES. Vol. 26. Heat pump design and application. In: Proceedings of winter annual meeting of the American society of mechanical engineers Atlanta, pp 33–38
10. Pons M, Grenier P (1986) A phenomenological adsorption equilibrium law extracted from experimental and theoretical considerations applied to the activated carbon + methanol pair. Carbon 24(5):615–625

Chapter 8
The Use of Municipality Water System for Building Cooling as an Alternative to Conventional Ground Source Heat Pump

Yigit Can Sezgin and Murat Celik

Abstract This paper investigates the usage of municipality water supply system for cooling a residential building. The study in this paper proposes that by using municipality water supply system, it is possible to reduce the cost of space cooling for residential units. Additionally, this system would have a lower initial investment cost in comparison with the conventional ground source heat pumps where specially designed piping system placed a few meters beneath the ground allows the usage of earth's stored heat. For a regular municipality water delivery system, the carried water circulates through kilometers of piping already located underground which allows the water to exchange heat with the earth. Thus, it is possible to replace conventional heat pump systems and use regular tap water as a coolant for a residential unit. In order to achieve cooling, pipes can be installed on the walls and the ceiling of a house. The tap water could be circulated within these pipes. The study represented here shows that 5.48 kW of building cooling load could be handled with only 50 W of pump power for water circulation through the pipes located inside the walls and the ceiling of a proposed typical house.

Keywords Building air conditioning • Cooling load • Panel cooling • Indoor thermal comfort

8.1 Introduction

Panel heating and cooling systems are becoming more widespread. Today, in South Korea, 95 % of the buildings use panel heating systems [1]. In such a panel heating and cooling system, a refrigerant such as water is circulated inside the pipes which are installed inside floors, walls, or ceilings. Because the circulating fluid has lower temperature with respect to other heating systems or higher temperature with respect to other cooling systems, its efficiency is higher, and renewable energy

Y.C. Sezgin (✉) • M. Celik
Department of Mechanical Engineering, Bogazici University, Bebek, İstanbul 34342, Turkey
e-mail: yigit.sezgin@boun.edu.tr

I. Dincer et al. (eds.), *Progress in Clean Energy, Volume 1*,
DOI 10.1007/978-3-319-16709-1_8

sources may also be used in such systems. In addition to this, the waste energy sources may be used in cooling or panel heating systems.

The panel cooling differs from the conventional cooling systems, because there is not only heat transfer by convection but also heat transfer by radiation. Cooling by radiation helps to achieve higher total human comfort level, because in radiant cooling systems, it is possible to have more uniform indoor temperature and to control the mean radiant temperature (MRT) of the panel cooling. Furthermore in these systems, produced noise is relatively lower than other conventional cooling systems such as fan coil [2].

Typically, air source heat pumps (i.e., split air conditioners) are used for space cooling purposes. However, they use a significant amount of energy. In addition, due to nonuniform cooling of the conditioned space and because of excessive air circulation, these systems cause some health problems.

Alternatively, ground source heat pump systems are used for space cooling. The ground which is only a few meters deep stays at a fairly constant temperature throughout the year. In typical ground source heat pump systems, pipes which are buried a few meters under the soil extract heat from the ground. This heat is used to cool or heat a residential building. In ground source heat pump systems, a mixture of water and antifreeze or a refrigerant fluid circulates inside the buried pipes. Heat from the ground is transferred to this circulating fluid. This heat is later used in a heat pump.

In ground source heat pumps, there is a risk of the refrigerant running through the pipes to leak to the ground and to cause contamination of groundwater sources. In the proposed cooling system, regular tap water from the municipality is used. This water is pumped through the pipes which are installed inside the walls and the ceiling of a building for cooling purposes. This water is then given back to the municipality water circulation system. There is no water wasted for this system, just the cooling capacity of the tap water is utilized.

If pipes are installed inside the walls and the ceiling and municipality water is circulated through the pipes, a higher temperature cooling fluid could be used. With the same system, heating can also be accomplished. Because of more uniform cooling, higher total human comfort level can be achieved. Also, it does not cause high-speed air circulation. In addition, maintenance cost could be kept low [3].

Cooling a house using municipality water running through the pipes inside the walls is a new technique and seems to be a promising one. In this paper, relevant calculations and analysis have been made, and the study of using tap water as a cooling fluid is presented. In the first part of the paper, a case problem is introduced. In the second part, heat gain of the hypothetical house is calculated. After that, heat transfer from tap water to the house is evaluated. Then, the pressure drop is estimated, and according to the obtained values, the pump work is assessed. The results show that this system has several advantages over other cooling systems in terms of economical and ecological considerations.

8.2 Case Problem and Assumptions

In this section, a hypothetical house is conceived to study the concept of using municipality water for cooling purposes, and all the major assumptions are presented.

8.2.1 The Hypothetical House

In order to make an analysis of the proposed concept, a hypothetical house which is depicted in Fig. 8.1 is studied. The hypothetical house is assumed to be located in Istanbul, Turkey, and positioned in north–south direction. It is assumed that four people reside in this hypothetical house and that it has 100 m^2 (10 m × 10 m) floor area. Thus, the sidewalls have a total of 40 m length. The walls are assumed to be 3 m in height.

The sidewalls are assumed to consist of 8 cm outer plaster, 5 cm insulation material, 24 cm brick, and 3 cm inner plaster layers. The layers of the walls are shown in Fig. 8.2. The ceiling is assumed to consist of 12 cm insulation material, 12 cm concrete, and 3 cm inner plaster layers. The layers of the ceiling are shown in Fig. 8.3. It is assumed that the hypothetical house has 3.75 m^2 window area on each sidewall, resulting in a total window area of 15 m^2. Therefore, the total wall area is 105 m^2, and the total ceiling area is 100 m^2.

Fig. 8.1 The hypothetical house with municipality water cooling system

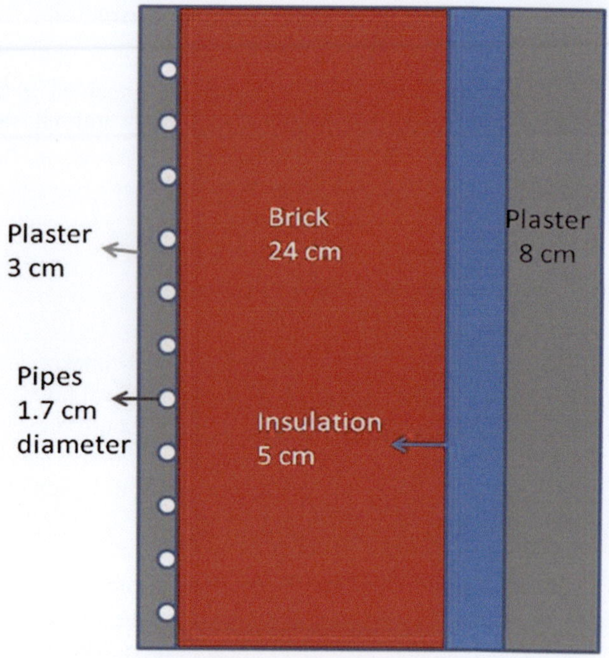

Fig. 8.2 Depiction of heat transfer from the walls

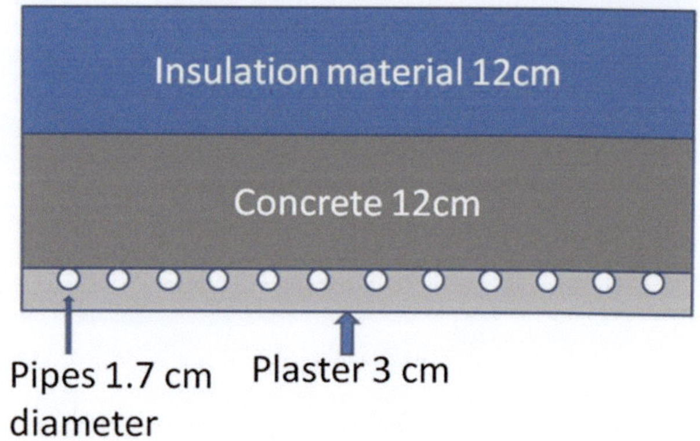

Fig. 8.3 Depiction of heat transfer from the ceiling

8.2.2 Assumptions

Istanbul's dry thermometer temperature is taken to be 33 °C for summer cooling load calculations [4]. The targeted inside temperature is taken to be 26 °C, which is the maximum desired value for summer room temperature [5]. It is assumed that the available tap water has a temperature of 20 °C for August in Istanbul. Also, it is assumed that the inside and outside temperature values are not changing in time. Heat transfer from the floor is not included in the analysis, because the temperature of the floor is assumed to be at the target room temperature.

8.3 Heat Transfer Analysis

In this section, the heat transfer analysis of the hypothetical house is presented. First, the cooling load of the hypothetical house is determined. Then, the heat transfer to water by convection and radiation is calculated, and the total heat loss from the pipes is evaluated. Later, the average water temperature through the pipes is determined, and the required pipe length is calculated.

8.3.1 Cooling Load Calculations

In this part, the cooling load for the hypothetical house is evaluated. There are heat transfer gains by conduction from the walls, windows, and roof. Also, there are radiation heat gain from the sun and heat gain from the inhabitants of the house, household devices, and ventilation through the windows. In order to calculate the heat gain from the walls, the equivalence increase in surface temperature is added, because of solar heating of the walls and roof. The accepted values are shown in Table 8.1 [6]. Then, the layer characteristics of the walls are specified. These values are tabulated in Table 8.2 [4]. It is also assumed that the walls have medium dark color and the ceiling has dark color.

By multiplying the thermal conductivity of the wall with the total area of the wall, total heat gain per degree can be calculated as 49.46 W/K. Then, by

Table 8.1 Equivalence increase in surface temperature because of the sun [6]

Surface type	Wall side			Flat roof
	East	South	West	
Dark color	5	3	5	11
Medium dark color	4	3	4	9
Light color	3	2	3	5

Table 8.2 Heat gain from walls

Wall materials	Thickness (m)	Thermal conductivity (W/mK)	Thermal resistance (m^2K/W)
R_1	–	–	0.13
Plaster	0.03	1	0.03
Brick	0.24	0.5	0.48
Insulation material	0.05	0.035	1.429
Plaster	0.08	0.35	0.023
R_e	–	–	0.04

Total thermal resistance = 2.132 m^2K/W; thermal conductivity = 0.471 W/m^2K

Table 8.3 Heat gain from the roof

Roof materials	Thickness (m)	Thermal conductivity (W/mK)	Thermal resistance (m^2K/W)
R_1	–	–	0.13
Plaster	0.02	1	0.02
Concrete	0.12	2.5	0.048
Insulation material	0.12	0.04	3
R_e	–	–	0.04

Total thermal resistance = 3.278 m^2K/W; thermal conductivity = 0.305 W/m^2K

Table 8.4 Heat gain from solar radiation in W/m^2 at 40° north latitude at various times

	08:00	12:00	16:00
West	50	50	500
East	500	50	50
South	50	200	50
North	50	50	50

multiplying this value with the temperature difference, the total heat gain from the walls can be calculated as 482.19 W.

After that, the heat gain from the windows is calculated. For double glass windows, heat loss per degree Kelvin is 36 W/K, and the total window area is 15 m^2. Therefore, the heat gain from the windows by conduction is found to be $36 \times 7 = 252$ W. Then the heat gain from the roof is calculated as presented in Table 8.3.

By multiplying the thermal conductivity of the roof with the total roof area, the total heat gain per degree can be calculated as 30.5 W/K. Then by multiplying this value with temperature difference, the total heat gain from the roof can be calculated as 549 W. Later, the heat gain from solar radiation is obtained from [7] as presented in Table 8.4.

In order to calculate the heat gain from radiation, first, the window areas and the directions are specified. The hypothetical house has windows of 3.75 m^2 on each wall. Using the solar radiation values presented in Table 8.4 for west, east, south, and north directions, it is found that the radiation heat gain from the windows is $Q = 3.75 \times 500 + 3.75 \times 50 \times 3 = 2{,}437.5$ W.

However, windows have a shadow factor which decreases the heat gain from radiation. Shadow factor for a double glass window is assumed to be 0.9. Then the net heat gain from radiation is $2{,}437.5 \times 0.9 = 2{,}193.75$ W.

There is also heat gain from the ventilation through the windows, because the outside air temperature is higher than the inside air temperature. For 1 person, 20 m^3/h fresh air is needed [7]. Thus, the total ventilation cooling load can be calculated: $Q_v = 7 \times n$ (person) $\times$ V (fresh air) [watt] [7], plugging in the assumed values, $Q_v = 7 \times 4 \times 20 = 560$ W. Here, V is the fresh air volume flow rate.

Then the heat gain due to the presence of inhabitants is calculated, since the dwellers are the heat source for the hypothetical house. It is assumed that for one person, there is 110 W heat gain. Therefore, the heat gain of the hypothetical house from people is $4 \times 110 = 440$ W [7].

Also it is assumed that there is no lighting load because the sunlight is sufficient for the considered times of the day. A heat gain of 1,000 W is assumed from household devices such as refrigerator, television, and computer [7]. After adding all these heat gains, the total heat gain of the hypothetical house is found to be 5.48 kW.

8.3.2 *Panel Cooling Calculations*

In order to cool the hypothetical house, the pipes are assumed to be installed inside the walls and ceiling as shown in Fig. 8.4. Tap water which is taken from the municipality water system will be circulated in these pipes and be given back to the municipality water system. To understand whether the cooling capacity of the tap water is sufficient, heat transfer to the walls has been calculated. First, heat transfer by convection and radiation is analyzed. Then, heat resistance of the panel has been evaluated. After that, the average water temperature is found, and the necessary pipe length has been determined. By knowing the pipe length and pipe properties, the pressure drop is calculated, and the required pump work is assessed.

Fig. 8.4 Sample wall cooling system (Yensis) [8]

8.3.2.1 Heat Transfer to the Water by Convection and Radiation

In order to calculate the heat transfer from the room to the wall, MRT method has been used. By using this method, the radiant heat flux is calculated using the following equation [2]:

$$q_r = 5 \times 10^{-8}\left[\left(T_p + 273.15\right)^4 - \left(T_{\mathrm{AUST}} + 273.15\right)^4\right] \tag{8.1}$$

where T_p is the effective panel surface temperature in °C and T_{AUST} is the area-weighted temperature of all indoor surfaces of the walls, ceiling, floor, windows, doors, etc. In the hypothetical house, T_p is assumed to be 23 °C, and T_{AUST} is assumed to be 26 °C. By substituting the values into the equation, q_r is found to be -15.82 W/m^2.

After that, cooling by convection is calculated. Natural convection heat flux between cooled ceiling surface and indoor air is calculated using the formula [2]:

$$q_{\text{c-ceiling}} = 2.42 \times \frac{\left|T_p - T_a\right|^{0.31}\left(T_p - T_a\right)}{D_e^{0.08}} \tag{8.2}$$

Natural convection heat flux between cooled wall panel surface and indoor air is calculated using the formula [2]:

$$q_{\text{c-wall}} = 1.87 \times \frac{\left|T_p - T_a\right|^{0.32}\left(T_p - T_a\right)}{H^{0.05}} \tag{8.3}$$

where T_p is 23 °C, T_a which is the designated dry-bulb indoor air temperature is 26 °C, D_e which is the equivalent diameter of the panel (4 × area/perimeter) is 10 m, and H which is the height of wall panel is 2.7 m. By using these values, $q_{\text{c-ceiling}}$ is found to be -8.48 W/m^2, and $q_{\text{c-wall}}$ is found to be -7.58 W/m^2.

8.3.2.2 Thermal Resistance of the Panel

In order to calculate the average water temperature, thermal resistance of active panel surface should be calculated. Thermal resistance of the panel affects the heat transfer to the pipes. The lower the thermal resistance, the higher the heat transfer. To calculate the thermal resistance of the panel, the following equation is used [2]:

$$r_u = r_t M + r_s M + r_p + r_c \tag{8.4}$$

where M is the spacing between adjacent tubes, r_t is the thermal resistance of the tube wall per unit tube spacing in a hydronic system, r_s is the thermal resistance between the tube and panel body per unit spacing between adjacent tubes, r_p is the thermal resistance of the panel body, r_c is the thermal resistance of active panel surface covers, and r_u is the characteristic (combined) panel's thermal resistance.

In the hypothetical house, the tubes are embedded in the plaster. When the tubes are embedded in the plaster, r_s may be neglected. Also in the sample house, there is only a painting on the surface; therefore, thermal resistance of the painting may be neglected.

The r_p of the panel may be calculated by using the following formula [2]:

$$r_p = \frac{x_p - \frac{D_o}{2}}{k_p} \tag{8.5}$$

where D_o is the outer diameter of the pipe, x_p is the distance between the center of the pipe and the inside surface of the wall, and k_p is the thermal conductivity of the plaster.

For the hypothetical house, x_p is 0.0305 m, D_o is 0.021 m, and k_p is 1.00 W/mK. Therefore, r_p is calculated as 0.009 m^{2}K/W. To calculate r_t for circular tubes, the following formula is used [2]:

$$r_t = \frac{\ln\left(\frac{D_o}{D_i}\right)}{2\pi k_t} \tag{8.6}$$

where D_i is the inner diameter of the pipe and k_t is the thermal conductivity of the pipe. For the hypothetical house, D_o is 0.021 m, D_i is 0.017 m, and k_t for the pipe made of PEX is 0.38 W/mK. Therefore, $r_t = 0.088$ m^{2}K/W. As a result $r_u = r_t M + r_p = 0.022$ mK/W for $M = 0.15$.

8.3.2.3 Panel Design

The average skin temperature of the tubing and the average circulating water temperature are the determining factors of the system. Panel surface temperature is determined to be 23 °C, and for this temperature value, the needed average water temperature should be calculated. Average water temperature depends on the thermal resistance of the panel, spacing between adjacent tubes, indoor design temperature, and panel surface temperature.

In order to find the average skin temperature of the tubing, the following equation is used [2]:

$$T_d \approx T_a + \frac{(T_p - T_a)M}{2W\eta} + q(r_p + r_c + r_s M) \tag{8.7}$$

where T_d is the average skin temperature of the tubing, q is the combined heat flux ($q = q_c + q_r$) on the panel surface, T_a is the air temperature and may be replaced by T_{AUST}, 2 W is the net spacing between tubing, and η is the fin efficiency. In order to calculate fin efficiency, the following equations are used [2]:

$$\eta = \frac{\tanh(fW)}{fW} \tag{8.8}$$

$$\eta \approx \frac{1}{fW} \quad \text{for } fW > 2 \tag{8.9}$$

where f is calculated using the following formula [2]:

$$f \approx \left[\frac{q}{m(T_p - T_a)\sum_{i=1}^{n} k_i x_i}\right]^{0.5} \tag{8.10}$$

where $m = 2 + r_c/2r_p$, n is the total number of different material layers including panel and surface covers, x_i is the characteristic thickness of each material layer, and k_i is the thermal conductivity of each layer.

In the hypothetical house, q is found to be -23.85 W/m^2 using the q_r and q_c values obtained in the section "Heat Transfer to the Water by Convection and Radiation," $\sum_{i=1}^{n} k_i x_i$ is obtained to be 0.00976 W/m, m is 2, f is 20.18, and W is 0.0645 m. With these, η is found to be 0.662, and T_d is found to be 21.15 °C. After, in order to find the average water temperature circulating through the pipes, the following formula is used [2]:

$$T_w = (q + q_b)Mr_t + T_d \tag{8.11}$$

where q_b is the back and perimeter heat losses. For the hypothetical house, if $q + q_b = -26.71$ W/m^2, then $T_w = 20.80$ °C, which means in order to meet the cooling need of the hypothetical house, the average water temperature circulating through the pipes should be 20.80 °C.

T_w value can also be determined using the design graph presented in Fig. 8.5, which provided the thermal resistance of the panel, the heat flux of the hypothetical house, and the appropriate geometrical pipe parameters.

8.3.3 *Heat Transferred to Water*

In order to determine the length of a pipe section, to be installed inside the walls and the ceiling, that satisfies the average circulating water temperature found in the previous section, first, the heat transferred to water is calculated using the following equation [2]:

$$q_w = 1{,}000\,\dot{m} c_p \Delta T \tag{8.12}$$

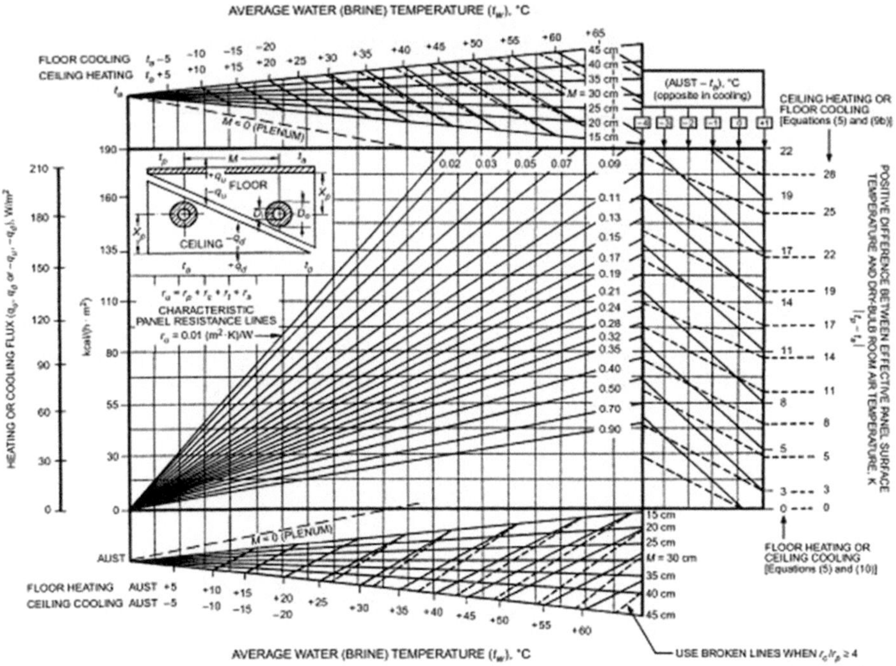

Fig. 8.5 Design graph for sensible heating and cooling with floor and ceiling panels [2]

where q_w is the heat transfer rate to water, $\dot{m}$ is the mass flow rate of the circulating water, c_p is the specific heat of water, and ΔT is the water temperature increase across one pipe section.

In the hypothetical house, it is assumed that municipality water temperature is 20 °C and the exit temperature of municipality water is 21.60 °C because average water temperature is 20.80 °C. Therefore, ΔT is equal to 1.60 °C.

Also it is assumed that the velocity of the water inside the pipes is 0.5 m/s to prevent noise, and it is known that the inside diameter of the pipe is 0.017 m. Therefore, the mass flow rate $\dot{m}$ is equal to 0.113 kg/s. It is assumed that c_p of water is 4.187 kJ/kgK. By plugging in the values to the equation, q_w is found to be 757 W. It means that the water can absorb 757 W while circulating at an average temperature of 20.80 °C. It is found in section "Panel Design" that the total heat transfer to the water pipes of the system is equal to 26.71 W/m^2. Therefore, at most, 28 m^2 of wall area is needed to provide this average water temperature. In the hypothetical house, 205 m^2 panel cooling is available, which means eight sections of pipe are needed.

8.4 Pressure Drop and Pump Work

Pressure drop in pipes affects the system efficiency: lower pressure drop means higher system efficiency, because it directly affects the needed pump work of the system. In order to achieve low pressure drop, due to its relatively small friction factor, PEX pipes are used. In order to calculate friction factor in pipes, the following equation is used [9]:

$$\frac{1}{\sqrt{f}} = -2.0\log\left(\frac{\varepsilon/D}{3.7} + \frac{2.51}{R_e\sqrt{f}}\right) \tag{8.13}$$

where f is the friction factor, ε/D is the relative roughness of the inside surface of the pipe, and Re is the Reynolds number. The Reynolds number is given by the following equation [9]:

$$R_e = \frac{\rho V D}{\mu} \tag{8.14}$$

where ρ is the density, V is the velocity, D is the diameter of the pipe, and μ is the dynamic viscosity. The density of water is taken to be 998.43 kg/m^3, the velocity of the flow is 0.5 m/s, the pipe inner diameter is 0.017 m, and the dynamic viscosity is 0.001 Pa·s. Then the Reynolds number is found to be 8,486, and thus the pipe flow is assumed to be turbulent. The relative roughness of the PEX pipe is taken to be 7.6×10^{-5}, and the friction factor, f, is calculated as 0.03238. The pressure drop of the water flow inside the pipe is then calculated using the following formula [9]:

$$\Delta P = f\frac{L}{D}\frac{\rho V^2}{2} \tag{8.15}$$

where L is the length of one section of the pipe. For the hypothetical house, L is calculated to be 198 m. Thus, the pressure drop in one such pipe section is 47.07 kPa.

In order to determine the system efficiency, the pump power input is calculated. In the hypothetical house, for one section of piping, the volume flow rate is equal to 0.113 L/s. In the hypothetical house, there are eight pipe sections; therefore, the total volume flow rate is equal to 0.904 L/s. Major head loss is evaluated to be 4.8 m. Minor head losses are neglected. With these values, the necessary pump work for the hypothetical house is found by using the following equation [10]:

$$\dot{W} = \gamma Q h \tag{8.16}$$

where γ is the specific weight of the water, Q is the mass flow rate, and h is the pump head. For the hypothetical house, the minimum required pump work is found to be 42.48 W.

8.5 Conclusions

In this paper, a study of using municipality water for building cooling has been presented. First, the cooling load of a hypothetical house is determined. The total cooling load is found as 5.48 kW. In order to meet this cooling load, the panel surface temperature is taken as 23 °C, and the heat transfer flux by convection and radiation is calculated. Panel thermal layers are specified, and the corresponding thermal resistance value is calculated. Then the distance between the adjacent pipes is specified as 0.15 m, and by using the obtained values, the average water temperature is determined as 20.80 °C. After that, the required lengths of the pipe sections are determined. Later the pressure drop in one section of the piping is evaluated, and the necessary pump power is calculated.

The obtained results show that the use of municipality water system for building cooling is a feasible and energy-saving system in comparison with other conventional cooling systems. The system would need less than 50 W of pump power to provide 5.48 kW of cooling for the proposed hypothetical house. In this system, cooling capacity of municipality water is used. Therefore, it decreases the energy consumption of cooling units and presents an environmental friendly method for space cooling. Since the system's circulating fluid is a regular water, there is no risk of contamination to the environment. The envisioned system would provide higher total human comfort level.

However, it is not possible to use this system for all buildings of a city. Therefore, there should be a permission from the municipality to use this system. However, this system can be used in governmental buildings such as hospitals, police stations, municipality service buildings, etc. Therefore, the proposed system is a promising one and deserves further research.

References

1. Karabay H, Arc M, Sandk M (2013) A numerical investigation of fluid flow and heat transfer inside a room for floor heating and wall heating systems. Energy Buildings 67:471–478
2. 2008 ASHRAE Handbook—HVAC Systems and Equipment (I-P): I-P Edition (includes CD in Dual Units). American Society of Heating, Refrigerating & Air-Conditioning Engineers, Incorporated
3. De Swardt CA, Meyer JP (2001) A performance comparison between an air-source and a ground source reversible heat pump. Int J Energy Res 25(10):899–910
4. Karakoc T (2001) Uygulamali TS825 ve kalorifer tesisati hesabi, Izocam A.S. (in Turkish)
5. Olesen B (2008) Radiant heating and cooling by embedded water-based systems. In: Proceedings of Congreso Climaplus 2011-I Congreso de Climatización Eficiente
6. Ozkol N (1988) Uygulamali sogutma teknigi, Makina Muhendisleri odasi, TMMOB Makina Muhendisleri Odasi (in Turkish)
7. Klima tesisati (2001) ISISAN yayinlari: Isisan calismalari dizisi, ISISAN (in Turkish)
8. http://www.yensis.com
9. Cengel Y (2003) Heat transfer: a practical approach, McGraw-Hill series in mechanical engineering. McGraw-Hill, New York
10. Munson B, Rothmayer A, Okiishi T (2012) Fundamentals of fluid mechanics, 7th edn. Wiley, Hoboken

Chapter 9
Study and Optimization of the Energy Profile of the Meat Industry in the Region of Alentejo, Portugal

Octávio Alves, Paulo Brito, Pedro Lopes, and Paula Reis

Abstract Today, the meat processing industry plays an important role in boosting the economy of the region of Alentejo, Portugal, which possesses a strong economic linkage to rural traditional activities. However, several problems have been detected relating to an efficient use of the energy in their productive processes, leading to excessive and unnecessary consumptions, increasing bills and therefore costs of final products, reducing company's competitiveness and causing increased emissions of greenhouse gases to the atmosphere.

The aim of this study is to characterize energy consumption of the meat processing industry in Alentejo, to understand where excesses are being committed and at the same time to identify opportunities for improvement of energy performance in the productive process. For this purpose data were collected through a field survey within a sample of companies and the respective findings were statistically analysed. Subsequently a study of the technical–economic impact of a set of efficiency measures was evaluated to reduce energy consumption and cost.

Results obtained showed that it is in the refrigeration systems and heat boilers that it is possible to achieve the largest reductions in energy consumptions and energy over costs, which may go up to 7 % (savings on consumption) and 19 % (savings on cost) through the implementation of efficiency measures with a payback period of less than 11 years.

Keywords Meat industry • Energy efficiency

O. Alves • P. Brito (✉) • P. Lopes
BioEnergia, C3i, Polytechnic Institute of Portalegre, Portalegre P-7300-110, Portugal
e-mail: pbrito@estgp.pt

P. Reis
NEISES, C3i, Polytechnic Institute of Portalegre, Portalegre P-7300-110, Portugal

I. Dincer et al. (eds.), *Progress in Clean Energy, Volume 1*,
DOI 10.1007/978-3-319-16709-1_9

Nomenclature

A	Surface area, m^2
a	Coefficient of the function for curve modelling, J/kg
b	Exponent of the function for curve modelling
C	Cost or investment, €
E	Annual active or reactive energy consumption, J or kVAr s
F	Luminous flux, lm
e	Specific energy consumption, J/kg
MEE	Measure of energy efficiency
m_{prod}	Total annual production, kg
n	Number of units of an equipment
P	Total electric power of equipment, W
R^2	Linear correlation coefficient

Greek Letters

η	Efficiency of equipment

Subscripts

a	Actual
cham	Cold chamber
cap	Capacitive
comp	Compressor of refrigeration system
ener	Energy consumption
ener u	Unit of energy consumed
i	Index of the equipment, company or class of reactive energy
ilum	Illumination of facilities
ilum cam	Internal illumination of cold chambers
ind	Inductive
ini	Initial
lamp	Existing lamps in the illumination system
o	Quantity optimized, after application of the measure
total	Total

9.1 Introduction

9.1.1 Relevance and Energy Use in the Meat Industry of Alentejo, Portugal

The region of Alentejo in Portugal is characterized by a growing economy which is closely linked to the agro-industrial sector, where the meat processing industry plays a predominant role. In 2012, this industry employed approximately 25 % of workers from the region and generated a turnover of EUR 286 million, the equivalent of 21 % of revenue within the food sector. Nineteen percent of companies belonging to the meat industry were engaged in animal slaughter while the remaining 81 % were engaged in the manufacturing of meat products: fresh goods for immediate consumption and processed products for canning (e.g., dry cured ham and sausages). Due to their extraordinary quality, these products have become a brand image of the region [1, 2].

Despite the importance of the meat industry, it has been found that productive processes do not operate efficiently in terms of energy consumption. The lack of thermal insulation in refrigeration pipework, the use of motors or lighting systems with low efficiencies and the lack of training of human resources are problems commonly found. It is estimated that about 6.5 % of the total energy consumed may be easily saved through the implementation of adequate corrective measures [3], reflected directly in the reduction of invoice costs and in the increase of competitiveness of companies.

Refrigeration systems make up 50 % of the total energy consumed inside such companies, showing therefore a great potential in terms of energy savings [4].

9.1.2 Framework of Present Study

With a view to improve energy usage in industry and also the environment, the European Commission set out a package of goals named “20-20-20” to be accomplished until 2020. These goals are:

- Twenty percent reduction in greenhouse gas emissions
- Twenty percent increase in the production of energy from renewable sources
- Twenty percent improvement in overall energy efficiency [5]

Previous goals were transported to Portuguese law through the National Action Plan of Energy Efficiency that stipulated two concrete targets:

- Reduction of 8 % of energy consumption in the industry.
- Obligation to reduce the energy consumed per kg of product between 4 and 6 % within 8 years and to maintain CO_2 emissions in all companies achieving energy consumptions higher than 8.370×10^{12} J/year; however, this measure can optionally be applied by other companies [6, 7]

To fulfil these targets, the same plan proposes the implementation of a series of practical measures of energy efficiency (MEE). Table 9.1 describes some of the available measures and the advantages of them, with a special focus on those that can be implemented in cold stores.

Table 9.1 Description and benefits of some MEE

No.	Measure	Description	Benefits
MEE1	Installation of variable speed drives in compressors and fans coupled to condensers of cold stores	Variation of torque and rotation speed of electric motor by changing the frequency of supply voltage. The equipment consists of a rectifier (converts AC voltage to DC), a filter (modulates the DC frequency wave) and an inverter (reconverts DC voltage to AC)	• Savings in consumption between 20 and 25 % • Eliminates peaks of starting currents • Control of power according to the needs [3]
MEE2	Installation of simulators of food temperature in cold chambers	Device that simulates the thermal inertia of food. The temperature probe inserted inside it reads the correct temperature to adequately control the functioning of the compressor instead of using the temperature of circulating air	• Savings in energy consumption between 10 and 30 % • Reduces the operation cycles of compressor, thus prolonging its useful life [8]
MEE3	Adjustment of internal illuminance of cold chambers	Reduction of illuminance to the recommended value of 200 lm/m^2 [9] through the deactivation of some lamps	• Reduction in energy consumption of lighting • Reduction of heat released by lamps and, consequently, the consumption of compressor
MEE4	Optimization of general lighting in the facilities	Replacement of bulbs with large energy consumptions (incandescent, halogen and discharge) by more economic bulbs (fluorescent and LED)	• Savings in consumption between 75 and 80 % compared with ancient bulb types • Minor maintenance interventions due to large durability of bulbs (10–25 times higher) [10]

(continued)

Table 9.1 (continued)

No.	Measure	Description	Benefits
MEE5	Adjustment of capacitor banks	Increases the power factor of the installation, inhibiting consumption or supply of reactive energy to the electric grid which is required for the generation of magnetic fields responsible for the operation of engines. This type of energy is usually penalized by power companies	• Reduces energy bill costs • Allows the use of electrical cables of lower sections • Mitigates the heating of cables [3, 11]
MEE6	Installation of biomass heat boilers	Replacement of conventional gas or diesel heat boilers by biomass heat boilers. Biomass is a new type of combustible obtained from plant and animal residues (pellets, wood)	• Reduced cost of biomass • Negligible CO_2 emissions • Ease of access to biomass [11]

9.1.3 Goals of Present Study

Considering the "potential" savings that energy efficiency can create in the meat processing industry through the reduction of energy consumption and costs with a negative impact on the sustainability of companies, and to contribute to the achievement of energy targets defined by the European Commission and other legislation, the present study comes with the following goals:

- To analyse the energy profile of meat industry in the region of Alentejo, Portugal
- To evaluate the impact of efficiency measures described above within companies, through the estimation of energy savings and associated costs

9.2 Methodology

The methodology adopted in each goal set for the study is summarized in Fig. 9.1.

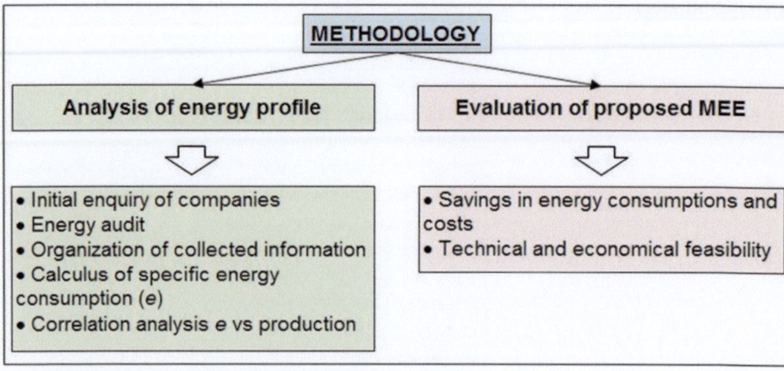

Fig. 9.1 Summary of the methodology used in each goal

9.2.1 Analysis of the Energy Profile of the Meat Processing Industry

The energy profile was determined through the study of a sample of six companies spread throughout the region, representing the universe of meat processing industry during the year of 2010. These companies have different dimensions, produce different types of products and they are identified by designations Ca1 to Ca6. Each step of the analysis identified in Fig. 9.1 is described below:

- *Initial enquiry of companies*: after a model of enquiry was prepared and approved, visits were performed in all six companies with a view to collect and retrieve the desired information. Categories and specific relevant variables that were raised are listed in Table 9.2.
- *Energy audit*: with the intention to validate the information collected in the six questionnaires, energy audits were carried out in two of the studied companies—Ca2 and Ca5—which are designated by pilot units. It was also intended to ascertain accurately the global energy needs and also consumptions of lighting and refrigeration systems.

  Measurements of electric energy consumptions were made by power quality analysers installed in switchboards during 1 week, and later they were extrapolated to the entire year of 2012. Additionally, energy costs and consumptions were extracted from invoices of electric energy, propane gas and diesel, as well as a survey of technical characteristics of existing heat boilers, lighting and components of refrigeration systems.

  Table 9.3 identifies the data that were retrieved in this step.

Table 9.2 Data collected in the initial enquiry of companies

Category	Variable
Refrigeration systems	• Number of cold chambers
	• Type of internal lighting
	• Power of internal lighting
	• Dimensions
Energy demands	• Total annual consumption
Production	• Activity type
	• Annual values of production

Table 9.3 Data collected during the energy audits

Category	Variable
Energy demands	• Global consumption of cold chambers
	• Lighting consumption inside cold chambers
	• Consumption of facility's lighting by bulb type
	• Consumption of reactive energy
	• Consumption of heat boilers
Equipment characteristics	• Type of lighting inside facilities
	• Lighting power inside facilities
	• Power of refrigeration system's components
	• Thermal power of heat boilers
	• Thermal fluid inside heat boilers
	• Maximum pressure of heat boilers

- *Organization of collected information*: data were converted to SI units by applying appropriate conversion factors and later presented in tables and graphs for comparison.
- *Calculation of specific energy consumption* (e): this indicator assesses the total consumption of energy required to produce 1 kg of product and was obtained for each company through Eq. (9.1) [7]:

$$e = E_{\text{total}}/m_{\text{prod}} \tag{9.1}$$

- *Correlation analysis between specific energy consumption* (e) *and level of production* (m_{prod}): according to collected data, the relationship between these variables is inversely proportional and so it was admitted that e is a function of m_{prod} through the estimating expression indicated by Eq. (9.2):

$$e = a \times \left(m_{\text{prod}}\right)^{\text{b}} \tag{9.2}$$

Since this is a model of non-linear statistical regression, Eq. (9.2) was transformed to a linear form by the application of natural logarithms to both members [12]. In this way, parameters a and b were determined by expressions of linear regression (Eqs. (9.3) and (9.4)), replacing in both of them data for e_i and $m_{\text{prod}\ i}$ collected in each company [13]:

$$b = \left[\Sigma\left(\ln\left(m_{\text{prod}i}\right) \times \ln(e_i)\right) - 1/6 \times \Sigma\left(\ln\left(m_{\text{prod}i}\right)\right) \times \Sigma(\ln(e_i))\right] / \left[\Sigma(\ln(e_i))^2 - 1/6 \times \left(\Sigma\left(\ln\left(m_{\text{prod}i}\right)\right)\right)^2\right] \tag{9.3}$$

$$a = \exp\left[1/6 \times \Sigma(\ln(e_i)) - \text{b} \times \Sigma\left(\ln\left(m_{\text{prod}\,i}\right)\right)\right] \tag{9.4}$$

The agreement between collected data and the regression curve was calculated by the linear correlation coefficient (R^2), as expressed by Eq. (9.5) [13]:

$$R^2 = \left[b \times \Sigma\left(\ln\left(m_{\text{prod}\,i}\right) \times \ln(e_i)\right) - 1/6 \times \Sigma\left(\ln\left(m_{\text{prod}\,i}\right)\right) \times \Sigma(\ln(e_i))\right] / \left[\Sigma(\ln(e_i)) - 1/6 \times \Sigma(\ln(e_i))\right]^2 \tag{9.5}$$

9.2.2 *Evaluation of the Impact of Proposed MEE in the Meat Processing Industry*

To test the reduction of energy consumption and the current energy cost it was simulated the implementation of the six MEEs that were previously described in the two pilot units.

The analysis of annual savings in both total energy consumptions and total energy costs is described below for each measure:

- *Installation of variable speed drives* (*MEE1*): the coupling of one drive in each compressor located in the refrigeration systems was admitted, according to its respective power. Considering that the actual consumption of compressors ($E_{\text{comp}\ a}$) is proportional to their total power (P_{comp}), Eq. (9.6) was applied to determine $E_{\text{comp}\ a}$:

$$E_{\text{comp}\,a} = P_{\text{comp}}/P_{\text{cham}} \times E_{\text{cham}\,a} \tag{9.6}$$

The optimized total energy consumption in each company ($E_{\text{total}\ o}$) was calculated by Eq. (9.7), assuming an energy saving of 25 % in actual consumption of compressors ($E_{\text{comp}\ a}$) [3, 14]:

$$E_{\text{toal}\,o} = E_{\text{total}\,a} - 0.25 \times E_{\text{comp}\,a} \tag{9.7}$$

The calculation of costs of optimized consumption and energy savings was done considering an average price for electric energy of 0.031 €/MJ [15]:

Table 9.4 Values used for lighting efficiency

Lighting type	Efficiency (lm/W)
Incandescent	13.5
Halogen	17
Tubular fluorescent	70
Compact fluorescent	60
Metal halide	92.5
Sodium vapor	105
LED	76

- *Installation of simulators of food temperature* (*MEE2*): in each cold chamber a simulator was installed to produce energy savings of about 10 % [8]. Thus, the optimized consumption for each company ($E_{\text{total } o}$) is given by Eq. (9.8):

$$E_{\text{total}\,o} = E_{\text{total}\,a} - 0.1 \times E_{\text{cham}\,a} \tag{9.8}$$

- *Adjustment of internal illuminance of cold chambers* (*MEE3*): when illuminance inside the chambers was higher than the recommended value of 200 lm/m^2 [9], lighting power was reduced to ensure such recommendation. By using the average of energy savings for each chamber, the optimized consumption for the company ($E_{\text{total } o}$) was determined through Eq. (9.9):

$$\begin{aligned} E_{\text{total}\,o} = {} & E_{\text{total }\,a} - E_{\text{ilum cham}\,a}/n_{\text{cham}} \\ & \times \Sigma(1 - 200 \times A_{\text{cham }i}/(\eta_{\text{ilum cham }i} \times P_{\text{ilum cham}\,i})) \end{aligned} \tag{9.9}$$

Values used for lighting efficiency ($\eta_{\text{ilum cham}}$) are listed in Table 9.4 [10].

- *Optimization of general lighting in the facilities* (*MEE4*): replacement of incandescent light bulbs by compact fluorescent lamps was tested, ensuring the maintenance of the existing luminous flux ($F_{\text{ilum }a}$). This was calculated by Eq. (9.10) and using values shown in Table 9.4. In this case, $P_{\text{ilum }a}$ refers to the power of current lighting that is intended to be replaced:

$$F_{\text{ilum }a} = \eta_{\text{ilum}} \times P_{\text{ilum }a} \tag{9.10}$$

The optimized total energy consumption of the company ($E_{\text{total }o}$) was calculated by Eq. (9.11), considering the relationship between the power of economic lamp ($P_{\text{ilum }o}$) and the power of current one ($P_{\text{ilum }a}$):

$$E_{\text{total }o} = E_{\text{total }a} - E_{\text{ilum }a} \times n_{\text{lamp}} \times (1 - P_{\text{ilum }o}/P_{\text{ilum }a}) \tag{9.11}$$

- *Adjustment of capacitor banks* (*MEE5*): reactive energy that can be eliminated through this measure was taken from invoices, and its cost was determined as a function of the power factor ($\cos(\varphi)$) recorded at each instant and also considering if the energy was inductive (consumed from the grid) or capacitive (supplied to the grid). Table 9.5 refers all prices applied in accordance with tariffs of medium voltage [15].

Table 9.5 Cost of reactive energy as a function of power factor

Reactive energy type	Class	Power factor range	Cost (€/(kVAr s))
Inductive	1	$0.93 < \cos(\varphi) \leq 0.96$	2.250×10^{-6}
	2	$0.89 < \cos(\varphi) \leq 0.93$	6.833×10^{-6}
	3	$\cos(\varphi) \leq 0.89$	2.050×10^{-5}
Capacitive	–	–	5.139×10^{-6}

The optimized energy cost ($C_{\text{ener } o}$) that was possible to get is expressed by Eq. (9.12) (E_{ind1}, E_{ind2}, E_{ind3} and E_{cap} are expressed in kVAr s):

$$C_{\text{ener } o} = C_{\text{ener } a} - \left(E_{\text{ind 1}} \times 2.250 \times 10^{-6} + E_{\text{ind 2}} \times 6.833 \times 10^{-6} + E_{\text{ind 3}} \times 2.050 \times 10^{-5} + E_{\text{cap}} \times 5.139 \times 10^{-6}\right) \quad (9.12)$$

- *Installation of biomass heat boilers* (*MEE6*): attending to the consumption of existing heat boilers the total cost of propane gas and pellets needed to generate the same amount of energy was determined, and thereafter the difference between both costs for comparison was calculated. Prices of propane gas and pellets correspond, respectively, to 0.058 €/MJ and 0.036 €/MJ, as obtained from local suppliers.

The assessment of economic feasibility was applied for all measures except MEE3 and MEE5, because they do not require the purchase of any additional equipment; only a technical human intervention is necessary. The indicators used for the analysis of the remaining measures are described below:

- *Initial cost of the measure* (C_{ini}): costs of required equipment were only considered and obtained from local suppliers
- *Payback period of the measure* (t_{ret}): the period to recover the investment, in years, was determined by Eq. (9.13), and the cost per unit of energy ($C_{\text{ener } u}$) was the same of electric energy (0.031 €/MJ) or pellets (0.036 €/MJ), depending on the measure under analysis [3]

$$t_{\text{ret}} = C_{\text{ini}} / \left(C_{\text{ener } u} \times (E_a - E_o)\right) \quad (9.13)$$

9.3 Results and Discussion

Tables 9.6 and 9.7 show, respectively, the data collected in the initial enquiry made to all six companies and during the energy audits carried out in the two pilot units.

Figure 9.2 shows the results for the specific energy consumption (e) in each company, while Fig. 9.3 presents the regression curve between that quantity and the

Table 9.6 Data obtained in the initial enquiry of companies (year 2010)

Company	Activity	Energy consumption (GJ)	Production (kg)
Ca1	Manufacturing of meat products	20.15	35,030
Ca2	Slaughter of animals	6,397.00	2,560,808
Ca3	Manufacturing of meat products	72.62	12,251
Ca4	Manufacturing of meat products	15.17	37,723
Ca5	Manufacturing of meat products	328.40	1,770,000
Ca6	Manufacturing of meat products	57.12	62,400

mass production ($e = a(m_{\text{prod}})^b$), as well as the adjustment points considering the following parameters: $a = 5,863,622\,\text{J/kg}$, $b = -0.158$ and $R\cdot = 0.079$.

Results indicate that company Ca3 had the highest specific energy consumption (5.928 MJ/kg), and therefore it is less efficient compared to others. In this way, the potential for improving energy efficiency is high because one would expect to record a specific energy consumption in the order of 0.5 to 1 MJ/kg, taking into account that manufactured products and mass production levels are similar to those observed in companies Ca1, Ca4 and Ca6.

On the other side of the scale stands company Ca5 with a specific energy consumption of 0.186 MJ/kg, and a possible justification for this low value may rely in its high production level. In fact, points drawn in Fig. 9.3 show that companies with large production levels tend to be more efficient (and with lower specific energy consumptions), due to the intensive series production (less time to manufacture one unit of product) and smaller starting cycles of machines, which are also responsible for higher consumptions (e.g., initial heating or cooling of components).

Figure 9.3 shows a weak correlation between the curve and the collected data ($R^2 < 8\ \%$) due to the remoteness of points that represent companies Ca3 (low energy efficiency) and Ca2 (with a distinct activity), which are unsuitable for the modelling. Removing these points it is possible to obtain values for parameters $a = 12631161\,\text{J/kg}$ and $b = -0.288$, which results in a correlation coefficient $R^2 = 64\,\%$ and, therefore, a better fitting of the curve to the remaining points.

Table 9.8 indicates quantities and reference prices (year 2014) considered for each equipment to be implemented in each MEE subjected to the analysis of economic feasibility within the two pilot units.

Table 9.9 specifies optimized energy consumptions and costs resulting from the application of each measure, energy savings, estimated initial investments and the underlying payback periods.

Table 9.10 shows energy savings achieved from the implementation of all six measures together, as well as the necessary investments and payback periods.

Figures 9.4 and 9.5 illustrate all savings achieved in total energy consumptions and related costs for each measure implemented in the two pilot units (values of 2012).

Table 9.7 Data obtained from the energy audits in the pilot units (year 2012)

Data		Pilot unit Ca2	Pilot unit Ca5
Total energy consumption (TJ)		7.124	8.206
Total energy cost (€)		305,913	293,985
Total production (kg)		5,957,464	2,066,288
Number of cold chambers		15	11
Lighting type inside cold chambers		Tubular fluorescent	Tubular fluorescent
Lighting power inside cold chambers with illuminance >200 lm/m^2 (W)		1,436	0
Energy consumption of lighting inside cold chambers (GJ)		0.804	0.419
Area of cold chambers with illuminance >200 lm/m^2 (m^2)		339.74	0
Power of cold chambers (W)		335,050	207,790
Number and power of cold chamber's compressors		1 × 18,500 W; 1 × 38,100 W; 1 × 55,000 W	2 × 2,450 W; 1 × 2,500 W; 1 × 2,700 W; 1 × 3,460 W; 1 × 4,500 W; 2 × 4,680 W; 1 × 5,300 W; 3 × 5,940 W; 1 × 9,100 W; 8 × 13,640 W
Energy consumption of cold chambers (TJ)		2.658	1.381
Number of heat boilers		2	1
Energy consumption of heat boilers (TJ)		2.624	0.764
Thermal fluid of heat boilers		Superheated vapor	Water
Thermal fluid flow (kg/s)		0.694	0.682
Maximum pressure of heat boiler (MPa)		1.3	1.8
Consumption of inductive energy class 1 (kVAr s)		269,593,200	181,850,400
Consumption of inductive energy class 2 (kVAr s)		73,296,000	142,286,400
Consumption of inductive energy class 3 (kVAr s)		856,800	33,796,800
Consumption of capacitive energy (kVAr s)		2,980,800	12,081,600
Typology of lamps of general lighting inside facilities to be improved (quantity/unit power)	Incandescent	9/60 W	2/60 W
	Halogen	1/250 W	–
	Metal halide	1/400 W	–
	Sodium vapor	32/70 W	–
Consumption of general lighting inside facilities to be improved (GJ)	Incandescent	0.277	0.281
	Halogen	3.089	–
	Metal halide	4.943	–
	Sodium vapor	27.680	–
Consumption of general lighting inside facilities (TJ)		0.170	0.103

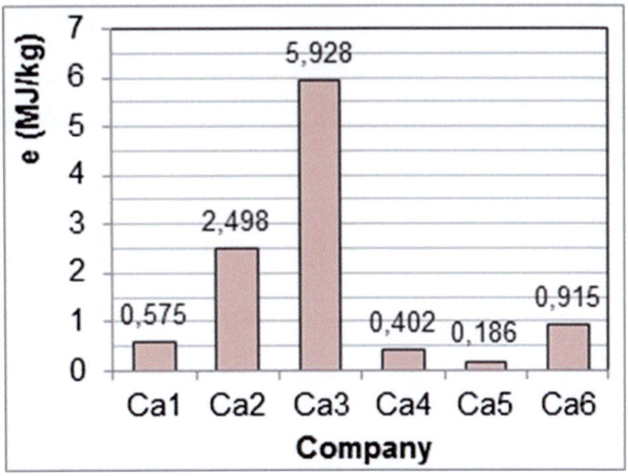

Fig. 9.2 Specific energy consumption for each company (2010)

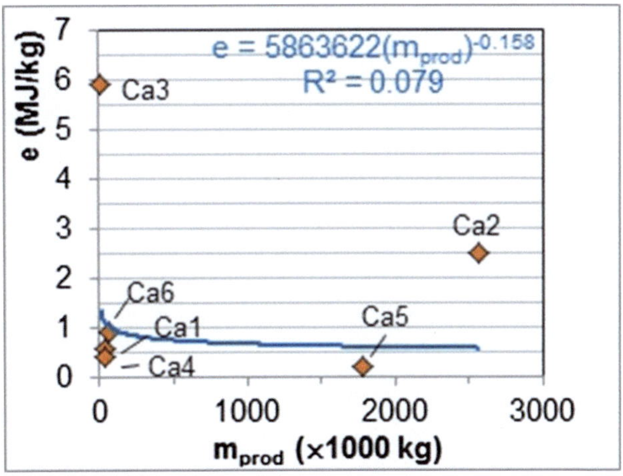

Fig. 9.3 Regression curve for specific energy consumption vs. production (2010)

According to the results, the most effective measures in terms of total energy savings were MEE1 (installation of variable speed drives) and MEE2 (use of temperature simulators), with results that averaged 3.9 % (Figs. 9.4 and 9.5). The same measures also provided interesting reductions in energy bill costs (average of 3 % or 8,996 €), although at a much lower degree than that observed in MEE6.

Considering that the consumption of refrigeration systems occupies a substantial part of total energy needs (37 % in Ca2 and 16 % in Ca5), the high influence of measures MEE1 and MEE2 is thus justified. This influence was greater in Ca2 due to the higher number of cold chambers and the higher energy consumptions registered by them.

Despite the higher initial investments (particularly in MEE1), the lower paybacks (<2 years) and the interesting savings obtained either in terms of energy or

Table 9.8 Equipment, quantities and reference prices used in the various measures (values of 2014)

Measure	Equipment	Unit price (€)	Quantity	
			Ca2	Ca5
MEE1	Variable speed drive 3,000 W	450	–	4
	Variable speed drive 4,000 W	500	–	1
	Variable speed drive 5,500 W	550	–	4
	Variable speed drive 7,500 W	700	–	3
	Variable speed drive 11,000 W	800	–	1
	Variable speed drive 15,000 W	1,000	–	8
	Variable speed drive 18,500 W	1,150	1	–
	Variable speed drive 55,000 W	3,000	5	–
MEE2	Food temperature simulator	150	18	23
MEE4	Compact fluorescent lamp (20 W)	6	9	2
MEE6	Biomass superheated steam boiler	300,000	1	–
	Biomass water boiler	70,000	–	1

cost lead to believe that measures MEE1 and MEE2 are the most recommended for both pilot units.

Measures associated with the improvement of lighting (MEE3 and MEE4) offered a negligible impact on the reduction of energy consumptions and associated costs (savings less than 0.1 %). In Ca5, measure MEE3 did not have any effect because illuminances of all cold chambers were always below the reference level (200 lm/m^2), although it is desirable that they remain close to such value, as far as is possible, to ensure a better visual comfort. The weak impact of both measures is justified by the low weight of lighting consumption in total energy needs (2 % in Ca2 and 1 % in Ca5) and by the awareness that both companies demonstrated in the use of more economic bulbs (especially the tubular fluorescent type), in most of existing spaces.

The lower economic and energy savings provided allied with greater payback periods (>2 years) suggest that measures MEE3 and MEE4 must not be a priority in terms of implementation. In Ca2, it was inclusively observed that the second one does not have economic feasibility because of its too high payback (almost 10 years) and because lighting is not frequently used. By this way, it seems to be preferable to maintain the existing type of lighting.

Measure MEE5 did not reveal a significant potential to reduce costs when compared to the remainder (MEE1, MEE2 and MEE6), which reached an average of 1,640 € (<1 % of total energy cost). Although both pilot units use capacitor banks to limit the reactive energy needs, the lack of maintenance or adjustment of such devices were found to be common practices, especially during the installation or replacement of an equipment with a higher inductive load (like transformers and electric motors). In this way, it is left here as an advice to draw up plans for periodic maintenance of capacitor banks.

Table 9.9 Energy and economic results provided by the application of each measure (values of 2012)

Measure	Company	Optimum consumption (TJ)	Energy saving (GJ)	Cost of optimum consumption (€)	Cost of energy saving (€)	Initial investment (€)	Payback (year)
MEE1	Ca2	6.651	473.0	291,456	14,457	16,150	1.1
	Ca5	7.906	300.4	284,804	9,181	15,400	1.7
MEE2	Ca2	6.858	265.8	209,610	8,124	2,700	0.3
	Ca5	8.068	138.2	246,580	4,222	3,450	0.8
MEE3	Ca2	7.124	0.151	305,908	5	–	–
	Ca5	8.206	0.000	293,985	0	–	–
MEE4	Ca2	7.117	6.947	305,701	212	54	9.6
	Ca5	8.206	0.184	293,979	6	12	2.1
MEE5	Ca2	–	–	304,771	1,142	–	–
	Ca5	–	–	291,848	2,137	–	–
MEE6	Ca2	–	–	248,674	57,239	600,000	10.5
	Ca5	–	–	277,309	16,676	70,000	4.2

Table 9.10 Energy and cost savings, initial investments and payback periods for the set of six measures (2012)

Company	Total energy saving (GJ)	Total cost of energy saving (€)	Total investment (€)	Payback (year)
Ca2	745.9	81,179	618,904	7.6
Ca5	438.8	32,222	88,862	2.8

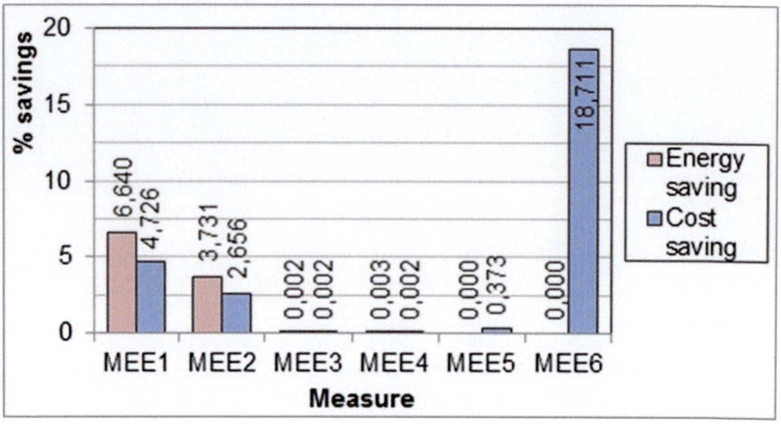

Fig. 9.4 Energy and economic savings of each measure in pilot unit Ca2 (values of 2012)

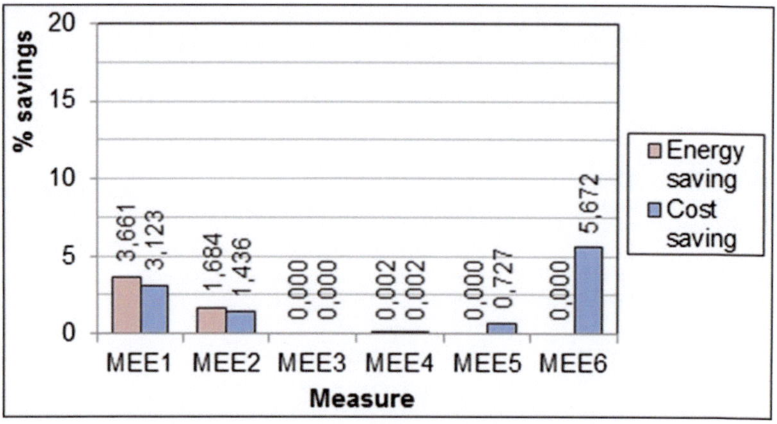

Fig. 9.5 Energy and economic savings of each measure in pilot unit Ca5 (values of 2012)

Measure MEE6 (installation of biomass heat boilers) holds the greatest potential in terms of savings in energy costs, which can go up to 18.7 % (57,239 €) in the case of company Ca2. This occurs because of the difference between prices of propane gas and pellets when generating the same amount of heat. Initial investment and payback are both very high (600,000 € and 10.5 years for Ca2, respectively), but given the depletion and consequent increase in the price of fossil fuels in the coming years, the ecological characteristics of biomass (non-toxic, without CO_2 emissions) and its ease of access, it is believed that the implementation of measure MEE6 holds a good interest.

The joint application of the six measures in both pilot units caused average reductions in energy consumptions and costs of 7.9 % and 18.7 %, respectively (Table 9.10). Due to the dilution of total investment by overall energy savings, a significant decrease in payback with an average variation of 2.2 years was observed.

It shall be pointed out that initial investments and paybacks may be slightly higher in reality, since just the investment in equipment acquisition was analysed; costs of installation and maintenance were not considered.

The reduction of specific energy consumption (e) recorded in 2012 from 1.189 to 1.071 MJ/kg (in the case of Ca2) and from 3.967 to 3.767 MJ/kg (in the case of Ca5) was also seen, which corresponded to variations of −9.92 and −5.04 %, respectively. The improvement was significantly higher in Ca2 due to the effect caused by measures MEE1 and MEE2 in refrigeration systems, as seen previously.

Although none of the pilot units were subjected to the obligation of reducing the specific energy consumption (e) because they do not have energy demands greater than 8.370×10^{12} J/year (as defined by the Portuguese National Action Plan of Energy Efficiency), it was found that the adoption of all measures enable them to fulfil the goal of reduction of that indicator between 4 and 6 % within 8 years. Considering the fact that energy consumptions of both pilot units are very close to the limit fixed at 8.370×10^{12} J/year, the early implementation of these or other new measures due to the likely growth of mass production and energy consumption in the coming years is recommended.

9.4 Conclusions

Despite the deviations found in some companies regarding the modelling curve of specific energy consumption vs. production, the developed correlation constitutes a gross reference base but still interesting to determine whether any company from the meat industry is above or below a level of optimum energy efficiency. The accuracy of the curve can be improved by extending the sample and by using more companies with different indices of energy performance.

The study of implementation of all MEE showed their importance in the reduction of energy consumption and costs inside companies from the meat industry, located in the region of Alentejo, Portugal. The potential for energy savings continues to be great, which has direct and indirect effects in environmental preservation through the mitigation of greenhouse gas emissions, particularly CO_2.

From the set of measures selected for the study, the ones which proved to be more effective were MEE1 (installation of variable speed drives), MEE2 (use of temperature simulators) and MEE6 (installation of biomass heat boilers), although the higher investments and payback periods may constitute a serious obstacle for their implementation. All total savings that were obtained, which averaged 7.9 % in energy consumption and 18.7 % in energy cost for the two pilot units that entered in the analysis, shall also be underlined.

In a region with a weak industrial and economic dynamism, the role that energy efficiency plays in its development becomes even more relevant by ensuring not only the sustainability of companies but also the availability of funds for the creation of innovative products, instead of applying them in unnecessary costs.

Acknowledgements This work was developed with data obtained from project "InovEnergy—Energy Efficiency in the Agroindustrial Sector", cofinanced by FEDER under the SIAC—COMPETE/QREN program.

References

1. INE—Instituto Nacional de Estatística, IP (2012) Portal do Instituto Nacional de Estatística. http://www.ine.pt. Accessed 5 Mar 2014
2. Turismo do Alentejo ERT (2014) Queijos, azeites, enchidos e presuntos. http://www.visitalentejo.pt/pt/o-alentejo/saboreie/queijos-azeites-enchidos-e-presuntos. Accessed 4 Mar 2014
3. Magueijo V, Fernandes MC, Nunes CP et al (2010) Medidas de eficiência energética aplicáveis à indústria portuguesa: um enquadramento tecnológico sucinto. ADENE—Agência para a Energia, Algés
4. Carbon Trust (2011) Refrigeration systems—guide to key energy saving opportunities. http://www.carbontrust.com/media/13055/ctg046_refrigeration_systems.pdf. Accessed 10 Mar 2014
5. European Commission (2007) The 2020 climate and energy package. http://ec.europa.eu/clima/policies/package/index_en.htm. Accessed 5 Mar 2014
6. CITEVE—Centro Tecnológico das Indústrias Têxtil e do Vestuário de Portugal (2012) Plano setorial de melhoria da eficiência energética em PME—setor agroalimentar. Instituto de Apoio às Pequenas e Médias Empresas e à Inovação, Lisboa
7. ADENE—Agência para a Energia (2014) Sistema de gestão dos consumos intensivos de energia. http://www.adene.pt/sgcie. Accessed 10 Mar 2014
8. Greenchoice Group (2014) Why monitoring air temperature to keep food cold? http://www.greenchoicegroup.com/energy-saving. Accessed 8 Mar 2014
9. ISO—International Organization for Standardization (2002) ISO 8995:2002—lighting of indoor work places. ISO—International Organization for Standardization, Geneva
10. U. S. Department of Energy (2010) Tips: lighting. http://energy.gov/energysaver/articles/tips-lighting. Accessed 8 Mar 2014
11. RECET—Associação dos Centros Tecnológicos de Portugal, CITEVE—Centro Tecnológico das Indústrias Têxtil e do Vestuário, CTCV—Centro Tecnológico da Cerâmica e do Vidro et al (2007) RENOVARE—Guia de Boas Práticas de Medidas de Utilização Racional de Energia e Energias Renováveis. RECET—Associação dos Centros Tecnológicos de Portugal/Fundación CARTIF, Vila Nova de Famalicão/Boecillo
12. Baker SL (2008) Non-linear regression. http://hspm.sph.sc.edu/Courses/J716/pdf/716-5%20Non-linear%20regression.pdf. Accessed 10 Mar 2014
13. Montgomery DC, Runger GC (1999) Applied statistics and probability for engineers. Wiley, New York
14. Almeida AT, Ferreira FJT (2005) Technical and economical considerations in the application of variable-speed drives with electric motor systems. IEEE T Ind Appl 41:188–199
15. ERSE—Entidade Reguladora dos Serviços Energéticos (2013) Tarifas transitórias de venda a clientes finais em portugal continental em 2014. http://www.erse.pt/pt/electricidade/tarifaseprecos/2014/Documents/PrecosTVCF%20PTCont_2014.pdf. Accessed 12 Mar 2014

Chapter 10
Energy and Exergy Analyses of Water Usage in Oil Sands Extraction and Upgrading Operations

M.K. Cohce, I. Dincer, and G.F. Naterer

Abstract Oil sands extraction and upgrading consume a significant amount of water, energy, as well as hydrogen to produce synthetic crude oil (SCO) from bitumen. This study examines the energy and exergy analysis of water usage for oil sands during extraction and upgrading. Steam and water supply to these processes are investigated by using Aspen Plus simulation software with consideration of heat transfer. These systems and their modifications are simulated and analyzed thermodynamically. The assessment involves the examination of several different factors and comparisons. The analysis includes energy and exergy flows with the amount of destroyed exergy during the processes and operations. The results provide new insight for the design, optimization, and modification of oil sands and upgrading processes.

Keywords Energy • Exergy • Efficiency • Oil extraction • Water usage

Nomenclature

$\dot{E}$	Energy flow rate, kJ/s
$\dot{\mathrm{E}}\mathrm{x}$	Exergy flow rate, kJ/s
ex	Specific exergy, kJ/kg
h	Specific enthalpy, kJ/kg
LHV	Lower heating value, MJ/kg
m_{i}	Inlet mass, kg

M.K. Cohce (✉) • I. Dincer
Faculty of Engineering and Applied Science, University of Ontario Institute of Technology, 2000 Simcoe Street North, Oshawa, ON, Canada L1H 7K4
e-mail: Mehmet.Cohce@uoit.ca; Ibrahim.Dincer@uoit.ca

G.F. Naterer
Faculty of Engineering and Applied Science, Memorial University, 240 Prince Phillip Drive, St. John's NL A1B 3X5
e-mail: Greg.Naterer@uoit.ca

I. Dincer et al. (eds.), *Progress in Clean Energy, Volume 1*,
DOI 10.1007/978-3-319-16709-1_10

m_o	Outlet mass, kg
P_o	Reference-environment pressure, kPa
Q	Heat, kJ
S	Entropy, kJ/K
T	Temperature, K
T_o	Reference-environment temperature, K
x	Exergy ratio

Subscripts

dest	Destroyed
en	Energy
gen	Generated
i,j	Index for components
in	Input
meth	Methane gas (CH_4)
out	Output
st	Steam
sys	System

Superscripts

ch	Chemical
ph	Physical

Acronyms

API	American Petroleum Institute
ARC	Alberta Research Council
bbl	Billion barrel
CONRAD	Canadian Oil Sands Network for Research and Development
COOL	Cooling
COS	Canadian Oil Sands
EIA	Energy Information Administration
GHG	Greenhouse gases
HE	Heat exchanger
IEA	International Energy Agency
Mbd	Million barrels per day
NEB	National Energy Board
PADD	Petroleum Administration for Defense District
SCO	Synthetic crude oil
SIRCA	Scientific and Industrial Research Council of Alberta

10.1 Introduction

There are large and eminent reserves of heavy oil, extra-heavy oil, and bitumen in Canada, Venezuela, Russia, the USA, and many other countries [1]. World oil reserves consist of a portion of conventional oil, 30 %; heavy oil, 15 %; and extra-heavy oil and bitumen, 55 % [2]. As shown in Fig. 10.1, the Canadian oil sands represent the world's third largest oil reserves [3]. These hydrocarbon resources are in the form of bitumen or oil sands (also called tar sands). Typically, oil sands contain about 75 % inorganic matter, 10 % bitumen, 10 % silt and clay, and 5 % water [4].

Water is necessary as a reactant while producing hydrogen (bitumen must be upgraded with an injection of hydrogen to produce SCO), where it also acts as a fluidizing agent in the form of steam to mix with oil sands during bitumen production. Oil sands operations in Alberta, Canada, are major consumers of freshwater from the Athabasca River [5]. Currently, 349 million cubic meters of freshwater from the Athabasca River per year is diverted, and this amount is expected to increase to more than 500 million cubic meters per year [6].

There are two major sources of water used during bitumen extraction—first, raw water imported from the Athabasca River and, second, oil sands process-affected water (OSPW), which refers to the water that has been in contact with oil sands or released from tailing deposits and reserved in tailing ponds [7]. Freshwater usage will change depending on the bitumen processing type, typically 2.41 units for mining and 0.84 units for in situ process [8]. The technologies used to mine, extract, and upgrade the bitumen to synthetic crude oil (SCO) make the product among the most environmentally costly sources of transport fuel in the world [9]. In addition, the recovery and upgrading of bitumen from the oil sands are high-energy-intensive activities, consuming large amounts of water, natural gas, electricity, transportation fuels, and hydrogen [10].

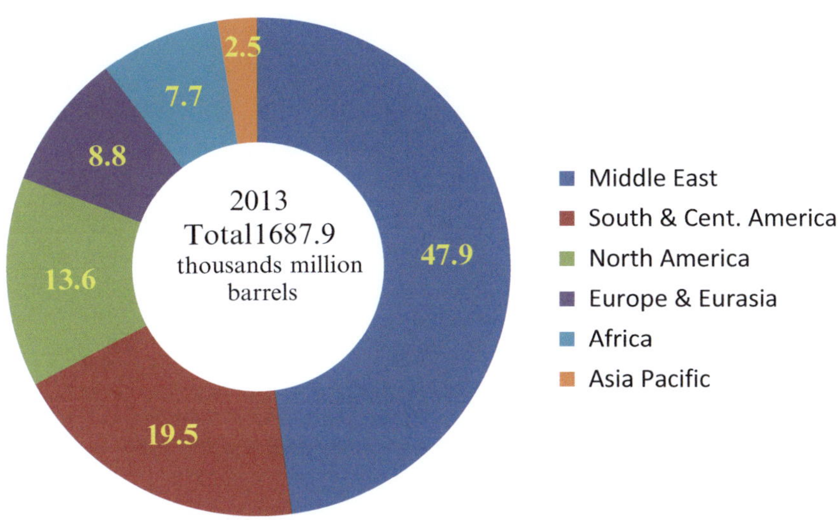

Fig. 10.1 Percentage distribution of proven oil reserves [3]

There are many techniques and technologies for bitumen extraction [11], such as surface mining (also called open-pit mining), steam-assisted gravity drainage (SAGD), cyclic steam simulation (CCS), toe-to-heel air injection (THAI), vapor extraction (VAPEX), and cold heavy oil production with sand (CHOPS). This study will focus on two of the most common techniques, SAGD and surface mining (SM), to investigate and evaluate the water usage during these processes with respect to the first and second law of thermodynamics.

10.2 Overall System Description

As shown in Fig. 10.2, the oil sands operations involve four main steps, sequentially: extraction, froth treatment, upgrading, and refining [12]. Open-pit mining is similar to many coal mining operations. Large shovels scoop the oil sands into giant trucks, which take it to crushers, where the large clumps of sand are broken down.

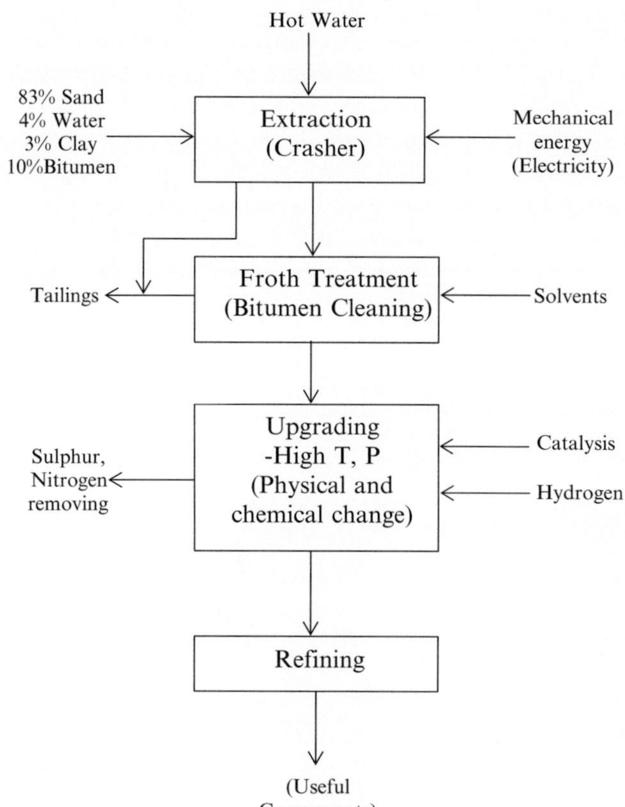

Fig. 10.2 Generalized oil sands operation scheme

The oil sands are then mixed with water (extraction) and transported by pipeline to a plant, where the bitumen is separated from the other components with further treatment, upgrading, and refining and is applied to produce SCO. This study will focus on two primary processes to extract bitumen, called ex situ (open-pit mining) and in situ (SAGD), respectively; a generic flow diagram of oil sands mining to bitumen upgrading is shown in Fig. 10.2.

10.2.1 *Surface Mining (Ex Situ) Process*

Surface mining is used when bitumen is close to the surface (within 250 ft) [13]. Figure 10.3 shows a block diagram for the water path during mining and upgrading. It displays where the most water spent during oil sands processes occurs, which in this study, water consumption is considered mainly during extraction and upgrading.

The total water consumption in the oil sands industry is very high in comparison with the water used for conventional oil production where each barrel of conventional oil requires about 0.1–0.3 barrels of water [14]. In contrast, for a barrel of bitumen obtained through surface mining, about 2–3 barrels of water are withdrawn from the Athabasca River, and around 85 % of the water used in the mining process is recycled [15].

10.2.2 *Steam-Assisted Gravity Drainage*

Also called an in situ process, in situ recovery techniques are used to extract deep deposits of bitumen without removing soil and materials above it (Fig. 10.4).

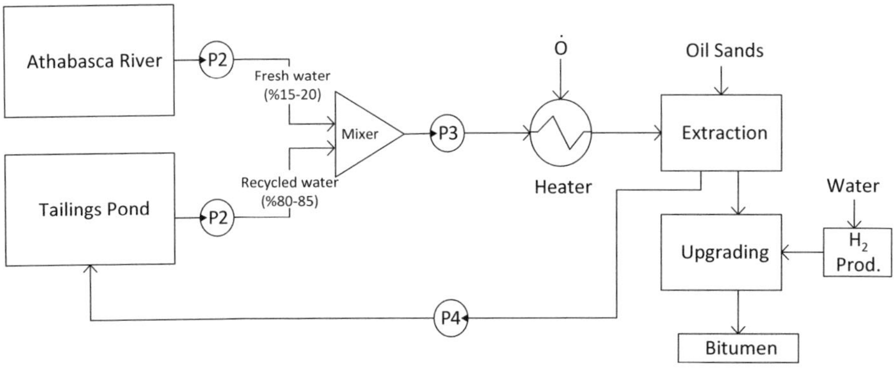

Fig. 10.3 Block diagram of water path during mining and upgrading operations

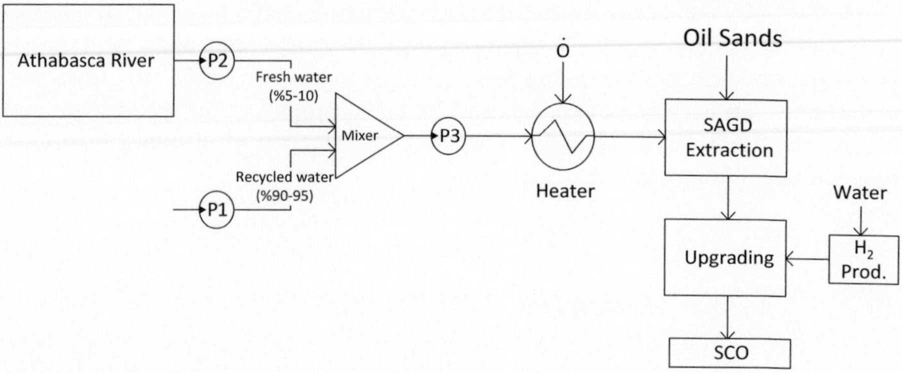

Fig. 10.4 Block diagram of water path during in situ and upgrading operations

Water is produced to steam to heat the bitumen underground and then pumped to the surface through wells. The majority of the oil sands exist more than 70 m (200 ft) [16] below the ground, which means it is excessively deep to be mined. These reserves can be recovered in situ (in place) by drilling wells. In situ drilling accounts for about 80 % of oil sands reserves [17]. Drilling methods disturb a small amount of land and do not require tailings ponds. Advanced technology is used to inject steam. Mostly this steam is produced by burning natural gas through an HRST to supply necessary heat into the reservoir. The heat warms the bitumen and reduces the viscosity so it can be pumped to the surface through recovery wells.

In situ technology has significantly reduced water consumption, resulting in only 0.9 barrels of water being used per barrel [14]. Furthermore, about 90–95 % of the water consumed in steam-assisted gravity drainage is recycled [15].

10.3 Water Requirements for Production of Hydrogen

As mentioned previously, hydrogen can be produced from steam methane reforming, coal gasification, water electrolysis, or thermochemical cycles like the Cu-Cl cycle. To compare these methods, the hydrogen production capacity must be compared on the same basis with respect to the heat source.

Figure 10.5 illustrates the required H_2O to produce 1 kmol H_2 and shows that in order to reduce the total amount of water used, SMR and Cu-Cl cycles are the most favorable processes. Less water is consumed with both of these aforementioned methods. Figure 10.5 also shows which hydrogen production techniques would be the finest to produce this amount of hydrogen to integrate the upgrading process with the aim of reducing consumed water.

For these processes, the total practical yield of hydrogen is normally within the range of 60–70 % [18, 19]. An average value of 65 % is assumed. The conversion to hydrogen is about 60 % and the thermal efficiency is 60–75 %, depending on the

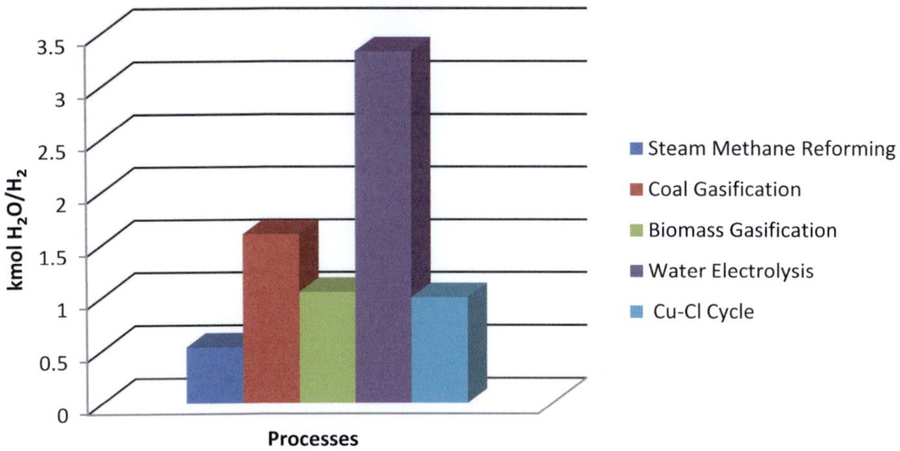

Fig. 10.5 Required H_2O to produce 1 kmol H_2 (data from [18])

form of carbon source and assuming a thermal efficiency of 60 %. Electrical energy required to split 1 mol of H_2O is 285.83 kJ/mol if the efficiency expected to be 100 % [20]. It is assumed in this study that natural gas is used for the electrical power generation, and the overall thermal efficiency of water electrolysis is 30 %.

In order to achieve the total bitumen production rate of 2.2×10^6 barrels per day, the required amount of hydrogen 7.26×10^6 is estimated. From Fig. 10.5, if the coal gasification is used to produce at the same amount of required hydrogen, the amount of necessary H_2O will be 12.1×10^6 barrels per day. From the same logic, the required water needs will be 264×10^6 barrels per day, if biomass gasification is chosen and if hydrogen comes from water electrolysis; it will be 8.1×10^6 barrels H_2O per day, respectively.

10.4 Thermodynamic Analysis

The mass and energy balances are evaluated with Aspen Plus. For a general steady-state process, the mass and energy balances, respectively, are written as

$$\sum_{i} \dot{m}_{i} = \sum_{o} \dot{m}_{o} \tag{10.1}$$

$$\sum_{i} \dot{E}_{in} = \sum_{o} \dot{E}_{out} \tag{10.2}$$

An overall exergy balance can be written for a steady-state process as follows:

$$\left(\sum \dot{\mathrm{E}}\mathrm{x}_i\right)_{\mathrm{in}} = \left(\sum \dot{\mathrm{E}}\mathrm{x}_i\right)_{\mathrm{out}} + \sum \dot{\mathrm{E}}\mathrm{x}_{\mathrm{dest}} \tag{10.3}$$

where during extraction processes

$$\left(\sum \dot{\mathrm{E}}\mathrm{x}_i\right)_{\mathrm{in}} = \dot{\mathrm{E}}\mathrm{x}_{\mathrm{recyled\ water}} + \dot{\mathrm{E}}\mathrm{x}_{\mathrm{fresh\ water}} + \dot{\mathrm{E}}\mathrm{x}_{\mathrm{oil\ sands}} \tag{10.4}$$

It is assumed that the content of the oil sands consists of 83 % sand, 4 % water, 3 % clay, and 10 % bitumen [4].

$$\left(\sum \dot{\mathrm{E}}\mathrm{x}_i\right)_{\mathrm{out}} = \dot{\mathrm{E}}\mathrm{x}_{\mathrm{st}} + \dot{\mathrm{E}}\mathrm{x}_{\mathrm{water}} + \dot{\mathrm{E}}\mathrm{x}_{\mathrm{bitumen}} + \dot{\mathrm{E}}\mathrm{x}_{\mathrm{sand}} \tag{10.5}$$

Both physical and chemical exergy inlet and outlet values are determined for the pumps and heaters and used to assess exergy destructions. The specific flow exergy associated with a specified state is expressed by the sum of specific physical and specific chemical exergy:

$$\mathrm{ex}_{\mathrm{prod}} = \mathrm{ex}^{\mathrm{ch}} + \mathrm{ex}^{\mathrm{ph}} \tag{10.6}$$

In this study, it has been assumed that there is no chemical reaction that occurs during water consumption; therefore, chemical exergy will not be changed ($\mathrm{ex}^{\mathrm{ch}} = 0$). In this particular study, all flows are considered to have only a physical exergy change. The physical exergy is defined as follows:

$$\mathrm{ex}^{\mathrm{ph}} = (h - h_{\mathrm{o}}) - T_0(s - s_{\mathrm{o}}) \tag{10.7}$$

A second way of finding total $\dot{\mathrm{E}}\mathrm{x}_{\mathrm{dest}}$ during extraction process can be calculated from using total $\dot{S}_{\mathrm{gen}}$ values. The entropy balance for a steady-flow reacting system can be written as

$$\sum \frac{\dot{Q}_j}{T_j} + \sum \dot{m}_i s_i - \sum \dot{m}_{\mathrm{o}} s_{\mathrm{o}} + \dot{S}_{\mathrm{gen}} = 0 \tag{10.8}$$

The exergy destroyed due to irreversibility can also be expressed as follows:

$$\dot{\mathrm{E}}\mathrm{x}_{\mathrm{dest}} = T_0 \dot{S}_{\mathrm{gen}} \tag{10.9}$$

The heat capacity of flows is determined using Aspen Plus property data and substituted into Eq. (10.8) to find generated entropy values ($\dot{S}_{\mathrm{gen}}$), which are used for the thermal exergy calculation in Eq. (10.9).

The exergy efficiency for component i may be written as

$$\psi_i = 1 - \frac{\left(\dot{\mathrm{E}}\mathrm{x}_{\mathrm{dest}}\right)_i}{\left(\dot{\mathrm{E}}\mathrm{x}_{\mathrm{in}}\right)_i} \tag{10.10}$$

where $(\dot{\mathrm{E}}\mathrm{x}_{\mathrm{dest}})_i$ and $(\dot{\mathrm{E}}\mathrm{x}_{\mathrm{in}})_i$, respectively, are the exergy destruction rate and the exergy input rate for component i.

Also, the ratio of exergy destruction, x_{dest}, for a component can be evaluated by dividing its exergy destruction by the total exergy provided to the system

$$x_{\mathrm{dest}} = \frac{\left(\dot{\mathrm{E}}\mathrm{x}_{\mathrm{dest}}\right)_i}{\left(\sum \dot{\mathrm{E}}\mathrm{x}_i\right)_{\mathrm{in}}} \tag{10.11}$$

where $(\dot{\mathrm{E}}\mathrm{x}_{\mathrm{dest}})_i$ is the exergy destruction for each component and $\left(\sum \dot{\mathrm{E}}\mathrm{x}_i\right)_{\mathrm{in}}$ is the exergy flow of all input material streams. These equations assumed steady state and no losses throughout the processes.

10.5 System Simulations

10.5.1 Mining Water Usage Simulation

It will be assumed that water is consumed during extraction, where oil sands and hot water mix to extract the bitumen. Second, during the upgrading process, a significant amount of hydrogen is needed by the system in order to upgrade the bitumen. Producing hydrogen will affect the total consumed water indirectly since also producing hydrogen needs water. For simplicity, hereafter the total amount of water consumption will be assumed to the sum of these two processes; in other words, the total water consumption is equal to water consumption during extraction plus the amount of required water during hydrogen production (Fig. 10.6).

10.5.2 In Situ Water Usage Simulation (SAGD)

Figure 10.7 shows Aspen Plus models for the refining section to establish how much water is needed during oil sands in situ operations. Water is used for mining, upgrading, conditioning, and during processes of hydrogen production from a steam methane reformer. Hydrogen is commonly produced by steam methane

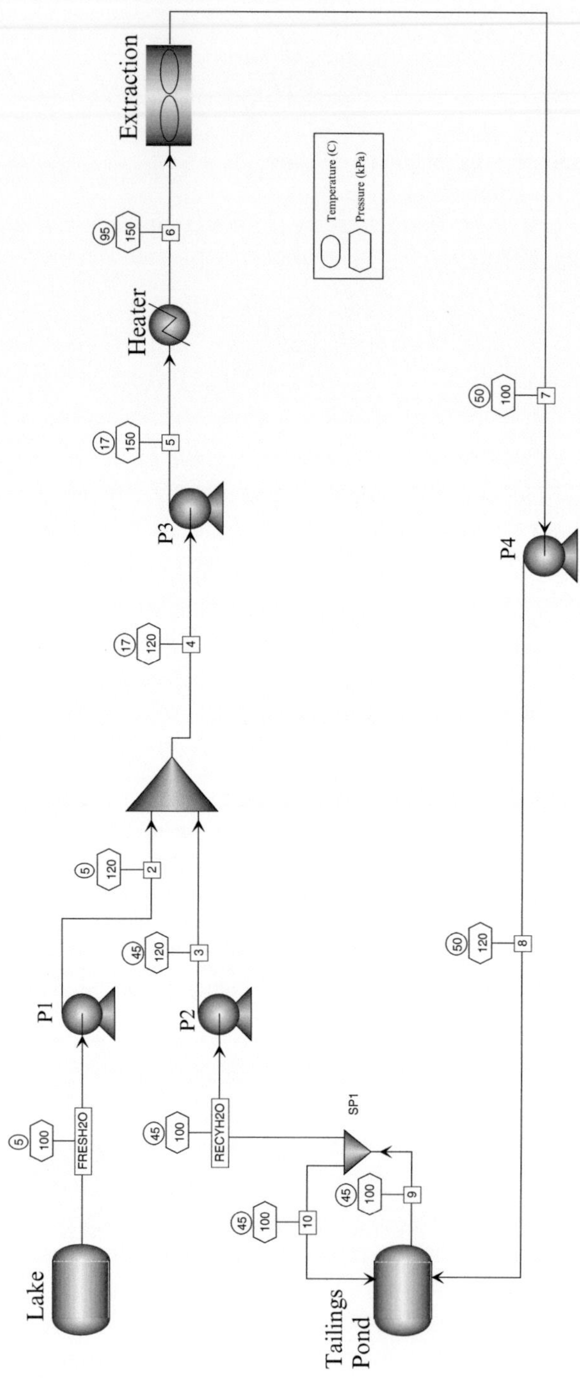

Fig. 10.6 Aspen model of water cycle during mining operation

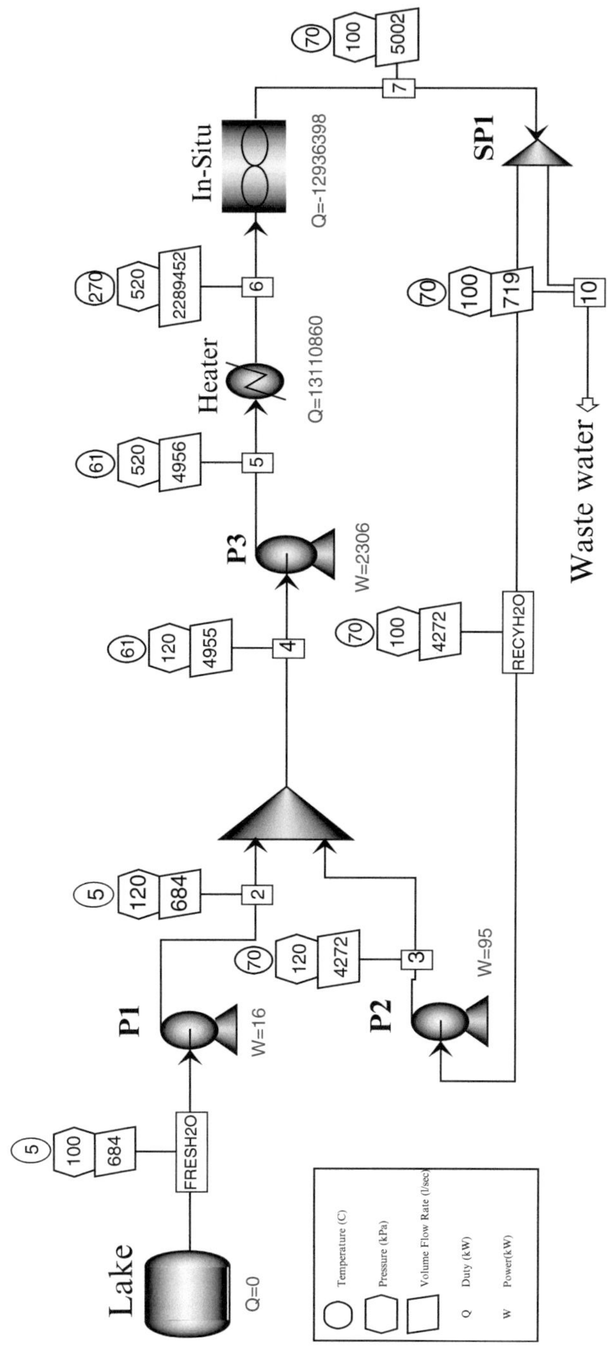

Fig. 10.7 Aspen model of water cycle during in situ process

reforming; however, in this case study, we will consider also other hydrogen production techniques to examine the amount of water used for each process for comparison purposes.

10.6 Results and Discussion

In Alberta, Canada, oil sands companies are currently diverting massive amounts of water from the Athabasca River, enough to satisfy the needs of a city of two million people [21]. Mining as well as in situ bitumen operations consume large volumes of water. Water requirements for oil sands projects range from 2.5–4.0 units of water for each unit of bitumen produced [22]. The primary challenge for process water is that no large-scale water treatment facilities exist near the oil sands. As a result, recycled percentage of process water stays very low.

Figure 10.8 shows that the heater (6 %) yields a large amount of exergy destruction. Since this heater is simulated as a heat recovery steam generator (HRSG), where phase change occurs, during this boiler process, the exergy destruction increases to a greater extent. If the hot inlet temperature stream passed through two or three more heat exchangers instead of one boiler process, the exergy destruction rate could be reduced. While this process would increase the equipment cost, it would allow the system to gradually reduce the hot inlet stream's temperature, producing less exergy destruction and a drastically decreased hot inlet temperature.

Table 10.1 shows water flow data for different points in the simulation. As a result, energy and exergy losses are occurring in the heating, mining, and tailing operations. The majority of exergy loss is in the tailing ponds since a large amount of

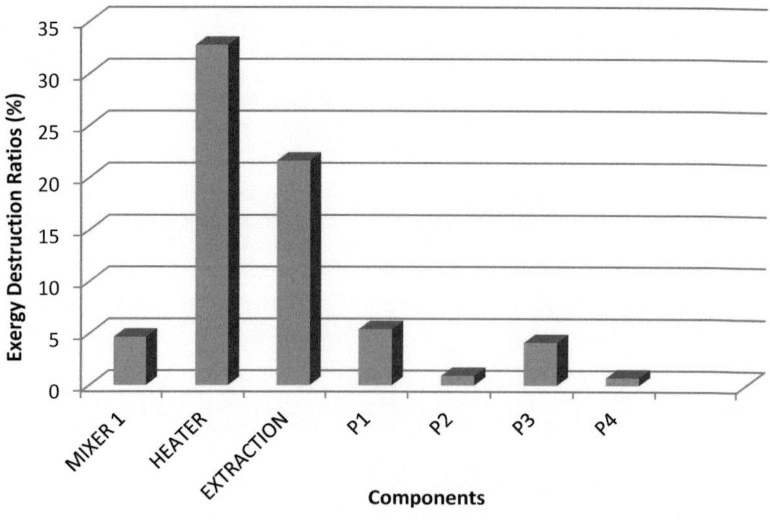

Fig. 10.8 Exergy destruction ratios of some components during surface mining processes

Table 10.1 Water flow data (mining)

Comp	Pressure (bar)	Temperature (°C)	Mass flow (kg/s)	Mole flow (kmol/s)	Volume flow (L/s)	Enthalpy (kJ/kg)	Entropy (kJ/kg K)	Exergy (kW)]	Destroyed exergy (kW)
P1	1.2	5	5,648	313	5,659	1,593	9.285	93,044	125
P2	1.2	45	8,797	488	8,789	15,779	8.777	93,044	104
P3	1.5	17	1,139	313	488	801	801	801	98
P4	1.5	50	20,527	5,648	8,797	14,448	14.52	14,449	345
Heater	1.5	95	21,065	5,576	9,028	14,601	146.01	15,641	12,559
Mining	1	60	28,535	28,535	28,049	28,391	283.91	28,427	8,035
Tailing	1	50	15,839	15,839	15,569	15,759	345.55	55,779	156,755

water (temperature around 50 °C) dumped to the atmosphere (average temperature around 15 °C) causes a significant amount of heat losses which is directly proportional with the amount of exergy loss, which means much of this water has a great work potential, yet it will be wasted. In addition, entropy values are increasing with heating, mining, and tailing operations. As the production of bitumen increases, so will the demand for water. As noted earlier, the EUB expected the production of bitumen from oil sands to more than double in the decade 2004–2014, which could see a comparable increase in the demand for water in northeast Alberta [21].

Figure 10.9 shows the exergy destruction ratios of some components during in situ processes; as it can be seen, the greatest exergy destruction occurs during in situ process and heating. The reason is that the amount of heat and pressure lost in these both processes cannot be reversed.

Exergy destruction ratios help establish the exergy destruction in any system as it identifies where the highest part of the total inlet exergy is destroyed. When analyzing Figs. 10.8 and 10.9, the mixing section consists of the lake and the tailing pond in the simulation. The results reflect the mixing exergy destruction ratio which is 4.7 %. The major exergy destruction ratio which is described in Eq. (10.11) in the thermodynamic analysis section was found to be 37 % which means that the total exergy of the system is declining after the heating and mixing (where fresh and cold water mix) processes. The extraction system has a large amount of exergy destruction, because in the extractor, oil sands and hot water mix. In addition, the SP1 and P1 exergy destruction ratios are 3.6 % and 4.2 %, respectively, while the other pumps have smaller exergy destruction ratios.

Table 10.2 illustrates water flow data for in situ simulation operation result. This process has a higher pressure and temperature; based on exergy phenomena, the higher the pressure and temperature difference, the greater the produced entropy; thus, the amount of produced entropy will be higher and destroyed exergy as well. However, the mass flow of water is less than the mining operation, so wasted water

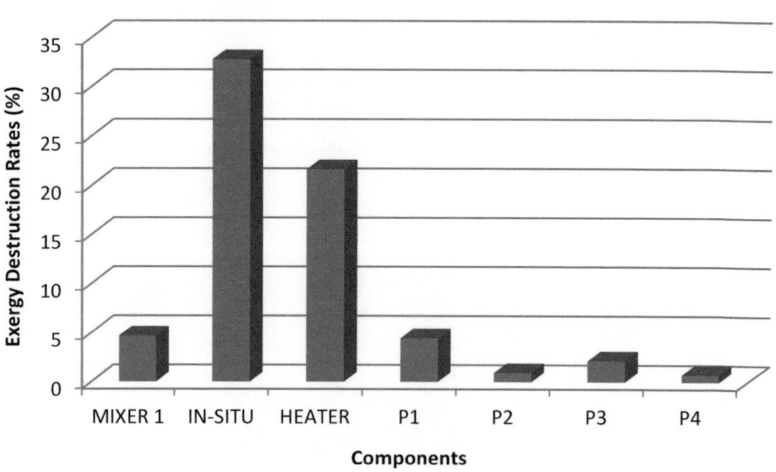

Fig. 10.9 Exergy destruction ratios of some components during in situ processes

Table 10.2 Water flow data (in situ)

Comp	Pressure (bar)	Temperature (°C)	Mass flow (kg/s)	Mole flow (kmol/s)	Volume flow (L/s)	Enthalpy (kJ/kg)	Entropy (kJ/kg K)	Exergy (kW)	Destroyed exergy (kW)
P1	1.2	5	693.1	38.5	694.0	15,677.4	8.47	83,044	143
P2	1.2	45	3,927.3	218.0	3,934	15,779	9.28	97,044	245
P3	1.5	17	4,620.4	256.5	4,629.3	15,877.4	8.47	100,675	456
P4	1.5	50	4,620.4	256.5	4,629.3	15,877.4	8.58	100,456	344
Heater	5.2	270	4,620.4	256.5	4,629.3	18,953.5	2.07	230,564	14,564
In situ	5	250	4,620.4	256.5	4,629.3	15,677.4	8.47	130,564	12,456

Table 10.3 Raw bitumen and crude oil production from oil sands [20]

Million barrels per day	2011	2015	2020	2025	2030
Mining	0.89	1.21	1.52	1.93	2.17
In situ	0.85	1.27	1.87	2.57	3.16
Total	*1.74*	*2.48*	*3.39*	*4.50*	*5.33*
Crude oil	*1.6*	*2.3*	*3.2*	*4.2*	*5.0*

and total destroyed exergy will be relatively lower than the mining operation. In 2003, Alberta's Environment Minister initiated a committee to find ways to reduce the oil and gas industry's consumption of freshwater. As part of the province's long-term water strategy, limits may be placed on the volume of freshwater that companies are allowed to use [23]. Table 10.3 displays total raw bitumen and crude oil production from oil sands, as it is clear that total production will increase with required water by time. It means produced energy and entropy will be increased as well.

In this study, there are two main water consumption processes that take place in the oil sands and bitumen production operations. The first process occurs during extraction or in situ processes where the hot water is mixing with oil sands. The second process is indirect water consumption, where the upgrading process needs a large amount of H_2 for the upgrading of bitumen to synthetic crude. Some hydrogen production methods consume less water than others to produce the same amount of H_2. In this study, water consumption investigation has also been used for different hydrogen production methods to determine which hydrogen production method is the best in terms of requiring less water. The natural gas needed to supply heat for the production of 2.2 million barrels bitumen per day is calculated as 25,603 MW (thermal energy) [24].

The required H_2O to produce 1 kmol H_2/h for the SMR, coal gasification, and biomass gasification is described previously. The goal is to reduce the amount of used water; thus, coal gasification seems to be the best option in terms of less water consumption; however, environmental concerns will be raised if the coal is used. So a second option which is SMR is more convenient for both environment and water usage concerns.

10.7 Conclusions

This study has examined the total water requirements of oil sands extraction and upgrading processes. In situ operations require much less water than mining. Approximately 1.0–1.25 GJ (280–350 kWh) of energy is needed to extract a barrel of bitumen and upgrade it to synthetic crude. Since a barrel of oil equivalent is about 6.117 GJ (1,699 kWh), this means it extracts about 5 or 6 times as much energy as consumed. Energy and exergy losses are an important factor in the ultimate magnitude of used water impacts, means the more water used, the more energy. In addition to energy perspective, there are many other reasons to minimize the amount of water used such as water availability, pollution, and security. Hence, reducing the water consumption and increasing the water return volume are a major challenge that must be resolved for oil sands development.

Acknowledgment The authors thankfully acknowledge the financial support provided by the Ontario Research Excellence Fund.

References

1. Clark B (2007) Working document of the NPC global oil & gas study; heavy oil, extra-heavy oil and bitumen unconventional oil. National Petroleum Council (NPC)
2. Macgregor DS (1996) Factors controlling the destruction or preservation of giant light oil fields. Pet Geosci 2:197–217
3. (2014) BP statistical review of world energy. www.BP.com
4. US natural gas: the role of unconventional gas. Energy Bull. http://www.energybulletin.net/node/44389
5. Mech M (2011) A comprehensive guide to the Alberta oil sands: understanding the environmental and human impacts. http://www.greenparty.ca/sites/greenparty.ca/files/attachments/
6. Dyer S, Moorhouse J, Laufenberg K, Powell R (2008) Under-mining the environment: the oil sands report card. World Wildlife Foundation and the Pembina Institute, Canada
7. Zubot W, MacKinnon MD, Chelme-Ayala P, Smith DW, El-Din MG (2012) Petroleum coke adsorption as a water management option for oil sands process-affected water. Sci Total Environ 427:364–372
8. Alberta Environment and Sustainable Resource Development. Oil sands information portal: oil sands operators, water use history. http://environment.alberta.ca/apps/OSIPDL/Dataset/Details/56
9. Draper D, Reed MG (2009) Our environment: a Canadian perspective, 4th edn. Nelson, Toronto
10. (2013) Canada's energy future—energy supply and demand projections to 2035. An Energy Market Assessment. ISSN 2292–1710
11. Veil JA, Quinn JJ (2010) Water issues associated with heavy oil production. US Department of Energy, National Energy Technology Laboratory, Morgantown
12. Alberta Chamber of Resources (2004) Oil sands technology roadmap: unlocking the potential. Alberta Chamber of Resources, Edmonton
13. Speight J (2011) An introduction to petroleum technology, economics, and politics. Scrivener Publishing, Salem. ISBN 9781118012994
14. Prebble P, Coxworth A, Simieritsch T, Dyer S, Huot M, and Walsh H (2009) Carbon copy: preventing oil sands fever in Saskatchewan. Oil sands fever series. Pembina Institute, Saskatchewan Environmental Society, Canadian Parks and Wilderness Society, Drayton Valley, pp 14–20
15. Humphries M (2008) North American oil sands: history of development, prospects for the future. Congressional Research Service, Washington, DC
16. Hershey PA (2012) Sustainable practices: concepts, methodologies, tools and applications. In: Sustainable systems and energy management, pp 184–201
17. Yeh S, Jordaan SM, Brandt AR, Turetsky MR, Spatari S, Keith DW (2010) Land use greenhouse gas emissions from conventional oil production and oil sands. Environ Sci Technol 44(22):8766–8772

18. NYSERDA (New York State Energy Research and Development Authority) (2010) Hydrogen fact sheet: hydrogen production steam methane reforming (SMR). Technical report-Clean Energy Initiative. http://www.getenergysmart
19. Bonaquist D (2010) Analysis of CO2 emissions, reductions, and capture for large-scale hydrogen production plants: a white paper. See also, http://www.praxair.com/praxair
20. Lewis MA, Masin JG, O'Hare PA (2009) Evaluation of alternative thermochemical cycles, part I: the methodology. Int J Hydrog Energy 34:4115–4124
21. Griffiths M, Taylor A, Woynillowicz D (2006) Troubled waters, troubling trends. The Pembina Institute, Drayton Valley
22. National Energy Board (2006) Canada's oil sands—opportunities and challenges to 2015: an update, p 17
23. Söderbergh B (2005) Canada's oil sands resources and its future impact on global oil supply. Masters of Science, Uppsala University, Uppsala
24. Wang ZL, Naterer GF (2010) Greenhouse gas reduction in oil sands upgrading and extraction operations with thermochemical hydrogen production. Int J Hydrog Energy 35:11816–11828

Chapter 11
Air Source Heat Pump Performance in Open, Semi-closed, and Closed Greenhouse Systems in British Columbia

İlhami Yıldız, Jin Yue, and Asena Cansu Yıldız

Abstract The specific objective of this study was to investigate regional spatio-temporal distributions and spatial correlations of energy and water consumption in open, semi-closed, and completely closed greenhouse systems in British Columbia. The findings showed that the energy and water consumptions were both spatially and system dependent. Latitude and longitude significantly predicted total energy and water consumptions. Semi-closed and open heat pump systems were the most energy conserving systems while closed system was the most water conserving. Semi-closed and closed systems at lower latitudes had more water consumptions compared to those at higher latitudes while open systems at higher latitudes had more compared to those at lower latitudes. Locations close to the southwest Pacific Coast had the lowest energy and water consumptions in all systems.

Keywords Heat pump • Air source • Energy consumption • Water consumption • Closed greenhouse • Semi-closed greenhouse • Heating • Cooling • Dehumidification • Spatiotemporal distribution • Spatial correlation • British Columbia

Nomenclature

C	Amount of water collection, kg/m^2
CHP	Closed heat pump greenhouse system
CON	Conventional greenhouse system
E	Amount of energy consumption, GJ/m^2

İ. Yıldız (✉) • J. Yue
Department of Engineering, Faculty of Agriculture, Dalhousie University, Truro-Bible Hill, NS, Canada B6L 3H9
e-mail: iyildiz@dal.ca

A.C. Yıldız
Department of Law, Queen Mary University of London, Mile End Road, London E1 4NS, UK

I. Dincer et al. (eds.), *Progress in Clean Energy, Volume 1*,
DOI 10.1007/978-3-319-16709-1_11

F	Statistical F test
OHP	Open heat pump greenhouse system
p	Significance level
R	Regression
SHP	Semi-closed heat pump greenhouse system
t	Statistical t test
W	Amount of water consumption, kg/m^2

Greek Letters

β Beta value

11.1 Introduction

One major factor hindering future expansion of greenhouse industry is the cost required for environmental control. Consequently, considerable effort is expended to conserve energy and look for alternative energy sources, especially environment-friendly renewable energy sources and technologies. Proper greenhouse and environmental management systems can significantly change the energy and moisture dynamics of greenhouse production systems. This study helps enhance the competitive position of British Columbia's (BC) agriculture and agri-food industry by introducing economically, environmentally, and socially sustainable technologies and management strategies. The investigated heat pump technology and operational modes (especially semi-closed and closed) bring increased yield, and energy and water-efficient solutions to the doorstep of BC's greenhouse producers. The overall goal of this study was to help reduce the load on power grid, demand for fossil fuels and irrigation water, and also supply CO_2 for the greenhouse production. The use of heat pumps for heating and cooling greenhouses makes it possible to use local and renewable energy sources while reducing or totally eliminating CO_2 emissions. Heat pumps also make an innovative confined greenhouse operation possible, which would totally eliminate the energy, CO_2, and water losses due to ventilation conserving all of these precious resources, and would make an insect-free greenhouse operation possible. The specific objective of this study was to investigate spatiotemporal performance distributions and spatial correlations of energy and water consumptions in open, semi-closed, and confined greenhouse systems in BC.

11.2 Methodology

This section focuses on the resources and procedures employed in this study, as well as the statistical analyses employed and data presentation.

11.2.1 Weather File

The Canadian Weather for Energy Calculations (CWEC) files created by concatenating 12 Typical Meteorological Months (TMY) selected from a database of 30 years of data were used for continuous full year analyses at 14 different locations. CWEC data sets were prepared under the direction of Environment Canada [1], and simulations were performed starting at the beginning of 1st day of January and ending at the end of 31st day of December for full 365-day simulations.

11.2.2 Greenhouse Characteristics and Operational Strategies

A 7.5 × 7.5 m double polyethylene greenhouse was used in this dynamic simulation study. The floor surface material was reflective mulch, and a cucumber crop was used with a full plant height of 2.0 m, a row spacing of 0.86 m, and a North–south orientation. Theoretical approach, simulation model assumptions and validation, and all the other characteristics of the model including the details of control systems and strategies were reported in another study [2]. Three conventional gas-fired furnaces (24,612 W of heat input each) provided heating with a multiposition proportional control, and an evaporative cooling system provided cooling in the conventional (CON) system. A furnace efficiency of 0.8 was assumed. In the heat pump systems, however, three 3-ton (based on system heat removal capacity) gas-fired heat pump units provided both heating and cooling. The greenhouse systems with heat pump units were operated as an open (OHP), semi-closed (SHP), and completely closed (CHP) system. One of the heat pump units in closed system operations was used as a dehumidifier while operating as a heater. The only difference from the original heating unit was circulating the inside greenhouse air through the outdoor coil instead of the outside air. This prevented moisture build up in closed system operations. An overhead plastic tube was used for the hot and the cold air distribution. No heat storage facility was used in this study. A variable shading system was used to reduce cooling loads during daytime and an aluminized (both sides) night curtain was used at night to reduce the heat loss due to long-wave radiation exchanges. In the open systems (CON, OHP), ventilation was provided by two fans, one with a fixed flow rate (0.005 $m^3/s{\cdot}m^2$) operating at all times, and the other one with a variable flow rate (0.08 $m^3/s{\cdot}m^2$ max). SHP system was operated as either OHP or CHP system at different times of the year for providing an optimal indoor environment with minimal energy consumption. Liquid CO_2 tanks were employed and a concentration of 380 ppm was maintained during the day in all four-greenhouse systems.

11.2.3 Data Analyses and Presentation

Annual total energy (heating and cooling) and water consumption (transpiration), and water collection were determined for each greenhouse system and location studied. A total of four systems (CON, OHP, SHP, and CHP) and 87 locations were studied. The two-way analysis of variance (ANOVA) (significance level of $p < 0.05$) was used to determine the effect of each independent variable, location and system, on the dependent variables (energy consumption, water consumption, and water collection). Tukey Simultaneous Test at a confidence level of 95 % was performed to compare the means within the two treatments (location and system) and determine statistically significant differences. For each system, energy and water consumption, and water collection by dehumidification units were also analyzed by multiple regressions, and spatial correlations were developed by using as regressors altitude, latitude, and longitude. With other variables held constant, the relationships and significance between the energy and water consumption and water collection and each of the regressors were analyzed. How well the regressors explain the variation in dependent variable, taken together, was assessed by the value of R^2. Whether the regressors, taken together, are significantly associated with the dependent variable were assessed by the statistic F in the standard ANOVA (significance level of $p < 0.05$). What relationship each regressor had with the dependent variable when all other regressors were held constant was assessed by looking at the regression coefficients. Whether the relationship of each regressor with the dependent variable was statistically significant or not, with all other regressors taken into account was answered by looking at the t values in the table of regression coefficients. To assess which regressor had the most effect on the dependent variable was addressed by using the *beta weights*. *Minitab 15* was used for the statistical analyses. Spatial distribution maps were generated by using *ArcGIS 10*.

11.3 Data Interpretation

This section covers the result and discussions for different greenhouse systems at different locations.

11.3.1 Energy Consumption, Water Consumption, and Water Collection

The two-way ANOVA $(p < 0.05)$ was used to determine the main effects of greenhouse location and system on total energy and water consumptions as well as water collection. There were significant main effects for the location and system

Table 11.1 Comparisons of means with respect to different locations using the Tukey Simultaneous Test at a 95 % confidence interval

Location	Mean annual energy use (GJ/m^2)	Mean annual water use (kg/m^2)	Mean annual water collection (kg/m^2)
Abbotsford	5.29 a	584.1 qr	109.7 wx
Comox	5.23 a	564.9 q	113.9 wx
Cranbrook	6.68 ab	638.8 qr	146.4 wx
Fort Nelson	8.74 c	616.1 qr	132.2 wx
Fort St John	7.85 bc	605.7 qr	125.6 wx
Kamloops	5.81 ab	685.0 r	169.4 x
Port Hardy	5.78 ab	473.1 p	87.4 w
Prince George	7.30 bc	549.9 q	117.7 wx
Prince Rupert	6.08 ab	468.5 p	100.8 wx
Sandspit	5.76 a	473.9 p	89.3 w
Smithers	7.18 b	565.4 q	122.3 w
Summerland	5.74 a	654.7 r	155.2 wx
Vancouver	5.31 a	545.4 pq	104.2 wx
Victoria	5.11 a	573.0 q	102.0 wx

Means in the same column not followed by the same letter are significantly different

on total energy consumption, $F(13, 39) = 8.70$ ($p < 0.05$), and $F(3, 39) = 129$ ($p < 0.05$), respectively (Tables 11.1 and 11.2).

Locations at lower latitudes had much less total energy consumption, and part of which was for cooling while northern locations had much less cooling requirements (Fig. 11.1). Table 11.2 shows that SHP and OHP systems had the lowest annual energy consumptions while CON system had the highest.

There were significant main effects for the location and system on water consumption as well, $F(13, 39) = 10.8$ ($p < 0.05$), and $F(3, 39) = 133$ ($p < 0.05$), respectively (Tables 11.1 and 11.2). The main effect for the system on water collection was also significant, $F(3, 39) = 288$ ($p < 0.05$); however, the main effect for the location on water collection was not, $F(13, 39) = 1.41$ ($p < 0.05$) (Tables 11.1 and 11.2).

Locations closer to the Pacific Coast had lower water consumption (transpiration) due to increased outside relative humidity levels, and hence reduced vapor pressure deficits. Water collection at these locations was also significantly lower than those at other locations. Even though the impact of latitude on water consumption was not statistically significant, semi-closed, and closed systems at lower latitudes had higher water consumption (transpiration) due to increased amount of available solar radiation (Fig. 11.2). Transpiration rates in the open systems at higher latitudes were relatively higher than those at lower latitudes as the vapor pressure deficits became larger due to cooler outside air. CHP systems had the lowest transpirations as much higher relative humidities were observed in these systems compared to those in the other systems. All the transpired water was collected by dehumidification system, and therefore, the overall annual water

Table 11.2 Comparisons of means with respect to different greenhouse systems using the Tukey Simultaneous Test at a 95 % confidence interval

System	Mean annual energy consumption (GJ/m^2)	Mean annual water consumption (kg/m^2)	Mean annual water collection (kg/m^2)
CON	9.30 a	635.2 p	0.0 w
OHP	4.61 b	679.3 q	5.0 w
SHP	4.40 b	578.3 r	81.3 x
CHP	6.79 c	392.6 s	392.6 y

Means in the same column not followed by the same letter are significantly different

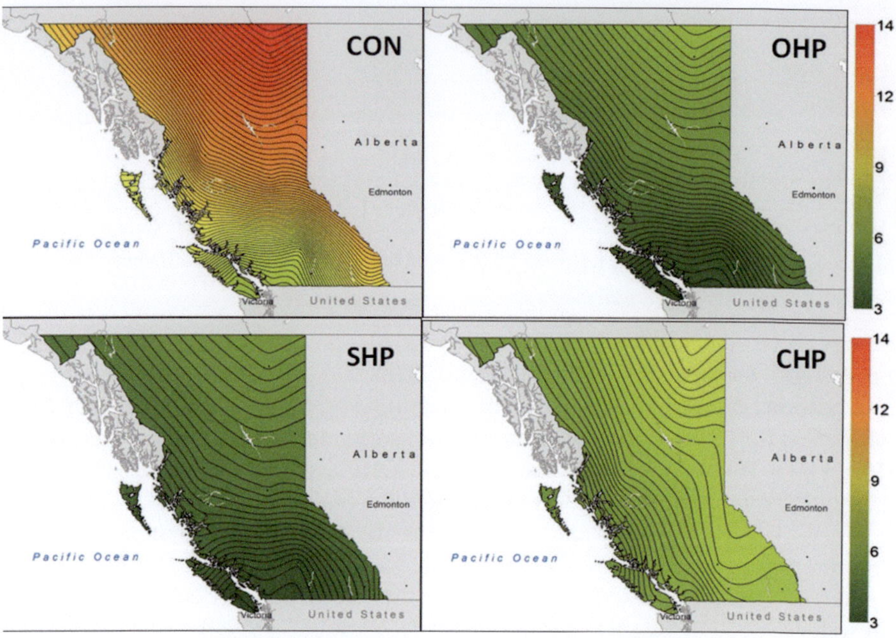

Fig. 11.1 Spatial distributions of annual total energy consumptions (GJ/m^2) in CON, OHP, SHP, and CHP systems in British Columbia (contour line intervals: 100 MJ/m^2)

consumption in CHP systems was essentially zero (Tables 11.1 and 11.2). CHP systems had relatively lower energy consumptions in colder months, but the consumption increased in warmer months due to increased cooling loads; as a result, the overall annual energy consumptions in CHP systems were much higher than those in OHP and SHP systems (Figs. 11.3 and 11.4). The energy consumption in SHP systems approached to those in closed (CHP) systems at higher latitudes (e.g., Fort Nelson) (Fig. 11.3) because SHP systems under such conditions operated more as a closed system from November through end of March. However, the energy consumption in SHP systems was exactly the same as in open (OHP)

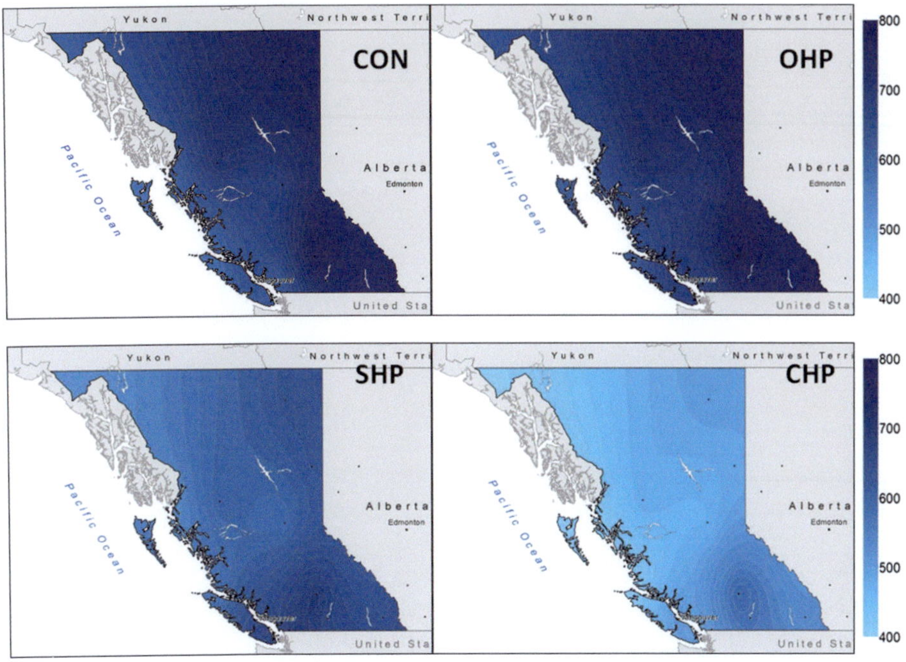

Fig. 11.2 Spatial distributions of annual amount of transpiration (kg/m^2) in CON, OHP, SHP, and CHP systems in British Columbia (contour line intervals: 15 kg/m^2)

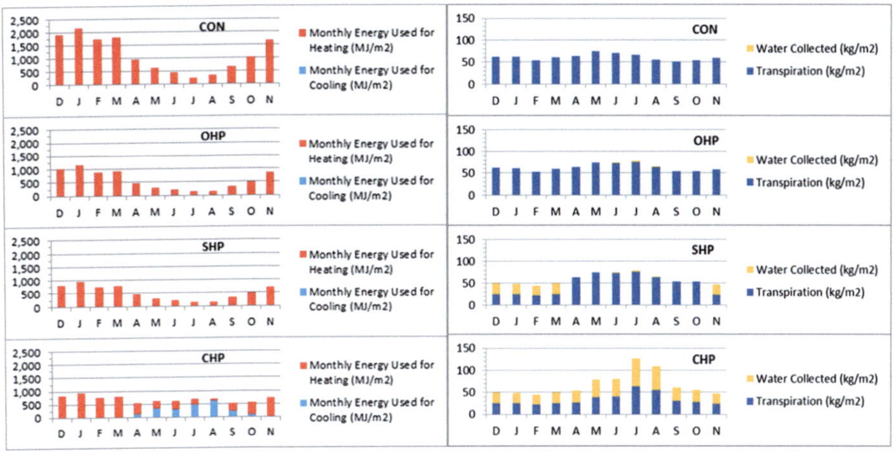

Fig. 11.3 Temporal distributions of energy consumption (*left*) and transpiration (*right*) in CON, OHP, SHP, and CHP systems at Fort Nelson, BC (58.83 N, 122.58 W, altitude: 382 m)

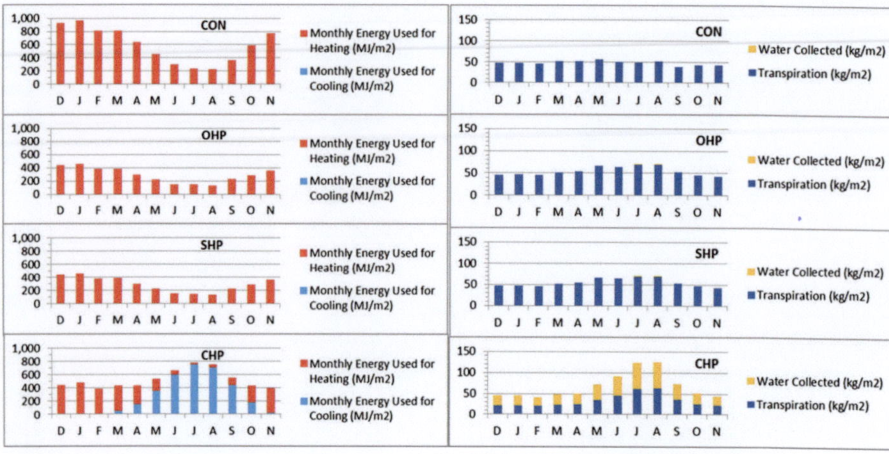

Fig. 11.4 Temporal distributions of energy consumption (*left*) and transpiration (*right*) in CON, OHP, SHP, and CHP systems at Victoria, BC (48.65 N, 123.43 W, altitude: 19 m)

systems at lower latitudes (e.g., Victoria) (Fig. 11.4) because SHP systems under such conditions operated exactly as an open (OHP) system throughout the year.

Further details of individual comparisons for both location and system were provided in Tables 11.1 and 11.2. And the individual system performances were provided below.

11.3.2 Conventional System

Amount of total energy consumption, water consumption (transpiration), and water collection on the dehumidifier were analyzed throughout the region by multiple regressions using as regressors latitude, longitude, and altitude. The regression for the total energy consumption was provided in Eq. (11.1).

$$E = -13.8 + 0.545 \times (\text{Latitude}) - 0.048 \times (\text{Longitude}) + 0.0020 \times (\text{Altitude}) \quad (11.1)$$

where E, the annual total energy consumption, was in GJ/m^2, altitude was in meters, latitude and longitude were both in degrees.

Latitude significantly predicted total energy consumption, $\beta = 0.83, t(10) = 10.1$ $(p < 0.05)$, and it was the most influential regressor for CON system. Latitude explained a significant proportion of variance in total energy consumption in CON systems as well, $R^2 = 0.96, F(3, 10) = 74.6 (p < 0.05)$, and the effects of regressors longitude and altitude were not significant.

The regression for water consumption was provided in Eq. (11.2).

$$W = 1659 + 8.02 \times (\text{Latitude}) - 11.8 \times (\text{Longitude}) + 0.0892 \times (\text{Altitude}) \quad (11.2)$$

where W, amount of water consumption, was in kg/m^2, altitude was in meters, latitude and longitude were both in degrees.

Longitude significantly predicted water consumption, $\beta = 0.59$, $t(12) = -2.23$ $(p < 0.05)$. Longitude also explained a significant proportion of variance in water consumption in CON systems, $R^2 = 0.76$, $F(3, 10) = 10.3$ $(p < 0.05)$. The effects of regressors latitude and altitude were not significant.

11.3.3 Open Heat Pump System

The regression for the total energy consumption was provided in Eq. (11.2).

$$E = -5.82 + 0.290 \times (\text{Latitude}) - 0.040 \times (\text{Longitude}) + 0.0008 \times (\text{Altitude}) \quad (11.3)$$

Latitude significantly predicted total energy consumption, $\beta = 0.87$, $t(10) = 10.4$ $(p < 0.05)$, and it was the most influential regressor for OHP system. Latitude explained a significant proportion of variance in total energy consumption in OHP systems as well, $R^2 = 0.96$, $F(3, 10) = 72.3$ $(p < 0.05)$, and the effects of regressors longitude and altitude were not significant.

The regression for water consumption was provided in Eq. (11.4).

$$W = 1849 + 4.78 \times (\text{Latitude}) - 11.5 \times (\text{Longitude}) + 0.0327 \times (\text{Altitude}) \quad (11.4)$$

Longitude significantly predicted water consumption, $\beta = 0.75$, $t(10) = -2.74$ $(p < 0.05)$. Longitude also explained a significant proportion of variance in water consumption, $R^2 = 0.74$, $F(3, 10) = 9.40$ $(p < 0.05)$. The effects of regressors latitude and altitude were not significant.

The effects of regressors latitude, longitude, and altitude on water collection, and the overall regression model were not significant and useful, $F(3, 10) = 1.77$ $(p > 0.05)$.

11.3.4 Semi-closed Heat Pump System

The regression for the total energy consumption was provided in Eq. (11.5).

$$E = -5.30 + 0.221 \times (\text{Latitude}) - 0.016 \times (\text{Longitude}) + 0.0007 \times (\text{Altitude}) \quad (11.5)$$

Latitude significantly predicted total energy consumption, $\beta = 0.86$, $t(10) = 9.11$ $(p < 0.05)$, and it was the most influential regressor. Latitude explained a significant proportion of variance in total energy consumption in SHP systems as well, $R^2 = 0.95$, $F(3, 10) = 57.0$ $(p < 0.05)$, and the effects of regressors longitude and altitude were not significant.

The regression for water consumption was provided in Eq. (11.6).

$$W = 2600 - 4.16 \times (\text{Latitude}) - 14.2 \times (\text{Longitude}) - 0.127 \times (\text{Altitude}) \quad (11.6)$$

Longitude significantly predicted water consumption, $\beta = 1.00$, $t(10) = -3.43$ $(p < 0.05)$. Longitude also explained a significant proportion of variance in water consumption in SHP systems, $R^2 = 0.70$, $F(3, 10) = 7.85$ $(p < 0.05)$. The effects of regressors latitude and altitude were not significant.

The regression for water collected by dehumidifier unit was provided in Eq. (11.7).

$$C = -488 + 5.20 \times (\text{Latitude}) + 2.12 \times (\text{Longitude}) + 0.115 \times (\text{Altitude}) \quad (11.7)$$

where C, amount of water collected on dehumidifier unit, was in kg/m^2, altitude was in meters, latitude and longitude were both in degrees.

Altitude did not have a statistically significant effect on water consumption; however, water consumption followed the increase in altitude as a result of the increased vapor pressure deficits at higher altitudes. As a result, altitude significantly predicted water collection, $\beta = 0.77$, $t(10) = 2.69$ $(p < 0.05)$. Altitude also explained a significant proportion of variance in water collection in SHP systems, $R^2 = 0.73$, $F(3, 10) = 8.90$ $(p < 0.05)$. The effects of regressors latitude and longitude were not significant.

11.3.5 Closed Heat Pump System

The regression for the total energy consumption was provided in Eq. (11.8).

$$E = 15.0 + 0.133 \times (\text{Latitude}) - 0.122 \times (\text{Longitude}) - 0.00002 \times (\text{Altitude}) \quad (11.8)$$

Latitude and longitude significantly predicted total energy consumption, $\beta = 0.59$, $t(10) = 3.17$ $(p < 0.05)$, and $\beta = 0.78$, $t(10) = -3.18$ $(p < 0.05)$, respectively; and longitude was the most influential regressor. Latitude and longitude both explained a significant proportion of variance in total energy consumption in CHP systems as well, $R^2 = 0.79$, $F(3, 10) = 12.3$ $(p < 0.05)$, and the effect of regressor altitude was not significant.

The regression for water consumption and collection was provided in the same Eq. (11.9) as the water collected was equal to the water consumption in CHP system.

$$W \text{ or } C = 2179 - 3.86 \times (\text{Latitude}) - 12.7 \times (\text{Longitude}) - 0.0386 \times (\text{Altitude}) \quad (11.9)$$

Longitude significantly predicted water consumption, $\beta = 0.82$, $t(10) = -2.42$ $(p < 0.05)$. Longitude also explained a significant proportion of variance in water consumption, $R^2 = 0.60$, $F(3, 10) = 4.96$ $(p < 0.05)$. The effects of regressors latitude and altitude were not significant in either case.

11.4 Concluding Remarks

The results in this study showed that the energy and water consumptions were both spatially and system dependent. Latitude and longitude significantly predicted total energy and water consumptions, respectively. Semi-closed and open heat pump systems proved to be the most energy conserving systems, while closed system was the most water conserving system. Highest energy consuming locations of semi-closed and open heat pump systems were still a lot more efficient than least energy consuming locations of conventional systems. Closed systems had the lowest energy consumptions in colder months, but the consumption increased in warmer months due to increased cooling loads hence the overall annual energy consumptions in closed systems were much higher than those in semi-closed systems. Locations on or close to the southwest Pacific Coast had the lowest energy consumptions in all systems. These locations had the lowest water consumptions (transpiration) as well due to increased outside relative humidity levels hence reduced vapor pressure deficits. Water consumptions in the open systems at higher latitudes were relatively higher than those at lower latitudes as the vapor pressure deficits became larger due to decreased outside air temperatures. Semi-closed and closed systems at lower latitudes had more water consumptions due to increased amount of available solar radiation. Semi-closed heat pump system transforms the non-energy-efficient, unfavorable regions under conventional systems into highly energy and water-efficient potential greenhouse regions all over British Columbia.

References

1. Siurna DL, D'Andrea LJ, Hollands KGT (1984) A Canadian representative meteorological year for solar system simulation. In: Proceedings of the 10th annual conference of the Solar Energy Society of Canada (SESCI '84), Calgary, 2–6 Aug 1984
2. Yildiz I, Stombaugh DP (2006) Dynamic modeling of microclimate and environmental control strategies in a greenhouse coupled with a heat pump system. Acta Horticult 718:331–340

Chapter 12
Air Source Heat Pump Performance in Open, Semi-closed, and Closed Greenhouse Systems in the Canadian Maritimes

İlhami Yıldız, Jin Yue, and Asena Cansu Yıldız

Abstract The specific objective of this study was to investigate regional spatio-temporal distributions and spatial correlations of energy and water consumption in open, semi-closed, and completely closed greenhouse systems in the Canadian Maritimes. The findings showed that the energy and water consumptions were both spatially and system dependent. Latitude, longitude, and altitude all significantly predicted total energy consumptions; however, none of these variables predicted water consumptions in the Canadian Maritimes. Semi-closed and open heat pump systems were the most energy conserving systems, while closed system was the most water conserving. Locations on or close to the south central Nova Scotia had the lowest energy consumptions in all systems, which is comparable to major greenhouse regions in southwestern British Columbia and Ontario. This study revealed potentially a new greenhouse region in the Canadian Maritimes, Nova Scotia, which currently does not have a significant greenhouse industry.

Keywords Heat pump • Air source • Energy consumption • Water consumption • Closed greenhouse • Semi-closed greenhouse • Heating • Cooling • Dehumidification • Spatiotemporal distribution • Spatial correlation • Canadian Maritimes

Nomenclature

C	Amount of water collection, kg/m^2
CHP	Closed heat pump greenhouse system
CON	Conventional greenhouse system
E	Amount of energy consumption, GJ/m^2

İ. Yıldız (✉) • J. Yue
Faculty of Agriculture, Department of Engineering, Dalhousie University,
Truro-Bible Hill, NS, Canada B6L 3H9
e-mail: iyildiz@dal.ca

A.C. Yıldız
Department of Law, Queen Mary University of London, Mile End Road, London E1 4NS, UK

I. Dincer et al. (eds.), *Progress in Clean Energy, Volume 1*,
DOI 10.1007/978-3-319-16709-1_12

F	Statistical F test
OHP	Open heat pump greenhouse system
p	Significance level
R	Regression
SHP	Semi-closed heat pump greenhouse system
t	Statistical t test
W	Amount of water consumption, kg/m^2

Greek Letter

β Beta value

12.1 Introduction

One major factor hindering future expansion of greenhouse industry is the cost required for environmental control. Consequently, considerable effort is expended to conserve energy and look for alternative energy sources, especially environment-friendly renewable energy sources and technologies. Proper greenhouse and environmental management systems can significantly change the energy and moisture dynamics of greenhouse production systems. This study helps enhance the competitive position of the Canadian Maritimes' agriculture and agri-food industry by introducing economically, environmentally, and socially sustainable technologies and management strategies. The investigated heat pump technology and operational modes (especially semi-closed and closed) bring increased yield, and energy and water-efficient solutions to the doorstep of the Canadian Maritimes' greenhouse producers. The overall goal of this study was to help reduce the load on power grid, demand for fossil fuels and irrigation water, and also supply CO_2 for the greenhouse production. The use of heat pumps for heating and cooling greenhouses makes it possible to use local and renewable energy sources while reducing or totally eliminating CO_2 emissions. Heat pumps also make an innovative confined greenhouse operation possible, which would totally eliminate the energy, CO_2, and water losses due to ventilation conserving all of these precious resources, and would make an insect-free greenhouse operation possible. The specific objective of this study was to investigate spatiotemporal performance distributions and spatial correlations of energy and water consumptions in open, semi-closed, and confined greenhouse systems in the Canadian Maritimes covering New Brunswick, Newfoundland, Nova Scotia, and Prince Edward Island as a single geographic region.

12.2 Methodology

This section focuses on the resources and procedures employed in this study, as well as the statistical analyses employed and data presentation.

12.2.1 *Weather File*

The Canadian Weather for Energy Calculations (CWEC) files created by concatenating 12 Typical Meteorological Months (TMY) selected from a database of 30 years of data were used for continuous full year analyses at 12 different locations. CWEC data sets were prepared under the direction of Environment Canada [1], and simulations were performed starting at the beginning of 1st day of January and ending at the end of 31st day of December for full 365-day simulations.

12.2.2 *Greenhouse Characteristics and Operational Strategies*

Theoretical approach, simulation model assumptions and validation, and all the other characteristics of the model including the details of control systems and strategies were reported in another study [2]. Full details of the greenhouse systems characteristics and operational strategies were reported by Yildiz et al. [3].

12.2.3 *Data Analyses and Presentation*

A total of four systems, conventional (CON), open heat pump (OHP), semi-closed heat pump (SHP), and closed heat pump (CHP), and a total of 12 locations were studied. Full details of the data analyses were reported by Yildiz et al. [3].

12.3 Data Interpretation

This section covers the result and discussions for different greenhouse systems at different locations.

12.3.1 Energy Consumption, Water Consumption, and Water Collection

The two-way ANOVA ($p < 0.05$) was used to determine the main effects of greenhouse location and system on total energy and water consumptions as well as water collection. There were significant main effects for the location and system on total energy consumption, $F(11, 33) = 4.26$ ($p < 0.05$), and $F(3, 33) = 885$ ($p < 0.05$), respectively (Tables 12.1 and 12.2). Locations at lower latitudes had less total energy consumption, and part of which was for cooling, while northern locations had much less cooling requirements (Fig. 12.1). Table 12.2 shows that SHP and OHP systems had the lowest annual energy consumptions, while CON system had the highest.

There were significant main effects for the location and system on water consumption as well, $F(11, 33) = 10.4$ ($p < 0.05$) and $F(3, 33) = 530$ ($p < 0.05$), respectively (Tables 12.1 and 12.2). The main effect for the system

Table 12.1 Comparisons of means with respect to different locations using the Tukey Simultaneous Test at a 95 % confidence interval

Location	Mean annual energy use (GJ/m^2)	Mean annual water use (kg/m^2)	Mean annual water collection (kg/m^2)
Charlottetown	6.92 a	531.8 q	124.0 w
Fredericton	7.06 a	558.8 pq	143.7 w
Gander	7.13 a	539.6 q	122.1 w
Greenwood	6.20 b	577.2 p	140.3 w
Miramichi	7.02 a	587.9 p	151.4 w
Moncton	6.69 ab	569.0 pq	149.6 w
Shearwater	6.34 b	536.6 q	125.3 w
Saint John	6.77 ab	550.0 pq	128.8 w
St. John's	7.13 a	485.5 r	115.9 w
Stephenville	6.77 ab	525.5 q	119.2 w
Sydney	6.73 ab	537.9 q	132.6 w
Truro	6.65 ab	537.2 q	127.3 w

Means in the same column not followed by the same letter are significantly different

Table 12.2 Comparisons of means with respect to different greenhouse systems using the Tukey Simultaneous Test at a 95 % confidence interval

System	Mean annual energy consumption (GJ/m^2)	Mean annual water consumption (kg/m^2)	Mean annual water collection (kg/m^2)
CON	10.10 a	650.0 p	0.0 w
OHP	5.13 b	705.6 q	3.9 w
SHP	4.82 b	554.0 r	117.1 x
CHP	7.11 c	472.4 s	472.4 y

Means in the same column not followed by the same letter are significantly different

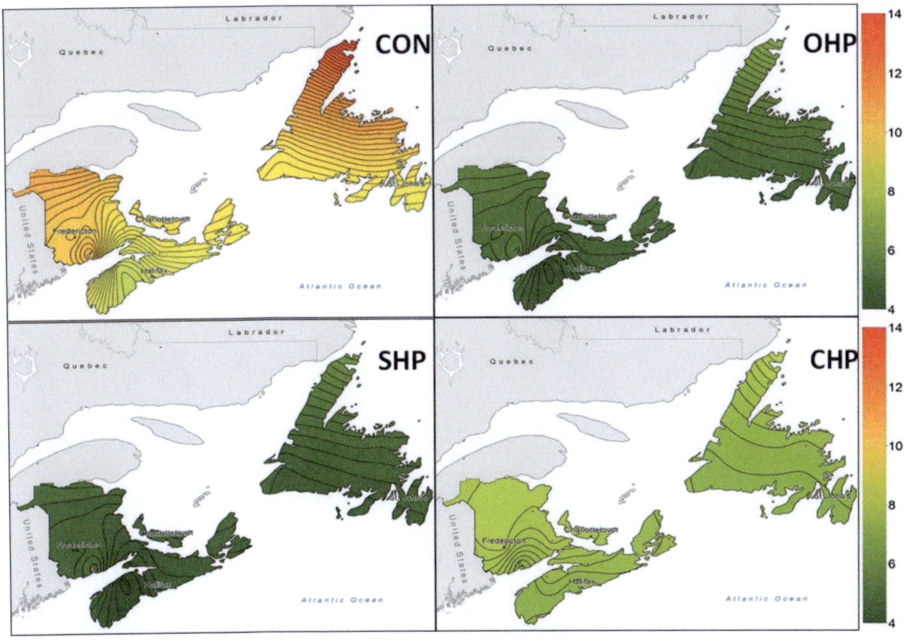

Fig. 12.1 Spatial distributions of annual total energy consumptions (GJ/m^2) in CON, OHP, SHP and CHP systems in the Maritimes (contour line intervals: 100 MJ/m^2)

on water collection was also significant, $F(3, 33) = 1050$ ($p < 0.05$); however, the main effect for the location on water collection was not, $F(11, 33) = 1.36$ ($p > 0.05$) (Tables 12.1 and 12.2). Locations at higher latitudes had more water consumption (transpiration) in open systems due to relatively colder and drier outside air, and CHP systems had higher transpiration rates at lower latitudes due to increased amount of available solar radiation (Fig. 12.2). In CHP systems, solar radiation was the only outside parameter affecting the amount of transpiration. CHP systems also had the lowest transpirations as much higher relative humidities were observed in these systems compared to those in the other systems. All the transpired water was collected by dehumidification system, and therefore, the overall annual water consumption in CHP systems was essentially zero (Tables 12.1 and 12.2).

CHP systems had relatively lower energy consumptions in colder months, but the consumption increased in warmer months due to increased cooling loads; as a result, the overall annual energy consumptions in CHP systems were much higher than OHP and SHP systems (Figs. 12.3 and 12.4). The energy consumption in SHP systems approached to those in CHP systems at higher latitudes (e.g., Gander, NF) (Fig. 12.3), because SHP systems under such conditions operated more as a closed system from November through end of March. However, the energy consumption in SHP systems approached to those in OHP systems at lower latitudes (e.g., Shearwater, NS) (Fig. 12.4) because SHP systems under such conditions operated more as an open system from April through end of November.

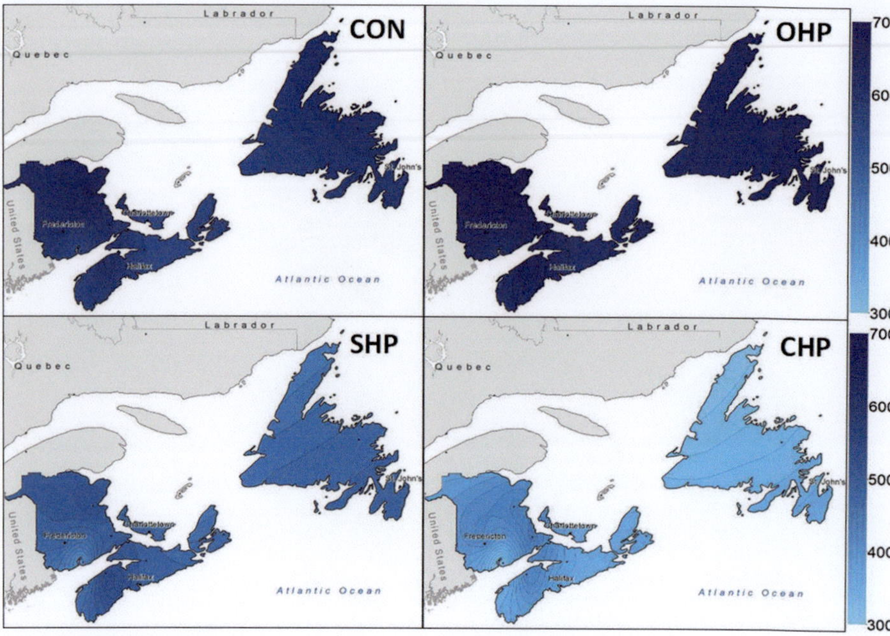

Fig. 12.2 Spatial distributions of annual amount of transpiration (kg/m^2) in CON, OHP, SHP, and CHP systems in the Maritimes (contour line intervals: 10 kg/m^2)

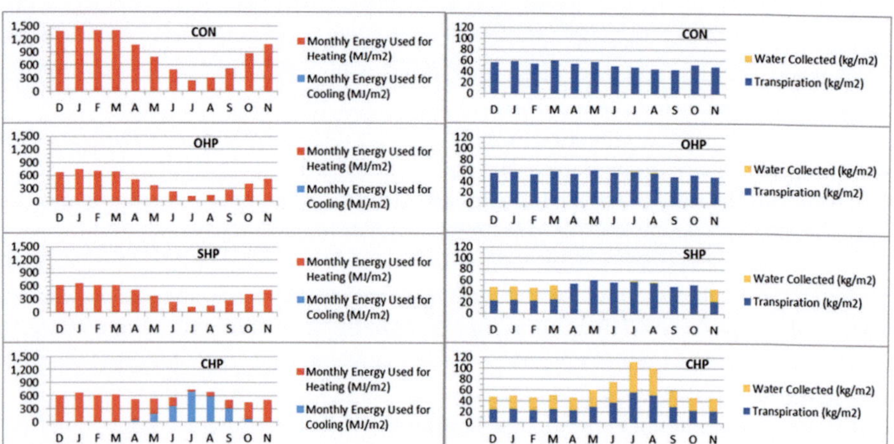

Fig. 12.3 Temporal distributions of energy consumption (*left*) and transpiration (*right*) in CON, OHP, SHP, and CHP systems at Gander, NF (48.95 N, 54.57 W, altitude: 151 m)

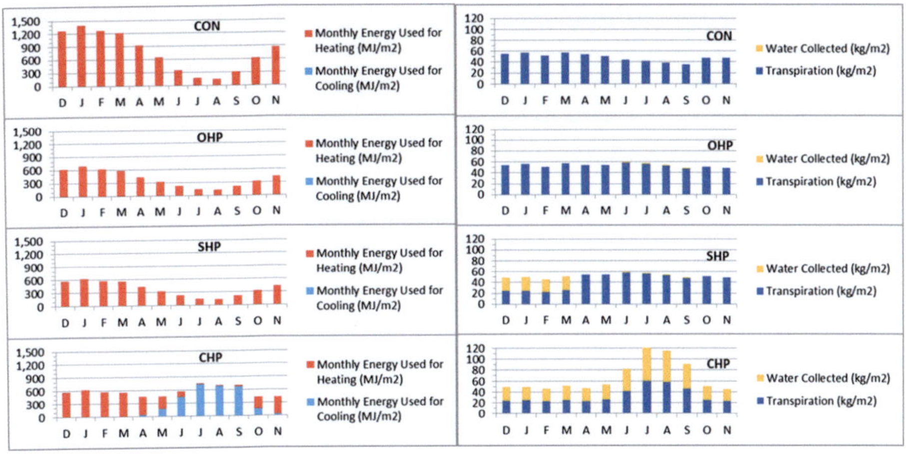

Fig. 12.4 Temporal distributions of energy consumption (*left*) and transpiration (*right*) in CON, OHP, SHP, and CHP systems at Shearwater, NS (44.63 N, 63.50 W, altitude: 51 m)

Further details of individual comparisons for both location and system were provided in Tables 12.1 and 12.2. And the individual system performances were provided below.

12.3.2 Conventional System

Amount of total energy consumption, water consumption (transpiration), and water collection on the dehumidifier were analyzed throughout the region by multiple regressions using as regressors latitude, longitude, and altitude. The regression for the total energy consumption in CON system was provided in Eq. (12.1).

$$E = -23.7 + 0.494^*(\text{Latitude}) + 0.166^*(\text{Longitude}) + 0.0106^*(\text{Altitude}) \quad (12.1)$$

where E, the annual total energy consumption, was in GJ/m^2, altitude was in meters, latitude and longitude were both in degrees.

Latitude, longitude, and altitude significantly predicted total energy consumption, $\beta = 1.10, t(8) = 4.64\,(p < 0.05), \beta = 1.23, t(8) = 4.08\,(p < 0.05), \beta = 0.85$, $t(8) = 3.76 \quad (p < 0.05)$, respectively; and longitude was the most influential regressor for CON system followed by latitude. Latitude, longitude, and altitude explained a significant proportion of variance in total energy consumption, $R^2 = 0.80, F(3, 8) = 10.7\ (p < 0.05)$.

None of the regressors significantly predicted water consumption or explained a significant proportion of variance in water consumption, $F(3, 8) = 0.91\,(p > 0.05)$.

12.3.3 Open Heat Pump System

The regression for the total energy consumption was provided in Eq. (12.2).

$$E = -7.84 + 0.187^{*}(\text{Latitude}) + 0.0649^{*}(\text{Longitude}) + 0.0046^{*}(\text{Altitude}) \quad (12.2)$$

Latitude, longitude, and altitude significantly predicted total energy consumption, $\beta = 0.94, t(8) = 2.99\,(p < 0.05), \beta = 1.08, t(8) = 2.71\,(p < 0.05), \beta = 0.82$, $t(8) = 2.76$ $(p < 0.05)$, respectively; and longitude was the most influential regressor for OHP system as well, followed by latitude. Latitude, longitude, and altitude explained a significant proportion of variance in total energy consumption, $R^2 = 0.65, F(3, 8) = 5.00\,(p < 0.05)$.

None of the regressors significantly predicted water consumption and collection or explained a significant proportion of variance in water consumption or collection in OHP system, $F(3, 8) = 2.10\,(p > 0.05)$.

12.3.4 Semi-closed Heat Pump System

The regression for the total energy consumption was provided in Eq. (12.3).

$$E = -5.21 + 0.140^{*}(\text{Latitude}) + 0.0531^{*}(\text{Longitude}) + 0.0046^{*}(\text{Altitude}) \quad (12.3)$$

Latitude, longitude, and altitude significantly predicted total energy consumption, $\beta = 0.79, t(8) = 2.54\,(p < 0.05), \beta = 0.99, t(8) = 2.52\,(p < 0.05), \beta = 0.92$, $t(8) = 3.13$ $(p < 0.05)$, respectively; longitude was the most influential regressor for SHP system as well, followed by altitude. Latitude, longitude, and altitude explained a significant proportion of variance in total energy consumption, $R^2 = 0.65, F(3, 8) = 5.10\,(p < 0.05)$.

None of the regressors significantly predicted water consumption and collection or explained a significant proportion of variance in water consumption or collection in SHP system, $F(3, 8) = 1.17$ $(p > 0.05)$ and $F(3, 8) = 1.48$ $(p > 0.05)$, respectively.

12.3.5 Closed Heat Pump System

None of the regressors significantly predicted energy consumption, water consumption, and collection or explained a significant proportion of variance in energy consumption, water consumption, or collection in CHP system, $F(3, 8) = 2.21$ $(p > 0.05)$, $F(3, 8) = 1.66$ $(p < 0.05)$, and $F(3, 8) = 1.65$ $(p > 0.05)$, respectively.

12.4 Concluding Remarks

The results in this study showed that the energy and water consumptions were both spatially and system dependent. Latitude, longitude, and altitude all significantly predicted total energy consumptions; however, none of these variables predicted water consumptions in the Canadian Maritimes. Semi-closed and OHP systems proved to be the most energy conserving systems, while closed system was the most water-conserving system. Highest energy consuming locations of semi-closed and OHP systems were still a lot more efficient than least energy consuming locations of conventional systems. Closed systems had the lowest energy consumptions in colder months, but the consumption increased in warmer months due to increased cooling loads, hence the overall annual energy consumptions in closed systems were much higher than those in semi-closed systems. Locations on or close to the south central Nova Scotia had the lowest energy consumptions in all systems, which is comparable to major greenhouse regions in southwestern British Columbia and Ontario. SHP system transforms the non-energy-efficient, unfavorable regions under conventional systems into highly energy and water-efficient potential greenhouse regions in the Canadian Maritimes, especially in Nova Scotia.

References

1. Siurna DL, D'Andrea LJ, Hollands KGT (1984) A Canadian representative meteorological year for solar system simulation. In: Proceedings of the 10th annual conference of the Solar Energy Society of Canada (SESCI '84), Calgary, 2–6 Aug 1984
2. Yildiz I, Stombaugh DP (2006) Dynamic modeling of microclimate and environmental control strategies in a greenhouse coupled with a heat pump system. Acta Horticult 718:331–340
3. Yildiz I, Yue J, Yildiz AC (2015) Air source heat pump performance in open, semi-closed and closed greenhouse systems in British Columbia. In: Dincer I et al. (eds) Progress in clean energy. Springer, New York

Chapter 13
Air Source Heat Pump Performance in Open, Semi-closed, and Closed Greenhouse Systems in Ontario

İlhami Yıldız, Jin Yue, and Asena Cansu Yıldız

Abstract The specific objective of this study was to investigate regional spatio-temporal distributions and spatial correlations of energy and water consumption in open, semi-closed, and completely closed greenhouse systems in Ontario. The findings showed that the energy and water consumptions were both spatially and system dependent. Latitude and longitude both significantly predicted total energy and water consumptions. Altitude was significant in conventional and open heat pump systems' energy consumptions. Semi-closed and open heat pump systems were the most energy conserving systems while closed system was the most water conserving. Semi-closed and closed systems at lower latitudes had more water consumptions. Open system had more water consumption at higher latitudes and longitudes compared to those at lower latitudes and longitudes, respectively. Locations in the southwest (a major greenhouse region in Ontario) had the lowest energy consumptions in all systems.

Keywords Heat pump • Air source • Energy consumption • Water consumption • Closed greenhouse • Semi-closed greenhouse • Heating • Cooling • Dehumidification • Spatiotemporal distribution • Spatial correlation • Ontario

Nomenclature

C	Amount of water collection, kg/m^2
CHP	Closed heat pump greenhouse system
CON	Conventional greenhouse system
E	Amount of energy consumption, GJ/m^2

İ. Yıldız (✉) • J. Yue
Department of Engineering, Faculty of Agriculture, Dalhousie University, Truro-Bible Hill, NS, Canada B6L 3H9
e-mail: iyildiz@dal.ca

A.C. Yıldız
Department of Law, Queen Mary University of London, Mile End Road, London E1 4NS, UK

I. Dincer et al. (eds.), *Progress in Clean Energy, Volume 1*,
DOI 10.1007/978-3-319-16709-1_13

F	Statistical F test
OHP	Open heat pump greenhouse system
p	Significance level
R	Regression
SHP	Semi-closed heat pump greenhouse system
t	Statistical t test
W	Amount of water consumption, kg/m^2

Greek Letters

β Beta value

13.1 Introduction

One major factor hindering future expansion of greenhouse industry is the cost required for environmental control. Consequently, considerable effort is expended to conserve energy and look for alternative energy sources, especially environment-friendly renewable energy sources and technologies. Proper greenhouse and environmental management systems can significantly change the energy and moisture dynamics of greenhouse production systems. This study helps enhance the competitive position of Ontario's agriculture and agri-food industry by introducing economically, environmentally, and socially sustainable technologies and management strategies. The investigated heat pump technology and operational modes (especially semi-closed and closed) bring increased yield, and energy and water-efficient solutions to the doorstep of Ontario's greenhouse producers. The overall goal of this study was to help reduce the load on power grid, demand for fossil fuels and irrigation water, and also supply CO_2 for the greenhouse production. The use of heat pumps for heating and cooling greenhouses makes it possible to use local and renewable energy sources while reducing or totally eliminating CO_2 emissions. Heat pumps also make an innovative confined greenhouse operation possible, which would totally eliminate the energy, CO_2, and water losses due to ventilation conserving all of these precious resources and would make an insect-free greenhouse operation possible. The specific objective of this study was to investigate spatio-temporal performance distributions and spatial correlations of energy and water consumptions in open, semi-closed, and confined greenhouse systems in Ontario.

13.2 Methodology

This section focuses on the resources and procedures employed in this study, as well as the statistical analyses employed and data presentation.

13.2.1 Weather File

The Canadian Weather for Energy Calculations (CWEC) files created by concatenating 12 Typical Meteorological Months (TMY) selected from a database of 30 years of data were used for continuous full year analyses at 16 different locations. CWEC data sets were prepared under the direction of Environment Canada [1], and simulations were performed starting at the beginning of 1st day of January and ending at the end of 31st day of December for full 365-day simulations.

13.2.2 Greenhouse Characteristics and Operational Strategies

Theoretical approach, simulation model assumptions and validation, and all the other characteristics of the model including the details of control systems and strategies were reported in another study [2]. Full details of the greenhouse systems characteristics and operational strategies were reported by Yildiz et al. [3].

13.2.3 Data Analyses and Presentation

A total of four systems, conventional (CON), open heat pump (OHP), semi-closed heat pump (SHP), and closed heat pump (CHP), and a total of 16 locations were studied. Full details of the data analyses were reported by Yildiz et al. [3].

13.3 Data Interpretation

This section covers the result and discussions for different greenhouse systems at different locations.

13.3.1 Energy Consumption, Water Consumption, and Water Collection

The two-way ANOVA $(p < 0.05)$ was used to determine the main effects of greenhouse location and system on total energy and water consumptions as well as water collection. There were significant main effects for the location and system on total energy consumption, $F(16, 45) = 8.16$ $(p < 0.05)$, and $F(3, 45) = 361$ $(p < 0.05)$, respectively (Tables 13.1 and 13.2). Locations at lower latitudes had much less total energy consumption, and part of which was for cooling while

Table 13.1 Comparisons of means with respect to different locations using the Tukey Simultaneous Test at a 95 % confidence interval

Location	Mean annual energy use (GJ/m^2)	Mean annual water use (kg/m^2)	Mean annual water collection (kg/m^2)
Kapuskasing	8.29 a	588.6 pq	153.8 w
Kenora	7.69 ab	627.4 p	161.1 w
Kingston	6.39 bc	571.4 q	149.6 w
London	6.30 bc	597.1 pq	157.5 w
Mount Forest	6.73 bc	585.2 pq	153.4 w
Muskoka	6.89 bc	591.2 pq	151.0 w
North Bay	7.42 ab	574.5 q	139.8 w
Ottawa	7.02 b	583.7 pq	151.5 w
Sault Ste Marie	7.04 b	572.3 q	139.7 w
Simcoe	6.23 bc	603.7 pq	154.0 w
Sioux Lookout	7.96 ab	625.8 p	156.6 w
Thunder Bay	7.65 ab	601.7 pq	142.8 w
Timmins	8.04 ab	591.1 pq	138.4 w
Toronto	6.43 bc	587.2 pq	152.9 w
Trenton	6.28 bc	604.8 pq	155.8 w
Windsor	5.76 c	622.4 p	168.8 w

Means in the same column not followed by the same letter are significantly different

Table 13.2 Comparisons of means with respect to different greenhouse systems using the Tukey Simultaneous Test at a 95 % confidence interval

System	Mean annual energy consumption (GJ/m^2)	Mean annual water consumption (kg/m^2)	Mean annual water collection (kg/m^2)
CON	10.25 a	650.0 p	0.0 w
OHP	5.22 b	705.6 q	9.9 w
SHP	4.79 b	554.0 r	124.6 x
CHP	7.75 c	472.4 s	472.4 y

Means in the same column not followed by the same letter are significantly different

northern locations had much less cooling requirements (Fig.13.1). Table 13.2 shows that SHP and OHP systems had the lowest annual energy consumptions, while CON system had the highest.

There were significant main effects for the location and system on water consumption as well, $F(11, 33) = 10.4$ $(p < 0.05)$ and $F(3, 33) = 530$ $(p < 0.05)$, respectively (Tables 13.1 and 13.2). The main effect for the system on water collection was also significant, $F(3, 33) = 1050$ $(p < 0.05)$; however, the main effect for the location on water collection was not, $F(11, 33) = 1.36$ $(p > 0.05)$ (Tables 13.1 and 13.2). Locations at higher latitudes had more water consumption (transpiration) in open systems due to relatively colder and drier outside air, and CHP systems had higher transpiration rates at lower latitudes due to increased amount of available solar radiation (Fig. 13.2). In CHP systems, solar radiation was the only outside parameter affecting the amount of transpiration. CHP systems

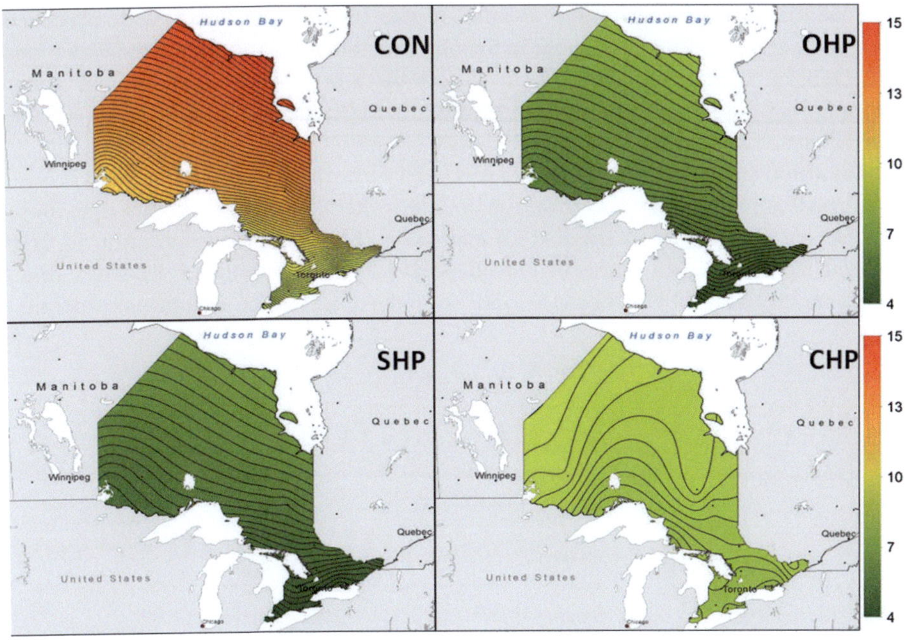

Fig. 13.1 Spatial distributions of annual total energy consumptions (GJ/m^2) in CON, OHP, SHP, and CHP systems in the Maritimes (contour line intervals: 100 MJ/m^2)

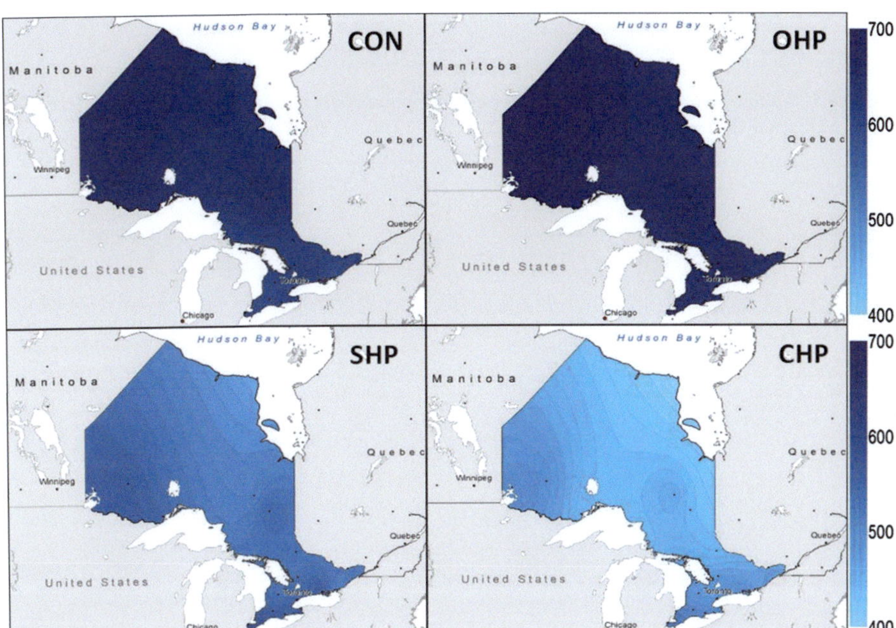

Fig. 13.2 Spatial distributions of annual amount of transpiration (kg/m^2) in CON, OHP, SHP, and CHP systems in Ontario (contour line intervals: 10 kg/m^2)

also had the lowest transpirations as much higher relative humidities were observed in these systems compared to those in the other systems. All the transpired water was collected by dehumidification system, and therefore, the overall annual water consumption in CHP systems was essentially zero (Tables 13.1 and 13.2).

CHP systems had relatively lower energy consumptions in colder months, but the consumption increased in warmer months due to increased cooling loads; therefore, the overall annual energy consumption in CHP systems was much higher than those in OHP and SHP systems (Figs. 13.3 and 13.4). The energy consumption in SHP systems approached to those in closed (CHP) systems at higher latitudes (e.g., Kapuskasing) (Fig. 13.3) because SHP systems under such conditions operated

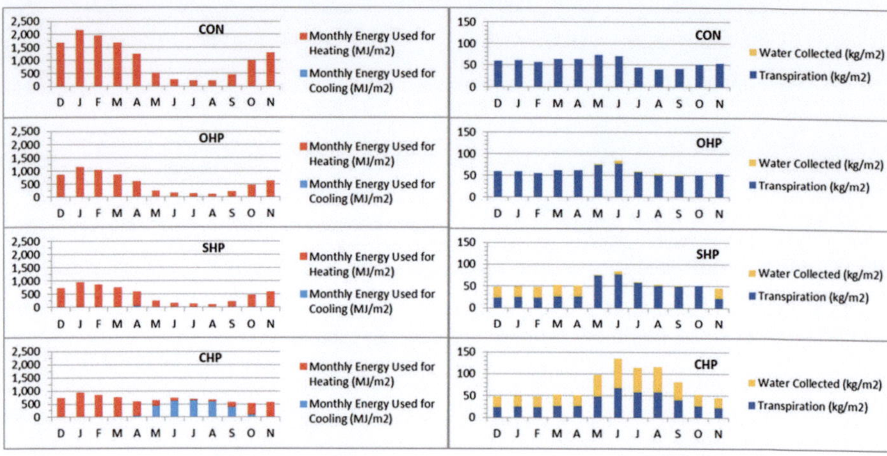

Fig. 13.3 Temporal distributions of energy consumption (*left*) and transpiration (*right*) in CON, OHP, SHP, and CHP systems at Kapuskasing, ON (49.40 N, 82.50 W, altitude: 227 m)

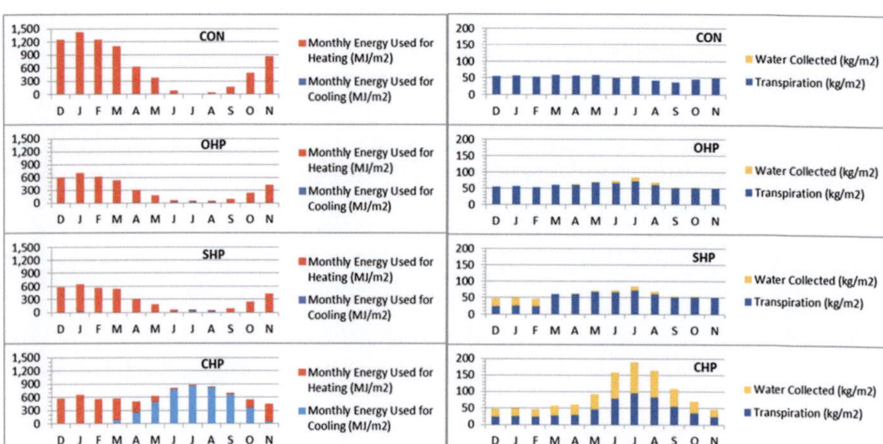

Fig. 13.4 Temporal distributions of energy consumption (*left*) and transpiration (*right*) in CON, OHP, SHP, and CHP systems at Windsor, ON (42.27 N, 82.97 W, altitude: 190 m)

more as a closed system from November through end of April. However, the energy consumption in SHP systems approached to those in open (OHP) systems at lower latitudes (e.g., Windsor) (Fig. 13.4) because SHP systems under such conditions operated more as an open system from March through end of November.

Further details of individual comparisons for both location and system were provided in Tables 13.1 and 13.2. And the individual system performances were provided below.

13.3.2 Conventional System

Amount of total energy consumption, water consumption (transpiration), and water collection on the dehumidifier were analyzed throughout the region by multiple regressions using as regressors latitude, longitude, and altitude.

The regression for the total energy consumption was provided in Eq. (13.1).

$$E = -10.5 + 0.623 \times (\text{Latitude}) + 0.101 \times (\text{Longitude}) + 0.0021 \times (\text{Altitude}) \quad (13.1)$$

where E, the annual total energy consumption, was in GJ/m^2, altitude was in meters, latitude and longitude were both in degrees.

Latitude significantly predicted total energy consumption, $\beta = 1.13, t(12) = 13.7$ ($p < 0.05$), and it was the most influential regressor for CON system. Longitude and altitude (moderately) as well significantly predicted total energy consumption, $\beta = 0.37,\ t(12) = -4.09\ \ (p < 0.05)$, and $\beta = 0.16,\ t(12) = 2.21\ \ (p < 0.05)$, respectively. Latitude and longitude explained a significant proportion of variance in total energy consumption as well, $R^2 = 0.96, F(3, 12) = 86.9$ ($p < 0.05$).

The regression for water consumption was provided in Eq. (13.2).

$$W = 161 + 8.01 \times (\text{Latitude}) + 1.47 \times (\text{Longitude}) + 0.0041 \times (\text{Altitude}) \quad (13.2)$$

where W, amount of water consumption, was in kg/m^2, altitude was in meters, latitude and longitude were both in degrees.

Latitude significantly predicted water consumption, $\beta = 0.73,\ t(12) = 5.21$ ($p < 0.05$). Latitude also explained a significant proportion of variance in water consumption in CON system, $R^2 = 0.87, F(3.12) = 27.8$ ($p < 0.05$). The effects of regressors longitude and altitude were not significant.

13.3.3 Open Heat Pump System

The regression for the total energy consumption was provided in Eq. (13.3).

$$E = -5.60 + 0.321 \times (\text{Latitude}) - 0.051 \times (\text{Longitude}) + 0.0011 \times (\text{Altitude}) \quad (13.3)$$

Latitude, longitude, and altitude significantly predicted total energy consumption, $\beta = 1.12,\ t(12) = 14.8\ \ (p < 0.05);\ \beta = 0.36,\ t(12) = -4.27\ \ (p < 0.05);$ and $\beta = 0.17,\ t(12) = 2.51\ \ (p < 0.05)$, respectively. Latitude, longitude, and altitude also explained a significant proportion of variance in total energy consumption, $R^2 = 0.96, F(3, 12) = 103\ (p < 0.05)$.

The regression for water consumption was provided in Eq. (13.4).

$$W = 415 + 2.53 \times (\text{Latitude}) + 2.14 \times (\text{Longitude}) + 0.0046 \times (\text{Altitude}) \quad (13.4)$$

Longitude significantly predicted water consumption, $\beta = 0.62,\ t(12) = 3.15$ $(p < 0.05)$. Longitude also explained a significant proportion of variance in water consumption, $R^2 = 0.79,\ F(3, 12) = 15.3\ \ (p < 0.05)$. The effects of regressors latitude and altitude were not significant.

The regression for water collected by dehumidifier unit was provided in Eq. (13.5) and the overall relationship was moderately significant.

$$C = 28.5 - 1.94 \times (\text{Latitude}) + 0.94 \times (\text{Longitude}) - 0.0291 \times (\text{Altitude}) \quad (13.5)$$

where C, amount of water collected by dehumidifier unit, was in kg/m^2, altitude was in meters, latitude and longitude were both in degrees.

Latitude and longitude significantly predicted water collection, $\beta = 0.74$, $t(12) = -2.63\ (p < 0.05)$, and $\beta = 0.72,\ t(12) = 2.34\ (p < 0.05)$, respectively. Latitude and longitude explained a significant proportion of variance in water collection in OHP system as well, $R^2 = 0.48,\ F(3, 12) = 3.75\ (p < 0.05)$. The effect of regressor altitude was not significant in OHP system.

13.3.4 Semi-closed Heat Pump System

The regression for the total energy consumption was provided in Eq. (13.6).

$$E = -3.32 + 0.247 \times (\text{Latitude}) - 0.041 \times (\text{Longitude}) + 0.0008 \times (\text{Altitude}) \quad (13.6)$$

Latitude and longitude significantly predicted total energy consumption, $\beta = 1.13,\ t(12) = 12.1\ \ (p < 0.05)$, and $\beta = 0.38,\ t(12) = -3.71\ \ (p < 0.05)$, respectively. Latitude and longitude also explained a significant proportion of variance in total energy consumption in SHP system, $R^2 = 0.94,\ F(3, 12) = 67.0$ $(p < 0.05)$, and the effect of regressor altitude was not significant.

The regression for water consumption was provided in Eq. (13.7).

$$W = 567 - 9.31 \times (\text{Latitude}) + 5.13 \times (\text{Longitude}) - 0.0354 \times (\text{Altitude}) \quad (13.7)$$

Latitude and longitude significantly predicted water consumption, $\beta = 0.97$, $t(12) = -4.12\ (p < 0.05)$, and $\beta = 1.07,\ t(12) = 4.16\ (p < 0.05)$, respectively.

Latitude and longitude also explained a significant proportion of variance in water consumption in SHP system, $R^2 = 0.64$, $F(3, 12) = 7.19$ $(p < 0.05)$. The effect of regressor altitude was not significant.

The regression for water collected by dehumidifier unit was provided in Eq. (13.8).

$$C = -33.4 + 5.54 \times (\text{Latitude}) - 1.20 \times (\text{Longitude}) + 0.0107 \times (\text{Altitude}) \quad (13.8)$$

Latitude significantly predicted water collection, $\beta = 1.07$, $t(12) = 5.29$ $(p < 0.05)$. Latitude also explained a significant proportion of variance in water collection in SHP system, $R^2 = 0.73$, $F(3, 12) = 11.0$ $(p < 0.05)$. The effects of regressors longitude and altitude were not significant.

13.3.5 *Closed Heat Pump System*

The regression for the total energy consumption was provided in Eq. (13.9).

$$E = 3.57 + 0.0910 \times (\text{Latitude}) - 0.0013 \times (\text{Longitude}) + 0.0005 \times (\text{Altitude}) \quad (13.9)$$

Latitude significantly predicted total energy consumption, $\beta = 0.74$, $t(12) = 3.21$ $(p < 0.05)$. Latitude explained a significant proportion of variance in total energy consumption CHP systems as well, $R^2 = 0.66$, $F(3, 12) = 7.62$ $(p < 0.05)$, and the effect of regressors longitude and altitude were not significant.

The regression for water consumption and collection was provided in the same Eq. (13.10) as the water collected was equal to the water consumption in CHP system.

$$W \text{ or } C = 681 - 12.9 \times (\text{Latitude}) + 4.68 \times (\text{Longitude}) - 0.0095 \times (\text{Altitude}) \quad (13.10)$$

Latitude and longitude significantly predicted water consumption, $\beta = 1.02$, $t(12) = -3.94$ $(p < 0.05)$, and $\beta = 0.74$, $t(12) = 2.62$ $(p < 0.05)$, respectively. Latitude and longitude also explained a significant proportion of variance in water consumption in CHP systems, $R^2 = 0.57$, $F(3, 12) = 5.23$ $(p < 0.05)$. The effect of regressor altitude was not significant in either case.

13.4 Concluding Remarks

This results in this study showed that the energy and water consumptions were both spatially and system dependent. Latitude and longitude both significantly predicted total energy and water consumptions. Altitude was significant in conventional and OHP systems' energy consumptions. Semi-closed and open heat pump systems proved to be the most energy conserving systems, while closed system was the most water conserving system. Highest energy consuming locations of semi-closed and open heat pump systems were still a lot more efficient than least energy consuming locations of conventional systems. Closed systems had the lowest energy consumptions in colder months, but the consumption increased in warmer months especially in southwestern Ontario due to increased cooling loads hence the overall annual energy consumptions in closed systems were much higher than those in semi-closed systems. Locations in the southwest (a major greenhouse region) had the lowest energy consumptions in all systems. Semi-closed and closed systems at lower latitudes had higher water consumption (transpiration) due to increased amount of available solar radiation. Since the vapor pressure deficits became larger due to colder and drier outside air, transpiration rates especially in open systems were relatively higher at higher latitudes and longitudes than those at lower latitudes and longitudes, respectively. Semi-closed heat pump system transforms the non-energy-efficient, unfavorable regions under conventional systems into highly energy and water-efficient potential greenhouse regions in Ontario.

References

1. Siurna DL, D'Andrea LJ, Hollands KGT (1984) A Canadian representative meteorological year for solar system simulation. In: Proceedings of the 10th annual conference of the Solar Energy Society of Canada (SESCI '84), Calgary, 2–6 Aug 1984
2. Yildiz I, Stombaugh DP (2006) Dynamic modeling of microclimate and environmental control strategies in a greenhouse coupled with a heat pump system. Acta Horticult 718:331–340
3. Yildiz I, Yue J, Yildiz AC (2015) Air source heat pump performance in open, semi-closed and closed greenhouse systems in British Columbia. In: Dincer I et al. (eds) Progress in clean energy. ©Springer Science+Business Media, New York

Chapter 14
Air Source Heat Pump Performance in Open, Semi-closed, and Closed Greenhouse Systems in the Canadian Prairies

İlhami Yıldız, Jin Yue, and Asena Cansu Yıldız

Abstract The specific objective of this study was to investigate regional spatiotemporal distributions and spatial correlations of energy and water consumption in open, semi-closed, and completely closed greenhouse systems in the Canadian Prairies. The findings showed that the energy and water consumptions were both spatially and system dependent. Both latitude and longitude significantly predicted total energy consumptions. Latitude significantly predicted water consumption in semi-closed and closed systems as well. Semi-closed system was the most energy conserving system, while closed system was the most water conserving. Locations at lower latitudes had higher water consumption due to increased amount of available solar radiation. Water consumptions in open systems at higher and lower latitudes approached to each other due to relatively increased transpiration rates at higher latitudes as the vapor pressure deficit became larger due to colder and drier outside air. Locations in the southwestern Prairies (Alberta, and partly Saskatchewan) had the lowest energy consumptions in all systems.

Keywords Heat pump • Air source • Energy consumption • Water consumption • Closed greenhouse • Semi-closed greenhouse • Heating • Cooling • Dehumidification • Spatiotemporal distribution • Spatial correlation • Alberta • Manitoba • Saskatchewan • Canadian Prairies

Nomenclature

C	Amount of water collection, kg/m^2
CHP	Closed heat pump greenhouse system
CON	Conventional greenhouse system

İ. Yıldız (✉) • J. Yue
Faculty of Agriculture, Department of Engineering, Dalhousie University, Truro-Bible Hill, NS, Canada B6L 3H9
e-mail: iyildiz@dal.ca

A.C. Yıldız
Department of Law, Queen Mary University of London, Mile End Road, London E1 4NS, UK

I. Dincer et al. (eds.), *Progress in Clean Energy, Volume 1*,
DOI 10.1007/978-3-319-16709-1_14

E	Amount of energy consumption, GJ/m^2
F	Statistical F test
OHP	Open heat pump greenhouse system
p	Significance level
R	Regression
SHP	Semi-closed heat pump greenhouse system
t	Statistical t test
W	Amount of water consumption, kg/m^2

Greek Letter

β Beta value

14.1 Introduction

One major factor hindering future expansion of greenhouse industry is the cost required for environmental control. Consequently, considerable effort is expended to conserve energy and look for alternative energy sources, especially environment-friendly renewable energy sources and technologies. Proper greenhouse and environmental management systems can significantly change the energy and moisture dynamics of greenhouse production systems. This study helps enhance the competitive position of the Canadian Prairies' agriculture and agri-food industry by introducing economically, environmentally, and socially sustainable technologies and management strategies. The investigated heat pump technology and operational modes (especially semi-closed and closed) bring increased yield, and energy and water-efficient solutions to the doorstep of the Canadian Prairies' greenhouse producers. The overall goal of this study was to help reduce the load on power grid, demand for fossil fuels and irrigation water, and also supply CO_2 for the greenhouse production. The use of heat pumps for heating and cooling greenhouses makes it possible to use local and renewable energy sources while reducing or totally eliminating CO_2 emissions. Heat pumps also make an innovative confined greenhouse operation possible, which would totally eliminate the energy, CO_2 and water losses due to ventilation conserving all of these precious resources, and would make an insect-free greenhouse operation possible. The specific objective of this study was to investigate spatiotemporal performance distributions and spatial correlations of energy and water consumptions in open, semi-closed, and confined greenhouse systems in the Canadian Prairies covering Alberta, Manitoba, and Saskatchewan.

14.2 Methodology

This section focuses on the resources and procedures employed in this study, as well as the statistical analyses employed and data presentation.

14.2.1 Weather File

The Canadian Weather for Energy Calculations (CWEC) files created by concatenating 12 Typical Meteorological Months (TMY) selected from a database of 30 years of data were used for continuous full year analyses at 16 different locations. CWEC data sets were prepared under the direction of Environment Canada [1], and simulations were performed starting at the beginning of 1st day of January and ending at the end of 31st day of December for full 365-day simulations.

14.2.2 Greenhouse Characteristics and Operational Strategies

Theoretical approach, simulation model assumptions and validation, and all the other characteristics of the model including the details of control systems and strategies were reported in another study [2]. Full details of the greenhouse systems characteristics and operational strategies were reported by Yildiz et al. [3].

14.2.3 Data Analyses and Presentation

A total of four systems, conventional (CON), open heat pump (OHP), semi-closed heat pump (SHP), and closed heat pump (CHP), and a total of 16 locations were studied. Full details of the data analyses were reported by Yildiz et al. [3].

14.3 Data Interpretation

This section covers the result and discussions for different greenhouse systems at different locations.

14.3.1 Energy Consumption, Water Consumption, and Water Collection

The two-way ANOVA ($p < 0.05$) was used to determine the main effects of greenhouse location and system on total energy and water consumptions as well as water collection. There were significant main effects for the location and system on total energy consumption, $F(20, 60) = 9.54$ ($p < 0.05$), and $F(3, 60) = 684$ ($p < 0.05$), respectively (Tables 14.1 and 14.2). Locations at lower latitudes had much less total energy consumption, and part of which was for cooling while northern locations had much less cooling requirements (Fig. 14.1). Table 14.2 shows that SHP system had the lowest annual energy consumptions, while CON system had the highest.

There were significant main effects for the location and system on water consumption as well, $F(20, 60) = 14.3$ ($p < 0.05$), and $F(3, 60) = 1287$ ($p < 0.05$), respectively (Tables 14.1 and 14.2). The main effect for the system on water collection was also significant, $F(3, 60) = 2234$ ($p < 0.05$); however, the main effect for the location on water collection was not, $F(20, 60) = 1.54$ ($p > 0.05$)

Table 14.1 Comparisons of means with respect to different locations using the Tukey Simultaneous Test at a 95 % confidence interval

Location	Annual energy use (GJ/m^2)	Annual water use (kg/m^2)	Annual water collection (kg/m^2)
Brandon	8.03 b	632.5 qr	155.2 wx
Calgary	7.15 ab	650.3 qr	142.3 wx
Cold Lake	8.02 b	626.5 q	151.9 wx
Edmonton	7.87 ab	574.9 p	126.2 w
Estevan	7.54 ab	662.6 r	174.3 x
Ft McMurray	8.30 ab	607.0 pq	135.4 w
G. Prairie	7.91 b	606.0 pq	129.5 w
La Ronge	8.55 b	620.4 q	145.8 wx
Lethbridge	6.83 a	691.8 rs	163.5 wx
Lynn Lake	9.60 c	597.4 pq	143.2 wx
Medicine Hat	6.85 a	703.1 s	165.7 wx
North Battle	8.07 b	630.6 qr	150.2 wx
Norway H.	8.72 bc	620.5 q	150.8 wx
Peace River	7.69 ab	626.7 q	133.6 w
Regina	7.76 ab	665.1 r	164.7 wx
Saskatoon	7.91 ab	652.2 qr	152.7 wx
Stony Rapids	9.91 c	582.4 p	135.0 w
Swift Current	7.56 ab	624.9 q	146.5 wx
Tha Pas	8.82 bc	576.0 p	143.1 wx
Thompson	9.36 bc	620.3 q	152.2 wx
Winnipeg	8.18 b	614.8 pq	159.8 wx

Means in the same column not followed by the same letter are significantly different

Table 14.2 Comparisons of means with respect to different greenhouse systems using the Tukey Simultaneous Test at a 95 % confidence interval

System	Mean annual energy consumption (GJ/m^2)	Mean annual water consumption (kg/m^2)	Mean annual water collection (kg/m^2)
CON	12.36 a	728.2 p	0.0 w
OHP	6.29 b	753.8 q	9.4 w
SHP	5.68 c	578.7 r	134.4 x
CHP	8.16 d	450.9 s	450.8 y

Means in the same column not followed by the same letter are significantly different

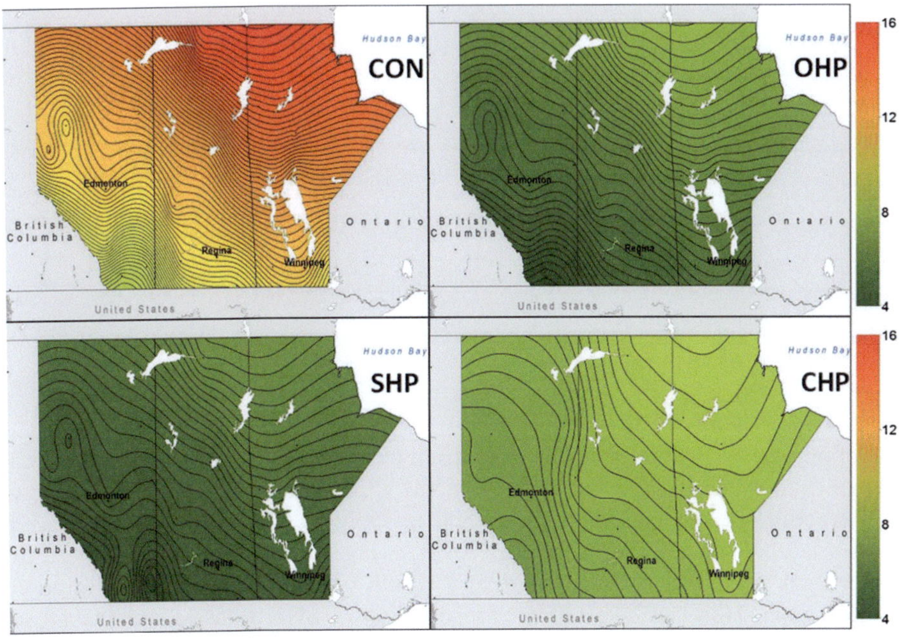

Fig. 14.1 Spatial distributions of annual total energy consumptions (GJ/m^2) in CON, OHP, SHP, and CHP systems in the Prairies (contour line intervals: 100 MJ/m^2)

(Tables 14.1 and 14.2). Locations at lower latitudes had more water consumption (transpiration) due to increased amount of available solar radiation (Fig. 14.2). Transpiration rates at higher and lower latitudes approached to each other in the open systems due to relatively increased transpiration rates at higher latitudes as the vapor pressure deficit became larger due to cooler outside air. CHP systems had the lowest transpiration as much higher relative humidities were observed in these systems compared to those in the other systems. All the transpired water was collected by the dehumidification system, and therefore, the overall annual water consumption in CHP systems were essentially zero (Tables 14.1 and 14.2).

CHP systems had relatively lower energy consumptions in colder months, but the consumptions increased in warmer months due to increased cooling loads;

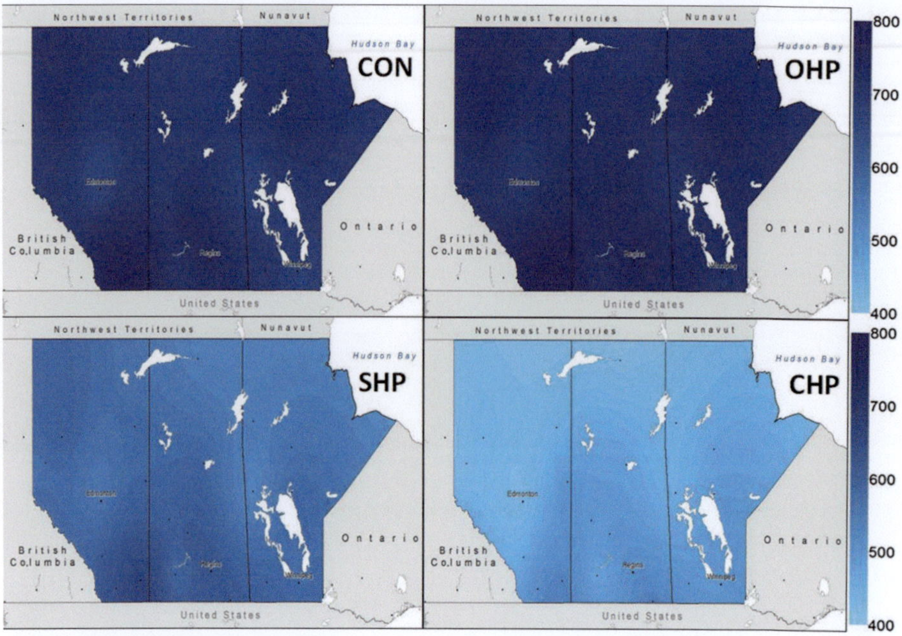

Fig. 14.2 Spatial distributions of annual amount of transpiration (kg/m^2) in CON, OHP, SHP, and CHP systems in the Prairies (contour line intervals: 10 kg/m^2)

therefore, the overall annual energy consumptions in CHP systems were much higher than those in OHP and SHP systems (Figs. 14.3 and 14.4). The energy consumption in SHP systems approached to those in closed (CHP) systems at higher latitudes (e.g., Stony Rapids, SASK) (Fig. 14.3) because SHP systems under such conditions operated more as a closed system from November through end of April. However, the energy consumption in SHP systems approached to those in open (OHP) systems at lower latitudes (e.g., Lethbridge, ALTA) (Fig. 14.4) because SHP systems under such conditions operated more as an open system from April through end of October.

Further details of individual comparisons for both location and system were provided in Tables 14.1 and 14.2. And the individual system performances were provided below.

14.3.2 Conventional System

Amount of total energy consumption, water consumption (transpiration), and water collection on the dehumidifier were analyzed throughout the region by multiple regressions using as regressors latitude, longitude, and altitude.

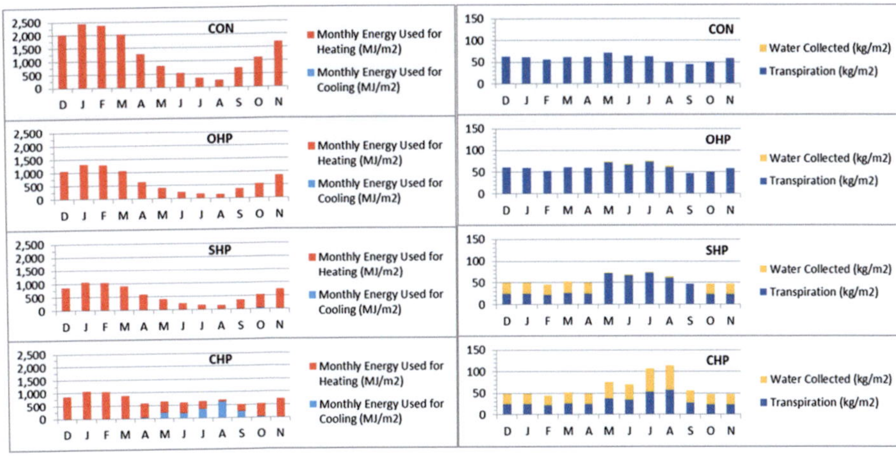

Fig. 14.3 Temporal distributions of energy consumption (*left*) and transpiration (*right*) in CON, OHP, SHP, and CHP systems at Stony Rapids, SASK (59.25 N, 105.83 W, altitude: 245 m)

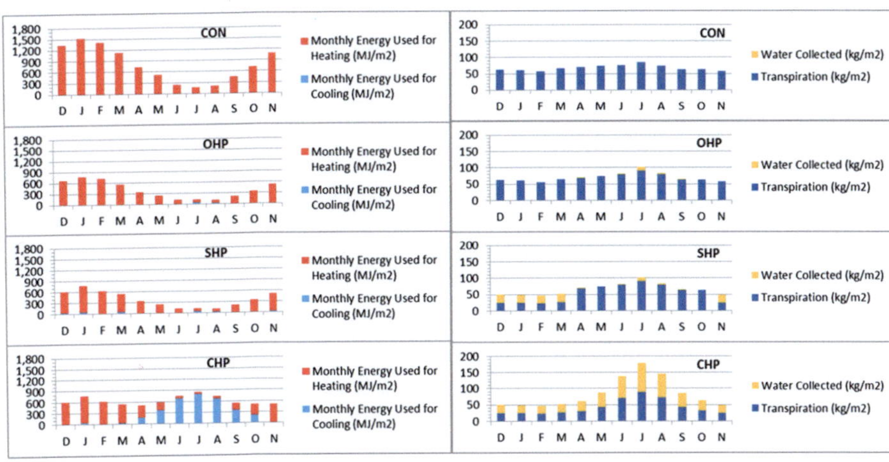

Fig. 14.4 Temporal distributions of energy consumption (*left*) and transpiration (*right*) in CON, OHP, SHP, and CHP systems at Lethbridge, ALTA (49.63 N, 112.80 W, altitude: 929 m)

The regression for the total energy consumption was provided in Eq. (14.1).

$$E = -10.5 + 0.623^{*}(\text{Latitude}) + 0.101^{*}(\text{Longitude}) + 0.0021^{*}(\text{Altitude}) \quad (14.1)$$

where E, the annual total energy consumption, was in GJ/m^2, altitude was in meters, latitude and longitude were both in degrees.

Latitude and longitude significantly predicted total energy consumption, $\beta = 0.95$, $t(17) = 9.30$ $(p < 0.05)$, and $\beta = 0.69$, $t(17) = -5.52$ $(p < 0.05)$, respectively; and latitude was the most influential regressor for CON system. Altitude's prediction of total energy consumption was not significant. Latitude and longitude explained a significant proportion of variance in total energy consumption as well, $R^2 = 0.94$, $F(3, 17) = 83.5$ ($p < 0.05$).

None of the regressors significantly predicted water consumption or explained a significant proportion of variance in water consumption, $F(3, 17) = 3.08 (p > 0.05)$.

14.3.3 Open Heat Pump System

The regression for the total energy consumption was provided in Eq. (14.2).

$$E = 1.99 + 0.260^*(\text{Latitude}) - 0.091^*(\text{Longitude}) + 0.0005^*(\text{Altitude}) \quad (14.2)$$

Latitude and longitude significantly predicted total energy consumption, $\beta = 0.91$, $t(17) = 9.93$ ($p < 0.05$), and $\beta = 0.71$, $t(17) = -6.19$ ($p < 0.05$), respectively. Latitude and longitude also explained a significant proportion of variance in total energy consumption, $R^2 = 0.95$, $F(3, 17) = 104$ ($p < 0.05$), and the effect of regressor altitude was not significant.

None of the regressors significantly predicted water consumption or explained a significant proportion of variance in water consumption in OHP system, $F(3, 17) = 3.44$ ($p > 0.05$).

The regression for water collected by dehumidifier unit was provided in Eq. (14.3) and the overall relationship was significant.

$$C = 137 - 2.70^*(\text{Latitude}) + 0.25^*(\text{Longitude}) - 0.0192^*(\text{Altitude}) \quad (14.3)$$

where C, amount of water collected by dehumidifier unit, was in kg/m^2, altitude was in meters, latitude and longitude were both in degrees.

Latitude significantly predicted water collection, $\beta = 0.99$, $t(17) = -3.84$ ($p < 0.05$). Latitude explained a significant proportion of variance in water collection as well, $R^2 = 0.59$, $F(3, 17) = 8.24$ ($p < 0.05$). The effects of regressors longitude and altitude were not significant.

14.3.4 Semi-closed Heat Pump System

The regression for the total energy consumption was provided in Eq. (14.4).

$$E = 2.23 + 0.200^*(\text{Latitude}) - 0.072^*(\text{Longitude}) + 0.0008^*(\text{Altitude}) \quad (14.4)$$

Latitude and longitude significantly predicted total energy consumption, $\beta = 1.00$, $t(17) = 6.61$ ($p < 0.05$), and $\beta = 0.79$, $t(17) = -4.18$ ($p < 0.05$), respectively. Latitude and longitude also explained a significant proportion of variance in total energy consumption, $R^2 = 0.86,\ F(3, 17) = 34.6\ \ (p < 0.05)$, and the effect of regressor altitude was not significant.

The regression for water consumption was provided in Eq. (14.5).

$$W = 860 - 11.8*(\text{Latitude}) + 3.43*(\text{Longitude}) - 0.0363*(\text{Altitude}) \quad (14.5)$$

where W, amount of water consumption (transpiration), was in kg/m^2, altitude was in meters, latitude and longitude were both in degrees.

Latitude significantly predicted water consumption, $\beta = 0.85,\ t(17) = -3.41\ (p < 0.05)$. Latitude also explained a significant proportion of variance in water consumption, $R^2 = 0.62,\ F(3, 17) = 9.35\ \ (p < 0.05)$. The effects of regressors longitude and altitude were not significant.

The regression for water collected by dehumidifier unit was provided in Eq. (14.6).

$$C = 224 + 1.89*(\text{Latitude}) - 1.88*(\text{Longitude}) + 0.0214*(\text{Altitude}) \quad (14.6)$$

Longitude significantly predicted water collection, $\beta = 0.99,\ t(17) = -2.77\ (p < 0.05)$. Longitude also explained a significant proportion of variance in water collection in SHP system, $R^2 = 0.49, F(3, 17) = 5.51\,(p < 0.05)$. The effects of regressors latitude and altitude were not significant.

14.3.5 Closed Heat Pump System

The regression for the total energy consumption was provided in Eq. (14.7).

$$E = 10.8 + 0.068*(\text{Latitude}) - 0.057*(\text{Longitude}) + 0.0002*(\text{Altitude}) \quad (14.7)$$

Latitude and longitude significantly predicted total energy consumption, $\beta = 0.44,\ t(17) = 4.03\ \ (p < 0.05)$, and $\beta = 0.82,\ t(17) = -6.05\ \ (p < 0.05)$, respectively. Latitude and longitude also explained a significant proportion of variance in total energy consumption, $R^2 = 0.93,\ F(3, 17) = 72.2\ \ (p < 0.05)$, and the effect of regressor altitude was not significant.

The regression for water consumption and collection was provided in the same Eq. (14.8) as the water collected was equal to the water consumption in CHP system.

$$W \text{ or } C = 1186 - 12.9*(\text{Latitude}) - 0.25*(\text{Longitude}) - 0.0421*(\text{Altitude}) \quad (14.8)$$

Latitude significantly predicted water consumption and collection, $\beta = 0.92,\ t(17) = -4.22\ \ (p < 0.05)$. Latitude also explained a significant proportion of variance in water consumption and collection in CHP system, $R^2 = 0.71, F(3, 12) = 13.8\ \ (p < 0.05)$. The effects of regressors longitude and altitude were not significant in either case.

14.4 Concluding Remarks

The results in this study showed that the energy and water consumptions were both spatially and system dependent. Both latitude and longitude significantly predicted total energy consumptions. Latitude significantly predicted water consumption (transpiration) in semi-closed and closed systems as well. Semi-closed system proved to be the most energy conserving system, while closed system was the most water conserving system. Highest energy consuming locations of semi-closed system were still a lot more efficient than least energy-consuming locations of conventional and closed systems. Closed systems had the lowest energy consumptions in colder months, but the consumption increased in warmer months due to increased cooling loads hence the overall annual energy consumptions in closed systems were much higher than those in semi-closed systems. Locations in the southwestern Prairies (Alberta, and partly Saskatchewan) had the lowest energy consumptions in all systems. Locations at lower latitudes had higher water consumption due to increased amount of available solar radiation. Transpiration rates at higher and lower latitudes approached to each other in the open systems due to relatively increased transpiration rates at higher latitudes as the vapor pressure deficit became larger due to colder and drier outside air. SHP system transforms the non-energy-efficient, unfavorable regions under conventional systems into highly energy and water-efficient potential greenhouse regions in the Canadian Prairies.

References

1. Siurna DL, D'Andrea LJ, Hollands KGT (1984) A Canadian representative meteorological year for solar system simulation. In: Proceedings of the 10th annual conference of the Solar Energy Society of Canada (SESCI '84), Calgary, 2–6 Aug 1984
2. Yildiz I, Stombaugh DP (2006) Dynamic modeling of microclimate and environmental control strategies in a greenhouse coupled with a heat pump system. Acta Horticult 718:331–340
3. Yildiz I, Yue J, Yildiz AC (2015) Air source heat pump performance in open, semi-closed and closed greenhouse systems in British Columbia. In: Dincer I et al. (eds) Progress in clean energy. Springer, New York

Chapter 15
Air Source Heat Pump Performance in Open, Semi-closed, and Closed Greenhouse Systems in Quebec and Labrador

İlhami Yıldız, Jin Yue, and Asena Cansu Yıldız

Abstract The specific objective of this study was to investigate regional spatiotemporal distributions and spatial correlations of energy and water consumption in open, semi-closed, and completely closed greenhouse systems in Quebec and Labrador. The findings showed that the energy and water consumptions were both spatially and system dependent. Latitude, and depending on the system longitude as well, significantly predicted total energy consumptions. Semi-closed and open heat pump systems were the most energy conserving systems while closed system was the most water conserving. Semi-closed and closed systems at lower latitudes had more water consumptions compared to those at higher latitudes. Water consumptions in open systems at higher latitudes approached to or were sometimes higher than those at lower latitudes as the vapor pressure deficit became larger due to cooler outside air. Southern Quebec had the lowest energy consumptions in all systems.

Keywords Heat pump • Air source • Energy consumption • Water consumption • Closed greenhouse • Semi-closed greenhouse • Heating • Cooling • Dehumidification • Spatiotemporal distribution • Spatial correlation • Labrador • Quebec

Nomenclature

C	Amount of water collection, kg/m^2
CHP	Closed heat pump greenhouse system
CON	Conventional greenhouse system

İ. Yıldız (✉) • J. Yue
Department of Engineering, Faculty of Agriculture, Dalhousie University,
Truro-Bible Hill, NS, Canada B6L 3H9
e-mail: iyildiz@dal.ca

A.C. Yıldız
Department of Law, Queen Mary University of London, Mile End Road, London E1 4NS, UK

I. Dincer et al. (eds.), *Progress in Clean Energy, Volume 1*,
DOI 10.1007/978-3-319-16709-1_15

E	Amount of energy consumption, GJ/m^2
F	Statistical *F* test
OHP	Open heat pump greenhouse system
p	Significance level
R	Regression
SHP	Semi-closed heat pump greenhouse system
t	Statistical *t* test
W	Amount of water consumption, kg/m^2

Greek Letters

β Beta value

15.1 Introduction

One major factor hindering future expansion of greenhouse industry is the cost required for environmental control. Consequently, considerable effort is expended to conserve energy and look for alternative energy sources, especially environment-friendly renewable energy sources and technologies. Proper greenhouse and environmental management systems can significantly change the energy and moisture dynamics of greenhouse production systems. This study helps enhance the competitive position of Quebec and Labrador's agriculture and agri-food industry by introducing economically, environmentally, and socially sustainable technologies and management strategies. The investigated heat pump technology and operational modes (especially semi-closed and closed) bring increased yield, and energy and water-efficient solutions to the doorstep of both Quebec and Labrador's greenhouse producers. The overall goal of this study was to help reduce the load on power grid, demand for fossil fuels and irrigation water, and also supply CO_2 for the greenhouse production. The use of heat pumps for heating and cooling greenhouses makes it possible to use local and renewable energy sources while reducing or totally eliminating CO_2 emissions. Heat pumps also make an innovative confined greenhouse operation possible, which would totally eliminate the energy, CO_2, and water losses due to ventilation conserving all of these precious resources, and would make an insect-free greenhouse operation possible. The specific objective of this study was to investigate spatiotemporal performance distributions and spatial correlations of energy and water consumptions in open, semi-closed, and confined greenhouse systems in Quebec and Labrador.

15.2 Methodology

This section focuses on the resources and procedures employed in this study, as well as the statistical analyses employed and data presentation.

15.2.1 Weather File

The Canadian Weather for Energy Calculations (CWEC) files created by concatenating 12 Typical Meteorological Months (TMY) selected from a database of 30 years of data were used for continuous full year analyses at 24 different locations. CWEC data sets were prepared under the direction of Environment Canada (Siurna et al. [1]), and simulations were performed starting at the beginning of 1st day of January and ending at the end of 31st day of December for full 365-day simulations.

15.2.2 Greenhouse Characteristics and Operational Strategies

Theoretical approach, simulation model assumptions and validation, and all the other characteristics of the model including the details of control systems and strategies were reported in another study [2]. Full details of the greenhouse systems characteristics and operational strategies were reported by Yildiz et al. [3].

15.2.3 Data Analyses and Presentation

A total of four systems, conventional (CON), open heat pump (OHP), semi-closed heat pump (SHP), and closed heat pump (CHP), and a total of 24 locations were studied. Full details of the data analyses were reported by Yildiz et al. [3].

15.3 Data Interpretation

This section covers the result and discussions for different greenhouse systems at different locations.

15.3.1 Energy Consumption, Water Consumption, and Water Collection

The two-way ANOVA $(p < 0.05)$ was used to determine the main effects of greenhouse location and system on total energy and water consumptions as well as water collection. There were significant main effects for the location and system on total energy consumption, $F(23, 69) = 6.18$ $(p < 0.05)$, and $F(3, 69) = 261$ $(p < 0.05)$, respectively (Tables 15.1 and 15.2). Locations at lower latitudes had less total energy consumption, and part of which was for cooling while northern locations had no cooling requirements (Fig. 15.1). SHP and OHP systems had the lowest annual energy consumptions while CON system had the highest.

Table 15.1 Comparisons of means with respect to different locations using the Tukey Simultaneous Test at a 95 % confidence interval

Location	Mean annual energy use (GJ/m^2)	Mean annual water use (kg/m^2)	Mean annual water collection (kg/m^2)
Bagotville	7.79 ab	573.0 qr	131.6 w
Baie Comeau	7.87 ab	549.7 q	130.0 w
Battle Harbour	8.87 b	534.0 pq	120.1 w
Cartwright	8.18 ab	528.9 pq	122.1 w
Goose Bay	8.28 ab	483.6 p	120.2 w
Gridstone Isl.	7.03 ab	519.0 pq	130.7 w
Kuujjuaq	10.9 c	511.6 pq	113.6 w
Kuujjuarapik	9.96 bc	507.4 pq	128.0 w
La Grande Riv.	9.60 bc	545.8 q	130.2 w
Lake Eon	9.59 bc	526.4 pq	123.8 w
Mont Joli	7.54 ab	567.9 qr	130.6 w
Montreal	6.98 ab	566.1 qr	151.3 w
Montreal Int.	7.34 ab	611.5 r	155.9 w
Nitchequon	10.0 bc	511.8 pq	120.0 w
Quebec	7.06 ab	587.4 qr	146.1 w
Riviere du L.	7.45 ab	560.2 qr	139.9 w
Roberval	7.78 ab	590.9 qr	139.7 w
Schefferville	9.98 bc	531.5 pq	121.0 w
Sept-Iles	8.41 ab	523.2 pq	115.0 w
Sherbrooke	7.19 ab	584.2 qr	144.7 w
St. Hubert	6.73 a	615.9 r	170.7 w
Ste. Agathe	7.48 ab	566.1 qr	137.6 w
Val d'Or	8.19 ab	580.1 qr	137.9 w
Wabush	9.34 bc	569.3 qr	130.8 w

Means in the same column not followed by the same letter are significantly different

Table 15.2 Comparisons of means with respect to different greenhouse systems using the Tukey Simultaneous Test at a 95 % confidence interval

System	Mean annual energy consumption (GJ/m^2)	Mean annual water consumption (kg/m^2)	Mean annual water collection (kg/m^2)
CON	12.89 a	649.5 p	0.0 w
OHP	6.67 b	684.3 q	3.5 w
SHP	5.94 b	492.3 r	147.0 x
CHP	7.75 c	381.5 s	381.5 y

Means in the same column not followed by the same letter are significantly different

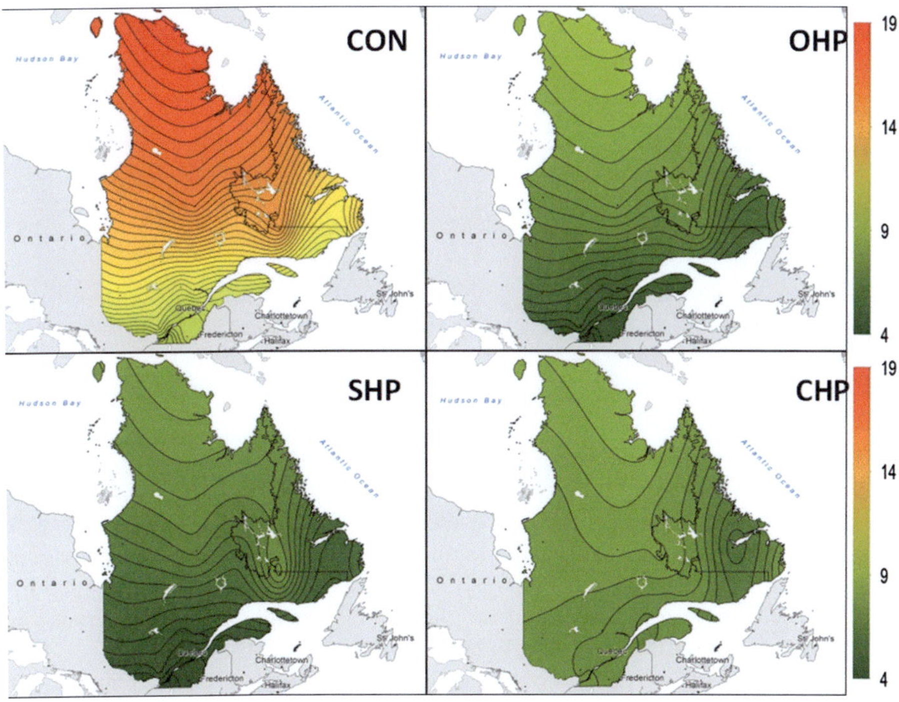

Fig. 15.1 Spatial distributions of annual total energy consumptions (GJ/m^2) in CON, OHP, SHP, and CHP systems in Quebec and Labrador (contour intervals: 250 MJ/m^2)

There were significant main effects for the location and system on water consumption as well, $F(23, 69) = 5.06$ ($p < 0.05$), and $F(3, 69) = 504$ ($p < 0.05$), respectively (Tables 15.1 and 15.2). The main effect for the system on water collection was also significant, $F(3, 69) = 626$ ($p < 0.05$); however, the main effect for the location on water collection was not, $F(23, 69) = 0.61$ ns (Tables 15.1 and 15.2). Semi-closed and closed systems at lower latitudes had more water consumption (transpiration) due to increased amount of available solar radiation (Fig. 15.2). In the open systems, transpiration rates at higher latitudes approached to or sometimes higher than those at lower latitudes as the vapor

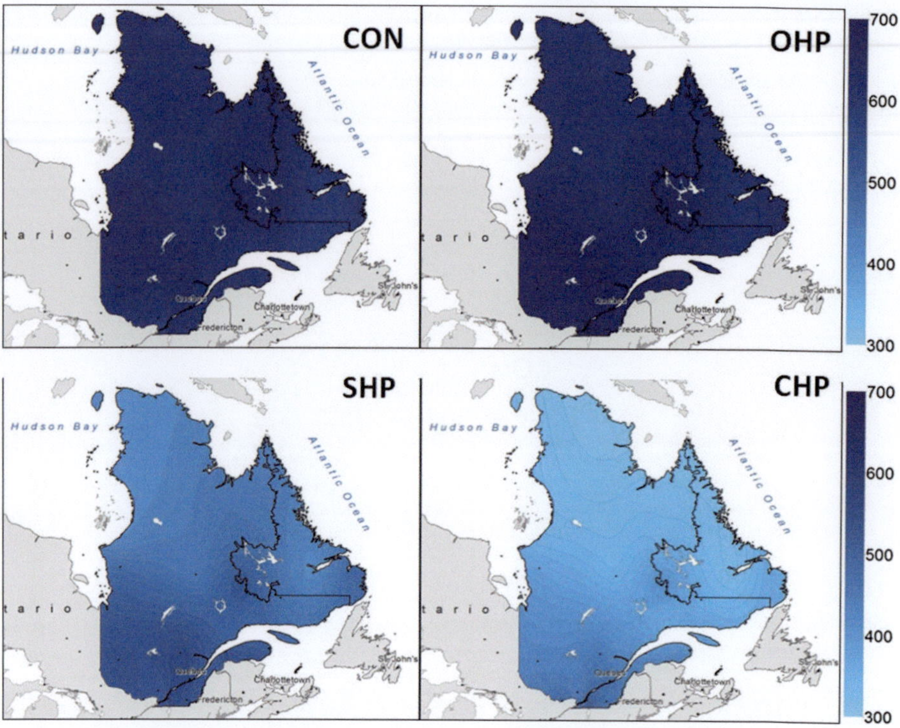

Fig. 15.2 Spatial distributions of annual amount of transpiration (kg/m^2) in CON, OHP, SHP, and CHP systems in Quebec and Labrador (contour intervals: 10 kg/m^2)

pressure deficit became larger due to cooler outside air. CHP systems had lowest transpiration as higher inside relative humidities were observed in closed systems compared to those of the other systems. All the transpired water was collected by dehumidification system, and therefore, the overall annual water consumption in CHP systems was essentially zero (Tables 15.1 and 15.2).

CHP systems had relatively lower energy consumptions in colder months, but the consumption increased in warmer months due to increased cooling loads; as a result, the overall annual energy consumptions in CHP systems were much higher than those in OHP and SHP systems (Figs. 15.3 and 15.4). The energy consumption in SHP systems approached to those in closed (CHP) systems at higher latitudes (e.g., Kuujjuaq, QC) (Fig. 15.3) because SHP systems under such conditions operated more as a closed system from October through end of April. However, the energy consumption in SHP systems approached to those in open (OHP) systems at lower latitudes (e.g., St. Hubert, QC) (Fig. 15.4) because SHP systems under such conditions operated more as an open system from April through end of October.

Further details of individual comparisons for both location and system were provided in Tables 15.1 and 15.2. And the individual system performances were provided below.

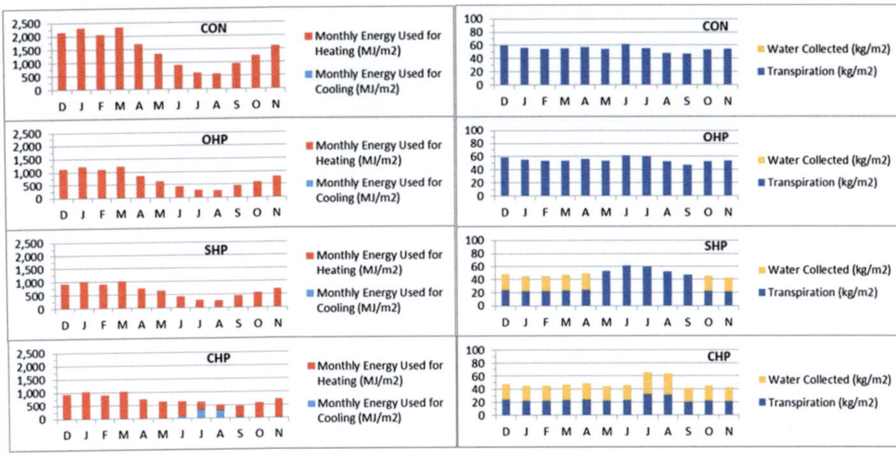

Fig. 15.3 Temporal distributions of energy consumption (*left*) and transpiration (*right*) in CON, OHP, SHP, and CHP systems at Kuujjuaq, QC (58.10 N, 68.42 W, altitude: 39 m)

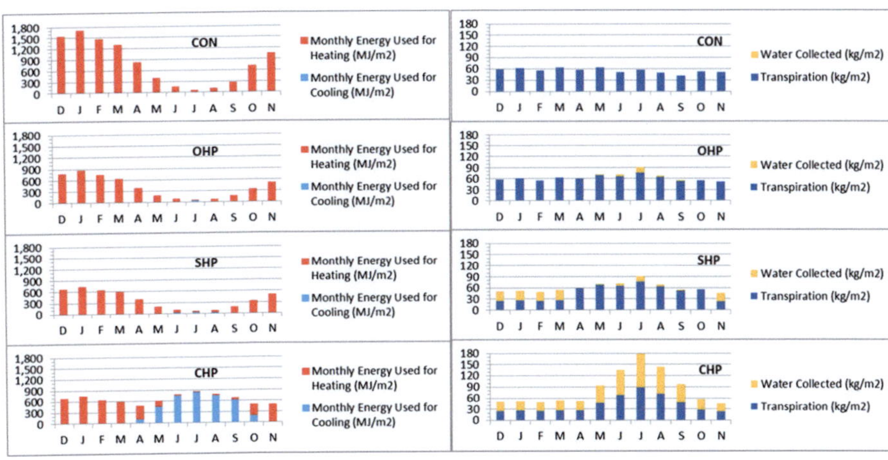

Fig. 15.4 Temporal distributions of energy consumption (*left*) and transpiration (*right*) in CON, OHP, SHP, and CHP systems at St. Hubert, QC (45.52 N, 73.42 W, altitude: 27 m)

15.3.2 Conventional System

Amount of total energy consumption, water consumption (transpiration), and water collection on the dehumidifier were analyzed throughout the region by multiple regressions using as regressors latitude, longitude, and altitude.

The regression for the total energy consumption was provided in Eq. (15.1).

$$E = -10.5 + 0.623 \times (\text{Latitude}) + 0.101 \times (\text{Longitude}) + 0.0021 \times (\text{Altitude}) \tag{15.1}$$

where E, the annual total energy consumption, was in GJ/m^2, altitude was in meters, latitude and longitude were both in degrees.

Latitude and longitude significantly predicted total energy consumption, $\beta = 0.95,\ t(17) = 9.30\ (p < 0.05)$, and $\beta = 0.69,\ t(17) = -5.52\ (p < 0.05)$, respectively; and latitude was the most influential regressor for CON system. Altitude's prediction of total energy consumption was not significant. Latitude and longitude explained a significant proportion of variance in total energy consumption as well, $R^2 = 0.94$, $F(3, 17) = 83.5\ (p < 0.05)$.

None of the regressors significantly predicted water consumption or explained a significant proportion of variance in water consumption, $F(3, 17) = 3.08\ (p > 0.05)$.

15.3.3 Open Heat Pump System

The regression for the total energy consumption was provided in Eq. (15.2).

$$E = -10.3 + 0.277 \times (\text{Latitude}) + 0.043 \times (\text{Longitude}) + 0.0010 \times (\text{Altitude}) \tag{15.2}$$

Latitude significantly predicted total energy consumption, $\beta = 0.87,\ t(20) = 7.41\ (p < 0.05)$. Latitude also explained a significant proportion of variance in total energy consumption in OHP systems, $R^2 = 0.76$, $F(3, 20) = 21.4\ (p < 0.05)$. The effects of regressors longitude and altitude were not significant in OHP system.

The regression for water consumption was provided in Eq. (15.3).

$$W = 730 - 2.92 \times (\text{Latitude}) + 1.42 \times (\text{Longitude}) + 0.0164 \times (\text{Altitude}) \tag{15.3}$$

where W, amount of water consumption, was in kg/m^2, altitude was in meters, latitude and longitude were both in degrees.

Latitude significantly predicted water consumption, $\beta = 0.45,\ t(20) = -2.49\ (p < 0.05)$. Latitude also explained a significant proportion of variance in water consumption in OHP systems, $R^2 = 0.44$, $F(3, 20) = 5.23\ (p < 0.05)$. The effects of regressors longitude and altitude were not significant.

The regression for water collected by dehumidifier unit was provided in Eq. (15.4) and the overall relationship was significant.

$$C = 27.7 - 0.82 \times (\text{Latitude}) + 0.26 \times (\text{Longitude}) - 0.0069 \times (\text{Altitude}) \tag{15.4}$$

where C, amount of water collected by dehumidifier unit, was in kg/m^2, altitude was in meters, latitude and longitude were both in degrees.

Latitude significantly predicted water collection, $\beta = 0.46$, $t(20) = -2.44$ ($p < 0.05$). Latitude also explained a significant proportion of variance in water collection in OHP systems, $R^2 = 0.39$, $F(3, 20) = 4.32$ ($p < 0.05$). The effects of regressors longitude and altitude were not significant in OHP system.

15.3.4 Semi-closed Heat Pump System

The regression for the total energy consumption was provided in Eq. (15.5).

$$E = -7.76 + 0.252 \times (\text{Latitude}) + 0.011 \times (\text{Longitude}) + 0.0016 \times (\text{Altitude}) \tag{15.5}$$

Latitude significantly predicted total energy consumption, $\beta = 0.89$, $t(20) = 13.0$ ($p < 0.05$), and it was the most influential regressor for SHP system. Altitude as well significantly predicted total energy consumption, $\beta = 0.29$, $t(20) = 4.39$ ($p < 0.05$). Latitude and altitude explained a significant proportion of variance in total energy consumption in SHP systems as well, $R^2 = 0.92$, $F(3, 20) = 76.0$ ($p < 0.05$), and the effect of regressor longitude was not significant.

The regression for water consumption was provided in Eq. (15.6).

$$W = 947 - 10.6 \times (\text{Latitude}) + 1.11 \times (\text{Longitude}) + 0.0013 \times (\text{Altitude}) \tag{15.6}$$

Latitude significantly predicted water consumption, $\beta = 0.77$, $t(20) = -5.70$ ($p < 0.05$). Latitude also explained a significant proportion of variance in water consumption in SHP systems, $R^2 = 0.68$, $F(3, 20) = 14.2$ ($p < 0.05$). The effects of regressors longitude and altitude were not significant.

The regression for water collected by dehumidifier unit was provided in Eq. (15.7).

$$C = -120 + 4.67 \times (\text{Latitude}) + 0.48 \times (\text{Longitude}) + 0.0020 \times (\text{Altitude}) \tag{15.7}$$

Latitude significantly predicted water collection, $\beta = 0.78$, $t(20) = 4.89$ ($p < 0.05$). Latitude also explained a significant proportion of variance in water collection in SHP systems, $R^2 = 0.56$, $F(3, 20) = 8.57$ ($p < 0.05$). The effects of regressors longitude and altitude were not significant.

15.3.5 Closed Heat Pump System

The regression for the total energy consumption was provided in Eq. (15.8).

$$E = 1.87 + 0.0624 \times (\text{Latitude}) + 0.0384 \times (\text{Longitude}) + 0.0006 \times (\text{Altitude}) \tag{15.8}$$

Latitude significantly predicted total energy consumption, $\beta = 0.52,\ t(20) = 3.07\ (p < 0.05)$. Longitude as well significantly predicted total energy consumption, $\beta = 0.53,\ t(20) = 3.14\ (p < 0.05)$. Latitude and longitude explained a significant proportion of variance in total energy consumption in CHP systems as well, $R^2 = 0.51,\ F(3, 20) = 6.81\ (p < 0.05)$, and the effect of regressor altitude was not significant.

The regression for water consumption and collection was provided in the same Eq. (15.9) as the water collected was equal to the water consumption in CHP system.

$$W \text{ or } C = 947 - 10.6 \times (\text{Latitude}) + 1.11 \times (\text{Longitude}) - 0.0013 \times (\text{Altitude}) \tag{15.9}$$

Latitude significantly predicted water consumption and collection, $\beta = 0.82$, $t(20) = -8.58\ (p < 0.05)$. Longitude as well significantly predicted water consumption and collection, $\beta = 0.20, t(20) = 2.14(p < 0.05)$. Latitude and longitude explained significant proportions of variance in water consumption and collection in CHP systems, $R^2 = 0.84, F(3, 20) = 35.2(p < 0.05)$, and the effect of regressor altitude was not significant in any case.

15.4 Concluding Remarks

The results in this study showed that the energy and water consumptions were both spatially and system dependent. Latitude, and depending on the system longitude as well, significantly predicted total energy consumptions. Semi-closed and OHP systems proved to be the most energy conserving systems, while closed system was the most water conserving system. Highest energy consuming locations of semi-closed and OHP systems were still a lot more efficient than least energy consuming locations of conventional systems. Closed systems had the lowest energy consumptions in colder months, but the consumption increased in warmer months especially in southern Quebec due to increased cooling loads; therefore, the overall annual energy consumptions in closed systems were much higher than those in semi-closed systems. Locations in southern Quebec had the lowest energy consumptions in all systems. Semi-closed and closed systems at lower latitudes had higher water consumption (transpiration) due to increased amount of available solar radiation. Water consumptions at higher latitudes in open systems approached to or were sometimes higher than those at lower latitudes as the vapor pressure deficit became larger due to cooler outside air. SHP system transforms the non-energy-efficient, unfavorable regions under conventional systems into highly energy and water-efficient potential greenhouse regions in Quebec and Labrador.

References

1. Siurna DL, D'Andrea LJ, Hollands KGT (1984) A Canadian representative meteorological year for solar system simulation. In: Proceedings of the 10th annual conference of the Solar Energy Society of Canada (SESCI '84), Calgary, 2–6 Aug 1984
2. Yildiz I, Stombaugh DP (2006) Dynamic modeling of microclimate and environmental control strategies in a greenhouse coupled with a heat pump system. Acta Horticult 718:331–340
3. Yildiz I, Yue J, Yildiz AC (2015) Air source heat pump performance in open, semi-closed and closed greenhouse systems in British Columbia. In: Dincer I et al. (eds) Progress in clean energy, ©Springer Science+Business Media New York

Chapter 16
Experimental Study of a Multilayer Active Magnetic Regenerator Refrigerator-Demonstrator

Younes Chiba, Osmann Sari, Arezki Smaïli, Cyril Mahmed, and Petri Nikkola

Abstract This work presents an experimental investigation on a reciprocating active magnetic regenerator (AMR) refrigerator-demonstrator operating near room temperature. La(Fe, Co)$_{13-x}$ Si$_x$ is used as a refrigerant material in the form of multilayer and water as heat transfer fluid to release and absorb heat at the hot and cold sources, respectively. Experimental tests using the multilayer regenerator with an applied magnetic field of 1.45 T have produced a temperature span of 15.6 K, without cooling load. The effects of the following cycle parameters, namely, utilisations, frequencies and mass flow rates, on the AMR cycle performance are presented and discussed. A parametric study was conducted on the behaviour of the temperature span to define the optimum demonstrator parameters to improve performance and validate the concept of multilayer and provide useful data for magnetic refrigerator design.

Keywords Multilayer magnetic materials • La(Fe, Co)$_{13-x}$ Si$_x$ • Active magnetic regenerator refrigerator-demonstrator • Thermal analysis

---

Y. Chiba
Department of Mechanical Engineering, Faculty of Technology,
University of Médéa, Médéa 26000, Algeria

Ecole Nationale Polytechnique, Laboratoire de Génie Mécanique et Développement,
P.B. 182, El-Harrach, Algiers 16200, Algeria

University of Applied Sciences of Western Switzerland, Avenue des sports 20,
Yverdon-Les-Bains CH 1400, Switzerland

O. Sari • C. Mahmed • P. Nikkola
University of Applied Sciences of Western Switzerland, Avenue des sports 20,
Yverdon-Les-Bains CH 1400, Switzerland

A. Smaïli (✉)
Ecole Nationale Polytechnique, Laboratoire de Génie Mécanique et Développement,
P.B. 182, El-Harrach, Algiers 16200, Algeria
e-mail: arezki.smaili@g.enp.edu.dz

I. Dincer et al. (eds.), *Progress in Clean Energy, Volume 1*,
DOI 10.1007/978-3-319-16709-1_16

Nomenclature

Variables

C_f Specific heat of fluid (J·(kg·K)$^{-1}$)
C_r Specific heat of magnetocaloric material (J·(kg·K)$^{-1}$)
$\dot{m}_f$ Mass flow rate (kg s^{-1})
m_r Mass of magnetocaloric material (kg)
U Utilisation factor

Greek Letters

τ_P Time of cycle (s)

Subscripts

f Fluid
r Regenerator
p Time period (s)

16.1 Introduction

Magnetic refrigeration system is considered actually as an attractive alternative for conventional vapour compression refrigeration systems; furthermore, all systems of magnetic refrigeration have no ozone depletion potential and no direct global warming potential. Magnetic refrigeration is a new technology for the production of cooling at room temperature based on the magnetocaloric effect [1–4]. This technique can be used to achieve extremely low temperatures as well as warm temperatures in the case of a heat pump. The magnetocaloric effect is defined in terms of an adiabatic change in temperature or isothermal change in magnetic entropy. This phenomenon is practically reversible for some magnetocaloric materials. Recently, many refrigerator-demonstrators that are linear and rotary, operating near room temperature, have been reported by Romero Gómez et al. [5], Yu et al. [6] and Gschneidner et al. [7]. Several works reported in the literature using multilayer include Legait et al. [8] using three materials $Pr_{0.65}Sr_{0.35}MnO_3$, La-Fe-Co-Si and gadolinium for four regenerator configurations, Tušek et al. [9] using seven and two materials layered based on different compounds of La-Fe-Co-Si and Engelbrecht et al. [10] using also two materials layered based on different compounds of La-Fe-Co-Si. Richard et al. [11] had described a reciprocating permanent magnet AMR using a layered regenerator consisting of two materials based on different alloys of gadolinium, while Zimm et al. [12] had described a rotary permanent magnet AMR using a layered regenerator for two materials. In 2006, Rowe and Tura [13] studied the impact of the rejection temperature and cooling power of three materials layered. Also, in 2006 Okamura [14] investigated the feasibility of using four materials layered based on different alloys of gadolinium. Arnold et al. [15] had described a reciprocating permanent magnet AMR using a layered regenerator consisting of two materials based on different alloys of gadolinium. The numerical model for predicting the efficiency of a multilayer has been presented by Engelbrecht et al. [16] and Aprea et al. [17]. Recently, in 2012, Egolf et al. [18] attempted to explain in detail the theory and concept of AMR device using multilayer.

In this paper, new compounds LaFeCoSi have been tested, characterised and used in our prototype in order to improve the efficiency of the machine. LaFeCoSi compounds are used in multilayer form in the regenerators at the direction of increasing the magnetocaloric effect. The performance characteristics of active magnetic refrigerator device are analysed for different cases taking into account the operational parameters of cycle, namely, utilisations, frequencies and mass flow rates. Some interesting cases are presented and discussed in this paper.

16.2 Experimental Apparatus

Figure 16.1 shows a schematic of an AMR device. The experimental apparatus is composed of two regenerators with LaFeCoSi plates, two heat exchangers and two permanent magnetic sources producing about 1.45 T. The regenerator is divided into two separated parts; each part contains LaFeCoSi flat plates with $e = 1$ mm thickness, $w = 8$ mm width and $l = 80$ mm length, corresponding to about 716.76 g of LaFeCoSi. The characteristic parameters of the machine are given in Table 16.1. The AMR cycle consists of four processes, namely, magnetisation and demagnetisation steps, application and removal of a magnetic field, and cold and hot blows (i.e. cooling and heating the circulating fluid). Taking into account the design and the scheme of operation of the demonstrator presented in Fig. 16.1, when the first part of the regenerator is magnetised, the second part is demagnetised which allows to compensate and reduce the magnetic forces in the magnetic refrigeration system, where each regenerator is divided into two separated parts. This new design decreases largely the magnetic forces. According to calculations, more than 90 % of the absorbed energy during magnetisation and demagnetisation can be saved [2, 19]. The calculations were confirmed by the measurement of the forces performed directly on the machine. The geometry and configuration of multilayer materials are presented in Fig. 16.2 with different transition temperatures.

16.3 Results and Discussions

The following equation can be used to calculate the various operational parameters of active magnetic refrigeration cycle:

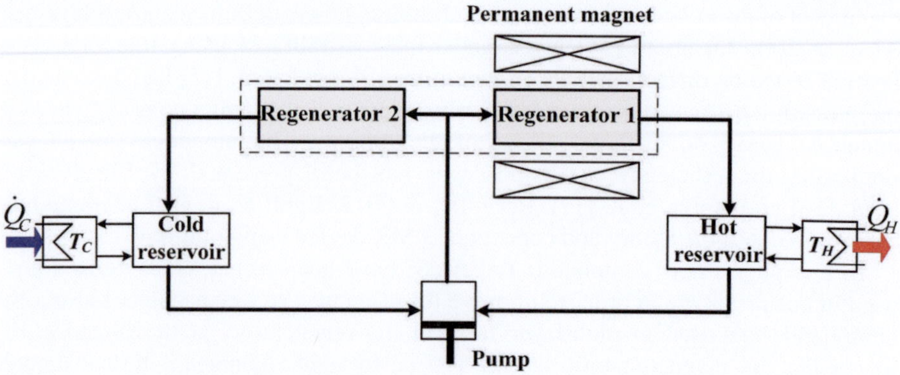

Fig. 16.1 Schematic diagram of the active magnetic refrigeration refrigerator setup

Table 16.1 Operating and technical parameters of demonstrator

Fluid	Water + 5 % Noxal
ΔT_{ad} LaFe(Co, Si)$_{13}$ (K)	2.4
Specific heat LaFe(Co, Si)$_{13}$ (J·(gK)$^{-1}$)	0.830
Fluid-specific heat (J·(gK)$^{-1}$)	4.18
Total mass of solid material (g)	716.76
Total mass of fluid (g)	45.6
Magnetocaloric material	La(Fe, Co)Si
Type of magnetic source	Permanent magnet (NdFeB)
Magnetic field (T)	1.45

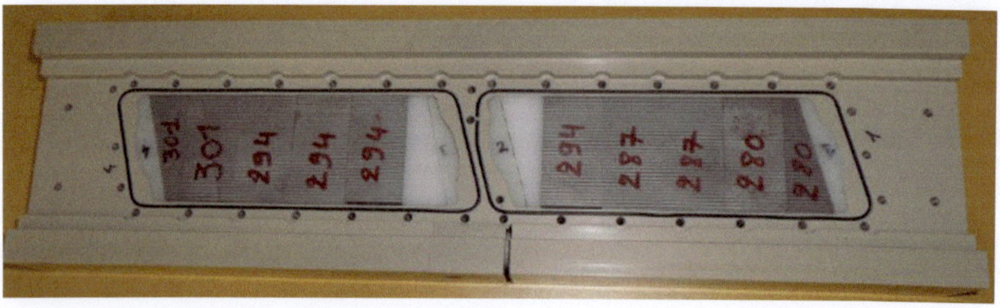

Fig. 16.2 Geometry and configuration of multilayer

$$U = \frac{m_f c_f}{m_r c_r} = \frac{\dot{m}_f c_f \tau_P}{m_r c_r} \tag{16.1}$$

The results presented in Fig. 16.3 show, in the apparatus operating range without load, the evolution of temperature span as functions of mass flow rate of heat transfer fluid and utilisation. As it can be seen, all the curves behave in the same trend: having maximum values around 25 g/s then decreasing with increasing mass

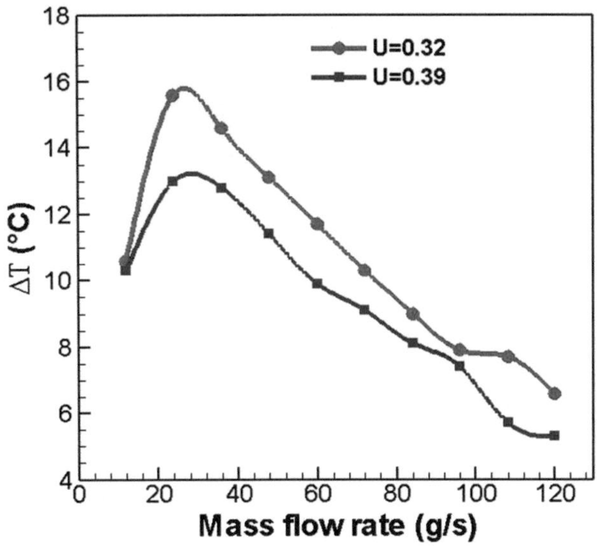

Fig. 16.3 Evolution of temperature span as function of mass flow rate for multilayer LaFe(Co, $Si)_{13}$ operating in ambient temperature of 20 °C

flow rate. Also, notice the positive impact of the decrease of utilisation on the temperature span.

Figure 16.4 shows the results of the evolution of pressure drop as a function of mass flow rate in demonstrator. As expected, the more the mass flow rate is increasing, the more the pressure drops.

Figure 16.5 shows the evolution of optimal temperature span as function of time for multilayer regenerator LaFe(Co, $Si)_{13}$ and Gd operating near room temperature, around 20 °C under operating conditions: $f = 0.24$ Hz, $U = 0.30$ and the magnetic field $B = 1.5$ T. As it can be seen, the temperature span as function of time exhibits first sharp variation and then constant trends (i.e. steady-state conditions) from about 200 and 1,400 s for Gd and multilayer, respectively. Also, differences up to 2.9 °C have been noted during steady-state conditions. These shifts might be attributed to positive effect of the feasibility of the multilayer refrigerant.

Figure 16.6 shows the results of the evolution of temperature span as function of utilisation for two different typical values of frequency obtained at optimum mass flow rate (24 g/s). Mainly, notice here the impact of frequency on the optimum operations of the demonstrator. Table 16.2 summarises the results.

Figure 16.7 shows the results of the evolution of temperature span as function of utilisation for three different typical values of mass flow rate. As it can be seen, these results confirm that the optimum mass flow rate is equal to 24 g/s for a given operation range of demonstrator.

Figure 16.8 shows the positive effect of increasing frequency on the cooling power for constant utilisation. However, notice that the increase in frequency has a negative impact on the cold temperature and vice versa for the cooling capacity.

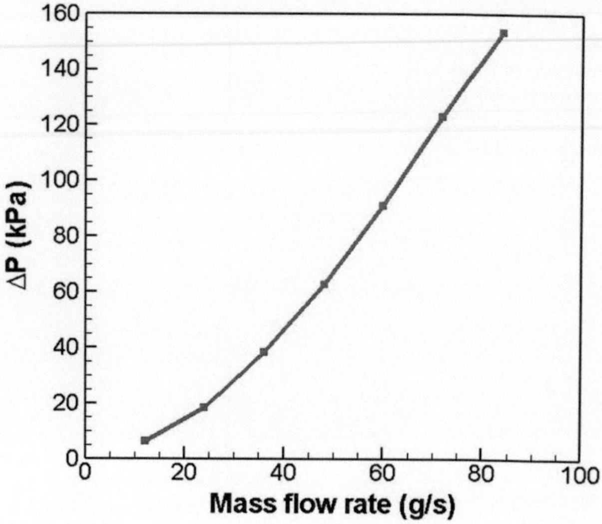

Fig. 16.4 Evolution of pressure drop as function of mass rate flow in demonstrator

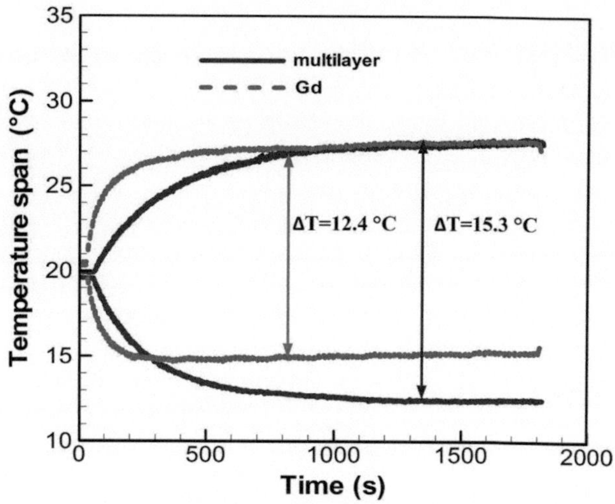

Fig. 16.5 Evolution of optimal temperature span as function of time for multilayer $LaFe(Co, Si)_{13}$ operating in ambient temperature of 20 °C

Indeed, this is because of the thermal losses which have an influence on the heat exchange between the fluid and the various layers of magnetocaloric materials chosen.

16.4 Conclusion

The alloys of $LaFe(Co, Si)_{13}$ were experimentally studied in an active magnetic regenerator demonstrator near room temperature. The multilayer regenerator was found to produce a higher temperature span and more cooling power than the

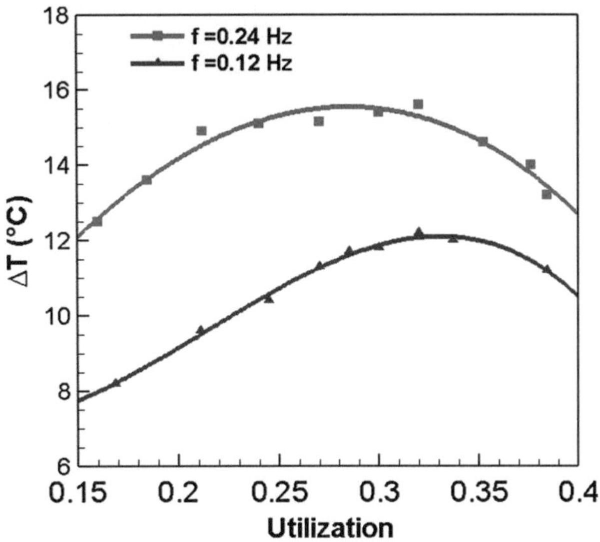

Fig. 16.6 Evolution of temperature span as function of utilisation for multilayer operating in ambient temperature of 20 °C

Table 16.2 Optimum operating parameters of demonstrator

f (Hz)	ΔT_{max} (°C)	$U_{optimal}$
0.24	15.6	0.28
0.12	12.2	0.32

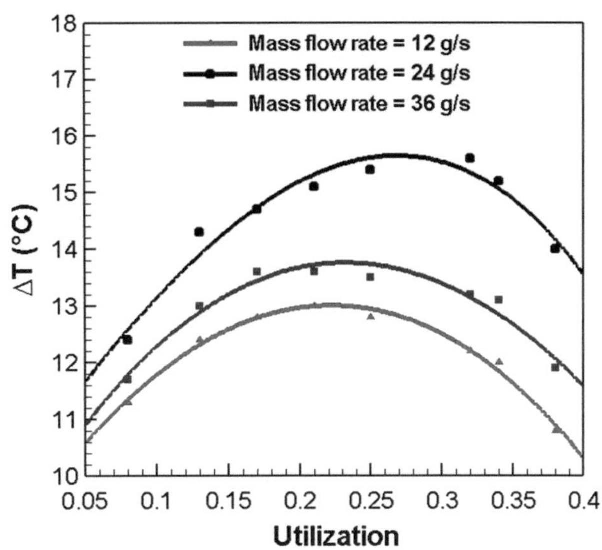

Fig. 16.7 Evolution of temperature span as function of utilisation for multilayer operating in ambient temperature of 20 °C

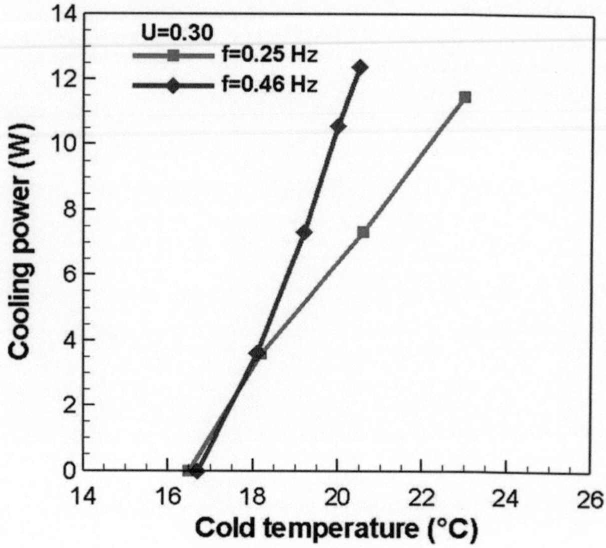

Fig. 16.8 Evolution of cooling power as function of cold temperature for frequency 0.26 and 0.48 Hz

single-material regenerator under certain operating conditions. The obtained experimental results using the demonstration unit enable to clarify the behaviour of the AMR cycle and the definition of the optimal operating parameters that are used to describe the performance of a magnetic refrigeration system.

Acknowledgments We further are grateful to the Direction Générale de l'Enseignement Supérieur du canton de Vaud, Switzerland, the Office fédéral de l'Energie (OFEN), Switzerland, INTERREG Iva France-Suisse programme, the Romande Energie RER, the Haute Ecole de Suisse Occidentale (HES-SO) and the Ministère de l'enseignement supérieur et la recherche scientifique of the Algerian government. We acknowledge their financial support.

References

1. Chiba Y, Smaïli A, Mahmed C, Balli M, Sari O (2014) Thermal investigations of an experimental active magnetic regenerative refrigerator operating near room temperature. Int J Refrig 37:36–42
2. Sari O, Balli M (2014) From conventional to magnetic refrigerator technology. Int J Refrig 37:8–15
3. Smaïli A, Ait-Ali S, Chahine R (2011) Performance predictions of a first stage magnetic hydrogen liquefier. Int J Hydrog Energy 36:4169–4177
4. Smaïli A, Chahine R (1998) Thermodynamic investigations of optimum active magnetic regenerators. Cryogenics 38:247–252
5. Romero Gómez J, Ferreiro Garcia R, Carbia Carril J, Romero Gómez M (2013) A review of room temperature linear reciprocating magnetic refrigerators. Renew Sustain Energy Rev 21:1–12
6. Yu B, Liu M, Egolf PW, Kitanovski A (2010) A review of magnetic refrigerator and heat pump prototypes built before 2010. Int J Refrig 33(6):1029–1060

7. Gschneidner KA Jr, Pecharski VK (2008) Thirty years of near room temperature magnetic cooling: where we are today and future prospects. Int J Refrig 31:945–961
8. Legait U, Guillou F, Kedous-Lebouc A, Hardy V, Almanza M (2014) An experimental comparison of four magnetocaloric regenerators using three different materials. Int J Refrig 37:147–155
9. Tušek J, Kitanovski A, Tomc U, Favero C, Poredoš A (2014) Experimental comparison of multi-layered La–Fe–Co–Si and single-layered Gd active magnetic regenerators for use in a room-temperature magnetic refrigerator. Int J Refrig 37:117–126
10. Engelbrecht K, Bahl CRH, Nielsen KK (2011) Experimental results for a magnetic refrigerator using three different types of magnetocaloric material regenerators. Int J Refrig 34:1132–1140
11. Richard MA, Rowe A, Chahine R (2004) Magnetic refrigeration: single and multi-material active magnetic regenerator experiments. J Appl Phys 95(4):2146–2150
12. Zimm C, Boeder A, Chell J, Sternberg A, Fujita A, Fujieda S, Fukamichi K (2006) Design and performance of a permanent-magnet rotary refrigerator. Int J Refrig 29:1302–1306
13. Rowe A, Tura A (2006) Experimental investigation of a three-material layered active magnetic regenerator. Int J Refrig 29:1286–1293
14. Okamura T, Yamada K, Hirano N, Nagaya S (2006) Performance of a room-temperature rotary magnetic refrigerator. Int J Refrig 29:1327–1331
15. Arnold DS, Tura A, Rowe A (2011) Experimental analysis of a two-material active magnetic regenerator. Int J Refrig 34:178–191
16. Engelbrecht KL, Nellis GF, Klein SA (2006) Predicting the performance of an active magnetic regenerator refrigerator used for space cooling and refrigeration. HVAC&R Res 12 (4):1077–1095
17. Aprea C, Greco A, Maiorino A (2011) A numerical analysis of an active magnetic regenerative refrigerant system with a multi-layer regenerator. Energy Convers Manag 52:97–107
18. Egolf PW, Vuarnoz D, Gravier L, Courret G (2012) The thermodynamics of devices with solid layered-bed magnetocaloric refrigerants. Int J Refrig 35:1506–1517
19. Balli M, Sari O, Mahmed C, Besson C, Bonhote P, Duc D, Forchelet J (2012) A pre-industrial magnetic cooling system for room temperature application. Appl Energy 98:556–561

Chapter 17
Evaluation of Thermal Properties of Refrigerant Clathrates with Additives

Sayem Zafar, Ibrahim Dincer, and Mohamed Gadalla

Abstract A modeling study is conducted to evaluate the heat transfer capabilities of novel refrigerant clathrate-based phase change materials with salts and nanoparticles as additives. The formation of refrigerant clathrates is studied for both active and passive cooling applications. In this regard, the refrigerants, e.g., R134a, R141b, and R32 clathrates are studied at different refrigerant mass fractions since the solubility of refrigerants, in water, change with change in temperature. The sodium chloride and magnesium nitrate hexahydrate are used as salt additives. The nanoparticles of pure aluminum, copper, and graphene are also studied to investigate the improvement in their thermal properties. Some empirical correlations are used to predict the thermal conductivities of refrigerant clathrates and the improvement with the addition of additives. The results show that an increase in refrigerant mass fraction lowers the thermal conductivity of the refrigerant clathrate but not extensively. The addition of salts results in a minor improvement in thermal conductivity. The inclusion of nanoparticles significantly improved the thermal conductivity of the phase change material. It is also obtained that adding the nanoparticles improves the thermal conductivity more than the salts. The specific heat capacity, however, was not generally improved by the nanoparticles as it depended on the additive used.

Keywords Thermal • Heat transfer • PCM • Refrigerant • Nanoparticle

Nomenclature

CFC	Chlorofluorocarbons
C_p	Specific heat capacity, J/kg K
D	Dipole moment, Debye

S. Zafar (✉) • I. Dincer
Faculty of Engineering and Applied Science, University of Ontario Institute of Technology, 2000 Simcoe Street North, Oshawa, ON, Canada L1H 7K4
e-mail: sayem.zafar@uoit.ca; ibrahim.dincer@uoit.ca

M. Gadalla
Department of Mechanical Engineering, American University of Sharjah, Sharjah, UAE
e-mail: mgadalla@aus.edu

I. Dincer et al. (eds.), *Progress in Clean Energy, Volume 1*,
DOI 10.1007/978-3-319-16709-1_17

f	Liquid fraction
HCFC	Hydro-chlorofluorocarbons
HFC	Hydrofluorocarbons
h	Volumetric enthalpy, J/m^3
k	Thermal conductivity coefficient, W/m K
M	Molar mass, g/mol
m	Mass, kg
PCM	Phase change material
Pr	Prandtl number
Q	Heat, J
T	Temperature, K
t	Time, s
TES	Thermal energy storage
x	Mass fraction

Greek Letters

α	Biot number
δ	Difference
ρ	Density, kg/m^3
λ	Specific latent heat of fusion, J/kg
β	Thermal diffusivity, m^2/s
φ	Volume fraction

Subscripts

b	Base fluid
c	Charging
cm	Critical
i	Initial
m	Mixture

17.1 Introduction

The ability to utilize energy was a great scientific achievement of the human recent past, and it continues to reform at a gradual pace. The recent global trend has changed from rudimentary energy production or rejection to precisely manage and absorb energy. The energy management is a challenge that needs to be dealt with to achieve the goal of sustainable growth. Thermal energy storage (TES) refers to the storage of heat by increasing or decreasing the temperature of a substance or by changing the phase of a substance [1]. TES is the temporary storage of high- or low-temperature energy for later use.

Refrigerant clathrate has recently appeared to be a promising way to store thermal energy for cooling applications. Clathrate is a solidified form of water that contains gas molecules in its molecular cavities [2, 3]. Clathrates form when water and gas combine under low temperatures and high pressures [4]. Since the phase change temperature of clathrate is above the freezing point of water, yet low enough to be used for comfort cooling, its use in air conditioning has been studied and found to be useful [5]. Refrigerant clathrates are considered more effective compared with other types of PCMs as they can be used through refrigerant loops and can be easily circulated [6, 7].

The clathrates have certain properties that make them a very good PCM for cold storage described as follows:

- They have high heat of fusion which means that they can store higher amount of energy.
- They have high density which means the storage size can be smaller per unit energy rate.
- The do not have added corrosive or toxicity, and hence normal refrigeration unit can be used to form clathrates of refrigerants.
- Since they require the mixture of water and refrigerant, it is cost effective and available in abundance.

Many chlorofluorocarbons (CFCs), hydrochlorofluorocarbons (HCFCs), and hydrofluorocarbons (HFCs) can form clathrates of refrigerant [8]. In order to be used for cold TES, an effective refrigerant clathrate should form at atmospheric pressure and temperature range between 278 and 285 K [9]. Several refrigerants form the clathrates, but only handful is commercialized.

Due to the stratospheric ozone layer depletion concerns, CFC clathrates are forbidden. This constrain leaves only the hydrochlorofluorocarbon and HFCs to be used for PCM. Refrigerant clathrates with R134a have been proposed as a PCM which demonstrated to be the most promising candidates for this goal [9]. Another refrigerant considered feasible for refrigerant clathrates is R141b due to its available engineering applications, low saturated vapor pressure, low cost, and low pressure character [10, 11]. Refrigerant R32 should also be of interest as it has low global warming potential, is accessible, and makes an effective PCM due to its high thermal conductivity, compared to its counterparts [12, 13].

Conventional PCMs, especially the ones based on refrigerant clathrates, have poor heat transfer properties. To further enhance the performance of refrigerant clathrates, to be effective PCMs, additives of different materials have been studied. For instance, adding calcium hypochlorite or benzenesulfonic acid sodium salt improved the cold energy storage capacity and the cold energy transfer rate of R141b-based clathrate [10]. For organic material, adding alcohol in R134a-based clathrate accelerates the cool storage rate and eliminates the floating clathrate during the hydration process [14].

To increase thermal conductivity, metallic nanoparticles have also been added to the existing heat transfer fluid. It has been reported that even a small fraction of nanoparticles of low thermal conductivity metallic oxides can favorably increase the thermal conductivity of pure substances, such as water [15, 16]. Even for organic compounds such as monoethylene glycol and paraffin fluids, copper oxide nanoparticles can improve the thermal conductivity [17]. The addition of pure copper nanoparticles in ethylene glycol increases the thermal conductivity by 40 % [18]. For the refrigerant hydrate, nanoparticles of copper are also studied which shows that the heat transfer increases with the addition of nanoparticles of copper [19].

Apart from thermal properties improvement, additives have other advantages to enhance the performance and usability of PCMs. The melting temperature of some refrigerant clathrates is a little higher than what is generally required for comfort, food, or electronics cooling. It is believed that this phase change temperature can be lowered by using additives such as salt, alcohol, and ethylene glycol to make clathrates more suitable for cooling applications [10]. This chapter describes a modeling study to investigate the effects of adding additives in the selected refrigerant clathrates on the enhancement of their thermal properties. Based on the preliminary results, the most appropriate refrigerant clathrates are proposed, and the use of suitable additives is also evaluated and discussed.

17.2 Analysis

To assess the heat to or from a PCM, an appropriate knowledge of heat of fusion and specific heat of both the phases is required. Predicting the behavior of phase change systems is difficult due to its inherent nonlinear nature at moving interfaces, for which displacement rate is controlled by the latent heat lost or absorbed at the boundary [20]. The governing equation for heat can be written as follows:

$$Q = (mC_P\delta T)_S + m\lambda + (mC_P\delta T)_l \tag{17.1}$$

where m is the mass of the PCM, C_P is the specific heat, T is the temperature, and λ is the heat of fusion of PCM. Subscripts s and l are for solid and liquid, respectively.

Here, the heat to or from a PCM depends on the heat of fusion, specific heat, and the change in temperature during sensible phase. Specific heat can be calculated by adding the product of the mass fraction and specific heat of the pure species present in the PCM:

$$C_{P,PCM} = x_1C_{P,1} + x_2C_{P,2} + x_3C_{P,3} \tag{17.2}$$

where $C_{P,i}$ is the heat capacity of species i and x is its mass fraction.

Another important parameter is the heat of fusion which can be described for a two-dimensional problem as [21]

$$\frac{\partial h}{\partial t} = \frac{\partial}{\partial x}\left(\beta \frac{\partial h_{\text{latent}}}{\partial x}\right) - \rho_L \lambda \frac{\partial f}{\partial t} \tag{17.3}$$

where "h" is the volumetric enthalpy, β is the thermal diffusivity, λ is the specific latent heat of fusion, "t" is the time, ρ is the density, and "f" is the liquid fraction described as follows:

$$f = \begin{cases} 0 & T < T_{\text{m}} \quad \text{solid,} \\ [0, 1] & T = T_{\text{m}} \quad \text{mushy,} \\ 1 & T > T_{\text{m}} \quad \text{liquid.} \end{cases} \tag{17.4}$$

The thermal diffusivity β is

$$\beta = \frac{k}{C_{\text{P}}\rho} \tag{17.5}$$

where "k" is the thermal conductivity coefficient and ρ is the density.

Substituting Eq. (17.5) into Eq. (17.3) yields the enthalpy of fusion with respect to thermal conductivity and density as follows:

$$\frac{\partial h}{\partial t} = \frac{\partial}{\partial x}\left[\left(\frac{k}{C_{\text{P}}\rho}\right)\frac{\partial h}{\partial x}\right] - \rho_L \lambda \frac{\partial f}{\partial t} \tag{17.6}$$

Note that PCM in solid phase has higher density as compared to liquid phase PCM while it is in the middle in the "mushy" region. To further simplify Eq. (17.6) to obtain a preliminary result, it could be assumed that the heat of fusion does not change with respect to time, hence treating it as steady state condition. With this assumption, the equation becomes

$$\rho_L \lambda \frac{\partial f}{\partial t} = \left(\frac{k}{C_{\text{P}}\rho}\right)\frac{\partial^2 h}{\partial x^2} \tag{17.7}$$

Assuming the change is linear with respect to time for "f" and distance for "h," Eq. (17.7) can be simplified to

$$\rho_L \lambda \frac{\Delta f}{\Delta t} = \left(\frac{k}{C_{\text{P}}\rho}\right)\frac{\partial^2 h}{\partial x^2} \tag{17.8}$$

Since it is difficult to predict the exact density during phase change, it is important to have the density eliminated with something that tends to stay constant during phase change. Unlike density, the mass of the material tends to stay the same so an expression can be arrived upon without density of the PCM.

Since "h" is the volumetric enthalpy, substituting mass per unit volume in place of density would yield the enthalpy value. The equation becomes

$$\rho_L \lambda \frac{\Delta f}{\Delta t} = \left(\frac{k}{C_{\mathrm{P}}m}\right)\frac{\partial^2 H}{\partial x^2} \tag{17.9}$$

where "m" is the mass and "H" is the volumetric enthalpy of fusion.

The thermal conductivity "k" for the refrigerant clathrate can be described as follows:

$$k = 0.4 \times \exp\left(\frac{T}{T_{\mathrm{cm}}} - 1\right)\left(\frac{D_1}{D_2}\right)^2 \times M^{0.5} x_1 x_2 \left(\frac{k_2}{M_2^{0.5}} - \frac{k_1}{M_1^{0.5}}\right) \tag{17.10}$$

where D is the dipole moment, x is the mass fraction, and M is the molar mass of the species.

Refrigerant and water properly mix to form the clathrate and both are in liquid phase when they are mixed. Adding solid additives, to improve thermal conductivity, will improve the thermal transport properties of the PCM. It requires a different set of equations to predict the thermodynamic properties of the PCM with solid additives, be it salts or nanoparticles.

The specific heat of the PCM containing solid particles can be described as [21]

$$\rho_{\mathrm{PCM}} \mathrm{Cp}_{\mathrm{PCM}} = (1 - \varphi)\rho_{\mathrm{f}} \mathrm{CP}_{\mathrm{f}} + \varphi \rho_{\mathrm{p}} \mathrm{CP}_{\mathrm{p}} \tag{17.11}$$

where $\mathrm{Cp_f}$ and $\mathrm{Cp_p}$ are the specific heats of a base fluid and nanoparticles, respectively. φ is the volume fraction of nanoparticles and ρ_{f} and ρ_{p} are the densities of a base fluid and nanoparticles, respectively. The bulk density of the PCM, ρ_{PCM}, can be described as

$$\rho_{\mathrm{PCM}} = (1 - \varphi)\rho_{\mathrm{f}} + \varphi \rho_{\mathrm{p}} \tag{17.12}$$

The thermal conductivity of the PCM having solid particles is not easy to calculate as it yields complex parameters upon which the values are based [18]. An effective model to present the improvement in thermal conductivity of fluids with nanoparticles is through single-phase Brownian model (SPBM) as follows [22]:

$$k = k_{\mathrm{c}}\left(1 + \frac{Re\,Pr}{4}\right)\frac{[k_{\mathrm{d}}(1+2\alpha) + 2k_{\mathrm{c}}] + 2\,V_{\mathrm{d}}[k_{\mathrm{d}}(1-\alpha) - k_{\mathrm{c}}]}{[k_{\mathrm{d}}(1+2\alpha) + 2k_{\mathrm{c}}] - V_{\mathrm{d}}[k_{\mathrm{d}}(1-\alpha) - k_{\mathrm{c}}]} \tag{17.13}$$

where the Biot number α is

$$\alpha = 2R_{\mathrm{b}}k_{\mathrm{c}}/d_{\mathrm{p}} \tag{17.14}$$

and the Prandtl number Pr is

$$Pr = \sqrt{\frac{Re}{Nu^2}} \tag{17.15}$$

The equation yields the thermal conductivity of the fluid with nanoparticles that incorporates the conduction contribution of the particles, particle–fluid thermal boundary resistance, and the convection contribution. Overall, Eq. (17.13) takes care of the localized convection due to Brownian motion as well, something that previous models failed to address. When the additives that are not nanoscale particles are used, the equation simplifies to only the static thermal conductivity as follows:

$$k = k_c \frac{[k_d(1+2\alpha)+2k_c]+2\,\varphi_d[k_d(1-\alpha)-k_c]}{[k_d(1+2\alpha)+2k_c]-\,\varphi_d[k_d(1-\alpha)-k_c]} \tag{17.16}$$

17.3 Results and Discussion

After the literature review, the three refrigerant candidates isolated are R141b, R134a, and R32. Several different additives are incorporated with the refrigerant clathrates to improve their thermal properties. Sodium chloride (NaCl) and magnesium nitrate hexahydrate ($Mg(NO_3)_2{\cdot}6H_2O$) are added in the refrigerant clathrates to evaluate the change in thermal conductivity and specific heat of the newly formed PCM. To develop PCMs with improved thermal conductivity, nanoparticles of highly conductive materials are added. Aluminum, copper, and graphene nanoparticles are selected as additives to study the improvement in the thermal conductivity of the clathrate based on the before-mentioned refrigerants. The thermal properties are determined for a variable fraction of refrigerants since their solubility changes with change in temperature [23, 24].

17.3.1 *Specific Heat*

The first thermal property to be calculated is the specific heat of the clathrate using Eq. (17.2). Specific heat capacity is the ability of the material to absorb the heat per unit change of its own temperature. It is desired to predict the specific heat as it helps determine the capacity of the PCM during the sensible heat storage temperature. The properties used to calculate the specific heat are listed in Table 17.1.

17.3.1.1 Effect of Salts on Specific Heat

The effect of salts on the specific heat capacity of the refrigerant clathrates is studied. It highlights the effect the salt has as an additive, on the refrigerant

Table 17.1 Material and their corresponding specific heat capacities used in the calculations

Material	Specific heat, J/kg K
R134a	850
R141b	1,160
R32	1,370
Water	4,200
Sodium chloride	880
Magnesium nitrate hexahydrate	4,700
Aluminum	9,100
Copper	3,900
Graphene	600

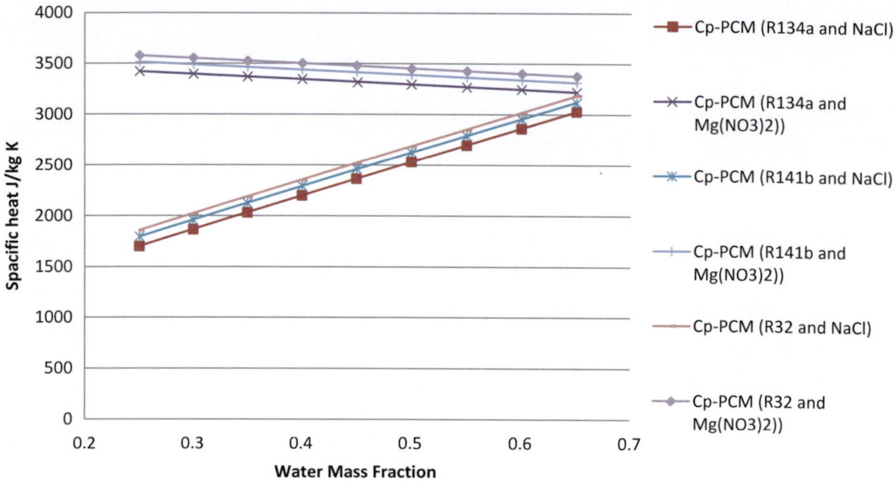

Fig. 17.1 Specific heat trend for different refrigerant clathrates at 30 %, with salts

clathrate's specific heat capacity. This section shows the overall specific heat capacity for each refrigerant-based clathrate with the two different salts studied.

Figure 17.1 shows the specific heat of R134a, R141b, and R32 clathrates with sodium chloride and magnesium nitrate hexahydrate. The results are shown for 30 % refrigerant clathrate. The graph shows the variation in specific heat with respect to water and salt mass fraction. It shows the linear increase in specific heat of sodium chloride-based PCM as the salt mass fraction decreases. This decrease is due to the lower specific heat of sodium chloride as compared to the clathrate. Clathrate based on magnesium nitrate hexahydrate shows linear increase in specific heat as the salt mass fraction increases due to its higher specific heat. The figure also shows that R32-based clathrates have the highest specific heat, followed by R141b and then R134a. This is simply due to the difference in their individual specific heats. It can be seen from the graph that using magnesium nitrate hexahydrate with the refrigerant clathrate yields significantly high specific heat at high salt mass fraction.

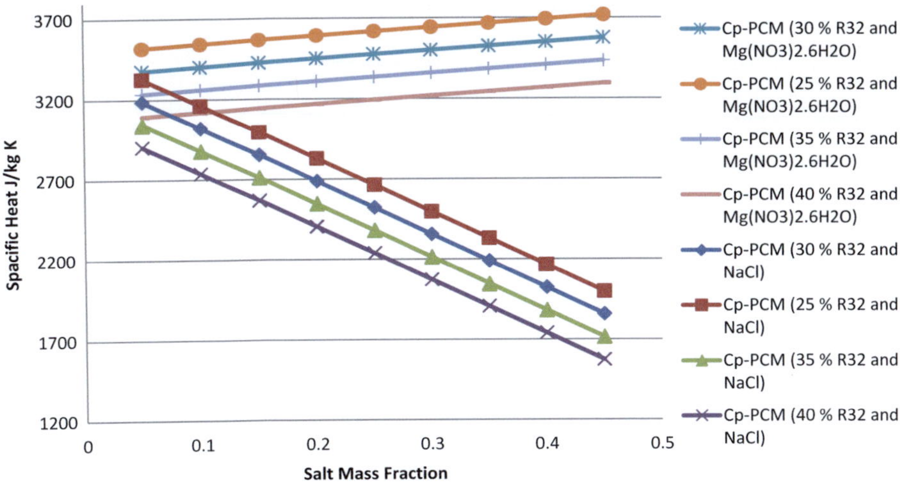

Fig. 17.2 R32 clathrate with salts at different refrigerant proportions

Figure 17.2 shows the change in specific heat of the PCM based on R32 refrigerant and salts. The variation is shown with respect to the change in salt and refrigerant mass fractions. The specific heat of the PCM increases with the increase in magnesium nitrate hexahydrate mass fraction and decreases with the increase in sodium chloride mass fraction. The increase in specific heat is due to the higher specific heat of magnesium nitrate hexahydrate whereas the decrease is due to the lower specific heat of sodium chloride, compared to the R32-based clathrate. The specific heat decreases with the increase in refrigerant mass fraction since the refrigerant has lower specific heat compared to water. Using sodium chloride as additive lowers the specific heat of the clathrate. As a TES material during sensible temperature period, sodium chloride would be a bad choice. Using magnesium nitrate hexahydrate would turn out to be a feasible choice as a sensible TES material. Adding salts in clathrate can improve the latent heat and vary the phase change temperature. These changes can only be observed by conducting the experiments.

Figure 17.3 shows the change in specific heat of the PCM based on R134a refrigerant and salts. The variation is shown with respect to the change in salt and refrigerant mass fractions. Similar to the other refrigerant-based PCMs, the specific heat increases with the increase in magnesium nitrate hexahydrate mass fraction and decreases with the increase in sodium chloride mass fraction. The specific heat decreases with the increase in refrigerant mass fraction since the refrigerant R134a has lower specific heat compared to water. Even for R134a-based PCMs, using magnesium nitrate hexahydrate would produce the PCM with better specific heat property as compared to sodium chloride.

Figure 17.4 shows the change in specific heat of the PCM based on R141b refrigerant and salts. The variation is shown with respect to the change in salt and refrigerant mass fractions. The specific heat trend is found to be the same as with the PCMs based

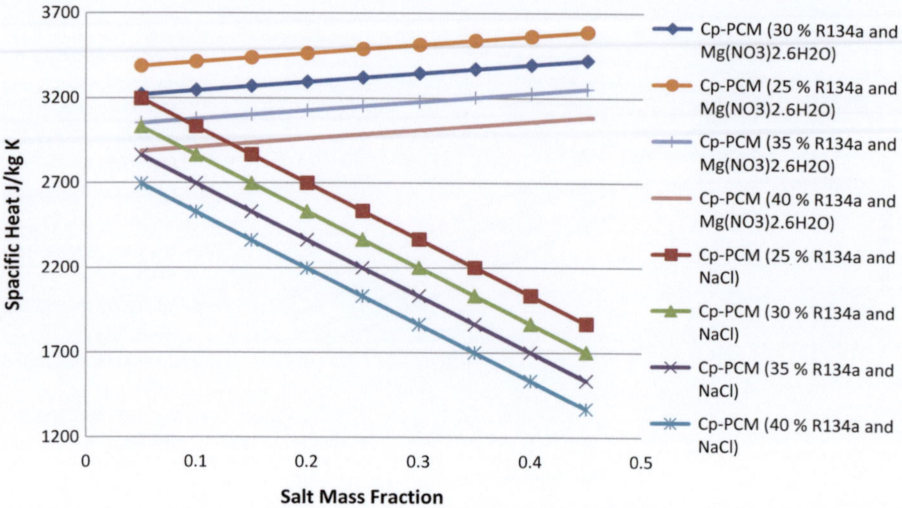

Fig. 17.3 R134a clathrate with salts at different refrigerant proportion

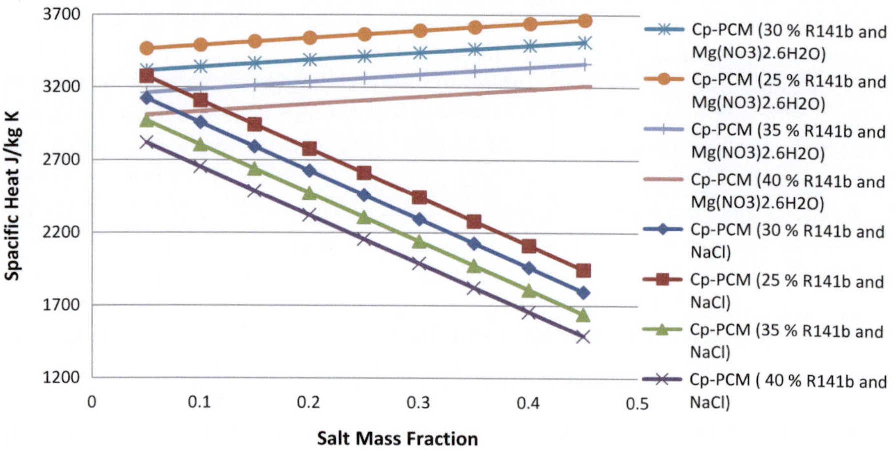

Fig. 17.4 R141b with salts at different refrigerant proportions

on R134a and R32. The specific heat increases with the increase in magnesium nitrate hexahydrate mass fraction and decreases with the increase in sodium chloride mass fraction. Once again, using magnesium nitrate hexahydrate would produce the PCM with better specific heat property as compared to sodium chloride.

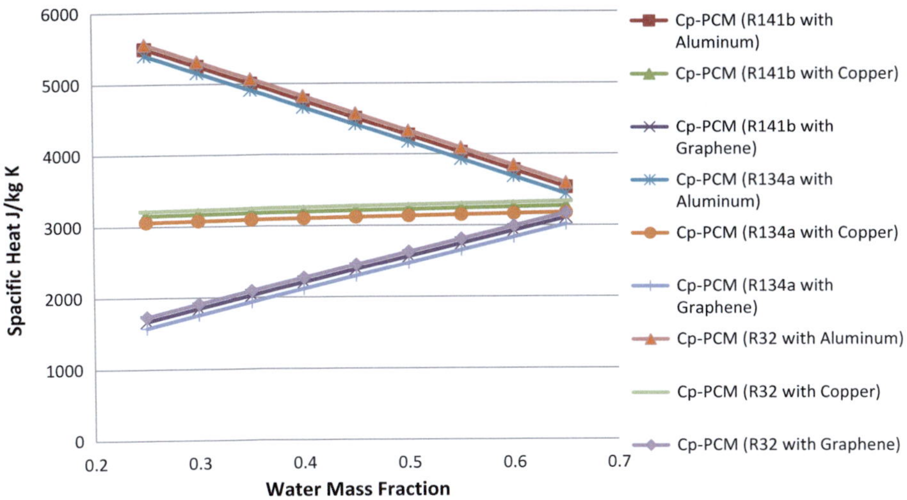

Fig. 17.5 Refrigerants at 30 % mass ratio clathrate with nanoparticles

17.3.1.2 Effect of Nanoparticles on Specific Heat

The effect of nanoparticles on the specific heat capacity of the refrigerant clathrates is also studied. This section presents the impact of each nanoparticle material on the refrigerant clathrate's specific heat. It also shows how much of an improvement can be achieved by using nanoparticles over salts.

Figure 17.5 shows the change in specific heat of the PCM based on studied refrigerants and nanoparticles. The variation is shown with respect to the change in water and nanoparticle mass fraction with 30 % refrigerant by mass. For PCM with graphene and copper nanoparticles, the specific heat decreases with the increase in nanoparticles' mass fraction. This trend is due to low specific heat of graphene and copper, compared to any of the studied refrigerant-based clathrates. For copper-based PCM, the decrease is not as steep as for graphene. PCM with aluminum nanoparticles, however, shows an increase in specific heat with the increase in mass fraction of aluminum. Among the refrigerants, R32 has the highest specific heat, followed by R141b and then R134a. The graph shows that PCM based on R32 with aluminum nanoparticles would serve effectively as a heat storage medium during sensible temperature region.

Figure 17.6 shows the change in specific heat of the PCM based on R32 and nanoparticles. The variation is shown with respect to the change in refrigerant percent mass fraction and nanoparticle mass fraction. As the nanoparticle mass fraction increases, the specific heat decreases for graphene and copper. For aluminum, the specific heat linearly increases. The specific heat is found to be higher when the refrigerant mass fraction is low. This is due to higher specific heat of water compared to R32. It can be seen from the graph presented in Fig. 17.6 that at low mass fraction of nanoparticles, refrigerant mass fraction has greater effect on the

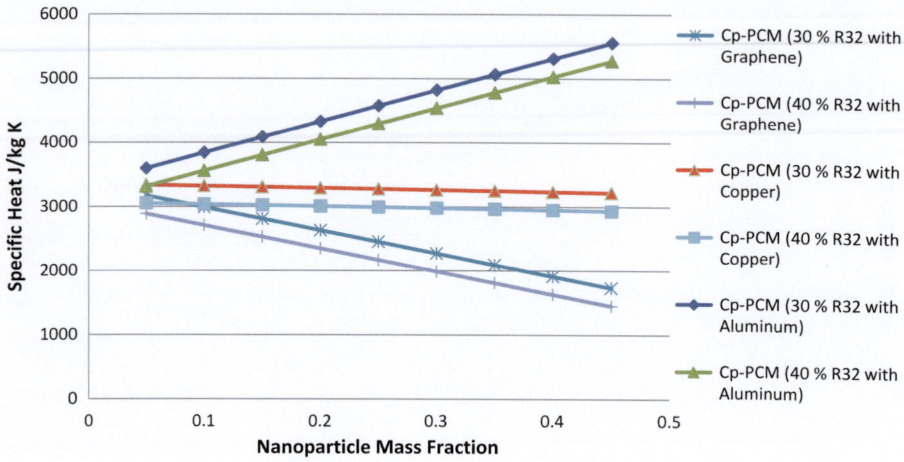

Fig. 17.6 R32 with nanoparticles at different refrigerant proportions

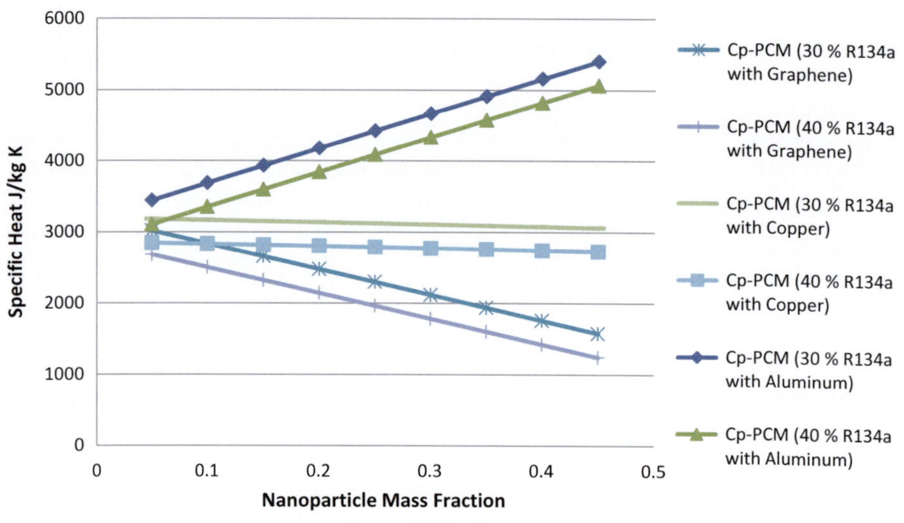

Fig. 17.7 R134a with nanoparticles at different refrigerant proportions

specific heat as compared to the nanoparticles' mass fraction. Figure 17.6 shows that aluminum with low R32 mass fraction appears to have the highest specific heat which increases with increase in nanoparticle mass fraction.

Figure 17.7 shows the change in specific heat of the PCM based on R134a and nanoparticles. The variation is shown with respect to the change in refrigerant percent mass fraction and nanoparticle mass fraction. As the nanoparticle mass fraction increases, the specific heat decreases for graphene and copper. For aluminum, the specific heat linearly increases with the increase in its mass fraction. The

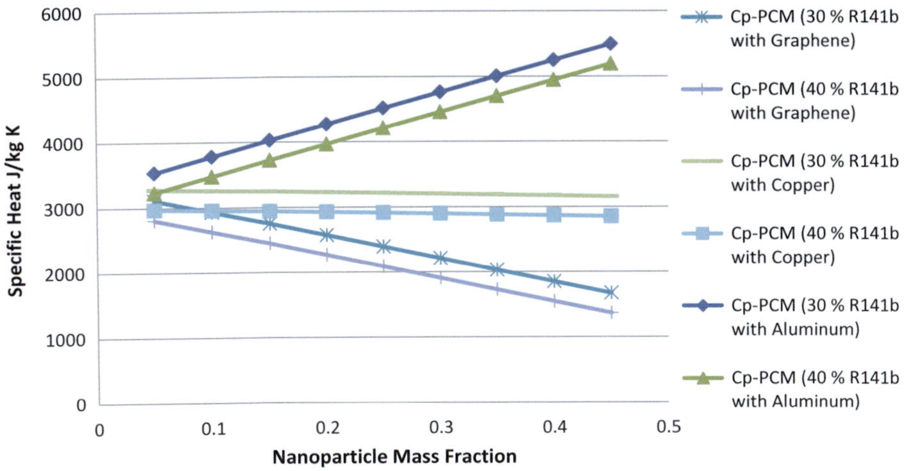

Fig. 17.8 R141b with nanoparticles at different refrigerant proportions

specific heat is again found to be higher when the refrigerant mass fraction is low due to higher specific heat of water compared to R134a. For low mass fraction of nanoparticles, the refrigerant mass fraction has greater effect on the specific heat as compared to the nanoparticles' mass fraction. Figure 17.7 shows that aluminum with low R134a mass fraction appears to have the highest specific heat which increases with increase in the nanoparticle mass fraction.

Figure 17.8 shows the change in specific heat of the PCM based on R141b and nanoparticles. The variation is shown with respect to the change in refrigerant percent mass fraction and nanoparticle mass fraction. The trend is found to be similar as with the R32- and R134a-based PCMs. As the nanoparticle mass fraction increases, the specific heat decreases for graphene and copper while for aluminum, the specific heat linearly increases. The specific heat is again found to be higher when the refrigerant mass fraction is low due to higher specific heat of water compared to R141b. Aluminum with low R141b mass fraction appears to have the highest specific heat which increases with increase in nanoparticle mass fraction.

17.3.2 Thermal Conductivity

Using Eq. (17.10), the thermal conductivity of the refrigerant clathrates is calculated. The solubility of the refrigerant in water changes with operating temperature; hence the thermal conductivity is determined for a variety of mass fractions over a range of operating temperatures. Properties used to calculate the thermal conductivity of the clathrate as taken from the literature sources [25–27] are listed in Table 17.2.

Figure 17.9 shows the change in thermal conductivity as the refrigerant mass fraction changes for R134a, R32, and R141b refrigerants. The thermal conductivity

Table 17.2 Parameters used to calculate the thermal conductivity of the refrigerant clathrates

	Water	R134a	R141b	R32
Critical temperature, T_c (K)	647	374	477	351
Thermal conductivity, k (W/m K)	0.58	0.092	0.093	0.155
Dipole moment, D	6.2 E−30 C m	4.85 E−30 C m	6.66 E−30 C m	2.8 E−30 C m
Molar mass, M (mol/g)	18	102	117	52

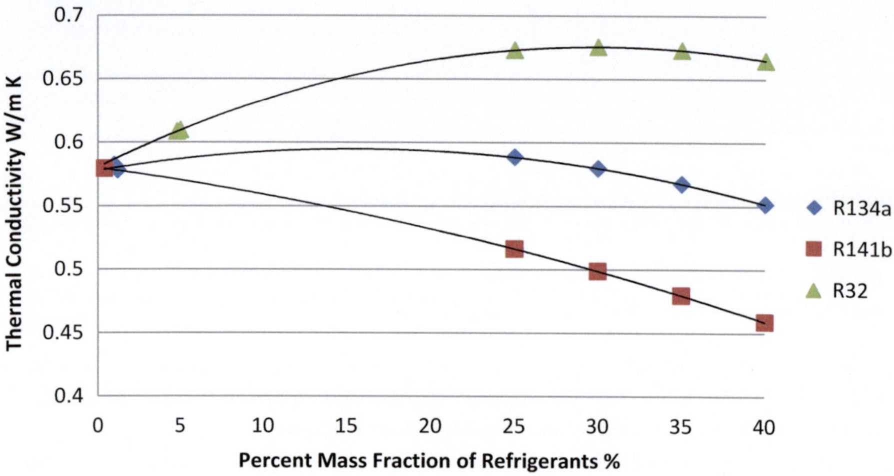

Fig. 17.9 Clathrate thermal conductivity with different refrigerants against change in refrigerant mass fraction

trend shown in Fig. 17.9 is for pure clathrate without any additive or nanoparticles. At low refrigerant mass fraction, the thermal conductivity does not vary much for the studied refrigerants. However, for high refrigerant mass fractions, distinct thermal conductivity starts to appear. Thermal conductivity for R141b-based clathrate is the lowest, followed by R134a. R32-based clathrate is found to have the highest thermal conductivity. This trend matches with the pure thermal conductivity of the discussed refrigerants. Graph in Fig. 17.9 also shows that the thermal conductivity linearly decreases as the mass fraction of R141b increases. Thermal conductivity for R134a-based clathrate also decreases as the mass fraction of R141b increases but the values are higher than R141b. For R32-based clathrate, the thermal conductivity increases and appears to stabilize near 30 % mass fraction of refrigerant. The parameters used to determine the thermal conductivities of the PCM are listed in Table 17.3.

Table 17.3 Parameters and their corresponding values used to calculate the thermal conductivities of the PCM

Parameter	Value
Thermal interface resistance—R_b	98 E−10 K m^2/W
Nanoparticle diameter—d_p	10 nm
Reynolds' number—*Re*	0.029
Nusselt number	1
Thermal conductivity—aluminum	210 W/m K
Thermal conductivity—copper	410 W/m K
Thermal conductivity—graphene	3,000 W/m K
Thermal conductivity—sodium chloride	6.5 W/m K
Thermal conductivity—magnesium nitrate hexahydrate	0.7 W/m K

17.3.2.1 Effect of Nanoparticles on Thermal Conductivity

The effect of nanoparticles, as additives, on thermal conductivity of the refrigerant clathrate is to be evaluated. The thermal conductivity improvement is important to establish as it helps select the most appropriate additive. The improvement in thermal conductivity results in reduced charging time and greater heat transfer rate during discharge. Using Eq. (17.13), the thermal conductivity of the PCM containing refrigerant clathrate with nanoparticles is calculated.

Figure 17.10 shows the variation in thermal conductivity of the PCM with different refrigerants and aluminum nanoparticles. The thermal conductivity is studied over a range of nanoparticle volume fraction. The figure shows the exponentially increasing trend in thermal conductivity of the PCM containing aluminum nanoparticles. The graph also shows that R32-based PCM produces the highest thermal conductivity due to the high thermal conductivity of the refrigerant. R141b-based PCM yields the lowest thermal conductivity.

Figure 17.11 shows the variation in thermal conductivity of the PCM with copper nanoparticles and different refrigerants. The thermal conductivity is studied over a range of nanoparticle volume fraction. It also shows the exponentially increasing trend in thermal conductivity of the PCM containing copper nanoparticles. The graph also shows that R32-based PCM produces the highest thermal conductivity due to the high thermal conductivity of the refrigerant. R141b-based PCM yields the lowest thermal conductivity.

Figure 17.12 shows the variation in thermal conductivity of the PCM with copper nanoparticles and different refrigerants. The thermal conductivity is studied over a range of nanoparticle volume fraction. The graph shows the exponentially increasing trend in thermal conductivity of the PCM containing copper nanoparticles. The graph also shows that R32-based PCM produces the highest thermal conductivity due to the high thermal conductivity of the refrigerant. R141b-based PCM yields the lowest thermal conductivity.

Figure 17.13 shows the improvement in thermal conductivity after using the selected nanoparticles for different refrigerant mass fractions. The graph also shows

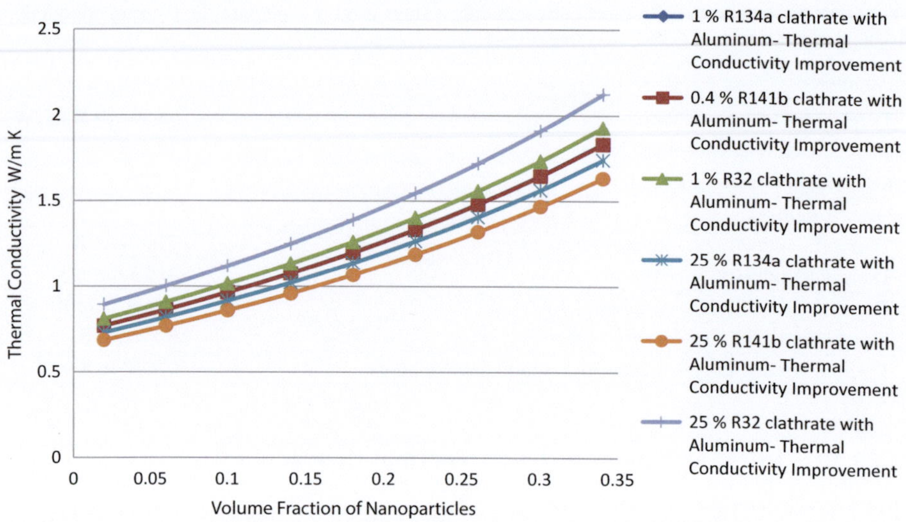

Fig. 17.10 Thermal conductivity of refrigerant clathrate with aluminum nanoparticles

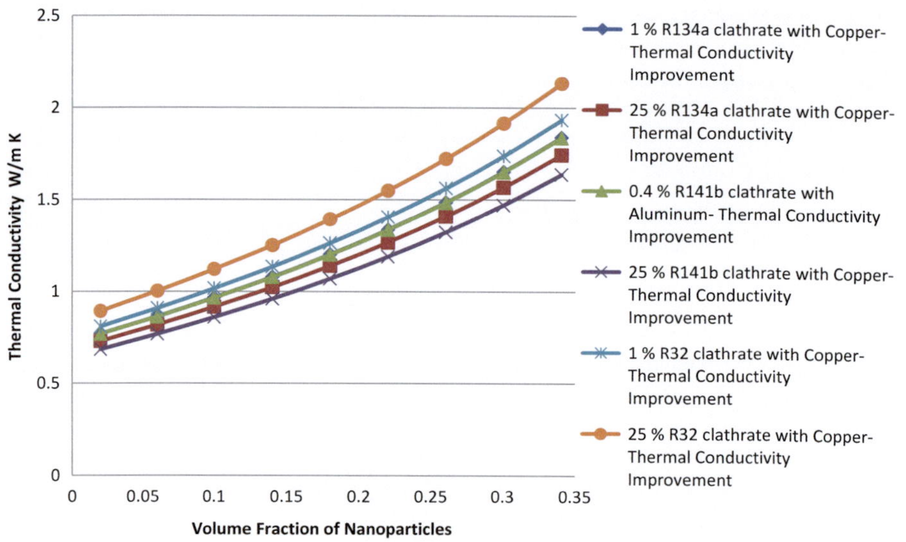

Fig. 17.11 Thermal conductivity of refrigerant clathrate with copper nanoparticles

the baseline thermal conductivity of R141b-based clathrate. The graph shows that the inclusion of nanoparticles improves the thermal conductivity of the PCM. Graphene improves the thermal conductivity the most, followed by copper and then aluminum. Although the thermal conductivity difference in pure species of nanoparticle is significant, their use as additives in refrigerant clathrate does not yield a significant increase in the thermal conductivity. For R141b clathrates, the difference in thermal

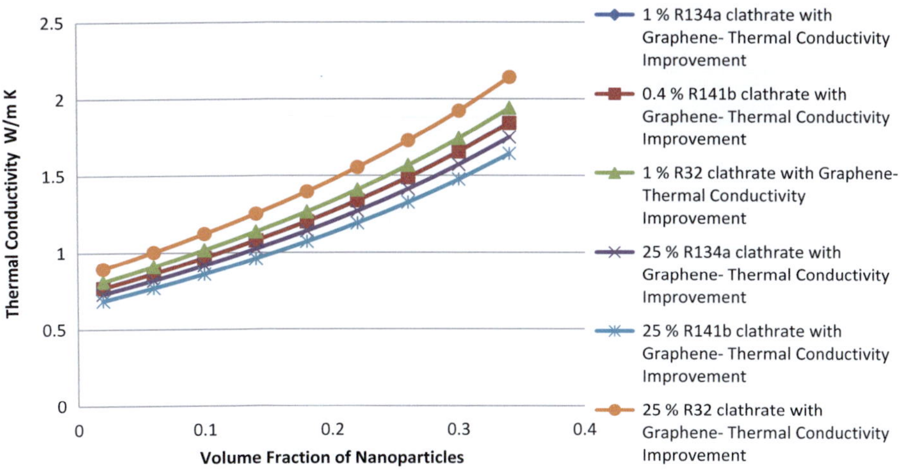

Fig. 17.12 Thermal conductivity of refrigerant clathrate with graphene nanoparticles

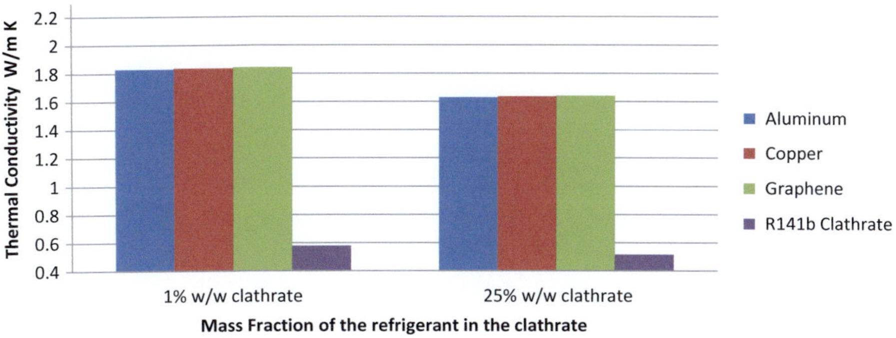

Fig. 17.13 Thermal conductivity comparison for different nanoparticles in R141b-based clathrate

conductivities is not large for different refrigerant mass fractions while low refrigerant mass fraction yields slightly higher thermal conductivity. The reason for improved thermal conductivity with small refrigerant fraction is the low thermal conductivity of R141b refrigerant.

Figure 17.14 shows the variation in thermal conductivity for PCM with aluminum, copper, and graphene nanoparticles for different refrigerant mass fractions. The graph also shows the baseline thermal conductivity of R32-based clathrate. The graph shows that the inclusion of nanoparticles improves the thermal conductivity of the PCM in which graphene improves the thermal conductivity the most, followed by copper and then aluminum. Although the thermal conductivity difference in pure species of nanoparticle is significant, their use as additives in refrigerant clathrate does not yield a significant increase in the thermal conductivity. For R32 clathrates, high refrigerant mass fraction yields slightly higher thermal conductivity for the PCM due to high thermal conductivity of R32 refrigerant.

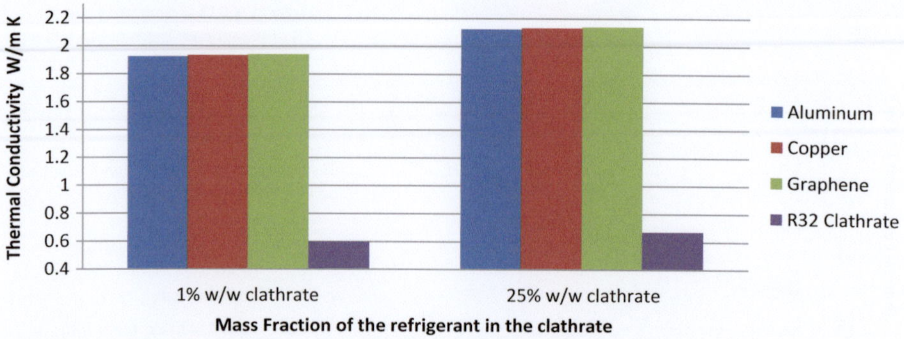

Fig. 17.14 Thermal conductivity comparison for different nanoparticles in R32-based clathrate

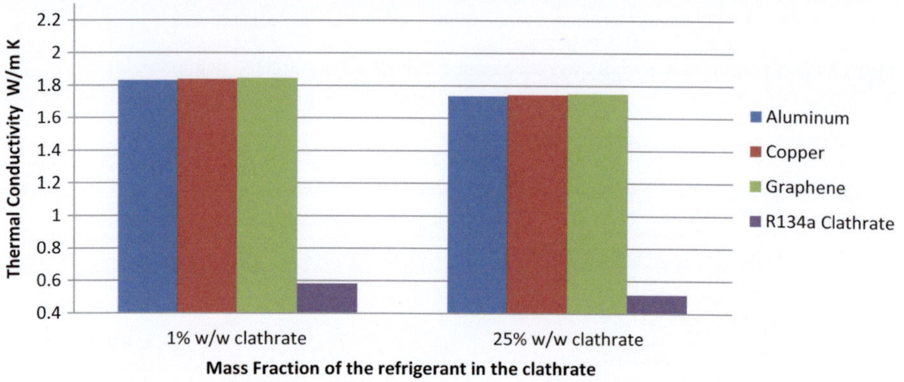

Fig. 17.15 Thermal conductivity comparison for different nanoparticles in R134a-based clathrate

Figure 17.15 shows the improvement in thermal conductivity after using the selected nanoparticles for different refrigerant mass fractions. The graph also shows the baseline thermal conductivity of R134a-based clathrate. The graph shows that the inclusion of nanoparticles in R134a-based clathrate improves the thermal conductivity of the PCM. Graphene improves the thermal conductivity the most, followed by copper and then aluminum. For R134a-based PCM, low refrigerant mass fraction yields slightly higher thermal conductivity. The reason for improved thermal conductivity with small refrigerant fraction is the low thermal conductivity of R134a refrigerant.

17.3.2.2 Effect of Salts on Thermal Conductivity

The effect of salts, as additives, on thermal conductivity is to be determined. The prediction of thermal conductivity for clathrates with salts is slightly different. For salt particles, the Brownian thermal conductivity part does not exist; hence the

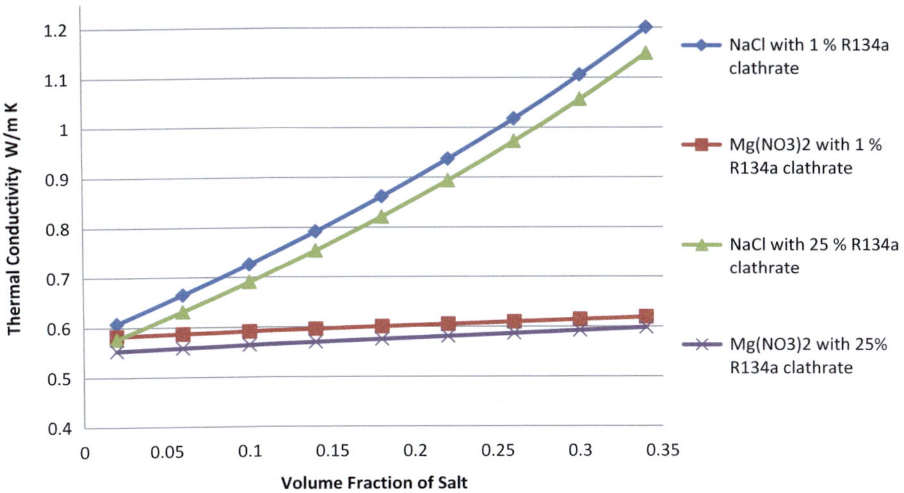

Fig. 17.16 Thermal conductivity of R134a clathrates with salts

equation does not incorporate that component. Using Eq. (17.16), the thermal conductivity of the PCM containing refrigerant clathrate with salts is calculated.

Figure 17.16 shows the thermal conductivity of R134a clathrate with salt additives against a wide range of salt volume fractions. The salts studied are sodium chloride and magnesium nitrate hexahydrate. Thermal conductivity of both the studied salts is higher; hence they both improved the thermal conductivity of the formed PCM. However, sodium chloride improved the thermal conductivity significantly more than magnesium nitrate hexahydrate. The thermal conductivity increases with the increase in volume fraction of salts.

Figure 17.17 shows the thermal conductivity of R141b clathrate with salt additives against a wide range of salt volume fractions. The salts studied are sodium chloride and magnesium nitrate hexahydrate. Thermal conductivity of both the studied salts is higher; hence they both improved the thermal conductivity of the formed PCM. However, sodium chloride improved the thermal conductivity significantly more than magnesium nitrate hexahydrate. The thermal conductivity rises with an increase in volume fraction of salts.

Figure 17.18 shows the thermal conductivity of R32 clathrate with salt additives against a wide range of salt volume fractions. The salts studied are sodium chloride and magnesium nitrate hexahydrate. Thermal conductivity of both the studied salts is higher; hence they both improved the thermal conductivity of the formed PCM. However, sodium chloride improved the thermal conductivity significantly more than magnesium nitrate hexahydrate. The thermal conductivity increases with the increase in volume fraction of salts.

Figure 17.19 shows the change in thermal conductivity with respect to the change in salt volume fraction. It shows that sodium chloride with R32 refrigerant

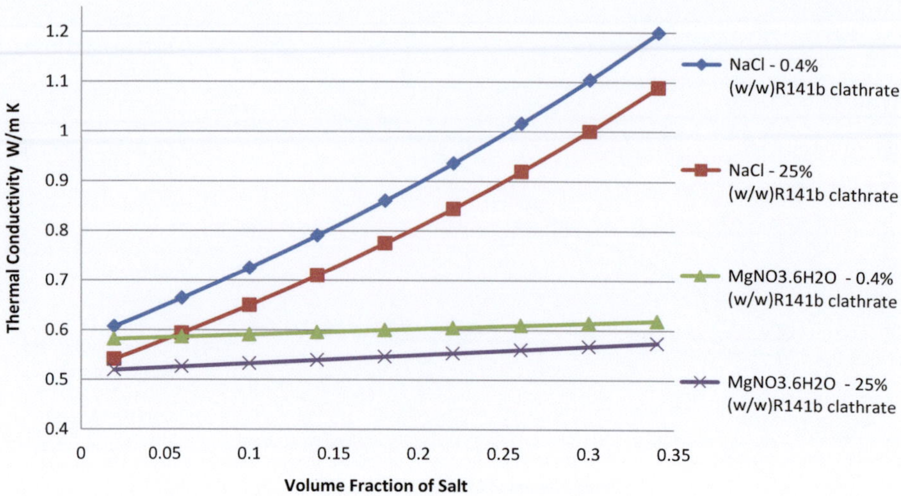

Fig. 17.17 Thermal conductivity of R141b clathrates with salts

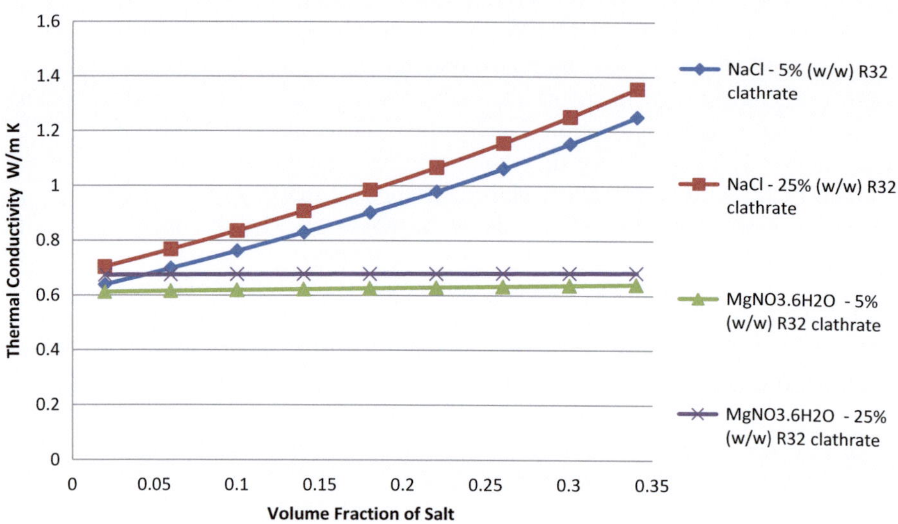

Fig. 17.18 Thermal conductivity of R32 clathrates with salts

has the highest thermal conductivity. The lowest thermal conductivity is with R141b and magnesium nitrate hexahydrate. Thermal conductivity with sodium chloride linearly increases with the increase in salt volume fraction. For magnesium nitrate hexahydrate, the thermal conductivity remains almost constant.

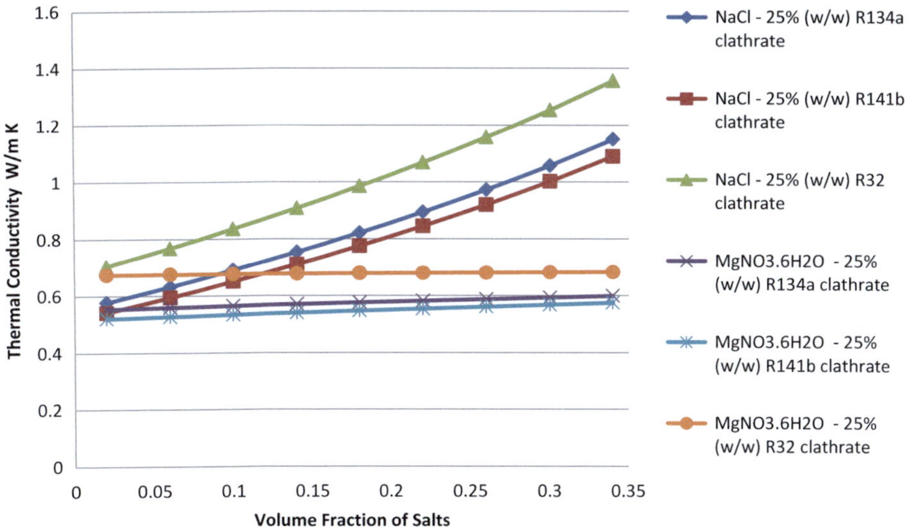

Fig. 17.19 Variation of thermal conductivity for different refrigerants for studied salts

17.4 Conclusions

This chapter looks at the possible candidates to be used as additives to improve the thermal properties of refrigerant-based clathrates. It also presents a model study to predict the thermal conductivities and specific heat capacities of the refrigerant clathrates with the possible additives. A theoretical model is presented in the chapter to determine the thermal conductivity and specific heat capacity. Thermal conductivity prediction is found to be based on empirical equations which vary for salts and nanoparticles. A parametric study is conducted to investigate the effect of mass fraction of refrigerant, salts, and nanoparticles on the specific heat capacity of the PCM. The parametric study for thermal conductivity is also conducted to see the impact of change in volume fraction of additives and mass fraction of refrigerants.

Among the refrigerant clathrates at 25 % refrigerant by mass, R32-based clathrates have been found to have the highest thermal conductivity which is 0.675 W/m K, followed by R134a at 0.59 W/m K and then R141b at 0.51 W/m K. With the addition of nanoparticles, the thermal conductivities of R32 clathrate improved to 2.14, 2.13, and 2.12 W/m K for graphene, copper and aluminum respectively. A similar trend is found with R134a- and R141b-based clathrates when nanoparticles are added but the thermal conductivities are lower than that of R32-based PCM. Although the difference in thermal conductivities of pure species of nanoparticles is significant, the predicted model does not show the same difference when used as additive. For salt additives, sodium chloride yielded a thermal conductivity of 1.35 W/m K while magnesium nitrate hexahydrate yielded 0.68 W/m K. The other thermal property that is analyzed is the specific heat capacity. R32 again has the best specific heat capacity followed by R141b and then R134a. Among salts, sodium chloride improved the specific heat capacity while magnesium nitrate hexahydrate lowered the specific heat of the PCM. Among nanoparticles, aluminum has the best specific heat followed by graphene and then copper. It can be safely concluded that adding metal refrigerants improve the thermal conductivity of refrigerant clathrates. The improvement in specific heat, however, depends solely on the specific heat capacity of the pure species.

References

1. Dincer I, Rosen MA (2002) Thermal energy storage—systems and applications. Wiley, Chichester
2. George A (1989) Hand book of thermal design. In: Guyer C (ed) Phase change thermal storage materials, chap 1. McGraw Hill, New York
3. Sloan ED (1990) Clathrate hydrates of natural gases. Marcel, New York
4. Nikbakhta F, Izadpanaha AA, Varaminianb F, Mohammadic AH (2012) Thermodynamic modeling of hydrate dissociation conditions for refrigerants R-134a, R-141b and R-152a. Int J Refrig 35:1914–1920
5. Mori YH, Isobe F (1991) A model for gas hydrate formation accompanying direct-contact evaporation of refrigerant drops in water. Int Commun Heat Mass Transfer 18:599–608
6. Bi YH, Guo TW, Zhu TY, Fan SS, Liang DQ, Zhang L (2004) Influence of volumetric-flow rate in the crystallizer on the gas-hydrate cool-storage process in a new gas-hydrate cool-storage system. Appl Energy 78:111–112
7. Inaba H (2000) New challenge in advanced thermal energy transportation using functionally thermal fluids. Int J Therm Sci 39:991–1003
8. Eslamimanesha A, Mohammadia AH, Richona D (2011) Thermodynamic model for predicting phase equilibria of simple clathrate hydrates of refrigerants. Chem Eng Sci 66:5439–5445
9. Guo KH, Shu BF, Zhang Y (1996) Transient behavior of energy charge–discharge and solid–liquid phase change in mixed gas-hydrate formation. In: Wang BX (ed) Heat transfer science and technology. Higher Education Press, Beijing, pp 728–733
10. Bi Y, Guo T, Zhu T, Zhang L, Chen L (2006) Influences of additives on the gas hydrate cool storage process in a new gas hydrate cool storage system. Energy Convers Manag 47:2974–2982
11. Bi Y, Guo T, Zhang L, Zhang H, Chen L (2009) Experimental study on cool release process of gas-hydrate with additives. Energy Build 41:120–124
12. Engineering ToolBox (2013) Refrigerants–ozone depletion (ODP) and global warming potential (GWP). http://www.engineeringtoolbox.com/Refrigerants-Environment-Properties-d_1220.html
13. Ro ST, Kim JY, Kim DS (1995) Thermal conductivity of R32 and its mixture with R134a. Int J Thermophys 16:1193–1201
14. Wua J, Wangb S (2012) Research on cool storage and release characteristics of R134a gas hydrate with additive. Energy Build 45:99–105
15. Murshed SMS, Leong KC, Yang C (2009) A combined model for the effective thermal conductivity of nanofluids. Appl Therm Eng 29:2477–2483
16. Duangthongsuk W, Wongwises S (2009) Measurement of temperature dependent thermal conductivity and viscosity of TiO_2-water nanofluids. Exp Therm Fluid Sci 33:706–714
17. Moghadassi AR, Hosseini SM, Henneke DE (2010) Effect of CuO nanoparticles in enhancing the thermal conductivities of monoethylene glycol and paraffin fluids. Ind Eng Chem Res 49:1900–1904

18. Eastman JA, Choi SUS, Li S, Yu W, Thompson LJ (2001) Anomalously increased effective thermal conductivities of ethylene glycol-based nanofluids containing copper nanoparticles. Appl Phys Lett 78:718–720
19. Lia J, Liangb D, Guob K, Wangc R, Fanb S (2006) Formation and dissociation of HFC134a gas hydrate in nano-copper suspension. Energy Convers Manag 47:201–210
20. Dutil Y, Rousse DR, Salah NB, Lassue S, Zalewski L (2011) A review on phase-change materials: mathematical modeling and simulations. Renew Sustain Energy Rev 15:112–130
21. Phelan PE, Bhattacharya P, Prasher RS (2005) Nanofluids for heat transfer applications. Annu Rev Heat Transf 14:255–275
22. Prasher R, Bhattacharya P, Phelan PE (2006) Brownian-motion-based convective-conductive model for the effective thermal conductivity of nanofluids. J Heat Transf 128:588–595
23. Hernandez O (2001) SIDS initial assessment report for 12th SIAM. UNEP Publications, Paris
24. DuPont™ Suva® refrigerants (2004) Mutual solubility of select HCFCs and HFCs and water. DuPont, USA
25. Poolen LJV, Holcom CD, Niesen VG (1997) Critical temperature and density from liquid-vapor coexistence data: application to refrigerants R32, R124, and R152a. Fluid Phase Equilib 129:105–111
26. Papadaki M, Schmitt M, Seitz A, Stephan K, Taxis B, Wakeham WA (1993) Thermal conductivity of R134a and R141b within the temperature range 240–307 K at the saturation vapor pressure. Int J Thermophys 14:173–181
27. Benedito JCC, Guedes RC, Pai-Panandikerb RS, Castrob CAN (2001) Hydrogen bonding and the dipole moment of hydrofluorocarbons by density functional theory. Phys Chem Chem Phys 3:4200–4207

Chapter 18
Refrigeration System Optimization for Drinking Water Production Through Atmospheric Air Dehumidification

Marco Bortolini, Mauro Gamberi, Alessandro Graziani, and Francesco Pilati

Abstract Drinking water availability is one of the emerging challenges of the twenty-first century. Different technologies are investigated as possible sources of water for the arid regions. Atmospheric water vapor processing is a developing approach whose aim is to cool air to condensate the water present in the atmospheric moisture. Air dehumidification allows obtaining pure drinking water for geographical regions far from sea, rivers, and lakes.

This chapter presents the optimization of a refrigeration system for drinking water production through atmospheric air dehumidification. The system uses a fan to force the air through a heat exchanger, in which it is cooled. The water vapor condensates on the cooled heat exchanger surfaces and it is collected by gravity in a tank.

The system's aim is to condensate the maximum water quantity achievable for every atmospheric air condition, represented by temperature, humidity, and pressure. Thus, a mathematical model is defined to determine the optimal atmospheric air flow that maximizes the condensed water production for every atmospheric air condition. Furthermore, to consider the atmospheric condition hourly profiles of the refrigeration system installation site, three air flow control strategies are proposed: hourly, monthly, and yearly. An experimental campaign is set up to validate the model. Experimental test results show that it accurately predicts the drinking water production (gap between −5.6 and +4.1 %). Finally, the case study of a refrigeration system installed in Dubai, United Arab Emirates, is presented to assess and compare the proposed three air flow control strategies.

M. Bortolini
Department of Industrial Engineering, University of Bologna,
Viale Risorgimento 2, Bologna 40136, Italy
e-mail: marco.bortolini3@unibo.it

M. Gamberi (✉) • A. Graziani • F. Pilati
Department of Management and Engineering, University of Padua,
Stradella San Nicola 3, Vicenza 36100, Italy
e-mail: mauro.gamberi@unipd.it; alessandro.graziani6@unibo.it; pilati@gest.unipd.it

I. Dincer et al. (eds.), *Progress in Clean Energy, Volume 1*,
DOI 10.1007/978-3-319-16709-1_18

Keywords Drinking water production • Atmospheric water vapor processing • Refrigeration system • Air flow optimization • Experimental campaign

Nomenclature

Latin Letters

AP	Inflow pipe section, m^2
c	Specific heat, J/kg°C
f	Conversion factor
F	Bypass factor
M	Mass, g
m	Molar mass, g/mol
n	Moles number, mol
P	Pressure, mbar
Q	Heat, W
$\dot{q}$	Condensed water flow, L/h
R	Universal gas constant, J/K mol
r	Evaporation latent heat, kJ/kg
RP	Refrigeration power, W
S	Air flow speed, m/s
T	Temperature, °C
V	Air and water vapor mixture volume
$\dot{v}$	Volumetric air flow, m^3/h

Greek Letters

ρ	Density, kg/m^3
ω	Absolute humidity, $g_{H_2O}/kg_{dry\ air}$

Subscripts

a	Atmospheric state
air	Air
c	Post-evaporator state
e	Heat exchanger exit state
H_2O	Water
l	Latent
s	Sensible

Superscripts

h	Hour
i	Month
*	Best value achievable
min	Minimum value admitted
tot	Total value
y	Year
Δ	Increment
0	Initial value

18.1 Introduction and Literature Review

The United Nations define the improvement of drinking water access as one of the Millennium Development Goals. Nowadays, about 768 million people around the world have no access to any source of drinking water [1], and by 2025 the number of people that will live in water-stressed countries is expected to rise up to three billion [2]. North Africa and the Middle East, in particular, have to face with severe water shortage. Fifteen percent of their population has no access to any source of clean and fresh safe water [3].

Several technologies, decentralized small-scale plants, are designed to produce drinking water in arid regions [4] as presented in the following list:

- Rainwater harvesting is the technique adopted to collect rainwater from rooftops or land surfaces and store it in natural or artificial reservoirs [5].
- Desalination removes salt and other minerals from saline water to produce drinking water [6]. Two are the processes traditionally exploited. Evaporation distils seawater using a heat source, whereas membrane process separates the drinking water from a saline concentrate [7].
- Atmospheric water vapor processing (AWVP) is the technique to extract water molecules from the atmospheric moisture, exploiting the water phase change from vapor to liquid [8]. Compared to the aforementioned technologies, AWVP represents a significant opportunity to produce drinking water even in regions far from natural water basins or affected by sporadic rainfalls [9]. The water contained in the earth atmospheric moisture is over 12.9 billion cubic meter [10, 11].

Considering AWVP, three are the processes traditionally adopted to produce drinking water from the atmospheric moisture:

- Atmospheric vapor concentration uses desiccants to extract water from air by means of vapor pressure reduction that determines a flow of water molecules toward the desiccant surface [12]. The desiccants are classified in absorbents and adsorbents [13]. The former change chemically or physically during the water molecule absorption, and the latter do not [14].

- Atmospheric controlled convection purpose is to induce air currents in tall tower structures from high altitude where the condensation occurs [15].
- Atmospheric moisture condensation forces an air flow to pass through surfaces cooled by a refrigeration system. The air temperature reduction below the dew point determines the atmospheric moisture condensation. Thus, the produced water is collected in tanks, and it is sterilized using ultraviolet irradiation to kill the microorganisms. Finally, the required minerals are added to obtain drinking water.

Several authors investigate the relation between the water production and the atmospheric air conditions, namely, temperature, pressure, and relative or absolute humidity, for atmospheric moisture condensation processes. Milani et al. [16] propose a relation between the atmospheric air conditions at the entrance of their thermoelectric cooler and the amount of condensable water. Scrivani and Bardi [17] evaluate the quantity of drinking water that can be obtained by a solar-powered system in Morocco, Jordan, and Lebanon. Nevertheless, both these authors do not investigate how the refrigeration system operating parameters affect the water production. Carrington and Liu [18] and Khalil [19] suggest that the air flow value significantly affects the condensed water quantity. Jradi et al. [20] investigate the relation between the air flow and the water production for a thermoelectric system installed in Lebanon. Habeebullah [21] estimates the condensed water quantity for different air flows for a refrigeration system installed in Saudi Arabia. However, the optimal air flow value suggested by both the authors is constant during the operating period. Thus, it does not consider the atmospheric air condition variations during the daily hours and the months.

This chapter proposes a mathematical model to determine the optimal air flow feeding a refrigeration system for AWVP that maximizes the condensed water production considering the atmospheric air conditions, namely, temperature, pressure, and relative or absolute humidity. Furthermore, to consider the atmospheric condition hourly profiles of the refrigeration system installation site, three air flow control strategies are presented: hourly, monthly, and yearly. The mathematical model is validated through an experimental campaign adopting a refrigeration system to produce drinking water and able to simulate different atmospheric air conditions. The validated model is used by the proposed control strategies to evaluate the water production of a refrigeration system installed in Dubai, United Arab Emirates. The three control strategies are compared to determine the most effective one.

According to the introduced topic and purposes, the remainder of this chapter is organized as follows: Sect. 18.2 proposes the AWVP mathematical model. Section 18.3 introduces the atmospheric air flow control strategies, while Sect. 18.4 presents the experimental campaign to validate the mathematical model, together with the refrigeration system and the test procedure. Section 18.5 illustrates the experimental campaign results and the optimal air flows determined by the AWVP model. Section 18.6 evaluates and compares the proposed air flow control strategies through the introduced case study. Finally, Sect. 18.7 presents the chapter conclusions and suggests further research opportunities.

18.2 AWVP Mathematical Model

The process to produce drinking water by means of atmospheric moisture condensation is described in this section, together with its mathematical formulation.

A volumetric flow $\dot{v}_a$ of air at atmospheric conditions defined by temperature, T_a; pressure, P_a; and absolute humidity, ω_a (or relative humidity, φ_a), is forced by a fan to pass through a cooled heat exchanger. The system refrigeration power, *RP*, is used to remove the sensible, Q_s, and the latent, Q_l, heat from the air flow Eq. (18.1).

$$RP = Q_s + Q_l \tag{18.1}$$

Q_s represents the refrigeration power component required to lower the air temperature from atmospheric conditions T_a to post-condensation one T_c Eq. (18.2), whereas Q_l accounts for the refrigeration power necessary for air dehumidification from ω_a to ω_c absolute humidity Eq. (18.3) and the related condensed water production $\dot{q}_c$ Eq. (18.4).

$$Q_s = \dot{v}_a \cdot \rho_{air} \cdot c_{air} \cdot (T_a - T_c) \tag{18.2}$$

$$Q_l = \dot{v}_a \cdot \rho_{air} \cdot r_{H_2O} \cdot (\omega_a - \omega_c) \tag{18.3}$$

$$\dot{q}_c = \dot{v}_a \cdot \rho_{air} \cdot (\omega_a - \omega_c) \tag{18.4}$$

The air at post-condensation state is in saturation condition. The relation between the temperature T_c and the absolute humidity ω_c of air in post-condensation state is determined exploiting Buck experimental equation [22]. Buck experimental Eq. (18.5), water vapor and dry air equation of state Eqs. (18.6) and (18.7), and Dalton's law Eq. (18.8) are necessary to determine the aforementioned relation:

$$P_{H_2O} = x \cdot (1 + A + B \cdot P_a) \cdot e^{\frac{y \cdot T_c}{z + T_c}} = D \cdot e^{\frac{y \cdot T_c}{z + T_c}} \tag{18.5}$$

$$P_{H_2O} \cdot V = n_{H_2O} \cdot R \cdot T_c = n_{H_2O} \cdot \frac{m_{H_2O}}{m_{H_2O}} \cdot R \cdot T_c = \frac{M_{H_2O}}{m_{H_2O}} \cdot R \cdot T_c \tag{18.6}$$

$$P_{air} \cdot V = n_{air} \cdot R \cdot T_c = n_{air} \cdot \frac{m_{air}}{m_{air}} \cdot R \cdot T_c = \frac{M_{air}}{m_{air}} \cdot R \cdot T_c \tag{18.7}$$

$$P_a = P_{H_2O} + P_{air} \tag{18.8}$$

Equations (18.6) and (18.7) are used to determine ω_c as a function of P_a and T_c Eq. (18.9). Combining Eq. (18.9) with Eq. (18.5) T_c is expressed as a function of ω_c and P_a Eq. (18.10). Furthermore, Eqs. (18.5)–(18.8) enable to determine the atmospheric air flow density ρ_{air} as a function of the atmospheric air conditions Eq. (18.11).

$$\omega_c = \frac{M_{H_2O}}{M_{air}} \cdot f = \frac{m_{H_2O}}{m_{air}} \cdot \frac{P_{H_2O}}{P_a - P_{H_2O}} \cdot f = \frac{m_{H_2O}}{m_{air}} \cdot \frac{D \cdot e^{\frac{y \cdot T_c}{z + T_c}}}{P_a - D \cdot e^{\frac{y \cdot T_c}{z + T_c}}} \cdot f \tag{18.9}$$

$$T_c = \frac{z \cdot \ln\left[\frac{\frac{\omega_c}{f} \cdot \frac{m_{air}}{m_{H_2O}} \cdot P_a}{\left(1 + \frac{\omega_c}{f} \cdot \frac{m_{air}}{m_{H_2O}}\right) \cdot D}\right]}{y - \ln\left[\frac{\frac{\omega_c}{f} \cdot \frac{m_{air}}{m_{H_2O}} \cdot P_a}{\left(1 + \frac{\omega_c}{f} \cdot \frac{m_{air}}{m_{H_2O}}\right) \cdot D}\right]} \tag{18.10}$$

$$\rho_{air} = \frac{P_a - \varphi_a \cdot D \cdot e^{\frac{y \cdot T_a}{z + T_a}}}{\frac{R}{m_{air}} \cdot T_a} \tag{18.11}$$

Combining Eqs. (18.1)–(18.4) and (18.10), (18.12) represents the relation between the air flow $\dot{v}_a$ and the post-condensation absolute humidity ω_c considering the system refrigeration power and the atmospheric air conditions. Equation 18.4 is rearranged to define ω_c as a function of $\dot{v}_a$, $\dot{q}_c$, and the atmospheric air conditions Eq. (18.13). Equation (18.14) is defined using Eq. (18.13) to substitute ω_c in Eq. (18.12). It represents the relation between the air flow $\dot{v}_a$ and the condensed water quantity $\dot{q}_c$ for a refrigeration system of refrigeration power *RP* operating at atmospheric air conditions T_a, P_a, and ω_a (or φ_a). Equation 18.15, similarly, relates the post-condensation air temperature T_c to the atmospheric air inflow $\dot{v}_a$.

$$RP = \dot{v}_a \cdot \rho_{air} \cdot c_{air} \cdot \left\{ T_a - \frac{z \cdot In\left[\frac{\frac{\omega_c}{f} \cdot \frac{m_{air}}{m_{H_2O}} \cdot P_a}{\left(1 + \frac{\omega_c}{f} \cdot \frac{m_{air}}{m_{H_2O}}\right) \cdot D}\right]}{y - In\left[\frac{\frac{\omega_c}{f} \cdot \frac{m_{air}}{m_{H_2O}} \cdot P_a}{\left(1 + \frac{\omega_c}{f} \cdot \frac{m_{air}}{m_{H_2O}}\right) \cdot D}\right]} \right\} + r_{H_2O} \cdot \dot{q}_c \tag{18.12}$$

$$\omega_c = \omega_a - \frac{\dot{q}_c}{\rho_{air} \cdot \dot{v}_a} \tag{18.13}$$

$$RP = \dot{v}_a \cdot \rho_{air} \cdot c_{air} \cdot \left\{ T_a - \frac{z \cdot In\left[\frac{\left(\omega_a - \frac{\dot{q}_c}{\rho_{air} \cdot \dot{v}_a}\right) \cdot \frac{m_{air}}{f \cdot m_{H_2O}} \cdot P_a}{\left(1 + \left(\omega_a - \frac{\dot{q}_c}{\rho_{air} \cdot \dot{v}_a}\right) \cdot \frac{m_{air}}{f \cdot m_{H_2O}}\right) \cdot D}\right]}{y - In\left[\frac{\left(\omega_a - \frac{\dot{q}_c}{\rho_{air} \cdot \dot{v}_a}\right) \cdot \frac{m_{air}}{f \cdot m_{H_2O}} \cdot P_a}{\left(1 + \left(\omega_a - \frac{\dot{q}_c}{\rho_{air} \cdot \dot{v}_a}\right) \cdot \frac{m_{air}}{f \cdot m_{H_2O}}\right) \cdot D}\right]} \right\} + r_{H_2O} \cdot \dot{q}_c \tag{18.14}$$

$$RP = \dot{v}_a \cdot \rho_{air} \cdot \left[c_{air} \cdot (T_a - T_c) + r_{H_2O} \cdot \left(\omega_a - \frac{m_{H_2O}}{m_{air}} \cdot \frac{D \cdot e^{\frac{y \cdot T_c}{z + T_c}}}{P_a - D \cdot e^{\frac{y \cdot T_c}{z + T_c}}} \cdot f \right) \right] \tag{18.15}$$

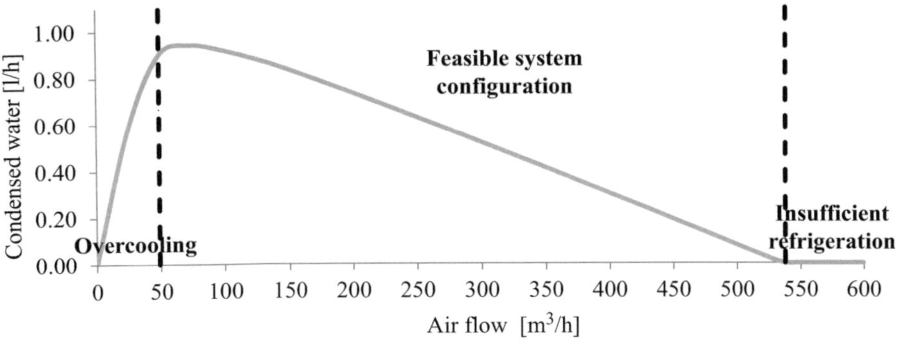

Fig. 18.1 Air flow $\dot{v}_a$—condensed water $\dot{q}_c$ relation Eq. (18.14) for $T_a = 32\ °C, P_a = 1,013$ mbar, $\varphi_a = 70\ \%$ and $RP = 1$ kW

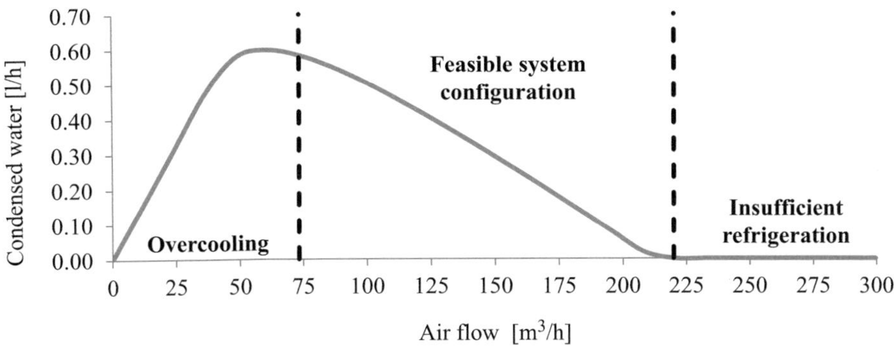

Fig. 18.2 Air flow $\dot{v}_a$—condensed water $\dot{q}_c$ relation Eq. (18.14) for $T_a = 32\ °C, P_a = 1,013$ mbar, $\varphi_a = 40\ \%$ and $RP = 1$ kW

Equation (18.14) enables to determine the optimal air flow that maximizes the condensed water production for every atmospheric air condition and system refrigeration power. Equation (18.15) estimates the air temperature at post-condensation state for the optimal air flow. Both Eqs. (18.14) and (18.15) are nonlinear; thus, an iterative procedure is required to solve them.

Figures 18.1 and 18.2 represent the trend of Eq. (18.14) for two sets of refrigeration power and atmospheric conditions distinguished by different humidity values, only. Equation (18.14) trend is distinguished by three areas: overcooling, insufficient refrigeration, and feasible system configuration. Low $\dot{v}_a$ do not represent a feasible refrigeration system configuration. As determined by Eq. (18.15), the overcooling refrigeration condensates the atmospheric moisture, but it determines frost formation at the heat exchanger due to T_c almost equal or lower than 0 °C. High $\dot{v}_a$ are distinguished by insufficient refrigeration. *RP* is not enough to start the dehumidification process. The heat removed by the air flow does not reduce the atmospheric air temperature below the dew point. Thus, Q_l is equal to 0 and no water is produced. Finally, the optimal $\dot{v}_a$ distinguished by the maximum $\dot{q}_c$ is selected among the feasible system configurations. For high relative humidity values, as shown in Fig. 18.1, the optimal $\dot{v}_a$ is not limited by the overcooling constraint. On the contrary, for low relative humidity, as shown in Fig. 18.2, $\dot{v}_a$ lower than the optimal value would increase the water production, but it does not represent a feasible system configuration.

Section 18.3 uses Eqs. (18.14) and (18.15) to define three air flow control strategies considering the atmospheric condition hourly profiles of a generic refrigeration system installation site. The definitions of the variables and the parameters of the equations proposed in this section are presented in the Nomenclature, whereas the values of the parameters are listed in the Appendix.

18.3 Air Flow Control Strategies

The refrigeration system installation site's most relevant features are the temperature, pressure, and absolute (or relative) humidity hourly profiles. Each month (i) is distinguished by the average temperature $T_a^{h,i}$, pressure $P_a^{h,i}$, and humidity $\omega_a^{h,i}$ (or $\varphi_a^{h,i}$) hourly (h) values. The refrigeration system processes the atmospheric air continuously, 24 h/day. This section proposes three air flow control strategies: hourly, monthly, and yearly.

The hourly control strategy determines the hourly optimal air flow $\dot{v}_a^{h,i,\mathrm{opt}}$ for each hour and month considering the corresponding values of temperature, pressure, and humidity to determine the best hourly water production $\dot{q}_c^{h,i,*}$. To determine $\dot{v}_a^{h,i,\mathrm{opt}}$ Eqs. (18.14) and (18.15) are used. The calculation of the optimal air flow for each hour and month ensures the maximum drinking water production achievable. However, this strategy requires a real-time acquisition of the atmospheric air conditions and an accurate control for the air flow fine regulation.

To overcome such disadvantages, the algorithm proposed in Fig. 18.3 presents the air flow optimization for the monthly control strategy. The purpose of this strategy is to determine the optimal air flow $\dot{v}_a^{i,\mathrm{opt}}$ for each month i, constant over the daily hours, that determines the best monthly water production $\dot{q}_c^{i,*}$. Thus, no real-time atmospheric condition acquisition is required, and the air flow can be manually set once a month. Considering the month i, the aforementioned algorithm initializes the air flow value $\dot{v}_a^i$ with $\dot{v}_a^0$ and uses Eqs. (18.14) and (18.15) to calculate the condensed water production $\dot{q}_c^{h,i}$ and the post-condensation temperature $T_c^{h,i}$ for each daily hour h. If the minimum post-condensation temperature constraint is satisfied for each hour, the monthly water production $\dot{q}_c^{i,\mathrm{tot}}$ is determined. This is compared to the maximum one achieved by the previous algorithm iterations $\dot{q}_c^{i,*}$. If $\dot{q}_c^{i,\mathrm{tot}}$ is greater than $\dot{q}_c^{i,*}$, $\dot{v}_a^i$ is considered the new air flow optimal value $\dot{v}_a^{i,\mathrm{opt}}$. $\dot{v}_a^i$ is increased by $\Delta\dot{v}_a$ and the algorithm iterates the aforementioned steps, while $\dot{q}_c^{i,\mathrm{tot}}$ is greater than zero. Monthly control strategy requires to evaluate each feasible $\dot{v}_a^i$ value from

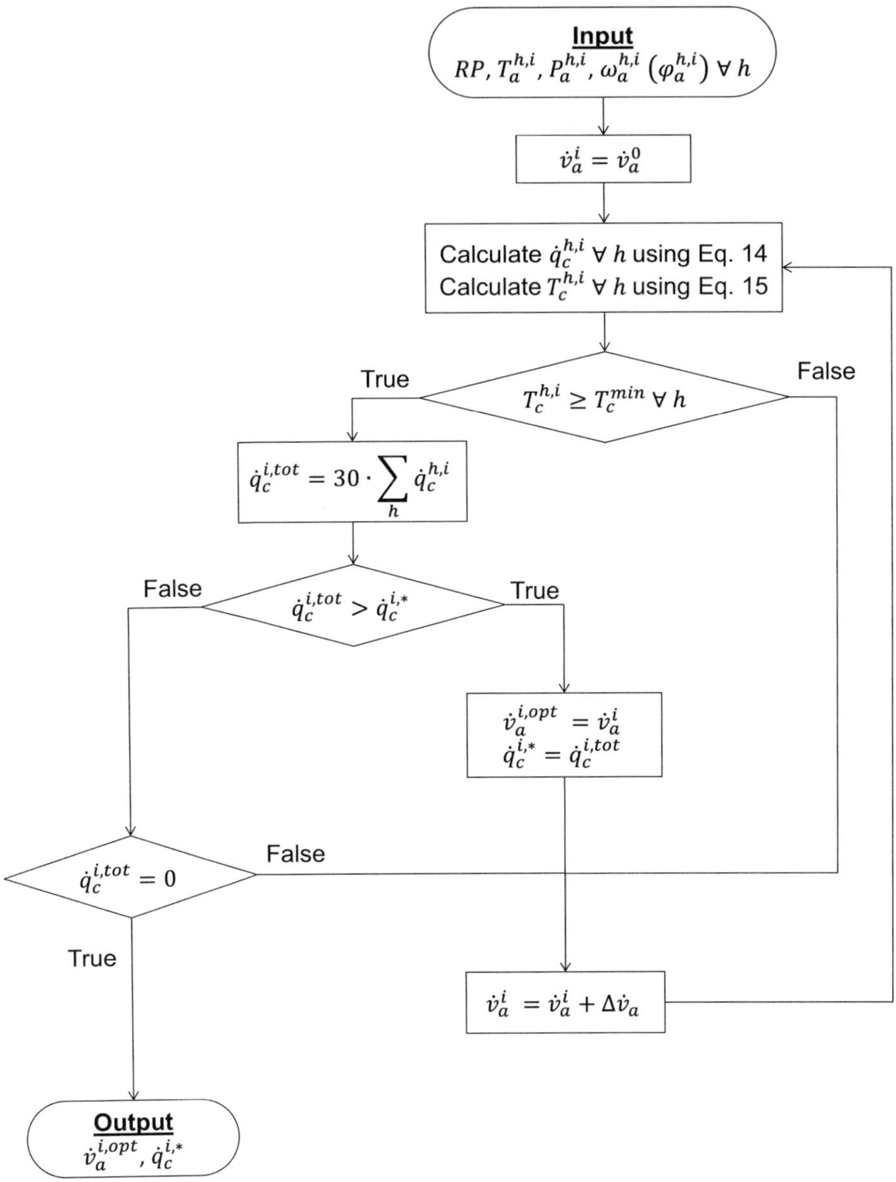

Fig. 18.3 Air flow optimization algorithm for monthly control strategy

$\dot{v}_a^0$ to the one distinguished by no water production to determine the monthly optimal air flow $\dot{v}_a^{i,\text{opt}}$ and the related monthly best water production $\dot{q}_c^{i,*}$.

The third air flow control strategy is the yearly one. It defines the yearly optimal air flow $\dot{v}_a^{y,\text{opt}}$, constant over the daily hours and the months, that determines the best yearly water production $\dot{q}_c^{y,*}$. Compared to the previous control strategies, the yearly

Table 18.1 Air flow control strategy features

	Hourly	Monthly	Yearly
Air flow value	Hourly optimal	Monthly optimal	Yearly optimal
Post-condensation temperature constraint	For the considered hour	For each hour of the month	For each hour and month of the year
Purpose	Determine the best hourly water production	Determine the best monthly water production	Determine the best yearly water production
Advantages	Maximum water production achievable	• No real-time atmospheric condition acquisition • Air flow manually set once a month	• No real-time atmospheric condition acquisition • Air flow set at system installation
Disadvantages	• Real-time atmospheric condition acquisition • Air flow accurate regulation	Small water production reduction	Huge water production reduction

one does not require any atmospheric air condition acquisition and air flow control over the hours and the months. The air flow is univocally set at the refrigeration system installation. The algorithm presented in Fig. 18.3 can be used to determine the yearly optimal air flow $\dot{v}_{\mathrm{a}}^{y,\mathrm{opt}}$ and the yearly best water production $\dot{q}_{\mathrm{c}}^{y,*}$.

Equations (18.14) and (18.15) have to calculate the condensed water production $\dot{q}_{\mathrm{c}}^{h,i}$ and the post-condensation temperature $T_{\mathrm{c}}^{h,i}$ for each daily hour h and month i. Thus, the minimum post-condensation temperature constraint has to be satisfied for each hour and month, to determine the yearly water production $\dot{q}_{\mathrm{c}}^{y,\mathrm{tot}}$. This is equal to the sum over i of the monthly water production $\dot{q}_{\mathrm{c}}^{i,\mathrm{tot}}$. The further algorithm steps do not require any correction. Thus, the yearly optimal air flow $\dot{v}_{\mathrm{a}}^{y,\mathrm{opt}}$ and the yearly best water production $\dot{q}_{\mathrm{c}}^{y,*}$ are determined.

Table 18.1 summarizes the three air flow control strategy features.

Section 18.4 presents the experimental campaign used to validate the AWVP mathematical model, together with the refrigeration system and the test procedure adopted.

18.4 Experimental Campaign

This section presents the experimental campaign focused on the validation of the AWVP mathematical model. Three are the experimental campaign-relevant features: the refrigeration system used for water condensation, the control system for input and output signal processing, and the test procedure. They are presented in the following.

18.4.1 Refrigeration System

A thermoelectric refrigeration system is designed and crafted for the experimental campaign. A centrifugal fan forces the air flow to pass through a plastic pipe of section AP where the air temperature and relative humidity are set to the simulated atmospheric conditions using a heater and a humidifier. Thus, the air flow passes through a vertical heat exchanger cooled by 20 thermoelectric packs mounted on two opposite sides of the heat exchanger. A thermoelectric pack is made of a couple of thermoelectric cells connected to a fan-cooled heat sink. The heat exchanger is thermally insulated through polyurethane films. The condensed water is collected by a funnel and stored in a bottle to avoid water evaporation. Figure 18.4a represents the refrigeration system diagram, while Fig. 18.4b is a picture of it. The refrigeration system component characteristics are listed in Table 18.2.

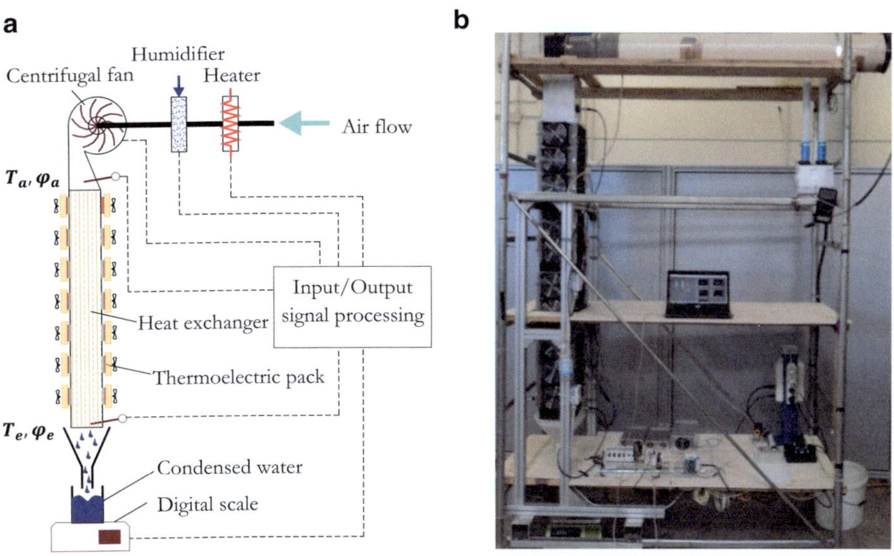

Fig. 18.4 (**a**) Refrigeration system diagram. (**b**) Refrigeration system picture

Table 18.2 Refrigeration system component characteristics

Component	Characteristics
Inflow pipe	Length 154 cm, circular section of 298.6 cm^2
Heater	Resistor, maximum thermal load 700 W
Humidifier	Ultrasonic humidifier, maximum vaporization capacity 1.2 L/h
Centrifugal fan	Power requirement 105 W, maximum air flow 320 m^3/h
Heat exchanger	Height 146 cm, section 160 × 160 mm
Thermoelectric cell	Area 40 × 40 mm. Specifications for 25 °C hot face temperature: maximum absorbable heat 72.0 W, maximum input current 8.5 A, maximum voltage 15.4 V, maximum temperature difference 65 °C

18.4.2 Control System

The validation of the AWVP mathematical model presented in Sect. 18.2 requires to simulate a particular set of atmospheric air conditions at the heat exchanger entrance. The air temperature T_a is measured by means of a PT100 resistance temperature detector with ± 0.3 °C accuracy, while the relative humidity φ_a is measured using a capacitive humidity sensor with ± 2 % accuracy. A real-time proportional-integral (PI) closed-loop controller uses these data to regulate the heater and the humidifier to set the temperature and humidity values. The air temperature T_e and relative humidity φ_e at the exit of the heat exchanger are measured by means of identical resistance temperature detector and capacitive humidity sensors. The air flow is regulated by the centrifugal fan, and its speed S is measured using an anemometer with ± 0.1 m/s accuracy, while the condensed water quantity is weighted by a digital scale with ± 1 g accuracy. The acquisition of the sensor signals, the data processing, and the PI controller are provided with a customized and easy-use real-time interface developed in the LabVIEW™ integrated development environment. Figure 18.5 shows the front panel of the monitoring and control tool.

18.4.3 Test Procedure

The aim of the experimental campaign is to measure the condensed water quantity and the post-condensation air temperature for several air flow values simulating a particular set of atmospheric air conditions. The following test procedures are used for this purpose:

- Air flow speed regulation to the required value
- Setup of the simulation atmospheric air conditions
- Condensed water quantity measurement at interval of 10 min
- Measurement of air conditions at heat exchanger exit
- Evaluation of air conditions in post-condensation state and bypass factor

Air conditions at the heat exchanger exit (T_e, φ_e) differ with the post-condensation values (T_c, φ_c). Part of the air flow is not affected by the heat exchange. Thus, its temperature and humidity do not vary. This percentage is defined bypass factor F. Most of the air flow $(1 - F)$ is distinguished by post-condensation conditions equal to T_c, φ_c. The air conditions measured at the heat exchanger exit T_e, φ_e are determined by the aforementioned air flow mix. Considering that $\varphi_c = 100\%$, T_c can be evaluated through the measurement of T_a, φ_a, T_e, φ_e. The temperature and humidity values of air at atmospheric, heat exchanger exit, and post-condensation conditions enable the bypass factor F evaluation:

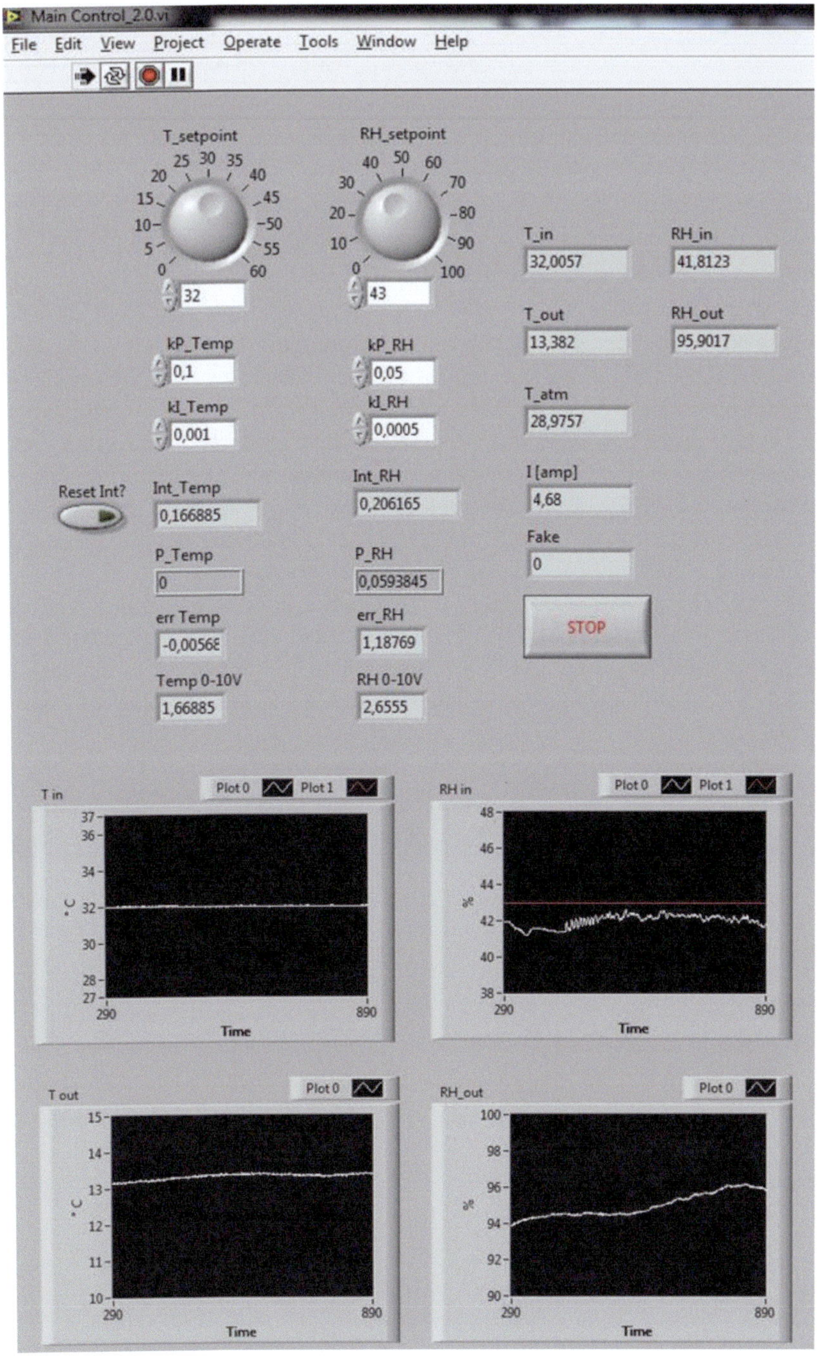

Fig. 18.5 Real-time monitoring and control tool

- Evaluation of the air flow through the following Eq. (18.16):

$$\dot{v}_a = S \cdot A \cdot (1 - F) \tag{18.16}$$

- Evaluation of the refrigeration power using Eqs. (18.1)–(18.3) and (18.11).

18.5 AWVP Model Validation

The validation of the AWVP mathematical model through the experimental campaign is presented in this section. Six different air flow values are tested by the refrigeration system: 57, 73, 88, 100, 115, and 127 m^3/h. The simulating atmospheric air conditions are $T_a = 32°C, P_a = 1{,}010$ mbar, and $\omega_a = 12.16\, g_{H_2O}/kg_{dry\ air}$ (or $\varphi_a = 39.5$ %). For each air flow value, the condensed water quantity and the post-condensation air temperature are measured. $\dot{v}_a - \dot{q}_c$ relation is presented in Fig. 18.6a, whereas Fig. 18.6b proposes $\dot{v}_a - T_c$ relation, both for the AWVP mathematical model and for the experimental campaign.

The AWVP model accurately predicts the condensed water and post-condensation temperature values. The difference between the $\dot{q}_c$ measured during the experimental campaign and the one estimated by the model is between +4.1 and −5.6 % for each tested air flow value, while the T_c error is between −0.3 and −0.5 %.

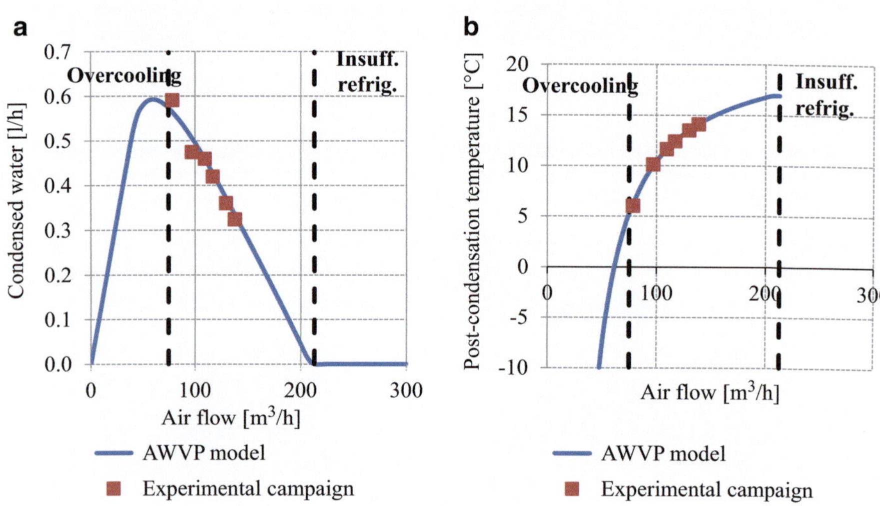

Fig. 18.6 (**a**) $\dot{v}_a - \dot{q}_c$ relation for $T_a = 32$ °C, $P_a = 1{,}010$ mbar, $\omega_a = 12.16\ g_{H_2O}/kg_{dry\ air}$ (or $\varphi_a = 39.5$ %) and $RP = 1$ kW. (**b**) $\dot{v}_a - T_c$ relation for $T_a = 32$ °C, $P_a = 1{,}010$ mbar, $\omega_a = 12.16\ g_{H_2O}/kg_{dry\,air}$ (or $\varphi_a = 39.5$ %) and $RP = 1$ kW

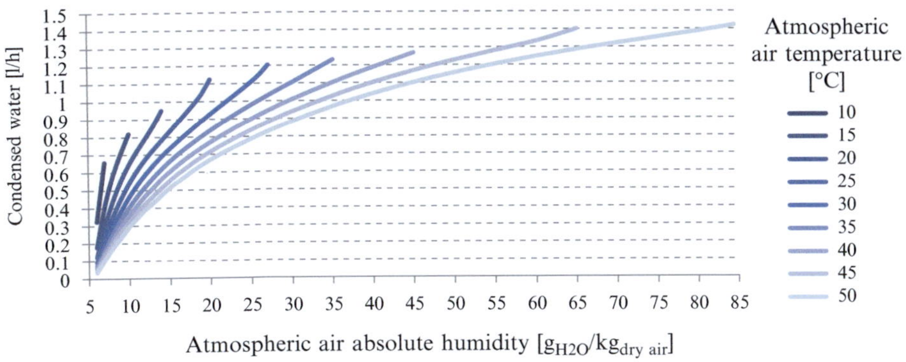

Fig. 18.7 Condensed water for every atmospheric air condition ($RP = 1$ kW, $P_a = 1{,}013$ mbar)

The validated AWVP mathematical model is used to determine the optimal air flow that maximizes the condensed water production for every atmospheric air condition between 10 and 50 °C for T_a and between 6 and 85 $g_{H_2O}/kg_{dry\ air}$ for ω_a considering $P_a = 1{,}013$ mbar and $RP = 1$ kW. This section presents the model results for the feasible $T_a - \omega_a$ combinations. As shown in Fig. 18.7, the higher the ω_a, the higher the maximum condensed water quantity, for each T_a value. On the contrary, lower T_a enables a greater water production for the same value of ω_a. For hot and humid atmospheric air conditions, distinguished by $T_a \geq 35$ °C and $\varphi_a \geq 70$ %, the maximum condensed water production varies between 1.00 and 1.42 L/h per kW of refrigeration power installed. Table 18.3 presents the optimal air flow values for every atmospheric condition. The value range is wide and included between 32 and 466 m^3/h. For each ω_a considered, the lower the T_a, the greater the optimal air flow value. As shown in Table 18.3, this relation is determined by Q_l/RP. The lower the T_a, the higher the Q_l/RP. Thus, high optimal air flow value enables to exploit this favorable condition and to increase the water production.

The validated AWVP model is used in Sect. 18.6 by the air flow control strategies presented in Sect. 18.3 to investigate the water production through atmospheric air dehumidification for a specific geographical location affected by water shortage. Furthermore, the three control strategies are compared to determine the most effective one.

18.6 Case Study and Results

This section presents the optimization of a refrigeration system for drinking water production installed in Dubai (the United Arab Emirates). The optimal air flow that determines the best condensed water production is determined for each air flow control strategy proposed in Sect. 18.3. Figures 18.8, 18.9, and 18.10 present the

Table 18.3 Optimal air flow and Q_l/RP for every atmospheric air condition ($RP = 1$ kW, $P_a = 1,013$ mbar)

$\omega_a\,[g_{H_2O}/kg_{dry\ air}]$	Optimal air flow (m^3/h) T_a (°C)									Q_l/RP (%) T_a (°C)								
	10	**15**	**20**	**25**	**30**	**35**	**40**	**45**	**50**	**10**	**15**	**20**	**25**	**30**	**35**	**40**	**45**	**50**
6	466	260	180	138	112	94	81	71	64	20	11	8	6	5	4	3	3	2
10		215	116	97	83	73	65	59	53		52	41	34	29	25	23	20	19
15				88	64	57	52	48	44				53	46	42	38	35	32
20				344	72	54	46	41	38				71	59	53	49	45	42
25					106	58	46	40	35					70	62	57	53	50
30						66	47	39	34						70	64	59	56
35						108	49	39	33						78	70	65	61
40							54	39	33							75	70	66
45							70	39	32							80	74	70
50								41	32								77	73
55								45	32								81	76
60								56	32								84	79
65								200	33								88	81
70									35									83
75									38									85
80									46									88
85									81									90

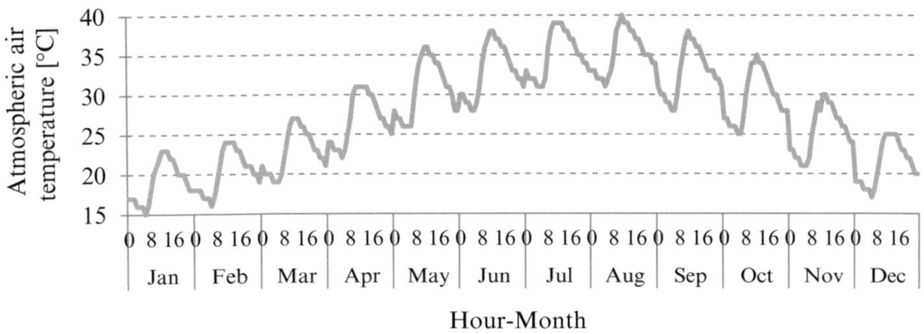

Fig. 18.8 Dubai atmospheric air temperature hourly profile of each month

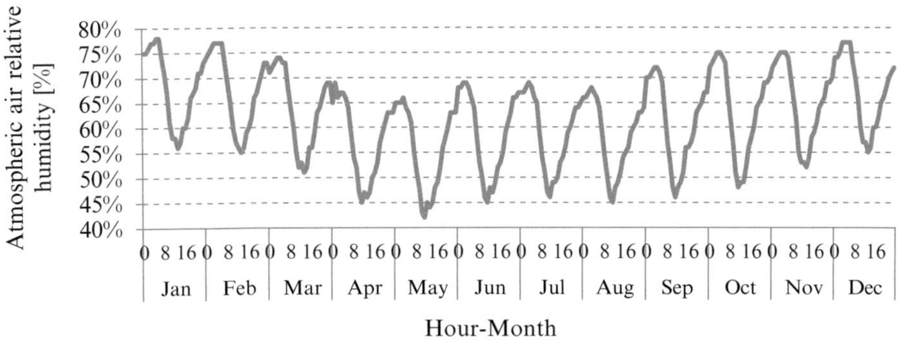

Fig. 18.9 Dubai atmospheric air relative humidity hourly profiles of each month

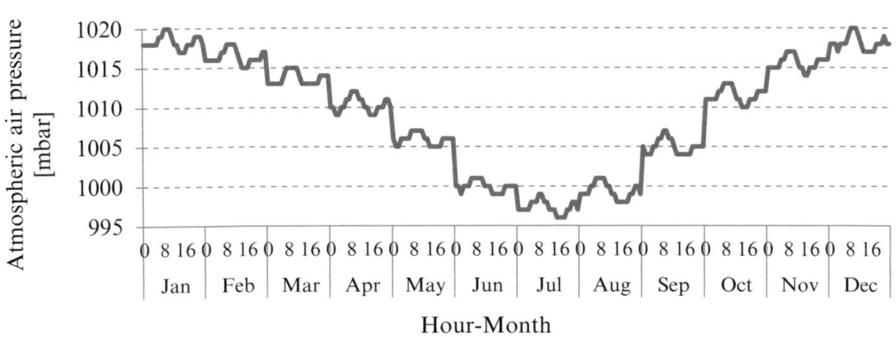

Fig. 18.10 Dubai atmospheric air pressure hourly profiles of each month

temperature $T_{\mathrm{a}}^{h,i}$, the relative humidity $\varphi_{\mathrm{a}}^{h,i}$, and the pressure $P_{\mathrm{a}}^{h,i}$ hourly profiles of each month [23].

The following results are presented per kW of refrigeration power installed ($RP = 1$ kW). The optimal air flow as well as the maximum condensed water

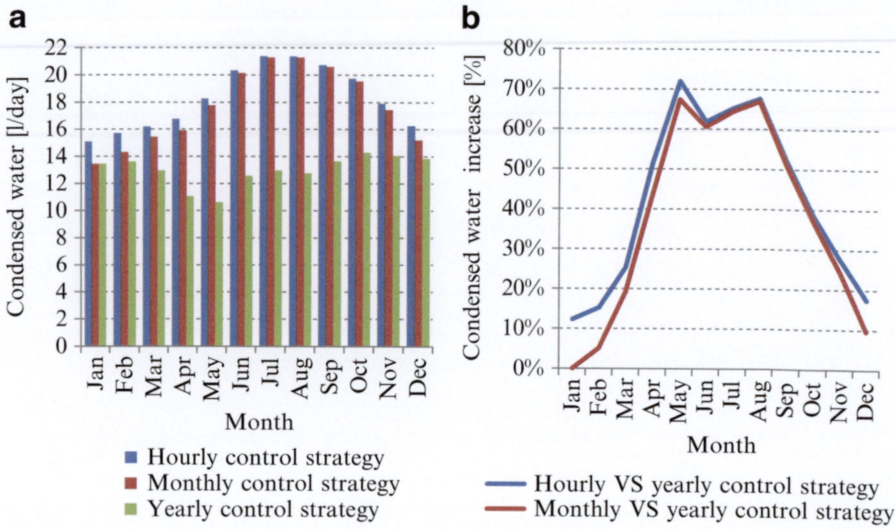

Fig. 18.11 (**a**) Daily condensed water production for each month and control strategy. (**b**) Condensed water increase for each month: hourly and monthly versus yearly control strategies

production of a refrigerator system distinguished by a refrigeration power *n* times *RP* are equal to n times the values presented in this section.

Figure 18.11a presents the daily condensed water production for each month and control strategy. Yearly control strategy guarantees from 10.6 L/day of condensed water in May to 14.3 L/day in October. Monthly control strategy is distinguished by a significant water production increase, that is, minimum in January (13.4 L/day) and maximum in July (21.3 L/day). Hourly control strategy guarantees the maximum water production achievable (as stated in Sect. 18.3), that is, equal to 15.1 L/day in January and 21.4 L/day in July. Compared to the yearly control strategy, hourly and monthly control strategies are distinguished by a significant water production increase. As shown in Fig. 18.11b, during the summer period (from May to September) the condensed water production of these strategies is 50 % or more greater than the yearly control strategy. Hourly and monthly control strategies are distinguished by similar water production values. Considering the entire year, hourly control strategy water production is 3.5 % greater than the monthly strategy one (219 versus 212 L/year).

The condensed water production difference among the air flow control strategies depends on the optimal air flows defined by these strategies. As shown in Fig. 18.12, for the winter months (from December to April) the monthly strategy optimal air flow, $\dot{v}_{\mathrm{a}}^{i,\mathrm{opt}}$, is equal to the maximum value over the daily hours of the optimal air flow defined by the hourly control strategy, $\max_h\left(\dot{v}_{\mathrm{a}}^{h,i,\mathrm{opt}}\right)$. Lower $\dot{v}_{\mathrm{a}}^{i,\mathrm{opt}}$ increase the water production. However, for certain hours, $T_{\mathrm{c}}^{h,i}$ would be lower than $T_{\mathrm{c}}^{\mathrm{min}}$, the lower bound to avoid frost formation. Thus, during the winter months $\dot{v}_{\mathrm{a}}^{i,\mathrm{opt}}$ is equal to the minimum value that satisfies the post-condensation temperature constraint for

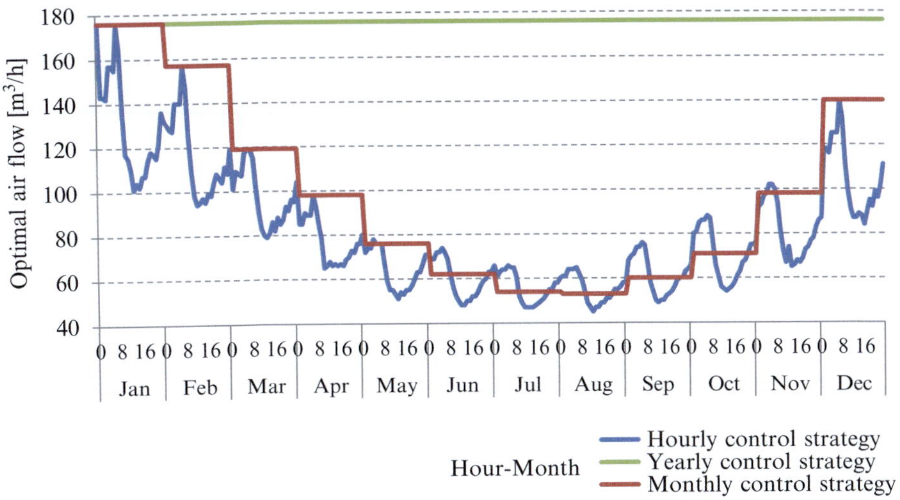

Fig. 18.12 Hourly, monthly, and yearly control strategy optimal air flows for each hour and month

each hour. This value is equal to $\max_h\left(\dot{v}_a^{h,i,\text{opt}}\right)$. During the summer period $T_a^{h,i}$ is higher; frost formation does not represent a constraint for $\dot{v}_a^{i,\text{opt}}$. Similarly, the yearly control strategy air flow optimal value, $\dot{v}_a^{y,\text{opt}}$, constant over the entire year, is the minimum air flow that satisfies post-condensation temperature constraint for each hour of each month. It is equal to the maximum value over the entire year of the optimal air flow defined by the hourly control strategy, $\max_{i,h}\left(\dot{v}_a^{h,i,\text{opt}}\right)$.

The limited increase in condensed water production that distinguishes the hourly control strategy compared to the monthly control strategy is overcome by the former strategy disadvantages, i.e., real-time atmospheric condition acquisition and air flow fine regulation. The monthly control strategy is suggested as the best strategy to adopt for the analyzed case study.

18.7 Conclusions

This chapter presents the optimization of a refrigeration system for drinking water production through atmospheric air dehumidification. An AWVP mathematical model is proposed to determine the optimal air flow entering the refrigeration system to maximize the condensed water production for every atmospheric air condition. Furthermore, to consider the atmospheric air condition hourly profiles of the refrigeration system installation site, three air flow control strategies are proposed: hourly, monthly, and yearly. Hourly control strategy defines the optimal air flow for each hour and month to determine the best hourly water production. Monthly control strategy estimates the optimal air flow for each month to determine the best monthly water production. Yearly control strategy evaluates the optimal air flow, constant for each hour and month, to determine the best yearly water production.

To validate the proposed AWVP mathematical model, an experimental campaign is set up. A thermoelectric refrigeration system is designed and crafted for the experimental analysis. A control system regulates a heater and a humidifier to simulate the required atmospheric air conditions. Several air flow values are tested, and the related condensed water production is measured. The AWVP mathematical model accurately predicts the drinking water production. The difference between the experimental campaign and the model values is between −5.6 and +4.1 %. The validated AWVP model is used to determine the optimal air flow for every atmospheric air condition between 10 and 50 °C for temperature and between 6 and 85 $g_{H_2O}/kg_{dry\ air}$ for absolute humidity.

Finally, to evaluate and compare the proposed air flow control strategies, the case study of a refrigeration system for AWVP installed in Dubai (the United Arab Emirates) is presented. Hourly and monthly control strategies ensure a significant water production increase compared to the yearly strategy, equal to 50 % or more during the summer period. Hourly and monthly control strategies are distinguished by similar water production values, equal to 219 and 212 L/year, respectively, per kW of refrigeration power installed. The limited water production increase that distinguishes the hourly control strategy compared to the monthly strategy is overcome by the former strategy disadvantages, i.e., real-time atmospheric condition acquisition and air flow fine regulation. The monthly control strategy is suggested as the one to adopt for the analyzed case study.

Further research has to improve the experimental campaign simulating the atmospheric air condition hourly profiles of a specific installation site and implementing the proposed air flow control strategies. Furthermore, the evaluation of the refrigeration system energy consumption per liter of drinking water produced is strongly recommended both for every atmospheric air condition and for the proposed air flow control strategies.

Appendix

AWVP mathematical model and control strategy parameter values

Parameter	Value	Unit of measure
A	0.0007	–
B	0.00000346	–
c_{air}	1,005	J/kg°C
f	1,000	g_{H_2O}/kg_{H_2O}
m_{air}	28.84	g/mol
m_{H_2O}	18.016	g/mol
R	8.3144	J/mol K

(continued)

r_{H_2O}	2,272	kJ/kg
T_c^{min}	5	°C
$\Delta\dot{v}_a$	1	m^3/h
$\dot{v}_a^0$	1	m^3/h
x	6.1121	–
y	17.123	–
z	234.95	–

References

1. United Nations (2013) The millennium development goals report 2013. United Nations, New York
2. Anbarasu T, Pavithra S (2011) Vapour compression refrigeration system generating fresh water from humidity in the air. In Proceedings of sustainable energy and intelligent systems (SEISCON 2011), Tamil Nadu, India, 20–22 July 2011
3. Miller JE (2003) Review of water resources and desalination technologies. Sandia National Laboratories—Unlimited Release, Albuquerque
4. Shanmugam G, Jawahar GS, Ravindran S (2004) Review on the uses of appropriate techniques for arid environment. In Proceedings of international conference on water resources and arid environment, Riyadh, Saudi Arabia, 5–8 Dec 2004
5. Helmreich B, Horn H (2009) Opportunities in rainwater harvesting. Desalination 248:118–124
6. Ghaffour N, Missimer TM, Amy GL (2013) Technical review and evaluation of the economics of water desalination: current and future challenges for better water supply sustainability. Desalination 309:197–207
7. Narayan GP, Sharqawy MH, Summers EK, Lienhard JH, Zubair SM, Antar MA (2010) The potential of solar-driven humidification–dehumidification desalination for small-scale decentralized water production. Renew Sustain Energy Rev 14:1187–1201
8. Wahlgren RV (2001) Atmospheric water vapour processor designs for potable water production: a review. Water Res 35:1–22
9. Mezher T, Fath H, Abbas Z, Khaled A (2011) Techno-economic assessment and environmental impacts of desalination technologies. Desalination 266:263–273
10. Gleick PH (2000) A look at twenty-first century water resources development. Water Int 25:127–138
11. Gleick PH (1996) Water resources. In: Schneider SH (ed) Encyclopedia of climate and weather. Oxford University Press, New York
12. Gad HE, Hamed AM, El-Sharkawy II (2001) Application of a solar desiccant/collector system for water recovery from atmospheric air. Renew Energy 22:541–556
13. Ji JG, Wang RZ, Li LX (2007) New composite adsorbent for solar-driven fresh water production from the atmosphere. Desalination 212:176–182
14. Abualhamayel HI, Gandhidasan P (1997) A method of obtaining fresh water from the humid atmosphere. Desalination 113:51–63
15. Starr VP (1972) Controlled atmospheric convection in an engineered structure. Nord Hydrol 3:1–21
16. Milani D, Abbas A, Vassallo A, Chiesa M, Bakri DA (2011) Evaluation of using thermoelectric coolers in a dehumidification system to generate freshwater from ambient air. Chem Eng Sci 66:2491–2501

17. Scrivani A, Bardi U (2008) A study of the use of solar concentrating plants for the atmospheric water vapour extraction from ambient air in the Middle East and Northern Africa region. Desalination 220:592–599
18. Carrington CG, Liu Q (1995) Calorimeter measurements of a heat pump dehumidifier: influence of evaporator air flow. Int J Energy Res 19:649–658
19. Khalil A (1993) Dehumidification of atmospheric air as a potential source of fresh water in the UAE. Desalination 93:587–596
20. Jradi M, Ghaddar N, Ghali K (2012) Experimental and theoretical study of an integrated thermoelectric–photovoltaic system for air dehumidification and fresh water production. Int J Energy Res 36:963–974
21. Habeebullah BA (2009) Potential use of evaporator coils for water extraction in hot and humid areas. Desalination 237:330–345
22. Buck AL (1981) New equations for computing vapor pressure and enhancement factor. J Appl Meteorol 20:1527–1532
23. WeatherSpark (2014) Average weather for Dubai, United Arab Emirates. https://weatherspark.com/#!dashboard;ws=32855. Accessed 15 March 2014

Chapter 19
Short-Term Forecasting of the Global Solar Irradiation Using the Fuzzy Modeling Technique: Case Study of Tamanrasset City, Algeria

Lyes Saad Saoud, Fayçal Rahmoune, Victor Tourtchine, and Kamel Baddari

Abstract In this chapter, the short-term forecasting of the global solar irradiation using the fuzzy modeling technique is proposed. The multi-input multi-output (MIMO) fuzzy models are used to predict the next 24 h ahead based on the mean values of the daily solar irradiation and the daily air temperature. The measured meteorological data of Tamanrasset City, Algeria (altitude, 1,362 m; latitude, 22°48 N; longitude, 05°26 E) is used, where the 2 years (2007–2008) are used for modeling and the year 2009 is used to validate the developed model. Several models are presented to test the feasibility and the performance of the fuzzy modeling technique for forecasting hourly solar irradiation in the MIMO strategy. Results obtained throughout this chapter show that the fuzzy modeling technique is suitable for a short-time forecasting of the solar irradiation.

Keywords Solar irradiation • Fuzzy modeling • Short-term forecasting • Algeria

Nomenclature

u_n	Input vector contains n elements
f	Nonlinear function
$\hat{y}$	Fuzzy model's output
i	Time index
A_j	Matrix with fuzzy sets
$\hat{y}_k$	kth local output
c	Number of cluster
b	Scalar vector

L. Saad Saoud (✉) • F. Rahmoune • V. Tourtchine • K. Baddari
Laboratory of Computer science, Modeling, Optimization and Electronic Systems (LIMOSE), Department of Physics, Faculty of sciences, University M'Hamed Bougara, Boumerdès 35000, Algeria
e-mail: saadsaoudl@umbb.dz

I. Dincer et al. (eds.), *Progress in Clean Energy, Volume 1*,
DOI 10.1007/978-3-319-16709-1_19

a_j	*j*th slope vector
μ_k	Overall degrees of the premise's membership
R_k	Rule of the fuzzy model
G_m	Daily global solar irradiation of the actual day, kJ/m^2
T_m	Daily air temperature of the actual day, °C
G_{jh}	Hourly global solar irradiation in the hour *j* of the next day, kJ/m^2
nRMSE	Normalized root mean squared error, %
R^2	Coefficient of determination, %
MAE	Mean absolute error, %
N	Number of samples
MIMO	Multi-input multi-output
MISO	Multi-input single-output

19.1 Introduction

Solar energy is a promising renewable energy source due to its pollution-free and inexhaustible nature. The most important parameter in the solar energy is its global irradiation, which is very important in the design of renewable and solar energy systems, particularly for the sizing of photovoltaic (PV) systems [1]. In literature, several models based on artificial intelligence to predict the global solar irradiation have been proposed [2–12], whereas the neural networks are widely used to forecast solar irradiation with good accuracy [9, 13–19]. For instance, Wang et al. [18] use the statistical feature parameters and the artificial neural network to forecast the short-term irradiance. The forecasting of global and direct solar irradiance using stochastic learning methods is investigated in [12]. The complex valued forecasting of the solar irradiation based on the complex valued neural network is proposed in [19]. Even though the fuzzy modeling proves its abilities in several fields, a little consideration is taken to use the fuzzy modeling technique to forecast the solar irradiation. Fuzzy modeling is a useful technique for the description of nonlinear systems [20] where nonlinear multiple linear models with fuzzy transitions approximate process behavior. It can be seen as logical models, which use "if-then" rules to establish qualitative relationships among the variables in the model [21]. On the other hand, in Algeria where the desert of Sahara occupied almost 80 % of its area, the solar energy exists with a large quantity, which makes it a primordial place for solar systems installation. Hence, models based on artificial intelligence are needed. In this chapter, the fuzzy modeling strategy is used to forecast the next 24 h based on the daily air temperature and/or the daily solar irradiation. Tamanrasset City, Algeria (altitude, 1,362 m; latitude, 22°48 N; longitude, 05°26 E), is used to validate the developed model.

19.2 Fuzzy Modeling

Fuzzy modeling and identification from measured data are effective tools for the approximation of nonlinear systems [21–23]. It is based on the clustering technique, in which several methods can be used [24]. Gustafson–Kessel clustering algorithm [25] promises a good approximation of the membership functions [26]. Based on this last one [25], Babuska et al. [27] identified a multi-input multi-output (MIMO) processes, and they have proved that such systems will be approximated by a collection of coupled multi-input single-output (MISO) discrete time fuzzy models. In our previous work [28, 29], the fuzzy modeling technique is used to model the static part in pH chemical reactor and predict the maximum power generated by the photovoltaic module, respectively.

The static MISO models used for the black box input–output are given:

$$\hat{y}(i) = f(u_n(i)) \tag{19.1}$$

where $u_n(i)$ is the input vector that contains n elements, f is a nonlinear function that maps the nonlinear real system that will be identified, $\hat{y}$ is the fuzzy model's output, and i is the time index.

The process can be approximated by a MISO static fuzzy model with rules of the following structure:

$$R_k : \text{If } u_1 \text{ is } A_1 \text{ and } u_2 \text{ is } A_2 \ \ldots\ u_n \text{ is } A_n \text{ then}$$

$$\hat{y}_k(i) = b + \sum_{j=1}^{n} a_j u_j(i-1) \tag{19.2}$$

A_j is a matrix with fuzzy sets, and $\hat{y}_k$ is the kth local output, with $k = 1,\ldots, c$ (c is the number of cluster). b is the scalar vector, and a_j is the jth slope vector. The overall degrees μ_k of the premise's membership of rule R_k can be calculated as

$$\mu_k = \min[A_1 \quad A_2 \quad \cdots \quad A_n] \tag{19.3}$$

The model output is calculated according to Eq. (19.4):

$$\hat{y} = \frac{\sum_{k=1}^{m} \mu_k(u)\hat{y}_k}{\sum_{k=1}^{m} \mu_k(u)} \tag{19.4}$$

In this work, the fuzzy models have one or two inputs, i.e., $u_n(i-1) = \left[T_m(i-1)^{\text{and/or}} G_m(i-1)\right]$, and 24 outputs which represent the 24 next hours, i.e., $y_n(i) = [G_{1h}(i), G_{2h}(i), \ldots, G_{24h}(i)]$, with $T_m(i-1)$ and $G_m(i-1)$ that represent the daily air temperature and the daily global solar irradiation of the actual day. $G_{1h}(i), G_{2h}(i), \ldots, G_{24h}(i)$ are the hourly global solar irradiation of the next day.

To have an idea about the prediction performance of the applied fuzzy models, the normalized root mean squared error (nRMSE) [13], the coefficient of determination (R^2), and the mean absolute error (MAE) [30] given below are taken like criteria for all examples (their units are in %), where N is the number of samples:

$$\text{nRMSE} = \frac{\sqrt{\frac{1}{N}\sum_{k=1}^{N} |y_k - \hat{y}_k|^2}}{\overline{y}} \tag{19.5}$$

$$R^2 = 1 - \frac{\sum_{k=1}^{N} |y_k - \hat{y}_k|^2}{\sum_{k=1}^{N} |y_k - \overline{y}|^2} \tag{19.6}$$

$$\text{MAE} = \frac{\sum_{k=1}^{N} |y_k - \hat{y}_k|}{N} \tag{19.7}$$

It should be noted that MAE in % is calculated by multiplying the obtained value by 100 and dividing the result by the maximum of the measured data.

19.3 Results and Discussions

In this part, the application of fuzzy modeling technique is presented. The measured meteorological data in 3 years (2007 through 2009) of Tamanrasset City, Algeria (altitude, 1,362 m; latitude, 22°48 N; longitude, 05°26 E) are used in this work. The two first years (2007–2008) are used to train the fuzzy models and the rest (2009) is for validation. First of all, the input vector contains one parameter (i.e., the air temperature or the daily solar irradiation) and the output vector contains the next 24 h. To have an idea about the influence of the number of clusters, different numbers are used. Table 19.1 presents the obtained results for this first strategy (i.e., using one input). When using the daily solar irradiation, we can see that using different values of clusters does not give a great improvement. And also increasing this number does not improve the results (i.e., 50 clusters give the worst results). Hence, when using just the daily solar irradiation to forecast the next 24 h, just three clusters are needed (MAE = 3.476 %, nRMSE = 33.18 %, and R^2 = 93.79 %). The same remark could be deduced when using the daily air temperature (i.e., changing the number of clusters does not produce almost any improvement). Using the daily solar irradiation gives better results than using the daily air temperature.

In the case of using two meteorological inputs (i.e., daily air temperature and daily solar irradiation) to forecast the next 24 h, the results given in Table 19.2 show the existence of small improvement (1 %). We can see also in the same Table 19.2 that using 3 or 7 clusters gives almost the same results and using 50 clusters produces the worst results. As a sample of these results, Fig. 19.1 shows those

Table 19.1 Obtained results using the fuzzy modeling technique in the case of using one meteorological input

Model's structure		Measure's criteria		
	Number of clusters "c"	MAE (%)	nRMSE (%)	R^2 (%)
$\{\hat{G}_{1h}(i), \hat{G}_{2h}(i), \ldots, \hat{G}_{24h}(i)\} = f(G_d(i-1))$	3	3.476	33.18	93.79
	5	3.491	33.33	93.73
	7	3.466	33.19	93.79
	11	3.480	33.28	93.75
	15	3.499	33.38	93.72
	20	3.522	33.76	93.57
	50	3.755	36.28	92.57
$\{\hat{G}_{1h}(i), \hat{G}_{2h}(i), \ldots, \hat{G}_{24h}(i)\} = f(T_d(i-1))$	3	4.217	38.45	91.66
	5	4.212	38.44	91.67
	7	4.225	38.53	91.62
	11	4.243	38.68	91.56
	15	4.274	39.01	91.42
	20	4.319	39.31	91.28
	50	4.480	40.91	90.56

Table 19.2 Obtained results using the fuzzy modeling technique in the case of using two meteorological inputs

Model's structure		Measure's criteria		
	Number of clusters "c"	MAE (%)	nRMSE (%)	R^2 (%)
$\{\hat{G}_{1h}(i), \hat{G}_{2h}(i), \ldots, \hat{G}_{24h}(i)\} = f(G_d(i-1), T_d(i-1))$	3	3.154	31.85	94.28
	5	3.138	32.04	94.21
	7	3.119	31.84	94.28
	11	3.145	31.98	94.23
	15	3.184	32.41	94.07
	20	3.240	32.92	93.88
	50	3.568	37.08	92.24

obtained for validation using 3 clusters (MAE = 3.154 %, nRMSE = 31.88 %, and $R^2 = 94.28$ %), and Fig. 19.2 presents the measured versus forecasted 24 h for the same part (i.e., validation part). It should be noted that the number of hours used to obtain these results is 8,736 h (i.e., 24*364).

To show better the obtained results, 2 months (January and May) from the validation part are chosen. Figure 19.3 presents the measured and the forecasted 24-h solar irradiation for the month of January, in which we can see the good results ($R^2 = 89.85$ %) even though the fuzzy model loses the same values in the cloudy days (i.e., good results for the sunny days). This disadvantage is not very important regarding Algeria in general and Tamanrasset City in particular, because it has

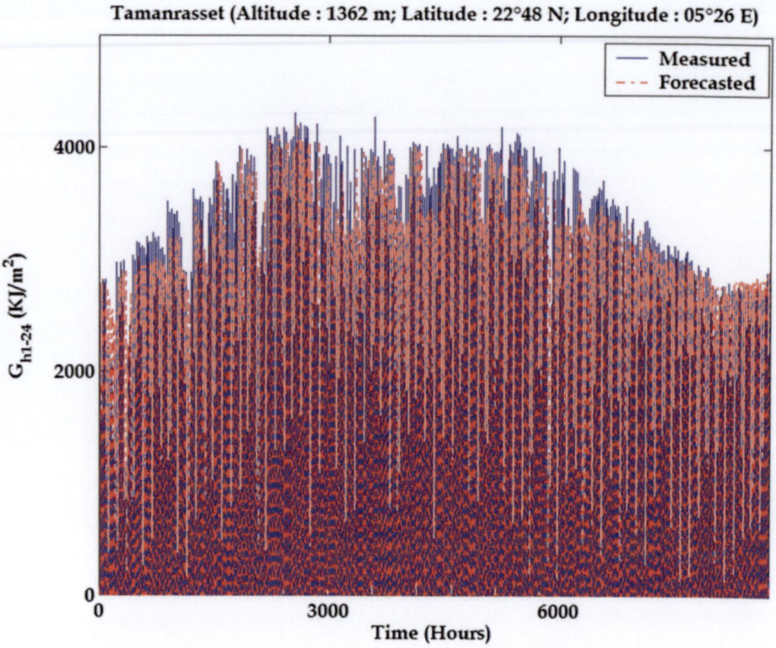

Fig 19.1 Measured and forecasted 24-h solar irradiation for the validation part (year 2009)

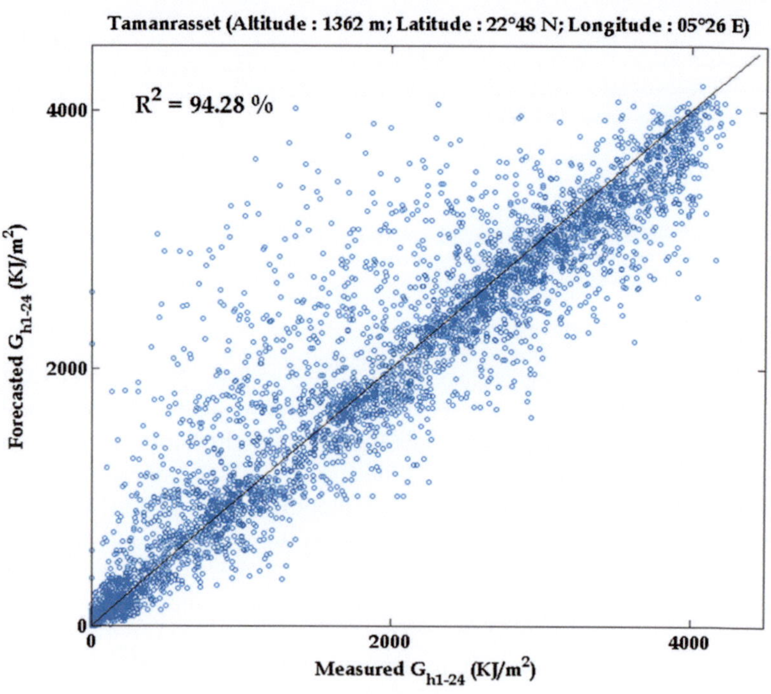

Fig 19.2 Measured versus forecasted 24-h solar irradiation for the validation part (year 2009)

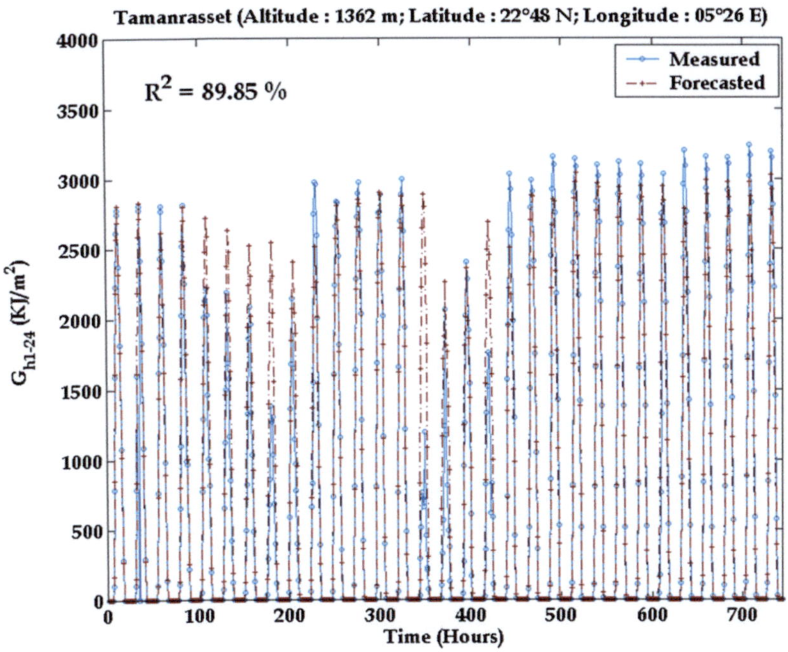

Fig 19.3 Measured and forecasted 24-h solar irradiation for the month of January, 2009

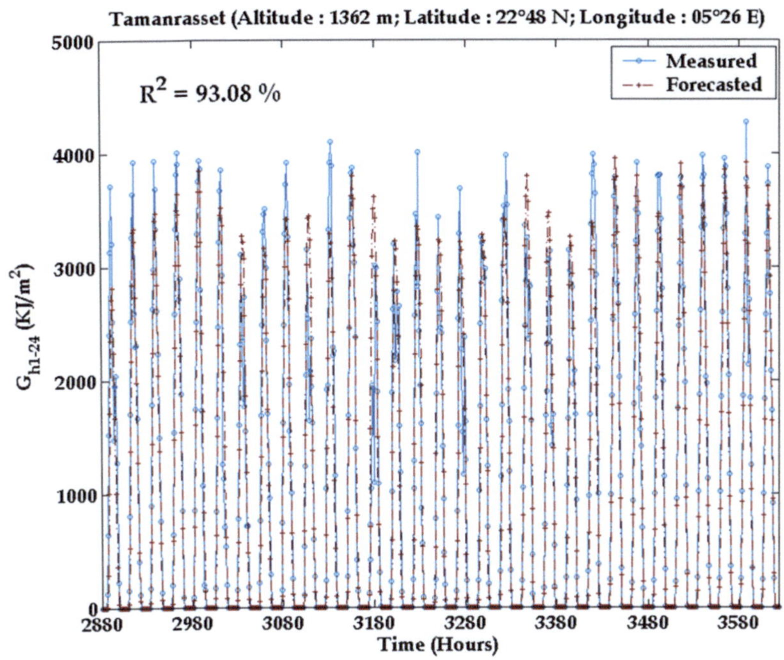

Fig 19.4 Measured and forecasted 24-h solar irradiation for the month of May, 2009

more sunny days than the cloudy days (i.e., more than 300 days are sunny in Algeria). The same remark could be deduced for the month of May (Fig. 19.4), in which we can see that the results are improved comparing to the month of January ($R^2 = 93.08$ %).

19.4 Conclusion

In this chapter, the short-term forecasting of the global solar irradiation using the fuzzy modeling technique is presented. The MIMO fuzzy models are used to predict the next 24 h ahead based on the mean values of the daily solar irradiation and/or the daily air temperature. In the case of using just one input, the daily solar irradiation gives better results compared to the use of the daily air temperature, where three clusters are sufficient. Using both parameters (the daily solar irradiation and the daily air temperature) adds a small improvement but not great (almost 1 %). According to the obtained results throughout this chapter, these remarks could be concluded: (1) the fuzzy modeling technique is suitable to forecast the short-term solar irradiation; (2) the daily solar irradiation or the daily air temperature or both of them could be used, but depending on the complexity and the cost, just the first parameter (i.e., the daily air temperature) is needed; (3) in all cases, just three (3) clusters are needed to construct the fuzzy models.

Acknowledgment The authors are grateful to the reviewers for their comments and recommendations. Also we thank the department's head of database in the National Office of Meteorology (NOM) of Algeria for providing the real dataset used throughout this chapter. This work is supported by the National Committee for Evaluation and Planning Unit of University Research, Ministry of Higher Education and Scientific Research, Algeria, under project number: J0200320130025.

References

1. Mellit A, Pavan AM (2010) A 24-h forecast of solar irradiance using artificial neural network: application for performance prediction of a grid-connected PV plant at Trieste, Italy. Sol Energy 84:807–821
2. Mellit A, Mekki H, Messai A, Kalogirou SA (2011) FPGA-based implementation of intelligent predictor for global solar irradiation, part I: theory and simulation. Expert Syst Appl 38:2668–2685
3. Solanki SK, Krivova NA, Wenzler T (2005) Irradiance models. Adv Space Res 35:376–383
4. Kalogirou S, Sencan A (2010) Artificial intelligence techniques in solar energy applications. In: Manyala R (ed) Solar collectors and panels, theory and applications. InTech, Rijeka
5. Winslow JC, Hunt ER Jr, Piper SC (2001) A globally applicable model of daily solar irradiance estimated from air temperature and precipitation data. Ecol Model 143:227–243
6. El-Sebaii AA, Al-Hazmi FS, Al-Ghamdi AA, Yaghmour SJ (2010) Global, direct and diffuse solar radiation on horizontal and tilted surfaces in Jeddah, Saudi Arabia. Appl Energy 87:568–576

7. Mefti A, Bouroubi MY, Adane A (2003) Generation of hourly solar radiation for inclined surfaces using monthly mean sunshine duration in Algeria. Energy Convers Manag 44:3125–3141
8. Dazhi Y, Jirutitijaroen P, Walsh WM (2012) Hourly solar irradiance time series forecasting using cloud cover index. Sol Energy 86:3531–3543
9. Wang F, Mi Z, Su S, Zhao H (2012) Short-term solar irradiance forecasting model based on artificial neural network using statistical feature parameters. Energies 5:1355–1370
10. Martin L, Zarzalejo LF, Polo J, Navarro A, Marchante R, Cony M (2010) Prediction of global solar irradiance based on time series analysis: application to solar thermal power plants energy production planning. Sol Energy 84:1772–1781
11. Gueymard CA (2009) Direct and indirect uncertainties in the prediction of tilted irradiance for solar engineering applications. Sol Energy 83:432–444
12. Marquez R, Coimbra CFM (2011) Forecasting of global and direct solar irradiance using stochastic learning methods, ground experiments and the NWS database. Sol Energy 85:746–756
13. Notton G, Paoli C, Vasileva S, Nivet ML, Canaletti J-L, Cristofari C (2012) Estimation of hourly global solar irradiation on tilted planes from horizontal one using artificial neural networks. Energy 39(1):166–179
14. Zervas PL, Sarimveis H, Palyvos JA, Markatos NCG (2008) Prediction of daily global solar irradiance on horizontal surfaces based on neural-network techniques. Renew Energy 33:1796–1803
15. Paoli C, Voyant C, Muselli M, Nivet M-L (2010) Forecasting of preprocessed daily solar radiation time series using neural networks. Sol Energy 84(12):2146–2160
16. Mellit A, Benghanem M, Kalogirou SA (2006) An adaptive wavelet-network model for forecasting daily total solar radiation. Appl Energy 83:705–722
17. Chen C, Duan S, Cai T, Liu B (2011) Online 24-h solar power forecasting based on weather type classification using artificial neural network. Sol Energy 85:2856–2870
18. Wang Z, Wang F, Su S (2011) Solar irradiance short-term prediction model based on BP neural network. In: ICSGCE 2011, Energy Procedia, no. 12, Chengdu, China, 27–30 Sept 2011, pp 488–494
19. Cao S, Cao J (2005) Forecast of solar irradiance using recurrent neural networks combined with wavelet analysis. Appl Therm Eng 25:161–172
20. Saad Saoud L, Rahmoune F, Tourtchine V, Baddari K (2013) Complex-valued forecasting of global solar irradiance. J Renew Sust Energy 5(4):043124–043145. http://dx.doi.org/10.1063/1.4818618
21. Roffel B, Betlem B (2006) Process dynamics and control, modeling for control and prediction. Wiley, Chichester
22. Babuska R (1998) Fuzzy modeling for control. Kluwer, Boston
23. Zhang H, Liu D (2006) Fuzzy modeling and fuzzy control. Birkhauser, Boston
24. Chafaa K, Ghanai M, Benmahammed K (2007) Fuzzy modelling using Kalman filter. IET Control Theory Appl 1(1):58–64
25. Gustafson DE, Kessel WC (1979) Fuzzy clustering with a fuzzy covariance matrix. In: Proceedings of IEEE CDC, San Diego, pp 761–766
26. Karaboga D, Ozturk C (2010) Fuzzy clustering with artificial bee colony algorithm. Sci Res Essays 5(14):1899–1902
27. Babuska R, Roubos JA, Verbruggen HB (1998) Identification of MIMO systems by input-output TS fuzzy models. In: The IEEE World Congress on Computational Intelligence, vol 1, 4–9 May 1998, pp 657–662
28. Saad Saoud L, Rahmoune F, Tourtchine V, Baddari K (2011) Modeling pH neutralization process using fuzzy dynamic neural units approaches. Int J Comput Appl 28(4):22–29
29. Saad Saoud L, Rahmoune F, Tourtchine V, Baddari K (2013) On the fuzzy modeling based estimation of maximum power generation from photovoltaic module. In: IEEE proceedings of the 3rd international conference on systems and control, Algiers, 29–31 Oct
30. Drossu R, Obradovic Z (1996) Rapid design of neural networks for time series prediction. IEEE Comput Sci Eng 3(2):78–89, ISSN: 1070-9924

Chapter 20
Improving WRF GHI Forecasts with Model Output Statistics

Burak Barutcu, Seyda Tilev Tanriover, Serim Sakarya, Selahattin Incecik, F. Mert Sayinta, Erhan Caliskan, Abdullah Kahraman, Bulent Aksoy, Ceyhan Kahya, and Sema Topcu

Abstract Solar energy applications need reliable forecasting of solar irradiance. In this study, we present an assessment of a short-term global horizontal irradiance forecasting system based on Advanced Research Weather Research and Forecasting (WRF-ARW) meteorological model and neural networks as a post-processing method to improve the skill of the system in a highly favorable location for the utilization of solar power in Turkey.

The WRF model was used to produce 1 month of 3 days ahead solar irradiance forecasts covering Southeastern Anatolia of Turkey with a horizontal resolution of 4 km.

Single-input single-output (SISO) and multi-input single-output (MISO) artificial neural networks (ANN) were used.

Furthermore, the overall results of the forecasting system were evaluated by means of statistical indicators: mean bias error, relative mean bias error, root mean square error, and relative root mean square error. The MISO ANN gives better results than the SISO ANN in terms of improving the model predictions, provided by WRF-ARW simulations for August 2011.

Keywords Model output statistics • Artificial neural networks • Weather research and forecasting • Turkey

---

B. Barutcu (✉)
Istanbul Technical University Energy Institute, ITU Ayazaga Campus, Maslak, Istanbul 34469, Turkey
e-mail: barutcub@itu.edu.tr

S.T. Tanriover • S. Sakarya • S. Incecik • F.M. Sayinta • E. Caliskan
A. Kahraman • C. Kahya • S. Topcu
Department of Meteorology, Istanbul Technical University, Maslak, Istanbul, Turkey

B. Aksoy
Turkish State Meteorological Service, Ankara, Turkey

I. Dincer et al. (eds.), *Progress in Clean Energy, Volume 1*,
DOI 10.1007/978-3-319-16709-1_20

20.1 Introduction

Short-term irradiance forecasting is an important issue in the field of solar energy for many applications. Numerical weather prediction (NWP) models have proven to be powerful tools for solar radiation forecasting. The use of numerical meteorological models in combination with statistical post-processing tools may have the potential to satisfy the requirements for up to 72 h ahead of irradiance forecast. In this study, we present an assessment of a short-term irradiance system based on the Advanced Research WRF (WRF-ARW) (V3) meteorological model and a post-processing method in order to improve the overall skills of the system for a full month simulation of the August 2011 over the Southeastern Anatolia Region (SEA) of Turkey.

20.2 Data and Methodology

20.2.1 Data

The study is carried out in SEA which is a highly favorable location for the utilization of solar power in Turkey (Fig. 20.1). SEA is located between the latitudes of 36°–38°N, covering about 7.5 million hectares of land which is approximately 10 % of the whole country. Global horizontal solar irradiation (GHI) data

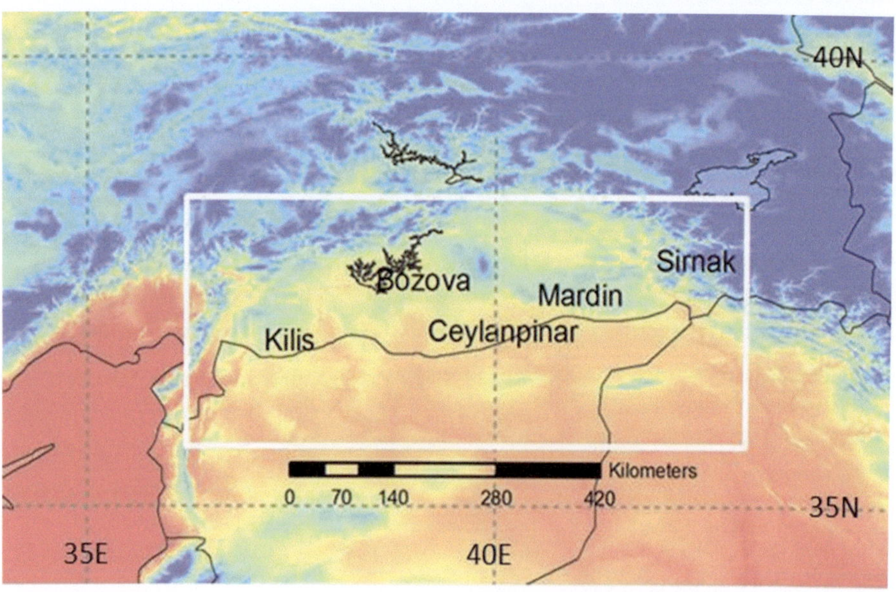

Fig. 20.1 GHI measurement sites in Southeastern Anatolia Region of Turkey

Table 20.1 Geographical elements of GHI measurement sites in Southeastern Anatolia Region

Site	Latitude (°N)	Longitude (°E)	Altitude (m)
Bozova	37.37	38.51	622
Ceylanpınar	36.84	40.03	360
Kilis	36.71	37.11	640
Mardin	37.31	40.73	1,040
Şırnak	37.52	42.45	1,350

Table 20.2 Parameterizations used in WRF model for the August 2011 simulations

Parameterization type	Scheme
Surface layer	Monin-Obukhov with Carlson-Boland (MM5 similarity)
Land surface model	Noah land surface model
Planetary boundary layer model	Yonsei University scheme
Atmospheric radiation	RRTMG
Cumulus	Kain-Fritsch (for domains 1 and 2, none for the innermost domains)
Microphysics	New Thompson scheme

were collected from five meteorological stations, representing the region of interest, which, namely, are Bozova, Ceylanpınar, Kilis, Mardin, and Şırnak.

To measure the GHI, Kipp & Zonen type pyranometers are used by the Turkish State Meteorological Service. The hourly observational GHI data from SEA are used for model verification. The stations are distributed over a complex terrain with station elevations ranging from 360 to 1,350 m, above sea level (Table 20.1). The full month of August 2011 is selected for the present work.

20.2.2 The WRF Model

The Weather Research and Forecasting (WRF) mesoscale meteorological model is applied to calculate the GHI over the study area. WRF is an Eulerian, 3-dimensional, non-hydrostatic mesoscale model with state-of-the-art parameterizations (radiation, planetary boundary layer, surface layer, land surface model, cumulus, and microphysics) in a massively parallel computing environment. Table 20.2 presents the parameterizations used in the WRF model run for this study.

In this study, the WRF model concerning the period of August 2011 was run on an initial regional grid of 36×36 km in space covering Europe. Inner domains following with 12×12 km and finally 4×4 km of horizontal resolution with 35 vertical layers over SEA of Turkey were used. The National Centers for Environmental Prediction (NCEP) Global Forecast System (GFS) data with 6 h temporal and 1° spatial resolution were used for initial and boundary conditions.

The GHI forecasts up to 72 h were performed using the WRF model. The conducted simulations were initialized from 00UTC of each day of August 2011. Two-way nesting option was selected for all nesting operations.

It is known that the mesoscale meteorological models can predict GHI without mean bias error (MBE) for clear sky conditions only. However, the bias was highly dependent on cloudy conditions and becomes strong in overcast conditions. Remund et al. [1] evaluated different NWP-based GHI forecasts in the USA, reporting relative root mean square error (rRMSE) values ranging from 20 to 40 % for a 24 h forecast horizon. Similar results were reported by Perez et al. [2], evaluating NWP-based irradiance forecasts in several places in the USA. Remund et al. [1] examined NWP biases compared to a single site and found that the next day (24 h), GHI forecasts of European Centre for Medium-Range Weather Forecast (ECMWF), a NWP model, and GFS have some MBE of 19 %. This MBE was found to be approximately constant for intraday (hour ahead) to 3 days ahead forecast horizons.

Lorenz et al. [3, 4] evaluated several NWP-based GHI forecasts in Europe, using statistical measures such as RMSE and MBE. In the chapter by Lorenz et al. [3], results showed rRMSE values of about 40 % for Central Europe and 30 % for Spain. Evaluating ECMWF's accuracy in Germany, Lorenz et al. [4] also showed that NWP MBE was largest for cloudy conditions with moderate clear sky indices, while forecasted clear conditions were relatively unbiased. They reported rRMSE values of about 35 % for single stations for 24 h horizon forecasts.

20.2.3 MOS and ANN

Post-processing techniques are usually applied to refine and improve the NWP model outputs. Model output statistics (MOS) is a post-processing technique used to objectively interpret numerical model output and produce site-specific forecasts. MOS relates observed meteorological parameters to appropriate variables (predictors) via a statistical approach. Hence, they can be used to reduce systematical forecast errors. There are several papers in literature about the MOS correction of GHI forecasts from NWP models [2, 5–7]. They indicate that the MOS application to the WRF GHI forecasts was successful in minimizing bias and reducing RMSE. Bofinger and Heilscher [7] used MOS locally with ECMWF GHI forecasts to create daily solar electricity predictions, accurate to 24.5 % rRMSE for averaged daily forecasts.

In this study, WRF model forecasts were used as the predictors. As an alternative to conventional approaches, artificial neural networks (ANNs) have been successfully applied for solar radiation estimation in the literature [5, 6, 8–10].

ANNs recognize patterns in the data and have been successfully applied to solar forecasting. Using training data, ANNs have been developed to reduce rRMSE of the GHI values. The innermost domain was used in the evaluation procedure. Furthermore, a group of post-processes were used: single-input single-output (SISO) ANN and multi-input single-output (MISO) ANN.

Feed-forward multilayer ANNs consist of three basic components: an input layer, one or more hidden layers, and an output layer. Due to the fact, there isn't any connection between the nodes in the same layer. The input layer of ANN transfers the information it receives from outside to the internal processing unit. Hidden layers are used for feature extraction during the flow of information. Output layer gives the system output. There are two major steps in the use of a feed-forward ANN, one is teaching and the other is recall [11, 12].

In this study, a group of ANNs were used to reduce the forecast errors. The ANN architectures which are applicable to this goal can be grouped under two main branches. In the first group which is called SISO ANNs, there is one node as input, one neuron as output layer, and a group of (e.g., 5, 7, etc.) neurons are applied as the number of neurons in their hidden layer. In the second group which is called MISO ANNs, a group of nodes (e.g., 3 or more) are set as input layer, and one neuron is applied to the output layer. In this group, embedding technique is used to improve the success, and two different selection criteria are used separately for determining the number of neurons in the hidden layers, one is building as a triangular architecture $(n+1)/2$ (where n is the number of nodes in the input layer) and the other is Hecht-Nielsen's $2n+1$ approximation [13] (which is based on Kolmogorov's proof [14]).

A series of trials were performed to define the embedding dimension of the MISO ANN. Figure 20.2 represents the chosen ANN structure that consists of three nodes as input layer, seven neurons for the hidden layer, and a single neuron as the output layer.

In both groups, two third of the signal is set for the training and the rest one third is used for the test. For the sake of comparison, five ANNs were created for each model in both architectures, and these ANNs were trained separately by Levenberg-

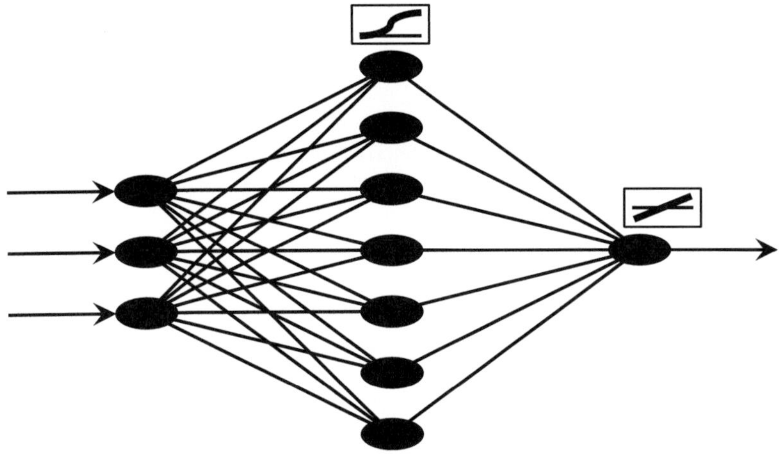

Fig. 20.2 The feed-forward MISO ANN architecture used to approximate the WRF outputs to the observation values. The ANN has logarithmic sigmoid activation functions in hidden layers and pure linear activation functions in output layers. Number of hidden layer neurons determined by Hecht-Nielsen's approximation as $2 \times 3 + 1 = 7$ neurons

Marquardt training algorithm for 500 steps. Logarithmic sigmoid function is used for all hidden layers (because solar irradiance is a nonnegative signal), and pure linear activation function is applied to all output layers. SISO ANNs could never achieve success levels as high as MISO architectures because MISO networks are designed by the embedding approach in prediction theory. Besides, SISOs do not use any information about the past values of a signal.

20.2.4 Adjustment of ANN

The best performed neural networks used to improve the WRF results are MISO ANNs. The logarithmic sigmoid activation function was used—as the irradiation data is a nonnegative signal—in the hidden layer. For the output layer, linear activation function proved to produce better results. The performance of ANNs depends on the weights and biases which are selected as small random numbers in the initialization step of ANNs. Therefore, starting with different weights and biases may yield different consequent performances. Keeping this in mind, five separate ANNs were trained for each station and their predictions for 24, 48, and 72 h. To be able to sufficiently diverge from the step until error reduction is no longer possible, all networks were trained for 500 steps with Levenberg-Marquardt algorithm, one of the best among quasi-Newtonian methods. The ANNs with the best performance for each prediction horizon were selected. As a result, 15 ANNs were established for five stations.

20.3 Results

In this study, an assessment of a short-term GHI forecasting system based on WRF GHI forecasts and a post-processing method, ANN for a full month of August 2011, was presented. The WRF model outputs and the ANN post-processes were compared by means of statistical error parameters, as shown in Table 20.3. These aforementioned parameters are as follows: mean bias error (MBE), relative mean bias error (rMBE), root mean square error (RMSE), and relative root mean square error (rRMSE).

In order to improve the solar irradiance forecasts, we employed the ANN methodology.

The statistical accuracy by the means of presented error values of verification of the WRF outputs to observations is shown in Table 20.3. The impact of ANN on RMSE and MBE or the performances of the ANNs for forecasts of 24, 48, and 72 h selected for this study are presented in Table 20.4. Tables 20.3 and 20.4 as a whole summarize the results for each station separately by showing the full August statistical measures, averaged over the stations for GHI.

Table 20.3 Error statistics of WRF simulations for August 2011 of Southeastern Anatolia Region in Turkey

Forecast horizon	Statistical parameters	WRF verifications				
		Bozova	Ceylanpınar	Kilis	Mardin	Şırnak
24 h	MBE (W/m^2)	26.704	47.109	32.665	31.922	34.715
	rMBE	0.094	0.171	0.113	0.107	0.116
	RMSE (W/m^2)	51.524	71.995	54.110	53.564	60.365
	rRMSE	0.182	0.262	0.190	0.180	0.202
48 h	MBE (W/m^2)	27.349	46.55	32.101	28.556	30.694
	rMBE	0.095	0.169	0.111	0.096	0.102
	RMSE (W/m^2)	51.535	71.672	51.664	55.693	56.099
	rRMSE	0.179	0.261	0.179	0.187	0.187
72 h	MBE (W/m^2)	26.662	45.814	31.517	29.226	30.950
	rMBE	0.094	0.169	0.110	0.100	0.105
	RMSE (W/m^2)	54.239	71.214	54.242	54.391	60.603
	rRMSE	0.191	0.264	0.191	0.186	0.206

Table 20.4 Error statistics of ANN post-processes and for August 2011 of Southeastern Anatolia Region in Turkey

Forecast horizon	Statistical parameters	WRF + ANN verifications				
		Bozova	Ceylanpınar	Kilis	Mardin	Şırnak
24 h	MBE (W/m^2)	−17.062	5.492	−18.899	−3.342	−6.182
	rMBE	−0.056	0.021	−0.062	−0.011	−0.021
	RMSE (W/m^2)	31.407	33.912	30.808	13.153	17.938
	rRMSE	0.103	0.127	0.101	0.045	0.061
48 h	MBE (W/m^2)	−19.468	6.712	−22.327	−3.377	−9.557
	rMBE	−0.064	0.025	−0.073	−0.012	−0.023
	RMSE (W/m^2)	45.099	35.522	33.020	14.739	23.010
	rRMSE	0.148	0.135	0.109	0.050	0.078
72 h	MBE (W/m^2)	−16.808	10.143	−19.896	−2.591	−5.089
	rMBE	−0.057	0.039	−0.066	−0.009	−0.017
	RMSE (W/m^2)	39.354	37.881	33.025	16.037	29.724
	rRMSE	0.133	0.146	0.110	0.055	0.100

The following are some regional conclusions based on the comparison of the results given in Tables 20.3 and 20.4:

- The regional accuracy of the WRF model outputs in rRMSE for SEA lies in the order of 20.32 % and is reduced to 8.74 % with ANN, for the first 24 h.
- Regarding the whole region, the rRMSE is reduced from 19.86 to 10.40 % and from 20.76 to 10.88 % for 48 and 72 h of forecast horizons, respectively.
- The RMSE reduction for August via ANN, regarding the 24 h WRF outputs (from 58.31 to 25.44 W/m^2), shows 56 % forecast improvement.

The quality of the forecast is also dependent on the location. The two best predicted sites with the lowest RMSE are places in Mardin and Şırnak with more sunny days than the others.

20.4 Concluded Remarks

In this study, The WRF model was used to produce 1 month of 3 days ahead solar irradiance forecasts covering Southeastern Anatolia of Turkey. Additionally, in order to improve the forecasts of GHI, we employed the ANN post-processing. The presented study gives the assessments of short-term GHI forecasting by WRF and a post-processing tool, ANN. The analysis was performed on a full month of August 2011 over SEA of Turkey. The evaluation focuses on the capability of the ANN methodology to improve the forecast of global horizontal solar irradiance by reducing the systematical error.

The MISO ANN presents better results than the SISO ANN in terms of improving the model predictions provided by WRF-ARW simulations. Besides, the ANN had a significant performance in August 2011.

On the basis of GHI measurements, the present study makes it possible to obtain more accurate predictions.

Acknowledgements This paper is funded by TUBITAK COST 111Y234 Project.

References

1. Remund J, Perez R, Lorenz E (2008) Comparison of solar radiation forecasts for the USA. In: Proceedings of the 23rd European PV conference, Valencia, pp 1–9
2. Perez R, Kivalov S, Schlemmer J, Hemker K Jr, Renné D, Hoff TE (2010) Validation of short and medium term operational solar radiation forecasts in the US. Sol Energy 84 (12):2161–2172
3. Lorenz E, Remund J, Müller SC, Traunmüller W, Steinmaurer G, Pozo D, Ruiz-Arias JA, Fanego VL, Ramirez L, Romeo MG, Kurz C, Pomares LM, Guerrero CG (2009) Benchmarking of different approaches to forecast solar irradiance. In: Proceedings of the 24th European photovoltaic solar energy conference, Hamburg, pp 4199–4208
4. Lorenz E, Hurka J, Heinemann D, Beyer HG (2009) Irradiance forecasting for the power prediction of grid-connected photovoltaic systems. IEEE J Sel Topics Appl Earth Observ 2 (1):2–10
5. Rincón A, Jorba O, Baldasano JM, Delle Monache L (2011) Assessment of short term irradiance forecasting based on post-processing tools applied on WRF meteorological simulations. In: State-of-the-Art Workshop, COST ES 1002: WIRE: Weather Intelligence for Renewable Energies. COST Action WIRE, Paris, 22–23 March 2011, pp 1–9
6. Diagne M, David M, Lauret P, Boland J, Schmutz N (2013) Review of solar irradiance forecasting methods and a proposition for small-scale insular grids. Renew Sustain Energy Rev 27:65–76

7. Bofinger S, Heilscher G (2006) Solar electricity forecast-approaches and first results. In: 21st European photovoltaic solar energy conference, no. 9, pp 4–8
8. Mathiesen P, Kleissl J (2011) Evaluation of numerical weather prediction for intra-day solar forecasting in the continental United States. Sol Energy 85(5):967–977
9. Rahimikhoob A (2010) Estimating global solar radiation using artificial neural network and air temperature data in a semi-arid environment. Renew Energy 35(9):2131–2135
10. Benghanem M, Mellit A, Alamri SN (2009) ANN-based modelling and estimation of daily global solar radiation data: a case study. Energy Convers Manag 50(7):1644–1655
11. Beale R, Jackson T (1998) Neural computing—an introduction. IOP, Bristol
12. Haykin S (1999) Neural networks: a comprehensive foundation. Prentice Hall, Upper Saddle River
13. Hecht-Nielsen R (1990) Neurocomputing. Addison-Wesley, Reading
14. Kolmogorov AN (1957) On the representation of continuous functions of many variables by superposition of continuous functions of one variable and addition. Dokl Akad Nauk SSSR 114:953–956. Translated in: Amer Math Soc Transl 28:55–59 (1963)

Chapter 21
CitInES: Design of a Decision Support Tool for Industrial Complexes and Cities

Sinem Nalbant, Eren Yaşar Çiçek, Fırat Uzman, Funda Çetin, Çağrı Savaşan, Laurent Fournie, and Sylvain Mouret

Abstract This paper focuses on the CitInES (City and Industry Energy System) project industrial use case, Tüpraş. The industrial use case includes plant modeling, historical replay, optimization, operation policy, and field testing. Reliable modeling was ensured by taking advantage of actual historical data. Missing or corrupted data were handled by data reconciliation. Optimization was carried out on historical operational data to develop optimized strategies which respect numerous actual technical constraints (design and safety constraints, efficiency curves, start-up costs). A guideline was created for operational management and another for electricity market bidding. During experimentation, the tool developed within the CitInES project handled complex utility systems and provided strategies for operating a power plant, enabled bidding on the Turkish electricity market, and answered feasibility studies for investment projects. The new production and bidding guidelines designed with the software reduced utility costs by 7 % without any new investment when compared to previous years.

Keywords Utility plant • Power plant • Scheduling • Optimization • Day-ahead electricity market • Bidding

Nomenclature

VHP	Very high pressure
HP	High pressure
MP	Medium pressure
LP	Low pressure
FO	Fuel oil

S. Nalbant (✉) • E.Y. Çiçek • F. Uzman • F. Çetin • Ç. Savaşan
Tüpraş—Turkish Petroleum Refineries Corporation, Güney Mah. Petrol Cad. No: 25
Körfez, Kocaeli 41780, Turkey
e-mail: sinem.nalbant@tupras.com.tr

L. Fournie • S. Mouret
Artelys, 12 rue du Quatre Septembre, Paris 75002, France

I. Dincer et al. (eds.), *Progress in Clean Energy, Volume 1*,
DOI 10.1007/978-3-319-16709-1_21

FG Fuel gas
NG Natural gas

21.1 Introduction

Industrial companies need to reinforce their competing capacities by reducing production costs in a sustainable manner. The oil and gas industry is a major consumer, as well as a producer, of energy [1]. Because of the large share of energy costs in the overall cost of refining operations, the industry has a strong financial motivation to improve energy efficiency. The refining industry is inherently very energy intensive. Several factors contribute to higher energy intensity in refining, offsetting part of the efficiency gains from new investments. More stringent oil product standards, such as low-sulfur fuels, increasing demand for lighter products, and heavier crude oil slates are forcing refiners to increase secondary processing and conversion of heavy residues. The introduction of carbon capture and storage at refineries while helping to offset increased emissions would significantly boost energy use as well.

The overall objective of CitInES is to design and develop two multi-scale multi-energy decision support tools to optimize the energy strategy of cities and large industrial complexes by defining sustainable, reliable, and cost-effective long-term energy plans. Demonstrations of the tools took place in two cities of Italy, Cesena and Bologna, and in Tüpraş, operating four oil refineries, with a total of 28.1 million tons annual crude oil processing capacity [2]. This paper discusses the industrial use case of Tüpraş, mainly focusing on the production and bidding guidelines and field testing.

Innovative energy system modeling and optimization algorithms have been designed for optimizing energy strategies through detailed simulations of local energy generation, storage, transport, distribution, and demand, including demand-side management and functionalities enabled by smart grid technologies [3]. All energy vectors (electricity, gas, heat), end uses (heating, air conditioning, lighting, transportation), and sectors (residential, industrial, tertiary, urban infrastructure) are used to draw a holistic map of the city/industry energy behavior [2].

The decision support tool used in the project is named Crystal Industry, an analysis software developed by a French project partner, Artelys [4, 5]. The motivation of the project partners was to better understand the dynamics and economics of the energy systems, to design new energy-efficient management strategies, and to evaluate the economic and environmental impacts of new investment projects. Crystal Industry is a decision support tool which models complex energy plants and operational constraints, generates optimized management strategies, and simulates current and new strategies to estimate their respective efficiency. It is designed as an off-line tool supporting energy management studies that require massive amount of data and complex mathematical/statistical functions. The main purpose of the Crystal Industry software is to study the behavior of the plant in operational conditions that are different from the current conditions (either due to the application of new management rules or the presence of new assets). In practice, two different types of studies can be performed, either management strategy studies or investment studies.

To answer the energy management needs of Tüpraş İzmit Refinery, the following functionalities were developed within Crystal Industry: detailed modeling of the physical and operational constraints of energy production assets and of their interactions, ability to replay historical strategies to enable rigorous comparison of optimized or personalized management strategies with the current operational rules, and study-oriented user interface specifically designed to support energy consultants or experts that perform complex energy management studies and make use of highly configurable simulation result views that enable detailed analysis and comparison of various strategies.

The CitInES project showed that the sole optimization of the operational management of turbine assets can lead up to a 7 % decrease of costs and CO_2 emissions. This joint decrease is due to the direct impact of the usage of combustible fuels on CO_2 emissions. Simple and practical management guidelines based on the optimal solution have been determined to help the refinery reach cost and emission reductions almost as high as would have been with the Crystal Industry optimizer. The guidelines were established for the purpose of managing Tüpraş İzmit Refinery Power Plant processes with optimum decisions and to conduct electricity trading from the market. The guideline helped the decision makers decide on which equipment to be switched on or not and the load of these assets, with respect to the steam and electricity demand of the refining operations. In addition, day-ahead import and export bids for the electricity market were formed with the help of the bidding guideline.

21.2 Demo Plant

Through distillation and chemical reactions, refineries convert crude oil into a variety of valuable fuels and lubricants, as well as feedstock for downstream process. Energy supply is very important for refineries since refining processes are highly energy dependent. The energy demand of İzmit Refinery is mostly covered by the power plants.

Tüpraş refining processes use energy in the forms of fuel, steam, and electricity. Refinery fuel gas (a by-product of refining processes), heavy fuel oil, and natural gas are the main fuels used in refineries. Produced steam is distributed through four pressure levels: very high pressure (VHP, 70 bar-g), high pressure (HP, 42 bar-g), medium pressure (MP, 13 bar-g), and low pressure (LP, 6 bar-g). VHP and HP are generated by utility boilers and the waste heat boiler of gas turbine. VHP and HP steam can be used by steam turbines to produce electricity or by letdown stations to downgrade steam shown in Fig. 21.1. VHP steam is not distributed to the refinery but downgraded by steam turbines or letdown stations; HP, MP, and LP steams are

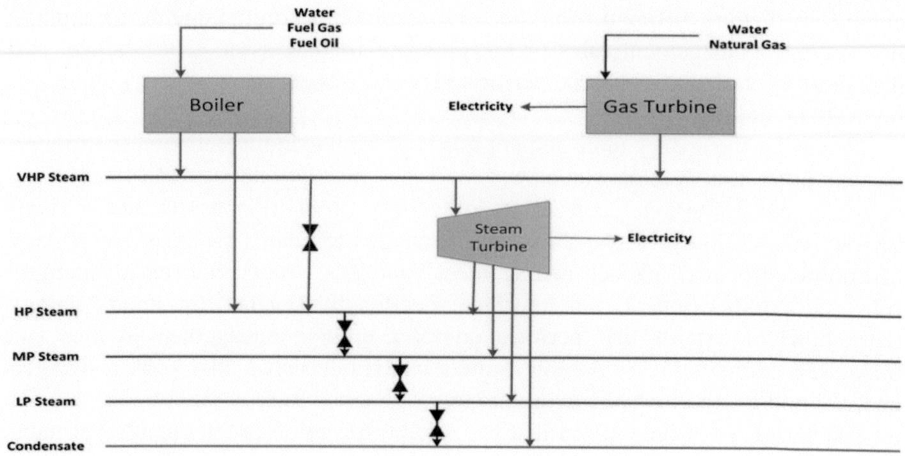

Fig. 21.1 Simplified scheme of power plant

sent to various stages of the refining processes. HP steam is used as the power for pumps and compressors. MP steam is provided to heat exchangers and to distillation columns as stripping steam. LP steam is used by heating services, including process heat exchangers, tank heating, pipeline tracing, and miscellaneous services.

Electricity is consumed mainly by the electric motor-driven pumps, compressors, and fans in air cooler instrumentation systems. The refinery provides its required electrical energy from four steam turbines, one gas turbine, and national grid operated by TEIAS (Turkish Electricity Transmission Corporation). In Tüpraş, natural gas (NG), fuel gas (FG), fuel oil (FO), and water are the raw materials of the power plant. The NG is used only by the gas turbine, whereas fuel gas and fuel oils are used in boiler assets. İzmit Refinery power plants contain eight utility boilers (four very high pressure boilers, four high pressure boilers), four steam turbines, a gas turbine, a waste heat boiler of gas turbine, and letdown stations. The total production capacity of all power plant assets is higher than overall refinery demand. Then it is important to run the right combination of equipment that can supply the demand while minimizing the total cost of energy production.

21.3 Guideline

The energy managers of Tüpraş have to both decide on the production plan of the power plants and enter biddings to the electricity market when it is necessary. The energy management group needed a user-friendly method to operate the power plant at the optimum parameters. In line with this requirement, at first, the power plant was modeled on Crystal Industry. Then power plant operations were optimized by a historical replay of the previous years. These optimum decisions were evaluated by Tüpraş and Artelys, and the guidelines for bidding and production were formed. By following these guidelines, suboptimum operational parameters that are very close to the optimum can be reached.

21.3.1 Bidding Guideline

The day-ahead Turkish market was used for the bidding guideline. When compared with the intraday (balancing) market, the day-ahead market allows an accurate prediction on the prices. This is because each actor is responsible for providing bids on his capacity to sell/buy from/to the grid and associated to their costs. The day-ahead bids are entered at noon each day for the market operations of the next day. The bids are resolved a few hours later. The day after the bidding, the accepted bids should be accommodated by the actors [6]. In the Tüpraş case, the accepted bids need to be coordinated with refinery demand. The bidding guideline was created to help the decision maker in this process.

Electricity sales bids were given by the suggestions of bidding guideline which took advantage of the expected unused capacity of electricity generating assets. When bids are accepted, assets run at higher and more efficient loads, which lead to a decrease in unit costs of energy. The asset, producing electricity with minimum cost, is the gas turbine.

Electricity buying bids are given lower than the unit cost of electricity production of the gas turbine which is dependent on the natural gas price. When a buy bid is accepted, refinery supplies part of its electricity demand from the market at a price lower than its production cost at the gas turbine. This situation rarely occurs since the market price is usually higher than the unit cost of the gas turbine and yields a rather low income.

The day-ahead bidding guideline accepts utility costs, asset capacities, forecasted hourly refinery demands for the next day, and risk margins as inputs. The forecast of the internal demand is the most critical input of the guideline. Indeed, if the refinery bids too much sell capacity and if the local demand happens to be higher than anticipated, it may not be possible to satisfy both demands, and the refinery will not meet its market engagements. Due to using forecasted demand, we take risk, and to mitigate this risk, the following risk margins are entered: a bid price margin and a safety ratio to maximum bid quantity that can be applied. Even though the risk margins decrease the risk taken, they also decrease the profits.

The main output of the bidding guideline is a bidding plan for the next day, which includes the sell and buy quantity as well as the associated price, for each hour of the day. The bidding guideline integrates one buy bid to be used when the gas turbine is no longer profitable compared to the market and four sell bids which are used to profit from available extra capacities for multiple units (gas turbine and other steam turbines).

For each bid, the results provided to the refinery operators are the optimal threshold price and the optimal quantity of electricity with respect to the entered risk margins. These bids are placed on the market, and the accepted bids together with the refinery demand should be supplied by the power plant the next day.

21.3.2 Production Guideline

The production guideline consists of 12 states in which each state lists the assets that should be active, min/max targets to be achieved within the state, and which assets are in load control mode. Also transitions are defined for each state to cope with potential events that will lead to a jump to another state. These transitions have a triggering event which is followed by an action and the destination state. To illustrate, as seen in Fig. 21.2, at state 3.1 when the condition, gas turbine reaches maximum, is triggered, the action is to change the VHP control mode and move to the destination state which is state 3.2.

By using the production guideline, managing the operation at optimum parameters has been quicker and efficient. At any time, the operator knows the state of the system (meaning which assets are running or not) and from this point on, he can use the guideline to decide on the changes to the operation policy in response to demand variation.

21.4 Field Experimentation

The bidding and production guidelines were applied to the power plant beginning in November 2013. Testing of both guidelines was performed separately for a period of 4 months. During the tests, the guidelines were adjusted according to the feedback from the operators and then were commissioned on March 2014.

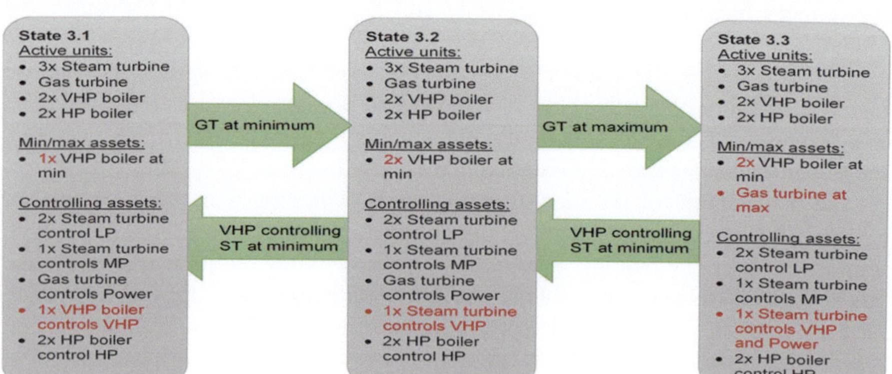

Fig. 21.2 Example for production states and transitions

During the testing phase, in December, the availability of the natural gas supply was limited by the government due to public demand for heating. Electricity prices go high in the winter since power generation in Turkey is highly dependent on natural gas as is the gas turbine. There are two possible actions against a shortage of natural gas: operating a gas turbine with fuel oil which is costly and more polluting than would be with natural gas and purchasing electricity from the grid instead of producing it internally [7]. According to the adjusted bidding guideline, VHP production from the boilers has increased, while gas turbine use has decreased. Buy bids were entered and accepted to make up for the limited power generation.

The production guideline developed by Artelys was applied in February 2014, and during this period, the gas turbine controlled the electricity demand. According to the production guideline, three steam turbines were used for electricity production and supplied steam at lower pressure levels to the refinery. Also two VHP boilers and two HP boilers were operated to provide steam to the plant and steam turbines. At the end of this period, the production, demand, and price data were collected and simulated in Crystal Industry for analysis.

Finally, the updated production and bidding guidelines were experimented simultaneously beginning in March 2014.

21.5 Results and Discussion

During the demonstration of the bidding guideline, from November 2013 to April 2014, total gross profit was obtained as 34,054.70 TL. As seen in Fig. 21.3, in December, monthly total gross profit is low due to natural gas crisis. In January, Tupraş İzmit Refinery underwent a planned shutdown process for general

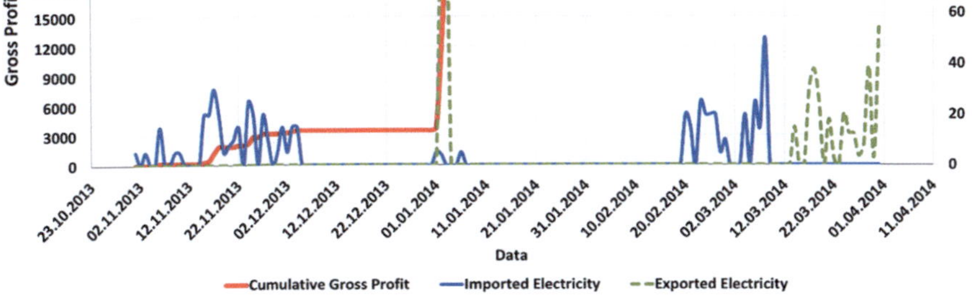

Fig. 21.3 Graph of cumulative gross profit vs. time

maintenance. Beginning with 6th of January, no electricity was imported or exported. Despite the shutdown, the gross profit was high because during the first days of the shutdown, the demand from the refinery was low, while the power plant continued its operation a while longer. The excess amount of electricity was exported with high income rates. In February, gross profit was also low like it was in December because of the shutdown.

In March, an additional 2.2 % cost reduction was achieved using the markets when compared to February 2014 experimentation results. Similar conclusions can be drawn with respect to CO_2 emissions. Furthermore, it should be noted that the energy management used a rather conservative approach, choosing to place market bids with lower capacities than what the guideline suggested. Had management placed bids at the top-end of the guideline, further gains could have been achieved. There is no doubt that the ideal production strategy without markets can be outperformed in the field with the help of a practical bidding guideline that utilizes the Turkish electricity market.

By applying the production guideline, refinery demand was met with the right combination of assets and their loads. According to refinery demand, assets were usually operated on state 3.2. At state 3.2, two VHP boilers run at minimum capacity due to safety considerations but still there is excess VHP production. By applying the management policy of the guideline, LP is only produced from G1 and G2 steam turbines, and MP demand is met from G4 steam turbine. With this combination, by increasing electricity production in steam turbines, the load of gas turbine and its waste heat boiler VHP production is decreased. A comparison of historical and optimal production strategies reveals that the guideline gives better operation instructions with lower production costs and CO_2 emissions and utility costs are reduced by 4 %.

When both production and bidding guidelines were experimented simultaneously beginning in March 2014, 7 % decrease in costs and CO_2 emissions were achieved.

21.6 Conclusion

The motivation to implement an energy efficiency program in a refinery is driven by three major factors: profit margins, asset management, and environmental regulation compliance. However, there are many constraints which make it difficult. Tüpraş experimentation aimed to improve the energy efficiency and reduce the environmental impact of the Tüpraş refinery in Izmit, Turkey, with implementation of new strategies developed. The key benefits gained by Tüpraş from CitInES project are mainly the capability to handle the complex utility system, new convenient and practical strategies to operate the assets, effective usage of the Turkish electricity market, and the ability to select the most appropriate investment projects. The CitInES project showed that the sole optimization of the operational management of boiler and turbine assets can lead up to a 7 % decrease of costs and CO_2

emissions. This joint decrease is due to the direct impact of the usage of combustible fuels on CO_2 emissions. Simple and practical management guidelines based on the optimal solution were determined and applied to help the refinery reach cost and emission reductions almost as high as what the theoretical optimized strategy suggests.

References

1. IPIECA (2007) Saving energy in the oil and gas industry
2. http://www.citines.com/. Accessed 1 Apr 2014
3. Nakata T (2004) Energy-economic models and the environment. Prog Energy Combust Sci 30 (4):417–475
4. http://www.citines.com/crystal-industry/presentation/. Accessed 1 Apr 2014
5. http://www.artelys.com/. Accessed 24 Mar 2014
6. Yucekaya A (2013) Bidding of price taker power generators in the deregulated Turkish power market. Renew Sustain Energy Rev 22:506–514
7. Stoppato A, Mirandola A et al (2012) On the operation strategy of steam power plants working at variable load: technical and economic issues. Energy 37(1):228–236

Chapter 22
Forecasting the Energy Consumption Using Neural Network Approach

Mohamed Bouabaz, Mourad Mordjaoui, Nabil Bouleknafet, and Badreddine Belghoul

Abstract This chapter presents the use of neural network for predicting energy consumption in buildings and their expenditure. Application of artificial intelligence by the use of neural networks to predict the energy consumption for heating rehabilitated buildings is underscored by the need to develop a generic model that can be used for prediction of the consumption of the energy in buildings. The model presented for the prediction of the energy consumption of natural gas has been developed on the basis of data obtained for the winter period. Alternatively, a comparative economic study was conducted. An average error of the training phase for the model was 2.4 %, while the test phase error was 3.2 %. This indicates that the neural network model is presented successfully to predict the energy consumption by using natural gas as clean energy for heating buildings

Keywords Artificial neural network • Optimization • Energy consumption

22.1 Introduction

The use of natural gas in residential housing for heating has a crucial impact in terms of costing. Previous studies have shown that the major proportion of the environmental impact of a residential building is due to the energy consumption for space heating. Heating and cooling systems differ in their source of energy, the type of appliance, and the distribution system; hence, they also vary in their

M. Bouabaz (✉) • N. Bouleknafet • B. Belghoul
Department of Civil Engineering, Faculty of Technology, LMGHU Laboratory, University 20 August 1955, Skikda, Algeria
e-mail: mbouabaz@hotmail.fr

M. Mordjaoui
Department of Electrical Engineering, Faculty of Technology, LRPCSI Laboratory, University 20 August 1955, Skikda, Algeria

I. Dincer et al. (eds.), *Progress in Clean Energy, Volume 1*,
DOI 10.1007/978-3-319-16709-1_22

environmental impacts. Energy efficiency is increased by investing in old buildings regarding the life cycle cost of and the old park housing [1, 2]. Artificial neural network can offer an alternative approach and is applied for complex problems such as modeling and optimization [3–5]. They are inspired from the biological structure of the human brain, which acquires knowledge through a learning process. A neural network is constructed by arranging several units in a number of layers. The output of a single layer provides the input of a subsequent layer and the strength of the output is determined by the connection weights between the processing units of two adjacent layers. The ability of neural network is to learn from examples to detect by themselves the relationships that link inputs to outputs. Artificial neural network are used in solving problems, where numerical solutions are hard to obtain. Few mathematical models have been developed.

22.2 Scope of the Work

The data used in the study for the development and testing were collected from the records of local agencies of energy distribution (Algeria), initiated for the critical period of the 2012. For the generalization of the model the data collected, cover three different regions for three types of residential housings.

22.3 Research Methodology

A research methodology was carried out to achieve the objective of the study. The data collected were selected and analyzed for best configuration of the model. Artificial neural network theories based on energy prediction were investigated. Model based on neural network for predicting and managing the thermal energy consumption was developed and tested on separate data. Regression analysis and back propagation neural network were used to develop energy consumption and cost expenditure models. NeuroSolution for excel was used for developing and testing the artificial neural network model. Regression analysis was used to describe the best performing neural network model in terms of their prediction performance. The accuracy of the model is expressed as follows:

$$\text{Accuracy} = \text{Mean error} \pm \text{Standard deviation} \tag{22.1}$$

$$\text{Mean error} = \frac{1}{n}\sum_{i=1}^{n}\left(\frac{\text{Predicted} - \text{Actual}}{\text{Actual}}\right) \times 100\,\% \tag{22.2}$$

22.3.1 *Application of Neural Network in Predicting the Energy Consumption*

Artificial Neural Network has been used by many researchers for its applicability as a tool for modeling and simulation in building energy consumptions over the past decade. Various neural network architectures have been used in modeling and optimization of energy consumption such as back propagation [6, 7], regarding its ability and preciseness, recurrent neural network, and generalized neural network are also among neural network models. Breekweg et al. [8] investigated a number of artificial neural network techniques in the development of a generalized method for building energy consumption. Radial basis function (RBF) was used for modeling. They suggested the necessity to test the developed ANN model with energy data from various buildings in order to guarantee the generalizing capacity of the model in terms of accuracy.

Melek et al. [9] developed a model using the power consumption of a central chiller plant using climate data for predicting the energy consumption based on artificial neural network method. Regarding their findings, an average absolute training error for the model was 9.7 % while the testing error was 10.0 %. They suggest that the model can successfully predict the particular air-conditioning energy consumption in a tropical climate.

The objective of the study aims to develop a model by application of artificial neural network using parametric data that can predict accurately. The model can be used in optimizing and managing the thermal energy consumption and the cost expenditure in old residential housings.

22.4 Development of Artificial Neural Network Model

22.4.1 *Model Input Output Data*

The Artificial Neural Network has been designed to contain an input layer of eight processing neurons corresponding to the eight input parameters and two output layer of one element each. The hidden layer with ten processing elements was selected after several trails for best configuration architecture during the training phase. Table 22.1 lists the input and output variables of the model development.

Figure 22.1 illustrates the architectural configuration, which consists of main layers.

22.4.2 *The Transfer Function*

The transfer function adopted for the model is the hyperbolic tangent sigmoid function which generates output values between -1 and $+1$ and describes the behavior of the network model. It is expressed as follows:

Table 22.1 The data used in the artificial neural network model development

No	Variables	Input	Output
1	Location	x1	
2	Construction type	x2	
3	Temp max	x3	
4	Temp mini	x4	
5	Humidity max	x5	
6	Humidity mini	x6	
7	Rainfall	x7	
8	Wind speed	x8	
9	Energy consumption (natural gas)		y1
10	Costs expenditure		y2

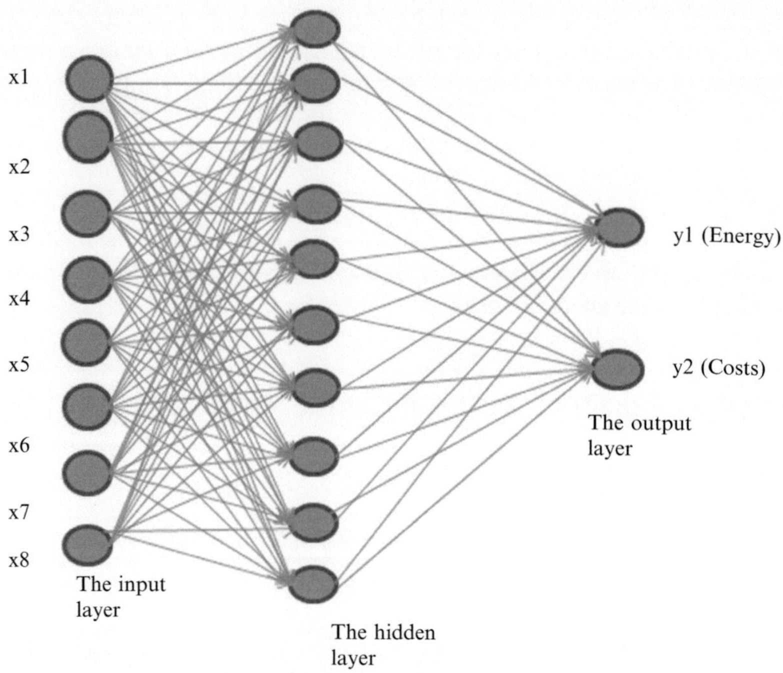

Fig. 22.1 The neural network model architecture

$$\text{Tansig}\,(x) = \frac{2}{1 + \exp^{(-2x)}} - 1 \tag{22.3}$$

The back propagation algorithm involves the gradual reduction of the error between model output and the output [10]. It develops the input to output by minimizing a mean square error.

The mean square error is expressed as follows:

$$\mathrm{Mse} = \frac{\sqrt{\sum_{i=1}^{n} (xi - Y(i))^2}}{n} \tag{22.4}$$

where n is the number of observations to be examined in the training phase, xi is the model output, and Y is the target. The mean square error is a good overall measure.

22.5 Results and Discussion

The results of the models at testing phase are presented in Table 22.2, according to the variation errors for each observation.

22.5.1 *Accuracy*

The accuracy of the cost model developed by artificial neural network is very accurate. It has been shown from the results that the model performs well and no significant difference could be discerned between the predicted and the actual values [11].

Table 22.2 Variation errors at tested phase

	Energy consumption			Costs		
No	Actual	Predicted	Error (%)	Actual	Predicted	Error (%)
1	402	425	5.721	1,641.36	1,645	0.221
2	276	260	−5.797	1,126.9	1,120	−0.612
3	307	300	−2.280	1,253.48	1,260	0.520
4	252	261	3.571	1,028.9	1,030	0.106
5	214	220	2.803	873.7	855	−2.140
6	985	980	−0.507	4,021.7	4,101	1.971
7	347	345	−0.576	1,416.8	1,425	0.578
8	390	395	1.282	1,592.4	1,601	0.540
9	260	268	3.076	1,061.6	1,108	4.370
10	927	925	−0.215	3,784.9	3,795	0.266
11	298	302	1.342	1,216.7	1,214	−0.221
12	478	470	−1.673	1,951.4	2,001	2.541
13	1,198	1,190	−0.667	4,891.4	4,870	−0.437
14	1,532	1,548	1.044	6,255	6,280	0.399
15	1,024	1,040	1.562	4,181	4,175	−0.143
16	1,112	1,140	2.5179	4,540	4,662	2.687
17	810	825	1.851	3,307	3,300	−0.211
18	3,679	3,670	−0.244	15,021	15,035	0.093
19	855	860	0.584	3,491	3,598	3.065
20	4,850	4,830	−0.412	19,802	19,823	0.106

Table 22.3 Expected accuracy at testing phase

Model	Maximum positive error (%)	Maximum negative error (%)	Mean error (%)	Standard deviation (%)	Accuracy range (%)
Energy	+5.721	−5.797	0.649	2.453	+3.102
					−1.804
Costs	+4.370	−2.140	0.685	1.506	+2.19
					−0.822

A level of accuracy is expected at the testing phase; this is to ensure validity of generated model. From the analysis extracted from Table 22.2, maximum expected negative error is −2.14 while maximum positive error is +4.37. This generated mean error of 0.685, with an absolute standard deviation of 1.506 and a percentage accuracy range within +2.19 to −0.822 for cost prediction.

For the energy consumption, the maximum negative error is −5.797 while maximum positive error is +5.721. The mean error is of 0.649, with an absolute standard deviation of 2.453 and a percentage accuracy range within +3.102 to −1.804.

Therefore, Table 22.3 contains the summary of error, standard deviation, and range of expected accuracy.

22.5.2 Regression Analysis

Regression analysis was used to ascertain the relationship between the predicted and actual values. The results of analysis are illustrated graphically. The statistics can be interpreted that the correlation coefficient is 0.99, indicating that, there is a good linear correlation between the actual and the predicted neural network values at tested phase. The results of linearly regressing are shown graphically in Figs. 22.2 and 22.3.

The regression models were compared graphically for closeness of fit and prediction performance based on the predicted versus the actual values.

The correlation between the actual and the predicted values of the ANN model for prediction of energy consumption and their costs yielded a good result in terms of accuracy as shown in Fig. 22.2a and b, respectively. The scatter plots in these figures at testing phase revealed that, for both models, the predictability was satisfactory and data points were well fitted over slope line.

22.5.3 The Models Meaning

The performance of the model is evaluated by using the percentage error. The percentage error of each model is shown graphically in Figs. 22.4 and 22.5 for the 20 testing samples.

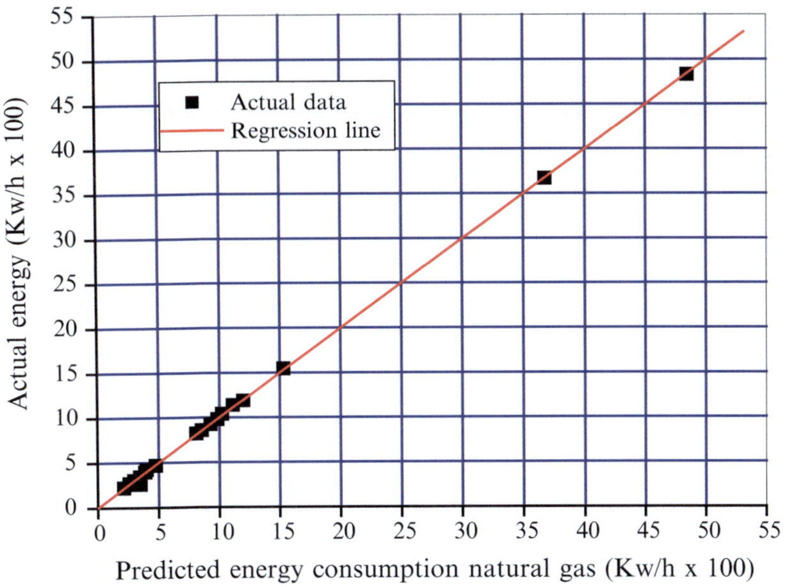

Fig. 22.2 Regression line of actual energy versus predicted energy consumption

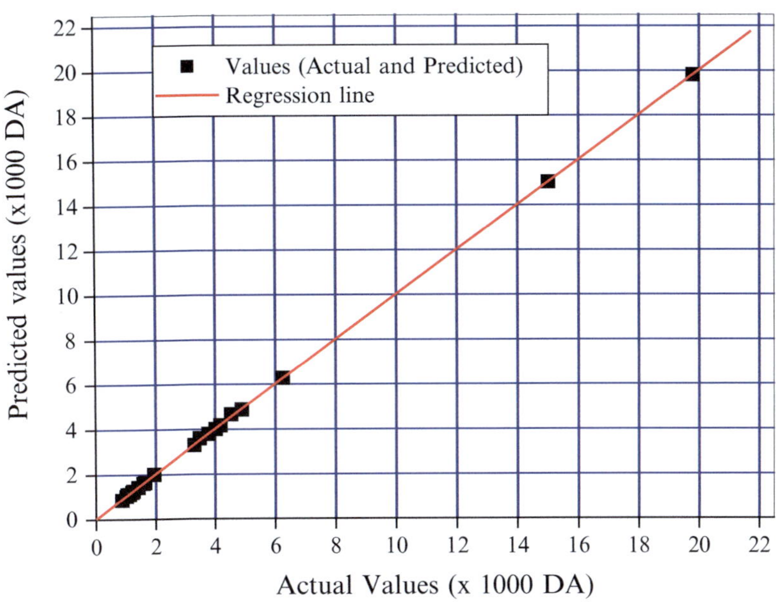

Fig. 22.3 Regression line of actual costs versus predicted values at testing phase

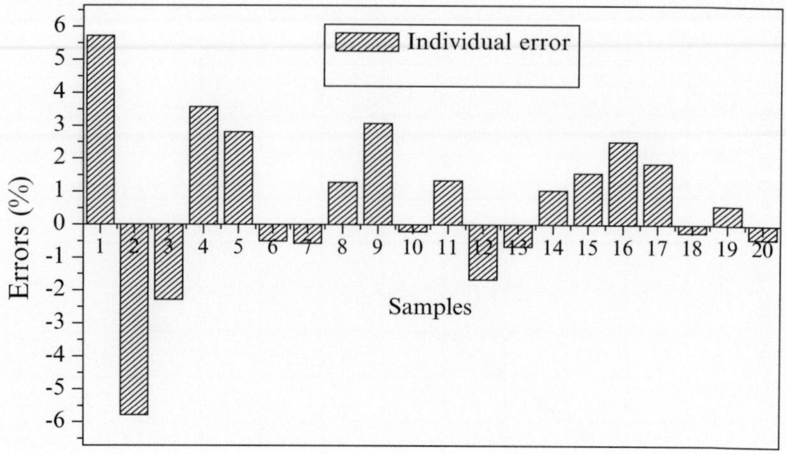

Fig. 22.4 Error variation at testing phase for energy consumption

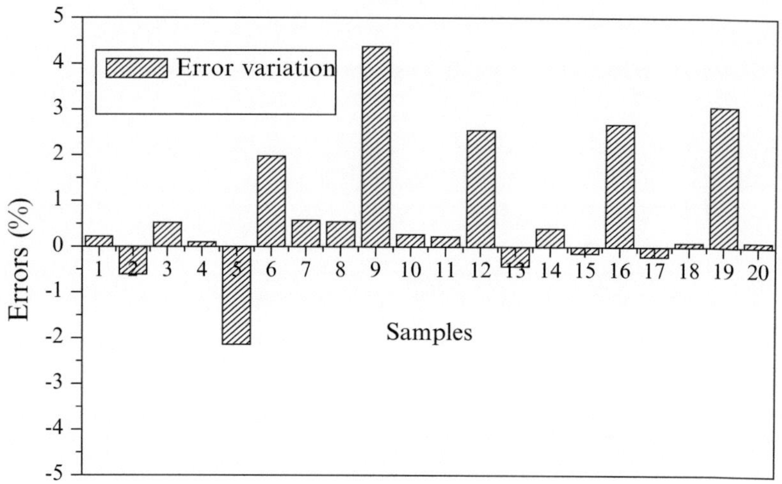

Fig. 22.5 Error variation at testing phase for costs prediction at testing phase

22.5.4 Sensitivity

Sensitivity analysis is a method used to determine the influence of each input parameters to output variable. Parameter sensitivity is performed as a series of tests showing which input parameter causes a change in the dynamic behavior of the structure. Sensitivity analysis can also indicate which parameter values are reasonable to use in the model. Therefore, the extraction of x2 reduces the accuracy of the model by 1 % which bears less significant than the other parameters as shown graphically in Fig. 22.6.

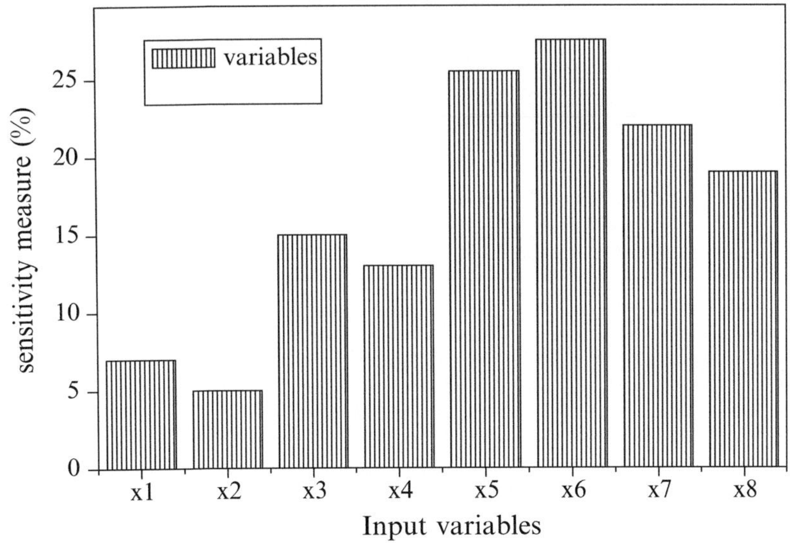

Fig. 22.6 Sensitivity measure about the mean

22.6 Conclusions

This chapter has proposed an artificial neural network model for forecasting energy consumption with its expenditure in terms of cost for a specific type of residential housings. The method has provided successful results regarding its effectiveness and preciseness. The main advantage of the proposed scheme is its simplicity for use. If a problem is particularly complicated, it may be broken down into a number of parts, each of which is then solved separately by its own network module. More often than not, the choice of connection scheme for a network is arbitrary. For some systems, however, the connection scheme is defined precisely by the problem under consideration. By the introduction of supervised and unsupervised training schemes, we may operate both prior to and during application of the network. It is recommended to further investigations of neural networking to other domains of energy consumption modeling techniques and their management, such as energy economics and consumption, by adopting appropriate methods of solving different classes of problems arising in energy research.

References

1. Entrop AG, Brouwers HJH, Reinders AHME (2010) Evaluation of energy performance indicators and financial aspects of energy saving techniques in residential real estate. Energy Build 42(5):618–629
2. Akande OK, Adebamowo MA (2010) Indoor thermal comfort for residential buildings in hot-dry climate in Nigeria. In: Proceedings of conference: adapting to change: new thinking on comfort Cumberland Lodge, Windsor, 2010

3. Emsley MW, Lowe DJ, Duff AR, Harding A, Hickson A (2002) Data modelling and the application of a neural network approach to the prediction of total costs. Constr Manag Econ 20:465–472
4. Bouabaz M, Hamami M (2008) A cost estimation model for repair bridges based on artificial neural networks. Am J Appl Sci 5(4):334–339
5. Ansett M, Kreider JF (1993) Application of neural networking models to predict energy use. ASHRAE Trans 99(1):505–517
6. Kalogirou SA (2001) Artificial neural networks in renewable energy systems applications: a review. Renew Sustain Energy Rev 5:373–401
7. Kreider JF, Claridge DE, Curtiss P, Haberl JS, Krarti M (1995) Building energy use prediction and system identification using recurrent networks. Trans ASME J Sol Energy Eng 117:161–166
8. Breekweg MRB, Gruber P, Ahmed O (2000) Development of generalized neural network model to detect faults in building energy performance. ASHRAE Trans 43(72):61–93
9. Melek Y, Sedat A (2005) Artificial neural networks application in building energy predictions and a case study for tropical climates. Int J Energy Res 29:891–901
10. Michelis-Tzanakou E (2011) Artificial neural networks: an overview. Netw Comput Neural Syst 22:208–230
11. Ashworth A, Skitmore RM (1982) Accuracy in estimating. Occasional paper no .27. Chartered Institute of Building, London

Chapter 23
Complex-Valued Wavelet Neural Network Prediction of the Daily Global Solar Irradiation of the Great Maghreb Region

Lyes Saad Saoud, Fayçal Rahmoune, Victor Tourtchine, and Kamel Baddari

Abstract In this chapter, the prediction of the daily global solar irradiation of the great Maghreb region using the complex-valued wavelet neural network (CVWNN) is presented. Both multi-input single output (MISO) and multi-input multi-output (MIMO) strategies are considered. The meteorological data of the capitals of the great Maghreb, which are Tripoli (Libya), Tunis (Tunisia), Algiers (Algeria), Rabat (Morocco), El Aaiun (Western Sahara), and Nouakchott (Mauritania), are used like samples from each country. To test the applicability and the feasibility of the CWNN to predict the daily global irradiation for the great Maghreb case, several models are presented. Results obtained throughout this chapter show that the CWN technique is suitable for prediction of the daily solar irradiation of the great Maghreb region.

Keywords Solar irradiation • Complex-valued wavelet neural networks • Forecasting • Great Maghreb

Nomenclature

CVWNN	Split complex-valued wavelet neural networks
n	Number of inputs
m	Number of neurons in the hidden layer
l	Number of output
X_n	Input vector
t_m	Translations of the hidden neurons

L. Saad Saoud (✉) • F. Rahmoune • V. Tourtchine • K. Baddari
Laboratory of Computer science, Modeling, Optimization and Electronic Systems (LIMOSE), Department of Physics, Faculty of sciences, University M'Hamed Bougara, Boumerdès 35000, Algeria
e-mail: saadsaoudl@umbb.dz

I. Dincer et al. (eds.), *Progress in Clean Energy, Volume 1*,
DOI 10.1007/978-3-319-16709-1_23

d_m	Dilations of the hidden neurons
$f_{\mathbb{C}}^1(.)$	Complex-valued wavelet used for the hidden layer
$f_{\mathbb{C}}^2(.)$	Complex-valued activation function used for the output layer
$j=\sqrt{-1}$	Imaginary unit
y_l	lth desired output
$\hat{y}_l$	lth predicted output
$t_{\mathbb{C}}(d)$	Complex-valued temporal index, $d = 1, \ldots, 365$
T_m	Daily air temperature, °C
H_m	Relative humidity, %
G_m	Daily global solar irradiation, kJ/m^2
T_d	Complex-valued daily air temperature
H_d	Complex-valued relative humidity
G_d	Complex-valued daily global solar irradiation
nRMSE	Normalized root mean squared error, %
R^2	Coefficient of determination, %
MAE	Mean absolute error, %
N	Number of samples
MIMO	Multi input multi output
MISO	Multi input single output

23.1 Introduction

Global solar irradiation is considered as the most important parameter in the design of renewable and solar energy systems, particularly for the sizing of Photovoltaic (PV) systems [1]. This parameter is influenced by the sunlight variation and other meteorological parameters which make it variable from location to another. As it is known, the great Maghreb has a great area occupied by the desert of Sahara which gives it the opportunity to be an advantageous place for solar systems installations. Due to the great nonlinearity of the solar irradiation model and the dependence between this last parameter with other meteorological and geo-astronomical data, many models based on computational techniques dedicated to predict the solar irradiation could be found in literature [2–12], whereas, the neural networks are widely used to forecast solar irradiation with good accuracy [13–19]. Complex-Valued Neural Networks (CVNNs) prove their abilities for the identification of nonlinear systems. They can outperform their real counterparts in many ways [20]. CVNNs give also the possibility of the simultaneous modeling and forecasting [21], where they have been used to forecast the wind's speed and direction simultaneously. The main motivation to use CVNNs is the faster convergence, reduction in learning parameters, and the ability to learn two dimensional motion of signal in complex-valued neural network [22]. The CVNNs are simply the generalization of the real-valued neural networks in the complex-valued domain, where all the parameters including weights, biases, inputs, and outputs could be complex variables. In real-valued strategy, the wavelet theory has found many applications in function approximation, numerical analysis, and signal processing [23]. Due to the similarity between wavelet decomposition and one-hidden-layer neural networks, the idea of combining both wavelets and neural networks has been proposed in various works [24–29]. This advantage motivates other researchers to move this strategy from the real-valued domain to the complex-valued one. For instance, Özbay et al. [30] proposed the split complex-valued wavelet artificial neural network, where Mexican hat and Haar wavelet functions have been used as activation functions. They showed that this architecture is suitable to classify Doppler signals. On the other hand, the split complex-valued neural networks have showed to be suitable and promising technique for forecasting the hourly and the daily solar irradiation [19]. In this chapter, split complex-valued wavelet artificial neural network [30] is applied to forecast the daily global solar irradiation of the great Maghreb region. The multi input single output (MISO) and the multi input multi output (MIMO) strategies are used. A comparison between the technique presented throughout this chapter and the split complex-valued neural networks is given.

23.2 Complex-Valued Wavelet Neural Networks

The split complex-valued wavelet neural networks (CVWNN) architecture is proposed in [30]. The main difference between the CVWNN and the traditional complex-valued neural networks is coming from the activation function. In the CVWNN architecture, one type of wavelet function is used. In other words, the CVWNN is simply complex-valued neural networks using a wavelet function as activation function. The complex Back-Propagation (BP) algorithm, which is the complex-valued version of the real-valued back propagation algorithm, is widely used to train the CVNNs [19, 31, 32]. This algorithm could be used easily to train the CVWNN [30].

Let us take the complex-valued wavelet neural network which has n inputs, m neurons in the hidden layer, and l outputs. The lth network's output could be calculated as follows:

$$\hat{y}_l(k) = f_{\mathrm{C}}^2\left(W_{l0}^2 + \sum_m W_{lm}^2 H_m\right) \tag{23.1}$$

And the output of each hidden neuron m is given like:

$$H_m = f_{\mathrm{C}}^1(U_m) \tag{23.2}$$

where $X_n = [y(k-1), y(k-2), \ldots, y(k-i), u(k-1), u(k-2), \ldots, u(k-q)]$, which contains $n = i + q$ complex-valued input. With n, m, i, l, and q are positive integers. U_m is given as follows:

$$U_m = \frac{t_m - \sum_n W^1_{mn} X_n}{d_m}$$

W^2_{l0} and W^2_{lm} are the biases and the weights from the hidden to the output layers, W^1_{mn} are the weights from the input to hidden, and t_m and d_m are translations and dilations of the hidden neurons.

$f^1_{\mathbb{C}}(.)$ is a complex-valued wavelet used as an activation function for the hidden layer.

According to the Liouville's theorem, in which the analytic and bounded functions on entire complex plane are constant, the complex function takes a great attention and several complex activation functions proposed in the literature [24]. In this chapter, the split Mexican Hat function is given as follows:

$$f^1_{\mathbb{C}}(U_m) = \left(1 - a(\mathrm{Re}(U_m))^2\right)\exp\left(-\frac{b}{2}(\mathrm{Re}(U_m))^2\right) + j\left(1 - a(\mathrm{Im}(U_m))^2\right)\exp\left(-\frac{b}{2}(\mathrm{Im}(U_m))^2\right) \quad (23.3)$$

where $z = x + jy$, $j = \sqrt{-1}$, a, and b are reals, in our case they are taken: $a = 2.7$ and $b = 3$.

$$f^2_{\mathbb{C}}(U_l) = \tanh(\mathrm{Re}(U_l)) + j\tanh(\mathrm{Im}(U_l)) \quad (23.4)$$

And $U_l = W^2_{l0} + \sum_m W^2_{lm} H_m$

The objective is to find the complex-valued parameters that minimize the cost function given as follows:

$$E(k) = \frac{1}{2}\sum_l |e_l(k)|^2 \quad (23.5)$$

where $e_l(k)$ is the error between the lth desired output $y_l(k)$ and the lth predicted output $\hat{y}_l(k)$ which is given as follows:

$$e_l(k) = y_l(k) - \hat{y}_l(k) \quad (23.6)$$

23.3 Data Preparation

In this study, the six cities of the great Maghreb (Fig. 23.1) are taken to implement the CVWNN technique. The miss of experimental data (or the difficulty to obtain them) of the whole region requires us to use satellite data. Satellite meteorological data (daily solar irradiation, daily air temperature, and daily relative humidity) are obtained from the official website of NASA [33] for the following cities:

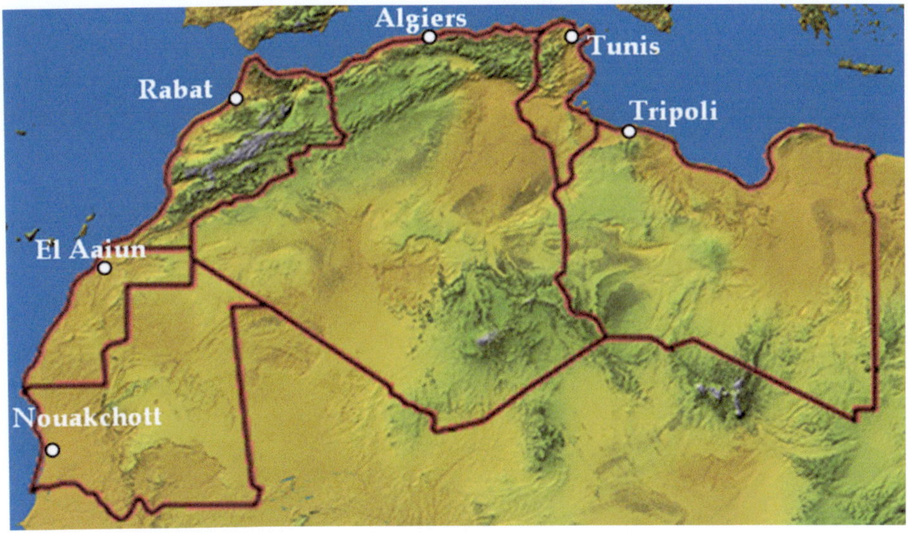

Fig. 23.1 The Great Maghreb with its capitals (modified form [34])

Tripoli, Libya: Latitude = 32°53 N, Longitude = 13°11 E
Tunis, Tunisia: Latitude = 36°49 N, Longitude = 10°10 E
Algiers, Algeria: Latitude = 36°45 N, Longitude = 3°3 E
Rabat, Morocco: Latitude = 34°1 N, Longitude = 6°50 W
El Aaiun, Western Sahara: Latitude = 27°8 N, Longitude = 13°13 W
Nouakchott, Mauritania: Latitude = 18°5, Longitude = 15°59 W

To make the data useful to the CVWNN, it should be transformed into the complex-valued domain. The same procedure given in [19] will be used in this work. Therefore, each year's day will be represented by an angle, with 1 year is assumed to be a circle (2π). In this case, the time index will be integrated into the data itself which produces one input that contains two simultaneous parameters (the meteorological data and the time index). The complex-valued temporal index $t_C(d)$ for the daily case is given by the following Eq. (23.7) [19]:

$$t_c(d) = \exp(j2\pi d/365) \tag{23.7}$$

where $d = 1, \ldots, 365$, represent the day number, with the first day is the 1st January.

By multiplying the obtained complex-valued time index with each meteorological parameter (T_m, H_m, G_m), the complex valued of: the daily air temperature T_d, relative humidity H_d, and the daily global solar irradiation G_d could be found, which are given by the following equations:

$$T_d = T_m \exp(j2\pi d/365) \tag{23.8}$$

$$H_d = H_m \exp(j2\pi d/365) \tag{23.9}$$

$$G_d = G_m \exp(j2\pi d/365) \tag{23.10}$$

The satellite data for the six cities (Tripoli, Tunis, Algiers, Rabat, El Aaiun, Nouakchott) are daily air temperature, relative humidity, and daily solar irradiation in the duration 01/01/2003 to 06/30/2005. Two years (2003–2004) are used to train the CVWNN and the rest (180 days) for its validation.

To have an idea about the prediction performance of the applied algorithm, the normalized Root Mean Squared Error (nRMSE) [13], the coefficient of determination (R^2), and the Mean Absolute Error (MAE) [35] given below are taken like criteria for all examples (its units are in %), where N is the number of samples:

$$\text{nRMSE} = \frac{\sqrt{\frac{1}{N}\sum_{k=1}^{N} |y_k - \hat{y}_k|^2}}{\overline{y}} \tag{23.11}$$

$$R^2 = 1 - \frac{\sum_{k=1}^{N} |y_k - \hat{y}_k|^2}{\sum_{k=1}^{N} |y_k - \overline{y}|^2} \tag{23.12}$$

$$\text{MAE} = \frac{\sum_{k=1}^{N} |y_k - \hat{y}_k|}{N} \tag{23.13}$$

It should be noted that MAE in % is calculated by multiplying the obtained value by 100 and dividing the result by the maximum of the measured data.

23.4 Results and Discussions

In this part, two strategies based on the CVWNN are presented. In the first time, the daily solar irradiation of the great Maghreb region is forecasted using the MISO strategy. The air temperature or/and the relative humidity are used to predicted 1 day ahead of the solar irradiation. These results are presented in Tables 23.1 and 23.2 for the case of predicting on day ahead of the daily solar irradiation prediction. It should be mentioned that we use the notation $<I \times H \times O>$ in all tables, where I, H, and O represent the number of neurons in input layer, the number of neurons in the hidden layer and the number of neurons in the output layer, respectively. According to the values of I, H, and O several architectures could be obtained.

We can see in Table 23.1, for the all cities, the temperature alone gives a good estimation results comparing to the relative humidity. Where the better results are obtained using air temperature for the city El Aaiun (MAE = 9.778 %

Table 23.1 Obtained results using CVWNN in the case of one meteorological input

Model's structure			Measure's criteria		
	$<I \times H \times O>$	Capitals	MAE (%)	nRMSE (%)	R^2 (%)
$\hat{G}_d(k) = f(T_d(k-1))$	$<1 \times 30 \times 1>$	Tripoli	11.419	24.44	91.19
		Tunis	12.950	30.73	87.27
		Algiers	15.892	33.55	85.22
		Rabat	12.532	22.80	92.22
		El Aaiun	9.778	17.67	95.02
		Nouakchott	13.755	22.69	91.62
$\hat{G}_d(k) = f(H_d(k-1))$	$<1 \times 30 \times 1>$	Tripoli	15.155	29.38	87.11
		Tunis	13.216	30.90	87.11
		Algiers	13.547	29.67	88.40
		Rabat	12.933	25.02	90.65
		El Aaiun	9.856	17.57	95.07
		Nouakchott	19.391	31.13	84.17

Table 23.2 Obtained results using CVWNN in the case of the two meteorological inputs

Model's structure			Measure's criteria		
	$<I \times H \times O>$	Capitals	MAE (%)	nRMSE (%)	R^2 (%)
$\hat{G}_d(k) = f(T_d(k-1), H_d(k-1))$	$<2 \times 30 \times 1>$	Tripoli	13.280	26.15	89.90
		Tunis	12.119	28.76	88.84
		Algiers	14.347	30.91	87.43
		Rabat	11.906	22.23	92.60
		El Aaiun	8.838	16.22	95.79
		Nouakchott	8.696	17.83	94.81

nRMSE = 17.67 % and R^2 = 95.02 %) compared to the other cities. The results for the city Algiers are the worst (MAE = 15.892 %, nRMSE = 33.55 %, and R^2 = 85.22 %). Using the relative humidity does not decrease a lot the results (almost the same results are obtained). In the case of using both the two meteorological parameters (Table 23.2), the results are a little improved for all the cities except Nouakchott city, where a great improvement are presented (MAE = 8.696 % nRMSE = 17.83 %, and R^2 = 94.81 %).

By introducing the past values of the daily solar irradiation (five values in this work), the results are improved (Table 23.3). We can see also in the same Table 23.3 that changing the number of neurons in the hidden layer does not produce a great improvement (almost 1 %).

In the MIMO case, two structures are presented for all cities. At the first time, the 25 past days of each month are used to forecast the rest 5 days in parallel (i.e., predict $\hat{G}_{26d}(k+1), \hat{G}_{27d}(k+1), \ldots, \hat{G}_{30d}(k+1)$ using $G_{1d}(k), G_{2d}(k), \ldots, G_{25d}(k)$ of each month). Three numbers on neurons (30, 50, and 100) in the hidden layer are used for all cities. The results of this strategy are given in Table 23.4.

Table 23.3 Obtained results using CVWNN in the case of the two meteorological inputs and delayed solar irradiation

Model's structure			Measure's criteria		
Forecasting $\hat{G_d}(k)$ using:	$<I \times H \times O>$	Capitals	MAE (%)	nRMSE (%)	R^2 (%)
$T_d(k-1), H_d(k-1), G_d(k-1), G_d(k-2)$ $G_d(k-3), G_d(k-4), G_d(k-5))$	$<7 \times 30 \times 1>$	Tripoli	10.837	23.15	91.98
	$<7 \times 100 \times 1>$		10.718	23.22	91.91
$T_d(k-1), H_d(k-1), G_d(k-1), G_d(k-2)$ $G_d(k-3), G_d(k-4), G_d(k-5))$	$<7 \times 30 \times 1>$	Tunis	11.798	28.45	89.01
	$<7 \times 100 \times 1>$		11.859	28.33	89.12
$T_d(k-1), H_d(k-1), G_d(k-1), G_d(k-2)$ $G_d(k-3), G_d(k-4), G_d(k-5))$	$<7 \times 30 \times 1>$	Algiers	12.210	26.96	90.30
	$<7 \times 100 \times 1>$		12.716	27.31	90.06
$T_d(k-1), H_d(k-1), G_d(k-1), G_d(k-2)$ $G_d(k-3), G_d(k-4), G_d(k-5))$	$<7 \times 30 \times 1>$	Rabat	11.820	21.86	92.80
	$<7 \times 100 \times 1>$		10.774	20.97	93.38
$T_d(k-1), H_d(k-1), G_d(k-1), G_d(k-2)$ $G_d(k-3), G_d(k-4), G_d(k-5))$	$<7 \times 30 \times 1>$	El Aaiun	10.062	17.83	94.79
	$<7 \times 100 \times 1>$		9.199	16.74	95.34
$T_d(k-1), H_d(k-1), G_d(k-1), G_d(k-2)$ $G_d(k-3), G_d(k-4), G_d(k-5))$	$<7 \times 30 \times 1>$	Nouakchott	10.193	18.31	94.46
	$<7 \times 100 \times 1>$		8.726	17.11	95.21

Table 23.4 Obtained results using the CVWNN for forecasting 5 days of each month

Model's structure			Measure's criteria		
Forecasting $\{\hat{G}_{26d}(k), \ldots, \hat{G}_{30d}(k)\}$ using	$<I \times H \times O>$	Capitals	MAE (%)	nRMSE (%)	R^2 (%)
$G_{1d}(k-1), G_{2d}(k-1), \ldots, G_{25d}(k-1)$	$<25 \times 30 \times 5>$	Tripoli	0.335	3.43	97.46
	$<25 \times 50 \times 5>$		0.578	5.40	95.91
	$<25 \times 100 \times 5>$		0.091	0.81	99.50
$G_{1d}(k-1), G_{2d}(k-1), \ldots, G_{25d}(k-1)$	$<25 \times 30 \times 5>$	Tunis	0.142	1.53	97.00
	$<25 \times 50 \times 5>$		0.244	3.10	96.68
	$<25 \times 100 \times 5>$		0.040	0.49	99.67
$G_{1d}(k-1), G_{2d}(k-1), \ldots, G_{25d}(k-1)$	$<25 \times 30 \times 5>$	Algiers	0.258	2.83	94.47
	$<25 \times 50 \times 5>$		0.579	6.21	95.10
	$<25 \times 100 \times 5>$		0.016	0.18	99.80
$G_{1d}(k-1), G_{2d}(k-1), \ldots, G_{25d}(k-1)$	$<25 \times 30 \times 5>$	Rabat	0.215	2.04	97.75
	$<25 \times 50 \times 5>$		0.104	0.93	99.05
	$<25 \times 100 \times 5>$		0.144	1.39	99.35
$G_{1d}(k-1), G_{2d}(k-1), \ldots, G_{25d}(k-1)$	$<25 \times 30 \times 5>$	El Aaiun	0.150	1.35	97.51
	$<25 \times 50 \times 5>$		0.379	3.24	98.58
	$<25 \times 100 \times 5>$		0.127	1.03	99.75
$G_{1d}(k-1), G_{2d}(k-1), \ldots, G_{25d}(k-1)$	$<25 \times 30 \times 5>$	Nouakchott	0.166	1.41	99.12
	$<25 \times 50 \times 5>$		0.057	0.44	99.62
	$<25 \times 100 \times 5>$		0.044	0.35	99.89

In this case, 100 neurons in the hidden layer give the better results for all cities, where the best of them are obtained for Nouakchott city (MAE = 0.044 %, nRMSE = 0.35 %, and $R^2 = 99.89$ %). As a sample of the obtained results for this case (i.e., forecasting the last 5 days of each month using the past 25 days), the results for the cities: Tripoli, Tunis, and Algiers are given in Figs. 23.2, 23.3, and 23.4. We can see in these figures the promising results either in the complex plane or in the temporal one.

Secondly, the same method (i.e., MIMO case) is used, where the last 15 days are forecasted using the 15 past days (i.e., predict $\hat{G}_{16d}(k+1), \hat{G}_{17d}(k+1), \ldots, \hat{G}_{30d}(k+1)$ using $G_{1d}(k), G_{2d}(k), \ldots, G_{15d}(k)$ of each month). These results are given in Table 23.5, in which we can see the good results. The cities Rabat, El Aaiun, and Nouakchott are chosen as examples for the case of prediction 15 days of each month using the past 15 days of each month. Results for the above selected cities are shown in Figs. 23.5, 23.6, and 23.7, in which the data are represented in both polar and temporal domains and the measured versus predicted outputs as well. We can see the high performance and quality given by the CVWNN in the MIMO case compared with MISO one.

Comparing the network used in this chapter (CVWNN) with the other strategies such as (CVNN [19]), we can see that both of them give a promising results. But and according to that the CVWNN has wavelet functions in its hidden layer, this network keeps the advantages of the traditional real-valued wavelet network and also works like the CVNN due to its output layer which is the split hyperbolic tangent. In the case of forecasting specially, the complex-valued wavelet neural network deals better than the CVNN for forecasting the daily solar irradiation in the region of the great Maghreb, but with more neurons in the hidden layer.

23.5 Conclusion

In this work, the complex-valued wavelet neural network technique is used to forecast the daily solar irradiation for the great Maghreb region. The CVWNN models have been validated using satellite data in the Great Maghreb, where the cities Tripoli (Libya), Tunis (Tunisia), Algiers (Algeria), Rabat (Morocco), El Aaiun (Western Sahara), and Nouakchott (Mauritania) are selected for the collection of data. The CVWNN proves its abilities to predict the daily solar irradiation in both cases (MISO and MIMO strategies) for all cities. Hence and according to the obtained results, we can say that the CWVNN is very suitable for forecasting the daily solar irradiation in this region.

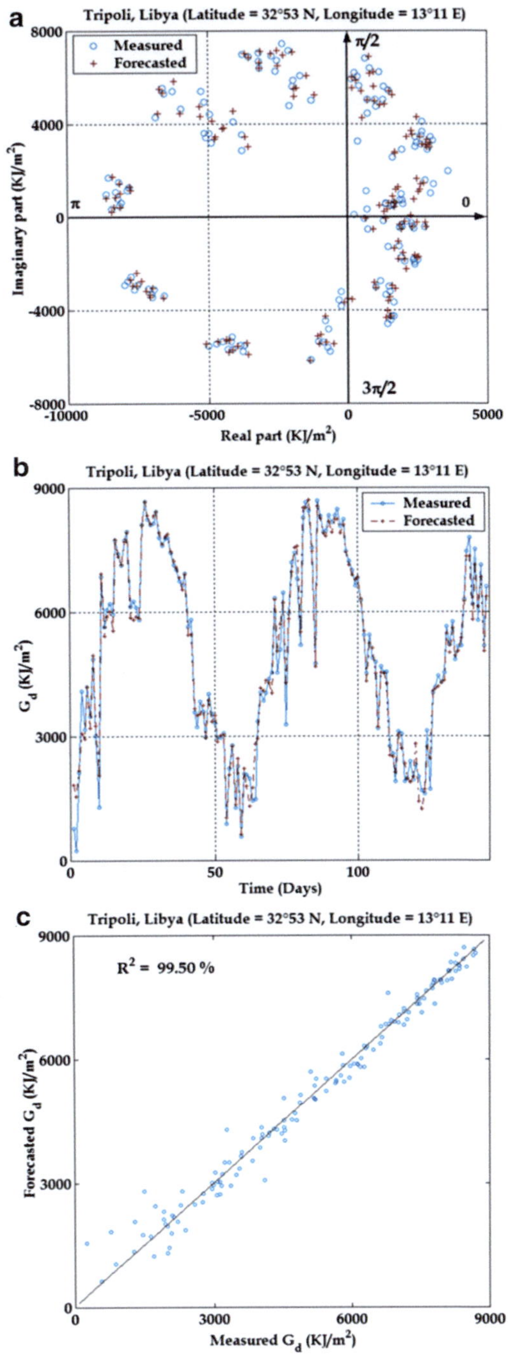

Fig. 23.2 Measured and predicted daily solar irradiation of 5 days ahead for the city Tripoli: (**a**) in the polar plane, (**b**) temporal, and (**c**) measured versus predicted

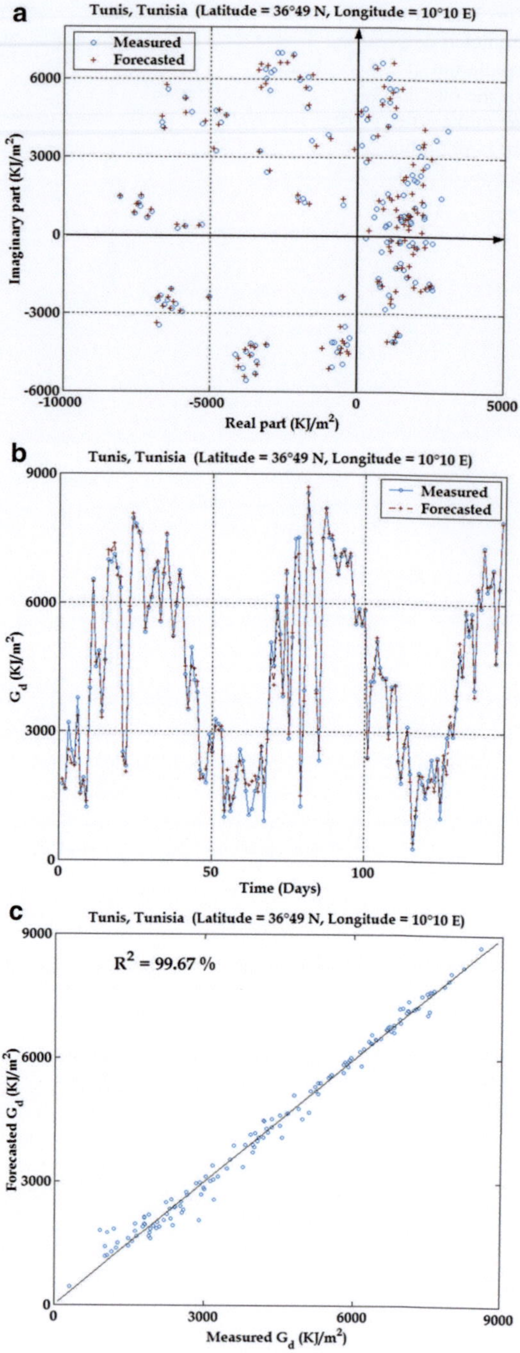

Fig. 23.3 Measured and predicted daily solar irradiation of 5 days ahead for the city Tunis: (**a**) in the polar plane, (**b**) temporal, and (**c**) measured versus predicted

Fig. 23.4 Measured and predicted daily solar irradiation of 5 days ahead for the city Algiers: (**a**) in the polar plane, (**b**) temporal, and (**c**) measured versus predicted

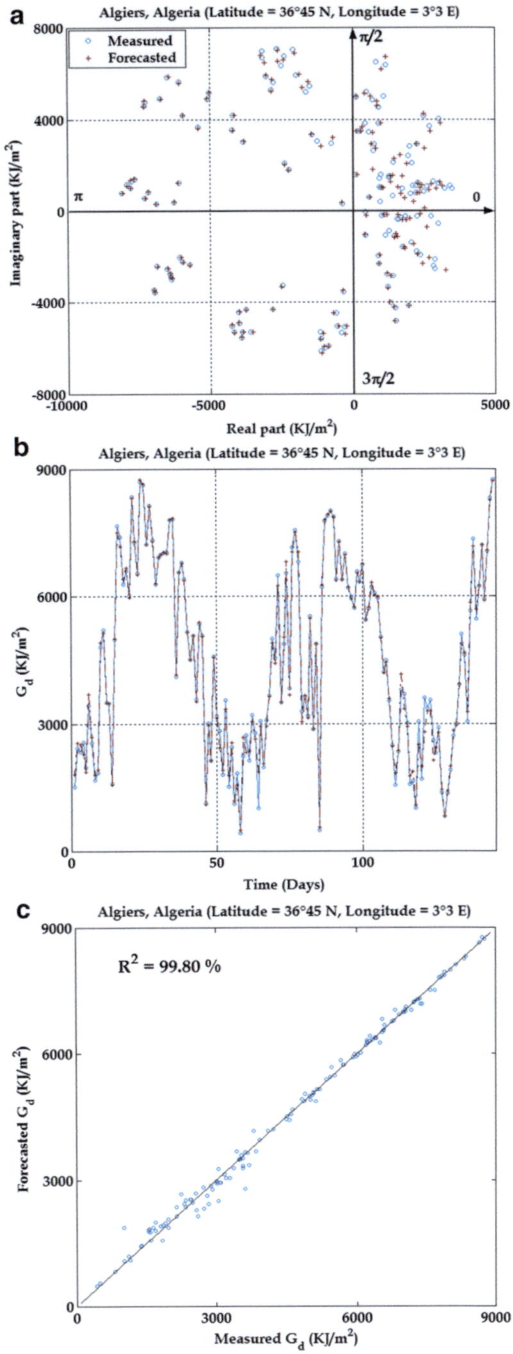

Table 23.5 Obtained results using the CVWNN for forecasting 15 days of each month

Model's structure			Measure's criteria		
Forecasting $\{\hat{G}_{16d}(k), \ldots, \hat{G}_{30d}(k)\}$ using	$<I \times H \times O>$	Capitals	MAE (%)	nRMSE (%)	R^2 (%)
$G_{1d}(k-1), G_{2d}(k-1), \ldots, G_{15d}(k-1)$	$<15 \times 30 \times 15>$	Tripoli	0.055	1.00	98.78
	$<15 \times 50 \times 15>$		0.106	1.64	98.92
	$<15 \times 100 \times 15>$		0.047	0.80	99.36
$G_{1d}(k-1), G_{2d}(k-1), \ldots, G_{15d}(k-1)$	$<15 \times 30 \times 15>$	Tunis	0.193	3.77	96.07
	$<15 \times 50 \times 15>$		0.152	3.14	97.73
	$<15 \times 100 \times 15>$		0.088	1.79	99.29
$G_{1d}(k-1), G_{2d}(k-1), \ldots, G_{15d}(k-1)$	$<15 \times 30 \times 15>$	Algiers	0.280	4.96	87.50
	$<15 \times 50 \times 15>$		0.089	1.70	98.49
	$<15 \times 100 \times 15>$		0.067	1.36	98.99
$G_{1d}(k-1), G_{2d}(k-1), \ldots, G_{15d}(k-1)$	$<15 \times 30 \times 15>$	Rabat	0.130	2.06	95.36
	$<15 \times 50 \times 15>$		0.060	1.08	98.79
	$<15 \times 100 \times 15>$		0.086	1.42	99.33
$G_{1d}(k-1), G_{2d}(k-1), \ldots, G_{15d}(k-1)$	$<15 \times 30 \times 15>$	El Aaiun	0.156	2.25	98.98
	$<15 \times 50 \times 15>$		0.064	0.90	99.46
	$<15 \times 100 \times 15>$		0.077	1.08	99.59
$G_{1d}(k-1), G_{2d}(k-1), \ldots, G_{15d}(k-1)$	$<15 \times 30 \times 15>$	Nouakchott	0.076	0.93	98.40
	$<15 \times 50 \times 15>$		0.042	0.54	98.65
	$<15 \times 100 \times 15>$		0.051	0.72	99.54

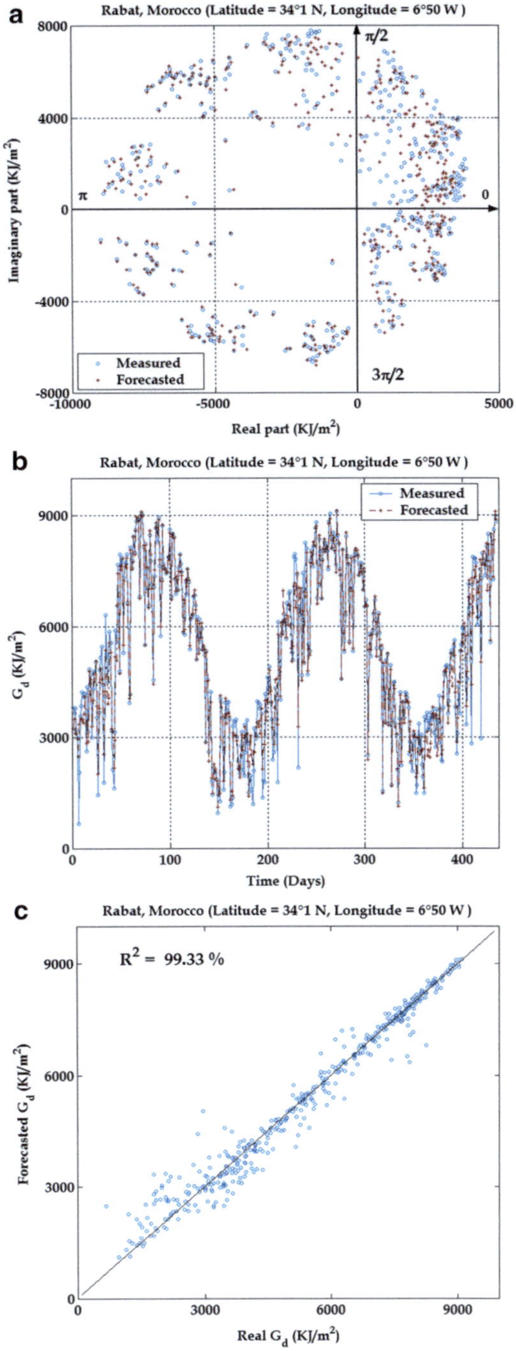

Fig. 23.5 Measured and predicted daily solar irradiation for 15 days in the city Rabat: (**a**) in the polar plane, (**b**) temporal, and (**c**) measured versus predicted

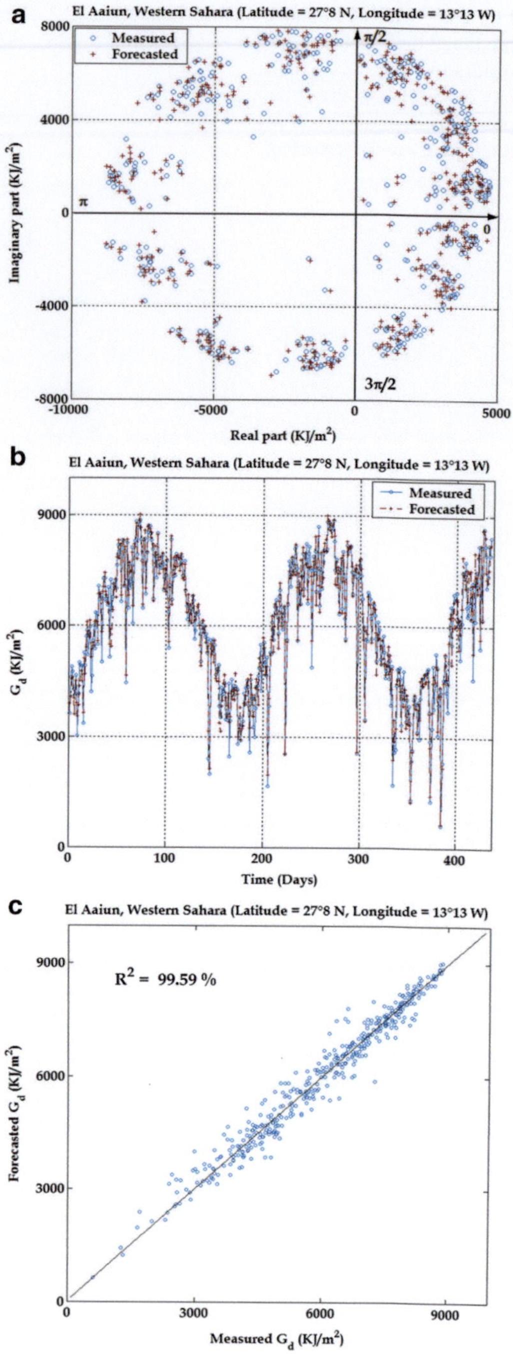

Fig. 23.6 Measured and predicted daily solar irradiation of 15 days ahead for the city El Aaiun: (**a**) in the polar plane, (**b**) temporal, and (**c**) measured versus predicted

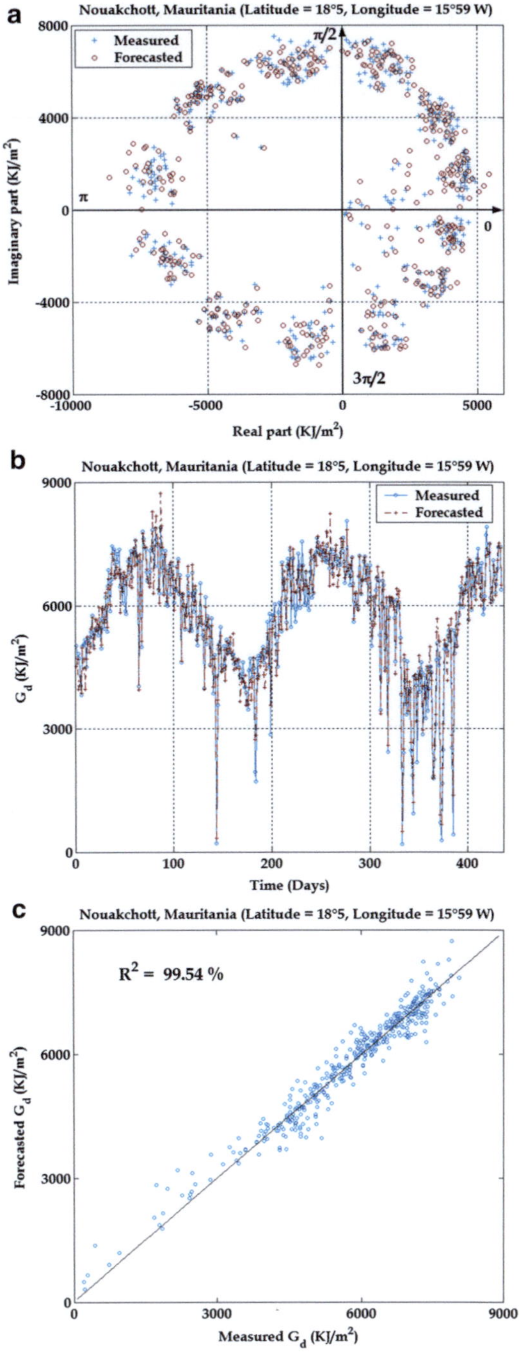

Fig. 23.7 Measured and predicted daily solar irradiation of 15 days ahead for the city Nouakchott: (**a**) in the polar plane, (**b**) temporal, and (**c**) measured versus predicted

References

1. Mellit A, Mekki H, Messai A, Kalogirou SA (2011) FPGA-based implementation of intelligent predictor for global solar irradiation, part I: theory and simulation. Expert Syst Appl 38:2668–2685
2. Mellit A, Pavan AM (2010) A 24-h forecast of solar irradiance using artificial neural network: application for performance prediction of a grid-connected PV plant at Trieste, Italy. Sol Energy 84:807–821
3. Solanki SK, Krivova NA, Wenzler T (2005) Irradiance models. Adv Space Res 35:376–383
4. Kalogirou S, Sencan A (2010) Artificial intelligence techniques in solar energy applications. In: Manyala R (ed) Solar collectors and panels, theory and applications. InTech, Rijeka
5. Winslow JC, Hunt ER Jr, Piper SC (2001) A globally applicable model of daily solar irradiance estimated from air temperature and precipitation data. Ecol Model 143:227–243
6. El-Sebaii AA, Al-Hazmi FS, Al-Ghamdi AA, Yaghmour SJ (2010) Global, direct and diffuse solar radiation on horizontal and tilted surfaces in Jeddah, Saudi Arabia. Appl Energy 87:568–576
7. Mefti A, Bouroubi MY, Adane A (2003) Generation of hourly solar radiation for inclined surfaces using monthly mean sunshine duration in Algeria. Energy Convers Manage 44:3125–3141
8. Dazhi Y, Jirutitijaroen P, Walsh WM (2012) Hourly solar irradiance time series forecasting using cloud cover index. Sol Energy 86:3531–3543
9. Mellit A, Eleuch H, Benghanem M, Elaoun C, Massi Pavan A (2010) An adaptive model for predicting of global, direct and diffuse hourly solar irradiance. Energy Convers Manage 51:771–782
10. Martin L, Zarzalejo LF, Polo J, Navarro A, Marchante R, Cony M (2010) Prediction of global solar irradiance based on time series analysis: application to solar thermal power plants energy production planning. Sol Energy 84:1772–1781
11. Gueymard CA (2009) Direct and indirect uncertainties in the prediction of tilted irradiance for solar engineering applications. Sol Energy 83:432–444
12. Marquez R, Coimbra CFM (2011) Forecasting of global and direct solar irradiance using stochastic learning methods, ground experiments and the NWS database. Sol Energy 85:746–756
13. Notton G, Paoli C, Vasileva S, Nivet ML, Canaletti J-L, Cristofari C (2012) Estimation of hourly global solar irradiation on tilted planes from horizontal one using artificial neural networks. Energy 39:166–179
14. Zervas PL, Sarimveis H, Palyvos JA, Markatos NCG (2008) Prediction of daily global solar irradiance on horizontal surfaces based on neural-network techniques. Renew Energy 33:1796–1803
15. Paoli C, Voyant C, Muselli M, Nivet M-L (2010) Forecasting of preprocessed daily solar radiation time series using neural networks. Sol Energy 84(12):2146–2160
16. Mellit A, Benghanem M, Kalogirou SA (2006) An adaptive wavelet-network model for forecasting daily total solar radiation. Appl Energy 83:705–722
17. Wang Z, Wang F, Su S (2011) Solar irradiance short-term prediction model based on BP neural network. In: ICSGCE 2011, 27–30 Sept 2011, Chengdu, China, Energy Procedia 12:488–494
18. Cao S, Cao J (2005) Forecast of solar irradiance using recurrent neural networks combined with wavelet analysis. Appl Therm Eng 25:161–172
19. Saad Saoud L, Rahmoune F, Tourtchine V, Baddari K (2013) Complex-valued forecasting of global solar irradiance. J Renew Sustain Energy 5(4):043124–043145
20. Rattan SSP, Hsieh WW (2005) Complex-valued neural networks for nonlinear complex principal component analysis. Neural Netw 18:61–69
21. Goh SL, Chen M, Popovic DH, Aihara K, Obradovic D, Mandic DP (2006) Complex-valued forecasting of wind profile. Renew Energy 31:1733–1750

22. Tripathi B, Chandra KB, Singh M, Kalra PK (2011) Complex generalized-mean neuron model and its applications. Appl Soft Comput 11:768–777
23. Saad Saoud L, Rahmoune F, Tourtchine V, Baddari K (2011) Wavelet network implementation on an inexpensive eight bit microcontroller. In: Kamanina NV (ed) Features of liquid crystal display materials and processes. Intech, Rijeka, pp 87–102
24. Zhang Q, Benveniste A (1992) Wavelet networks. IEEE Trans Neural Netw 3(6):889–898
25. Pati YC, Krishnaprasad PS (1993) Analysis and synthesis of feedforward neural networks using discrete affine wavelet transformations. IEEE Trans Neural Netw 4:73–85
26. Hong J (1992) Identification of stable systems by wavelet transform and artificial neural networks. Ph.D. dissertation, University of Pittsburgh, Pittsburgh
27. Tsatsanis MK, Giannakis GB (1993) Time-varying system identification and model validation using wavelets. IEEE Trans Signal Process 41:3512–3523
28. Kreinovich V, Sirisaengtaksin O, Cabrera S (1994) Wavelet neural networks are asymptotically optimal approximators for functions of one variable. In: Proceeding IEEE international conference on neural networks, Orlando, pp 299–304
29. Delyon B, Juditskyand A, Benveniste A (1995) Accuracy analysis for wavelet approximations. IEEE Trans Neural Netw 6:332–348
30. Özbay Y, Kara S, Latifoğlu F, Ceylan R, Ceylan M (2007) Complex-valued wavelet artificial neural network for Doppler signals classifying. Artif Intell Med 40(2):143–156
31. Nitta T (1997) An extension of the back-propagation algorithm to complex numbers. Neural Netw 10(8):1391–1415
32. Özdemir N, Iskender BB, Özgür NY (2011) Complex valued neural network with Möbius activation function. Comm Nonlinear Sci Numer Simulat 16:4698–4703
33. Educypedia (2013) http://educypedia.karadimov.info/education/atlastopics.htm. Acceded August 2013
34. NASA Database (2013) https://eosweb.larc.nasa.gov/cgi-bin/sse/. Acceded Mar 2013
35. Drossu R, Obradovic Z (1996) Rapid design of neural networks for time series prediction. IEEE Comput Sci Eng 3(2):78–89, ISSN 1070-9924

Chapter 24
Neural Modeling Adsorption of Copper, Chromium, Nickel, and Lead from Aqueous Solution by Natural Wastes

Samia Rebouh, Mounir Bouhedda, Salah Hanini, and Abdenour Djellal

Abstract An artificial neural network (ANN) is used to model the static adsorption of copper, chromium, lead, and nickel by natural wastes, which are respectively charred cereal waste, Mediterranean biomass (Posidonia oceanica (L) DELILE), activated carbon as well as olive kernel and pulp. This intelligent model is used to predict and estimate the amount of adsorbed metal per mass unit of adsorbent or the yield percentage of the adsorption. The results obtained using multilayer neural network shows its effectiveness in predicting the experimental results. The relative error is 0.2 mg/g for charred cereal waste/Copper, Biomass/Chromium, and 1.9 % for the combinations activated carbon/Lead, olive kernel/Nickel, and olive pulp/Nickel, respectively. Furthermore, the same artificial neural network is exploited to predict the effect of some operating parameters (pH, temperature, initial metal concentration, contact time, agitation speed, ionic strength, and adsorbent weight) that affect the static adsorption of these metals by several types of adsorbents.

Keywords Artificial neural networks • Prediction • Static adsorption • Heavy metals • Adsorbents

Nomenclature

AC	Activated carbon
ANN	Artificial neural network
b_i	Bias of the ith node
BM	Mediterranean biomass

S. Rebouh (✉) • S. Hanini • A. Djellal
Laboratory of Biomaterials and Transport Phenomena (LBMPT), University of Medea, Avenue de l'ALN, Ain Dheb, Medea 26000, Algeria
e-mail: samrebouh@yahoo.fr

M. Bouhedda
Laboratory of Advanced Electronic Systems (LSEA), University of Medea, Nouveau pôle urbain, Medea 26000, Algeria

I. Dincer et al. (eds.), *Progress in Clean Energy, Volume 1*,
DOI 10.1007/978-3-319-16709-1_24

C_{in}	Initial concentration, mg/g
Cu	Copper
I_s	Inonic strength, mol m^3
J	Jacobian matrix of weights and biases derivate
L	ANN output dimension
M	Number of iterations
Ni	Nickel
OK	Olive kernel
OP	Olive pulp
Pb	Lead
s	Neuron inputs sum
S_a	Agitation speed, round per minute (rpm)
T	Temperature, °C
t_c	Contact time
w_{ij}	Connection weight between node and node j
WCC	Waste charred cereal
W_d	Adsorbent weight
y_i	ith desired output
z_i	ith neuron output

Greek Letters

μ	Learning coefficient
ε_i	Output error of the ith node

Subscripts

i	Node and output index
j	Node index

24.1 Introduction

The pollution affects all compartments of the environment: water, air, and soil. It is the result of the diffusion of organic and inorganic contaminants. Climate change, the loss of plant and animal species are the first signs. The water pollution took the great part of environmental concerns, the fact that water resources are limited and that economic development creates pollution problems where the volumes of wastewater generated by the various sectors become increasingly important.

The main sources of water contamination are domestic and industrial wastewater, agricultural production, air pollutants, old waste dumps, and the use of hazardous substances in water. Many pollutants, in particular, heavy metals may harm the health and the future of the living beings. The metal elements are, in various forms, always present in the environment, and they are necessary or essential to living beings [1]. However, at high concentration, they exhibit a more or less strong toxicity. The presence of heavy metals in the environment results from natural causes and human activities. It is a particular problem because they are not biodegradable [2].

These heavy metals do not all have the same risks because of their different effects on organisms and their chemical, physicochemical and biological properties. Their toxicity is highly variable and their impact on environment is very different [2].

The conventional methods used for the effluents treatment [3–7] including heavy metals removal, such as precipitation, redox, ion exchange, filtration, membrane processes, and evaporation are toxic, inefficient, and too expensive. They are applied for solutions at low concentration [8–12]. A good alternative to these methods is the use of original biopolymers such as industrial and agricultural wastes [13–20], which have physicochemical properties to capture heavy metals by adsorption mechanism.

Adsorption as a separation process has been widely used in environmental chemistry because of its relatively low cost, simple design, and capacity for adsorbing a board range of pollutants at low concentration [21]. Many biosorbents have been tested for heavy metals removal because of their availability, low cost, and easy obtaining. They have been investigated as potential adsorbents to remove this kind of pollutants. The removal of copper ions from aqueous solution by hazelnut shell has been predicted [22]. The use of activated carbon prepared from peanut shells for hexavalent chromium adsorption has been studied [23]. The adsorption of Cd (II), Zn (II), Cr (III), and Cr (IV) from aqueous solutions using hazelnut has been reported [24]. Removal of heavy metals ions by banana pith has been investigated [24]. The adsorption of Cu (II) from synthetic solutions using Irish sphagnum peat moss has been studied [25]. The removal of heavy metal ions from aqueous solutions by means of rice bran, soybean, and cottonseed hulls has been investigated [26]. The use of shells of lentil, wheat, and rice for Cu (II) removal from aqueous solutions has been studied [27]. The adsorption of Cu (II) from water using decaying Tamarix gallica leaves has been reported [28].

Many agricultural by-products have proved to be good low cost adsorbents for the removal of both copper and chromium from water. Lignocellulosic residues include wood residues (sawdust and paper mill discards) and agricultural residues include lignocellulosic agrowastes (sugarcane bagasse, wheat bran, wheat straw, corn stoves, etc.). Lignocellulosic wastes have an adsorption capacity comparable to other natural sorbents, beside, they have the advantage of very low or no cost, great availability, simple operational process through their use as biosorbents, an added value is provided to products that otherwise would be considered as a waste [29].

The removal capacity of a biosorbent presenting nonlinearities with the operating parameters has to be described accurately in order to get an effective prediction model. To get the model, the designer has to follow two ways. The first one is referring to the mathematical modelling where he has to use the knowledge of chemistry, biology, and other sciences to describe an equation which is extremely difficult for the cases treated in this work. The second way requires the use of the experimental data to identify relationship between output and inputs. This method is used in the cases where the process involves extremely complex physical phenomena, or exhibits strong nonlinearities and dependence on various parameters. This method matches exactly with the treated problem in this work.

Obtaining a mathematical model for a system can be rather complex and time consuming. This fact has led the researches to exploit the intelligent techniques like ANNs for modelling complex systems using the input–output data sets.

Artificial neural networks (ANNs) can be advantageous because they provide an adjusting mechanism to get at the end, an intelligent model for the prediction of the adsorption amount.

In this work, three different interests were brought together to provide maximum gains and benefits:

- Environmental interest:
  - Removal of heavy metals that are toxic components
  - Use of natural and biodegradable materials such as activated charcoal, olive waste, algae, and grains waste to solve problems of harmful toxic waste in the environment
- Economic interest: Valorization of biomass as an abundant and inexpensive renewable energy
- Optimization and prediction of adsorption amount: furnished by using a reliable and precise intelligent model

In the present work, an ANN is synthesized to predict and estimate the amount of different metals adsorbed per mass unit of adsorbent using Neural Network Toolbox of MATLAB® where an experimental database is collected and exploited to provide a large learning database for the established ANN model.

24.2 Artificial Neural Networks

24.2.1 ANN Basics

As mentioned above, ANNs appear usually in literature describing nonlinear complex system building [30]. For this purpose, the most often used ANN architecture is the feedforward network with one or more hidden layers using Levenberg–Marquardt learning algorithm. Some related basic relationships are reported in the following; further detailed information about the feedforward ANN can be found in [31].

In the feedforward ANN, a typical ANN element (node) transfer function is the hyperbolic tangent. In this case, the output z_i of the ith node is computed as (24.1).

$$z_i = \frac{e^{\alpha s} - e^{-\alpha s}}{e^{\alpha s} + e^{-\alpha s}} \tag{24.1}$$

where α is the coefficient that defines the steepness of the sigmoidal transfer function and the expression of s is given by (24.2)

$$s = \sum_i w_{ij} x_{ij} - b_i \tag{24.2}$$

where x_{ij} is the input of the ith node given by jth node of the previous layer; w_{ij} is the connection weight of the ith node with the jth node of the previous layer; and b_i is the ith node bias.

The use of ANN implies three separate phases:

- *The learning phase*, in which the ANN is forced to furnish the desired outputs corresponding to a determined inputs (learning set); in this phase, the ANN learning level is verified by means of suitable performance indexes computed with reference to data (test data) that are different but coherent with the data of the learning set.
- *The validation phase*, in which the ANN generalization capability is verified by means of data (validation test) completely different from the data used in the previous phase.
- *The production phase*, in which the ANN is capable of providing the outputs required that correspond to any input.

In the learning phase, Levenberg–Marquardt algorithm is used to modify w_{ij} and b_i values by means of an adaptive process which minimizes the output node errors (supervised learning) on the basis of the relationships given by (24.3) and (24.4).

$$w_{ij}(k) = w_{ij}(k-1) - \left[\mathbf{J}^{\mathrm{T}}\mathbf{J} + \mu\mathbf{I}\right]^{-1}\mathbf{J}^{\mathrm{T}}\varepsilon_i(k-1) \tag{24.3}$$

$$b_i(k) = b_i(k-1) - \left[\mathbf{J}^{\mathrm{T}}\mathbf{J} + \mu\mathbf{I}\right]^{-1}\mathbf{J}^{\mathrm{T}}\varepsilon_i(k-1) \tag{24.4}$$

where $\mathbf{J}$ is the Jacobian matrix that contains first derivatives of the network errors with respect to the weights and biases, k is the current iteration of the learning algorithm, ε_i is the output error of the ith node and μ is the learning coefficient.

The ANN learning progress is monitored by means of suitable monitor index which is the Mean Square Error (MSE), where the expression is given by (24.5) [32].

$$\mathrm{MSE} = \frac{1}{M}\sum_{k=1}^{M}\sum_{i=1}^{L}\left(y_i^k - z_i^k\right)^2 \tag{24.5}$$

where M the number of iterations of Levenberg–Marquardt algorithm, L is the output layer dimension, and y_i is the ith desired outputs of the ANN.

In the validation and production phases, the ANN outputs are computed by mean of simple sums and products, whose number is related only to the chosen ANN architecture.

24.2.2 ANN Learning

The adopted learning type is the supervised training. First, a multilayer structure of the ANN is chosen with giving random values to the synaptic weights (w_{ij}), where both the inputs and the desired output are provided. The ANN then processes the inputs and compares its resulting output against the desired output (Fig. 24.1). Errors are then propagated back through the system, causing the system to adjust the weights of the ANN. This process occurs over and over as the weights are continually tweaked. The set of data which enables the training is called the "training set." During the training of a network, the same set of data is processed many times using Levenberg–Marquardt algorithm as the connection weights are ever refined.

24.3 Methods

24.3.1 Synthesis Methodology

The synthesis of the artificial neural networks for static adsorption modeling is illustrated on the flowchart presented in Fig. 24.2. In order to establish a reliable and representative database four steps are followed:

- Database collection
- Selection of the neural model's inputs and output
- Database preparation
- Data format

Fig. 24.1 ANN supervised learning

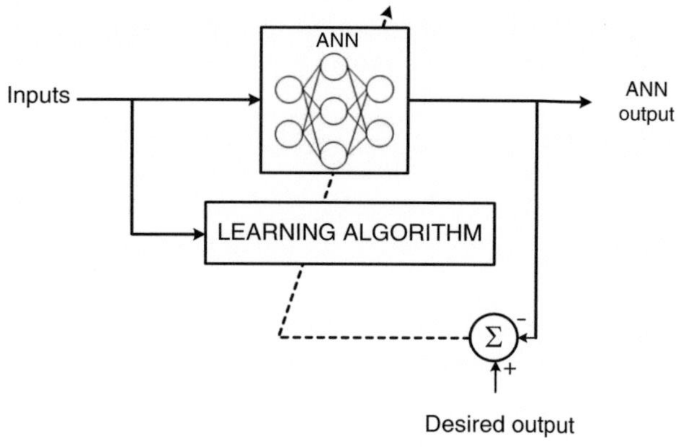

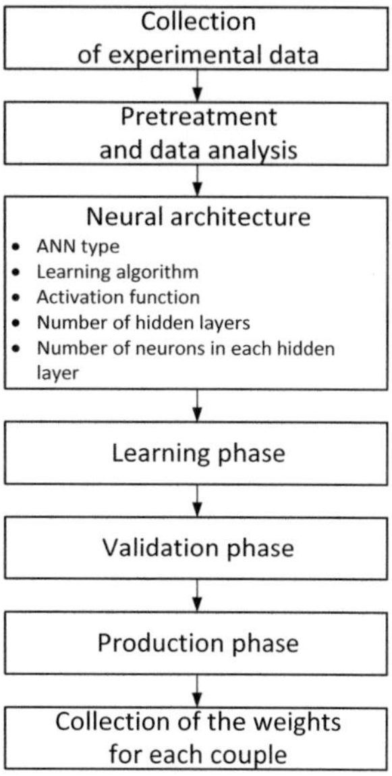

Fig. 24.2 Flowchart of the ANN synthesis

24.3.2 Database Collection

Data concerning metals adsorbed quantities or yields for five adsorbent/heavy metal couples are grouped after being taken from previous experimental studies. The five different selected couples are:

- Couple 1: Waste charred cereal/Copper (WCC/Cu) [33]
- Couple 2: Mediterranean biomass (Posidonia Oceanica (L) DELILE)/Chromium (MB/Cr) [34]
- Couple 3: Activated carbon/Lead (AC/Pb) [35]
- Couple 4: Olive kernel/Nickel (OK/Ni) [36]
- Couple 5: Olive pulp/Nickel (OP/Ni) [37]

24.3.3 Selection of the Neural Model's Inputs and Outputs

The various factors which can affect the biosorption are agitation speed, pH, temperature, sorption time, sorbent quantity, ionic strength, sorbent/metal ratio, and initial metal-ion concentration [33–37]. The inputs are the parameters that affect adsorption with relevant effects on yield or on the adsorption capacity of heavy metals. Some variables are used as the input vectors to train the network whereas the others are kept constant at their optimum values. However, the outputs are adsorption yield (%) and adsorption capacity (mg/g).

24.3.4 Preparing and Formatting Data

Like any ANN modelization, three phases are followed: Learning, validation, and production phase. In each phase, an amount of data is taken from the learning database:

- 80 % data for learning
- 10 % data for the validation
- 10 % for the production

24.3.5 Partial Database Normalization

Normalization is an important step in the process of compiling the data. It becomes in most cases, a requirement for the input data, the fact that often transfer functions of bounded sigmoid-type are used in static models. In this study, a sub-base separate data have been used, corresponding to adsorbent/metal couples. The numerical values were normalized in order to improve the optimization to get the output values within the interval [–0.9, 0.9].

24.3.6 Adopted ANN Structure

The basic structure of a multilayer perceptron is retained at three layers: an input layer, a hidden layer, and an output layer. The used learning algorithm is Levenberg–Marquardt.

The used activation functions are hyperbolic tangent for the neurons of the hidden layer and linear function for the neurons of the output layer.

One ANN is used for couple 1, 2, and 3 and another one is used for the two remaining couples. Each ANN has its appropriate inputs: pH, temperature (T), initial metal concentration (C_{in}), contact time (t_c), agitation speed (S_a), ionic strength (I_s), and adsorbent weight (W_g) and additional input which is used to indicate the concerned couple. More details about the structure and learning phase results of the two trained ANNs are given in Table 24.1. The architectures of the two used ANNs are illustrated in Fig. 24.3.

For simulation and validation of the network, eight experiments kept for the test phase are considered for each couple; for each experiment, the values of the variables are selected randomly within the selected ranges, and the results are given in Table 24.2. This step ensures the adoptability validation of the network for inputs which are out of the training data set.

Table 24.1 ANNs learning results

ANN	Couples	Inputs	Output	Number of hidden layers	Number of neurons in the hidden layer	Number of learning data	Maximum epochs	MSE
ANN_1	WCC/Cu	Couple, T, pH, C_{in}, W_a, t_c	Adsorption yield (%)	01	05	246	10,000	2.67×10^{-3}
	MB/Cr							
ANN_2	AC/L	Couple, T, pH, C_{in}, W_a, t_c, I_s, S_a	Adsorption capacity (mg/g)	01	06	120	8,000	6.88×10^{-6}
	OK/Ni							
	OP/Ni							

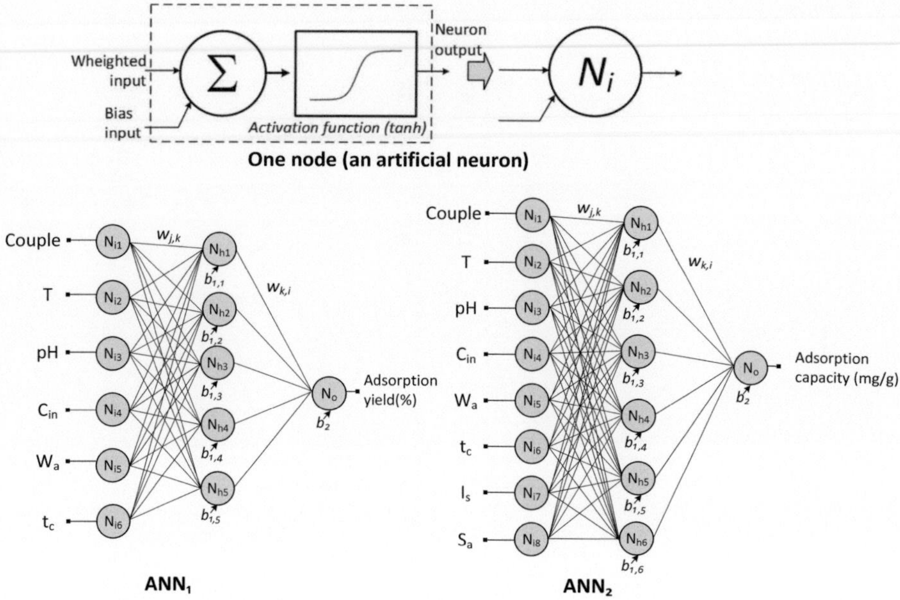

Fig. 24.3 Architecture of ANN_1 and ANN_2

24.4 Results and Discussion

The realized experimental design is made with the purpose of exploring the effect of interaction of the eight parameters which are namely pH, temperature, initial metal concentration, contact time, agitation speed, ionic strength, and adsorbent weight by implementing the permutations of variables within their selected range.

The performance of the ANNs upon training with the test data may be revealed by the closely followed trends between the experimental and the neural predicted patterns in Fig. 24.4. The ANN is found to be very efficient in predicting the amount of adsorption within the range of data contained in the training set.

Figure 24.5 represents comparison between ANN simulated and experimental output data (12 sets for ANN_1 and 45 sets for ANN_2). The maximum relative error is 0.2 mg/g for ANN_1 and 1.9 % for ANN_2. The predicted adsorption amount efficiencies of obtained ANNs results are as close as with training and testing data given to the network.

Table 24.2 Simulation and validation data—adsorption of different metals by different adsorbents

Experimental run	Couple	T (°C)	pH	C_{in} (mg/g)	t_c (mn)	I_s (mol m^3)	S_a (rpm)	W_a (g)	Experimental amount	Simulated amount	Relative error	Percentage relative error
1	1	20	6	–	3	0	100	1.00	4.711 mg/g	4.640 mg/g	−0.071	−1.518
2	1	20	6	–	10	0	100	1.00	7.324 mg/g	7.221 mg/g	−0.103	−1.408
3	1	20	6	–	20	0	100	1.00	8.308 mg/g	8.211 mg/g	−0.097	−1.171
4	1	20	6	–	90	0	100	1.00	7.568 mg/g	7.611 mg/g	0.043	0.570
5	1	20	6	–	5	0	600	1.00	7.834 mg/g	7.772 mg/g	−0.062	−0.794
6	1	20	6	–	20	0	600	1.00	8.347 mg/g	8.479 mg/g	0.132	1.576
7	1	20	6	–	60	0	600	1.00	9.161 mg/g	9.079 mg/g	−0.082	−0.900
8	1	20	6	–	120	0	600	1.00	8.885 mg/g	9.072 mg/g	0.187	2.108
9	2	30	2	5	10	0	0	1.00	0.097 mg/g	0.101 mg/g	0.004	3.965
10	2	30	2	10	10	0	0	1.00	0.158 mg/g	0.161 mg/g	0.003	1.723
11	2	30	2	25	10	0	0	1.00	0.910 mg/g	0.912 mg/g	0.002	0.266
12	2	30	2	50	10	0	0	1.00	1.941 mg/g	1.922 mg/g	−0.019	−1.000
13	2	10	2	20	10	0	0	1.00	0.124 mg/g	0.123 mg/g	−0.001	−0.943
14	2	20	2	20	10	0	0	1.00	0.827 mg/g	0.838 mg/g	0.011	1.327
15	2	30	2	20	10	0	0	1.00	0.936 mg/g	0.909 mg/g	−0.027	−2.886
16	2	50	2	20	10	0	0	1.00	0.548 mg/g	0.551 mg/g	0.003	0.510
17	3	25	7	10	15	200	0	1.00	54.04 %	52.907 %	−1.133	−2.097
18	3	25	7	10	15	200	0	1.00	42.79 %	41.329 %	−1.461	−3.414
19	3	25	7	10	15	200	0	1.00	35.50 %	36.843 %	1.343	3.783
20	3	25	7	10	15	200	0	1.00	22.80 %	21.712 %	−1.088	−4.773
21	3	25	2.5	10	15	0	0	1.00	40.04 %	39.564 %	−0.476	−1.190
22	3	25	4	10	15	0	0	1.00	42.80 %	44.249 %	1.449	3.385
23	3	25	7	10	15	0	0	1.00	54.04 %	54.837 %	0.797	1.476
24	3	25	9	10	15	0	0	1.00	50.74 %	50.571 %	−0.169	−0.333

(continued)

Table 24.2 (continued)

Experimental run	Couple	T (°C)	pH	C_{in} (mg/g)	t_c (mn)	I_s (mol m^3)	S_a (rpm)	W_a (g)	Experimental amount	Simulated amount	Relative error	Percentage relative error
25	4	25	3.5	10	30	0	0	0.10	53.18 %	52.167 %	−1.013	−1.905
26	4	25	3.5	30	30	0	0	0.10	39.27 %	38.424 %	−0.846	−2.153
27	4	25	3.5	50	30	0	0	0.10	22.18 %	22.687 %	0.507	2.284
28	4	25	3.5	10	30	0	0	0.55	53.18 %	52.109 %	−1.071	−2.013
29	4	25	3.5	10	30	0	0	1.00	50.84 %	50.139 %	−0.701	−1.379
30	4	25	3.5	10	30	0	0	0.10	35.52 %	35.584 %	0.064	0.181
31	4	25	3.5	10	10	0	0	0.10	23.00 %	22.827 %	−0.173	−0.753
32	4	25	3.5	10	30	0	0	0.10	53.18 %	53.988 %	0.808	1.519
33	5	25	3.5	10	30	0	0	0.10	91.16 %	91.410 %	0.250	0.275
34	5	25	3.5	30	30	0	0	0.10	63.24 %	62.210 %	−1.030	−1.629
35	5	25	3.5	50	30	0	0	0.10	42.10 %	41.465 %	−0.635	−1.508
36	5	25	3.5	10	30	0	0	0.55	91.16 %	89.690 %	−1.470	−1.613
37	5	25	3.5	10	30	0	0	0.55	77.67 %	77.172 %	−0.498	−0.642
38	5	25	3.5	10	30	0	0	1.00	73.21 %	74.696 %	1.486	2.029
39	5	25	3.5	10	10	0	0	0.10	21.86 %	21.768 %	−0.092	−0.420
40	5	25	3.5	10	30	0	0	0.10	91.16 %	90.071 %	−1.089	−1.194

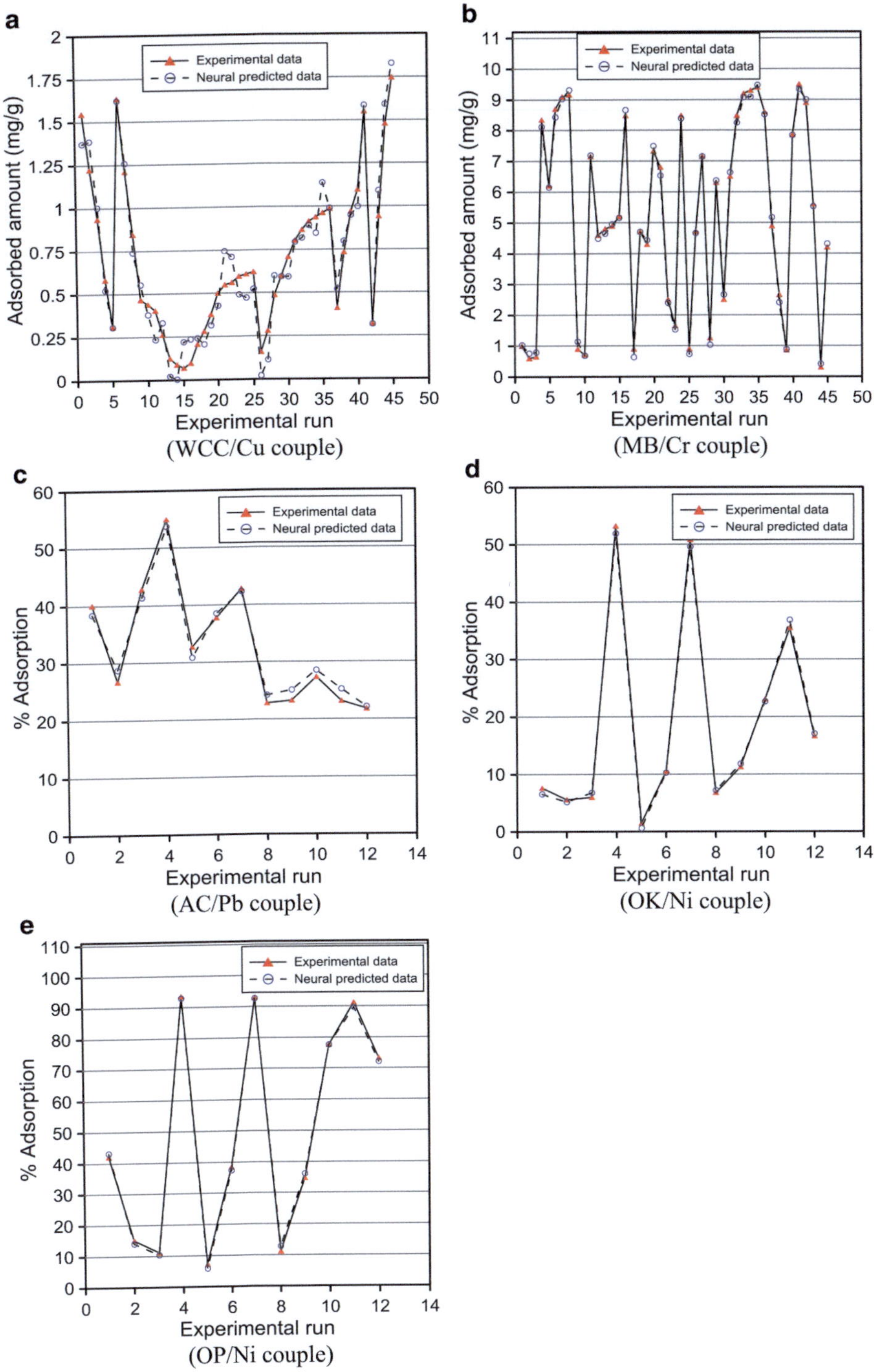

Fig. 24.4 Experimental and neural predicted adsorption amount

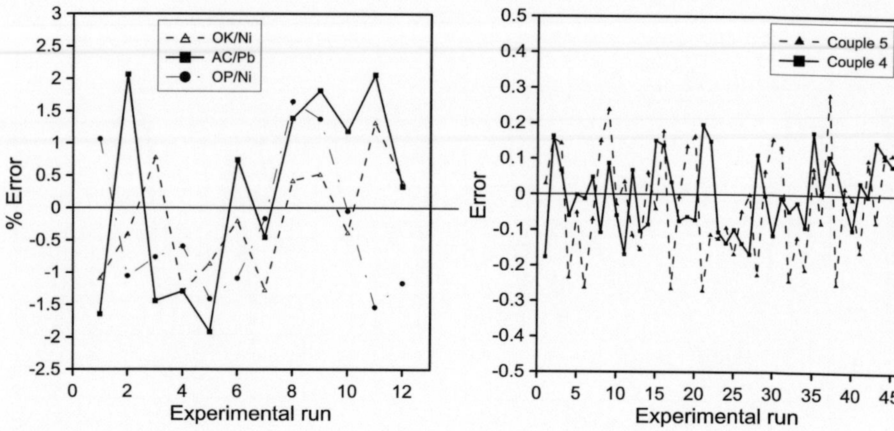

Fig. 24.5 Relative error between experimental and neural predicted adsorption amount

24.5 Conclusion

In the present study, a recurrent neural network has been designed and demonstrated to predict the adsorption amount of Cu, Pb, Cr, and Ni by charred cereal waste, biomass, activated carbon, olive kernel, and olive pulp.

The obtained results using two multilayer neural networks show the effectiveness in predicting the experimental results for the metal adsorption by the different natural adsorbents. This can be observed through the error values found which are 0.2 mg/g are for the ANN_1 for WCC/Cu, BM/Cr couples, and 1.9 % for the ANN_2 for AC/Pb, OK/Ni, and OP/Ni couples, respectively.

Also, artificial neural networks can be considered as an effective supplement for the conventional and complicated models in the prediction of bioprocesses parameters.

References

1. Mohand D, Pittman CU, Seele P (2007) Arsenic removal from water/wastewater using adsorbents—a critical review. J Hazard Mater 142(1–2):1–53
2. Nacim R, Tahir S (2001) Removal of Pb(II) from aqueous/acidic solutions by using bentonite as an adsorbent. Water Res 33(11):3982–3986
3. Boren A, Bowen DE, Santos FR (2003) Understanding biotechnology, 1st edn. Prentice Hall, Upper Saddle River
4. Ullmann (2002) Ullmann's encyclopedia of industrial chemistry. Wiley, Weinheim
5. Vogel H (1997) Fermentation and biochemical engineering handbook principles, process design, and equipment, 2nd edn. Noyes, Westwood
6. Meinck M (1982) Les eaux résiduaires industrielles. Masson & Cie, Paris
7. Hendricks D (2010) Fundamentals of water treatment unit processes: physical, chemical, and biological. CRC Press, Boca Raton

8. Adebajo MO, Frost RL, Kloprogge JT, Carmody O, Kokot S (2003) Porous materials for oil spill cleanup: a review of synthesis and absorbing properties. J Porous Mater 1:159–170
9. Babel S, Kurniawan TA (2003) Low-cost adsorbents for heavy metals uptake from contaminated water: a review. J Hazard Mater B97:219–243
10. Bailey SE, Olin TJ, Bricka RM, Adrian DD (1999) A review of potentially low-cost sorbents for heavy metals. Water Res 33(11):2469–2479
11. Bal Y, Bal KE, Lallam A (2003) Removal of Bi (III) and Zn (II) by nonliving Streptomyces rimosus biomass from nitric solutions. Eur J Miner Process Environ Prot 3(1):42–48
12. Kong B, Tang B, Liu X, Zeng X, Duan H, Luo S, Wei W (2009) Kinetic and equilibrium studies for the adsorption process of cadmium (II) and copper (II) onto Pseudomonas aeruginosa using square wave anodic stripping voltammetry method. J Hazard Mater 167 (1):455–460
13. Angelova N, Henkler D (1999) Rationalizing the design of polymeric biomaterials. Trends Biotechnol 17:409–420
14. Cardenas G, Orlando P, Edelio T (2001) Synthesis and applications of chitosan mercaptanes as heavy metal retention agent. Int J Biol Macromol 28:167–174
15. Chiron N, Guilet R, Deydier E (2003) Adsorption of Cu(II) and Pb(II) onto a grafted silica: isotherms and kinetic models. Water Res 37:3079–3086
16. Dupont L, Guillon E, Bouanda J, Dumonceau J, Laplincourt M (2002) EXAFS and XANES studies of retention of copper and lead by a lignocellulosic biomaterial. Environ Sci Technol 36:5062–5066
17. Dias MA, Castro HF, Pimentel F, Gomes NCM, Rosa CA, Linardi VR (2000) Removal of heavy metals from stainless steel effluents by waste biomass from Brazilian alcoholic beverage production. World J Microbiol Biotechnol 16:107–108
18. Fenga D, Aldrich C (2004) Adsorption of heavy metals by biomaterials derived from the marine alga Ecklonia maxima. Hydrometallurgy 73:1–10
19. Dakiky M, Khamis M, Manassra A, Mereb M (2002) Selective adsorption of chromium (VI) in industrial wastewater using low-cost abundantly available adsorbents. Adv Environ Res 6:533–540
20. Kadirvelu K, Thamaraiselvi K, Namasivayam C (2001) Removal of heavy metals from industrial wastewaters by adsorption onto activated carbon prepared from an agricultural solid waste. Bioresour Technol 76:63–65
21. Dong A, Xie J, Wang W, Yu L, Liu Q, Yin Y (2010) A novel method for amino starch preparation and its adsorption for Cu(II) and Cr(VI). J Hazard Mater 181(1–3):448–454
22. Meneses M, Llobet J, Granero S, Schuhmacher M, Domingo J (1999) Monitoring metals in the vicinity of a municipal waste incinerator: temporal variation in soils and vegetation. Sci Total Environ 226(2–3):157–164
23. Al-Othman Z, Ali R, Mu N (2012) Hexavalent chromium removal from aqueous medium by activated carbon prepared from peanut shell: adsorption kinetics, equilibrium and thermodynamic studies. Chem Eng J 184:238–247
24. Low KS, Lee CK, Leo AC (1995) Removal of metals from electroplating wastes using banana pith. Bioresour Technol 51(2–3):227–231
25. Ho YS, McKay G (2003) Sorption of dyes and copper ions onto biosorbents. Process Biochem 38(7):1047–1061
26. Marshall WE, Wartelle LH, Boler DE, Johns M, Toles CA (1999) Toles enhanced metal adsorption by soybean hulls modified with citric acid. Bioresour Technol 69(3):263–268
27. Aydın H, Bulut Y, Yerlikaya Ç (2008) Removal of copper (II) from aqueous solution by adsorption onto low-cost adsorbents. J Environ Manage 87(1):37–45
28. Zaggout FR (2001) Removal of copper from water by decaying Tamarix gallica leaves. Asian J Chem 13(2):639–650
29. Qiang Y, Xinshu Z, Qiong W, Wei Q, Xuesong T, Zhenhong Y (2012) Hydrolysis of sweet sorghum bagasse and eucalyptus wood chips with liquid hot water. Bioresour Technol 116:220–225

30. Darsey JA (2014) Neural networks in chemical and physical systems. World Scientific, Singapore
31. Schalkoff RJ (1997) Artificial neural networks. McGraw-Hill, New York
32. Hassoun M (2003) Fundamentals of artificial neural networks. A Bradford Book, MIT Press, Cambridge
33. Jeon C (2011) Removal of copper ion using rice hulls. J Ind Eng Chem 17(3):517–520
34. Ncibi MC (2008) Étude de la biosorption du chrome (VI) par une biomasse méditerranéenne: Posidonia oceanica (L.) delile. J Water Sci 21(4):441–449
35. Boulkrah H (2008) Etude comparative de l'adsorption des ions de plomb sur différents adsorbants. Théses et mémoires Faculté des sciences Mémoires de Magistère : Chimie Option:Pollution Chimique et Environnement, Skikda
36. Fiola N, Villaescusaa I, Martínezb M, Mirallesb N, Pochc J, Serarolsc J (2006) Sorption of Pb (II), Ni(II), Cu(II) and Cd(II) from aqueous solution by olive stone waste. Sep Purif Technol 50 (1):132–140
37. Gharaibeh S, Abu-el-sha'r WY, Al-Kofahi M (1998) Removal of selected heavy metals from aqueous solutions using processed solid residue of olive mill products. Water Res 32 (2):498–502

Chapter 25
Computational Study of the Effects of Heat Generating Finned Annular Pipe on the Conjugate Heat Transfer

Sofiane Touahri and Toufik Boufendi

Abstract In this work, we numerically study the effects of heat generating fins in an annular pipe on the three-dimensional conjugate heat transfer: conduction in the solid and mixed convection in the working fluid. The horizontal annular pipe is equipped by longitudinal attached fins on internal surface of outer cylinder. The external pipe and the fins are heated by an electrical current passing through their small thickness. The number of longitudinal fins studied is: 2 vertical, 4 and 8 fins. The convection in the fluid domain is conjugated to thermal conduction in the pipes and fins solid thickness. The physical properties of the fluid are thermal dependent. The heat losses from the external pipe surface to the surrounding ambient are considered. The model equations of continuity, momenta, and energy are numerically solved by a finite volume method with a second-order spatiotemporal discretization. The obtained results showed that the axial Nusselt number increases with the increasing of number and height of fins. The participation of fins located in the lower part of the tube on the improvement of heat transfer is higher than the participation of the upper fins.

Keywords Conjugate heat transfer • Mixed convection • Annulus • Fins • Numerical simulation

Nomenclature

D	Diameter (m)
g	Gravitational acceleration (m/s^2)
G	Volumetric heat source (W/m^3)
h	Heat transfer coefficient (W/m^2°C)
k	Thermal conductivity (W/m°C)

S. Touahri • T. Boufendi (✉)
Energy Physics Laboratory, Department of Physics, Faculty of Exact Science, Constantine 1 University, Constantine, Algeria
e-mail: boufendit@yahoo.fr

I. Dincer et al. (eds.), *Progress in Clean Energy, Volume 1*,
DOI 10.1007/978-3-319-16709-1_25

L Length (m)
Nu Nusselt number
P Pressure (N/m^2)
Pr Prandtl number
r Radial coordinate (m)
Ra Rayleigh number
Re Reynolds number
t Time, s
T Temperature (°K)
U Radial velocity component (m/s)
V Axial velocity component (m/s)
W Angular velocity component (m/s)
z Axial coordinate (m)

Greek Letters

α Thermal diffusivity, (m^2/s)
β Thermal expansion coefficient, (1/°K)
ε Emissivity coefficient
μ Dynamic viscosity, (kg·m/s)
θ Angular coordinate, (rad)
τ Viscous stress, (N/m^2)
ρ Density, (kg/m^3)

Subscripts

b Bulk
i, o Reference to the inner and outer surface of the duct, respectively
m Mean
r, θ, z Reference to the radial, angular and axial direction, respectively
0 Duct inlet
∞ Ambient air away from the outer wall

Superscripts

* Nondimensional

25.1 Introduction

The study of internal mixed convection flows was the subject of much theoretical (analytical, numerical) and experimental research because the problem is dependent on several control parameters, such as the Reynolds, the Grashof and the Prandtl numbers, the geometrical parameter such as the inclination angle and the entry thermal length and finally the thermal boundary conditions such as the heat flux imposed on the duct wall or the imposed temperature. In addition to fundamental aspects of these phenomena, their industrial applications are numerous and diverse, such as compact heat exchanger, solar collectors, and cooling components electronic. When the convection in the fluid and conduction in the solid are simultaneously considered, we say that the heat transfer is conjugated as is indicated in the thermal transfer book, Incropera et al. [1]. Finned tubes are often used in many engineering sectors for extend the contact surface between the tube wall and the fluid and improve the heat transfer; the researchers have studied the problem of optimizing the shape and geometry of attached fins in order to increase heat transfer effectiveness. Many investigations, both experimental and numerical, have been conducted for different kinds of internally finned tubes. Patankar et al. [2] presented an analytical model for fully developed turbulent air flow in internally finned tubes and annuli. In their study, the longitudinal attached fins in the inner wall were considered. With thermal boundary conditions such as the constant heat flux at the inner surface, the results of this study concern the heat transfer and the pressure drop coefficients. The obtained results are presented as determining the Nusselt numbers and coefficients of charge loss. Agrawal et al. [3] studied numerically the effect of varying the geometric parameters and the Reynolds number on the pressure drop and the heat transfer. This study was conducted for the case of longitudinal fins attached to the inside of an annulus. In the numerical work of Farinas et al. [4], the authors studied the laminar mixed convection in an annulus with inner fins for two, four, and sixteen fins. The inner wall is hot while the outer wall is cold. The Grashof number varied from 10^2 to 10^4. The conservation equations are solved by the finite difference method. The results are presented for the air with Rayleigh numbers varied from 10^3 to 10^6 for different configurations of fins (fine, rounded, or divergent) and different lengths of blades ($L = 0.25$, 0.5 and 0.75). The results are presented as isotherms graphs, velocity fields, and the variation of Nusselt numbers. The heat transfer is enhanced to a rounded configuration of the fins. Similar studies were also treated experimentally by Wei-Mon Yan et al. [5], Yu et al. [6] and Wang et al. [7].

In the present work, we have studied numerically the heat transfer by mixed convection in annulus between two concentric cylinders. Longitudinal fins are attached in the inner wall of the outer cylinder. The mixed convection is conjugated with the thermal conduction in the pipes and fins walls. The physical properties of the fluid are thermal dependent and heat losses to the outside environment are taken into account while the inner cylinder is adiabatic at its inner wall. The objective of our work is to study the improvement of heat transfer in the annulus using 2, 4, and 8 longitudinal fins generating heat.

25.2 The Geometry and the Mathematical Model

Figure 25.1 illustrates the geometry of the problem studied. Two long horizontal concentric cylinders having a length $L = 1$ m, the inner tube with an internal diameter $D_{1i} = 0.46$ cm, and an outer diameter $D_{1o} = 0.5$ cm, while the outer tube with an internal diameter $D_{2i} = 0.96$ cm and an outer diameter $D_{2o} = 1$ cm. Longitudinal fins are attached to the inner wall of the outer cylinder. At the entrance, the flow has an average axial velocity equal to 9.88 10^{-2} m/s and a constant temperature of 15 °C. The Reynolds and the Prandtl numbers are equal to 399.02 and 8.082, respectively. The nondimensional fluid viscosity and thermal conductivity variations with temperature are represented by the functions $\mu^*(T^*)$ and $K^*(T^*)$ obtained by smooth fittings of the tabulated values cited by Baehr et al. [8].

This problem, which combines, simultaneously, the two heat transfer modes, convection in the fluid and conduction in the solid, belongs to the class of conjugate heat transfer problems. It will be modeled by the following dimensionless conservation equations of the mass, momenta, and energy with the appropriate boundary conditions, [9].

$$\begin{array}{ll}\text{At } t^* = 0, & U^* = V^* = W^* = T^* = 0 \\ \text{At}: t^* > 0 & \end{array} \tag{25.1}$$

25.2.1 Mass Conservation Equation

$$\frac{1}{r^*}\frac{\partial}{\partial r^*}\left(r^* V_r^*\right) + \frac{1}{r^*}\frac{\partial V_\theta^*}{\partial \theta} + \frac{\partial V_z^*}{\partial z^*} = 0 \tag{25.2}$$

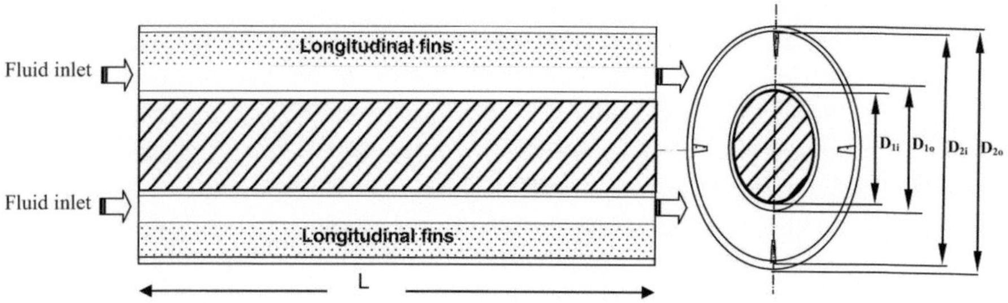

Fig. 25.1 Geometry and dimensions $\left(D_{1i}^* = 1, D_{1e}^* = 1.09, D_{2i}^* = 2.09, D_{2e}^* = 2.17, L^* = 217.39\right)$

25.2.2 Radial Momentum Conservation Equation

$$\frac{\partial V_r^*}{\partial t^*}+\frac{1}{r^*}\frac{\partial}{\partial r^*}\left(r^*V_r^*V_r^*\right)+\frac{1}{r^*}\frac{\partial}{\partial\theta}\left(V_\theta^*V_r^*\right)+\frac{\partial}{\partial z^*}\left(V_z^*V_r^*\right)-\frac{V_\theta^{*}2}{r^*}$$
$$=-\frac{\partial P^*}{\partial r^*}+\frac{Gr_0^*}{Re_0^2}\cos\theta\ T^*+\frac{1}{Re_0}\left[\frac{1}{r^*}\frac{\partial}{\partial r^*}\left(r^*\tau_{rr}^*\right)+\frac{1}{r^*}\frac{\partial}{\partial\theta}\left(\tau_{r\theta}^*\right)-\frac{\tau_{\theta\theta}^*}{r^*}+\frac{\partial}{\partial z^*}\left(\tau_{rz}^*\right)\right] \tag{25.3}$$

25.2.3 Angular Momentum Conservation Equation

$$\frac{\partial V_\theta^*}{\partial t^*}+\frac{1}{r^*}\frac{\partial}{\partial r^*}\left(r^*V_r^*V_\theta^*\right)+\frac{1}{r^*}\frac{\partial}{\partial\theta}\left(V_\theta^*V_\theta^*\right)+\frac{\partial}{\partial z^*}\left(V_z^*V_\theta^*\right)+\frac{V_r^*V_\theta^*}{r^*}$$
$$=-\frac{1}{r^*}\frac{\partial P^*}{\partial\theta}-\frac{Gr_0^*}{Re_0^2}\sin\theta\ T^*+\frac{1}{Re_0}\left[\frac{1}{r^{*2}}\frac{\partial}{\partial r^*}\left(r^{*2}\tau_{\theta r}^*\right)+\frac{1}{r^*}\frac{\partial}{\partial\theta}\left(\tau_{\theta\theta}^*\right)+\frac{\partial}{\partial z^*}\left(\tau_{\theta z}^*\right)\right] \tag{25.4}$$

25.2.4 Axial Momentum Conservation Equation

$$\frac{\partial V_z^*}{\partial t^*}+\frac{1}{r^*}\frac{\partial}{\partial r^*}\left(r^*V_r^*V_z^*\right)+\frac{1}{r^*}\frac{\partial}{\partial\theta}\left(V_\theta^*V_z^*\right)+\frac{\partial}{\partial z^*}\left(V_z^*V_z^*\right)$$
$$=-\frac{\partial P^*}{\partial z^*}+\frac{1}{Re_0}\left[\frac{1}{r^*}\frac{\partial}{\partial r^*}\left(r^*\tau_{rz}^*\right)+\frac{1}{r^*}\frac{\partial}{\partial\theta}\left(\tau_{\theta z}^*\right)+\frac{\partial}{\partial z^*}\left(\tau_{zz}^*\right)\right] \tag{25.5}$$

25.2.5 Energy Conservation Equation

$$\frac{\partial T^*}{\partial t^*}+\frac{1}{r^*}\frac{\partial}{\partial r^*}\left(r^*V_r^*T^*\right)+\frac{1}{r^*}\frac{\partial}{\partial\theta}\left(V_\theta^*T^*\right)+\frac{\partial}{\partial z^*}\left(V_z^*T^*\right)$$
$$=G^*-\frac{1}{Re_0Pr_0}\left[\frac{1}{r^*}\frac{\partial}{\partial r^*}\left(r^*q_r^*\right)+\frac{1}{r^*}\frac{\partial}{\partial\theta}\left(q_\theta^*\right)+\frac{\partial}{\partial z^*}\left(q_z^*\right)\right] \tag{25.6}$$

where $G^*=\begin{cases}K_s^*/(Re_0Pr_0) & \text{in the solid domain}\\ 0 & \text{in the fluid domain}\end{cases}$

The components of the viscous stress tensor are:

$$\tau_{rr}^* = 2\mu^* \frac{\partial V_r^*}{\partial r^*},\ \tau_{r\theta}^* = \tau_{\theta r}^* = \mu^* \left[r^* \frac{\partial}{\partial r^*}\left(\frac{V_\theta^*}{r^*}\right) + \frac{1}{r^*}\frac{\partial V_r^*}{\partial \theta}\right],\ \tau_{\theta\theta}^* = 2\mu^* \left[\frac{1}{r^*}\frac{\partial V_\theta^*}{\partial \theta} + \frac{V_r^*}{r^*}\right]$$

$$\tau_{\theta z}^* = \tau_{z\theta}^* = \mu^* \left[\frac{\partial V_\theta^*}{\partial z^*} + \frac{1}{r^*}\frac{\partial V_z^*}{\partial \theta}\right],\ \tau_{zz}^* = 2\mu^* \frac{\partial V_z^*}{\partial z^*},\ \tau_{zr}^* = \tau_{rz}^* = \mu^* \left[\frac{\partial V_z^*}{\partial r^*} + \frac{\partial V_r^*}{\partial z^*}\right] \quad (25.7)$$

and the heat flux are : $q_r^* = K^* \frac{\partial T^*}{\partial r^*},\quad q_\theta^* = \frac{K^*}{r^*}\frac{\partial T^*}{\partial \theta^*},$ and $q_z^* = K^* \frac{\partial T^*}{\partial z^*}$ (25.8)

The dimensionless numbers that characterize the problem of mixed convection are evaluated at the entrance fluid temperature. With the following boundary conditions:

At the annulus entrance: $z^* = 0$

$$\text{In the fluid domain}:\quad V_r^* = V_\theta^* = T^* = 0,\quad V_z^* = 1 \quad (25.9)$$

$$\text{In the solid domain}:\quad V_r^* = V_\theta^* = V_z^* = T^* = 0 \quad (25.10)$$

At the annulus exit: $z^* = 217.39$

$$\text{In the fluid domain}:\quad \frac{\partial V_r^*}{\partial z^*} = \frac{\partial V_\theta^*}{\partial z^*} = \frac{\partial V_z^*}{\partial z^*} = \frac{\partial}{\partial z^*}\left(K^* \frac{\partial T^*}{\partial z^*}\right) = 0 \quad (25.11)$$

$$\text{In the solid domain}:\quad V_r^* = V_\theta^* = V_z^* = \frac{\partial}{\partial z^*}\left(K^* \frac{\partial T^*}{\partial z^*}\right) = 0 \quad (25.12)$$

$$\text{At the inside wall of internal pipe,}\quad r^* = 0.5\quad V_r^* = V_\theta^* = V_z^* = 0\quad et\quad \frac{\partial T^*}{\partial r^*} = 0 \quad (25.13)$$

At the outer wall of external pipe: $r^* = 1.087$

The non-slip condition is imposed and the radial conductive heat flux is equal to the sum of the heat fluxes of the radiation and natural convection losses.

$$r^* = 1.087 \text{ for } 0 \le \theta \le 2\pi \text{ and}$$

$$0 \le z^* \le 217.39 \begin{cases} V_r^* = V_\theta^* = V_z^* = 0 \\ -K^* \dfrac{\partial T^*}{\partial r^*} = \dfrac{(h_r + h_c)D_i}{K_0} T^* \end{cases} \quad (25.14)$$

Where the radiative heat transfer is: $h_r = \varepsilon\sigma\left(T^2 + T_\infty^2\right)\ (T + T_\infty)$

The emissivity of the outer wall ε is arbitrarily chosen to 0.9 while h_c is derived from the correlation of Churchill et al. [10] valid for all Pr and for Rayleigh numbers in the range $10^{-6} \le Ra \le 10^9$

$$Nu = [h_c D_i / K_{air}]$$
$$= \left[0.6 + \left(0.387 Ra^{1/6} / \left(1 + (0.559/\mathrm{Pr}_{air})^{9/16}\right)^{8/27}\right)\right]^2 \quad (25.15)$$

with the local numbers of Rayleigh and Prandtl are defined by $Ra = \frac{g\beta[T(R_o,\theta,z)T_\infty]+D_o^3}{\alpha_{air}\nu_{air}}$, $\mathrm{Pr}_{air} = \nu_{air}/\alpha_{air}$.

In this expressions, the thermophysical properties of the ambient air are evaluated at the local film temperature, as: $T_{film} = [T(R_0,\theta,z) + T_\infty]/2$.

In addition, the functions $\mu^*(T^*)$ and $K^*(T^*)$ were obtained by smooth fitting of the tabulated values cited by [8], as follows:

$$\mu^*(T^*) = 0.23087 + 0.78727\exp(-T^*/0.11386) \quad (25.16)$$

$$K^*(T^*) = 1.00111 + 0.80477T^* - 1.06002T^{*2} \quad (25.17)$$

In this conjugate problem, the solid is treated as a fluid with an extremely large value of dynamic viscosity equal to 10^{30}. This allows you to cancel the velocity components in the solid domain, and hence heat transfer will occur only by conduction. This procedure is discussed in Patankar [11].

25.2.6 *The Nusselt Number*

At the pipes wall interface $(r^* = r^*_{2i} = 1.0435)$, the local Nusselt number is defined as:

$$Nu(\theta, z^*) = \frac{h(\theta, z^*)D_i}{K_0} = \left[\frac{(K^*\partial T^*/\partial r^*)|_{r^*=r^*_{2i}}}{T^*(0.5,\theta,z^*) - T^*_m(z^*)}\right] \quad (25.18)$$

At the fins wall interface $(\theta = \theta_{fin})$, the local Nusselt number is defined as:

$$Nu(r^*, z^*) = \frac{h(r^*, z^*)D_h}{K_0} = \left[\frac{(K^*/r^*)(\partial T^*/\partial\theta)|_{\theta=\theta_{fin}}}{T^*(r^*,\theta_{fin},z^*) - T^*_m(z^*)}\right] \quad (25.19)$$

25.3 The Numerical Procedure

The coupled nonlinear system of model equations, Eqs. (25.1)–(25.6), with their boundary conditions Eqs. (25.9)–(25.17) are discretized using the finite volume method, well described by [11] and Minkowycz et al. [12]. In this study, all discretization schemes are of second order precision. The temporal discretization of the derivative terms follows the backward Euler scheme whereas the convective and the nonlinear terms follow the Adams–Bashforth scheme whose the truncature error is of Δt^{*2}. The spatial dicretization of the diffusive terms and the pressure gradient follows the fully implicit of the central difference scheme whose the truncatures errors are of Δr^{*2}, $\Delta\theta^2$ and Δz^{*2} orders. The systems of the linearized algebraic equations obtained are solved sequentially by following the SIMPLER algorithm, [11]. The iterative solution is achieved by the line by line sweeping method using the Thomas algorithm in the radial and axial directions and the tri-diagonal cyclic algorithm in the angular direction. With suitable step time of $\Delta t^* = 10^{-3}$ and 5.10^{-4} the time marching is continued until the steady state is reached. The convergence is confirmed by the satisfaction of the global mass and energy balances as well as the temporal invariance of the local and average-dependent variables. In the r^*, θ, z^* directions, the numerical grid used is $52 \times 88 \times 162$ nodes in radial, angular, and axial directions successively corresponding to a total of 189,540 nodes. In the radial direction, five nodes are located in each small solid thickness. The numerical code used in this study has been thoroughly tested and validated by comparison with other results published in previous works. Full details of this validation can be found in the papers published by Boufendi et al. [13] and Touahri et al. [14, 15]. One of these validations is the comparison of our results with those of Nazrul et al. [16] who studied the laminar mixed convection heat transfer in a horizontal concentric annulus using air and water as the working fluid. The thermal boundary condition chosen is that of uniform heat flux at the inner wall and an adiabatic outer wall. The controlling parameters of the problem are: $D_2/D_1 = 2$, $Re = 200$, $Pr = 0.7$, and $Gr = 0$. In Fig. 25.2, we illustrate the axial evolution of the circumferentially averaged Nusselt number. It is seen that there is a good agreement between our results and theirs.

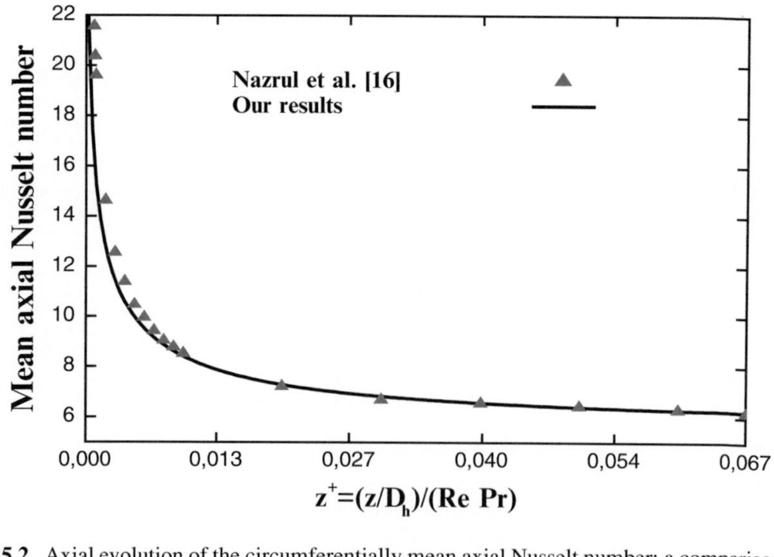

Fig. 25.2 Axial evolution of the circumferentially mean axial Nusselt number; a comparison with the results of Nazrul et al. [16]

25.4 Results and Discussions

Extensive numerical calculus where conducted for the conjugate heat transfer, laminar forced, and mixed convection in the working fluid and conduction in the solid, for the horizontal annular duct. All the results presented in this chapter were calculated for Reynolds number, $Re = 399.02$, the Prandtl number, $Pr = 8.082$ (distilled water), and the Grashof number is equal to 12,801 while the fin height H^* is equal to 0.12. For the brevity of the chapter, the results obtained for another fin height ($H^* = 0.24$) are not presented. The dynamic and thermal fields are represented in judiciously axial stations chosen.

25.4.1 Development of the Hydrodynamic Flow

25.4.1.1 The Flow Field

The obtained flow for the studied cases is characterized by a main flow along the axial direction and a secondary flow influenced by the density variation with temperature, which occurs in the plane ($r^* - \theta$). These flows are presented for eight longitudinal fins.

In Fig. 25.3, we present the secondary flow vectors at the annulus exit ($Z^* = 217.39$). The transverse flow is explained as follows: the hot fluid moves along the inner wall of the external cylinder from the bottom $(\theta = \pi)$ to the top $(\theta = 0)$, this movement is blocked by the walls of the longitudinal fins. After, a portion of the fluid continues moving upwardly under the effect of the thermal buoyancy force while another part is conveyed downwardly with the relatively cold fluid which descends near of the inner cylinder. The transverse flow in the $\left(r^* - \theta\right)$ plane is represented by counter rotating cells; the cells number is proportional to longitudinal fins number used. The vertical plane passing through the angles $(\theta = 0)$ and $(\theta = \pi)$ is a plane of symmetry. Regarding axial flow, this latter is influenced by the generation of the secondary flow which causes an angular variation explained as follows: the thermal viscosity is inversely proportional to the fluid temperature and the axial velocity increases with the decrease of viscosity, automatically we will have an axial velocity relatively high in the upper part of the annulus where the fluid temperature is greater than that of the lower portion. In Fig. 25.4, we represent an illustration of the variation of the axial velocity in the exit of annulus ($Z^* = 217.39$).

25.4.2 The Development of the Thermal Field

The angular distribution of the temperature field in the presence of eight longitudinal fins at the annulus exit is shown in Fig. 25.5. The maximum fluid temperature is equal to 0.5356 situated at the top of the annulus at $r^* = 1.0435$, $\theta = 0$, and

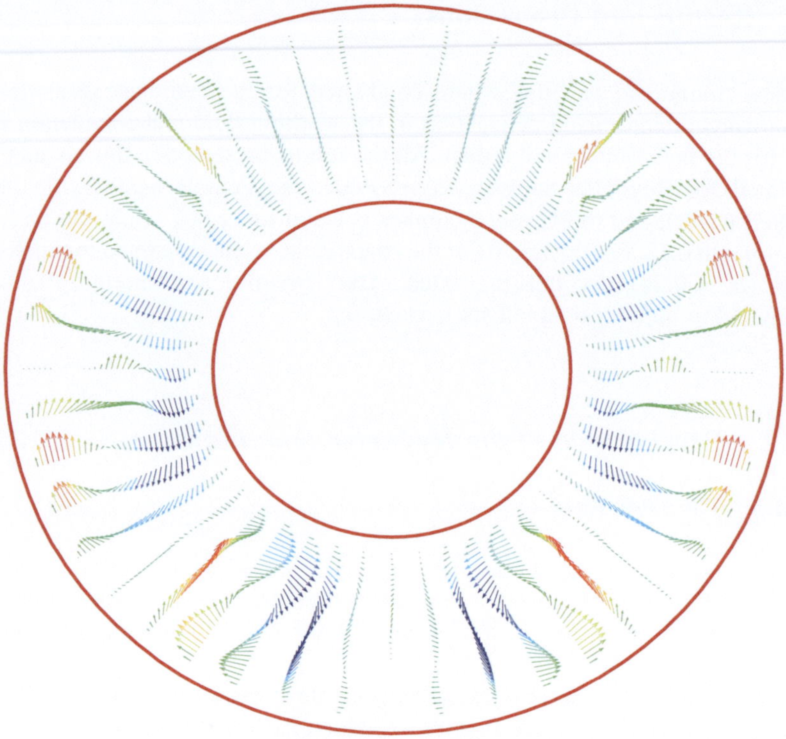

Z*=217.39

Fig. 25.3 Development of the secondary flow at the annulus exit

$z^* = 217.39$. In this axial position, the minimum temperature of the fluid is in the lower part of annulus at $\theta = \pi$ and $r^* = 0.6450$. In Fig. 25.6, we represent the variation of the axial temperature at the end of each fin at $r^* = 0.8169$. It is clear that the temperature of the vertical fin placed at $\theta = 0$ is highest, followed by fin placed at $\theta = \pi/4$, $\pi/2$, $3\pi/4$, and π.

25.4.3 Evolution of the Nusselt Number

The variation of the Nusselt number at the interface of the outer cylinder is illustrated in Fig. 25.7. The local Nusselt number at the interface (external pipe-fins) is zero. Apart from these azimuthal positions, the local Nusselt number takes a minimum value at the top of the cylindrical interface and a maximum value at the bottom of the cylindrical interface. At the exit of annulus, the local Nusselt number of the outer cylinder take a maximum value equal to 48.02 at $\theta = 2.8203$.

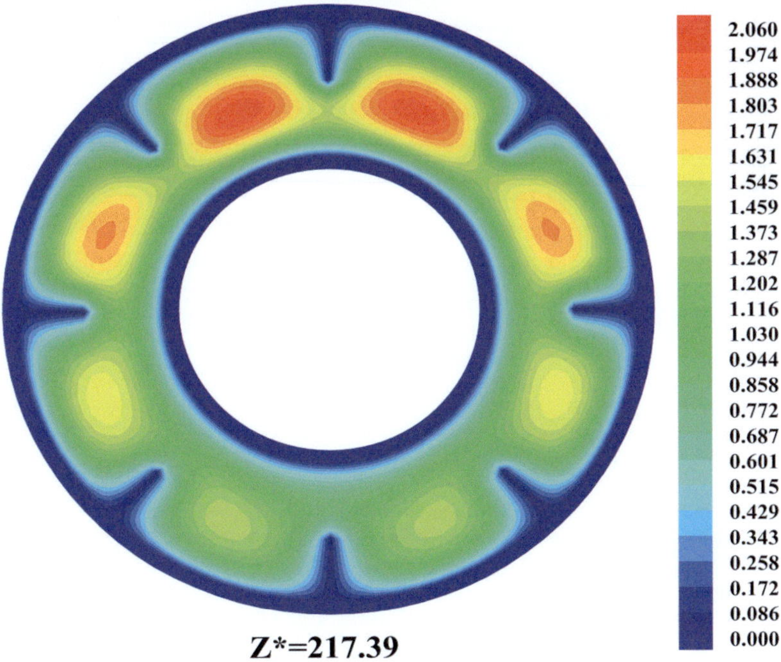

Fig. 25.4 Axial velocity profile at the annulus exit

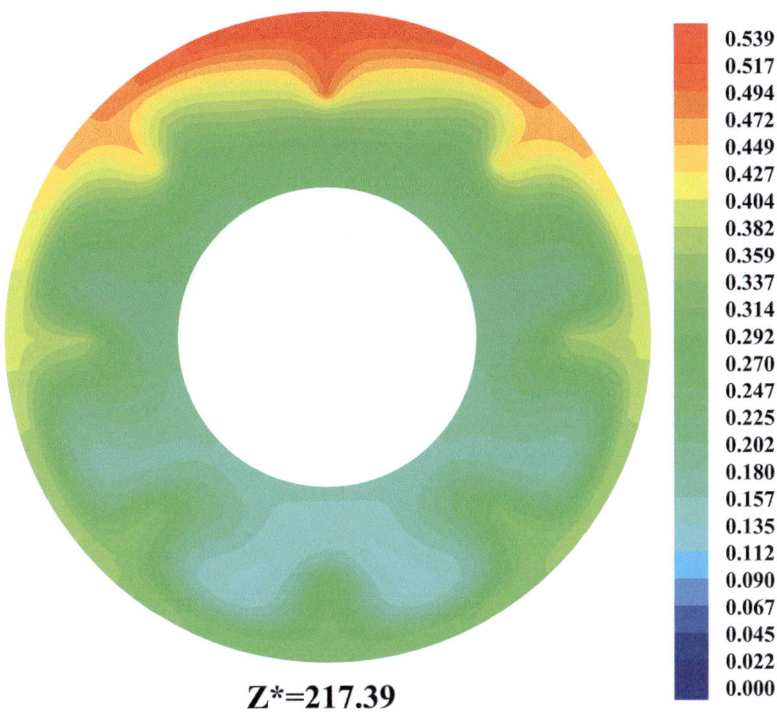

Fig. 25.5 Temperature field at the exit of annulus

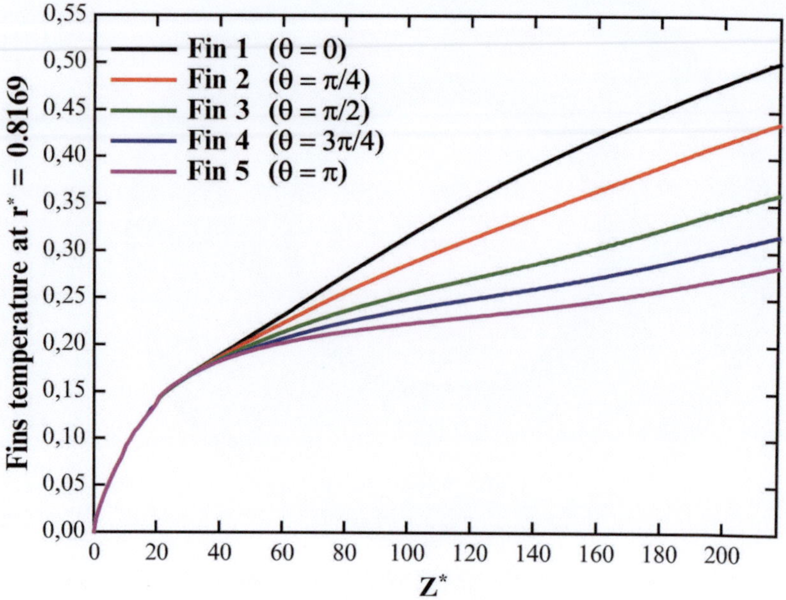

Fig. 25.6 Axial temperature variation of the fins at $r^* = 0.8169$

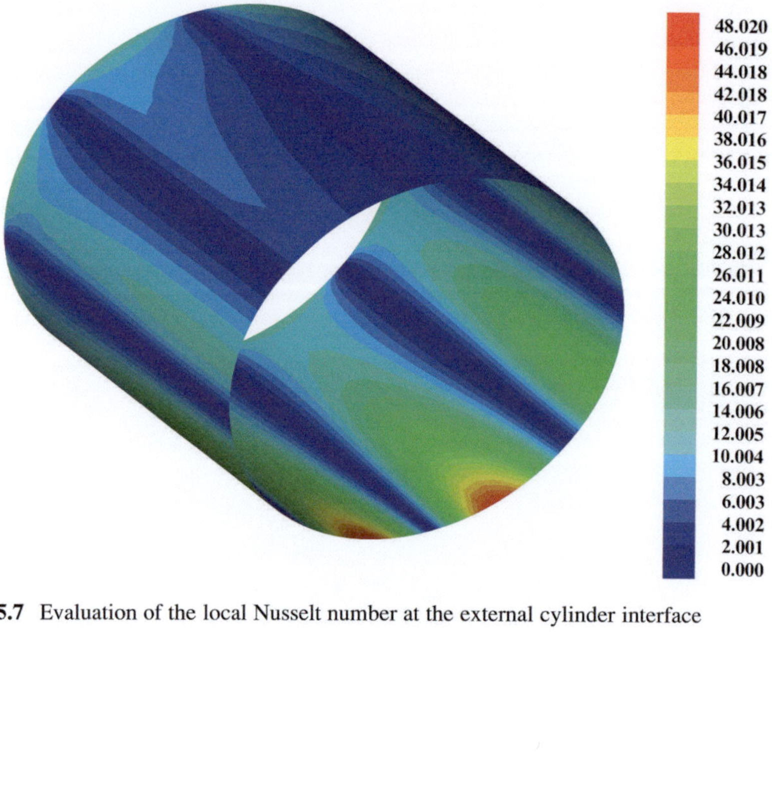

Fig. 25.7 Evaluation of the local Nusselt number at the external cylinder interface

Fig. 25.8 Evaluation of the local Nusselt number at the fin interface located at ($\theta = \pi$)

The local Nusselt number of the longitudinal fin placed at ($\theta = \pi$) is shown in Fig. 25.8. This number takes a maximum value equal to 130.00 at $z^* = 217.39$ and $r^* = 0.8169$.

25.5 Conclusion

This study considers the numerical simulation of the three dimensional mixed convection heat transfer in annulus equipped by longitudinal fins. The pipe and the fins are heated by an electrical current passing through its small thickness. The results show that the increase in the number of fins increases the axial Nusselt number especially when the flow is fully developed. The increase in the height of the fin from 0.12 to 0.24 improves the average Nusselt number from 41.89 to 60.09 in cases of eight fins. In this case, the maximum heat rate transferred to the fluid, is equal to 4.039, and is positioned on the fin located at ($\theta = 0$, $z^* = 93.07$) while the maximum Nusselt number equal to 130.00 is positioned on the fin located at ($\theta = \pi$, $z^* = 217.39$).

References

1. Incropera FP, DeWitt DP, Bergman TL, Lavine AS (2012) Foundations of heat transfer, 6th edn. Wiley
2. Patankar SV, Ivanovic M, Sparrow EM (1979) Analysis of turbulent flow and heat transfer in internally finned tubes and annuli. J Heat Transf 101:29–37
3. Agrawal AK, Sengupta S (1990) Laminar flow and heat transfer in a finned tube annulus. Int J Heat Fluid Flow 11(1):54–59
4. Farinas MI, Garon A, Saint-Lous K (1997) Study of heat transfer in a horizontal cylinder with fins. Revue Générale de Thermique 36:398–410

5. Yan WM, Sheen PJ (2000) Heat transfer and friction characteristics of Fin and tube heat exchangers. Int J Heat Mass Transf 43:1651–1659
6. Yu B, Nie JH, Wang QW, Tao WQ (1999) Experimental study on the pressure drop and heat transfer characteristics of tubes with internal longitudinal fins. Heat Mass Transf 35:65–73
7. Wang CC, Fu WL, Chang CT (1997) Heat transfer and friction characteristics of typical wavy Fin and tube heat exchanger. Exp Therm Fluid Sci 14:174–186
8. Baehr HD, Stephan K (1998) Heat and mass transfer. Springer, Berlin
9. Bejan A (2013) Convection heat transfer, 4th edn. Wiley, New York
10. Churchill SW, Chu HS (1975) Correlating equation for laminar and turbulent free convection from a horizontal cylinder. Int J Heat Mass Transf 18:1049–1053
11. Patankar S (1980) Numerical heat transfer and fluid flow. McGraw-Hill, New-York
12. Minkowycz WJ, Sparrow EM, Murthy JY (2006) Handbook of numerical heat transfer, 2nd edn. Wiley, New York
13. Boufendi T, Afrid M (2005) Three dimensional conjugate conduction-mixed convection with variable fluid properties in a heated horizontal pipe. Revue des Energies Renouvelables 8(1):1–18
14. Touahri S, Boufendi T (2012) Numerical study of the conjugate heat transfer in a horizontal pipe heated by joulean effect. Therm Science 16(1):53–67
15. Touahri S, Boufendi T (2014) In: Dincer I, Çolpan CÖ, Kızılkan Ö, Acar C, Hamut HS, Ezan MA, Özbilen A. (Eds.) Proceedings of the 13th International Conference on Clean Energy, Istanbul, Turkey, 08–12 June, pp. 3183–3190.
16. Nazrul I, Gaitonde UN, Sharma GK (2001) Mixed convection heat transfer in the entrance region of horizontal annuli. Int J Heat Mass Transf 44(11):2107–2120

Chapter 26
Exergy-Based Design and Analysis of Heat Exchanger Networks

S. Aghahosseini and I. Dincer

Abstract In this chapter, a new methodology for heat exchanger networks synthesis, extending traditional pinch technology to include exergy-destruction cost, is described and employed for practical applications. In the proposed approach, the cost rate of exergy destruction substitutes the utility cost in a trade-off between the operating and capital costs in the conventional pinch analysis to determine the optimum minimum approach temperature, ΔT_{min}, in the energy assessment of heat exchanger networks (HENs). It is demonstrated that the balanced composite curves can be used directly in the calculation of the exergy destruction, heat recovery, and heat-transfer area for a specified minimum approach temperature. Moreover, it is showed that the grand composite curve can be utilized for determination of the optimal external utility allocation. The inclusion of thermal exergy destruction into pinch analysis decreases the optimum value of network ΔT_{min}, due to increase in the system operating costs by considering both the external utility requirements and internal exergy destruction. The results of this assessment provide a better and more realistic utilization of external utilities considers the required capital investment. Two case studies are presented to show how to apply the proposed methodology effectively and practically. Case studies are analyzed using Aspen energy analyzer software.

Keywords Heat exchanger networks • Exergy-destruction cost • Pinch analysis • Heat-transfer area

Nomenclature

ΔT	Temperature difference, °K
ΔT_{min}	Minimum approach temperature, °K
A	Heat-transfer area (m^2)

S. Aghahosseini (✉) • I. Dincer
Clean Energy Research Laboratory, Faculty of Engineering and Applied Science, University of Ontario Institute of Technology, 2000 Simcoe Street North, Oshawa, ON, Canada, L1H 74K
e-mail: Seyedali.Aghahosseini@uoit.ca

I. Dincer et al. (eds.), *Progress in Clean Energy, Volume 1*,
DOI 10.1007/978-3-319-16709-1_26

T_0	Surrounding temperature (°K)
q	Heat flow (kW)
T	Temperature (°K)
h	File heat-transfer coefficient ($kW \cdot m^{-2} \cdot K^{-1}$)
$\dot{E}x$	Exergy flow (kW)
cx	Cost per unit of exergy ($\$ \cdot kW^{-1}$)
$\dot{K}$	Cost rate ($\$ \cdot s^{-1}$)
CC	Capital cost of a heat exchanger ($)
a	The installation cost set coefficient of a shell and tube heat exchanger
b, c	The duty/area-related cost set coefficients of a shell and tube heat exchanger
N_{shell}	The number of heat exchanger shells in a heat exchanger
OC	Operating cost ($\$ \cdot year^{-1}$)
C	Annual utility cost ($\$ \cdot kW^{-1} \cdot year^{-1}$)
A_f	The annualization factor ($year^{-1}$)
ROR	The rate of return, percent of capital
PL	Plant lifetime (year)

Subscripts

LMTD	Log mean temperature difference
H	Hot
C	Cold
des	Destruction
net	Network
hu	Hot utility
cu	Cold utility
min	Minimum
in	Inlet
out	Outlet
f	Fuel
CI	Capital investment
OM	Operation and maintenance

26.1 Introduction

Pinch analysis is an efficient approach that enables many industries increase their profitability reducing their utility consumption, greenhouse gas (GHG) emissions, and waste generation [1]. The heat exchanger networks synthesis using pinch analysis is a mature technology for energy integration in energy industries, which has already been well established and applied effectively in numerous design and retrofit projects for about three decades [2–4]. Pinch technology provides the ability to identify the most economical way of maximizing heat recovery within process streams and minimizing requirements for external utilities (e.g., steam, fuels, and cooling water) [5].

In most previously published analyses, the optimum value of the minimum approach temperature, ΔT_{min}, is calculated based on trade-offs between external energy demand, utility cost, and heat-transfer area required within a heat exchanger network, and capital costs. Except the beneficial results from a reduction in utility requirements, there is no clear reflection of correlating heat-transfer loss to the HENs synthesis in conventional pinch analysis. In fact, based on thermodynamics, the effect of entropy generation, which varies considerably with the temperature level at which it occurs, is neglected [6, 7]. Accordingly, it is more reasonable to consider exergy destruction rather than energy demand as the annual operating cost in a trade-off with capital costs to determine the optimal minimum approach temperature in HENs synthesis.

Many studies have been carried out on the concept of linking exergy calculation with economic parameters called exergoeconomic analysis [8, 9]. The exergoeconomic analysis of a single heat exchanger has been already presented [10] and some papers have been published in the application of the exergy concept into HENs synthesis [11, 12]. The combination of pinch analysis with the exergy load distribution method is studied analytically [13, 14]. The exergy composite curves are proposed elsewhere to focus on the opportunity of using heat pumps in energy conversion systems and optimal integration of utility systems using Mixed Integer Linear Programming (MILP) [15]. However, calculation of the exergy destruction in HENs analysis and the trade-off between exergy-destruction cost and the capital expense are not considered practically.

In this chapter, it is demonstrated that the balanced composite curves can be used directly for calculation of the exergy destruction, heat recovery, and heat-transfer areas having a specified minimum approach temperature. The application of the proposed exergy-pinch methodology is examined for two different case studies using Aspen energy analyzer software and results are compared with conventional pinch methodology. It is revealed that the inclusion of heat exergy destruction into the pinch analysis decreases the optimum value of network ΔT_{min} due to an increase in the influence of operating cost by considering both the external utility requirements and internal heat exergy destruction. It is also showed that the greater the driving force, the higher exergy destruction and conversely, the smaller the size of heat exchangers in a network. It is concluded that the new approach allows determination of more realistic energy targets for hot and cold utilities by trade-offs between exergy-destruction cost, instead of external utility costs, and the heat-transfer area investment.

26.2 Conventional Pinch Analysis

Pinch technology is a systematic analysis of heat recovery within industrial heat exchanger networks and trade-offs between the networks' operating and capital costs for optimum selection of the heat-transfer minimum approach temperature.

In the pinch analysis, process stream data is depicted like energy flows as a function of temperature versus heat load. These data are integrated for all process streams to create composite curves for all hot streams releasing heat and for all cold streams requiring heat. The most constrained region for the internal heat transfer corresponds to the minimum vertical distance between two composite curves, ΔT_{min}, and is referred to as the pinch temperature. Practically, in the pinch analysis, the heat exchangers that are located between streams with a temperature above and below the pinch point are often replaced with alternative heat exchangers to make the process satisfy its energy target [16].

The composite curves include necessary information to predict the network's required heat-transfer area. The optimal minimum temperature difference between the curves is calculated by an economic trade-off between external energy demand and the required heat-transfer area. The minimum approach temperature, ΔT_{min}, has substantial effect on the amount of heat recovery and external utility requirements. When the hot and cold curves intersect, there is no driving force for heat transfer in the process. It means infinite heat-transfer area is needed and consequently, infinite capital cost is required. In contrast, an increase in the minimum temperature difference throughout the process, results in decreasing the heat-transfer area. Moreover, utility costs increase when ΔT_{min} increases. So, the cost analysis of HENs indicates specific temperature difference for all individual heat exchangers that is equal or more than the optimal ΔT_{min}.

In order to predict the network's required total heat-transfer area from the composite curves, utility streams must be incorporated with process streams to form the balanced composite curves [16]. The balanced composite curves are divided into vertical enthalpy intervals as shown in Fig. 26.1 considering individual stream with specific heat-transfer film coefficient and countercurrent heat transfer.

The total area required for a HEN is equal to the summation of the calculated area for each enthalpy interval as presented in the following equation [17, 18],

$$A_{\text{net}} = \sum_k \frac{1}{\Delta T_{\text{LMTD,k}}} \left[\sum_i^{\text{HStream}} \frac{\dot{q}_{i,k}}{h_i} + \sum_j^{\text{CStream}} \frac{\dot{q}_{j,k}}{h_j} \right] \quad (26.1)$$

where $\dot{q}_{i,k}$ and $\dot{q}_{j,k}$ are stream duties on hot stream i and cold stream j in enthalpy interval k, h_{i}, and h_{j} are film heat-transfer coefficients for hot stream i and cold stream j, respectively, including wall and fouling resistances, and $\Delta T_{\text{LMTD,k}}$ is a logarithmic mean temperature difference for interval k. If film heat-transfer coefficients change significantly from stream to stream, the precise minimum area requirements must be calculated using linear programming [19]. Although Eq. (26.1) is not 100 % accurate, but the estimated targeted area which is required for energy and area trade-offs, and also to evaluate alternative flowsheet options, provides significant reduction in calculation complexity with the acceptance of small penalty [16].

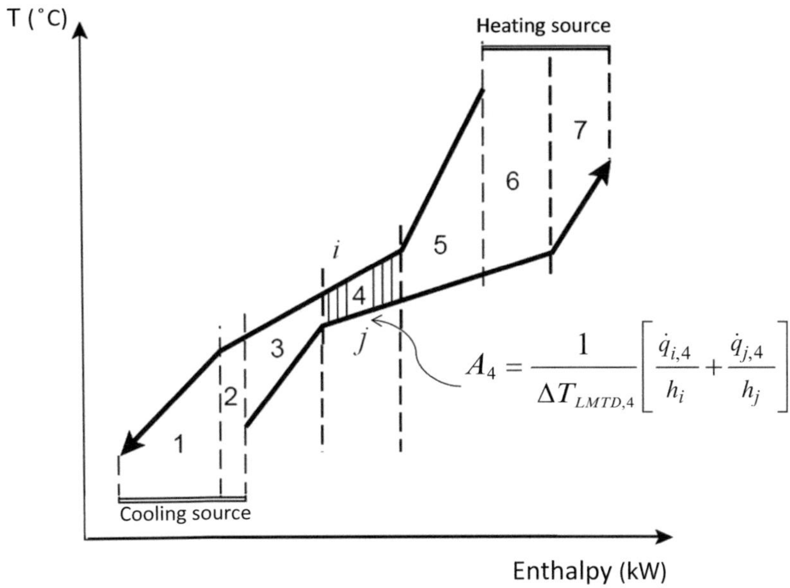

Fig. 26.1 Composite curves divided into intervals for heat-transfer area calculations

26.3 Integrated Exergy-Pinch Analysis

Heat-transfer process is the inevitable source of entropy generation mainly related to energy degradation due to an internal temperature drop. As a result, looking only to the energy consumption in HENs synthesis is not a correct approach. In the process of heat transfer, exergy is destroyed due to irreversibilities and it grows as the degree of irreversibility increases. However, according to the first law of thermodynamics, energy is conserved but the second law of thermodynamics indicates that energy quality decreases. Therefore, taking the exergy into account in HENs analysis enables better and more realistic energy targeting for different pinch temperature differences. Figure 26.2 depicts exergy-destruction concept for heat-transfer processes [6].

Based on the second law of thermodynamics, the exergy is defined as the maximum obtainable work from a process that brings a system into equilibrium with its surrounding environment [20]. Exergy is the energy quantity that is accessible to be used. Consequently, when a process and its surroundings reach equilibrium, exergy is zero. Therefore, exergy is interpreted as the energy quality that depends on the state of both the system and environment. Although energy is conserved, however, exergy is destroyed due to irreversible phenomena. In the heat-transfer process, the maximum possible conversion of heat to useful work, or heat exergy content, depends on the temperatures at which heat is available and is rejected. In fact, the exergy destruction actuates the process of heat transfer. Therefore, it is more logical to consider the amount of exergy being destroyed

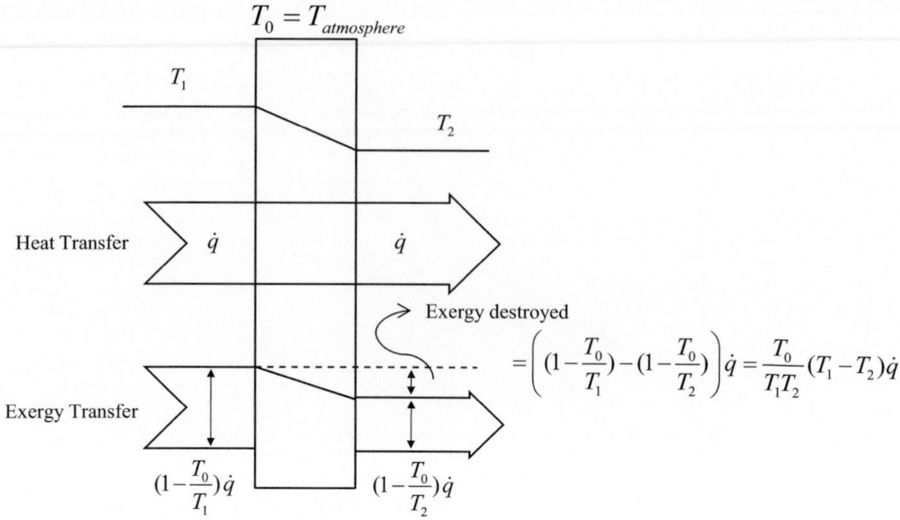

Fig. 26.2 Representation of exergy destruction in a heat-transfer process

rather than the amount of external energy being used as the annual operating cost in a trade off with total network area cost in order to determine the optimal minimum approach temperature, $\Delta T_{\min}$. The amount of heat is transferred from point 1 with the thermodynamic temperature T_1 to the point 2 with the thermodynamic temperature T_2 is always accompanied by exergy destruction which can be calculated by the following equation [6],

$$\dot{E}x_{dest} = \dot{E}x_1 - \dot{E}x_2 = \left(\left(1 - \frac{T_0}{T_1}\right) - \left(1 - \frac{T_0}{T_1}\right)\right)\dot{q} = \frac{T_0}{T_1 T_2}(T_1 - T_2)\dot{q} \quad (26.2)$$

In the calculation of exergy destruction during the process of heat transfer the temperatures of process streams are continuously changing. When the temperature–enthalpy relation is linear, process streams have constant specific heat capacities where using the log mean temperature difference results in an accurate calculations. If the temperature–enthalpy relation is nonlinear, a linearization strategy should be applied. Moreover, it is necessary to mention that the amount of exergy destruction for HENs is directly related to the specified $\Delta T_{\min}$, but it is independent of how the network is configured as long as the utility consumption is constant.

In the proposed methodology, the balanced composite curves that are divided into enthalpy intervals, obtained from pinch analysis, are used to calculate the network exergy destruction. As illustrated in Fig. 26.3, the points A and B on the hot and cold composite curves are selected and exergy destruction of the heat transfer between them is calculated using differential elements. Since T_A and T_B are being changed continuously with the line correlating temperature and enthalpy, there are different curve slopes due to different physical properties of process streams.

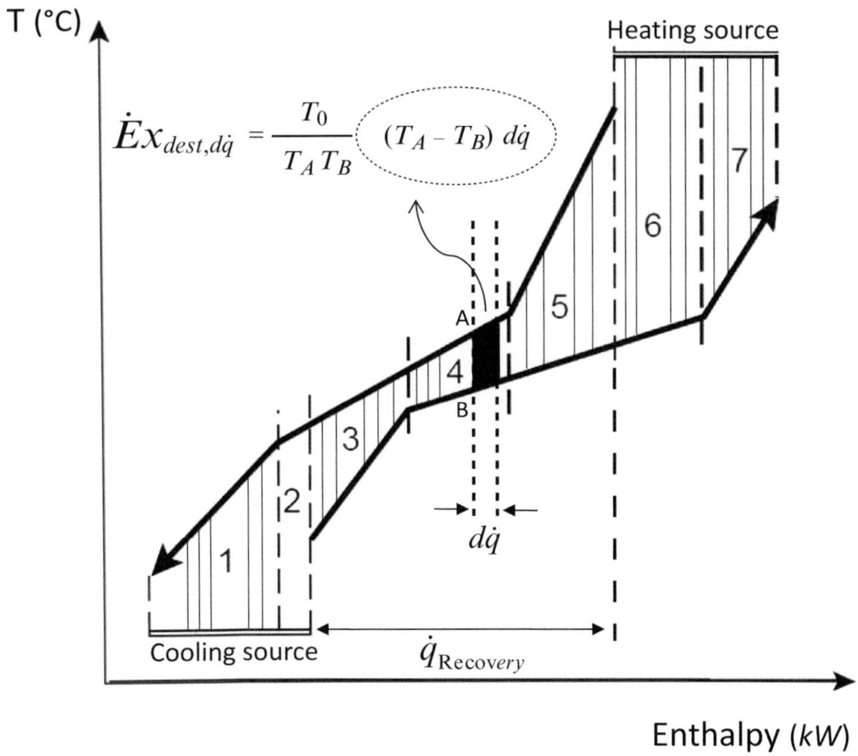

Fig. 26.3 Calculation of exergy destruction in balanced composite curves

First, the enthalpy value corresponding to the each interval is calculated. Subsequently, the number of integration sections is determined according to the number of enthalpy intervals. As a result, the exergy destruction for the network which is depicted in Fig. 26.3 could be calculated using the following equation,

$$\dot{E}x_{\mathrm{dest,net}} = \int_{0}^{\dot{q}_1} \frac{T_0}{T_{\mathrm{A}}T_{\mathrm{B}}}(T_{\mathrm{A}} - T_{\mathrm{B}})d\dot{q} + \int_{\dot{q}_1}^{\dot{q}_2} \frac{T_0}{T_{\mathrm{A}}T_{\mathrm{B}}}(T_{\mathrm{A}} - T_{\mathrm{B}})d\dot{q} + \ldots + \int_{\dot{q}_6}^{\dot{q}_7} \frac{T_0}{T_{\mathrm{A}}T_{\mathrm{B}}}(T_{\mathrm{A}} - T_{\mathrm{B}})d\dot{q} \tag{26.3}$$

The inclusion of exergy destruction calculations in the pinch analysis provides a practical and more realistic estimation of the optimum pinch temperature difference in trade-off between required HEN area and operating costs. In order to calculate the cost of exergy destruction for a system operating at a steady-state condition, cost balances should be formulated for each component separately. The cost balance for each component shows that the sum of cost rates for all output exergy streams is equal to the sum of cost rates of all input exergy streams plus the cost associated to capital investment and operating and maintenance costs [21]. The cost balance for a single adiabatic heat exchanger is presented in the below equation,

$$\sum_{\text{in}} \text{cx}_{\text{in}}\dot{\text{E}}\text{x}_{\text{in}} = \sum_{\text{out}} \text{cx}_{\text{out}}\dot{\text{E}}\text{x}_{\text{out}} + \dot{K}_{\text{CI}} + \dot{K}_{\text{OM}} \quad (26.4)$$

where cx_{in} and cx_{out} denote average cost per unit of exergy in dollar per kW for inlet and outlet streams respectively, $\dot{\text{E}}\text{x}$ is the exergy rate, $\dot{K}_{\text{CI}}$ is the coat rate of capital investment, and $\dot{K}_{\text{OM}}$ is the cost rate associated with operating and maintenance. It is assumed that the costs per exergy unit of all entering streams for a component are known from the component they exit or from the purchase cost which is commercially available [21]. Consequently, the unknown cost rates for a heat exchanger in the network are calculated using a cost balance from its predecessor.

The cost rate of exergy destruction in a heat exchanger, assuming that the purpose is to heat the cold stream and also the unit cost of fuel (hot stream) is independent of the exergy destruction, is calculated with the following equation [22],

$$\dot{K}_{\text{des}} = \text{cx}_{\text{f}}\dot{\text{E}}\text{x}_{\text{des}} \quad (26.5)$$

where $\dot{K}_{\text{des}}$ is the cost of exergy destruction, cx_{f} is the unit cost of hot stream as a fuel that is assumed to be constant for inlet and outlet, and $\dot{\text{E}}\text{x}_{\text{des}}$ is the rate of exergy destruction.

For heat exchanger networks, the cost of exergy destruction could be calculated for each enthalpy interval through the composite curves for a specific minimum approach temperature. In some research papers, the cost of exergy destruction is estimated by the multiplication of the unit price of electricity, operating hours, and the amount of exergy destruction [23, 24]. The reason behind this calculation is that it could be theoretically assumed that the electrical energy is converted completely to useful work. It means that the annual cost of exergy destruction substitutes the energy cost of hot and cold utilities, as the operating costs, and is applied in a trade-off with the cost of HEN heat-transfer area, as the network capital cost, to calculate the optimal ΔT_{min}.

The economic parameters used to calculate the cost of the heat exchanger network are categorized into capital costs, operating costs, and total annualized costs (TAC). The capital costs are the fixed costs to purchase and install the heat exchangers which is calculated for shell and tube heat exchangers using the following equation [25],

$$\mathrm{CC} = a + b\left(\frac{A}{N_{\mathrm{shell}}}\right)^{C} \times N_{\mathrm{shell}} \tag{26.6}$$

where CC is the capital costs of a heat exchanger, *a*, *b*, and *c* are the installation and the duty/area-related cost set coefficients of the heat exchanger, respectively. *A* is the heat-transfer area, and N_{shell} is the number of heat exchanger shells in the heat exchanger. The value of these parameters is based on the shell and tube type exchanger with carbon steel as the construction material.

The operating costs are a time-dependent expenses that represent the energy costs to run the equipment. The operating costs are dependent on the calculated energy targets in the heat exchanger networks and calculated as below,

$$\mathrm{OC} = \sum \left(C_{\mathrm{hu}} \times \dot{q}_{\mathrm{hu,min}}\right) + \sum \left(C_{\mathrm{cu}} \times \dot{q}_{\mathrm{cu,min}}\right) \tag{26.7}$$

where OC is the operating costs per year, C_{hu} and C_{cu} are the annual utility cost for hot and cold utilities, $\dot{q}_{\mathrm{hu,min}}$ and $\dot{q}_{\mathrm{cu,min}}$ are the heat flows for energy target of hot and cold utilities, respectively.

The TAC consists of both the capital and operating costs associated with the heat exchangers in the HENs. The below equation is used to calculate the TAC,

$$\mathrm{TAC} = A_{\mathrm{f}} \times \sum \mathrm{CC} + \mathrm{OC} \tag{26.8}$$

where A_{f} is the annualization factor which accounts for the depreciation of capital costs in a heat exchanger network and is represented with the following equation [25]. It should be noted that the capital costs and operating costs of a heat exchanger network do not have the same units.

$$A_{\mathrm{f}} = \frac{\left(1 + \frac{\mathrm{ROR}}{100}\right)^{\mathrm{PL}}}{\mathrm{PL}} \tag{26.9}$$

where ROR is the rate of return in percent of capital and PL is the plant lifetime. TAC is usually considered to be minimized in the optimal minimum approach temperature.

It is feasible to decrease the heat-transfer area by increasing the minimum temperature difference. Higher ΔT_{min} results in greater degrees of irreversibility and accordingly higher exergy destruction. Hence, there is an optimum point in a trade-off between exergy-destruction cost and heat exchanger capital expense. Figure 26.4a shows the conventional capital and energy trade-off using pinch analysis and Fig. 26.4b illustrates the proposed substituting trade-off between the network exergy-destruction costs and capital costs.

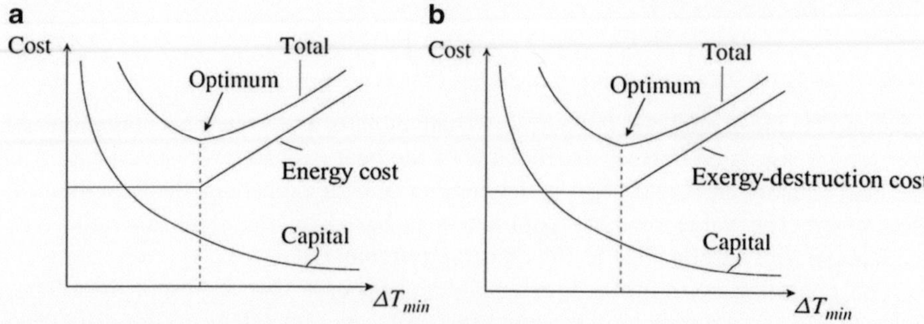

Fig. 26.4 Optimum setting for ΔT_{min} by trade-off between (**a**) energy costs, (**b**) exergy-destruction costs and network heat-transfer area (capital costs)

Table 26.1 Data and information used in Case Study I

Stream	Supply temp. (°C)	Target temp. (°C)	Heat capacity flow rate (kW/°C)	Heat load (kW)
Hot stream—1	160	93	8.7	588.9
Hot stream—2	249	138	10.55	1,171
Cold stream—1	60	160	7.62	762
Cold stream—2	116	260	6.08	875.5
Cooling water	30	50		
Hot oil	230	270		
Heat exchanger capital cost (US$)		10,000 + 800 (Area/Shells)^0.8*Shells[a]		
Electricity cost (US$)		0.05/kWh		
Rate of Return (ROR)		40 %		
Hours of operation		8,500 h/year		
Plant life		15 years		

[a]Based on Aspen Energy analyzer software

26.4 Case Study I

The first case study was initially presented by Linnhoff and Flower [4, 26] which has two hot and two cold streams. The streams and utility information with cost data are listed in Table 26.1.

The network is simulated in Aspen energy analyzer and the exergy-destruction costs, network capital costs, and the TAC are calculated over the range of ΔT_{min} from 1 to 20 °C. The rate of return is assumed 40 %. The plant lifetime and the hours of operation are considered to be 15 years and 8,500 hours per year, respectively. Figure 26.5 illustrates the trade-off between capital costs and the cost rate of exergy destruction. This figure indicates the optimal ΔT_{min} is at 6 °C that is lower from the optimal ΔT_{min} of 9 °C determined in the references. A lower minimum approach temperature saves exergy with relatively larger equipment. It is demonstrated that the exergy-destruction costs increase with the increase of ΔT_{min}.

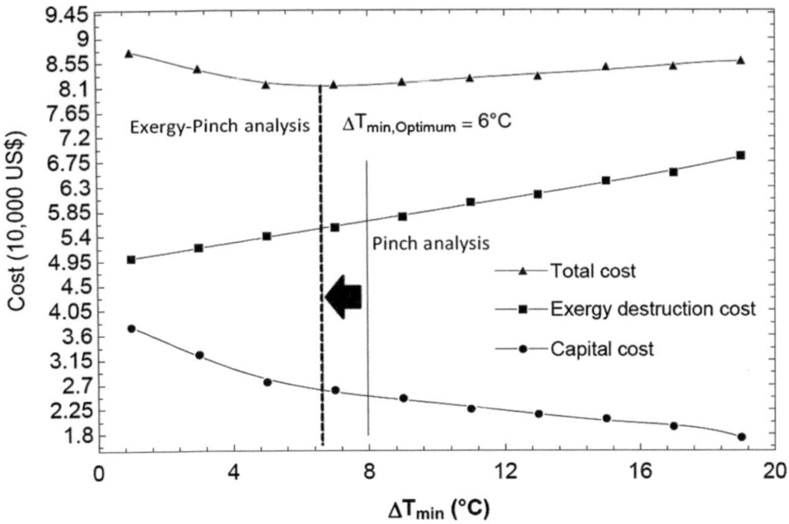

Fig. 26.5 Optimum $\Delta T_{\min}$ calculation for the network minimum total cost, Case Study I

Furthermore, Fig. 26.5 shows that the cost of exergy destruction decreases with increases in the capital costs. The depicted trend is similar to the energy cost analysis. However, exergy destruction analysis takes into account the energy degradation. Consequently, in line with sustainable development strategy, integrated exergy-pinch analysis offers an important way to reduce resource depletion which in a real-world application.

26.5 Case Study II

In the second case study, the chemical process flowsheet, as shown in Fig. 26.6, is considered to examine the application of the integrated exergy-pinch analysis [16]. The streams and utility information with cost data, as used, are listed in Table 26.2. The flowsheet consists of two main parts, such as one separation column and four heating and cooling units. The rate of return is assumed to be 40 %. The plant lifetime and the hours of operation are again considered to be 15 years and 8,500 hours per year, respectively. Despite the fact that the heat capacities are assumed constant, the nonlinear temperature–enthalpy behavior could have occurred by a series of linear segments [27]. Based on the conventional pinch analysis and the considered optimum $\Delta T_{\min} = 10°\text{C}$, retrofit design corresponding to the flowsheet in Fig. 26.6 achieves the target of the minimum 7.5 MW heating utility and minimum 10 MW cooling utility with the total of 51.5 MW of internal process heat recovery [16]. Figure 26.7 shows the corresponding grid diagram of the retrofit design flowsheet of Fig. 26.6.

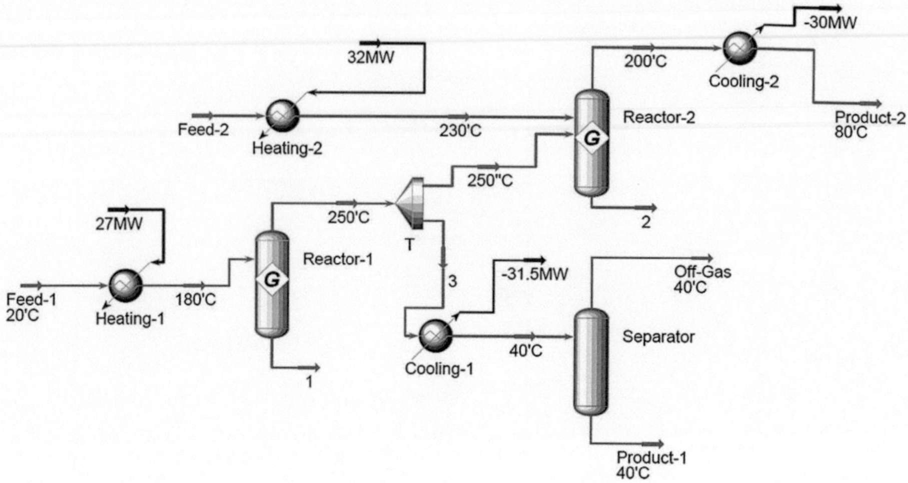

Fig. 26.6 Process flowsheet for Case Study II

Table 26.2 Data and information used in Case Study II

Stream	Supply temp. (°C)	Target temp. (°C)	Heat capacity flow rate (MW/°K)	Film heat-transfer coefficient (MW/m^2 · °K)
Feed—1	20	180	0.2	0.0006
Product—1	250	40	0.15	0.001
Feed—2	140	230	0.3	0.0008
Product—2	200	80	0.25	0.0008
Cooling water	20	30	1	0.001
Steam	240	239	7.5	0.003
Heat exchanger capital cost (US$)		10,000 + 800 (Area/Shells)^0.8*Shells[a]		
Electricity cost (US$)		0.05/kWh		
Rate of Return (ROR)		40 %		
Hours of operation		8,500 h/year		
Plant life		15 years		

[a]Based on Aspen Energy analyzer software

The heat exchanger network and corresponding balanced composite curves are simulated in Aspen energy analyzer software and used to calculate the network heat-transfer area as a capital costs and the exergy-destruction costs as a substitute for the external utilities costs to determine the network optimal $\Delta T_{\min}$. Figure 26.8 shows the balanced composite curves for Case Study II.

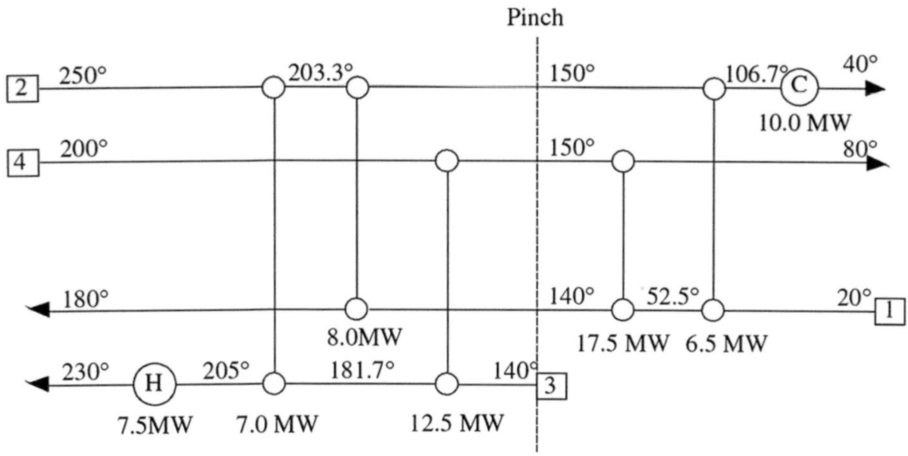

Fig. 26.7 Grid diagram of retrofit design for Case Study II

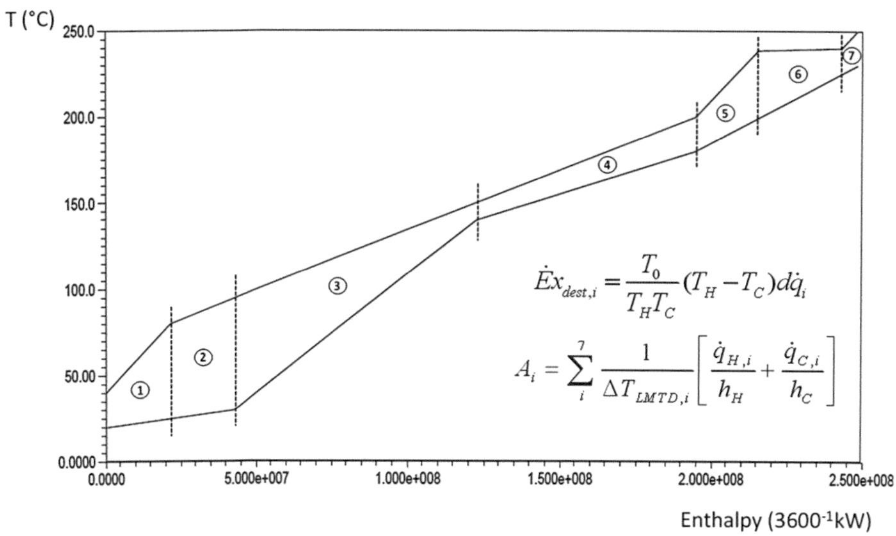

Fig. 26.8 Exergy destruction and heat-transfer area calculation for Case Study II

The network exergy-destruction costs, the required capital investment, and the network's TAC are calculated over a range of $\Delta T_{\min}$ from 1 to 15 °C and results are given in Fig. 26.9. This figure indicates that the optimal $\Delta T_{\min}$ is at 8 °C which is lower from $\Delta T_{\min}$ of 10 °C in the reference study [16]. It is demonstrated that a lower minimum approach temperature, which could save exergy with a relatively larger heat-transfer area, provides the opportunity of better resource utilization for the network.

Figure 26.10 depicts Grand Composite Curve (GCC) of the heat exchanger network for Case Study II. This graph is mainly analyzed for an optimal utility

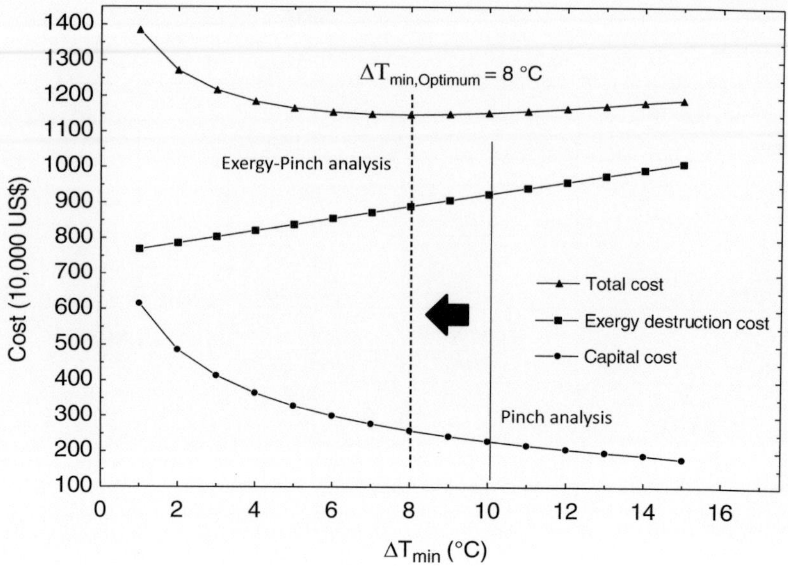

Fig. 26.9 Optimum $\Delta T_{\min}$ calculation for the network minimum total cost, Case Study II

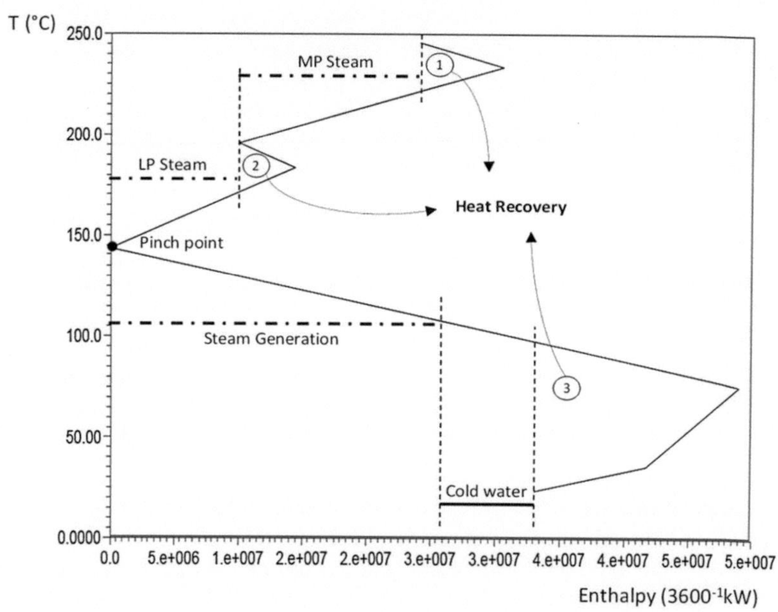

Fig. 26.10 Grand composite curve for optimum utility allocation, Case Study II

selection. The zero heat flow is the heat recovery pinch point and the numbered areas represent the heat recovery through process-to-process heat transfer. The profile of the GCC shows residual heating and cooling requirements after heat recovery in the heat exchanger networks.

The objective of optimum resource allocation using GCC is to maximize the use of the cheapest utilities. It is based on the assumption that the least expensive utilities are the hottest cold utilities and the coldest hot utilities. It means that a utility is considered cheaper based on temperature not based on cost in order to minimize the amount of exergy destruction in the system. Consequently, a hot utility at a lower temperature considered cheaper than a hot utility at a higher temperature. The most common hot utility is steam and is usually available at several levels, the Medium Pressure (MP) steam and Low Pressure (LP) steam.

Figure 26.10 shows that 2,778 kW LP steam at 180 °C and 4,722 kW MP steam at 240 °C become the minimum external heat requirements to satisfy the hot water utility demand of the process flowsheet in this case study. It is also demonstrated that there is an opportunity of 8,334 kW LP steam generation at 120 °C. It is also noted that the allocation of the external hot and cold utilities in the system using GCC results in employing the cheapest resource with a minimum exergy destruction.

26.6 Conclusions

In the proposed methodology for the heat exchanger network analysis, the cost rate of exergy destruction substitutes the utility cost in a trade-off between the operating and capital costs in the conventional pinch analysis to determine the optimum minimum approach temperature (ΔT_{min}). It is demonstrated that the balanced composite curves can directly be used for the calculation of exergy destruction, heat recovery, and heat-transfer area for a specified minimum approach temperature. It is shown that GCC can be employed for an optimal external utility allocation. Two case studies are presented as examples for application of the proposed methodology to demonstrate its advantages over the conventional pinch analysis. It is shown that the combination of heat exergy destruction into the pinch analysis decreases the optimum value of network ΔT_{min}. It is also shown that the greater the driving force, the higher exergy destruction and conversely, the smaller the size of heat exchangers in the network. The proposed methodology allows determination of realistic energy targets for hot and cold utilities in the network by trade-offs between exergy-destruction costs, instead of external utility costs, and heat-transfer area investment.

References

1. Azapagic A (1999) Life cycle assessment and its application to process selection, design and optimization. Chem Eng J 73(1):1–21
2. Linnhoff B et al (1982) User guide on process integration for the efficient Use of energy. Pergamon, Oxford
3. Gundersen T, Naess L (1988) The synthesis of cost optimal heat exchanger networks, an industrial review of the state of the art. Comput Chem Eng 12(6):503–530
4. Linnhoff B, Flower JR (1978) Synthesis of heat exchanger networks. AIChE J 24(4):633–654
5. Linnhoff B (1993) Pinch analysis—a state of the art overview. Trans Inst Chem Eng 71 (A):503–522
6. Çengel YA, Boles MA (2006) Thermodynamics: an engineering approach, 5th edn. McGraw-Hill, New York
7. Dincer I, Rosen M (2007) Exergy, energy, environment and sustainable development, 1st edn. Elsevier, Oxford
8. Fani M, Mozafari AA, Farhanieh B (2009) Coordination of process integration and exergoeconomic methodology for analysis and optimization of a pulp and paper mill. Sharif University of Technology. Trans B Mech Eng 16(4):301–312
9. Bakan K, Dincer I, Rosen MA (2008) Exergoeconomic analysis of glycol cold thermal energy storage systems. Int J Energ Res 32(3):215–225
10. Can A, Buyruk E, Eryener D (2002) Exergoeconomic analysis of condenser type heat exchangers. Exergy An Int J 2(2):113–118
11. Li ZH, Hua B (2000) Modeling and optimizing for heat exchanger networks synthesis based on expert system and exergo-economic objective function. Comput Chem Eng 24 (2–7):1223–1228
12. Jiang N, Li L (2011) Heat exchanger network integration considering flow exergy loss. Chem Eng Technol 34(12):1997–2004
13. Sorin M, Paris J (1997) Combined exergy and pinch approach to process analysis. Comput Chem Eng 21(Suppl):S23–S28
14. Sorin M, Paris J (1999) Integrated exergy load distribution method and pinch analysis. Comput Chem Eng 23:497–507
15. Marechal F, Favart D (2005) Combined exergy and pinch analysis for the optimal integration of energy conversion technologies. In: 18th International conference on efficiency, cost, optimization, simulation and environmental impact of energy systems, pp. 177–184
16. Smith R (2005) Chemical process design and integration. Wiley, England
17. Townsend DW, Linnhoff B (1984) Surface area targets for heat exchanger networks. Institution of Chemical Engineers Annual Research Meeting, Bath
18. Linnhoff B, Ahmad S (1990) Cost optimum heat exchanger networks-1. Minimum energy and capital using simple models for capital cost. Comput Chem Eng 14:729
19. Ahmad S, Linnhoff B, Smith R (1990) Cost optimum heat exchanger networks- 2. Targets and design for detailed capital cost models. Comput Chem Eng 14:751
20. Dincer I, Çengel YA (2001) Energy, entropy, and exergy concepts and their roles in thermal engineering. Entropy 3:116–149
21. Bejan A, Tsatsaronis G, Moran M (1991) Thermal design and optimization. Wiley, New York
22. Tsatsaronis G, Moran M (1997) Exergy-aided cost minimization. Energy Convers Manage 38 (15–17):1535–1542
23. Rosen MA, Dincer I (2003) Exergoeconomic analysis of power plants operating on various fuels. Appl Therm Eng 23:643–658
24. Rosen MA, Dincer I (2008) A concise review of exergy-based economic methods. EE'08 Proceedings of the 3rd IASME/WSEAS international conference on energy & environment. Wisconsin, pp. 136–144

25. Aspen energy analyzer reference guide (2009) Aspen Technology, Burlington
26. Linnhoff B, Flower JR (1978) Synthesis of heat exchanger networks: II. Evolutionary generation of networks with various criteria of optimality. AIChE J 24:642
27. Van Reisen JLB et al (1995) The placement of two-stream and multistream heat exchangers in an existing network through path analysis. Comput Chem Eng 19(Suppl):143–148

Chapter 27
Flow Velocity Prediction for Heat Exchanger for Fuel Cell by Test and Analysis

Seonhwa Kim, Jonghyek Kim, and Jinheok Jeong

Abstract In fuel cell systems, the heat exchanger is one of the most important pieces of equipment. and needs to be of robust design and optimization. This study is about the serpentine type heat exchanger with a rectangular shell. It has a tube bundle with a lot of pass and is operated for two-phase service. It is essential to estimate the exact velocity value in order to predict the effect of flow-induced vibration and heat transfer characteristics. It is difficult to estimate the velocity for special heat exchangers due to the variable temperature gradient on each tube. This chapter presents the methodology for the velocity prediction. By comparison with real hot test and the results of analysis, results, the reliability of the 1D calculation could be verified. Also this chapter presents the detail values that are the scalar values for the heat exchanger. Then the velocity profile can be obtained as a vector value on each tubes as well as their physical properties.

Keywords Heat Exchanger • Heat Transfer • CFD • Porous medium

Nomenclature

E	Total energy per unit mass
f_r	Body force due to rotation
f_g	Body force due to gravity
f_p	Porous media body force
f_u	User-defined body force
f_ω	Vortices confinement specific force
f_L	Lorentz force or Laplace force
P	Fluid pressure
P_v	Viscous (linear) resistance tensor (β)
P_i	Inertial (quadratic) resistance tensor (α)
$\mathbf{q}''$	Heat flux vector

S. Kim (✉) • J. Kim • J. Jeong
Fuel Cell Division, POSCO Energy Inc, #13-9 Jukcheon-ri, Heunghae-eup, Buk-gu, Pohang-si, Gyeongsangbuk-do 791-941, Korea
e-mail: makgang11@gmail.com

I. Dincer et al. (eds.), *Progress in Clean Energy, Volume 1*,
DOI 10.1007/978-3-319-16709-1_27

S	Modulus of the mean strain rate tensor
S_k, $S\varepsilon$	User-specified source term
T	Viscous stress tensor
v	Velocity
v_g	Gas velocity vector
$\nabla \cdot v$	Velocity divergence
ε_0	Ambient turbulence
γ_M	Fluid turbulence
ρ	Density

27.1 Introduction

The heat exchanger is used widely in industry for the purpose of energy regeneration. Likewise, it is used in molten carbonate fuel cell systems for plant balance. The primary purpose of the heat exchanger is to supply suitable temperature fuel for reforming a reaction. Thus the heat exchanger is an important piece of equipment in a fuel cell system. In general, the flow velocity is simply calculated by the correlation between flow rate and hydro diameter. But in a heat exchanger, in order to estimate the heat transfer characteristics and the force from the induced fluid flow, the velocity should be calculated in more detail due to the complex structure, temperature distribution, and pressure. In addition, because the fluid service used is a two-phase flow, the temperature distribution depends on heat flux regarding latent heat. Also the heat exchanger of interest is a kind of evaporator that contains a number of serpentine-type tubes [1].

27.2 Approach

This chapter presents a methodology to estimate flow velocity more exactly as shown in the procedure in Fig. 27.1. At first, the heat exchanger was designed by HTRI with known geometry information which is a kind of program capable of thermal rating including a number of physics for fluid flow and heat transfer characteristics for a general heat exchanger. Because the thermal rating model is a preliminary design and based on fundamental theory it should be used with real hot testing data [2]. Also the model should be updated using the test results in this chapter. The temperature and pressure distributions are obtained from the model, as well as the flow velocity in each tube pass. At that time, it is assumed that the distribution value is the same at each row of tubes. Therefore computational fluid dynamics should be performed to obtain the exact flow velocity at each row for the tube bundle. Koh and Colony have modeled the microstructure as a porous medium [3]. Later Tien and Kuo, and Kim extended their work in order to analyze the heat transfer phenomenon in a microchannel heat sink [4–6]. Using the porous medium approach, they evaluated

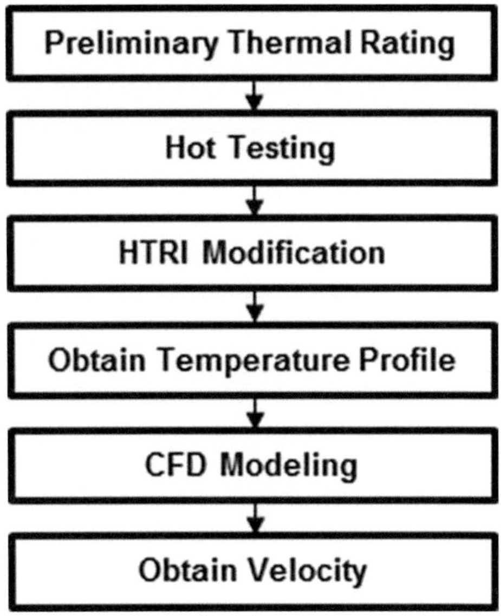

Fig. 27.1 Proposed analysis flow

the effects of geometric parameters and physical properties without tedious numerical computations. Similarly, Srinivasan et al. studied fluid flow and heat transfer through spirally fluted tubes using a porous substrate approach [7]. The model divided the flow domain into two regions, the flute region and the core region, with the flutes being modeled as a porous substrate. There are several assumptions for fin tube modeling for the CFD in this study. The first is that the fin tubes could be modeled as porous media. The second is that the cold side service fluid could be modeled by single phase, because the temperature value could be adjusted by the heat sink on the porous media. Because the cold side service fluid is a two-phase flow with natural gas and liquid water for the reforming reaction, it is not linear between the increasing temperature and heat flux.

27.3 Results and Discussion

In order to analyze the thermal rating, the heat exchanger's geometric information is essential for tube row, pass, tube form, pitch, clearance to the wall, and fin geometric information as well as material. By the thermal rating, the temperature distribution for the hot and cold sides for this heat exchanger can be calculated. The stream properties are similar to those shown in Fig. 27.2 and Table 27.1, which were calculated by the gas state equation of Peng-Robinson [8].

The HTRI model used has an estimated base on the experiment in this study. According to the structure of the heat exchanger, the hot side service fluid flows to

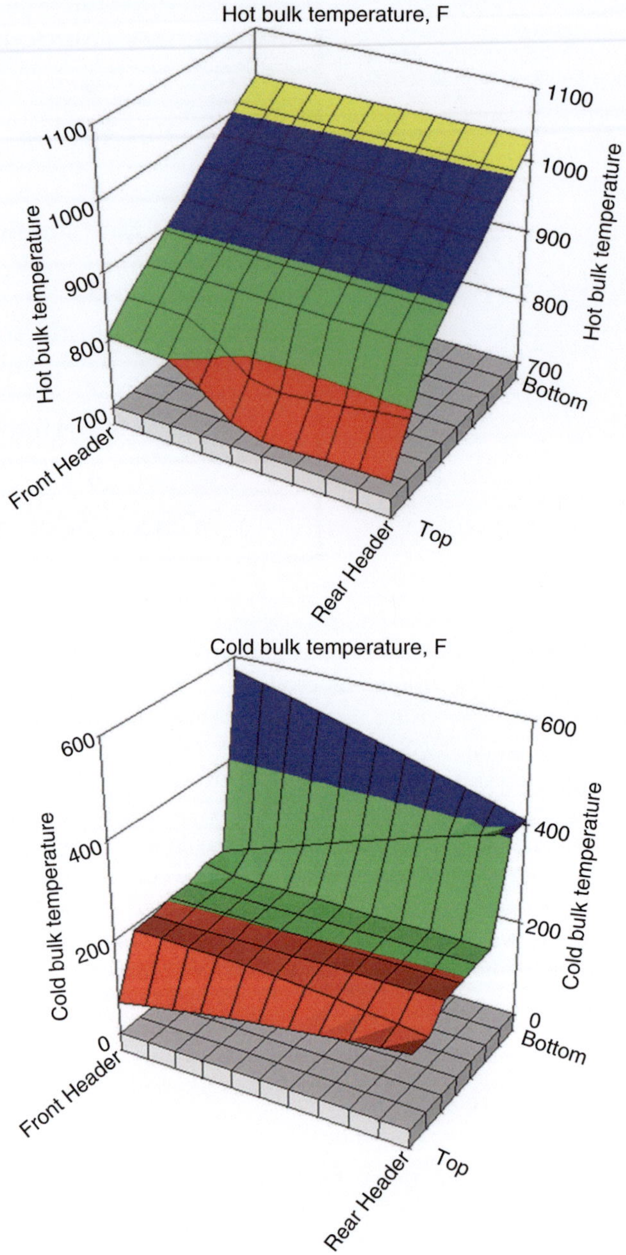

Fig. 27.2 Temperature distribution

Table 27.1 Physics for cold side fluid of liquid

Temp (°C)	Enthalpy (kJ/kg)	Vapor fraction	Density (kg/m³)	Viscosity (mN-s/m²)	Heat capacity (kJ/kg-C)	Conductivity (W/m-C)
10	−1,140.64	0.33	1.08	0.01	2.14	0.03
18	−1,105.11	0.33	1.04	0.01	2.16	0.03
45	−962.99	0.35	0.96	0.01	2.20	0.04
50	−927.46	0.36	0.94	0.01	2.21	0.04
70	−714.29	0.42	0.89	0.01	2.20	0.04
96	398.96	0.87	0.84	0.01	2.08	0.03
97	552.92	0.93	0.84	0.01	2.07	0.03
98	706.88	1.00	0.84	0.01	2.06	0.03
345	1,256.14	1.00	0.50	0.02	2.40	0.06
593	1,897.08	1.00	0.36	0.03	2.78	0.10

the bottom from the top. The entrance to the cold side is the front header to the top and the discharging nozzle is located on the bottom header in the same direction on this heat exchanger. In Fig. 27.3, the heat flux can also be seen as not a constant value. To be specific, the tube side fluid service is a two-phase flow with vapor and liquid coexisting. Table 27.4 shows the boiling regime for the inside of the tube. The cold stream would be boiled with stratified flow in passes 1 and 2. The film boiling from pass 3 to 6 can be seen in this table. Figure 27.3 shows the vapor fraction of the fluid in the tube; especially the dry-out in the sixth pass can be seen in this figure (Tables 27.2, 27.3, and 27.4).

In this study, computational fluid dynamics was utilized to carry out a commercial application for use in STAR-CCM+. Inside the equipment during operation due to the evaporation of water from the pipeline in order to reflect the change in temperature, the phenomenon reflects HTRI analysis of the temperature distribution inside the tubes at that point about the CFD model to follow the user field function applied using a heat sink. The flow analysis model used in this study is the steady-state model is analyzed for the experiment. In addition, as the heat exchanger characteristic, each fluid temperature varies with the position of the volume and pressure variation due to temperature changes to be considered. In this study, the gas model was applied to ideal gas. The speed of the flow inside the heat exchanger is subsonic; the humidifier assembly fin inside the tube is caused by myriad flow resistance elements present. In order to simulate such a turbulent environment, the realizable k-epsilon model is used in this chapter. The model transport equation for ε and k are as follows [2].

$$\frac{d}{dt}\int_V \rho k dV + \int_{\mathbf{A}} \rho k\left(v - v_g\right) \cdot da = \int \left(\mu + \frac{\mu_t}{\sigma_k}\right) \nabla k \cdot da + \int [G_k + G_b - \rho((\varepsilon - \varepsilon_0) + \gamma_M) + S_k] dV \tag{27.1}$$

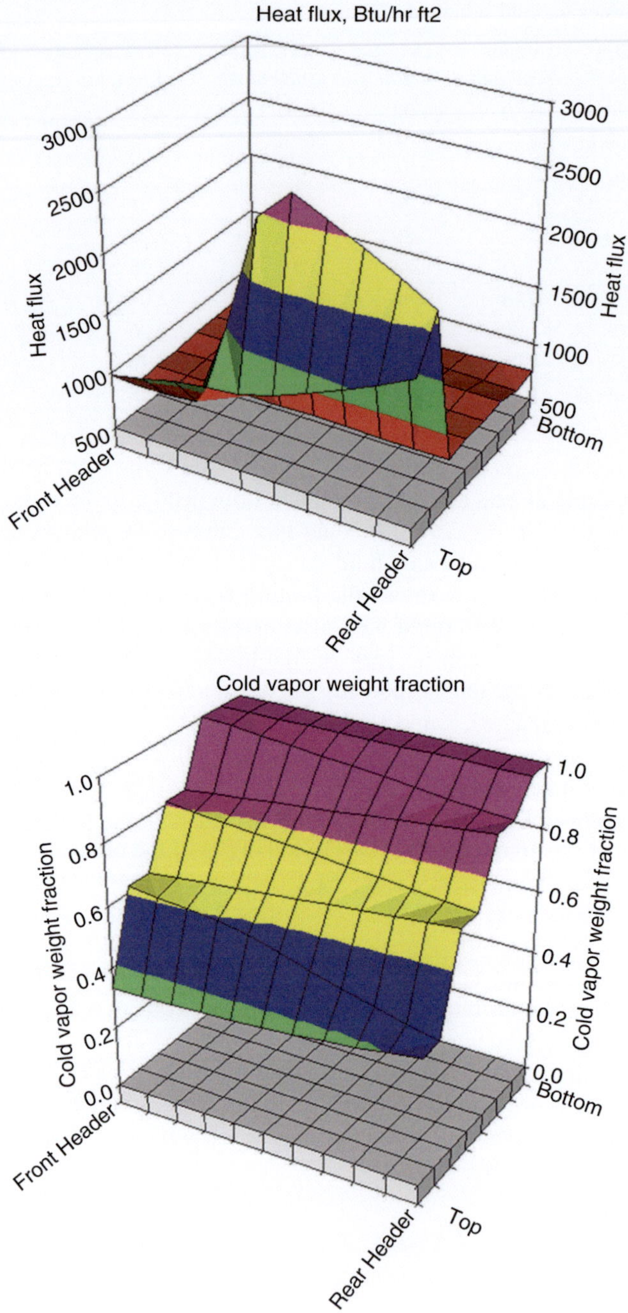

Fig. 27.3 Heat flux distribution and vapor fraction

Table 27.2 Physics for cold side fluid of vapor

Temp (°C)	Density (kg/m³)	Viscosity (mN-s/m²)	Heat capacity (kJ/kg-C)	Conductivity (W/m-C)	Surface tension (mN/m)	Critical pressure (kPa)
10	1,000.16	1.31	4.23	0.58	74.22	22,054.80
18	998.09	1.04	4.22	0.60	72.95	22,054.85
45	988.93	0.59	4.22	0.64	68.71	22,054.92
50	986.83	0.54	4.22	0.64	67.88	22,054.93
70	977.20	0.40	4.24	0.66	64.47	22,054.95
96	961.32	0.29	4.29	0.68	59.63	22,054.98
97	960.53	0.29	4.30	0.68	59.40	22,054.98
98	959.85	0.29	4.30	0.68	59.21	22,054.98

Table 27.3 Physics for hot side fluid of vapor

Temp (°C)	Enthalpy (kJ/kg)	Vapor fraction	Density (kg/m³)	Viscosity (mN-s/m²)	Heat capacity (kJ/kg-C)	Conductivity (W/m-C)
593	997.37	1.00	0.418	0.038	1.260	0.065
501	882.12	1.00	0.468	0.035	1.233	0.058
490	869.32	1.00	0.474	0.034	1.230	0.057
406	766.88	1.00	0.533	0.031	1.205	0.051
396	754.07	1.00	0.541	0.031	1.202	0.050
310	651.63	1.00	0.622	0.028	1.178	0.044
288	626.03	1.00	0.646	0.027	1.172	0.042

Table 27.4 Boiling regime for each tube pass

Tube pass	Boiling regime
1	Wavy stratified
2	Wavy stratified
3	Film
4	Film
5	Film
6	Mist
7	Sensible
8	Sensible

$$\frac{d}{dt}\int_V \rho\varepsilon dV + \int_A \rho\varepsilon\left(v - v_g\right)\cdot da = \int\left(\mu + \frac{\mu_t}{\sigma_k}\right)\nabla\varepsilon k \cdot da + \int\left[C_{\varepsilon 1}S\varepsilon + \frac{\varepsilon}{k}C_{\varepsilon 1}C_{\varepsilon 3}G_b - \frac{\varepsilon}{k+\sqrt{v\varepsilon}}C_{\varepsilon 2}\rho(\varepsilon - \varepsilon_0) + S_\varepsilon\right]dV \tag{27.2}$$

$$\frac{\partial}{\partial t}\int_V \boldsymbol{W}\chi dV + \oint [\boldsymbol{F} - \boldsymbol{G}] \cdot d\mathbf{a} = \int_V \boldsymbol{H} dV \tag{27.3}$$

$$\boldsymbol{W} = \begin{bmatrix} \rho \\ \rho\mathbf{v} \\ \rho E \end{bmatrix}, \qquad \boldsymbol{F} = \begin{bmatrix} \rho(\mathbf{v} - \mathbf{v}_{\mathrm{g}}) \\ \rho(\mathbf{v} - \mathbf{v}_{\mathrm{g}}) \otimes \mathbf{v} + p\mathbf{I} \\ \rho(\mathbf{v} - \mathbf{v}_{\mathrm{g}})H + p\mathbf{v}_{\mathrm{g}} \end{bmatrix}, \qquad \boldsymbol{G} = \begin{bmatrix} 0 \\ \mathbf{T} \\ \mathbf{T} \cdot \mathbf{v} + \mathbf{q}'' \end{bmatrix},$$

$$\boldsymbol{H} = \begin{bmatrix} S_{\mathrm{u}} \\ \boldsymbol{f}_{\mathrm{r}} + \boldsymbol{f}_{\mathrm{g}} + \boldsymbol{f}_{\mathrm{p}} + \boldsymbol{f}_{\mathrm{u}} + \boldsymbol{f}_{\omega} + \boldsymbol{f}_{\mathrm{L}} \\ S_{\mathrm{u}} \end{bmatrix}$$

The boundary layer of the fluid model with a realizable K-epsilon two layer, all wall treatment model is used for the boundary layer of the fluid model due to the size and elements of mesh. Because the gas flows into the heat exchanger, the more accurate interpretation of the various types of gas for multicomponent gas was simulated using the model. In this chapter, the research also shows that no gas is assumed to be a reaction between species. Also this simulation is analyzed by the coupled solver, because the solver could calculate the combined fluid volume between fluid flow and energy as an ideal gas model. Equation (27.3) presents the solver for the coupled flow model.

Due to the complex geometry of the finned tube and the conjugated heat transfer between the fluid and the fins, the conventional energy equation cannot be solved analytically. A so-called porous medium approach has been chosen for modeling fluid flow and heat transfer with a heat transfer augmentation device. The fin tube could be modeled as shown in the following figure which has a fin density of 275.6 per meter and fin thickness of 0.787 mm, so that a dense grid mesh is essential for finite volume in order to perform fine analysis. But that is adverse for calculation time and reliability of results, thus it is necessary to reduce the number of grid mesh. By using the original shape information for the fin tube, the local FVM model can be generated.

$$\frac{\Delta P}{L} = -(P_i|v| + P_v)v \tag{27.4}$$

This model is applied to an arbitrary flow rate, and then the pressure drop values for each direction can be obtained. A theoretical value for the pressure drop per unit length of the porous media is as follows. Ultimately, this study shows the relationship between pressure drop and flow rate is obtained; this second-order polynomial interpolation of the viscous resistance and inertial resistance coefficient is calculated (Figs. 27.4 and 27.5).

As shown in the figure, the x-axis direction is defined as the axial fin tube, the y-axis and z-axis are defined as the radial direction of the fin tube. In this study, the radial flow resistance is assumed to be the same. This spiral fin tube in the axial direction of the flow flows along the fin, whereas the radial flow flows along the gap

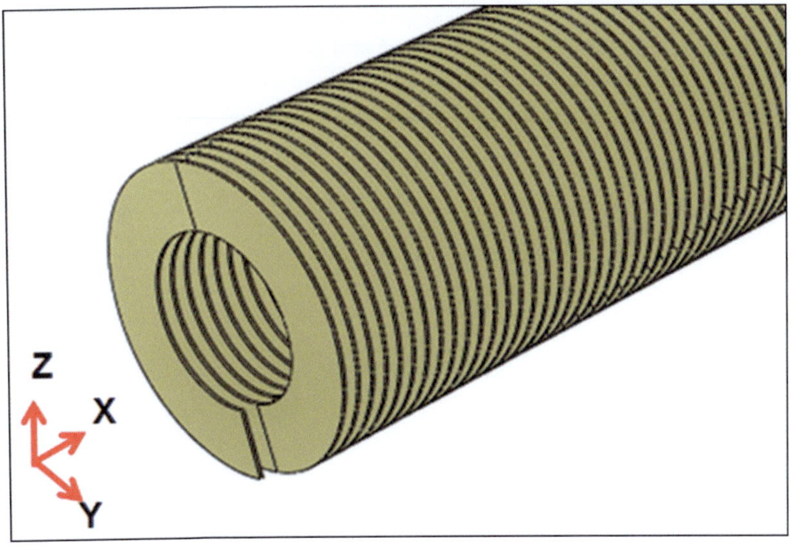

Fig. 27.4 Local fin tube modeling

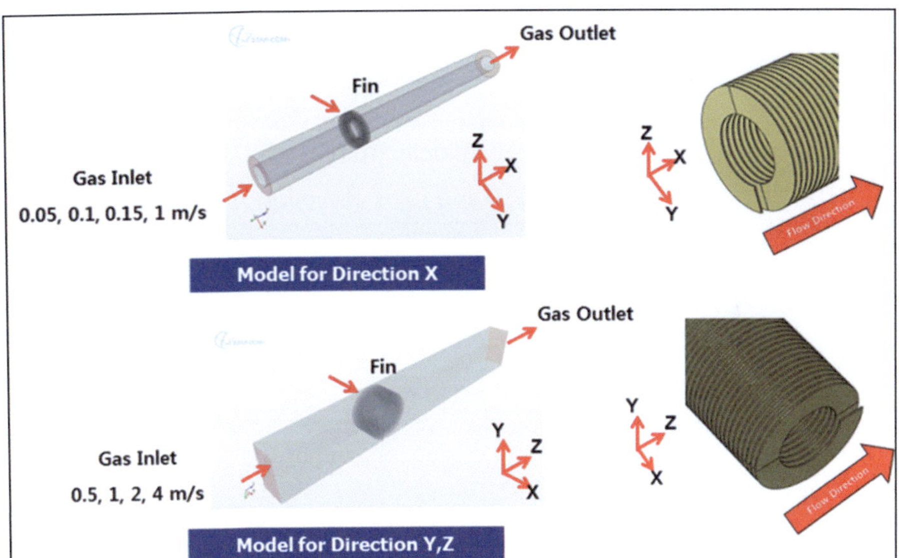

Fig. 27.5 Boundary condition for local modeling

between the junctions of the fin. Because of these flow characteristics, the velocity values are given as 0.05, 0.10, 0.15, and 1 m/s for the x direction and 0.5, 1.0, 2.0, and 4.0 m/s for the y direction. The values for the inertial resistance and viscous resistance could be obtained from the interpolated equation in this simulation which are 699116.1 kg/m^4, 114795.6 kg/m^3 s, respectively, for the X direction. Also the α, β values of the Y direction are 48.3 kg/m^4 and 63.1 kg/m^3 s. These values would then be used in the global CFD model for this heat exchanger (Fig. 27.6).

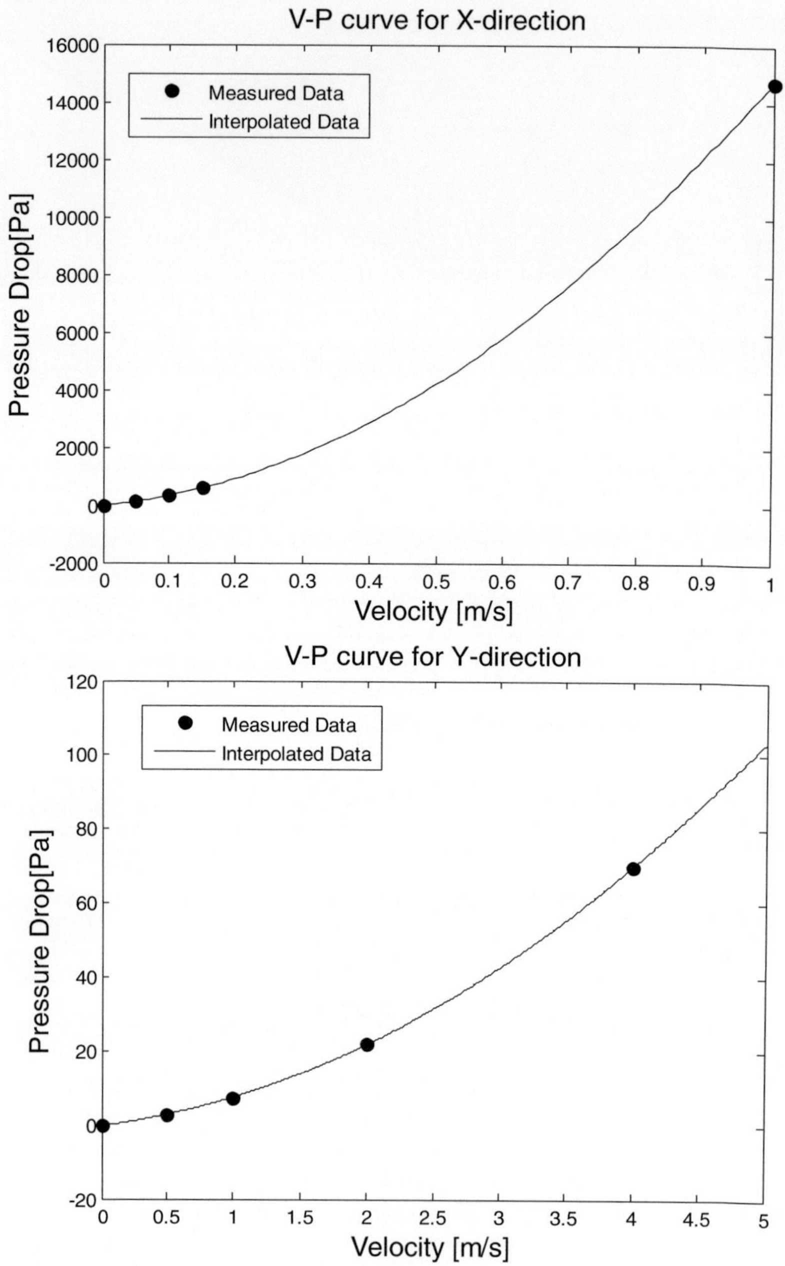

Fig. 27.6 P–V curve for Z direction and Y direction

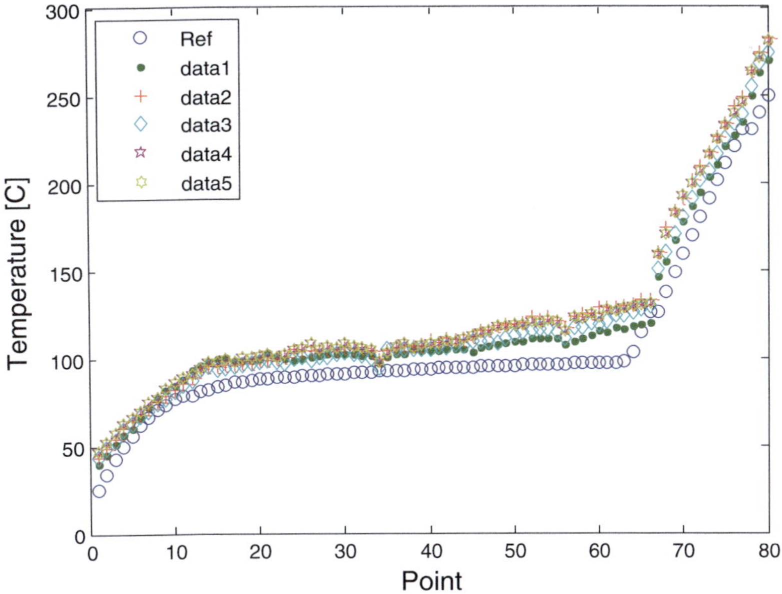

Fig. 27.7 Temperature profile of CFD results with heat sink

The mixture flow inside the tube in this heat exchanger is a two-phase flow containing liquid water and natural gas. So the heat flux and temperature are nonlinear because of the latent heat of the water. In this study, a heat sink was applied to the tube elements in order to consider the phenomenon of water evaporation. Also by adjusting the heat sink value against the thermal rating results, this simulation could be more reliable in real operation. Figure 27.7 shows the value of the heat sink tube and location of the applied values, as well as the cold side temperature distribution for several points.

The heat sink value is determined through iterated calculation as shown in Fig 27.7 to come to similar results against existing data as well as the values which are varied depending on their location. Although the temperature could be different against that of the layer for the tube bundle in the real model, the thermal rating results show the same. Therefore to study the temperature value on several positions in a row is significant. The inlet temperature is 25 °C and the outlet temperature is 318 °C according to the HTRI results. The temperature is increase excessively from point 1 to 2, imperceptibly from point 3 to 6, and excessively from point 7 to 8 due to a typical characteristic for the multiphase flow. In comparison between the results of CFD and thermal rating results, the averaged standard deviation is about 6 °C. The predicted flow velocities are as shown in Fig. 27.8.

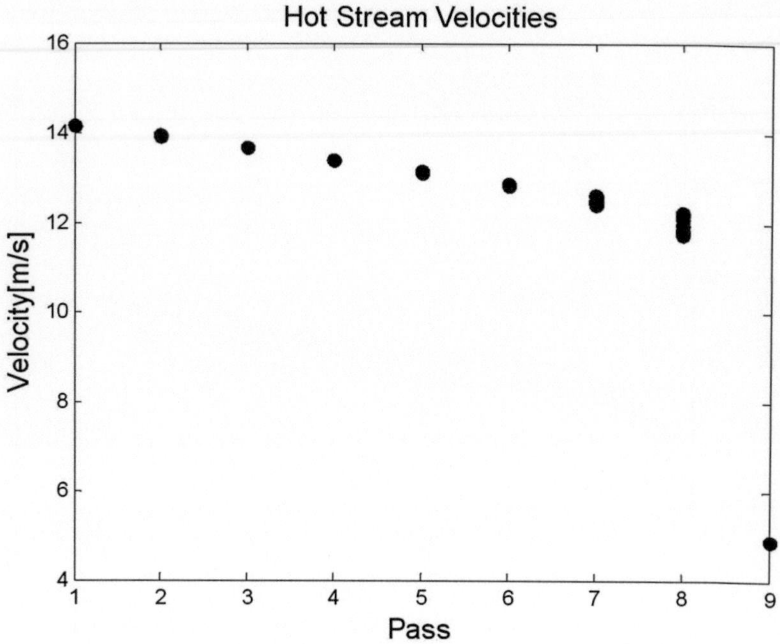

Fig. 27.8 Flow velocities of distribution for hot stream

27.4 Conclusion

Reliable thermal rating data indicate that, in this study, the temperature of the heat exchanger tubes and other physical values were calculated. The finite volume model was designed to analyze by CFD for the fin tube from the actual operation of the heat exchanger geometry. In the local model of the finite volume, the viscous resistance and inertial resistance were estimated by using the correlation between flow velocities and pressure drop. In the global FVM model, these values were applied on the fin tube modeling as a porous medium. Also in order to obtain a reliable model, a heat sink was modeled onto the tube and its value was adjusted according to the thermal rating results. For this procedure, the profile of temperature and velocities could be estimated.

References

1. Smith JM, Van Ness HC, Abbott MM (2007) Introduction to chemical engineering thermodynamics, 7th edn. McGraw-Hill, New York
2. CD-adapco, User guide STAR-CCM+
3. Koh JCY, Colony R (1986) Heat transfer of microstructures for integrated circuits. Heat Mass Transf 13:89–98

4. Tien CL and Kuo SM (1987) Analysis of forced convection in microstructures for electronic system cooling. In: Proceedings of the international symposium on cooling technology for electronic equipment, Honolulu, pp 217–226
5. Kim SJ, Kim D (1999) Forced convection in microstructures for electronic equipment cooling. ASME J Heat Transfer 121:639–645
6. Kim SJ, Kim D (2000) On the local thermal equilibrium in micro channel heat sinks. Heat Mass Transf 43:1735–1748
7. Srinivasan V, Vafai K, Christensen RN (1994) Analysis of heat transfer and fluid flow through a spirally fluted tube using a porous substrate approach. ASME J Heat Transf 116:543–551
8. Abudour AM, Mohammad SA, Robinson RL Jr, Gasem KAM (2013) Volume translated Peng-Robinson equation of state for liquid densities of diverse binary mixtures Original Research Article. Fluid Phase Equilib 349:37–55

Chapter 28
Preparation and Characterization of Nanoencapsulated *n*-Nonadecane for Convective Heat Transfer

Semahat Barlak, Ali Karaipekli, O. Nuri Sara, and Sinan Yapici

Abstract In this study, polyurethane nanocapsules containing phase change material were synthesized by interfacial polycondensation polymerization method. Toluene-2,4-diisocyanate (TDI) and diethylenetriamine (DETA) were chosen as monomers. *n*-Nonadecane was employed as a core material. The properties of nanocapsules were characterized by DSC, FT-IR, and SEM. The results show that the nanocapsules were synthesized successfully and that the phase change temperature was about 29.6 °C, and the latent heat of fusion was about 82 Jg^{-1}. The particle size was found to range from 100 to 340 nm.

Keywords Nanoencapsulation • PCM • Nanofluid • *n*-Nonadecane • Heat transfer

28.1 Introduction

Phase change materials (PCMs) absorb or release energy during phase change depending on the phase transition and have been considered for various applications such as thermal energy systems, buildings, textile, etc. [1–5]. The solid–liquid phase change is commonly used, and many organic and inorganic materials are used as a concept of solid–liquid PCM [5–10]. Although the PCMs can be used directly, they are usually enclosed in a medium to protect them from the effect of external environments; therefore, encapsulation is an important process for PCM applications. The capsules are classified as macrocapsule, microcapsule, and nanocapsule according to their particle size [11, 12]. The shell materials of capsules can be different with respect to production method, but polymer shell is commonly used [12]. Various techniques are available for microencapsulation of PCMs, which

S. Barlak • A. Karaipekli • O.N. Sara (✉)
Faculty of Engineering, Department of Chemical Engineering,
Çankiri Karatekin University, Çankiri 18100, Turkey
e-mail: onurisara@karatekin.edu.tr

S. Yapici
Faculty of Engineering, Department of Chemical Engineering,
Atatürk University, Erzurum 25240, Turkey

I. Dincer et al. (eds.), *Progress in Clean Energy, Volume 1*,
DOI 10.1007/978-3-319-16709-1_28

are mainly classified as physical and chemical techniques. Chemical techniques include interface polymerization, emulsion polymerization, coacervation, etc. [12]. A number of researchers investigated microencapsulation of PCMs for the purpose of using thermal storage and heat transfer systems, and the detailed reviews of these studies were given by several researchers [9, 12–14].

With technological development, the nanoencapsulation technology has been applied to encapsulate PCMs for the use in thermal systems. Nanocapsules provide better properties such as a lower crushing rate by pumping, higher thermal stability, lower sedimentation, higher thermal coefficient, large surface-to-volume ratio, and the ability to disperse uniformly within the liquid [15, 16]. Nanoencapsulation of various PCMs, such as paraffin wax by miniemulsion method [17], *n*-octadecane by in situ polymerization [18], *n*-dodecanol by miniemulsion polymerization [19], methyl stearate by dispersion and suspension polymerization [15], and *n*-octadecane by miniemulsion in situ polymerization [20], has been investigated. A study of polymer encapsulated paraffin nanoparticles suspended in water as a working fluid for jet impingement system was investigated by [16]. They reported that when compared to water, the slurry with 28 % particle volume fraction enhanced heat transfer coefficient by up to 50 % for jet impingement. Although there are several studies on microencapsulation of PCMs, the studies on the nanoencapsulation of PCMs and their use in thermal system as suspended particles in a base fluid are limited. Therefore, the further investigation is required. In this study, the nanocapsules consist of *n*-nonadecane as the core material and polyurethane as shell material was synthesized by interfacial polycondensation polymerization to use in a convective heat transfer system as suspended particles in a base heat transfer fluid such as water and ethylene glycol. This paper is limited only with preparation and characterization of nanocapsules. The morphology, structure, and thermal properties of nanocapsules were characterized by SEM, FT-IR, and DSC.

28.2 Experimental

28.2.1 Materials

Toluene-2,4-diisocyanate (TDI) (98 wt%, TCI) and diethylenetriamine (DETA) (98 wt% Merck) are used as monomers. *n*-Nonadecane (m.p., 28–32 °C; 99 wt% Alfa Aesar) was chosen as core material (PCM), and cyclohexane (99 wt% Sigma-Aldrich) and distilled water are used as oil phase and aqueous phase, respectively. NP-10 (98 wt% Sigma-Aldrich) is used as emulsifier. No additional purification to the used chemicals is applied.

28.2.2 Nanocapsule Synthesis

Preparation of nanocapsules was performed by interfacial polymerization method, which is a popular method [21]. Before encapsulation, *n*-nonadecane as the core material was dissolved in a solution containing cyclohexane and TDI, and this solution was added to the aqueous surfactant solution, which was previously prepared with NP-10. The mixture was emulsified by magnetic stirring at a rate of 1,000 rpm to form an oil/water emulsion. After stirring for 5 min, the aqueous solution of DETA was poured into the emulsion systems, and the mixture was heated to 60 °C. The interfacial polymerization reaction took place between TDI and DETA at oil/water interface. Similar process was used for encapsulation of different materials by [22, 21]. After about 2 h, the mixture was filtered, washed, and dried at 40 °C in an oven dryer for about 24 h. This process was carried out with different TDI/DETA ratio (w/w), different amounts of *n*-nonadecane, and surfactants.

28.2.3 Characterization of Nanocapsules

The morphology and size of the capsules were observed by scanning electron microscope (SEM). The diameter of the capsules was determined with commercial software from SEM photographs. The average diameter, d_{m}, was calculated according to the following equation:

$$d_{\mathrm{m}} = \frac{1}{N}\sum_{i=1}^{N} d_i \tag{28.1}$$

where d_i is the measured diameter of the capsules and N is the number of the capsules, which are about 200 for this study. The thermal properties of nanocapsules were determined using a differential scanning calorimeter (DSC, Setaram DSC 131 evo). The heating and cooling rate of 10 °C/min was used between 5 and 60 °C. The core content of nanocapsules, the amount of *n*-nonadecane that was capsulated, was calculated using the following equation [23]:

$$\%\ n\text{-nonadecane content}\,(\mathrm{Wt}) = \frac{\Delta H_{\mathrm{NEPCM}}}{\Delta H_{\mathrm{PCM}}} \times 100 \tag{28.2}$$

where H_{NEPCM} is the heat of fusion of nanocapsules containing *n*-nonadecane in Jg^{-1} and H_{PCM} is the heat of fusion of pure *n*-nonadecane in Jg^{-1}. The Fourier transform IR spectrum (FT-IR) of the nanocapsules was obtained to identify the structure of the shell polymer.

28.3 Results and Discussions

The detail of reaction scheme of shell formation process is given by [24]. The polyurethane shell is formed by reaction of the amine groups (–NH_2) of DETA with isocyanate groups (–NCO) of TDI at the interface. Interfacial polymerization of TDI and DETA occurs rapidly at room temperature. On the other hand, TDI monomers can be hydrolyzed slowly at the interface to produce –NH_2 groups, which react with TDI to form the shell of the polyurethane nanocapsules. The latter interfacial reaction occurs on the oil side of the interface.

28.3.1 Determination of Reactive TDI/DETA Ratio

In order to study the influence of TDI/DETA ratio on polymerization, nanocapsules were synthesized using a constant amount of TDI and different amounts of DETA at stirring rate of 750 rpm. Cyclohexane was completely removed during drying process after polymerization, and the nanocapsules consisting of polyurethane and nonadecane were weighted to determine their dry weights. The dry weight graph and photograph views of nanocapsules synthesized at different amount of DETA are shown in Fig. 28.1. As seen from the Fig. 28.1a, the dry weight of the nanocapsules increased linearly as DETA amount increased from 0 to 0.8 g, but only a little increase was observed as DETA weight increased from 0.8 up to 2 g. This means that excess DETA cannot react once TDI is depleted at the interface. On the other hand, some parts of the obtained products have an image like wax as DETA amount increased over 1.0 (in Fig. 28.1b). This result is because of the coagulation of capsules with excess addition of DETA. From Fig. 28.1, it can be concluded that the mass ratio of TDI to DETA is about 3 to 1.

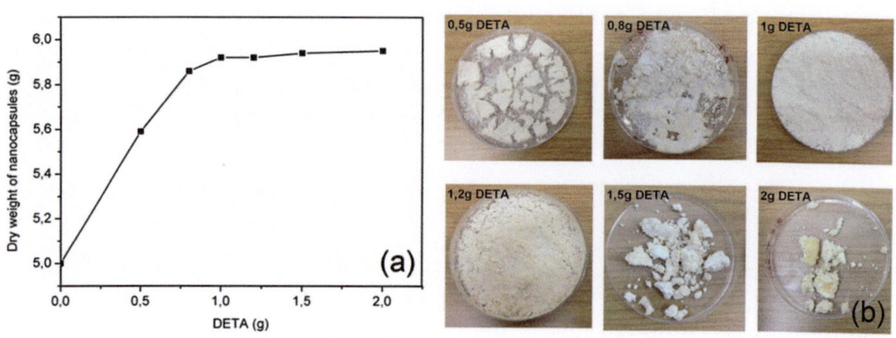

Fig. 28.1 (**a**) Dry weight graph and (**b**) photograph images of synthesized nanocapsules at various TDI/DETA molar ratios

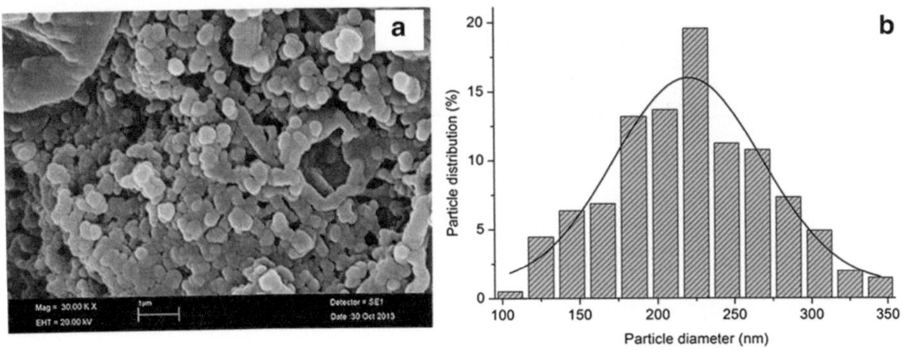

Fig. 28.2 SEM image and particle size distribution graph of the PU/nonadecane nanocapsules

28.3.2 *Morphologies and Particle Size Distribution of Nanocapsules*

Surface morphology of the obtained nanocapsules was revealed by SEM. Figure 28.2 shows the SEM image and particle size distribution graph of synthesized PU/nonadecane nanocapsules. As can be seen from Fig. 28.2a, the nanocapsules have smooth surface, spherical profile, and a tendency of agglomeration. The agglomeration of the nanocapsules is due to the presence of isocyanate (–NCO) groups that were not reacted completely to form polyurethane shell. In addition, these unreacted (–NCO) groups might react with air during air-drying of nanocapsules and effect the agglomeration of the nanocapsules.

Figure 28.2b indicates the particle size distribution of the produced nanocapsules. As seen from this figure, the particle sizes of the nanocapsules are distributed in the range of 100–345 nm, and the mean diameter of the nanocapsules calculated by Eq. (28.1) is 213 nm. The percentage of the nanocapsules around the mean particle size is about 50 %. The irregularity in the size distribution of the nanocapsules may be a result of the stirring rate. However, the diameter of the prepared nanocapsules is suitable to prepare nanofluids containing PCM nanocapsules.

28.3.3 *FT-IR Analysis of Shell Material and PU/Nonadecane Nanocapsules*

FT-IR analysis enables to determine the chemical structure of the polymer shell and the content of the nanocapsules. Also, it is necessary to understand the real reaction degree of TDI. The FT-IR spectra of TDI, empty PU nanocapsules, *n*-nonadecane, and PU/nonadecane nanocapsules are shown in Fig. 28.3. In FT-IR spectrum of TDI,

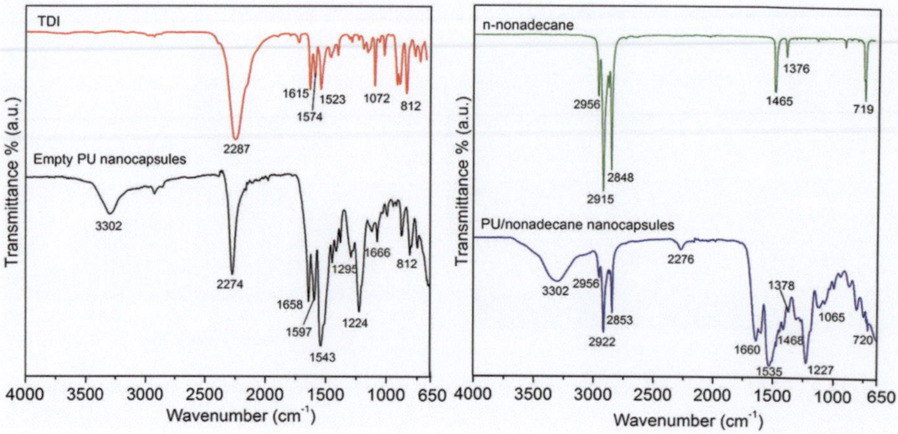

Fig. 28.3 FT-IR spectra of TDI, empty PU nanocapsules, *n*-nonadecane, and PU/nonadecane nanocapsules

a characteristic absorption bands belonging to the isocyanate (–NCO) group of TDI were observed at 2,287 and 1,523–1,574 cm^{-1}. As can be seen from Fig. 28.3, the empty PU nanocapsules have many characteristic peaks at 812, 1,224, 1,295, 1,543, 1,597, 1,643, 2,274, and 3,302 cm^{-1}. This FT-IR spectrum indicated that the absorption bands at 2,274 and 1,543–1,597 cm^{-1} came from the unreacted –NCO group of TDI. We can conclude that the nanocapsules still have unreacted isocyanate groups that are located in the terminal end of PU polymer. The peak at 3,302 cm^{-1} from N–H stretching vibration and the peak at 1,533 cm^{-1} from N–H bending vibration show the formation of urethane bond. Furthermore, the peaks at 1,643 cm^{-1} from carbonyl –CO group of PU and peak at 1,066 cm^{-1} of ester (C–O–C) confirmed the urethane formation in the nanocapsules. These results confirm that the shell structure of nanocapsules is polyurethane. On the other hand, in the *n*-nonadecane spectrum, the peaks at 2,956, 2,915, and 2,848 cm^{-1} are C–H stretching peaks, and the peaks at 1,465, 1,376, and 719 cm^{-1} are characteristic for *n*-alkanes.

Comparing the FT-IR spectra of *n*-nonadecane and PU/nonadecane nanocapsules, it is clearly seen that the FT-IR spectrum of the nanocapsules consists of both the peaks of *n*-nonadecane and PU. In addition, the peak positions in the spectrum of nanocapsules slightly deviate from the positions in their pure spectra. For instance, the peak at 1,465 cm^{-1} in the *n*-nonadecane spectrum shifts to 1,468 cm^{-1} in the nanocapsules spectrum. These specific peaks observed in the spectra of the PU/nonadecane nanocapsules indicate the existence of nonadecane in the PU nanocapsules.

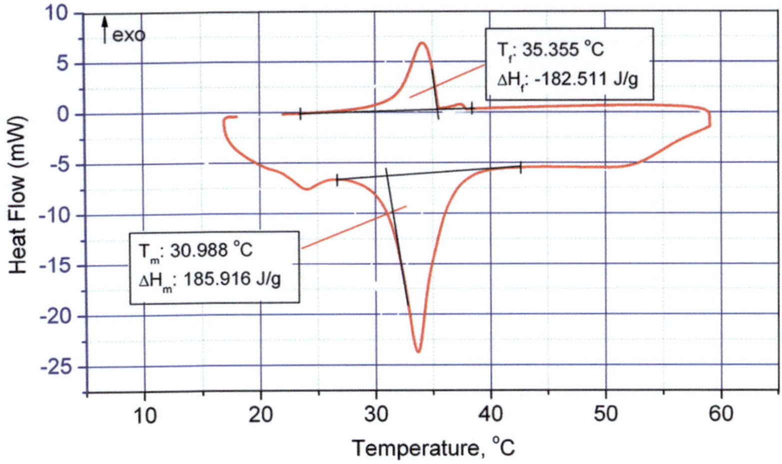

Fig. 28.4 DSC thermogram of *n*-nonadecane

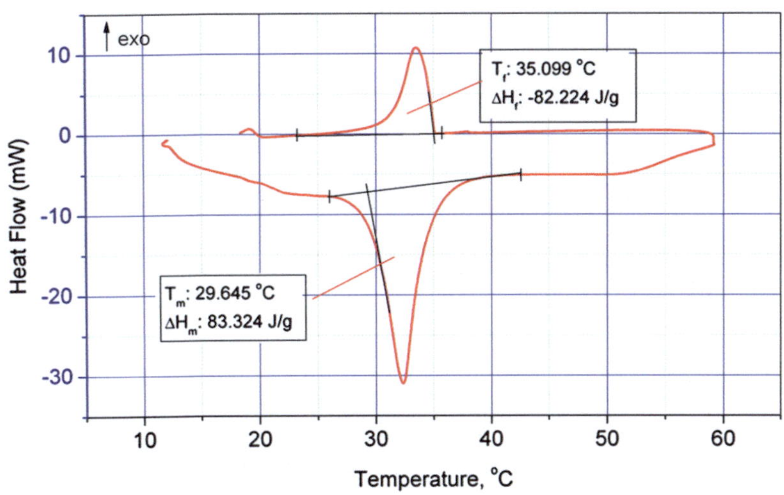

Fig. 28.5 DSC thermogram of PU/nonadecane nanocapsules

28.3.4 Thermal Properties of Nanocapsules

DSC thermograms of *n*-nonadecane and PU/nonadecane nanocapsules are shown in Figs. 28.4 and 28.5, respectively. Thermal properties obtained from the thermograms indicate that PU/nonadecane nanocapsules melt at 29.6 °C and freeze at 35.1 °C, while pure nonadecane has a melting point of 30.9 °C and a freezing point of 35.4 °C. The latent heats of melting and freezing of PU/nonadecane nanocapsules were measured as 83.3 and 82.2 Jg^{-1}, respectively. The encapsulation ratio of *n*-nonadecane was calculated as 45 wt% with using Eq. (28.2).

28.4 Conclusion

In this study, PU nanocapsules containing *n*-nonadecane were successfully prepared with an average diameter of 213 nm by the stirring rate of around 1,000 rpm. FT-IR results confirmed that the nanocapsules were fabricated by interfacial polymerization between TDI and DETA, and *n*-nonadecane was successfully encapsulated within the polymer shell. SEM analysis indicated that the prepared nanoPCMs have smooth and relatively spherical profiles. The temperatures and latent heats of melting and freezing of PU/nonadecane nanocapsules were determined using DSC analysis method. The results also showed that the nanocapsules consisted of an average of 45 wt% nonadecane. Based on all these results, it can be concluded that the fabricated nanocapsulated PCMs have a great potential to increase thermal energy storage capacity of heat transfer fluids.

Acknowledgment The authors are grateful to The Scientific and Technological Research Council of Turkey (TÜBİTAK-MAG Grant number: 113 M133) and Çankırı Karatekin University (Grant number: BAP 2011/25) for the financial support of this research. The SEM analysis of samples was done at Erciyes University, Technology Research and Application Center.

References

1. Schossig P, Henning HM, Gschwander S, Haussmann T (2005) Micro-encapsulated phase-change materials integrated into construction materials. Sol Energy Mater Sol Cells 89:297–306
2. Alay S, Gode F, Alkan C (2011) Synthesis and thermal properties of poly(n-butyl acrylate)/n-hexadecane microcapsules using different cross-linkers and their application to textile fabrics. J Appl Polym Sci 120:2821–2829
3. Parameshwaran R, Kalaiselvam S, Harikrishnan S, Elayaperumal A (2012) Sustainable thermal energy storage technologies for buildings: a review. Renew Sustain Energy Rev 16:2394–2433
4. Li G, Hwang Y, Radermacher R, Chun H-H (2013) Review of cold storage materials for subzero applications. Energy 51:1–17
5. Sarier N, Onder E (2012) Organic phase change materials and their textile applications: an overview. Thermochim Acta 540:7–60
6. Tyagi VV, Kaushik SC, Tyagi SK, Akiyama T (2011) Development of phase change materials based microencapsulated technology for buildings: a review. Renew Sust Energy Rev 15:1373–1391
7. Sharma A, Tyagi VV, Chen CR, Buddhi D (2009) Review on thermal energy storage with phase change materials and applications. Renew Sust Energy Rev 13:318–345

8. Zalba B, Marin JM, Cabeza LF, Mehling H (2003) Review on thermal energy storage with phase change: materials, heat transfer analysis and applications. Appl Therm Eng 23:251–283
9. Agyenim F, Hewitt N, Eames P, Smyth M (2010) A review of materials, heat transfer and phase change problem formulation for latent heat thermal energy storage systems (LHTESS). Renew Sust Energy Rev 14:615–628
10. Farid MM, Khudhair AM, Razack SAK, Al-Hallaj S (2004) A review on phase change energy storage: materials and applications. Energy Convers Manag 45:1597–1615
11. Kim EY, Do Kim H (2005) Preparation and properties of microencapsulated octadecane with waterborne polyurethane. J Appl Polym Sci 96:1596–1604
12. Dubey R, Shami TC, Rao KUB (2009) Microencapsulation technology and applications. Def Sci J 59:82–95
13. Regin AF, Solanki SC, Saini JS (2008) Heat transfer characteristics of thermal energy storage system using PCM capsules: a review. Renew Sustain Energy Rev 12:2438–2458
14. Zhao CY, Zhang GH (2011) Review on microencapsulated phase change materials (MEPCMs): fabrication, characterization and applications. Renew Sustain Energy Rev 15:3813–3832
15. Cheng F, Wie Y, Zhang Y, Wang F, Shen T, Zong C (2013) Preparation and characterization of phase-change material nanocapsules with amphiphilic polyurethane synthesized by 3-allyloxy-1,2-propanediol. J Appl Polym Sci 130:1879–1889
16. Wu W, Bostanci H, Chow LC, Ding SJ, Hong Y, Su M, Kizito JP, Gschwender L, Snyder CE (2011) Jet impingement and spray cooling using slurry of nanoencapsulated phase change materials. Int J Heat Mass Transf 54:2715–2723
17. de Cortazar MG, Rodriguez R (2013) Thermal storage nanocapsules by miniemulsion polymerization. J Appl Polym Sci 127:5059–5064
18. Zhang XX, Fan YF, Tao XM, Yick KL (2004) Fabrication and properties of microcapsules and nanocapsules containing n-octadecane. Mater Chem Phys 88:300–307
19. Chen Z-H, Yu F, Zeng X-R, Zhang Z-G (2012) Preparation, characterization and thermal properties of nanocapsules containing phase change material n-dodecanol by miniemulsion polymerization with polymerizable emulsifier. Appl Energy 91:7–12
20. Fang YT, Kuang SY, Gao XN, Mang ZG (2008) Preparation and characterization of novel nanoencapsulated phase change materials. Energy Convers Manag 49:3704–3707
21. Chen L, Xu LL, Shang HB, Zhang ZB (2009) Microencapsulation of butyl stearate as a phase change material by interfacial polycondensation in a polyurea system. Energy Convers Manag 50:723–729
22. Cho JS, Kwon A, Cho CG (2002) Microencapsulation of octadecane as a phase-change material by interfacial polymerization in an emulsion system. Colloid Polym Sci 280:260–266
23. Alkan C, Sari A, Karaipekli A, Uzun O (2009) Preparation, characterization, and thermal properties of microencapsulated phase change material for thermal energy storage. Sol Energy Mater Sol Cells 93:143–147
24. Su JF, Wang LX, Ren L, Huang Z, Meng XW (2006) Preparation and characterization of polyurethane microcapsules containing n-octadecane with styrene-maleic anhydride as a surfactant by interfacial polycondensation. J Appl Polym Sci 102:4996–5006

Chapter 29
Numerical Simulation of Heat Transfer to TiO_2-Water Nanofluid Flow in a Double-Tube Counter Flow Heat Exchanger

C.S. Oon, H. Nordin, A. Al-Shamma'a, S.N. Kazi, A. Badarudin, and B.T. Chew

Abstract Recently, the means of improvement of heat transfer has been rapidly studied. One of the methods that enhance the heat transfer is by changing the heat exchanging fluids. The poor heat transfer coefficient of common fluids compared to the most solids becomes the primary obstacle to design high compactness and effectiveness of heat exchanger. The primary objective of this chapter is to conduct the study of the heat transfer between the water and nanofluid. Both of the fluids were flowed in the horizontal counter flow heat exchanger under the turbulent flow condition. The flow velocity of the fluids varied with Re between 4,000 and 18,000. Literature review states that the heat transfer coefficient of nanofluid is higher than the water by about 6–11 %. Heat transfer to the nanofluid and water is investigated using a computer fluid dynamics software. Ten percent heat transfer augmentation is observed utilizing nanofluid as heat exchanging fluid compared to water. The results also showed the enhancement of the Reynolds number increases the heat transfer to the nanofluid studied in this investigation.

Keywords Numerical simulation • Heat transfer • TiO_2 • Nanofluid • Heat exchanger

---

C.S. Oon (✉)
School of Built Environment, Liverpool John Moores University, Byrom Street, Liverpool L3 3AF, UK

Department of Mechanical Engineering, Faculty of Engineering, University of Malaya, Kuala Lumpur 50603, Malaysia
e-mail: oonsean2280@yahoo.com

H. Nordin • S.N. Kazi • A. Badarudin • B.T. Chew
Department of Mechanical Engineering, Faculty of Engineering, University of Malaya, Kuala Lumpur 50603, Malaysia

A. Al-Shamma'a
School of Built Environment, Liverpool John Moores University, Byrom Street, Liverpool L3 3AF, UK

I. Dincer et al. (eds.), *Progress in Clean Energy, Volume 1*,
DOI 10.1007/978-3-319-16709-1_29

Nomenclature

Pr	Prandtl number
Re	Reynolds number
d	Diameter, m
f	Friction factor
k	Thermal conductivity, W/m^3 K
Q	Heat transfer rate, W/m^2
Cp	Heat capacity, kJ/kg °C
T	Temperature, °C
Nu	Nusselt number
h_x	Heat transfer coefficient
Pe	Peclet number

Greek Letters

ø	Volume fraction
μ	Viscosity, kg/ms
ε	Surface roughness, m
ρ	Density, kg/m^3
f	Friction factor

29.1 Introduction

Nowadays, the usage of Computational Fluid Dynamic (CFD) has been widely used in industrial and nonindustrial areas. CFD is the software for the analysis system involving fluid flow, heat transfer, and associated phenomena by means of computer-based simulation [1]. One of the examples of CFD usage is in the field of turbo machinery where the CFD is used to simulate the flow inside rotating passage, diffuser, etc. and it is due to the cost for conducting the simulation is lower than performing an experiment. Back in the 1960s, the aerospace industries used CFD software to design, conduct research and development (R&D), and manufacture the body of the aircraft.

Present research has been carried out to study the heat transfer characteristic in counter flow heat exchanger. Armaly et al. had experimented flow separation with the laser-Doppler anemometer to define the quantitatively variation of separation length with Reynolds number [2]. In the experiment, both side of the channel wall showed multiple regions of separation downstream of backward facing step. The separation and reattachment region has strong influence on the heat transfer characteristic.

Heat transfer enhancement can also be achieved by enhancing the thermo physical properties of the fluid itself. Today, the study regarding improving the thermal conductivities of fluids is by using the nanofluid where small solid particles are suspended in the fluid [3]. Recently, a lot of researches were conducted by using nanofluids to investigate the heat transfer characteristic with various geometries [4–6]. Most of the studies of heat transfer have been focusing on ducts and circular pipe flow whereas a little research has conducted in annular passage [7, 8].

This chapter is focused on simulation for the nanofluid and showed the validity of the Nusselt number of previous investigation such as the work of Duangthongsuk and Wongwises [9]. Nanofluid can be defined as the suspension of ultrafine particles in a conventional base fluid, where it can increase the heat transfer characteristics of the original fluid [10]. Moreover, the nanofluid is suitable for practical applications as their use incurs little or no penalty in pressure drop because the particle of the nanofluid is ultrafine. Hence, the nanofluid looks like more to be single-phase rather than a solid–liquid mixture. Generally, the nanoparticles commonly available are: aluminum oxide (Al_2O_3), copper (Cu), Titanium oxide (TiO_2), copper oxide (CuO), gold (Au), silver (AgO), silica nanoparticles, and carbon nanotube [10]. The base fluid could be the water, acetone, decene, and ethylene glycol. The nanoparticle can be produced by several processes such as gas condensation, mechanical attrition, or chemical precipitation techniques [11]. There are two factors why the nanoparticle has been selected for making nanofluids which are capable of increasing the heat transfer. Firstly, the nanoparticles in the fluid can increase the thermal conductivity of the fluids. Secondly, the chaotic movement of the nanoparticles in the fluid causes the turbulence in the fluid that increases the energy exchange process. Many experiments have been conducted using the nanofluids, where it showed much higher thermal conductivity. In the experiment that carried out by Wen and Ding, they proposed the nanoparticles that enhanced the thermal conductivity of the fluid [12]. This is because the nanoparticles in the fluid have changed the properties of the base fluid such as water, oil, and ethylene glycol to a better suspension of higher thermal conductivity. In addition, an experiment carried out by Lee et al. where the CuO and Al_2O_3 have been used to enhance the heat transfer of the fluid. They proposed that the enhancement of the heat transfer is caused by the chaotic movement of the nanoparticles in the fluid. Moreover, they also suggested that the enhancement of heat transfer is also affected by the size and shape of the particles.

Das et al. have done the analysis of turbulent flow and heat transfer of different nanofluids; CuO, Al_2O_3, and SiO_2 in ethylene glycol and water mixture at constant heat flux in circular tube [13]. Through their research, they found that the smaller the particle, the higher the viscosity and Nusselt number. Other than that, the higher the volume concentration of nanofluids and Reynolds number, the heat transfer coefficient will increase. However, the increasing of volume concentration of nanofluids and Reynolds number caused the pressure loss.

On the other hand, there were several reports about inconsistency of nanofluids behavior in terms of heat transfer coefficient. Pak and Cho reported that for the forced convective heat transfer, the Nusselt number for the Al_2O_3-water and TiO_2-water nanofluids is increased with the volume fraction of suspended nanoparticles and Reynolds number [14]. Moreover, Yang et al. report the similar observation for the graphite nanofluids in laminar flow regime [15].

29.2 Methodology

The numerical simulation has been conducted to verify the accuracy of the heat transfer experimentally. Figure 29.1 shows the pipe drawn and meshed by using the CFD software. The inner diameter of the copper pipe is set at 8.13 mm and the outer diameter is 9.2 mm. The thickness of the copper pipe is about 0.7 mm. The temperature of the water in the copper pipe is 40 °C, and the flow rate is 3 L/min. On the other hand, the outer pipe is made up of PVC, where the nanofluid flows through outer pipe. The dimension of the inner diameter of PVC pipe is 27.2 mm whereas the outer diameter is 33.9 mm. The nanofluid flows in the counter direction of the water with 27 °C. The annular passage is considered in the simulation, where the Reynolds number varied between 4,000 and 18,000. Figure 29.1 shows the schematic diagram of the counter flow heat exchanging equipment.

29.2.1 Computational Fluid Dynamics Simulation

The aim of this research is to investigate the heat transfer, typically Nusselt number and understand the phenomena of flow in annular pipe. Thus, the finite volume-based flow solver of computational dynamics software (FLUENT) is chosen in this investigation. The geometry is drawn in the CFD software. Moreover, the K-epsilon viscous model has been chosen for the iteration based on the energy equation and Reynolds averaged Navier Stokes equations. The flow condition in the pipe between the base fluid (i.e., water) and the nanofluid is assumed to be in thermal equilibrium and no slip occurs between them [16].

In the simulations, the initial information of the study such as geometry, dimension, input and initial values, dependent and independent variables of the study have been determined. Then, based on these data gained in case study and according to literature review the geometry considered for the simulation can be simplified. This case is simplified by using 2-dimensional pipe and the pipe is cut into half to make it

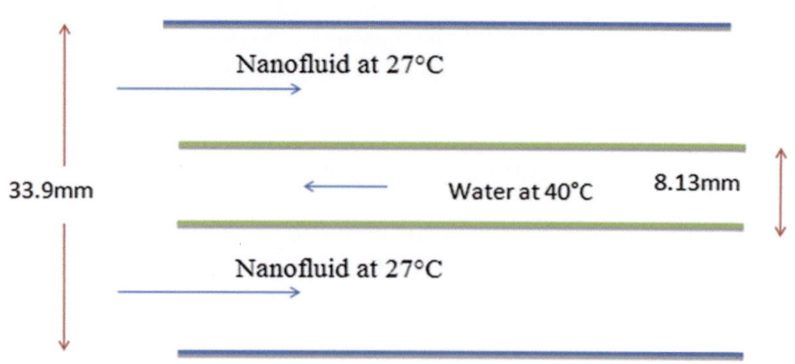

Fig. 29.1 Counter flow heat exchanger drawing

symmetry. Fluid flow in a physical domain is governed by the laws of conservation of mass and momentum.

Upon meshing, the governing equations including some partial differential equations are derived and discretized by one of the existing methods such as finite element, infinite element, and finite volume to be converted into algebraic equations. In addition, the boundary conditions, material properties, and initial values were obtained to solve the iteration. These steps are called preprocessing steps. Once the preprocessing part is completed, governing equations are solved during several iterative algorithms. The measure of convergence error and the problem initialization are specified before running the solution algorithms.

After the preprocessing step, the post-processing step takes place where the result is obtained. During post-processing step, all the data including temperature, velocity, and pressure can be obtained. The accuracy of the results is then verified. Furthermore, the Nusselt number is then calculated from the temperature data.

In this research, TiO_2 nanoparticles mixed with the water by 0.2 % were used to investigate the heat transfer characteristic with different variables such as Reynolds number and temperature of the flowing nanofluid and mass flow rate. The simulation conditions used in this study are as follows:

- The Reynolds number of the nanofluid varied from 4,000 to 18,000
- The temperature of the nanofluid is 27 °C
- The mass flow rates of the hot water are 3 LPM and 4.5 LPM
- The temperature of the hot water is 40 °C

29.2.2 *Mathematical Model*

The thermal physical properties of the nanofluid can be calculated utilizing the following equations.

Pak and Cho calculated the density by using Eq. (29.1) [14]:

$$\rho_{\mathrm{nf}} = \phi\rho_{\mathrm{p}} + (1 - \phi) \cdot \rho_{\mathrm{w}} \tag{29.1}$$

where ρ_{nf} the volume fraction of nanoparticles is, ρ_{p} is the density of the nanoparticles, and ρ_{w} is the density of the base fluid.

Donald and Stephen calculated the viscosity by using well-known Einstein equation (29.2) which is applicable to spherical particles in volume fractions of less than 5 % [17].

$$\mu_{\mathrm{nf}} = (1 + 2.5\phi) \cdot \mu_{\mathrm{w}} \tag{29.2}$$

The μ_{nf} is referred to the viscosity of the nanofluid, and the μ_{w} referred to the base fluid which is water.

Thermal conductivity of the nanofluid is obtained using Eq. (29.3) as suggested by Yu and Choi [18].

$$k_{nf} = \left[k_p + 2k_w + 2(k_p - k_w)(1+\beta)^2\phi\right] / \left[k_p + 2k_w - (k_p - k_w)(1+\beta)^2\phi\right] \tag{29.3}$$

where k_{nf} is the thermal conductivity of the nanofluid, k_p is the thermal conductivity of the nanoparticles, k_w is the thermal conductivity of the base fluid. Moreover, the Nusselt number is calculated from Gnielinski equation and Colebrook equation and then compared to the simulation results. The Gnielinski equation (29.4) and Colebrook equation (29.5) are as follow [19, 20]:

$$Nu = [(f/8)(Re - 1000)Pr] / \left[1 + 12.7(f/8)^{0.2}\left(Pr^{1/2} - 1\right)\right] \tag{29.4}$$

The Nu is the Nusselt number, Re is the Reynolds number, Pr is the Prandtl number, and f is the friction factor.

Next, the Colebrook equation is defined as:

$$F^{(-1/2)} = -2\,\mathrm{Log}\left[(\varepsilon/3.7D) + \left(2.51/Re \cdot f^{0.5}\right)\right] \tag{29.5}$$

The temperature of the inlet and outlet of copper tube were calculated to obtain the heat transfer rate from the water to the nanofluid through the copper tube. The heat transfer rate is calculated using the Eq. (29.6) which is suggested by Duangthongsuk and Wongwises [21].

$$Q_w = m \cdot \mathrm{C}p_w(T_{in} - T_{out})_w \tag{29.6}$$

The Q_w is the heat transfer rate for the hot water, and m is the mass flow rate of the water.

Next, the local heat transfer is calculated using the convection heat flux as used by Oon et al. [22]:

$$h_x = q_c/T_b \tag{29.7}$$

where q_c is the heat transfer from water to nanofluid, and the T_b is the temperature different between the outlet temperature and the inlet temperature. The value of the Nusselt number is calculated from the equation:

$$Nu = h_x \cdot d/k_{nf} \tag{29.8}$$

where the value of d is the hydraulic diameter, and the k_{nf} is the thermal conductivity of the water.

Then, Pak and Cho and Xuan and Li correlations are utilized to predict and compare the Nusselt number of TiO_2 nanofluid. The Pak and Cho correlation is shown in Eq. (29.9) [14, 23]:

$$Nu_{\mathrm{nf}} = 0.021 \cdot Re_{\mathrm{nf}}^{0.8} Pr_{\mathrm{nf}}^{0.5} \tag{29.9}$$

The Xuan and Li correlation is shown in Eq. (29.10):

$$Nu_{\mathrm{nf}} = 0.0059\left(1 + 7.6286\,\phi^{0.6886} Pe_{\mathrm{nf}}^{0.001}\right) Re_{\mathrm{nf}}^{0.9238} Pr_{\mathrm{nf}}^{0.4} \tag{29.10}$$

where Pe_{nf} is the Peclet number. Table 29.1 shows the thermophysical properties of the water, nanoparticles, and the nanofluid calculated from the equations above.

29.3 Results and Discussion

Figure 29.2 shows the heat transfer coefficient increases with the increase of Reynolds number. The data obtained from simulation is compared with the Gnielinski equation for water in turbulent flow. The simulation result obtained is slightly higher than the Gnielinski equation partly due to turbulent flow. The value

Table 29.1 Thermal physical properties

Property	Fluid phase (water)	TiO_2 particles	TiO_2-water (0.2 %)
Density, ρ (kg/m^3)	998.2	4,250	1,004.7
Heat capacity, Cp (J/kg K)	4,182	686.2	4,175
Thermal conductivity, k (W/m K)	0.6	8.9538	1.0677

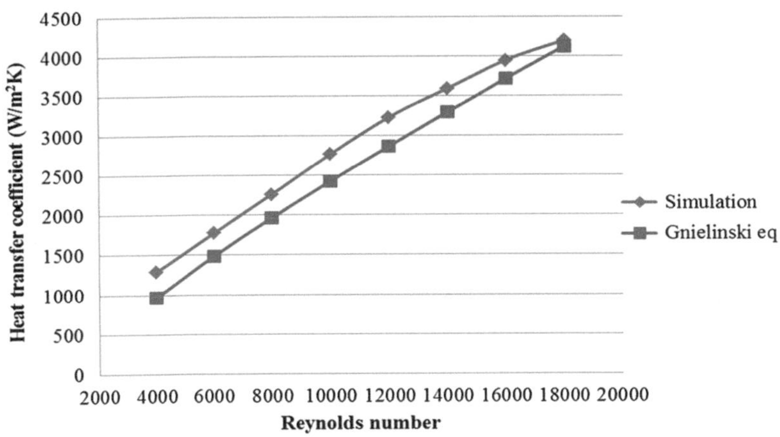

Fig. 29.2 The comparison between the heat transfer coefficients against Reynolds number for water

obtained from the simulation is compared with the Gnielinski equation. Hence, the same parameter used in the computational study can be extended for simulation of TiO_2-water nanofluid.

Figure 29.3 shows the simulation values of Nusselt number of TiO_2-water nanofluids which are closer towards Pak and Cho correlation than the Xuan and Li correlation. This is simply because Xuan and Li correlation was established from the data of Cu-water nanofluids while the Pak and Cho correlation was established from the data of TiO_2-water nanofluids.

Figure 29.4 shows the results for different variables such as change of mass flow rate of hot water from 3 LPM to 4.5 LPM for the 0.2 % TiO_2-water nanofluid.

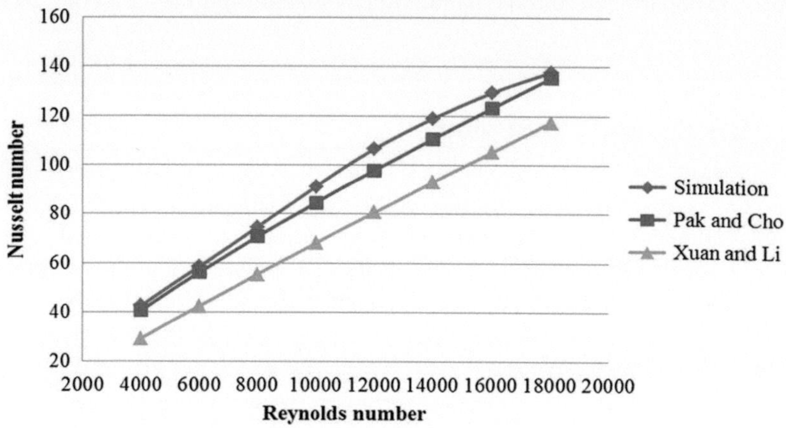

Fig. 29.3 The comparison of Nusselt number between the TiO_2

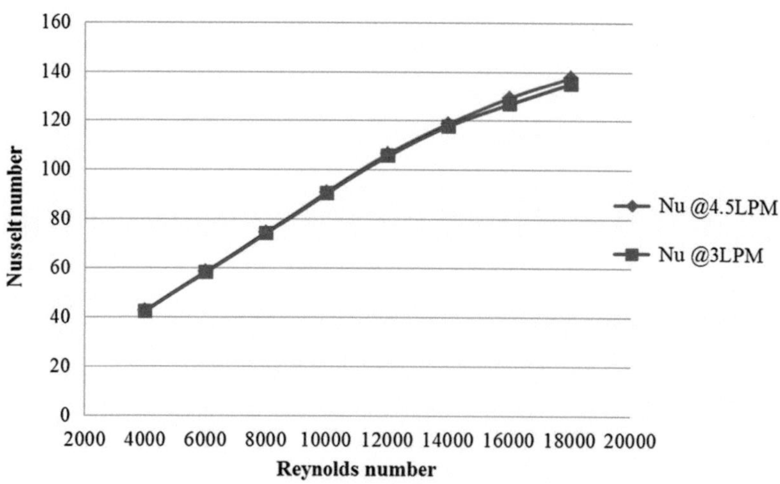

Fig. 29.4 The comparison of Nusselt number of TiO_2 between 3 LPM and 4.5 LPM water flow rate

It is clearly seen that the 4.5 LPM flow rate shows higher heat transfer compared to 3 LPM flow rate. The result also shows that the heat transfer coefficient of the nanofluid increases with an increase in the mass flow rate of the hot water.

29.4 Conclusion

The numerical simulation has shown that the increasing in Reynolds number also causes the heat transfer to increase. The higher Reynolds number contributes to greater heat transfer from the water to the nanofluid. In addition, the numerical simulation results agree with the Pak and Cho correlation as both of the results are in compliance. Finally, with the development of computational software it has become easier to provide a fare and comparable result as in the present case.

Acknowledgment Thanks to School of Built Environment and Graduate School, Liverpool John Moores University who provide the funding. The authors also gratefully acknowledge High Impact Research Grant UM.C/625/1/HIR/MOHE/ENG/46 and the UMRG RP012B-13AET University of Malaya, Malaysia for support to conduct this research work.

References

1. Versteeg HK, Malalasekera W (2007) An introduction to computational fluid dynamics. Pearson, Glasgow
2. Armaly BF, Durst F, Pereira JCF, Schonung B (1983) Experimental and theoretical investigation of backward facing step flow. J Fluid Mech 127:473–496
3. Bianco V, Chiacchio F, Manca O, Nardini S (2009) Numerical investigation of nanofluids forced convection in circular tube. Appl Therm Eng 29:3632–3642
4. Aminossadati SM, Ghasemi B (2009) Natural convection cooling of a localised heat source at the bottom of a nanofluid-filled enclosure. Eur J Mech B Fluids 28:630–640
5. Oon CS et al. (2013) Numerical investigation of heat transfer to fully developed turbulent air flow in a concentric pipe. In: Fifth international conference on computational intelligence, modelling and simulation (CIMSim), pp 288–293
6. Oon CS, Togun H, Kazi SN, Badarudin A, Sadeghinezhad E (2013) Computational simulation of heat transfer to separation fluid flow in an annular passage. Int Commun Heat Mass Transf 46:92–96
7. Oon CS, Togun H, Kazi SN, Badarudin A, Zubir MNM, Sadeghinezhad E (2012) Numerical simulation of heat transfer to separation air flow in an annular pipe. Int Commun Heat Mass Transf 39:1176–1180
8. Oon CS, Badarudin A, Kazi SN, Fadhli M (2014) Simulation of heat transfer to turbulent nanofluid flow in an annular passage. Adv Mater Res 925:625–629
9. Duangthongsuk W, Wongwises S (2009) Heat transfer enhancement and pressure drop characteristics of TiO_2—water nanofluid in a double-tube counter flow heat exchanger. Int J Heat Mass Transf 52:2059–2067
10. Trisaksri V, Wongwises S (2007) Critical review of heat transfer characteristics of nanofluids. Renew Sustain Energy Rev 11:512–523

11. Lee S, Choi SUS, Li S, Eastman JA (1999) Measuring thermal conductivity of fluids containing oxide nanoparticles. J Heat Transf 121:280–289
12. Wen D, Ding Y (2004) Experimental investigation into convective heat transfer of nanofluids at the entrance region under laminar flow conditions. Int J Heat Mass Transf 47:5181–5188
13. Das SK, Putra N, Roetzel W (2003) Pool boiling characteristics of nano-fluids. Int J Heat Mass Transf 46:851–862
14. Pak BC, Cho YI (1998) Hydrodynamic and heat transfer study of dispersed fluids with submicron metallic oxide particles. Exp Heat Transf 11:151–170
15. Yang Y, Zhang ZG, Grulke EA, Anderson WB, Wu G (2005) Heat transfer properties of nanoparticle-in-fluid dispersions (nanofluids) in laminar flow. Int J Heat Mass Transf 48:1107–1116
16. Abu-Nada E (2008) Application of nanofluids for heat transfer enhancement of separated flows encountered in a backward facing step. Int J Heat Fluid Flow 29:242–249
17. Donald AD, Stephen LP (1999) Theory of multicomponent fluids, vol 135, Applied mathematical sciences. Springer, New York
18. Yu W, Choi SUS (2003) The role of interfacial layers in the enhanced thermal conductivity of nanofluids: a renovated maxwell model. J Nanoparticle Res 5:167–171
19. Gnielinski V (1975) New equation for heat and mass transfer in the turbulent flow in pipes and channel. Forsch Ingenieurwes 41:8–16
20. Colebrook CF (1939) Turbulent flow in pipes with particular reference to the transition region between the smooth and rough pipe laws. J ICE 11:135–156
21. Daungthongsuk W, Wongwises S (2007) A critical review of convective heat transfer of nanofluids. Renew Sustain Energy Rev 11:797–817
22. Oon CS, Al-Shamma'a A, Kazi SN, Chew BT, Badarudin A, Sadeghinezhad E (2014) Simulation of heat transfer to separation air flow in a concentric pipe. Int Commun Heat Mass Transf 57:48–52
23. Xuan Y, Li Q (2003) Investigation on convective heat transfer and flow features of nanofluids. J Heat Transf 125:151–155

Chapter 30
The Effects of Oil Palm Shell Volume Fractions on Thermal Conductivity for Insulation Concrete

Eravan Serri, M. Zailan Suleman, and M. Azree Othuman

Abstract Industrial waste has been considered as an option in the initiative to promote green and sustainable construction. Oil palm shell (OPS) is one of the industrial wastes produced from the processing of palm oil and its ability to be used as a lightweight aggregate in concrete mixes has been tested. OPS has a high porosity content, which means that it is a good heat insulation material in concrete. This study focuses on the OPS volume fraction in concrete and its effects on the thermal insulation concrete. The volume fractions used are 30, 32, 34, 36, and 38 % from concrete density. Density decreases with the increase of volume fraction. The highest reduction from air dry to oven density is 13 %, which is obtained from a volume fraction of 34 % OPS. The volume fraction affected compressive strength and thermal conductivity. All mixes achieved for the requirement for load-bearing strength based on compressive strength were obtained. The highest strength was 22 Mpa by volume fraction, 30 % used. Volume fractions used are within the range of the semi-structure which has the same capacity as thermal insulation materials, below 0.75 W/m K (according to RILEM requirements) except for the volume fraction, 30 %. The thermal properties increased according to the increase in density except for the specific heat result, and they have a strong relationship within the thermal conductivity and compressive strength results. Thus the OPS lightweight concrete (OPSLC) capacity as a heat insulation material is proven and it can reduce energy use in buildings.

Keywords Density • Ultrasonic pulse velocity (UPV) • Mechanical strength • Thermal properties

E. Serri (✉)
Department of Building Technology, Housing Building Planning School,
Universiti Sains Malaysia, 1700 Minden, Penang, Malaysia

University Colleague Sabah Foundations, Kota Kinabalu, Sabah 88867, Malaysia
e-mail: eravan06@yahoo.com

M.Z. Suleman • M.A. Othuman
Department of Building Technology, Housing Building Planning School,
Universiti Sains Malaysia, 1700 Minden, Penang, Malaysia

I. Dincer et al. (eds.), *Progress in Clean Energy, Volume 1*,
DOI 10.1007/978-3-319-16709-1_30

Nomenclature

k Thermal conductivity, W/m K
fcu Compressive strength, Mpa

Subscripts

OPS Oil palm shell
OPSLC Oil palm shell lightweight concrete
UPV Ultrasonic pulse velocity
Sp Superplasticizer
SSD Saturated dry density

30.1 Introduction

Concrete is the main material in the construction industry and is used throughout the world. The high demand for concrete will affect natural resources as ingredients in concrete. Because of the wide usage of this material, many studies are investigating the properties of these materials. One type of concrete of interest to researchers is lightweight concrete. And to produce lightweight concrete, the use of lightweight aggregate is one method that can be implemented. Lightweight aggregate often is made of waste materials that occur as by-products from other industries, and hence their use is ecologically desirable. Growing popular as solid waste material in concrete is oil palm shell (OPS). Resulting from the processing of oil palm, OPS contributes 5.5 % as solid waste from overall oil palm manufacture [1] and by using OPS as aggregate made from waste materials occurring as by-products from other industries, their use is ecologically desirable [2]. OPS is traditionally used as solid fuel for steam boilers to run turbines for electricity [3], cover the surface of the roads in the plantation area [1], dandified into briquettes [4], converted to bio-oil by using a pyrolysis process for biomass energy [5], and for the production of charcoal and activated carbon. For more than 20 years, researchers have investigated [6] the potential of OPS as a structural lightweight aggregate. The highest compressive strength obtained by Shafigh et al. [3] was about 53 MPa by using crushed OPS. The tensile/compressive strength ratio of OPSLC is lower than normal concrete, but it is comparable with the lightweight concrete aggregate concrete made with an artificial lightweight aggregate of an equivalent grade. The bond strength of OPSLC was found to be about 2.3–3.9 times higher compared to the design bond strength as recommended by BS 8110 [7]. Thus the bond stress of the OPS lightweight concrete (OPSLC) showed that it satisfied the bond criterion as per the code's requirement.

Other advantages of OPS as a lightweight aggregate are high porosity content (Fig. 30.1) and Okpala [8] reported that the porosity of OPS at about 37 %. The nature of porosity is one of the important factors in the selection of thermal insulation

Fig. 30.1 *Right*: OPS shape, *left*: porosity inside the shell

material . Thus, the porosity of a lightweight concrete is a function of the "self-porosity" of the aggregate (i.e., the porosity of the aggregate as a material), the shape-factor and surface quality of the aggregates, and also of the porosity of the whole concrete mass, which is generally used to reduce the unit volume of concrete, and is known to have excellent heat interception and noise absorption properties due to the shapes of porous elements. The aim is to utilize this porosity to the best advantage in obtaining good insulation properties, while also achieving the required strength.

That advantage gives OPS the same low thermal conductivity value as other lightweight aggregates such as clinker, expanded vermiculite, expanded slate, and expanded clay. The previous study stated that using lightweight aggregate can be good thermal insulation in lightweight concrete [9, 10], but thermal conductivity of concrete will increase with increasing cement content [9]. To reduce the cement content, Lee et al. [11] investigated by introducing entrained air into a lightweight aggregate; concrete mix and aggregate normally constitute about 70–80 % by volume of Portland cement concrete [11]. The fineness of the pores will give better insulation properties and also, most important, moisture content (increases in moisture by mass linearly will increase thermal conductivity 42 % [12]). Due to that condition, this study investigates the effect of volume fraction OPS to lightweight concrete on thermal insulation concrete requirements and mechanical strength. The use of OPS as insulation material in concrete is a new area of research not yet explored.

30.2 Material and Experimental Program

Ordinary Portland cement ASTM type I was used as binder, with specific gravity of 3.10 and Blaice specific surface area of 3,510 cm^2/g; and local sand with specific gravity, fines modulus, water absorption and maximum grain size of 2.67 %,

2.66 %, 0.95 %, and 2.36 mm., respectively. Superplasticizer was used in the range 1 % of cement volume. Potable water was used. Old OPS was used as coarse aggregate, implying they were discarded approximately for half a year in a palm oil mill yard. Old OPS does not have fiber and has less oil coating, which results in a better bond within the OPS surface and mortar. OPS in deferent shape and sizes were used and,a stone crushing machine was used to crush the OPS for their crushed shape. The physical properties of OPS used in this study are shown in Table 30.1. Due to the high water absorption of OPS, it was washed and kept in a saturated dry (SSD) condition before mixing.

30.2.1 Mix Proportion

The mix design for lightweight aggregate concrete was used for semi-structural with thermal insulation capacity purposes. The mix design was based on the absolute volume method with five different volume fractions of OPS. The mix design details are shown in Table 30.2.

Table 30.1 The physical properties of OPS

OPS shape	Crushed
Specific gravity	1.22
Water absorption (%)	
24 h	18.73
Bulk density	626.4 kg/m^3
Fines modulus	5.82
Thermal conductivity	0.17 W/m K
Grading (% by weight passing sieve size stated)	
14 mm	100
10 mm	99.9
6.3 mm	35.98
5 mm	14.19
2.3 mm	0.6

Table 30.2 Mix proportion

	OPS		Cement	Sand	Water	Sp
Mix	Crushed	Volume (%)	(kg/m^3)	(kg/m^3)	(kg/m^3)	(kg/m^3)
C-30	537	30	400	720.75	160.16	40
C-32	581	32	375	675.26	150.06	38
C-34	618	34	350	631.46	140.33	35
C-36	647	36	325	584.91	129.98	33
C-38	678	38	300	540.62	120.14	30

30.2.2 Testing of Specimens

The specimens were tested based on the following testing standard. The density test was carried out according to the BS EN 12390-7:2009, the specimens are weighted in the air and suspended in water, as they are removed from the curing or exposure conditions.

The compressive strength test was done in accordance with BS 12390-3:2009, and three 100-mm cubes were tested, each at 7, 28, 90, and 180 days. At the age of 28 days, the splitting tensile strength, modulus of elasticity, and flexural strength were measured in accordance with ASTM C496/C496-11, ASTM C469-10, and ASTM C78-10. Ultrasonic pulse velocity was measured according to BS EN 12504-4:2004.

The thermal properties of specimens at room temperature were measured with a Hot-Disk probe TPS2500. The system is based on the transient plane source technology. This method is based on recording the temperature rise of a plane source heating the surrounding materials to be measured. The basic concept of the corresponding sensor is the fact that the conducting pattern is used both as a heat source and as a temperature sensor [13].

All specimens were prepared according to ASTM C332-99 [14] and the testing met the ISO/DIS 22007-2.2 standard. A sensor chosen with radius 9 mm, consisting of a very fine nickel double spiral covered with two thin layers of electrically insulating material, is placed between two specimens 75 ϕ and 45 mm thickness. The two sample pieces were prepared with a section of flat surface each in order to obtain a contact surface with as thin air layers as possible.

30.3 Result and Discussion

30.3.1 Density

Based on Table 30.3, the dry density is 165–211 kg/m^3 lower than air dry density. Generally the air dry density for OPSLC is within 1,725–2,025 kg/m and only uses 10–15 % volume fraction of OPS. In this study, mix C-36 and C-38 are of lower density than previous studies because the volume fraction of OPS used are highest

Table 30.3 Density and ultrasonic pulse velocity (UPV)

Mix	UPV (km/s)	Density (kg/m^3) Air	Density (kg/m^3) Dry
C-30	3.174	1,815	1,649
C-32	3.043	1,812	1,640
C-34	2.974	1,783	1,584
C-36	2.91	1,678	1,452
C-38	2.83	1,600	1,424

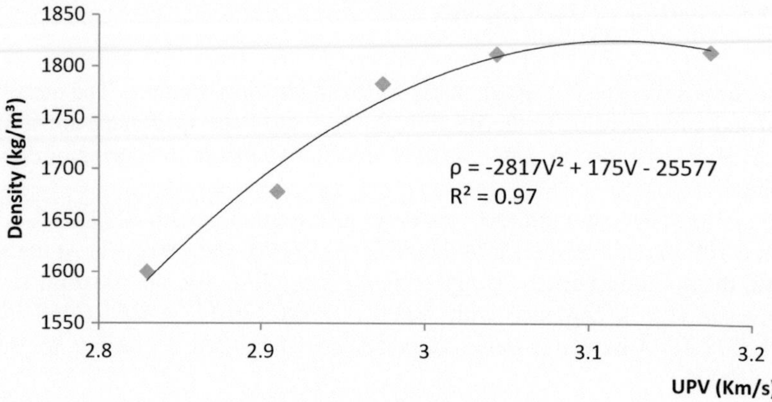

Fig. 30.2 Relationship density and ultrasonic pulse velocity (UPV)

which is more than 34 % of the total weight. The lowest air density obtained by mix C-38, 1,600 kg/m, may be due to more voids rate on concrete. The density is also contributed from the specific gravity of OPS; more volume of the OPS will reduce the bulk density of concrete. The highest reducing air to dry density was obtained by mix C-38, which is 26 % due to high porosity content. For ultrasonic pulse velocity, it was observed that there was an increase in pulse velocity values inversely proportional to the amount of volume fractions. The smaller volume fraction will give a better UPV value. Only mix C-30 and mix C-32 have moderate quality [15] and the rest have a low quality of concrete. In addition, the UPV value of concrete may also be greatly affected by the perfection of compaction, and U. Jonson [6] reported the proportion of OPS did not cause much variation in workability. The compactness of the microstructures would definitely be affected. At the same time, the porous cellular structure and OPS itself have also partially contributed to the low UPV value. The strong polynomial relationship within the density and UPV value obtained are shown in Fig. 30.2. A lower density will produce a weak quality of concrete.

30.3.2 Mechanical Strength

Previous researchers used a volume fraction of not more than 15 % because they designed for structural lightweight strength, but the present study investigates a higher volume fraction up to 38 % of the total volume of concrete. The lower volume fraction will produce high mechanical strength. Based on Table 30.2, C-30 has the highest compressive strength; 22 Mpa and C-38 had the lowest compressive strength which is 12 Mpa. However, all mixes comply with structural requirement strength according to the ASTM and BS standard which is more than 11 Mpa. For flexural strength, the values are 8–10 % from the compressive strength for all

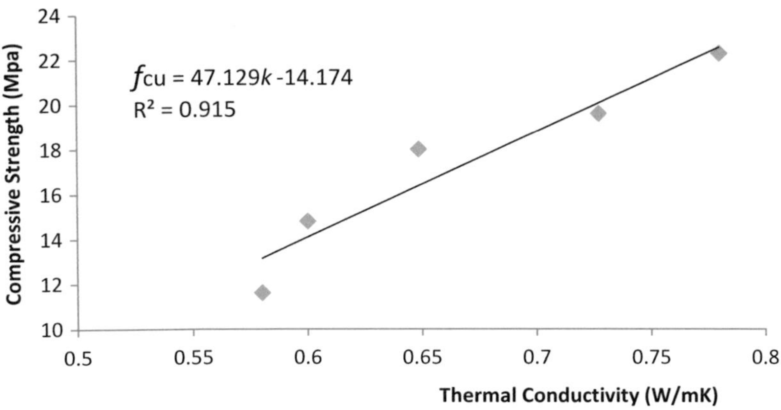

Fig. 30.3 Relationship compressive strength and thermal conductivity

Table 30.4 Mechanical strength

Mix	Compressive strength (MPa)	Flexural strength (MPa)	Tensile strength (MPa)
C-30	22	2.05	1.35
C-32	20	1.79	1.29
C-34	18	1.48	1.13
C-36	15	1.23	1.04
C-38	12	1.05	0.87

mixes, the same as reported by Shafigh et al. [3]. For high strength OPS the flexural strength is approximately 10 % from compressive strength. The tensile strength is in the range 0.87–1.35 Mpa (Table 30.4). It can be observed that the splitting tensile strength increased with increasing the compressive strength. Compared to the previous [6] study, the range of splitting tensile is 2.0–2.21 Mpa. This study has the low splitting tensile due to the bond within the cement paste aggregate slightly weaker because more volume fraction of OPS was used. It can be concluded that the mechanical strength of this study is lower than previous studies that design for structural purpose, but all mix designs are archived for the structural standard in order to achieve insulation concrete with load-bearing strength. Figure 30.3 shows the strong relationship within compressive strength and thermal conductivity; $R^2 = 0.915$.

30.3.3 Thermal Properties

The thermal conductivity (k) of the study ranged from 0.58 to 0.78 W/m K. The increasing of the volume fraction of OPS in concrete significantly contributes to the thermal conductivity value because the thermal conductivity of aggregate will influence the thermal conductivity of concrete. Mix C-38 has the highest OPS content, therefore the thermal conductivity value showed the lowest compared with other mixes. The thermal conductivity also increases with increasing cement content in concrete because cement has high thermal conductivity compared to OPS [16, 17]. The mix design of this study is based on the absolute volume method, thus more volume fraction of OPS will reduce the volume of cement and vice versa. Except for mix C-30, all mix designs were accepted with semi-structure insulation concrete according to the RILEM requirement, which is below 0.75 W/m K. They have a strong relationship within compressive strength and thermal conductivity as shown in Fig. 30.3 and are written as

$$\mathrm{fcu} = 47.129k - 14.174 \quad \left(R^2 = 0.92\right) \tag{30.1}$$

For specific heat value, the results show inconsistency because the distribution of aggregate is not uniform in concrete. The specific heat influence of distribution of OPS is shown in Fig. 30.4, especially for the transient plane source method used to measure specific heat value [18]. The sensor only measures in the specific area surrounding the sensor, therefore different samples will have different distributions and sizes of OPS also will influence the result. The thermal diffusivity value is the same as the thermal conductivity trend, which is influenced by the volume fraction OPS used (Table 30.5).

Fig. 30.4 Distribution of OPS in concrete (volume fraction from left: 30, 32, 34, 36 and 38 %)

Table 30.5 Thermal properties

Mix	Thermal conductivity (W/m K)	Specific heat (MJ/m^3K)	Thermal diffusivity (mm^2/s)
C-30	0.78	1.25	0.57
C-32	0.73	1.61	0.48
C-34	0.65	1.42	0.45
C-36	0.6	1.98	0.31
C-38	0.58	1.71	0.34

30.4 Conclusions

The following conclusions can be made regarding mix design with different volume fractions of OPS used.

1. The lowest density with load-bearing strength obtained by mix contains volume fraction 38 % of OPS; more volume fraction will decrease the density and strength.
2. There is a strong relationship within the density and ultrasonic pulse velocity value. A high volume fraction will give a low pulse velocity value due to more air void and porosity content in the concrete.
3. All mechanical strength of mixes achieved the structural capacity in compliance with the ASTM and BS standards.
4. Except the mix that used volume fraction 30 % of OPS, all mix designs are in the range for semi-structure insulation concrete with RILEM standard for which the thermal conductivity is lower than 0.75 W/m K.
5. By using the transient plane source method, specific heat OPS concrete will give not consistencies reading base on distributions and sizes of aggregate. The increase of energy required is due to moisture content of the aggregate.
6. Thermal conductivity and thermal diffusivity results have the same indicator, which is increased according the density of OPS concrete.

Acknowledgment The author would like to thank the Universiti Sains Malaysia for awarding the Research University grant (1001/PPBGN/814213) to carry out the present research.

References

1. Abullah N, Sulaiman F, Gerhouser H, Shariff A (2011) An outlook of Malaysian energy, oil palm industry and its utilization of waste as useful resources. Biomass Bioenergy 35:3775–3786
2. Pankhurst RNW (1993) Construction. In: Clarke JL (ed) Structural lightweight aggregate concrete. Chapman & Hall, London; chapter 4
3. Shafigh P, Zamin Jumaat M, Mahmud H, Johnsom Alengaram U (2011) A new method of producing high strength oil palm shell lightweight concrete. Mater Des 32:4839–4843
4. Husain Z, Zainac Z, Abdullah Z (2002) Briqutting of palm fibre and shell from the processing of palm nuts to palm oil. Biomass Bioenergy 22:505–509
5. Sahu JN, Abnisa F, Daud WMA, Husin WMW (2011) Utilization possibilities of palm shell as a source of biomass energy in Malaysia by producing bio-oil in pyrolysis process. Biomass Bioenergy 35:1863–1872
6. Johnson Alengaram U, Al Muhhit BA, bin Jumaat MZ (2013) Utilization of oil palm kernel as lightweight aggregate in concrete—a review. Construct Build Mater 38:161–172
7. Teo DCL, Mannan MA, Kurian VJ (2006) Structural concrete using oil palm shell (OPS) as lightweight aggregate. Turk J Eng Environ Sci 30:251–257
8. Okpala DC (1990) Palm kernel shell as lightweight aggregate in concrete. Build Environ 25:291–296

9. Demirbo˘ga R, Gul R (2003) Thermal conductivity and compressive strength of expanded perlite aggregate concrete with mineral admixtures. Energy Build 35:1155–1159
10. Al-sibahy A, Edward R (2012) Mechanical and thermal properties of novel lightweight concrete mixtures containing recycled glass and metakaolin. Construct Build Mater 31:157–167
11. Lee HK, Kim HK, Jeon JH (2012) Workability and mechanical, acoustic and thermal properties of lightweight aggregate concrete with volume of entrained air. Construct Build Mater 29:193–200
12. Narayanan N, Ramamurthy K (2000) Structure and properties of aerated concrete: a review. Cem Concr Compos 22:321–329
13. Bouguerra A, Laurent JP, Goual MS, Queneudec M (1997) The measurement of the thermal conductivity of solid aggregate using the transient plane source technique. J Phys D Appl Phys 30:2900–2904
14. ASTM 332-99. Standard specification for lightweight aggregates for insulating concrete. ASTM, West Conshohocken
15. Solis-Carcano R, Moreno EI (2008) Evaluation of concrete made with crushed lime stone aggregate based on ultrasonic pulse velocity. Construct Build Mater 30:1225–1231
16. Serri E, Zailan Suleiman M, Md Azree OM (2014) The effect of oil palm shell aggregate shape on the thermal properties and density of concrete. Adv Mater Res 935:172–175
17. Serri E, Zailan Suleiman M, Md Azree OM (2014) Thermal properties of oil palm shell lightweight concrete with different mix design. Jurnal Teknologi 70(1):155–159
18. Demirboğa R, Gül R (2013) The effects of expanded perlite aggregate, silica fume and fly ash on the thermal conductivity of lightweight concrete. Cem Concr Res 33:723–727

Chapter 31
Double Diffusion Effects on Entropy Generation in Convective Heat and Mass Transfer

Zeroual Aouachria, Djamel Haddad, and F. Benzemit

Abstract The aim of the present study is to examine the Dufour influence and Soret effects on entropy generation in convective heat and mass transfer for the case of a binary gas mixture with a single diffusive species in a square cavity. We numerically investigate the thermosolutal convection in a square cavity, where a binary mixture of incompressible fluid is confined under the conditions of both thermal and solutal gradients. Furthermore, we analyze the effect of Dufour and Soret on entropy generation. We find that, according to Grashof number values, the entropy generation could be mainly due to heat transfer or to fluid friction irreversibility and that the Soret and Dufour effects have great impact on the production of entropy.

Keywords Natural convection • Entropy generation • Numerical methods • Double diffusion • Square cavity

Nomenclature

Symbols

T	Temperature of the fluid, K
T_1	Temperature of hot wall, K
T_2	Temperature of the cold wall, K
T_0	Average temperature of fluid, K
C	Molar concentration, mol L^{-1}
C_0	Average concentration, mol L^{-1}

Z. Aouachria (✉) • F. Benzemit
Laboratory (LPEA), Faculty of Science, University of Batna, Batna, Algeria
e-mail: aouachria2001@gmail.com

D. Haddad
Laboratory (LESEI), Faculty of Engineering, University of Batna, Batna, Algeria

I. Dincer et al. (eds.), *Progress in Clean Energy, Volume 1*,
DOI 10.1007/978-3-319-16709-1_31

D	Mass diffusivity, $m^2\ s^{-1}$
D_F	Dufour coefficient, $m^2\ s^{-1}\ K^{-1}$
D_T	Soret coefficient, $m^2\ s^{-1}\ K^{-1}$
P	Dimensionless pressure
p	Pressure, Pa
u, v	Velocity components, $m\ s^{-1}$
U, V	Velocity dimensionless components
x, y	Dimensional Cartesian coordinates, m
X, Y	Dimensionless Cartesian coordinates
V_0	Inlet fluid velocity, $m\ s^{-1}$
S	Entropy production, $J\ m^{-3}\ s^{-1}\ K^{-1}$
t	Dimensional time, s
R	Thermodynamic constant of fluid, $J\ kg^{-1}\ K^{-1}$
α	Thermal diffusivity, $m^2\ s^{-1}$
K	Thermal conductivity, $W\ m^{-1}\ K^{-1}$
ρ_0	Mass density, $kg\ m^{-3}$
ν	Kinetic viscosity, $m^2\ s^{-1}$
μ	Dynamic viscosity, Pa s
β_T	Thermal expansion coefficient, K^{-1}
μ_{SP}	Mass chemical potential, $J\ kg^{-1}$
g	Gravity acceleration, $m\ s^{-2}$
β_S	Solutal expansion coefficient, $m^3\ mol^{-1}$
C_P	Specific heat at constant pressure, $J\ kg^{-1}\ K^{-1}$
θ	Dimensionless temperature
τ	Dimensionless time
ξ	Dimensionless concentration
ϕ	Dependent variable
Γ	Coefficient of the diffusion

Formulae

Thermal Grashof number	$Gr_{\mathrm{Th}} = \frac{g\beta_{\mathrm{T}}(T_1-T_2)L^3}{\nu^2}$
Solutal Grashof number	$Gr_{\mathrm{S}} = \frac{g\beta_{\mathrm{S}}\cdot(T_1-T_2)L^3}{\nu^2}$
Lewis number	$L_e = \frac{\alpha}{D}$
Buoyancy number	$N = \frac{Gr_S}{Gr_T}$
Dufour parameter	$K_F = \left[\left(\frac{W\cdot D_F}{\alpha}\right)\cdot\left(\frac{\beta_{\mathrm{T}}}{\beta_{\mathrm{S}}}\right)\right]$
Soret parameter	$K_T = \frac{[D_T\cdot C\cdot(1-C)T]}{D}$
Positive thermodynamic parameter	$W = \left[\frac{C_0\cdot T_0}{C_P}\right]\left(\frac{\partial\mu_{SP}}{\partial C}\right)$
Prandtl number	$Pr = \frac{\nu}{\alpha}$

31.1 Introduction

The phenomena of heat and mass transfer through natural convection with double diffusion are due to the simultaneous presence of temperature and concentration gradients. These gradients cause nonuniform distribution of the density of the mixture which in turn causes the convective motion under the effect of gravity. In most situations, whether it is in nature or in industry, the fluid studied consists of two or more components. So the generated natural flows are called thermosolutal natural convection and are characterized by the irreversibility phenomena. The latter that are expressed by entropy generation are of important interest during the design of any thermodynamic system. Due to its fundamental and practical importance (oceanography, metallurgy, geophysics, and air pollution), the double diffusive convection is essentially characterized by the phenomena of heat and mass transfer accompanying the entropy production. In addition, when the process of heat and mass transfer occurs simultaneously, they interfere and produce the so-called cross-diffusion effects. So one witnesses a coupling of heat flow induced by a concentration gradient and the flow of matter induced by a temperature gradient, which are respectively known by the names of the effects of Dufour and Soret and which have been neglected in most studies.

Many works concerning entropy generation have been carried out [1–4]. The efficient use of energy is the overarching objective in the design of a system or process, which may be achieved by minimizing the entropy production (known as energy destruction). Many works have analyzed the production of entropy in terms of external and internal irreversibilities. In general, the origin of irreversibility is related to the phenomena of heat and mass transfer, chemical reactions, and turbulence in fluid flow [5].

The analysis on the entropy generation has become a very useful tool for thermal design of thermodynamic systems in order to save energy. The various sources of irreversibility are responsible for entropy production such as the transfer of heat and mass, magnetic field effect, and viscous dissipation effect. Many studies have been published on the entropy production. Weaver and Viskanta [6] indicated that the effects of Soret and Dufour have an important role when the differences in temperature and concentration are large and when the difference in molecular weight between two elements dispersing in the mixture is significant. Their study was applied in a square cavity, and they showed that the contribution of the Soret effect to the total mass flux can reach up to 10–15 %. The energy flow caused by the Dufour effect can be comparable to that obtained by pure conduction. Haddad and Al Nimr [7] conducted a study on the generation of entropy in a simple rectangular micro channel with a boundary of constant heat flux. They numerically modeled the micro channel to find velocity distributions and the temperature along the flow direction in the channel of the coolant. Magherbi et al. [8] studied the influence of the angle of inclination of the cavity relative to the horizon, the thermal Grashof number, and the ratio of buoyancy on the production of total entropy. The location of the irreversibility due to heat transfer, mass transfer, and fluid friction is discussed for three angles of inclination for a fixed thermal number of Grashof. The results showed that the total entropy generation increases with the thermal number of Grashof and the ratio of buoyancy to moderate numbers of Lewis. Locally, the irreversibility due to heat and mass transfer is almost identical and located at the bottom and top walls respectively heated and cooled. Hidouri et al. [9] studied the influence of the Soret effect on the production of entropy in thermosolutal natural convection. They showed that the entropy production is at minimum value for a ratio of buoyancy in the vicinity of -1 and for a thermal Grashof number greater than 10^4.

Since the efficient use of energy is the main objective in the design of a system or a technical process, this can be obtained by the minimum destruction of energy, that is to say, the minimization of the production of entropy which influences Dufour and Soret effects significantly. Indeed, the Dufour effect tends to increase the entropy production, while the Soret effect tends to decrease it especially for the thermal Grashof number. Alam [10] and Aouachria et al. [11] examined the problem of a regular free convective flow over a vertical semi-infinite flat plate with uninterrupted movement in a porous medium taking into account the effects of Dufour and Soret. Such effects in the convective flow along a flat plate embedded in a porous medium with variable Darcy suction were examined taking into account the temperature and concentration at the plate as functions of time [12–14]. In our study, we propose the analysis of the entropy production in double diffusion convection in the case of a square cavity taking into account the Soret and Dufour effects. The aim of our work is the study of natural double diffusion convection and the Dufour and Soret effects on the production of entropy in a thermodynamic system consisting of a binary mixture of gases (air-CO_2) located in a semiadiabatic square cavity.

31.2 Mathematical Formulation

31.2.1 Problem Generation

The analysis of thermal convection is based on the application of three basic laws of general physics (conservation of mass, momentum, energy and species), taking into account the Dufour and Soret effects. The equations we study are derived from the conservation laws and cannot be solved analytically, forcing one to look for other methods to solve them, which should be associated with simplifying assumptions that may be adequate to solve these types of equations. So we consider a two-dimensional square cavity of length L, where a binary mixture of incompressible fluid is confined under the conditions of both thermal and solutal gradients. As shown in Fig. 31.1, the generated convective motion is assumed to be two-dimensional and unsteady.

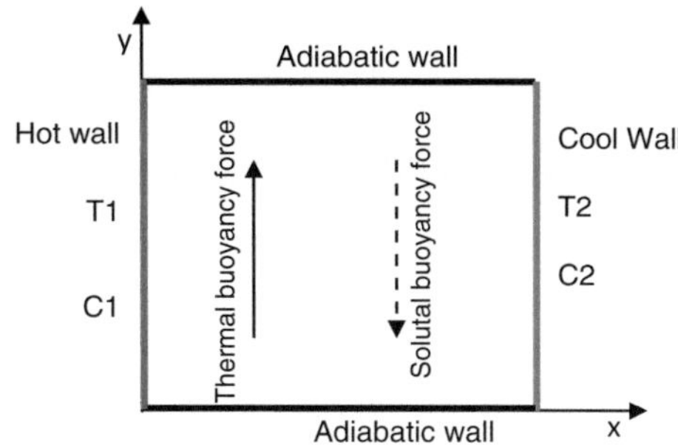

Fig. 31.1 Schematic view of 2D square cavity

The different physical–chemical parameters are assumed constant, except for the applied volume force (ρg) for which we use Boussinesq approximation as the temperature gradient is not sufficiently large. We also neglect the effect of chemical reactions. This volume force is given by the expression

$$\rho(T, C) = \rho_0[(1 - \beta_T(T - T_0) - \beta_S(C - C_0))] \tag{31.1}$$

where $\beta_T = -\frac{1}{\rho_0}\left(\frac{\partial \rho}{\partial T}\right)$, $\beta_T > 0$ and $\beta_S = -\frac{1}{\rho_0}\left(\frac{\partial \rho}{\partial C}\right)$, $\beta_s > 0$ represent, respectively, thermal and solutal expansion coefficients of the fluid solution; $\rho_0 = \rho(T_0, C_0), T_0$, C_0 are the density, the temperature, and the reference concentration, respectively.

31.2.2 *Mathematical Modeling*

Under the conditions mentioned above, we can establish the various equations necessary to solve this problem as follows. From the non-dimensionalization of the dependent and independent variables, we use characteristic quantities that are constant throughout the flow field and temperature. These quantities are L, V_0, T_1, T_2, and P:

$$X = \frac{x}{L},\ Y = \frac{y}{L},\ \nu_0 = \frac{\alpha}{L},\ U = \frac{u}{\nu_0},\ V = \frac{v}{\nu_0},\ \tau = t\frac{\nu_0}{L},\ p = \frac{p}{\rho \nu_0},$$
$$\theta = \left(\frac{T - T_0}{T_1 - T_2}\right),\ \xi = \left(\frac{C - C_0}{C_1 - C_2}\right) \tag{31.2}$$

Using Cartesian coordinates and defining the following dimensionless variables, we obtain

$$\frac{\partial U}{\partial X} + \frac{\partial V}{\partial Y} = 0 \tag{31.3}$$

$$\frac{\partial U}{\partial \tau}+\frac{\partial U\cdot U}{\partial X}+\frac{\partial V\cdot U}{\partial Y}=-\frac{\partial P}{\partial X}+Pr\left(\frac{\partial^2 U}{\partial X^2}+\frac{\partial^2 U}{\partial Y^2}\right) \tag{31.4}$$

$$\frac{\partial V}{\partial \tau}+\frac{\partial U\cdot V}{\partial X}+\frac{\partial V\cdot V}{\partial Y}=-\frac{\partial P}{\partial Y}+Pr\left(\frac{\partial^2 V}{\partial X^2}+\frac{\partial^2 V}{\partial Y^2}\right)+Pr^2\cdot Gr_{\mathrm{T}} \cdot(\theta+N\cdot\xi) \tag{31.5}$$

$$\frac{\partial \theta}{\partial \tau}+\frac{\partial U\cdot \theta}{\partial X}+\frac{\partial V\cdot \theta}{\partial Y}=\left(\frac{\partial^2 \theta}{\partial X^2}+\frac{\partial^2 \theta}{\partial Y^2}\right)+K_F\left(\frac{\partial^2 \xi}{\partial X^2}+\frac{\partial^2 \xi}{\partial Y^2}\right) \tag{31.6}$$

$$\frac{\partial \xi}{\partial \tau}+\frac{\partial U\cdot \xi}{\partial X}+\frac{\partial V\cdot \xi}{\partial Y}=\frac{1}{Le}\left(\frac{\partial^2 \xi}{\partial X^2}+\frac{\partial^2 \xi}{\partial Y^2}\right)+\frac{1}{Le}\frac{\partial}{\partial X}\left(\frac{K_T}{\theta}\frac{\partial \theta}{\partial X}\right)+\frac{1}{Le}\cdot\frac{\partial}{\partial Y}\left(\frac{K_T}{\theta}\frac{\partial \theta}{\partial Y}\right) \tag{31.7}$$

where the thermal diffusivity of the mixture is $\alpha=\frac{k}{\rho C_P}$ and the thermodynamic parameter $W=\frac{TC_0}{C_P}\left(\frac{\partial\mu}{\partial C}\right)$, with $\frac{K_T}{T}=C_1C_2\frac{D_T}{D}$. In general, the ratio of the thermal diffusion coefficient D_T and the mass diffusivity D is defined as the Soret coefficient S_T, in which, sometimes, the product $C_1C_2=C_1(1-C_1)$ is included to the extent that the concentration variations are assumed small compared to the rest of the mixture. The associated boundary conditions to this problem are

$$\begin{array}{ll}
\text{for } x=0 \text{ and } 0\le y\le 1 & \theta=1,\ \xi=1,\ u=v=0\\
\text{for } x=1 \text{ and } 0\le y\le 1 & \theta=0,\ \xi=0,\ u=v=0\\
\text{for } y=0, \text{ and } y=1 \text{ and } \forall x & \left(\frac{\partial\theta}{\partial y}\right)=0,\ \left(\frac{\partial\xi}{\partial y}\right)=0,\ u=v=0\\
\multicolumn{2}{l}{\text{at } t=0,\ \theta=0.5,\ \xi=0,\ u=v=0 \text{ and } P=0 \text{ at any point in the cavity}}
\end{array} \tag{31.8}$$

This system of equations can be written in compact form as

$$\frac{\partial\rho\phi}{\partial t}+\frac{\partial\rho U\phi}{\partial x}+\frac{\partial\rho V\phi}{\partial y}=\frac{\partial}{\partial x}\left(\Gamma\frac{\partial\phi}{\partial x}\right)+\frac{\partial}{\partial y}\left(\Gamma\frac{\partial\phi}{\partial y}\right)+S_p\cdot\phi+S_u \tag{31.9}$$

where S_p and S_u are the source terms, Γ is the diffusion coefficient, and ϕ indicates any variables as U, θ, ξ, etc.

31.2.3 Modeling of the Entropy Generation

The existence of thermal and diffusive gradients, between the active walls of the cavity, places the fluid in a nonequilibrium state which causes the generation of entropy in the system. According to local thermodynamic equilibrium with the linear transport theory, the generation of the local entropy is given by

$$S_T = \frac{k(\text{grad}\,T^2)}{T_0^2} - \frac{1}{T}\sum \vec{J}_{\alpha i}\text{grad}\,\mu_i + \frac{\vec{\tau} : \overline{\overline{\text{grad}}}}{T}\vec{v}_d \tag{31.10}$$

$J_{\alpha i}$ and μ_i are the mass diffusion flux of spice i in phase and chemical potential, respectively.

As can be seen from Eq. (31.10), the right-hand side of this equation represents three terms: (1) irreversibility due to heat transfer, (2) irreversibility due to mass transfer, and (3) irreversibility due to the frictions in the fluid. The dimensionless expression for the production of local entropy, for the case of a perfect binary mixture gas in double diffusion by convection, is given by

$$S_{\mathrm{t}} = \left(\frac{\partial \theta}{\partial X}\right)^2 + \left(\frac{\partial \theta}{\partial Y}\right)^2 \tag{31.11}$$

$$S_v = \lambda_1\left[2\left(\left(\frac{\partial U}{\partial X}\right)^2 + \left(\frac{\partial V}{\partial Y}\right)^2\right) + \left(\frac{\partial U}{\partial Y} + \frac{\partial V}{\partial X}\right)^2\right] \tag{31.12}$$

$$S_{\xi\cdot\xi} = \lambda_2\left[\left(\frac{\partial \xi}{\partial X}\right)^2 + \left(\frac{\partial \xi}{\partial Y}\right)^2\right] \tag{31.13}$$

$$S_{\theta\cdot\xi} = \lambda_3\left[\left(\frac{\partial \theta}{\partial X}\right)\left(\frac{\partial \xi}{\partial X}\right) + \left(\frac{\partial \theta}{\partial Y}\right)\left(\frac{\partial \xi}{\partial Y}\right)\right] \tag{31.14}$$

S_{t}, S_{v}, $S_{\xi\cdot\xi}$, and $S_{\theta\cdot\xi}$ are respectively related to the irreversible thermal gradients, the effects of viscosity, the sum of the concentration gradients, and the sum of a combination product of the temperature and concentration gradients. The latter two are the local production of entropy due to the mass distribution (S_{d})

$$S_{\mathrm{d}} = S_{\xi,\xi} + S_{\theta,\xi} \tag{31.15}$$

λ_1, λ_3, and λ_2 are the irreversibility distribution ratios, and they are given by

$$\lambda_1 = \left[\left(\frac{\mu \cdot T_0}{K}\right)\right] \cdot \left[\frac{\alpha}{L} \cdot (T_1 - T_2)\right]^2,$$
$$\lambda_2 = \left[\left(\frac{\rho_0 \cdot D \cdot R \cdot T_0^2}{K \cdot C_0}\right)\right] \cdot \left[\left(\frac{C_1 - C_2}{T_1 - T_2}\right)\right]^2, \quad (31.16)$$
$$\lambda_3 = \left[\left(\frac{\rho_0 \cdot D \cdot R \cdot T_0}{K}\right)\right] \cdot \left[\left(\frac{C_1 - C_2}{T_1 - T_2}\right)\right]$$

31.3 Results and Discussion

We have developed a program in FORTRAN language which solves different equations governing this phenomenon. Parameters and physical properties of the thermodynamic system used for the calculations are shown in Table 31.1. The analysis is carried out for different values of Soret and Dufour numbers assigned in the Table 31.2. We must note that the results we present below concern only the stationary part of the phenomenon that will level after a while for the condition's initial fixed time.

31.3.1 Velocity Field Analysis

Figure 31.2a–d illustrates the change in the modulus of the flow rate and its components. From these profiles, we can note a few key points. The flow generated in the vessel is a rotational flow in the area adjacent to the walls of the latter and which is carried around a fixed central mass.

31.3.2 Change Profile of Velocity Magnitude Depending on k_T *for Different Values of* K_F

Figure 31.3 shows the evolution of the modulus of the velocity of flow and its components. From these profiles, some essential points can be deduced. The flow generated in the vessel is a rotational flow in the area adjacent to the walls, which takes place around a fixed central point. Increasing the Dufour parameter relatively decreases the velocity field. Indeed, Fig. 31.3c shows two ranges of Dufour parameter which have a different effect on the profile of the magnitude velocity. For a given value of K_T, the increasing K_F increases the speed modulus in the area close to the two nonadiabatic walls.

Table 31.1 Parameters and physical properties of the thermodynamic system

Parameters	*Le*	*Pr*	Gr_T	λ_1	λ_2	λ_3
Values	1.2	0.71	$7 \cdot 10^4$	10^{-4}	0.5	10^{-2}

Table 31.2 Different values of Soret and Dufour numbers

K_T	0	0.2	0.4	0.6	0.9	1.0
K_F	0	0.25	0.5	0.6	0.75	1.25

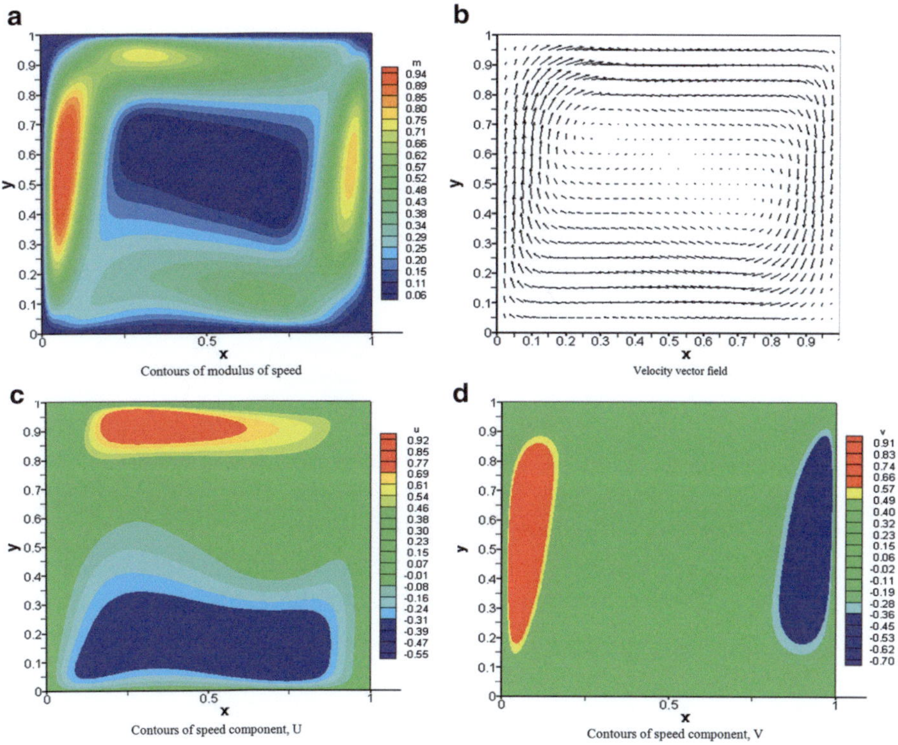

Fig. 31.2 Field of velocities. (**a**) Contours of modulus of speed. (**b**) Velocity vector field. (**c**) Contours of speed component, *U*. (**d**) Contours of speed component, *V*

31.3.3 Analysis of Thermal and Solutal Fields

Figure 31.4a shows the thermal and solutal contours. We observe that the temperature profile is well pronounced in the vicinity of the hot wall by moving more toward the top adiabatic wall but slightly in the direction of the thermal gradient. This explains why the Soret effect is more prominent compared to the thermal diffusion. This is in agreement with and explains the evolution of the solutal field represented in Fig. 31.3.

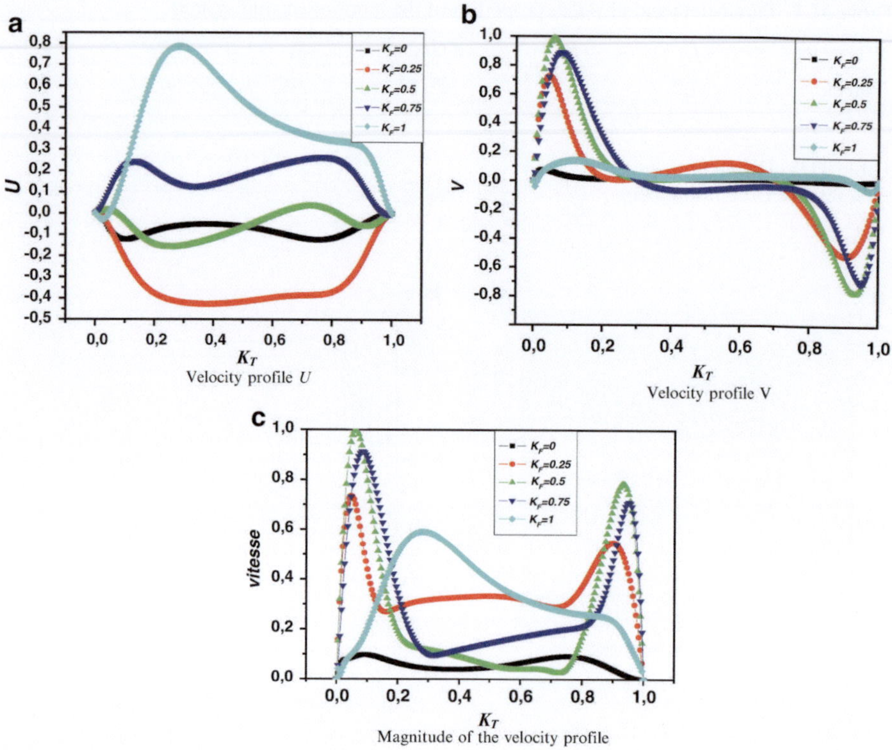

Fig. 31.3 Evolution of the speed profile depending on k_T, for different values of K_F. (**a**) Velocity profile U. (**b**) Velocity profile V. (**c**) Magnitude of the velocity profile

31.3.4 Change Profile of Solutal and Thermal Fields in Terms of k_T for Different Values of K_F

In Fig. 31.5, we include the thermal and solutal profiles for different values of Soret and Dufour numbers. It is noted in Fig. 31.4a that the temperature evaluation is carried out in the same direction of change in the number of Dufour, while it decreases with increasing Soret number. This explains the molecular diffusion motivates well this aspect and is well nuanced in the vicinities of the two walls. However, it is practically zero at the center of the tank. This phenomenon is analogous to changing the solutal field; see Fig. 31.4b.

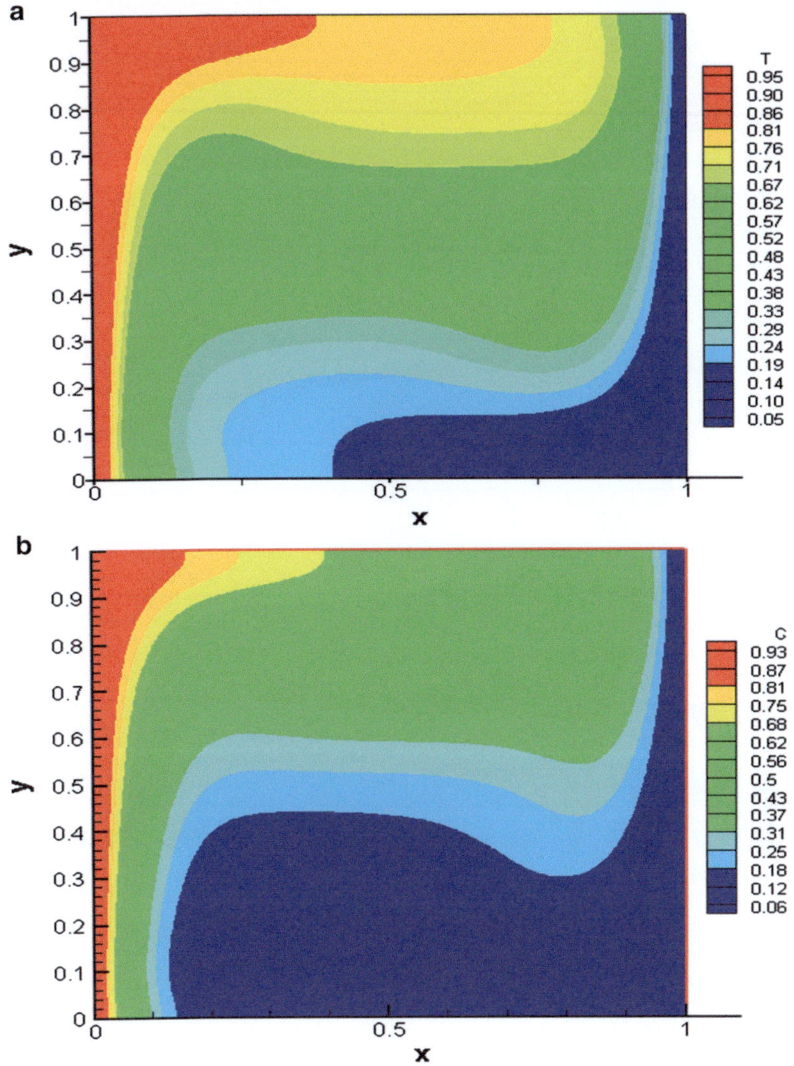

Fig. 31.4 Thermal and solutal fields. (**a**) Iso-contours of the temperature and (**b**) iso-contours of the concentration

31.3.5 Entropy Production of Origin: Thermal, Viscous, Solutal, and Total Entropy in Function of $\mathbf{k_T}$ *for Different Values of* $\mathbf{K_F}$

Figure 31.6 shows the iso-contours of the entropy production due to thermal, viscous, and solutal effects. The entropy production of thermal origin occurs in specific zones where it is most important near the most active and cold zones

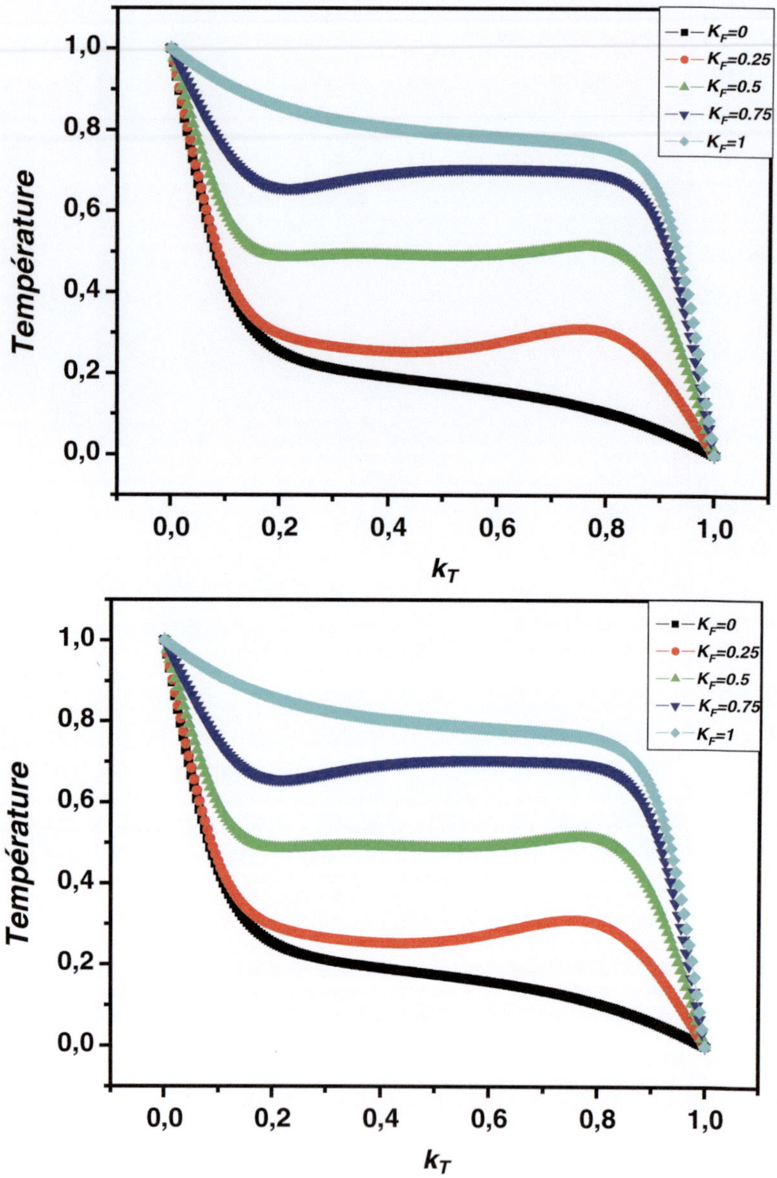

Fig. 31.5 Thermal and solutal fields versus K_T for different values of K_D

(Fig. 31.6a). The iso-contours of the entropy generation due to the viscous effect of the fluid which is under the influence of heat and solutal gradients are presented in Fig. 31.6b. We note that it appears in specific areas. In fact, its maximum locates in the upper region of the cavity and close to the hot wall. This area is characterized by the large viscous effect of the fluid acceleration. More acceleration is high more

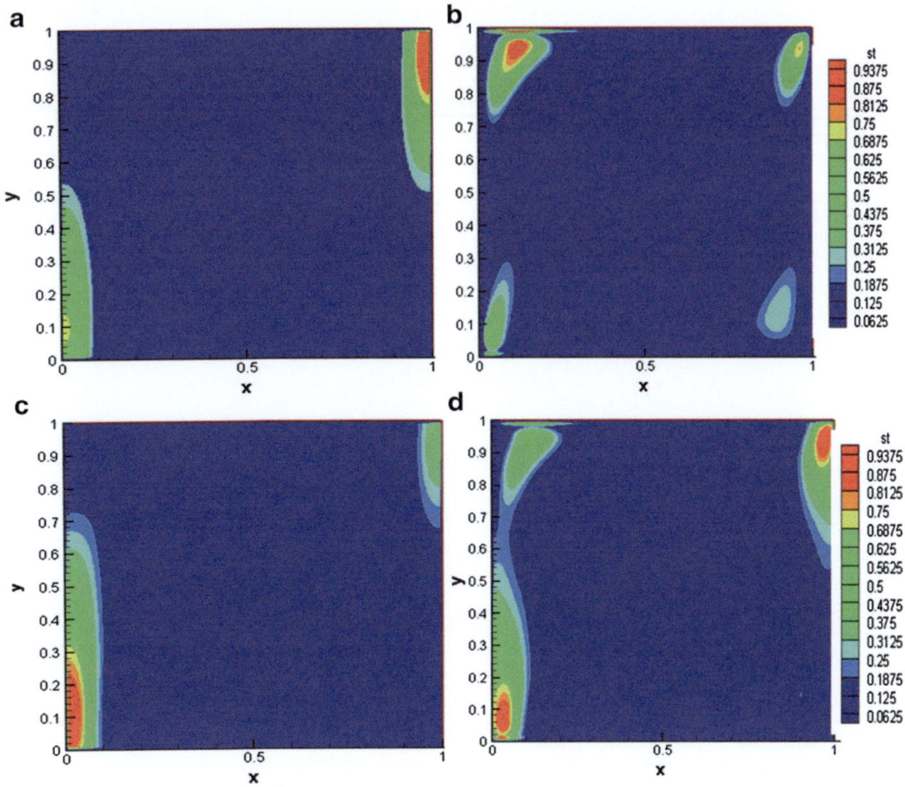

Fig. 31.6 Iso-contours of entropy generation by thermal, viscous, and solutal effects and the total entropy

entropy production is great. This production is related to the molecular diffusion which is accentuated in the lower left area of the cavity. We also note a production of entropy, but relatively low in the diagonally opposite the latter area. The total entropy production of different entropies is shown in Fig. 31.6d; its iso-contours have the appearance that they do not reflect the logical summation of positive quantities. At the local level, entropy generation is located on the lower corner of the heated side and upper corner of the cooled side of the enclosure. This can be understood as a paradox but does take into account the coupling effects that have antagonistic effects.

We have drawn, in Fig. 31.7, the variation in the entropy due to thermodynamic irreversibilities depending on the number of Soret, for different values of Dufour. The analysis of the profile shows two areas on which the influence of these two parameters is much nuanced. These areas are localized near the two nonadiabatic walls (K_T = [0 to 0.9] and [0.8 to 1]). The increase of Dufour number increases the entropy near the cold wall, but it decreases on the hot wall; see Fig. 31.7a.

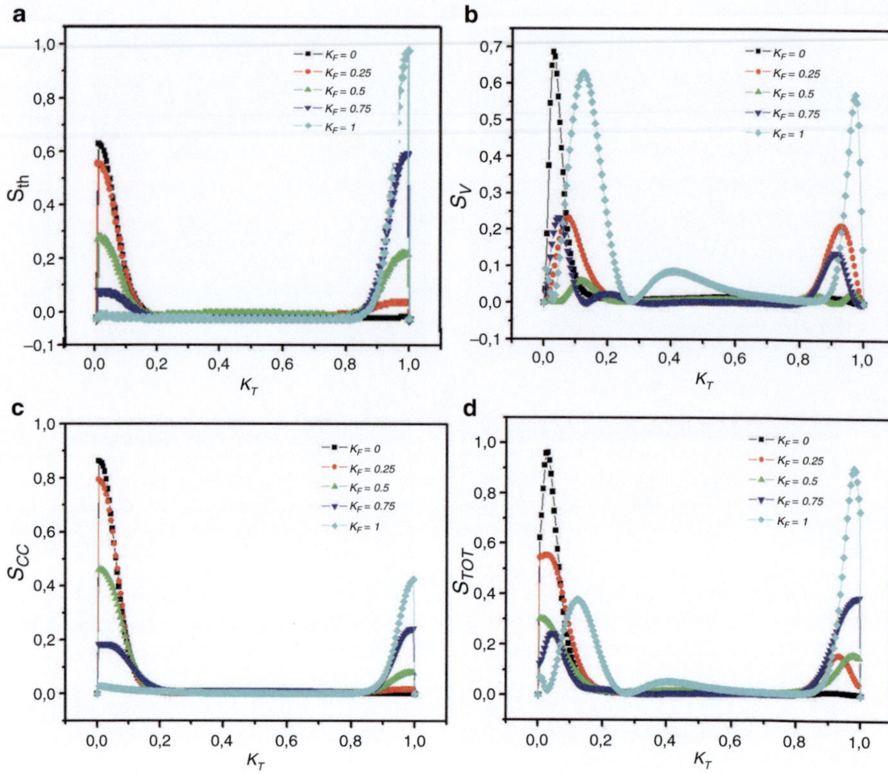

Fig. 31.7 Entropy generation by thermal, viscous, and solutal effects and the total entropy versus K_T for different values of K_D

The analysis of the profile shows that in the interval [0.25 to 0.87], the effect of K_T and K_F on viscous entropy is absent. When Soret varies in the range [0 to 0.25] and [0.87 to 1], the increased number of Dufour, viscous entropy decreases, whereas when it is in the range [0–0.5], it actually increases near the hot wall and conversely near the cold wall. This effect was reversed when it varies in the range [0.5–1]. The similar impacts occur on thermal and solutal fields and are also shown on entropy production profiles; see Fig. 31.7c. The total entropy increases with the parameter Dufour, near the cold wall, but it decreases near from the hot wall, with its increase. It should be noted that there is no influence in the central region of the cavity; see Fig. 31.7d.

31.4 Conclusion

According to what is presented, it was concluded that the Soret and Dufour effects have a significant impact on the entropy production. In addition, they have reverse influences. The Dufour effect always causes entropy production but for small numbers of Grashof, while the Soret effect appears more in the case of relatively large numbers of Grashof. At the local level, entropy generation is located on the lower corner of the heated side and upper corner of the cooled side of the enclosure.

References

1. Bejan A (1979) A study of entropy generation in fundamental convective heat transfer. J Heat Transfer 1:718–725
2. Poulikakos D, Bejan A (1982) Fin geometry for minimum entropy generation in forced convection. J Heat Transfer 104:616–623
3. Sahin AZ (2000) Entropy generation in turbulent liquid flow through a smooth duct subjected to constant wall temperature. Int J Heat Mass Transfer 43:1469–1478
4. Bouabid M, Magherbi M, Hidouri N, Ben Brahim A (2011) Entropy generation at natural convection in an inclined rectangular cavity. Entropy 13:1020–1033. doi:10.3390/e13051020
5. Magherbi M, Ben Brahim A (2007) Influence des effets de diffusion croisés sur la production d'entropie en convection par double diffusion. Int J Exergy 4(3):1–5
6. Weaver RV, Viskanta J (2005) Natural convection due to horizontal temperature and concentration—2. Species interdiffusion, Soret and Dufour effects. Far East J Appl Math 25:179–197
7. Haddad O, Al Nimr M (2004) Entropy generation due to laminar incompressible forced convection flow through parallel-plates microchannel. Entropy 6(5):413–426
8. Magherbi M, Abbassi H, Hidour N, Ben Brahim A (2006) Second law analysis in convective heat and mass transfer. Entropy 8:1–17
9. Hidouri N, Magherbi M, Ben Brahim A (2007) Influence of thermodiffusion effect on entropy generation in thermosolutal convection. Far East J Appl Math 25:179–197
10. Alam MS (2006) Dufour and Soret effects study free convection and mass transfer flow past a semi-infinite vertical porous plate in a porous medium. Int J Appl Mech Eng 11(3):535–545
11. Aouachria Z, Rouichi F, Haddad D (2012) Double diffusion effects on convection in flow on vertical plate imbedded in a porous media. Front Heat Mass Transfer 3:023004. www.ThermalFluidsCentral.org. doi:10.5098/hmt.v3.2.30042012
12. Dimia AM (2011) Contribution à l'étude thermodynamique des diagrammes de phases binaires: étude de l'influence de la température sur les enthalpies et les entropies de mélange. Thèse de magistère, Université el hadj Lakhdar
13. Aouachria Z (2007) Cours de thermodynamique des phénomènes irréversibles. Université el hadj Lakhdar Batna
14. Polttier N (2007) Physique statistique hors d'équilibre Processus irréversibles linéaires. Cnrs éditions éd. 2007, Imprimé France

Chapter 32
Three-Dimensional Numerical Study of the Heat Transfer on The Planar Solid Oxide Fuel Cell: Joules Effect

Youcef Sahli, Bariza Zitouni, Hocine Ben Moussa, and Hafsia Abdenebi

Abstract The aim of this work is to analyze the three-dimensional temperature fields in a planar solid oxide fuel cell (SOFC) single cell with different geometric configurations: supported anode, electrolyte, or cathode (SA, SE, and SC). The temperature distribution is determined by taking into account only the largest heat source due to ohmic overpotential loss resulting from the Joule effect. The temperature values are obtained using a program in FORTRAN language which is based on the method of three-dimensional finite difference. The three-dimensional numerical study result analysis shows the localization of the highest temperature value at the SOFC component's specific area: cathode (C), electrolyte (E), anode (A), and interconnector.

Keywords SOFC • Temperature • Heat source • Three-dimensional • FORTRAN

Nomenclature

Roman Letters

i Current density, A/m^2
L_x Width of the cell along the axis (X), μm
L_y Thickness of the cell along the axis (Y), μm

Y. Sahli (✉)
Unité de Recherche en Energies Renouvelables en Milieu Saharien, URERMS,
Centre de Développement des Energies Renouvelables, CDER 01000, Adrar, Algeria

Department of Mechanical, Faculty of Technology, University Hadj Lakhder, Batna, Algeria
e-mail: sahli.sofc@gmail.com; H2SOFC@gmail.com

B. Zitouni
Department of Food Technology, Institute of Veterinary Sciences and Agricultural Sciences, University Hadj Lakhder, Batna, Algeria

Department of Mechanical, Faculty of Technology, University Hadj Lakhder, Batna, Algeria

H. Ben Moussa • H. Abdenebi
Department of Mechanical, Faculty of Technology, University Hadj Lakhder, Batna, Algeria

I. Dincer et al. (eds.), *Progress in Clean Energy, Volume 1*,
DOI 10.1007/978-3-319-16709-1_32

L_z Length of the cell along the axis (Z), μm
S Heat source, W/m^3
T Temperature, K

Greek Letters

λ Thermal conductivity, W/m K
σ Electric conductivity, 1/Ω m

Subscripts

an Anode
cat Cathode
elec Electrolyte
in Interconnector
max Maximum
ohm Ohmic

Abbreviations

SA Supported anode
SC Supported cathode
SE Supported electrolyte

32.1 Introduction

Several previous works were done about this subject. Kakaça et al. [1] publish a large literature review about planar or tubular SOFC physical models. They used 120 references. It contains electrochemical models, gas dynamics, and mass and heat transfer. The literature review conducted by the authors includes all heat transfer modes.

Inui et al. [2] have examined in detail the influence of the hydrogen and carbon monoxide ratio at the planar SOFC performance using three-dimensional numerical simulation code. Chaisantikulwat et al. [3] studied also presented a three-dimensional model of planar SOFC-SA. It provides the polarization curve (potential/current density), mole fractions fields, velocity, temperature, species concentrations, and the current distribution at the cell. The model is based on a conductive heat transfer in the solid part (electrolyte and interconnector) and a heat transfer by convection and conduction at the porous part (electrodes). They use an effective thermal conductivity. The documents authors [3, 4], are indicated that the studies are carried out by unsteady-state models.

Concerning the heat source type, Yang et al. [4] described a three-dimensional steady simulation tool, which neglect the radiative transfer for the planar SOFC-SA. The heat source is a three-source sum: the source due to electrochemical reactions, activation overvoltage, and ohmic source. The described tool is intended to simulate more parameters. The temperature distributions, species changes, and current densities are calculated under different conditions. Ho et al. [5] studied a planar SOFC with supported anode (SA) with an unsteady three-dimensional numerical model. It contains the equations describing the mass, species, and energy transport of the chemical and electrochemical processes. The model ignores also the radiation heat transfer. The heat source is similar except the chemical heat source. Equation solving is made by the STAR-CD software. The authors Khaleel et al. [6] proposed a three-dimensional tool simulation for the planar SOFC-SA. The tool combines a finite element analysis code with a robust electrochemical module internally developed. It is based on the characteristics obtained experimentally. It calculates the current density distribution, heat production, and the species (fuel and oxidant) concentrations.

Sun and Ou [7] studied and evaluated the channel design influence on a planar SOFC (ES) single cell at a three-dimensional and unsteady model that neglected the radiation transfer. They have compared the results of three different oxidant compositions: 100 % O_2, 50 % N_2, and 50 % air O_2. Also, a comparison was made between the temperature conditions, adiabatic or isothermal, applied to the outer limit of the SOFC.

Wang et al. [8] presented a three-dimensional mathematical model to simulate the electrochemical characteristics and the multispecies heat transfer in (SOFC) planar steady state. The equations governing continuity, mass, energy, and the electric charge conservation are solved simultaneously by the finite volume method. Polarization activation, Joules effect, and concentration overpotential are considered the main heat sources.

At the present work which is based on our previous works [9–13], our interest is focused on the collection and analysis of distributions and temperature profiles in the case of a three-dimensional geometry which recognizes only the Joule effect, the heat source present in all the SOFC components, cathode, electrolyte, anode, channels, and interconnectors.

32.2 Mathematical Modeling

32.2.1 Area Calculation

The computational domain is limited to a SOFC single cell that consists of two gas flow channels, anode and cathode, the electrolyte, and the interconnectors, Fig. 32.1.

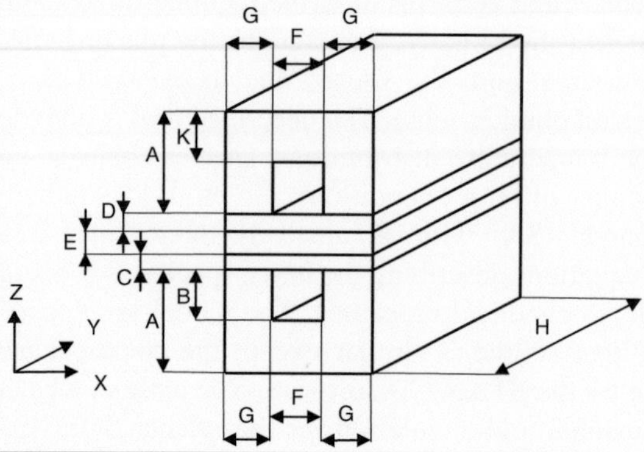

	AS (μm)	ES (μm)	CS (μm)
A	3000	3000	3000
B	2000	2000	2000
C	50	50	2000
D	700	50	150
E	10	180	40
F	2000	2000	2000
G	1000	1000	1000
H	30000	30000	30000
K	1000	1000	1000

Fig. 32.1 SOFC single-cell dimension

32.2.2 *Simplifying Assumptions*

It is clear that the assumption's introduction is necessary to simplify the three-dimensional equation system and make the problem less difficult. In this work, the following assumptions quote:

- A three-dimensional model and stationary computing.
- A continuous medium.
- Components are homogeneous and isotropic.
- The channel entrance fluid temperature is constant.
- The gas electrical conductivity is negligible.
- The channel flow velocity is assumed very low.
- Heat transfer by radiation is neglected.
- A constant current density for all the SOFC components.

32.2.3 *Heat Transfer Equations*

Energy is transported by conduction in the cell, the effects of the Joule effect losses are taken into account by an additional term in the source of the energy balance equation, the three-dimensional equation of energy transport is given by:

$$\frac{\partial}{\partial x}\left\{\lambda\frac{\partial T}{\partial x}\right\}+\frac{\partial}{\partial y}\left\{\lambda\frac{\partial T}{\partial y}\right\}+\frac{\partial}{\partial z}\left\{\lambda\frac{\partial T}{\partial z}\right\}+S=0 \tag{32.1}$$

Table 32.1 Electrical conductivity of the SOFC components

Components	σ [Ω^{-1} cm^{-1}]	References
Electrolyte	$\sigma_{ele} = 3.34 \times 10^4 \exp(-10,300/T)$	[3, 5, 10, 12–19]
Cathode	$\sigma_{cat} = \frac{4.2\times10^7}{T}\exp(-1,200/T)$	[10, 12, 13, 15]
Anode	$\sigma_{an} = \frac{9.5\times10^7}{T}\exp(-1,150/T)$	[3, 10, 12–16, 18, 19]
Interconnector	$\sigma_{in} = \frac{9.3\times10^5}{T}\exp(-1,100/T)$	[13, 15]

32.2.4 *The Heat Source Term*

The heat source that exists in the solid parts of the SOFC is due to ohmic losses and is expressed as follows:

$$S = S_{\text{Ohm, in}} = S_{\text{Ohm, elec}} = S_{\text{Ohm, an}} = S_{\text{Ohm, ca}} = \frac{i^2}{\sigma} \tag{32.2}$$

The heat source due to the Joule effect (ohmic losses) depends on the ratio of the current divided by the square of the electrical conductivity of each point of the solid cell. The electric conductivity varies from a material to another (Table 32.1).

Energy is transported by convection and conduction at the anode and the cathode. The gas velocities in the SOFC channels are very small. Therefore, the convective term is neglected, and then we consider the conduction mode transfer at gas channels, which gives the following equation of energy conservation:

$$\frac{\partial}{\partial x}\left\{\lambda\frac{\partial T}{\partial x}\right\} + \frac{\partial}{\partial y}\left\{\lambda\frac{\partial T}{\partial y}\right\} + \frac{\partial}{\partial z}\left\{\lambda\frac{\partial T}{\partial z}\right\} = 0 \tag{32.3}$$

32.2.5 *Boundary Conditions*

On all external surfaces and all directions, the boundary conditions are of Neumann type, except at the entrance channel where we take a constant temperature with the Dirichlet conditions that are as shown in Fig. 32.2 and Table 32.2.

32.2.6 *Computation Program*

The equations are partial second-order differential type (parabolic) with constant coefficients. They have nonlinear source terms. Due to the complexity of these equations, they cannot be solved analytically. However, they can be solved numerically. The finite difference method is used. The various data used are listed in Tables 32.2 and 32.3.

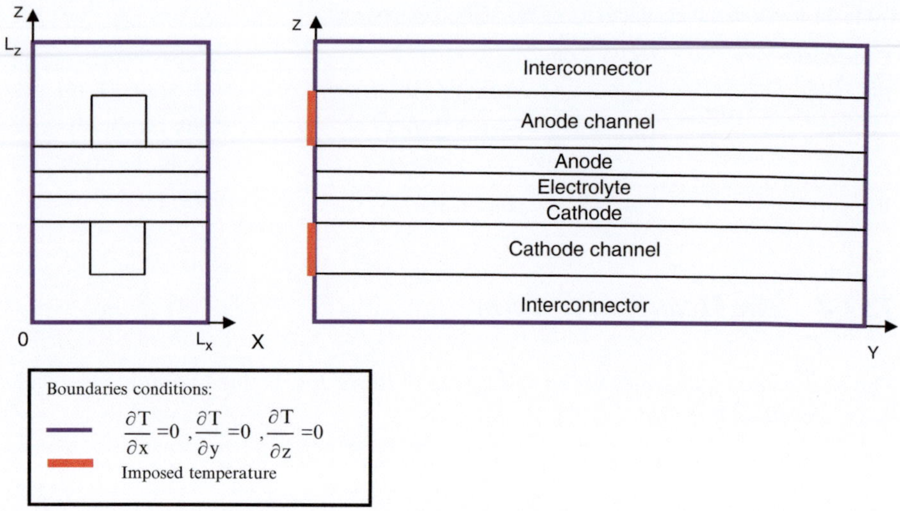

Fig. 32.2 Boundary conditions of a SOFC elementary cell

Table 32.2 Boundary conditions

	Axis OX		Axis OY		Axis OZ	
Components	$(0, y, z)$	(Lx, y, z)	$(x, 0, z)$	(x, Ly, z)	$(x, y, 0)$	(x, y, Lz)
Electrolyte	$\frac{\partial \Phi}{\partial x}=0$	$\frac{\partial \Phi}{\partial x}=0$	$\frac{\partial \Phi}{\partial y}=0$	$\frac{\partial \Phi}{\partial y}=0$	$\frac{\partial \Phi}{\partial z}=0$	$\frac{\partial \Phi}{\partial z}=0$
Cathode	$\frac{\partial \Phi}{\partial x}=0$	$\frac{\partial \Phi}{\partial x}=0$	$\frac{\partial \Phi}{\partial y}=0$	$\frac{\partial \Phi}{\partial y}=0$	$\frac{\partial \Phi}{\partial z}=0$	$\frac{\partial \Phi}{\partial z}=0$
Anode	$\frac{\partial \Phi}{\partial x}=0$	$\frac{\partial \Phi}{\partial x}=0$	$\frac{\partial \Phi}{\partial y}=0$	$\frac{\partial \Phi}{\partial y}=0$	$\frac{\partial \Phi}{\partial z}=0$	$\frac{\partial \Phi}{\partial z}=0$
Interconnector	$\frac{\partial \Phi}{\partial x}=0$	$\frac{\partial \Phi}{\partial x}=0$	$\frac{\partial \Phi}{\partial y}=0$	$\frac{\partial \Phi}{\partial y}=0$	$\frac{\partial \Phi}{\partial z}=0$	$\frac{\partial \Phi}{\partial z}=0$
Anode channel	$\frac{\partial \Phi}{\partial x}=0$	$\frac{\partial \Phi}{\partial x}=0$	$T=T_{H_2}=C^{ste}$	$\frac{\partial \Phi}{\partial y}=0$	$\frac{\partial \Phi}{\partial z}=0$	$\frac{\partial \Phi}{\partial z}=0$
Cathode channel	$\frac{\partial \Phi}{\partial x}=0$	$\frac{\partial \Phi}{\partial x}=0$	$T=T_{air}=C^{ste}$	$\frac{\partial \Phi}{\partial y}=0$	$\frac{\partial \Phi}{\partial z}=0$	$\frac{\partial \Phi}{\partial z}=0$

Table 32.3 Thermal conductivity values

Components	Thermal conductivity λ (W m^{-1} K^{-1})	Material	References
Anode	5.84	Ni-YSZ	[2, 20–22]
Cathode	4	$La_{1-x}Sr_xMnO_3$	[5, 9, 13, 21, 23–26]
Electrolyte	2.16	YSZ	[2, 20–22]
Interconnector	6	$LaCrO_3$	[1, 27]

32.3 Results and Discussion

The temperature values obtained at a three-dimensional SOFC are obtained by inlet gas channel temperature values equal to 1,023 K [3, 5, 14, 17] and an imposed current density equal to 20,000 A/m^2 [5, 12, 17, 28, 29]. The sizes are taken from

[14], for a cell having a supported anode (SA). The case with a supported electrolyte (SE) is referred to [27]. Finally, [5, 17, 30, 31] are as a reference for a SOFC with supported cathode (SC).

As first results, the cells having an anode supported or electrolyte supported or cathode supported exhibit almost the same shape of the field, that is to say, the maximum value of the temperature is the output of the SOFC cell regardless of its geometric configuration (case of heat production).

From the results obtained, it is noted that the elevation of the temperature values is important in case of a cell having an SC relative to the other geometric configurations; it is greater than 1,220 K (Fig. 32.5).

Based on the inlet temperature value of the two gases, it is noted that the increasing maximum values of the cell temperature that have SC are displayed by a temperature difference of about 180 ten units (Fig. 32.5). The temperature value evolution compared with the inlet temperature is almost for 180 K for SE cell configuration (Fig. 32.4). The low temperature elevation value is noted at the SOFC SA. It is in order of 18 K (Fig. 32.3). The maximum values of the temperature are situated in different locations in the input of the stack. According to each geometric configuration, the temperature maximum values and their locations are shown in Table 32.4.

Moreover, it is noted that the upper portion of the interconnector (anode portion) is hotter relative to the lower portion of the interconnector (cathode portion) for the SA and SE (Figs. 32.3 and 32.4); besides in the cell by SC (Fig. 32.5), is the totally opposed to cases presented by (Figs. 32.3 and 32.4). It means that the cathode portion is hotter than the anode portion (Fig. 32.5). The temperature increase for each configuration in the direction (OY) and positions along the axis (OX) ($x = 0$ and $x = 0.5x_{max}$) to the output of the stack is indicated in Fig. 32.6.

Note that the temperature variation value range according to the (OY) direction at positions ($x = 0$ and $x = 0.5x_{max}$) located at the outlet of the cell is nearly of the order of five units for every SOFC geometric configuration (Fig. 32.6).

At the SA configuration, the cell middle is totally hotter than the terminals (left and right). Also, it is observed that the anode middle part is hotter than the terminals (left and right) situated in the same part of the cell. The contrary is presented in the other cases (Fig. 32.6).

As for the case of an SE geometric configuration, Fig. 32.3 shows that the cell lower part (cathode, cathode and interconnector channel) situated at the stack middle is hotter than the terminals (left and right) situated in the cell same part.

Finally, in the case of a cell with SC, Fig. 32.6 shows that the lower stack part (cathode, cathode and interconnector channel) is hotter than the upper part (anode, anode and interconnector channel). Contrary, in other cases, the lower stack part (cathode, interconnector and cathode channel) is cooler than the upper stack part (anode, anode and interconnector channel). It has to be observed that the middle of that cell is fully colder than its terminals (left and right), while the middle of the lower part of the cell is cooler than the wings (left and right) situated in the same stack portion which gives us perfectly the opposite in the other cases (Fig. 32.6).

According to the curves shown in Fig. 32.6, cells which possess a SE or an SA, show clearly that the maximum difference of the temperature values on the axis

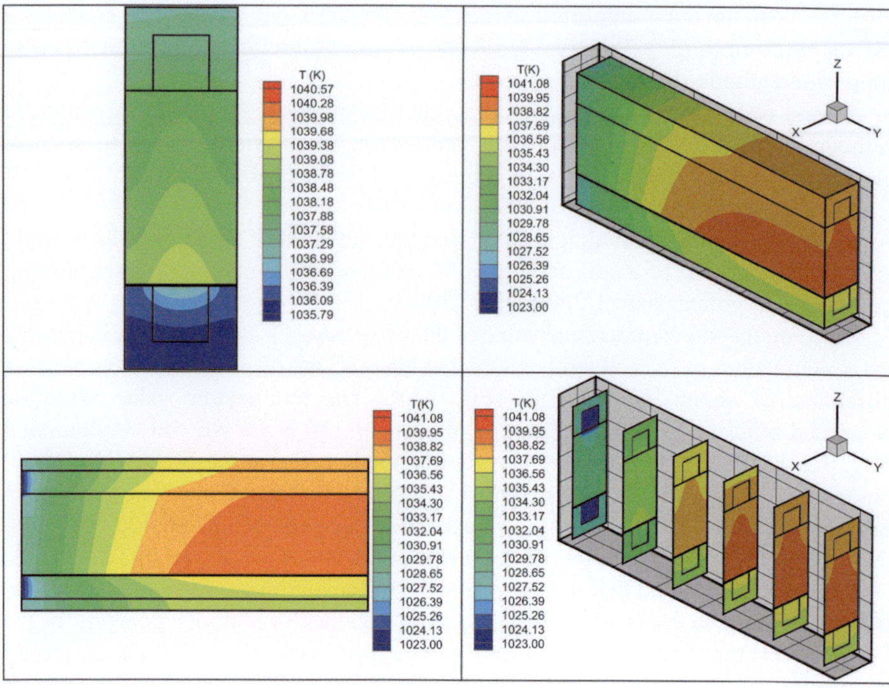

Fig. 32.3 Temperature field in a cell with SA

Table 32.4 Maximum temperature at the cell output according to each geometric configuration

Cell type	T_{max} (K)	Location of T_{max}
SA	1,041.08	The electrolyte and the anode
SE	1,199.10	The upper part of the electrolyte and the lower part of the anode
SC	1,220.47	The electrolyte and the upper part of the cathode

(OY) to the cell output, is situated in the cell middle, the opposite case is logically shown in the case of a cell with cathode supported, that is to say, in the case of SOFC-SC, the maximum difference of the temperature values on the axis (OY) to the cell output, is situated on both sides, left and right, of the cell.

Figure 32.7 shows the temperature evolution for each configuration according to the axis (OZ) from the input to the output of the stack for the position ($y = 0.5y_{max}$ and $x = 0.5x_{max}$).

The temperature increase is remarkable at the cell beginning in all cell configuration types which is related to the gas inlet temperature value. At the cell entrance, in the case of an SA, a cell with the value of the temperature remains almost constant.

Then, starting from the last values of the temperature, the increase is hyperbolical. According to an SA configuration as shown in Fig. 32.7, at the distance from 17 mm from the entrance, the temperature reaches almost its final value at the output of the cell (no temperature increase values in the direction (OZ), after all

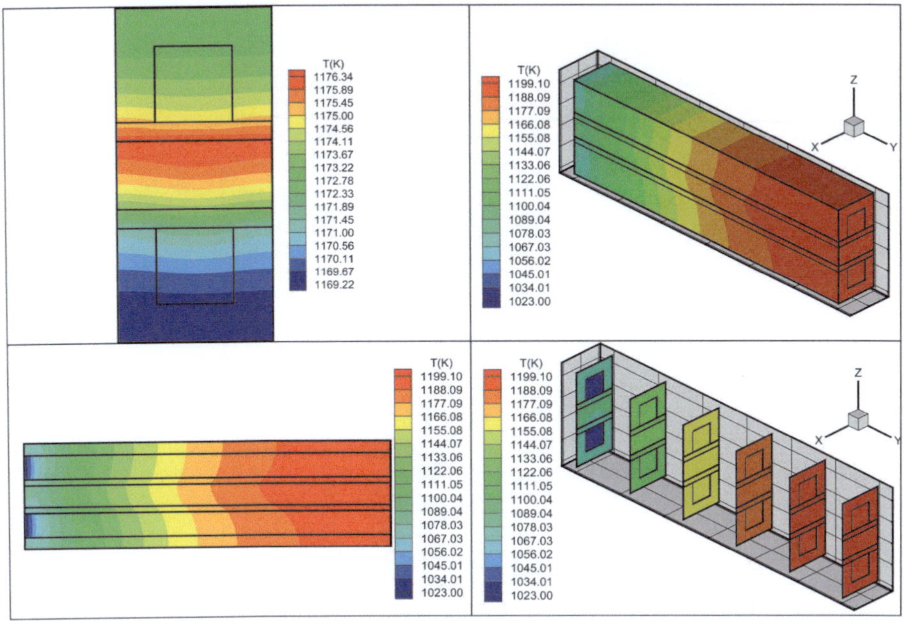

Fig. 32.4 Temperature field in a cell with SE

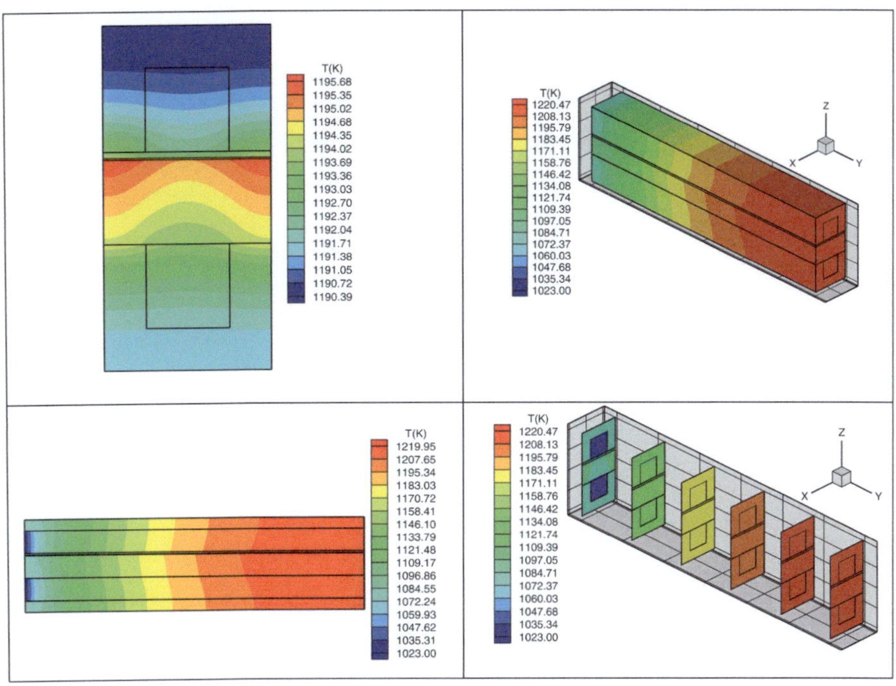

Fig. 32.5 Temperature field in a cell with SC

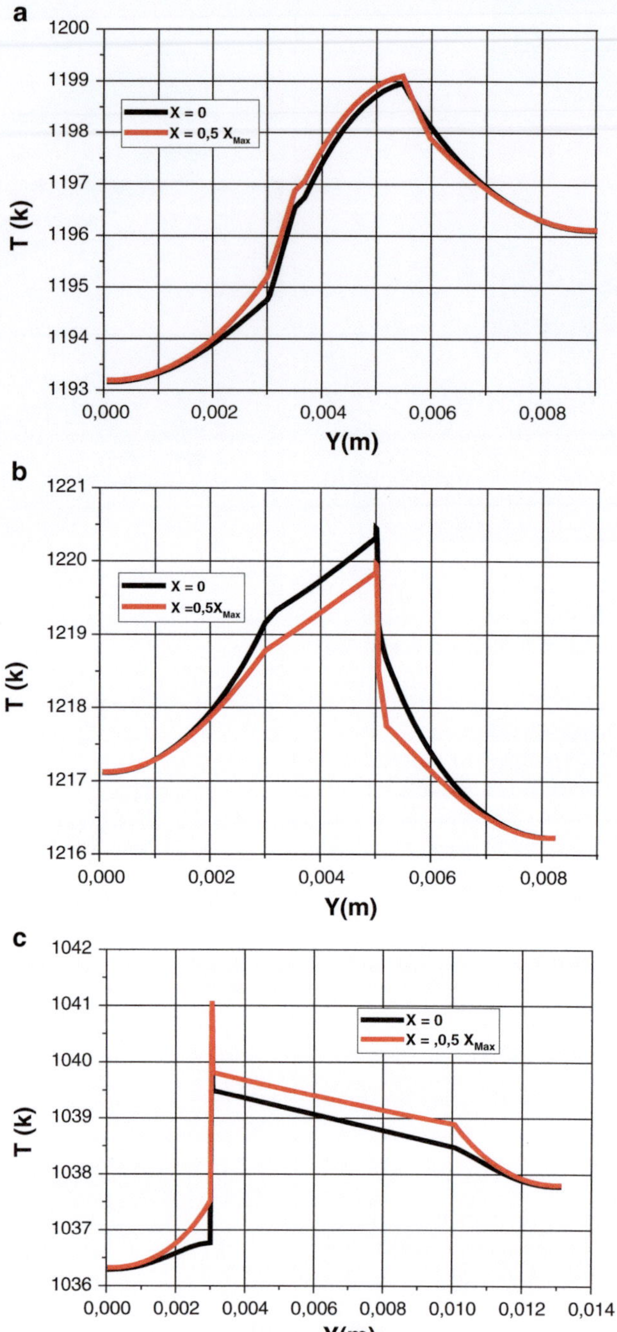

Fig. 32.6 Temperature evolution along the axis (OY) at ($x = 0$, $x = 0.5x_{\max}$) to the stack output and for several configurations. SE, SC, SA

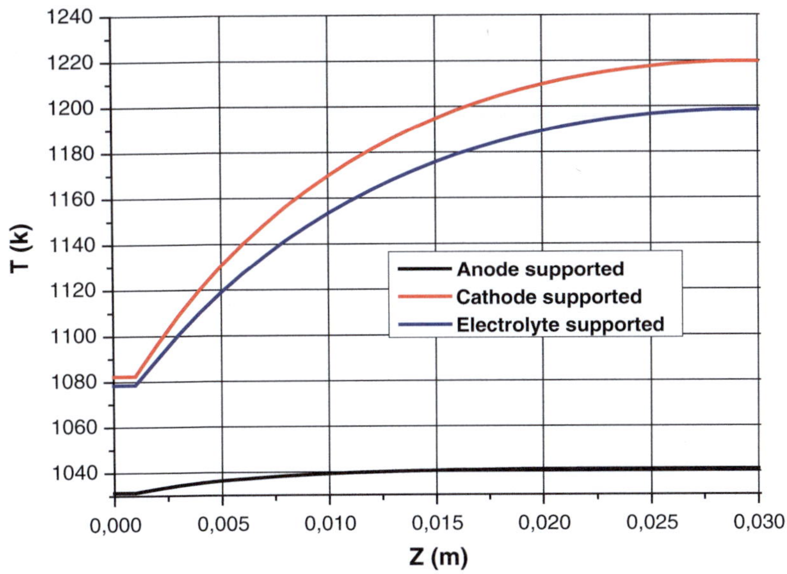

Fig. 32.7 Temperature evolution along the axis (OZ) at ($y = 0.5y_{max}$, $x = 0.5x_{max}$) to the stack output and for several configurations

superior distances of 17 mm). For the other geometric configurations, the temperature approaches its final value beyond a length of 3 in.

32.4 Conclusions

This heat transfer study evaluates and locates the values of maximum and minimum temperatures in an elementary SOFC by a three-dimensional and stationary model that recognizes only the Joule effect as a heat source.

The results show that the temperature elevation values are the largest in the case of a cell having a SC relative to other geometries. The low values are that of the cell having an SA. The highest temperature values are located in the electrolyte and at the upper part of the wings to the cathode of the cell in case of a cell having an SC. Contrarily, their localizations are at the electrolyte and at the anode lower part at the cell middle in the other case.

Based on the obtained result analysis, it appeared that the model studied for the SOFC stack, planar type, allowed us to understand the geometric configuration effect at fields and profile temperature values in all the elementary planar cell SOFC parts.

References

1. Kakaça S, Pramuanjaroenkijb A, Zhoub XY (2007) A review of numerical modeling of solid oxide fuel cells. Int J Hydrogen Energy 32:761–786
2. Inui Y, Urata A, Ito N, Nakajima T, Tanaka T (2006) Performance simulation of planar SOFC using mixed hydrogen and carbon monoxide gases as fuel. Energy Convers Manag 47:1738–1747
3. Chaisantikulwat A, Diaz-Goano C, Meadows ES (2008) Dynamic modelling and control of planar anode-supported solid oxide fuel cell. Comput Chem Eng 32:2365–2381
4. Yang Y, Wang G, Zhang H, Xia W (2007) Computational analysis of thermo-fluid and electrochemical characteristics of MOLB-type SOFC stacks. J Power Sources 173:233–239
5. Ho TX, Kosinski P, Hoffmann AC, Vik A (2009) Numerical analysis of a planar anode-supported SOFC with composite electrodes. Int J Hydrogen Energy 34:3488–3499
6. Khaleel MA, Lin Z, Singh P, Surdoval W, Collin D (2004) A finite element analysis modeling tool for solid oxide fuel cell development: coupled electrochemistry, thermal and flow analysis in MARC. J Power Sources 130:136–148
7. Sun CL, Ou HC (2008) Numerical characterization of a microscale solid-oxide fuel cell. J Power Sources 185:363–373
8. Wang G, Yang Y, Zhang H, Xia W (2007) 3-D model of thermo-fluid and electrochemical for planar SOFC. J Power Sources 167:398–405
9. Zitouni B, Ben-Moussa H, Oulmi K, Saighi S, Chetehouna K (2009) Temperature field, H_2 and H_2O mass transfer in SOFC single cell: electrode and electrolyte thickness effects. J Hydrogen Energy 34:5032–5039
10. Abdenebi H, Zitouni B, Haddad D, Ben-Moussa H, George MA, Abdessemed S (2011) SOFC fuel cell heat production: analysis. Energy Procedia 6:643–650
11. Ben-Moussa H, Zitouni B, Oulmi K, Mahmah B, Belhamel M, Mandin P (2009) Hydrogen consumption and power density in a co-flow planar SOFC. J Hydrogen Energy 34:5022–5031
12. Haddad D, Abdenebi H, Zitouni B, Ben-Moussa H, Oulmi K (2013) Thermal field in SOFC fed by hydrogen: inlet gases temperature effect. J Hydrogen Energy 38:8575–8583
13. Oulmi K, Zitouni B, Ben-Moussa H, Abdenebi H, Andreadi GM (2011) Total polarization effect on the location of maximum temperature value in planar SOFC. J Hydrogen Energy 36:4236–4243
14. Chan SH, Ho HK, Tian Y (2002) Modelling of simple hybrid solid oxide fuel cell and gas turbine power plant. J Power Sources 109:111–120
15. Ni M, Leung MKH, Leung DYC (2007) Parametric study of solid oxide fuel cell performance. Energy Convers Manag 48:1525–1535
16. Chung TD, Hong WT, Chyou YP, Yu DD, Lin KF, Lee CH (2008) Efficiency analyses of solid oxide fuel cell power plant systems. Appl Therm Eng 28:933–941
17. Aguiar P, Adjiman CS, Brandon NP (2004) Anode-supported intermediate temperature direct internal reforming solid oxide fuel cell. I: Model-based steady-state performance. J Power Sources 138:120–136
18. Zhang X, Li G, Li J, Feng Z (2007) Numerical study on electric characteristics of solid oxide fuel cells. Energy Convers Manag 48:977–989
19. Zinovik I, Poulikakos D (2009) Modeling the temperature field in the reforming anode of a button-shaped solid oxide fuel cell. Electrochim Acta 54:6234–6243
20. Chyou YP, Chung TD, Chen JS, Shie RF (2005) Integrated thermal engineering analyses with heat transfer at periphery of planar solid oxide fuel cell. J Power Sources 139:126–140
21. Ramakrishna PA, Yang S, Sohn CH (2006) Innovative design to improve the power density of a solid oxide fuel cell. J Power Sources 158:378–384
22. Damm DL, Fedorov AG (2006) Reduced-order transient thermal modeling for SOFC heating and cooling. J Power Sources 159:956–967

23. Mahcene H, Ben-Moussa H, Bouguettaia H, Bechki D, Babay S, Meftah MS (2011) Study of species, temperature distributions and the solid oxide fuel cells performance in a 2-D model. J Hydrogen Energy 36:4244–4252
24. Otaa T, Koyamaa M, Wena C, Yamadab K, Takahashia H (2003) Object-based modeling of SOFC system: dynamic behavior of micro-tube SOFC. J Power Sources 118:430–439
25. Petruzzi L, Cocchi S, Fineschi F (2003) A global thermo-electrochemical model for SOFC systems design and engineering. J Power Sources 118:96–107
26. Suwanwarangkul R, Croiset E, Pritzker MD, Fowler MW, Douglas PL, Entchev E (2006) Mechanistic modelling of a cathode-supported tubular solid oxide fuel cell. J Power Sources 154:74–85
27. Sangtongkitcharoen W, Vivanpatarakij S, Laosiripojana N, Arpornwichanop A, Assabumrungrat S (2008) Performance analysis of methanol-fueled solid oxide fuel cell system incorporated with palladium membrane reactor. Chem Eng J 138:436–441
28. Wongchanapai S, Iwai H, Saito M, Yoshida H (2012) Selection of suitable operating conditions for planar anode-supported direct-internal-reforming solid-oxide fuel cell. J Power Sources 204:14–24
29. Wang Y, Yoshiba F, Watanabe T, Weng S (2007) Numerical analysis of electrochemical characteristics and heat/species transport for planar porous-electrode-supported SOFC. J Power Sources 170:101–110
30. Zhou L, Cheng M, Yi B, Dong Y, Cong Y, Yang W (2008) Performance of an anode-supported tubular solid oxide fuel cell (SOFC) under pressurized conditions. Electrochim Acta 53:5195–5198
31. Wang L, Zhang H, Weng S (2008) Modeling and simulation of solid oxide fuel cell based on the volume–resistance characteristic modeling technique. J Power Sources 177:579–589

Chapter 33
Modeling of Heat Transfer in the PEMFC: Velocity Inlet and Current Density Effect

Djamel Haddad, Kafia Oulmi, Hocine Benmoussa, Zeroual Aouachria, and Sahli Youcef

Abstract The purpose of this work is to present a two-dimensional transient model of the gas flow in the fuel cell (PEMFC). The model includes various conservation equations such movement and energy equations. The governing equations were resolved by the finite volume method. The objective of this work is to know the maximum temperature and its location in PEMFC and to determine the performance conditions of the fuel cell under the current density and velocity inlet effect. The polarization curve obtained numerically is compared with much numerical work. The numerical results show the regime flow effect and the nature of porous middle on the gas distribution in the membrane electrode assembly (MEA).

Keywords Temperature • Flow laminar • Finite volume method • PEMFC

Nomenclature

C_P	Specific heat capacity, J/kg K
C_K	Molar concentration, mol/m^3
D_K	Effective diffusivity of species K, m^2/s
E_{Nernst}	Ideal potential, V
e_{m}	Membrane thickness, m
F	Faraday constant, 96,487 C/mol
I	Current density, A/cm^2
i_{Max}	Limit current density, A/cm^2
J_{a}	Transfer current anode, A/cm^3
J_{c}	Transfer current cathode, A/cm^3
P	Pressure, Pa

D. Haddad (✉) • H. Benmoussa • S. Youcef
Laboratory (LESEI), Faculty of Engineering, University of Batna, Batna, Algeria
e-mail: djamel_hd2@yahoo.fr

K. Oulmi • Z. Aouachria
Laboratory (LPEA), Faculty of Science, University of Batna, Batna, Algeria

I. Dincer et al. (eds.), *Progress in Clean Energy, Volume 1*,
DOI 10.1007/978-3-319-16709-1_33

R Gas constant, J/mol K
S_C Source terms in the species equation
S_T Source terms in the species equation
T Temperature, K
t Time, s
U Velocity vector, m/s
V_{Cell} Real potential, V

Greek Letters

ρ Density of the gas, kg/m^3
ε Porosity
η_{act} Activation polarization of the anode and the cathode, V
η_{ohm} Ohmic polarization, V
η_{conc} Concentration polarization of the anode and the cathode, V
μ Dynamic viscosity, kg/m s

Subscripts

a Anode
c Cathode
m Membrane

33.1 Introduction

Proton exchange membrane fuel cells (PEMFCs) are efficient and environmentally clean electrical generators that are being developed for both stationary and mobile applications. One of the fuel cell interests is that the temperatures are one more low level than in turbines or combustion engines [1–3]. Over the past decade, a significant research effort of PEMFC has been focused on fuel cell structure design, the development of better catalyst, the better structure of catalyst layer and gas diffusion layer, and the development of membrane with high performance, thermal, and water management [4]. Within the literature, there are many types of fuel cell models. In some cases, these models are aimed at examining specific phenomena, such as water management and flooding. In general, it is possible to categorize all models as belonging to one of two categories: partial cell models or complete cell models. The one-dimensional model by Genevey et al. [5] is an example of the partial cell models that include only the membrane. In this work, transport of protons, dissolved water, and energy are modeled for this element. The work of Wang and Wang [6] is an example of the complete cell models that include all or nearly all of the components of the fuel cell. In this case, the transport of species and protons is included in a two-dimensional, down-the-channel geometry.

An evaluation of temperature gradient in PEMFC is important for its functioning and change in the conductivity of the membrane. For this aim of the present work was to develop a two-dimensional model transient that describes the flow laminar of the reactants in the fuel cell. The model includes various conservation equations such movement and energy equations.

The model is used to present the transition from the continued medium (channel) toward the porous medium (MEA) as well as the current density and velocity inlet effect on the temperature field.

33.2 Mathematical Model

Figure 33.1 schematically shows a PEMFC fuel cell divided into seven subregions: the anode gas channel, gas diffusion anode, anode catalyst layer, membrane, cathode catalyst layer, gas diffusion cathode, and cathode gas channel. The present model considers the anode channel feed consisting of hydrogen and water vapor, whereas humidified oxygen is fed into the cathode channel. Hydrogen oxidation and oxygen reduction reactions are considered to occur only within the active catalyst layers which can be described by

Hydrogen oxidation reactions:

$$H_2 \rightarrow 2H^+ + 2e^-$$

Oxygen reduction reactions:

$$2H^+ + 2e^- + \frac{1}{2}O_2 \rightarrow H_2O$$

In this model, the mathematical equations of transport, the source terms, and the closure relations are presented. The shape of the equations can change for one

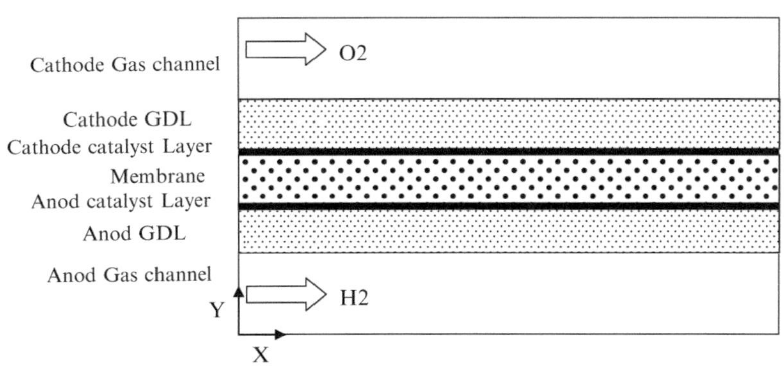

Fig. 33.1 Schema of proton exchange membrane fuel cell (PEMFC)

Table 33.1 The source terms and the diffusion coefficient for each element of the PEMFC

	S_U	S_V	μ^{eff}	S_T	α^{eff}
Gas channel	0	0	μ	0	α
Gas diffuser	$-\varepsilon_{\text{d}}\frac{\mu}{k_{\text{d}}}U$	$-\varepsilon_{\text{d}}\frac{\mu}{k_{\text{d}}}V$	$r \cdot \mu$	0	k_{eff}/C_P
Catalyst layer	$-\varepsilon_{\text{c}}\frac{\mu}{k_{\text{c}}}U$	$-\varepsilon_{\text{c}}\frac{\mu}{k_{\text{c}}}V$	$r \cdot \mu$	$i \cdot \eta_{\text{act}}/(\rho vcp)$	k_{eff}/C_P
Membrane	$-\varepsilon_{\text{m}}\frac{\mu}{k_{\text{m}}}U$	$-\varepsilon_{\text{m}}\frac{\mu}{k_{\text{m}}}V$	$r \cdot \mu$	$i \cdot \eta_{\text{ohm}}/\rho \cdot cp$	k_{eff}/C_P

component to another of the fuel cell; in other words, every element possesses its own equation. The gas channels are considered as a continuous middle; the flow regime laminar is considered and is governed by the Navier–Stokes equation. The diffusion layer, the catalyst layer, and the membrane are porous middle; the effect of the strength exercised by the solid on the fluid (i.e., Darcy's term) is characterized by a new additional term in the movement equation [4, 6] (Table 33.1):

- Continuity equation:

$$\frac{\partial \rho}{\partial t} + \frac{\partial \rho U}{\partial x} + \frac{\partial \rho V}{\partial y} = 0 \tag{33.1}$$

- Momentum equation:

$$\frac{\partial \rho U}{\partial t} + U\frac{\partial \rho U}{\partial x} + V\frac{\partial \rho U}{\partial y} = -\frac{\partial P}{\partial x} + \frac{\partial}{\partial x}\left(\mu^{\text{eff}}\frac{\partial U}{\partial x}\right) + \frac{\partial}{\partial y}\left(\mu^{\text{eff}}\frac{\partial U}{\partial y}\right) + S_U \tag{33.2}$$

$$\frac{\partial \rho V}{\partial t} + U\frac{\partial \rho V}{\partial x} + V\frac{\partial \rho V}{\partial y} = -\frac{\partial P}{\partial y} + \frac{\partial}{\partial x}\left(\mu^{\text{eff}}\frac{\partial V}{\partial x}\right) + \frac{\partial}{\partial y}\left(\mu^{\text{eff}}\frac{\partial V}{\partial y}\right) + S_V \tag{33.3}$$

- Energy equation:

$$\frac{\partial T}{\partial t} + U\frac{\partial T}{\partial x} + V\frac{\partial T}{\partial y} = \frac{\partial}{\partial x}\left(\alpha^{\text{eff}}\frac{\partial T}{\partial x}\right) + \frac{\partial}{\partial y}\left(\alpha^{\text{eff}}\frac{\partial T}{\partial y}\right) + S_T \tag{33.4}$$

- The correction factor of the porous middle r in the equation of impulse has for expression [6, 7]:

$$r = 2.25\frac{(1-\varepsilon)^2}{\varepsilon^2} \tag{33.5}$$

- The proton conductivity in the membrane phase has been correlated [1, 7]:

$$\sigma_m = \exp\left(1{,}268 \cdot \left(\frac{1}{303} - \frac{1}{T}\right)\right) \cdot (0.5139 \cdot \lambda - 0.326) \tag{33.6}$$

where the water content in the membrane, λ_{H_2O/SO_3^-}, depends on the water molar concentration $C^m_{H_2O}$, according to the following fit of the experimental data [7, 8]:

$$\lambda_{H_2O/SO_3} = \frac{C^{\mathrm{m}}_{H_2O}}{\frac{\rho^{\mathrm{m}}_{\mathrm{sec}}}{M^{\mathrm{m}}} - 0.025 \cdot C^{\mathrm{m}}_{H_2O}} \tag{33.7}$$

where M^{m} is the molecular coefficient of the membrane and $\rho^{\mathrm{m}}_{\mathrm{sec}}$ the density of the dry membrane.

Ohmic losses in the electrolyte obey Ohm's law [9]:

$$\eta_{\mathrm{ohm}} = \frac{i}{\sigma_{\mathrm{m}}} \tag{33.8}$$

Knowing that the activation polarization increases with current density, its expression can be written as [10]

$$\eta_{\mathrm{act}} = -[\xi_1 + \xi_2 \cdot T + \xi_3 \cdot T \cdot \ln(C_{O_2}) + \xi_4 \cdot T \cdot \ln(i)] \tag{33.9}$$

where i is the current density and ξ_1, ξ_2, ξ_3, and ξ_4 are parametric coefficients for each fuel cell model:

$$\xi_1 = -0.948,\ \ \xi_2 = 0.00286 + 0.0002 \cdot \ln(A) + \left(4.3 \cdot 10^{-5}\right)\ln(C_{H_2}),$$
$$\xi_3 = 7.6 \cdot 10^{-5},\ \ \xi_4 = -1.93 \cdot 10^{-4}$$

where C_{O_2} is the oxygen concentration at the interface of the cathode catalytic surface and C_{H_2} is the hydrogen concentration at the interface of the anode catalytic surface; they are expressed by Henry's law as

$$C_{O_2} = \frac{P^*_{\mathrm{c}}}{5.08 \cdot 10^6 \cdot \exp\left(-\frac{498}{T}\right)} \tag{33.10}$$

$$C_{H_2} = \frac{P^*_{\mathrm{a}}}{1.09 \cdot 10^6 \cdot \exp\left(\frac{77}{T}\right)} \tag{33.11}$$

The expression of the Nernst equation according to Maher [11]

$$E_{\mathrm{Nernest}} = 1.229 - 0.85 \cdot 10^{-3} \cdot (T - 298.15) + 4.31 \cdot 10^{-5} \cdot T \cdot \left[\ln\left(P^*_{\mathrm{a}}\right) + \frac{1}{2}\ln\left(P^*_{\mathrm{c}}\right)\right] \tag{33.12}$$

The real potential of a fuel cell decreases compared to the Nernst potential, because of polarization phenomena, which are activation polarization, ohmic polarization, and concentration polarization:

$$V_{\mathrm{cell}} = E_{\mathrm{Nernest}} - \eta_{\mathrm{act}} - \eta_{\mathrm{ohm}} - \eta_{\mathrm{conc}} \tag{33.13}$$

where η_{conc} is the concentration polarization [12]:

$$\eta_{\mathrm{con}} = -\frac{R \cdot T}{2 \cdot F} \ln\left(1 - \frac{i}{i_{\max}}\right) \tag{33.14}$$

33.3 Numerical Method

The solution to the governing equations is performed by employing a finite volume scheme with the model domain divided into a number of cells as control volumes. The governing equations are numerically integrated over each of these computational cells or control volumes (Table 33.2).

The method exploits a collocated cell-centered variable arrangement with the local or cell-averaged values of the physical quantities evaluated and stored at each cell center (Fig. 33.2). The governing equations can be expressed in the form of a generalized convection–diffusion type of transport equation:

Table 33.2 The boundary conditions

	V	U	T	P
I	$V=0$	$U=U_{0,O_2}$	$T=T_{O_2}$	$P=P_{O_2}$
II	$V=0$	$U=U_{0,H_2}$	$T=T_{H_2}$	$P=P_{H_2}$
III	$V=0$	$U=0$	$\frac{\partial T}{\partial x}=0$	$\frac{\partial P}{\partial x}=0$
IV	$V=0$	$U=0$	$\frac{\partial T}{\partial y}=0$	$\frac{\partial P}{\partial y}=0$
V	$V=0$	$\frac{\partial U}{\partial x}=0$	$\frac{\partial T}{\partial x}=0$	$\frac{\partial P}{\partial x}=0$

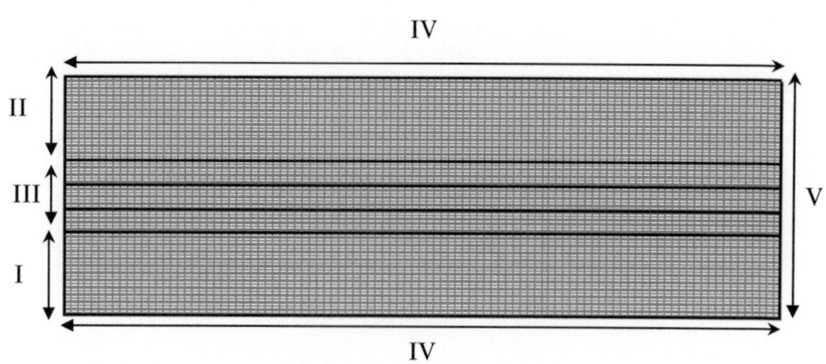

Fig. 33.2 Presentation of the mesh and the boundary conditions

$$\frac{\partial(\rho\phi)}{\partial t}+\frac{\partial(\rho u\phi)}{\partial x}+\frac{\partial(\rho v\phi)}{\partial y}=\frac{\partial}{\partial x}\left(\Gamma\frac{\partial\phi}{\partial x}\right)+\frac{\partial}{\partial y}\left(\Gamma\frac{\partial\phi}{\partial y}\right)+b\cdot\phi+c \quad (33.15)$$

The hybrid differencing scheme is chosen here on the grounds of its stability as the preferred method for the treatment of convection terms, so here we quote the implicit/hybrid difference form of the transient convection–diffusion equations [14, 15].

33.4 Results and Discussions

The mathematical model with various limit conditions illustrates the transport phenomena which occur in the fuel cell. The discretization of the equations governing the heat transfer phenomena in all parts of the fuel cell was made by the finite volume method. We supposed that the flow regime is laminar. The results representing the transient behavior of the fuel cell PEMFC are presented in this study. The current density and velocity inlet is among the important parameters which have a big influence on the functioning of the fuel cell. A computer code was developed in this direction to obtain the temperature profiles.

- *Effect of current density*
  Figures 33.4 and 33.5 show the variation of the temperature field as a function of current density and under the effect of ohmic source. The results also show that there is proportionality between the increase in current density and temperature field. The absence of the electrochemical reaction ($I=0$ cm^2) shows a small scatter of the temperature field, and maximum temperature is located in this case the input channels. The presence of the electrochemical reaction, for higher density currents, leads to a distribution of temperature field away from the input channels of the MEA. The increase in current density gives a higher maximum temperature in the membrane.

- *Effect of velocity inlet*
  The effect of velocity inlet in the fields of temperature is shown as contours and profiles (Figs. 33.6 and 33.7). These figures show that increasing the velocity inlet, the heat is transported from input to output of the fuel cell. The temperature increase is localized in the side downstream of the membrane.

  On the polarization curve Fig. 33.8 obtained numerically, there is a decrease in voltage when the current density increases due to the different polarizations (ohmic, activation, concentration). The power density shows a maximum for an optimum current density. When the velocity inlet is increasing, the power density increases. Model validation involves the comparison of model results with much numerical work.

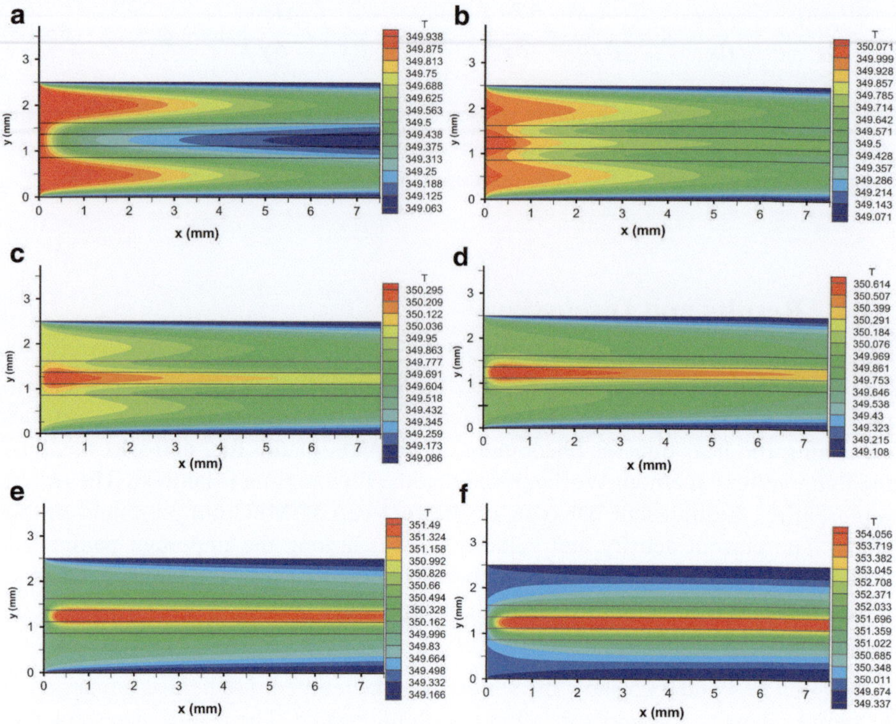

Fig. 33.4 Temperature fields in the PEMFC: current density effect. (**a**) $i = 0\ \mathrm{A/cm^2}$, (**b**) $i = 0.4\ \mathrm{A/cm^2}$, (**c**) $i = 0.5\ \mathrm{A/cm^2}$, (**d**) $i = 0.6\ \mathrm{A/cm^2}$, (**e**) $i = 0.8\ \mathrm{A/cm^2}$, and (**f**) $i = 1\ \mathrm{A/cm^2}$

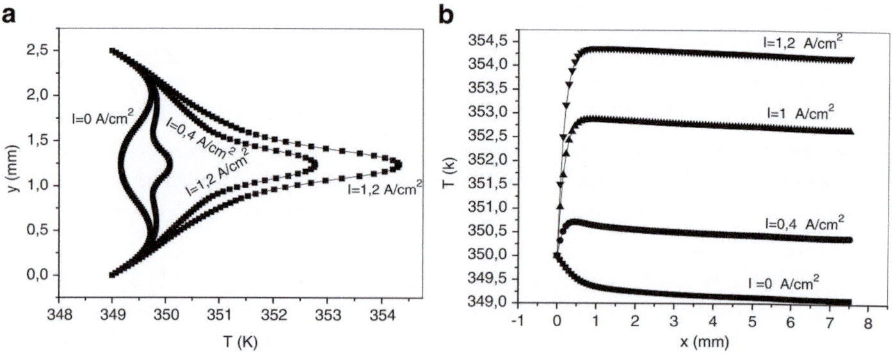

Fig. 33.5 Temperature profile for different current densities. (**a**) In the PEMFC. (**b**) In the membrane

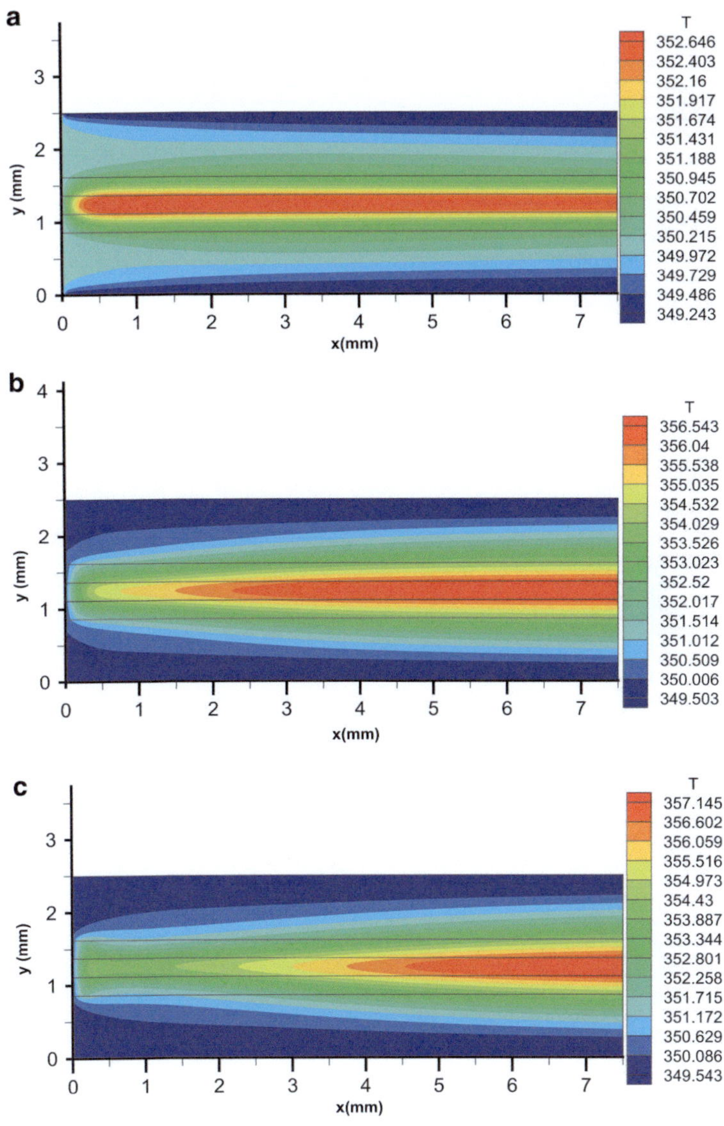

Fig. 33.6 Temperature fields in the PEMFC: velocity inlet effect. (**a**) $V_0 = 0.16\ \text{m/s}$, (**b**) $V_0 = 0.64\ \text{m/s}$, and (**c**) $V_0 = 1.6\ \text{m/s}$

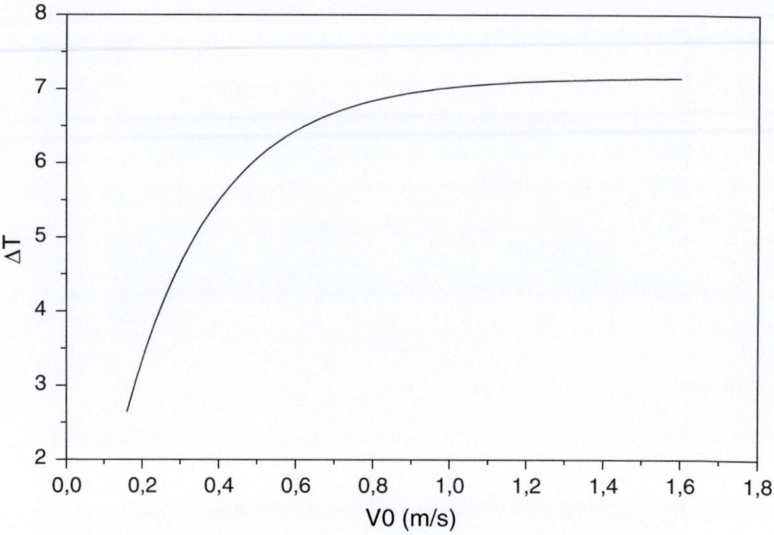

Fig. 33.7 Maximum temperature variation for different velocity inlets

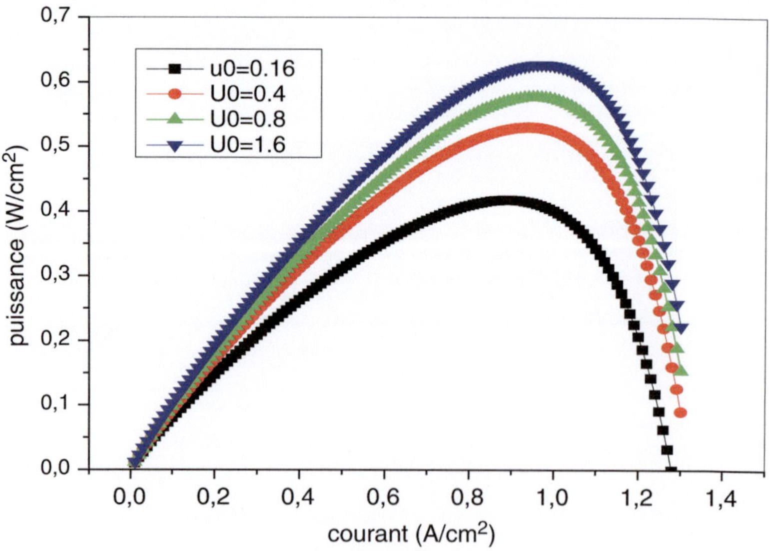

Fig. 33.8 Polarization curves: velocity inlet effect

33.5 Conclusion

Two-dimensional model of the transfer phenomena, heat in the fuel cell, was presented. The flow regime is considered laminar in the channel and the MEA. The results obtained are:

The increase in current density gives a higher maximum temperature in the membrane.

The increase of the velocity inlet leads to the transport of heat from input to output of the fuel cell. When the velocity inlet is increasing, the power density increases.

References

1. Springer TE, Zawodzinski TA, Gottesfeld S (1991) Polymer electrolyte fuel cell model. J Electrochem Soc 138:2334–2341
2. Broka K, Ekdunge P (1999) Modelling the PEM fuel cell cathode. J Appl Electrochem 27:281–289
3. Sui PC, Chen LD, Seaba JP, Wariishi Y (1999) Modeling and optimization of a PEMFC catalyst layer. In: SAE congress, pp 61–70
4. Natarajan D, Nguyen TV (2003) Three-dimensional effects of liquid water flooding in the cathode of a PEM fuel cell. J Power Sources 115:66–80
5. Genevey D, von Spakovsky MR, Ellis MW, Nelson DJ, Olsommer B, Topin F, Montel N, Siegel NP (2002) Transient model of the heat, mass and charge transfer as well as electrochemistry in the catalyst layer. In: PEMFC international mechanical engineering congress and exposition—IMECE'2002, ASME IMECE
6. Wang ZH, Wang CY (2000) Two-phase flow and transport in the interdigitated air cathode of proton exchange membrane fuel cells. Proc ASME Heat Transfer Div 366:27–33
7. Haddad D, Oulmi K, Benmoussa H, Aouachria Z, Bourmada N (2013) Transport phenomena effect on the performance of proton exchange membrane fuel cell (PEMFC). Int J Hydrogen Energy 38:8550–8556
8. Haddad D, Abdenebi H, Zitouni B, Benmoussa H, Oulmi K (2013) Thermal field in SOFC fed by hydrogen: inlet gases temperature effect. Int J Hydrogen Energy 38:8575–8583
9. Bard J, Faulkner E (1980) Électrochimie, principes, méthodes et application. Masson, Paris
10. Mikkola M (2004) Experimental studies on polymer electrolyte membrane fuel cell stacks. Thesis, Helsinki University of Department of Engineering Physics and Mathematic, Finland
11. Sadiq Al-Baghdadi MAR (2005) Modelling of proton exchange membrane fuel cell performance based on semi-empirical equations. Renew Energy 30:1587–1599
12. Chahine R, Laurencelle F, Hamelin J, Agbossou K, Fournier M, Bose TK et al (2001) Characterization of a Ballard MK5-E proton exchange membrane fuel cell stack. Fuel Cells 1:66–71
13. Versteeg HK, Malalasekera W (1995) An introduction to computational fluid dynamics the finite volume method. Longman Scientific & Technical, Harlow, ISBN 0-470
14. Natarajan D, Van Nguyen T (2001) A two-dimensional, two-phase multi component, transient model for the cathode of a proton exchange membrane fuel cell using conventional gas distributors. J Electrochem Soc 148:1324–1335
15. Rhee GH, Sung HJ (2000) Numerical prediction of locally forced turbulent separated and reattaching flow. Fluid Dyn Res 26:421–436

Chapter 34
Modeling the Structure Based on GaAsNBi/GaAs for Solar Cell

A. Aissat, A. Djili, S. Zelazel, and J.P. Vilcot

Abstract This work focuses on modeling and simulation of the structure based on GaAsNBi/GaAs for photovoltaic application. Indeed, the incorporation of a small composition nitrogen N <5 % and bismuth Bi <12 % induces the splitting of the conduction band into two bands and the valence band, respectively. Under the effect of this splitting, there is reduction of the bandgap which is very interesting for increasing the absorption. For $x = 1$ % and $y = 10$ %, the bandgap energy is $E_g = 0.68$ eV and the absorption coefficient equal to 1.15×10^6 cm^{-1} with a yield around 20 %. The use of structures based on new materials allows us to improve the performance of the optical conversion.

Keywords News materials • GaAsNBi/GaAs • Solar cell • Optoelectronics

Nomenclature

N	Nitrogen
Bi	Bismuth
x	Nitrogen concentration
y	Bismuth concentration
E_g	Bandgap
a_e	Lattice parameter of epitaxial layer
a_s	Lattice parameter of substrate
$a_\perp$	Perpendicular lattice parameter
C_{ij}	Elastic stiffness constants
ε_{xx}	Parallel strain
ε_{zz}	Perpendicular strain

A. Aissat (✉) • A. Djili • S. Zelazel
LATSI Laboratory, Faculty of Technology, University of Blida 1, Blida, Algeria
e-mail: sakre23@yahoo.fr

J.P. Vilcot
Institut d'Electronique, de Microélectronique et de Nanotechnologie (IEMN), UMR CNRS 8520, Université des Sciences et Technologies de Lille 1, Avenue Poincaré, BP 60069, Villeneuve d'Ascq 59652, France

I. Dincer et al. (eds.), *Progress in Clean Energy, Volume 1*,
DOI 10.1007/978-3-319-16709-1_34

$E_g^{unstrain}$	Unstrain bandgap
C	Bowing coefficient
E_g^{strain}	Strain bandgap
ΔE_C^{hy}	Hydrostatic conduction energy
$\Delta E_v^{hy}{}_{av}$	Average hydrostatic valence sub-band energy
ΔE_{hh}^{sh}	Shear valence energy of heavy holes
α	Absorption coefficient
α_0	Initial absorption coefficient
E	Incident photon energy
J	Current density
V	Voltage
P	Power
ε	Strain
J_{ph}	Photocurrent density
V_{co}	Open-circuit voltage
FF	Form factor
η	Efficiency

34.1 Introduction

Today, photovoltaic electricity production is widely dominated by silicon with 95 % of installed photovoltaic panels (monocrystalline, polycrystalline, and thin film) [1]. However, these commercial panels suffer from a low average yield, about 13 % (between 12 and 18 % for monocrystalline, between 12 and 15 % for polycrystalline, and between 6 and 9 % for thin films) [2]. Hence, many researches focus on increasing the efficiency of silicon panels (nanostructures, thin layers, and plasmatic) and alternative materials with a higher yield (CdTe,CIGS, GaAs, etc.). Generally, these new materials are deposited in thin layers (a few hundreds of nanometers) but have always interesting yields thus reducing production costs. Among the new thin film materials considered for photovoltaics, GaAsN alloys, due to the giant bowing coefficient [3], provide the rare opportunity among the III–V systems to simultaneously lower both the lattice constant and the bandgap. However, due to the large size mismatch between nitrogen and arsenic, the growth of high-quality GaAsN on GaAs substrate is difficult. In order to overcome this problem, the coalloying approach is proposed. By substituting large atom such as indium on the cation site or antimony on the anion site, this new alloy can be made lattice matched to GaAs and has a bandgap close to (1 eV). Most of the previous efforts focused on the growth of InGaAsN on a GaAs substrate. However, it appears that when the nitrogen concentration increases, the alloy quality deteriorates very fast [4, 5]. Furthermore, experimental observations show that the presence of indium does not affect significantly the incorporation of nitrogen in GaAs [6]. For over two decades, III–V-Bi alloys have attracted interest for potential applications in the infrared spectral region, because of the belief that alloying conventional III–V semiconductors with the semi-metallic III-Bi compounds could lead to low-bandgap materials like GaAsNBi alloy [7].

34.2 Theoretical Approach

$GaAs_{1-x}N_x$ has been studied for more than a decade due to its potential applications for optoelectronic devices. Because of giant bowing coefficient, $GaAs_{1-x}N_x$ alloy simultaneously lowers both the lattice constant and the bandgap. However, due to the large size mismatch between nitrogen and arsenic, the growth of high-quality $GaAs_{1-x}N_x$ on GaAs substrate is difficult. In order to overcome this problem, a coalloying approach has been proposed. By substituting a large-type atom like bismuth, the new alloy GaAsNBi can be made lattice matched to GaAs while mitigating the undesirable effects produced by nitrogen [8]. The coalloying of bismuth (Bi) with nitrogen (N) in GaAs can significantly lower the nitrogen concentration required to reduce the alloy bandgap. The strain compensation between the small-sized N and the large-sized Bi also reduces the alloy formation energies [8]. Several parameters of GaAsNBi alloy can be determined by linear interpolation between the binary parameters of the constituent semiconductor, this is called interpolation Vegard law, and it is given by the following expression for the quaternary $GaAs_{1-x-y}N_xBi_y$ [9].

$$P(x, y) = (1 - x - y) \cdot P_{\mathrm{GaAs}} + x \cdot P_{\mathrm{GaN}} + y \cdot P_{\mathrm{GaBi}} \qquad (34.1)$$

where $P(x, y)$ is the parameter to be determined.

34.2.1 Deformation

The epitaxy of detuned materials became well controlled, especially to the progress in growth techniques. Thus, the lattice matching is not an imperative; moreover the strain effects are exploited in many high-performance electronic and optoelectronic components. Indeed, the use of heterostructure strain adjusts the emission wavelength, by changing the energy positions of the valence band and conduction band. Two separate cases should be considered when studying a heterostructure:

- Adapted lattice parameter is the most favorable case where all materials are epitaxial lattice matching with the substrate, and therefore no dislocation cannot exist.
- Pseudomorphic: the epitaxial materials have different parameters, but not too far. During the growth of a semiconductor layer for which the lattice parameter is a_e different from that of the substrate a_s, the support material imposes its lattice parameter in the plane of layer. Therefore, the epitaxial layer becomes the seat of compressive strain if $(a_e > a_s)$ or tensile if $(a_e < a_s)$, thus resulting a tetragonal deformation of the lattice (Fig. 34.1).

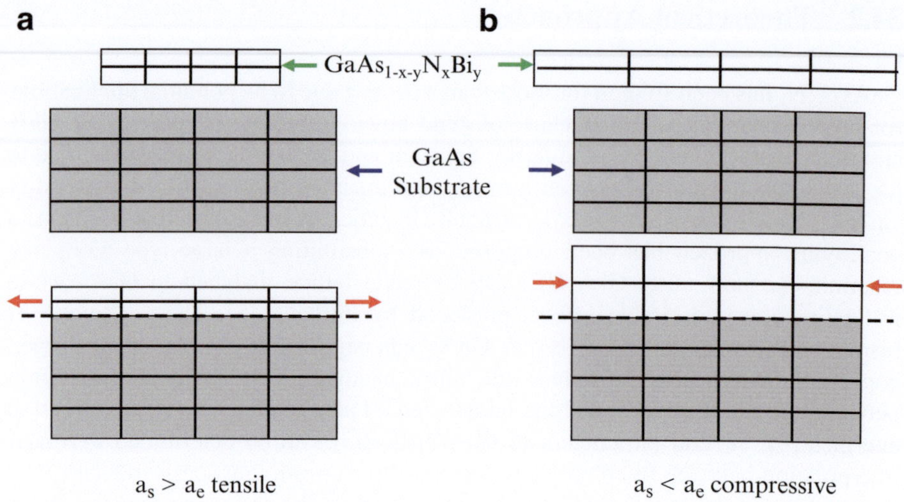

Fig. 34.1 Illustration of the two types of strain: (**a**) tensile strain and (**b**) compressive strain

Considering a semiconductor layer pseudomorphically grown on a (0 0 1) oriented substrate, the adaptation of the lattice parameter of the epitaxial layer with the substrate in the plane of the layer leads to the following strains [10]:

– In the plane of layer,

$$\varepsilon_{xx} = \varepsilon_{yy} = \frac{a_s - a_e}{a_e} = \varepsilon_{//} \tag{34.2}$$

– In the direction of growth,

$$\varepsilon_{zz} = \frac{a_{\perp} - a_e}{a_e} = \varepsilon_{\perp} \tag{34.3}$$

$$\varepsilon_{\perp} = -2\frac{C_{12}}{C_{11}}\varepsilon_{//} \tag{34.4}$$

The necessary condition for a good heteroepitaxial is obviously that the two materials have the same crystal structure. When the lattice parameters are different, the material constituting the large thickness layer imposes its lattice to another, at least in the vicinity of the interface. This leads to the existence in the thin material, a biaxial strain in the plane of the layers [11]. To describe the evolution of the bandgap energy of the unstrained bulk material of $GaAs_{1-x-y}N_xBi_y$ depending on the nitrogen and bismuth composition, we used linear interpolation law which is described by Chuang [11]:

Table 34.1 Material parameters

E_g (GaAs)	E_g (GaN)	E_g (GaBi)	C_{GaAsN}	C_{GaAsBi}	$C_{GaAsNBi}$
1.42 eV	3.36 eV	−1.45 eV	20 eV	6.1 eV	16.2 eV

$$E_g^{unstrain}(x, y) = (1 - x - y) \cdot E_{g(GaAs)} + x \cdot E_{g(GaN)} + y \cdot E_{g(GaBi)} - C_{GaAsN}x(1 - x) - C_{GaAsBi}y(1 - x) - C_{GaAsNBi}xy \quad (34.5)$$

where C is the bowing coefficient.

The bandgap energy of the binary (GaAs, GaN, GaBi) and the bowing coefficient C of GaAsN, GaAsBi, and GaAsNBi are given in Table 34.1.

Starting from a certain percentage of nitrogen and bismuth, the conduction band and the valence band are divided into two sub-bands. The increase of the nitrogen and bismuth concentration causes the splitting of the sub-bands from each other. The calculation of strain bandgap is expressed by Cuminal [12]:

$$E_g^{strain} = E_{g(x,y)}^{unstrain} + \Delta E_c^{hy} - \Delta E_{v,av}^{hy} - \Delta E_{hh}^{sh} \quad (34.6)$$

where ΔE_c^{hy}, $\Delta E_{v,av}^{hy}$ are the energy offset of gravity centers of the conduction band and the valence band ΔE_{hh}^{sh} is the energy offset induced by the shear strain for the heavy holes.

34.2.2 *Absorption*

The electrons in the valence band of a semiconductor can absorb photons whose energy is greater than the energy of the bandgap E_g and then move to the conduction band. The absorption coefficient $\alpha(E)$, for a photon energy E greater than the energy of the bandgap, is given by the following relationship [13]:

$$\alpha(E) = \alpha_0 \frac{\sqrt{E - E_g(x, y)}}{E} \quad (34.7)$$

where α_0 is constant.

34.3 Results and Discussion

Figure 34.2 illustrates the variation of lattice mismatch in the plane of layer ε_{xx} and in the direction of growth ε_{zz} of $GaAs_{1-x-y}N_xBi_y$/GaAs structure. It is observed that the strain increases with increasing concentration of bismuth, while it decreases with the increase of the nitrogen concentration.

Figure 34.3 displays the variation of strain gap of GaAsNBi on GaAs as function of bismuth and nitrogen compositions. It is shown that the quaternary alloys cover a wide energy range from 0.2 to 1.2 eV, that is to say, covering the near- and mid-infrared for concentrations of bismuth up to 12 % and nitrogen up to 5 %, for such a range, the strain of GaAsNBi/GaAs is within ±1.5 %. This small strain of GaAsNBi/GaAs is due to the strain compensation between GaAsN and GaAsBi. We can find couples x and y that give us the most optimal bandgap energy with an acceptable strain.

Figure 34.4 shows the evolution of the absorption coefficient as a function of nitrogen and bismuth concentration for fixed photon energy of 1.05 eV; we find that for a bandgap energy very close to the photon energy, the absorption is relatively low because only the electrons at the limit of the valence band can interact with photons. We note that increased concentrations of nitrogen and bismuth increase the absorption coefficient of GaAsNBi structure. This study allows us to find the optimal structure to improve conversion of light into electricity. Figure 34.5 shows the variation of the current density and power depending on the voltage. If the concentration of bismuth is increased from 0 to 10 %, with $x = 4$ % of nitrogen concentration, the photocurrent density increases by 12 %; however the open-circuit voltage decreases by 4 %. Similarly, if we set the bismuth concentration at 10 % and the concentration of nitrogen is varied from 0 to 4 %, the photocurrent density increases by 58 %, but the open-circuit voltage decreases by 15 %. From this study we determine the optimal structure. It is also noted that both concentrations of bismuth and nitrogen increase the maximum power

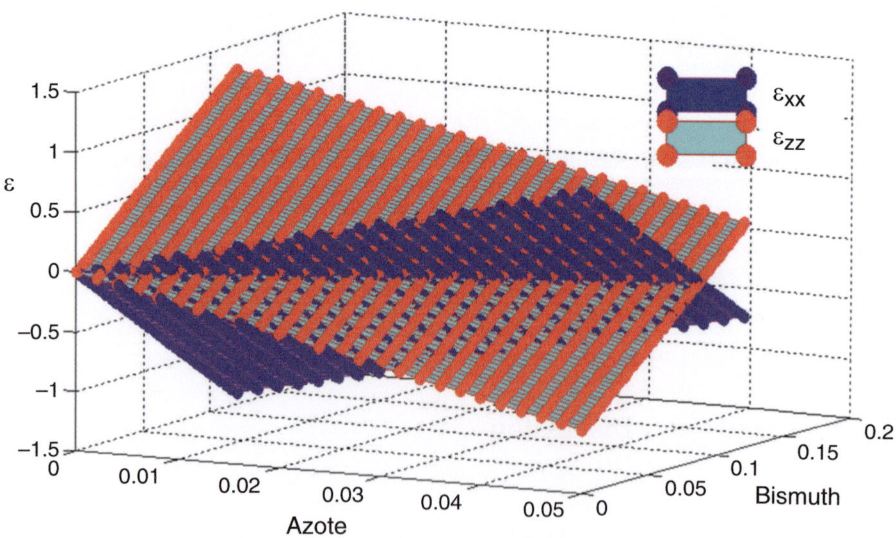

Fig. 34.2 Variation of lattice mismatch as a function of nitrogen and bismuth concentration of $GaAs_{1-x-y}N_xBi_y$/GaAs structure

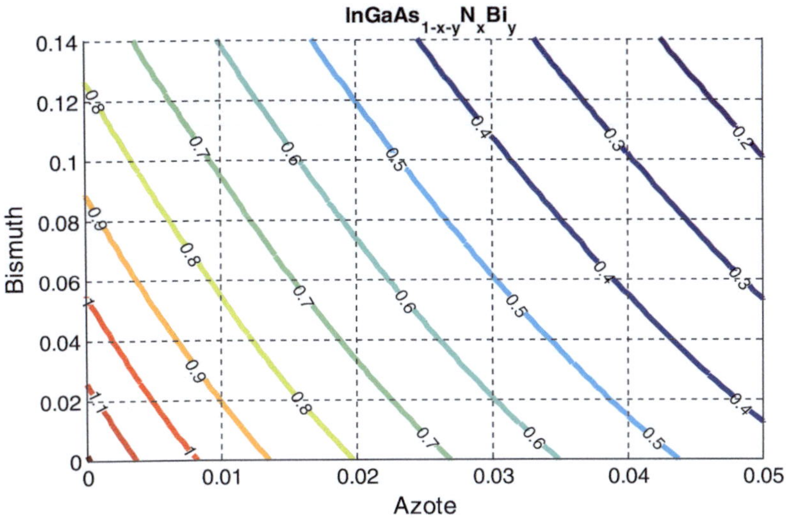

Fig. 34.3 Variation of the strained bandgap energy as a function of nitrogen and bismuth concentration of $GaAs_{1-x-y}N_xBi_y$/GaAs structure

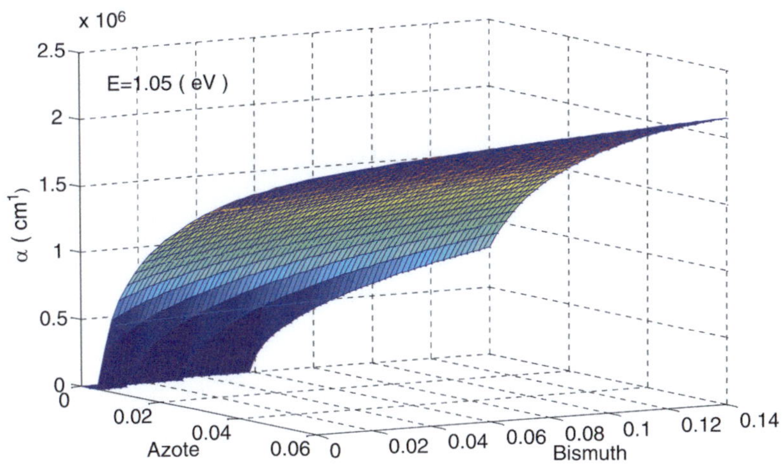

Fig. 34.4 Variation of the absorption coefficient as a function of the nitrogen and bismuth concentration of $GaAs_{1-x-y}N_xBi_y$/GaAs structure for incident photon energy equal to 1.05 eV

delivered by the cell structure created by the $GaAs_{1-y-x}N_xBi_y$/GaAs. Table 34.2 shows the result simulation of a structure based on $GaAs_{1-x-y}N_xBi_y$/GaAs.

When these nitrogen and bismuth concentrations increase, the bandgap decreases and, on the other side, the absorption coefficient increases. We observe also the increase in the photocurrent density, the form factor, and the conversion efficiency η. This study allows us to determine reliable structure.

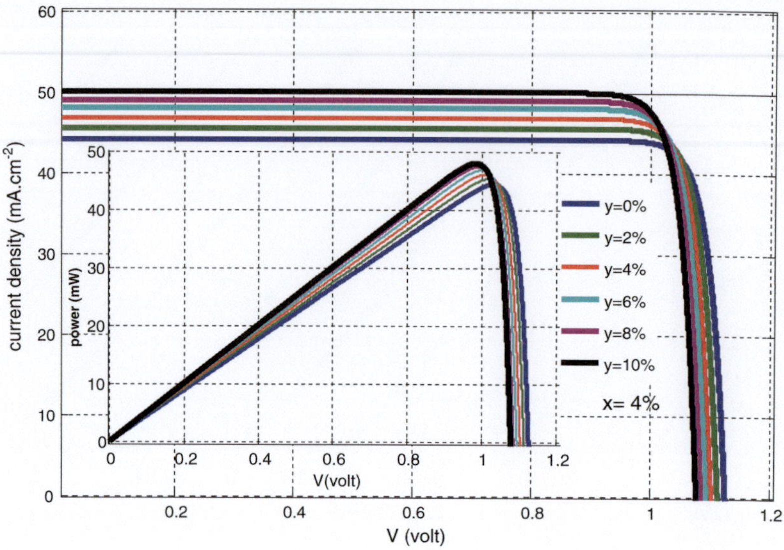

Fig. 34.5 Characteristics $J = f(V)$ and $P = f(V)$ of $GaAs_{1-x-y}N_xBi_y$/GaAs structure

Table 34.2 Representation of simulation results of the structure based on $GaAs_{1-x-y}N_xBi_y$/GaAs

	x (%)	y (%)	□ (%)	E_g (eV)	α $(cm^{-1}) \cdot 10^6$	J_{ph} (mA/cm^2)	V_{co} (V)	FF (%)	η (%)
Structure of GaAsNBi/GaAs	1	1	0.3	0.3	1.33	27.41	1.25	0.52	16.28
	1	6	0.5	0.78	1.43	30.62	1.23	0.54	18.33
	1	8	0.74	0.73	1.46	31.85	1.22	0.54	19.13
	1	10	0.97	0.68	1.49	32.90	1.21	0.55	19.82
	2	1	−0.3	0.76	1.50	33.93	1.20	0.56	20.51
	2	4	0.06	0.68	1.58	36.28	1.18	0.57	22.79
	2	10	0.77	0.54	1.67	39.55	1.16	0.59	24.39

34.4 Conclusion

In this study we have presented the influence of the concentration of nitrogen N and bismuth Bi on the different physical parameters of the structure based on $GaAs_{1-x-y}N_xBi_y$ on GaAs substrate. The nitrogen concentration affects the conduction band and gives us a splitting of this band. By against the bismuth concentration affects the valence sub-bands and gives us a splitting of the two sub-bands of heavy and light holes. Indeed, the incorporation of nitrogen and bismuth in GaAs decreases considerably the energy of the bandgap. Thus, it is noted that the strain affects the structure. The absorption coefficient increases with increasing concentrations of both bismuth and nitrogen. This work allows us to improve the absorption coefficient and the efficiency of a structure based on new materials.

References

1. Palais O (2009) Cellules photovoltaïques: la filière silicium cristallin aujourd'hui et demain. Mater Tech 97:241–245
2. Green MA, Emery K, King DL, Hishikawa Y, Warta W (2006) Solar cell efficiency tables (version 28). Prog Photovolt Res Appl 14:455–461
3. Geisz JF, Friedman DJ, Olson JM, Kurtz SR, Keyes BM (1998) Photocurrent of 1 eV GaInNAs lattice-matched to GaAs. J Cryst Growth 195(1–4):401–408
4. Kurtz SR, Allerman AA, Jones ED, Gee JM, Banas JJ, Hammons BE (1999) InGaAsN solar cells with 1.0 eV band gap, lattice matched to GaAs. Appl Phys Lett 74(5):729–732
5. Wei S-H, Zunger A (1996) Giant and composition-dependent optical bowing coefficient InGaAsN alloys. Phys Rev Lett 76(5):664–667
6. Harmand JC, Ungaro G, Largeau L, Le Roux G (2000) Comparison of nitrogen incorporation in molecular-beam epitaxy of GaAsN, GaInAsN, and GaAsSbN. Appl Phys Lett 77 (16):2482–2485
7. Berding MA, Sher A, Chen AB, Miller WE (1988) Structural properties of bismuth-bearing semiconductor alloys. J Appl Phys 63(1):107–116
8. Wei P, Tixier S, Chicoine M, Francoeur S, Mascarenhas A, Tiedje T, Schiettekatte F (2004) Ion beam characterization of $GaAs_{1-x-y}N_xBi_y$ epitaxial layers. Nucl Instrum Methods Phys Res Sect B 219:671–675
9. Misiewicz J, Kudrawiec R, Gladysiewicz M, Harris JS (2008) Electro-modulation spectroscopy of GaInNAsSb/GaAs quantum wells: the conduction band offset and the electron effective mass issues. Mater Sci 105:163–179
10. Van de Walle CG (1989) Band lineups and deformation potentials in the model solid theory. Phys Rev B 39(3):1871–1883
11. Chuang SL (1995) Physics of optoelectronic devices, Wiley series in pure and applied optics. Wiley, New York
12. Cuminal Y (1997) Réalisation et etude de diodes lazers à base de GaSb émettant vers 2.3μm pour application à l'analyse de gaz. Université Montpellier II
13. Casey HC Jr, Sell DD, Wecht KW (1975) Concentration dependence of the absorption coefficient for n-type and p-type GaAs between 1.3 and 1.6 eV. J Appl Phys 46:250

Chapter 35
Phenomenological Study of a Cylindrical Solar Water Heater: Critical Analysis of the Mathematical Model

Omar Bait, Mohamed Si-Ameur, and Abdelaziz Benmoussa

Abstract This study presents a mathematical model for simulating the transient processes of cylindrical solar water heater. The model simulates the solar collector system including the glass cover, the copper coil, and the working fluid. The equations are based on the energy balance for each part of the solar system; the physical parameters are identified promptly to make the model accessible; constant thermophysical properties are used. The differential equations were solved using the fourth-order Runge–Kutta method. A computer code has been conceived in this study using the Fortran 6.6 platform. Numerical simulations have been carried out to explore the temperature profiles of each part of the present configuration, in order to estimate its efficiency in two different situations: collector with and without gap layer. We have analyzed the effect of the instationary term in this context. A quasi-stationary state shifted due to air gap and temperature variation of the working fluid and the absorber. The results have been confronted to experimental comparison measurements available in the open literature.

Keywords Solar energy • Flat-plat collectors • Solar water heater

Nomenclature

A_{col}	Collector area, m^2
c_p	Specific heat capacity, $J\ kg^{-1}\ K^{-1}$
C_b	The bond conductance, $J\ kg^{-1}\ K^{-1}$
D_{ab}	Diameter of the absorber, m
$D_{ab\text{-}i}$	Interior diameter of the absorber, m
D_g	Glass cover cylinder diameter, m
dx	Spatial coordinate
F'	Collector efficiency factor

O. Bait (✉) • M. Si-Ameur • A. Benmoussa
Department of Mechanical Engineering, Faculty of Technology,
LESEI Laboratory, University of Batna, 05000 Batna, Algeria
e-mail: obait80@yahoo.fr

I. Dincer et al. (eds.), *Progress in Clean Energy, Volume 1*,
DOI 10.1007/978-3-319-16709-1_35

F''	Collector flow factor
F_R	Collector heat removal factor
Gr	Grashof number
I_t	Total insolation on a horizontal surface, W m^{-2}
k_{air}	Heat conductivity, W kg^{-1} K^{-1}
L	Length, m
$\dot{m}$	Mass flow rate of water, kg s^{-1}
Nu	Nusselt number
Pr	Prandtl number
Q_u	Useful energy gained, J
h_c	Convection heat transfer coefficient between the absorber and the glass cover, W m^{-2} K^{-1}
$h_{r,ab\text{-}g}$	Radiation heat transfer coefficient from the absorber to the glass cover, W m^{-2} K^{-1}
$h_{r,g\text{-}am}$	Radiation heat transfer coefficient for the cover glass to the ambient, W m^{-2} K^{-1}
h_{wind}	Heat transfer coefficient due to the wind, W m^{-2}
R	Tilt correction factor
S	Absorbed solar radiation, W m^{-2}
S_{ab}	Absorber surface tube, m^2
$S_{ab\text{-}i}$	Interior surface absorber tube, m^2
S_a	Air gap surface, m^2
S_g	Glass cover surface tube, m^2
T_{am}	Ambient temperature, K
T_{mean}	Mean temperature of air and transparent cover, K
U_L	Overall heat loss coefficient, W m^{-2} K^{-1}
V_{wind}	Wind velocity, m s^{-1}
W	Spacing tube, m

Greek Letters

α	Thermal diffusivity, m^2 s^{-1}
α_{ab}	Absorptivity of copper tube
β	Tilt angle
δ	Declination
δ_{ab}	Absorber thickness, m
δ_a	Air gap thickness, m
δ_g	Glass cover thickness, m
ΔT	Temperature increase of water between inlet and exit from the heater, K
η	Instantaneous efficiency
Φ	Latitude angle
ν	Kinematic viscosity, m^2 s^{-1}
ρ	Density, kg m^{-3}

σ	Stefan–Boltzmann constant, W m^{-2} K^{-4}
$(\tau\alpha)$	Effective transmittance–absorption coefficient
τ_g	Transmissivity of transparent cover
ρ_g	Reflectance of transparent cover
ω	Hour angle

Subscripts

a	Air gap
ab	Absorber
am	Ambient
c	Convection
ext	Exterior
f	Fluid
fi	Fluid inlet
fo	Fluid outlet
g	Glass cover
i	Inner
int	Interior
out	Outer
r	Radiation

35.1 Introduction

Solar energy is one of the crucial energy sources recognized by cleanliness and efficiency. Indeed, it provides a promising solution which becomes an alternative to other dangerous sources such as fossil fuels, which produce harmful gas emissions and therefore are not viable due to their dwindling reserves. One of the most important applications of solar energy is solar radiation conversion into heat utilized in water heating systems. A new generation of preheating systems is representing in the apparition of a cylindrical solar water heater. The system has a simple design, which initially consists of a spiral copper coil housed in a cylindrical transparent tube made from glass. The extremities of the transparent tube were sealed with circular transparent or opaque cylindrical plates, in which holes have been made to allow the spiral copper tube to pass through. The copper coil tube is painted black and works as a collector to incident solar energy. On the other hand, thermocouples have been used between the copper coil tube inlet and outlet in order to measure the temperature of water. In the following context, scientists have been attracted to conduct intensively several works on solar water heating systems. The aim is to improve their efficiency. Taherian et al. [1], have treated by mean of simulation a solar heater, thermosyphon-collector-type. Among the conducted parameters we find, the cover, the absorber temperatures and the efficiency. Reached results are encouragement with experiments. Whereas, an analysis has been performed by Saroja et al. [2] for a solar heater cylindrical type. The set equations were solved numerically. It is found that experiment and numerical procedures have shown a concordance. The work of Nahar and Malhotra [3] is based on both design and test of a cylindrical solar heater in order to examine its performance. The system has shown a good ability to heat up water. Experimental investigation of a cylindrical solar collector has been conducted by Al-Madani [4]. It is shown that the system has 41.8 % of efficiency. Ogueke et al. [5] have conducted experimentally a cylindrical solar water collector. Test results reach to three obtained efficiencies of 53.99 %, 56.21 %, and 57.09 % corresponding, respectively, to the mass rates of 7.2, 10.8, and 3.6 kg/h.

In the present chapter, we analyze by means of numerical simulations the performances of the new type of solar collector (as it is designed by the use of SolidWorks software). All numerical computations have thoroughly been performed for a location in Batna (35°33′N, 6°11′E), Algeria.

35.2 Mathematical Model of a Cylindrical Solar Collector System

In this work, we present a mathematical model describing the cylindrical solar collector system inspired from Saleh AM [6], which is adapted to our case, by considering that the thermophysical properties are constant in different parts. The analyzed control volume of cylindrical solar collector contains four nodes (glass cover, air gap, absorber (copper coil), and the fluid) describing our model. The global system presented in Fig. 35.1 was designed by the authors using SolidWorks software as tool of conception. Figure 35.2 represents a cross section of the system which shows different parts inside the cylindrical solar water heater system.

35.3 Mathematical Analysis of Each Part

35.3.1 The Glass Cover

Considering the constant properties of the glass, the governing equation can be obtained from an energy balance in a differential volume of thickness δ_g. The heat transfers considered in glass cover part are by convection between the cover and the ambient and by radiation from the sun and the absorber (Fig. 35.3):

$$cp_g\rho_g V_g \frac{dT_g}{dt} = S_g\left(h_{\text{wind}}\left(T_{\text{am}} - T_g\right) + h_{\text{r, g-am}}\left(T_{\text{ab}} - T_g\right) + h_c\left(T_a - T_g\right) + \alpha_g I_t\right) \tag{35.1}$$

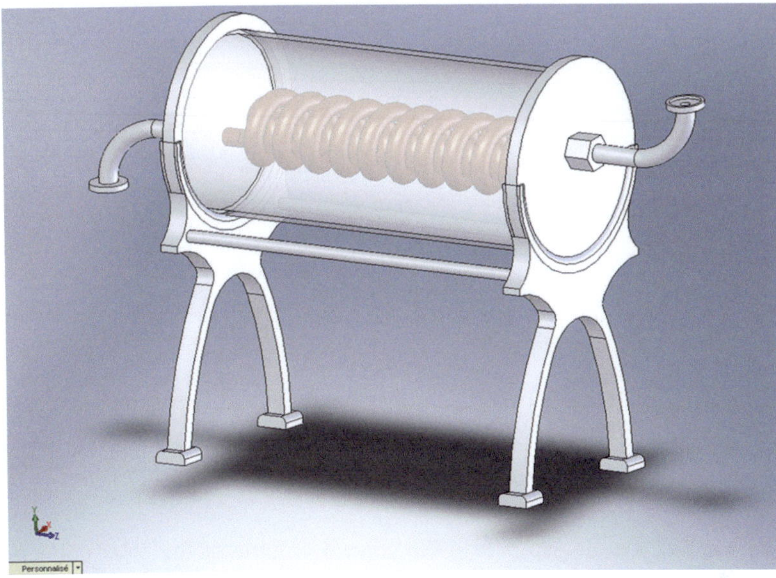

Fig. 35.1 Cylindrical solar water heater designed on SolidWorks

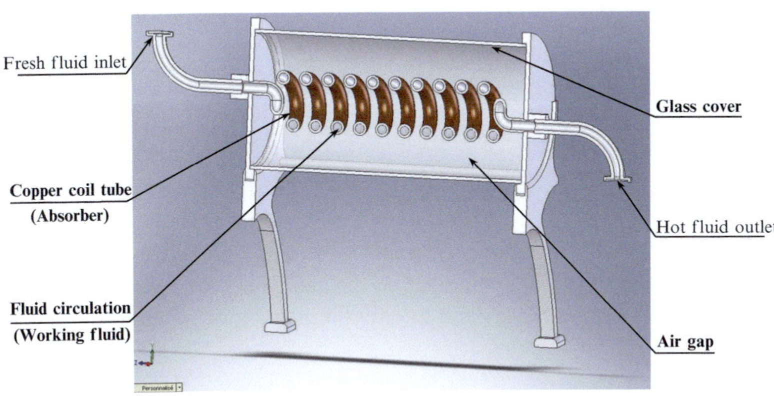

Fig. 35.2 Sketch of the four nodes analyzed in the cylindrical solar collector model

Fig. 35.3 Heat transfer in the glass cover

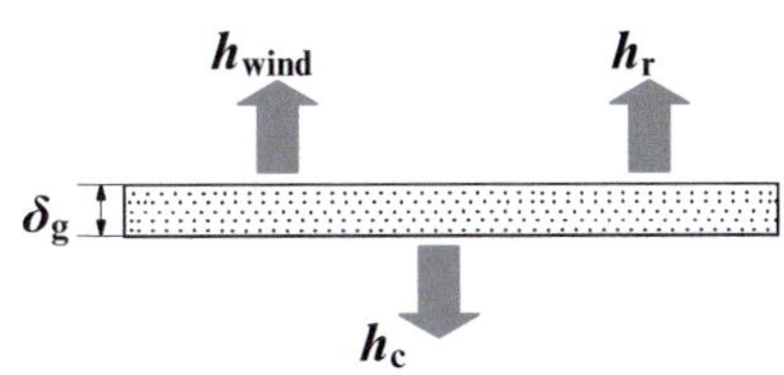

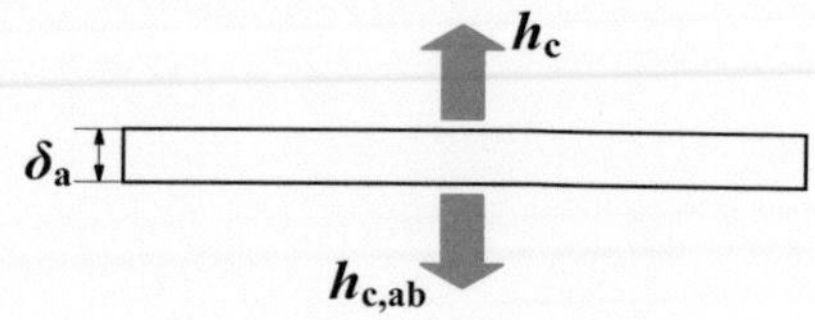

Fig. 35.4 Heat transfer in the air gap between cover and absorber

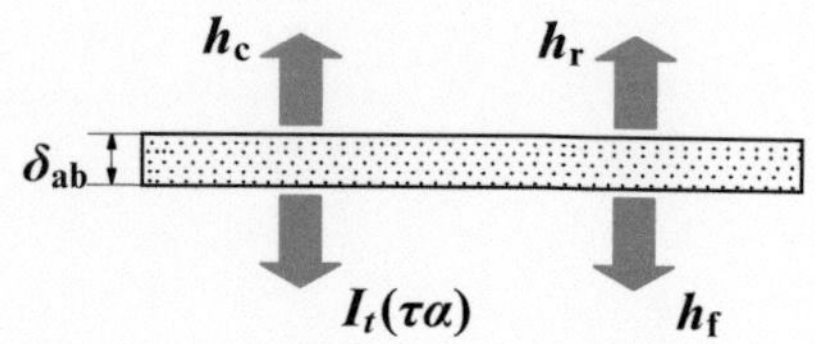

Fig. 35.5 Heat transfer in the absorber

35.3.2 The Air Gap Between the Cover and the Absorber

In the analysis of the air gap part in the control volume of the solar collector, we also consider that the thermophysical properties of the air are constant. The heat transfer into the air gap thickness is by convection between air gap and the glass cover housing side and the absorber on the other side (Fig. 35.4):

$$c p_a \rho_a V_a \frac{dT_a}{dt} = S_a \left(h_c \left(T_g - T_a \right) + h_{c,\,ab} \left(T_{ab} - T_a \right) \right) \tag{35.2}$$

35.3.3 The Absorber

The heat energy balance for the absorber in a differential volume of thickness δ_{ab} is shown in Fig. 35.5. Considering the thermophysical properties of the absorber material as constant, the solar irradiance on the absorber part, the radiation between the absorber and the glass cover, and the heat transfer by convection with the fluid flow, we write

$$\begin{aligned} c p_{ab} \rho_{ab} V_{ab} \frac{dT_{ab}}{dt} &= S_{ab} \left((\tau\alpha) I_t + h_{r,\,ab\text{-}g} \left(T_g - T_{ab} \right) + h_c \left(T_a - T_{ab} \right) \right) \\ &\quad + \pi D_{ab\text{-}i} L_{ab} h_f \left(T_f - T_{ab} \right) \end{aligned} \tag{35.3}$$

35.3.4 The Working Fluid

Figure 35.6 shows the energy balance in a control volume of the working fluid in a part of the solar tube collector. Considering the heat energy absorption by the absorber tube from the incident solar radiation into the fluid control volume, we can write

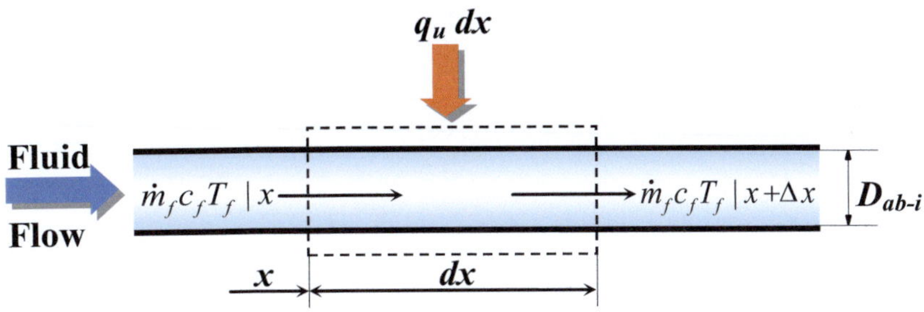

Fig. 35.6 Energy balance in a control volume of the working fluid in a part of the solar tube collector

$$cp_{\mathrm{f}}\rho_{\mathrm{f}}S_{\text{ab-i}}\frac{dT_{\mathrm{f}}}{dt} = \pi D_{\text{ab-i}}h_{\mathrm{f}}(T_{\mathrm{ab}} - T_{\mathrm{f}}) - \dot{m}_{\mathrm{f}}cp_{\mathrm{f}}\frac{dT_{\mathrm{f}}}{dx} \tag{35.4}$$

Therefore, the temperature distribution in flow direction can be expressed using the energy balance on the fluid working through the tube of length dx as [7]

$$\dot{m}_{\mathrm{f}}cp_{\mathrm{f}}T_{\mathrm{f}}|_x - \dot{m}_{\mathrm{f}}cp_{\mathrm{f}}T_{\mathrm{f}}|_{x+dx} + dxq_u = 0$$

thus

$$\dot{m}_{\mathrm{f}}cp_{\mathrm{f}}\frac{dT_{\mathrm{f}}}{dx} = WF'[S - U_{\mathrm{L}}(T_{\mathrm{f}} - T_{\mathrm{am}})] = WF'[(\tau\alpha)I_{\mathrm{t}} - U_{\mathrm{L}}(T_{\mathrm{f}} - T_{\mathrm{am}})] \tag{35.5}$$

The integration of Eq. (35.5) leads to find the outlet temperature T_{fo}. Using the boundary conditions at $T_{\mathrm{f}}\,(x=0)=T_{\mathrm{fi}}$:

$$\frac{T_{\mathrm{fo}} - T_{\mathrm{am}} - (\tau\alpha)I_{\mathrm{t}}/U_{\mathrm{L}}}{T_{\mathrm{fi}} - T_{\mathrm{am}} - (\tau\alpha)I_{\mathrm{t}}/U_{\mathrm{L}}} = \exp\left(-A_{\mathrm{col}}U_{\mathrm{L}}F'/\dot{m}_{\mathrm{f}}cp_{\mathrm{f}}\right) \tag{35.6}$$

The new form of Eq. (35.4) can be written as

$$cp_{\mathrm{f}}\rho_{\mathrm{f}}S_{\text{ab-i}}\frac{dT_{\mathrm{f}}}{dt} = \pi D_{\text{ab-i}}h_{\mathrm{f}}(T_{\mathrm{ab}} - T_{\mathrm{f}}) - WF'[(\tau\alpha)I_{\mathrm{t}} - U_{\mathrm{L}}(T_{\mathrm{f}} - T_{\mathrm{am}})] \tag{35.7}$$

The effective transmittance–absorption coefficient is given by

$$(\tau\alpha) = \tau_{\mathrm{g}}\alpha_{\mathrm{ab}}\sum_{n=0}^{\infty}\left[(1-\alpha_{\mathrm{ab}})\rho_{\mathrm{g}}\right]^n = \left(\tau_{\mathrm{g}}\alpha_{\mathrm{ab}}\right)/\left(1-(1-\alpha_{\mathrm{ab}})\rho_{\mathrm{g}}\right) \tag{35.8}$$

35.4 Performance of the System

A measure of collector performance is the collection efficiency, defined as the ratio of the useful gain over some specified time period to the incident solar energy over the same time period:

$$\eta = Q_u/(A_{col}I_t) \tag{35.9}$$

35.5 Collector Heat Removal Factor

It is convenient to define a quantity that relates the actual useful energy gain of a collector to the useful gain if the whole collector surface were at fluid inlet temperature. The efficiency factor F' is given as

$$F' = \frac{(1/U_L)}{W((1/(U_L[D_{ab\text{-}out} + (W - D_{ab\text{-}out})F])) + (1/C_b) + (1/(\pi D_{ab\text{-}i} h_f)))} \tag{35.10}$$

In our case, we assumed that C_b is zero, because no metal plate or bond exists:

$$F'' = F_R/F' = (\dot{m}cp)/(A_{col}U_L)\left(1 - \exp\left(-\left(A_{col}U_L F'\right)/\dot{m}cp\right)\right) \tag{35.11}$$

The quantity F_R is defined as the ratio of the actual heat transfer to the maximum possible heat transfer.

35.6 Useful Gain Energy Q_u

The product of the collector heat removal factor times this maximum possible useful energy gain is equal to the actual useful energy gain Q_u [7]:

$$Q_u = A_{col}q_u = A_{col}F_R[S - U_L(T_f - T_{am})] \tag{35.12}$$

where the absorbed solar radiation S can be calculated from

$$S = I_t R(\tau\alpha) \tag{35.13}$$

and R is calculated from

$$R = \frac{\cos(\phi - \beta)\cos\delta\cos\omega + \sin(\phi - \beta)\sin\delta}{\cos\phi\cos\delta\cos\omega + \sin\phi\sin\delta} \tag{35.14}$$

Due to the cylindrical shape of the system, it does not need to be inclined to the horizontal. Thus, the tilt correction factor R is unity [5].

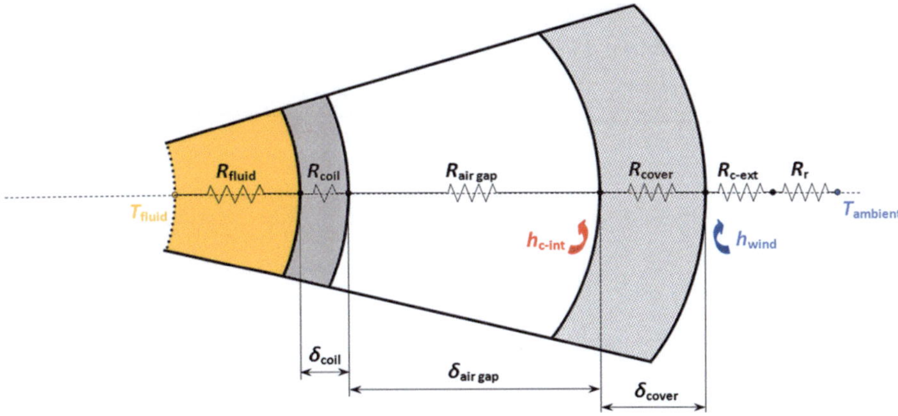

Fig. 35.7 Thermal resistance network

35.7 Overall Coefficient of Heat Losses U_L

The thermal resistance network of the system is shown in Fig. 35.7. It consists of four resistances: resistance due to radiation, to convection outside the glass cover, to convection inside the glass cover and outside the absorber, and finally the resistance due to convection from the absorber to the fluid (water). The overall heat transfer coefficient U_L is given in Eq. (35.15). We assume that the shape of the copper coil tube (absorber) approximates to a cylinder which is concentric to the transparent cylindrical housing. Losses U_e and U_b from the ends of the cylindrical housing are assumed to be negligible due to the poor thermal conductivity of the material of the cylindrical housing. The overall coefficient of heat losses can be calculated from [5, 8]

$$U_L = \left(\left(1/\left(h_c + h_{r,\text{ ab-g}}\right)\right) + \left(1/\left(h_{\text{wind}} + h_{r,\text{ g-am}}\right)\right)\right)^{-1} + U_e + U_b \qquad (35.15)$$

35.8 Calculation of Heat Transfer Coefficients

The determination of the overall coefficient of heat losses U_L requires calculating heat transfer coefficients of each part of the solar water heater, starting with the following.

35.8.1 Convection Heat Transfer Coefficient h_c

The convection heat transfer coefficient between the absorber and the glass cover is [1]

$$h_{\mathrm{c}} = (k_{\mathrm{air}} Nu)/\delta_{\mathrm{a}} \tag{35.16}$$

while the Nusselt number is calculated using the following correlation [9]:

$$Nu = 0.55055(Gr \times Pr)^{0.25} \tag{35.17}$$

And Grashof number is calculated as

$$Gr = \left(g\beta\Delta T L^3\right)/\nu^2 \tag{35.18}$$

35.8.2 Radiation Heat Transfer Coefficient $h_{r,ab\text{-}g}$

The radiation coefficient from the absorber to the glass cover is [7]

$$h_{\mathrm{r,\,ab\text{-}g}} = \sigma\left(T_{\mathrm{ab}}^2 + T_{\mathrm{g}}^2\right)\left(T_{\mathrm{ab}} + T_{\mathrm{g}}\right)/\varepsilon_{\mathrm{eff}} \tag{35.19}$$

The effective emittance between the absorber and the glass cover will be

$$\varepsilon_{\mathrm{eff}} = (1/\varepsilon_{\mathrm{ab}}) + \left(1/\varepsilon_{\mathrm{g}}\right) - 1 \tag{35.20}$$

35.8.3 Convection Heat Transfer Coefficient h_{wind}

The coefficient h_{wind} is obtained from the correlation for flow over horizontal cylinders [10]:

$$h_{\mathrm{wind}} = (4.22 - 0.00257 T_{\mathrm{mean}})\left(V_{\mathrm{wind}}^{0.633}/D_{\mathrm{g\text{-}out}}{}^{0.367}\right) \tag{35.21}$$

T_{mean} is the mean temperature of air and transparent glass cover given as

$$T_{\mathrm{mean}} = \left(T_{\mathrm{am}} + T_{\mathrm{g}}\right)/2 \tag{35.22}$$

35.8.4 Radiation Heat Transfer Coefficient $h_{r,g\text{-}am}$

Also, the radiation coefficient for the glass cover to the ambient is [7]

$$h_{r,\ g\text{-}am} = \varepsilon_g \sigma \left(T_g^2 + T_s^2 \right) \left(T_g + T_s \right) \quad (35.23)$$

where T_s is the sky or sunny temperature given as

$$T_s = 0.0552 T_{am}^{1.5} \quad (35.24)$$

35.9 Results and Discussion

A study of a cylindrical solar water heater was carried out numerically under the meteorological conditions of Batna City, Algeria. From Fig. 35.8, which represents the variations of global solar radiative through 24 h for horizontal surface and in a clear day, it can be seen that in the morning (6:00 a.m.), insolation is generally low, but it increases with sunrise to achieve its maximum in midday which reveals that the peak of the curve shape depends on the latitude and time of year, and it keeps decreasing until the sunset.

Table 35.2 represents experimental measurements taken from the weather station of Batna Airport, which shows the monthly variation of the global insolation (in hours) (source: The Meteorological National Office). Figure 35.9 shows the average solar insolation between the years 1995 and 2004.

This part is based on the results of two works that have been done experimentally by Ogueke et al. [5] and Al-Madani [4]. The objective was to study the performance of a solar water heater, cylindrical type.

The former considered that the convection is present inside the collector, namely, between the collector itself and the absorber, in which water is heated

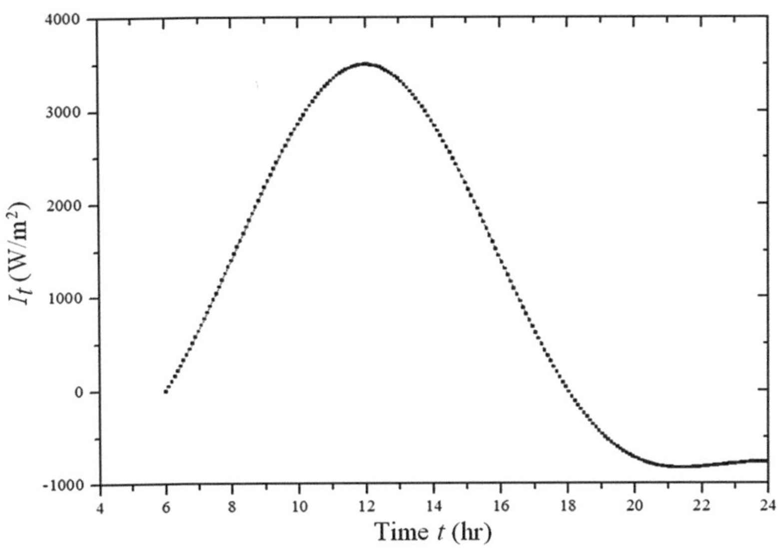

Fig. 35.8 Total solar radiation on horizontal surface

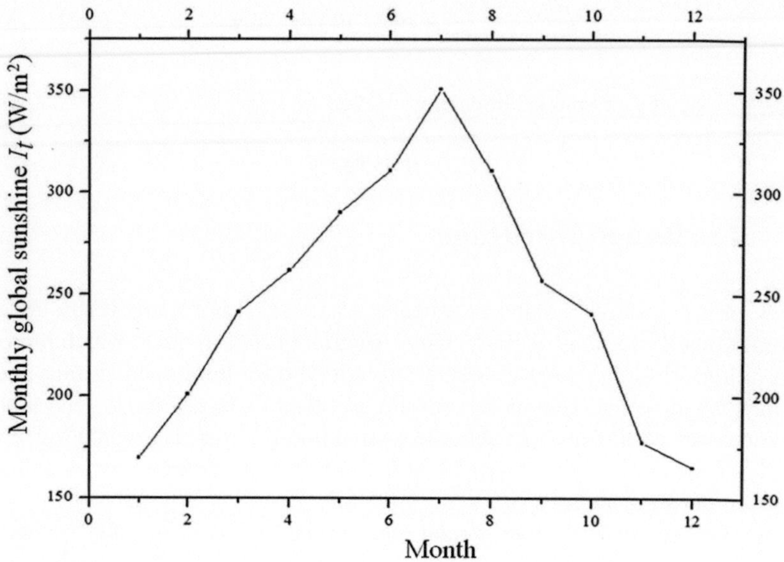

Fig. 35.9 Monthly average insolation for Batna City (period: 1995–2004)

Table 35.1 Experimental data

Components	Absorptivity	Specific heat	Emissivity	Density	Flow velocity
Glass cover	0.005	720	0.88	2,500	–
Absorber	0.36	385	0.1	8,795	–
Water	–	4,190	–	1,000	–
Air	–	–	–	–	1.5

using solar energy. The latter considered removing air flowing inside the collector to prevent water vapor created by heated water in the absorber. For these reasons, we have treated two cases:

– The first is a system working under normal conditions.
– The second is a collector emptied using the vacuum operation.

The set of first-order differential equations governing the system presented previously is solved by the fourth-order Runge–Kutta method. The physical parameters used for the numerical calculation and experimental investigation are given in Table 35.1. Figure 35.10 shows the temperature profiles of the glass cover, the air gap, the working fluid, and the absorber. It reveals that the variation of the temperature for each part of the cylindrical solar water heater is strongly dependent on the insolation. The temperature profiles of the system appear to follow the pattern of the insolation intensity on the entire day as shown in Fig. 35.8, increasing and decreasing in response to the changing weather conditions, focusing that the maximum temperatures occur usually at the times when the insolation is highest

Table 35.2 Monthly global sunshine duration (in hours), Batna Airport Weather Station, period: 1995–2004

Months/years	1995	1996	1997	1998	1999	2000	2001	2002	2003	2004	Average
January	174.6	161.5	140.3	187.3	151.3	198.4	181.2	192.1	131.7	178.6	169.7
February	240.0	145.4	224.3	201.1	170.7	214.1	215.3	220.6	166.6	211.7	201.0
March	238.6	228.3	288.2	251.8	244.5	247.7	214.9	255.6	227.7	217.4	241.5
April	282.0	206.0	258.0	268.5	299.9	272.5	271.7	275.4	245.4	239.7	261.9
May	329.0	299.5	309.7	267.3	297.9	241.8	292.3	309.6	293.6	259.2	290.0
June	285.2	298.2	330.4	328.2	230.6	306.1	343.9	344.1	338.4	302.4	310.7
July	375.2	347.4	348.3	390.2	346.9	366.2	340.3	325.1	328.0	342.2	351.0
August	320.3	311.1	320.0	314.0	287.5	336.0	318.9	271.3	314.3	312.9	310.6
September	248.6	260.9	233.2	269.5	270.8	249.5	270.4	283.9	229.2	252.0	256.8
October	256.4	261.2	233.2	241.7	239.8	221.9	268.1	272.3	186.2	224.7	240.6
November	196.3	206.1	173.2	170.6	180.9	194.5	183.0	148.3	182.1	139.6	177.5
December	162.6	176.5	183.9	178.2	152.4	193.9	171.0	168.1	153.5	112.9	165.3
Annual	3,108.8	2,902.1	3,042.7	3,068.4	2,873.2	3,042.6	3,071.0	3,066.4	2,796.7	2,793.3	2,976.5

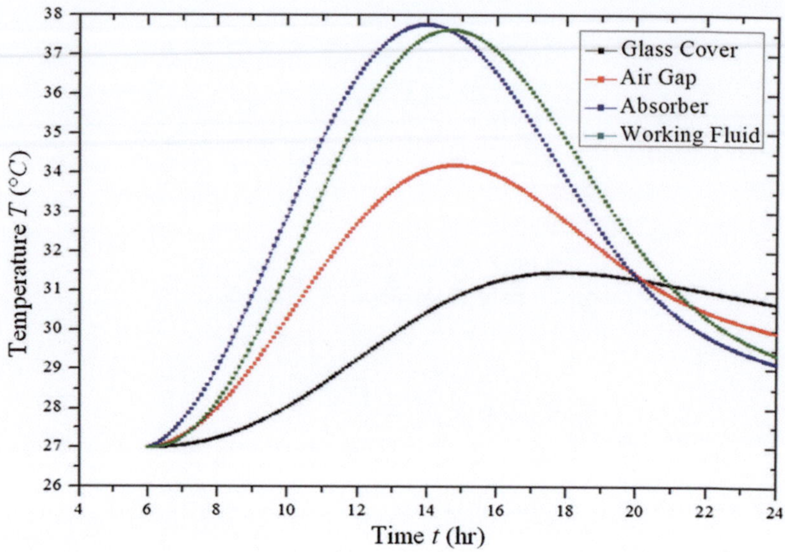

Fig. 35.10 Influence of convection on temperature profiles for different parts of cylindrical solar collector

(12:00 a.m.). Considering the convection (air gap) inside the system, two cases can be studied: the first one with air gap thickness of 25 mm leads to a maximum value of absorber temperature equal to 37.62 °C. The second shows that there is no influence of stationarity.

Figure 35.11 shows a significant increase of the absorber and fluid temperatures without considering convection effect. It reveals that the maximum temperature of the working fluid is 49.40 °C, which means that a shift of stationarity has taken place, implying the importance of unsteady term in modeling a solar collector without considering convection effect.

We have shown that the effect of convection results in a draconic decrease in fluid temperature. However, quasi-stationary state occurs almost at true noon, and without considering convection effect, it will be shifted to 4 p.m. It would be wise to make a techno-economic study to justify the use of both devices. The increase in temperature with the presence of convection needs a study of the thickness of the air layer in the system.

In Fig. 35.12, two values of air gap thickness were identified. In the case of thin air layer equal to 5 mm, the temperature of the fluid increases and achieves 44 °C, which confirms convection influence besides that of working conditions. The case of 25 mm air layer thickness has been dealt with in Fig. 35.10.

Figure 35.13 shows the instantaneous efficiency of the system versus the variable $(T_{fi} - T_{am})/I_t$. Here, we can analyze Eqs. (35.9) and (35.12), knowing that the product $F_R(\tau\alpha)$ determines the way energy is absorbed, whereas $F_R U_L$ determines how energy is lost. The efficiency of the cylindrical solar water heater versus time is

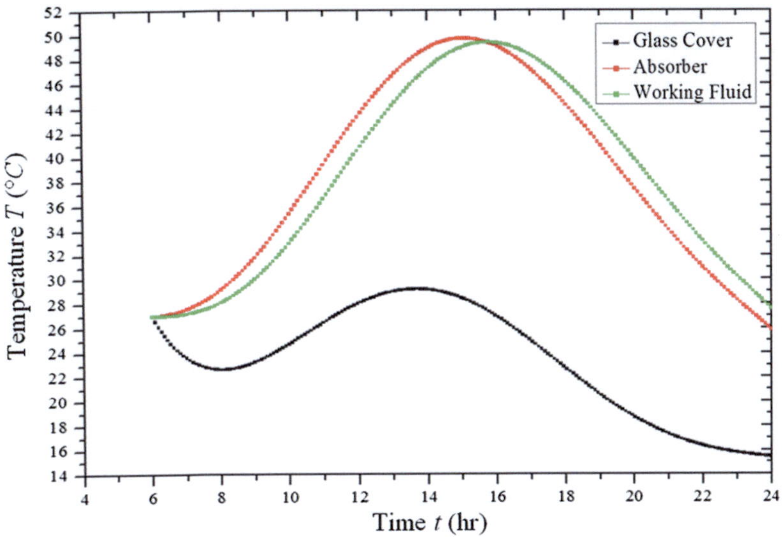

Fig. 35.11 Influence of the absence of convection inside the collector system

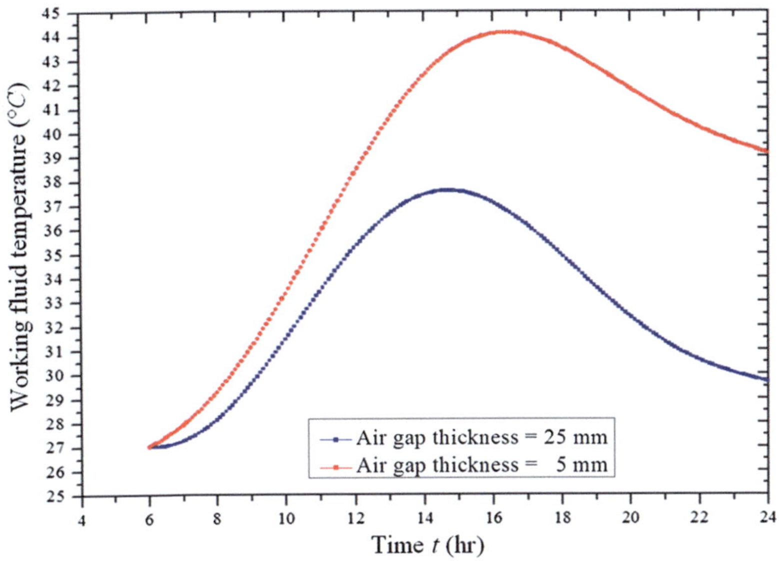

Fig. 35.12 Working fluid temperature profiles for different values of air gap thickness

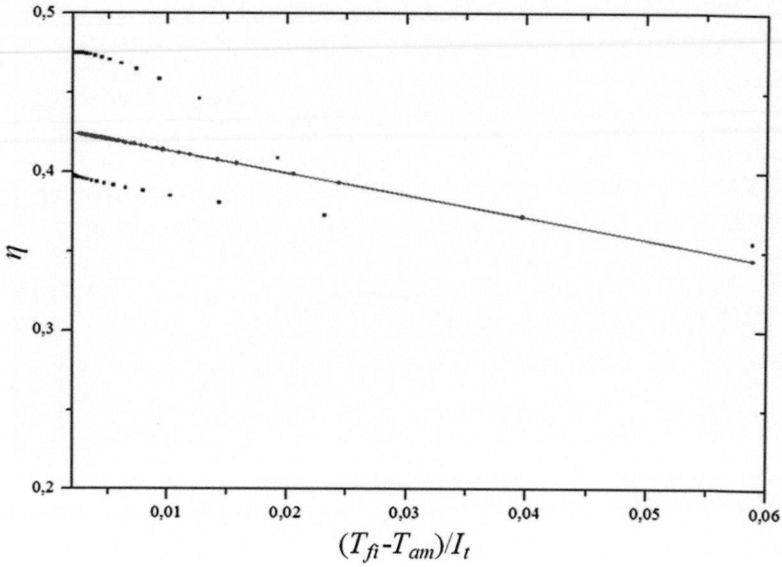

Fig. 35.13 Instantaneous efficiency η of the solar collector versus $(T_{\mathrm{fi}} - T_{\mathrm{am}})/I_{\mathrm{t}}$ for mass flow rate: 0.003 kg/s

approximately steady and its maximum value is 42 %, which is a good result compared to that obtained by Al-Madani [4] and Ogueke et al. [5]. A remarkable point in Eq. (35.9) is that the instantaneous efficiency of the system decreases when the ratio of temperature to incident radiation increases.

35.10 Conclusions

A major advantage of solar energy is its nonpolluting nature. Several objectives are achieved using solar energy to minimize emissions of fossil fuels and reduce the environment damaging. A new type of solar collectors, namely, cylindrical solar water heater, can be used as a preheating system that can reduce the amount of fossil fuel or electrical energy required to generate hot water. It is known that the cylindrical shaped solar water heater does not require to be inclined to the horizontal at a value that is equal to the location's local latitude. Besides having the advantage of taking a small area, it does not need to be directed to the sun due to its circular shape. Also, the device has a good capability of converting solar energy to heat which is used for heating water. Finally, cylindrical solar water heater has a simple geometry which consists of two essential elements: a cylindrical housing and a coil tube.

Our numerical study of a cylindrical solar water heater is carried out on the basis of energy balance analysis in the different constituents of the device. Thereby, our finding may be drawn as:

1. The mathematical model adapted in this context has been shown to work adequately, as the issued numerical simulation results are in good agreement with experimental measurements.
2. The parametric analysis regarding the air gap thickness indicates a considerable alteration of temperature levels comparing to a vacuum situation.
3. The quasi-stationary state for the working fluid has been observed to shift forward as the air gap is explicitly inside this solar system.

We have now undertaken a supplement study based on technical and economic investigation to analyze the thermo-sensibility of such a system with and without vacuum situations.

References

1. Taherian H, Rezania A, Sadeghi S, Ganji DD (2011) Experimental validation of dynamic simulation of the flat plate collector in a closed thermosyphon solar water heater. Energy Convers Manag 52:301–307
2. Saroja S, Nithiarasu P, Seetharamu KN (1997) Transient analysis of a cylindrical solar water heater. Energy Convers Manag 38:1833–1840
3. Nahar NM, Malhotra KS (1984) Year round performance of a cylindrical solar water heater. Energy Convers Manag 24:277–280
4. Al-Madani H (2006) The performance of a cylindrical solar water heater. Renew Energy 31:1751–1763
5. Ogueke NV, Njoku MC, Anyanwu EE (2009) Design, construction, and testing of a cylindrical solar water heater. Int J Energy Clean Environ 10(1–4):57–72
6. Saleh AM (2012) Modeling of flat-plate solar collector operation in transient states. Purdue University, Fort Wayne
7. Duffie JA, Beckman WA (1991) Solar engineering of thermal processes. Wiley, New York
8. Ammari HD (2003) A mathematical model of thermal performance of a solar air heater with slats. Renew Energy 28:1597–1615
9. Holman JP (2002) Heat transfer. McGraw-Hill, New York
10. ASHRAE (2005) Handbook fundamental. American Society of Heating, Refrigerating and Air-Conditioning Engineers, Atlanta

Chapter 36
Assessment of Turbulence Models for Aerodynamic Performance Analysis of a Commercial Horizontal Axis Wind Turbine

Mojtaba Tahani

Abstract In this chapter, the results of three-dimensional computational fluid dynamics (CFD) finite volume simulations of airflow around a commercial Vestas V80 Horizontal Axis Wind Turbine (HAWT), with a rated output power of 2 MW, are presented. The grid used in the simulations consists of two main parts, i.e., unstructured mesh rotating with blades and structured hexahedral stationary one for the external domain. Several cases with different free stream velocities (and different tip speed ratios and mean pitch angles) are studied, employing four different turbulence models: $k-\omega$ SST, $\overline{v}^2-f$, $k-\varepsilon$ RNG and Spalart–Allmaras one-equation, in order to examine their ability to predict the output generated power of HAWTs. The investigation outcomes are compared with each other and existing experimental result given in previous studies. It is shown that the numerical results are in acceptable agreement with experiments. Regarding assumptions during simulations, more sensible output power values are obtained through $k-\varepsilon$ RNG and $\overline{v}^2-f$ models. In addition, maximum value of power coefficient occurs at more accurate associated wind speed using $\overline{v}^2-f$ model. The simulations provide useful guidelines to design more efficient large commercial wind turbines.

Keywords Aerodynamic • Turbulence modeling • Wind turbine • CFD

Nomenclature

A_S	Swept area (m^2)
BEM	Blade element momentum
C_D	Drag coefficient
CFD	Computational fluid dynamic

M. Tahani (✉)
Faculty of New Sciences and Technologies, University of Tehran, Tehran, Iran
e-mail: m.tahani@ut.ac.ir

I. Dincer et al. (eds.), *Progress in Clean Energy, Volume 1*,
DOI 10.1007/978-3-319-16709-1_36

C_m	Moment coefficient
C_P	Power coefficient
HAWT	Horizontal Axis Wind Turbine
I	Turbulence intensity (%)
k	Turbulent kinetic energy (m^2/s^2)
LES	Large eddy simulation
p	Pressure (N m^2)
Re	Reynolds number
RNG	Renormalization group
SA	Spalart–Allmaras
SST	Shear stress transport
U_∞	Velocity at infinity (m/s)
V	Local velocity magnitude (m/s)
y^+	Distance to wall in viscous units

Greek Letters

ε	Turbulent energy dissipation rate (m^2/s^3)
λ	Tip speed ratio
μ	Dynamic viscosity (N s/m^2)
μ_t	Turbulent viscosity (N s/m^2)
Ω	Angular velocity (rad/s)
ω	Specific rate of turbulent energy dissipation (s^{-1})
ρ	Density (kg/m^3)
τ	Stress tensor (N/m^2)
$\varnothing$	Normal components of the pressure-strain

Subscripts and superscripts

Eff	Effective
i, j, k	Space subscripts
t	Time subscript
T	Transpose of a matrix
v	Viscous

36.1 Introduction

Wind turbines play significant role in providing economic power to satisfy today's huge demand of sustainable and clean forms of energy. Nowadays, a major concern of the wind turbine designers is their power efficiency and overall performance. As a result, it is useful to investigate the air flow behavior passing through and behind the turbine. Several studies have been done in recent years, each of which has focused on a particular region of the flow. The blades of an HAWT should be designed considering many important factors such as aerodynamic efficiency, loaded forces and shear/normal stress, total mass, final cost, and their balance with design parameters associated with internal and external shape [1]. Thus, CFD approaches could be used in order to analyze complexity of flow field around a wind turbine. Hartwanger and Horvat obtained power, lift, drag, and moment coefficients for an isolated wind turbine using CFX software. They also used 3D CFD results, obtained through 2D analysis of blade sections, to predict values for actuator disk induction factors [2]. Later, Galdamez et al. performed simulations for a Vestas V39 wind turbine using CFD models, blade element momentum (BEM) theory, and some algorithms for optimization, which is expected to lead to optimization of its winglet design [3]. The BEM theory is actually a two-dimensional approach extended to the third dimension and uses corrections taken from correlations obtained through measurements or CFD simulations to capture 3D phenomena. BEM theory must be used only when the blades have uniform circulation [1], as Tangler states that the BEM method basically under-predicts the overall performance of a wind turbine, however, overpredicts the peak power value [4]. The main reason for such an important assumption is the high computational cost of CFD methods. Virtually, all the modern rotors for HAWTs that exist today were designed using the BEM method. However, as a result of simplifications in derivation, this method could have many shortcomings [1]. Vaz et al. [5] have presented a mathematical model for wind turbine based on BEM theory. In this study in addition of axial and tangential induction factor in rotor plane, the effect of wake behind the rotor plane is considered by applying axial and tangential induction factor at the rotor wake. Esfahanian et al. [6] used a mixed CFD and BEM analysis for aerodynamic simulation of an NREL Phase II wind turbine. In this study, a CFD analysis is implemented to calculate the aerodynamic coefficients, then simulating a 3D flow through the wind turbine by the use of BEM theory. Sezer-Uzol and Long chose PUMA2 as their solver, which used the 4-stage Runge–Kutta numerical time integration method and Roe's numerical flux scheme in the computations. It is shown that considerable separation happens at higher wind speeds (15 m/s) even with zero yaw angles [7]. Zhou and Chow investigated properties of stable boundary layer (SBL) for wind energy using LES method and reconstruction turbulence model approach [8]. This is important for lifetime and performance of wind turbines caused by fatigue [8]. Chatelain et al. studied wake behavior behind isolated HAWTs using Large Eddy Scale (LES), vortex particle-mesh coupling method to analyze unsteady output power of the turbines, generator performance, and to understand their aerodynamic characteristics. Besides they investigated the possible effects of type of used turbulent inflow [9]. Among the most recent works, Jeon and Lee performed simulations using vortex lattice method to study unsteady aerodynamics of an offshore floating wind turbine (NREL 5 MW OFWT). Their results showed that when the turbine is running at low speed inflow, considerable turbulent wakes behind the turbine, which cannot be predicted by BEM [10]. Also Lanzafame et al. employed a modified correlation-based transitional model to simulate the flow around the turbine and compared the results to outcomes of 1D model, which is based on BEM theory, and SST $k - \omega$ method [11]. Wang et al. [12] studied blade-tower interaction of NREL Phase VI wind turbine in upwind configuration by using the open source Open FOAM tools coupled with arbitrary mesh interface (AMI) method. Their results show that the wind turbine tower has little effect on the whole aerodynamic performance of an upwind wind turbine, while the rotating rotor will induce an obvious cyclic drop in the front pressure of the tower.

In the present work, it is focused on four turbulence models and their precision on results for 22 cases of different wind speeds, ranging from cut-in to cut-out values, and associated mean pitch angles of blades. Each one of employed methods contains a package of rules and assumptions for specific regions of flow, including near the wall, far field, etc. Also the study constitutes an instructive exercise in wind turbines, and opens the way to understand the complexities of airflow behind the HAWTs better.

36.2 Governing Equations

In the averaging of steady-state flows, the conservation equations can be solved based on the average Reynolds values or the time-averaging approach; however, the most common method of modeling turbulent flows is the time-averaging method. Using this approach for the case of steady and compressible fluid flows, the general forms of the continuity and momentum equations could be expressed as relations (36.1) and (36.2), respectively.

$$\nabla \cdot \left(\rho \vec{V}\right) = 0 \tag{36.1}$$

$$\nabla \cdot \left(\rho \vec{V} \vec{V}\right) = -\nabla p + \nabla \cdot (\bar{\bar{\tau}}) \tag{36.2}$$

$$\bar{\bar{\tau}} = \mu \left[\left(\nabla \vec{V} + \nabla \vec{V}^{\mathrm{T}} \right) - \frac{2}{3} \nabla \cdot \vec{V} I \right] \tag{36.3}$$

Stress tensor is calculated by means of following equation:

In this numerical solution, the finite volume method has been used for the discretization of equations, but the analysis of the geometry has been based on the finite element approach. So, the geometrical flexibility of the finite element method can be used; however, the equations are dealt with in the form of finite volume [13].

36.3 Turbulent Models

Four turbulence models, which are used in this study, are briefly discussed in this section.

36.3.1 *Standard Spalart–Allmaras One-Equation Model (SA)*

In this section, detailed information is given on the equations for standard forms of the Spalart–Allmaras turbulence model. The form of the model are given here is a linear eddy viscosity model. Linear models use the Boussinesq assumption:

$$\tau_{ij} = 2\mu_t\left(S_{ij} - \frac{1}{3}\frac{\partial u_k}{\partial x_k}\delta_{ij}\right) - \frac{2}{3}\rho k\delta_{ij} \tag{36.4}$$

where the last term is generally ignored for Spalart–Allmaras because k is not readily available (the term is sometimes ignored for non-supersonic speed flows for other models as well). The one-equation model is given by the following equation:

$$\begin{aligned} u_j\frac{\partial \hat{v}}{\partial x_j} &= c_{b1}(1 - f_{t2})\hat{S}\,\hat{v} - \left[c_{w1}f_w - \frac{c_{b1}}{\kappa^2}f_{t2}\right]\left(\frac{\hat{v}}{d}\right)^2 \\ &+ \frac{1}{\sigma}\left[\frac{\partial}{\partial x_j}\left((v + \hat{v})\frac{\partial \hat{v}}{\partial x_j}\right) + c_{b2}\frac{\partial \hat{v}}{\partial x_i}\frac{\partial \hat{v}}{\partial x_i}\right] \end{aligned} \tag{36.5}$$

And the turbulent eddy viscosity is computed from:

$$\mu_t = \rho\hat{v}f_{v1} \tag{36.6}$$

where

$$f_{v1} = \frac{\chi^3}{\chi^3 + c_{v1}^3}, \quad \chi = \frac{\hat{v}}{v} \tag{36.7}$$

And ρ is the density, $v = \mu/\rho$ is the molecular kinematic viscosity and μ is the molecular dynamic viscosity.

36.3.2 $k-\varepsilon$ RNG

$k-\varepsilon$ RNG is the complementary model for standard $k-\varepsilon$ model. This model was introduced by Yakhot and Orszag [14]. The standard model is presented mainly for regions with high Reynolds numbers, whereas by considering Eq. (36.8) in RNG model to determine the turbulent viscosity value in order to take low-*Re* regions into account as well.

$$d\left(\frac{\rho^2 k}{\sqrt{\varepsilon\mu}}\right) = 1.72\frac{\hat{\nu}}{\sqrt{\hat{\nu}^3 - 1 + C_\nu}}d\hat{\nu} \tag{36.8}$$

where ν and the constant value, C_ν are defined as follows:

$$\nu = \frac{\mu_{\text{eff}}}{\mu} \quad \text{and} \quad C_\nu \approx 100 \tag{36.9}$$

36.3.3 $k-\omega$ SST

In the case of $k-\omega$ shear stress transport (SST) model, a differential equation for specific rate of dissipation (ω) is solved except for dissipation rate of turbulent energy (ε). Menter introduced SST model for $k-\omega$ to integrate exact and strong $k-\omega$ equations near the wall region, with independent $k-\varepsilon$ equations in the far field. Turbulent viscosity in this model is calculated from below equation [15].

$$\mu_{\text{T}} = \frac{\rho k}{\omega}\frac{1}{\max\left[\frac{1}{\alpha^*}, \frac{SF_2}{a_1\omega}\right]} \tag{36.10}$$

where α^*, S, F_2, and a_1 can be defined separately.

36.3.4 $\overline{\upsilon}^2 - f$ (Four Equations)

Here, the equations for the Reynolds stress tensor components is replaced by a transport equation for the value $\overline{\upsilon}^2$ and an elliptic equation for a scalar function f related to the energy distribution in the equation for $\overline{\upsilon}^2$. This model is similar to standard $k-\varepsilon$ model, with the transport equations for k and ε being solved as follows [16]:

$$\frac{Dk}{Dt} = \frac{\partial k}{\partial t} + U_j \nabla^j k = P_k - \varepsilon + \nabla^j\left[\left(\nu + \frac{\nu_t}{\sigma_k}\right)\nabla_j k\right] \tag{36.11}$$

$$\frac{D\varepsilon}{Dt} = \frac{\partial \varepsilon}{\partial t} + U_j \nabla^j \varepsilon = \frac{C_{\varepsilon 1} P_k - C_{\varepsilon 2} \varepsilon}{T} + \nabla^j \left[\left(\nu + \frac{\nu_t}{\sigma_\varepsilon} \right) \nabla_j \varepsilon \right] \tag{36.12}$$

where $C_{\varepsilon 1}$ and $C_{\varepsilon 2}$ are related to the creation and the destruction of dissipation accordingly, P_k is the production of the turbulent energy.

The transport equation for $\overline{\upsilon}^2$ is as follows:

$$\frac{D\overline{\upsilon}^2}{Dt} = \frac{\partial \overline{\upsilon}^2}{\partial t} + U_j \nabla^j \overline{\upsilon}^2 = kf - \overline{\upsilon}^2 \frac{\varepsilon}{k} + \nabla^j \left[\left(\nu + \frac{\nu_t}{\sigma_k} \right) \nabla_j \overline{\upsilon}^2 \right] \tag{36.13}$$

In Eq. (36.13), k_f is defined as

$$k_f = \varnothing_{22} - \boldsymbol{\varepsilon}_{22} + \frac{\overline{\upsilon}^2}{k} \varepsilon$$

where $\varnothing_{22}$ and ε_{22} are normal components of the pressure-strain and dissipation tensors to the wall, respectively. Also the following elliptic equation is solved for f:

$$L^2 \nabla^2 f - f = \frac{1}{T}(C_1 - 1) \left[\frac{\overline{\upsilon}^2}{k} - \frac{2}{3} \right] - C_2 \frac{P_k}{k} \tag{36.14}$$

where

$$T = \max \left[\frac{k}{\varepsilon}, 6 \sqrt{\frac{\nu}{\varepsilon}} \right], \quad L = C_L \max \left[\frac{k^{\frac{3}{2}}}{\varepsilon}, \frac{C_\eta \nu^{\frac{3}{4}}}{\varepsilon^{\frac{1}{4}}} \right] \tag{36.15}$$

The turbulent viscosity is defined as $\nu_t = C_\mu \overline{\upsilon}^2 T$ and coefficients of the model are as follows [17]:

$$\begin{aligned} &C_\mu = 0.19, \quad \sigma_k = 1, \quad \sigma_\varepsilon = 1.3, \quad C_{\varepsilon 1} = 1.4 \left[1 + 0.045 \sqrt{\frac{k}{\overline{\upsilon}^2}} \right], \quad C_{\varepsilon 2} = 1.9, \\ &C_1 = 1.4\; C_2 = 0.3, \quad C_L = 0.3, \quad C_\eta = 70 \end{aligned} \tag{36.16}$$

In Eq. (36.14), anisotropy is considered and in homogeneous flow ($\nabla^2 f = 0$), the classical model for $\varnothing_{22}$ is recovered.

In all turbulent models above, inlet boundary conditions could be determined based on direct specification of k, ε, or ω or based on specification of turbulent intensity along with length scale, hydraulic diameter, or viscosity ratio. Selection of each choice is dependent on flow field physical properties, surroundings, and known values. Providing turbulent intensity and one of other three parameters at inlet, values of k, ε, and ω are calculated with separate relations.

36.4 Numerical Calculation Setup

The grid used in the solution consists of around 650,000 cells including both tetrahedral and hexahedral types, getting smaller near the blades due to boundary layer and rotating effects. A cross-section of the generated mesh is given in Figs. 36.1 and 36.2. The far-field distance in domain is chosen great enough (ten times characteristic length which is blade length), so that the simulation gives reliable results. The inner domain surrounds the turbine blades and hub having

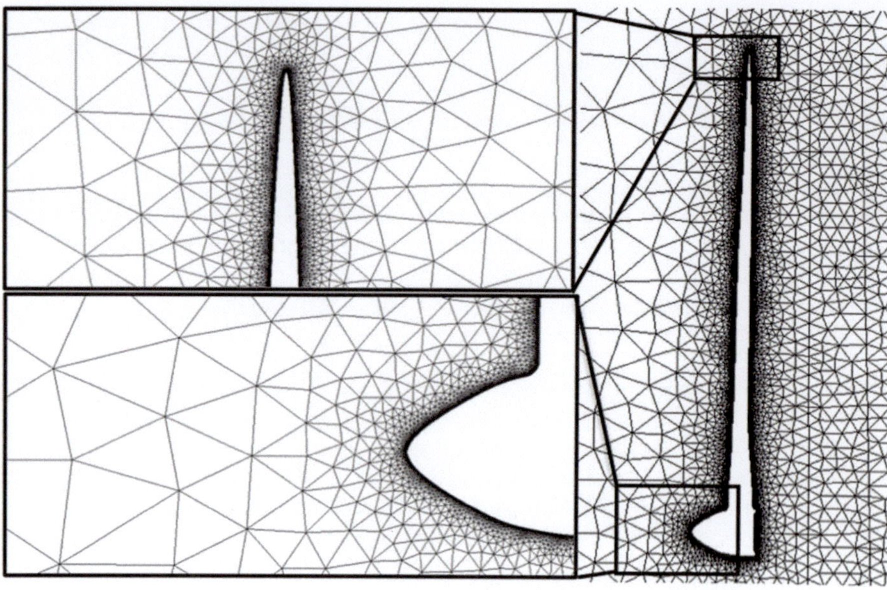

Fig. 36.1 Unstructured mesh construction of rotating parts on normal cut plane

Fig. 36.2 Fluid domain for CFD model, including rotating unstructured and stationary structured parts

2–6 m offset from the blades surface, while the outer domain includes the rest of it of hexahedral type. The tetrahedral-type domain is actually sliding in the complement hexahedral region to capture the rotating effects generated by high angular velocity. Since the free stream velocity is known, boundary condition at inlet is taken as velocity-inlet normal to the turbine in the x-direction. On the other hand, despite the inlet, pressure is known for outlet instead of velocity, which is the atmosphere condition. As a result, gauge pressure is set for outlet. It is to mention that turbulence level can't be definitely set, for it totally depends on environmental condition in which the turbine is running. Moreover, since the path passed by air to meet the blades is long enough, turbulence intensity of flow is damped enough. However, an initial estimation of 4–12 %, for working speed range of turbine, will be fine as discussed in [18].

36.5 Grid Study

Obviously, no numerical simulation is reliable without a grid study. The first rough mesh, generated at the first step often contains cells especially near wall or in boundary layer region, which is not capable of capturing all gradients effects or phenomena happening in separation region. So by slightly changing the mesh, the results may take new values. Therefore, the mesh must be treated in a couple of steps to get results, which become independent of mesh size. The first mesh generated in this study contains around 650,000 cells, more of which are in cylindrical sliding region. Then, the mesh was treated in five stages leading the mesh size to reach 3,500,000 cells. During five level of treatment, it was seen that the results don't change considerably, which shows independency is already achieved. Consequently, the first grid was chosen because of simplicity in computational cost. Table 36.1 shows the details about number of elements and variation of moment coefficient (C_m) during grid study process for case 7, RNG method (as a sample case). It must be mentioned that solution residuals values in whole domain were chosen as the convergence criteria in order to make sure that solution is converged all across the domain. This was evaluated by checking residual contours.

Table 36.1 Variation of C_m versus number of mesh elements for case 7

Number of elements	C_m about rotor axis
649,358	0.05767
1,203,423	0.05753
1,923,117	0.05747
2,602,863	0.05745
3,489,261	0.05743

36.6 Turbulent Model Investigation

Referring to brochure data given in [19], the cut-in and cut-out speeds for Vestas V80 HAWT are, respectively, 4 m/s and 25 m/s. Taking these two values as minimum and maximum of free stream velocity range, minimum and maximum values of Reynolds number can be calculated, based on blade length, as 1.1e+7 and 6.7e+7. Even the minimum value is higher than critical Reynolds number for transition to outer turbulent flow. Hence, one can assume the flow fully turbulent. Thanks to $k - \varepsilon$ RNG and $k - \omega$ SST methodologies for both near-wall and far distance regions, surely all turbulent and boundary layer phenomena, including separation, viscous layer interactions, and wakes are investigated in the simulations, using these models. To get results for a wide performance range of turbine, 22 cases were considered, the variant parameters of which are free stream velocity and mean pitch angle of blades.

36.7 Wall y^+ Number

To fully capture the fluid flow behavior in boundary layer region, we need to correctly set Δy distance as the first cell height. According to a correlation for Δy value, we have the following equation for our cell-based method:

$$\Delta y = 2L\sqrt{74}y^{+}Re^{-\frac{13}{14}} \tag{36.17}$$

where y^+ is the dimensionless distance from the wall. As mentioned before, the Reynolds number is around 6.7e+7. Therefore, having above equation and taking y^+ as 30, Δy becomes about 0.0026 m (2.6 mm). Setting this value in mesh generation, we can make sure all interactions in the boundary layer are fully captured.

36.8 Output Power Calculation

To calculate the output power of the turbine, moment coefficient value should be used. Having angular velocity and moment about rotating axis of turbine power delivered by turbine can be defined. The torque T can be calculated by the following equation:

$$T = \frac{1}{2}\rho \cdot A_{\mathrm{S}} \cdot U_{\infty}^{2} \cdot R \cdot C_{\mathrm{m}} \tag{36.18}$$

Table 36.2 Brochure data for Vestas V80 [19]

Power regulation	Pitch regulated with variable speed
Rated power	2,000 kW
Cut-in wind speed	4 m/s
Rated wind speed	16 m/s
Cut-out wind speed	25 m/s
Rotor diameter	80 m
Swept area	5,027 m^2
Blade length	39 m

where ρ and U_∞ are density and velocity of free stream, respectively, C_m (moment coefficient) is obtained directly through developed CFD code, A_S is circular area swept by rotating blades and R is blade radius in SI units. Assuming standard condition we use the value of 1.225 kg m^{-3} for air density. Other parameters, provided by the manufacturer [19], are given in Table 36.2.

One important characteristic of any wind turbine is the tip speed ratio, λ, which means at what wind speed, turbine will rotate at specific angular velocity. It is defined by following equation:

$$\lambda = \frac{\Omega \cdot R}{U_\infty} \tag{36.19}$$

where R is blade radius and Ω is turbine angular velocity in SI units. Tip speed ratio values applied in this study are calculated from angular velocity plot given by [18]. The corresponding curve is given in Fig. 36.3. Taking account these values in our calculations, data set in approach are given in Table 36.3. The most important factor used to evaluate overall performance of a wind turbine is its power coefficient, which is defined as:

$$C_\mathrm{P} = \frac{\text{Actual power}}{\frac{1}{2}\rho A_\mathrm{S} U_\infty^3} \tag{36.20}$$

36.9 Solution Results and Discussion

The corresponding results for four mentioned CFD methods are given in Fig. 36.4. Clearly the $k-\varepsilon$ RNG has given bigger moment coefficient and output power values for all free stream velocities. So did the $\overline{\upsilon}^2 - f$ method, but with less deviation from manufacturer's brochure data, while $k-\omega$ SST and Spalart–Allmaras models have under-predicted the power output values after around infinity velocity of 7 m/s. It is to be noted that the wind turbine geometry simulated in this study is not the same as experimented by the manufacturer; mainly because of three reasons: (1) the Vestas V80 turbine geometry used here doesn't include its nacelle

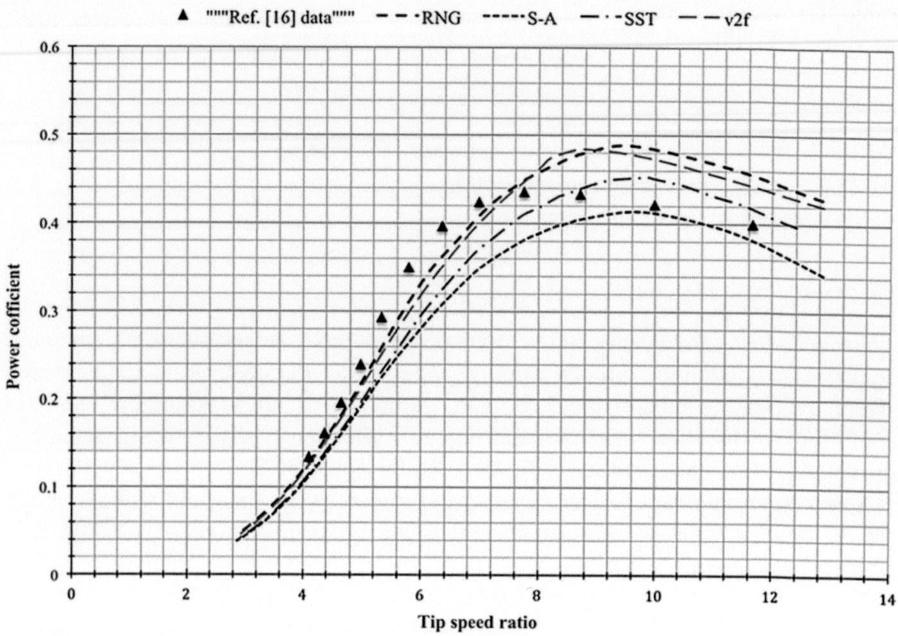

Fig. 36.3 Rotor speed and pitch angle both as a function of wind speed [20]

Table 36.3 Cases with associated data and results

Case	U_∞ (m/s)	Ω (rad/s)	λ	Pitch angle (deg)
1	4	1.319	12.865	−1.75
2	5	1.299	10.128	−1
3	6	1.361	8.849	0
4	7	1.508	8.402	0.8
5	8	1.676	8.168	1
6	9	1.843	7.987	1.25
7	10	1.885	7.351	1.75
8	11	1.885	6.683	−0.5
9	12	1.885	6.126	−3
10	13	1.885	5.655	−6.5
11	14	1.885	5.251	−10.5
12	15	1.885	4.901	−12
13	16	1.885	4.595	−13.75
14	17	1.885	4.324	−15.8
15	18	1.885	4.084	−17.5
16	19	1.885	3.869	−18.75
17	20	1.885	3.676	−19
18	21	1.885	3.501	−19
19	22	1.885	3.342	−19

(continued)

Table 36.3 (continued)

Case	U_∞ (m/s)	Ω (rad/s)	λ	Pitch angle (deg)
20	23	1.885	3.196	−19
21	24	1.885	3.063	−19
22	25	1.885	2.941	−19
Case	$k-\omega$ SST	$k-\varepsilon$ RNG	SA	$\overline{\upsilon^2}-f$
1	78.5	84.9	67.7	83.6
2	188.4	215.1	151	167.2
3	291.2	336.6	261.3	318.8
4	491.1	499.2	304.1	512.2
5	697.9	722.8	652.4	747.4
6	935.5	1,000.3	917.7	1,063.6
7	1,158.6	1,338.8	1,096.3	1,306.6
8	1,408.2	1,626.2	1,306.6	1,549.6
9	1,561.3	1,827.3	1,483.5	1,803.1
10	1,704.7	2,003.1	1,643.2	1,931.2
11	1,813.8	2,095.4	1,737.8	2,040.9
12	1,853.2	2,078.6	1,795.3	2,137.6
13	1,882.9	2,218.6	1,837.1	2,164.2
14	1,898.3	2,246.5	1,852.8	2,192.5
15	1,900.2	2,255.7	1,853.4	2,216
16	1,909.4	2,278.9	1,860.8	2,229.1
17	1,912.8	2,293.5	1,867.4	2,231.7
18	1,917	2,302.7	1,866	2,236.9
19	1,917.6	2,318.3	1,868.5	2,238.4
20	1,921.3	2,319.4	1,868.7	2,239.5
21	1,928.5	2,324.5	1,869.8	2,242.2
22	1,932.7	2,326.6	1,873.7	2,244.7

and tower, so their effect on the air flow is neglected; (2) these turbines are mounted at hub height of around 70–80 m above the ground [19], while we have assumed the rotating blades free in the air, so this height is not considered in the simulations; (3) wind turbines are usually installed together in farms in big quantities resulting in reduction of overall performance of turbines, mainly because of wakes produced by adjacent turbines. But current study doesn't take this issue in account, so that the results are optimistic to some extent.

Therefore considering above assumptions, roughly we can expect that the results should show some over prediction of power values. The $k-\varepsilon$ RNG has overpredicted values with about 22–50 % deviation, least of which corresponds to rated power of turbine, while maximum error is for cut-in velocities of turbine. However, $\overline{\upsilon^2}-f$ results show smaller deviation and can be labeled as more "realistic" model among the other approaches regarding assumptions of the problem.

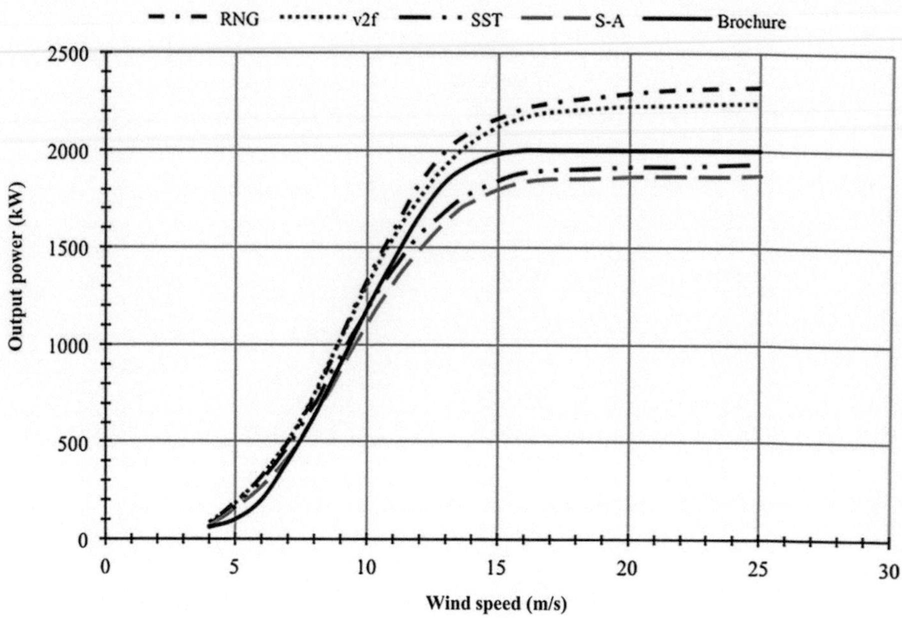

Fig. 36.4 Power curves of four used CFD methods compared with brochure data

In addition to power curve ($P - U_\infty$ diagram), another important plot, which is never disregarded by wind turbine designers, is the variation of power coefficient with tip speed ratio because we are always looking for a quasi-constant C_P values over various operating wind velocities [20]. The corresponding plots extracted from simulation results are given in Fig. 36.5 as well as data obtained in [20]. As shown in Fig. 36.5, the peak value of pressure coefficient (0.452) is well predicted by $k - \omega$ SST model in comparison with that of experiments, which is about 0.436. This could be because of the model's intrinsic assumption, that is, the hybrid SST model solves simplified Navier–Stokes equations for low Reynolds regime near the wall, while it uses high Reynolds assumption for region outside the boundary layer.

Given the pathlines, colored by velocity magnitude, and velocity vectors obviously showing mild separation around blade tip in Fig. 36.6, as a future work it may be useful to mount winglet-like end plates on blade tips to prevent separation. By employing such plates, clearly the flow would pass the tip area smoother, hence with fewer disturbances.

The pressure coefficient distribution on the surface and leading edge of the wind turbine's blades is illustrated in Fig. 36.7. As indicated, the pressure coefficient values increase on the blade as nearing to the leading edge. This is due to stagnation near the leading edge; the fluid particles almost stop when they meet the first contact area on blades, which means a sudden fall in velocity magnitude. The change in pressure is accounted for by the Bernoulli equation, which means for

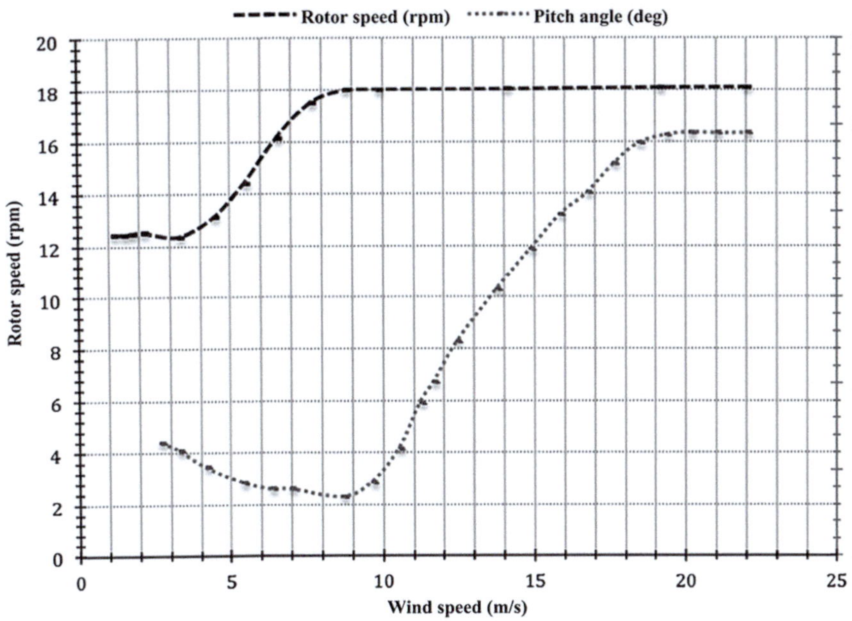

Fig. 36.5 Power coefficient versus tip speed ratio plot of four used CFD models and trend given in [18]

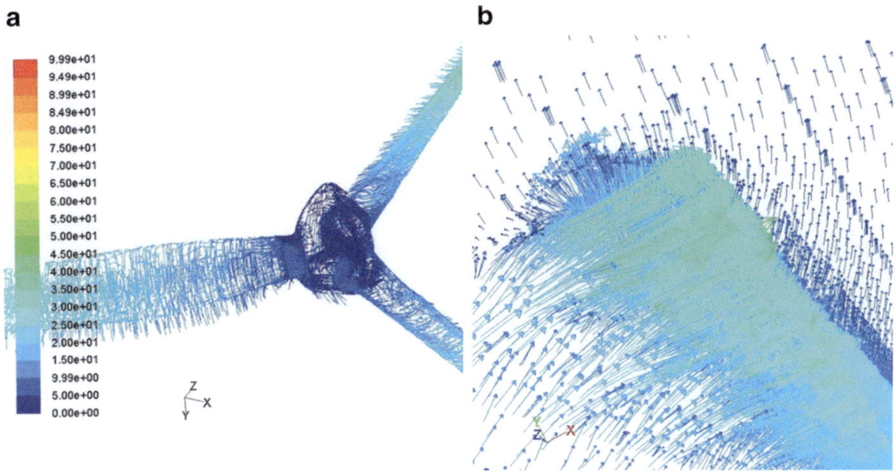

Fig. 36.6 (**a**) Pathlines by velocity magnitude and (**b**) velocity vectors showing mild separation around blade tip

horizontal fluid flow, a decrease in the velocity of flow will result in an increase in static pressure. This fact could be of help to designers to consider such a magnificent load on the leading edge to prevent destructive phenomena such as fatigue or creeping in a more effective way.

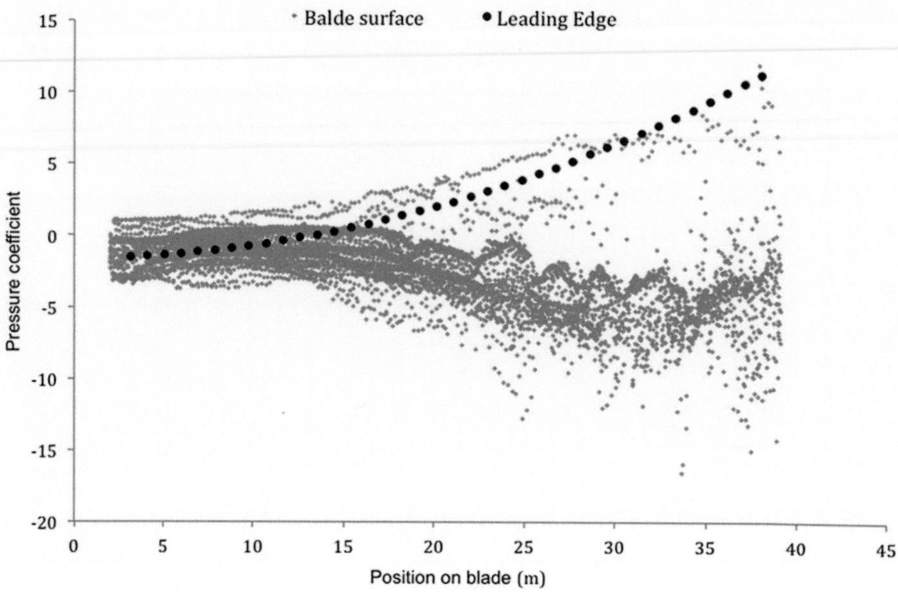

Fig. 36.7 Pressure coefficient distribution along blade surface and the leading edge

Finally, as discussed earlier, the closest results for generated output power, in comparison with the experiments, are related to $k - \varepsilon$ RNG turbulence method. However, for the power coefficient all employed methods have given acceptable outcomes for output power.

36.10 Conclusion

The main objective of this study was to numerically investigate the turbulent airflow passed a commercial HAWT using four different turbulence models with least simplifying considerations to get the most possible precise results, and then comparing gained results with those of experimental methods as well as brochure data provided by the manufacturer. The significant outstanding difference between current study and the others is inclusion of boundary layer and near-wall effects in the simulations, which gives more realistic and accurate results. It was demonstrated that results of performed simulations were in good agreement with experimental data presented previously. The closest results, for generated output power, compared to experiments are associated with the $k - \varepsilon$ RNG turbulence method. In addition, due to the mild separation occurring at the blades tips, it is suggested to mount winglet-like end plates to reduce the induced drag and improve the overall performance of the turbine.

References

1. Widjanarko SM (2010) Steady blade element momentum code for wind turbine design validation tool. Internship Vestas Wind System A/S, Twente University
2. Hartwagner D, Horvat A (2008) 3D modeling of a wind turbine using CFD. In: NAFEMS conference
3. Galdamez RG, Ferguson DM, Gutierrez JR (2011) Design optimization of winglets for wind turbine rotor blades report. Thesis, Florida International University
4. Tangler J (2002) The nebulous art of using wind-tunnel airfoil data for predicting rotor performance. NREL/CP-500-31243, National Renewable Energy Laboratory, CO
5. Vaz JRP, Pinho JT, Mesquita ALA (2011) An extension of BEM method applied to horizontal-axis wind turbine design. J Renew Energy 36:1734–1740
6. Esfahanian V, Salavati Pour A, Harsini I, Haghani A, Pasandeh A, Shahbazi A, Ahmadi G (2013) Numerical analysis of flow field around NREL Phase II wind turbine by a hybrid CFD/BEM method. J Wind Eng Ind Aerodyn 120:29–36
7. Sezer-Uzol N, Long LN (2006) 3-D time-accurate CFD simulations of wind turbine rotor flow fields. AIAA. Paper No. 2006-0394
8. Zhou B, Chow FK (2012) Turbulence modeling for the stable atmospheric boundary layer and implications for wind energy. J Flow Turbul Combust 88:255–277. doi:10.1007/s10494-011-9359-7
9. Chathelain P, Backaert S, Winkelmans G, Kern S (2013) Large eddy simulation of wind turbine wakes. Flow Turbul Combust 1(3):587–605. doi:10.1007/s10494-013-9474-8
10. Jeon M, Lee S (2014) Unsteady aerodynamics of offshore floating wind turbines in platform pitching motion using vortex lattice method. J Renew Energy 65:207–212
11. Lanzafame R, Mauro S, Messina M (2013) Wind turbine CFD modeling using a correlation-based transitional model. J Renew Energy 52:31–39
12. Wang Q, Zhou H, Wan D (2012) Numerical simulation of wind turbine blade-tower interaction. J Mar Sci 11:321–327
13. Shojaeefard MH, Tahani M, Ehghaghi MB, Fallahian MA, Beglari M (2012) Numerical study of the effects of some geometric characteristics of a centrifugal pump impeller that pumps a viscous fluid. Comput Fluids 60:61–70
14. Yakhot V, Orszag SA (1986) Renormalization group analysis of turbulence: I. Basic theory. J Sci Comput 1(1):1–51
15. Menter FR (1994) Two-equation eddy-viscosity turbulence models for engineering applications. AIAA J 32(8):1598–1605
16. Laurence DR, Uribe JC, Utyuzhnikov SV (2004) A robust formulation of the $\overline{v}^2 - f$ model. J Flow Turbul Combust 73:169–185
17. Durbin PA, Pettersson Reif BA (2001) Statistical theory and modeling for turbulent flows. Wiley, Chichester
18. Hansen KS, Barthelmie RJ, Jensen LE, Sommer A (2012) The impact of turbulence intensity and atmospheric stability on power deficits due to wind turbine wakes at horns rev wind farm. J Wind Energy 15:183–196. doi:10.1002/we.512
19. http://www.vestas.com. Accessed 29 Sept 2013
20. Wood D (2011) Small wind turbines, analysis, design, and application. Springer, London

Chapter 37
Fabrication and Characterization of $Zn_{1-x}Mg_xO$ Films for Photovoltaic Application

Olcay Gencyilmaz, Ferhunde Atay, and Idris Akyuz

Abstract $Zn_{1-x}Mg_xO$ (ZMO) films were prepared by the ultrasonic spray pyrolysis technique and effects of doping on structural, optical, electrical, and surface properties were examined. The film structures were studied by X-ray diffraction. X-ray diffraction patterns of the films showed that the ZMO films exhibited hexagonal wurtzite crystal structure with a preferred orientation along (0 0 2) direction. Texture coefficient, grain size values, and lattice constants were calculated. The optical properties like transmission, reflection, and absorption were investigated with UV–Vis spectrophotometer. The optical measurements reveal a shift in absorption edge and optical band gap of ZMO films changed with Mg content. Optical parameters (refractive index, extinction coefficient) and thicknesses of the films were investigated by spectroscopic ellipsometry (SE). Surface and electrical properties of the films were studied using atomic force microscopy and four-probe technique, respectively. After all investigations, it is concluded that the ZMO thin films can be used in photovoltaic solar cells as window materials and cell efficiencies can be increased using different content rates.

Keywords Spray pyrolysis • Spectroscopic ellipsometry • Atomic force microscopy • Four-probe technique

Nomenclature

λ	Wavelength, nm
2θ	Diffraction angle, degree
B_n	SE model parameter, $(nm)^2$
C_n	SE model parameter, $(nm)^4$
E_g	Band gap, eV

O. Gencyilmaz (✉)
Physics Department, Cankiri Karatekin University, Cankiri 18100, Turkey
e-mail: eren_o@hotmail.com

F. Atay • I. Akyuz
Physics Department, Eskisehir Osmangazi University, Eskisehir 26480, Turkey

I. Dincer et al. (eds.), *Progress in Clean Energy, Volume 1*,
DOI 10.1007/978-3-319-16709-1_37

d Thickness, nm
h Planck constant, eV s
α Absorption coefficient, m^{-1}
ν Frequency, sn^{-1}
ρ Resistivity, Ω cm
σ Conductivity, $(\Omega\ cm)^{-1}$
ϕ Quality factor, Ω^{-1}
a Lattice parameter, Å
c Lattice parameter, Å
D Grain size, nm

37.1 Introduction

ZnO is one of the most important members of Transparent Conductive Oxide (TCO) materials [1–4], and it is a very interesting material for many different applications in both microelectronic and optoelectronic devices. ZnO thin film is an n-type semiconductor with an optical gap of about 3.3 eV, and it exhibits a high transmittance in the visible region of the radiation spectra [5–7]. Moreover, ZnO thin films used in transparent conductive contacts, solar cells, laser diodes, ultraviolet lasers, thin film transistors, optoelectronic and piezoelectric applications to surface acoustic wave devices [8, 9]. Recently, ternary $Zn_{1-x}Mg_xO$ (ZMO) with wider band gap than ZnO ($E_g = 3.37$ eV) has received much attention due to its potential applications in ultraviolet optoelectronic devices [10, 11]. ZMO thin films have emerged as one of the important compound semiconductors due to high exciton binding energy and its tunable band gap from 3.37 to 6.7 eV depending upon the Mg content.

Different deposition techniques have been widely used to produce ZnO and ZMO thin films. However, seeking the most reliable and economic deposition technique is the main goal. The most intensively studied techniques include chemical vapor deposition (CVD) [12], sol–gel method [13], thermal evaporation [14], RF magnetron sputtering [15], and spray pyrolysis [16, 17]. Among these, spray pyrolysis is one of the most widely used methods. Spray pyrolysis has been developed as a powerful tool to prepare various kinds of thin films such as metal oxides, superconducting materials, and nanophase materials. In comparison with other chemical deposition techniques, spray pyrolysis has several advantages such as high purity, simplicity, safety, low cost of the apparatus and raw materials, reproducible, excellent control of chemical uniformity, and stoichiometry in multicomponent system. The other advantage of the spray pyrolysis method is that it can be adapted easily for production of large-area films.

In this study, ZMO thin films were deposited by spray pyrolysis on glass substrates. The influence of Mg contribution rates on the structural, optical, and electrical properties of ZnO thin films was investigated. A detailed study of changes of physical properties in the film would be interesting and can provide useful information related to these properties, which are very important for better understanding and improvement of the quality of the film and for the fabrication of ZnO-based devices.

37.2 Results and Discussion

37.2.1 *Experimental*

ZnO and $Zn_{1-x}Mg_xO$ (ZMO) thin films of various Mg doped concentrations (1 %, 3 %, 5 %) were grown at 350 ± 5 °C on glass substrates by ultrasonic spray pyrolysis. 0.1 M zinc acetate dihydrate [$Zn(CH_3COO)_2 \cdot 2H_2O$] and 0.1 M magnesium acetate [$Mg(CH_3COO)_2$] were used as starting precursors. The glass substrates were ultrasonically cleaned in acetone and rinsed in deionized water. The nozzle substrate separation used was 30 cm. The solution flow rate was kept at 5 mL min^{-1} and controlled by a follow meter. Totally, 100 mL of solution was used and sprayed for 20 min.

The crystal quality of the ZnO and ZMO thin films was analyzed by XRD in θ–2θ geometry. Cu Kα source ($\lambda = 1.5406$ nm) was used, and the scanning range was between $2\theta = 20°$ and 80°. Optical properties in the wavelength range of 300–900 nm were studied by a UV–Vis spectrometer. The film thickness, refractive index, and extinction coefficient were measured by spectroscopic ellipsometry (SE) with a He–Ne laser as the light source. The resistivity of the film was measured by a four-point probe resistivity method at room temperature. The surface morphology was characterized by atomic force microscopy (AFM).

37.2.2 *Structural Properties*

The crystal structures of the ZnO and ZMO thin films were investigated by XRD. The normalized XRD patterns of the ZMO films prepared under different Mg concentration are presented in Fig. 37.1. The X-ray pattern (Fig. 37.1) depicted that all the films were polycrystalline in nature with hexagonal wurtzite structure. The (0 0 2) peaks are stronger than other peaks, indicating that all the samples have a wurtzite structure and are preferentially oriented along the c-axis direction perpendicular to the substrate surface. Besides the change in the (0 0 2) peak position, there is also a change in the intensity of (0 0 2) peaks of the samples. For samples ZnO, ZMO1, ZMO3, and ZMO5, there is no significant change in the intensity of (0 0 2) peaks. As the Mg concentration is increased, another small diffraction peaks corresponding to the (1 0 2), (1 0 3), (1 1 2), (1 1 0) plane of ZnO becomes visible. Also, the (1 0 1) peak is enhanced as the Mg concentration increases. Diffraction peaks from other phase such as MgO are not detected, indicating that the ZMO thin films are composed of a single-phase wurtzite ZnO.

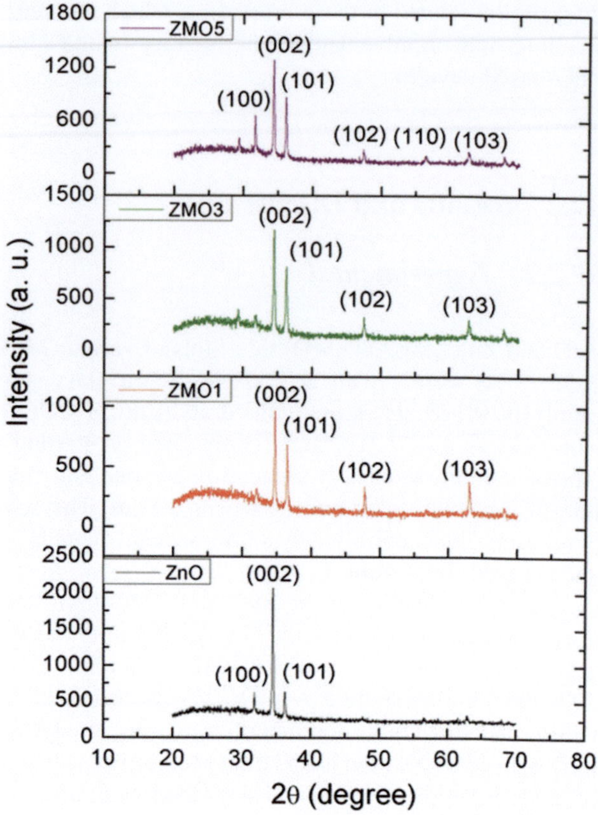

Fig. 37.1 XRD patterns of ZnO and ZMO thin films prepared under different Mg concentration

Table 37.1 Lattice parameters (*a* and *c*), texture coefficient (TC) values, and the grain size (*D*) for ZMO thin films

Sample no.	TC (0 0 2)	*D* (nm)	*a* (Å)	*c* (Å)
ZMO	2.89	33	3.22	5.16
ZMO1	2.88	38	3.25	5.20
ZMO3	2.81	40	3.24	5.21
ZMO5	2.16	44	3.24	5.20

Since the Mg content of the ZMO films is small, the ZMO wurtzite structure is not changed. However, when Mg is doped in ZnO, the (0 0 2) peak shifts towards the big angle direction. Because the radius of Mg^{2+} is smaller than that of Zn^{2+}, when Mg doping concentration is relatively high, it will lead to large lattice distortion in ZnO. The similar phenomena have also been reported by others [18–24].

In order to investigate a possibility of the preferred orientation, the Harris analysis was performed [25, 26]. The calculated texture coefficient (TC) values and the grain size (*D*) are given in Table 37.1. It was determined that all films have three TC values bigger than one. Also, lattice parameters for all samples have been determined for the orientations with highest TC value and compared with the ones in ASTM (American Society of Testing Materials) cards. These values are given in Table 37.1. A good agreement is found between these two values.

37.2.3 Optical Properties

PHE-102 spectroscopic ellipsometer (250–2,300 nm) was used to determine the ψ parameters, refractive indices (n), extinction coefficient (k), and thicknesses (d) of the films. Cauchy–Urbach dispersion model was used to fit the experimental ψ parameters. This model is a modified type of Cauchy model. In the Cauchy–Urbach dispersion model, the refractive index $n(\lambda)$ and the extinction coefficient $k(\lambda)$ as a function of the wavelength are given by,

$$n(\lambda) = A + \frac{B}{\lambda^2} + \frac{C}{\lambda^4} \tag{37.1}$$

$$k(\lambda) = \alpha \exp \beta \left(1{,}240 \left(\frac{1}{\lambda} - \frac{1}{\gamma} \right) \right) \tag{37.2}$$

where A, B, C, α, β, and γ are model parameters [27]. The incident angle is an important factor for the films having depolarization effect. Three different incident angles (50°, 60°, and 70°) were tried to take the measurements. The best angle was determined to be 70° using experimental Ψ spectra. Later, the elipsometric parameters (A, B, C,α, β, and γ) related to the Cauchy–Urbach model were defined. Spectroscopic ellipsometry (SE) spectra of ZMO films are shown in Fig. 37.2. Thicknesses and model parameters are given in Table 37.2.

Ψ spectra are found a good fit between model and experimental data in Fig. 37.2. However, there are some small deviations on Ψ values. This deviation is probably due to the depolarizing effect of roughness, grain boundaries, and morphologies of

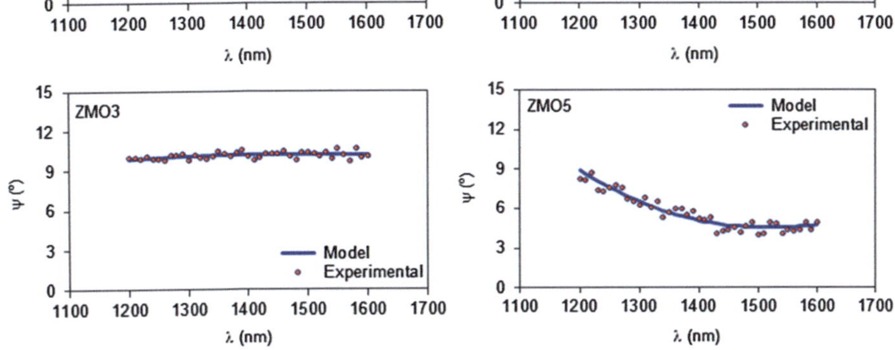

Fig. 37.2 SE spectra of ZMO films

Table 37.2 Thicknesses, model parameters, and band gaps of ZnMO films

Sample no.	d (nm)	A_n	B_n ($\times 10^{-2}$) (nm)^2	C_n ($\times 10^{-3}$) (nm)^4	A_k	B_k (eV)$^{-1}$	E_g (eV)
ZnO	258	2.21	0.08	0.02	0.01	1.51	3.20
ZMO1	268	2.11	0.01	0.03	0.03	1.43	3.06
ZMO3	273	1.72	0.09	0.02	0.01	1.56	3.32
ZMO5	271	2.08	0.01	0.09	0.03	1.47	3.35

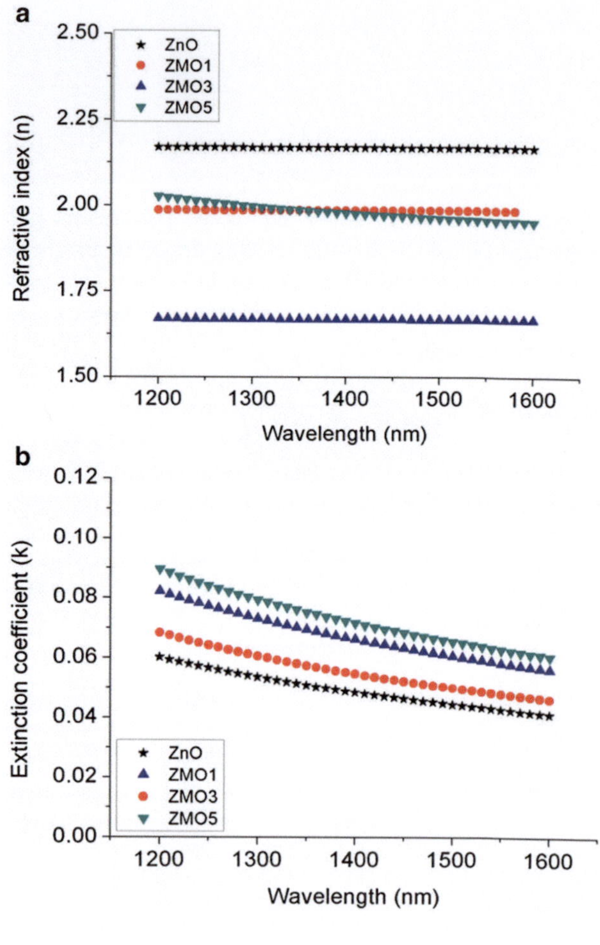

Fig. 37.3 (**a**) Refractive index (n) and (**b**) extinction coefficient (k) spectra of ZMO films

the films which affect the experimental data. Backside reflection of glass substrates may also cause deviations on fitted values.

Figure 37.3a and b shows the changes of refractive index and extinction coefficient of the films with different Mg content in the wavelength range of 1,200–1,600 nm. It is clear that Mg content has an important effect on refractive index and extinction coefficient values. As can be seen, the refractive index of the ZMO films

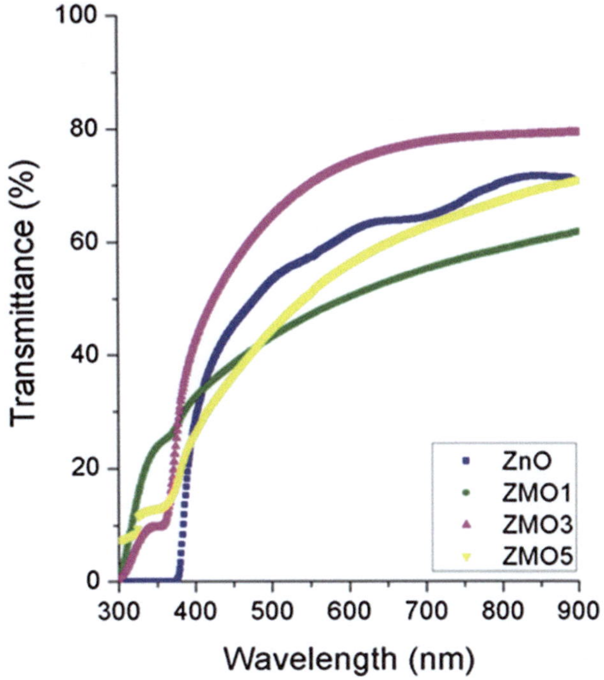

Fig. 37.4 Transmission spectra of ZMO films

decreases with Mg-doped in the wavelength range from 1,200 to 1,600 nm, but extinction coefficient values increases.

In addition, the extinction coefficient decreases when the wavelength is larger than 1,200 nm. We think that there may be three reasons for this variation: (1) it is well known that Mg impurity doped into ZnO films can act as effective n-type donors to generate free carriers. (2) With the increasing Mg content, the carrier concentration in the ZMO films increases. The increase of the carrier concentration results in the decrease of the refractive index. Therefore, we can control the refractive index of the ZMO films by varying the Mg content, which is important for the applications in designing integrated photovoltaic application.

The transmission spectra of ZMO films were recorded as depicted in Fig. 37.4. The optical transmission is around 75 % in the visible region (400–700 nm). From Fig. 37.4 it is clear that all the films have sharp absorption edges in the wavelength region between 360 and 380 nm. These absorption edges shifted to shorter wavelengths (blue shifted) when Mg was incorporated into ZnO films; we think that these blue shifts may have been caused by doping-induced film degradation.

Also, we measured the thickness of the films by spectroscopic ellipsometry. The sharp step required for measuring the thickness in this method was obtained by etching and the corresponding step height gives the actual thickness of the film. This measured thickness was then used to calculate the absorption coefficient and

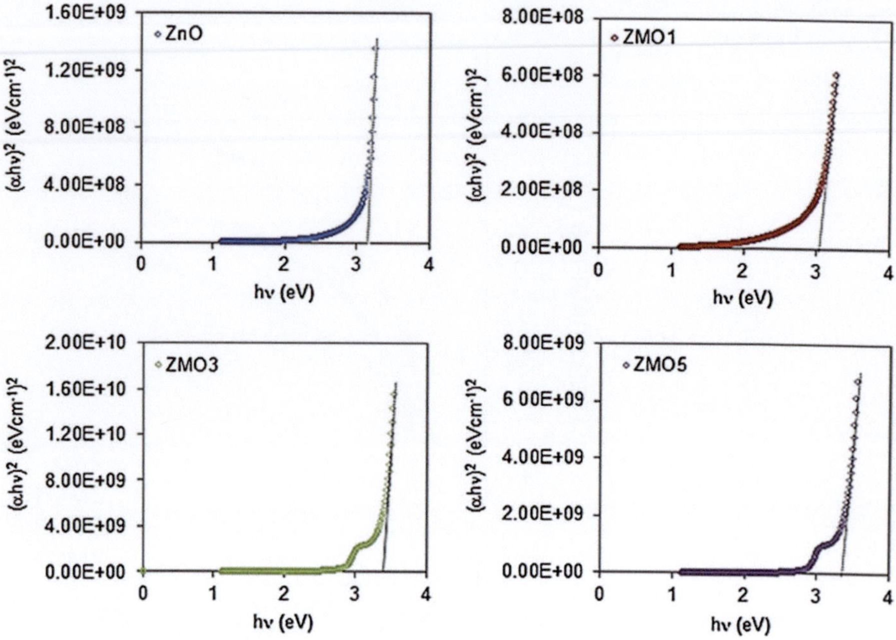

Fig. 37.5 Optical band gap energy estimation of ZMO films

band gap values of the films were determined by optical method and we used the Tauc relationship [28] as follows:

$$\alpha h\nu = A\left(h\nu - E_g\right)^n \tag{37.3}$$

where α is the absorption coefficient, A the constant, h the Planck's constant, n the photon frequency, E_g the optical band gap, and n the 1/2 for direct band gap semiconductors.

Since $E_g = h\nu$ when $(\alpha h\nu)^2 = 0$, an extrapolation of the linear region of the plot of $(\alpha h\nu)^2$ versus photon energy ($h\nu$) on the x-axis gives the value of the optical band gap E_g. Effect of Mg content on the band gap of ZMO films were investigated as shown in Fig. 37.5. From Fig. 37.5, we notice that the band gap of the ZMO films first increases and then decreases, reaching a maximum values 3.35 eV at Mg content 5 %. The shift in the band gap towards higher value of band gap energy suggest that most of Mg atoms may replace the Zn lattice sites in the crystal structure and remaining Mg may have been incorporated at the interstitial sites and grain boundaries.

37.2.4 Surface Properties

An AFM was used to measure the surface roughness of the films over (5 μm × 5 μm) area in noncontact mode. The three-dimensional (3D) images of AFM micrographs

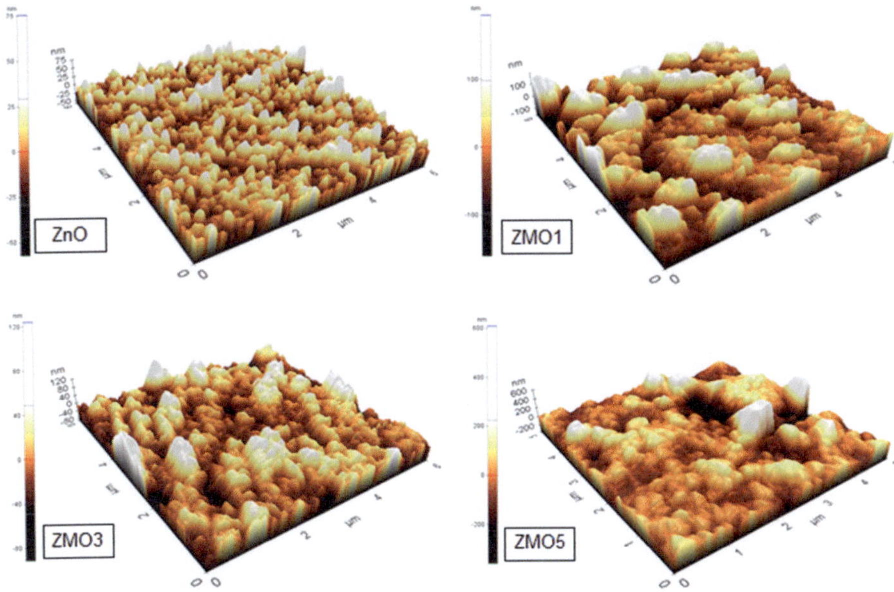

Fig. 37.6 ZMO films surface topography

Table 37.3 Electrical resistivity and roughness values of ZMO films

Sample no.	$\rho \times 10^6$ (Ω cm)	R_q (nm)	R_a (nm)	R_{pv} (nm)
ZnO	1.21	14	11	132
ZMO1	3.96	49	39	354
ZMO3	6.68	24	19	215
ZMO5	8.09	54	44	304

are given in Fig. 37.6. Also, root mean square (R_q), average (R_l), and peak-valley (R_{pv}) roughness values of the films have been determined by XEI version 1.7.1 software and these values are given in Table 37.3. ZnO films a partly granular texture and island-growth mechanism is dominant when compared to others. For ZMO films have a smoother surface texture showing a layer-by-layer growth mechanism. As expected, the surface roughness increases as a result of grains increasing caused by the Mg content. Consequently, all films have a reasonable roughness value which is a desired characteristic for opto-electronic applications.

37.2.5 *Electrical Properties*

Electrical resistivity of ZMO films were determined by a four-point probe setup. Electrical resistivity values of ZMO films obtained on different Mg content are

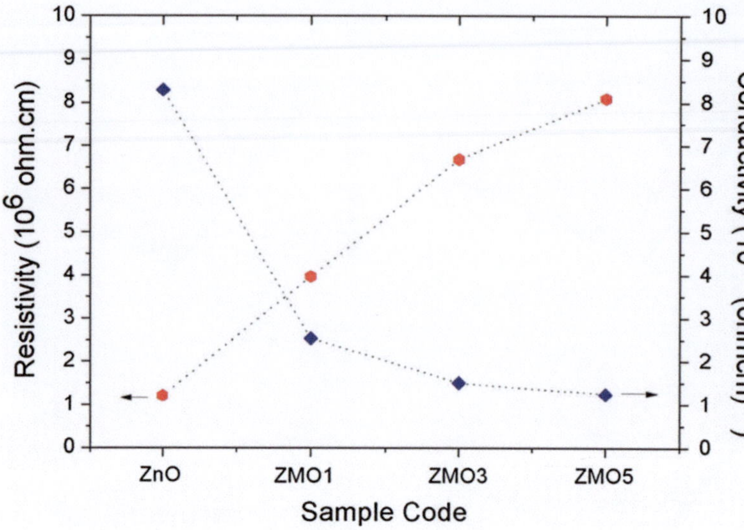

Fig. 37.7 Resistivities and conductivity values of ZMO films

given in Table 37.3. Figure 37.7 shows the resistivities and conductivity values of the ZMO films with different Mg content. It is clear that the resistivity of the ZnO films increases with increasing of Mg concentration. As we know that the valences of Zn^{2+} and Mg^{2+} are the same, and the substitution of Mg^{2+} for the group II cation of Zn should not generate or consume carriers of ZnO. However, electrical properties of the ZnO thin films change after Mg doping. This presumably results from the substitution of Mg ions with a smaller ionic radius for Zn sites [29]. As a result, the resistivity of the ZnO thin films increases after Mg is doped. We think that with the help of XRD results, which results in lattice distortion to some extent. Also, the ZMO films may have less oxygen vacancies and this situation will lower the carrier concentration and make the films have higher resistivity than ZnO films. Because oxygen helps to decrease the number of excess Mg carriers, which also contributes to the increasing resistivity. On the other hand, there is also the possibility that a spot of Mg at the grain boundaries may produce electrical barriers, increasing scattering of the carriers, and thus increase the resistivity. The most important characteristics for transparent conductive films in solar cell applications are low resistivity and high transparency. For this reason, we try to investigate a quality factor (ϕ) for the films including average transmittance (at 300–900 nm) and sheet resistance (R_s) values using Haacke's figure of merit $\left(\phi = T^{10}/R_s\right)$ [30]. The ϕ values of all films were calculated as 2.80×10^{-11}, 1.75×10^{-11}, 2.18×10^{-11}, and 3.95×10^{-11} $(\Omega)^{-1}$, respectively. Figure 37.8 shows quality factor values of ZMO films. It is seen that the sample with 5 % Mg concentration has a high-quality factor.

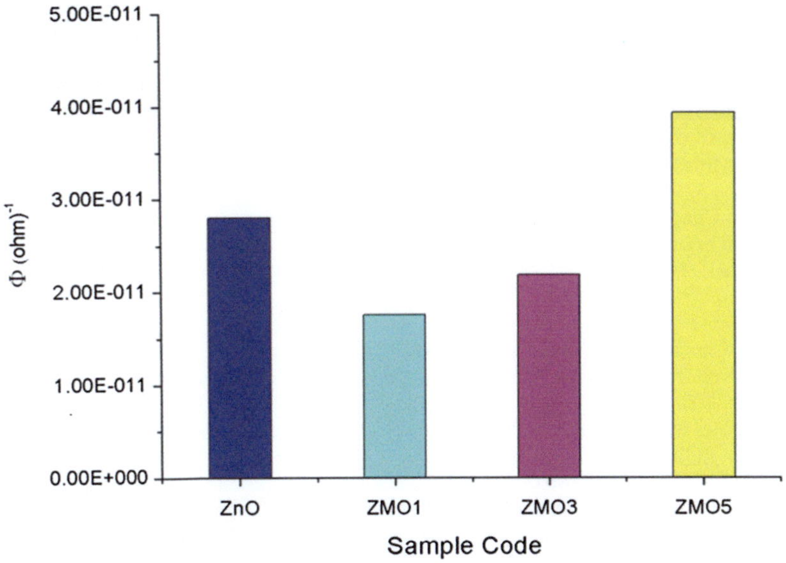

Fig. 37.8 Quality factor values of ZMO films

37.3 Conclusions

In this work, ZnO and ZMO films have been successfully prepared on glass substrates using ultrasonic spray pyrolysis. We have studied the structural, optical, electrical, and surface properties of ZMO films prepared under different Mg content rates. The results show that, the low cost of the production system is an advantage for potential applications of ZMO films. Also, XRD patterns reveal that the ZnO and ZMO films are composed entirely of the wurtzite ZnO structure. ZMO3 films caused the transmittance values to increase which is a desired development for transparent conducting oxide industry and solar cells. AFM analysis revealed that roughness values of ZMO films increases with increasing Mg content. The resistivity increases from 1.21×10^6 Ω cm for the ZnO films to approximately 8.09×10^6 Ω cm for ZMO films. The variations in optical absorption edges and band gap energies of ZMO films were also investigated. The band gap values have been obtained from 3.20 to 3.35 eV for ZMO films. These properties demonstrate that the ZMO3 films are an ideal material for the window layer of solar cells and also have potential application in UV detectors.

References

1. Tang ZK, Wong GKL, Yu P, Kawasaki M, Ohtomo A, Koinuma H, Segawa Y (1998) Room temperature ultraviolet laser emission from self-assembled ZnO micro crystallite thin films. Appl Phys Lett 72:3270
2. Bagnall DM, Chen YF et al (1997) Optically pumped lasing of ZnO at room temperature. Appl Phys Lett 70:2230
3. Major S, Chopra KL (1988) Indium-doped zinc oxide films as transparent electrodes for solar cells. Sol Energy Mater 17:319
4. Chopra KL, Major S, Panday DK (1983) Transparent conductors a status review. Thin Solid Films 102:1
5. Liu J, Xia H, Xue D, Lu L (2009) Double-shelled nanocapsules of V_2O_5-based composites as high-performance anode and cathode materials for Li ion batteries. J Am Chem Soc 131:12086
6. Yan C, Xue D (2008) Formation of Nb_2O_5 nanotube arrays through phase transformation. Adv Mater 20:1055
7. Xia H, Xiao W, Lai MO, Lu L (2009) Facile synthesis of novel nanostructured MnO_2 thin films and their application in supercapacitors. Funct Mater Lett 2:13
8. Jiang DX, Cao L, Liu W, Su G, Qu H, Sun YG, Dong BH (2009) Synthesis and luminescence properties of core/shell ZnS:Mn/ZnO nanoparticles. Nanoscale Res Lett 4:78
9. Hwang DK, Jeong MC, Myoung JM (2004) Effects of deposition temperature on the properties of $Zn_{1-x}Mg_xO$ thin films. Appl Surf Sci 225:217
10. Kaidashev E, Lorenz M, Von Wenckstern H, Rahm A, Semmelhack HC, Han KH, Benndorf G, Bundesmann C, Hochmuth H, Grundmann M (2003) High electronmobility of epitaxial ZnO thin films on c-plane sapphire grown by multisteppulsed-laser deposition. Appl Phys Lett 82:3901
11. Bonasewicz P, Hirschwald W, Neumann G (1986) Influence of surface processes on electrical, photochemical, and thermodynamical properties of zinc oxide films. J Electrochem Soc 133:2270
12. Lin CW, Cheng TY, Chang L, Juang JY (2004) Chemical vapor deposition of zinc oxide thin films on Y_2O_3/Si substrates. Phys Status Solidi C 1:851
13. Ohyama M, Kozuka H, Yoko T, Sakka S (1996) Preparation of ZnO thin films with preferential orientation by sol–gel method. J Ceram Soc 104:296
14. Dinescu M, Verardi P (1996) ZnO thin films deposition by laser ablation of Zn target in oxygen reactive atmosphere. Appl Surf Sci 106:149
15. Jayaraj MK (2002) Transparent conducting zinc oxide thin film prepared by off-axis RF magnetron sputtering. Bull Mater Sci 25:227
16. Wu X, Cotts TJ, Mulligan WP (1997) Properties of transparent conducting oxides formed from CdO and ZnO alloyed with SnO_2 and In_2O_3. J Vac Sci Technol A 15:1057
17. Subramanyam TK, Uthanna S, Sinivasulu Naidu B (1998) Preparation and characterization of CdO films deposited by dc magnetron reactive sputtering. Mater Lett 35:214
18. Kumar S, Gupte V, Screenivas K (2006) Structural and optical properties of magnetron sputtered $Mg_xZn_{1-x}O$ thin films. J Phys Condens Matter 18:3343
19. Ghosh R, Basak D (2009) The effect of growth ambient on the structural and optical properties of $Mg_xZn_{1-x}O$ thin films. Appl Surf Sci 255:7238
20. Kumar P, Singh JP, Kumar Y, Gaur A, Malik HK, Asokan K (2012) Investigation of phase segregation in $Zn_{1-x}Mg_xO$ systems. Curr Appl Phys 12:1166
21. Tsukazaki A, Ohtomo A, Nakano M, Kawasaki M (2008) Photoinduced insulator-to-metal transition in ZnO/$Mg_{0.15}Zn_{0.85}O$ heterostructures. Appl Surf Sci 92:052105
22. Chu S, Zhao J, Zuo Z, Kong J, Li L, Liu J (2011) Enhanced output power using MgZnO/ZnO/MgZnO double heterostructure in ZnO homojunction light emitting diode. J Appl Phys 109:123110
23. Benelmadjat H, Boudine B, Halimi O, Sebais M (2009) Fabrication and characterization of pure and Sn/Sb-doped ZnO thin films deposited by sol–gel method. Opt Laser Technol 41:630

24. Liu YC, Chen YW, Shao CL, Lu SX (2006) Structural, optical and photoelectric properties of ZnO:In and $Mg_xZn_{1-x}O$ nanofilms prepared by sol–gel method. Sol–Gel Sci Technol 39:57
25. Barrett CS, Massalski TB (1980) Structure of metals. Pergamon, Oxford, p 204
26. Bilgin V, Kose S, Atay F, Akyuz I (2005) The effect of Sn concentration on some physical properties of zinc oxide films prepared by ultrasonic spray pyrolysis. J Mater Sci 40:1909
27. Atay F, Bilgin V, Akyuz I, Kose S (2003) The effect of In doping on some physical properties of CdS films. Mater Sci Semicond Process 6:197
28. Girtan M, Folcher G (2003) Structural and optical properties of indium oxide thin films prepared by an ultrasonic spray CVD process. Surf Coat Technol 172:242
29. Han JP, Senos AMR, Mantas PQ (2002) Varistor behaviour of Mn-doped ZnO ceramics. J Eur Ceram Soc 22:1653
30. Haacke G (1976) New figure of merit for transparent conductors. J Appl Phys 47:4086

Chapter 38
Activated Zeolites and Heteropolyacids Have Efficient Catalysts for Synthesis Without Use of Organic Solvent at Room Temperature

Mohamed Hammadi, Hadjila Dokari, Didier Villemin, and Nassima Benferrah

Abstract The Thiele–Winter reaction is of interest for synthesis of triacetoxyaromatic precursors of hydroquinones.

Solid acids such as heteropolyacids and activated zeolites have an efficient catalyst in acetoxylation reaction of quinones without the use of organic solvent at room temperature. Hydroquinones and substituted hydroquinones was easily oxidized at room temperature in quinones by using (Pc[Co]/K10) and air (1 atm). We have also tested this type of reaction in naphthoquinone series. Many naphthoquinones are natural products with interesting biological properties. The catalytic system (Pc[Co]/K10) in the presence of pure oxygen easily oxidizes the 1,5-dihydroxynaphthalene in Juglone (yield 87 %).

Keywords Activated zeolites • Heteropolyacids • Without solvent • Room temperature • Catalytic system (Pc[Co]/K10)

Nomenclature

(Pc[Co]/K10)	Phthalocyanine (Co) supported on clay K10
Ho	Index hammett
$H_3PMo_{12}O_{40}$	Molybdophosphoric acid
$ClSO_3H$	Chlorosulfonic acid
$H_3PW_{12}O_{40}$	12 Tunsto phosphoric acid
H_2SO_4	Sulfuric acid
$(CH_3)_2(CO)_2O$	Acetic anhydride
Rt	Room temperature
Mp	Melting points

M. Hammadi (✉) • H. Dokari • N. Benferrah
Faculty of Sciences, University of Boumerdes, Boumerdes 35000, Algeria
e-mail: mohammedhammadi@yahoo.fr

D. Villemin
Ensicaen, LCMT—UMR 6507 CNRS, University of Caen, Caen 14050, France

I. Dincer et al. (eds.), *Progress in Clean Energy, Volume 1*,
DOI 10.1007/978-3-319-16709-1_38

NMR	Nuclear magnetic resonance
IR	Infrared spectroscopy
UV	Ultraviolet–visible spectrophotometry
MO	Microwave irradiation
$COCl_2$	Cobalt chloride
H	Hour
p	Power
T	Temperature
t	Time
Λ	Wavelengths
ε_λ	Molar extinction coefficient
$H_3SiW_{12}O_{40}$	Tungstosilicic acid

38.1 Introduction

The founding principles of green chemistry are mainly based on:

- The removal, if possible, of pollutant solvents that do not respect the environment. Nowadays, the use of inorganic solid acid catalysts in dry reactions (without organic solvent) offers many possibilities in the field of organic synthesis. Indeed they provide a simplification of the experimental protocols, easy recycling, and thereby easy protection of the environment.
- The improvement in the energy efficiency that beneficially affects the economy and the environment that must be considered and should be minimized by the development of synthetic methods at room temperature conditions.

One of the main problems in the chemical industry is the search for tolerable procedures for chemical production. There is a growing demand for the development of selective, efficient, and environmentally appropriate synthetic methods.

Organic solvents are not only expensive, but are often flammable, toxic, and hazardous. Generally the use of solvent is considered to be necessary in organic reactions.

The hydroxynaphthoquinones [1] are well-known substances for their chemical and biologic properties [2]. The hydroxynaphthoquinones including Phthiocol [3], Lawsone [4, 5], Juglone [6, 7], Menadione, and Atovaquone [8] have gained much interest due to their presence in natural products and their pharmacological properties as antitumoral, antiprotozoal, anti-inflammatory, antiviral, and antifungal [9, 10].

We have used this methodology to prepare the acetoxylation reactions of quinones and naphthoquinones acid-catalyzed by $ClSO_3H$ ($Ho = -13.80$) and heteropolyacids $H_3PW_{12}O_{40}$ ($Ho = -13.6$), in the absence of solvent at room temperature.

Our attention has focused on the use of metallic phthalocyanines supported in the oxidative reactions of naphthotriols prepared from the Thiele–Winter reaction in order to synthesize hydroxynaphthoquinone natural products having antibiotic

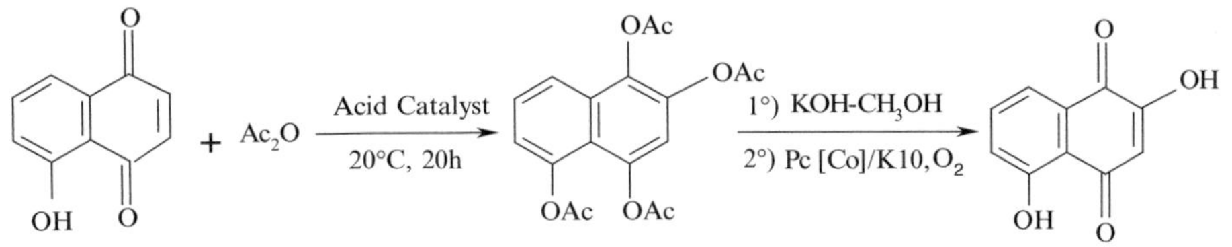

Fig. 38.1 Synthesis of (**3b**) from Juglone

properties. The acetoxylation reaction of Juglone, Menadione, and Lawsone with acetic anhydride $(CH3)_2(CO)_2O$ catalyzed by sulfuric acid H_2SO_4 [11, 12] (Ho = −11.94), at room temperature is slow with low yields especially in the presence of substituents on the naphthoquinone as electron donors. Sulfuric acid (H_2SO_4) was replaced by chlorozincique acid [13] and tetrafluoroboric acid [1] with low yields. To overcome this problem where the acidity of the medium thus protonation of the quinone is the decisive step, we opted for the use of liquid hyperacid more acidic than sulfuric acid in homogeneous conditions such as chlorosulfonic acid: $ClSO_3H$ [14] (Ho = −13.80) and solid super acid as heterogeneous conditions in the heteropolyacid: $H_3PW_{12}O_{40}$ (Ho = −13.6).

In our work we are particularly interested in the synthesis of 2,5-dihydroxy-1,4-naphthoquinone (**3b**). This synthesis involves first the oxidation of 1,5-dihydroxynaphthalene in product (**3**) and its transformation into tetracetoxy (**3a**) by the Thiele–Winter reaction. The saponification of (**3a**) followed by catalytic oxidation in situ of naphthotriols is shown in Fig. 38.1.

38.2 Experimental Procedure

Melting points (m.p.) were determined with a Kofler hot apparatus and are uncorrected. Proton NMR spectra (PMR) were determined on Brucker AC 250 (250 MHz, $CDCl_3$, Me_4Si).

The IR spectra were recorded as KBr pellets on Perkin Elmer 16 PC FT-IR spectrometer. UV–visible spectra (λ_{max} log(ε)) were obtained with spectrophotometer Perkin-Elmer Lamda 15.

Microwave irradiation was carried out with a commercial microwave oven (Toschiba ER 7620) at 2,450 MHz. and with resonance cavity TEo13, joined to a generator MES 73-800 of microwaves. Mass spectra were carried out on a Nermag Riber R10; TLC analyses were performed by using Kieselgel Schleicher and Shull F 1500 Ls 254 and Merck 60F 254. The grinding of products was carried out on analytical grinder A 10 of Janke and Kenkel-IKA Labortechnik.

38.2.1 Phthalocyanine(Co) Supported on K10

Procedure:

38.2.1.1 Preparation of Montmorillonite K10 Exchanged by Co^{2+}

In a 250-mL flask, the Montmorillonite K10 (20 g) was added to a solution of metallic salt $CoCl_2$ (0.2 mol) dissolved in 100 mL of distilled water. The reaction al mixture was stirred for 24 h at room temperature. The suspension was washed twice with distilled water then centrifuged. The Montmorillonite exchanged by Co^{2+} was washed with methanol and recentrifuged. The solid was dried for 24 h in vacuum then finely ground. The final product was a clear beige color.

38.2.1.2 Phthalocyanine Intercalated in the Montmorillonite K10

A solution of phthalonitrile (20 mmol; 2.56 g) dissolved in 20 mL of dichloromethane was added to the solid Montmorillonite K10 (5 g) exchanged with some metallic cations (Co^{2+}). After contact for 2 h, the remaining liquid was evaporated under reduced pressure. The activation of the solid under microwave irradiation (*MO*): $p =$ power, $t =$ time of irradiation) was carried out in a resonance cavity. After cooling, the solid was successively washed with water, acetone (20 mL), and then with dichloromethane (20 mL). The solid was dried under reduced pressure and the extracted with acetonitrile as a solvent using a Soxhlet for 8 h. Catalysts were characterized by FT-IR; electronic spectra of metalated phthalocyanine intercalated into Montmorillonite were very close to those observed with pure metalated phthalocyanine, but the bands were shifted.

(**Pc[CO] Supported on K10**): Purple solid, m.p. >300 °C MO; microwaves irradiation Resonance Cavity. ($p = 630$ W, $t = 10$ min); $C_{32}H_{16}CoN_8$; yields: 85 % UV–visible λ_{max} log(ε)/(1-chloronaphthalene) nm: 669.5 (4.42); 642.1 (3.88); 604.2 (3.78); 580.1 (3.22); IR (KBr) cm^{-1}: 1636, 1522, 1400, 1044, 870, 796, 756, 532, 525, 466.

38.2.2 Preparation of Zeolites Exchanged H^+

Procedure:
Ten grams of zeolite (ZSM5 or DAY) are placed in a 50-mL flask and then treated with a solution of NH_4Cl (0.1 M) previously prepared with deionized water. The stoppered flask is allowed to stand overnight at room temperature. The solid residue is recovered by filtration and then activated in a muffle furnace at 500 °C for 24 h. The zeolite in H^+ (H^+-ZSM5 or H^+-DAY) is stored in a dry container.

38.2.3 Preparation of Hydroquinones

Procedure:
These were prepared by the Elbs reaction of phenols. In a typical experiment, the phenol (100 mL) was dissolved in 200 mL of a 10 % solution of sodium hydroxide under stirring at room temperature. A saturated aqueous solution of sodium persulfate (100 mL); 23.8 g) was slowly and carefully added for 3 h. At the end of the addition, stirring was maintained at room temperature. The mixture stood overnight. The solution was acidified with Congo red and was extracted twice with ether. The aqueous layer was treated with an excess of hydrochloric acid and warmed in a steam bath for 30 min. After cooling, it was first dried on magnesium sulphate and finally distilled to afford the corresponding hydroquinone.

2,6-Dimethoxyhydroquinone (**1c**): Prepared from 2,6-dimethoxyphenol (104 mmol; 16.21 g) and sodium persulfate (10 mmol; 23.8 g) brown solid, yield: 69 %, m.p.: 149 °C, $C_8H_{10}O_4$, calcd: H, 5.92 %; C, 56.47 %; found: H, 5.85 %; C, 56.38 %; IR (KBr) cm^{-1}: 3420 (γ OH), 3062 (γ CH, 2952, 2852 (γ OCH_3), 1696, 1646, 1594, 1482, 1322, 1260, 1220, 1108, 878 (δ CH, H arom), 668, 605; ^1H NMR ($CDCl_3$): δ 2.22 (s, 1H, CH_3); 6.45–6.72 (m, 3H, H arom); 7.18 (s, 1H, OH); 7.48 (s, 1H, OH).

38.2.4 Oxidation to Quinones

Procedure:
In a typical experiment, a current of air is passed through a U tube fitted with a filter flask. The tube contains the supported phthalocyanine (0.1 g) in suspension in a solution of hydroquinone (5 mmol) dissolved in 10 mL of a mixture of dioxane/water (60:40) for 6 h at room temperature. The oxidation of hydroquinone is followed by TLC. After disappearance of the hydroquinone, the solution is filtered and evaporated under vacuum. The quinones are purified by chromatography on silica gel or by sublimation. All products are known compounds and identified by comparison of their physical and spectral data with those of authentic samples. The UV spectra of quinones were carried out in a water–dioxane mixture.

2-Methyl-1,4-benzoquinone (**2a**): Brown solid, yield: 95 %, m.p.: 72–73°C (lit. 73 °C)20 $C_7H_6O_2$, calcd: H, 4.95 %; C, 68.85 %; found: H, 4.87 %; C, 68.73 %; UV–visible λ_{max} log(ε) (dioxane/H_2O) nm: 278.2 (4.15), 312.3 (2.75), 422.6 (1.31); ^1H NMR ($CDCl_3$): δ 2.01 (s, 1H, CH_3); 6.23 (s, 1H); 6.52 (m, 1H), 6.85 (d, 2H); IR (KBr) cm^{-1}: 1689 (γ C=O).

2,6-Dimethyl-1,4-benzoquinone (**2b**): Brown solid, yield: 90 %, m.p.: 73–74 °C (lit. 71–73 °C)21 $C_8H_8O_2$, calcd: H, 5.92 %; C, 70.58 %; found: H, 6.01 %; C, 70.50 %; UV–visible λ_{max} log(ε) (dioxane/H_2O) nm: 270.6 (4.29), 318.1 (2.52), 414.3 (1.25); IR (KBr) cm^{-1}: 1685 (γ C=O).

5-Hydroxy-1,4-naphthoquinone: (Juglone) (**3**): Red solid, yield: 87 %, m.p.: 159–160 °C (lit. 161 °C)21 $C_{10}H_6O_3$, calcd: H, 3.47 %; C, 68.97 %; found: H, 3.58 %; C, 68.92 %; UV–visible λ_{max} log(ε) (dioxane/H_2O) nm: 249.3 (4.21), 331.1 (3.18), 422.1 (2.98); ^{1}H NMR ($CDCl_3$): δ 6.76 (s, 2H, CH=); 7.25–7.45 (m, 3H, H arom.); 7.77 (s, 1H, OH); IR (KBr) cm^{-1} : 3241 (γ OH), 1683 (γ C=O), 1587, 1170, 884, 536.

2-Hydroxy-1,4-naphthoquinone : Lawsone (**4**): Orange solid, yield: 79 %, m.p.: 194 °C (lit. 191–194 °C)21 $C_{10}H_6O_3$, calcd: H, 3.47 %; C, 68.97 %; found: H, 3.53 %; C, 68.92 %; UV–visible λ_{max} log(ε) (dioxane/H_2O) nm: 276.3 (4.48), 334.3 (3.28); ^{1}H NMR ($CDCl_3$): δ 6.37 (s, 1H, OH); 7.26 (s, 1H, H arom.); 7.25–8.14 (m, 4H, H arom.); IR (KBr) cm^{-1}: 3166 (γ OH), 1676 (γ C=O), 1592, 1170, 874, 668, 536.

38.2.5 Acetoxylation Reaction

Procedure:

38.2.5.1 Acetoxylation Reaction Under Homogeneous Conditions (Liquid Acids)

In a typical experiment, 10 mmol of Juglone were dissolved in acetic anhydre (0.2 mol), 5 mmol of catalyst were added, and the mixture was stirred at room temperature. After 20 h the mixture was again stirred at room temperature. The solution was poured on ice and after the ice melted the mixture was filtered and the solid 1,2,4,5-tetracetoxynaphthalene, produce (**3a**) was crystallized in ethanol.

38.2.5.2 Acetoxylation Reaction Under Heterogeneous Conditions (Solid Acids)

In a typical experiment, 10 mmol of Juglone were dissolved in acetic anhydre (0.2 mol; 21.64 g).

One gram of catalyst was added and the mixture was stirred at room temperature. After 48 h, the products were separated by chromatography on a silica column eluted successfully with ethyl acetate/cyclohexane (10/90). The different fractions gave, respectively, a yellow solid (unreacted quinone) and a beige solid (1,2,4,5-tetracetoxynaphthalene) (**3a**). It must be noted that traces of products of diacetylation could be detected by thin-layer chromatography (TLC).

We have used acid solids (see below for the procedure): zeolites and heteropolyacides have been tested. The results are reported in Table 38.2.

1,2,4,5-Tétracetoxynaphthalene (**3a**): Marron solid, yield: 63 %, m.p.: 137–138 °C, $C_{18}H_{16}O_8$, ^1H NMR ($CDCl_3$): δ 2.33 (s, 3H, $OCOCH_3$); 2.35 (s, 3H, $OCOCH_3$); 2.40 (s, 3H, $OCOCH_3$); 2.42 (s, 3H, OCOCH3); 7.68–8.07 (m, 4H, H arom.); IR (KBr) cm^{-1}: 3009 (γ CH arom.), 1774 (γ $OCOCH_3$), 1762, 1719.

1,2,3,4-Tétracetoxynaphthalene (**4a**): Beige solid, yield 66 %, m.p.: 140–143 °C, $C_{18}H_{16}O_8$, ^1H NMR ($CDCl_3$): δ 2.39 (s, 6H, $OCOCH_3$); 2.44 (s, 6H, $OCOCH_3$); 7.46–8.14 (m, 4H, H arom.); IR (KBr) cm^{-1} : 3012 (γ CH arom.), 1760 (γ $OCOCH_3$), 1718; MS *m/z* (%); 360 (M^+; 24.46), 318 (3.41), 276 (16.38), 192 (100), 146 (11.95), 105 (23.55).

2-Methyl-1,3,4-triacetoxynaphthalene (**5a**): Beige solid, yield: 59 %, m.p.: 148–150 °C, $C_{17}H_{16}O_8$, ^1H NMR ($CDCl_3$): δ 2.16 (s, 3H, CH_3); 2.37 (s, 3H, $OCOCH_3$); 2.45 (s, 3H, $OCOCH_3$)7, 2.49 (s, 3H, $OCOCH_3$) 7.48–7.76 (m, 4H, H arom.); IR (KBr) cm^{-1}: 3000 (γ CH arom.), 1774 (γ $OCOCH_3$), 1760; MS *m/z* (%); 316 (M^+; 39.13), 274 (12.50), 233 (17.38), 174 (6.63), 115 (2.88).

38.2.6 Saponification and Oxidation In Situ

Procedure:
In a typical experiment, a current of air is passed through a U tube fitted with a filter flask. The tube contains the (Pc(Mn)/K10 (0.1 g) in suspension in a solution of product (**4**) (0.6 g) dissolved in CH_3OH (20 mL) and KOH (2 g) for 6 h at room temperature. After filtration and evaporation of the methanol, the product (**5**) is crystallized.

2,5-Dihydroxy-1,4-naphthoquinone (**3b**): Yellow solid, yield: 89 %, m.p.: 210 °C, $C_{10}H_6O_4$, UV–visible λ_{max} log(ε) (dioxane/H_2O) nm: 279.3 (4.68), 331.3 (3.08); ^1H NMR ($CDCl_3$): δ 6.39 (s, 1H, OH); 6.40 (s, 1H, OH); 7.18–8.01 (m, 5H, H arom.); IR (KBr) cm^{-1}: 3168 (γ OH), 1679 (γ C=O), 1591, 1176, 881, 544.

2,3-Dihydroxy-1,4-naphthoquinone (**4b**): Beige clear solid, yield: 91 %, m.p.: 201 °C, $C_{10}H_6O_4$, UV–visible λ_{max} log(ε) (dioxane/H_2O) nm: 289.3 (4.77), 339.3 (3.27); ^1H NMR ($CDCl_3$): δ 6.41 (s, 1H, OH); 6.42 (s, 1H, OH); 7.31–8.11 (d, 4H, H arom.); IR (KBr) cm^{-1}: 3172 (γ OH), 1678 (γ C=O), 1589, 1174, 878, 536.

2-Hydroxy-3-methyl-1,4-naphthoquinone (**5b**): Yellow solid, yield: 92 %, m.p.: 173–174 °C (ether + hexane), $C_{11}H_8O_3$ calcd: H, 4.29 %; C, 70.21; found: H, 4.35 %; C, 70.08 %, UV–visible λ_{max} log(ε) (dioxane/H_2O) nm: 271.3 (4.44), 346.6 (3.21); ^1H NMR ($CDCl_3$): δ 2.08 (s, 1H, CH_3); 6.42 (s, 1H, OH); 7.29 (s, 1H, OH); 7.5–8.1 (m, 4H, H arom); IR (KBr) cm^{-1}: 3330 (γ OH), 1655 (γ C=O), 1589, 1174, 878, 536.

38.3 Results and Discussion

38.3.1 Oxidation of Phenols and Hydroquinones into Quinones

The phenol oxidation products are used for the synthesis of natural products such as vitamins and their intermediates. Concerning the oxidation of phenols or hydroquinones into quinones, numerous oxidants were described in the literature such as silver oxide [15] and ceric ammonium nitrate [16].

In a preliminary study, we were inspired by the Elbs reaction [17] to access hydroquinones by oxidation of phenols by sodium persulfate ($Na_2S_2O_8$).We have tested the oxidation of hydroquinones into quinones by air and catalyzed a metalated phthalocyanine [18] on a sheet of a Montmorillonite K10. Hydroquinones and substituted hydroquinones were easily oxidized at room temperature in quinones by using (Pc[Co]/K10) and air (1 atm), as indicated in Fig. 38.2.

Results obtained from different phenols are reported in Table 38.1.

The direct oxidation of phenols with oxygen in the presence of supported metalled phthalocyanines as oxidant generally gave quinones at room temperature but the reaction is very slow and not selective.

We have also tested this type of reaction in a naphthoquinone series. Many naphthoquinones are natural products with interesting biological properties. The catalytic system (Pc[Co]/K10) in the presence of pure oxygen easily oxidizes the 1,5-dihydroxynaphthalene in Juglone (yield 87 %) . (This latter is used as an alimentary stain for soft drinks.) The total oxidation was about 6 h at room temperature. Results obtained from different phenols are reported in Table 38.2.

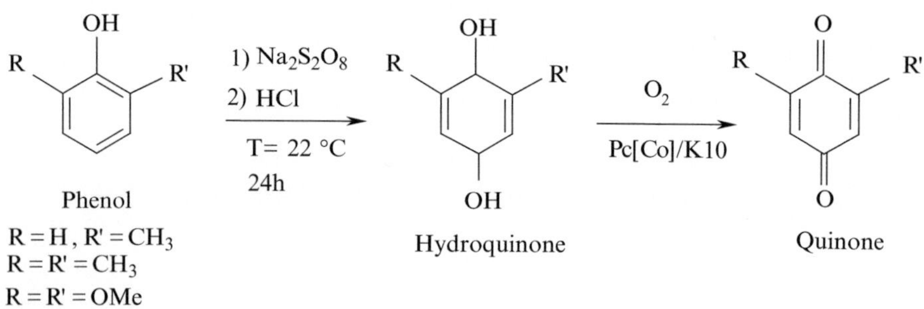

Fig. 38.2 Oxidation of hydroquinones into quinones

Table 38.1 Oxidation of hydroquinones into quinone with supported metalated phthalocyanines

Phenols	Hydroquinones	Oxidation catalysts	Quinones	M.p. (°C)	Yield (%)
2-Methylphenol	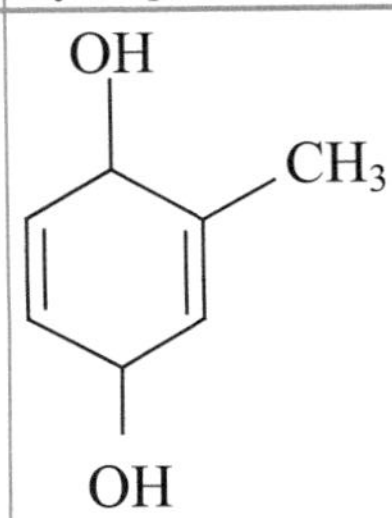 (1a)	Pc[Co]/K10	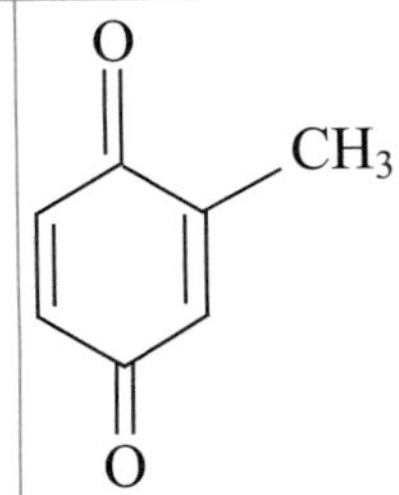 (2a)	72	95
2,6-Dimethylphenol	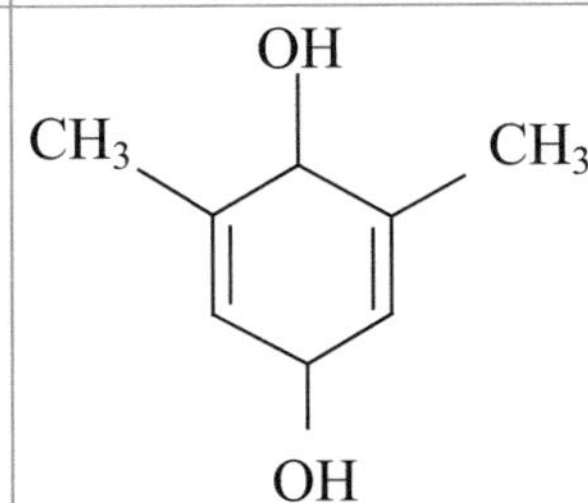 (1b)	Pc[Co]/K10	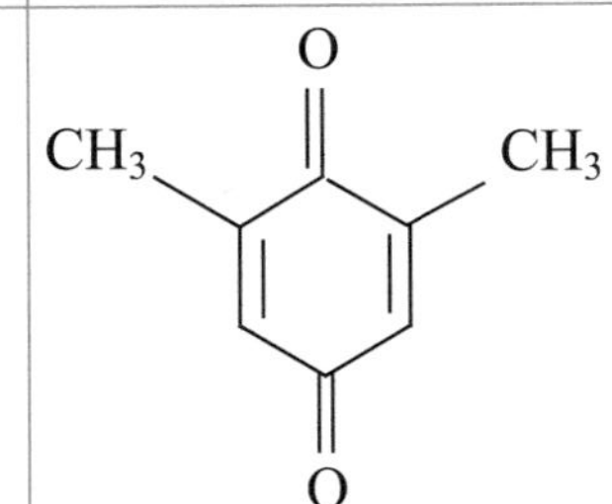 (2b)	69	90
2,6-Dimethoxyphenol	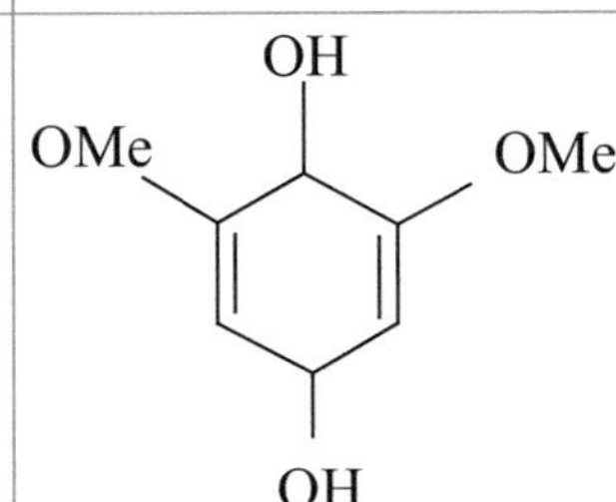 (1c)	Pc[Co]/K10	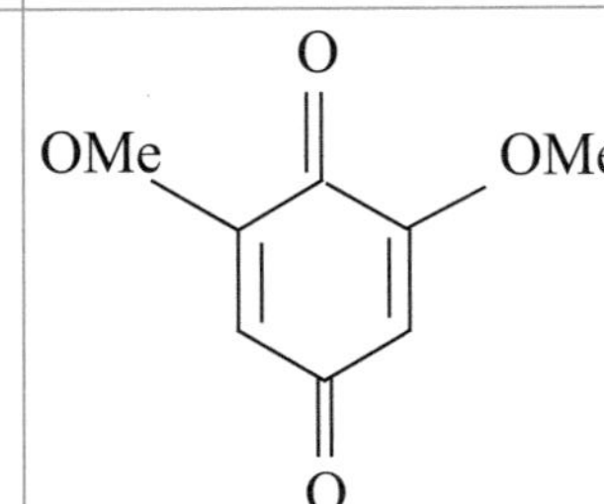 (2c)	252	89

Table 38.2 Oxidation of hydroxynaphthalene

Hydroxynaphthalene	Oxidation catalysts	Naphthoquinone	M.p. (°C)	UV–visible λ_{max} (nm)	Yield (%)
1,5-Dihydroxynaphthalene	Pc [Co]/K10	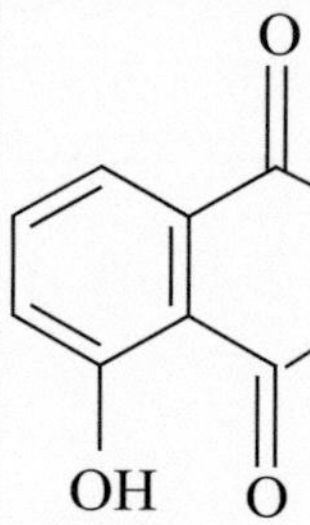 Juglone (3)	159	331.1	87
1,3-Dihydroxynaphthalene	Pc [Co]/K10	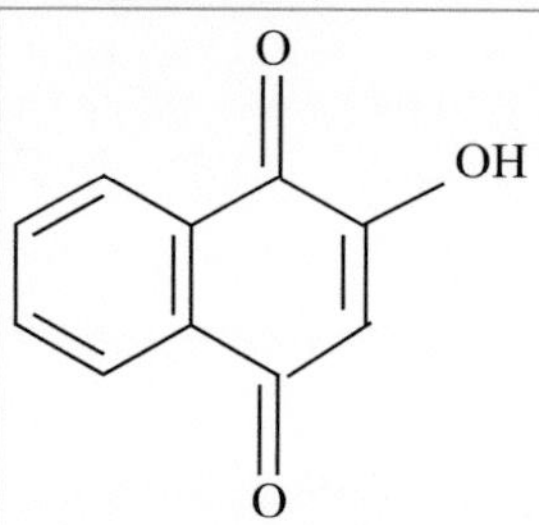 Lawsone (4)	197	334.3	79
2-Methyl-1,4-dihydroxynaphthalene	Pc [Co]/K10	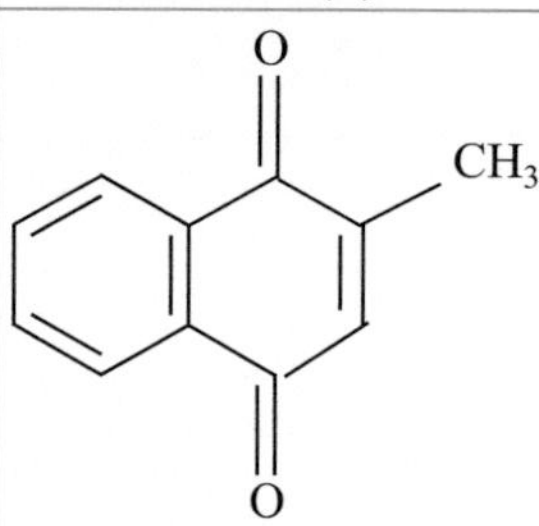 Menadione (5)	103	332.5	80

38.3.2 *Thiele–Winter Reaction*

The 5-hydroxynaphthoquinone (**3**) treated with acetic anhydride in the presence of catalyst converts into the 1,2,4,5-tetracetoxynaphthalene (**3a**) (yield 63 %).

We have studied this reaction with strong acids. Chlorosulfonique acid (Ho = −13.80) and perchloric acid [19–21] (Ho = −10.31) led to the desired product.

The reaction with solid super acids shows a high Brônsted acidity and was lower than with the homogeny phase.

Heteropolyacids (HPAs also called polyoxometalates) are catalysts of very great interest [22, 23]. The display has a very high Brônsted acidity close to super acids. We decided to test three heteropolyacids ($H_3PW_{12}O_{40}$ (PW), $H_4SiW_{12}O_{40}$ (SiW), and $H_3PMo_{12}O_{40}$ (PMO). The (SiW) provided the product with a yield of 63 % whereas a yield of 61 % was achieved with the (PW). It has been noticed that the (SiW) appeared a better catalyst than the (PW). This could be because of the higher ratio of the number of protons in (SiW) [24].

Zeolites with a high Si–Al ratio (15–20) are known to posses strongly acidic sites 10. Two acid zeolites (H+ -ZSM5 and H+ -DAY) were also tested successfully. Zeolites can be reused without loss of activity when the reaction is carried out at room temperature. No poisoning by tar formation was observed with ZSM5 or DAY. With solid catalysts such as zeolite, the Juglone is too wide to enter the pores of ZSM5, and, the reaction takes place on the zeolite surface.

We have noticed that no product (**3a**) was obtained with the $H_3PMo_{12}O_{40}$ (PMO). The reaction medium color changed from green to deep blue, proving the reduction of PMO. The result with (PMO) is not surprising owing to its high redox potential resulting from the presence of the molybdenum atom. Results of acetoxylation of Juglone by acetic anhydride obtained from different acids catalysts are reported in Table 38.3.

The kinetics of the reaction was monitored hourly by visible spectroscopy ultra-purple. The kinetics of Menadione transformation (bleaching) was monitored at 333 nm, which is the maximum point of Menadione (**5**) (the vitamin K3 carrier). In Fig. 38.3 we indicate changes in the UV–visible spectrum in the presence of $ClSO_3H$. The kinetic of Menadione transformation (bleaching) depends on the acidity of the catalyst. It seems that the limiting step of the Thiele–Winter reaction is the protonation of the quinine.

The 1,2,4,5-tetracetoxynaphthalene (**3a**) was easily transformed into (**3b**) by saponification basic and oxidation with potassium hydroxide (KOH) in methanol (CH_3OH) in the presence of air at room temperature. The yield of 2,5-dihydroxy-1,4-naphthoquinone from (**3a**) is good (89 %). The results obtained from the two-step sequence (triacetoxyaromatic and saponification–oxidation in situ) are reported in Table 38.4.

Table 38.3 Acetoxylation of Juglone at 20 °C

Catalyst	Aspect	Hammet: Ho	Experimental	M.p. (°C)	IR vester (cm^{-1})	Yield (%)
$ClSO_3H$	Liquid	−13.80	20 °C, 20 h	148	1,773	55
H^+ -DAY	Solid	–	20 °C, 48 h	149	1,772	56
H^+ -ZSM5	Solid	–	20 °C, 48 h	148	1,776	54
$H_3PW_{12}O_{40}$	Solid	−13.6	20 °C, 48 h	149	1,773	61
$H_3PMo_{12}O_{40}$	Solid	–	20 °C, 48 h	–	–	–
$H_4SiW_{12}O_{40}$	Solid	–	20 °C, 48 h	152	1,774	63

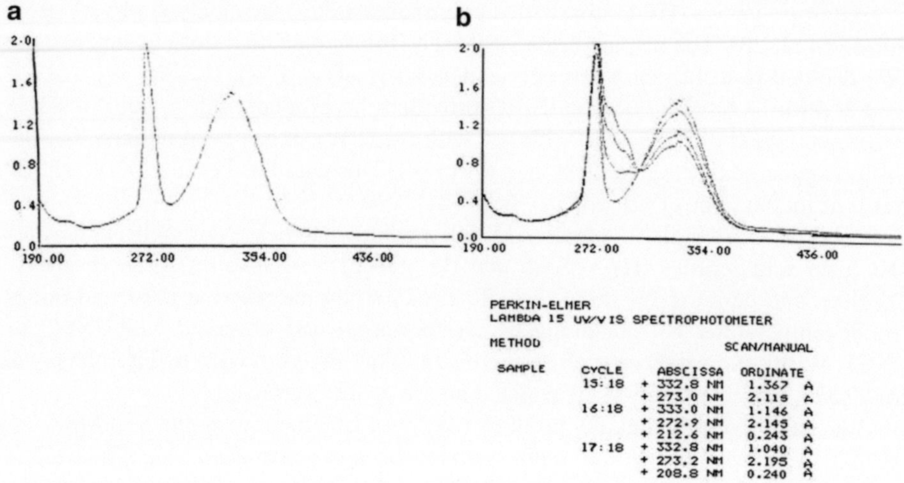

Fig. 38.3 The evolution of UV–visible spectrum in the presence of $ClSO_3H$. (**a**) Ménadione + acetic anhydride [temps: $t = 0$] and (**b**) Menadione + acetic anhydride + $ClSO_3H$

38.4 Conclusion

In conclusion it appears that chlorosulfonique acid is the most effective acid catalyst found for the acetoxylation reaction of Juglone under homogeneous conditions. Zeolites (DAY-H+) heteropolyacids ($H_3PMo_{12}O_{40}$) can easily catalyze the Thiele–Winter reaction. This work represents the first example of catalysis of this reaction by solid acids. The sequence of the two steps (acetoxylation, saponification–oxidation) constitutes a new synthesis of hydroxynaphthoquinones from hydroquinones.

Table 38.4 Solid acid as catalyst for oxidation by molecular oxygen

(1) Thiele–Winter: triacetoxyaromatic		(2) [Saponification–oxidation] in situ	
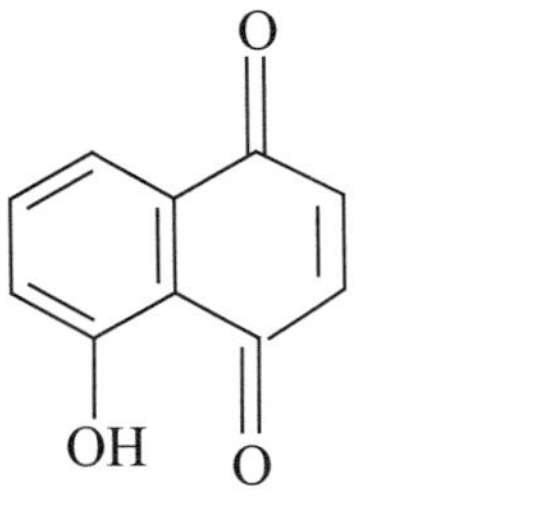 (3)	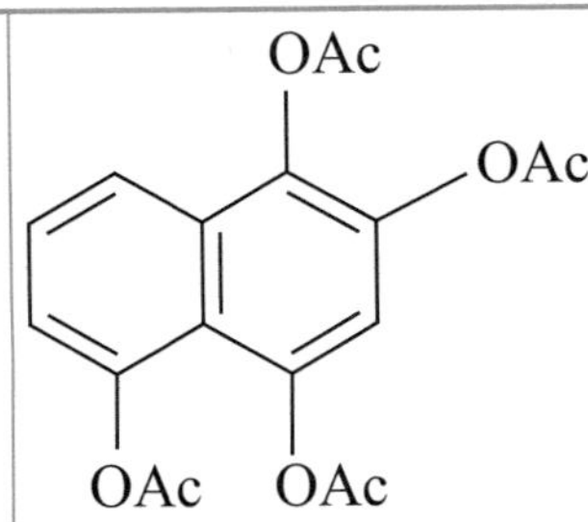 (3a)	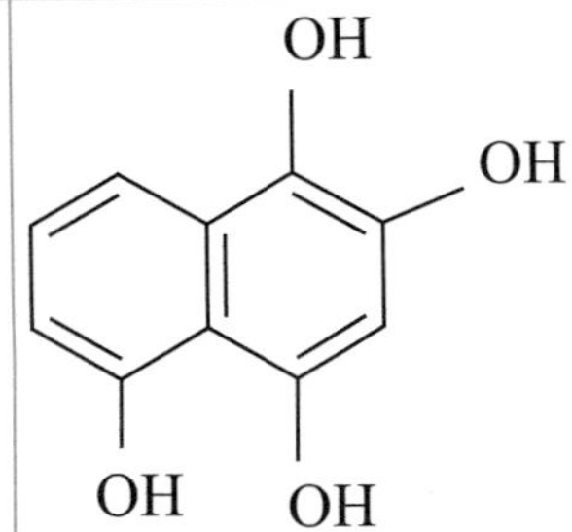	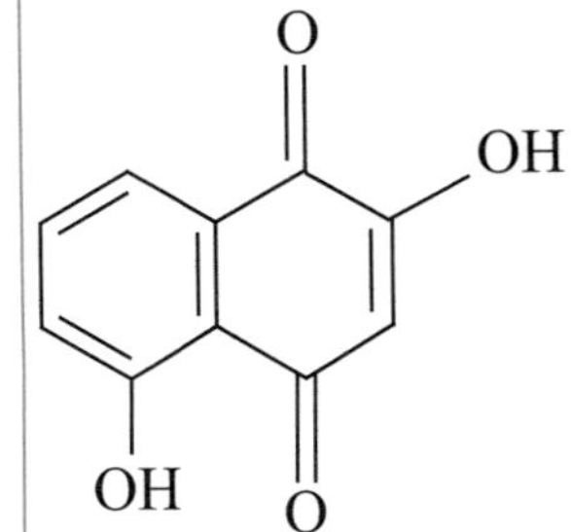 (3b)
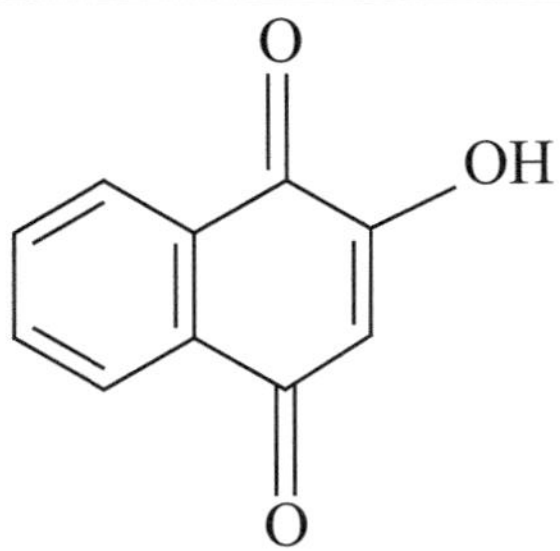 (4)	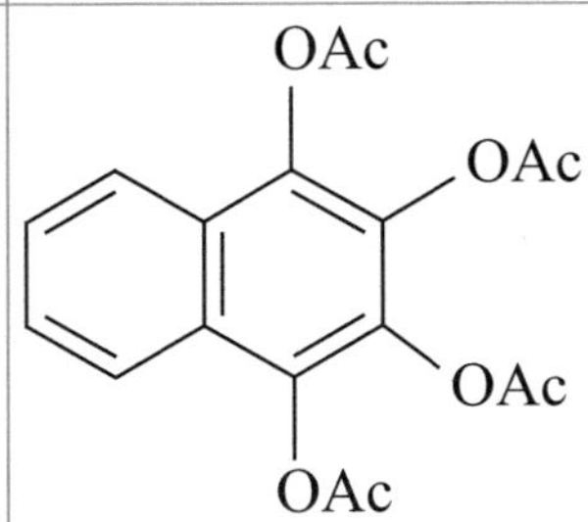 (4a)	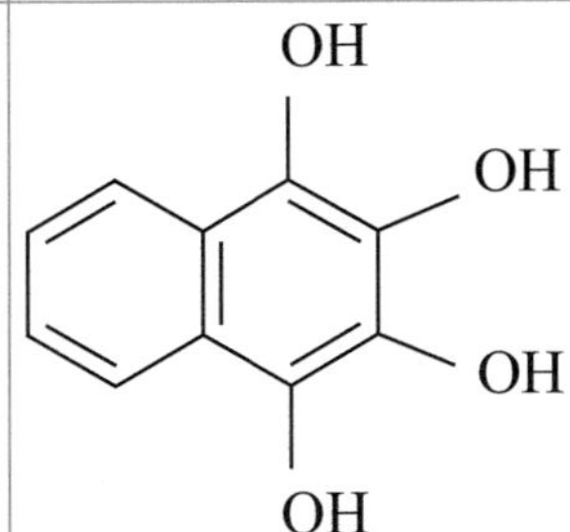	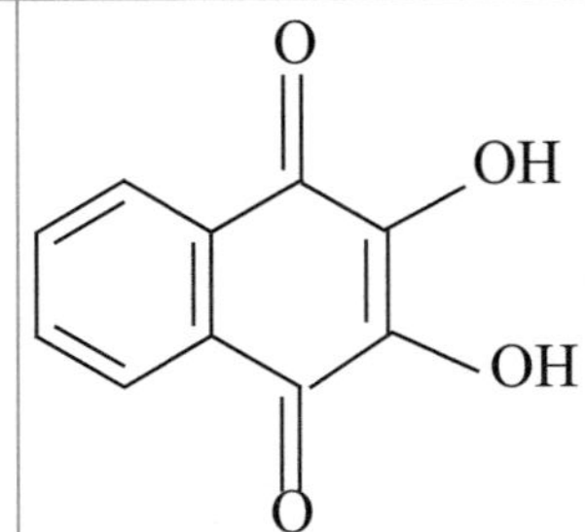 (4b)

(continued)

Table 38.4 (continued)

(1) Thiele–Winter: triacetoxyaromatic		(2) [Saponification–oxidation] in situ	
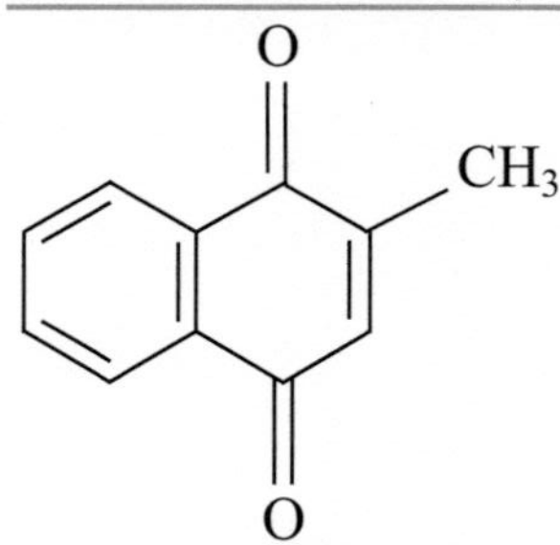 (5)	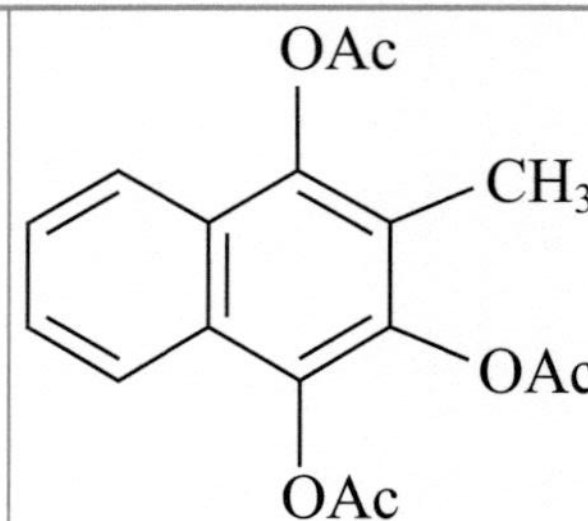 (5a)	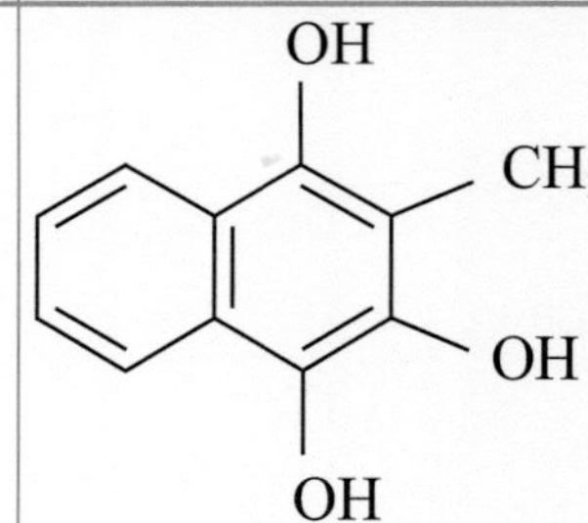	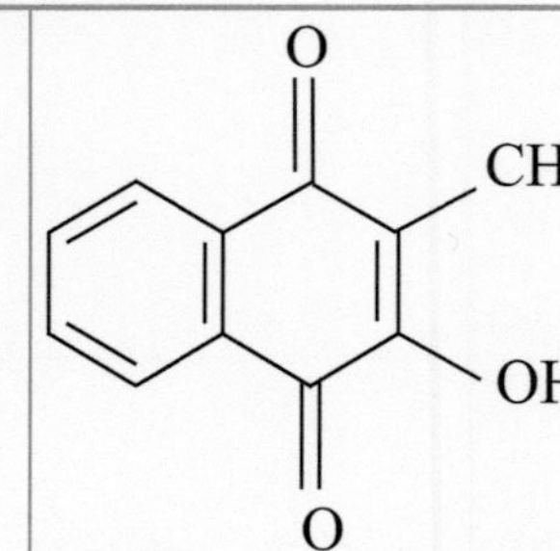 (5b)

Synthesis of hydroxynaphthoquinones

References

1. McOmie JFW, Blatchey JM (1972) Thiele–Winter acetoxylation of quinones. Org React 9:199
2. Spyroudis S (2000) Hydroxyquinones: synthesis and reactivity. Molecules 5:1291
3. Lira AM, Araújo AAS, Basílio IDJ, Santos BLL, Santana DP, Macedo RO (2007) Compatibility studies of lapachol with pharmaceutical excipients for the development of topical formulations. Thermochim Acta 457:1
4. Sauriasari R, Wang DH, Takemura Y, Tsutsui K, Masuoka N, Sano K, Horita M, Wang BL, Ogino K (2007) Cytotoxicity of lawsone and cytoprotective activity of antioxidants in catalase mutant Escherichia coli. Toxicology 235(1–2):103–111
5. Windholz M (1993) The Merck index, 12th edn. Merck, Rahway
6. Combes R (1907) Procédé de préparation et de purification des dérivés oxyanthrquinoniques et oxynaphtoquinoniques en général, du juglonet de l'émodine en particulier. Bull Soc Chim France 4* série. I, pp 800–816
7. Soderquist CJ (1973) Juglone and allelopathy. J Chem Educ 50:782–783
8. Kessl JJ, Moskalev NV, Gribble GW, Nasr M, Meshnicr SR, Trumpower BL (2007) Parameters determining the relative efficacy of hydroxy-naphthoquinone inhibitors of the cytochrome bc1 complex. Biochim Biophys Acta 1767(4):319–326
9. Teimouri MB, Khavasi HR (2007) One-pot three-component regioselective synthesis of linear naphtho[2,3-b]-furan-4,9-diones. Tetrahedron 63:10269–10275
10. Valente C, Moreira R, Guedes RC, Iley J, Jaffar M, Douglas KT (2007) The 1,4-naphthoquinone scaffold in the design of cysteine protease inhibitors. Bioorg Med Chem 15(15):5340–5350
11. Thiele J (1898) Ueber die Einwirkung von Essigsäure-anhydrid auf Chinon und auf Dibenzoylstyrol. Ber Dtsch 31:1247–1249
12. Thiele J, Winter E (1900) Ueber die Einwirkung von Essigsäureanhydrid und Schwefelsäure auf Chinone. Ann Chem 311:341–352
13. Burton H, Praill PFG (1952) Acylation reactions catalysed by strong acids. Part VI. A comparison of zinc chloride and perchloric acid as catalysts for the Thiele acetylation of quinines. J Chem Soc 755–759
14. Gillspie R (1969) J Can Chem Educ 4:9
15. Michie JK, Miller JA (1981) Phosphorus trichloride as catalyst in the preparation of 1,1-diacetates from aldehydes. Synthesis 10:824–825
16. Kochhar KS, Bal BS, Deshpande RP, Rajadhyaksha SN, Pinnick HWJ (1983) Protecting groups in organic synthesis. Part 8. Conversion of aldehydes into geminal diacetates. J Org Chem 48:1765–1767
17. Elbs K (1993) Ueber nitrohydrochinon. J Prakt Chem 48:179
18. Villemin D (1996) Synthesis and encapsulated metal intercalated phtalocyanure without solvent under microwave irradiation. Actes of 3th Colloque Franco-Magrébin pp. 611
19. Hull NF, Conant JB (1927) A study of superacid solutions the use of the chloranil electrode in glacial acetic acid and the strength of certain weak bases. J Am Chem Soc 49:3047–3061
20. Olah GA, Pradash GKS, Sommer J (1985) Superacids. Wiley, New York
21. Olah GA (1993) Superelectrophiles. Angew Chem Int Ed Engl 32:767
22. Izumi Y, Urabe K, Onaka M (1992) Zeolite, clay and heteropolyacid, in organic reactions, vol 100. VCH, Weinheim, p 120
23. Dupont P, Lefebvre F (1996) Esterification of propanoic acid by butanol and 2-ethylhexanol catalyzed by heteropolyacids pure or supported on carbon. J Mol Catal A Chem 114 (1–3):299–307
24. Izumi YK, Matsuo K, Urabe K (1983) Efficient homogenous acid catalysis of heteropoly acid and its characterisation through ether cleavage reactions. J Mol Catal 18:299–314

Chapter 39
Utilization of Iron Oxides in Chemical Looping Combustion Processes

Nesibe Dilmaç, Omer Faruk Dilmaç, Osman Nuri Şara, and Sedat Yörük

Abstract Fossil fuel combustion is the major source of CO_2 emissions. There are a number of techniques that can be used to separate CO_2 from the rest of the flue gas. The main disadvantage of these techniques is the large amount of energy that is required for the separation, which means a relative reduction of the overall efficiency of a power plant. Chemical looping combustion (CLC) is a new developing clean-combustion technology that has the advantage of inherent CO_2 separation ability. Thus, storage-ready, concentrated CO_2 can be obtained without the use of above-mentioned energy-intensive separation operations. This process involves two separate reactors, i.e., an air and a fuel reactor, and an oxygen carrier material which circulates between the reactors in order to extract and transport oxygen from air to fuel for many cycles. It is essential for oxygen carrier to be reactive, cheap to prepare, and durable over many cycles of reduction and oxidation at high reaction temperatures. Because of its good thermodynamical properties, physical strength, low cost, and low toxicity, iron oxide is a good candidate to be an oxygen carrier in CLC. This study will present a brief summary of developments obtained via CLC processes utilizing iron oxides as oxygen carrier.

Keywords Chemical looping combustion • Oxygen carrier • CO_2 emission • CO_2 capture

39.1 Introduction

Chemical looping combustion (CLC) is a combustion technique where the greenhouse gas CO_2 is inherently separated during combustion [1] via dividing traditional combustion route into two steps which occur simultaneously at two different

N. Dilmaç (✉) • O.F. Dilmaç • O.N. Şara
Department of Chemical Engineering, Faculty of Engineering, Çankırı Karatekin University, Çankırı 18100, Turkey
e-mail: ndilmac@karatekin.edu.tr

S. Yörük
Department of Chemical Engineering, Faculty of Engineering, Atatürk University, Erzurum 25400, Turkey

I. Dincer et al. (eds.), *Progress in Clean Energy, Volume 1*,
DOI 10.1007/978-3-319-16709-1_39

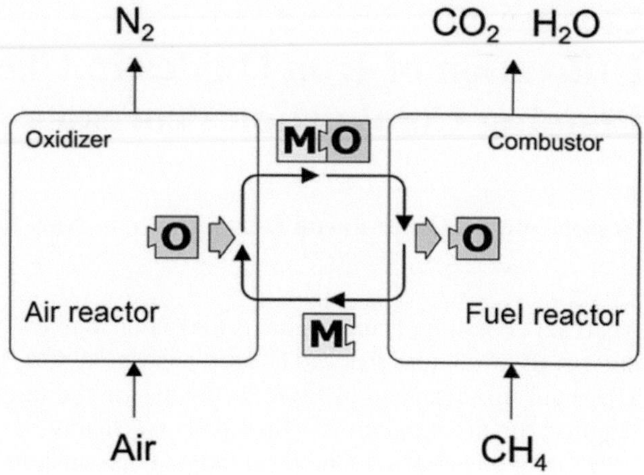

Fig. 39.1 Chemical looping combustion concept [2]

reactors; an air and a fuel reactor as seen in Fig. 39.1 and given in (39.1)–(39.3). (The letter “M” represents a metal and MO represents its oxide.)

Reduction: mostly endothermic reaction (occurring at fuel reactor)

$$CH_4 + 4MO \rightarrow CO_2 + 2H_2O \ + 4M \quad (39.1)$$

Oxidation: exothermic reaction (occurring at air reactor)

$$M + (1/2)\, O_2 \rightarrow MO \quad (39.2)$$

Net/conventional combustion reaction

$$CH_4 + 2O_2 \rightarrow CO_2 + 2H_2O \quad (39.3)$$

Although the CLC concept was first offered as a way to produce CO_2 from fossil fuels [3], soon after it was realized that power station efficiency could be enhanced by that way possibly due to the enhanced reversibility of redox reactions given in (39.1) and (39.2) compared to single traditional combustion reaction given in (39.3) which the release of a fuel’s energy occurs in a highly irreversible manner [4]. Beside the increased efficiency advantage, in recent years interest has been shown on CLC as a carbon capture technique.

The total amount of the heat evolved from whole process is same as that for normal combustion, where the oxygen is in direct contact with the fuel. However, the advantage with this system, compared to normal combustion is that the CO_2 and H_2O are inherently separated from the rest of the flue gases and no energy is expended for this separation. Thus, compared to other techniques for CO_2 separation, CLC is potentially much cheaper [1] and the main driving force for research in CLC is the convenient CO_2 capture properties of this concept.

A successful operation of CLC process strongly depends on the effective performance of the oxygen carrier particles that are usually composed of primary metal oxide, support, and promoter/doping agents [5]. The metal oxides in the oxygen carriers provide lattice oxygen to oxidize the carbonaceous fuel into CO_2 adequately and the supports improve the resistance of oxygen carrier to fragmentation, abrasion, and sintering. Thus, regardless of which reactor arrangement is used, the oxygen carrier is of crucial importance on CLC researches [6]. There are some important criteria for the selection of oxygen carrier particles:

- First, the thermodynamic equilibrium for the reaction with the fuel has to be favorable in order to achieve high fuel conversion to CO_2 and H_2O.
- Also the rate of oxidation and reduction has to be sufficiently fast. Otherwise, the amount of the oxygen carrier needed for circulation between reactors would be too large.
- Moreover, the oxygen transport capacity of the carrier must be sufficient in order to avoid circulation of large amounts of carrier between reactors.
- Since the proposed reactor systems mostly consist of fluidized beds, the particles need to have a low tendency for fragmentation and attrition. It is also vital they do not agglomerate under real reaction conditions.
- The deactivation of the oxygen carrier is another factor to be analyzed. For example, sulfur compounds in the fuel gases could react with the oxygen carrier resulting in deactivation of the particles because of formation of metal sulfides or sulfates. Thus, the oxygen carrier must be resistant to deactivation.
- Cost of the carrier must be acceptable and proper and it must be safe in terms of health and environmental aspects [7].

By the contribution of researchers from worldwide, over than 700 different solid materials (natural or synthetically produced by several methods) suggested as oxygen carrier have been tested with the focus on redox reactivity, recyclability, and attrition behavior up to date. The most promising metal/metal oxide couples are: Ni/NiO, Cu/CuO, and Fe_3O_4/Fe_2O_3.

There are pros and cons associated by each couple. NiO has high reactivity but thermodynamic limitation and concern of toxicity. CuO has exothermic characteristic but low melting point and possible defluidization as well as potential secondary pollution. Comparing to NiO and CuO, Fe_2O_3 has a striking economical strength for its lower price, environmental benignity without potential pollution, and higher melting points [8]. Certainly, the performance of oxygen carrier can be improved by making mixed metal oxides. Since the manufacturing cost of the oxygen carriers based on a combination of metal oxides and/or nonmetal oxides is high and the preparation of the composite particles involves a complex trial and error procedure depending on types of metal oxide and support, weight percentage and synthesis method, oxygen carrier researches have been focused on finding extensive and inexpensive sources in recent years. One of the most promising and abundant resource for CLC oxygen carrier is natural iron ore.

Wide variety of researches on iron-based oxygen carriers have been carried out in Chalmers University of Technology from Sweden and Instituto de Carboquimica from Spain.

This study will present a brief summary of developments obtained via CLC processes utilizing iron oxides as oxygen carrier.

39.2 Iron Oxide as an Oxygen Carrier

Mattisson et al. [9] investigated the reactivity of natural iron ore (Carajas hematite) with an average diameter of 200 μm as an oxygen carrier in a fixed bed quartz reactor. Iron oxide was exposed to repeated cycles of air and methane at 950 °C, with the outlet gas concentrations measured. It was observed for all experiments where the reduction was investigated that the outlet concentration of methane, carbon dioxide, and carbon monoxide showed similar graphs which implies the durability of the carrier. Also high conversion of methane to carbon dioxide and water for six cycles and fast oxidation rates were observed. According to the findings, the authors offered a dual circulating fluidized bed setup for pilot CLC application using iron oxide as oxygen carrier.

In another study, Mattisson et al. [1] investigated the oxygen carrying performance of synthesized particles composed of 40–80 wt% Fe_2O_3, together with Al_2O_3, ZrO_2, TiO_2, or $MgAl_2O_4$ which was prepared by freeze granulation in a fluidized bed reactor of quartz at 950 °C. Particles were exposed to alternating atmospheres of CH_4 and O_2 containing gas mixtures which simulate the cyclic conditions of a CLC system and the particles were tested in this manner for 6–26 cycles. The authors observed that the type of inert used and the sintering temperature have a significant effect on the reactivity and the strength of the carrier. The crushing strength has a tendency to increase as a function of sintering temperature and the reactivity decreases as the sintering temperature increases. It is concluded that the carrier whose active phase is Fe_2O_3, supported by $MgAl_2O_4$ and sintered at 950 °C showed the highest initial reactivity.

In a study carried out by Abad et al. [10] in Chalmers University of Technology, an iron-based oxygen carrier was used in a continuously operating laboratory CLC unit, consisting of two interconnected fluidized beds. Natural gas or syngas was used as fuel and the thermal power was between 100 and 300 W. Tests were performed at four temperatures: 1,073, 1,123, 1,173, and 1,223 K. The prototype was successfully operated for all tests and stable conditions were maintained during the combustion. The combustion efficiency of syngas was high, about 99 % for all experimental conditions. However, in the combustion tests with natural gas there was unconverted methane in the exit flue gases. Higher temperature and lower fuel flow increase the combustion efficiency. No signs of agglomeration and mass loss were detected and the crushing strength of the oxygen carrier particles did not change significantly.

The use of iron oxide as an oxygen carrier has been investigated in another study carried out by He et al. [11]. Particles composed of 80 wt% Fe_2O_3 together with Al_2O_3 as binder have been prepared by impregnation method. The reactivity of the oxygen carrier particles has been studied in 20 cycle reduction and oxidation tests in a TGA reactor. It is observed that the reaction rates became slightly higher with the number of cyclic reactions increasing. The SEM analysis confirmed that the pore size inside the particle had been enlarged by the thermal stress during the reaction, which was favorable for diffusion of the gaseous reactants into particles.

The experimental results suggested that the Fe_2O_3/Al_2O_3 oxygen carrier was a promising candidate for a CLC system.

Mayer et al. [12] compared oxygen carrying performances of micaceous iron oxide and ilmenite with respect to syngas conversion in a 10 kW_{th} bubbling fluidized bed reactor. Both carriers were alternatively reduced and oxidized at 900 °C. The conversion of syngas with micaceous iron ore was higher than that with ilmenite. The authors concluded that micaceous iron oxide could be a promising new oxygen carrier for direct coal combustion unit.

A study performed by Song et al. [13] presents a continuous operation chemical looping process with a natural Australian hematite. In that study, a 1 kW_{th} prototype unit consisted of a spouted fluidized bed as the fuel reactor, and a fast fluidized bed as the air reactor was employed. The unit was also equipped with a coal feeding device in order to transfer coal into the fuel reactor at a constant rate. The oxygen from air was transferred to the fuel by the solid hematite that circulated between the interconnected fluidized bed reactors. Experimental results indicated that the Australian hematite showed a stable reactivity and resistance to agglomeration–attrition. At a fuel reactor temperature of 950 °C, a little CH_4 was measured and there were neither hydrocarbons heavier than CH_4 nor tars in the exit of the fuel reactor. The carbon conversion efficiency was about 81.2 %, and the loss rate of the hematite oxygen carrier due to attrition was about 0.0625 %/h. XRD results showed that the active phase, i.e., Fe_2O_3 of the hematite oxygen carrier was mostly reduced to Fe_3O_4 phase by coal gasification products in the fuel reactor, and only small part of oxygen carrier was further reduced to FeO. No tendency of decreased reactivity of the hematite oxygen carrier was observed during 10 h of operation. It is concluded that Australian hematite had a good behavior as an oxygen carrier, suitable for use in CLC with coal.

In same study, Song et al. [13] investigated the reduction behavior of hematite oxygen carrier in a thermal gravimetric analyzer using H_2. The solid reduction products were characterized by XRD and SEM-EDS. The results show that there is a maximum reaction rate when the conversion is 0.11. In this stage, Fe_2O_3, i.e., the active phase was converted to Fe_3O_4 and the reduction reaction was easy to happen. Then, there is a decrease of reaction rate. When the conversion of hematite oxygen carrier was 0.178, the solid reduction products were composed of Fe_3O_4 and FeO. When the conversion is 0.477, Fe_3O_4, FeO, and Fe were all found in the reduction products. SEM-EDS results show that the grains at the surface of hematite oxygen carrier are gathered and grown up, and the particle volume shrinkage and sintering effect are observed, especially in the region rich in Fe element. However, a relative stable structure was seen in the region rich in SiO_2 or Al_2O_3 contents. Further, compared with the results based on interconnected beds, the sintering of hematite was suppressed due to a good inert material distribution, which kept a good long-term reactivity of hematite oxygen carrier.

Table 39.1 [14] summarizes a comparison among most popular oxygen carrier metal/metaloxide couples. The most reactive oxides, i.e., CuO and NiO are the most expensive ones. For NiO and CuO, there are also health and environmental safety issues which cannot be neglected. Furthermore, NiO has a thermodynamic

Table 39.1 Qualitative estimation of the active oxides [14]

	Fe_2O_3/Fe_3O_4	Mn_3O_4/MnO	CuO/Cu	NiO/Ni
Oxygen transfer capacity, R_o	0.03	0.07	0.20	0.21
Reactivity	←Decreasing increasing→			
Cost	←Decreasing increasing→			
Health and environmental aspects	+	+	–	–
Thermodynamics	+	+	+	Cannot fully convert fuels to CO_2 and H_2O, beyond a maximum conversion of 99.5 %
Methane yield at atmospheric pressure and 1,000 °C	1.0000	1.0000	1.0000	0.9883
Melting point	+	+	1,085 °C	+

restriction; it can convert fuels fully to CO_2 and H_2O beyond a maximum conversion of 99–99.5 %.

Reduction of all oxides with methane is endothermic, except CuO. This can be evaluated as an advantage for CuO, since it reduces the particle circulation needed maintain fuel reactor temperature. On the other hand, Cu has the disadvantage of a low melting temperature which causes defluidization problems in reactors. Despite low reactivity and oxygen transfer capacity, other properties make iron oxide one of the most proper oxygen carriers for CLC processes.

39.3 Conclusions

Because of its good thermodynamical properties, physical strength, low cost and low toxicity, iron oxide is a good candidate to be an oxygen carrier in CLC. Results showed that SiO_2 naturally exists as an inert material in the iron ore, does not react with the active phase of the ore, i.e., Fe_2O_3, suppresses sintering of the oxygen carrier particles at high CLC operation temperatures, and enhances durability of the carrier [13]. Furthermore, in consideration of using solid fuels, it is important to use cheap and natural oxygen carriers, since there will probably be some loss of bed material while discharging ash from the system [12].

Acknowledgement We appreciate the support of TÜBİTAK (Project No: 113 M548) for this study.

References

1. Mattisson T, Johansson M, Lyngfelt A (2004) Multicycle reduction and oxidation of different types of iron oxide particles—application to chemical looping combustion. Energy Fuel 18:628–637
2. Ryu HJ, Bae DH, Jin GT (2003) Effect of temperature on reduction reactivity of oxygen carrier particles in a fixed bed chemical-looping combustor. Korean J Chem Eng 20(5):960–966
3. Lewis W, Gilliland E, Sweeney M (1951) Gasification of carbon. Chem Eng Prog 47:251–256
4. Richter HJ, Knoche KF (1983) Reversibility of combustion processes. Efficiency and costing—second law analysis of processes. ACS Symp Ser 235:71–85
5. Fan LS, Li F (2010) Chemical looping technology and its fossil energy conversion applications. Ind Eng Chem Res 49:10200–10211
6. Schwebel GL, Leion H, Krumm W (2012) Comparison of natural ilmenites as oxygen carriers in chemical-looping combustion and influence of water gas shift reaction on gas composition. Chem Eng Res Des 90:1351–1390
7. Abad A, Mattisson T, Lyngfelt A, Ryden M (2006) Chemical-looping combustion in a 300 W continuously operating reactor system using a manganese-based oxygen carrier. Fuel 85:1174–1185
8. Wang S, Wang G, Jiang F, Luo M, Li F (2009) Chemical looping combustion of coke oven gas by using Fe_2O_3/CuO with $Mg_2Al_2O_4$ as oxygen carrier. Energy Environ Sci 3:1353–1360
9. Mattisson T, Lyngfelt A, Cho P (2001) The use of iron oxide as an oxygen carrier in CLC of methane with inherent separation of CO_2. Fuel 80:1953–1962

10. Abad A, Mattisson T, Lyngfelt A, Johansson M (2007) The use of iron oxide as oxygen carrier in a chemical looping reactor. Fuel 86:1021–1035
11. He F, Wang H, Dai Y (2007) Application of Fe_2O_3/Al_2O_3 composite particles as oxygen carrier of chemical looping combustion. J Nat Gas Chem 16:155–161
12. Mayer F, Bidwe AR, Schopf A, Taheri K, Zieba M (2013) Comparison of a new micaceous iron oxide and ilmenite as oxygen carrier for chemical looping combustion with respect to syngas conversion. Appl Energy 113:1863–1868
13. Song T, Shen L, Xiao J, Gao Z, Gu H, Zhang S (2011) Characterization of hematite oxygen carrier in CLC at high reduction temperature. J Fuel Chem Technol 39(8):567–574
14. Lyngfelt A, Johansson M, Mattisson T (2008) In: Proceedings of the ninth international conference on circulating fluidized beds (CFB-9), Hamburg, Germany

Chapter 40
On the Catalytic Activity of Palladium Nanoparticles-Based Anodes Towards Formic Acid Electro-oxidation: Effect of Electrodeposition Potential

Islam M. Al-Akraa, Ahmad M. Mohammad, Mohamed S. El-Deab, and Bahgat E. El-Anadouli

Abstract In this investigation, the catalytic activity of palladium nanoparticles (PdNPs)-modified glassy carbon (GC) (simply noted as PdNPs/GC) electrodes towards the formic acid electro-oxidation (FAO) was investigated. The deposition of PdNPs on the GC substrate was carried out by a potentiostatic technique at different potentials and the corresponding influence on the particles size and crystal structure of PdNPs as well as the catalytic activity towards FAO was studied. Scanning electron microscopy (SEM) demonstrated the deposition of PdNPs in spherical shapes and the average particle size of PdNPs deposited at a potential of 0 V vs. Ag/AgCl/KCl(sat.) was the smallest (ca. 8 nm) in comparison to other cases, where the deposition proceeded at higher potentials. The electrochemical measurements agreed consistently with this, where the highest surface area of PdNPs was calculated similarly for the deposition carried out at 0 V vs. Ag/AgCl/KCl(sat.). Interestingly, the X-ray diffraction (XRD) analysis revealed a similar dependency of the PdNPs crystal structure on their particle size and distribution. The deposition of PdNPs at 0 V vs. Ag/AgCl/KCl(sat.) seemed exhibiting the best crystallinity. From the electrocatalytic point of view, the activity of the PdNPs/GC electrode towards FAO decreased with the deposition potential

---

I.M. Al-Akraa (✉)
Department of Chemical Engineering, Faculty of Engineering, The British University in Egypt, Cairo 11837, Egypt
e-mail: islam.ahmed@bue.edu.eg

A.M. Mohammad • M.S. El-Deab
Department of Chemical Engineering, Faculty of Engineering, The British University in Egypt, Cairo 11837, Egypt

Department of Chemistry, Faculty of Science, Cairo University, Cairo 12613, Egypt
e-mail: ammohammad@cu.edu.eg; msaada68@yahoo.com

B.E. El-Anadouli
Department of Chemistry, Faculty of Science, Cairo University, Cairo 12613, Egypt
e-mail: bahgat30@yahoo.com

I. Dincer et al. (eds.), *Progress in Clean Energy, Volume 1*,
DOI 10.1007/978-3-319-16709-1_40

of PdNPs, which influenced consequently the particle size, shape, and/or crystallographic orientation of PdNPs.

Keywords Palladium nanoparticles • Electrodeposition • Formic acid • Fuel cells • Electrocatalysis • Particle size

40.1 Introduction

Research in fuel cells is intensively growing with the global desire to obtain efficient, incessant, and green energy sources [1, 2]. In this regard, the direct formic acid fuel cells (DFAFCs) have shown superiority over the traditional hydrogen and direct methanol fuel cells (DMFCs) in providing electricity for portable electronic devices [3–5]. The hydrogen fuel cells (HFCs) have long been investigated but the commercialization was restricted by difficulties associated with hydrogen storage and transportation. On the other hand, the DMFCs appeared promising as a consequence of the ease handling (methanol is a liquid fuel) and the high theoretical energy density (approximately 4.9 kWh L^{-1}) of methanol [6]. However, unfortunately, the DMFCs endured an inherent toxicity and a slow oxidation kinetics of methanol as well as the high crossover of methanol through Nafion-based membranes [7]. This has actually drawn attention to find a more convenient fuel for proton exchange membrane fuel cells (PEMFCs). Formic acid (FA) appeared promising in this regard, where it exhibited a smaller crossover flux through Nafion membrane than methanol [8, 9], which, interestingly, allow using high concentrated fuel solutions and thinner membranes in DFAFCs. This is highly desirable for the design of compact portable power systems. Furthermore, DFAFCs have a higher theoretical open-circuit potential (1.40 V) than that of HFCs (1.23 V) and DMFCs (1.21 V) [6, 7]. Nevertheless, DFAFCs experience a severe drawback where the catalytic activity of the Pt anodes (that typically used in DFAFCs), on which FA electro-oxidation (FAO) proceeds, ceases with time. This results from the adsorption of poisoning CO intermediate resulting from the "non-faradaic" dissociation of FA. This ultimately deteriorates the overall performance of DFAFC [3, 4]. Typically on Pt-based materials, FAO proceeds in two parallel pathways; the direct (desirable—involving the dehydrogenation of FA to CO_2 at a low anodic potential), and the indirect (undesirable—involving the chemical dehydration of FA with the release of poisonous CO that next be oxidized by platinum hydroxide at higher potentials (see Eq. (40.1)). Unfortunately, the adsorption of CO on Pt surface in the low potential domain (or the surface poisoning with CO) induces a catalytic deactivation for the electrode, which ultimately impedes the direct route of FAO.

$$\mathrm{Pt + HCOOH} \begin{cases} \nearrow \ \mathrm{Pt + CO_2 + 2H^+ + 2e^-} \quad \text{Direct dehydrogenation} \\ \searrow \ \mathrm{Pt\text{-}CO + H_2O \longrightarrow Pt + CO_2 + 2H^+ + 2e^-} \quad \text{Indirect dehydration} \end{cases} \tag{40.1}$$

To overcome the electrode's poisoning with CO and to prevent the catalytic deterioration of the electrode's activity, the development of new efficient and stable anodic catalysts for FAO will be necessary. Actually, Pd represents the reference catalyst for FAO where it provides even a better catalytic enhancement than Pt-based materials [10–12]. However, unfortunately, this catalytic activity deteriorates rapidly as a consequence of the poor stability of Pd catalyst [13]. That is why attention was shifted towards Pt-based materials regardless the lower catalytic activity provided. If we could solve the catalytic deactivation of Pd surfaces, it will be much interesting than Pt. The reaction mechanism of FAO on Pd proceeds exclusively via the dehydrogenation pathway (Eq. (40.1)) to produce carbon dioxide (CO_2) with very minor poisoning by CO [13, 14].

$$Pd + HCOOH \rightarrow Pd + CO_2 + 2H^+ + 2e^- \qquad (40.2)$$

The Pd catalyst was previously prepared by several approaches such as the dip coating [15], spraying [16], painting [17], sputtering, and electroplating (electrodeposition) [18–20]. Among these techniques, the electrodeposition of Pd catalyst onto a carbon substrate has attracted a particular attention due to the ease of preparation, suitability for special-shaped electrodes and low cost requirement. However, several studies indicated that the potential of electrodeposition may affect the particle size, geometry, and distribution of the Pd catalyst on the electrode surface, which will definitely change the electrochemical surface area and influence the electrocatalytic performance [21, 22].

With the revolution in nano-manufacturing, the use of palladium nanoparticles (PdNPs)-modified catalyst for FAO will be promising in twofold; the materials' saving and, hopefully, overcoming the catalytic deterioration of Pd. However, initially we need to investigate the influence of deposition potential on the particle size distribution and the catalytic activity of PdNPs towards FAO.

In this study, PdNPs will be electrodeposited by a potentiostatic method at different potentials onto GC electrodes to investigate how the deposition potential influences the PdNPs' particle size, crystal structure, and electrocatalytic activity towards FAO. The electrochemical measurements, field-emission scanning electron microscopy (FE-SEM), and X-ray diffraction (XRD) spectroscopy are all combined to reveal the surface morphology, composition, and crystal structure of PdNPs and to understand the origin of enhancement in the catalytic activity of the catalyst.

40.2 Experimental

Typically cleaned glassy carbon electrode (GC, $d = 3.0$ mm) was used as the working electrode. A spiral Pt wire and Ag/AgCl/KCl(sat) were used as the counter and reference electrodes, respectively. All potentials in this study are referenced to Ag/AgCl/KCl(sat). The electrodeposition of PdNPs on the bare GC was carried out in 0.1 M H_2SO_4 solution containing 1.0 mM $Pd(CH_3COO)_2$ (Merck KGaA) via a constant potential electrolysis technique by holding the potential at 0, 0.05, 0.10, 0.15, and 0.20 V corresponding to overpotentials (η) of –0.80, –0.75, –0.70, –0.65, and –0.60 V, respectively. Overpotentials stated here are relative to the standard potential for Pd^{2+} reduction to Pd^0 which is 0.715 V vs. Ag/AgCl/KCl(sat). All of the chemicals used in this investigation were of analytical grade and used without further purification.

The electrochemical measurements were performed at room temperature (25 ± 1 °C) in a conventional two-compartment three-electrode glass cell using BioLogic SAS potentiostat operated with EC-lab® software. A field emission scanning electron microscope (FE-SEM, QUANTA FEG 250) was employed to evaluate the electrode's morphology. The X-ray diffraction (XRD, PANalytical, X'Pert PRO) operated with Cu target ($\lambda = 1.54$ Å) was used to identify the crystallographic structure of PdNPs. The electrocatalytic activity of the modified electrodes towards FAO was examined in a solution of 0.3 M FA at pH of 3.5 (the pH was adjusted by adding a proper amount of NaOH).

40.3 Results and Discussion

40.3.1 Electrochemical and Materials Characterization

Figure 40.1 shows the cyclic voltammogram (CV) obtained at GC electrode in 0.1 M H_2SO_4 containing 1.0 mM $Pd(CH_3COO)_2$ solution at a scan rate of 20 mV s^{-1}. The potential scan started from 1.05 V to more negative potentials. The absence of the cathodic current at potentials more positive than 0.55 V reveals that the under potential deposition (UPD) in this system has not yet reached and that the electro-crystallization started in the overpotential deposition (OPD) region. This indicates a weak deposit–substrate interaction and recommends the Volmer–Weber growth mechanism for the deposition process [23, 24]. This growth mechanism operates on electrodes with low surface energy as has been observed for the electrodeposition of metals on a GC substrate [24]. Looking inside, a single well-defined cathodic peak (peak a) at ca. 0.2 V was observed and assigned to the Pd deposition. At more cathodic potentials ($E = -0.35$ V), the current increased intensively due to the hydrogen adsorption and evolution. On the reverse (anodic direction) sweep, peak (b) represents the hydrogen desorption and/or oxidation. The backward and the forward scans intersect and a nucleation loop is observed at a potential of ca. 0.3 V. The appearance of such a loop indicates that the deposition of Pd on Pd is easier than on GC [25]. At more positive potential ($E = 0.7$ V), the oxidation peak (c) assigns the dissolution of deposited Pd.

The data of Fig. 40.1 was employed to electrodeposit PdNPs onto the GC electrode in 0.1 M H_2SO_4 containing 1.0 mM $Pd(CH_3COO)_2$ solution via a constant potential electrolysis technique. Figure 40.2a shows the current transients of PdNPs electrodeposition onto GC electrode for 10 s at different applied potentials ($E = 0$,

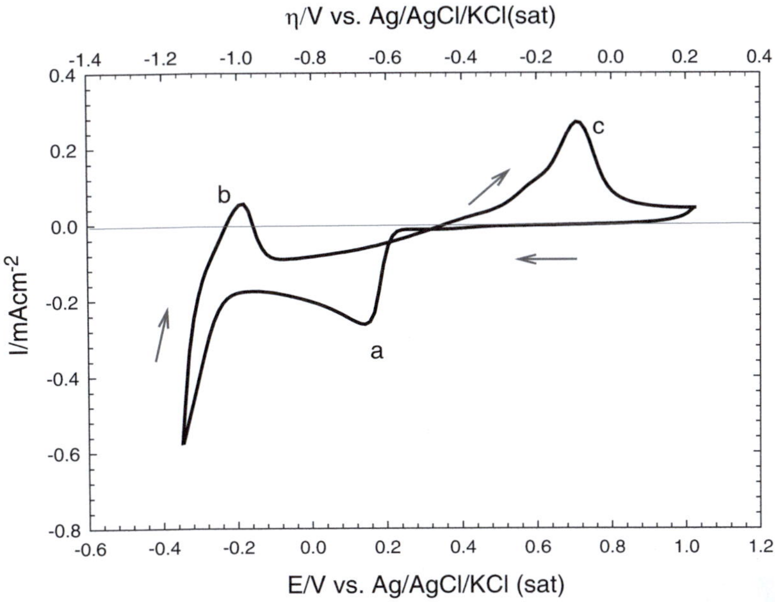

Fig. 40.1 CV obtained at GC electrode in N_2-saturated 0.1 M H_2SO_4 solution containing 1.0 mM $Pd(CH_3COO)_2$. Potential scan rate: 20 mV s^{-1}

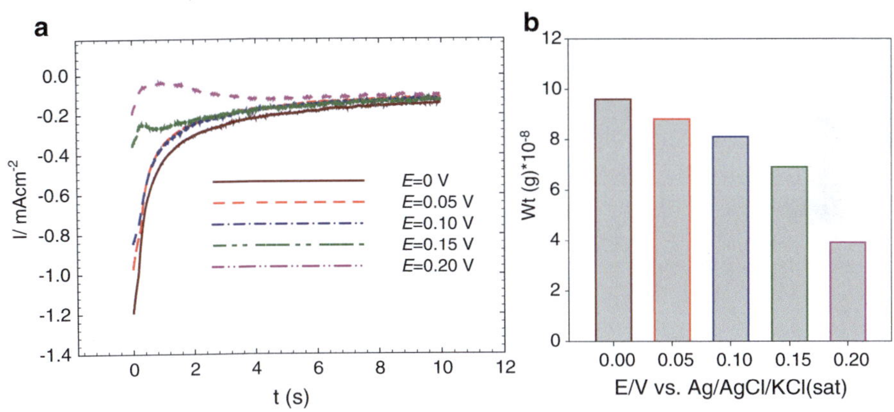

Fig. 40.2 (**a**) Current transients obtained at GC electrode in N_2-saturated 0.1 M H_2SO_4 solution containing 1.0 mM $Pd(CH_3COO)_2$ at different applied potentials ($E=0$, 0.05, 0.10, 0.15, and 0.20 V), and (**b**) variation of deposited mass of PdNPs with applied electrodeposition potential

0.05, 0.10, 0.15, and 0.20 V). As observed in Fig. 40.2, the charge of deposition decreases with the electrodeposition potential, which will definitely stimulate a consequent decrease in the mass of the deposited PdNPs, as calculated from Faraday's law of electrolysis (see Fig. 40.2b). Actually, not only the mass, but also the particle size, geometry, crystallinity, and electrocatalytic activity of PdNPs

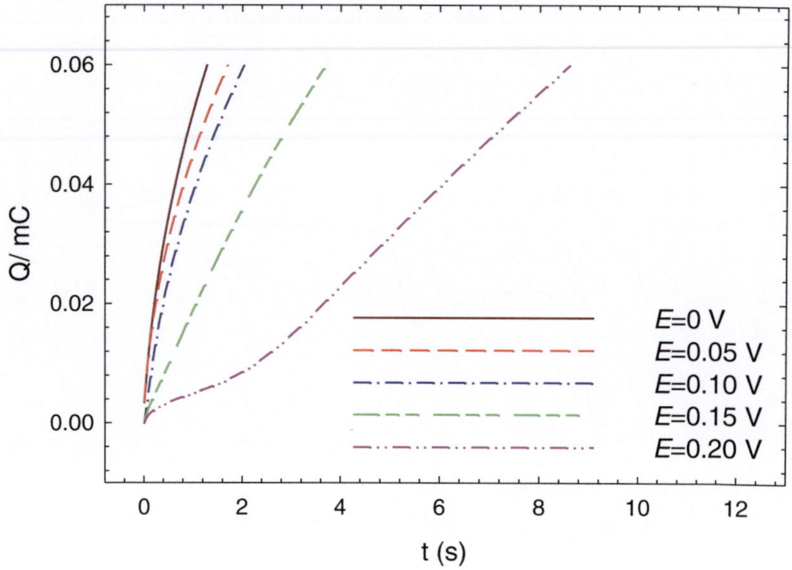

Fig. 40.3 Charge transients obtained at GC electrode in N_2-saturated 0.1 M H_2SO_4 solution containing 1.0 mM $Pd(CH_3COO)_2$ at different applied potentials ($E = 0$, 0.05, 0.10, 0.15, and 0.20 V)

may be influenced by the deposition potential. However, in order to understand precisely this influence, we need to fix the mass of PdNPs deposited at all potentials. This can be achieved if the deposition charge (Q) of PdNPs was kept constant regardless the deposition potential.

Figure 40.3 shows the charge transients of PdNPs electrodeposited onto the GC electrode at different potentials ($E = 0$, 0.05, 0.10, 0.15, and 0.20 V), where Q is kept constant at 60 μC.

As Fig. 40.3 depicts, a deposition time of (1.3, 1.7, 2.1, 3.7, and 8.6 s) was required to pass 60 μC for the deposition of PdNPs with a theoretical net mass of 3.3×10^{-8} g. Now, after avoiding the influence of the catalyst mass, we can investigate the dependence of the particle size, crystallographic structure, and electrocatalytic activity of PdNPs on the deposition potential.

Figure 40.4 shows the characteristic CVs at PdNPs/GC electrode at different applied potentials ($E = 0$, 0.05, 0.10, 0.15, and 0.20 V) applying only a charge of 60 μC. The characteristic behavior of Pd is clearly shown; the oxidation of Pd, which extends over a wide range of potential, is coupled with the oxide reduction peak at ca. 0.50 V. This couple corresponds to the solid-state surface redox transition (SSSRT) involving Pd/PdO. In addition, well-defined peaks for the hydrogen adsorption/desorption are shown in the potential range from 0.0 to −0.2 V. Interestingly, the active real surface area of PdNPs (calculated utilizing the oxide reduction peak at ca. 0.50 V) decreased with the deposition potential, which infers a difference in the particle size and distribution, as long as the mass is

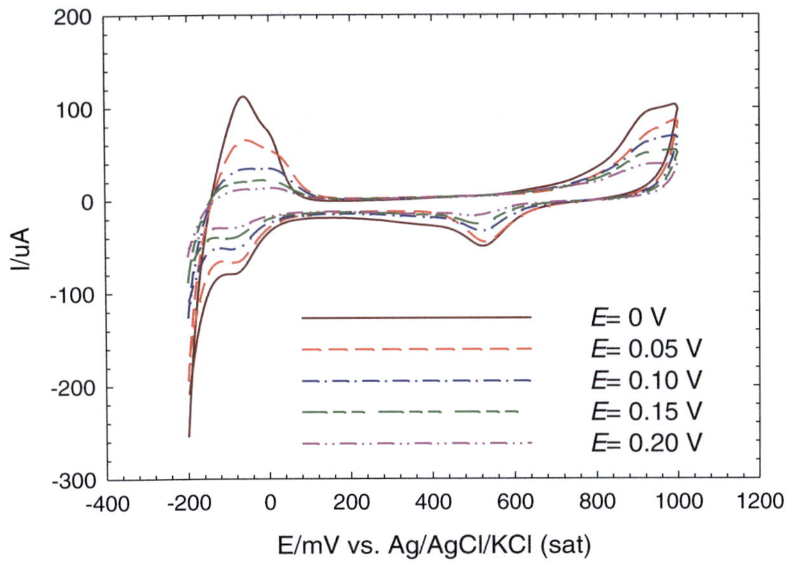

Fig. 40.4 CVs obtained at PdNPs/GC electrode in N_2-saturated 0.5 M H_2SO_4. A same charge of 60 μC is applied at all electrodeposition potentials. Potential scan rate: 100 mV s^{-1}

fixed. A recent report has actually indicated the dependence of the catalyst's particle size on the deposition potential [26].

Interestingly, the deposition of PdNPs at 0 V provided the highest real surface area among the other deposition potentials (see Fig. 40.4). And consequently, we expect the smallest particle size for PdNPs deposited under this potential. This is the role of microscopy to support the assumption. The FE-SEM micrographs in Fig. 40.5 show the corresponding morphology for the set of PdNPs catalysts deposited at 0, 0.10, and 0.20 V. The PdNPs were deposited at 0, 0.10, and 0.20 V in spherical nanoparticles of average particle sizes of ca. 8, 36, and 48 nm, respectively. This agrees very much with our expectation, recommending an increase of average particle size with the increase of the deposition potential. The particles' distribution of PdNPs deposited at 0 V was much more homogeneous if compared to other deposition potentials. It seems with this deposition potential (0 V), the number of individual starting nuclei for PdNPs deposition was higher assisting the uniform distribution of small particles. On the other hand, the deposition of PdNPs at higher potentials favored the deposition of fewer number of PdNPs' nuclei but assisted their rapid growth to end with low density but coarser particles [25].

The influence of the deposition potential on the particle's size of PdNPs may induce a further change in the crystal structure of PdNPs. Recent experiments explained the possibility of changing the lattice parameters of nanoparticles with their particle size, and the dependency became more significant at extremely small particle sizes [27]. The XRD technique was next employ to verify this dependence.

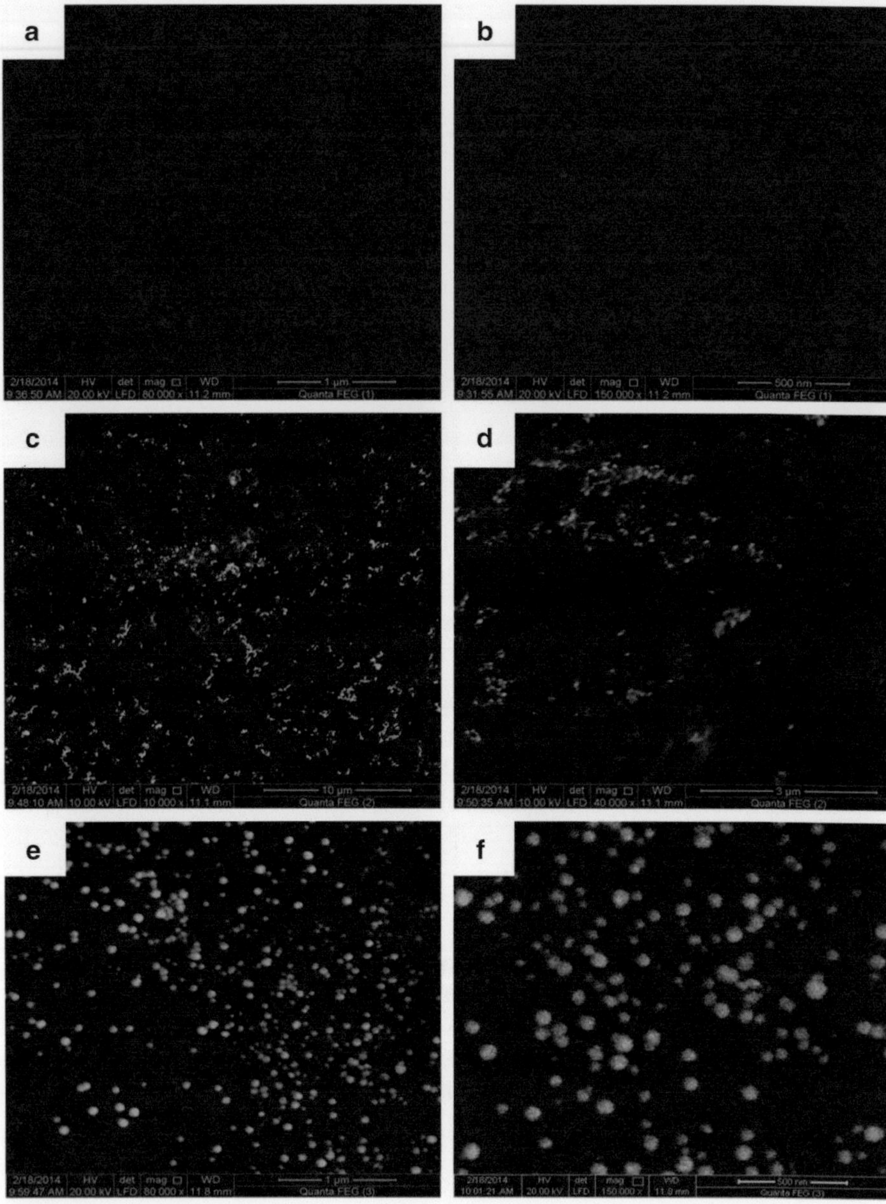

Fig. 40.5 FE-SEM micrographs of PdNPs/GC electrode. The electrodeposition of PdNPs was carried out at (**a**, **b**) 0 V, (**c**, **d**) 0.10 V, and (**e**, **f**) 0.20 V keeping the deposition charge constant at 60 μC

Figure 40.6 depicts the XRD pattern of PdNPs (8, 36, and 48 nm) that were electrodeposited at 0, 0.10, and 0.20 V, respectively. In Fig. 40.6a, the XRD pattern of PdNPs (8 nm) shows several peaks ca. 24°, 38°, 44°, 65°, and 78°. These peaks correspond, respectively, to the (0 0 2) plane of C, (1 1 1), (2 0 0), (2 2 0), and (3 1 1)

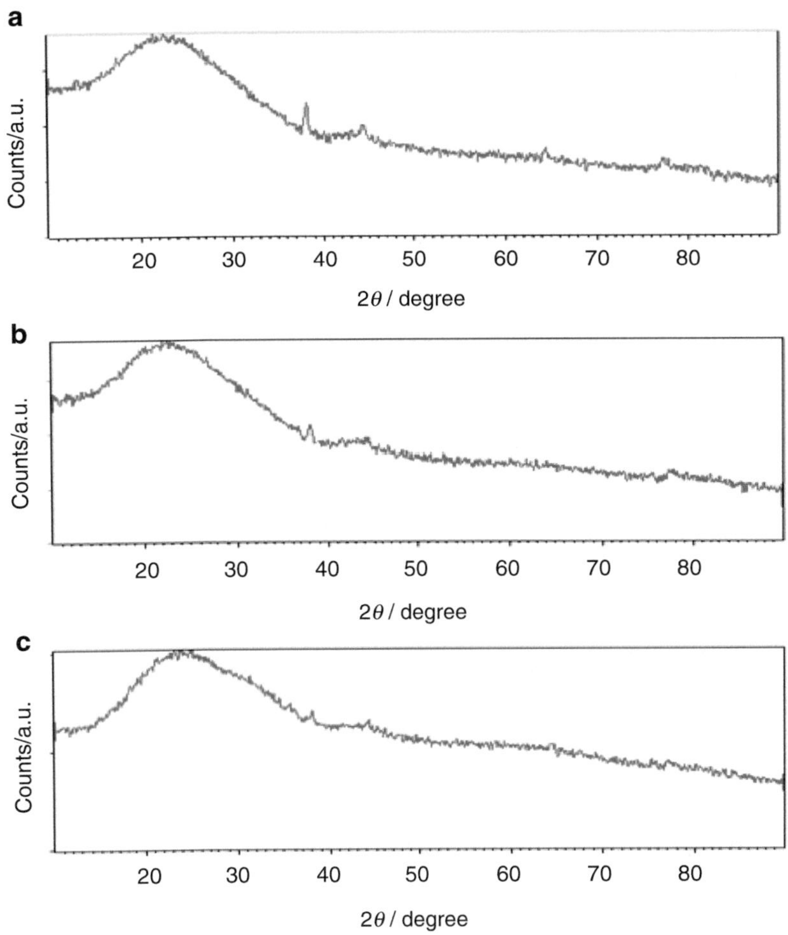

Fig. 40.6 XRD pattern of PdNPs/GC electrode. The electrodeposition of PdNPs was carried out at (**a**) 0 V, (**b**) 0.10 V, and (**c**) 0.20 V keeping the deposition charge constant at 60 μC

planes of Pd face-centered cubic (fcc) lattice (JCPDS standard 05-0681 (Pd)) [21, 28]. However, the intensity of these peaks for PdNPs deposited at 0.10 (36 nm) and 0.20 (48 nm) V was a little bit broad and shorter (Fig. 40.6b and c). This may happen with a certain degree of distortion in the crystal lattice parameters that was associated with the change in the average particle size [29]. The diffraction angles of the lattice planes in Fig. 40.6 were shifted as well to lower values with the decrease in the particle size, which reveals the expansion of the Pd–Pd interatomic distance [28]. The lattice constant of PdNPs, in contrast to other metal nanoparticles, increased with the decrease of particle size, perhaps due to a structural change or incorporation of oxygen, carbon, or hydrogen into the palladium lattice [28]. Table 40.1 summarizes the differences in diffraction angles, interplanar spacing of (1 1 1) plane, and lattice constants of PdNPs deposited at different potentials.

Table 40.1 Differences in diffraction angle (2θ), interplanar spacing (d) of (1 1 1) plane, and lattice constant (a) with PdNPs deposition potential (E), overpotential (η), and PdNPs average size

E (V)	η (V)	Average particle size (nm)	2θ of (1 1 1) (degree)	d (Å)	a (Å)
0	–0.80	8	38.17	2.357	4.082
0.10	–0.70	36	38.23	2.354	4.078
0.20	–0.60	48	38.39	2.344	4.059

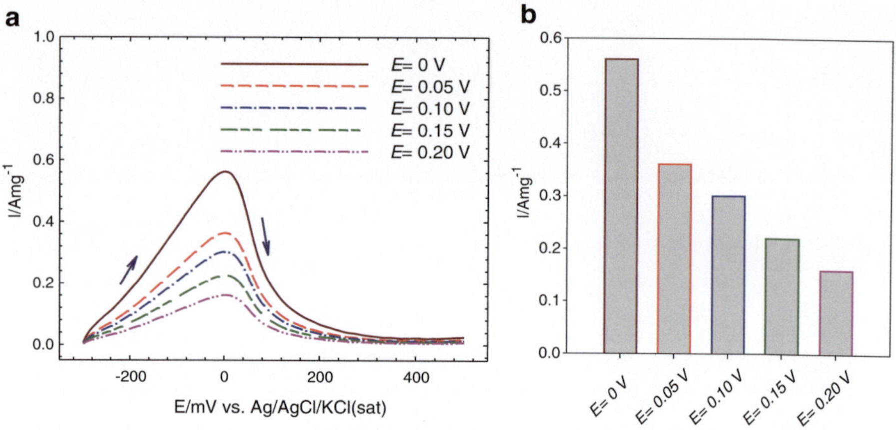

Fig. 40.7 (**a**) LSVs obtained at PdNPs/GC electrode with the same applied charge of 60 μC in N_2-saturated 0.3 M FA (pH 3.5). Potential scan rate: 100 mV s^{-1}, and (**b**) variation of electrocatalytic activity of PdNPs (in terms of oxidation peak specific current) towards FAO with electrodeposition potential

40.3.2 Formic Acid Electro-Oxidation

Figure 40.7a shows the linear sweep voltammograms (LSVs) of FAO at the PdNPs/GC electrode in an aqueous solution of 0.3 M formic acid (pH 3.5), where the deposition of PdNPs was carried out at potentials ($E = 0$, 0.05, 0.10, 0.15, and 0.20 V). A pronounced single oxidation peak is observed at ca. 0 V, which assigns the direct oxidation of FA to CO_2.

The electrocatalytic activity of metal nanoparticles is expected to depend on their size, shape, and crystal structure [30–32]. In our case of PdNPs, the oxidation peak current can be used as a probe to measure the electrocatalytic activity towards FAO. Investigation of these peaks revealed a decrease in the catalytic activity of the PdNPs/GC electrode towards FAO with the increase in deposition potential of PdNPs or with the increase of the average particle size of PdNPs (see Fig. 40.7b). Interestingly, the specific current of the oxidation peak for PdNPs deposited at 0 V was almost 3.5 times larger than that deposited at 0.20 V. As we mentioned previously, the deposition at 0 V resulted in PdNPs with the smallest particle size (8 nm), the highest active real surface area, and the most perfect crystal structure.

Therefore, this deposition condition (0 V) is expected to offer more active sites for the adsorption of FA, which ultimately will lead to a better enhancement in the electrocatalytic activity towards FAO.

40.4 Conclusions

The deposition of PdNPs on the GC electrode was carried out by a potentiostatic technique at different potentials. Electrochemical, SEM, and XRD investigations confirmed that the deposition of PdNPs at a potential of 0 V provided the smallest particle sizes (ca. 8 nm), the largest active surface area, and the most perfect crystal structure for PdNPs. From another view, the electrocatalytic activity of the PdNPs/GC electrode towards FAO decreased with the deposition potential of PdNPs, and the best enhancement was achieved when the deposition was achieved at 0 V. This is definitely attributed to a consequent change in the particle size, distribution, and/or crystallographic orientation of PdNPs.

References

1. Abdullah AM, Okajima T, Mohammad AM et al (2007) Temperature gradients measurements within a segmented H_2/air PEM fuel cell. J Power Sources 172:209–214
2. Abdullah AM, Mohammad AM, Okajima T et al (2009) Effect of load, temperature and humidity on the pH of the water drained out from H_2/air polymer electrolyte membrane fuel cells. J Power Sources 190:264–270
3. Al-Akraa IM, Mohammad AM, El-Deab MS et al (2011) Electrooxidation of formic acid at platinum–gold nanoparticle-modified electrodes. Chem Lett 40:1374–1375
4. Al-Akraa IM, Mohammad AM, El-Deab MS et al (2012) Development of tailor-designed gold-platinum nanoparticles binary catalysts for efficient formic acid electrooxidation. Int J Electrochem Sci 7:3939–3946
5. Zhang S, Shao Y, Yin G et al (2010) Facile synthesis of PtAu alloy nanoparticles with high activity for formic acid oxidation. J Power Sources 195:1103–1106
6. Yu X, Pickup PG (2008) Recent advances in direct formic acid fuel cells (DFAFC). J Power Sources 182:124–132
7. Demirci UB (2007) Direct liquid-feed fuel cells: thermodynamic and environmental concerns. J Power Sources 169:239–246
8. Wang X, Hu J-M, Hsing I-M (2004) Electrochemical investigation of formic acid electro-oxidation and its crossover through a Nafion® membrane. J Electroanal Chem 562:73–80
9. Rhee YW, Ha SY, Masel RI (2003) Crossover of formic acid through Nafion® membranes. J Power Sources 117:35–38
10. Rice C, Ha S, Masel RI et al (2003) Catalysts for direct formic acid fuel cells. J Power Sources 115:229–235
11. Jung WS, Han J, Ha S (2007) Analysis of palladium-based anode electrode using electrochemical impedance spectra in direct formic acid fuel cells. J Power Sources 173:53–59
12. Han SD, Choi JH, Noh SY et al (2009) Performance characterization of direct formic acid fuel cell using porous carbon-supported palladium anode catalysts. Korean J Chem Eng 26:1040–1046

13. Baik SM, Han J, Kim J et al (2011) Effect of deactivation and reactivation of palladium anode catalyst on performance of direct formic acid fuel cell (DFAFC. Int J Hydrogen Energy 36:14719–14724
14. Choi J-H, Jeong K-J, Dong Y et al (2006) Electro-oxidation of methanol and formic acid on PtRu and PtAu for direct liquid fuel cells. J Power Sources 163:71–75
15. Jayashree RS, Gances L, Choban ER et al (2005) Air-breathing laminar flow-based microfluidic fuel cell. J Am Chem Soc 127:16758–16759
16. Gago AS, Morales-Acosta D, Arriaga LG et al (2011) Carbon supported ruthenium chalcogenide as cathode catalyst in a microfluidic formic acid fuel cell. J Power Sources 196:1324–1328
17. Brushett FR, Jayashree RS, Zhou W-P et al (2009) Investigation of fuel and media flexible laminar flow-based fuel cells. Electrochim Acta 54:7099–7105
18. Kjeang E, Michel R, Harrington DA et al (2008) An alkaline microfluidic fuel cell based on formate and hypochlorite bleach. Electrochim Acta 54:698–705
19. Kjeang E, Brolo AG, Harrington DA et al (2007) Hydrogen peroxide as an oxidant for microfluidic fuel cells fuel cells and energy conversion. J Electrochem Soc 154:B1220–B1226
20. Lopez-Montesinos PO, Yossakda N, Schmidt A et al (2011) Design, fabrication, and characterization of a planar, silicon-based, monolithically integrated micro laminar flow fuel cell with a bridge-shaped microchannel cross-section. J Power Sources 196:4638–4645
21. Zhang B, Ye D, Li J et al (2012) Electrodeposition of Pd catalyst layer on graphite rod electrodes for direct formic acid oxidation. J Power Sources 214:277–284
22. Kim H, Subramanian NP, Popov BN (2004) Preparation of PEM fuel cell electrodes using pulse electrodeposition. J Power Sources 138:14–24
23. Alvarez AE, Salinas DR (2010) Formation of Cu/Pd bimetallic crystals by electrochemical deposition. Electrochim Acta 55:3714–3720
24. Komsiyska L, Staikov G (2008) Electrocrystallization of Au nanoparticles on glassy carbon from $HClO_4$ solution containing $[AuCl4]^-$. Electrochim Acta 54:168–172
25. Rezaei M, Tabaian SH, Haghshenas DF (2012) Nucleation and growth of Pd nanoparticles during electrocrystallization on pencil graphite. Electrochim Acta 59:360–366
26. Rezaei M, Tabaian SH, Haghshenas DF (2012) A kinetic description of pd electrodeposition under mixed control of charge transfer and diffusion. J Electroanal Chem 687:95–101
27. Qi WH, Wang MP (2005) Size and shape dependent lattice parameters of metallic nanoparticles. J Nanopart Res 7:51–57
28. Teranishi T, Miyake M (1998) Size control palladium nanoparticles and their crystal structures. Chem Mater 10:594–600
29. Mohanlal SK, Sanjeeviraja C (1986) An X-ray diffraction study of the size effect on a crystalline solid solution of semiconductor system $Ge_{0.2}Si_{0.8}$. Cryst Res Technol 21:K77–K80
30. Kalimuthu P, John SA (2008) Size dependent electrocatalytic activity of gold nanoparticles immobilized onto three dimensional sol–gel network. J Electroanal Chem 617:164–170
31. Hoshi N, Nakamura M, Kida K (2007) Structural effects on the oxidation of formic acid on the high index planes of palladium. Electrochem Commun 9:279–282
32. Zhang L, Sui Q, Tang T et al (2013) Surfactant-free palladium nanodendrite assemblies with enhanced electrocatalytic performance for formic acid oxidation. Electrochem Commun 32:43–46

Chapter 41
The Effect of Temperature and Initial Concentration on Synthesis of Ammonia Borane

Derya Öncel Özgür and Göksel Özkan

Abstract In this work, production of ammonium borane (NH_3BH_3, abbreviated as AB) was studied. In synthesis works, the effect of reaction temperature and reactive inlet material concentrations on the conversion into AB were examined. Synthesis of ammonium borane was obtained from the reaction of $NaBH_4$ and $(NH_4)_2SO_4$ in THF. As a result of parametrical experiments, it has been determined that product conversion yield is increased from 95.3 to 98.2 % and reaction period is decreased by up to 2 h and product purity value is increased up to 100 % when reaction temperature is increased from 25 to 50 °C. It has been determined that when increased amount of reactive materials is compared with stoichiometric amounts, product conversion increased but limiting agent in the reaction is determined to be $NaBH_4$. According to FTIR, XRD, and ^{11}B NMR analyses results, AB may be synthesized up to purity phase.

Keywords Ammonia borane • Hydrogen storage • High purity sythesis

41.1 Introduction

With a high chemical energy density and zero emissions when produced from renewable resources, hydrogen is set to become a major fuel of the future, bridging the gap between intermittent renewable and rapidly depleting fossil fuels. It can be used across a wide range of applications, from portable electronics to energy distribution systems within the grid and transportation. One of the important problems related to hydrogen energy is the improvement of appropriate hydrogen storing materials and the enablement of use in practice [1]. It is seen to be a good chemical hydrogen storing material because AB among the boron compounds has the capacity to store high hydrogen. AB is a boron nitrogen hydride compound whose peculiarity is to be able to store very high hydrogen content (up to 19 % by weight) in a safe and easy transportable manner [2]. For this reason, it has attracted

D.Ö. Özgür (✉) • G. Özkan
Department of Chemical Engineering, Gazi University, Ankara 06570, Turkey
e-mail: deryaoncel@gazi.edu.tr; gozkan@gazi.edu.tr

I. Dincer et al. (eds.), *Progress in Clean Energy, Volume 1*,
DOI 10.1007/978-3-319-16709-1_41

a lot of attention as a source of hydrogen due to the increasing demand for this element as fuel or reducing agent. Therefore, this study focuses on the AB production and characterization.

41.2 Experimental

NH_3BH_3 was synthesized by the reaction of $NaBH_4$ with $(NH_4)_2SO_4$ in the presence of THF as the solvent as given by Eq. (41.1). Experiments were performed in a 3-neck round-bottom flask fitted with septum inlet, a reflux and thermometer. $NaBH_4$ and $(NH_4)_2SO_4$ were added to a 500 mL flask. THF was transferred into the flask, and the contents were stirred. Constant reaction temperature was achieved as a result of a hotplate combined with heater control system. The synthesis of AB was carried out at different temperature in the range of 25–50 °C using different inlet mole ratio as $NaBH_4/(NH_4)_2SO_4$ mole ratio in the range between 1 and 4. Upon completion, the reaction mixture was cooled to room temperature and filtered. The solvent was removed from the filtrate through vacuum distillation. The resulting ammonia borane was dissolved in diethyl ether and filtered to remove any insoluble material. Purity of the product ammonia borane was assessed by means of X-ray powder diffraction, ^{11}B NMR spectroscopy, infrared spectroscopy, and dehydrogenation reaction.

$$2NaBH_4 + (NH_4)_2SO_4 \xrightarrow{THF} 2NH_3BH_3 + Na_2SO_4 + 2H_2 \quad (41.1)$$

The experimental procedure consisted of five key steps: (1) Reaction, (2) Filtration, (3) Vacuum Distillation, (4) Extraction, and (5) Drying. The flow chart of NH_3BH_3 synthesis is shown in Fig. 41.1

41.3 Results and Discussion

In this work, the synthesis and characterization of AB were studied. In synthesis works, the effect of temperature and initial molar ratio as $NaBH_4/(NH_4)_2SO_4$ on the conversion into AB are investigated.

41.3.1 *Effects of Temperature*

Experiments were carried out in the range of 25–50 °C. Also, the reaction at temperatures above 50 °C was not tried as the close to THF boiling point. As a result of parametrical experiments, it has been determined that product conversion

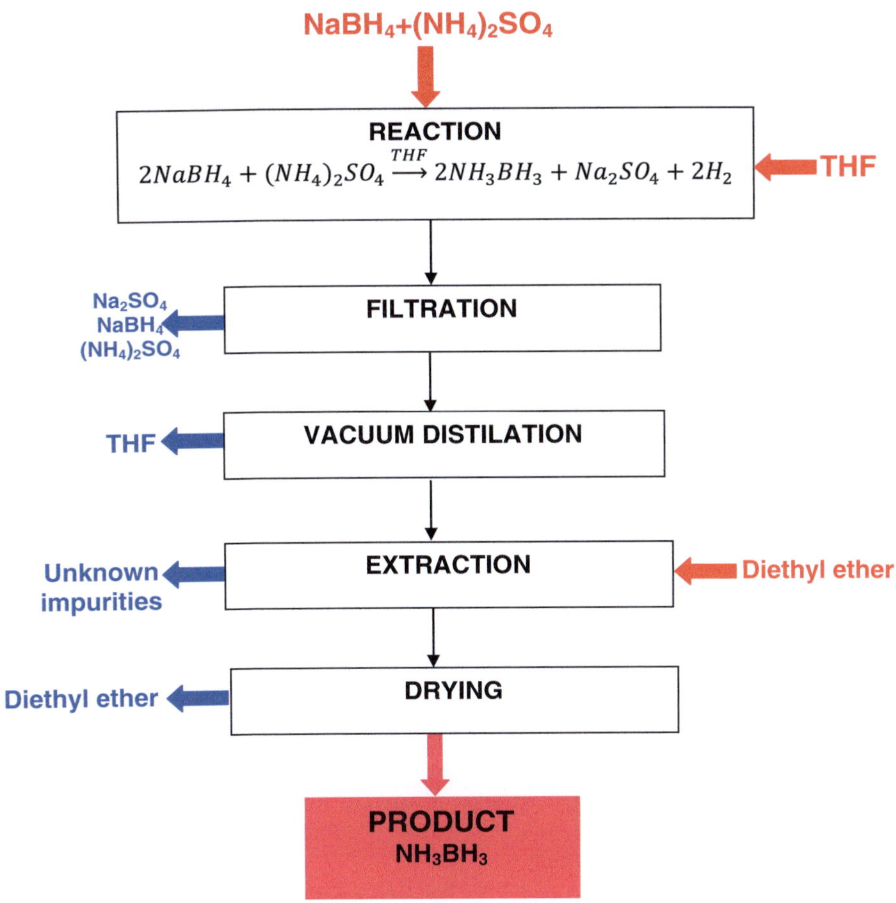

Fig. 41.1 Process flow chart for AB synthesis

yield is increased from 95.3 to 98.2 % and reaction period is decreased by up to 2 h and product purity value is increased up to 100 % when reaction temperature is increased from 25 to 50 °C (Fig. 41.2)

41.3.2 Effects of Initial Molar Ratio

The synthesis of AB was carried out at different initial $NaBH_4/(NH_4)_2SO_4$ molar ratio in the range between 1 and 4. Figure 41.3 shows the effect of input molar ratio on product yield. We undertook synthesis experiments at 40 °C and 200 mL THF. When input molar ratio equals to 1, it means that $(NH_4)_2SO_4$ is excess amount in the reaction medium and $NaBH_4$ is limiting compound. When input molar ratio equals

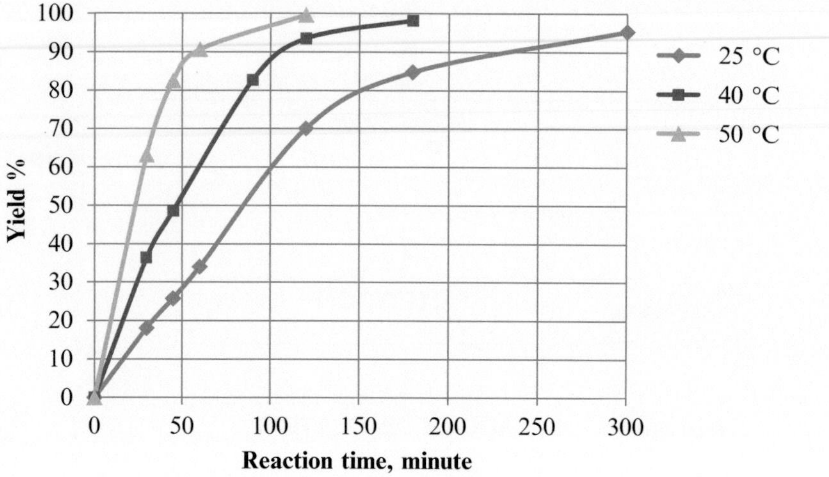

Fig. 41.2 The effect of the temperature on AB product yield

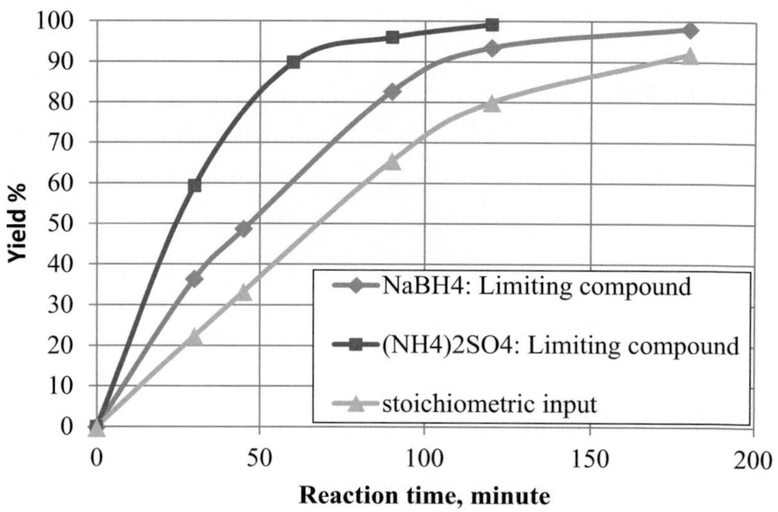

Fig. 41.3 The effect of feed molar ratio as $NaBH_4/(NH_4)_2SO_4$ on AB product yield

to 4, it means that $NaBH_4$ is excess amount in the reaction medium and $(NH_4)_2SO_4$ is limiting compound. It has been shown that when increased amount of reactive materials is compared with stoichiometric amounts (input molar ratio is 2), product conversion increased but limiting agent in the reaction is determined to be $NaBH_4$.

41.3.3 *Characterization of AB*

Product AB was characterized by tree different technique: XRD, FTIR, and ^{11}B NMR. The FT-IR spectrum of this compound is shown in Fig. 41.4. FT-IR result demonstrates that the produced AB had the same chemical structure as NH_3BH_3.

Figure 41.5 shows the XRD patterns of product AB. Produced AB was compared with the reference AB has characteristic peaks. This result showed that product was precisely AB.

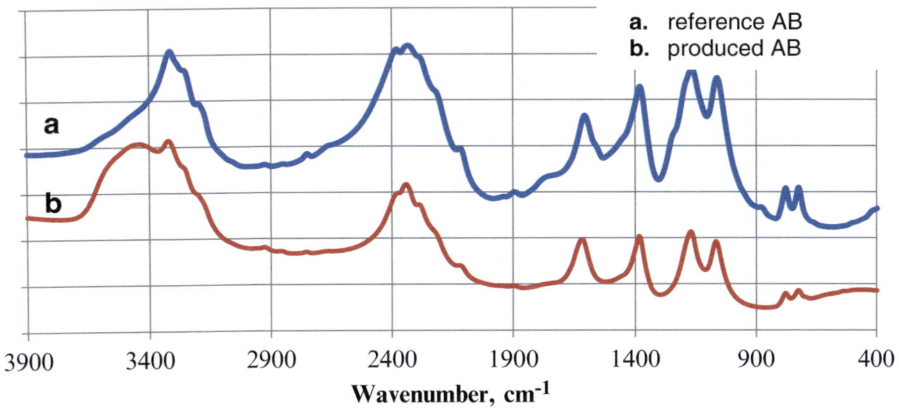

Fig. 41.4 FT-IR spectrum of product AB

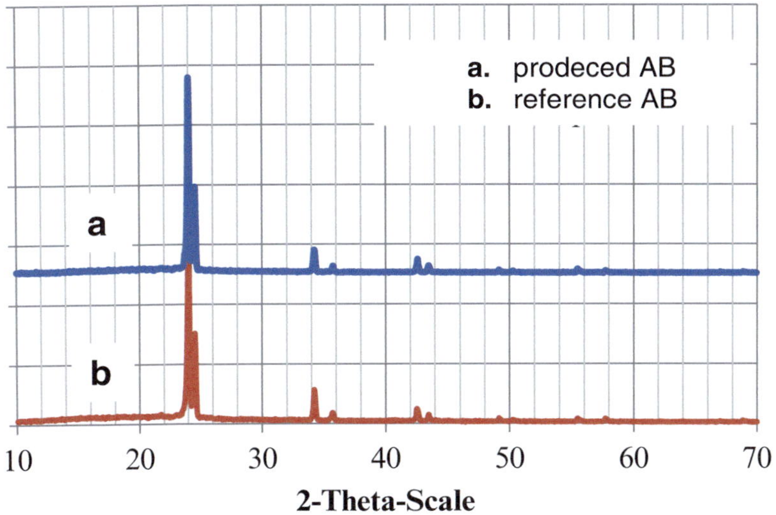

Fig. 41.5 XRD pattern of product AB

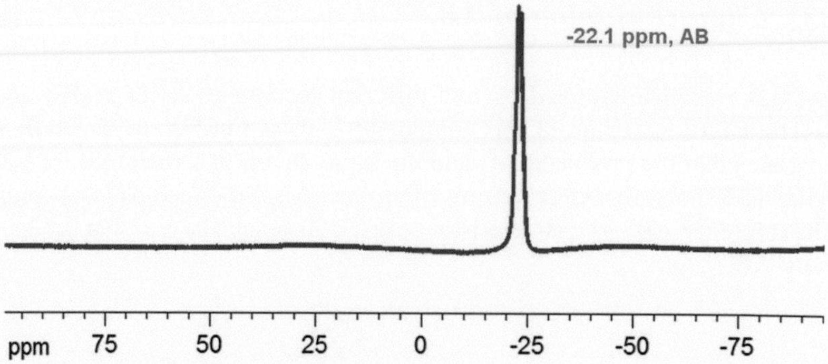

Fig. 41.6 ^{11}B NMR spectrum of product AB

Figure 41.6 shows the ^{11}B NMR spectra of product AB. According to the NMR results, chemical shift value was similar with literature value (−22.1 ppm) and demonstrated that synthesis product was AB [3].

Acknowledgements The authors express their gratitude to BOREN for supporting the project 2011. Ç0294

References

1. Aslan Ö (2007) Hidrojen Ekonomisine Doğru. İstanbul Ticaret Üniversitesi Sosyal Bilimler Dergisi 11:283–298
2. Umegaki T, Yan J, Zhang X, Shioyama H, Kuriyama N, Xu Q (2009) Boron-and nitrogen-based chemical hydrogen storage materials. Int J Hydrog Energy 3:2303–2311
3. Ramachandran PV, Gagare PD (2007) Preparation of ammonia borane in high yield and purity, methanolysis and regeneration. Inorg Chem 46:7810–7817

Chapter 42
Electrocatalysis of Formic Acid Electro-Oxidation at Platinum Nanoparticles Modified Surfaces with Nickel and Cobalt Oxides Nanostructures

Gumaa A. El-Nagar, Ahmad M. Mohammad, Mohamed S. El-Deab, and Bahgat E. El-Anadouli

Abstract The present study proposes a novel promising binary catalyst for formic acid electro-oxidation (FAO); the main anodic reaction in direct formic acid fuel cells (DFAFCs). The catalyst is basically composed of two metal oxides of nickel and cobalt nanostructures (i.e., NiOx and CoOx) assembled onto a platinum nanoparticles (PtNPs)−modified glassy carbon (Pt/GC) electrode. Actually, FAO proceeds at bare Pt surfaces in two parallel routes; one of them is desirable (called direct or hydrogenation) and occurred at a low potential domain (I_p^d). While, the other (undesirable) involves the dehydration of formic acid (FA) at low potential domain to produce a poisoning intermediate (CO), which next be oxidized (indirect, I_p^{ind}) at a higher potential domain after the platinum surface becomes hydroxylated. Unfortunately, the peak current ratio (I_p^d/I_p^{ind}) of the two oxidation routes, which monitors the degree of the catalytic enhancement and the poisoning level, stands for bare Pt surfaces at a low value (less than 0.2). Interestingly, this ratio increased significantly as a result of the further modification of the Pt/GC electrode with NiOx $\left(I_p^d/I_p^{ind} = 3\right)$, CoOx $\left(I_p^d/I_p^{ind} = 4\right)$ and a binary mixture of both $\left(I_p^d/I_p^{ind} = 15\right)$. This highlights the essential role of these in promoting the direct FAO, presumably via a mediation process that ultimately improved the oxidation kinetics or through a catalytic enhancement for the oxidation of the poisoning CO at the low potential domain of the direct FAO. The effect of the deposition order of NiOx and CoOx on the catalytic activity was addressed and fount influencing. The addition of CoOx to

G.A. El-Nagar • B.E. El-Anadouli (✉)
Department of Chemistry, Faculty of Science, Cairo University, Cairo 12613, Egypt
e-mail: gumaa@sci.cu.edu.eg; bahgat30@yahoo.com

A.M. Mohammad • M.S. El-Deab
Department of Chemistry, Faculty of Science, Cairo University, Cairo 12613, Egypt

Department of Chemical Engineering, Faculty of Engineering,
The British University in Egypt, Cairo 11837, Egypt
e-mail: ammohammad@cu.edu.eg; msaada68@yahoo.com

I. Dincer et al. (eds.), *Progress in Clean Energy, Volume 1*,
DOI 10.1007/978-3-319-16709-1_42

the catalyst was really important, particularly in improving the catalytic stability of the catalyst towards a long-term continuous electrolysis experiment, which actually imitates the real industrial applications.

Keywords Nanostructured • DFAFCs • Nano-CoOx • Electrocatalysis

42.1 Introduction

Fuel cell (FC) technology is hastily growing as a consequence of the global need to clean, eco-friendly, and efficient energy sources. They are expected to replace the fossil fuel-based energy sources to meet the growing infinite needs of electric power in industry and daily-life activities. Clean Energy Patent Growth Index shows growth in all clean energy technology patents. More than 1,000 FC patents issued in 2012. Samsung has designed prototype portable direct methanol fuel cell (DMFC) systems that can directly power a notebook/cell phone or recharge the battery in the computer to extend the operating time. Honda motor company Ltd. has commercially launched their Honda Clarity fuel cell electric vehicle (FCEV) which delivers 60 % driving energy efficiency compared to <20 % for a compact gasoline vehicle. Bloom energy has developed an SOFC with a power range between 100 and 500 kW for electricity generation. In this regard, the direct formic acid fuel cells (DFAFCs), in which the formic acid (FA) electro-oxidation (FAO) is the principal anodic reaction, have been proved attractive even more than the traditional hydrogen (HFCs) and methanol (MFCs) fuel cells. The eco-friendly nature (nontoxic, nonexplosive, and nonflammable at room temperature) of FA as a liquid fuel (i.e., can easily and safely be stored and transported) in addition to its less crossover via the Nafion membrane, separating the anodic and cathodic parts in DFAFCs, were behind this attractiveness for polymer electrolyte membrane (PEM) fuel cells [1–7]. However, one of the main problems in DFAFCs is the development anodic catalysts with high activity and stability towards FAO [2, 4, 7]. For long time, platinum (Pt) has served as one of the best candidates for FAO. However, the high sensitivity of Pt to poisoning with CO (produced as a result of the "non-faradaic" dissociation of FA) could lead to a reduction of the Pt active sites for FAO, which ultimately delays the reaction kinetics and deteriorate the catalytic activity [8–10]. We wish to emphasize here that the poisoning CO species can be oxidatively removed from the Pt surface according to the Langmuir–Hinshelwood model after being the Pt surface hydroxylated at a higher positive potential. This motivated us to think about alternatives to fulfill the enrichment of the Pt surface with oxygen entities (e.g., transition metal oxides) to induce the oxidative removal of the poisoning CO intermediate catalytically at lower potentials. One of these alternatives depended on modifying the Pt surface with transition metal oxides which enable the electrochemical dissociation of water at potentials more negative than that of bare Pt surfaces [8, 9]. Alternatively, just using Pt alloys such as Pt-Au, Pt-Bi, Pt-Ni, and Pt-Mn could retard the CO poisoning and improve the catalytic activity towards FAO [8–14]. In fact, the Ni/Co alloys have long been investigated for several electrochemical reactions due to their unique strength and high heat conductivity, in addition to the catalytic synergism that always appears in their coexistence [15].

In the present work, a Pt-based anodic catalyst modified with NiOx and CoOx nanostructures is recommended for FAO. In slightly acidic medium has been studied by cyclic voltammetry in FA solution with (pH 3.5). The catalyst' layers were deposited electrochemically sequentially on Pt nanoparticles–modified GC substrate and the influence of the deposition sequence on the catalytic efficiency and stability was investigated.

42.2 Experimental

Glassy carbon electrode GC ($d = 1.6$ mm) was served as the working electrode. A Ag/AgCl/KCl (sat) and a spiral Pt wire were used as reference and counter electrodes, respectively. Conventional pretreatment methods were applied to clean the GC electrode as described previously [8, 9].

42.2.1 Preparation of the Modified Electrodes

The deposition of PtNPs and nickel oxide nanoparticles on the GC electrode was reported previously in detail [8, 9]. Typically, the electrodeposition of PtNPs was achieved in 0.2 M H_2SO_4 containing 1.0 mM H_2PtCl_6 solution using potential step electrolysis from 1 to 0.1 V vs. Ag/AgCl/KCl(sat) for 300 s. However, the electrode's modification with nano-NiOx was achieved in two sequential steps. The first involved the electrodeposition of metallic nickel onto the working electrode from an aqueous solution of 0.1 M acetate buffer solution (ABS, pH = 4.0) containing 1 mM $Ni(NO_3)_2$ by a constant potential electrolysis at −1 V vs. Ag/AgCl/KCl (sat.) for 60 s. Next, the electrodeposited Ni was passivated (oxidized) in 0.1 M phosphate buffer solution (PBS, pH = 7) by cycling the potential between −0.5 and 1 V vs. Ag/AgCl/KCl (sat) for 10 cycles at 200 mV/s. On the other hand, the deposition of CoOx took place in 0.1 M PBS (pH = 7.0) containing 1 mM $CoCl_2$, and the potential was cycled from 1.2 V to −1.1 V vs. Ag/AgCl/KCl (sat.) at 100 m Vs^{-1}. The surface coverage (θ) of the NiOx on the Pt/GC electrode was controlled by the deposition time. The values of θ are listed in Table 42.1 for various electrodes. The real surface areas of the unmodified (A_{bare}) and metal oxide (MOx) modified (A_{mod}) Pt/GC electrodes were estimated from Fig. 42.1 based on the charge associated with the reduction of the surface Pt-oxide monolayer (at ca. −250 mV in 0.5 M KOH utilizing the reported values of 420 μC/cm^2 [16]) and the corresponding values are shown in Table 42.1.

Table 42.1 Summary of the electrochemical data obtained for the electrodes employed in this investigation

Electrodes	Real area/cm^2	Surface coverage (θ, %)	I_p^d/mA cm^{-2}	I_p^{int}/mA cm^{-2}	I_p^d/I_p^{ind}	Onset potential/ mV	I_p^d/I_b
Pt/GC	0.099	0.0	1.5	2	0.7	30	0.2
NiOx/Pt/GC	0.025	50	8.5	2	4.0	−122	0.7
CoOx/Pt/GC	0.041	50	4	0.5	8	−164	0.95
NiO*x*-CoOx/ Pt/GC	0.053	40	10	0.2	50	−179	1.05
CoO*x*-NiOx/ Pt/GC	0.029	50	5	0.2	25	−100	1.09

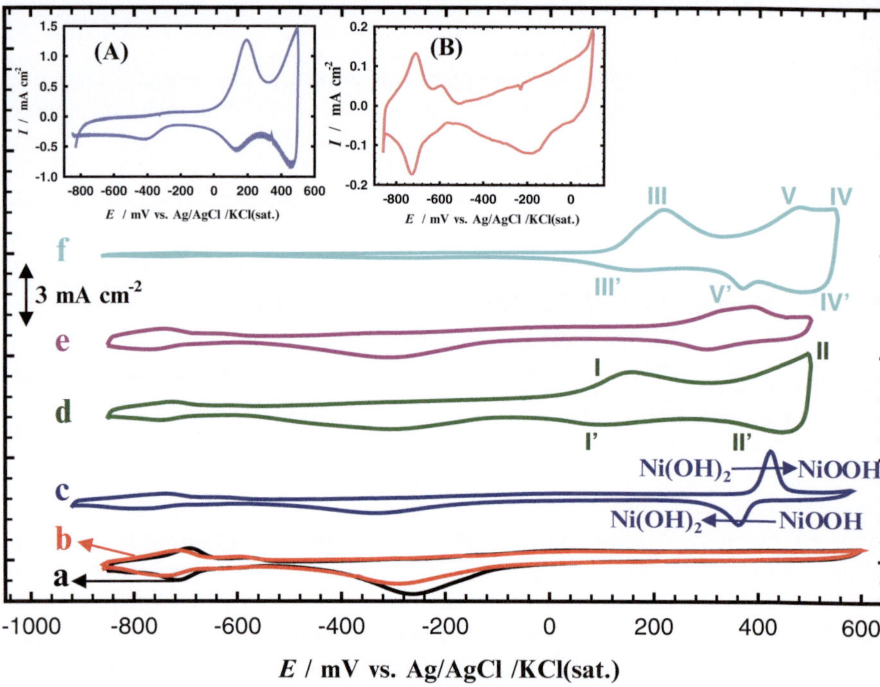

Fig. 42.1 CVs obtained at (**a**) bare Pt, (**b**) Pt/GC, (**c**) NiOx/Pt/GC, (**d**) CoOx/PtGC, (**e**) NiOx/CoOx/Pt/GC, and (**f**) CoOx/NiOx/Pt/GC electrodes in 0.5 M KOH at a scan rate of 100 mV s^{-1}. *Inset* (*A*) CV of CoOx/GC and (*B*) zoom of curve f

42.2.2 Measurements

The electrocatalytic activity of the Pt/GC-modified electrodes with (CoOx and NiOx nanostructures) towards FAO was examined in an aqueous solution of 0.3 M formic acid (pH = 3.5). The pH was adjusted by adding a proper amount of NaOH.

As, the use of highly acidic solutions would diminish the stability of nickel oxide (albeit at slow kinetics), the current study is conducted at slightly acidic pH, which lies within the stability domain of nickel oxide [17]. Moreover, at this pH an appreciable amount of FA is ionized to formate anion (about one third). This would enhance the ionic conductivity in the solution, thus reduces the resistance polarization, in addition to compressing the thickness of the diffusion layer. Cyclic voltammetry (CV) was performed in a conventional two-compartment three-electrode glass cell. All measurements were performed at room temperature (25 ± 1 °C) using an EG&G potentiostat (model 273A) operated with Echem 270 software. A field emission scanning electron microscope, FE-SEM, (QUANTA FEG 250) coupled with an energy dispersive X-ray spectrometer (EDS) unit was employed to evaluate the electrode's morphology and surface composition. Current densities were calculated on the basis of the geometric surface area of the working electrodes.

42.2.3 Electrode's Abbreviation

The electrocatalytic activities of several electrodes are compared in order to reveal the best deposition order for PtNPs, CoOx, and NiOx. This includes the Pt/GC, NiOx/Pt/GC, CoOx/Pt/GC, CoOx/NiOx/Pt/GC, and NiOx/CoOx/Pt/GC electrodes, where Pt in the electrode's assembly abbreviates PtNPs. The deposition sequence is exactly matching the order indicated in the electrode's name. For example, for the CoOx/NiOx/Pt/GC electrode, PtNPs were directly deposited onto the GC substrate and NiOx were next deposited onto the Pt/GC electrode and finally CoOx was deposited onto the NiOx/Pt/GC electrode.

42.3 Results and Discussion

42.3.1 Electrochemical and Morphological Characterization

Fortunately, the electrochemical methods of characterization are powerful and sensitive to very low amounts of the catalyst' ingredients (e.g., Pt, Ni, and Co) and can, moreover, distinguish firmly between all of them. Interestingly, for the bare Pt (Fig. 42.1a) and the Pt/GC (Fig. 42.1b) electrodes, the typical response characterizing a clean poly-Pt substrate in alkaline medium appeared. That is, the Pt oxidation, extending over a wide range of potential, is coupled with the reduction peak at ca. −0.28 V. This couple corresponds to the solid-state surface redox transition (SSSRT) involving Pt/PtO. In addition, the typical peaks for the hydrogen adsorption–desorption at Pt substrate ($H_{ads/des}$) are shown in the potential range from −0.86 to −0.6 V. Interestingly, upon modifying the Pt/GC electrode with

NiO*x* (Fig. 42.1c), a noticeable decrease in the intensity of the reduction peak of PtO (at ca. −0.28 V) is observed along with a decrease in the current intensity of the $H_{ads/des}$ peaks. Moreover, a new redox couple appeared at ca. 0.4 V, which is assigned to the $Ni(OH)_2$/NiOOH transformation, (note that $Ni(OH)_2$ is formed at the surface of NiOx during the CV measurements in KOH [8, 9, 18]:

$$Ni(OH)_2 + OH^- \Leftrightarrow NiOOH + H_2O + e^- \tag{42.1}$$

The surface coverage (θ) of NiOx in NiOx/Pt/GC electrode was estimated to be about 50 % of the Pt surface; revealing the partial exposure of PtNPs to the electrolyte (as evident from the appearance of the PtO reduction peak, albeit at lower intensity in the presence of NiOx). Similarly, Fig. 42.1d shows the CV response obtained at CoOx/Pt/GC electrode (with θ of CoOx = 50 %) a visible decrease in the intensity of the reduction peak of PtO along with a decrease in the current intensity of the $H_{ads/des}$ peaks in comparison to the unmodified Pt/GC electrode. Figure 42.1d also depicts the appearance of two new redox peak couples at ca. 0.18 and 0.50 V. Those peaks are assigned to the successive oxidation (in the anodic sweep) and reduction (in the cathodic sweep) of several intermediates of CoOx in alkaline medium, e.g., the formation of $Co(OH)_2$, Co_3O_4, CoOOH, and CoO_2 [19] in addition to a reduction peak at ca. −0.4 V which is assigned to the GC substrate. A similar behavior was observed at the CoO*x*/GC electrode (inset a of Fig. 42.1) with the appearance of two redox peak couples of cobalt oxide/hydroxide transformation according to the following equations [20, 21]:

$$Co_3O_4 + H_2O + OH^- \Leftrightarrow 3CoOOH + e^- \tag{42.2}$$

$$CoOOH + OH^- \Leftrightarrow CoO_2 + H_2O + e^- \tag{42.3}$$

The electrodeposition of NiO*x* into the CoO*x*/Pt/GC electrode (NiOx-CoOx/Pt/GC) was next sought. The formation of a NiCoOx nano-composite (e.g., $NiCoO_2$ and/or $NiCo_2O_4$) on the Pt/GC electrode's surface is presumed. This is inferred from the broadening and negative shift of the potential corresponding to the Ni (II)/Ni (III) transformation [22]. This shift may be due to the overlapping of the waves of Ni (II)/Ni (III) and Co (III)/Co (IV) couples. It has been reported that the anodic peak of the formation of higher valence oxides/hydroxides of cobalt appears between 0.42 and 0.5 V [18, 20, 21] which agrees with the current measurements. The obtained pair of current peaks corresponds to the surface faradaic reactions where the proton diffusion and charge transfer are both contributing to the total electrode reaction. Equation (42.1) describes the oxidation reaction of the lower oxidation state of Ni, i.e., NiO and/or $Ni(OH)_2$ to the higher oxidation state, e.g., NiOOH in which nickel hydroxide can form at the surface of nickel oxide during cycling in NaOH solution [8, 18]. Equations (42.2) and (42.3) correspond to an analogous transformation of the lower oxidation state of Co(II) (oxide and/or hydroxide) to the higher oxidation state (CoOOH) [20, 21].

However, when the deposition order is reversed, i.e., cobalt oxide (CoOx) is electrodeposited onto the NiOx/Pt/GC (i.e., CoOx/NiOx/Pt/GC, Fig. 42.1f), three redox peak couples were observed namely III, IV, and V in the anodic-going potential sweep and III′, IV′, and V′ in the cathodic-going potential sweep. Peaks III and IV, located at 0.19 and 0.55 V; were attributed to the oxidative transformation of cobalt oxides (i.e., $Co(OH)_2$, Co_3O_4, CoOOH, CoO_2) [20, 21] and the corresponding cathodic peaks III′ and IV′ located at 0.15 and 0.52 V were assigned to the corresponding reduction processes. While the redox peak couple V/V′ observed at ca. 0.4 V is assigned to the $Ni(OH)_2$/NiOOH redox pair [8, 9, 18].

In order to investigate the formation and growth of the cobalt oxide and/or nickel oxide nanostructures, different parts of the electrode surface were examined by field emission scanning electron microscope (FE-SEM). Figure 42.2a shows FE-SEM micrograph of NiOx/Pt/GC. As seen, flower-like Pt and Ni structures are obtained with average particle size of 80 nm. On the other hand, Fig. 42.2b shows a highly porous structure of CoOx/Pt/GC (flower-like shape) with an average size of less

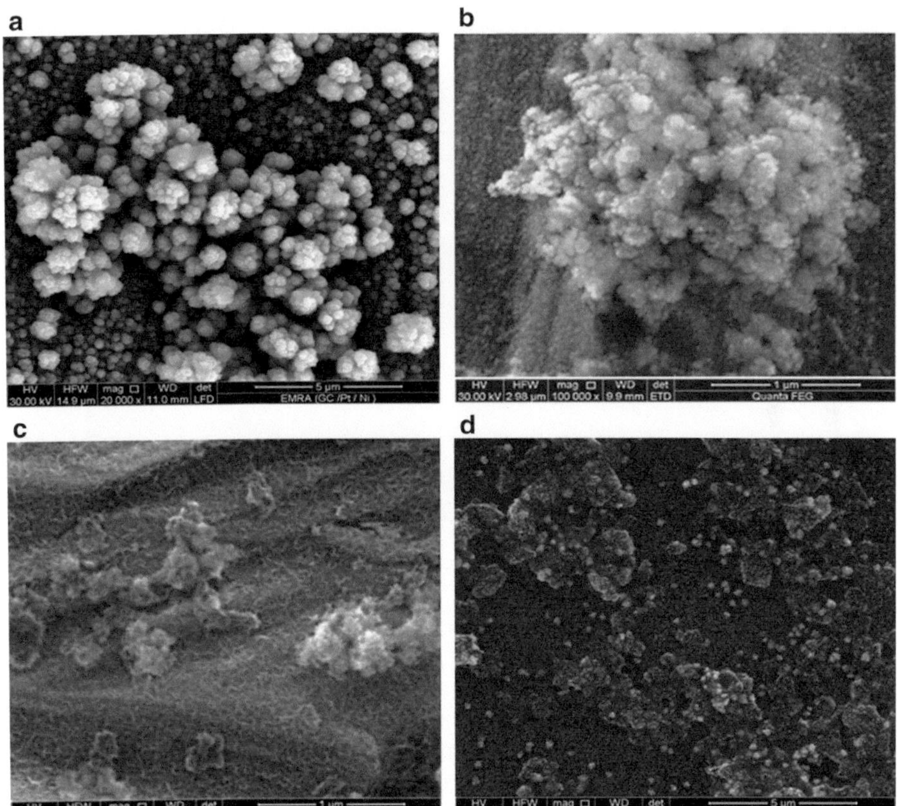

Fig. 42.2 FE-SEM micrographs obtained for (**a**) NiOx/Pt/GC, (**b**) CoO*x*/Pt/GC, (**c**) NiOx/CoOx/Pt/GC, and (**d**) CoOx/NiOx/Pt/GC electrodes

than 100 nm together with small particles (round-shaped) with an average particle size of 50 nm. While, Fig. 42.2c shows an SEM image of NiOx/CoOx/Pt/GC electrode (with CoOx firstly deposited). Inspection of this figure reveals that the CoOx and NiOx are electrodeposited in a porous texture (with an average size of ca. 25 nm) onto the Pt/GC electrode. This porous nano-texture homogeneously covers the entire surface of the electrode, which enables the accessibility of the solution species to the underlying substrate. Figure 42.2c, d discloses the influence of the deposition order on the morphology of deposited nanoparticles.

42.3.2 Electrocatalytic Activity Towards Formic Acid (FA) Oxidation

The electrocatalytic behavior of the various modified PtNPs-based electrodes with NiOx and/or CoOx towards FA oxidation was addressed by measuring the CVs in an aqueous solution of 0.3 M FA (pH = 3.5) as shown in Fig. 42.3. Note that the use of highly acidic solutions is avoided in this investigation as it would diminish the stability of nickel and cobalt oxides (albeit at slow kinetics). Instead a slightly acidic solution was used, with pH within the stability domain of nickel and cobalt

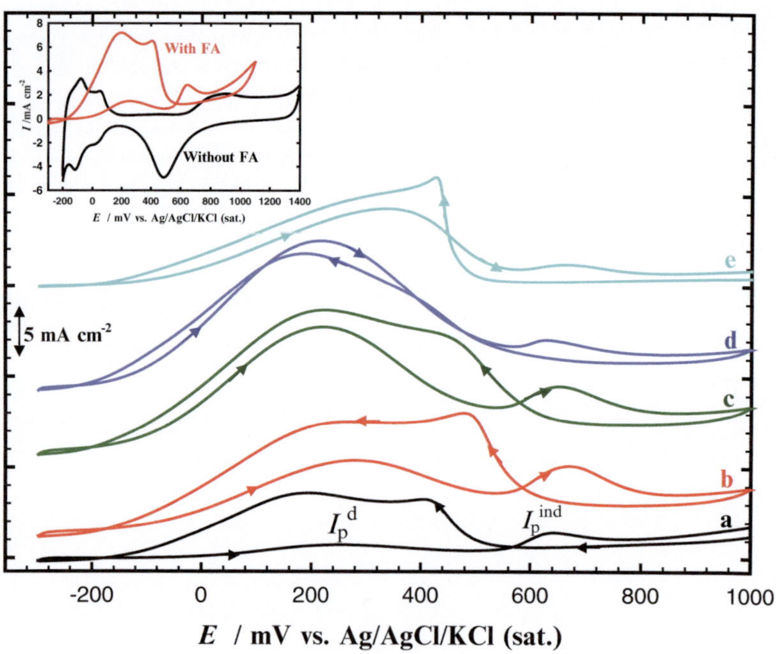

Fig. 42.3 CVs response obtained at (**a**) unmodified Pt/GC, (**b**) NiOx/Pt/GC, (**c**) CoOx/Pt/GC, (**d**) NiOx/CoOx/Pt/GC, and (**e**) CoOx/NiOx/Pt/GC electrodes in 0.3 M FA (pH~3.5) at a scan rate of 100 mV s^{-1}. Inset CV of (a) Pt/GC electrode in 0.5 M H_2SO_4 in the absence and (b) in the presence of 0.3 M FA

oxides. At this pH, an appreciable amount of FA is ionized to formate anion (about one third). This would enhance the ionic conductivity in the solution by reducing the polarization resistance, and in addition would help in compressing the thickness of the diffusion layer. It has been widely accepted in the literature that FA in the forward scan is oxidized to CO_2 via a dual path mechanism on Pt, which involves a reactive intermediate (dehydrogenation pathway) and CO_{ad} as a poisoning species (dehydration pathway). In one pathway, direct oxidation of FA is said to occur (dehydrogenation pathway) whereas in the other pathway, FA is oxidized to CO (dehydration pathway), which in turns must be removed by activating water. While the net reaction is the same in both of the two pathways, the water dissociation reaction is rather difficult. The OH_{ad} formed by the dissociation of water molecules on Pt surface aids in removing the adsorbed surface poison CO_{ad} by oxidizing it, but indeed this process is very intricate, as a higher potential is required for water activation (>0.5 V) on Pt surfaces. Consequently, the electrode surface will be blocked by large amounts of CO_{ad} species thereby hindering further adsorption of other FA molecules on the electrode surface. This drawback will make only a few number of FA molecules to be oxidized as the electrode poison CO_{ad} remains on the electrode surface for a long time occupying active catalyst sites thereby reducing the overall activity of FA oxidation. Hence, the rate of FA oxidation primarily depends on the amount of CO_{ad} removed from the electrode surface during the anodic-going potential scan.

In the backward scan, the reduction of Pt oxide to Pt occurs:

$$\mathrm{PtOx} + \mathrm{xH}^+ + \mathrm{x}e^- \longrightarrow \mathrm{Pt} + \mathrm{xH_2O} \tag{42.4}$$

Moreover, an oxidation peak has been observed which is attributed to the oxidation of the adsorbed carbonaceous species (i.e., CO) to CO_2:

$$\mathrm{Pt}\ldots\mathrm{CO_{ad}} + \mathrm{H_2O} \Rightarrow \mathrm{Pt} + \mathrm{CO_2} + 2\mathrm{H}^+ + 2e^- \tag{42.5}$$

In Fig. 42.3, the electrocatalytic activity of Pt/GC, NiOx/Pt/GC, CoOx/Pt/GC, NiOx/CoOx/Pt/GC, and CoOx/NiOx/Pt/GC electrodes towards FAO was evaluated by comparing the CVs measured in 0.3 M FA (pH = 3.5). In Fig. 42.3a, two oxidation peaks (with current intensities I_p^d and I_p^{ind}) were observed at ca. 0.25 V and 0.65 V, respectively, in the anodic sweep for the Pt/GC electrode. The first peak (I_p^d) is assigned to the direct oxidation of formic acid (dehydrogenation pathway) to CO_2, which shows a lower intensity compared with the second peak indicating a noticeable retardation of the direct FAO due to poisoning of the surface. This deactivation may arise from the accumulation of poisoning reaction intermediates (i.e., CO) and/or the adsorption of spectator species such as OH_{ad}.

At high potentials, the oxidation of the adsorbed CO starts at ca. 0.5 V, after the hydroxylation of the Pt surface. Note that CO_{ad} is formed as a result of the *non-faradaic* dissociation of FA. During the backward scan, similar oxidation peaks to those of the forward direction have appeared and were assigned to the same reactions but with a higher catalytic activity. This increase in the catalytic activity was not surprising as the surface has become relatively free from most of the poisoning CO and the Pt surface has mostly become hydroxylated $Pt - OH_{ad}$ [8, 9, 23–25]. Upon modifying the Pt/GC electrode with NiO*x* (Fig. 42.3b) or CoOx (Fig. 42.3c), a significant increase in the first peak intensity, I_p^d, concurrently with a noticeable depression of the second oxidation peak intensity, I_p^{ind}, were observed. This indicates that the direct pathway is enhanced and less poisoning intermediate (CO) is produced. The depression of I_p^{ind} can be explained in such a way that the presence of metal oxides could successfully retard the formation of CO_{ads}. Additionally, one might suggest that metal oxides stimulated a catalytic enhancement for CO oxidation at low potential. The ratio of the two oxidation peak currents (in the anodic sweep direction) reflects the preferential oxidation pathway of FA at a particular electrode. The ratio between the two oxidation peaks, I_p^d/I_p^{ind}, increases from 0.7 for Pt/GC up to about 4.0 and 8.0 for NiOx/Pt/GC (Fig. 42.3b) and CoOx/Pt/GC (Fig. 42.3c) electrodes, respectively. This reflects that the preferential pathway of FAO at the metal oxide (NiOx, CoOx)-modified electrodes is the dehydrogenation (direct) pathway. A closer examination of the two CVs shown in the inset depicts that the hydrogen adsorption/desorption patterns disappear in the presence of FA (inset of Fig. 42.3) compared to the FA-free solution. This implies two points: (i) FA is effectively adsorbed at the Pt surface in this potential region and disables the H adsorption and (ii) the amount of charge consumed for FA and CO oxidation (both are 2-electron transfer reaction) is surprisingly much less than that of the H desorption (1-electron transfer reaction), implying that the FA and CO adsorption consume more than one Pt atom per adsorbed molecule of each. It should be noted that, for low concentration of FA, the hydrogen adsorption/desorption peaks are still visible showing that FA does not prevent completely the hydrogen adsorption. Interestingly, the electrodeposition of both CoO*x* and NiOx onto Pt/GC electrode, NiOx/CoOx/Pt/GC (Fig. 42.3c) resulted in an outstanding enhancement of the direct peak current (I_p^d) with the depression of I_p^{ind}. The ratio between the two oxidation peaks, I_p^d/I_p^{ind}, increases from 0.7 for Pt/GC electrode up to 50 for NiOx/CoOx/Pt/GC (Fig. 42.3c) electrode (which is ca. 70 times higher). This reflects that FAO shifts exclusively towards the dehydrogenation pathway and has a high CO tolerance.

The effect of the nickel and cobalt oxides on the electrooxidation reaction of FA can be satisfactorily explained on the basis of at least one of the following: (a) the bifunctional mechanism, which should consider adsorption properties of CO and OH surface species, (b) the electronic interaction between the surface modifier elements (Co and/or Ni) and Pt, and/or (c) the possibility that CoOx and NiOx stimulated a catalytic enhancement for CO oxidation at low potential [12, 14]. According to the bifunctional theory, an efficient catalyst favors CO adsorption on Pt and OH_{ad} formation takes place on the second metal (Co or Ni). Hence, the binary combination yields the best overall activity for FAO. Therefore, CO adsorption mainly occurs on Pt, while OH_{ad} species easily interact with Ni and Co surfaces. Based on the above deduction, the ratio of the forward peak current (I_p^d) to the backward peak current (I_b) reflects the ratio of the amount of FA oxidized to CO_2 relative to the amount of CO. Hence, the ratio I_p^d/I_b can be used to describe the catalyst tolerance to carbonaceous species accumulation. Basically, a higher I_p^d/I_b value represents a relatively complete oxidation of FA, producing CO_2, while a low $I_p{}^d/I_b$ ratio indicates poor oxidation of FA to CO_2 during the anodic scan and excessive accumulation of carbonaceous residues on the catalyst surface [8–10].

In other words, this ratio essentially reflects the fraction of the catalyst surface that is not poisoned by CO_{ad} and can be used to measure the catalyst tolerance to CO poisoning. This ratio (I_p^d/I_b) increased from ca. 0.2 at Pt/GC electrode (Table 42.1, Fig. 42.3a) to ca. 0.7 for NiOx/PtGC (Table 42.1 and Fig. 42.3b), 0.95 for CoOx/Pt/GC (Fig. 42.3c), and to ca. 1.0 NiOx/CoOx/Pt/GC electrode and its *mirror image* electrode with NiOx electrodeposited first (Fig. 42.5d, e). The ratio of I_p^d/I_b at the modified electrodes is higher than that observed at the Pt/GC electrode, indicating that more intermediate carbonaceous species were oxidized to CO_2 in the forward scan at the modified electrodes with NiOx and/or CoOx oxides; that is, less amount of CO was produced in the forward scan at the modified electrodes than at the unmodified PtNPs-based surface. This modification could stimulate a higher degree of reversibility for the FAO, a feature that is always desirable in fuel cells manufacturing. We believe that NiOx and CoOx furnished a crucial role as catalytic mediators (through a reversible Ni(II)/Ni(III), Co (III)/Co (IV) redox system), which facilitated either the direct FAO (to CO_2) or the oxidation of the poisoning intermediate (CO) at a reasonable lower anodic potential. The ratio I_p^d/I_p^{ind} increased from 0.7 for Pt/GC to 4.0 for NiOx/Pt/GC ($\theta = 50$ %), 8.0 for CoOx/Pt/GC ($\theta = 50$ %), 21 for CoOx/NiOx/PtGC ($\theta = 50$ %), and 29 for NiOx/CoOx/Pt/GC ($\theta = 40$ %) electrodes, which reflects the superiority of binary-modified PtNPs-based electrode with NiOx *and* CoOx to unmodified Pt/GC or the single-oxide-modified Pt/GC electrodes for FAO. The increase in I_p^d/I_p^{ind} ratio indicates improvement in catalytic activity of the modified electrode towards FAO; presumably via lowering the CO adsorption and favoring the direct oxidation path. Table 42.1 shows that the current of FAO in forward scan are significantly enhanced after the electrodeposition of NiOx, CoOx, or a binary mix of both oxides. The later catalysts exhibit high I_p^d/I_p^{ind} ratios by at least 5 times higher than the unmodified Pt/GC electrode. Such high current ratio represents that most of the intermediate carbonaceous species can be oxidized to CO_2 in the forward scan on binary-modified catalysts. A further investigation was done to find out the mechanistic approach of the electrocatalytic enhancement at binary-modified Pt/GC with CoOx and NiOx. The influence of the scan rate (v) on the electrooxidation of FA at the NiOx/CoOx/Pt/GC was investigated (Fig. 42.4). As shown in Fig. 42.4a, the dependence of the voltammetric profile with the scan rate is different in the forward and backward scans. Whereas the backward scan is almost insensitive to the scan rate (only small changes in the maximum current obtained), the currents recorded in the forward scan increase as the scan rate

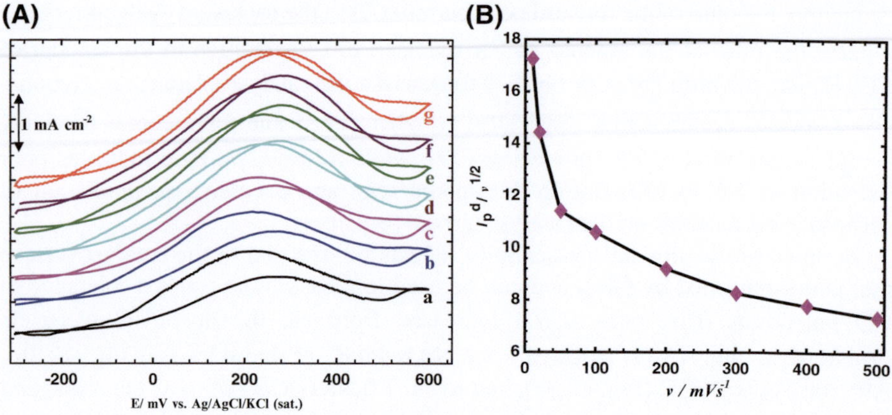

Fig. 42.4 (**A**) CVs for FOA at NiOx/CoOx/Pt/GC electrode ($\theta \approx 50$ %) in 0.3 M formic acid (pH 3.45) at potential scan rate, ν, of (*a*) 10, (*b*) 20, (*c*) 50, (*d*) 100, (*e*) 200, (*f*) 300, (*g*) 500 mV s^{-1}. (**B**) The variation of $(I_p/\nu^{1/2})$ with ν

increases. This figure shows that I_p^d increases with ν in a rather nonlinear fashion. Figure 42.4b shows the variation of $I_p^d/V^{1/2}$ ratio with ν. As can be seen, $I_p^d/V^{1/2}$ decreases with increasing ν, which is a characteristic feature of catalytic reactions. In order to verify the catalytic enhancement of FAO in the presence of CoOx and/or NiOx nanostructures, CO was adsorbed at open circuit potential at Pt/GC, NiOx/Pt/GC, and NiOx/CoOx/Pt/GC electrodes, and then oxidized in the CO-free electrolyte (0.5 M Na_2SO_4, pH = 3.5). The voltammograms of CO stripping corrected for the background current are presented in Fig. 42.5. The oxidation of CO was sought to estimate quantitatively its amount. Enhancement of the FAO at Pt by the addition of the metal oxide(s) can be caused by the favored dehydrogenation reaction (third-body and/or electronic effect), or by the efficient removal of CO_{ads} formed in the dehydration path (bifunctional effect). Closer look at CO stripping peaks at Pt/GC (Fig. 42.5a) and NiOx/Pt/GC (Fig. 42.5b) electrodes, the same amount of CO adsorbed at the two electrodes, but the onset potential at the NiOx/Pt/GC electrode shifted to more negative potential (ca., −140 mV) which means that NiOx did not resist the adsorption of CO at Pt surface, but catalyzed its oxidative removal at lower potential (bifunctional effect). The same behavior was obtained at CoOx/Pt/GC (data not shown). On the other hand, the electrodeposition of CoOx and NiOx at Pt/GC electrode (NiOx/CoOx/Pt/GC) resulted in a slight negative shift in the CO stripping peak (ca., 250 mV, Fig. 42.5c) and that means the observed significant enhancement of FAO at this electrode was mainly due to the catalysis of CO oxidation (to CO_2) at low potential which is called bifunctional effect.

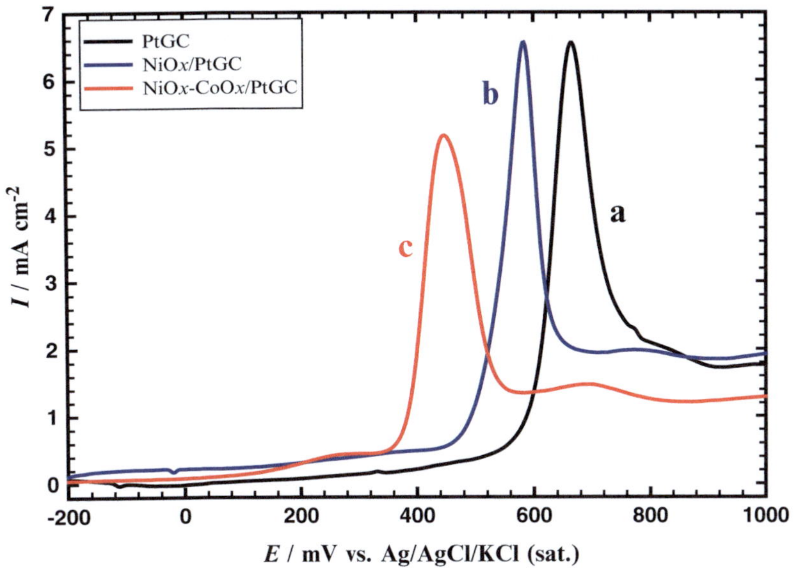

Fig. 42.5 Oxidative CO stripping at (**a**) Pt/GC, (**b**) NiOx/Pt/GC, and (**c**) NiOx/CoOx/Pt/GC electrodes in 0.5 M Na_2SO_4 measured at 50 mV s^{-1}

42.3.3 Stability of Binary Catalysts

To investigate the stability of the proposed catalysts, current transients (*i-t* curves) were measured for FAO at the proposed electrodes at a constant potential of 0.3 V. Figure 42.6 depicts that NiOx/CoOx/Pt/GC electrode supports a higher oxidation current (ca. 83, 12 and 3 times) than that obtained at Pt/GC, NiOx/Pt/GC, and CoOx/Pt/GC electrodes, respectively. This level of enhancement could still be observed up to 15 h of continuous electrolysis. This again demonstrates the preference of the direct oxidation path of FA at the NiOx/CoOx/Pt/GC electrode and its high CO tolerance. However, a gradual decrease of current was observed during a continuous electrolysis for 15 h at NiOx and/or CoOx-modified Pt/GC electrodes. It should be mentioned here that the initial decay in current density at the three electrodes might originate from the accumulation of CO_{ads} on the electrodes' surface to different extents. The partial dissolution and/or detachment of NiOx and/or CoOx away from the electrode' surface might also be contributing. A similar observation has previously been reported for NiOx-modified electrodes towards the oxygen evolution reaction. However and fortunately, the addition of a minute amount of CoOx to the modified catalyst could extremely improve the stability which can explain by the addition of $Co(OH)_2$ may increase the charge-acceptance of Nickel electrodes and prevent the formation of another form of nickel oxide (γ-NiOOH), which can cause the swelling of Ni electrodes.

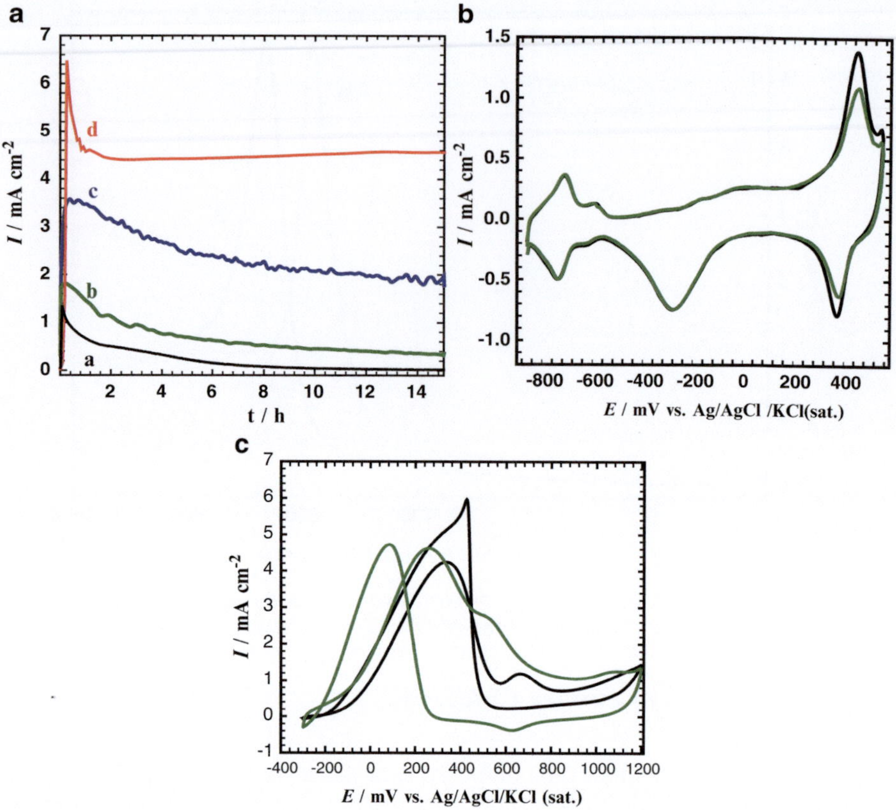

Fig. 42.6 (**a**) the current transients obtained at (*a*) Pt/GC, (*b*) NiO*x*/Pt/GC, (*c*) CoO*x*/Pt/GC, and (*d*) NiO*x*-CoO*x*/Pt/GC in 0.3 M FA solution (pH = 3.5) at a constant potential of +0.3 V, (**b**) CVs response obtained at NiO*x*-CoO*x*/Pt/GC electrodes (*a*) before and (*b*) after ageing for 2 h in FA solution with pH 3.5 at + 0.3 V in 0.5 M KOH with scan rate 100 mV s^{-1}, and (**c**) CVs response of FA oxidation obtained at NiO*x*-CoO*x*/Pt/GC electrode (*a*) before, and (*b*) after ageing for 15 h in FA solution with (pH 3.5) with scan rate 100 mV s^{-1}

Nevertheless, Fig. 42.6a still infers a better catalytic enhancement for FAO at NiOx/CoOx/Pt/GC electrode than on NiOx/Pt/GC electrode. After a period of 15 h, the current densities are 0.4 and 4.5 mA cm^{-2} for NiO*x*/Pt/GC and NiOx/CoOx/Pt/GC electrodes, respectively (i.e., 11 times larger). This again demonstrates the preference of the direct oxidation path of FA at the NiOx/CoOx/Pt/GC electrode and its high CO tolerance. In an attempt to understand the origin of the gradual decrease in oxidation current at NiOx/CoOx/Pt/GC electrode towards FAO, the characteristic CVs have been measured (see Fig. 42.6b) in 0.5 M KOH before (curve a) and after ageing for 15 h (curve b) of continuous electrolysis in 0.3 M FA (pH = 3.5) at a constant potential of 0.3 V. This figure depicted no change in the real surface area of PtNPs, as revealed by comparing the peak current intensities at ca. −0.37 V corresponding to the reduction of Pt oxide layer. This, interestingly, reflects a good mechanical stability and a good adhesion of both NiOx and CoOx oxides on the Pt/GC electrode. Nevertheless, a significant decrease in the current intensities of the redox couples corresponding to the $Ni(OH)_2$/NiOOH and CoOOH/CoO_2 transformations were observed after ageing for 2 h, and then remained without any change (see curve b). Explaining this behavior is a bit tricky, as the decrease in the current intensities of the redox couples corresponding to the Ni $(OH)_2$/NiOOH and CoOOH/CoO_2 transformations were not associated with a similar increase in the real surface area of PtNPs. If this was the case, one might argue that the depletion of oxides from the Pt surface not occurred.

On the other hand, the decrease in the peak currents of the $Ni(OH)_2$/NiOOH and CoOOH/CoO_2 transformations can be attributed to a phase change which is not active for FAO at this potential domain. Therefore, we believe that the reason behind the current decay observed in Fig. 42.6a (curve d) at the early stage of electrolysis is twofold: the Pt surface poisoning with CO and the deactivation of the catalytic mediator. While after 2 h the peak couple of transformations of Ni and Co oxides remain unchanged and that explained why the *i-t* is curve of NiOx/CoOx/Pt/GC did not decrease after 2 h. Figure 42.6c shows the corresponding CVs of FAO at binary modified Pt/GC electrode with CoOx (first electrodeposited) and NiOx nanostructures (i.e., NiOx/CoOx/Pt/GC electrode) before and after ageing in FA solution for 15 h at 0.3 V. It's clear from this figure that the ratio I_p^d/I_b, increased from 0.9 to 1.0 and I_p^{ind} completely disappeared after ageing for 15 in FA solution ($pH = 3.5$). This demonstrates the preference of the direct oxidation path of FA at the binary-modified electrode, its high CO tolerance after ageing, and its high stability.

42.3.4 *Origin of the Electrocatalytic Enhancement*

It has been reported that FAO proceeds in the forward scan to CO_2 via a dual path mechanism on Pt, which involves a reactive intermediate (dehydrogenation pathway) and CO_{ad} as a poisoning species (dehydration pathway). The latter is frequently assigned to a *nonfaradaic* dehydration of FA, which can effectively block the surface and thus hinders the formation of OH_{ad} required to keep the catalyst in an active state. It has been reported that the oxidation of small organic molecules as well as the oxygen evolution reaction (OER) are greatly enhanced at nickel oxide (β-NiOOH)-modified electrodes. However, Pt (albeit a crucial component) does not support the direct oxidation of FA at a reasonable rate. Thus, Co*Ox* and Ni*Ox* are believed to act as a catalytic mediators (possibly through a reversible Ni(II)/Ni(III) and Co(III)/Co(IV) redox systems, see Eqs. (42.6a) and (42.6b), respectively, below) which facilitates the charge transfer during the direct oxidation of FA to CO_2 (see Eqs. (42.7a) and (42.7b). On parallel, Co*Ox* and NiOx might facilitate the oxidation of the poisoning CO intermediate at a reasonably low potential (Fig. 42.5 and see Eqs. (42.8a) and (42.8b)). Recent reports provided arguments supporting the catalytic enhancement of CO oxidation (to CO_2) by manganese oxide nanostructures in similar media. In this context, the following set of equations is proposed as a plausible explanation of the observed enhancing role of nano-MOx (M = Co or Ni) through a reversible transformation between lower and higher oxidation states of metal oxides as follows:

$$\mathrm{Ni(OH)_2} \leftrightarrow \mathrm{NiOOH} + \mathrm{H}^+ + e^- \qquad (42.6a)$$

$$\mathrm{CoOOH} \Leftrightarrow \mathrm{CoO_2} + \mathrm{H}^+ + e^- \qquad (42.6b)$$

Thermodynamically, the above reaction is feasible under the prevailing conditions of pH and potential. The electrogenerated NiOOH and CoOOH species react with FA acid (and formate anions) as follows:

$$\mathrm{Pt}\ldots\mathrm{HCOOH_{ad}} + \mathrm{NiOOH} \rightarrow \mathrm{Pt_{free}} + \mathrm{CO_2} + \mathrm{Ni(OH)_2} + \mathrm{H}^+ + e^- \qquad (42.7a)$$

$$\mathrm{Pt}\ldots\mathrm{HCOOH_{ad}} + \mathrm{CoO_2} \rightarrow \mathrm{Pt_{free}} + \mathrm{CO_2} + \mathrm{CoOOH} + \mathrm{H}^+ + e^- \qquad (42.7a')$$

$$\mathrm{Pt}\ldots\mathrm{HCOO^-_{ads}} + \mathrm{NiOOH} \rightarrow \mathrm{Pt_{free}} + \mathrm{CO_2} + \mathrm{Ni(OH)_2} + e^- \qquad (42.7b)$$

$$\mathrm{Pt}\ldots\mathrm{HCOO^-_{ads}} + \mathrm{CoO_2} \rightarrow \mathrm{Pt_{free}} + \mathrm{CO_2} + \mathrm{CoOOH} + e^- \qquad (42.7b')$$

and/or

$$\mathrm{Pt}\ldots\mathrm{CO_{ads}} + \mathrm{NiOOH} + \mathrm{H_2O} \rightarrow \mathrm{Pt_{free}} + \mathrm{Ni(OH)_2} + \mathrm{CO_2} + \mathrm{H}^+ + e^- \qquad (42.8a)$$

$$\mathrm{Pt}\ldots\mathrm{CO_{ads}} + \mathrm{CoO_2} + \mathrm{H_2O} \rightarrow \mathrm{Pt_{free}} + \mathrm{CoOOH} + \mathrm{CO_2} + \mathrm{H}^+ + e^- \qquad (42.8b)$$

Addition of Eqs. (42.6a) and (42.6b) to either Eqs. (42.7a) or (42.7a′) gives, respectively:

$$\mathrm{Pt}\ldots\mathrm{HCOOH_{ad}} \rightarrow \mathrm{Pt_{free}} + \mathrm{CO_2} + 2\mathrm{H}^+ + 2e^- \qquad (42.9a)$$

Addition of Eqs. (42.6a) and (42.6b) to either Eqs. (42.7b) or (42.7b′) gives, respectively:

$$\mathrm{Pt}\ldots\mathrm{HCOO}^-_{ad} \rightarrow \mathrm{Pt_{free}} + \mathrm{CO_2} + \mathrm{H}^+ + 2e^- \qquad (42.9b)$$

Also, addition of Eqs. (42.6a) and (42.6b) to either Eqs. (42.8a) or (42.8b) gives, respectively:

$$\mathrm{Pt}\ldots\mathrm{CO_{ads}} + \mathrm{H_2O} \rightarrow \mathrm{Pt_{free}} + \mathrm{CO_2} + 2\mathrm{H}^+ + 2e^- \qquad (42.10)$$

Equations (42.9a) and (42.9b) displays that the observed oxidation current is solely supported by the oxidation of FA, and nickel and cobalt oxides function as catalytic mediators. Moreover, Reactions (42.7a), (42.7b), and (42.8a) indicate the regeneration of $Ni(OH)_2$, while the reaction (42.7a′), (42.7b′), and (42.8b) indicate the regeneration of CoOOH which are believed to start new cycles of catalytic media through the above set of equations. In all cases, NiOOH and CoOOH help the retrieval of Pt active sites for the further oxidation to proceed.

42.4 Conclusions

A novel binary metal oxide composed of CoOx and/or NiO*x*-modified Pt catalyst was developed for the direct electro-oxidation of FA. This modification resulted in a superb enhancement of the direct oxidation pathway of FA to CO_2. The ratio I_p^d/I_p^{ind} increased up to 75 times upon modifying the Pt substrate with a CoOx (first electrodeposited) and NiOx. This reflects that the direct dehydrogenation pathway has become preferential for the FAO. NiOx (in the β-NiOOH phase) and cobalt oxide (in the CoOOH phase) are believed to mediate the oxidation scheme of FA in such a way that facilitate the charge transfer and/or remove the poisoning CO. The prepared catalyst exhibits satisfactory stability and reproducibility for 15 h of continues electrolysis, which makes it attractive as anode in DFAFCs and the similar applications.

References

1. Baik SM, Kim J, Han J, Kwon Y (2011) Performance improvement in direct formic acid fuel cells (DFAFCs) using metal catalyst prepared by dual mode spraying. Int J Hydrog Energy 36:12583–12590
2. Cai W, Liang L, Zhang Y, Xing W, Liu C (2013) Real contribution of formic acid in direct formic acid fuel cell: investigation of origin and guiding for micro structure design. Int J Hydrog Energy 38:212–218
3. Hong P, Luo F, Liao S, Zeng J (2011) Effects of Pt/C, Pd/C and PdPt/C anode catalysts on the performance and stability of air breathing direct formic acid fuel cells. Int J Hydrog Energy 36:8518–8524
4. Rhee Y-W, Ha SY, Masel RI (2003) Crossover of formic acid through Nafion® membranes. J Power Sources 117:35–38
5. Rice C, Ha S, Masel RI, Waszczuk P, Wieckowski A, Barnard T (2002) Direct formic acid fuel cells. J Power Sources 111:83–89
6. Tsujiguchi T, Iwakami T, Hirano S, Nakagawa N (2014) Water transport characteristics of the passive direct formic acid fuel cell. J Power Sources 250:266–273
7. Yu X, Pickup PG (2008) Recent advances in direct formic acid fuel cells (DFAFC). J Power Sources 124:124–132
8. El-Nagar GA, Mohammad AM, El-Deab MS, El-Anadouli BE (2012) Facilitated electro-oxidation of formic acid at nickel oxide nanoparticles modified electrodes. J Electrochem Soc 159:F249–F254
9. El-Nagar GA, Mohammad AM, El-Deab MS, El-Anadouli BE (2013) Electrocatalysis by design: enhanced electrooxidation of formic acid at platinum nanoparticles–nickel oxide nanoparticles binary catalysts. Electrochim Acta 94:62–71

10. El-Deab MS, Kibler LA, Kolb DM (2009) Enhanced electro-oxidation of formic acid at manganese oxide single crystalline nanorod-modified Pt electrodes. Electrochem Commun 11:776–778
11. Habibi B, Delnavaz N (2011) Carbon–ceramic supported bimetallic Pt–Ni nanoparticles as an electrocatalyst for oxidation of formic acid. Int J Hydrog Energy 36:9581–9590
12. Obradović MD, Rogan JR, Babić BM, Tripković AV, Gautam AS, Radmilović VR et al (2012) Formic acid oxidation on Pt–Au nanoparticles: relation between the catalyst activity and the poisoning rate. J Power Sources 197:72–79
13. Tripković AV, Popović KD, Stevanović RM, Socha R, Kowal A (2006) Activity of a PtBi alloy in the electrochemical oxidation of formic acid. Electrochem Commun 8:1492–1498
14. Zhang S, Shao Y, Yin G, Lin Y (2010) Facile synthesis of PtAu alloy nanoparticles with high activity for formic acid oxidation. J Power Sources 195:1103–1106
15. Rafailović LD, Karnthaler HP, Trišović T, Minić DM (2010) Microstructure and mechanical properties of disperse Ni–Co alloys electrodeposited on Cu substrates. Mater Chem Phys 120:409–416
16. Trasatti S, Petrii O (1992) Real surface area measurements in electrochemistry. J Electroanal Chem 327:353–376
17. Deltombe E, De Zoubov N, Pourbaix M (1966) Atlas of electrochemical equilibria in aqueous solutions, vol 168. Pergamon Press, Oxford, pp 343–349
18. Giovanelli D, Lawrence NS, Jiang L, Jones TGJ, Compton RG (2003) Electrochemical determination of sulphide at nickel electrodes in alkaline media: a new electrochemical sensor. Sens Actuators B 88:320–328
19. Song Y, He Z, Zhu H, Hou H, Wang L (2011) Electrochemical and electrocatalytic properties of cobalt nanoparticles deposited on graphene modified glassy carbon electrode: application to some amino acids detection. Electrochim Acta 58:757–763
20. Švegl F, Orel B, Grabec-Švegl I, Kaučič V (2000) Characterization of spinel Co_3O_4 and Li-doped Co_3O_4 thin film electrocatalysts prepared by the sol–gel route. Electrochim Acta 45:4359–4371
21. Ding Y, Wang Y, Su L, Bellagamba M, Zhang H, Lei Y (2010) Electrospun Co_3O_4 nanofibers for sensitive and selective glucose detection. Biosens Bioelectron 26:542–548
22. Oshitani M, Yufu H, Takashima K, Tsuji S, Matsumaru Y (1989) Development of a pasted nickel electrode with high active material utilization. J Electrochem Soc 136:1590–1593
23. Joo J, Uchida T, Cuesta A, Koper MTM, Osawa M (2013) Importance of acid–base equilibrium in electrocatalytic oxidation of formic acid on platinum. J Am Chem Soc 135:9991–9994
24. Joo J, Uchida T, Cuesta A, Koper MTM, Osawa M (2014) The effect of pH on the electrocatalytic oxidation of formic acid/formate on platinum: a mechanistic study by surface-enhanced infrared spectroscopy coupled with cyclic voltammetry. Electrochim Acta 129:127–136
25. Samjeské G, Miki A, Ye S, Osawa M (2006) Mechanistic study of electrocatalytic oxidation of formic acid at platinum in acidic solution by time-resolved surface-enhanced infrared absorption spectroscopy. J Phys Chem B 110:16559–16566

Chapter 43
Electro-Oxidation of Formic Acid, Glucose, and Methanol at Nickel Oxide Nanoparticle Modified Platinum Electrodes

Sayed M. El-Refaei, Gumaa A. El-Nagar, Ahmad M. Mohammad, and Bahgat E. El-Anadouli

Abstract The current study presents a comparison for the electro-oxidation of formic acid (FA), glucose (GL), and methanol (ME) at nickel oxide nanoparticles (NiOx) modified electrodes. The modification with NiOx was pursed onto a bare glassy carbon (GC) and Pt-modified (Pt/GC) electrodes electrochemically, and the catalytic activity was measured in 0.3 M NaOH. Cyclic voltammetry (CV), scanning electron microscopy (SEM), and energy dispersive X-ray spectroscopy (EDX) are all used to provide a concrete characterization of the prepared electrodes. A catalytic enhancement of GL oxidation (GLO) and ME oxidation (MEO) was observed at the NiOx-modified GC (NiOx/GC) electrode, while the same electrode did not show any activity towards FA oxidation (FAO), revealing that FAO is substrate dependent. On the other hand, assembling NiOx onto the Pt/GC electrode assisted in improving the catalytic activity of all reactions (GLO, MEO, and FAO). The catalytic enhancement observed at the NiOx/Pt/GC electrode for GLO, MEO, and FAO was not only confined in the large increase of the oxidation current but also in a negative shift in the onset potential of the oxidation reaction. We believe NiOx could successfully play an essential role in this catalytic enhancement, presumably via participation in these reactions in a way facilitating the charge transfer or providing the oxygen atmosphere necessary for promoting an oxidative removal for unwanted poisoning species.

Keywords Electrocatalysis • Nickel oxide nanoparticles • Platinum nanoparticles • Fuel cells

S.M. El-Refaei • G.A. El-Nagar • B.E. El-Anadouli (✉)
Department of Chemistry, Faculty of Science, Cairo University, Cairo 12613, Egypt
e-mail: selrefaei@sci.cu.edu.eg; elnagar087@yahoo.com; Bahgat30@yahoo.com

A.M. Mohammad
Department of Chemistry, Faculty of Science, Cairo University, Cairo 12613, Egypt

Department of Chemical Engineering, Faculty of Engineering,
The British University in Egypt, Cairo 11837, Egypt
e-mail: ammohammad@cu.edu.eg

I. Dincer et al. (eds.), *Progress in Clean Energy, Volume 1*,
DOI 10.1007/978-3-319-16709-1_43

43.1 Introduction

Fuel cells (FCs) are systems for conversion of energy on a continuous regime. The smallest and easiest fuel for these devices is hydrogen. However, the use of hydrogen presents some significant drawbacks related to its explosive and flammable nature, which makes its storage and transportation problematic [1]. Therefore, different other small organic molecules have been proposed as fuels to substitute hydrogen. Among these fuels, formic acid (FA), glucose (GL), and methanol (ME) have been shown interesting oxidation efficiency. Actually, these organic molecules are commercially available and environmentally safe, which invokes no problem in transportation and storage [2–4].

Typically, these fuels are injected to the anodic part of FCs and oxidized at certain anodic catalysts of reasonable catalytic activity. Platinum represents the most widely used anodic catalyst for these anodic processes, particularly in the low temperature regime [5, 6]. However, the generation of strongly adsorbed poisoning intermediates (e.g., CO_{ad} and other carbonaceous materials), which are generally produced during the oxidation process, can ultimately hinder the oxidation process and deteriorate the catalytic activity of the catalyst, which ultimately leads to a severe failure of FCs as a whole [7].

Recently, Pt–M-based binary catalysts, where M = Ru, Ir, Mo, Os, or Ni are recommended to enhance the catalytic activity towards the oxidation of these fuels by eliminating or inhibiting the CO poisoning effect [3, 4, 8–10]. Effort is expanded to develop new ternary or quaternary alloy catalysts for these oxidation processes [11].

This study compares the catalytic activity of a new catalyst developed by electrodepositing crystallographically oriented nickel oxide nanoparticles (NiOx) on Pt surfaces [12] towards FA oxidation (FAO), GL oxidation (GLO), and ME oxidation (MEO) in alkaline media. NiOx is believed to provide the oxygen species to the adsorbed poisoning intermediates to enhance their oxidative removal at low potential.

43.2 Experimental

The working electrode was a glassy carbon (GC) rod ($d = 3.0$ mm) sealed in a Teflon jacket leaving an exposed geometric surface area of 0.07 cm^2. A spiral Pt wire and an Ag/AgCl/KCl (sat.) were used as the counter and the reference electrodes, respectively. Conventional methods were applied for cleaning the working electrode [3]. Platinum nanoparticles (PtNPs) were electrodeposited on GC (next noted as Pt/GC) electrodes from an acidic solution of 0.1 M H_2SO_4 containing 2.0 mM H_2PtCl_6 using potential step electrolysis from 1 to 0.1 V vs. Ag/AgCl/KCl (sat.) for 500 s. This resulted in the deposition of 4.3 μg of Pt (as estimated from the charge of the *i–t* curve recorded during the potentiostatic electrolysis). On the other hand, the modification of the GC and Pt/GC electrodes with NiOx (next noted as NiOx/GC and NiOx/Pt/GC, respectively) was achieved in two steps. First, Ni was electrodeposited from an aqueous 0.1 M acetate buffer solution (ABS, pH = 4.0) containing 1 mM $Ni(NO_3)_2$ by constant potential electrolysis at −1 V vs. Ag/AgCl/KCl (sat.) for 120 s. Next, an electrochemical passivation step for Ni was carried out in 0.1 M phosphate buffer solution (PBS, pH = 7) to allow the formation of NiOx [3, 12].

Cyclic voltammetric (CV) measurements were performed in a conventional two-compartment three-electrode glass cell. All measurements were performed at room temperature (25 ± 1 °C) using an EG&G potentiostat (model 273A) operated with Echem 270 software. A field emission scanning electron microscope (FE-SEM, QUANTA FEG 250) coupled with an energy dispersive X-ray spectrometer (EDS) unit was employed to evaluate the electrode's morphology. The electrocatalytic activity of the NiOx/GC and NiOx/Pt/GC electrodes towards FAO, GLO, and MEO was examined in a solution of 0.3 M NaOH containing 0.3 M FA, 40 mM glucose, and 0.3 M methanol, respectively. All chemicals were used without further purification. Current densities were calculated on the basis of the real surface area. The surface coverage θ of NiOx on the Pt/GC electrode was estimated from the decrease of the peak current intensity at ca. 0.4 V corresponding to the reduction of the Pt surface oxide formed during the anodic scan [13, 14] (see later Fig. 43.2d).

43.3 Results and Discussion

43.3.1 Materials and Electrochemical characterization

The morphological characterization of the modified electrodes was disclosed by FE-SEM imaging. Figure 43.1a shows the typical FE-SEM micrograph of NiOx/GC electrode after regular passivation (the inset depicts the morphology of the same electrode before passivation). It looks the electrodeposition of metallic Ni has occurred in a dendritic nanostructure with an average particle size of 80 nm that partially covers the surface of the GC electrode. After the electrochemical passivation, the average particle size of NiOx increased to ca. 120 nm, which infers the aggregation of Ni particles during the passivation. Figure 43.1b, showing the SEM micrograph for Pt/GC electrode, depicts round-shape Pt nanoparticles with reasonably uniform size and density distribution and a particle size of ca. 100 nm. At the NiOx/Pt/GC electrode (Fig. 43.1c), a flower-like Pt and Ni nanostructures are obtained with a significantly larger average particle size compared with Fig. 43.1a, b. The EDX spectrum for the NiOx/Pt/GC electrode was disclosed in Fig. 43.1d. It indicates that both Pt and Ni coexist at the surface of the modified GC electrode, which is helpful in understanding the catalytic activity and poisoning.

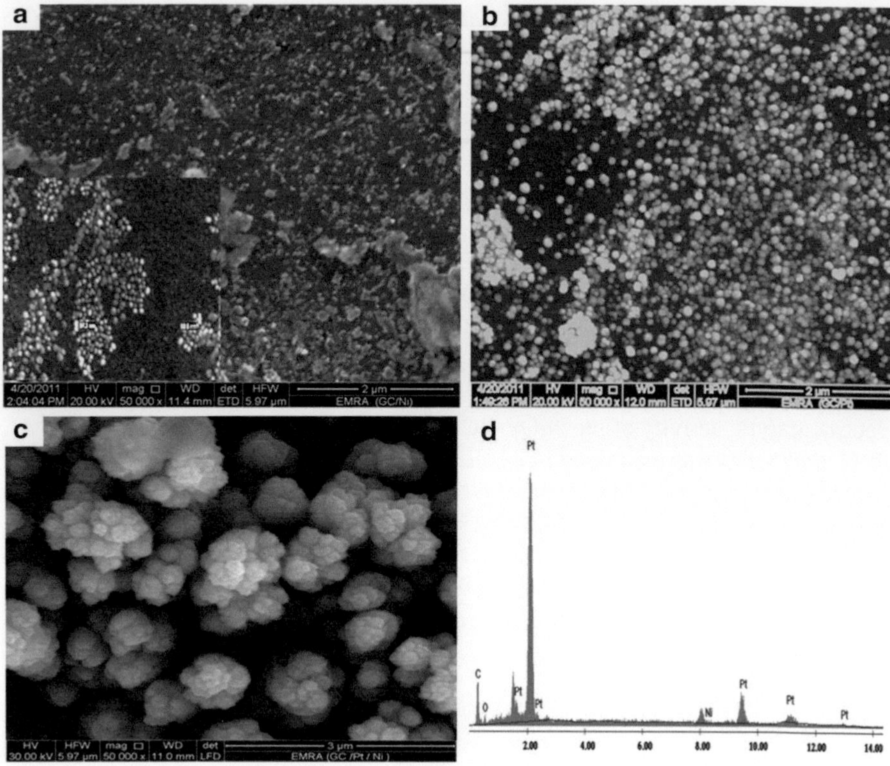

Fig. 43.1 FE-SEM micrographs obtained for NiOx/GC (*inset*, Ni/GC) (**a**), Pt/GC (**b**), NiOx/Pt/GC (**c**). EDX chart for the NiOx/Pt/GC (**d**)

Figure 43.2a–d shows CVs for GC (a), NiOx/GC (b), Pt/GC (c), NiOx/Pt/GC (d) electrodes, respectively, in 0.3 M NaOH at a scan rate of 200 mV s^{-1}. Figure 43.2a depicts the absence of any electrochemical response at the bare GC electrode under the operating conditions of potential and pH. While, at the NiOx/GC electrode (Figure 43.2b), a well-defined redox waves appeared around 0.4 V vs. Ag/AgCl/KCl(sat.) corresponding to a surface confined Ni(II)/Ni(III) reversible transformation, which stands as a characteristic feature for NiOx deposition [12, 15]. At the Pt/GC electrode (Fig. 43.2c), the characteristic response of polycrystalline Pt surface is obtained, inferring the successful deposition of PtNPs. Typically, a broad oxidation peak for the Pt oxide in the potential range of ca. −100 to 600 mV vs. Ag/AgCl/KCl (sat.) coupled with a single reduction peak centered at ca. −0.3 V vs. Ag/AgCl/KCl appeared. In addition, well-defined peaks of the hydrogen adsorption–desorption ($H_{ads}/_{des}$) appeared at potential negative to −0.5 V vs. Ag/AgCl/KCl [3].

Interestingly, the modification of the Pt/GC electrode with NiOx (Fig. 43.2d) resulted in a noticeable decrease in the intensity of the Pt oxide reduction peak (at ca. −0.34 V) along with a decrease in the current intensity of the ($H_{ads}/_{des}$) peaks

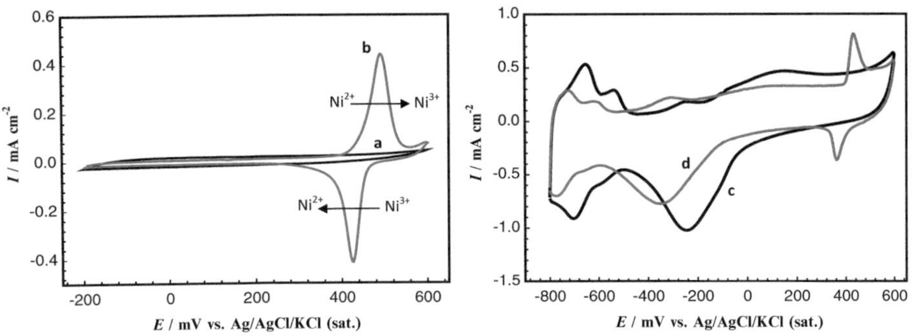

Fig. 43.2 CVs of GC (**a**), NiOx/GC (**b**), Pt/GC (**c**), and NiOx/Pt/GC (**d**) electrodes in 0.5 M NaOH solution at a scan rate 200 mV s^{-1}

(in the potential region from −0.6 to −0.9 V vs. Ag/AgCl/KCl (sat)) and hence, the accessible surface area of Pt is decreased (surface coverage of NiOx ≈ 25 %). In addition, a redox pair appears at ca. 0.45 V corresponding to the Ni(II)/Ni(III) reversible transformation [3, 12].

43.3.2 *Electrocatalytic activity towards FAO, GLO, and MEO*

The electrocatalytic activity of the various modified GC electrodes towards FAO, GLO, and MEO is addressed by measuring CVs in 0.3 M NaOH solution containing a definite concentration of each fuel at a scan rate 200 mV s^{-1}. Figure 43.3 shows the CVs obtained at the GC (a, c, and e) and NiOx/GC (b, d, and f) for FAO, GLO, and MEO, respectively. Basically, there was no electrochemical activity for all fuels at the bare GC electrode within the applied conditions of pH and potential (Fig. 43.3a, c, and e). The NiOx/GC electrode did not show any significant catalytic activity towards FAO, except increasing the charging current due to the presence of NiOx at the GC surface (Fig. 43.3b). This obviously depicts that FAO is a substrate-dependent reaction [16]. On the other hand, the same NiOx/GC electrode exhibited a reasonable catalytic activity towards GLO if compared to the bare GC electrode (Fig. 43.3d). NiOx is believed to act as a catalytic mediator in GLO in such a way facilitating the reaction kinetics [15, 17]. We believe that the adsorbed GL molecules at the surface of the electrode are oxidized at a high potential coincident with the Ni(II)/Ni(III) transformation. This process leads to reduce the number of active sites for glucose adsorption that, along with the poisoning effect of the products or intermediates of the reaction, tends to decrease the overall rate of glucose oxidation. Thus, the anodic current passes through a maximum in the forward scan.

It is worth mentioning that the electrocatalytic GLO occurs not only in the forward scan but also in the backward scan. Interestingly, in the backward scan,

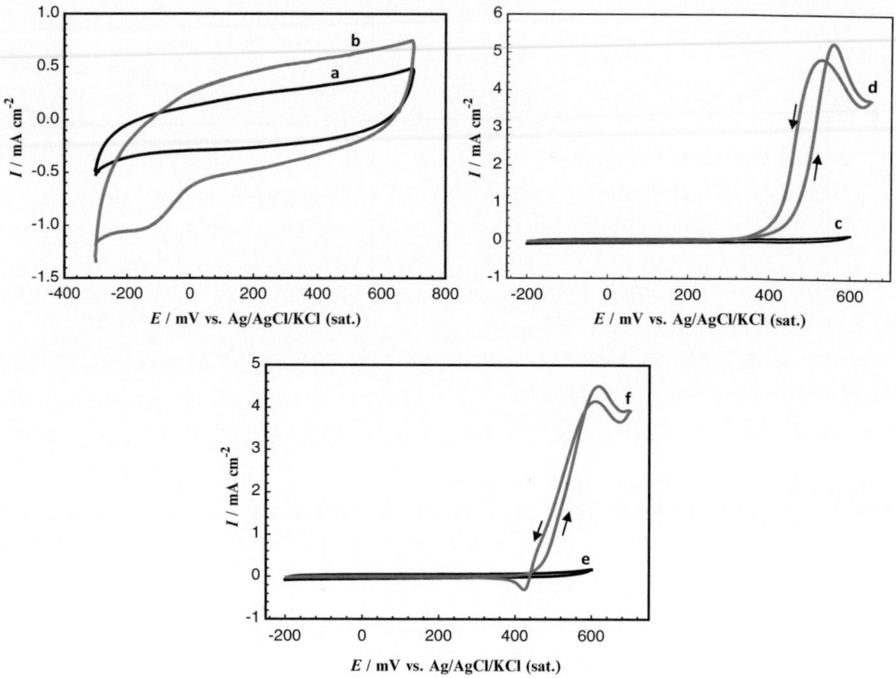

Fig. 43.3 (**a–f**) CVs for GC (**a**, **c**, **e**),NiOx/GC (**b**, **d**, **f**) electrodes in 0.3 M NaOH containing 0.3 M FA m, 40 mM glucose, and 0.3 M methanol oxidation measured at 200 mV s^{-1}, respectively

the oxidation current continues increasing to pass by a maximum due to the regeneration of active sites occupied by poisoning intermediates and product such as gluconolactone [15, 18] to participate again in GLO. Hence, the mechanism of GLO at the NiOx/GC electrode can be described as

$$\mathrm{Ni(II)} \Rightarrow \mathrm{Ni(III)} + e^{-}$$
$$\mathrm{Ni(III)} + \mathrm{glucose} \Rightarrow \mathrm{Ni(II)} + \mathrm{gluconolactone}$$

Figure 43.3f shows the electrocatalytic response of NiOx/GC towards methanol oxidation. As has happened for GLO, the electrocatalytic activity towards MOR has increased at the NiOx/GC electrode, in agreement with previous results [17, 19]. The mechanism of MEO at NiOx/GC electrode can be described similarly, presuming that NiOx could act as a catalytic mediator to facilitate the oxidation kinetics. However, a minor reduction peak was observed during the reverse scan of MEO at NiOx/GC, which may be attributed to reduction of remaining NiOOH [19].

Figure 43.4 shows the CVs obtained at Pt/GC (a, c, and e) and NiOx/Pt/GC (b, d, and f) for FAO, GLO, and MEO, respectively, in 0.3 M NaOH solution containing definite concentration for each fuel at 200 mV s^{-1}. Figure 43.4a shows the CVs responses for FAO at Pt/GC electrode, which appeared similar to that observed at bulk polycrystalline Pt substrates. In the forward scan, two oxidation peaks

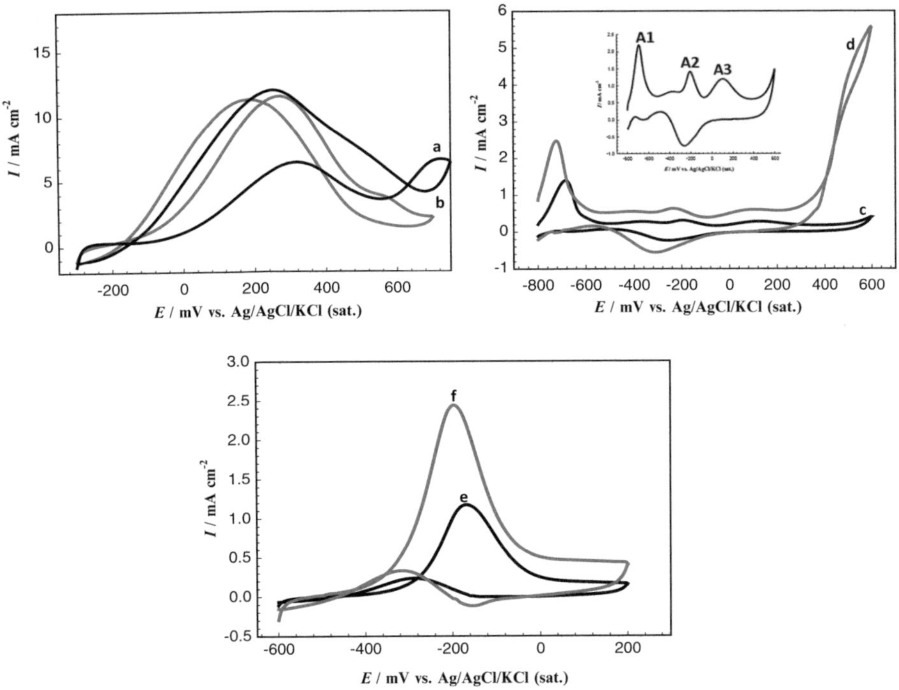

Fig. 43.4 (**a–f**) CVs for Pt/GC (**a**, **c**, **e**) and NiOx/Pt/GC (**b**, **d**, **f**) electrodes in 0.3 M NaOH containing 0.3 M FA m, 40 mM glucose, and 0.3 M methanol oxidation measured at 200 mV s^{-1}, respectively. *Inset*, is the curve c for glucose oxidation at Pt/GC electrode to clear the oxidation process

appeared; the first (I_p^d, at ca. 0.25 V) is assigned to the direct FAO to CO_2 (dehydrogenation pathway), while the second (dehydration pathway $\left(I_p^{ind}\right)$, at ca. 0.65 V) to the oxidation of the adsorbed poisoning CO intermediate (produced from non-faradaic dissociation of FA) to CO_2 [16, 20]. Oxidation current is also observed during the backward scan with a higher current (I_b). This is simply because the retrieval of the Pt active sites from the poisoning CO, which is accelerated after being the Pt surface mostly hydroxylated Pt–OH_{ad} [3, 16].

On the other hand, at the NiOx/Pt/GC electrode (Fig. 43.4b), several interesting features appears:

(I) A significant enhancement of I_p^d with a concurrent depression of I_p^{ind}. This may be attributed to the impact of NiOx in retarding the formation of CO and/or accelerating its oxidative removal at low potential [16, 20].

(II) The I_p^d/I_b ratio approaches unity in a similar fashion to the ideal catalyst (Pd) for FAO. This indicates that less amount of CO is produced in the forward scan at the NiOx/Pt/GC electrode than at the Pt/GC electrode [3, 21].

The mechanism of FAO at the NiOx/Pt/GC was discussed in detail previously [3, 16].

Figure 43.4c shows the CV responses for glucose oxidation at Pt/GC electrode in 0.3 M NaOH containing 40 mM glucose at a scan rate 200 mV s^{-1}, and the oxidation behavior appears clearly in the inset of this figure. Inspection of this figure (inset) reveals that there are three clear oxidation peaks A1, A2, and A3 for GLO on the forward scan at the Pt/GC electrode in agreement with previous results [2, 22]. The peak A1 is due to the chemisorption and dehydrogenation of GL in the hydrogen region with the removal of the first hydrogen atom (considered the rate determining step). The dehydrogenated intermediate formed by electro-oxidation at A1 leads to formation of gluconate by successive steps. The peak A2 appears at Pt-OH_{ads} adsorbed catalyst surface by direct glucose oxidation from the bulk to lactone (double layer region), which on hydrolysis produces gluconate. The peak A3 is obtained on already oxidized catalyst surface and may be due to oxidation of adsorbed residues [2, 22].

On the other hand, at NiOx/Pt/GC electrode, catalytic activity is enhanced (Fig. 43.4d). That is, the peak A1 appears at a more negative potential than Pt/GC electrode. In addition, the peak current of the peaks (A1, A2, and A3) at NiOx/Pt/GC electrode is substantially higher than that of the Pt/GC electrode. Another peak appears on the case of NiOx/Pt/GC on the potential range higher than 400 mV vs. Ag/AgCl/KCl (sat.). This peak is attributed to GLO at NiOx only (no contribution from Pt) in parallel with the Ni(II)/Ni(III) transformation, as observed at NiOx/GC electrode (Fig. 43.3d). The enhanced electrocatalytic activity for NiOx/Pt/GC electrode towards GLO may be attributed to the role of NiOx in providing the oxygen species necessary for the oxidative removal of poisoning species at lower potential compared to the unmodified Pt electrode.

Figure 43.4e shows the CV responses for methanol oxidation at Pt/GC electrode at a scan rate 200 mV s^{-1}. Analysis of this figure reveals the appearance of an anodic peak (I^{f}_{ap}) at ca. −190 mV vs. Ag/AgCl/KCl(sat.) at the Pt/GC electrode, during the forward scan, with a gradual decrease of the peak current, at potentials larger than ca. −190 mV. This decrease in the peak current may be attributed to the gradual formation of Pt oxide (see Fig. 43.2c) which is inactive towards methanol oxidation. Another anodic peak (I^{b}_{ap}) appears during the backward scan, which was assigned to the oxidation of methanol and/or methanol residues (e.g., CO) at the freshly reduced PtO surface [23, 24]. Interestingly, the modification of the Pt/GC electrode with NiOx has greatly enhanced its catalytic activity towards methanol oxidation (Fig. 43.4f). That is, the onset potential of methanol oxidation shifts by about ca. 100 mV to the negative direction of potential compared to that observed at the Pt/GC electrode concurrently with a significant increase in the peak current (I_p). The relative peak current density of I^{f}_{ap}/I^{b}_{ap} can be used as an index to describe the tolerance of the electrode against the carbonaceous species poisoning [24]. A large ratio means a good tolerance of the anode against the poisoning species. This ratio increased from 4.875 at the Pt/GC electrode to 6.528 at the NiOx/Pt/GC electrode. This catalytic enhancement points to a crucial role of NiOx in the oxidation of methanol. The enhanced activity of NiOx/Pt/GC for MEO in alkaline media can be attributed to the synergistic role of Pt and nickel oxides/hydroxides in the catalytic enhancement. The poisoned Pt nanoparticles surface with CO_{ad} can be regenerated via the reaction of surface CO with nickel oxides/hydroxides on the catalyst surface [25, 26].

43.4 Conclusion

Nickel oxide (NiOx)-modified GC and Pt/GC electrodes were prepared electrochemically and tested towards FA, Gl, and ME oxidation in 0.3 M NaOH solutions. The NiOx/GC electrode exhibited a catalytic activity towards glucose and methanol oxidation but did not show any activity towards FA oxidation. However, the NiOx/Pt/GC electrode showed a superb activity towards the oxidation of FA, Gl, and ME in comparison to Pt/GC electrode. Generally, the superb catalytic activity of NiOx/Pt/GC electrode is attributed to the role of NiOx, which could provide the necessary oxygen species to the catalyst surface to accelerate the oxidative removal of poisoning materials at lower potential.

References

1. Sharaf OZ, Orhan MF (2014) An overview of fuel cell technology: fundamentals and applications. Renew Sustain Energy Rev 32:810–853
2. Basu D, Basu S (2012) Performance studies of Pd–Pt and Pt–Pd–Au catalyst for electro-oxidation of glucose in direct glucose fuel cell. Int J Hydrogen Energy 37:4678–4684
3. El-Nagar GA, Mohammad AM, El-Deab MS, El-Anadouli BE (2013) Electrocatalysis by design: enhanced electrooxidation of formic acid at platinum nanoparticles–nickel oxide nanoparticles binary catalysts. Electrochim Acta 94:62–71
4. Freitas RG, Antunes EP, Pereira EC (2009) CO and methanol electrooxidation on Pt/Ir/Pt multilayers electrodes. Electrochim Acta 54:1999–2003
5. Miki A, Ye S, Senzaki T, Osawa M (2004) Surface-enhanced infrared study of catalytic electrooxidation of formaldehyde, methyl formate, and dimethoxymethane on platinum electrodes in acidic solution. J Electroanal Chem 563:23–31
6. Beden B, Léger J-M, Lamy C (1992) Electrocatalytic oxidation of oxygenated aliphatic organic compounds at noble metal electrodes. In: Bockris JOM, Conway BE, White R (eds) Modern aspects of electrochemistry. Springer, Philadelphia, pp 97–264
7. Iwasita T, Nart FC, Lopez B, Vielstich W (1992) On the study of adsorbed species at platinum from methanol, formic acid and reduced carbon dioxide via in situ FT-ir spectroscopy. Electrochim Acta 37:2361–2367
8. Radmilovic V, Gasteiger HA, Ross PN (1995) Structure and chemical composition of a supported Pt-Ru electrocatalyst for methanol oxidation. J Catal 154:98–106
9. Zhang H, Wang Y, Fachini ER, Cabrera CR (1999) Electrochemically codeposited platinum/molybdenum oxide electrode for catalytic oxidation of methanol in acid solution. Electrochem Solid-State Lett 2:437–439
10. Huang J, Yang H, Huang Q, Tang Y, Lu T, Akins DL (2004) Methanol oxidation on carbon-supported Pt-Os bimetallic nanoparticle electrocatalysts. J Electrochem Soc 151:A1810–A1815

11. Gurau B, Viswanathan R, Liu R, Lafrenz TJ, Ley KL, Smotkin ES et al (1998) Structural and electrochemical characterization of binary, ternary, and quaternary platinum alloy catalysts for methanol electro-oxidation 1. J Phys Chem B 102:9997–10003
12. Sadiek IM, Mohammad AM, El-Shakre ME, El-Deab MS (2012) Electrocatalytic activity of nickel oxide nanoparticles-modified electrodes: optimization of the loading level and operating pH towards the oxygen evolution reaction. Int J Hydrogen Energy 37:68–77
13. Trasatti S, Petrii OA (1992) Real surface area measurements in electrochemistry. J Electroanal Chem 327:353–376
14. El-Deab MS (2010) Electrocatalysis by nanoparticles: oxidation of formic acid at manganese oxide nanorods-modified Pt planar and nanohole-arrays. J Adv Res 1:87–93
15. El-Refaei SM, Awad MI, El-Anadouli BE, Saleh MM (2013) Electrocatalytic glucose oxidation at binary catalyst of nickel and manganese oxides nanoparticles modified glassy carbon electrode: optimization of the loading level and order of deposition. Electrochim Acta 92:460–467
16. El-Nagar GA, Mohammad AM, El-Deab MS, El-Anadouli BE (2012) Facilitated electro-oxidation of formic acid at nickel oxide nanoparticles modified electrodes. J Electrochem Soc 159:F249–F254
17. Fleischmann M, Korinek K, Pletcher D (1971) The oxidation of organic compounds at a nickel anode in alkaline solution. J Electroanal Chem Interfacial Electrochem 31:39–49
18. Jafarian M, Forouzandeh F, Danaee I, Gobal F, Mahjani MG (2009) Electrocatalytic oxidation of glucose on Ni and NiCu alloy modified glassy carbon electrode. J Solid State Electrochem 13:1171–1179
19. El-Shafei AA (1999) Electrocatalytic oxidation of methanol at a nickel hydroxide/glassy carbon modified electrode in alkaline medium. J Electroanal Chem 471:89–95
20. Masud J, Alam MT, Miah MR, Okajima T, Ohsaka T (2011) Enhanced electrooxidation of formic acid at Ta_2O_5-modified Pt electrode. Electrochem Commun 13:86–89
21. Habibi B, Delnavaz N (2011) Carbon–ceramic supported bimetallic Pt–Ni nanoparticles as an electrocatalyst for oxidation of formic acid. Int J Hydrog Energy 36:9581–9590
22. Basu D, Basu S (2011) Synthesis and characterization of Pt–Au/C catalyst for glucose electro-oxidation for the application in direct glucose fuel cell. Int J Hydrog Energy 36:14923–14929
23. Wang J, Shi R, Guo X, Xi J, Zhao J, Song C et al (2014) Highly active Pt-on-Au catalysts for methanol oxidation in alkaline media involving a synergistic interaction between Pt and Au. Electrochim Acta 123:309–316
24. Masud J, Alam MT, Awaludin Z, El-Deab MS, Okajima T, Ohsaka T (2012) Electrocatalytic oxidation of methanol at tantalum oxide-modified Pt electrodes. J Power Sources 220:399–404
25. Fu X-Z, Liang Y, Chen S-P, Lin J-D, Liao D-W (2009) Pt-rich shell coated Ni nanoparticles as catalysts for methanol electro-oxidation in alkaline media. Catal Commun 10:1893–1897
26. Xiong L, Yang X, Xu M, Xu Y, Wu D (2013) Pt–Ni alloy nanoparticles supported on multiwalled carbon nanotubes for methanol oxidation in alkaline media. J Solid State Electrochem 17:805–810

Chapter 44
Hydrotreatment of High-Acidity Vegetable Oil-Heavy Gas Oil Mixtures over a CoMo Catalyst

A. Vonortas, A. Zerva, K. Philippopoulos, and N. Papayannakos

Abstract The effect of free fatty acids (FFAs) on the hydrodesulfurization (HDS) reaction of an atmospheric heavy gas oil over a commercial CoMo/γ-Al_2O_3 catalyst was studied. Co-hydroprocessing of all liquid feed mixtures was carried out in a trickle-bed bench-scale three-phase stainless steel reactor at typical HDS conditions (310–350 °C reaction temperature and 33 bar reaction pressure). FFA content in the liquid feed varied from 0 to 20 % wt. Measurements concerning the final sulfur concentration, saponification number and acidity number of the final liquid product, the overall hydrogen consumption (HCON) for the liquid feed, and the composition of the gaseous phase after the hydrotreatment were obtained. The effect of the vegetable oil molecules on the HDS of gas oil was evaluated. Saponification and acidity number analysis showed full conversion (hydrodeoxygenation, HDO) of the heavy oxygenated molecules at all applied conditions. From the gaseous phase analysis, the products from the HDO of the carboxylic acids were determined. During the experimental tests, catalyst's activity was measured and deactivation was found to be negligible.

Abbreviations

FFAs	Free fatty acids
HDS	Hydrodesulfurization
HCON	Hydrogen consumption
HDO	Hydrodeoxygenation
RPO	Refined palm oil
HDT	Hydrotreatment
WGS	Water-gas shift
WHSV	Weight hourly space velocity
DMDS	Dimethyl disulfide

A. Vonortas • A. Zerva • K. Philippopoulos • N. Papayannakos (✉)
School of Chemical Engineering, National Technical University of Athens,
Heroon Polytechniou 9, Zografos, Athens 15780, Greece
e-mail: npap@central.ntua.gr

I. Dincer et al. (eds.), *Progress in Clean Energy, Volume 1*,
DOI 10.1007/978-3-319-16709-1_44

44.1 Introduction

The dwindling fossil fuel reserves and the current legislation concerning the use of renewable resources as fuels, especially in the transportation sector, lead to the increased interest in the research and development of new technologies for the production of liquid fuels which are not dependent on crude oil. The European Union has set the objective of 20 % substitution of conventional fuels with renewable ones by the year 2020 [1]. Moreover, several environmental concerns have emerged from the intensive use of petroleum and its derivatives making the use of alternative sources imperative. Vegetable oils and fats are an abundant source for producing biofuels resembling petroleum-derived diesel.

Vegetable oils and their derivatives have been investigated in the last decades with the purpose of producing fuels fully compatible with petroleum-derived diesel. Biodiesel was the first biofuel introduced in the market that could be used neat or in blends with diesel in diesel engines due to its similar properties to conventional petro-diesel and lower SOx and CO emissions from its combustion [2]. Though its production is relatively simple, its application is accompanied by certain drawbacks mainly related to its high oxygen content [3] and its conventional production process, which requires vegetable oils and fats with low acidity, methanol, and a continuously consumed base as catalyst [4].

An alternative route for the production of bio-components resembling diesel is hydrotreatment (HDT) of vegetable oils and their derivatives. This is a promising process that can partially substitute diesel fuel. Although HDT of vegetable oils is a more complex process than their transesterification, the necessary infrastructure and hydrogen already exist in oil refineries making this new technology even more attractive. The researcher's interest has been focused on shedding light on all different aspects concerning this new technology. Research work has already been carried out with both neat vegetable oils and blends with petro-diesel to determine optimal operating conditions [5], suitable catalysts [6], reaction mechanisms [7], catalyst deactivation [8], and final fuel properties [9, 10]. It was found that vegetable oils can be converted into useful hydrocarbons at milder reaction conditions (temperature, pressure) than gas oil's HDS. Huber [5] found that in mixtures of sunflower oil and heavy vacuum oil, the former was totally converted, while the latter reached only 41 % HDS conversion at 350 °C, indicating that even lower temperatures are adequate for triglyceride full HDO.

The goal of this work is to study the co-hydroprocessing of mixtures composed of an atmospheric heavy gas oil and FFAs using a bimetallic CoMo/γ-Al_2O_3 catalyst under typical HDS conditions. The main points are the study of the effect of the presence of the oxygenated and acidic molecules on the HDS performance of the three-phase reactor and the estimation of the influence of various parameters, like temperature, vegetable oil/gas oil ratio in the feed, and H_2S presence on the production of gaseous products from the FFA cleavage. For the estimation of the H_2S impact on the selectivity of the HDO reactions, a set of experiments was conducted using blends of a low-sulfur-content diesel and FFAs.

44.2 Experimental

Experiments were carried out in an integral, isothermal, bench-scale, tubular reactor. The detailed description of the experimental apparatus is presented in a previous communication [11]. A part of the experiments was performed for determining the effect of the presence of the FFAs on final HDS conversion, while another part concerned the investigation of gaseous phase composition after the hydrotreatment process. The experiments were carried out at three different reaction temperatures (310, 330, and 350 °C) and a constant total pressure of 33 bar that corresponds to typical mild HDS conditions. The gaseous feed was hydrogen of industrial purity provided by Air Liquide, while the liquid feed was neat atmospheric heavy gas oil, blends of this gas oil with FFAs (2.5, 5, and 10 %) and with an RPO, and finally blends of diesel with FFAs. 40 g of a commercial CoMo/γ-Al_2O_3 was used, diluted with 60 g of fine SiC inert particles to avoid liquid maldistribution over the solid catalyst and wall effects. The catalyst sulfidation was performed in situ, using a mixture of DMDS and light gas oil with 2 % wt. sulfur concentration. Gas flow at the reactor inlet was kept constant at approximately 20 Nl/h. The liquid flow rates tested were 28, 40, and 56 g/h. Gas oil and ultra-desulfurized diesel used in this work were provided by Hellenic Petroleum Inc. The analysis of the fatty acid composition for both vegetable oils was provided by Minerva Co. that also donated the vegetable oils. All liquid feed analyses are presented elsewhere [12]. The sulfur content of the final product was determined using an ANTEK 9000 Series sulfur analyzer. Ester bond conversion was determined from saponification number analysis (ASTM D 94-07) and was proved to be practically total (above 99 %) for all the experimental conditions applied in this study. Final acidity of the samples derived from the hydrotreatment of feeds containing acid vegetable oil was measured [13], and it was found that full conversion of the acid molecules had been achieved at all experimental conditions. Gas phase analysis was performed using a Hewlett Packard HP 6890 Series GC with a 5973 Mass selective detector. The analysis indicated the presence of CO and CO_2 due to FFA cleavage and water-gas shift (WGS) reaction, C_3H_8 from the HDO of the triglyceride molecules, and CH_4 from the methanation reaction occurring among some of the gas phase species. The partial pressure of H_2S in the gas phase, when present, was calculated by its mass balance and the conversion of organic sulfur bearing molecules in gas oil. The catalytic activity during the experimental tests was determined by repetition of a standard experiment.

44.3 Results and Discussion

44.3.1 Final Sulfur Concentration and HCON

The final sulfur concentration in ppm for all the four different feeds tested (0, 2.5, 5, and 10 % FFAs in gas oil) is given in Figs. 44.1, 44.2, 44.3, and 44.4 as a function of WHSV, keeping constant the reaction temperature.

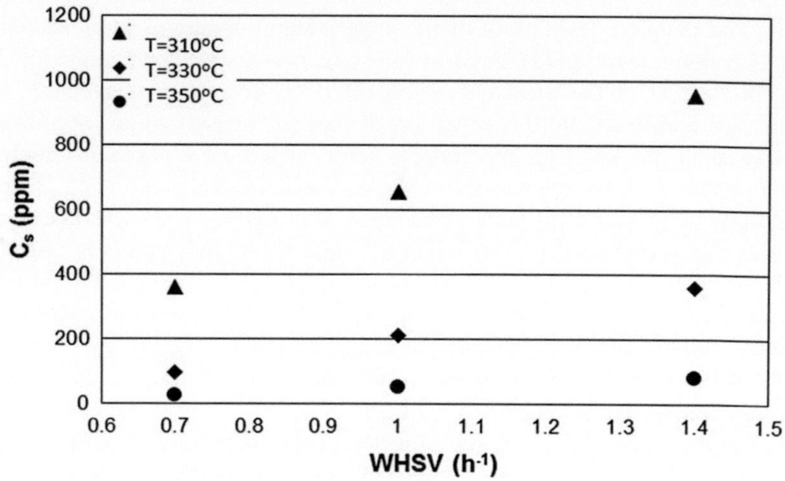

Fig. 44.1 Sulfur concentration with varying WHSV for neat gas oil for all three temperatures

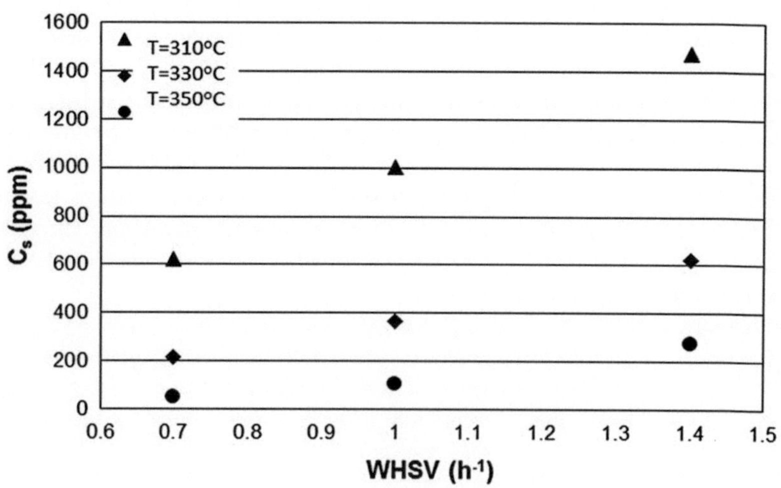

Fig. 44.2 Sulfur concentration with varying WHSV for 2.5 % wt. FFAs in gas oil for all three temperatures

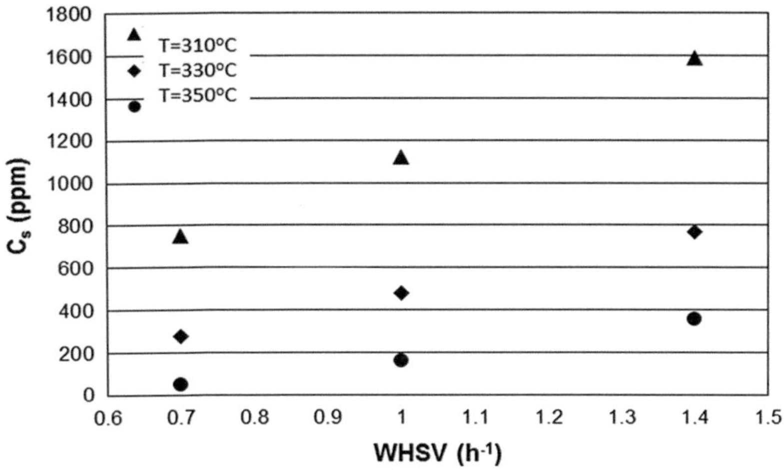

Fig. 44.3 Sulfur concentration with varying WHSV for 5 % wt. FFAs in gas oil for all three temperatures

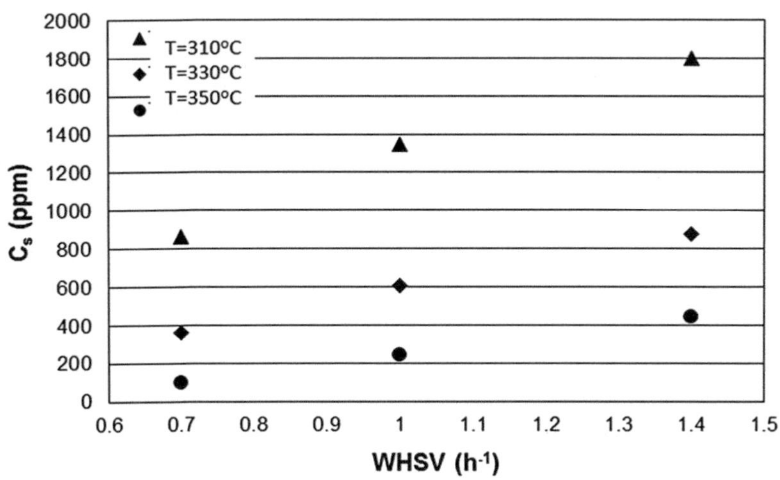

Fig. 44.4 Sulfur concentration with varying WHSV for 10 % wt. FFAs in gas oil for all three temperatures

Comparing the sulfur concentration for the same temperature and WHSV, it is obvious that as FFA content in the feed increases, sulfur concentration of the final product also increases. This increase becomes less steep at contents more than 5 %. This fact implies that FFAs have a strong impact on HDS reaction rates, an impact more severe at lower contents and less severe at higher contents. By carefully looking at the experimental points, it is clear that oxygenated molecule effect on

HDS is more evident as temperature increases (350 °C points showing bigger inhibition by FFAs or other oxygenated species than 330 or 310 °C). It should be noted that no significant catalyst deactivation concerning the HDS reaction was measured during the experimental runs.

HCON in Nl/Kg for all the four different feeds tested (0, 2.5, 5, and 10 % FFAs in gas oil) is given in Figs. 44.5, 44.6, 44.7, and 44.8 as a function of WHSV, keeping constant the temperature.

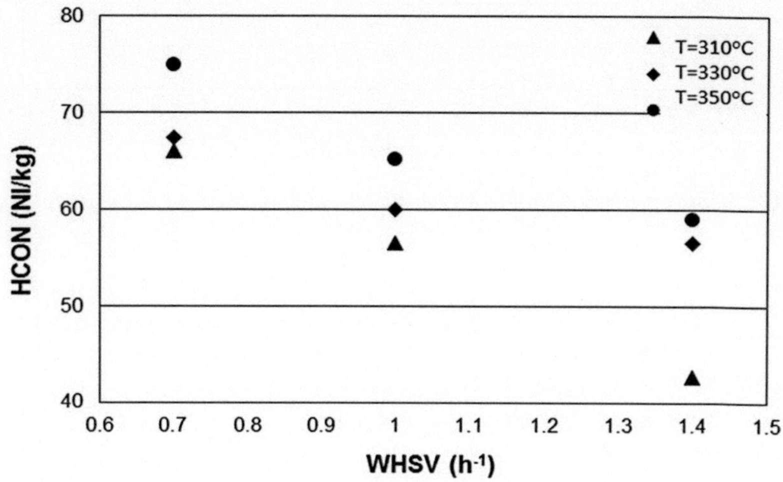

Fig. 44.5 HCON with varying WHSV for neat gas oil for all three temperatures

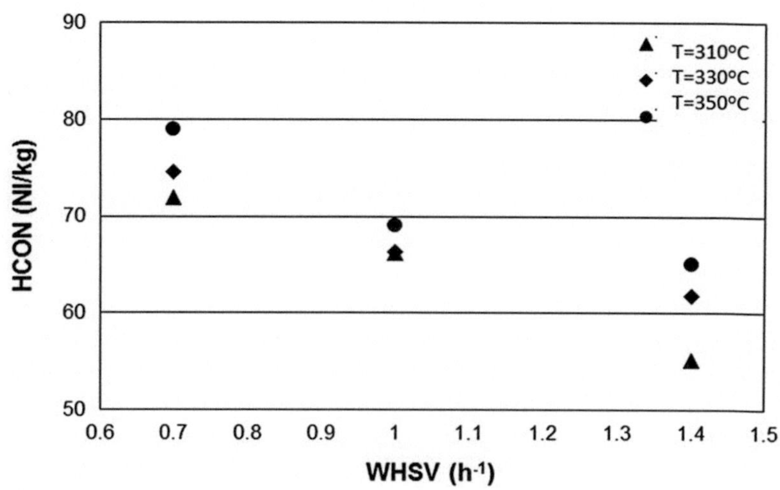

Fig. 44.6 HCON with varying WHSV for 2.5 % wt. FFAs in gas oil for all three temperatures

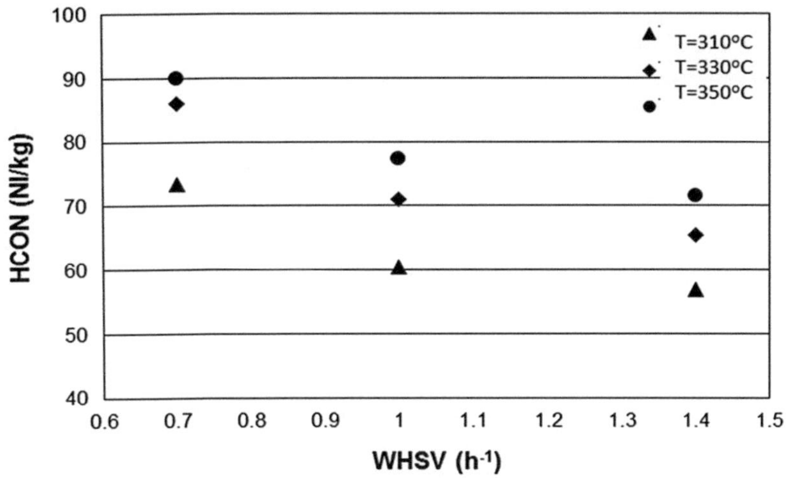

Fig. 44.7 HCON with varying WHSV for 5 % wt. FFAs in gas oil for all three temperatures

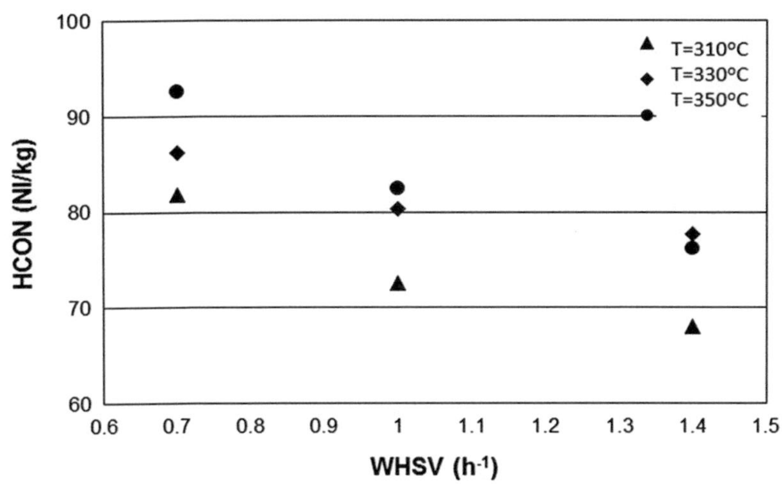

Fig. 44.8 HCON with varying WHSV for 10 % wt. FFAs in gas oil for all three temperatures

HCON is a crucial parameter for process economics during hydrotreatment processes. Because vegetable oils or their derivatives, apart from the removal of oxygen atoms which involves hydrogen, have a large number of unsaturated bonds between adjacent carbon atoms, their overall HDO requires large amounts of hydrogen. So, it is of utmost importance to determine the increased demands in hydrogen when vegetable oils are hydroprocessed along with petroleum fractions. From Figs. 44.5, 44.6, 44.7, and 44.8, it is obvious that hydrogen consumption increases with increasing FFA content in the initial feed, by increasing reaction temperature or by increasing the average residence time of the liquid molecules. For the two elevated temperatures, 330 and 350 °C, HCON is almost similar for all different contents, while 310 °C shows less HCON.

44.3.2 Gas Phase Analysis

The gas stream at the outlet of the reactor was analyzed and CO, CO_2, CH_4, and C_3H_8 (the latter only when RPO was fed to the reactor) were found to be present. Both carbon oxides result from carboxylic acid HDO, while methane is produced via the methanation reaction occurring at the prevailing experimental conditions from either of them. The determination of HDO reaction's selectivity was based upon the gaseous species amounts and is an important aspect of HDO process. It is widely known that HDO of triglycerides or carboxylic acids follows three distinct reaction pathways that are decarboxylation, decarbonylation, and deoxygenation. These three paths consume different amounts of hydrogen and yield different products. Decarboxylation has the lowest hydrogen consumption while deoxygenation the highest. Decarboxylation path yields CO_2 as by-product, decarbonylation CO and one molecule of water, and finally deoxygenation two molecules of water. While decarboxylation is the desired path according to its low hydrogen consumption, it yields paraffinic molecules with one carbon atom less than the original ester or acid, like decarbonylation path, thus resulting in final molecules with lower molecular weight. Complete knowledge of the HDO reaction's selectivity is therefore crucial for designing new processes or tailoring catalytic properties.

Because of the WGS reaction's equilibrium at the prevailing conditions, CO and CO_2 amounts cannot be attributed to either decarboxylation or decarbonylation path, and its amounts are not included herein.

Methane production varies with the FFA content in the initial feed tested. CH_4 amount increases with increasing vegetable oil content due to more available carbon oxides through the hydrodeoxygenation reactions. While the two vegetable oil feeds in the same matrix (gas oil) show a similar trend, which means that the type of the maternal oxygenated molecule does not play any important role for methanation reaction, when FFAs were mixed with desulfurized diesel, methane's production decreased by almost 25 %. This fact shows that H_2S, which is absent in the latter case, plays a significant role in methanation reaction. The same tendency has been observed for the three different temperatures tested in this work.

44.4 Conclusions

In this study, a commercial CoMo/γ-Al_2O_3 catalyst was used for the study of hydrotreatment of neat gas oil and its mixtures with FFAs (2.5, 5, 10, and 20 % wt.) in a micro-pilot hydrotreatment reactor at typical HDS conditions. FFA molecules or oxygenated products are found to have an adverse impact on HDS reaction, an impact that is steeper in lower contents and higher at contents more than 5 %. HCON measurements revealed that the presence of oxygenated molecules increased the hydrogen required for the hydrotreatment process. Saponification and acidity number analyses showed that full conversion is achieved at the experimental conditions tested. Gas phase analysis showed the existence of oxygenated by-products of HDO reaction along with hydrogen sulfide. CH_4 is produced from either of the carbon oxides and H_2S showed to influence its production. Catalyst activity appeared to be constant for the whole experimentation time, and oxygenated molecules do not promote catalyst deactivation.

Acknowledgment This research has been cofinanced by the European Union (European Social Fund—ESF) and Greek national funds through the Operational Program "Education and Lifelong Learning" of the National Strategic Reference Framework (NSRF)—Research Funding Program: Thales, investing in knowledge society through the European Social Fund.

References

1. The promotion of the use of biofuels or related other renewable fuels for transport, EU Directive 2001/0265(COD), 25 Feb 2003
2. Nabi N, Akhter S, Shahadat MZ (2006) Improvement of engine emissions with conventional diesel fuel and diesel–biodiesel blends. Bioresour Technol 97:372–378
3. Jain S, Sharma MP (2010) Stability of biodiesel and its blends: a review. Renew Sustain Energy Rev 14(2):667–678
4. Nair P (2007) UOP/Eni EcofiningTM process refining renewable feedstocks. In: International Symposium on Biofuels, September 25–26, New Delhi, India
5. Huber WG, O'Connor P, Corma A (2007) Processing biomass in conventional oil refineries: production of high quality diesel by hydrotreating vegetable oils in heavy vacuum oil mixtures. Appl Catal Gen 329:120–129
6. Bambang V, Han JY, Kim SK, Hong S-A, Kim YJ, Lim JS, Shu Y-W, Seong-Geun O, Kim J (2012) Production of renewable diesel by hydroprocessing of soybean oil: effect of catalysts. Fuel 94:578–585
7. Donnis B, Egeberg RG, Blom P, Knudsen KG (2009) Hydroprocessing of bio-oils and oxygenates to hydrocarbons. Understanding the reaction routes. Top Catal 52:229–240
8. Kubička D, Horáček J (2011) Deactivation of HDS catalysts in deoxygenation of vegetable oils. Appl Catal A-Gen 394(1–2):9–17
9. Simaček P, Kubička D (2010) Hydrocracking of petroleum vacuum distillate containing rapeseed oil: evaluation of diesel fuel. Fuel 89:1508–1513
10. Simaček P, Kubička D, Sebor G, Pospisil M (2010) Fuel properties of hydroprocessed rapeseed oil. Fuel 89:611–615
11. Vonortas A, Templis C, Papayannakos N (2012) Effect of palm oil content on deep hydrodesulphurization of gasoil-palm oil mixtures. Energ Fuel 26:3856–3863
12. Vonortas A, Kubička D, Papayannakos N (2014) Catalytic co-hydroprocessing of gasoil-palm oil/AVO mixtures over a NiMo/γ-Al_2O_3 catalyst. Fuel 116:49–55
13. Pasias S, Barakos N, Alexopoulos C, Papayannakos N (2006) Heterogeneously catalyzed esterification of FFAs in vegetable oils. Chem Eng Technol 29(11):1365–1371

Chapter 45
Carbon Sequestration by Direct Seeding of Wheat in Setif High Plains (North East Of Algeria)

Mohamed Fenni, Kamel Nadjem, and Abdelhamid Mekhlouf

Abstract Direct seeding is an important element of conservation agriculture; it contributes to environmental conservation and to sustainable agricultural production. Conservation agriculture increases carbon concentrations in the topsoil. It can also reduce the amount of fossil fuel consumed in intensive tillage and by other farm operations, and thus decrease the rate of CO_2 build-up in the atmosphere. Today, the global area of no-till farming is more than 100 million hectares. Setif high plains (north east of Algeria) is characterized by a semi arid climate with a long-term average annual precipitation ranging between 300 and 400 mm, the dry-farming system is commonly used, it is based on cereal/sheep production, the straw is used as main feed for livestock. In this region, direct seeding of cereals has recently been adopted, there is almost 9 years. The objective of this study was to quantify the carbon sequestered in the field by direct seeding of wheat. The experiment was conducted at the Setif ITGC experimental site during the 2010/2011 cropping season. The results indicate that dry straw left by the durum wheat varieties varies between 343 and 537 g/m^2. It varies between 251 and 643 g/m^2 for bread wheat genotypes. Thus, the average amount of carbon sequestered each year is 260 gC/m^2 and 257 gC/m^2 successively for durum and bread wheat where the genotypes MBB and WAHA have high values of GY and C for durum wheat and RMADA have high values of GY and C for bread wheat. The direct seeding of wheat, relatively to conventional till, can reduce the use of fuel by 50–70 %, machinery requirements by 60 %, resulting in a decrease in production costs. Thus, a combination of the economic and environmental benefits through reduced labor requirements, time savings, reduced machinery and fuel savings, direct seeding has universal appeal. Indirect measures of social benefits as society enjoys a better quality of life from environmental quality enhancement will be difficult to

M. Fenni (✉)
Laboratory of Valorisation of Biological and Natural Resources, Faculty of Life and Natural Sciences, University Ferhat Abbas Setif 1, Sétif 19000, Algeria
e-mail: fennimodz@yahoo.fr

K. Nadjem • A. Mekhlouf
Agronomy Department, Faculty of Life and Natural Sciences, University Ferhat Abbas Setif 1, Sétif 19000, Algeria

I. Dincer et al. (eds.), *Progress in Clean Energy, Volume 1*,
DOI 10.1007/978-3-319-16709-1_45

quantify. Conservation agriculture, working in harmony with nature, enables the protection of nonrenewable natural resources and their preservation for future generations can be beneficial for feeding and greening the world.

Keywords Carbon sequestration • Conservation agriculture • Direct seeding • Wheat • Global warming

Nomenclature

C	Carbon
CT	Conventional tillage
GHG	Greenhouse gases
GY	Grain yield
ITGC-ARS	Institute of Field Crop Agricultural Research Station of Setif
MAP	Monoammonium phosphate
Mt C	Million metric tons of carbon
NT	No-tillage
SOC	Soil organic carbon
STR	Straw yield

45.1 Introduction

Increasing concerns about global climate change, driven by rising atmospheric concentrations of greenhouse gases (GHG), have enhanced the interest in soil C sequestration as one of the strategies to offset anthropogenic CO_2 emissions. Soils can function either as a source or a sink for atmospheric GHG depending on land use and soil management. Appropriate management can enable agricultural soils to be a net sink for sequestering atmospheric CO_2 and other GHG [1–3]. Now, it is advocated that sequestration of Carbon (C) in terrestrial ecosystems, including in agricultural soils, might be used to offset some of the emissions of CO_2 from burning fossil fuels [4]. Emissions of CO_2 from agriculture are generated from three sources (1) machinery used for cultivating the land (2) production and application of fertilizers and pesticides, and oxidation of the Soil Organic Carbon (SOC) following soil disturbance. The amount of soil that is disturbed, in turn causing decomposition and oxidation of SOC, is largely dependent on the tillage practices used. The amount of fertilizers and pesticides applied varies among crop types, crop rotations, and tillage practices [5]. Currently, agriculture and other forms of land use contribute 32 % to the world's GHG emissions [6]. Lal et al. [7] estimate that changes in global agricultural practices could sequester over 200 million metric tons of C (Mt C) per year; indeed, changes in agronomic practices in the United States are thought to have the potential to offset nearly 10 % of its total carbon emissions [8]. In the context of rising atmospheric CO_2, no-tillage

(NT) has also been advocated as a way of slowing the raise in atmospheric CO_2 by sequestering additional organic carbon in the soil [7]. Additional carbon sequestration in turn has been advocated as strategy by which agriculture could mitigate climate change [9, 10]. Replacing conventional tillage (CT) with NT generally results in net sequestration of SOC [11, 12]. NT is the only type of conservation tillage that appears to bring about carbon benefits [13]. The combined effect of NT and straw incorporation on the accumulation of SOC was reported to be greater than the effect of either tillage reduction or straw incorporation alone [14]. Integrated crop residue management promotes carbon sequestration and has large potentials of reversing the net carbon flows from the atmosphere to the biosphere [15]. The management of crop residues may improve crop yields and land resilience against drought and other hazards while at the same time protecting and stimulating the biological functions of the soil [16]. Paustian et al. [2] compared 39 paired tillage experiments, ranging in duration from 5 to 20 years, and estimated that NT resulted in an average soil C increase of 285 g m^2 with respect to CT.

Conservation tillage systems have a higher SOC concentration in surface layer than CT [17] because the decomposition of plant residues is slower under conservation tillage due to reduced soil residue contact and incorporating residue in conventional tillage into the plow layer, and therefore, increasing soil and crop residue contact. The latter increases the exposure of SOC in inter- and intra-aggregate zones to microorganisms allowing a rapid decomposition [18]. In Setif high plains, the main cropping system is the dry-farming, it's based on cereal/sheep production, and the straw is used as main feed for livestock. This cropping system, such as practiced on the Setif high plains, degrades the structure of the tilled horizon and reduces its organic matter content [19].

The present study was conducted to reveal the contribution of no tillage in carbon sequestration in Setif high plains (north east of Algeria), the objective of this study was to estimate the amount of carbon sequestrated by straw of some wheat genotypes (durum wheat and bread wheat) under no tillage.

45.2 Materials and Methods

The experiment was located on Setif high plains of Algeria; it was established during the 2010/2011 cropping season at the Institute of Field Crop Agricultural Research Station of Setif (ITGC-ARS Setif, 36°08'N, 5°20'E, elevation 962 m above sea level). The soil of experimental station was classified as a steppic soil [20] (pH = 7.37, organic matter = 2.6 %). Eight wheat cultivars including four (04) durum cultivars (Zairi, Mohammed ben bachir (Mbb), Boussalam, and Waha) and four (04) bread cultivars (Acsad 901, Hiddab 1220, Wifak, and Rmada) were used in this study. The experiment was conducted under rainfed conditions in a randomized complete design with three replications. The sowing was done on 28 November 2010 with zero-tillage drill (Semeato S/A, Brazil) at a seed rate of 129 kg ha^{-1}. Monoammonium phosphate (MAP) and urea were used as source of phosphorus and nitrogen, respectively; plots were fertilized with 80 kg ha^{-1} of MAP at sowing and 120 kg ha^{-1} of urea 46 % at 21 March 2011. Percent residue cover was estimated by the line-transect method [21]. This involved stretching a 7.5 m measuring tape diagonally at about 45° angle across the crop rows and counting the number of the 15 cm marks along the tape that intercepted a piece of crop residue. The percent residue cover for the experimental area was then obtained by multiplying this count by 2. We found that 30 % of the soil surface was covered by crop residues. At harvest, one row, 1 m long, was harvested manually from each plot for recording grain yield (GY, g m^{-2}), straw yield (STR, g m^{-2}), and estimation of carbon in straw using the following formulae (C, g m^{-2}) = Dry matter/1.72.

The data collected from this study was analyzed statistically following appropriate methods as analysis of variance (ANOVA) using the SAS 9.00 software package [22]. Means separation was performed by Fisher's (protected) least significant differences (LSD) when the ANOVA results indicated significant effects at the 0.05 probability level.

45.3 Results and Discussion

45.3.1 Grain Yield and Straw Yields

GY analysis of variance showed significant differences for durum wheat genotypes and highly significant differences for bread wheat genotypes. In addition, no differences were shown between durum wheat genotypes for STR and very highly significant differences between bread wheat genotypes for STR (Table 45.1). Means values of GY and STR as shown in Table 45.3.

45.3.2 Carbon in Straw

Carbon analysis of variance showed no significant differences for durum wheat and highly significant differences between bread wheat genotypes (Table 45.2). For durum wheat genotypes, means value of C (g/m^2) varied between 199.49 and

Table 45.1 Analysis of variance of GY and STR for tested wheat genotypes

		Mean of square			
		GY		STR	
Source of variation	DF	Durum wheat	Bread wheat	Durum wheat	Bread wheat
Genotype	3	17,997.2*	70,921.85**	20,000^{ns}	59,811***
Error	8	3,220.85	5,922.43	6,213.96	4,607
CV%	/	14.31	16.3	17.55	15.33

DF degree of freedom, ns, *, **, and ***: not significant, significant, highly significant, and very highly significant at $P < 0.05$, $P < 0.01$, and $P < 0.001$, respectively, *CV* coefficient of variation

Table 45.2 Analysis of variance of carbon in straw for all tested wheat genotypes

Source of variation	DF	Mean of square	
		Durum wheat	Bread wheat
Genotype	3	$6{,}084^{ns}$	$23{,}973^{**}$
Error	8	1,088	2,002
CV%	/	14.3	16.3

DF degree of freedom, ns, *, **, and ***: not significant, significant, highly significant, and very highly significant at $P < 0.05$, $P < 0.01$, and $P < 0.001$, respectively, *CV* coefficient of variation

Table 45.3 Means values of GY, STR, and carbon of all genotypes

	Genotypes	Grain yield (g/m^2)	Straw yield (g/m^2)	Carbon in straw (g/m^2)
Durum wheat	MBB	423.52	537.25	311.19
	WAHA	494.12	478.43	278.15
	BOUSSALAM	333.33	437.25	254.21
	ZAIRI	335.29	343.13	199.49
Bread wheat	RMADA	674.05	643.13	373.91
	ACSAD 901	533.33	535.29	311.21
	WIFAK	419.6	341.17	198.35
	HIDAB 1220	288.23	250.98	145.91

311.19 for ZAIRI and MBB, respectively (Table 45.3). For bread wheat genotypes, means value of C (g/m^2) are 145.91, 198.35, 311.21, and 373.91 for HIDAB 1220, WIFAK, ACSAD 901, and RMADA (Table 45.3).

Agricultural ecosystems represent 11 % of the earth's land surface and include some of the most productive and carbon-rich soils. As a result, they play a significant role in the storage and release of C within the terrestrial carbon cycle [23]. Soil carbon is one of the most important factors that promote soil fertility, pest control, soil-water moisture, and farm productivity.

The management of crop residues may improve crop yields and land resilience against drought and other hazards while at the same time protecting and stimulating the biological functions of the soil [16]. The Kyoto protocol plans that the net emission of greenhouse gases can be reduced or by decreasing the rate in which they are emitted towards the atmosphere or by increasing the rate in which gases are removed from the atmosphere through wells. The agricultural soil is considered the most important reservoirs of carbon of the planet and their potential of detention can be spread. They thus supply a forward-looking solution to mitigate the increasing concentration of atmospheric CO_2. Conservation agriculture aims at sustainable and profitable cropping systems; its objective is the conciliation of the productivity and the environmental preservation. At Setif high plains the dry-farming is the cropping system dominating for purpose to secure the yields of cereal through the preservation of the water in the soil profile. But several studies showed that the dry-farming arrived at its limits especially those negative effects on the pedological

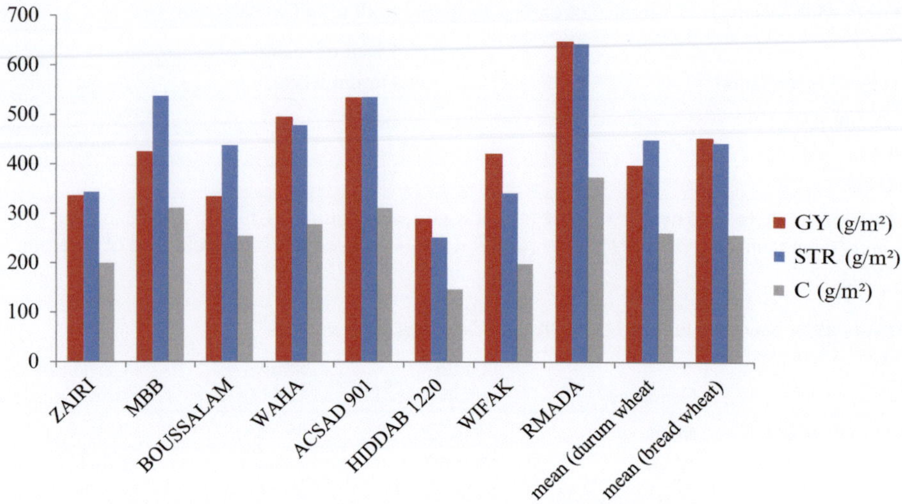

Fig. 45.1 Means values of GY, STR, and C of all tested genotypes

plan (decrease of the rate of organic matter and low fertility) and no increase of yields. With the technique of no tillage we try to improve the yields and to protect the environment especially the improvement of the rates of the organic matter in soil. Through this study we try to combine both objectives of the no tillage simultaneously at Setif high plains. Through this study we try to combine both simultaneously at Setif high plains. That is we look for the genotype which gives a high yield and leaves an important straw quantity on soil, and automatically a high quantity of carbon which will be held in the soil. For the durum wheat genotypes, MBB and WAHA have the highest means values of grain yield and the quantity of carbon in the straw, for bread wheat genotypes RMADA have the highest means values of GY and C in straw Fig. 45.1.

45.4 Conclusion

Carbon sequestration in soil organic matter will have a direct positive impact on soil quality and fertility. In Setif high plains north-eastern of Algeria, cropping system must adjust to help mitigate the effects of climate-warming gases lies in management of soil to increase organic content; the impact of no-tillage practices on carbon sequestration has been of great interest in recent years. In Setif high plain, the dry-farming is the dominant cropping system where half of the surface intended for cereals stay every year in fallow. In case the no tillage will be totally adopted in the region of the Setif high plain, that means that the surface intended for cereals will enhance twofold who leads automatically to the reduction of the fallow. And the increase of the yields and consequently the increase of the quantity of carbon held in soil in the region.

References

1. Lal R, Kimble JM (1997) Conservation tillage for carbon sequestration. Nutr Cycl Agroecosyst 49:243–253
2. Paustian K, Andren O, Janzen HH, Lal R, Smith P, Tian G, Tiessen H, Van Noordwijk M, Woomer PL (1997) Agricultural soils as a sink to mitigate CO2 emissions. Soil Use and Management 13:230–244
3. West TO, Post WM (2002) Soil organic carbon sequestration rates by tillage and crop rotation: a global data analysis. Soil Sci Soc Am J 66:1930–1946
4. IPCC (2000) Land use, land-use change, and forestry. Cambridge University Press, New York
5. Tristram OW, Marland G (2002) A synthesis of carbon sequestration, carbon emissions, and net carbon flux in agriculture: comparing tillage practices in the United States. Agric Ecosyst Environ 91:217–232
6. IPCC (2007) Climate change impacts, adaptation and vulnerability, Report of Working Group II to the Fourth Assessment Report of the Intergovernmental Panel on Climate Change, Cambridge University Press, Cambridge
7. Lal R, Kimble LM, Follett RF, Cole CV (1998) The potential of U.S. cropland to sequester C and mitigate the greenhouse effect. Ann Arbor Press, Chelsea, MI
8. FAO (2001) Food and fuel in a warmer world, News & Highlights, FAO Newsletter
9. Paustian K, Cole CV, Sauerbeck D, Sampson N (1998) CO_2 mitigation by agriculture: an overview. Clim Change 40:135–162
10. Lal R (2004) Soil carbon sequestration to mitigate climate change. Geoderma 123:1–22
11. Lal R, Follett RF, Kimble JM, Cole CV (1999) Management of U.S. cropland to sequester carbon in soil. J Soil Water Conserv 54:374–381
12. Robertson GP, Paul E, Harwood RR (2000) Greenhouse gases in intensive agriculture: contributions of individual gases to the radiative forcing of the atmosphere. Science 289:1922–1925
13. Uri ND (2001) The potential impact of conservation practices in US agriculture on global climate change. J Sustain Agric 18:109–131
14. Kushwaha CP, Tripathi SK, Singh KP (2001) Soil organic matter and water-stable aggregate under different tillage and residue conditions in a tropical dryland agroecosystem. Appl Soil Ecol 16:229–241
15. Marland G, Obersteiner M, Schlamadinger B (2007) The carbon benefits of fuels and forests. Science 318:1066–1068
16. Unger PW, Langdale GW, Papendick RI (1988) Role of crop residues in improving water conservation and use. In: Hargrove WL (ed) Cropping strategies for efficient use of water and nitrogen. ASA, Madison
17. Hutchinson JJ, Campbell CA, Desjardins RL (2007) Some perspectives on carbon sequestration in agriculture. Agric Meteorol 142:288–302
18. Al-Kaisi MM, Yin XH (2005) Tillage and crop residue effects on soil carbon and carbon dioxide emission in corn–soybean rotations. J Environ Qual 34:437–445
19. Kribaa M, Hallaire S, Curmi J (2001) Effects of tillage methods on soil hydraulic conductivity and durum wheat grain yield in semi-arid area. Soil Tillage Res 37:17–28
20. Kribaa M, Hallaire S, Curmi J, Lahmar R (2001) Effect of various cultivation methods on the structure and hydraulic properties of a soil in semi-arid climate. Soil Tillage Res 60:43–53
21. Richards BK, Walter MF, Much RE (1984) Variation in line transect measurements of crop residue cover. J Soil Water Conserv 39:60–61
22. SAS Institue (2002) SAS/STAT user's guide, version 9.00. SAS Institute, Inc., Cary, NC
23. Lal R, Bruce JP (1999) The potential of world cropland soils to sequester C and mitigate the greenhouse effect. Environ Sci Policy 2:77–185

Chapter 46
Chemical Activation of a Sewage Sludge for Elimination of Cationic die (Rhodamine-B) From Aqueous Solution

Meriem Zamouche, Sihem Arris, and Mossaab Bencheikh LeHocine

Abstract Sewage sludge, which was collected from drying beds of the municipal wastewater treatment station in Algeria, was used to eliminate cationic dye (Rhodamine B) from aqueous solution. Sludge was chemically treated by different reagents and at different conditions in order to optimize the best conditions giving a maximum sorption of dye. Sulfuric acid at 0.1 mol/L was the efficient reagent to activated sewage sludge. The effect of ratio and temperature impregnation was evaluated; the results show that the impregnation ratio which gives a good coverage of the sorbent surface by acids is 25 at ambient temperature.

Keywords Sewage sludge • Chemical activation • Dye • Impregnation ratio

Nomenclature

q_e	Uptake of dye (mg/g)
C_0	Initial dye concentrations (mg/L)
C_e	Final dye concentrations (mg/L)
V	Volume of solution (L)
W	Sorbent weight (g)

Subscripts

RS	Raw sludge
AS	Acids activated sewage sludge
BS	Bases activated sewage sludge

M. Zamouche • S. Arris (✉) • M.B. LeHocine
Laboratory of the Engineering and the Processes of Environment (LIPE), Faculty of Pharmaceutical Engineering Processes, University of Constantine 3, Constantine 25000, Algeria
e-mail: zamouche_meriem@yahoo.fr; arris_s@yahoo.fr; mossaabbb@yahoo.fr

I. Dincer et al. (eds.), *Progress in Clean Energy, Volume 1*,
DOI 10.1007/978-3-319-16709-1_46

AS_N	Sewage sludge activated by HNO_3
AS_S	Sewage sludge activated by H_2SO_4
AS_P	Sewage sludge activated by H_3PO_4
BSn	Sewage sludge activated by NaOH
BS_{CO3}	Sewage sludge activated by Na_2CO_3
BS_{HCO3}	Sewage sludge by $NaHCO_3$
BS_{SO4}	Sewage sludge by Na_2SO_4
RhB	Rhodamine B
IR	Impregnation ratio

46.1 Introduction

The adsorption technique has been used by several authors for the removal of various pollutants, such as dyes and heavy metals, from aqueous solution. Since the range of pollutants are various and varied, the adsorbent used are also more numerous and different. This is due to the use of by-products and recovery of certain wastes as adsorbents. Thus, the availability of by-product and waste from different genres expands the range of choice of sorbent to be used. In the literature, we find that several authors have used sewage sludge as a by-product, for purification of waste water by adsorption. The ways of activation of the sludge differ from one author to another. This mainly depends on the nature of pollutant to be removed. In the study carried by Caner et al. [1], the anaerobic sludge sampled from anaerobic digester of a municipal waste water treatment was perfectly used as adsorbent dye blue burazol ED. For improving the performance of sewage sludge for a large game of pollutants, sludge can follow different treatments, by chemical or physical activation. Ju et al. [2] have treated sludge chemically by three chemical reagents: sulfuric acid, hydrochloric acid and the NaOH, the sorption results of cationic dye Rhodamine B, show that the treatment with the acid is more effective than that by the base, while sulfuric acid gives the greater capacity of sorption. Once again, the sewage sludge was chemically activated by nitric acid in the study ported by Junxiong et al. [3]; the treatment efficiency was tested on the sorption of methylene blue (MB) and Reactive Red 4 (CE4); the characterization of the activated sludge shows that the surface sorbent has become complex; and anionic, carboxylic, and phosphonate functional groups were identified as binding site for the methylene blue, while the amine groups have been identified as the site adsorption of RR4. However, we focused our research on the optimization of the parameters of the activation of sewage sludge, the choice of activating reagent, and its concentration are among the most important parameters of the chemical activation, without even neglected the effect of ratio and temperature impregnation which may influence the effectiveness of the activation. A good impregnation of sorbent ensures a good distribution of functional groups on the surface of the sorbent. However, the rate and temperature of impregnation were also studied and the sorbate selected was the cationic dye Rhodamine B.

46.2 Materials and Methods

46.2.1 Preparation of Chemical Activated Sludge

The sewage sludge samples in this study were collected from drying beds of the municipal biological wastewater treatment station of IBN ZIED, Constantine, Algeria. Sewage sludge was dried at 105 °C, crushed and sieved to obtain a particle size $\leq$315 nm. The crushed raw sludge (RS) was prepared by chemical activation using different chemical reagents such as acids: H_3PO_4, H_2SO_4, and HNO_3, and bases: NaOH, Na_2CO_3, $NaHCO_3$, and Na_2SO_4. First, sludge was mixed with activating agent solution with a ratio of 1:25 (weight of crushed sludge/volume of activating agent) at an ambient temperature, for 24 h in a continuously mixed reactor. After that, the chemical activated sludge was separated from chemical agent by centrifugation (sigma 2–16) at 4,000 rpm for 5 min. The previously activated sludge was washed several times with distilled water in order to eliminate the excess of chemical activated agent, centrifuged and oven dried at 105 °C. Finally, the dried chemical activated sludge was subsequently crushed and sieved into a uniform size of equal or less than 315 nm. We noted by AS and BS the acids and bases activated sewage sludge, respectively, and by AS_N, AS_S, AS_P, BSn, BS_{CO3}, BS_{HCO3}, and BS_{SO4} for those activated by HNO_3, H_2SO_4, H_3PO_4, NaOH, Na_2CO_3, $NaHCO_3$, and Na_2SO_4, respectively. The chemical activation can be influenced by different parameters, the ratio and temperature impregnation, and the concentration of the chemical reagent. These parameters may produce sorbent with different characterization and surface structure. For this reason, the concentration of activated acids ranging from 0.01 to 2 mol/L was tested on the RS, keeping the same chemical activation conditions. The effect of the impregnation ratio was evaluated; however, the same amount of sludge was impregnated with different sulfuric acid volumes (50, 100, 200, 300, 400, and 600 mL). The impregnation temperature was tested from 10 to 40 °C including the ambient temperature.

46.2.2 Adsorbate

Rhodamine B (RhB) dye which was supplied from Sigma-Aldrich was used to prepare solutions. The chemical structure of this dye is shown in Fig. 46.1. The concentrations of Rhodamine B dye were determined by using UV–Vis spectrophotometer (Shimadzu mini 1240) set at a wavelength of 554 nm.

46.2.3 The Adsorption Procedure

The effect of chemical activation was carried on the sorption of Rhodamine B in batch system. For this, a 400 mL of RhB was stirred with 4 g of activated sludge at

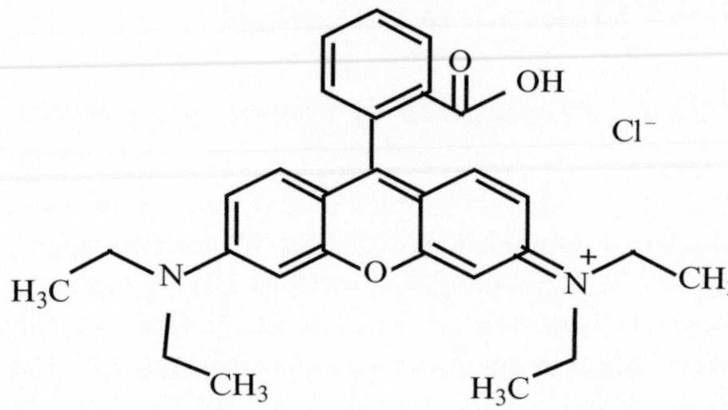

Fig. 46.1 Structure of Rhodamine B

300 rpm, by maintaining a constant temperature of 25 °C until equilibrium was reached. At predetermined time intervals, a volume 0.5 mL of solution was sampled, centrifuged at 4,000 rpm for 5 min, and the supernatant was used for the determination of final concentration of RhB. The dye uptake q_e (mg/g) was determined as follows:

$$q_e = \frac{(C_0 - C_e)V}{W} \tag{46.1}$$

where C_0 and C_e are the initial and final dye concentrations (mg/L), respectively, V is the volume of solution (L), and W is the sorbent weight (g).

46.3 Results and Discussion

46.3.1 Effect of Chemical Activation by Acids and Bases

To study the effect of acids and bases, the chemical activation reagents were used at the same concentration of 0.1 mol/L. The figure below shows the effect of activation by acids and bases. From this figure we can note that:

The activation of RS by acids (AS) is beneficial because the sorption capacity is higher than the sludge without activation. However, the BS treatment was not efficacy for any types of bases reagent used. Among the acids used, sulfuric acid gives the greater percent removal of dye compared to the other acids used; this result was obtained by different authors [4–6].

Chemical activation is intended to introduce the active site on the surface of the sorbent. However, activation of sorbent by bases is enabled by introducing a basic functional surface, and the activation with acids enables the formation of acidic functional surface character. Thus as the RhB is cationic, it is clear that the sorption on activated sludge by acid (AS) is favored. Since the acidic functional surface exists in the origin on the surface of RS, other functions are created; therefore, the

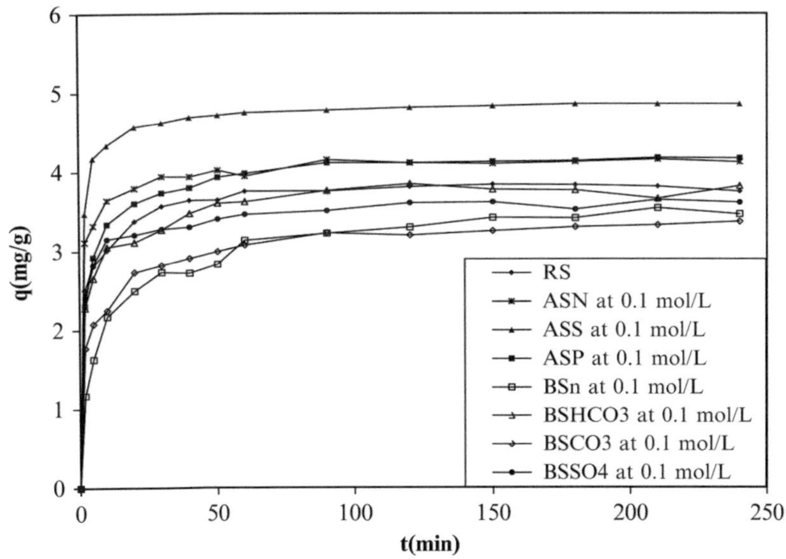

Fig. 46.2 Effect of treatment of sludge by acids and bases on the sorption of Rhodamine B

cationic dye is increasingly removed. The decrease in sorption capacity of the activated sludge by bases is expected because it is possible that the repulsion forces between the basic surface functions and the cationic dye are formed (Fig. 46.2).

46.3.2 Effect of Acids Concentration

To investigate the effect of acid concentration on the RS, only the three acids studies are used as chemical activation reagent. The concentration varied for each acid between 0.01 and 2 mol/L; the results obtained are presented in the following graphs:

As shown in Figs. 46.3, 46.4, and 46.5, between the three acids used, sulfuric acid revealed the best acids for activating the sludge for all the concentration studies, with the exception at the concentration of 2 mol/L. The biggest amount sorbed of AS_S is given by sulfuric acid at a concentration of 0.1 mol/L.

Since sulfuric acid at the concentration of 0.1 mol/L was found to be the best chemical activation reagent of RS, it was used at the same concentration for studies of the effects of ratio and temperature impregnation.

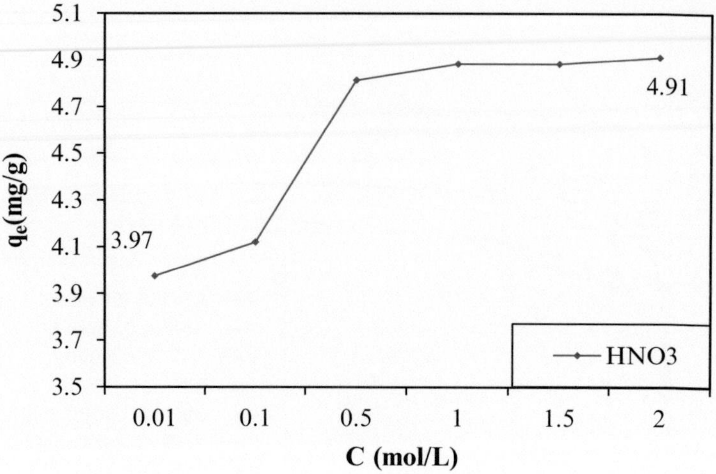

Fig. 46.3 Evolution of the sorption capacity of AS_N as function of the HNO_3 concentration

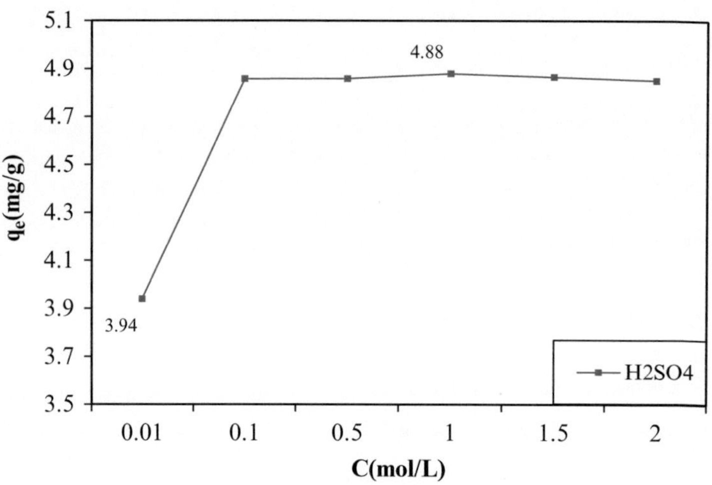

Fig. 46.4 Evolution of the sorption capacity of AS_S as function of the H_2SO_4 concentration

46.3.3 *Effect of Impregnation Ratio*

Chemical activation has a direct relation with the functional surfaces introducing on the surface of sorbent, the amount and concentration of the functional surface, depends on the mode and manner of activation. A good impregnation of sorbent can promote the enrichment of the sorbent surface by a functional surface. The impregnation ratio was tested by maintaining the same mass of RS and varying the volume

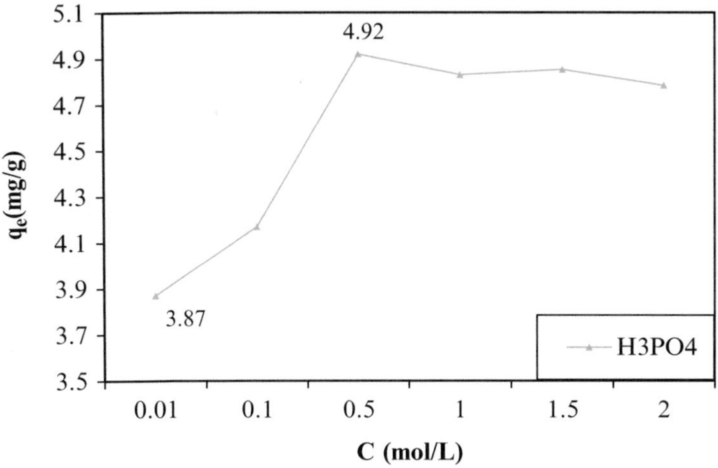

Fig. 46.5 Evolution of the sorption capacity of AS_P as function of the H_3PO_4 concentration

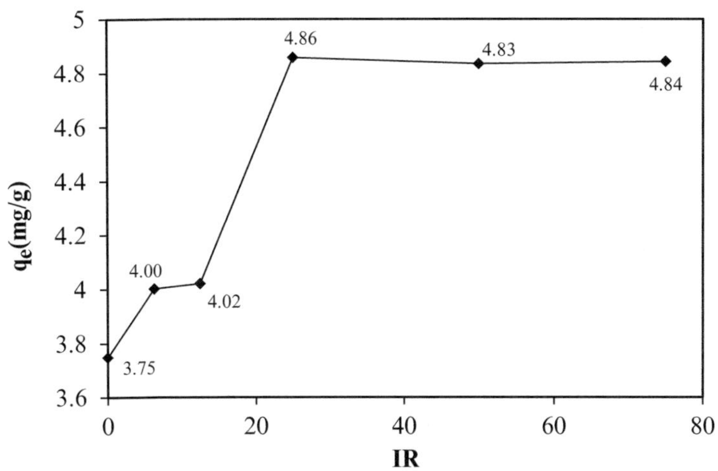

Fig. 46.6 Effect of impregnation ratio on the sorption capacity of AS_S

of sulfuric acid for obtaining impregnation ratio of 6.25, 12.5, 25, 50, and 75. Figure 46.6 shows that with an impregnation ratio (IR) of 25 we have obtained a very high sorption capacity, which means that at this IR we provide a good impregnation of the surface sorbent. At 6.25 and 12.5 IR, the sludge surface wasn't well impregnated. Above IR of 25, there is no more functional surface formed confirmed.

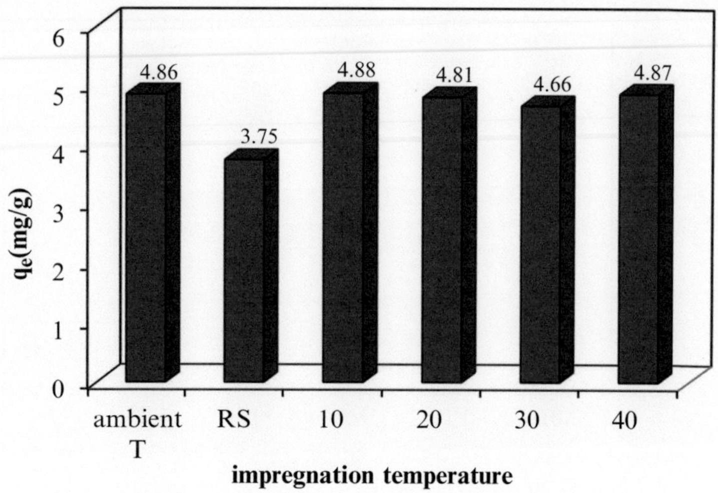

Fig. 46.7 Effect of impregnation temperature on the sorption capacity of AS_S

46.3.4 Effect of Impregnation Temperature

The effect of the temperature impregnation is studied for the temperatures (10, 20, 30, 40), by maintaining an IR equal to 25.

According to Fig. 46.7, it is clear that the impregnation temperature has no remarkable effect on the sorption capacity of AS_S, and the impregnation at ambient temperature gives a maximum amount of dye sorption.

46.4 Conclusion

The sewage sludge chemically treated by H_2SO_4 at 0.1 mol/L has to be found a good novel sorbent of cationic dye Rhodamine B from aqueous solution, the impregnation ratio which gives a maximum capacity of sorption of AS_S was 25 at an ambient temperature.

References

1. Caner N, Kiran I, Ilhan S (2009) Cansu Filik Iscen, Isotherm and kinetic studies of Burazol Blue ED dye biosorption by dried anaerobic sludge. J Hazard Mater 165:279–284
2. Ju DJ, Byun IG, Park JJ, Lee CH, Ahn GH, Park TJ (2008) Biosorption of a reactive dye (Rhodamine-B) from an aqueous solution using dried biomass of activated sludge. Bioresour Technol 99:7971–7975

3. Junxiong CAI, Longzhe CUI, Yanxin WANG, Chengfu LIU (2009) Effect of functional groups on sludge for biosorption of reactive dyes. J Environ Sci 21:534–538
4. Martin MJ, Artola A, Balaguer MD, Rigola M (2003) Activated carbons developed from surplus sewage sludge for the removal of dyes from dilute aqueous solutions. Chem Eng J 94:231–239
5. Kargi F, Ozmihci S (2005) Comparison of adsorption performances of powdered activated sludge and powdered activated carbon for removal of turquoise blue dyestuff. Process Biochem 40:2539–2544
6. Otero M, Rozada F, Calvo LF, Garcia AI, Moran A (2003) Elimination of organic water pollutants using adsorbents obtained from sewage sludge. Dyes Pigment 57:55–65

Chapter 47
Preferential CO Oxidation Over Ru/Al_2O_3-Coated Metal Monolith Catalyst for Small-Scale Fuel Processor

Kee Young Koo, Hyun Ji Eom, Un Ho Jung, and Wang Lai Yoon

Abstract Ru/Al_2O_3-coated FeCralloy monolith catalyst was applied for the preferential CO oxidation (PrOx) to reduce CO concentration less than 10 ppm in reforming process. FeCralloy monolith was pre-calcined at 900 °C after electrochemical surface treatment which results in the formation of uniform Al_2O_3 layer on the metal substrate. Pre-calcined monolith was coated with 10 wt% Al_2O_3 sol and followed by 1.2 wt% Ru/Al_2O_3 catalyst washcoating. The highly dispersed 1.2 wt% Ru/Al_2O_3 catalyst was prepared by the deposition-precipitation method using 5 wt% NaOH solution as a precipitant. The characterization as to surface area, metal dispersion, and reduction temperature of catalysts were analyzed by BET, CO-chemisorption and H_2-TPR. PrOx test was performed with GHSV of 5,000–30,000 h^{-1} at 100–200 °C. The $\lambda(2[O_2]/[CO])$ was adjusted between 1 and 2 and the effect of H_2O and CO_2 was examined at $\lambda = 2$. As a result, the metal dispersion of Ru coated on the FeCralloy monolith is higher than that of commercial pellet catalyst with the shape of sphere. The monolith catalyst shows higher CO conversion and CO_2 selectivity than the commercial catalyst due to the enhancement of thermal conductivity and the maximization of available Ru active site on the metal substrate. In addition, monolith catalyst has a superior tolerance to H_2O and CO_2. From this study, it is found that the Ru/Al_2O_3-coated monolith catalyst shows robust catalytic activity with 100 % CO conversion and 50 % CO_2 selectivity under 0.61 % CO, 0.61 % O_2, 59 % H_2, 19 % H_2O, 16 % CO_2, N_2 balance at GHSV = 5,000 h^{-1} from 100 to 160 °C in PrOx.

Keywords Preferential CO oxidation (PrOx) • Ru/Al_2O_3 • Deposition-precipitation • Monolith catalyst

---

K.Y. Koo • H.J. Eom • U.H. Jung • W.L. Yoon (✉)
Division of New and Renewable Energy Research, Hydrogen Laboratory, Korea Institute of Energy Research (KIER), 152, Gajeong-ro, Yuseong-gu, Daejeon 305-343, Republic of Korea
e-mail: wlyoon@kier.re.kr

I. Dincer et al. (eds.), *Progress in Clean Energy, Volume 1*,
DOI 10.1007/978-3-319-16709-1_47

47.1 Introduction

The Pt catalyst in the proton exchange membrane fuel cell (PMEFC) can be easily poisoned by irreversible CO adsorption. To prevent poisoning of the Pt catalyst, hydrogen fuel with a CO concentration of 10 ppm or less is needed [1–4]. Hydrogen is mostly produced by steam reforming of natural gas, but the reformed gas contains 5–15 % CO, and thus a CO removal process is required. The CO concentration of the reformed gas can be lowered to around 0.5–1 % through the water gas shift (WGS) reaction. However, the CO concentration must be maintained at around 10 ppm or lower in order for the Pt catalyst to operate stably, and this always requires an additional CO removal process. Generally, the preferential CO oxidation (PrOx) reaction is widely used as the most economical and effective CO removal process [1, 2, 4, 5]. The PrOx reaction is a fast exothermic reaction in which CO oxidation and H_2 oxidation occur competitively, and it makes it difficult to control the operating temperature [6]. Moreover, as the reaction temperature increases, side reactions such as H_2 oxidation, r-WGS, and methanation lower the CO conversion and CO_2 selectivity, so control of the reaction temperature is important. Therefore, a PrOx catalyst with a high CO conversion and CO_2 selectivity in a wide low temperature range of 100–200 °C, as well as a high tolerance to the H_2O and CO_2 present in the reformed gas is needed to effectively remove CO.

In general chemical processes including the PrOx reaction, a packed bed reactor with ceramic pellet catalysts has been used. However, the conventional packed bed reactor has a number of drawbacks including its low thermal conductivity, slow response characteristics, pressure drop, and channeling [2, 3, 7–9]. In particular, the PrOx reaction is a very fast catalytic surface reaction, and thus the diffusion of reactant gas to the catalyst surface becomes the rate limiting step. Therefore, only the active sites exposed on the support surface are involved in the reaction, lowering the utilization efficiency of the catalysts. As such, it entails the problem of increasing reactor volume as large amounts of pellet catalyst are required to obtain sufficient conversion [10]. Moreover, the temperature control in the catalyst bed is difficult due to the formation of localized hot spot and heat transfer limitation.

To overcome such problems, metal structured catalysts such as metal monolith have been studied as an alternative [1, 2, 7]. Since a metal structured catalyst provides a large geometric reaction surface area to reaction volume (S/V) and increases the heat transfer rate per unit volume of process flow (US/V), it enables a compact reactor design [11]. Moreover, the radial concentration and temperature gradient as well as pressure drop can be minimized by the well-behaved uniform flow of the metal structures through the regular channel. According to the previous reports concerning metal structured catalysts for PrOx reaction, CuO/CeO_2, Au, and Pt have been used as the active metals [1, 2, 7, 8, 12–14]. Snytnikov et al. [13] fabricated the microchannel reactor by washcoating the Cu/CeO_{2-x} catalyst on the stainless steel plate. In the PrOx reaction, the CO concentration was reduced from 1.5 vol.% to 10 ppm or less at $[O_2]/[CO] = 1.5$, WHSV = 240 L/g h, and temperature range of 230–240 °C, although the same H_2O and CO_2 concentrations as the actual reforming process existed. Laguna et al. [7] compared the catalyst activity of the CuO_x/CeO_2 powder catalyst and CuO/CeO_2 catalysts which were washcoated onto Fecralloy microchannel reactor in PrOx reaction. They explained that the microchannel reactor showed better CO conversion than the powder catalysts because effective heat removal restrained the H_2 oxidation and r-WGS reactions. There have also been other studies which used metal structured catalysts to which precious metals such as Pt and Au were deposited for the PrOx reaction [2, 12, 14]. Martínez et al. [14] compared the performance of powder catalysts with the monolith catalysts which were prepared by coating $Au\text{-}Al_2O_3$ and $Au\text{-}CeO_2$ on an austenitic stainless steel (AISI 304) monolith in the PrOx reaction. The results showed that the monolith catalysts had a superior CO conversion compared to the powder catalysts. Moreover, the 1 wt% $Au\text{-}CeO_2$/AISI 304 monolith catalyst showed the most outstanding catalyst activity even in the presence of H_2O because of the synergy effect between the oxidation capability of gold atoms and the redox property of the CeO_2 in the $Au\text{-}CeO_2$ catalysts. According to Sirijaruphan et al.'s report [12], the metal structured catalysts fabricated by washcoating the Fe-promoted $Pt/\gamma\text{-}Al_2O_3$ powder catalyst on the FeCrAlY metal foam showed more stable catalyst activity and higher CO selectivity than the powder catalysts in the PrOx reaction. However, as far as we know, few studies have used a metal structured catalyst coated with Ru/Al_2O_3 catalyst with outstanding CO oxidation activity at a low temperature.

In this study, we have investigated the characteristics of the PrOx reaction of the $Ru/\alpha\text{-}Al_2O_3$ washcoated FeCralloy monolith catalysts and the effect of H_2O and CO_2 addition as an important reaction variable in the PrOx reaction of actual reforming process. Moreover, we compared the catalyst activity and CO_2 selectivity at a low reaction temperature (100–200 °C) of the metal structured catalyst and an Ru/Al_2O_3 commercial pellet catalyst as a reference catalyst.

47.2 Experimental

47.2.1 Catalyst Preparation

FeCralloy (Fe:72.8, Cr:22, Al:5, Y:0.1, Zr:0.1 wt%) monolith used as a metal structured carrier was manufactured by rolling a flat metal strip and corrugated strip into the shape of a cylinder. The diameter of the cylinder was 22 mm, the height was 20 mm, and the cell density was 690 cpi (cells/in^2). The metal monolith was rinsed with acetone in ultrasonic bath for 30 min before the surface was treated electrochemically. After the preliminary treatment of the metal surface with proprietary method [10], it was calcined at 900 °C for 6 h. Pre-calcined monolith was repeatedly washcoated by 10 wt% Al_2O_3 sol solution until a sufficient amount of Al_2O_3 sol was loaded to the metal monolith. After which, the Al_2O_3-coated monolith was calcined again at 500 °C for 6 h. Al_2O_3 solution was prepared by

Fig. 47.1 Photos of (a) Ru/Al_2O_3-coated monolith catalyst and (b) commercial pellet catalysts

dissolving boehmite (DISPERAL P2, SASOL) in distilled water. Moreover, the viscosity of the sol solution was maintained constant by keeping the pH of the solution at 4.5. The Al_2O_3-coated monolith was washcoated by 1.2 wt% Ru/α-Al_2O_3 catalyst slurry, which was the mixture of 1.2 wt% Ru/α-Al_2O_3 catalyst powder, boehmite, and distilled water in ratios of 1:1:0.5, respectively. The washcoated monolith was dried at 150 °C for 1 h. The monolith catalyst was washcoated repeatedly up to the loaded catalyst amount of 1.0 g. 1.2 wt% Ru/α-Al_2O_3 catalyst was prepared by deposition-precipitation method using the α-Al_2O_3 and Ru nitrosyl nitrate solution ($Ru(NO)(NO_3)_x(OH)_y$, Aldrich) as a support and Ru metal precursor, respectively. α-Al_2O_3 support was prepared by the calcinations of γ-Al_2O_3 (SASOL) at 1,200 °C for 6 h. The pH of the solution was adjusted at 6.5 with 5 wt% NaOH (D.S.P. GR Reagent) solution as a precipitation agent [5]. The BET surface area, metal dispersion, and metal surface area of 1.2 wt% Ru/α-Al_2O_3 powder were 30.0 m^2/g, 91.1 %, and 3.0 m^2/g, respectively. For comparison, the 0.4 wt% Ru/α-Al_2O_3 commercial pellet catalyst was used as a reference catalyst. The commercial pellet catalyst was a spherical shape with a size of 1.5–2 mm. Figure 47.1 shows photos of 1.2 wt% Ru/α-Al_2O_3-coated monolith catalyst and 0.4 wt% Ru/α-Al_2O_3 commercial pellet catalyst.

47.2.2 Characterization

The surface area of catalysts was measured by BET (Belsorp max, BEL Japan Inc.) through the nitrogen adsorption at −196 °C. In the case of monolith catalyst, the coated samples were cut into a specific size to analyze the physicochemical properties. Temperature programmed reduction (TPR, BEL CAT B, BEL JAPAN Inc.) was carried out to measure the reduction temperature and H_2 consumption of catalysts. The sample of 0.1 g was loaded into a U-shape quartz reactor and heated by an electrical furnace at heating rate of 10 °C/min from room temperature up to

300 °C under 10 % H_2/Ar gas. H_2 consumption was estimated by integrating the peak area of the reduction profile. The metal dispersion and Ru particle size were measured by CO-chemisorption (BEL METAL-3, BEL JAPAN Inc.). The sample of 0.05 g in a U-shape quartz reactor was reduced with H_2 flow at 200 °C for 2 h and a CO pulse (10 % CO/He) was injected into the catalyst at 50 °C. The adsorbed CO amount obtained by assuming the adsorption stoichiometry of CO/Ru = 1 was used to estimate the Ru dispersion and particle size.

47.2.3 *Catalytic Tests*

To test the activity of the catalyst, the monolith catalysts were placed on the center of 1″ Incornel reactor; the thermocouple was positioned at the center of the bottom surface of the monolith to control the catalyst bed temperature. The monolith catalyst and pellet catalyst were reduced at 200 °C for 2 h under 10 % H_2/N_2 before the reaction. The PrOx reaction was performed at reaction temperature of 100–200 °C. The reactant feed was 0.61 % CO, 0.46–0.61 % O_2, 59 % H_2, N_2 balance, and the gas hourly space velocity (GHSV) was 5,000–30,000 h^{-1}. The effect of H_2O and CO_2 addition on the catalytic activity was investigated under 0.61 % CO, 0.61 % O_2, 59 % H_2, 19 % H_2O, 16 % CO_2, and N_2 balance. The $\lambda(2[O_2]/[CO])$ was adjusted between 1.5 and 2. The effluent was passed through a trap to condensate residual water, and the composition of gases was analyzed by online Micro-GC (Agilent 3000) equipped with a TCD detector. The Ru amount of commercial catalyst was filled equally into the amount of Ru metal coated over monolith catalyst for the comparison of catalytic activity in PrOx reaction. The CO conversion, O_2 conversion, CO_2 selectivity, and CH_4 yield are defined as follows:

$$\text{CO conversion } (\%) = \frac{[\text{CO}]_{\text{in}} - [\text{CO}]_{\text{out}}}{[\text{CO}]_{\text{in}}} \times 100 \tag{47.1}$$

$$\text{O}_2 \text{ conversion } (\%) = \frac{[\text{O}_2]_{in} - [\text{O}_2]_{\text{out}}}{[\text{O}_2]_{\text{in}}} \times 100 \tag{47.2}$$

$$\text{CO}_2 \text{ selectivity } (\%) = \frac{0.5 \times \{[\text{CO}]_{\text{in}} - [\text{CO}]_{\text{out}} - [\text{CH}_4]_{\text{out}}\}}{[\text{O}_2]_{\text{in}} - [\text{O}_2]_{\text{out}}} \times 100 \tag{47.3}$$

$$\text{CH}_4 \text{ yield } (\%) = \frac{[\text{CH}_4]_{\text{out}}}{[\text{CO}]_{\text{in}}} \times 100 \tag{47.4}$$

47.3 Results and Discussion

47.3.1 Catalyst Characterization

Table 47.1 summarizes the BET surface area, Ru metal dispersion, and particle size of the Ru/Al_2O_3-coated monolith catalyst and Ru/Al_2O_3 pellet catalyst. The BET surface area (2.7 m^2/g) of the pellet catalyst was smaller than the surface area (7.0 m^2/g) of the monolith catalyst. The measurement of metal dispersion and Ru particle size using CO-chemisorption analysis revealed that the metal dispersions of the monolith catalyst and pellet catalyst were 43.5 % and 19.6 %, respectively, whereas their Ru particle sizes were 3.1 nm and 6.8 nm, respectively. It was confirmed that small Ru metal particles were highly dispersed on the surface of the Ru/Al_2O_3-coated monolith catalyst. In general, highly dispersed Ru active metal improves CO oxidative activity in the PrOx reaction at low temperature [15, 16]. Therefore, the monolith catalyst with highly dispersed Ru is expected to show better CO oxidative activity than the commercial pellet catalyst at a low temperature.

Figure 47.2 presents the result of the TPR analysis to measure the reduction temperature of the monolith and pellet catalysts. Both catalysts were completely

Table 47.1 Characteristics of the Ru/Al_2O_3-coated monolith catalyst and Ru/Al_2O_3 pellet catalyst

Catalyst	BET (m^2/g)[a]	Metal dispersion (%)[b]	Ru particle size (nm)[b]
Monolith	7.0	43.5	3.1
Pellet	2.7	19.6	6.8

[a]Estimated from N_2 adsorption at −196 °C
[b]Estimated from CO-chemisorption at 50 °C

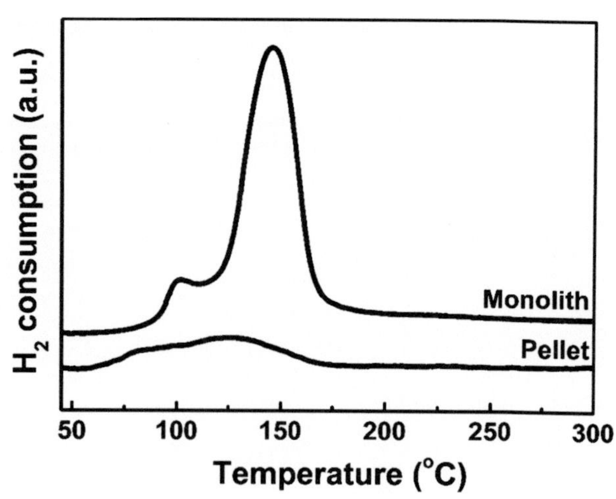

Fig. 47.2 H_2-TPR patterns of Ru/Al_2O_3-coated monolith catalyst and Ru/Al_2O_3 pellet catalyst

reduced at low temperature below 200 °C. This result is in agreement with the previous reports that all supported Ru catalysts were fully reduced at 250 °C in H_2 gas [17]. The reduction peak of the pellet catalyst was shown in a broad range of 65 and 175 °C. In the case of the monolith catalyst, two reduction peaks were observed at 100 and 150 °C. The reduction peak detected at below 100 °C was caused by the reduction of the free RuO_2 particle, whereas the TPR peak around 160 °C was attributed to the interaction between the RuOx and the support [18]. It indicates that the highly dispersed Ru catalysts are coated on the surface of metal monolith. The comparison of H_2 consumption estimated by integrating the area of the reduction peak found the H_2 consumption of the monolith catalyst to be much greater than that of the pellet catalyst. This means that the monolith catalyst has more reduced Ru active metal than the pellet catalyst. Such TPR result is consistent with the H_2-chemisorption result. In the case of the fast exothermic reaction on the catalyst surface such as PrOx reaction, the monolith catalyst, which has a large amount of Ru exposed on the support surface, has more active sites participating in the reaction; thus, it is expected to have better catalyst activity than the pellet catalyst.

47.3.2 Comparison of Monolith Catalyst and Pellet Catalyst

Figure 47.3 shows the performance of the Ru/Al_2O_3 pellet catalyst and Ru/Al_2O_3-coated monolith catalyst at different PrOx reaction temperatures. The reaction gas composition was 0.61 % CO, 0.46 % O_2, 59 % H_2, and N_2 as balance at GHSV = 15,000 h^{-1}. The loading amount of Ru metal in the monolith catalyst, and the pellet catalyst was the same. The CO conversion and O_2 conversion of the monolith catalyst were higher than those of the pellet catalyst throughout the reaction temperature range of 100–200 °C. The CO conversion of the monolith catalyst was as high as 95.9 % at low temperature of 120 °C, whereas the CO conversion of the pellet catalyst was low with 71.7 %. The commercial catalyst showed the highest CO conversion of 82.5 % at 140 °C. The reasons for the metal monolith catalyst showing higher CO oxidation reactivity at low temperature than the ceramic pellet catalyst were high metal dispersion, low-temperature reducibility, and good thermal conductivity of the metal monolith. Likewise, the CO_2 selectivity of the monolith catalyst was higher than the pellet catalyst. Note, however, that the CO_2 selectivity of the monolith catalyst nominally decreases at 200 °C, and such is attributed to methanation [19]. Although the methanation reaction is also the CO removal reaction, it consumes the hydrogen in the reformed gas; thus, lowering the hydrogen production efficiency. Therefore, the Ru catalyst must be designed to have stable and high CO oxidative activity at temperature of 200 °C or lower [5]. In other words, monolith catalyst with higher CO conversion and CO_2 selectivity than those of the pellet catalyst is more suitable for the PrOx reaction, which requires the catalyst to have outstanding CO oxidative activity at low temperature. The CO oxidative activity of the monolith catalyst is higher because it is better for the highly dispersed deposition of the Ru active metal, and it has more active metals involved in

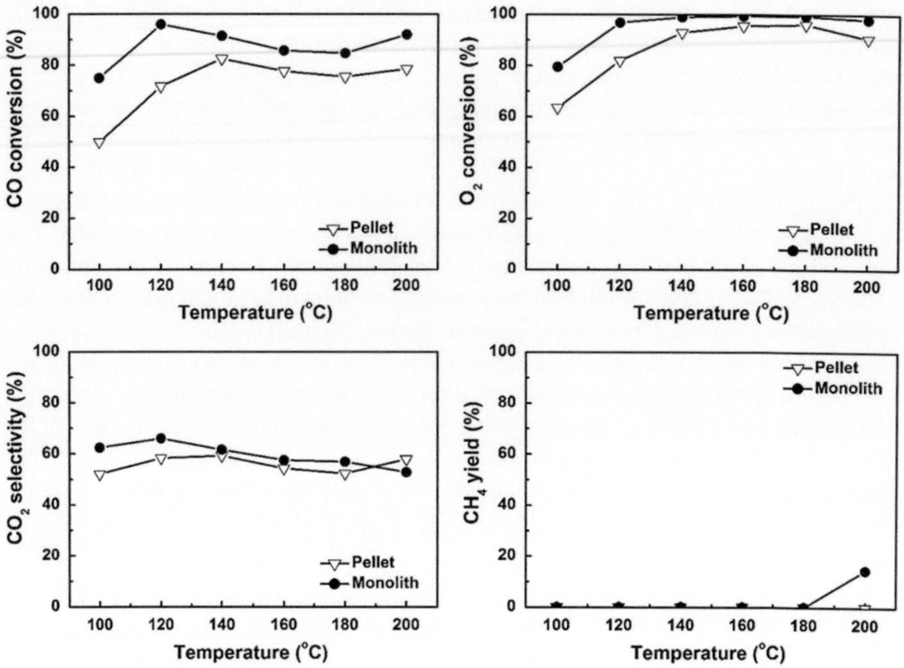

Fig. 47.3 Comparison of monolith catalyst and pellet catalyst with reaction temperatures in PrOx. (*Inverted filled triangle*): Ru/Al_2O_3 commercial pellet catalyst, (*filled circle*): Ru/Al_2O_3-coated monolith catalyst (Reaction condition: 0.61 % CO, 0.46 % O_2, 59 % H_2, N_2 as balance, GHSV = 15,000 h^{-1})

the reaction on the catalyst surface since it provides a higher ratio of geometric reaction surface area to reaction volume [2, 11]. Such reaction result is consistent with the characteristics of catalysts as described above.

Figure 47.4 presents the catalytic activity of the monolith catalyst and pellet catalyst with various GHSV in PrOx reaction. As GHSV increases from 5,000 to 30,000 h^{-1}, both CO conversion and CO_2 selectivity decreased. At high GHSV of 30,000 h^{-1}, the CO conversion of the monolith catalyst and pellet catalyst was 78.4 % and 66.4 %, respectively, indicating that the CO conversion of the monolith catalyst was higher. The monolith catalyst showed higher CO oxidative activity in PrOx reaction unlike the pellet catalyst because of better heat and mass transfer. In other words, the low CO oxidative activity of the pellet catalyst is attributed to the heat and mass transfer limitation [7, 10]. In detail, not all deposited Ru active metals of the pellet catalyst were involved in the reaction because of the low effectiveness factor, and temperature control was difficult since the heat generated by the fast exothermic reaction was not completely removed [7]. On the other hand, the Ru/Al_2O_3-coated monolith catalyst exhibited higher CO conversion and CO_2 selectivity in the PrOx reaction because of more active metals and higher thermal conductivity.

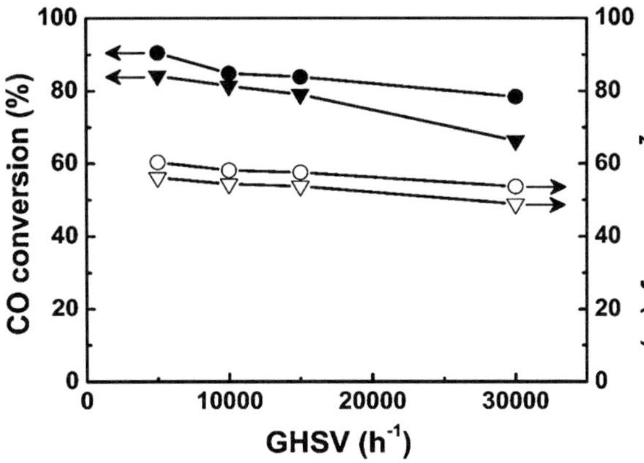

Fig. 47.4 Comparison of CO conversion and CO_2 selectivity with GHSV over Ru/Al_2O_3-coated monolith catalyst and Ru/Al_2O_3 commercial pellet catalyst in PrOx. (*Inverted filled triangle*): pellet catalyst, (*filled circle*): monolith catalyst (Reaction condition: 0.61 % CO, 0.46 % O_2, 59 % H_2, N_2 as balance, T = 160 °C)

47.3.3 *Effect of H_2O and CO_2 Addition*

In the reforming process, the reformed gas generated by the WGS reaction contains both H_2O and CO_2. Such impurities degrade the catalyst activity by blockage of the active sites of the PrOx catalyst or promoting the r-WGS as a side reaction [16, 19, 20]. Therefore, the PrOx catalysts applied in the reforming process require high resistances to impurities such as H_2O and CO_2. Figure 47.5 shows the effect of H_2O and CO_2 addition on the CO oxidative activity of the Ru/Al_2O_3-coated monolith catalyst and pellet catalyst in PrOx reaction. The composition of the reactant gas was 0.61 % CO, 0.61 % O_2, 59 % H_2, 19 % H_2O, 16 % CO_2, and N_2 as balance; the PrOx reaction was performed at GHSV = 15,000 h^{-1}. When H_2O and CO_2 were not present, the CO conversion of the monolith catalyst reached 100 % at temperature of 120 °C or above. When both H_2O and CO_2 were added together, however, CO conversion decreased throughout the reaction temperature range of 100–200 °C. As mentioned above, the blockage of catalyst active sites by the competitive absorption of H_2O and CO_2 impurities as well as the r-WGS reaction lowered the CO oxidative activity in PrOx reaction. Interestingly, adding CO_2 only did not affect the CO conversion at all; adding H_2O only decreased the CO conversion at temperature of 160 °C or above. This negative effect of H_2O addition is attributed to the blockage of catalyst active sites by the adsorbed H_2O and the formation of CO–H_2O complexes on the surface [3, 5].

When there were no H_2O and CO_2, the CO conversion of the Ru/Al_2O_3 commercial pellet catalyst reached 100 % at 140 °C. Unlike the Ru/ Al_2O_3-coated

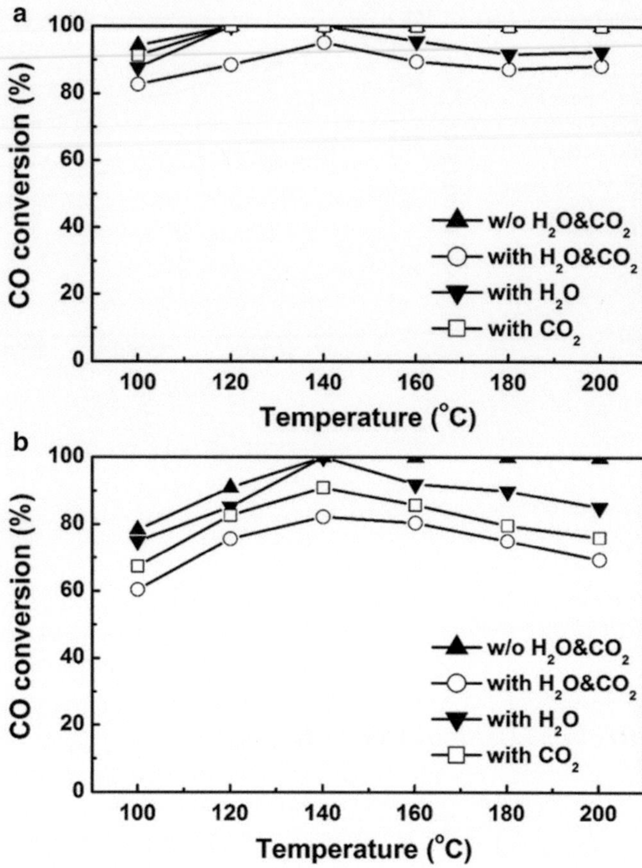

Fig. 47.5 Effect of H_2O and CO_2 addition on CO conversion of (**a**) Ru/Al_2O_3-coated monolith catalyst and (**b**) Ru/Al_2O_3 commercial pellet catalyst in PrOx reaction (Reaction condition: 0.61 % CO, 0.61 % O_2, 59 % H_2, 19 % H_2O, 16 % CO_2, N_2 as balance, GHSV = 15,000 h^{-1})

monolith catalyst, however, its CO conversion greatly decreased when any H_2O or CO_2 existed. In particular, when only CO_2 was added, the CO oxidative activity of the pellet catalyst greatly decreased, unlike the monolith catalyst whose CO conversion did not change at all. The decrease of CO conversion as the reaction temperature increases from 140 °C is attributed to the r-WGS as a side reaction. It is evident that the monolith catalyst has higher resistance to H_2O and CO_2 than the pellet catalyst. This is due to the fact that the highly dispersed Ru catalyst which is deposited on the monolith surface participates in the reaction.

47.3.4 Effect of $\lambda(2[O_2]/[CO])$

$\lambda(2[O_2]/[CO])$ is one of the most important reaction variables in the PrOx reaction. Since CO oxidation and H_2 oxidation occur competitively in a PrOx reaction, excess oxygen must be fed to remove CO completely. Figure 47.6 presents the performance of Ru/Al_2O_3-coated monolith catalyst at $\lambda = 1.5$ and $\lambda = 2$. At reaction temperature of 100–200 °C, higher CO conversion was observed at $\lambda = 2$ than at $\lambda = 1.5$. At low temperature below 120 °C, the difference in CO conversion at $\lambda = 1.5$ and $\lambda = 2$ was less than 4 %. There was little difference in catalyst activity with λ changed since the H_2 oxidation does not occur in the temperature range of 120 °C or below. At $\lambda = 1.5$, the CO conversion decreased as the reaction temperature increased, but O_2 conversion reached 100 % at 140 °C or above. It demonstrated that the H_2 oxidation as a side reaction occurred competitively [7]. This is in line with the decrease of CO_2 selectivity. At $\lambda = 1.5$ with lower oxygen amount, the decrease of CO_2 selectivity was clearer due to the competitively occurring H_2 oxidation as temperature increased from 120 to 200 °C. As the reaction temperature increased, the CO_2 selectivity slightly decreased at $\lambda = 2$. That is because CO in the

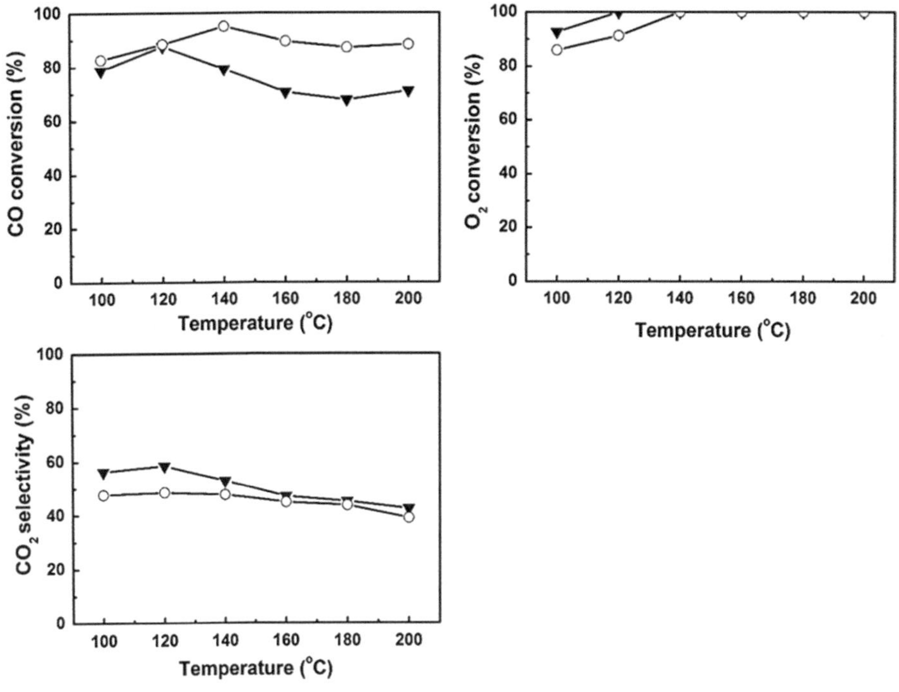

Fig. 47.6 Effect of $\lambda(2[O_2]/[CO])$ on catalytic activity of Ru/Al_2O_3-coated monolith catalyst in PrOx reaction. (*Open circle*): $\lambda = 2$ (*Inverted filled triangle*): $\lambda = 1.5$ (Reaction condition: 0.61 % CO, 0.46–0.61 % O_2, 59 % H_2, 19 % H_2O, 16 % CO_2, N_2 as balance, GHSV = 15,000 h^{-1})

reactant was completely removed by the excess oxygen supply. As described above, the increase in CO conversion and decrease of CO_2 selectivity at 200 °C are attributed to methanation.

47.3.5 *Operating Temperature Window*

Figure 47.7 shows the catalytic performance of 1.2 wt% Ru/α-Al_2O_3-coated monolith catalyst under 0.61 % CO, 0.61 % O_2, 59 % H_2, 19 % H_2O, 16 % CO_2, N_2 balance

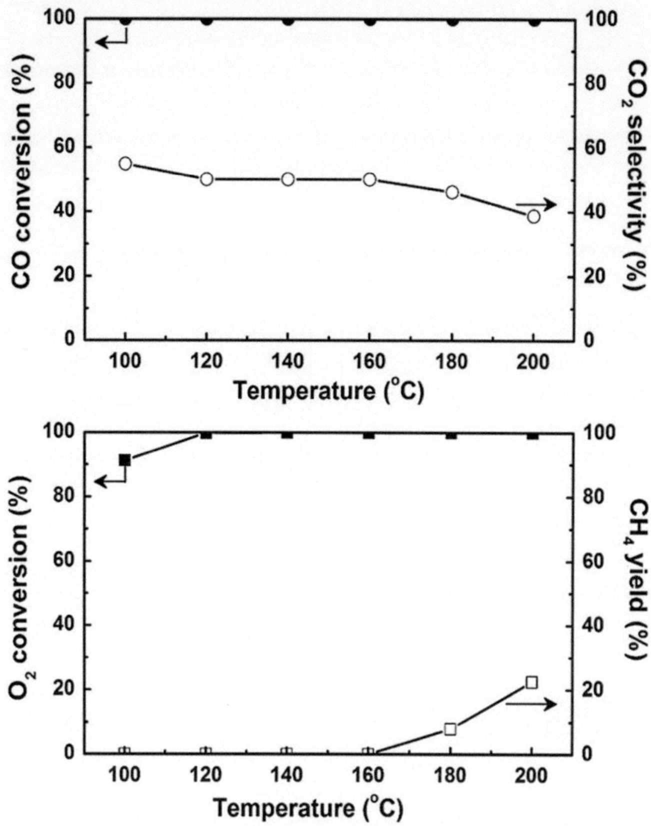

Fig. 47.7 Catalytic performance of Ru/Al_2O_3-coated monolith catalyst in PrOx reaction (Reaction condition: 0.61 % CO, 0.61 % O_2, 59 % H_2, 19 % H_2O, 16 % CO_2, N_2 as balance, GHSV = 5,000 h^{-1})

at GHSV = 5,000 h^{-1} with $\lambda = 2$. CO conversion and CO_2 selectivity in the range of reaction temperature between 100 and 200 °C reached at 100 % and 50 %, respectively. In addition, the O_2 conversion corresponds to the CO conversion except for that at 100 °C. Although O_2 conversion is lower than 100 %, CO_2 selectivity is greater than 50 wt% due to the complete CO oxidation. It is confirmed that CO_2 selectivity decreases at high temperature of over 180 °C because of the methanation reaction. Therefore, the 1.2 wt% Ru/α-Al_2O_3-coated monolith catalyst showed a good CO oxidation activity and CO_2 selectivity within a wide temperature window from 100 and 160 °C under the reaction condition of H_2O and CO_2 existence.

47.4 Conclusions

The FeCralloy monolith catalyst washcoated with 1.2 wt% Ru/Al_2O_3 catalyst was applied for the PrOx reaction to reduce CO concentration less than 10 ppm in reforming process. The Ru/Al_2O_3-coated monolith catalyst showed better CO conversion and CO_2 selectivity and superior tolerance of H_2O and CO_2 than Ru/Al_2O_3 commercial pellet catalyst with shape of sphere in PrOx. It was due to the fact that the CO oxidative activity of monolith catalyst was improved by the highly dispersed Ru catalyst on the surface of metal monolith and good thermal conductivity. In addition, monolith catalyst showed robust catalytic activity with 100 % CO conversion and 50 % CO_2 selectivity under 0.61 % CO, 0.61 % O_2, 59 % H_2, 19 % H_2O, 16 % CO_2, N_2 balance at GHSV = 5,000 h^{-1} from 100 to 160 °C in PrOx. Therefore, 1.2 wt% Ru/Al_2O_3-coated monolith catalyst is considered a promising catalyst with outstanding catalytic activity and high tolerance to H_2O and CO_2 applicable to the PrOx reaction at low temperature.

References

1. Zeng SH, Liu Y, Wang YQ (2007) CuO-CeO_2/Al_2O_3/FeCrAl monolithic catalysts prepared by sol-pyrolysis method for preferential oxidation of carbon monoxide. Catal Lett 117:119–125
2. Yu X, Li H, Tu S-T, Yan J, Wang Z (2011) Pt–Co catalyst-coated channel plate reactor for preferential CO oxidation. Int J Hydrog Energy 36:3778–3788
3. Zhang Q, Shore L, Farrauto RJ (2012) Selective CO oxidation over a commercial PROX monolith catalyst for hydrogen fuel cell applications. Int J Hydrog Energy 37:10874–10880
4. Ahluwalia RK, Zhang Q, Chmielewski DJ, Lauzze KC, Inbody MA (2005) Performance of CO preferential oxidation reactor with noble-metal catalyst coated on ceramic monolith for on-board fuel processing applications. Catal Today 99:271–283
5. Eom HJ, Koo KY, Jung UH, Rhee YW, Yoon WL (2010) Preparation of Highly Dispersed Ru/α-Al_2O_3 catalyst for preferential CO Oxidation. Trans Korean Hydrog New Energ Soc 21:390–397
6. Kim KH, Koo KY, Jung UH, Yoon WL (2012) Preferential CO oxidation over Ce-Promoted Pt/γ-Al_2O_3 catalyst. Trans Korean Hydrog New Energ Soc 23:640–646

7. Laguna OH, Domínguez MI, Oraá S, Navajas A, Arzamendi G, Gandía LM, Centeno MA, Montes M, Odriozola JA (2013) Influence of the O_2/CO ratio and the presence of H_2O and CO_2 in the feed-stream during the preferential oxidation of CO (PROX) over a CuO_x/CeO_2-coated microchannel reactor. Catal Today 203:182–187
8. Potemkin DI, Snytnikov PV, Belyaev VD, Sobyanin VA (2011) Preferential CO oxidation over Cu/CeO_{2-x} catalyst: internal mass transport limitation. Chem Eng J 176–177:165–171
9. Patcas FC, Garrido GI, Kraushaar-Czarnetzki B (2007) CO oxidation over structured carriers: a comparison of ceramic foams, honeycombs and beads. Chem Eng Sci 62:3984–3990
10. Koo KY, Joo H, Jung UH, Choi EJ, You S, Yoon WL (2011) Novel surface pretreatment for metal structured catalyst. Catal Today 164:52–57
11. Sirijaruphan A, Goodwin JG Jr, Rice RW, Wei D, Butcher KR, Roberts GW, Spivey JJ (2005) Metal foam supported Pt catalysts for the selective oxidation of CO in hydrogen. Appl Catal A Gen 281:1–9
12. Sirijaruphan A, Goodwin JG Jr, Rice RW, Wei D, Butcher KR, Roberts GW, Spivey JJ (2005) Effect of metal foam supports on the selective oxidation of CO on Fe-promoted $Pt/\gamma\text{-}Al_2O_3$. Appl Catal A Gen 281:11–18
13. Snytnikov PV, Potemkin DI, Rebrov EV, Sobyanin VA, Hessel V, Schouten JC (2010) Design, scale-out, and operation of a microchannel reactor with a Cu/CeO_{2-x} catalytic coating for preferential CO oxidation. Chem Eng J 160:923–929
14. Martínez LM, Frías DM, Centeno MA, Paúl A, Montes M, Odriozola JA (2008) Preparation of Au-CeO_2 and Au-Al_2O_3/AISI 304 austenitic stainless steel monoliths and their performance in the catalytic oxidation of CO. Chem Eng J 136:390–397
15. Kim YH, Park ED, Lee HC, Lee D (2009) Selective CO removal in a H_2-rich stream over supported Ru catalysts for the polymer electrolyte membrane fuel cell (PEMFC). Appl Catal A Gen 366:363–369
16. Kim YH, Park ED (2010) The effect of the crystalline phase of alumina on the selective CO oxidation in a hydrogen-rich stream over Ru/Al_2O_3. Appl Catal B Environ 96:41–50
17. Hadjiivanov K, Lavalley JC, Lamotte J, Maugé F, Saint-Just J, Che M (1998) FTIR study of CO interaction with Ru/TiO_2 catalysts. J Catal 176:415–425
18. Li D, Ichikuni N, Shimazu S, Uematsu T (1998) Catalytic properties of sprayed Ru/Al_2O_3 and promoter effects of alkali metals in CO_2 hydrogenation. Appl Catal A Gen 172:351–358
19. Niu T, Shen LM, Liu Y (2013) Preparation of meso-macroporous α-alumina using carbon nanotube as the template for the mesopore and their application to the preferential oxidation of CO in H_2-rich gases. J Porous Mater 20:789–798
20. Snytnikov PV, Sobyanin VA, Belyaev VD, Tsyrulnikov PG, Shitova NB, Shlyapin DA (2003) Selective oxidation of carbon monoxide in excess hydrogen over Pt-, Ru- and Pd-supported catalysts. Appl Catal A Gen 239:149–156

Chapter 48
Equilibrium and Kinetic Studies of Adsorption of Cd (II), Zn(II), and Cu(II) from Aqueous Solution into Cereal By-Products

S. Arris, F. Belaib, M. Bencheikh Lehocine, and H.-A. Meniai

Abstract The present study concerns the test of natural adsorbent obtained from a local cereal by-product as adsorbent for the removal of heavy metals such as Cu, Zn, and Cd ions from aqueous solutions. The solid support was used after calcinations at a temperature of 600 °C in exclusion of the air. The study was an opportunity to investigate the adsorption kinetics where equilibrium was reached after 30 min for the three metallic pollutants. The kinetic was of pseudo-second-order and controlled by intra-particle diffusion phenomenon. The isotherm of adsorption was also examined and showed a type C from Langmuir classification. The effect of important parameters such as the initial concentration and the contact time was also considered. The results showed a high retention capacity of the cereal by-product adsorbent, where yield values exceeding 90 % was reached for an initial concentration of 10 mg/L, at 20 °C, a mean size diameter of 0.1 mm, a mixing velocity of 600 rpm, a solid–liquid ratio of 10 g/L, a pH between 3 and 6, and a contact time of 2 h.

Keywords Removal • Heavy metals • Sorption • Low cost material

Nomenclature

K_1	Constant of pseudo-first-order model
K_2	Constant of pseudo-second-order model
K_{int}	Constant of intra-particle diffusion model
m	Mass of sorbent in g
n	Constant of Freundlich model

S. Arris (✉) • F. Belaib • M. Bencheikh Lehocine • H.-A. Meniai
Laboratory of the engineering and the processes of environment (LIPE),
Department of Environmental Engineering, Faculty of Pharmaceutical Process Engineering,
University Constantine 3, Constantine 25000, Algeria
e-mail: arris_s@yahoo.fr

I. Dincer et al. (eds.), *Progress in Clean Energy, Volume 1*,
DOI 10.1007/978-3-319-16709-1_48

q_t	Sorption capacity at time t (mg/g)
q_e	Sorption capacity at equilibrium (mg/g)
q_{max}	Maximale sorption capacity (mg/g)
v	Volume of solution in L
A, B	Constants of Temkin model
C_0	Initial metal concentration in mg/L
C_t	Metal concentration at time t in mg/L
K_L	Langmuir constant
K_F	Freundlich constant
P	Yield (%)
R^2	Correlation factor
R_L	Separation factor
T	Temperature °K
t	Temps (min)
α	Initial adsorption rate in mg g^{-1} min^{-1}
β	Desorption constant in g mg^{-1}

48.1 Introduction

Heavy metals are toxic to aquatic flora and fauna even in relatively low concentrations. Some metals can be assimilated, stored, and concentrated by organisms [1]. Industries, including mining and electroplating, discharge aqueous effluents containing high levels of such heavy metals as uranium, cadmium, mercury, and copper. Untreated effluents have an adverse impact on the environment [2]. Cadmium is very toxic, which can cause serious damage to the kidneys and bones. It is best known for its association with itai–itai disease [3]. Cadmium is classified as a soft acid. Cadmium ions have little tendency to hydrolyze at pH values below 8, but above 11 all cadmium exists as its hydroxo-complex [4]. In fresh water at pH 6–8, Cd (II) predominates. $CdOH^+$, $Cd(OH)_2$, $Cd(OH)_2$, $Cd(OH)^{-3}$, and $Cd(OH)_2^{-4}$ also exist depending upon the solution pH. The chloro-complexes $CdCl^+$, $CdCl_2$, and $CdCl^{-3}$ predominate in sea water and Cd (II) is present in very small amounts [4]. Free cadmium ions are highly toxic to plants and animals.

Cadmium accumulates in humans causing erythrocyte destruction, nausea, salivation, diarrhea, and muscular cramps, renal degradation, chronic pulmonary problems, and skeletal deformity [1]. The drinking water guideline recommended by World Health Organization and American Water Works Association (AWWA) is 0.005 mg Cd/ L. In many ground waters that contain bicarbonate/carbonate anions, the aqueous speciation of cadmium includes several complexes with bicarbonate/carbonate. Cadmium carbonate can be a solubility control for some high alkaline environments that contain high cadmium contamination. Copper usually occurs in nature as oxides and sulfides. In acidic environments, free aqueous Cu^{2+} dominates. At pH 6–8, the predominant species are Cu^{2+}, $Cu(OH)_2$, $CuHCO^{+3}$, $CuCO_3$, and $CuOH^+$, while at pH >10 the major species are $Cu(OH)_2^{-4}$ and Cu $(OH)^{-3}$ [4]. Mining wastes and acid mine drainage contribute significant quantities of dissolved copper to effluent streams. Other sources are fertilizer manufacturing, petroleum refining, paints and pigments, steelworks, foundries, electroplating and electrical equipment, brass, etc. Copper is an essential element but acute doses cause metabolic disorders. Inhalation of copper produces symptoms similar to those of silicosis and allergic contact dermatitis. Chronic copper poisoning cause hemolytic anemia, neurological abnormalities, and corneal opacity [4].

The reduction of the pollutant to an acceptable level is necessary when toxic metals are present in the aquatic system. Adsorption and ion exchange processes are the most useful methods to remove them. These methods explore the availability of different kinds of adsorbents associated with convenient procedures for obtaining high efficiency. A large number of different adsorbent materials containing a variety of attached chemical functional groups have been reported for this purpose. For instance, activated carbon is the most popular material; however, its high cost restricts its large-scale use. In recent years, special attention has been focused on the use of natural adsorbents as an alternative to replace the conventional adsorbents, based on both the environmental and the economical points of view [5]. Natural materials that are available in large quantities, or certain waste products from industrial or agricultural operations, may have potential as inexpensive sorbents. Due to their low cost, when these materials the end of their life time, they can be disposed of without expensive regeneration. The abundance and availability of agricultural by-products make them good sources of raw materials for natural sorbents.

48.2 Material and Methods

48.2.1 Sorbent

The cereal by-product was washed with distillated water and then dried at 110 °C in for 3 h. The carbon was obtained by cereal by-product carbonization at 600 °C during 1 h in exclusion of air in electrical furnace (HEARAEUS D-6450) HANAU/ Germany). After, the solid obtained was crushed in a crusher (FRITSCH industry. 86580 IdarOberstein). Finally, the solid material obtained is stored in desiccators for use.

48.2.2 Solution

A stock solution of Cu, Cd, and Zn were prepared (1,000 mg/L) by dissolving 1 g (AR grade) in 1.0 L of deionized water. The stock solution was diluted with deionized water to obtain the desired concentration range of Cu, Cd, and Zn solutions.

The pH was adjusted using HCl 0.1 M or NaOH 0.1 M solutions. All the used chemicals were of analytical grade.

48.2.3 Batch Biosorption Experiments

Batch adsorption experiments were carried out by shaking 0.5 g of carbonized by-product with 50 mL of metal solution of known concentration in Erlenmeyer flasks at specified temperatures in a thermostated water bath mechanical shaker. The flasks were removed after the desired contact time. A definite volume (5 mL) of the solution was withdrawn and quickly filtered through Whatmann No.40 filter paper. After, the filtrates collected were analyzed using atomic absorption (Absorption Atom is Spectrophotometer Spectra AA-20 plus). Finally, we have determinate the capacity of adsorption [6].

Effect of parameters such as the time of contact, the initial concentration, was studied. The amount of considered metal adsorbed q_t was calculated using the following equation [7]:

$$q_t = ((C_0 - C_t)V)/m \qquad (48.1)$$

where q_t is the amount of metal ions adsorbed on the biomass (mg/g), C_0: initial metal concentration in mg/L,

C_t: metal concentration at time t, m: mass of carbon in g, V: Volume of metal solution in L [8].

The percentage of removed Cr (VI) ions (R %) in solution was calculated using Eq. (48.2):

$$P(\%) = ((C_0 - C_t)/C_0) \times 100 \qquad (48.2)$$

48.3 Results and Discussion

48.3.1 Impact of Contacting Time and Kinetic Study

The studied phenomenon is the transfer of contaminants from the liquid phase to the solid phase, without any reaction in appearance. Kinetic experiment was conducted to decide time needed to reach adsorption equilibrium. The study consists of contacting, in batch, the pollutant with cereal by-product at different time, and the concentration of metallic ions in liquid phase was measured. The results for the three pollutants studied are shown in the Figs. 48.1, 48.2, and 48.3.

Equilibrium contact time was found to be 90 min for Cu (II), Zn (II), and Cd (II). The removal efficiency at these contact times was 90 %, 87 %, and 70 %, respectively. Figures 48.1, 48.2, and 48.3 show that biosorption of Cu, Zn and Cd were rapid within the first 30 min of contact time due to vacant sites on biosorbent. The rate of adsorption of metals gradually decreased with increase in contact time due to decrease of sorption sites. For three cations, q changes rapidly in the first, then we can see a plate indicating saturation.

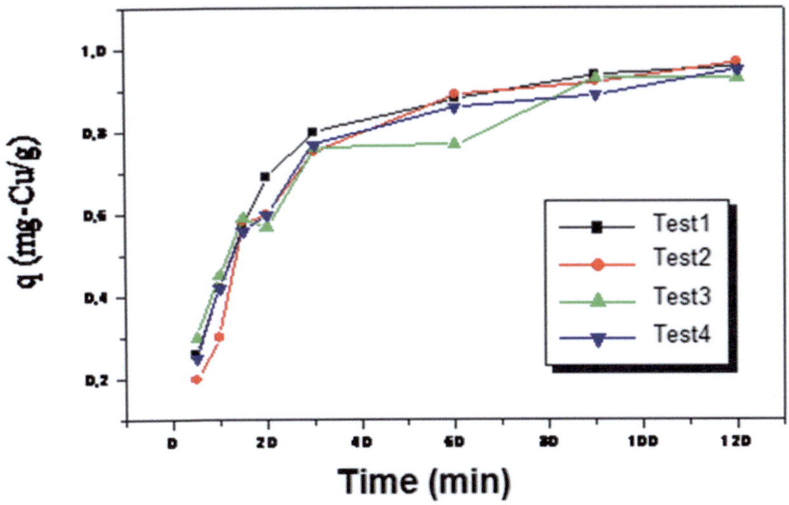

Fig. 48.1 Kinetic of Copper adsorption $v = 600$ tr/min, $r = 10$ g/L, $d < 0.1$ mm

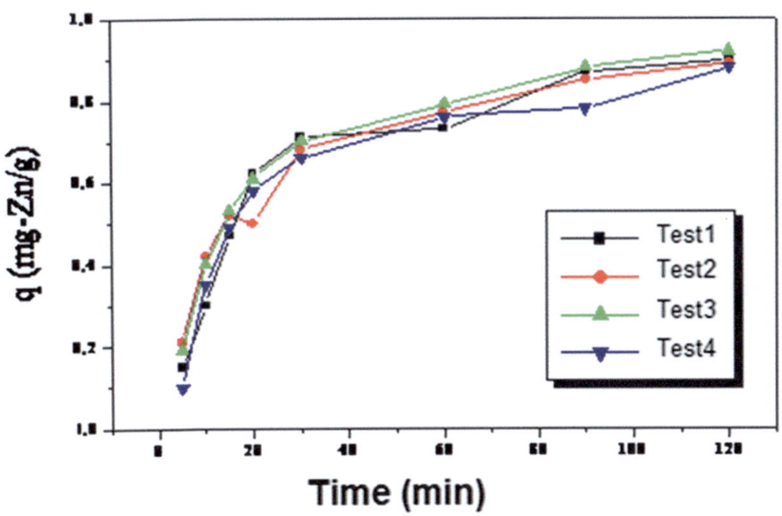

Fig. 48.2 Kinetic of Zinc adsorption $C_0 = 10$ mg/L, pH6, 20 °C, $C_0 = 10$ mg/L, pH6, 20 °C, $v = 600$ tr/min, $r = 10$ g/L, $d < 0.1$ mm

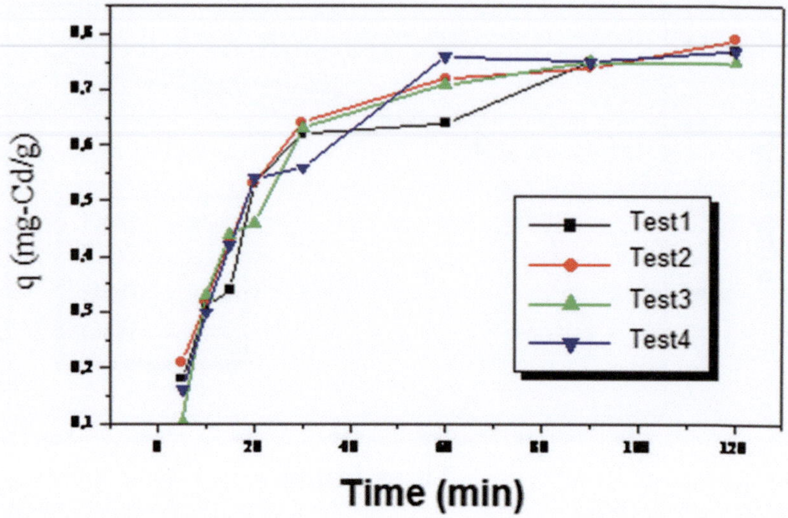

Fig. 48.3 Kinetic of Cadmium adsorption Conditions: $C_0 = 10$ mg/L, pH6, 20 °C, $v = 600$ tr/min, $r = 10$ g/L, $d < 0.1$ mm

Kinetic investigation was conducted by testing experimental data on pseudo-first, pseudo-second-order, Elovich and intra-particle diffusion kinetic models.

48.3.2 Lagergen Model

The pseudo-first-order kinetic model was proposed by Lagergren [9] and [10]. The integral form of the model was generally expressed as follows:

$$\mathrm{Log}(q_e - q_t) = \log q_e - \frac{K_1}{2.303}t \tag{48.3}$$

where q_e is the adsorption amounts at equilibrium (mg/g); q_t: the adsorption amounts at time t (mg/g); t: *time* (min), and K_1: the rate constant in the pseudo-first-order adsorption process (min-1). The constants were determined experimentally by plotting log ($q_e - q_t$) versus t Figs. 48.4, 48.5, and 48.6.

48.3.3 Pseudo-Second-Order Model

The pseudo-second-order kinetic model, proposed by Y. S. Ho and McKay [11], is based on the assumption that the adsorption follows second-order chemisorption. The linear form can be written as:

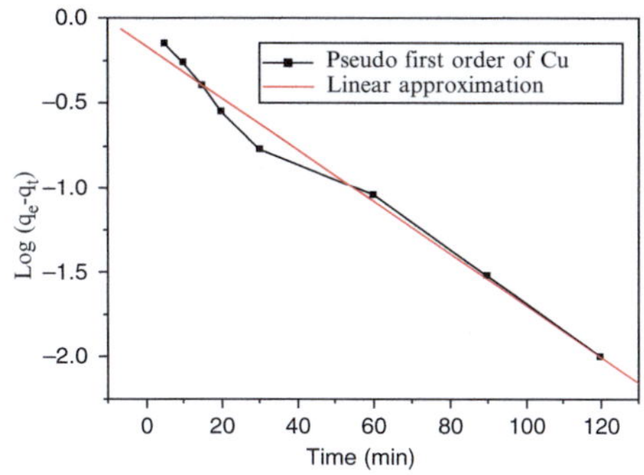

Fig. 48.4 Pseudo-first-order kinetic model of Cu sorption

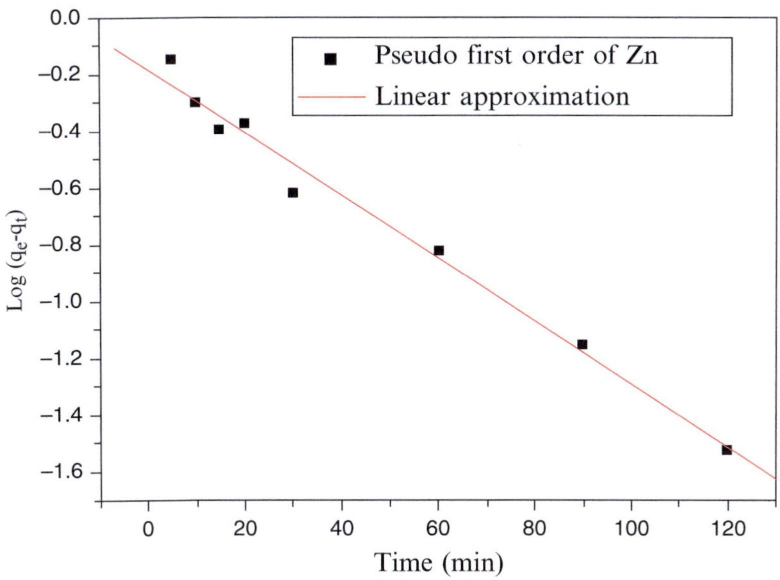

Fig. 48.5 Pseudo-first-order kinetic model of Zn sorption

$$\frac{t}{q_t} - \left(1/K_2 q_e^2\right) + \left(1/q_e^2\right)t \qquad (48.4)$$

where K_2 is the rate constant of adsorption (g/mg/min), q_e: the adsorbed metal amount (mg/g). By plotting a curve of t/q_t against t, q_e and K_2 can be evaluated Figs. 48.7, 48.8, and 48.9.

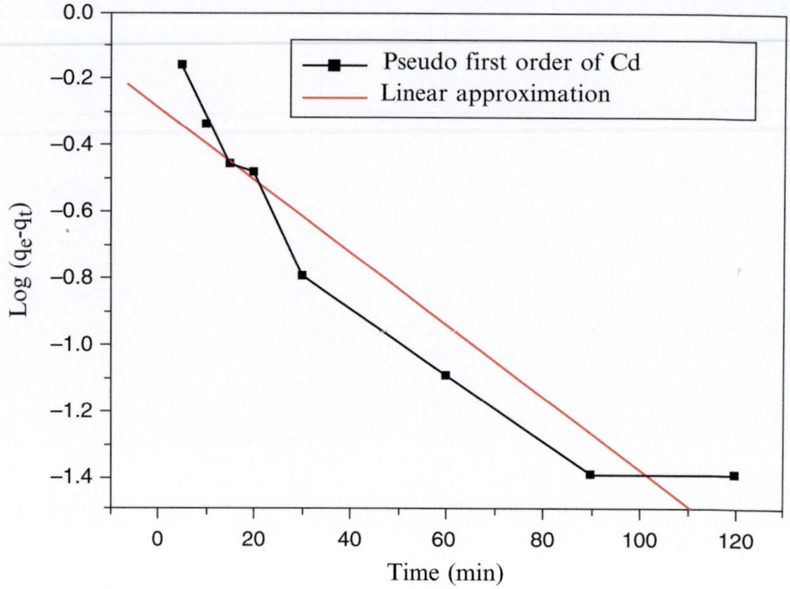

Fig. 48.6 Pseudo-first-order kinetic model of Cd sorption

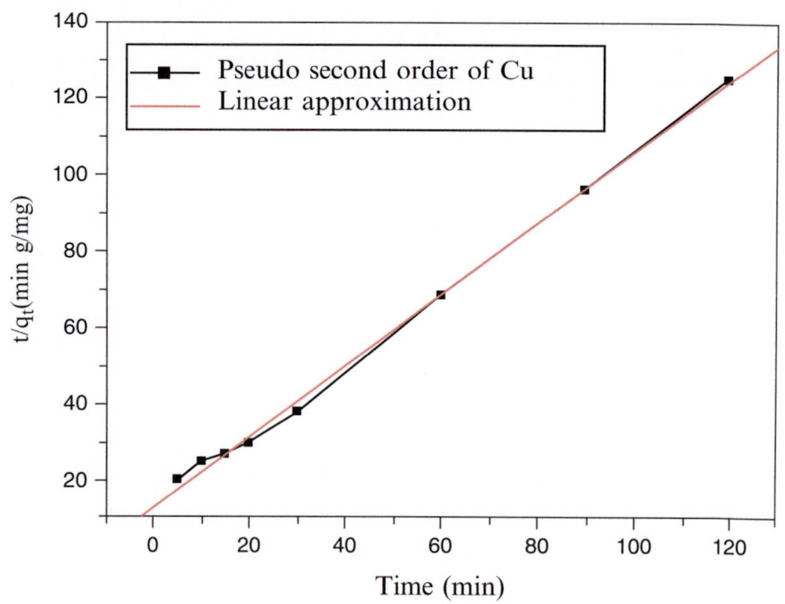

Fig. 48.7 Pseudo-second-order kinetic model of Cu sorption

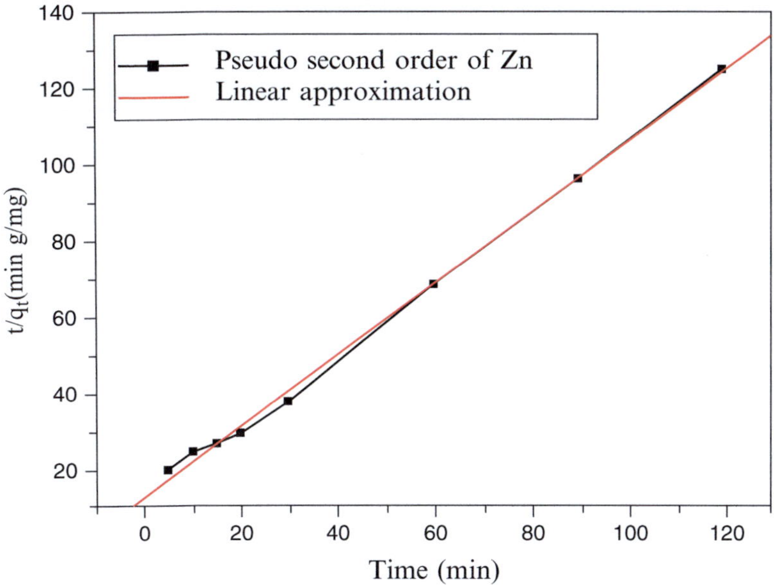

Fig. 48.8 Pseudo-second-order kinetic model of Zn sorption

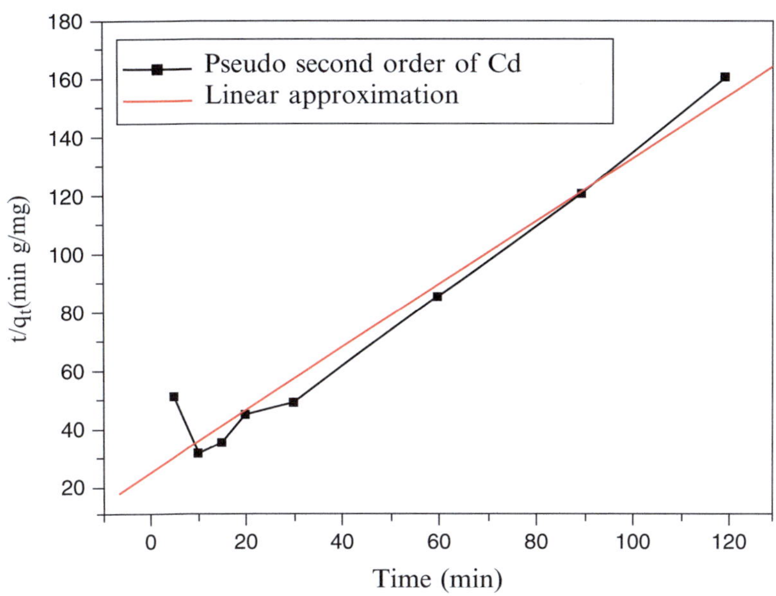

Fig. 48.9 Pseudo-second-order kinetic model of Cd sorption

48.3.4 *Intra-Particle Diffusion Model*

To investigate the internal diffusion mechanism during the adsorption of metal ions onto carbonized cereal by-product, the intra-particle diffusion equation has been used, considering that adsorption is usually controlled by an external film resistance and/or mass transfer is controlled by internal or intra-particle diffusion [12].

$$q_t = K_{int} t^{1/2} \tag{48.5}$$

where

q_t: the adsorbed metal amount in (mg/g),
K_{int}: the intra-particle diffusion rate constant (mg/g/min$^{1/2}$).
The coefficient K_{int} is determined from the initial linear slope of q_t versus time$^{0.5}$.

According to Weber and Morris (1963), if the intra-particle diffusion is the rate-limiting step in the adsorption process, the graph of q_t vs. $t^{0.5}$ should yield a straight line passing through the origin. McKay and Allen (1980) suggested that three linear sections on the plot q_t vs. $t^{0.5}$ can be identified. That means that two or three steps can occur. The first portion represents external surface adsorption or an instantaneous adsorption stage. The second portion is a gradual adsorption stage, where the intra-particle diffusion is the controlling factor. The third portion is a final equilibrium stage where the intra-particle diffusion starts to decelerate due to extremely low solute concentrations in the solution [13] Figs. 48.10, 48.11, and 48.12.

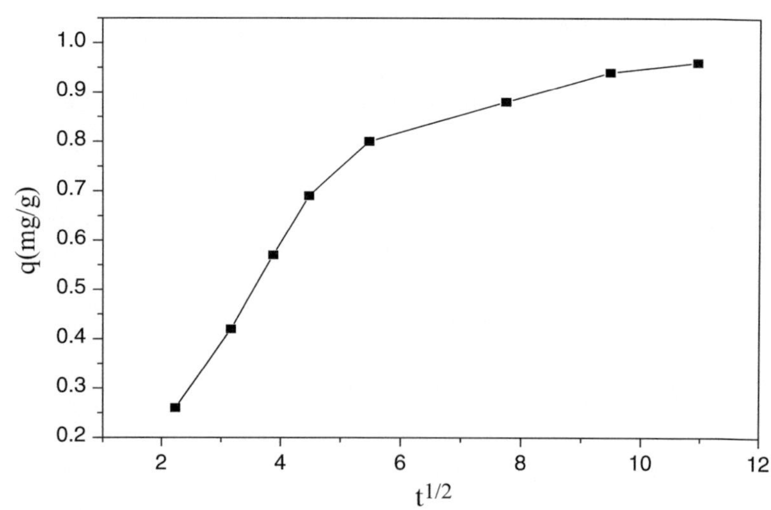

Fig. 48.10 Diffusional kinetic model of Cu sorption

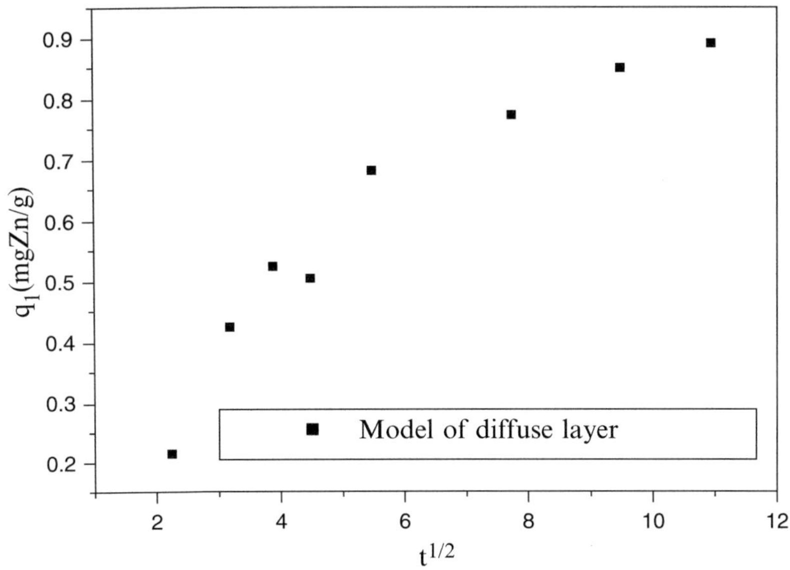

Fig. 48.11 Diffusional kinetic model of Zn sorption

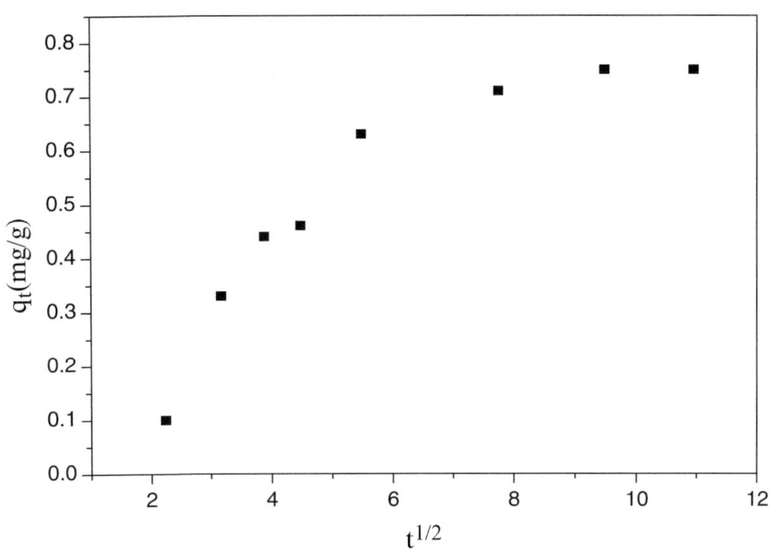

Fig. 48.12 Diffusional kinetic model of Cd sorption

48.3.5 Elovich Kinetic

The Elovich model equation is generally expressed as (Chien and Clayton, 1980; Sparks, 1986):

$$dq_t/dt = \alpha \exp(-\beta q_t) \tag{48.6}$$

where

α: initial adsorption rate $(mg \cdot g^{-1} \cdot min^{-1})$
β: desorption constant $(g \cdot mg^{-1})$ during any one experiment.
To simplify the Elovich equation, Chien and Clayton (1980) assumed $\alpha\beta t >> t$ and by applying the boundary conditions $q_t = 0$ at $t = 0$ and $q_t = q_t$ at $t = t$. Eq. (48.6) becomes:

$$q_t = 1/\beta(\ln(\alpha\beta) + 1/\beta \ln t \tag{48.7}$$

Elovich equation is commonly used to determine the kinetics of chemisorption of gases onto heterogeneous solids, and is quite restricted, as it only describes a limiting property ultimately reached by the kinetic curve [14]. In addition, the Elovich equation has also been used to describe the adsorption of pollutants from aqueous solutions in recent years.

If metal adsorption fits the Elovich model, a plot of q_t vs. ln(t) should yield a linear relationship with a slope of (1/β) and an intercept of (1/β) ln(αβ). Figures 48.13, 48.14, and 48.15 shows a plot of linearization form of Elovich model

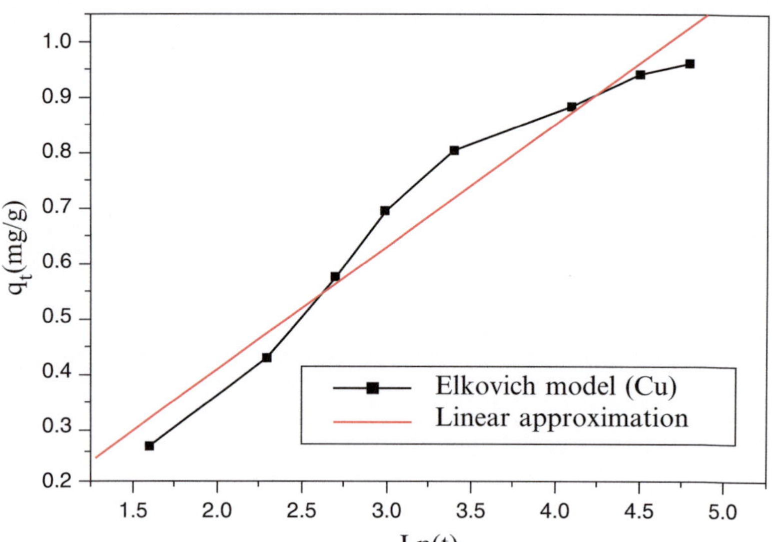

Fig. 48.13 Elovich kinetic model of Cu sorption

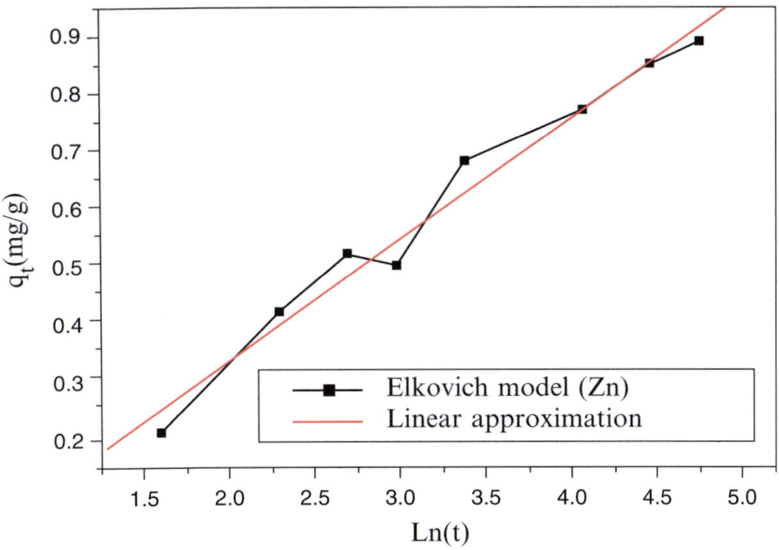

Fig. 48.14 Elovich kinetic model of Zn sorption

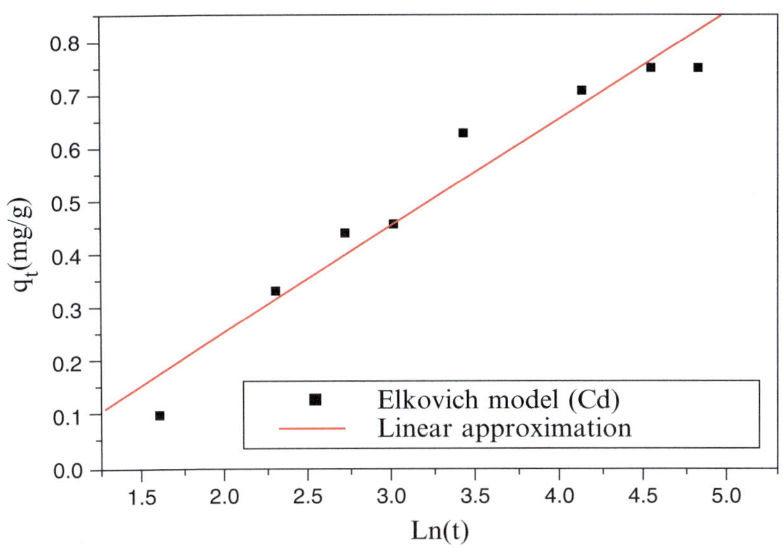

Fig. 48.15 Elkovich kinetic model of Cd sorption

at all concentrations studied. The slopes and intercepts of plots of q_t versus lnt were used to determine the constant β and the initial adsorption rate α.

From the linear regression of the previous models, we can see that graphically experimental data are conforming to the empirical pseudo-second-order model.

Table 48.1 Kinetic constants of Cu, Zn, and Cd

		Copper	Zinc	Cadmium
Pseudo-first-order model	$K_1(\text{min}^{-1})$	0.03111	0.02556	0.02372
	q_e(mg/g)	0.67928	0.65688	0.51604
	Correlation factor R^2	−0.99293	−0.9924	−0.95317
Pseudo-second-order model	$K_2(\text{g mg}^{-1}\text{ min}^{-1})$	0.26347	0.24392	0.22252
	q_e(mg/g)	1.07609	1.00855	0.92159
	Correlation factor R^2	0.99897	0.99833	0.97935
Diffusional model	k_{int}	0.17205	0.14034	0.15452
	Correlation factor R^2	0.9931	0.97654	0.97775
	D(cm^2/s)	$\leq 2.6\ 10^{-9}$	$\leq 2.6\ 10^{-9}$	$\leq 2.6\ 10^{-9}$
Elkovich model	α	0.17852	0.13998	0.09747
	β	4.44820	4.79547	4.92125
	Correlation factor R^2	−0.99244	0.98803	0.97055

And this is valid for three inorganic pollutants considered. This appears on the different values of the correlation factor. Low correlation coefficient values obtained for the pseudo-first-order model indicated that sorption does not occur exclusively on one site per ion.

The experimental value of the holding capacity at equilibrium q_e (q_e(Cu) = 1.092, q_e(Zn) = 1.046, and q_e(Cd) = 0.882 mg/g) is almost equal to the amount of the pseudo-second-order model (q_e(Cu) = 1.076, q_e(Zn) = 1.0085, and q_e(Cd) = 0.9215 mg/g). So we can conclude that the adsorption kinetics of copper, zinc, and cadmium is consistent with a pseudo-second-order kinetics. This suggests that the rate-limiting step may be the chemical adsorption not the mass transport limitation. The values of q_e, K_2, and R^2 are listed in Table 48.1. All the R^2 values are closer to "one," confirming the applicability of the pseudo-second-order equation.

According to the Weber model, the plot of uptake (q) versus the square root of time should be linear if intra-particle diffusion is involved in the adsorption process, and if these lines pass through the origin then intra-particle diffusion is the rate-controlling step. For the diffusional model, we calculated the slope of the straight part of the curve. This slope represents the rate constant k_{int}, the coefficient of diffusion D ($D = 0.03\ r_0^2/t^{1/2}$) for the three elements is the same because the equilibrium is reached after 30 min of contact for all three pollutants. This value are $\leq 2.603\ 10^{-9}$ cm s^{-1} ($r_0 \leq 0.0125$ cm); this indicates that the adsorption process can be controlled by the diffusional model. This result is consistent with the conclusion of Lecheng Lei and al.

From the table it can be seen, that the cereal by-product has affinity for the three ions in the following order: Copper > Zinc > Cadmium. This result is in agreement with the findings of Prashant Srivastava study on the competition adsorption of heavy metals on kaolinite.

48.3.6 Impact on Initial Concentration

In order to study the impact of the initial concentration of the pollutant on the retention phenomenon, we considered the follow values 1, 2, 3, 4, 5, 6, 7, 8, 9, 10, 11, 12, 15, 17, 20, and 25 mg/L. The results obtained for the three mineral elements are presented in Figs. 48.16, 48.17, and 48.18.

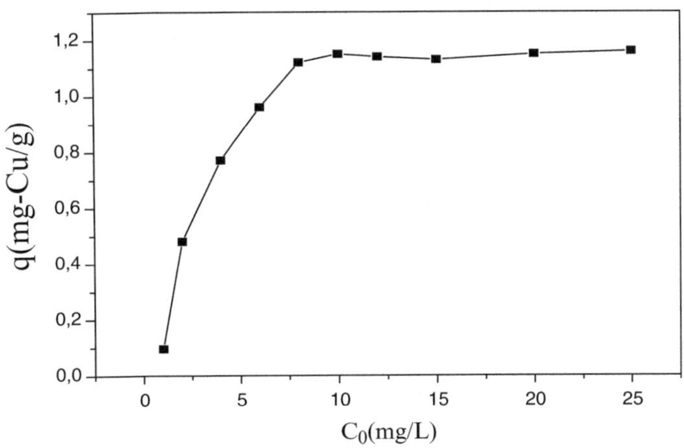

Fig. 48.16 Impact of initial concentration of Cu $t = 120$ min, pH6, 20 °C, $v = 600$ tr/min, $r = 10$ g/L, $d < 0.1$ mm

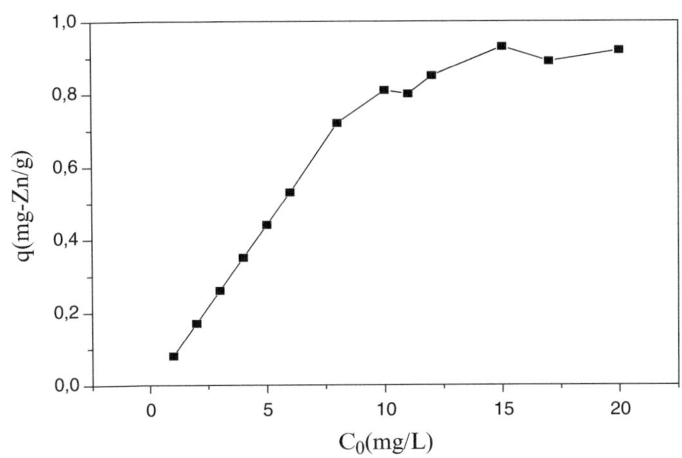

Fig. 48.17 Impact of initial concentration of Zn $t = 120$ min, pH6, 20 °C, $v = 600$ tr/min, $r = 10$ g/L, $d < 0.1$ mm

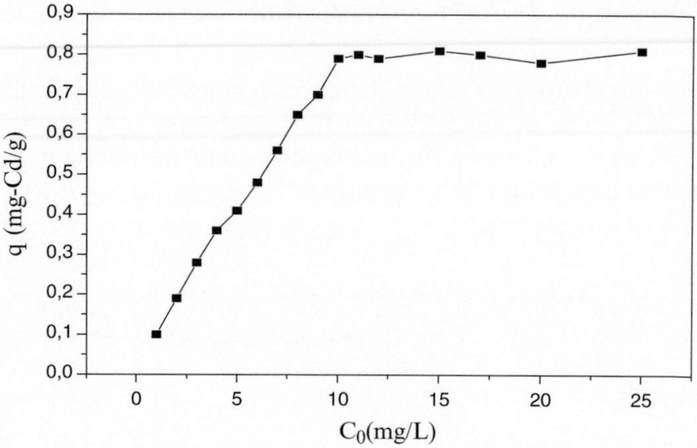

Fig. 48.18 Impact of initial concentration of Cd conditions: $t = 120$ min, pH6, 20 °C, $v = 600$ tr/min, $r = 10$ g/L, $d < 0.1$ mm

The curves show a relatively rapid increase in retention capacity at low concentration (1–10 mg/L). The fixation capacity continues to increase with increasing initial concentration indicating the existence of attraction forces between the exchange surface and metal cations. Then, a balance plate starts to appear as the initial value of the concentration reached 10 mg/L. This plateau may reflect saturation of active sites involved in the adsorption process.

48.3.7 Adsorption Isotherm

Study of adsorption isotherm helps to find the maximum adsorption capacity of the biomaterial. The maximum adsorption capacity of the biomaterial for Cu (II), Cd (II), and Zn (II) were investigated at the various concentrations from 1 to 25 mg/L of the respective metal ions. Figures 48.19, 48.20, and 48.21 represent plots of the adsorption amount (q_e mg/g) versus the equilibrium metal ions concentration in the solution (C_e mg/L).

Three equilibrium isotherm equations were used to find relation between equilibrium concentrations of adsorbate in liquid phase and in solid phase. These isotherms are:

48.3.7.1 Langmuir Isotherm

The Langmuir equation assumes that: the surface consists of adsorption sites, all adsorbed species interact only with a site and not with each other and, the adsorption is limited to monolayer. It is then assumed that once a metal ion occupies a site,

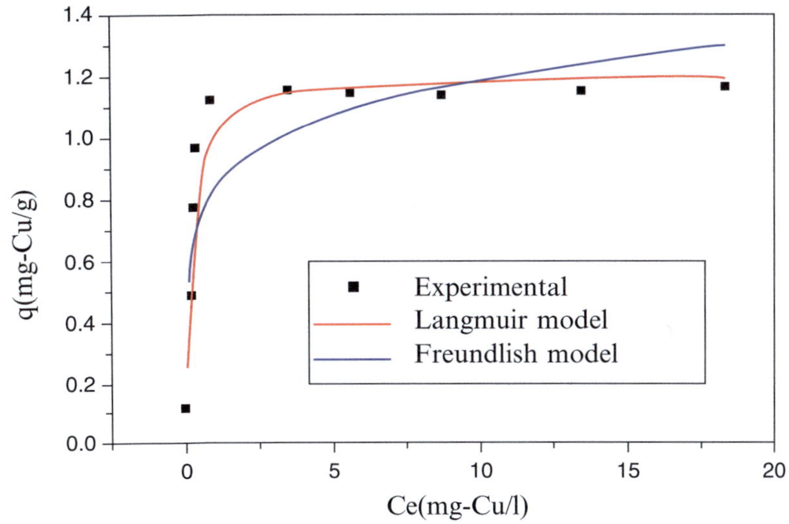

Fig. 48.19 Langmuir and Freundlich isotherm model of Cu

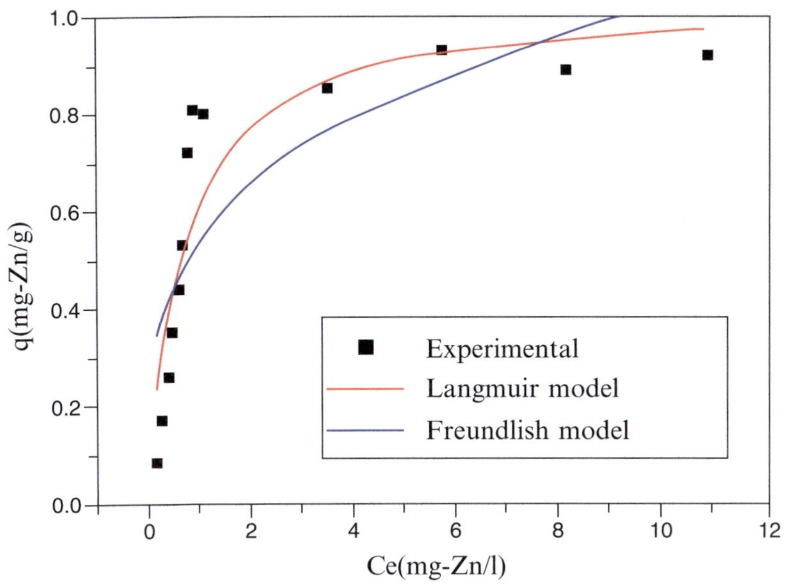

Fig. 48.20 Langmuir and Freundlich isotherm model of Zn

no further sorption can take place. This suggested that all binding sites are energetically equivalent, and there is no migration or interaction between the adsorbed ions on the surface area [15].

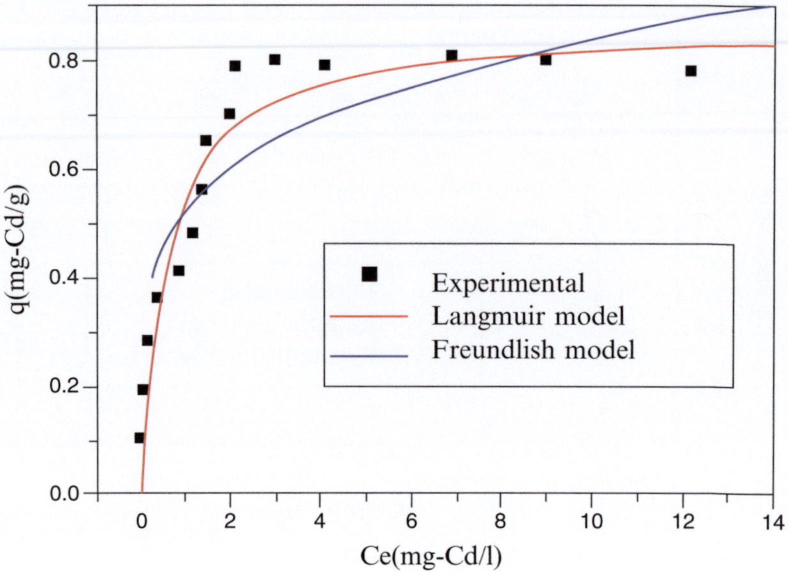

Fig. 48.21 Langmuir and Freundlich isotherm model of Cd

$$1/q_e = (1/K_L q_{max}) \cdot (1/C_e) + (1/q_{max}) \tag{48.8}$$

where

q_e: sorption capacity (mg/g) of ion sorbed,
C_e: the equilibrium concentration of ion in solution (mg/L),
q_{max}: the maximum amount of ion which can be taken up by biosorbent (mg/g),
K_L: the Langmuir constant (Lmg^{-1}).

48.3.7.2 Separation Factor (R_L)

A dimensionless constant, separation factor (R_L) can be used to predict whether a sorption system is favorable or unfavorable in batch adsorption process. R_L was calculated from Langmuir isotherm with the following equation [16].

$$R_L = 1(1 + K_L C_i) \tag{48.9}$$

where

K_L: the constant from Langmuir equation,
C_i: the initial concentration (mg/L).
The parameter, R_L, indicates the shape of the isotherm and the nature of the sorption process as given below:
$R_L > 1 =$ unfavorable isotherm.

$R_L = 1 =$ linear isotherm.
$R_L = 0 =$ irreversible isotherm.
$0 < R_L < 1 =$ favorable isotherm

The values of R_L for Cu(II), Zn(II), and Cd(II) were calculated at 10 mg/L and at 25 mg/L, and given in table. From the values it can be noted that adsorption process is favorable on low as well as high concentrations [12].

48.3.7.3 Freundlich Isotherm

The Freundlich isotherm is applicable to nonideal adsorption on heterogeneous surfaces. The general form of this model is:

$$q_e = K_F C^{1/n} \tag{48.10}$$

This can be linearized by taking logarithm of both sides of the equation to give:

$$\log q_e = \log K_F + 1/n \log C_e \tag{48.11}$$

where
K_F and $1/n$ are the Freundlich constants,
q_e: the sorption capacity (mg/g) of ion sorbed.

48.3.7.4 Temkin Isotherm

The Temkin isotherm assumes that the heat of adsorption for all the molecules in the layer decreases linearly with coverage due to adsorbent–adsorbate interactions, and that the adsorption is characterized by a uniform distribution of the binding energies, up to some maximum binding energy [3, 17]. The linear form of the Temkin isotherm is represented as:

$$q_e = (\mathrm{RT}/b)\ln \mathrm{A} + = (\mathrm{RT}/b)\ln C_e \tag{48.12}$$

where constant B1 = RT/b, which is related to the heat of adsorption, R is the universal gas constant (J/mol/K), T is the temperature (°K), b is the variation of adsorption energy (J/mol), and KT is the equilibrium binding constant (L/mg) corresponding to the maximum binding energy.

The adsorption increases with an increase in equilibrium concentration at low metal ions concentration and tends to approach constant value for each metal ion at their higher concentration, suggesting that these metal ions are adsorbed onto the biomaterial according to Langmuir adsorption. The adsorption of metal ions is

Table 48.2 Constants of Langmuir, Freundlich, and Temkin isotherm model of Cu, Zn, and Cd

Elements		Cuivre	Zinc	Cadmium
Langmuir model $q_e = a \cdot b \cdot C_e/(1 + b \cdot C_e)$	a	1.20625	1.03764	0.86541
	b	5.15936	1.46274	1.72114
	R_L	0.162–0.007	0.406–0.026	0.367–0.022
	R^2	0.92294	0.84085	0.92523
Freundlich model $q_e = a \cdot C_e^n$	a	0.83613	0.54344	0.52033
	n	0.15121	0.27612	0.20723
	R^2	0.64132	0.66734	0.77175
Temkin model $q_e = \text{B} \ln A + \text{B} \ln C_e$	a	259.233	15.2878	47.2111
	b	0.15443	0.20726	0.13703
	R^2	0.85973	0.8794	0.92527

eventually limited by the fixed numbers of active sites and a resulting plateau can be observed, which is well described by Langmuir parameter qmax [18].

As can be seen from the isotherms in Figs. 48.19, 48.20, and 48.21 and regression coefficients R^2 in Table 48.2, the Langmuir model shows the best fit compared to Freundlich and Temkin models. The Freundlich constants K_F and $1/n$ are calculated. The n value between 1 and 10 indicates good sorption efficiency but all the values obtained for copper, zinc, and cadmium: 0.1512, 0.2761, and 0.2072, respectively, are not included in this interval. The Langmuir model fitted better to experimental data in comparison to the Temkin model as represented from the value of its constants, q_{max} and R^2.

The conformity of the biosorption data to the Langmuir isotherm could be interpreted as an indication of a homogeny adsorption process, leading to monolayer binding. In addition all of the R_L values are between 0 and 1, which indicates that the biosorbent is a favorable adsorbent for the removal of Cu (II), Cd (II), and Zn (II) ions from aqueous solution Figs. 48.22, 48.23, and 48.24.

The value of the maximum sorption capacity of each metal is 1.092, 1.046, and 0.882 mg/g for copper, zinc, and cadmium, respectively. They reach the unit; except for Cadmium (This explains the superiority of the correlation factor). This is almost impossible for the Temkin model, which excludes low and high recoveries (valid in intermediate conditions). Sorption phenomena follow the Langmuir model, where the theoretical capacity of retention reached unit: 1.2062, 1.0376, and 0.8654 mg/g for copper, zinc, and cadmium.

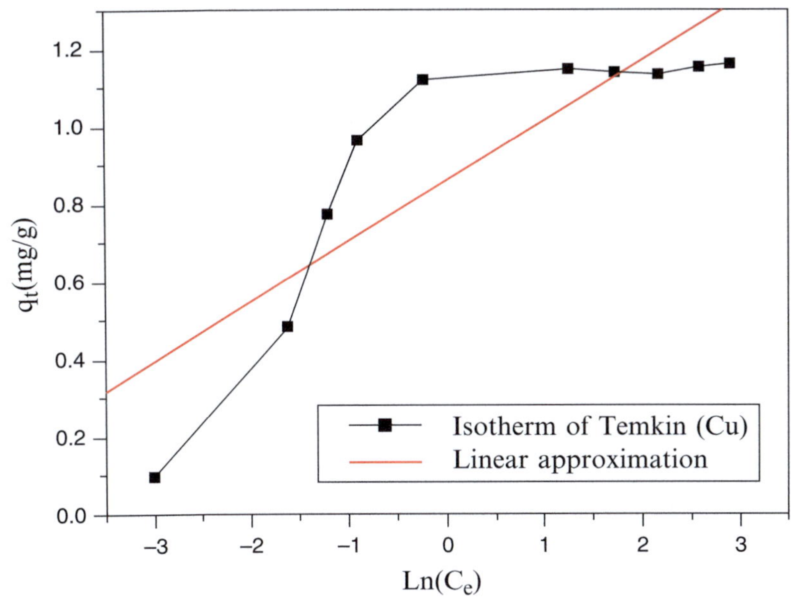

Fig. 48.22 Temkin isotherm model of Copper

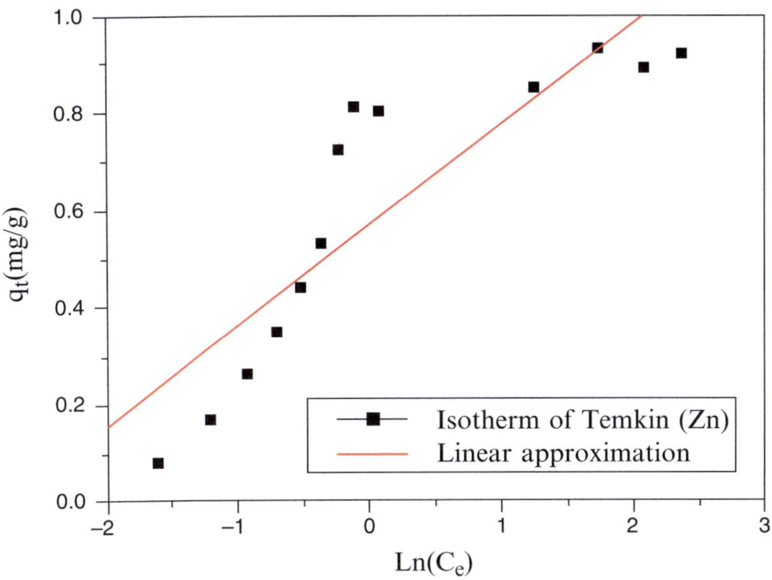

Fig. 48.23 Temkin isotherm model of Zinc

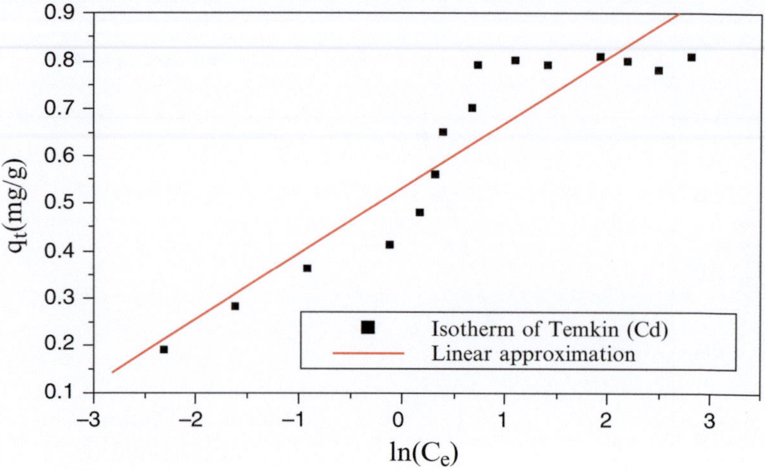

Fig. 48.24 Temkin isotherm model of Cadmium

48.4 Conclusions

The present study demonstrated that carbonized cereal by-product is an effective biosorbent for the removal of Cu (II), Zn (II), and Cd (II) ions from aqueous solutions by adsorption. The uptake of metal ions seemed to be quite rapid and the experimental data obeyed the pseudo-second-order model well. This is clearly confirmed by the values of correlation factors corresponding to each metal R^2: 0.99, 0.99, and 0.98 for Cu, Zn, and Cd, respectively. Also, the equilibrium data fit well to the Langmuir isotherm model. For a concentration of 10 mg/L, a temperature of 20 °C, particle size less than 0.1 mm, a stirring speed of 600 tr/min, a ratio of 0.5 g adsorbent/50 mL of solution (10 g/L), $3 < \text{pH} < 6$, the sorption yield reaches 90 %.

The results also show that carbonized cereal by-product has an affinity towards three elements in the following order: Copper > Zinc > Cadmium.

Based on the obtained results, it is believed that the application of the biosorption process using carbonized cereal by-product for the purification of industrial waste water containing Cu (II), Zn (II), and Cd (II) ions can be achieved in a bioreactor design or large-scale batch biosorption systems.

References

1. WHO, World Health Organization, Geneva Guidelines for Drinking Water Quality (1984) Toxicological profile for chromium. US Department of Health and Human Services Public Health Service Agency of Toxic Substances and Disease Registry, September, 2000
2. Yu MH (2005) Environmental toxicology- biological and health effects of pollutants, 2nd edn. CRC, Boca Raton

3. Manaham SE (2000) Environmental chemistry, 7th edn. CRC, Boca Raton
4. Mohan D, Singh KP (2002) Single- and multi-component adsorption of cadmium and zinc using activated carbon derived from bagasse- an agricultural waste. Water Res 36:2304–2318
5. Ghosh D, Bhattachacharyyra KG (2002) Adsorption of methylene blue on kaolinite. J Appl Clay Sci 20:295–300
6. LeHocine MB, Arris S, Meniai A-H, Morcellet M, Bacquet M, Martel M, Mansri A (2003) Study and identification of retention process of heavy metals by adsorption on agricultural by-products, chemical engineering technology. Eng Life Sci 3(2003): 9, Wiley, Germany
7. Zamouche M, Arris S, LeHocine MB, Meniai A-H (2014) Removal of Rhodamine B from water by cedar cone: effect of calcinations and chemical activation. Int J Hydrog Energy 39 (2014):1523–1531
8. Zou WH, Han RP, Chen ZZ (2006) Kinetic study of adsorption of Cu (II) and Pb (II) from aqueous solutions using manganese oxide coated zeolite in batch mode. Colloids Surf A Physicochem Eng Asp 279:238–246
9. Ho YS, McKay G (1998) Kinetic model for lead (II) sorption onto peat. Adsorpt Sci Technol 16:243–255
10. Ho YS, McKay G (1999) The sorption lead (II) ions on peat. Water Res 33:578–584
11. Ho YS, Porter JF, McKay G (2002) Equilibrium isotherm studies for the sorption of divalent metal ions onto peat: copper, nickel and lead single component systems. Water Air Soil Pollut 141:1–33
12. Ho YS (2003) Removal of copper ions from aqueous solution by tree fern. Water Res 37:2323–2330
13. Rengaraj S, Moon SH (2002) Kinetics of adsorption of Co(II) removal from water and wastewater by ion exchange. Water Res 36:1783–1793
14. Ruthven DM (1984) Principle of adsorption and desorption process. Wiley, New York
15. Subramanyam B, Das A (2009) Linearized and non-linearized isotherm models comparative study on adsorption of aqueous phenol solution in soil. Environ Sci Tech 6:633–640
16. Vasanthand K, Sivaneson S (2006) Equilibrium data, isotherm parameters and process design for partial and complete isotherm of methylene blue onto activated carbon. J Hazard Mater 137:237–244
17. Yabe MJS, Oliveira E (2003) Heavy metals removal in industrial effluents by sequential adsorbent treatment. Adv Environ Res 7:263–272
18. Orumwense FFO (1996) Removal of lead from water by adsorption on a kaolinitic clay. J Chem Tech Biotechnol 65:363–369
19. Ünlü N, Ersoz M (2006) Adsorption characteristics of heavy metal ions onto a low cost biopolymeric sorbent from aqueous solutions. J Hazard Mater 136(2):272–280

Chapter 49
Response Surface Modeling and Optimization of Chromium (VI) Removal from Aqueous Solution Using Date Stones Activated Carbon in Batch Process

F. Kaouah, S. Boumaza, T. Berrama, L. Brahmi, and Z. Bendjama

Abstract This chapter discusses response surface methodology (RSM) as an efficient approach for predictive model building and optimization of chromium adsorption on developed activated carbon. In this work, the application of RSM is presented for optimizing the removal of Cr (VI) ions from aqua solutions using activated carbon as adsorbent. All experiments were performed according to statistical designs in order to develop the predictive regression models used for optimization. The optimization of adsorption of chromium on activated carbon was carried out to ensure high adsorption efficiency at low adsorbent dose and high initial concentration of Cr (VI). While the goal of adsorption of chromium optimization was to improve adsorption conditions in batch process, i.e., to minimize the adsorbent dose and to increase the initial concentration of Cr (VI). In the adsorption experiments, a laboratory developed date stones activated carbon made of chemical activation (phosphoric acid) was used. A 2^4 full factorial central composite design experimental design was employed. Analysis of variance (ANOVA) showed a high coefficient of determination value and satisfactory prediction second-order regression model was derived. Maximum chromium removal efficiency was predicted and experimentally validated.

Keywords Adsorption • Activated carbon • Modeling • Optimization • Response surface method • Wastewater treatment

F. Kaouah (✉) • S. Boumaza • T. Berrama • L. Brahmi • Z. Bendjama
Laboratoire des sciences de génie des procédés industriels—Faculté de Génie Mécanique et de Génie des Procédés, USTHB, BP 32, El-Alia, Bab-Ezzouar 16111, Alger (Algérie)
e-mail: faridakaouah@yahoo.fr

I. Dincer et al. (eds.), *Progress in Clean Energy, Volume 1*,
DOI 10.1007/978-3-319-16709-1_49

Nomenclature

C_0 Initial concentrations of chromium solution (mg/L)
C_e Equilibrium concentrations of chromium solution (mg/L)
n Number of variables (dimensionless)
N Total number of tests (dimensionless)
X_1 Adsorbent dose (g/L)
X_2 Temperature (°C)
X_3 pH (dimensionless)
X_4 Initial feed concentrations of chromium (mg/L)
Xi Variables of action called factors
Y Response of the system adsorption of chromium (%)

49.1 Introduction

Heavy metal ions such as chromium, lead, cadmium, mercury, nickel, zinc, and copper are nonbiodegradable can be toxic and carcinogenic even at very low concentrations, and hence, usually pose a serious threat to the environmental and public health. The traditional methods for the treatment of chromium and other toxic heavy metal contaminated in wastewaters include complexation, chemical oxidation or reduction, solvent extraction, chemical precipitation, reverse osmosis, ion exchange, filtration, and membrane have disadvantages including incomplete metal removal, high consumption of reagent and energy, low selectivity, high capital and operational cost, and generation of secondary wastes that are difficult to be disposed off [1]. For these reasons, cost-effective alternative technologies for the treatment of metals contaminated waste streams are needed.

According to the ranking of metal interested priorities referred by Volesky [2], Cr (VI) is one of the most interesting heavy metal for removal and/or recovery considering the combination of environmental risk and reserve depletion. This metal and its compounds are widely used in many industries such as metal finishing, dyes, pigments, inks, glass, ceramics, and certain glues. It is also used in chromium tanning, textile, dyeing, and wood-preserving industries. The effluent from these industries contain hexavalent chromium, Cr(VI), at concentrations ranging from tens to hundreds of mg/L. Cr (VI)is considered by the IARC (International Agency for Research on Cancer) as a powerful carcinogenic agent that modifies the DNA transcription process causing important chromosomic aberrations. On the other hand, the presence of Cr (VI) in water causes significant environmental problems [3]. The U.S. Environmental Protection Agency recommends that the levels of Cr (VI) in drinking water should be less than 0.1 mg/L. It is therefore, essential to remove Cr (VI) from wastewater before disposal. Several methods are utilized to remove chromium from the industrial wastewater. These methods include reduction followed by chemical precipitation, ion exchange, reduction, electrochemical precipitation, solvent extraction membrane separation, reverse osmosis, and biosorption

[4]. These methods are cost-intensive and are unaffordable for a large-scale treatment of wastewater that is rich in Cr (VI). Adsorption using the activated carbon is an effective method for the treatment of industrial effluents contaminated with Cr (VI) and quite popular as compared to the other methods. The cost associated with the commercial activated carbon is very high which makes the adsorption process expensive. The cost of adsorption technology application can be reduced, if the adsorbent is inexpensive. Therefore, evaluation of biomass is getting increased attention in all over the world as it is renewable, widely available, cheap, and environmental friendly [5]. There are a number of biomass sources, such as forest residues, low grade plants, agricultural residues, and municipal solid wastes, which can be utilized for activated carbon precursor [6]. The adsorption of solutes from aqueous solution by any adsorbent is affected by several parameters such as the initial concentration of adsorbate, temperature, adsorbent dosage, and pH. This classical method does not depict the combined effect of all process parameters, is time consuming, and requires a number of experiments to determine optimum levels, which may be unreliable. These limitations of a classical method can be eliminated by optimizing all the process parameters collectively by statistical experimental design such as Response Surface Methodology (RSM) [7, 8].

This study reports the combined effect of four process parameters, such as adsorbent dose, pH, temperature, and adsorbat concentration, on the removal of chromium from synthetic wastewater by activated carbon as adsorbent prepared from date stones with chemical activation phosphoric acid using Central Composite Face-Centered Experimental Design in RSM.

49.2 Materials and Methods

49.2.1 Preparation of Chromium Solution and Analytical Methods

All the chemicals used are analytical grade. A stock solution of Cr (VI) is prepared by dissolving 2.8287 g of 99.9 % potassium dichromate ($K_2Cr_2O_7$) in 1,000 mL of solution. This solution is diluted as required to obtain the standard solutions containing 20–60 mg/L of Cr (VI). The solution pH is adjusted in the range of 2–10 by adding 0.1 N HCl or 0.1 N NaOH solutions as per the requirement and is measured by a pH meter. The concentration of free Cr (VI) ions in the effluent is determined spectrophotometrically by developing a purple-violet color with 1, 5-diphenyl carbazide in acidic solution as a complexing agent at 540 nm [9].

49.2.2 Activated Carbon Preparation

Date stones used for preparation of activated carbon was obtained locally. The precursor was first washed with distilled water and was then dried overnight at

105 °C. Chemical activation method using phosphoric acid was used to activate the raw materials. 20 g of raw materials was impregnated by certain amount of 60 wt% concentration phosphoric acid with occasional stirring. The amount of phosphoric acid solution used was adjusted to give an impregnation ratio of 3:1. After impregnation, the solution was filtered to take the residual acid. Subsequently, impregnated samples were dried in air oven at 105 °C for 48 h. Activation of phosphoric impregnated precursor was carried out at 600 °C with a carbonization time of 1 h under a nitrogen flow rate of 150 cm^3/min at a heating rate of 10 °C/min. After activation, the samples were cooled to room temperature under nitrogen flow and were washed sequentially several times with hot distilled water in order to remove any unconverted activating agent from carbonaceous material. The washing of the sample was continued until the pH of the filtrate was found to be 6–7. Finally, the samples were oven dried at 105 °C for 24 h and then stored in plastic containers.

49.2.3 *Batch Adsorption Experiment*

The experiments were performed in a thermal shaker at controlled temperature for a period of 2 h at 200 rpm using Erlenmeyer flask containing 100 mL of the aqueous solution of different initial concentration (20–60 mg/L) at different temperature (10–50 °C), at different weights (0.15–0.55 g), and at solution pH (2–10). The batch process was used so that there is no need of volume correction. Samples were taken out at regular interval and the residual concentration in solution was analyzed using spectrophotometer UV-Visible after filtering the adsorbent with Whatman filter paper to make it carbon free.

By knowing the chromium concentration at initial concentrations and equilibrium concentrations, the efficiency of adsorption of chromium by activated carbon can be calculated using the following equation for efficiency of adsorption chromium:

$$\text{Adsorption}(\%) = \frac{C_0 - C_e}{C_0} \times 100 \tag{49.1}$$

49.2.4 *Response Surface Methodology*

Response surface methodology (RSM) is a statistical method that uses quantitative data from appropriate experiments to determine regression model equations and operating conditions [10]. A standard RSM design called central composite design (CCD) was applied in this work to study the variables for adsorption of chromium from aqueous solution using prepared activated carbon in a batch process. This method is suitable for fitting a quadratic surface, and it helps to optimize the effective parameters with a minimum number of experiments, as well as to analyze the interaction between the parameters [11]. Generally, the CCD consists of a 2n factorial runs with 2n axial runs and n_c center runs (six replicates). The dependant

Table 49.1 Experimental range and levels of independent variables for adsorption of Cr (VI)

Variables	Symbol	$-\alpha$	-1	0	+1	$+\alpha$
Adsorbent dose (g/L)	X_1	1.5	2.5	3.5	4.5	5.5
Temperature (°C)	X_2	10	20	30	40	50
pH	X_3	2	4	6	8	10
Initial concentration (mg/L)	X_4	20	30	40	50	60

variables selected for this study were adsorbent dose (X_1, g/L), temperature (X_2, °C), pH (X_3), and initial concentration of chromium (X_4, mg/L). These four variables together with their respective ranges were chosen based on some preliminary studies. The range of the factors and levels, which were varied according to the experimental design, are given in Table 49.1.

This method requires factorial runs (2^n), axial runs ($2n$), and center runs (6), where n is the number of parameters [12–14]. Therefore, the number of experiments required for this study is 30 Eq. (49.2):

$$N = 2^n + 2n + 6 \tag{49.2}$$

Center runs include six replications which are performed by setting all factors at their midpoints to estimate the residual error and the reproducibility of the data. The axial points are located at ($\pm$ α, 0, 0), (0, $\pm$ α, 0), and (0, 0, $\pm$ α), where α is the distance of the axial point from center and makes the design rotatable. In this study, α value was fixed at 2 (rotatable). The experimental sequence was randomized in order to minimize the effects of the uncontrolled factor. The response was used to develop an empirical model which correlated the response to the adsorption of chromium from aqueous solution using prepared activated carbon in batch process variables using a second-degree polynomial equation such as Eq. (49.3):

$$Y = b_0 + \sum_{i=1}^{n} b_i X_i + \sum_{i=1}^{n} b_{ii} X_i^2 + \sum_{i=1}^{n-1} \sum_{j=i+1}^{n} b_{ij} X_i X_j \tag{49.3}$$

where Y is the predicted response, b_o the constant coefficient, bi the linear coefficients, $b_{i,j}$ the interaction coefficients, $b_{i,i}$ the quadratic coefficients, and $x_{i,}$,x_j are the coded values of the adsorption of chromium on prepared activated carbon variables.

49.3 Results and Discussion

49.3.1 Development of Regression Model Equation

Experiments were planned to obtain a quadratic model consisting of 2^4 trials plus a star configuration ($\alpha = \pm 2$) and there replicates at the center point. The design of

Table 49.2 Experimental design

Run	Coded level of variables X_1	X_2	X_3	X_4	Adsorption Y (%)
1	−1	−1	−1	−1	35.55
2	1	−1	−1	−1	64.63
3	−1	1	−1	−1	51.45
4	1	1	−1	−1	69.74
5	−1	−1	1	−1	9.45
6	1	−1	1	−1	48.85
7	−1	1	1	−1	19.05
8	1	1	1	−1	50.20
9	−1	−1	−1	1	34.05
10	1	−1	−1	1	38.37
11	−1	1	−1	1	67.42
12	1	1	−1	1	43.17
13	−1	−1	1	1	28.95
14	1	−1	1	1	45.85
15	−1	1	1	1	27.05
16	1	1	1	1	49.05
17	−2	0	0	0	38.30
18	2	0	0	0	84.43
19	0	−2	0	0	21.33
20	0	2	0	0	40.80
21	0	0	−2	0	83.14
22	0	0	2	0	31.33
23	0	0	0	−2	86.15
24	0	0	0	2	61.56
25	0	0	0	0	80.70
26	0	0	0	0	80.69
27	0	0	0	0	80.69
28	0	0	0	0	80.69
29	0	0	0	0	80.69
30	0	0	0	0	80.70

this experiment is given in Table 49.2, together with the experimental results. Regression analysis was performed to fit the response function of adsorption of chromium (%). The model expressed by Eq. (49.3), where the variables take their coded values, represents adsorption (Y) as a function of adsorbent dose (X_1), temperature (X_2), pH (X_3), and initial concentration (X_4). The final empirical model in terms of coded factors for adsorption of chromium (Y) is given in Eq. (49.4):

$$Y = 80.69 + 9.55X_1 + 4.60X_2 - 9.56X_3 - 2.67X_4 - 2.65X_1X_2 + 5.12X_1X_3 - 6.18X_1X_4 - 2.93X_2X_3 + 0.47X_2X_4 + 3.86X_3X_4 - 7.03X_1^2 - 14.60X_2^2 - 8.07X_3^2 - 3.91X_4^2 \quad (49.4)$$

The coefficient with one factor represents the effect of the particular factor, while the coefficients with two factors and those with second-order terms represent the interaction between two factors and quadratic effect, respectively. The positive sign in front of the terms indicates synergistic effect, whereas negative sign indicates antagonistic effect. The quality of the model developed was evaluated based on the correlation coefficient, R^2 value. In fact, the model developed seems to be the best at high R^2 statistics which is closer to unity as it will give predicted value closer to the actual value for the responses. In this experiment, the R^2 value for Eq. (49.4) was 0.901. This indicated that 90.1 % of the total variation in the chromium removal was attributed to the experimental variables studied.

49.3.2 Analysis of Variance

The significance and adequacy of the model were further justified through analysis of variance (ANOVA). The ANOVA of the regression model demonstrates that the model is highly significant as evident from the calculated F-value and a very low probability value. If the value of Prob. > F less than 0.05, the model terms are considered as significant [15]. The ANOVA for the quadratic model for chromium removal of date stones activated carbon is listed in Table 49.3. The model F-value of 9.84 and Prob. > F of less than 0.0001 implied that this model was significant. In these cases, X_1, X_2, X_3, X_1^2, X_2^2, X_3^2 and X_1X_4 factors were significant model terms where as X_4, X_4^2, X_1X_2, X_1X_3, X_2X_3, X_2X_4, and X_3X_4 were insignificant to the response. From the statistical results obtained, it was shown that the above model was adequate to predict the chromium removal within the range of variables studied. In addition, Fig. 49.1 shows the predicted values versus the experimental values for chromium removal. It can be seen that the model developed was successful in capturing the correlation between the variables for adsorption of chromium to the response when the predicted values obtained were quite close to the experimental values.

49.3.3 Chromium Removal

To investigate the effects of the four factors on the adsorption of chromium, the response surface methodology was used, and three-dimensional plots were drawn. Based on the ANOVA results obtained, adsorbent dose, pH, and temperature were found to have significant effects on the adsorption of chromium, with adsorbent

Table 49.3 Analysis of variance (ANOVA) for response surface quadratic model for adsorption of Cr (VI)

Source	Degree of freedom (DF)	Sum of squares	Mean square	F-Value	Prob. > F
Model	14	13,921.57	994.40	9.84	<0.0001[a]
*X*1	1	2,216.64	2,216.64	21.94	0.00029[a]
*X*2	1	493.86	493.86	4.89	0.04299[a]
*X*3	1	2,224.34	2,224.34	22.02	0.00029[a]
*X*4	1	163.75	163.75	1.62	0.22236
*X*1***X*2	1	121.06	121.06	1.20	0.29095
*X*1***X*3	1	405.12	405.12	4.01	0.06365
*X*2***X*3	1	128.99	128.99	1.28	0.27624
*X*1***X*4	1	593.53	593.53	5.87	0.02847[a]
*X*2***X*4	1	2.26	2.26	0.02	0.88316
*X*3***X*4	1	226.58	226.58	2.24	0.15499
*X*1***X*1	1	1,350.53	1,350.53	13.37	0.00234[a]
*X*2***X*2	1	5,840.25	5,840.25	57.81	<0.0001[a]
*X*3***X*3	1	1,777.21	1,777.21	17.59	0.00078[a]
*X*4***X*4	1	416.01	416.01	4.12	0.06057
Residual	15	1,515.45	101.03	–	–

[a]Significant

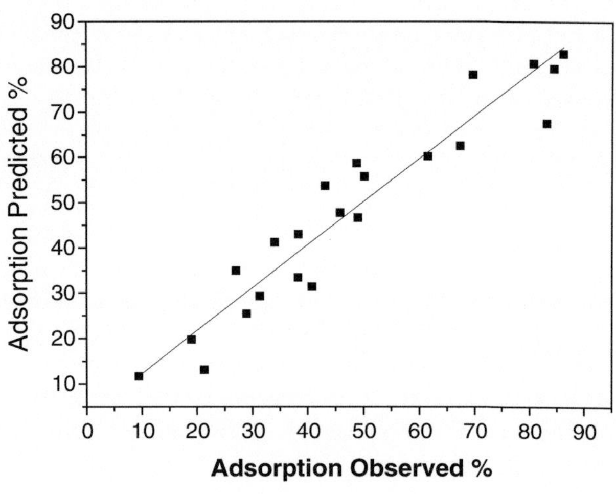

Fig. 49.1 The observed and predicted plot of adsorption of chromium (VI)

dose and pH imposing the greatest effect on adsorption of chromium. Initial feed concentration of Cr (VI) on the other hand imposed the least effect on the response. The quadratic effects of initial feed concentration of Cr (VI) as well as the interaction effects between X_1X_2, X_1X_3, X_2X_3, X_2X_4, and X_3X_4 were considered moderate. The adsorption of chromium percent response surface graphs is shown in Figs. 49.2, 49.3, and 49.4.

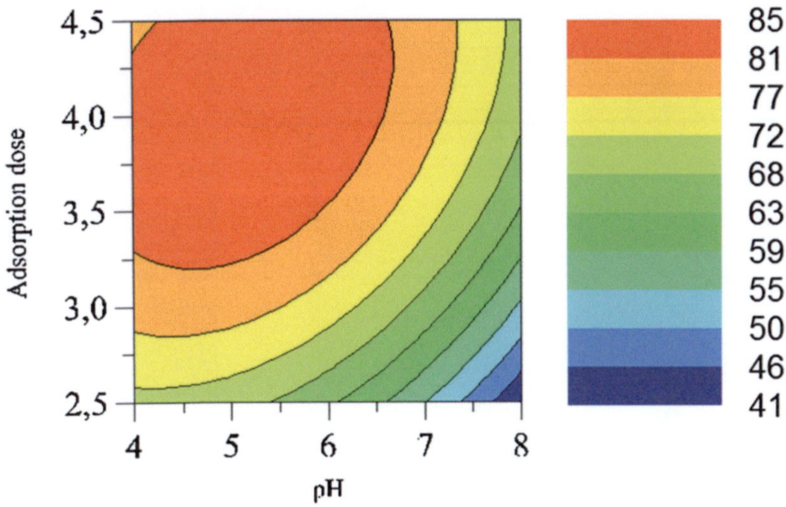

Fig. 49.2 The combined effect of adsorbent dose and pH on adsorption on Cr (VI) at constant temperature (30 °C) and initial concentration of Cr (VI) (40 mg/L)

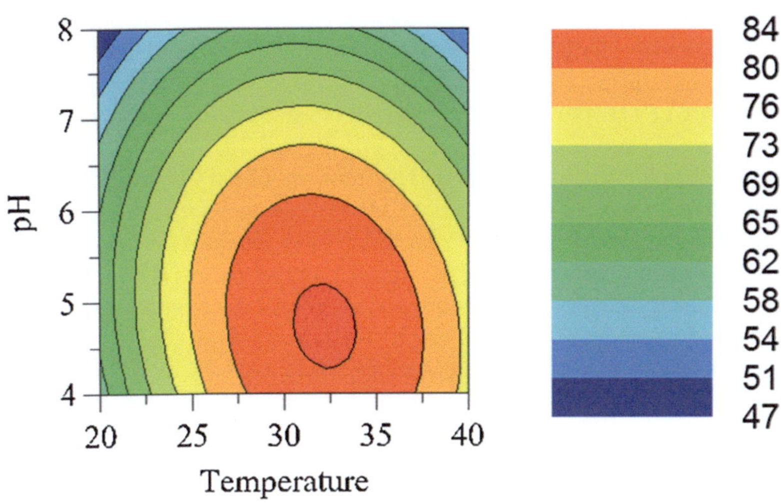

Fig. 49.3 The combined effect of temperature and pH on adsorption on Cr (VI) at constant adsorbent dose (3, 5 g/L) and initial concentration of Cr (VI) (40 mg/L)

Figure 49.2 shows the three-dimensional response surfaces which were constructed to show the interaction effects of the adsorbent dose and pH on the Cr (VI) removal. For this plot, the temperature and initial concentration were fixed at zero level (T = 30 °C, $C_0 = 40$ mg/L). It can be seen from the figure that initially the percentage removal increases very sharply with the increase in adsorbent dosage but beyond a certain value 3.25–4 g, the percentage removal reaches almost

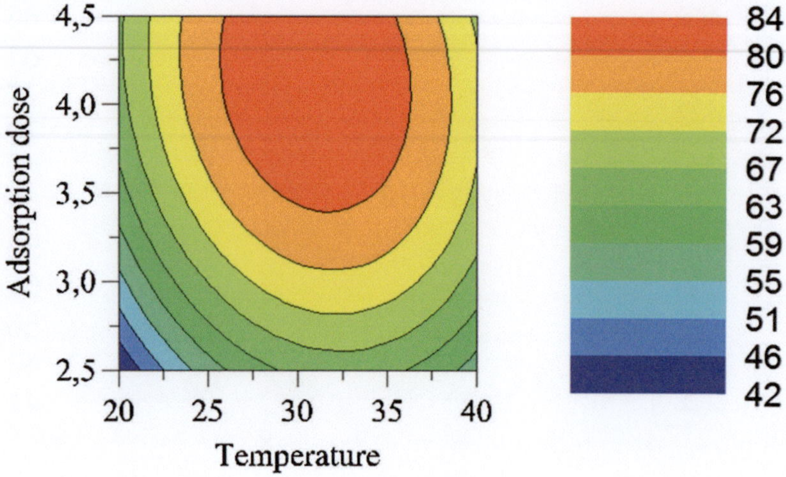

Fig. 49.4 The combined effect of temperature and adsorbent dose on adsorption on Cr (VI) at constant pH (6) and initial concentration of Cr (VI) (40 mg/L)

a constant value. A maximum adsorption of chromium > 85 % was determined. The results obtained were in agreement with the work done by Sahu and al. [16].

The interactive effect of temperature and pH of Cr (VI) of the solution on percent adsorption of Cr (VI) onto activated carbon at constant initial feed concentration of Cr (VI) (40 mg/L) and adsorbent dose 3.5 g/L is shown in Fig. 49.3. Increasing the temperature from 20 to 35 °C facilitated the removal of Cr(VI) ions. It is clear from this figure that the percent adsorption of Cr (VI) decreases with increase in pH from 4.0 to 8. A maximum adsorption of chromium >84 % was determined at constant initial feed concentration of Cr (VI) (30 mg/L) and adsorbent dose (3, 5 g/L).

The combined effect of temperature and adsorbent dose on adsorption of chromium (VI) at constant pH (6) and initial concentration of Cr (VI) (40 mg/L) is shown in Fig. 49.4, the three-dimensional response surfaces. A maximum adsorption of chromium >82 % was determined at constant pH (6) and adsorbent dose (3. 5 g/L).

49.3.4 Optimization Analysis

One of the main aims of this study was to find the optimum process parameters to maximize the adsorption of chromium from the developed mathematical model equations. The experimental conditions with the function of desirability were applied using JMP8.

In the optimization analysis, the target criteria was set as maximum values for the Cr (VI) removal Y while the values of the four variables were set in the ranges

Table 49.4 Model validation

Adsorbent dose (g/L)	Temperature (°C)	pH	Initial concentration (mg/L)	Adsorption of Cr (VI) (%)	
				Predicted	Experimental
4.39	31	4.86	30	90.48	88.42

being studied. The experimental conditions with the highest desirability (0.985) were selected to be verified. The predicted and experimental results of Cr (VI) removal obtained at optimum conditions are listed in Table 49.4. The optimum adsorption conditions was obtained using adsorbent dose, temperature, pH, and initial feed concentration of Cr(VI) of 4.39 g/L, 31 °C, and 4.86, 30 mg/L, respectively. It was observed that the experimental value obtained was in good agreement with the value predicted from the model, with relatively small error between the predicted and the actual value, which was only 2.28 %.

49.4 Conclusion

Response surface methodology was successfully used to investigate the effects of adsorbent doses (ranging 2.5–4.5 g/L), temperatures (ranging 20–40 °C), pH (ranging 4–8), and initial feed concentrations of Cr (VI) (ranging 30–50 mg/L) on the adsorption of chromium. The regression analysis, statistical significance, and response surface were done using JMP8 for predicting the responses in all experimental regions. Models were developed to correlate the adsorption variables to the responses. Through analysis of the response surfaces derived from the models, adsorbent dose, pH, and temperature were found to have the most significant effect on adsorption of chromium. Process optimization was carried out and the experimental values obtained for the adsorption of chromium are found to agree satisfactorily with the values predicted by the models. The optimal adsorption of chromium was obtained as adsorbent dose, temperature, pH, and initial concentration of Cr (VI) of the Cr (VI) solution and these were found to be 4.39 g/L, 31 °C, 4.86, and 30 mg/L, respectively, resulting in 90.48 % of adsorption of chromium.

References

1. Aksu Z, Gonen F, Demircan Z (2002) Biosorption of Chromium (VI) ions by mowital B3OH resin immobilized activated sludge in a packed bed: comparison with granular activated carbon. Process Biochem 38:175–186
2. Volesky B (2001) Detoxification of metal-bearing effluents: biosorption for the next century. Hydrometallurgy 59:203–216
3. Wang XS, Li ZZ, Tao SR (2009) Removal of chromium (VI) from aqueous solution using walnut hull. J Environ Manage 90:721–729

4. Gupta S, Babu BV (2009) Utilization of waste product (Tamarind seeds) for the removal of Cr(VI) from aqueous solutions: equilibrium, kinetics, and regeneration studies. J Environ Manage 90(10):3013–22
5. Karagoz S, Tay T, Ucar T, Erdem M (2008) Activated carbons from waste biomass by sulfuric acid activation and their use on methylene blue adsorption. Bioresour Technol 99:6214–6222
6. Mohamed FS, Khater WA, Mostafa MR (2006) Characterization and phenols sorptive properties of carbons activated by sulphuric acid. Chem Eng J 116:47–52
7. Mahalik K, Sahu JN, Patwardhan AV, Meikap BC (2010) Statistical modelling and optimization of hydrolysis of urea to generate ammonia for flue gas conditioning. J Hazard Mater 182 (1–3):603–610
8. Sahu JN, Acharya J, Meikap BC (2010) Optimization of production conditions for activated carbon from Tamarind wood by zinc chloride using response surface methodology. Bioresour Technol 10:1974–1982
9. APHA (1985) Standard methods for the examination of water and wastewater, 16th edn. APHA, AWWA, WPCF, Washington, DC
10. Alam MZ, Muyibi SA, Toramae J (2007) Statistical optimization of adsorption processes for removal of 2,4-dichlorophenol by activated carbon derived from oil palm empty fruit bunches. J Environ Sci 19:674–677
11. Azargohar R, Dalai AK (2005) Production of activated carbon from Luscar char: experimental and modeling studies. Micropor Mesopor Mater 85:219–225
12. Cronje KJ, Chetty K, Carsky M, Sahu JN, Meikap BC (2010) Optimization of chromium (VI) sorption potential using developed activated carbon from sugarcane bagasse with chemical activation by zinc chloride. Desalination 275:76–284
13. Azargohar R, Dalai AK (2008) Steam and KOH activation of biochar: experimental and modeling studies. Micropor Mesopor Mater 110:413–421
14. Lazic ZR (2004) Design of experiments in chemical engineering. Wiley, Weinheim
15. Sahu JN, Patwardhan AV, Meikap BC (2010) Optimization for production of ammonia from urea in a semi-batch reactor for safe feedstock in power plants: experimental and statistical studies. Clean Soil, Air Water 38(5–6):533–542
16. Sahu JN, Jyotikusum A, Meikap BC (2009) Response surface modeling and optimization of chromium(VI) removal from aqueous solution using Tamarind wood activated carbon in batch process. J Hazard Mater 172:818–825

Chapter 50
Carbon Dioxide Adsorption on Coconut Shell Biochar

Wan Azlina Wan Ab Karim Ghani, Nur Zalikha Rebitanim, Mohamad Amran Mohd Salleh, and Azil Bahari Alias

Abstract Biochar has been acknowledged for its unique property which makes it potential candidates as adsorbent for carbon dioxide (CO_2) in the flue gas system. In this study, the properties of raw coconut shell biochar (CSB) and amine treated coconut shell biochar (ACSB) are being compared. Physiochemical characterization has been performed to characterize the biochar properties. Fourier transform infrared spectroscopy (FTIR) and scanning electron microscopy (SEM) were used to evaluate the functional groups and surface morphology of the biochar. Thermogravimetric analyzer (TGA) was used to discover the thermal properties, reactivity during adsorption. During the adsorption study, it was observed that amine treated coconut shell gasified at 800 °C gave the highest adsorption of 35.57 mg CO_2/g sorbent at temperature of 30 °C. Nitrogen functionalities and the basicity of samples increased after the amine treatment and is said to assist the adsorption capacity.

Keywords Biochar • Carbon dioxide adsorption • Amine • Thermogravimetric • Flue gas

---

W.A.W.A.K. Ghani (✉) • N.Z. Rebitanim
Department of Chemical and Environmental Engineering, Faculty of Engineering, Universiti Putra Malaysia, Serdang, Selangor 43400, Malaysia
e-mail: wanazlina@upm.edu.my

M.A.M. Salleh
Department of Chemical and Environmental Engineering, Faculty of Engineering, Universiti Putra Malaysia, Serdang, Selangor 43400, Malaysia

Faculty of Engineering, Institute of Advance Technology, Universiti Putra Malaysia, Serdang, Selangor 43400, Malaysia

A.B. Alias
Faculty of Chemical Engineering, Universiti Technology MARA Malaysia, Shah Alam, Selangor 43450, Malaysia

I. Dincer et al. (eds.), *Progress in Clean Energy, Volume 1*,
DOI 10.1007/978-3-319-16709-1_50

50.1 Introduction

Extensive studies are currently being performed to reduce the discharge of CO_2 to the atmosphere. Countless organizations are finding alternatives to reduce the CO_2 concentration and considering methods to control the emissions of this greenhouse gases. The energy generated from the combustion of fossil fuels is one of the major sources of the greenhouse gases [1]. In Malaysia, CO_2 emission has increased rapidly since 1980, and this number is expected to increase massively in years to come. The amount of CO_2 emission in 1980 shows an emission of 25 million metric tons of CO_2 and increased to 160 million metric tons of CO_2 by 2006 which contributes to 540 % increase from the initial amount in 1980 [2]. Monoethanolamine (MEA) solvent is widely used in the flue gas which involves introducing the gas stream to an aqueous amine solution which reacts with the CO_2 in the gas by an acid–base neutralization reaction. A problem that the industry is facing with the usage of amine in the flue gas system is the degradation of the amine as it is being recycled continuously. Other than the by-products created by degradation of amine which reduces the adsorption capacity of CO_2, amine also accelerates the corrosion process in machines involved [3]. One promising method to handle the carbon dioxide uptake is by using activated carbon which has adsorptive properties such as microporous structure and high surface area [4]. Other than these criterions, the surface chemistry also plays a role in the effectiveness of the adsorption process [5]. In this chapter, the treatment of biochar is focused on MEA treatment to incorporate nitrogen component to the biochar structure.

50.2 Materials and Methodology

50.2.1 Preparation of Coconut Shell Biochar

Coconut shell (CS) biochar adsorbent used was derived from a lab scale air blown gasifier reactor operated at 800 °C to generate energy. The samples are grounded to diameter of 500 μm and dried overnight at 90 °C to remove any moisture present. The biochar is separated into two parts of raw and chemically treated sample for the CO_2 adsorption capacity study. The samples were treated with Monoethanolamine (MEA), and the sample was stirred for 20 min. The treated sample was then dried at 100 °C for 24 h [6]. The raw biochar sample is identified as CS while the amine treated sample is regarded as N-CS.

50.2.2 Adsorbent Characterization

Ultimate analysis was performed using CHNS elemental analyzer to determine the carbon, hydrogen, nitrogen, and sulfur contents in the adsorbent. Brunauer–Emmett–Teller (BET) surface area of the biochar was analyzed using an automatic Quantacome AS1WinTM—Automated Gas Sorption Data Analyzer. Mettler Toledo TGA/SDTA851 (USA) Thermal Gravimetric Analyzer (TGA) was used to check the stability of the sample. Fourier Transform Infrared Spectroscopy (FTIR) is used to identify the chemical composition of char by recording the infrared spectrum of respective samples. The compositions of the sample can be recognized by analyzing the spectra with the relative intensities of a functional group.

50.2.3 Isothermal Adsorption Tests

The isothermal carbon dioxide (CO_2) adsorption analyses were carried out at temperature 30 °C to evaluate the adsorption capacity of the char sample using EXSTAR 6,000 Thermal Gravimetric Analyzer (TGA). It is used to measure reactivity of carbonaceous material with CO_2. Around 10 mg of sample is placed in crucible before being placed in the conveyer. The sample is then heated up to desired temperature of 30 and 70 °C in 100 ml/min nitrogen flow. The sample is held at this temperature until the weight of sample is stabled (10–20 min). Gas is then switched to CO_2 at 100 mL/min to measure CO_2 adsorption and change back to nitrogen flow of 100 mL/min for desorption test.

50.3 Results and Discussion

50.3.1 Characterization of Coconut Shell Biochar

The difference characteristics of two samples can be seen from the results of the ultimate analysis in Table 50.1. The BET surface area of the raw sample differs tremendously after the chemical treatment while the nitrogen content of the treated char shows an increase of 77.7 %. This is due to the nitrogen component that has been incorporated to the structure of the char by amine treatment [7].

From the N_2 adsorption isotherm at −196 °C of the raw and treated biochar in Fig. 50.1 shows that the biochar illustrate nitrogen adsorption isotherms of type I, characteristics of microporous materials which have relatively small external surface whereby the limiting uptake are mainly controlled by the micropore volume rather than internal surface area [8]. The treated biochar is giving lower adsorption

Table 50.1 Chemical characteristics of the coconut shell (CS) biochar sorbents

No	Sample	Surface area (m^2/g)	Ultimate analysis (wt%), dry basis				pH
			C	H	N	O[a]	
1	CS	171.956	80.592	0.821	0.265	18.322	5.86
2	N-CS	10.335	62.21	1.727	1.19	34.873	7.95

[a]Calculated by difference

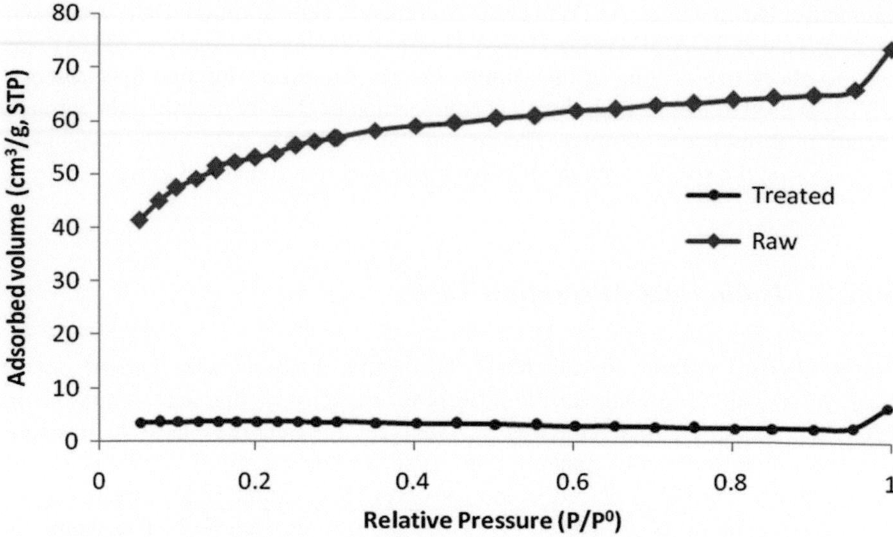

Fig. 50.1 N_2 adsorption isotherm at −196 °C of the raw and treated biochar

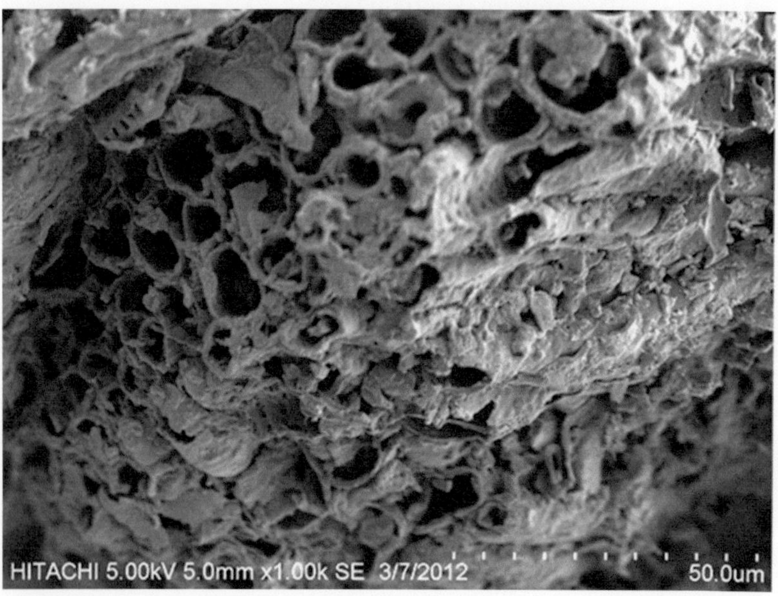

Fig. 50.2 SEM image of coconut shell char produced at 800 °C

of nitrogen as compared to raw biochar sample. This may be due to the decrement of surface area from 171.956 to 10.335 m^2/g because of pores blockage which inhibits the adsorption of CO_2.

Microscopic observation shows the particle character and structure dissimilarity of the raw and treated sample. Figure 50.2 shows the gasification effects of the

Fig. 50.3 SEM image of amine treated coconut shell char

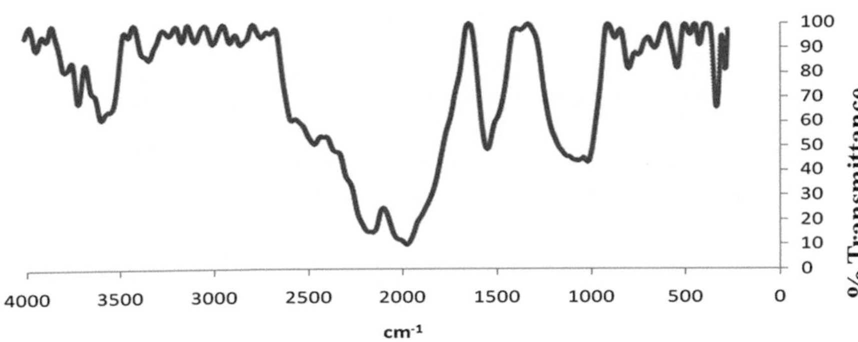

Fig. 50.4 FTIR spectra of coconut shell (CS) biochar gasified at 800 °C

biochar production provide pore openings which are characterized as randomly distributed and heterogeneous in size. The particles are irregular in shape and some agglomerations are observed. In Fig. 50.3 the effects of amine treated sample shows closest packing of the biochar grains with fewer cavities and voids. Small pores are observed among amorphous agglomerations in the pores which may be due to decomposition of carbon after the amine treatment. This may justify the decrement of the surface area from the BET results of the treated sample where the pores are blocked.

The coconut shell char in Fig. 50.4 showed a band of hydroxyl group above 3,000 cm^{-1} indication alcoholic or phenolic components. The most characteristics

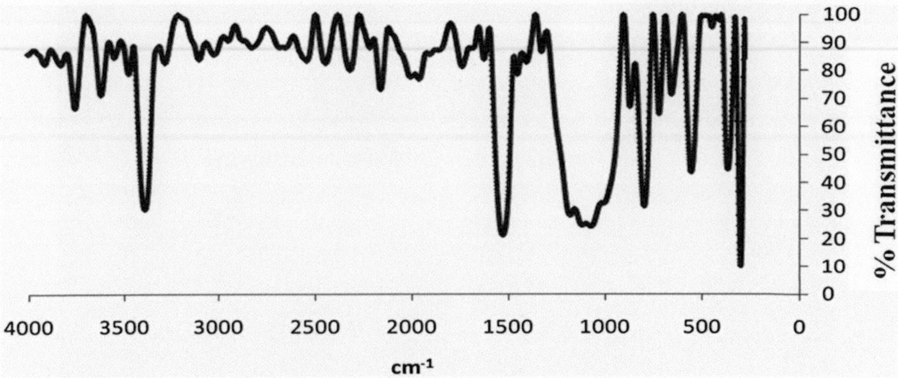

Fig. 50.5 FTIR spectra of amine treated coconut shell (N-CS) biochar

bands of lignin can be seen at 1,513 and 1,597 m^{-1} indicating aromatic ring vibrations and between 1,470 and 1,460 cm^{-1} which signify the CH deformation and aromatic ring vibrations [9]. A broad band between 1,000 and 1,500 cm^{-1} can be attributed to carboxyl-carbonates structures. The intensity of the band increases after amine treatment of the char as shown in Fig. 50.5. The presence of the basic oxides helps to explain the basicity of biochar and may be related to its ability as carbon dioxide adsorbent [10]. Broad and complex band within 3,200–3,700 cm^{-1} may be due to existence of surface hydroxylic groups and chemisorbed water. The increase of intensity of the treated char at this range as compared to the raw biochar may be due to the addition of chemisorbed water during amine treatment [11]. Both raw and coconut shell biochar shows a sharp peak near 1,600 cm^{-1} in which the treated sample giving higher peak after treatment, indicating appearance of quinine group in the sample. The increasing intensity of quinine group in the treated sample may assist to increase the CO_2 adsorption of the adsorbent. The increment intensity at 1,600 cm^{-1} can be attributed to cyclic amide functionality that may explain the effectiveness of the nitrogen incorporation onto the biochar carbon surface [5].

50.3.2 Assessment of Carbon Dioxide Adsorption–Desorption Using Thermo Gravimetric Analyzer Method

Figures 50.6 and 50.7 show the isothermal adsorption of the raw and treated coconut shell biochar at two adsorption temperatures of 30 and 70 °C. The experimental work was performed by using TGA flowed with CO_2 at 100 ml/min. The samples were first flown through nitrogen to remove any moisture present and then flow with CO_2 to study the adsorption capacity of the biochar. The highest

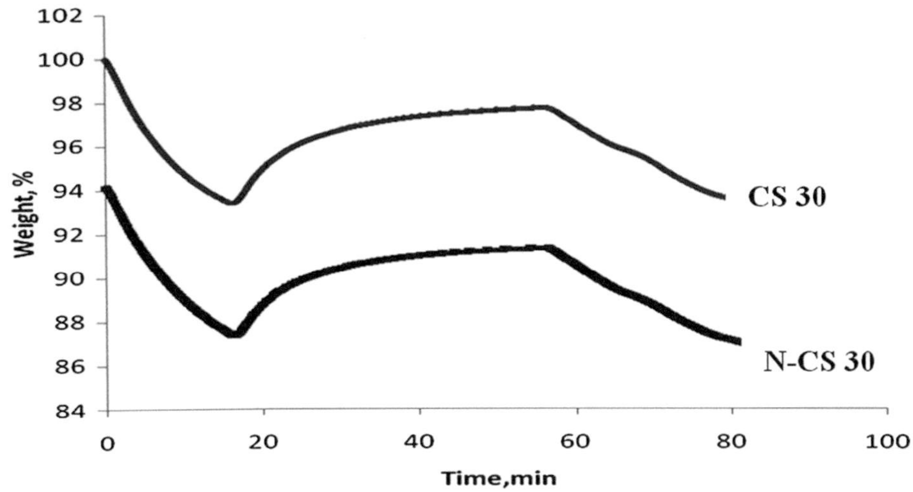

Fig. 50.6 The CO_2 adsorption and desorption curve of coconut shell char (CS 30) and treated char (N-CS 30) at 30 °C

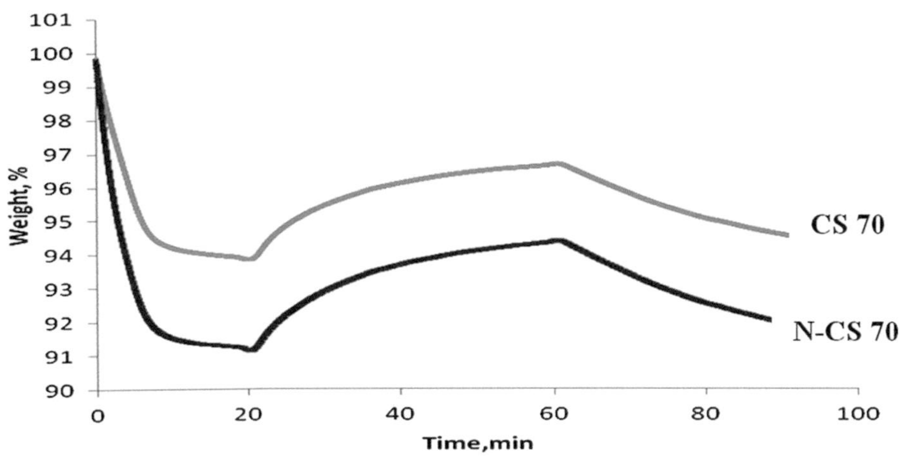

Fig. 50.7 The CO_2 adsorption and desorption curve of coconut shell char (CS 70) and treated char (N-CS 30) at 70 °C

adsorption capacity of 46.387 mg CO_2/g sorbent was acquired from the raw coconut shell biochar at adsorption temperature of 30 °C as shown in Table 50.2. This high adsorption capacity may be due to the domination of physisorption process in which the adsorption is highly dependent on the surface area of the biochar. Since the surface area of raw biochar is higher than treated biochar of 171.956–10.335 m^2/g accordingly, the raw char with higher surface area is able to sequester higher volume of CO_2 [12].

Table 50.2 Adsorption capacity of raw and treated coconut shell biochar at 30 °C and 70 °C

Adsorbents	Temperature (°C)	Surface area (m^2/g)	Adsorption capacity ($mgCO_2/g$ sorbent)	Pore volume (cm^3/g)
CS 30	30	171.956	46.387	0.031
N-CS 30	30	10.335	45.576	0.006
CS 70	70	171.956	30.114	0.031
N-CS 70	70	10.335	35.496	0.006

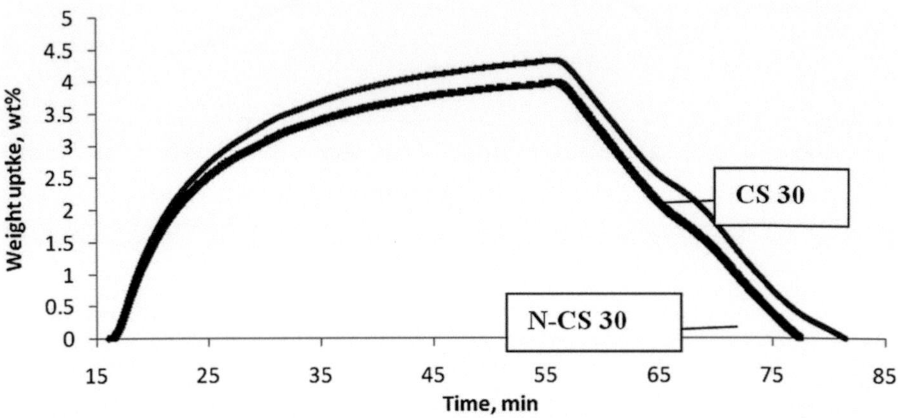

Fig. 50.8 The CO_2 capture capacities of coconut shell char (CS 30) and amine treated char (N-CS 30) at isothermal of 30 °C

For adsorption at temperature of 70 °C, the dominant process is mainly chemisorption whereby amine treated coconut shell biochar is giving higher adsorption of 35.496 mg CO_2/g sorbent as compared to raw which gives 30.114 mg CO_2/g sorbent. CO_2 is an acidic gas therefore by incorporating amine (with high value of basicity) onto biochar will assist in the adsorption capacity of sample through the attraction of the acid–base properties [13].

The CO_2 adsorption and desorption potential of the prepared char adsorbents were measured in Fig. 50.8 for isothermal of 30 °C. The percentage weight uptake of CO_2 shows a value of 4.3 % for the raw CS biochar as compared to treated biochar which gives a value of 3.98 % CO_2 uptake. Amount of adsorption is higher for the raw sample as compared to the amine treated sample. The CO_2 capture at 30 °C may be due to the physisorption which takes place within the pore structure. The physisorption behaviour is directly proportional to the surface area. For physisorption adsorption behaviour, the attraction between adsorbent and adsorbate exist by the formation of intermolecular electrostatic forces. As observed in this study, pore blockages during amine treatment for treated samples has reduced its surface area and hence inhibited the intermolecular interaction on its surface. In Fig. 50.9, the percentage of CO_2 adsorption is 2.7 % for raw sample and 3.2 % for treated sample at temperature of 70 °C. The low adsorption of CO_2 compared to

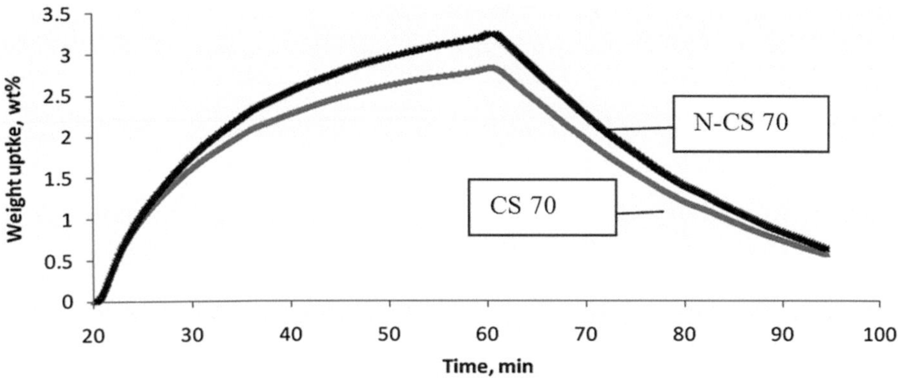

Fig. 50.9 The CO_2 capture capacities of coconut shell char (CS 70) and amine treated char (N-CS 70) at isothermal of 70 °C

Table 50.3 Adsorption and desorption behavior of raw and treated coconut shell biochar in CO_2 flow

Adsorbents	Temperature (°C)	Adsorption capacity (mg/g sorbents)	Desorption capacity (%)
CS	30	46.4	100
N-CS	30	45.6	100
CS	70	30.1	78.9
N-CS	70	35.5	79.7

adsorption at lower temperature is due to the subtle contact of the carbon support with CO_2 at higher temperature [14].

In Table 50.3, it is shown that the adsorption capacity of the raw CS decreases from 46.4 to 30.1 mg/g adsorbent as temperature from 30 to 70 °C. When temperature increases, the molecule diffusion rate increases which resulted in unsteady gas adsorption and desorption on the carbon surfaces due to higher energy provided [15]. The adsorption of CO_2 enhances with amine filling onto the sample as compared to the raw biochar with adsorption capacity from 46.4 to 45.6 mg/g adsorbent at temperature 30 °C and from 30.1 to 35.5 mg/g adsorbent at temperature 70 °C. This may be due to the high nitrogen content incorporated onto the sample which increases the basicity of the adsorbent [16].

Table 50.4 shows the comparison of the CO_2 adsorption onto raw and amine treated carbon samples from other experimental works. Most of the carbon exhibit decrement of surface area after amine treatment which may be due to the incorporation of nitrogen functionalities onto the carbon surface. The adsorption at temperature of 25–30 °C gives higher CO_2 uptake as compared to adsorption at higher temperature. The physisorption process that takes place at lower temperature is assisted by high surface area in which giving higher CO_2 sequestration. The effectiveness of the biochar used in this study is seen to be compatible when compared with other sorbents and commercial activated carbons.

Table 50.4 Comparison of adsorption from other studies

Sample	Treatment	Raw BET surface area (m^2/g)	Treated BET surface area (m^2/g)	CO_2 adsorption capacity ($mgCO_2/g$ adsorbent)	References
Fly carbon	Amine	75	241	Temperature 30 °C	[6]
				MEA: 68.6	
				Temperature 70 °C	
				MEA: 49.8	
Anthracites	Polyethylenimine (PEI)	1,000	<1	Temperature 75 °C	[16]
				Raw: 16.05	
				PEI: 26.30	
Palm shell activated carbon (AC)	Polyethylenimine (PEI)	941	1,052	Temperature 25 °C	[17]
				Raw: 3.56	
				PEI: 3.26	
Coconut shell biochar	Monoethanolamine (MEA)	171.956	10.335	Temperature 30 °C	This study
				Raw: 46.387	
				MEA: 45.576	
				Temperature 70 °C	
				Raw: 30.114	
				MEA: 35.496	

50.4 Conclusions

Gasification residues of coconut shell biochar provides CO_2 capture capacity of 46.387 mg CO_2/g adsorbent and 30.114 mg CO_2/g adsorbent at temperature of 30 and 70 °C, accordingly. The high uptake of CO_2 is mainly due to the physisorption process that takes place at low temperature in which the adsorption capacity is highly associated to the high values of surface area and pore volume. Amine treated biochar gives an advantage of higher CO_2 capture capacity at higher temperature in which 30.114 mg CO_2/g adsorbent and 35.496 mg CO_2/g adsorbent were adsorbed at temperature of 30 and 70 °C, accordingly. The higher CO_2 uptake adhered by the treated biochar is due to the basicity of the carbon surface and appearance of functional groups such as quinine and cyclic amide onto the sample that assist in the CO_2 uptake. The study performed showed that coconut shell biochar has the potential to be used as CO_2 sequester in flue gas system in efforts to reduce greenhouse gases emissions.

Acknowledgement The authors express their gratitude to the Ministry of Science, Technology and Innovation (MOSTI) of Malaysia (SF 03-01-04-SF1440/54506709) for financial support.

References

1. Siriwardane RV, Ming-Shing S, Edward PF, James AP (2001) Adsorption of CO_2 on molecular sieves and activated carbon. Energy Fuel 15:279–284
2. Tick HO (2010) Carbon capture and storage potential in coal-fired plant in Malaysia—a review. Renew Sust Energ Rev 14:2697–2709
3. Brian RS, Richard RA, Curt MW. Carbonation of Finnish magnesium silicates for CO_2 sequestration. https://www.netl.doe.gov/publications/journals/vol1/Papers/32-39.pdf. Accessed on January 2012
4. Aygun A, Yenisoy-Karakas S, Duman I (2003) Production of granular activated carbon from fruit stones and nutshells and evaluation of their physical, chemical and adsorption properties. Microporous Mesoporous Mater 66:189–195
5. Christian LM, Kelly RB, James E, Kenneth LF (2001) High temperature ammonia treatment of activated carbon for enhancement of CO_2 adsorption. Carbon 39:1809–1820
6. Mercedes MM-V, Zhe L, Yinzhi Z, Zhong T (2008) Sorbents for CO_2 capture from high carbon fly ashes. Waste Manag 28:2320–2328
7. Plaza MG, Pevida C, Arias B, Fermoso J, Arenillas A, Rubiera F, Pis JJ (2008) Application of thermogravimetric analysis to the evaluation of aminated solid sorbents for CO_2 capture. J Therm Anal Calorim 92(2):601–606
8. IUPAC (1985) Reporting physisorption data for Gas/solid systems with special reference to the determination of surface area and porosity. Pure Appl Chem 57:603–619
9. Ramesh KS, Jan BW, Vicki LB, Xuehao L, Chan WG, Mohammad RH (2004) Characterization of chars from pyrolysis of lignin. Fuel 83:1469–1482
10. Plaza MG, Pevida C, Arias B, Fermosa J, Casal MD, Martin CF, Rubierra F, Pis JJ (2009) Development of low-cost biomass-based adsorbents for postcombustion CO_2 capture. Fuel 88:2442–2447
11. Wenzhong S, Shouchun Z, Peimang J, Yihong L (2010) Surface chemistry of pyrolyzed starch carbons on adsorption of ammonia and carbon disulfide. Colloids Surf A Physicochem Eng Asp 356:16–20
12. Mohamed SS, Wan MAWD, Amirhossein H, Arash A-N (2011) Ammonia modification of activated carbon to enhance carbon dioxide adsorption: effect of pre-oxidation. J Appl Surf Sci 257:3936–3942
13. Dong-Il J, Soo JP (2001) Influence of amine grafting on carbon dioxide adsorption behaviors of activated carbons. Bull Kor Chem Soc 32:3377–3379
14. Xu XC, Song C, Andresen JM, Miller BG, Scaroni AW (2002) Preparation and characterization of novel CO_2 "molecular basket" adsorbents based on polymer-modified mesoporous molecular sieve MCM-41. Energy Fuel 16:1463–1469
15. Mercedes MM-V, Zhong T, Yinzhi Z (2005) CO_2 capture by activated and impregnated anthracites. Fuel Process Technol 86:1487–1502
16. Plaza MG, Pevida C, Arias B, Fermoso J, Rubiera F, Pis JJ (2011) Microporous phenol–formaldehyde resin-based adsorbents for pre-combustion CO_2capture. Energy Procedia 1:1107–1113
17. Mohamed KA, Wan MAWD, Chun YY, Donni A (2008) Adsorption capacities of carbon dioxide, oxygen, nitrogen and methane on carbon molecular basket derived from polyethyleneimine impregnation on microporous palm shell activated carbon. Sep Purif Technol 62:609–613

Chapter 51
Metal Dispersion and Interaction with the Supports in the Coke Production Over Ethanol Steam Reforming Catalysts

Gianguido Ramis, Ilenia Rossetti, Elisabetta Finocchio, Matteo Compagnoni, Michela Signoretto, and Alessandro Di Michele

Abstract Two families of Ni/TiO_2 catalysts were prepared for use in the steam reforming of ethanol. The catalytic performances, in terms of both H_2 productivity and stability towards coking and sintering, were related to the physico-chemical properties of the catalysts.

The samples were prepared either by synthesis of the support by precipitation and subsequent impregnation with the active phase, or by direct synthesis through Flame Pyrolysis. Many techniques have been used to assess the physico-chemical properties of both the fresh and spent catalysts. The samples showed different textural, structural and morphological properties, as well as different reducibility and thermal resistance, depending on the preparation method and support.

The performance of the titania-supported catalysts were found very dependent on the preparation procedure, and we may conclude that operation at 625 °C can be satisfactory from all the points of view of activity, productivity and C balance, allowing to limit the heat input to the reactor with respect to operation at 750 °C.

Keywords Ethanol steam reforming • Coke deposition • Ni dispersion • Flame pyrolysis

---

G. Ramis (✉) • E. Finocchio
Dip. Ingegneria Chimica, Civile e Ambientale, Università di Genova and INSTM Unit, P.le Kennedy 1, Genova 16129, Italy
e-mail: gianguidoramis@unige.it; Elisabetta.Finocchio@unige.it

I. Rossetti • M. Compagnoni
Dip. Chimica, Università degli Studi di Milano, via C. Golgi 19, Milano 20133, Italy

CNR-ISTM and INSTM Unit, Milano Università, via C. Golgi 19, Milano 20133, Italy
e-mail: ilenia.rossetti@unimi.it

M. Signoretto
Dip. di Scienze Molecolari e Nanosistemi and INSTM Unit, Università Ca' Foscari Venezia, Calle Larga S. Marta, 2137, Venezia, Italy

A. Di Michele
Dip. di Fisica, Università degli Studi di Perugia, Via Pascoli, Perugia 06123, Italy

I. Dincer et al. (eds.), *Progress in Clean Energy, Volume 1*,
DOI 10.1007/978-3-319-16709-1_51

51.1 Introduction

The steam reforming of ethanol (ESR) is gaining growing attention for the possibility to produce hydrogen from renewable sources. Demonstrative projects aim to a proof of concept of the application of this process to raw materials, which do not compete with the food and feed chain. Particular attention has been devoted to increase ethanol conversion and to maximise hydrogen productivity by depressing side reactions.

The reaction usually proceeds through direct ethanol decomposition or its dehydrogenation to acetaldehyde, followed by reforming [1]. Unfortunately, possible side reactions may induce the formation of coke over catalyst surface, often leading to its deactivation. Surface acidity, due to the support nature or to the active phase itself, may induce ethanol dehydration and the subsequent polymerisation of ethylene. This leads to the possible formation of graphitic layers over catalyst surface, sometimes blocking the active phase. In other cases, coke accumulation mainly occurs on support surface only, but this may anyway reflect on activity and selectivity, since the support is involved in steam activation and in supplying activated oxydril species. Other possible reactions can provoke carbon formation. For instance, the thermal decomposition of methane (one of the possible reaction products), occurring at high temperature, or CO disproportion (Boudouard reaction), exothermal and thus favoured at relatively low temperature.

Some coking mechanisms involve the active phase itself, as in the case of Ni- and Co-based catalysts. Both metals are indeed able to promote the formation of carbon deposits in the form of nanotubes, as well documented for Ni in the steam reforming of methane, ethanol and glycerol [2–9]. It has been widely discussed in most of that references that the tendency to form carbon nanotubes is lower when the active phase is well dispersed and stabilised by a strong metal–support interaction. Steam reforming of ethanol may be satisfactorily achieved at much lower temperature with respect to the SR of methane. Indeed, some catalytic formulations demonstrated sufficiently active to completely convert ethanol below 600 °C. However, at such low temperature, different coke forming mechanisms are active, while the coke gasification rate may be insufficient, leading to the accumulation of carbonaceous deposits, raising challenges for catalyst stability. Nevertheless, the possibility to operate at much lower temperature with respect to the steam reforming of natural gas would be attractive to limit the thermal input to the reactor, with all the positive consequences (fuel consumption, materials, safety, etc.). The study of the factors influencing catalyst stability is therefore particularly interesting for this aspect of practical importance. In our previous investigations, we compared different supports for Ni, e.g. SiO_2, ZrO_2 and TiO_2. The latter in particular showed interesting redox properties, which may in principle help cleaning the surface from accumulated C. Furthermore, we recently evidenced the impact of the preparation procedure and above all calcination temperature on the dispersion and strength of interaction with the support in Ni/TiO_2 catalysts prepared by impregnation. It was found that Ni differently dispersed over the support depending on the crystal lattice of the latter (anatase or rutile). Calcination at high temperature (800 °C) allowed to obtain, together with more sintered particles, very small Ni nanoparticles upon activation, likely due to the reduction of an ilmenite ($NiTiO_3$) phase forming only after calcination of such high temperature. A high calcination temperature also allowed strengthening the metal–support interaction, particularly in the case of titania [10–12] thus favouring the thermal resistance of Ni particles during high temperature operation in ESR. Both these aspects positively affect activity and resistance to coking.

Therefore, in the present work we compared Ni/TiO_2 catalysts prepared by impregnation and calcined at 800 °C, with a comparative sample prepared by flame pyrolysis. This technique allows the one-pot preparation of nanoparticles by flash calcination at high temperature (1,200–1,500 °C depending on the operating conditions) for a few milliseconds. It has been adopted here to check the effect of this innovative preparation procedure on Ni dispersion and strength of interaction with the support and in turn on catalyst performance for ESR.

51.2 Materials

Two families of Titania-based catalysts have been prepared through conventional precipitation and Flame Pyrolysis (FP), resulting in the formulations reported in Table 51.1, together with the corresponding notations. As for the conventional precipitation method, the titania support was prepared as follow: 20 g of $TiOSO_4$ $xH_2SO_4\,xH_2O$ (Sigma Aldrich, purity synthesis grade) were dissolved in 300 mL of distilled water at room temperature. NaOH (Carlo Erba, 9 M) or NH_3 (Riedel-de Haën, 9 M) was added dropwise until the system reached a pH of 5.5. The samples were labelled T-S or T-A, respectively. The precipitate was aged at 60 °C for 20 h,

Table 51.1 Main physical-chemical properties of the samples prepared

Sample	Preparation method	Ni loading (wt%)[a]	SSA (m^2/g)[b]	mL O_2 (0 °C)/g_{Ni}	A (m^2/g_{Ni})[c]	d (nm)[c]	T_{max} (°C)[d]
T-A	TiO_2 precipitated with NH_3, calcined at 800 °C	7.3	7	0.61	2	–	650
T-S	TiO_2 precipitated with NaOH, calcined at 800 °C	6.7	4	No chemisorption	–	–	700
T-FP	Prepared by FP	8.2	63	14.50	51	11	370
							480
							600

[a]From atomic absorption analysis
[b]SSA = specific surface area, from BET model
[c]Average metallic surface area and Ni particle size, as determined from O_2 chemisorption data
[d]Temperature of the maximum of reduction peaks from TPR analysis

then repeatedly washed with distilled water and then dried overnight at 110 °C. The active phase was added to each support by incipient wetness impregnation with an aqueous solution of the metallic precursor ($Ni(NO_3)_2{\cdot}6H_2O$, Sigma Aldrich, purity ≥98.5 %), in the proper concentration in order to obtain 10 wt% Ni loading. The catalyst was dried overnight at 110 °C and then calcined at 800 °C for 4 h. A comparative sample was prepared in nanopowder form by means of an FP apparatus [13, 14]. This technique allows the continuous and one-step synthesis of oxides, single or mixed. It is based on a specially designed burner fed with oxygen and an organic solution of the precursors, with the solvent acting as fuel for the flame. FP-synthesised samples usually show good phase purity, once the procedure is optimised for each material, along with nanometer-size particles and hence very high surface area (up to 250 m^2/g). In addition, the high temperature of the flame in principle should also ensure thermal stability, provided that a solvent with sufficiently high combustion enthalpy is chosen [15, 16]. The active metal (Ni) has been here directly incorporated during the support synthesis. The FP sample (T-FP) was synthesised using a solution of Titanium(IV)-isopropylate (Aldrich, pur. 97 %) in xylene, with a 0.67 M concentration referred to TiO_2. Ni was added to such mother solution by dissolving Ni(II) acetate (Aldrich, pur. 98 %) in propionic acid (Aldrich, pur. 97 %), with 10 wt% metal loading and a 1:1 vol/vol solution of the two solvents. The solutions were fed to the burner nozzle with a flow rate of 2.2 mL/min and a 1.5 bar pressure drop across the nozzle, which was also fed with 5 L/min of O_2.

51.3 Results and Discussion

51.3.1 Characterisation

The N_2 adsorption/desorption isotherms were analysed by a Micromeritics, ASAP 2000 Analyser to assess surface area and the pores size distribution.

The high temperature calcination caused for samples T-S and T-A the collapse of the porous structure of the oxides, thus obtaining a substantially non-porous material. By contrast, the FP prepared sample was characterised by higher surface area (ca. 60 m^2/g) with a certain contribution of (small) mesopores.

The latter sample was also characterised by an adsorption feature at high relative pressure, suggesting the contribution of a macroporous system. Indeed, the FP technique leads to dense nanoparticles where the porosity, if present, is related to the formation of agglomerates of particles. Therefore, in this case only interparticle porosity was found and the pore distribution showed very broad.

The TPR technique was employed to identify the different Ni oxide species present in the catalysts according to the reduction temperatures. Moreover, this technique allows to evaluate the interaction strength between the active phase and the support, thus supporting the evaluation of a strong metal–support interaction (SMSI) effect.

In this set of experiments, the catalyst was placed in a quartz reactor and heated by 10 °C/min from r.t. to 800 °C in a 5 % H_2/Ar mixed gas stream flowing at 40 mL/min. Characteristic peak temperature of every sample are collected in Table 51.1.

The TPR of both the catalysts calcined at 800 °C show NiO reduction peaks at high temperatures (>600 °C), higher for sample T-S than T-A [9]. It is well known that the interaction between metal and support increases with calcination temperature. The present TPR profiles thus suggest stronger metal–support interaction for sample T-S than T-A. This is mainly ascribed to the formation of $NiTiO_3$, as confirmed by XRD analysis. This may be of outmost importance in order to stabilise Ni in the most dispersed form, leading to the formation of small metal particles upon activation, as better deepened elsewhere [9].

For the FP prepared sample, a TPR–TPO–TPR cycle was carried out [6]. Indeed, the preparation procedure induced a partial incorporation of Ni into the support, possibly leading to a mixed oxide phase. It may be supposed that some reconstruction of the oxide may occur during metal reduction. In a parallel way, Ni that is initially well dispersed in the support, after a first reduction process may arrange in clusters, is likely characterised by a different reducibility.

The first TPR run of T-FP showed a series of broad and overlapping peaks spanning over a very wide temperature range (300–700 °C) ascribable to the reduction of free surface NiO, bulk NiO and NiO more and more deeply interacting with the TiO_2 surface, as previously discussed, as well as to $NiTiO_3$. By contrast, the second TPR run showed a much more uniform pattern, with only one peak centred around 600 °C. This indicates the presence of a uniform type of Ni species, likely $NiTiO_3$. This feature also indicated the irreversibility of the process with transformation of the "free" NiO particles into species strongly bound to the support, maybe as a mixed oxide.

By comparing the three samples, T-S showed the less reducible, thus characterised by Ni species very strongly interacting with the support. The patterns of T-A and T-FP (second run) were instead very similar.

O_2 chemisorption (Table 51.1) carried out on samples T-A revealed poor oxygen uptake, with respect to catalyst T-FP, characterised by a much higher Ni dispersion. However, dispersion of the latter sample dropped to values comparable to T-A when aging was carried out in conditions mimicking activation.

Sample T-S did not evidence any oxygen uptake. This, combined with TPR which concluded a stronger metal–support interaction for T-S than T-A, may be related with the SMSI effect [10–12]. Indeed, in addition to the low dispersion which characterises also sample T-A, the strong interaction strength with the support of sample T-S may further modify the electronic state of the metal definitely inhibiting its interaction with O_2.

XRD analyses were performed in order to identify the different phases present in the samples. Rietveld refinement was also done to quantify the content of each phase.

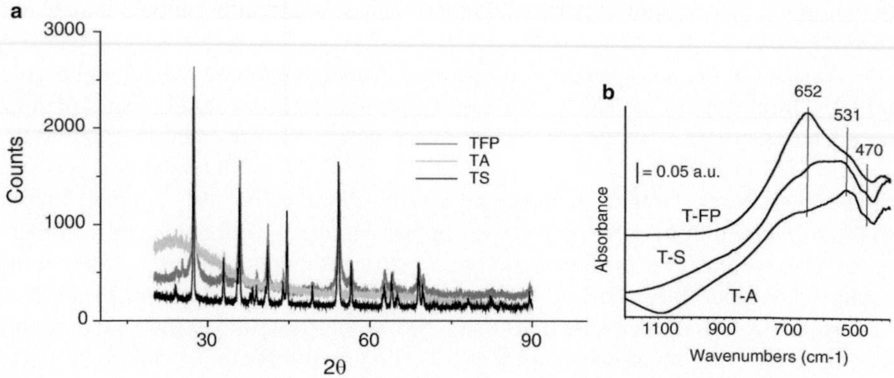

Fig. 51.1 XRD diffractograms (**a**) and FT-IR skeletal spectra (**b**) of the studied samples

By comparing the diffractograms (Fig. 51.1a) of the titania-supported samples with literature data, one may conclude that all the samples T-A and T-S were constituted by rutile, plus ca. 15–17 wt% of ilmenite ($NiTiO_3$), as confirmed by the TPR profiles. After activation of the samples, an XRD profile compatible with metallic Ni supported on rutile was always obtained.

The average Ni crystal size showed higher for samples T-A and T-S than for the T-FP one, confirming chemisorption data. This confirms the higher Ni-dispersion of the latter, in spite of the very high calcination temperature attained in the flame.

In Fig. 51.1, skeletal FT-IR spectra of the Ti-based catalysts diluted in KBr (1 % w/w) are also reported. As mentioned before, for samples T-A and T-S calcination at high temperature allows the formation of a titanate phase, characterised by peaks at 530 and 410 cm^{-1} (band at 320 cm^{-1} in the FAR-IR region, not reported) in agreement with literature data [17]. The shoulder at 690 cm^{-1} can be due to the rutile phase formation (also detected by XRD) whereas a weak shoulder centred at 470 cm^{-1} characterises residual titania anatase phase.

SEM micrographs of the catalyst T-FP revealed a rather uniform array of nanoparticles (not reported). EDS analysis also confirmed the Ni loading with respect to atomic absorption and repeated analyses in different zones demonstrated a uniform distribution of the active phase. The latter conclusion has been also supported by several maps revealing an even incorporation of Ni into the oxide matrix.

A more detailed view on sample morphology and particle size was obtained by TEM analysis (Philips 208 Transmission Electron Microscope) (Fig. 51.2a), which showed that the T-FP sample was characterised by dense 5–20 nm spheres.

By contrast, sample T-S showed bigger particles (>200 nm), but much more uniform in size than T-A.

In the latter, indeed big aggregates and small particles coexisted as shown in Fig. 51.2. This may imply a higher thermal resistance of the T-S sample and explain the lower crystallinity evidenced by XRD for catalyst T-A (Fig. 51.1).

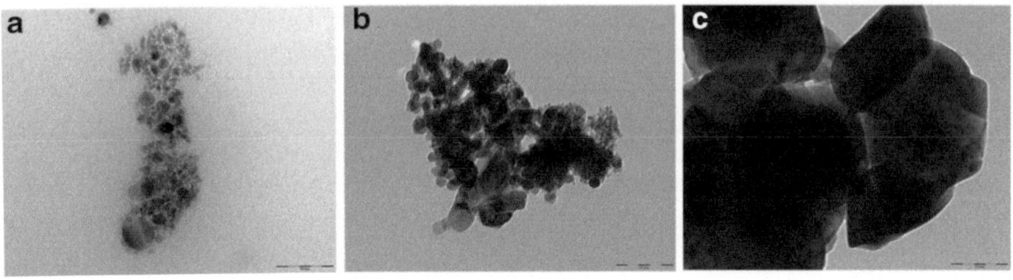

Fig. 51.2 TEM micrographs of the samples: (**a**) T-FP; (**b**) T-A; (**c**) T-S. Marker size 100 nm

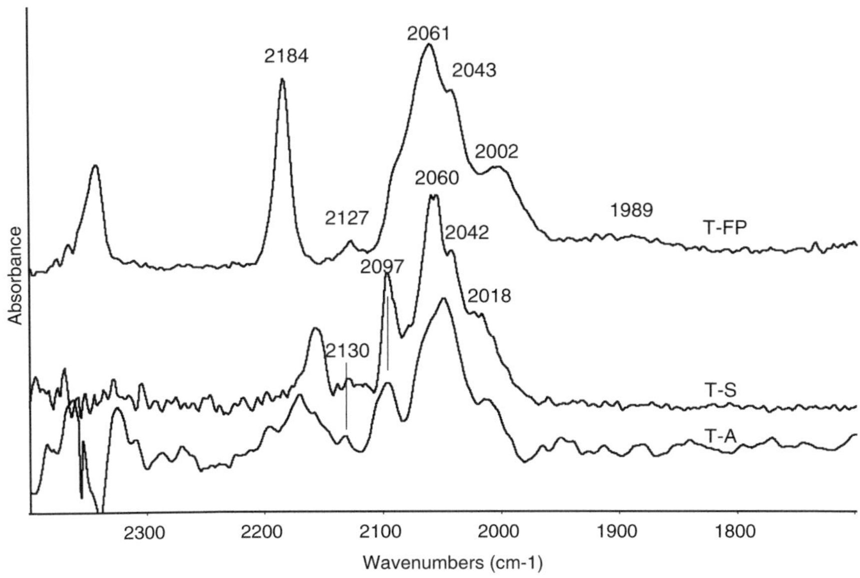

Fig. 51.3 FT-IR spectra of surface species arising from CO adsorption over reduced catalysts (liquid nitrogen temperature). The activated surface spectra have been subtracted

51.3.2 FT-IR Analysis of Catalysts Surface

The analysis of surface sites has been carried out over self-supporting disks of pure powder catalysts (30 mg average weight) placed in the IR cell connected to a conventional gas manipulation apparatus. After a reduction step in hydrogen at 773 K and thermal treatment in high vacuum, the spectrum of T-FP catalyst shows a clean surface almost completely dehydroxilated.

Low temperature CO adsorption over the same catalyst gives rise to several IR bands in the C–O carbonyl spectral region (Fig. 51.3, high CO coverage). As widely reported in the literature and well summarised by Hadjiivanov et al. [18] the absorption at 2,184 cm^{-1} (HF band) characterises acidic Ti cations (Ti^{4+}) acting as Lewis sites of different strength. Another weak component at 2,139 cm^{-1}, together with the shoulder at 2,090 cm^{-1}, indicates the presence of gem dicarbonyls species over dispersed Ni^{+} ions. This band disappears only after evacuation at room temperature, in agreement with literature findings that reported Ni^{+}–CO complexes being much more stable than Ni^{2+}–CO complexes [19, 20]. In the region, 2,000–2,100 cm^{-1} (LF bands) several bands can be detected: a main band at 2,061 cm^{-1} with shoulders at 2,043 cm^{-1} and 2,002 cm^{-1}, showing similar resistance to outgassing, was registered. These features can be associated with CO linearly adsorbed on Ni^{0} small crystals (main band at 2,061 cm^{-1}), but also to polycarbonyl species, unstable after outgassing. Beside, an additional broad and weak band is centred at 1,989 cm^{-1}. This band corresponds to bridging CO over larger Ni metal particles.

No bands associated with the presence of residual Ni^{2+}–CO species of NiO or nickel aluminate are detected following the reduction treatment: thus, Ni is broadly reduced at the catalyst surface, although highly dispersed. On the other side, the heterogeneity of the Ni species is in agreement with TPR results, pointing out the presence of at least three kinds of nickel species: Ni metal particles, Ni metal clusters strongly interacting with the support, Ni ions having different reducibility.

As for sample T-S, IR absorptions of OH stretching mode in the high frequency region of the spectrum are strong and ill-defined, possibly due to the low specific surface area of the sample. Low temperature CO adsorption has been carried out in the same conditions applied for T-FP sample. Several bands are detected in the C–O carbonyl spectral region: at 2,156 cm^{-1} (CO interacting with OH groups, possibly overlapped with CO adsorbed over Ni ions, and completely disappearing following outgassing at low temperature), at 2,130 (weak), together with the band at 2,095 cm^{-1} (symmetric/asymmetric stretching modes of poly-carbonyl species, likely gem dicarbonyl species coordinated over Ni^{+}). Bands at 2,060, and possibly 2,040 cm^{-1}, are assigned to poly carbonyl species on Ni^{0} crystals, as well as the band at 2,020 (shoulder) whose lower frequency suggests the assignation to larger metal Ni^{0} particles. No bands due to CO coordinated over exposed Ti centres can be detected (about 2,190 cm^{-1}), and this can be due to the collapse of surface area in these samples and also to the formation of a titanate phase (as pointed out by XRD and IR results).

CO adsorption over T-A catalyst leads to the detection of very similar features: a band at 2,160 cm^{-1} (H-bound PN), two bands at 2,138 and 2,090 cm^{-1} (polycarbonyl species, namely gem dicarbonyls coordinated over Ni^{+} ions), few bands below 2,060 cm^{-1} assigned to CO coordinated over Ni metal particles and to polycarbonyl species $Ni^{0}(CO)_x$, too. In these spectra, the last component seems to be predominant in comparison with sample T-FP, suggesting an increased fraction of highly dispersed Ni metal cluster or atomically dispersed Ni^{0}, strongly interacting with the surface, together with a fraction of Ni ionic species. This is in agreement with the TPR data indicating for these samples the main reduction peaks at temperatures above 600 °C, higher than T-FP sample.

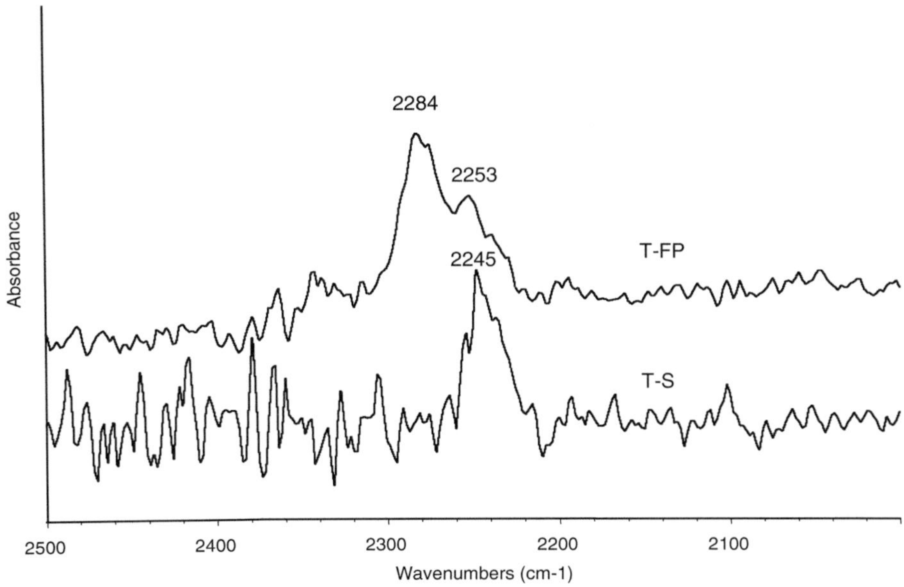

Fig. 51.4 FT-IR spectra of surface species arising from PN adsorption and following outgassing at room temperature over T-catalysts. The activated surface spectrum has been subtracted

Moreover, bands due to CO coordinated on Ti ions are not clearly detected in the T-S and T-A spectra (possibly a weak absorption at 2,170 cm^{-1} can be observable in the spectrum of sample T-A).

PN adsorption at room temperature has also been studied in order to evaluate the acidity of surface sites (Fig. 51.4).

Over oxidic surfaces usually PN adsorption gives rise to two families of signals in the CN stretching region: band in the range 2,300–2,270 cm^{-1} (PN coordinated over Lewis sites) and bands about 2,250 cm^{-1} (PN H-bound to hydroxyl groups). In fact, following PN adsorption on T-FP sample, two complex bands are detected in the CN stretching region at 2,278 and 2,247 cm^{-1} (with a shoulder at lower frequency). The former is due to PN interacting with medium strength Lewis sites (Ti ions), while the latter, strongly decreasing following outgassing at room temperature, is due to PN H-bound to residual exposed OH groups. The low relative intensity of this band is in agreement with the low degree of hydroxylation of this surface. Correspondingly, only a weak negative band can be detected in the OH stretching region of the subtraction spectrum.

On T-S sample, nitrile adsorption leads to the detection of just one band, due to H-bound species, disappearing following outgassing at room temperature and, correspondingly, no bands due to PN interacting with Lewis centres are reported. This is in agreement with results from CO adsorption. Spectra arising from PN adsorption over T-A sample, completely consistent with T-S sample are not reported here, due to the even higher noise level and the very weak band intensities.

Possibly, the disappearing of the Ti centres (i.e. Ti ions are not detected by CO or PN adsorption) is to be related to the collapse of the surface area or to the "shielding" effect of the Ni to the exposed Ti ions in the titanate structure.

51.3.3 ESR Catalytic Tests

Activity test were performed by means of a micropilot plant constituted by an Incoloy 800 continuous downflow reactor (i.d. 0.9 cm, length 40 cm), heated by an electric oven. The reactor temperature was controlled by a Eurotherm 3204 TIC.

The catalysts were pressed, ground and sieved into 0.15–0.25 mm particles and ca. 0.5 g were then loaded into the reactor after dilution 1:3 (vol/vol) with SiC of the same particle size.

Catalyst activation was accomplished by feeding 50 cm^3/min of a 20 % H_2 in N_2 gas mixture, while heating by 10 °C/min up to 800 °C, then kept for 1 h. During activity testing 0.017 cm^3/min of a 3:1 (mol/mol) H_2O:CH_3CH_2OH liquid mixture was fed to the reactor by means of a Hitachi, mod. L7100, HPLC pump, added with 56 cm^3/min of N_2, used as internal standard, and 174 cm^3/min of He. Such dilution of the feed stream was calibrated so to keep the reactants mixture in the vapour phase even at zero conversion at the reactor outlet.

The activity tests were carried out at atmospheric pressure, GHSV = 1,750 h^{-1} (referred to the ethanol + water gaseous mixture) at 500, 625 and 750 °C.

The analysis of the out-flowing gas was carried out by a gaschromatograph (Agilent, mod. 7980) equipped with two columns connected in series (MS and Poraplot Q) with a thermal conductivity detector (TCD), properly calibrated for the detection of ethanol, acetaldehyde, acetic acid, acetone, water, ethylene, CO, CO_2, H_2.

Material balance on C-containing products was checked to quantify coke deposition. Repeated analyses of the effluent gas were carried out every hour and the whole duration of every test at each temperature was ca. 8 h. The raw data, expressed as mol/min of each species out-flowing from the reactor, have been elaborated as detailed [7].

The tests for ethanol SR have been carried out on each catalyst and the results have been summarised in Table 51.2. If not else specified, the data represent the average performance over the last 4–8 h-on-stream. Anyway, the qualitative inspection of the full data set at every temperature allowed to assess the stability of the sample.

A blank test was carried out at 500 and 750 °C with quartz beads and SiC, but without any catalyst. At 750 °C ethanol conversion was rather high, indeed after starting values higher that 80 % it attested on ca. 50 %, due to thermal activation of the substrate, in accordance with literature data [21]. However, the main products at the reactor outlet were acetaldehyde ($S_{CH_3CHO} = 60\%$) and ethylene ($S_{CH_2CH_2} = 30\%$). Poor ethanol decomposition to CO, no CH_4 or CO_2

Table 51.2 Activity data at different reaction temperature

	Blank	T-A	T-S	T-FP
500 °C				
CO/CO_2	0.00 ± 0.00	8.7 ± 0.2	1.32 ± 0.14	0.62 ± 0.04
C balance (%)	91 ± 2	70 ± 6	76.7 ± 1.1	82 ± 8
Conv. EtOH (%)	13 ± 5	89 ± 9	99.5 ± 1.0	93 ± 8
H_2 productivity (mol/min kg_{cat})	0.00 ± 0.00	0.91 ± 0.13	0.84 ± 0.02	0.90 ± 0.06
CH_3CHO sel. (%)	24 ± 10	2.7 ± 0.6	0.2 ± 0.3	2.5 ± 0.9
CH_4 sel. (%)	0	9.9 ± 1.0	7.1 ± 0.3	3.5 ± 0.3
CH_2CH_2 sel. (%)	0	6 ± 3	0.00 ± 0.00	–
625 °C				
CO/CO_2	–	4.8 ± 0.3	3.0 ± 0.4	1.26 ± 0.07
C balance (%)	–	88 ± 3	94.3 ± 1.8	99 ± 3
Conv. EtOH (%)	–	83 ± 3	100 ± 0	100 ± 0
H_2 productivity (mol/min kg_{cat})	–	0.76 ± 0.08	0.99 ± 0.05	1.23 ± 0.03
CH_3CHO sel. (%)	–	15.7 ± 1.1	0.00 ± 0.00	1.7 ± 0.4
CH_4 sel. (%)	–	8.0 ± 0.3	3.51 ± 0.5	0.9 ± 0.2
CH_2CH_2 sel. (%)	–	–	0.00 ± 0.00	0.00 ± 0.00
750 °C				
CO/CO_2	Only CO	17 ± 2	1.82 ± 0.11	1.71 ± 0.14
C balance (%)	103 ± 3	103 ± 8	96 ± 5	100 ± 3
Conv. EtOH (%)	54 ± 3	100 ± 0	100 ± 0	100 ± 0
H_2 productivity (mol/min kg_{cat})	0.061 ± 0.006	1.42 ± 0.13	1.22 ± 0.06	1.40 ± 0.02
CH_3CHO sel. (%)	62 ± 4	0.00 ± 0.00	0.00 ± 0.00	0.00 ± 0.00
CH_4 sel. (%)	0.00 ± 0.00	10 ± 4	1.9 ± 0.4	0.00 ± 0.00
CH_2CH_2 sel. (%)	31 ± 12	0.00 ± 0.00	0.00 ± 0.00	0.00 ± 0.00

$P = 1$ atm, $H_2O/CH_3CH_2OH = 3$ (mol/mol); GHSV = 1,750 h^{-1}

was achieved and no evidence of coke deposition was desumed due to 100 % C balance. Therefore, at 750 °C thermal activation proved sufficient to promote both the dehydrogenation (mostly) and dehydration reactions, but no further conversion of the products may be obtained.

By lowering reaction temperature to 500 °C ca. 15 % ethanol conversion was observed, but with C balance closing only to 91 % due to coke deposition on the filling material.

In this case, acetaldehyde was the main product and no evidence of ethylene was found at the reactor outlet, likely due to its full polymerisation to form coke.

Sample T-A tested at 750 °C showed interesting values of carbon balance and hydrogen productivity, though it was characterised by some drawbacks.

At first, it needed a longer induction period before reaching stable performance. In such period, the carbon balance was oscillating. Furthermore, some unreformed methane was present even at this high temperature ($S_{CH_4} = 10\%$) and the CO/CO_2 ratio was much higher for other samples. This indicates a lower activity for the

WGS reaction, which is however not fully compatible with the high H_2 yield observed.

On the other hand, water conversion was very high in comparison to CO removal, so that coke deposed during the first hours-on-stream (carbon balance < 100 %) was gasified and additional hydrogen came from carbon gasification by steam. Very high values of the CO/CO_2 ratio may be also compatible with a considerable activity for the dry reforming of ethanol.

At 500 °C the C balance was much lower than the blank test, and some ethylene was also detected. Ethanol conversion was not optimal and the selectivity to the main by-products non-negligible ($S_{CH_4} = 10\%$, $S_{CH_2CH_2} = 6\%$ and $S_{CH_3CHO} = 2.7\%$).

By increasing the temperature to 625 °C carbon balance, H_2 productivity and ethanol conversion remained unacceptably low in comparison to other samples, indicating some irreversible deactivation of the sample. Furthermore, non-negligible selectivity to undesired products was observed ($S_{CH_4} = 8\%$ and $S_{CH_3CHO} = 16\%$), which is usually nil at this temperature for the best performing catalysts.

Therefore, deactivation mainly occurred during the test at 500 °C for this sample, likely due to severe coking, and it was not completely reversible at 625 °C, at the contrary of other samples that showed good performance at 625 °C even if at 500 °C their C balance was not optimal.

Indeed, coke may form after dehydration of the substrate on acidic sites of the support followed by polymerisation. If only acid sites are blocked by carbon deposition, the system may still reach a stable steady-state performance after deactivation of such sites [22]. This may in part influence selectivity because of the role of the support in activation of water, but if coking of the support surface sites is not too extensive stable performance may occur.

In other cases, coking may interest also the metal active phase, which becomes covered or encapsulated by carbon, with blockage of the active sites [23]. Ni is indeed particularly active for the formation of carbon nanotubes, especially when its particle size is big. Evidence of nanotubes formation has been obtained in previous investigations [6–9, 24, 25], and here (*vide infra*).

Sample T-S was better performing than the parent T-A sample, though not yet completely satisfactory. Indeed, at 750 °C the carbon balance was acceptable and CO/CO_2 ratio was lower for T-A. However, the H_2 productivity was lower and some methane was still present ($S_{CH_4} = 1.9\%$). At 625 °C the carbon balance diminished to ca. 94 %, even if it was much higher (up to 120 %) during the first hours-on-stream due to the gasification of most of the coke deposits cumulated during the previous testing at 500 °C.

Methane selectivity increased to 3.5 % at 625 °C and to 7.1 % at 500 °C, when some acetaldehyde also appeared ($S_{CH_3CHO} = 0.2\%$). At the latter reaction temperature, the C balance was ca. 77 % and a bit unstable for the first testing hours and the hydrogen productivity was lower for other samples, in spite of the very good ethanol conversion.

Sample T-FP showed stable at 750 °C with good C balance, full ethanol conversion, no by-products and high hydrogen productivity. At 625 °C, the carbon balance was higher than 100 % for the first h-on-stream due to the gasification of coke deposed at 500 °C, then it stabilised to ca. 98 %. H_2 productivity was a bit lower than expected, also due to the presence of some by-products ($S_{CH_4} = 1.0\,\%$ and $S_{CH_3CHO} = 1.8\,\%$). Nevertheless, at 500 °C the sample showed unstable.

The carbon balance value reported in Table 51.1 may be misleading since this parameter increased monotonously from ca. 63 to ca. 92 %, meanwhile a decrease of ethanol conversion was observed. Ethanol conversion remained stable at 100 % for the first 4 h-on-stream, but it was evidently correlated to a huge coking activity.

Together with the increase of carbon balance, finally stabilised above 80 %, the progressive decrease of ethanol and water conversions occurred, indicating catalyst deactivation. As soon as conversion started decreasing, we also observed an increase of selectivity to by-products. For example S_{CH_4} increased to 3.5 %, whereas acetaldehyde was absent during the first h-on-stream, then rose to 2.5 %. At difference with sample T-A, such deactivation process was likely due to reversible coking, as testified by the good performance during the subsequent testing at 625 °C. Moreover, catalyst T-FP led to the lowest CO/CO_2 ratio (high activity for the WGS reaction) and to the highest H_2 productivity. However, at difference with some silica- and zirconia-supported catalysts [7, 8] the present titania-supported ones showed satisfactory performance only if tested at high temperature, mainly due to coking activity at low temperature.

Similar conclusions were drawn by Denis et al. [26], who compared differently supported Ni catalysts. Ethanol conversion was higher for the TiO_2-supported catalysts in the medium-high temperature range.

Low selectivity to C_2 has been also reported, but one of the highest ones to CH_4 (ca. 40 and 15 mol% at 500 °C and 600 °C, respectively, at difference with the present results), with additional coking problems.

51.3.4 Characterisation of the Spent Catalysts

The spent catalysts have been characterised by FE-SEM and TEM (Fig. 51.5). Evidence of coking is obtained for sample T-A during testing at 500 °C, according to its lowest C balance. Severe deactivation occurred for this sample, with decreasing ethanol conversion and increasing selectivity to by-products with time-on-stream. The increase of reaction temperature to 625 °C did not allow to completely gasify coke, as testified by the huge amount of C filaments still present on the spent sample. As a result, conversion and hydrogen productivity remained unacceptably low. TEM showed that carbon nanotubes (multiwalled) formed extensively on sample T-A, sometimes covering the whole particle and unacceptably depressing activity. The formation of C nanotubes also occurred over sample T-S, which on the contrary was not irreversibly deactivated being coking much more limited (Fig. 51.5). Therefore, increasing temperature to 625 °C is sufficient to recover full ethanol conversion and satisfactory selectivity. Some nanotubes were also present on sample T-FP.

In Fig. 51.5, one may also observe that some Ni sintering occurred, more evident for sample T-A than T-S, according to the reducibility scale and, thus, on the strength of the metal–support interaction.

The formation of C nanotubes and in general C deposition has been also confirmed by Raman analysis, where the typical D and G bands are present with decreasing intensity when passing from sample T-A to T-FP and T-S (Fig. 51.6).

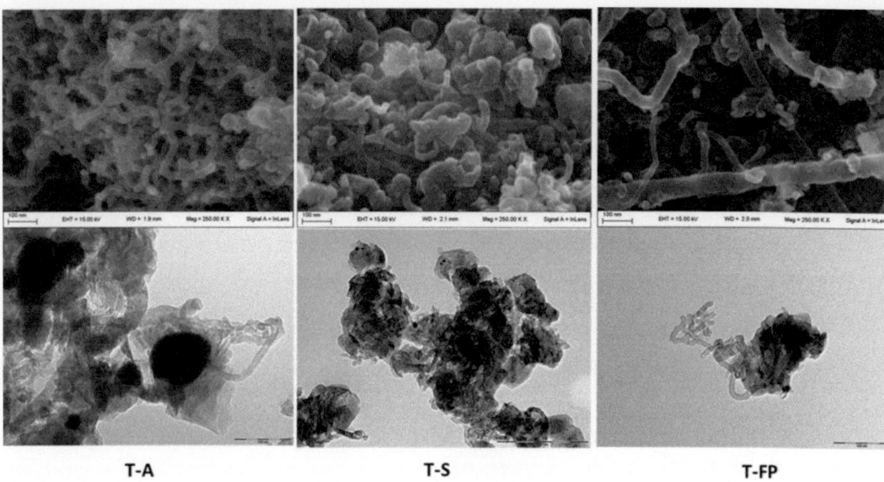

Fig. 51.5 TEM micrographs of spent samples. Marker size 100 nm

Fig. 51.6 Raman spectra of spent samples

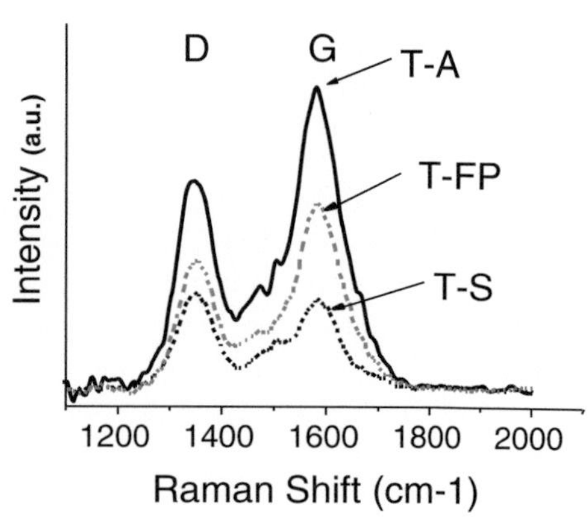

51.4 Conclusions

The performance of the titania-supported catalysts was found very dependent on the preparation procedure. In a previous investigation [9, 25], they demonstrated not active when calcined at 500 °C, likely due to surface reconstruction of the active phase and the support. Some better results have been achieved by calcination at higher temperature (i.e. at 800 °C as in the present case), leading to a stronger metal–support interaction and to a better dispersion of the active phase, as clearly demonstrated by data from TPR and CO adsorption. In principle, this should provide a better resistance towards the formation of carbon nanotubes during testing at 500 °C. Therefore, a new FP preparation procedure was adopted, leading to the formation of a mixed oxide, evolving after reduction to a very dispersed supported Ni active phase.

By looking at Tables 51.1 and 51.2, one may observe that Ni dispersion decreased from the FP sample to those prepared by precipitation (decreasing crystal size and metal surface area). Accordingly, carbon balance at 500 °C increased in the order T-A < T-S < T-FP (Table 51.2).

By comparing the different reaction conditions, it is possible to conclude that at 750 °C the differences among the samples flatten and the system approaches the equilibrium behaviour. When tested at 625 °C, the catalysts start to differentiate from one another.

At 500 °C no fully satisfactory catalyst has been found among this series, mainly due to carbon loss. Further improvements of catalyst formulation are needed, together with an optimisation of the reaction conditions (for example, increasing the water/ethanol feeding ratio or modifying contact time).

Support interaction with water and its acidity may play a role in the activation of the reactants and in catalyst resistance towards coking, strong acid sites being responsible for ethanol dehydration and subsequent ethylene polymerisation. Lewis acidity was predominant in the case of the present titania-supported samples, with medium strength acid sites. In every case, the absence of strong acid sites prevented from severe coking of the catalyst from this point of view. No preferred nature of acid sites (Lewis or Brønsted) may be invoked to explain coke formation, especially during testing at 500 °C, when the phenomenon was more relevant, but mainly ascribed to the formation of carbon nanotubes over the Ni active phase. For instance, sample FP exhibiting exposed Ti sites was not characterised by the highest coking rate. Therefore, provided that strong surface acidity is ruled out, no tight relationship between acidity and coking may be drawn, as already well pointed out by Liguras et al. [27].

By considering all these points, one may conclude that operation at 625 °C can be satisfactory from all the points of view of activity, productivity and C balance, allowing to limit the heat input to the reactor with respect to operation at 750 °C. According to the present results, an interesting candidate for ESR at 625 °C is sample T-FP.

According to pilot projects on stationary PEM fuel cell systems, characterised by 5 $kW_{electric}$ power size and efficiency variable from 0.35 to 0.41, H_2 consumption has been reported between 3.5 and 4.1 Nm^3/h [28]. According to the highest productivity here reported at each temperature we may estimate that the production of 4 Nm^3/h of H_2 requires ca. 2.1 kg of the T-FP catalyst operated at 750 °C and 2.4 kg at 625 °C. Therefore, there is no real advantage in operating at 750 °C since H_2 productivity does not increase appreciably (a higher reforming activity for by-products is counterbalanced by an unfavourable thermodynamics of the WGS reaction), and therefore similar installation costs for the catalyst are required, while evident drawbacks are higher variable costs for reactor heating and for purification from CO. Furthermore, in order to limit coking rate operating at 625 °C seems sufficient to guarantee stable catalyst performance, provided that a sufficiently SMSI is achieved during preparation, avoiding metal sintering and coking.

Acknowledgements The authors are indebted with Regione Lombardia and the Consortium for Material Science and Technology (INSTM) for financial support. The characterisation was partly supported by H2FC European Infrastructure Project (Integrating European Infrastructure to support science and development of Hydrogen and Fuel Cell Technologies towards European Strategy for Sustainable Competitive and Secure Energy), project reference 284522.

References

1. Fatsikostas AN, Verykios XE (2004) Reaction network of steam reforming of ethanol over Ni-based catalysts. J Catal 225:439–452
2. Centi G, Perathoner S (2009) Opportunities and prospects in the chemical recycling of carbon dioxide to fuels. Catal Today 148:191–205
3. Gonzalez-Dela Cruz VM, Holagado JP, Pereniguez R, Caballero A (2008) Morphology changes induced by strong metal-support interaction on a Ni-ceria catalytic system. J Catal 257:307–314
4. Christensen KO, Chen D, Lodeng R, Holmen A (2006) Effect of supports and Ni crystal size on carbon formation and sintering during steam methane reforming. Appl Catal Gen 314:9–22
5. Chen D, Christensen KO, Ochoa-Fernandez E, Yu Z, Totdal B, Latorre N, Monzón A, Holmen A (2005) Synthesis of carbon nanofibers: effects of Ni crystal size during methane decomposition. J Catal 229:82–96
6. Rossetti I, Gallo A, Dal Santo V, Bianchi CL, Nichele V, Signoretto M, Finocchio E, Ramis G, Garbarino G, Di Michele A (2013) Ni catalysts supported over TiO_2, SiO_2 and ZrO_2 for the steam reforming of glycerol. ChemCatChem 5:294–306
7. Rossetti I, Biffi C, Bianchi C, Nichele V, Signoretto M, Menegazzo F, Finocchio E, Ramis G, Di Michele A (2012) Ni/SiO_2 and Ni/ZrO_2 catalysts for steam reforming of ethanol. Appl Catal Environ 384:117–118
8. Rossetti I, Lasso J, Nichele V, Signoretto M, Finocchio E, Ramis G, Di Michele A (2014) Silica and zirconia supported catalysts for the low-temperature ethanol steam reforming. Appl Catal Environ 150–151:257–267
9. Rossetti I, Lasso J, Finocchio E, Ramis G, Nichele V, Signoretto M, Di Michele A (2014) TiO_2-supported catalysts for the steam reforming of ethanol. Appl Catal Gen 477:42–53
10. Tauster SJ, Fung SC, Garten RL (1978) Strong metal-support interactions. Group 8 noble metals supported on TiO_2. J Am Chem Soc 100:170–175
11. Tauster SJ (1987) Strong metal-support interactions. Account Chem Res 20:389–394

12. Subramanian S (1992) Temperature-programmed reduction of platinum group metals catalysts. Platin Met Rev 36:98–103
13. Chiarello GL, Rossetti I, Forni L (2005) Flame-spray pyrolysis preparation of perovskites for methane catalytic combustion. J Catal 236:251–261
14. Chiarello GL, Rossetti I, Lopinto P, Migliavacca G, Forni L (2006) Preparation by flame-spray pyrolysis of $ABO_{3\pm\delta}$ catalysts for the flameless combustion of methane. Catal Today 117:549–553
15. Chiarello GL, Rossetti I, Lopinto P, Migliavacca G, Forni L (2007) Solvent nature effect in preparation of perovskites by flame pyrolysis. 1. Carboxylic acids. Appl Catal Environ 72:218–226
16. Chiarello GL, Rossetti I, Lopinto P, Migliavacca G, Forni L (2007) Solvent nature effect in preparation of perovskites by flame pyrolysis. 2. Alcohols and alcohols + propionic acid mixture. Appl Catal Environ 72:227–232
17. Baraton MI, Busca G, Prieto MC, Ricchiardi G, Sanchez Escribano V (1994) On the vibrational spectra and structure of $FeCrO_3$ and of the ilmenite-type compounds $CoTiO_3$ and $NiTiO_3$. J Solid State Chem 112:9–14
18. Hadjiivanov KI, Vayssilov GN (2002) CO as a probe molecule in the characterization of oxide surfaces and zeolites. In: Gates BC, Knozinger H (eds) Advances in catalysis, vol 47. Academic, Burlington
19. Garbarino G, Sanchez Escribano V, Finocchio E, Busca G (2012) Steam reforming of phenol–ethanol mixture over 5% Ni/Al_2O_3. Appl Catal Environ 113–114:281–289
20. Garbarino G, Finocchio E, Lagazzo A, Valsamakis I, Riani P, Sanchez Escribano V, Busca G (2014) Steam reforming of ethanol-phenol mixture on Ni/Al_2O_3: effect of magnesium and boron on catalytic activity in the presence and absence of sulphur. Appl Catal Environ 147:813–826
21. Barattini L, Ramis G, Resini C, Busca G, Sisani M, Costantino U (2009) Reaction path of ethanol and acetic acid steam reforming over Ni–Zn–Al catalysts. Flow reactor studies. Chem Eng J 153(1–3):43–49
22. Le Valant A, Garron A, Bion N, Duprez D, Epron F (2011) Effect of higher alcohols on the performances of a 1% $Rh/MgAl_2O_4/Al_2O_3$ catalyst for hydrogen production by crude bioethanol steam reforming. Int J Hydrogen Energy 36:311–318
23. Wang F, Li Y, Cai W, Zhan E, Mu X, Shen W (2009) Ethanol steam reforming over Ni and Ni-Cu catalysts. Catal Today 146:31–36
24. Finocchio E, Rossetti I, Ramis G (2013) Redox properties of Co and Cu-based catalysts for the steam reforming of ethanol. Int J Hydrogen Energy 38:3213–3225
25. Nichele V, Signoretto M, Menegazzo F, Rossetti I, Cruciani G (2014) Hydrogen production by ethanol steam reforming: effect of the synthesis parameters on the activity of Ni/TiO_2 catalysts. Int J Hydrogen Energy 39:4252–4258
26. Denis A, Grzegorczyk W, Gac W, Machocki A (2008) Steam reforming of ethanol over Ni/support catalysts for generation of hydrogen for fuel cell applications. Catal Today 137:453–459
27. Liguras D, Kondarides D, Verykios X (2003) Production of hydrogen for fuel cells by steam reforming of ethanol over supported noble metal catalysts. Appl Catal Environ 43:345–354
28. Rossetti I, Biffi C, Tantardini GF, Raimondi M, Vitto E, Alberti D (2012) 5 kWe + 5 kWt reformer-PEMFC energy generator from bioethanol first data on the fuel processor from a demonstrative project. Int J Hydrogen Energy 37:8499–8504

Chapter 52
Biodegradation of *o*-Cresol by a *Pseudomonas* spp

Adh'ya-eddine Hamitouche, Zoubida Bendjama, Abdeltif Amrane, and Farida Kaouah

Abstract The *o*-cresol is one of the VOCs toxic compounds, it is used as pharmaceutical intermediates, reagents, custom synthesis and as a solvent, disinfectant and chemical intermediate for a wide variety of products including resins, paints and textiles. Efficient treatment technologies are required to reduce *o*-cresol concentration in wastewater to acceptable levels because *o*-cresol is hazardous even at low concentration. Biological treatment methods have been researched to treat *o*-cresol in wastewater. Several studies have shown that *o*-cresol can aerobically degraded by a wide variety of microorganisms.

In this study, a series of experiments were performed to examine the effects of the mineral medium composition and the pH on *o*-cresol removal. In this purpose, *o*-cresol biodegradation was carried out in a batch reactor containing mixed bacteria; the temperature (30 °C), the stirring velocity (200 r/min), the KH_2PO_4 concentration (1.5 g/L), the K_2HPO_4 concentration (2 g/L), and *o*-cresol concentration (100 mg/L) were kept constants. The initial pH was varied in the range 5–9 and the mineral components were tested in the following concentration ranges: 0–2 g/L for nitrogen sources (NH_4Cl, KNO_3, and NH_4NO_3), 0–0.5 g/L for NaCl, and 0–0.2 g/L for $MgSO_4$. Their effects on *o*-cresol biodegradation and specific growth rate were

A.-e. Hamitouche (✉)
Laboratoire des Sciences du Génie des Procédés Industriels, Faculté de Génie Mécanique et de Génie des Procédés, Université des Sciences et de la Technologie Houari Boumediene, BP 32, El-Alia, Bab ezzouar, Alger 16111, Algeria

Centre de Recherche scientifique et technique en Analyses Physico-Chimiques, BP 248, CRAPC, Alger, Algeria
e-mail: ahamitouche2@yahoo.fr

Z. Bendjama • F. Kaouah
Laboratoire des Sciences du Génie des Procédés Industriels, Faculté de Génie Mécanique et de Génie des Procédés, Université des Sciences et de la Technologie Houari Boumediene, BP 32, El-Alia, Bab ezzouar, Alger 16111, Algeria

A. Amrane
Ecole Nationale Supérieure de Chimie de Rennes, Université Rennes 1, CNRS, UMR 6226, Avenue du Général Leclerc, CS 50837, Rennes Cedex 7 35708, France

Université européenne de Bretagne, Rennes 35000, France

I. Dincer et al. (eds.), *Progress in Clean Energy, Volume 1*,
DOI 10.1007/978-3-319-16709-1_52

examined. The shorter biodegradation time of *o*-cresol was 35.6 h for NH_4Cl, NaCl, and $MgSO_4$ concentrations of 1 g/L, 0.3 g/L, and 0.1 g/L, respectively. Maximum specific growth rate (0.33 h^{-1}) and total *o*-cresol removal (99.99 %) were recorded for an optimal pH value of 8.

Keywords Biodegradation • *o*-Cresol • *Pseudomonas* spp. • Kinetic

52.1 Introduction

Volatile organic compounds (VOCs) released into the environment result in environmental damage as pollutants. It is strongly advised not to let the chemical enter into the environment because it persists in the environment. Special attention should be given to water, since many industrial processes have used highly volatile compounds as starting materials, as intermediates or solvent and some of them end up in process waters.

The removal of VOCs from industrial water effluents is of growing concern and there is an urgent need to develop strategies to deal with VOC-contaminated wastewaters. Although several techniques are available for treatment of VOCs, in recent years most of the hazardous VOCs entering process waters are biologically treated due to the environmental friendly nature of biotechnology [1–3].

Therefore, biodegradation of VOCs by pure or mixed cultures of microorganisms has been extensively studied.

As chemicals, pharmaceutical intermediates, reagents, custom synthesis, and as a solvent, disinfectant and chemical intermediate for a wide variety of products including resins, paints and textiles, *o*-cresol is a toxic one of VOCs and it is used. Efficient treatment technologies are required to reduce *o*-cresol concentration in wastewater to acceptable levels because *o*-cresol is hazardous even at low concentration.

Biological treatment methods have been researched to treat *o*-cresol in wastewater. Several studies have shown that *o*-cresol can aerobically degraded by a wide variety of microorganisms [4–6]. Bacteria are a class of microorganisms actively involved in the degradation of organic pollutants from contaminated sites. A number of bacterial species are known to degrade phenolic compounds. Most of them, showing high biodegradation efficiency, are isolated from contaminated soil or sediments [7]. The identified organisms belonged to several genera, like *Pseudomonas*, as well as *Agrobacterium*, *Bacillus* [8], *Burkholderia*, *Sphingomonas* [7], *Rhodococcus* [9] species, and mixed culture [10, 11].

The main goal of this chapter was to investigate the biodegradation of *o*-cresol by mixed bacteria; the effect of pH and the mineral medium composition were optimized to determine the best conditions for *o*-cresol removal, especially the mineral salts supplementation and the optimal mineral nitrogen concentration.

52.2 Materials and Methods

52.2.1 Microorganisms Cultivation

The mixed bacteria of Pseudomonas genus (*Pseudomonas* aeruginosa, *Pseudomonas* putida, *Pseudomonas* fluorescens) used in this work were obtained from activated sludge from the hazardous wastewater station of Boumerdès (Algeria). The stock cultures were stored at 4 °C. The mixed bacteria were activated for 24 h at 30 °C in the nutrient medium (NB) containing (g/L): peptone, 15, yeast extract, 3, sodium chloride, 6, and (D+)-glucose, 1.

After 24 h, when cells were grown, the biomass was harvested by centrifugation. The microorganisms collected after centrifugation (3,000 rpm) for 30 min were suspended in NaCl 0.5 % and recentrifuged. After the third washing, the microorganisms collected after centrifugation were resuspended in NaCl 0.5 % to determine the concentration of the mixed bacteria. This solution (mixed bacteria and NaCl 0.5 %) was analyzed by measuring OD at 600 nm using a Vis spectrophotometer (HACH DR2800); the OD value was then converted to dry cell mass using a dry weight calibration curve. The dry cell mass density (g/L) was found to follow the following regression equation $\times$ (g/L) = 1.044 $\times$ OD_{600}.

Specific growth rate was determined in the exponential growth phase [12–15]. For each flask, the specific growth value was determined from linear semilogarithmic plot of cell concentration versus time during the exponential growth phase, namely when specific growth rate became nearly constant [16].

52.2.2 Biodegradation Experiments

As the OD value of adapted cells reached 2.7–2.9, an aliquot of the culture was centrifuged at 3,000 rpm for 30 min.

To wash the biomass, it was resuspended in NaCl 0.5 % and centrifuged. The cells (1 mL) were then transferred and inoculated in Erlenmeyer flasks (250 mL) to yield an initial OD of 0.078, and containing 100 mL of medium containing nitrogen source (NH_4Cl, KNO_3, and NH_4NO_3) and the following mineral salt supplementation (MSS), namely KH_2PO_4, KH_2PO_4, and $MgSO_4$ at the required concentrations, and 100 mg/L of *o*-cresol. The cells were cultivated at 30 °C and 200 rpm.

Samples were withdrawn at suitable time intervals, and the concentration of cells was deduced from optical density measurement and *o*-cresol was measured as described below.

52.2.3 Residual o-Cresol Determination

The residual of *o*-cresol was estimated using a Vis spectrophotometer (specord 210 plus, Analytik-Jena) and was calculated by absorbance at 271 nm.

52.3 Results and Discussion

52.3.1 Optimal Salt Concentrations

The effect of mineral salt supplementation of culture medium on *o*-cresol degradation was shown for instance for NaCl (Table 52.1). The Fig. 52.1 shows that the lag phase was at least 35.6 h in the range of NaCl concentrations tested; total *o*-cresol removal (100 mg/L) was recorded in less than 39.25 h (Table 52.1).

Table 52.1 shows that the effect of culture medium components ([NaCl] and [$MgSO_4$]) on influential parameters on biodegradation of *o*-cresol. The specific growth rate increase with increasing of NaCl concentration. The maximum value of specific growth rate was 0.31 h^{-1} recorded for 0.3 g/L NaCl. This amount was in agreement with the mineral supplementation considered by other workers, since Luo et al. [18] and Nakano et al. [19] supplemented with 0.1 and 0.21 g/L NaCl to biodegrade phenol, while 0.4 g/L was used by Zilouei and al [20] to biodegrade chlorophenols (2-chlorophenol, 4-chlorophenol, 2, 4-dichlorophenol, and 2, 4, 6-trichlorophenol).

At 30 °C and pH 7, a series of degradation batch test at different $MgSO_4$ concentration were conducted with mixed culture. The effect of $MgSO_4$ concentration on *o*-cresol degradation was observed on Table 52.1 and Fig. 52.2.

Table 52.1 Effect of [NaCl] and [$MgSO_4$] concentrations on influential parameters on biodegradation of *o*-cresol

Parameters (g/L)	Value of parameters	Specific growth rate (h^{-1})	Time of lag phase (h)	% Removal of *o*-cresol	Time of total *o*-cresol removal (h)
NaCl	0	0.093	35.6	100	39.25
	0.05	0.12	35.6	100	39.25
	0.1	0.141	35.6	100	39.25
	0.2	0.216	35.6	100	39.25
	0.3	*0.31*	35.6	100	39.25
	0.4	0.181	35.6	100	39.25
	0.5	0.15	35.6	100	39.25
$MgSO_4$	0.05	0.209	41	100	45.25
	0.1	*0.31*	35.6	100	39.25
	0.15	0.247	37.7	100	41.25
	0.2	0.157	40	100	43.25

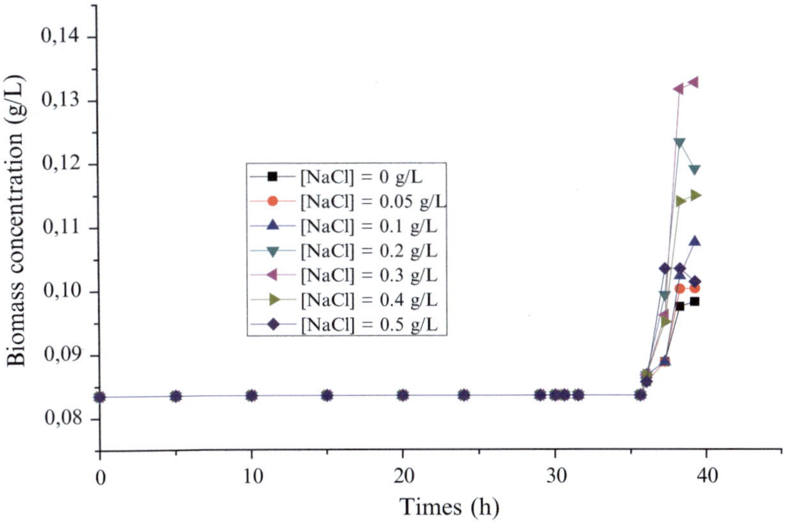

Fig. 52.1 Evolution of biomass concentration versus times for different value of NaCl concentration

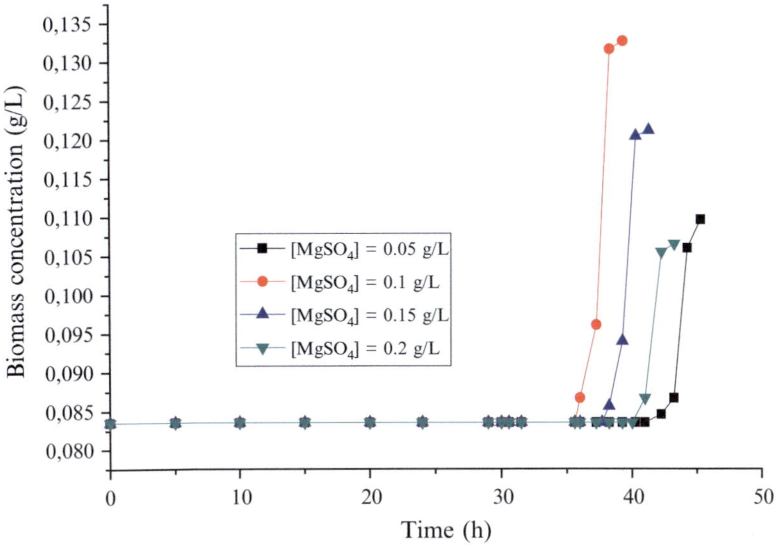

Fig. 52.2 Evolution of biomass concentration versus times for different value of $MgSO_4$ concentration

The Table 52.1 shows that a mixed bacterium degrades a 100 mg/L of *o*-cresol within 45.25 h, the shorter time is equal to 39.25 h when $MgSO_4$ concentration is equal to 0.1 g/L.

The Fig. 52.2 shows that a time of lag phase and maximum biomass concentration varied from 35.6 to 41 h and 0.106 g/L to 0.132 g/L, respectively.

The evolution of specific growth rate versus $MgSO_4$ concentration (Table 52.1) shows that the maximum value equal to 0.31 h^{-1} when $MgSO_4$ concentration is equal to 0.1 g/L, the same concentration of $MgSO_4$ in mineral salt medium was used by Zhao et al. and Santos VLd et al. [21, 22] to biodegrade phenol.

52.3.2 Optimal Nitrogen Source Concentration

A series of biodegradation batch test at different nitrogen concentration sources were conducted with *Pseudomonas* spp. mixture to see the effect of the three nitrogen sources selected on biodegradation of *o*-cresol. The results were regrouped in Table 52.2 and Figs. 52.3, 52.4, and 52.5.

Table 52.2 indicated the effect of nitrogen sources concentration on *o*-cresol degradation. A 100 g/L of *o*-cresol was entirely degraded within 48.25 h; the time shorter for *o*-cresol biodegradation is equal to 39.25 h when the optimal concentration is equal to 1 g/L.

The evolution of cell concentration shows that a lag phase is equal between 35.6 and 41 h for different nitrogen sources concentration.

Table 52.2 shows that the effect of evolution of concentration nitrogen sources on specific growth rate, the maximum value of specific growth rate is equal to 0.31 h^{-1}, 0.283 h^{-1}, and 0.299 h^{-1} when NH_4Cl, KNO_3, and NH_4NO_3, respectively, is equal to 1 g/L.

Table 52.2 Effect of nitrogen sources on the specific growth rate

Parameters (g/L)	Value of parameters	Specific growth rate (h^{-1})	Time of lag phase (h)	% Removal of *o*-cresol	Time of total *o*-cresol removal (h)
NH_4Cl	0.25	0.156	39	100	48.25
	0.5	0.258	38	100	46.25
	1	*0.31*	35.6	100	39.25
	1.5	0.198	40	100	45.25
	2	0.156	41	100	44.25
KNO_3	0.25	0.191	41	100	48.5
	0.5	0.251	41	100	48.5
	1	*0.283*	36	100	39.25
	1.5	0.228	39	100	45.25
	2	0.181	40	100	44.25
NH_4NO_3	0.25	0.197	41	100	48.25
	0.5	0.271	41	100	48.25
	1	*0.299*	36.25	100	39.25
	1.5	0.230	38	100	45.25
	2	0.206	38	100	44.25

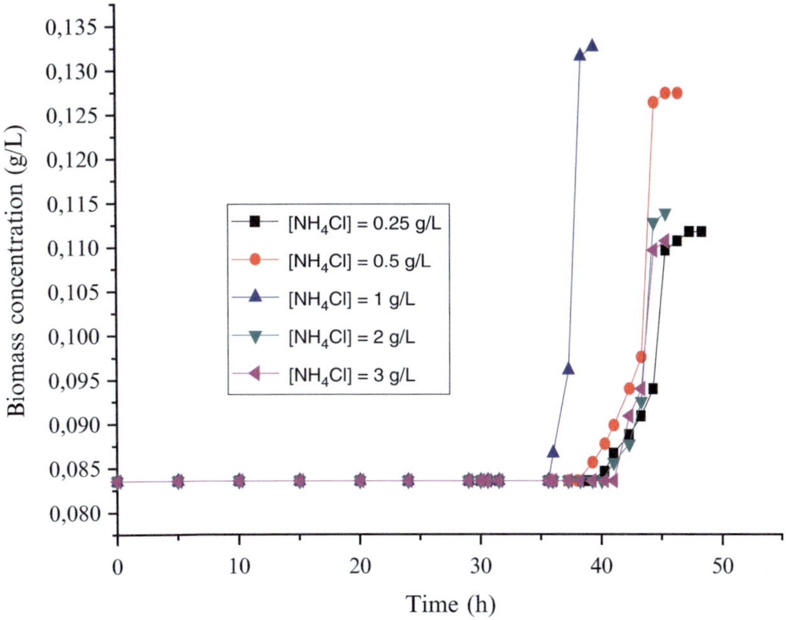

Fig. 52.3 Evolution of biomass concentration versus time for different value of NH_4Cl concentration

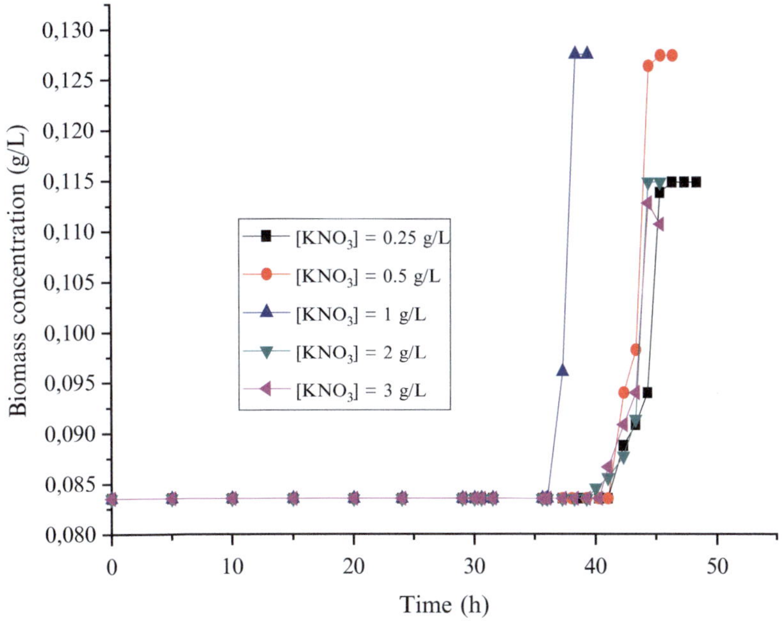

Fig. 52.4 Evolution of biomass concentration versus time for different KNO_3 concentration

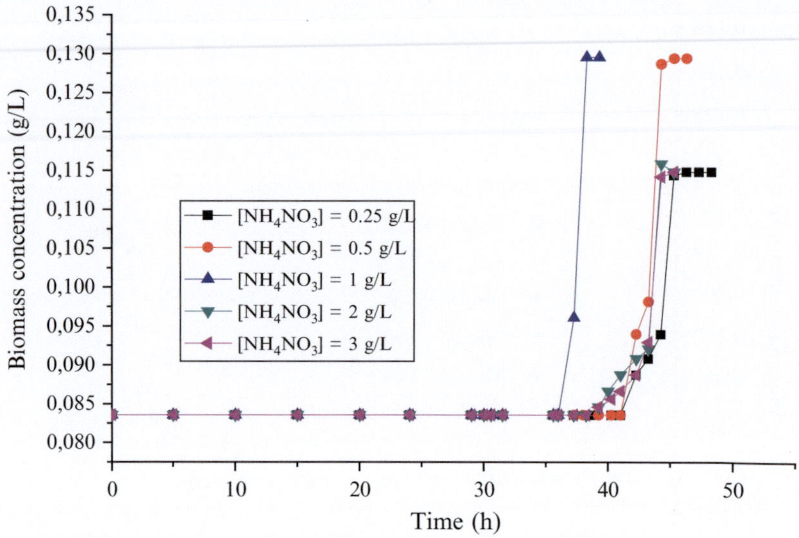

Fig. 52.5 Evolution of biomass concentration versus time for different NH_4NO_3 concentration

52.3.3 Optimal pH

Batch cultures of mixed bacteria were conducted in optimal condition of salt and nitrogen concentration containing initial *o*-cresol concentration equal to 100 mg/L and initial pH ranging from 5 to 9.

The evolution of biomass concentration versus time for different value of initial pH was represented in Figs 52.6, 52.7, and 52.8.

An *o*-cresol (100 mg/L) was entirely degraded within 48.25 h; the shorter time for *o*-cresol biodegradation is equal to 39.25 h when pH is equal to 7.

The evolution of cell concentration shows that a lag phase is equal to 35.6–41 h for different pH value.

Figure 52.9 shows that the effect of pH on specific growth rate. The maximum value of specific growth rate is equal to 0.33 h^{-1}, 0.314 h^{-1}, and 0.329 h^{-1} for NH_4Cl, KNO_3, and NH_4NO_3, respectively, when pH is equal to 8. The study of Kuo-Ling Ho et al. [2], Xinghui Qiu et al. [23], and Wang Ying et al. [24] obtained the same optimal pH with a phenol and p-nitrophenol.

The optimal values of specific growth rate are very higher than those found in the literatures, where the specific growth rate of mixed culture obtained has been reported to be in the range of 0.13–0.36 h^{-1} [16, 17, 25–29].

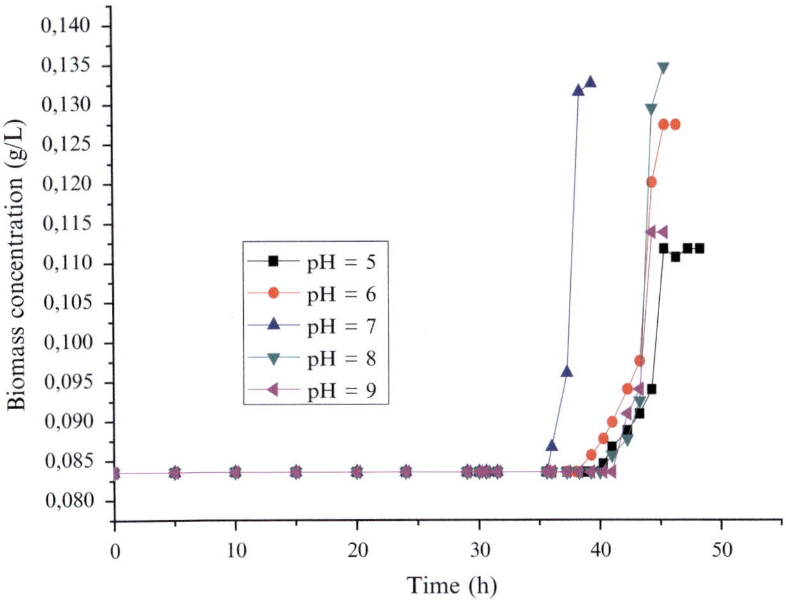

Fig. 52.6 Evolution of biomass concentration versus time for different initial pH value and NH_4Cl as nitrogen source

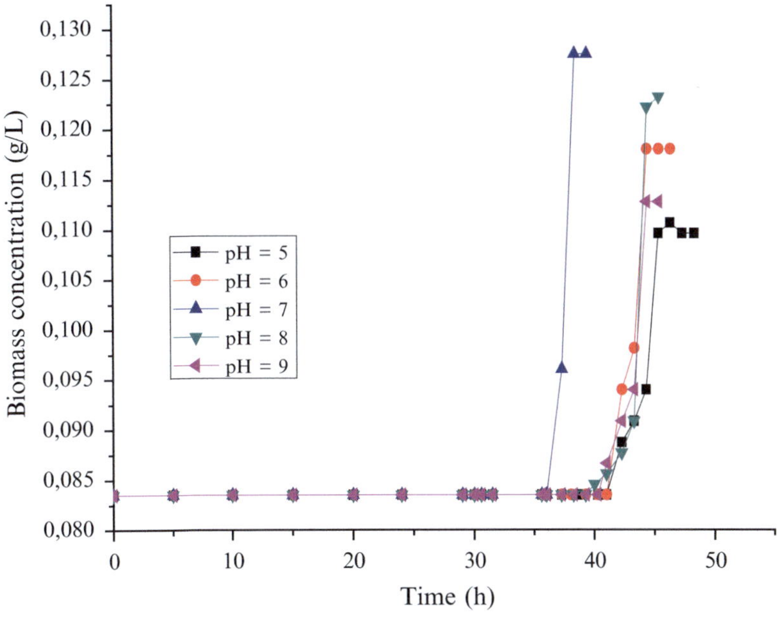

Fig. 52.7 Evolution of biomass concentration versus time for different initial pH value and KNO_3 as nitrogen source

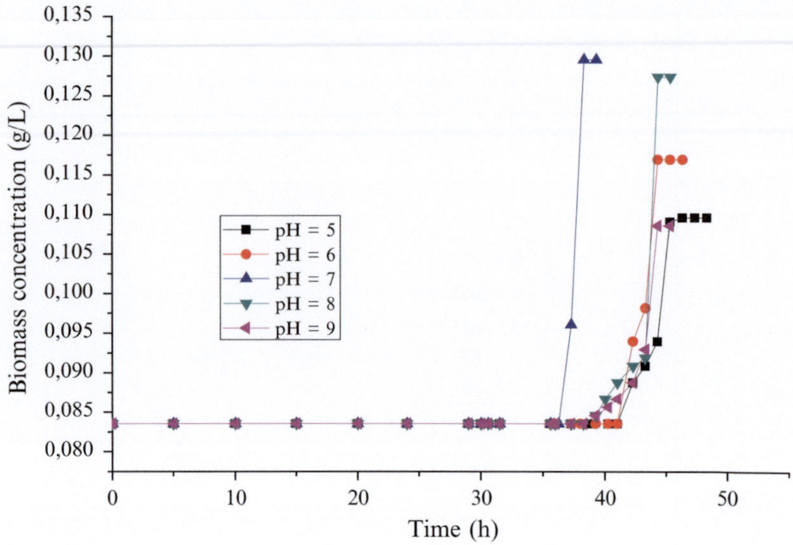

Fig. 52.8 Evolution of biomass concentration versus time for different initial pH value and NH_4NO_3 as nitrogen source

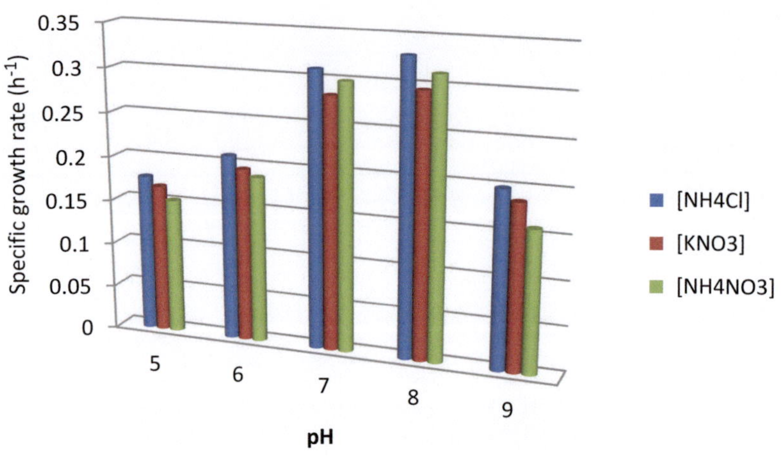

Fig. 52.9 Evolution of specific growth rate versus pH for different nitrogen sources

52.4 Conclusion

Growth kinetics of mixed bacteria and their degradation nature for phenol were investigated.

The following conclusions were drawn:

– As the mixed culture have high biodegradation activity where specific growth rate grow up until 0.33 h^{-1}, with a time of biodegradation equal to 40.27 h.

- A 100 mg/L of *o*-cresol are completely degrade for each batch experience and for different parameter studied, where time of biodegradation range is to 40.27 until 53.15 h.
- The optimal mineral medium concentrations (g/L) are equal to 0.3, 0.1, and 1 for NaCl, $MgSO_4$, and nitrogen sources, respectively
- The optimal value of pH is equal to 8.

References

1. Sa CSA, Boaventura RAR (2001) Biodegradation of phenol by pseudomonas putida DSM 548in a trickling bed reactor. Biochem Eng J 9:211–219
2. Ho K-L, Lin B, Chen Y-Y, Lee D-J (2009) Biodegradation of phenol using Corynebacterium sp. DJ1 aerobic granules. Bioresour Technol 100:5051–5055
3. Sittig M (1997) How to remove pollutants and toxic materials from air and water, a practical guide. Noyes Data Corporation, Park Ridge
4. Nemerow NL (1978) Industrial water pollution, origin, characteristics and treatment. Addison-Wesley, London
5. Patterson JW (1985) Industrial wastewater treatment technology, 2nd edn. Butterworths, Stoneham
6. Berkowitz JB (1988) Standard handbook of hazardous water treatment and disposal—hazardous waste recovery processes. McGraw-Hill, New York (Chapter 6)
7. Haritash AK, Kaushik CP (2009) Biodegradation aspects of polycyclic aromatic hydrocarbons (PAHs): a review. J Hazard Mater 169:1–15
8. Gurujeyalakshmi G, Oriel P (1989) Isolation of phenol degrading Bacillus stearothermophilus and partial characterisation of the phenol hydroxylase. Appl Environ Microbiol 55:500–502
9. Margesin R, Fonteyne PA, Redl B (2005) Low-temperature biodegradation of high amounts of phenol by Rhodococcus spp and basidiomycetous yeasts. Res Microbiol 156:68–75
10. Hamitouche A, Amrane A, Bendjama Z, Kaouah F (2010) Effect of the ammonium chloride concentration on the mineral medium composition — Biodegradation of phenol by a microbial consortium. Int J Environ Res 4(4):849–854
11. Hamitouche A, Amrane A, Bendjama Z, Kaouah F (2011) Phenol biodegradation by mixed culture in batch reactor—Optimization of the mineral medium composition. Desalin. Water Treat. 25:20–24
12. Dagley S, Gibson DT (1965) The bacterial degradation of catechol. Biochem J 95:466–474
13. Chiam HF, Harris IJ (1982) A model for non inhibitory microbial growth. Biotechnol Bioeng 24:37–55
14. Stanier RY, Palleroni NJ, Doudoroff M (1966) The aerobic Pseudomonasa taxonomic study. J Gen Microbiol 43:159–275
15. Worden RM, Donaldson TL (1987) Dynamics of a biological fixed film for phenol degradation in a fluidized-bed bioreactor. Biotechnol Bioeng 30:398–412
16. D'Adamo PD, Rozich AF, Gaudy AF Jr (1984) Analysis of growth data with inhibitory carbon sources. Biotechnol Bioeng 26:397–402
17. Yang RD, Humphrey AE (1975) Dynamic and steady state studies of phenol biodegradation in pure and mixed cultures. Biotechnol Bioeng 17:1211–1235
18. Luo H, Liu G, Zhang R, Jin S (2009) Phenol degradation in microbial fuel cells. Chem Eng J 147(2–3):259–264
19. Nakano Y, Nishijima W, Soto E, Okada M (1999) Relationship between growth rate of phenol utilizing bacteria and the toxic effect of metabolic intermediates of trichloroethylene (TCE). Water Res 33(4):1085–1089

20. Zilouei H, Guieysse B, Mattiasson B (2006) Biological degradation of chlorophenols in packed-bed bioreactors using mixed bacterial consortia. Process Biochem 41(5):1083–1089
21. Santos VLd, de Souza Monteiro A, Danúbia Telles B, Marcelo Matos S (2009) Phenol degradation by Aureobasidium pullulans FE13 isolated from industrial effluents. J Hazard Mater 161(2–3):1413–1420
22. Zhao G, Zhou L, Li Y, Liu X, Ren X, Liu X (2009) Enhancement of phenol degradation using immobilized microorganisms and organic modified montmorillonite in a two-phase partitioning bioreactor. J Hazard Mater 169:402–410
23. Qiu X, Ping W, Zhang H, Li M, Yan Z (2009) Isolation and characterization of Arthobacter sp. HY2 capable of degrading a high concentration of p-nitrophenol. Bioresour Technol 100:5243–5248
24. Wang Y, Tian Y, Han B, Zhao HB, Bi JN, Cai BL (2007) Biodegradation of phenol by free and immobilized Acinetobacter sp. Strain PD12. J Environ Sci 19:222–225
25. Pawlowsky U, Howell JA (1973) Mixed culture biooxidation of phenol, Determination of kinetic parameters. Biotechnol Bioeng 15:889–896
26. Hill GA, Robinson CW (1975) Substrate inhibition kinetics: phenol degradation by Pseudomonas putida. Biotechnol Bioeng 17:1599–1615
27. Sokol W (1987) Oxidation of an inhibitory substrate by washed cells (oxidation of phenol by Pseudomonas putida. Biotechnol Bioeng 30:921–927
28. Allsop PJ, Chisti Y, Moo-Young M, Sullivan GR (1993) Dynamic of phenol degradation by Pseudomonas putida. Biotechnol Bioeng 41:572–580
29. Rozich AF, Colvin RJ (1986) Effects of glucose on phenol biodegradation by heterogeneous populations. Biotechnol Bioeng 28:965–971

Chapter 53
Bioadditive Synthesis from Glycerol by Esterification Using Catalytic Chitosan Membrane

Derya Unlu and Nilufer Durmaz Hilmioglu

Abstract Glycerol is the main by-product in biodiesel production. The increasing production of biodiesel has resulted in significant amount of glycerol deposition. Esterification of glycerol with acetic acid is a kind of method to consume excess glycerol. The valuable chemicals like that bioadditives are produced in this method. $Zr(SO_4)_2{\cdot}4H_2O$ loaded chitosan catalytic membrane was prepared by the solution casting method and used as catalyst for esterification of glycerol and acetic acid. The reactions were performed in a batch reactor and in a pervaporation catalytic membrane reactor (PVCMR). The structure of the membranes was investigated by means of Fourier transform infrared spectroscopy (FTIR) and scanning electron microscopy (SEM). The results of esterification of acetic acid and glycerol show that the conversion was slightly improved with the $Zr(SO_4)_2{\cdot}4H_2O$ content in the membrane at a fixed chitosan content. The reaction rate constant increased with increasing acetic acid/glycerol molar ratio, amount of catalyst loading in the catalytic membrane and reaction temperature. Higher conversion values were obtained by PVCMR.

Keywords Catalytic membrane • Fuel bioadditive • Global warming • Glycerol • Green processes

Nomenclature

N_A	The mole of acetic acid at any time, mole
N_{A0}	The initial mole of acetic acid, mole
X	Conversion of acetic acid, %

D. Unlu (✉) • N.D. Hilmioglu
Department of Chemical Engineering, Kocaeli University, Kocaeli 41380, Turkey
e-mail: derya.unlu@kocaeli.edu.tr; niluferh@kocaeli.edu.tr

I. Dincer et al. (eds.), *Progress in Clean Energy, Volume 1*,
DOI 10.1007/978-3-319-16709-1_53

53.1 Introduction

Global warming is the biggest problem at the beginning of this century. Renewable energy sources reduce the impact of CO_2 emission. Therefore, renewable fuels have gained considerable attention alternative to fossil fuels [1]. Biodiesel is ordinarily obtained by transesterification of vegetable oils with short chain alcohol [2]. Comparing on traditional fuels derived from crude oil: Biodiesel is nature, renewable, biodegradable, harmless, and fundamentally free of sulfurous and aromatic species, reduces exhaust and has high ignition quality and a high cetane number [3].

Glycerol is the by-product of the biodiesel production by transesterification of oil and alcohol. The increasing demand and production of biodiesel has resulted in the accumulation of huge amounts of glycerol. Approximately glycerol is produced by 10 wt% of total biodiesel synthesis. Due to the increase of glycerol amount, a price decline has been observed [4]. Glycerol can be used in food additives and beverages, surfactants, cosmetic ingredients, and pharmaceutical applications. However, it is not possible to deplete all glycerol manufactured by the biodiesel industry by those markets. The great attempts occur to develop alternative techniques for obtaining high value-added products from glycerol [5].

Glycerol as promising values has been envisaged for the transportation sectors and biorefinery industries to convert it into fuel additive. Glycerol could play an important role as the main component in the development of incorporating biorefinery industries along with cleanness of environmental circuit. The fuel additives produced from glycerol are compatible with vehicle engine and economic and environmental friendly also. There are different methods such as decomposition to propane diols and for the achievements of another oxidation product for the production of invaluable chemicals from glycerol [5].

The esterification of glycerol with acetic acid can be a good alternative application for using glycerol by-product of biodiesel. Monoacetin (MAG), diacetin (DAG), and triacetin (TAG) are the main products of esterification reaction of glycerol with acetic acid. Applications of MAG, DAG, and TAG are in the cryogenic industries, the synthesis of biodegradable polymers and uses from cosmetics to food additives [3]. Additionally, DAG and TAG are used as fuel bioadditives for liquid fuels because they can improve cold and viscosity properties, octane rating, and reduce CO_2 emissions [6].

Bioadditives are generally produced by esterification of glycerol with acetic acid under base catalysis conditions. The homogeneous strong acids are very effective for the esterification reaction. But these catalysts have some problems, such as catalyst neutralization, catalyst removal from reaction mixture filtration or recovery, and corrosion. Therefore, it is important to replace the liquid acids with environmental-friendly solid acid catalysts in order to become the processes into "green processes" [3, 7]. Preparation of functional catalytic membranes, which has strong acid groups like that of solid acid catalysts, is an alternative method to overcome these problems [8]. Besides that the esterification of glycerol with acetic acid is an equilibrium-limited reaction. These reactions have a problem such as lower conversion and product purification. Pervaporation is a method which is

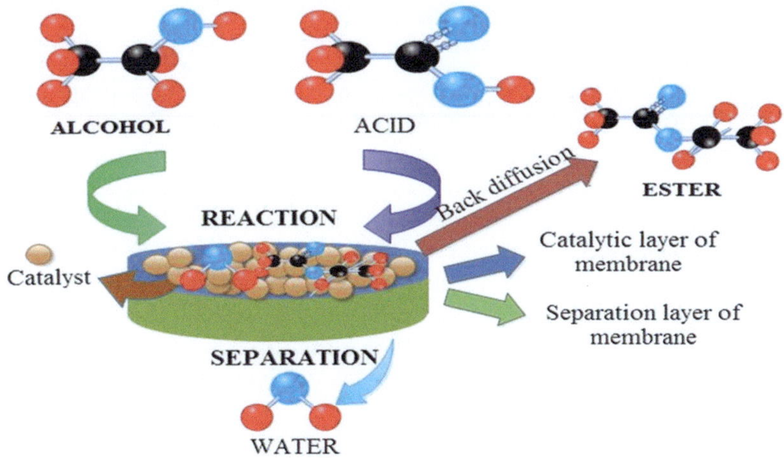

Fig. 53.1 The reaction mechanism in a composite catalytic membrane [12]

developed to overcome these problems. In this study, pervaporation catalytic membrane reactor (PVCMR) is used for the esterification of acetic acid and glycerol. PVCMR is an energy-intensive process which is occurred with combination of reaction and separation system [9, 10]. The reaction takes place in the catalytic area of membrane and conversion of reaction is increased due to the selective removal of product by the separation area of the membrane [11]. The reaction mechanism in a composite catalytic membrane is schematically shown in Fig. 53.1.

Esterification of glycerol with acetic acid by catalytic membrane is an innovative process. In this study, a $Zr(SO_4)_2 \cdot 4H_2O$ loaded chitosan membrane was prepared and used as a heterogeneous catalyst in the esterification of the glycerol with acetic acid for producing biofuel additives. The catalytic membranes were tested at different temperatures, the amount of catalysts and molar feed ratio of reaction in order to obtain maximum conversion for the esterification. The reaction was carried out in mild conditions, such as lower temperature and lower pressure for energy saving.

53.2 Experimental

53.2.1 Materials

Chitosan (high molecular weight polymer) was received from Sigma Aldrich. Acetic acid (AAc) and glycerol (GOH) were purchased Merck Chemicals. Crosslinking agents such as sulfuric acid, acetone, and sodium hydroxide were obtained from Merck Chemicals. The catalyst zirconium sulfate tetrahydrate was supplied by Alfa Aesar.

53.2.2 Membrane Preparation

For the preparation of the catalytic membrane, chitosan was dissolved in aqueous acetic acid solution, stirred for 24 h. After that polymer solution was casted on a plate for drying. Later, $Zr(SO_4)_2{\cdot}4H_2O$ was added to diluted chitosan solution for preparing the catalytic layer of membrane. After catalysts are dispersed homogeneously, catalytic chitosan solution poured onto the separation layer [13]. The dried chitosan membrane was treated with NaOH in a 50 % v/v aqueous ethanol solution to make insoluble in water. The membrane was dried and cross-linked with H_2SO_4 in 50 % aqueous acetone solution during 5 min. The cross-linked chitosan membranes were washed with distillate water to eliminate residuals and then dried at ambient conditions [14]. The thickness of the catalytic chitosan membrane was 100 μm.

53.2.3 Membrane Characterization

53.2.3.1 Fourier Transform Infrared Spectroscopy

The Fourier transform infrared spectroscopy (FTIR) spectra of uncross-linked and cross-linked chitosan membrane by sulfuric acid were scanned in the range 600–4,000 cm^{-1} using Perkin Elmer Spectrum 100 FTIR spectrophotometer.

53.2.3.2 Scanning Electron Microscopy

The structure of the catalytic membranes was characterized by scanning electron microscopy (SEM). Surface and cross-section images of catalytic membrane were viewed by SEM analysis. SEM samples were prepared by breaking with liquid nitrogen, and then the surface of the membrane was covered with gold.

53.2.4 Batch Reactor Experiments

Esterification of acetic acid with glycerol was carried out in a batch reactor. Reactants were mixed in a glass reactor which was connected with a reflux condenser. The scheme of the batch reactor process is given in Fig. 53.2.

Experiments were carried out with different temperature, initial molar ratio (AAc to GOH), and catalyst amount. The reaction mixture was stirred at 350 rpm. Reactants which were AAc and GOH were placed into the reactor. The functional catalytic membrane was cut into pieces and used as heterogenous catalyst. The different ratio of catalyst loaded catalytic membrane was added to the reactor. The liquid sample was taken from the reactor every other hour for

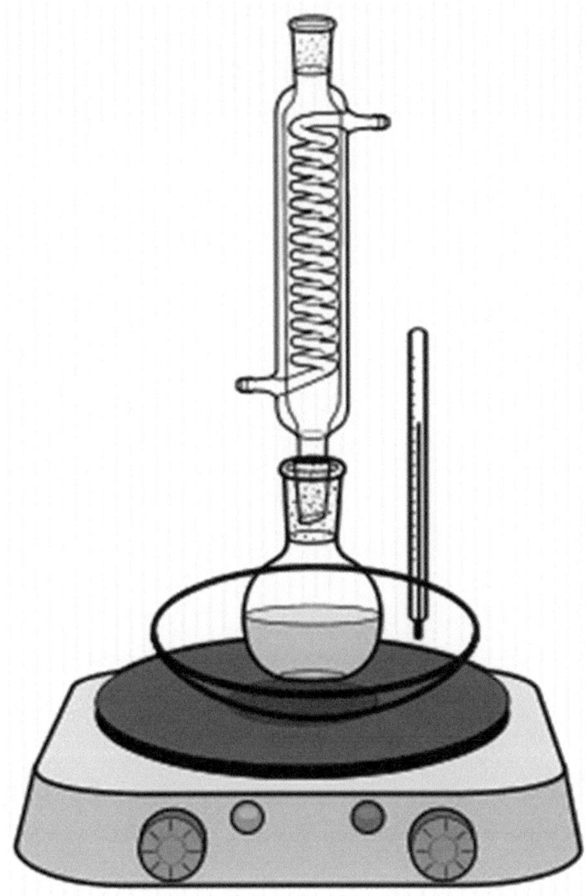

Fig. 53.2 Batch reactor setup

conversion analysis. The sample was titrated with a standard NaOH solution; the reaction was followed for 360 min. Unconverted acetic acid was determined by 0.1 M NaOH. Conversion was calculated by Eq. (53.1);

$$x = \frac{N_{A0} - N_A}{N_{A0}} \tag{53.1}$$

N_{A0} is the initial mole of acetic acid, N_A is the mole of acetic acid at any time [15].

53.2.5 PVCMR Experiments

Different reaction temperature was used in the experiments. The performance of the catalyst coated pervaporation membrane was determined in the esterification reaction. Acetic acid and glycerol was a liquid feed mixture. The reversible reaction

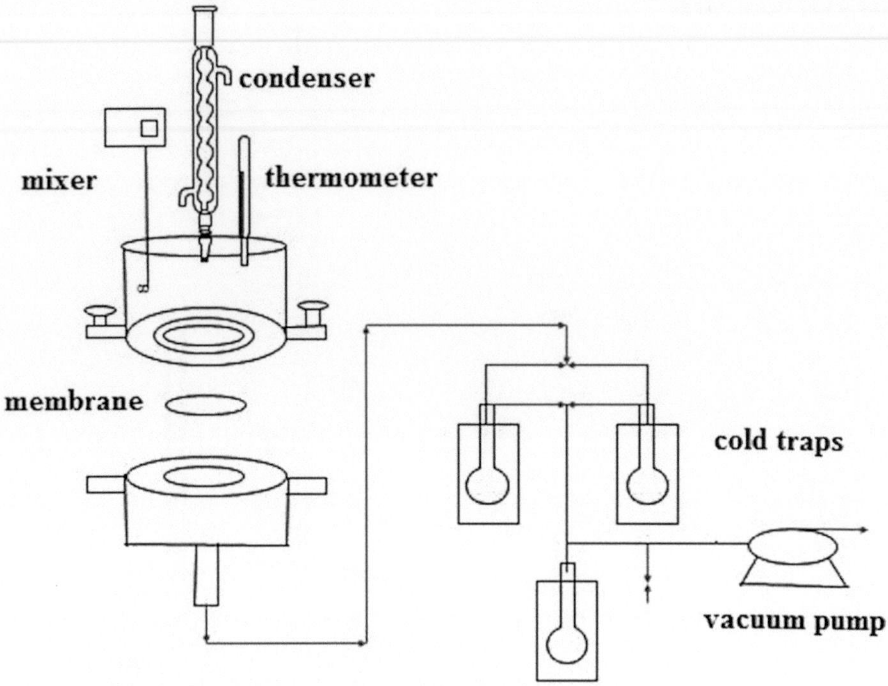

Fig. 53.3 PVCMR unit [16]

takes place into a catalytic layer of membrane. Reactive membrane $Zr(SO_4)_2{\cdot}4H_2O$ coated chitosan was used at membrane chamber. Reactants were sent to the catalytic layer of the composite membrane by a feed pump. The membrane has both selective and catalytic functions. Conversion was determined with acetic acid titration and calculated by Eq. (53.1). The scheme of PVCMR is represented in Fig. 53.3.

53.3 Results and Discussion

53.3.1 Membrane Characterization

53.3.1.1 FTIR of Chitosan Membrane

FTIR spectra of uncross-linked and sulfuric acid cross-linked chitosan are shown in Fig. 53.4. The characteristic peak of NH_2 is observed in 1,578 cm^{-1} is shown in Fig. 53.4. After cross-linking, the peak of NH_2 has been seen at 1,637 cm^{-1}. Groups of NH_2 in chitosan chain are protonated with groups of H^+ in H_2SO_4. Free hydroxyl group is observed in 3,255 cm^{-1} [17, 18].

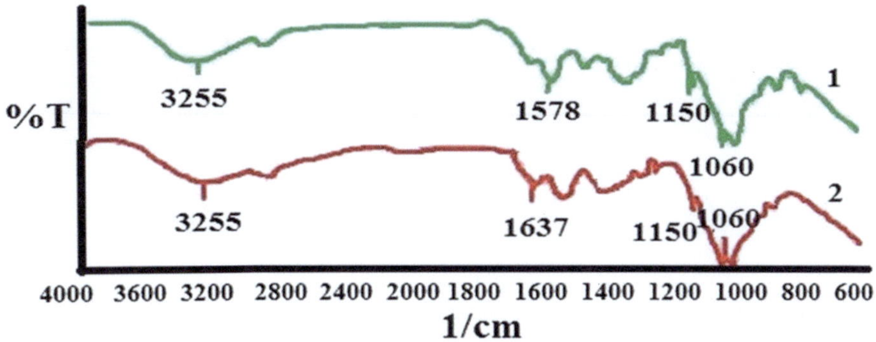

Fig. 53.4 FTIR spectra of uncross-linking and cross-linking chitosan membrane

53.3.1.2 SEM of Catalytic Chitosan Membrane

$Zr(SO_4)_2{\cdot}4H_2O$ loaded chitosan catalytic membrane is shown Fig. 53.5a, b. The cross-section image of the membranes shows a good interaction between chitosan and catalytic layer. The surface view of the membranes is shown that $Zr(SO_4)_2{\cdot}4H_2O$ catalysts distributions are homogenous on the surface of the membrane.

53.3.2 Results of Batch Reactor Experiments

The experiments were performed in a batch reactor. The esterification reaction between acetic acid and glycerol was carried out by $Zr(SO_4)_2{\cdot}4H_2O$ loaded catalytic membrane. Temperatures, amount of catalyst, and molar feed ratio of reaction were examined and interpreted batch reactor performance.

53.3.2.1 Effect of Temperature

Temperature is an important factor that influences reaction kinetics. The esterification reaction was performed with catalytic membrane at three different temperature (70, 80, 90 °C). Figure 53.6 shows that the variation of conversion of acetic acid with time for different temperatures at initial molar ratio ($M = 3{:}1$) in the presence of catalytic membrane as catalyst. Catalyst amount is 0.1 g in the membrane.

Figure 53.6 shows that when the temperature was increased, conversion increased in the batch reactor. The conversion of AAc was increased by increasing temperature because temperature has an important effect for kinetics of esterification reaction. An increase in temperature provided accelerations of esterification reaction. Reaction rate constant is a function of process temperature. Rate constant

Fig. 53.5 SEM photos of catalytic membrane (**a**) surface (**b**) cross section

is increased by temperature and also reaction rate is increased. When reactants temperature is risen, kinetic energy of particle is increased, so rate is increased [19].

Conversion was obtained as 58 % in the batch reactor at 90 °C in 6 h. The conversion of AAc increased from about 33 to 58 % with an increase in temperature from 70 to 90 °C in 6 h. Conversion was continuously increased at batch reactor for 6 h. It shows that the higher temperature gives the better conversion of acetic acid. 70 and 80 °C are chosen for investigate to effect of initial molar ratio. When the temperature is chosen, boiling point of reactants is taken into account.

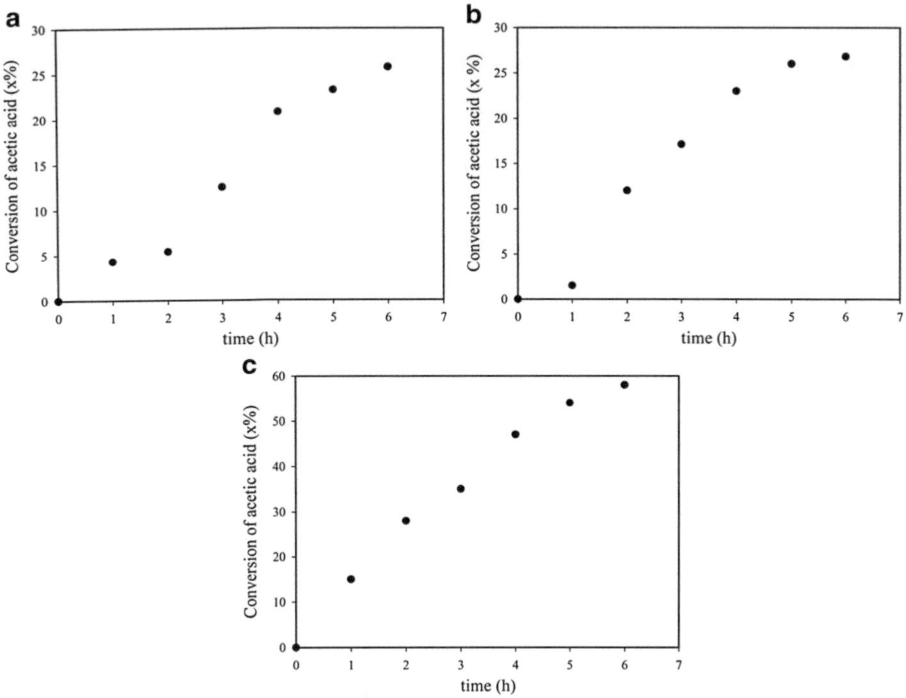

Fig. 53.6 Effect of temperature on acetic acid conversion (**a**) 70 °C (**b**) 80 °C (**c**) 90 °C ($M = 3{:}1$, amount of catalyst: 0.1 g)

53.3.2.2 Effect of Initial Molar Ratio (Acid/Alcohol)

The effect of glycerol to AAc initial molar ratio has been investigated by changing the acid to alcohol molar ratio 3:1 and 6:1. Temperature and catalyst amount were taken, respectively, as 70 °C and 80 °C and 0.1 g. In Fig. 53.7, the conversion of the AAc has been plotted as a function of time.

The initial molar ratio is an important factor in an equilibrium reaction such as esterification reactions. The use of excess reactant is an alternative way to increase the conversion of reversible reaction. According to Le Chatelier principle, the use of excess reactant is shifting the reaction equilibrium to product. Also, this method is increased to possibility contact of the other reactant due to conversion of acetic acid [20].

The highest conversion was obtained as 38 % for $M = 6{:}1$ at 6 h in the batch reactor. But significant difference wasn't observed between $M = 3{:}1$ and $M = 6{:}1$. $M = 3{:}1$ and 80 °C are preferred as mild conditions for analyzing the effect of catalyst amount.

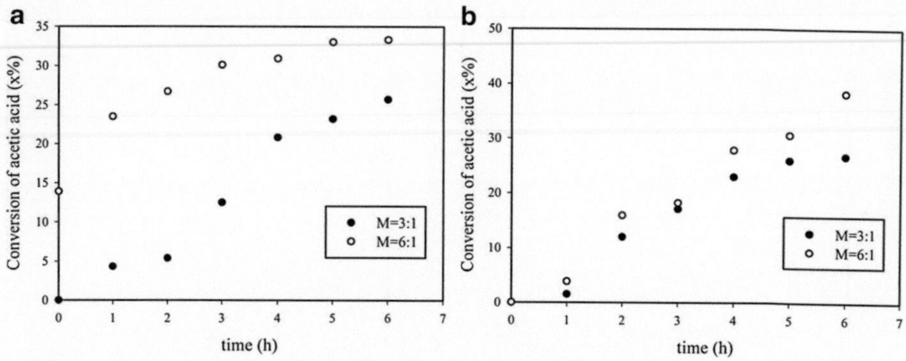

Fig. 53.7 Effect of initial molar ratio on acetic acid conversion. (**a**) 70 °C, $M = 3{:}1$ and $M = 6{:}1$ (**b**) 80 °C, $M = 3{:}1$ and $M = 6{:}1$ (amount of catalyst: 0.1 g)

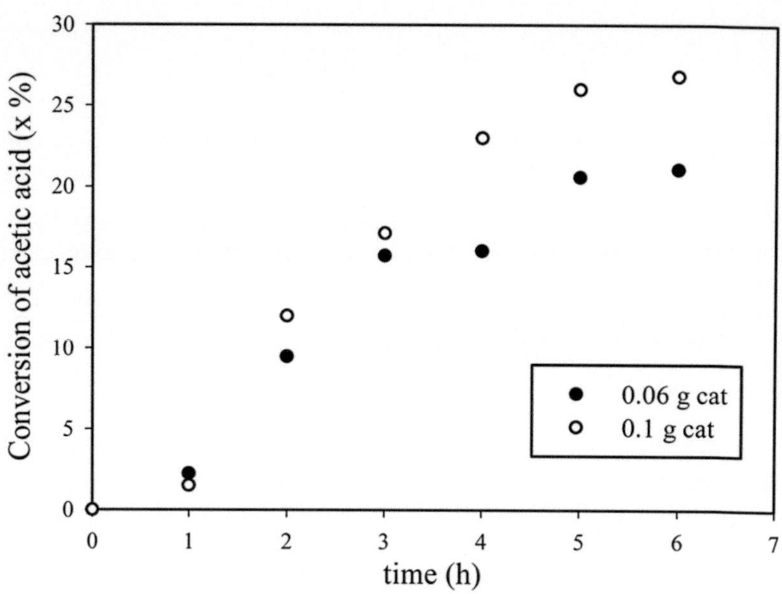

Fig. 53.8 Effect of catalyst amount on conversion (80 °C, $M = 3{:}1$)

53.3.2.3 Effect of Catalyst Amount

The effect of catalyst loading on the conversion of acetic acid is represented in Fig. 53.8. Catalyst loading was changed from 0.06 to 0.1 g. Fixed value of the molar feed ratio was 3:1 and the reaction was carried out at 80 °C.

Figure 53.8 represents that the catalyst concentration increases, and the reaction rate increases. An increase in the catalyst amount is related to be formed products in the reactor [21]. Increasing catalyst amount in the catalytic membrane, more reactants reached acid sites of catalyst in the catalytic membrane easily [22]. Thus, the time which is necessary to reach the reaction equilibrium was reduced.

For batch reactor, it's not appropriate to use more than 0.1 g $Zr(SO_4)_2 \cdot 4H_2O$ catalyst because the use of more than 0.1 g catalyst damage membrane structure.

Conversion was calculated as 26 % at the end of 6 h at the batch reactor.

53.3.3 PVCMR Experiments

The best reaction conditions (initial molar ratio and amount of catalyst) are determined by the batch reactor. The effect of reaction temperature on conversion of acetic acid was performed by PVCMR at specified conditions. The effect of temperature has been investigated by temperatures 70, 80, and 90 °C. In Fig. 53.9, the conversion of the AAc has been plotted as a function of time.

Figure 53.9 shows that when the temperature was increased, conversion value increased in PVCMR. Also the conversion in PVCMR was higher than batch reactor. In PVCMR, hydrophilic composite chitosan membranes were used to remove water which occurred as a result of the reaction. So reaction equilibrium

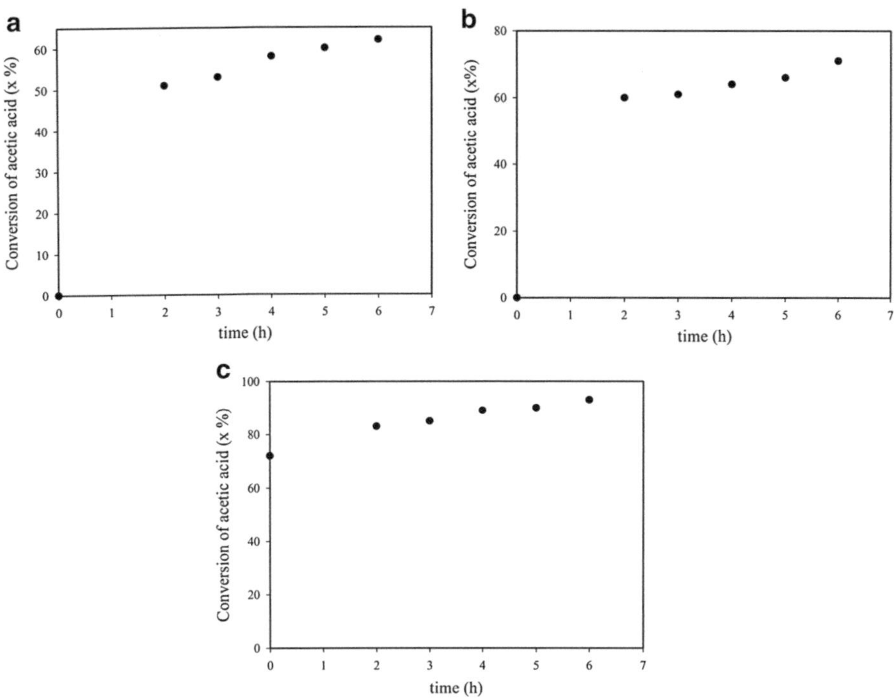

Fig. 53.9 Effect of temperature on acetic acid conversion (**a**) 70 °C (**b**) 80 °C (**c**) 90 °C ($M = 3{:}1$, amount of catalyst: 0.1 g)

was changed to products side and esterification reaction rate was increased. Since thermal motion of polymer chains increased with increasing temperature, transport of selective component increased across the free volume of membrane. Thus, conversion increased by removing of water [23].

Conversion was obtained as 58 % in batch reactor at 90 °C in 6 h. The conversion of AAc increased from about 33 to 58 % with an increase in temperature from 70 to 90 °C in 6 h. In PVCMR, the maximum conversion of AAc was obtained 93 % at 90 °C in 6 h. When PVCMR was used, time which was required to reach equilibrium was decreased. PVCMR had higher conversion than batch conversion at the same time. Conversion was continuously increased at batch reactor for 6 h. PVCMR was efficient both increasing of conversion and decreasing of the time.

53.4 Conclusions

$Zr(SO_4)_2{\cdot}4H_2O$ loaded chitosan was used as a catalyst in the esterification of acetic acid with glycerol. Chitosan is a natural product, so it has the advantage as green catalysis. The catalysts were fixed by cross-linking of membrane. Catalytic chitosan membrane can be used again and again. So recycle and reuse of catalyst can be easy.

Effects of temperature, catalyst loading amount, and initial molar ratio were investigated on reaction performance.

In this study, higher conversion values were obtained by using PVCMR for esterification of acetic acid and glycerol. PVCMR was more effective than the batch reactor to increase conversion. The conversions obtained in PVCMR were higher than the conversions achieved in batch reactor, due to the selective removal of water from the reaction mixture by pervaporation through the catalytic composite chitosan membrane. Because membrane had hydrophilic property, permeate flux and selectivity of water were higher than other components.

Catalyst in the membrane can be used repeatedly in batch reactor and PVCMR. There is no need to any process for separation or cleaning of the catalyst. Reusability of the catalyst without any additional process is very important from the point of cost saving. Batch reactor and PVCMR is an alternative process to conventional process; therefore, it can be used as energy and environment-friendly technology.

References

1. Gonçalves LC, Pinto BP, Silva JC, Mota CJA (2008) Acetylation of glycerol catalyzed by different solid acids. Catal Today 133–135:673–677
2. Trejda M, Stawicka K, Dubinska A, Ziolek M (2012) Development of niobium containing acidic catalysts for glycerol esterification. Catal Today 187:129–134

3. Rodríguez DI, Adriany C, Gaigneaux EM (2011) Glycerol acetylation on sulphated zirconia in mild conditions. Catal Today 167:56–63
4. Balaraju M, Nikhitha P, Jagadeeswaraiah K, Srilatha K, Prasad PS, Lingaiah N (2010) Acetylation of glycerol to synthesize bioadditives over niobic acid supported tungstophosphoric acid catalysts. Fuel Process Technol 91:249–253
5. Gonçalves CE, Laier LO, Cardoso AL, Silva MJ (2012) Bioadditive synthesis from H3PW12O40-catalyzed glycerol esterification with HOAc under mild reaction conditions. Fuel Process Technol 102:46–52
6. Zhou L, Al-Zaini E, Adesina AA (2013) Catalytic characteristics and parameters optimization of the glycerol acetylation over solid acid catalysts. Fuel 103:617–625
7. Ferreira P, Fonseca IM, Ramos AM, Vital J, Castanheiro JE (2009) Esterification of glycerol with acetic acid over dodecamolybdophosphoric acid encaged in USY zeolite. Catal Commun 10:481–484
8. Castanheiro JE, Ramos AM, Fonseca IM, Vital J (2006) Esterification of acetic acid by isoamylic alcohol over catalytic membranes of poly(vinyl alcohol) containing sulfonic acid groups. Appl Catal A Gen 311:17–23
9. Assabumrungrata S, Phongpatthanapanicha J, Praserthdama P, Tagawab T, Gotob S (2003) Theoretical study on the synthesis of methyl acetate from methanol and acetic acid in pervaporation membrane reactors: effect of continuous-flow modes. Chem Eng J 95:57–65
10. Park BG, Tsotsis TT (2004) Models and experiments with pervaporation membrane reactors integrated with an adsorbent system. Chem Eng Process 43:1171–1180
11. Peters TA (2006) Catalytic pervaporation membranes for close integration of reaction and separation. PhD Thesis, Technische Universiteit Eindhoven, Eindhoven
12. Unlu D, Hilmioglu ND (2014) Pervaporation catalytic membrane reactor study for the production of ethyl acetate using $Zr(SO_4)_2{\cdot}4H_2O$ coated chitosan membrane. J Chem Technol Biot Doi:10.1002/jctb.4546
13. Peters TA, Benes NE, Keurentjes JTF (2007) Preparation of Amberlyst-coated pervaporation membranes and their application in the esterification of acetic acid and butanol. Appl Catal A Gen 317:113–119
14. Won W, Feng X, Lawless D (2003) Separation of dimethyl carbonate/methanol/water mixtures by pervaporation using crosslinked chitosan membranes. Sep Purif Technol 31:129–140
15. Benedict DJ, Parulekar S, Tsai S (2006) Pervaporation-assisted esterification of lactic and succinic acids with downstream ester recovery. J Membr Sci 281:435–445
16. Ünlü D (2012) Investigation of the esterification reaction for the ethyl acetate production using pervaporation catalytic membrane reactor. Master Thesis, Kocaeli University, Graduate School of Natural and Applied Sciences, Kocaeli
17. Cui Z, Xiang Y, Si J, Yang M, Zhang Q, Zhang T (2008) Ionic interactions between sulfuric acid and chitosan membranes. Carbohydr Polym 73:111–116
18. Mulder MHV (1991) Thermodynamic principles of pervaporation. In: Huang RYM (ed) Pervaporation membrane separation processes. Elsevier, Amsterdam
19. Delgado P, Sanz MT, Beltran S (2009) Pervaporation of the quaternary mixture present during the esterification of lactic acid with ethanol. J Membr Sci 332:113–120
20. Çimenler Ü (2011) Investigation of the esterification reaction for the production of butyl propionate in pervaporation membrane reactor using Amberlite Ir-120 catalyst. Master Thesis, Yıldız Technical University, Graduate School of Natural and Applied Sciences, İstanbul
21. Zou Y, Tong Z, LIU K, Feng X (2010) Modeling of esterification in a batch reactor coupled with pervaporation for production of n-butyl acetate. Chin J Catal 31:999–1005
22. Ma J, Zhang M, Lu L, Yin X, Chen J, Jiang Z (2009) Intensifying esterification reaction between lactic acid and ethanol by pervaporation dehydration using chitosan–TEOS hybrid membranes. Chem Eng J 155:800–809
23. Kwang-Je K, Sang-Hee P, Won-Wook S, Sang-Jın M (2001) Pervaporation separation of aqueous organic mixtures through sulfated zirconia-poly(vinyl alcohol) membrane. J Appl Polym Sci 79:1450–1455

Chapter 54
Biosorption of Pharmaceutical Pollutant

Hakima Cherifi, Bentahar Fatiha, and Hanini Salah

Abstract The adsorption of penicillin onto dried biomass was studied as a function of initial penicillin concentration and temperature. The maximum penicillin adsorption yield was obtained at the temperature of 30 °C; the equilibrium uptake increased with increasing initial penicillin concentration. The Freundlich and Langmuir adsorption models have been used for the mathematical description of the adsorption equilibrium. The results show that experimental data fit perfectly the Langmuir model. The pseudo-first-order and pseudo-second-order kinetic models were applied to test the experimental data. The latter provided the best correlation of the experimental data compared to the pseudo-first-order model.

Keywords Adsorption • Pleurotus mutilus • Equilibrium • Kinetic

Nomenclature

C	Unadsorbed penicillin concentration in solution at any time, mg/L
C_e	Unadsorbed penicillin concentration in solution at equilibrium, mg/L
C_0	Initial penicillin concentration, mg/L
k	Adsorption equilibrium constant, mg/L
k_f	Freundlich constant
n_f	Freundlich adsorption constant
m	Mass of biomass, g
q	Adsorbed penicillin quantity per gram of adsorbed at any time, mg/g
q_e	Adsorbed penicillin quantity per gram of adsorbed at equilibrium, mg/g
Q_0	Maximum amount of penicillin per gram of biomass to form a complete monolayer on the surface, mg/g

H. Cherifi (✉) • H. Salah
Laboratoire des Biomatériaux et des Phénomènes de Transferts,
Université de Yahia Fares, Médéa 26000, Algéria
e-mail: ha_cherifi@yahoo.fr

B. Fatiha
Laboratoire des Phénomènes de Transfert, Université des Sciences et Technologie
Houari Boumedien, BP 32 EL Alia, Bab Ezzouar 16111, Algéria

I. Dincer et al. (eds.), *Progress in Clean Energy, Volume 1*,
DOI 10.1007/978-3-319-16709-1_54

V Volume of the penicillin solution, L
R^2 Regression correlation coefficient
T Temperature, K, °C

54.1 Introduction

Removal of hazardous compounds from industrial effluents is one of the growing needs of the present time. Recently, the attention of many researchers working in the environmental field was focused on the presence in the environment (and more specifically in waters) of pharmaceuticals as a new class of pollutants [1–3]. Different sources can be indicated to explain the appearance of these xenobiotics in waters and soils [4, 5]. Antibiotics are among the environmental micropollutants being responsible for disturbing the wastewater treatment processes and the microbial ecology of surface waters. At present, we may consider to be safe by expecting antibiotics to be present in trace concentrations. But in the future, they will cause a pollution that will reach a level that could destroy the ecological equilibrium [6]. Many pharmaceuticals are only partially removed during biological processes in sewage treatment plants (STP) with their consequent release into surface waters [6–8]. Therefore, extensive researches would be necessary to assess the environmental risk involved in exposing of antibiotics to the environment and for the removal of antibiotics from the waste water [9, 6]. Various techniques like coagulation, adsorption, chemical oxidation, and froth floatation have been used for the removal of organics as well as inorganics from wastewaters. Amongst these, adsorption is considered to be the most potential one due to its high efficiency and the ability to separate wide range of chemical compounds [10]. Adsorption is a well established and powerful technique for treating domestic and industrial effluents, where the activated carbon is the most widely used adsorbent. Recently, the use of less expensive and efficient biosorbents for the removal of antibiotics to replace activated carbon is of significant interest.

In the present work, the attention is focused on penicillin, a broad-spectrum antibiotic, and one of the most widely used antibiotics in Algeria in human and veterinary medicine. Its presence in STP effluents has been often documented [8, 11].

As a substitute for powdered activated carbon, the adsorbent used in this study is a waste of a drug produced by fermentation called Pleurotus mutilus biomass. The potential use of dried Pleurotus mutilus biomass for removal of penicillin from aqueous solution was studied. We examined the kinetic of static adsorption of the penicillin on dried biomass. The influence of some essential parameters, namely, initial penicillin concentration, contact time, and temperature on the static adsorption of penicillin on this biomass has been investigated.

Batch adsorption models, based on the assumption of the pseudo-first-order and pseudo-second-order mechanism were applied to study the kinetics of the adsorption.

54.2 Theory

The equilibrium study on adsorption has provided information on the capacity of the adsorbent. An adsorption isotherm is characterized by some constant values, which express the surface properties and affinity of the adsorbent and can also be used to compare the adsorptive capacities of the adsorbent for different pollutants. This kind of adsorption isotherm generally fits the Langmuir or Freundlich model [12, 13].

54.2.1 *Langmuir Model*

The Langmuir model assumes the presence of a finite number of binding sites, homogeneously distributed over the adsorbent surface, presenting the same affinity for sorption of a single layer, and having no interaction between adsorbed species. The well-known Langmuir expression [14] is presented as follows Eq. (54.1):

$$q_e = \frac{Q_0 k C_e}{1 + k C_e} \tag{54.1}$$

where q_e is the amount of penicillin bound per gram of biomass at equilibrium (mg/g), C_e (mg/L) the residual penicillin concentration in solution after binding, Q_0 (mg/g) the maximum amount of penicillin per unit weight of biomass to form a complete monolayer on the surface bound at high C_e, and k (L/mg) a constant related to the affinity of the binding sites. A plot of $1/q_e$ versus $1/C_e$ indicates a straight line of slope $1/kQ_0$ and an intercept $1/Q_0$.

54.2.2 *Freundlich Model*

The Freundlich model as stated in Eq. (54.2) is an empirical equation based on adsorption on a heterogeneous surface suggesting that binding sites are not equivalent and/or independent [15].

$$q_e = k_f C_e^{\frac{1}{n_f}} \tag{54.2}$$

where k_f and n_f are Freundlich parameters relating to the adsorption capacity and adsorption intensity, respectively. From the linear plot of ln q_e versus ln C_e, k_f and $1/n_f$ values can be determined.

54.3 Materials and Methods

54.3.1 Adsorbent

Pleurotus mutilus biomass is a result of industrial fermentation residue of an antibiotic (pleuromutilin) for veterinary use produced at the Antibiotic Complex of Médéa, Algeria (SAIDAL) used as biosorbents. It has been used in dry form after several washes with distilled water and after chemical treatments with HCl. The biomass was oven dried at 105 °C for 24 h, then crushed and sifted on a standard sifting and crushing.

54.3.2 Chemicals

A stock solution was prepared by dissolving 0.5 g of penicillin of analytical reagent grade in 500 mL of distilled water. The test solutions were prepared by diluting some stock solution up to the desired concentration. The ranges of concentrations of penicillin prepared from stock solution varied between the values of 30 and 100 mg/L.

54.3.3 Adsorption Studies

This method involved agitating (150 rpm) 250 mL Erlenmeyer flasks containing 1 g biomass and 100 mL of penicillin solutions at desired concentration, granulometry, and temperature. Samples (5 mL) were taken before mixing the biomass and penicillin solution and at predetermined time intervals for determining the residual penicillin concentration in the medium. Before analysis, samples were filtered and supernatant fluid analyzed for the remaining penicillin. The adsorption equilibrium was reached after 50 min.

54.3.4 Analysis

The concentration of residual penicillin in the adsorption media was analyzed using an HPLC (SHIMADZU LC-10A).

54.4 Results and Discussion

Adsorption data for the uptake of penicillin onto biomass was investigated at various initial concentration and temperature. The results are given as the units of adsorbed penicillin quantity per gram of adsorbent at time t (q, mg/g) and at equilibrium (q_e, mg/g), unadsorbed penicillin concentration in solution at time t (C, mg/g), and at equilibrium (C_e, mg/g).

The adsorbed quantity of aqueous solution (q) per unit mass of the solid support is calculated by the following formula (54.3):

$$q_e = (C_0 - C_e)\frac{V}{m} \tag{54.3}$$

54.4.1 Effect of Contact Time

In order to determine the time of contact necessary for the establishment of adsorption equilibrium, the absorbed quantity of aqueous solution on biomass is measured as a function of the time of contact corresponding to adsorption equilibrium. The variation of the quantity adsorbed versus the time of contact is shown in Fig. 54.1.

Adsorption experiments were carried out for 2 h, and it was observed that the amount of adsorbed penicillin increased linearly with time at the beginning of adsorption. Then, it reaches saturation at the equilibrium time. The large amount of penicillin was removed from the solution within the first 60 min of contact time and the equilibrium was established after 90 min at all the temperatures studied.

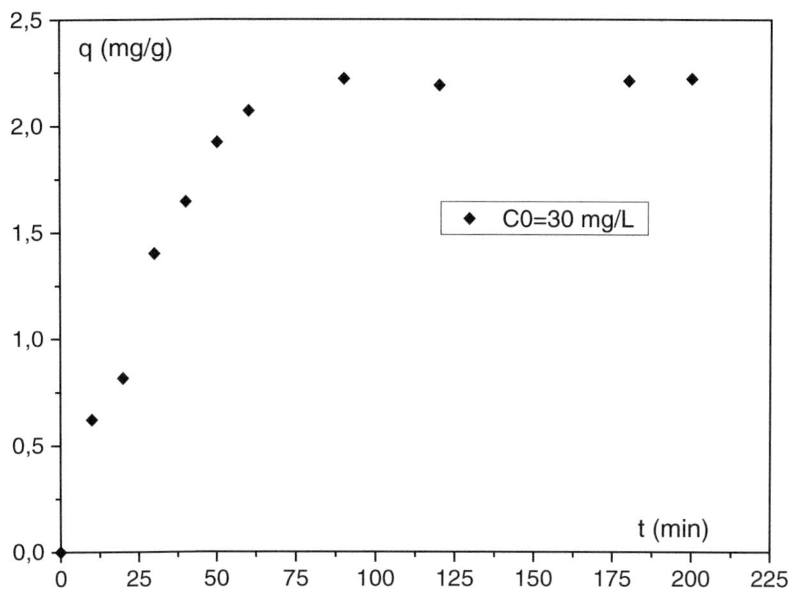

Fig. 54.1 Effect of contact time on adsorption at given condition at 25 °C, $V = 100$ mL, $m_s = 1$ g, $T = 23$ °C, stirring velocity = 500 rpm, and $0.05 < d < 0.15$ mm

54.4.2 Effect of the Initial Concentration of Penicillin

To determine the impact of the initial concentration on the adsorption of penicillin in aqueous solution in contact with *Pleurotus mutilus* biomass, different initial concentrations C_0 (30, 50, 80, and 100 mg/L) has been examined. The results of these experiments are displayed in Fig. 54.2, which shows that the sorption capacity of the sorbent increases from 2 to 5.5 mg/g with increasing initial penicillin concentration from 30 to 100 mg/L. The increase of loading capacity of *Pleurotus mutilus* biomass with increasing initial penicillin concentration may also be due to higher interactions between penicillin and *Pleurotus mutilus* biomass. As biomass offered a finite number of binding sites, penicillin sorption showed a saturation trend at higher initial penicillin concentration.

54.4.3 Effect of Temperature on Adsorption Equilibrium

Effect of temperature on the equilibrium penicillin sorption capacity of *Pleurotus mutilus* biomass was investigated in the temperature range of 25–45 °C at an initial penicillin concentration of 30 mg/L.

As seen from Fig. 54.3, the effect of temperature on the adsorption equilibrium was not significant. However, for higher temperature at 30 °C, the adsorption

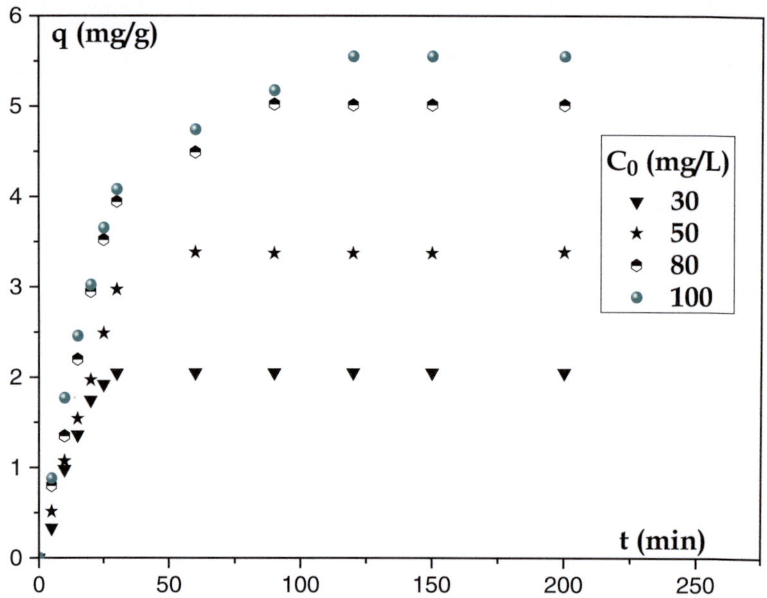

Fig. 54.2 The variation of adsorption capacity with adsorption time at various C_0 of penicillin at $T = 25$ °C, stirring velocity = 500 rpm, and $0.05 < d < 0.15$ mm

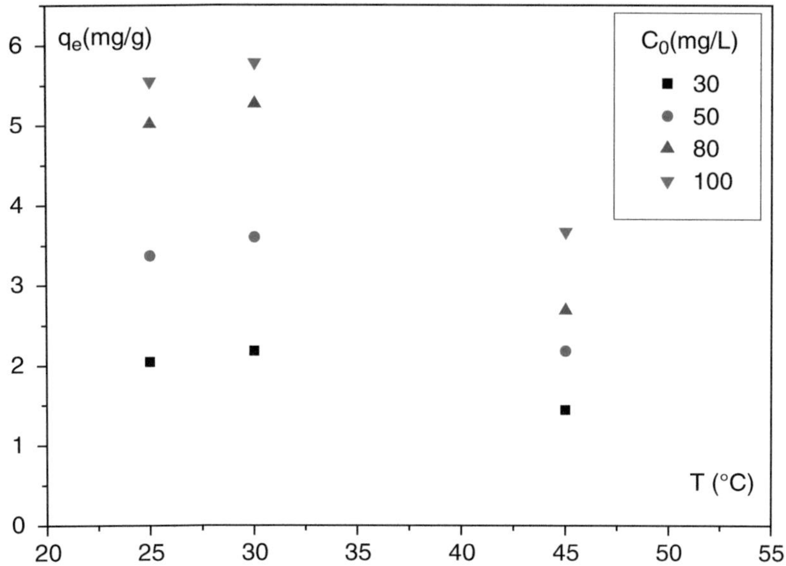

Fig. 54.3 Effect of temperature on adsorption equilibrium at various initial penicillin concentrations

Table 54.1 Langmuir and Freundlich isotherm model constants and correlation coefficients for adsorption of penicillin onto dried *Pleurotus mutilus* biomass

	Isotherm Langmuir model			*Isotherm Freundlich* model		
T (°C)	Q_0 (mg/g)	*k* (l/mg)	R^2	k_f (mg/g (l/mg)$^{1/n}$)	$1/n_f$	R^2
25	12.769	0.0204	0.9831	0.4973	0.6579	0.9409
30	11.543	0.0204	0.9753	0.6721	0.5991	0.9215
45	12.619	0.0081	0.9741	0.1356	0.8676	0.9572

equilibrium decreased with increased temperature. This result can be attributed to change in surface characteristics and activity of dried *Pleurotus mutilus* biomass, indicating the exothermic nature of the adsorption reaction. A similar observation was reported by the sorption of penicillin G onto *Rhizopus arrhizus* biomass [8].

54.4.4 Adsorption Isotherms

In this study, the adsorption equilibrium of penicillin on dried biomass was modelled using Langmuir and Freundlich isotherms. The Langmuir and Freundlich adsorption constants evaluated from the isotherms are given in Table 54.1 with the correlation coefficients.

As seen from this table, although correlation coefficients of both equations are considerably very significant at all temperatures, the Langmuir model exhibited better fit to the adsorption data ($R^2 > 0.97$) than the Freundlich model. The maximum capacity, Q_0 determined from the Langmuir equation defines the total capacity of the adsorbent for penicillin. The other Langmuir constant, k, indicates the affinity for the binding of penicillin. The higher value of k found at 30 and 25 °C showed the strong bonding of penicillin to the biomass at these temperatures.

The applicability of these isotherm models to the adsorption of penicillin on *Pleurotus mutilus* biomass was examined by judging the correlation coefficients R^2 values.

54.4.5 Adsorption Kinetics

Two simplified kinetic models were adopted to examine the mechanism of the adsorption process. First, the kinetics of adsorption was analyzed by the pseudo-first-order equation given by Langergren and Svenska, 1898 [16] as:

$$\mathrm{Log}(q_e - q_t) = \mathrm{Log}\, q_e - \frac{k_1}{2.303} t \tag{54.4}$$

where q_e and q_t are the amounts of penicillin adsorbed (mg/g) at equilibrium and at time t (min), respectively, and k_1 (min^{-1}) is the rate constant of adsorption. Values of k_1 at 25 °C were calculated from the plots of $\mathrm{Log}(q_e - q_t)$ versus t (Fig. 54.4) for

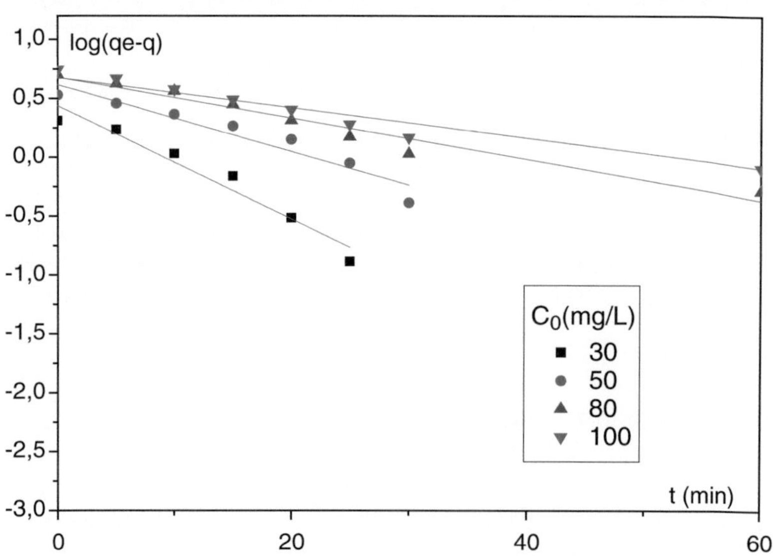

Fig. 54.4 Pseudo-first-order kinetics for adsorption of penicillin onto biomass at 23 °C

different initial concentrations of penicillin. It was observed from Fig. 54.4 that the pseudo-first-order model fits well ($R^2 > 0.955$), indicating the applicability of this model to describe the adsorption process of penicillin onto the biomass (Table 54.2).

On the other hand, the pseudo-second-order equation based on equilibrium adsorption [17] is expressed as:

$$\frac{t}{q_t} = \frac{1}{k_2 q_e^2} + \frac{1}{q_e} t \tag{54.5}$$

where k_2 (g/mg min) is the rate constant of second-order adsorption. The linear plot of t/q_t versus t at 25 °C, as shown in Fig. 54.5, yielded R^2 values that were greater than 0.956 for all penicillin concentrations. It also showed a good agreement between the experimental and the calculated q_e values (Table 54.2), the model can be applied for the entire data (Table 54.2).

54.4.6 Validity of Kinetic Models

The adsorption kinetics of penicillin onto the prepared biomass was verified at different initial concentrations. The validity of each model was determined by the sum of squared errors (SSE, %) given by:

$$\text{SSE} = \sqrt{\frac{\sum \left(q_{e,\exp} - q_{e,\text{cal}}\right)^2}{N}} \tag{54.6}$$

where N is the number of data points.

The lower the value of SSE indicates the better a fit is. Table 54.2 lists the SSE values obtained for the two kinetic models studied. It was found that the pseudo-first-order kinetic model yielded the lowest SSE values. This proves that the adsorption of penicillin onto the biomass could be best described by the pseudo-first-order kinetic model.

54.5 Conclusions

In this study, the batch adsorption of penicillin on Pleurotus mutilus biomass was investigated in function of temperature and initial penicillin concentration. The results showed that the solution temperature played an important role in the determination of the uptake capacities of sorbents. The optimum value of temperature was determined as 30 °C for Pleurotus mutilus biomass. The equilibrium uptake increased with increasing initial penicillin concentration. The results

Table 54.2 Kinetic parameters for the adsorption of penicillin by dried biomass

Penicillin/biomass PM'		Pseudo-first-order kinetics				Pseudo-second-order kinetics			
C_0 (mg/L)	q_{exp} (mg/g)	$q_{e,cal}$ (mg/g)	K_1 (min^{-1})	R^2	SSE (%)	$q_{e,cal}$ (mg/g)	K_2 (g/mg min)	R^2	SSE (%)
5	0.39	0.4108	0.0239	0.972	0.021	0.4860	0.0504	0.956	0.096
15	1.13	1.1356	0.0314	0.979	0.005	1.3106	0.0315	0.985	0.180
20	1.48	1.4610	0.0357	0.993	0.002	1.6510	0.0339	0.992	0.171
30	2.22	2.5006	0.0385	0.955	0.280	2.5720	0.0155	0.980	0.352
50	2.80	2.8768	0.0226	0.983	0.076	3.3054	0.0097	0.994	0.505

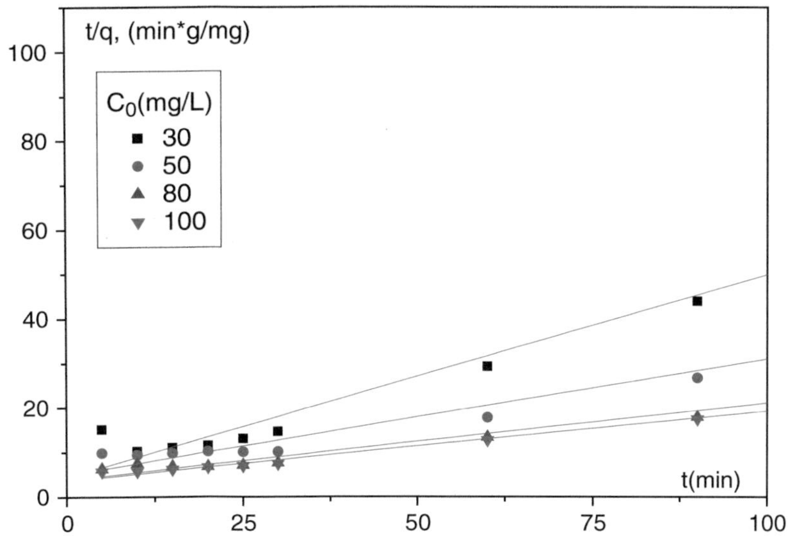

Fig. 54.5 Pseudo-second-order kinetics for adsorption of penicillin onto biomass at 23 °C

obtained for penicillin-biomass system showed that the adsorption equilibrium data fitted the Langmuir model very well at all temperatures studied. The pseudo-first-order and pseudo-second-order kinetic models were applied to test the experimental data. The latter provided the best correlation of the experimental data compared to the pseudo-first-order model.

References

1. Holm JH, Rugge K, Bjerg PL, Christensen TH (1995) Occurrence and distribution of pharmaceutical organic compounds in the groundwater down gradient of a landfill (Griendsted, Denmark). Environ Sci Technol 29:1415–1420
2. Ternes TA (1998) Occurrence of drugs in German sewage treatment plants and rivers. Water Res 32:3245–3260
3. Daughton CG, Ternes TA (1999) Pharmaceuticals and personal care products in the environment: agents of subtle change? Environ Health Perspect 107:907–938
4. Fukahori S, Fujiwara T, Ito R, Funamizu N (2011) pH-dependent adsorption of sulfa drugs on high silica zeolite: modeling and kinetic study. Desalination 275:237–242
5. Heberer T (2002) Occurrence, fate and removal of pharmaceutical residues in the aquatic environment: a review of recent research data. Toxicol Lett 131:5–17
6. Heberer T, Stan HJ (1996) Occurrence of polar organic contaminants in Berlin drinking water. Vom Wasser 86:19–31
7. Roberto A, Marisa C, Raffaele M, Nicklas P (2005) Antibiotic removal from wastewaters: the ozonation of amoxicillin. J Hazard Mater 122:243–250
8. Aksu Z, Tunç O (2005) Application of biosorption for penicillin G removal: comparison with activated carbon. Process Biochem 40:831–847

9. Hirsch R, Ternes T, Haberer K, Kratz KL (1999) Occurrence of antibiotics in the aquatic environment. Sci Total Environ 225:109–118
10. Mittal A, Krishnan L, Gupta VK (2005) Use of waste materials-Bottom Ash and De-Oiled Soya, as potential adsorbents for the removal of Amaranth from aqueous solutions. J Hazard Mater 117:171–178
11. Cherifi H, Hanini S, Bentahar F (2009) Adsorption of phenol from wastewater using vegetal cords as a new adsorbent. Desalination 244:177–187
12. Smith JM III (ed) (1981) Chemical engineering kinetics. McGraw-Hill, Singapore
13. Montanher SF, Oliveira EA, Rollemberg MC (2005) Removal of metal ions from aqueous solutions by sorption onto rice bran. J Hazard Mater 117:207–211
14. Langmuir I (1918) The constitution and fundamental properties of solids and liquids. J Am Chem Soc 40:1361–1403
15. Freundlich HMF (1926) Colloid and capillary chemistry. Methuen, London
16. Langergren S, Svenska BK (1898) Zur theorie der sogenannten adsorption geloester stoffe. Veternskapsakad Handlingar 24:1–39
17. Ho YS, Mckay G (1998) Sorption of dye from aqueous solution by peat. Chem Eng J 70:115–124

Chapter 55
Switchable Magnetic Properties of Hydrogenated Metal Alloys

Ioseb Ratishvili and Natela Namoradze

Article is dedicated to Lili, Ayfer, and T. Nejat Veziroglu.

Abstract Hydrogen detection becomes, generally, an actual problem as "H-technologies" are widely presented in the different areas of industry. On following these tendencies, we have made an attempt to consider possibilities of investigation of switchable magnetic properties of hydrogen-containing dilute paramagnetic metal alloys, on using information obtained in previous considerations of metal hydrides and of localized magnetic moments in metal alloys.

Keywords Metal-hydrogen interstitial alloys • Magnetic impurities in nonmagnetic metals • Fermi-level of conduction electrons

55.1 Introduction

It is well known that "hydrogen-based technologies" little-by-little become more and more popular in the world (see for e.g., [1]). But it is known as well that hydrogen–air mixtures at some particular hydrogen concentrations are very explosive and oblige us to be very careful and to control the content of free hydrogen in

---

I. Ratishvili (✉)
Andronikashvili Institute of Physics, I. Javakhishvili Tbilisi State University,
6 Tamarashvili Str, Tbilisi 0177, Georgia
e-mail: ratisoso@gmail.com

N. Namoradze
Chavchanidze Institute of Cybernetics, Georgian Technical University,
5 Sandro Euli Str, Tbilisi 0183, Georgia

I. Dincer et al. (eds.), *Progress in Clean Energy, Volume 1*,
DOI 10.1007/978-3-319-16709-1_55

the environment. There are known today a sufficiently great number of different hydrogen detectors. All these are based on the registration of physical and/or chemical property changes of different hydrogen-absorbing materials. We know, particularly, that hydrogenated metals change their volume, their lattice, and electronic properties revealed in different measurements (see for e.g., [2]). The most impressive phenomenon is the optical property changes registered in hydrogenated rare-earth metals and alloys [3–5]. Particularly, the trivalent lanthanum metal on hydrogenating it up to LaH_3-state becomes a nonmetal (an insulator or a semiconductor). Correspondingly, nevertheless LaH_2 (as pure lanthanum) reveals the metallic mirror properties, the trihydride LaH_3 is transparent for the sunlight [6]. This qualitative explanation of the optical phenomenon was supported by the NMR-measurements [7] performed on lanthanum hydrides within the concentration range LaH_2–LaH_3 (see below).

Taking in mind the above-mentioned influence of interstitial hydrogen atoms on the conduction band population of the host metal, we intend to analyze another interesting effect related to the "concentration of conduction electrons." We imply the influence of hydrogenation processes on the formation of *localized magnetic moments* in the dilute alloys of paramagnetic impurities in nonmagnetic metals.

Below we shall recall briefly experimental data on localized impurity states in alloys and particularly on localized magnetic moments in the dilute alloys of paramagnetic ions in nonmagnetic metals. Then, we shall try to estimate the role of hydrogenation process on these metallic systems. We intend to avoid any sophisticated mathematical schemes and concentrate our attention on the physics of considered effects.

55.2 Localized Electronic States in Alloys

Let us consider dilute substitution alloys of metal elements. It is assumed usually that valence electrons of all atoms (those of host metal atoms and impurity atoms) form an unfiled unified conduction band and electro-conductivity of the system depends mainly on the density of states at the Fermi-level of the electronic states. However, cited in [8] experimental investigations of the residual (impurity-induced) resistivity measurements performed on aluminum- and copper-based metal alloys had shown that this simplest picture has to be corrected (see Fig. 55.1a, b). Both figures indicate nontrivial dependence of the residual resistivity on the number of electrons in impurity atoms.

These results had stimulated Friedel to propose a conception of localized electronic states formed in dilute alloys. He had assumed [8] that if the valences of impurity atoms differ from that of the host metal atoms then the impurities will be playing the role of attractive (or repulsive) centers which provide formation of spatially sufficiently delocalized (but still concentrated near the impurity sites) specific electronic states. These impurity states are characterized by the wave functions of the same symmetry which they would have in the isolated impurity

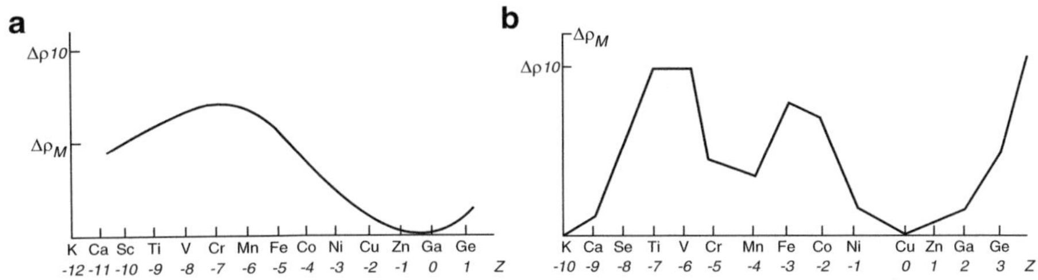

Fig. 55.1 Residual resistivity (in [μΩ-cm per percent]) for series of transition element and other impurities in a given host metal. (*Z* denotes the "valence difference of impurity and host metal atoms" [8]) (**a**) Dissolution of impurity atoms in aluminum, (**b**) dissolution of impurity atoms in copper

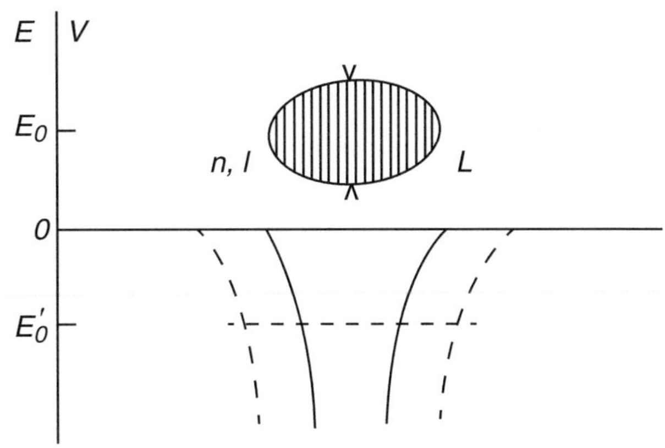

Fig. 55.2 Localized virtual (*n*, *l*)-level of conduction electrons scattered at the impurity potential *V*(*r*). The localized level energy is E_0 and width *L*. (E_0' is the energy of imaginary level responding the perturbation potential indicated by the *dashed cone* [8])

atom (i.e., they have *s*-, *p*-, *d*-, or *f*-symmetry) and an energy level *E*, which will be enlarged $(E \rightarrow E \pm \Delta E)$ due to the interference of the impurity states with the conduction band electronic states of the host metal. These spatially localized impurity-induced states provide additional resonance scattering of conduction electrons. The effectiveness of the resonance character of this scattering depends on: (a) location of the Fermi-level with respect to the impurity-level, (b) the density of conduction electron states at the Fermi-level, (c) the width $L = 2\Delta E$ of the localized impurity-induced state, and (d) the spin-structure of this impurity state.

The above model can be illustrated by Figs. 55.2 and 55.3.

Figure 55.2 illustrates formation of the "virtual impurity-state": an additional positive charge of impurity atom forms a potential which attracts negative particles. If this potential would be sufficiently strong (see dashed curves), then below the bottom of the conduction band there would be formed a bound state with an energy E_0' (dashed horizontal line), but if this potential is less effective, then the corresponding energy level E_0 will be located into the conduction band. In the latter case, the impurity electronic state interacts with the conduction electron states and provides enlargement of the energy level. Finally, the spatially localized

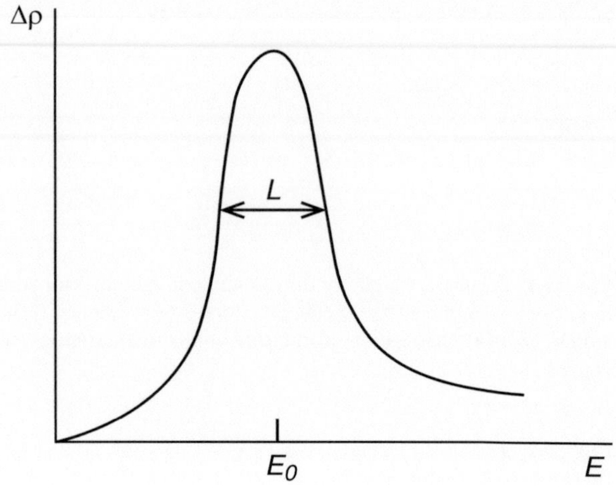

Fig. 55.3 Additional electro-resistivity caused by the virtual electronic state (the impurity energy level is E_0 and its width is L [8])

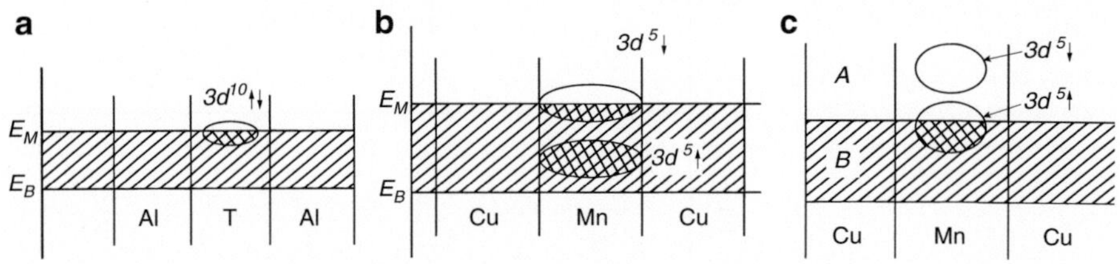

Fig. 55.4 Location of the enlarged localized virtual states with respect to Fermi-level. (**a**) Impurity atoms (denoted as T) dissolved in Al; (**b**, **c**) impurity atoms dissolved in Cu (see text)

"impurity-induced electronic state," which causes additional diffraction of conduction electrons, we characterize by some energy level E_0 and the width L. It is obvious that such local perturbations in the conduction band will provide additional resonance diffraction of charged particles manifested clearly in residual resistivity. On the energy scale, the density of states of the enlarged impurity-level is not constant. It is approximated usually by a bell-shaped curve resembling that shown in Fig. 55.3, and maximum of residual resistivity is assumed to occur when the Fermi-level coincides with the bell-curve peak.

It is well known that in a metallic system only band-electrons located directly at the Fermi-level are responsible for the charge transport. It is obvious as well that in the case of different "impurity—host-lattice" pairs the localized states will be placed at different positions with respect to the Fermi-level of the host metal. It follows that the series of impurities along the row of Periodic Table will provide residual resistivity data shown in Fig. 55.1.

Some examples of different localization of impurity virtual states with respect to the conduction band of the host metal are presented in Fig. 55.4a–c [8, 9]. (Figure 55.4b, c reproduce resembling situations: Fig. 55.4b is given by Friedel [8] and Fig. 55.4c—by Kittel [9].)

Let us reproduce comments of Kittel [9] on these experimental data.

Transition element impurities in aluminum. When Ti, at the beginning of the first transition group, is dissolved in Al, we expect its *d*-shell to be relatively unstable compared with the *d*-shell of Ni at the end of the group. When going continuously from Ti to Ni added to Al, the *d*-shell states of the impurity should descend and cross the Fermi-level of Al and give rise to virtual bound *d*-states. The residual resistivity of these impurities in Al is plotted in Fig. 55.1a, and show a large broad peak around Cr. The peak is believed to be caused by resonance scattering of electrons at the Fermi-level of Al when this level crosses the broadened *d*-shell levels of the impurity.

Transition element impurities in copper. The exchange and particularly the coulomb interactions within a *d*-shell tend to split it into two halves of opposite spin directions. In the free atom, the first five *d*-electrons have parallel spins, in conformity with Hund's rule; the sixth and subsequent *d*-electrons have their spin antiparallel to the direction of the spins of first five electrons. In the alloy, this spin splitting is maintained if the associated energy is larger than the width of the virtual level and larger than the distance from the Fermi-level to the virtual level. The numerical criterion is discussed in detail by Anderson [10]. Not all transition element impurities are magnetic in copper.

The exchange splitting explains the double peak (Fig. 55.1b) of the residual resistivity in the series from Ti to Ni dissolved in Cu, Ag, or Au. Peak *A* corresponds to the emptying of the upper half *A* of the *d*-shell, peak *B* to the lower half *B* with opposite spin direction" [9]. In review [9] in the "impurity charge - residual resistivity" dependence, $\Delta\rho(Z)$ (shown in Fig. 55.1b) the maxima are named, respectively, as maximum A (centered at Fe-Cr) and maximum B (centered at Ti).

Must be noted that experimental data reproduced in Fig. 55.1 were taken directly from [8]. Same dependences were presented in [9] in a slightly modified form, but for our qualitative consideration this difference seems to be negligible.

The above comments of Kittel clearly describe the physics of the residual resistivity measurements: there is emphasized the dominate role of the localized, impurity atom induced, electronic virtual states formed within the conduction band of the host material.

55.3 Localized Magnetic Moments in Alloys

Conception of the impurity-induced localized electronic states in metal alloys was formulated on using experimental investigations of a series of transition element impurities introduced in the *given host metals*. The model of localized magnetic moments formed on impurity paramagnetic ions dissolved in nonmagnetic metals was formulated on taking into account the idea of localized impurity levels split into "spin-up" and "spin-down" sub-levels (which are differently occupied by particles as in Fig. 55.4b, c), and on the basis of experiments dealing with the alloys of a *given paramagnetic ion* in a series of solvent metals and alloys. In Fig. 55.5, we

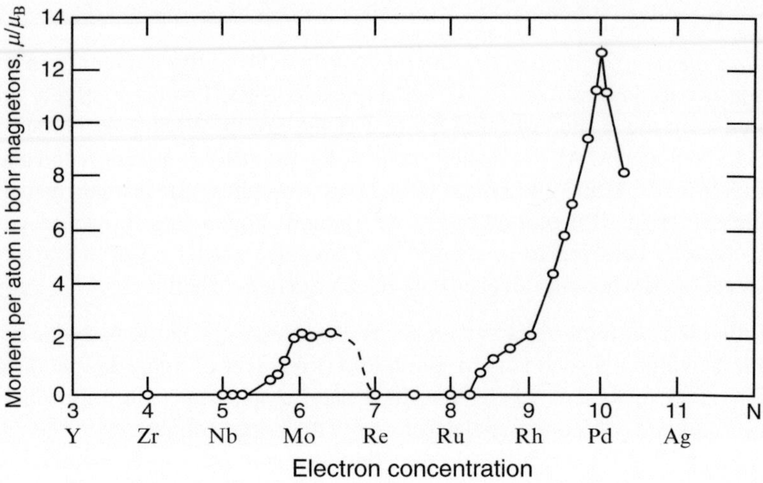

Fig. 55.5 Magnetic moments on impurity iron atoms dissolved in a series of 4*d*-alloys [9, 11]

have presented experimental data [11] of the measurements of magnetic moments associated with iron atoms dissolved in nonmagnetic host-metals (particularly, in 4*d*-metals and their alloys).

In the already mentioned article [10], the mathematical scheme which illustrates conditions for formation of spatially localized impurity states within the energy range of conduction electron band is described. The final stage of this consideration is represented by a pair of self-consistent equations which determine the equilibrium self-consistent values of additional electronic "spin-up" and "spin-down" densities ($\langle n_{d\uparrow}\rangle$ and $\langle n_{d\downarrow}\rangle$, respectively) localized at the impurity sites [9, 10]

$$\langle n_{d\uparrow}\rangle = \pi^{-1}\cot^{-1}\left[\left(E - E_F + U\langle n_{d\downarrow}\rangle\right)/\Delta\right],$$
$$\langle n_{d\downarrow}\rangle = \pi^{-1}\cot^{-1}\left[\left(E - E_F + U\langle n_{d\uparrow}\rangle\right)/\Delta\right]$$

In these expressions, E is the unperturbed energy of the d-state on an impurity atom, U is the coulomb repulsive energy between electrons in $d\uparrow$ and $d\downarrow$ states, E_F is the Fermi-level of the alloy, and Δ is the width of impurity energy levels. Energy parameters involved in above expressions were estimated in [10]. There were formulated numerical conditions allowing us to make rough predictions concerning magnetic behavior of alloys based on different solvent-solute pairs.

Localized impurity states in dilute alloys were considered by other investigators as well (see for e.g., [12, 13]). Formation of the so-called giant magnetic moments in Fe–Pd alloys (of the order 10 μ_B per iron atom) was considered, particularly, in [14, 15]. These anomalously great moments per impurity atoms were ascribed to Fe-ions induced polarization of surrounding Pd-sites.

55.4 Hydrogenated Magnetic Materials

Influence of hydrogen atoms on magnetic materials was investigated very intensively in previous decades. There are two aspects within the problem named "Hydrogen in magnetic materials." First one includes a set of investigations aimed to understand—what additional specific conditions are superimposed upon hydrogen atoms introduced in crystals with a magnetic subsystem, while the second unites investigations aimed to determine—what modifications of the crystalline magnetic subsystems are induced by interstitial hydrogen atoms. First problem was discussed in [16] and the second—in [17]. We pay more attention to the second one.

In [17], author had collected many empirical results based upon experimental investigations of magnetic materials absorbing hydrogen atoms. One of the expected effects is significant influence of inserted interstitial atoms upon the direct and indirect exchange interactions between the magnetic ion moments. In the first case, modifications of atom–atom interactions are associated usually with changes of interatomic distances caused by inserted hydrogen atoms, and in the second—by the influence of H-atoms on the electronic conduction band of the host material.

Influence of hydrogen on magnetic structure and other physical properties of rare-earth-based compounds were intensively investigated during the last few decades. In order to illustrate the investigation aims and main results obtained in this direction, we point to the set of corresponding articles published by one group of scientists from Laboratoire de Chimie Métallurgique et Spectroscopie des Terres Rares (CNRS) [18–23]. They have investigated intensively different properties of magnetic alloys based on the rare-earth and/or Fe-group metals and paid main attention to register relations between the crystalline structure and behavior of the magnetic subsystem. Particularly, different phases of the given $MH(D)_x$ compounds ($M \equiv RFe_2$, RMn_2, RNi_5; R—rare-earth element), as well as their stability regions, corresponding unit cell dimensions, magnetic structures, etc. were described. It follows that light atoms introduced in the metal lattice act at the magnetic subsystem mainly by distorting the host lattice and by introducing additional indirect interatomic bonds, modifying the spin–spin interaction conditions.

It must be stressed that sometimes interpretation of experimental data is quite difficult. For example, in Fig. 55.6a, b (reproduced from [17]) results of hydrogenation of magnetic compounds Y_6Mn_{23} and Th_6Mn_{23} are presented. We see that on adding H-atoms to these compounds the hydrides, $Y_6Mn_{23}H_{25}$ and $Th_6Mn_{23}H_{30}$, are formed. Note that the second compound being practically nonmagnetic, becomes strongly magnetic on hydrogenation, while the first one being magnetic without hydrogen, becomes practically nonmagnetic on hydrogen loading. Note as well that both compounds, Y_6Mn_{23} and Th_6Mn_{23}, are isostructural intermetallic compounds [17].

Hydrogen influence on magnetic subsystems of rare-earth metals was analyzed in [24].

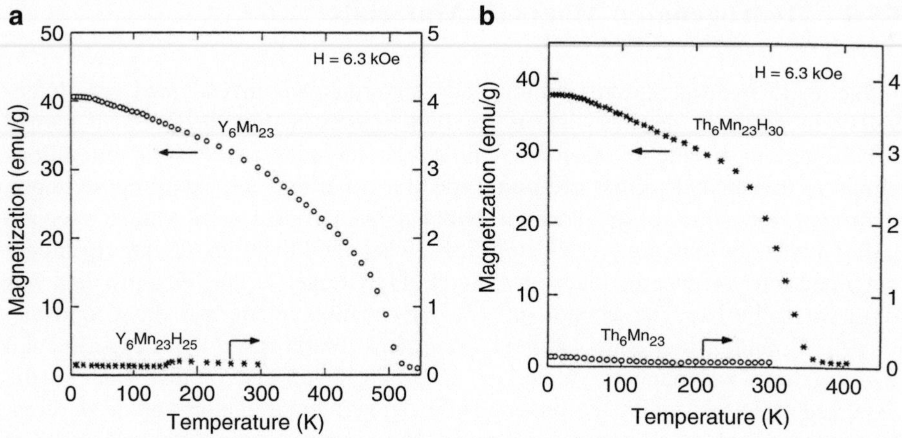

Fig. 55.6 Temperature dependence of the magnetization of pure and hydrogenated binary alloys [17] (**a**) Y_6Mn_{23} and $Y_6Mn_{23}H_{25}$, (**b**) Th_6Mn_{23} and $Th_6Mn_{23}H_{30}$

55.5 Hydrogen Influence on the Host Metal Conduction Band

We have seen above that hydrogen atoms inserted as interstitial centers in metal lattices modify the unit cells of the host materials and in this way influence on the direct and indirect atom–atom interactions. These lattice modifications are revealed particularly in changes of mechanical, magnetic, and electronic properties of the host materials. But we have mentioned already that hydrogen impurities influence as well upon the conduction band electrons of the solvent material. Each H-atom introduced in the metal lattice can give an electron to the conduction band, but provides an additional electronic level which can be occupied by a pair of electrons (in "spin-up" and "spin-down" states). If the hydrogen electronic level is well below the Fermi-level of the host metal, then it will attract conduction electrons from higher conduction states and will diminish the number of "free conduction electrons." If all electrons in the conduction band are transferred to the hydrogen 1*s*-states, then the hydrogenated compound will be transferred from metallic to nonmetallic (transparent) state. Fig. 55.7 [6] shows just this optical effect. The described model of rare-earth hydrides behavior supported by NMR experiments [7] is shown in Fig. 55.8, where results of spin-lattice relaxation times measurements in RE-H_2–RE-H_3 system are presented. This metal-insulator transition was analyzed in a number of theoretical investigations. Results of corresponding experimental investigations were collected, particularly, in [25, 26].

Results of NMR experiments performed on lanthanum hydrides: measurements of the electronic input in the spin-lattice relaxation rate (left-hand-side scale) and measurements of the Knight-shifts on lanthanum ion nucleus (right-hand-side scale) are presented. Both physical quantities are related to the electronic density

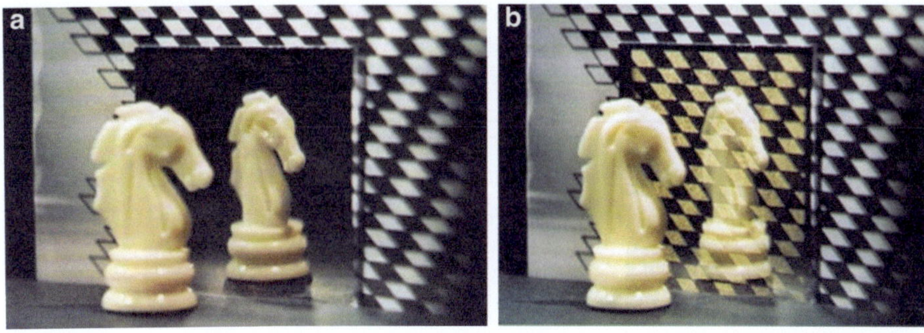

Fig. 55.7 Hydrogen-induced metal-insulator transition in rare-earth (yttrium, lanthanum) hydrides [6] (**a**) Metallic ("hydrogen-poor," mirror) state, (**b**) Nonmetallic ("hydrogen-rich," half-transparent) state

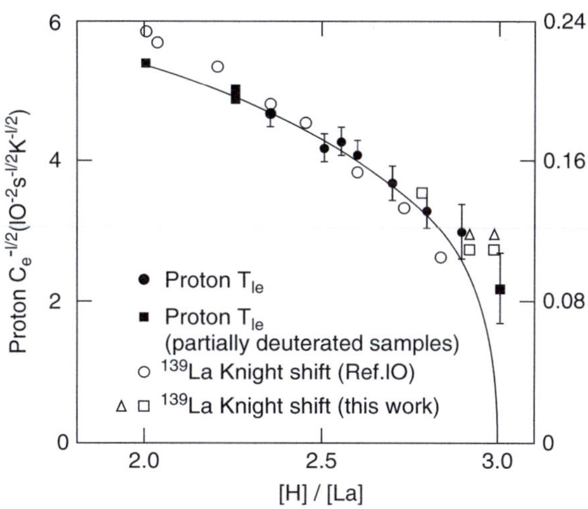

Fig. 55.8 Hydrogen-induced reduction of the conduction band filling [7]

of states in the conduction band. We see from these experiments that in hydrogen-rich compounds the density of conduction electrons decrease nonlinearly (the solid curve describing the set of experimental points is represented as proportional to $N(E_{\mathrm{F}})^{\sim}(3-x)^{1/3}$, where x is the hydrogen/metal atomic ratio [7]).

Interesting information on experimental investigations of metal-insulator transitions in rare-earth metal hydrides was collected, particularly, in [25, 26].

55.6 Expected Hydrogen Influence on Impurity Magnetic States in Dilute Alloys

We have already mentioned above that *d*-metal atoms dissolved in the lattice of normal metals form electronic states localized near impurity sites. These electronic states provide additional resistivity which depends on different characteristics of the localized state, among which the dominant role plays location of the localized energy level E_0 with respect to the Fermi-level E_F of the compound: when the difference $|E_0 - E_F|$ is minimal, the residual resistivity is maximal (see Figs. 55.1 and 55.4).

It was mentioned as well that if localized impurity state is split into pair of "spin-up" and "spin-down" states, then location of the Fermi-level between this "sub-levels" induces existence of unpaired spin-moment at the impurity site.

In performed experimental investigations, particularly in [11], the Fermi-level was displaced along the energy scale by *introducing in the lattice elements with valences different from that of the host atoms.*

We propose to shift the Fermi-level of the host lattice by inserting hydrogen atoms in interstitial sites. This procedure can be realized only in the case of host metals which easily absorb hydrogen and form stable hydrides (deuterides). These are niobium, palladium, rare-earth elements, and some others. As these metals form dilute alloys with paramagnetic ions, their magnetic properties will depend on hydrogen-induced location of the Fermi-level: if without hydrogen the Fermi-level will be placed outside of the paramagnetic impurity "spin-up" and "spin-down" levels—the system will be nonmagnetic, but if at adding hydrogen atoms the Fermi-level will be displaced and the "spin-up" and "spin-down" levels will occupy positions on both side of Fermi-energy, then the system will reveal a paramagnetic susceptibility (which can be registered). In other words, such hydrogen-absorbing dilute alloy of paramagnetic ions can be used as hydrogen-sensitive indicator.

55.7 Concluding Remarks

We have collected some experimental results and theoretical models in order to illustrate existing possibilities to obtain a system with variable magnetic properties on using the hydrogen-absorbing properties of the host nonmagnetic metals containing some paramagnetic impurity ions. As it can be seen from the cited examples, an adequate theoretical conception can be constructed after performing preliminary experimental investigations. Particularly, we need to know in what degree interstitial hydrogen atoms influence on the host lattice, on the conduction band, and on the "conduction band–paramagnetic impurity" interactions; what influence will have on the considered system application of external strong magnetic and/or electric fields, etc. This information is really needful in order to estimate possibilities of the given systems to be used as hydrogen detectors working in different external conditions.

Acknowledgements The manuscript was prepared within the frames of financial supports from Shota Rustaveli National Science Foundation (Georgia) and Science and Technology Center in Ukraine (STCU). Authors sincerely thank Zaza Rostomashivili for his help with illustrations.

References

1. Special issue (2002) The energy challenge. Phys Today
2. Alefeld G, Völkl J (eds) (1978) Hydrogen in metals, vol I and II. Springer, Berlin
3. Huiberts JI (1995) On the road to dirty metallic atomic hydrogen. Thesis, Vrije Universiteit, Amsterdam
4. Ngene P, Radeva T, Slaman M, Westerwaal RJ, Schreuders H, Dam B (2014) Seeing hydrogen in colors: low-cost and highly sensitive eye readable hydrogen detectors. Adv Funct Mater 24:2374–2382
5. Gordon Research Conference (GRC) on Hydrogen-metal systems, Lucca (Barga), 14–19 July 2013
6. Huiberts JH, Griessen R, Rector JH, Wijngaarden RJ, Dekker JP, de Groot DG, Koeman NJ (1996) Yttrium and lanthanum hydride films with switchable optical properties. Nature 380:231–234
7. Barnes RG, Chang C-T, Belhoul M, Torgeson DR, Schoenberger RJ, Beaudry BJ, Seymour EPW (1991) Magnetic resonance determination of the electronic density of states and the metal—non-metal transition in the LaH_2-LaH_3 system. J Less Common Met 172–174:411–418
8. Friedel J (1958) On the electronic structure of transition and refractory metals and their alloys. J Phys Radium 19(6):573
9. Kittel C (1966) Quantum theory of solids. Wiley, New York, Ch. 18
10. Anderson PW (1961) Localized magnetic states in metals. Phys Rev 124:41
11. Clogstone AM, Matthias B, Peter M, Williams H, Correnzwit E, Sherwood R (1962) Localized magnetic moments of iron atoms dissolved in different transition metal alloys". Phys Rev 125:541
12. Wolff PA (1961) Localized magnetic moments in metals. Phys Rev 124:1030
13. Clogston AM (1962) Impurity atoms states in metals. Phys Rev 125:439
14. Ratishvili IG (1967) Influence of magnetic localized states on magnetic susceptibility of solid solutions. JETPh 53:1353–1361
15. Ratishvili IG (1967) Distribution of hyperfine fields in the magnetic localized states. FMM 23:1013–1018 (in Russian)
16. Kronmuller H (1978) Magnetic aftereffects of hydrogen isotopes in ferromagnetic metals and alloys. In: Alefeld G, Völkl J (eds) Hydrogen in metals, vol I. Springer, Berlin, pp 289–320
17. Wallace WE (1978) Magnetic properties of metal hydrides and hydrogenated intermetallic compounds. In: Alefeld G, Völkl J (eds) Hydrogen in metals, vol I. Springer, Berlin, pp 169–195
18. Paul-Boncour V et al (1997) Structural and magnetic study of new YFe_2D_x compounds ($0<x<3.5$). J Alloys Compd 253–254:272–274
19. Paul-Boncour V et al (2003) Neutron diffraction study of ZrM_2D_x deuterides (M = Fe, Co). J Alloys Compd 356–357:69–72
20. Paul-Boncour V et al (2003) Influence of deuterium absorption on structural and magnetic properties of $ErFe_2$. J Alloys Compd 356–357:195–199
21. Paul-Boncour V et al (2005) Electronic and crystallographic properties of the system $CeY_2Ni_9D_x$ ($0 < x < 8.7$) measured by X-ray absorption spectroscopy. J Alloys Compd 404–406:60–63
22. Paul-Boncour V et al (2005) Origin of the first-order magnetostructural transition in $YFe_2D_{4.2}$. J Alloys Compd 404–406:355–359
23. Paul-Boncour V et al (2006) Homogeneity range and order-disorder transitions in $R_{1-x}Ni_2$ Laves phase compounds. Intermetallics 14(5):483–490
24. Vajda P (2005) Current progress in the magnetism of rare-earth-hydrogen systems. J Alloys Compd 404–406:131–135
25. Vajda P (1995) Hydrogen ordering and metal-semiconductor transitions in superstoichiometric rare-earth dihydrides. J Alloys Compd 231:170–175
26. Vajda P (2004) Hydrogen in rare-earths: a wealth of structural and electronic phenomena. Solid State Ionics 168:271–279

Chapter 56
Effect of Thickness on Photocatalytic Activity of TiO_2 Thin Films

F. Bensouici, M. Bououdina, A. Iratni, M. Toubane, and R. Tala-Ighil

Abstract TiO_2 thin films were deposited on glass substrates by hydrolysis and condensation of tetraethyl orthotitanate [TEOT], and the pH of the solution is controlled by the addition of concentrated HNO_3. The effect of thickness of the films on structural, optical, and photocatalytic properties has been investigated. The phase transformation temperature was determined by DTA-TGA; the nature of the phases and the crystallite size of TiO_2 are derived from X-ray diffraction patterns. From the transmittance spectra of the layers, the refractive index, and thickness were calculated using the envelope method. The results show that the films crystallize into anatase phase only with grains size in the range of 17 nm. The films showed high transmittance in the visible (93 %) and opaque in the UV. Finally, the as-prepared TiO_2 films exhibit high efficiency in the photocatalytic degradation of MB and RdB.

Keywords TiO_2 • Anatase • Photocatalysis • Sol-gel • Thin films

Nomenclature

TiO_2	Titanium dioxide
$Ti(OC_2H_5)_4$	Tetraethylorthotitanate
HNO_3	Nitric acid
H_2O	Water
XRD	X-ray diffraction

F. Bensouici (✉) • A. Iratni • M. Toubane
URMPE Unite, Department of Physics, UMBB University, Boumerdes 35000, Algeria
e-mail: fbensouici@yahoo.fr

M. Bououdina
Department of Physics, College of Science, University of Bahrain, Sakhir, Kingdom of Bahrain

Nanotechnology Centre, University of Bahrain, Sakhir, Kingdom of Bahrain

R. Tala-Ighil
URMPE Unite, Department of Physics, Institute of electrical and electronic engineering UMBB University, Boumerdes 35000, Algeria

I. Dincer et al. (eds.), *Progress in Clean Energy, Volume 1*,
DOI 10.1007/978-3-319-16709-1_56

UV/VIS/NIR	Ultraviolet/visible/near infrared
MB	Methylene blue
RhB	Rhodamine B
D	Crystallite size (Å)
(hkl)	Miller indices
Δ_{hkl}	Full width at half maximum (FWHM) of (hkl) diffraction peak
Δ_{inst}	FWHM corresponding the instrumental contribution
θ	Bragg angle (rad)
$T_{(hkl)}$	Texture coefficient of the (hkl) plane
α	Absorption coefficient
T	Transmittance (%)
E_g	Band gap (eV)
h	Plank constant (6.626176×10^{-34})
n_λ	Refractive index
n_s	Refractive index of substrate
C_0	The initial concentration of dye solution (mol/L)
C_t	The concentration of dye at different time interval under UV irradiation (mol/L)
k_{app}	Reaction rate constant (min^{-1})

56.1 Introduction

Titanium dioxide is a semiconductor material having a large number of applications such as solar energy conversion [1], photocatalysts [2], gas sensor [3], photochromic devices [4], optoelectronics [5], and ceramic membrane [6]. This is mainly due to its favorable properties like non-toxicity, chemical inertness, and stability over a wide pH range under irradiation conditions. Since Fujishima and Honda discovered the photocatalytic splitting of water on TiO_2 electrodes in 1972 [7], TiO_2 became the favorite semiconductor for heterogeneous photocatalysis thus attracted much attention of researchers in particular with the continuous increase of environmental problems. As we know TiO_2 crystallizes in three phase: anatase, rutile and brookite. It has been reported that the rutile phase is the most stable whereas the anatase phase is more active in photocatalysis. The band gap ($E_g > 3.25$ eV) of anatase is excited by UV light ($\lambda < 388$ nm) to allow electrons transition from the valence band to the conduction band [8]. Therefore, this limits the use of sunlight or visible light as an irradiation source in photocatalytic reactions on TiO_2 [9].

One of the most important characters of the photocatalytic reaction is the adsorption of photons to generate electron-hole pairs. These pairs production depend strongly on the films' thickness which influences afterwards the photocatalytic properties [10, 11]. Therefore, the surface properties of the film is deeply influenced by the thickness, it may be attributed to the roughness which provides more reaction sites for photocatalysis.

It has been reported that most of the physical and photocatalytic properties of thin films depend enormously on the preparation technique. During the last two decades, several methods have been developed to prepare TiO_2 thin films such as: electron beam evaporation [12], chemical vapor deposition [13, 14], physical vapor deposition [15], atmospheric pressure chemical vapor deposition [16], and the sol–gel method [17]. Additionally, the sol–gel dip-coating process has been identified as one of the most appropriate techniques to prepare vitreous or polycrystalline films with excellent homogeneity, purity, and uniformity on any type of substrates.

In this work, a simple method to prepare TiO_2 thin films by sol–gel dip-coating technique was adopted. The aim was to investigate the effect of deposited layers thickness on structural and optical properties thereby their influence on photocatalytic degradation of Rhodamine B (RhB) and Methylene Blue (MB). The films were analyzed by X-ray diffraction (XRD), Raman spectroscopy, and UV–Vis spectrophotometry.

56.2 Experimental Part

56.2.1 Solution Preparation

The precursor solutions for TiO_2 films were prepared using tetraethylorthotitanate, ethanol, water, and HNO_3 acid. The tetraethylorthotitanate $Ti(OC_2H_5)_4$ (99 %, Merck) is dissolved in ethanol, then the resulted mixture was injected into a prepared mixture containing distilled water, ethanol, and different amounts of nitric acid HNO_3. Ethanol (Aldrich 98 %) was chosen as a solvent because of its high volatility leading to a fast elimination during the drying phase. After several attempts, the adopted molar proportions [Ti]/[H_2O]/[acid]/ethanol were 1.00/0.55/0.33/21.18. The solutions were stirred for 1 h and further aged at room temperature under atmospheric pressure. The final solution became transparent and light yellow.

56.2.2 Films Deposition

Soda-lime glasses (40 × 20 × 1 mm) were used as substrates. TiO_2 films deposition was carried out by dip-coating method, consisting in immersing the substrate and then withdrawing it from a Teflon pot containing the sol. The withdrawal speed was 1 mm s^{-1} and the thickness of the layer was controlled indirectly by the number of dipping. Subsequently, the coated sample was heated at 400 °C for 10 min in order to remove the solvent and residual organics. In addition, a second longer heat treatment was performed at 450 °C (10 °C/min) for 1 h in air to increase the film density.

56.2.3 Films Characterizations

TG/DTA measurements were carried out using NETZSCH STA 409 PC/PG apparatus with heating rate 10 ° C/min. in an inert atmosphere. The crystalline phase of TiO_2 films was determined by XRD method in thin film mode using bruker D8 Advance $(\theta, 2\theta)$ diffractometer equipped with Cu-Kα radiation ($\lambda = 1.5418$ Å) and confirmed by Raman Spectroscopy measurements. Optical transmittance and absorbance spectral were measured at normal incidence in UV–Vis spectral region (190–1,100 nm) with a Lambda EZ 210 double beam spectrophotometer. PlasCalc-2000 UV/Vis/NIR used to determine the spectra of the lamp UV used in the photocatalytic test.

56.2.4 Photocatalysis Measurements

The photocatalytic activity was measured in a homemade system. It was quantified by means of the bleaching of water both diluted Methylene blue (MB) and Rhodamine B solutions. For preparing these solutions, 35 mL of Methylene blue (MB) solution at an initial concentration of 2.5×10^{-5} mol/L and Rhodamine B at an initial concentration of 2.08×10^{-5} mol/L introduced in a glass cells. Four samples are immersed into the cell containing the Methylene blue (MB) solution (cell + solution + sample) at an angle of 70 °C to ensure the surface is opposite to the radius of the UV lamp and the immersed surface is about 4×2 cm^2. Subsequently, the cells are exposed to UV-light at different exposure times. AG 15T8 germicidal lamp of 15 W was used as UV source; the spectral of this lamp is represented in Fig. 56.1. The distance between the cell and UV lamp was fixed to 7 cm.

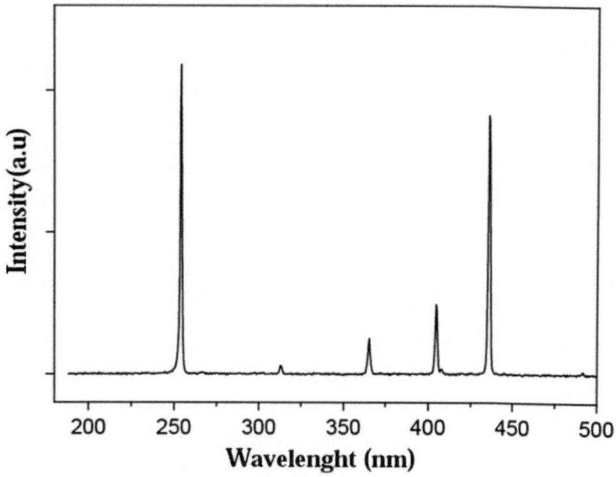

Fig. 56.1 Specter of AG 15 T8 germicidal lamp

56.3 Results and Discussion

The thermogravimetric (TG) curve shows that the mass of the xerogel TiO_2 decreases with increasing heating temperature and became stable after 400 °C, see Fig. 56.2a. This mass loss with temperature is accompanied by both endothermic and exothermic peaks. Between 25 and 300 °C residual water evaporation, ethanol decomposition and burning of organic compounds took place corresponding to the endothermic peak. The exothermic peak, centered around 400 °C, corresponds to the crystallization of TiO_2. To determine the evolution of the structure as a function of annealing time, an isotherm at 450 °C versus time was plotted in Fig. 56.2b. The results showed that annealing time higher than 40 min is enough for TiO_2 crystallization. It is also observed that annealing temperature equal or higher than 400 °C for one hour would be largely sufficient for the entire formation of TiO_2.

Figure 56.3 shows a typical Raman spectrum of the as-prepared films with different thickness after heat treatment at 450 °C for 60 min. It can be noticed that no peaks can be detected for the film with one layer only. After the deposition of two layers, only the main peak located at 125 cm^{-1} can be easily observed. Afterwards, four main peaks located at 125, 399, 517, and 636 cm^{-1} appeared in the case of film with four layers, which is attributed to the formation of pure anatase phase, in agreement with the reported data in the literature [18, 19].

Moreover, it can be noticed that the peak intensity increases with the thickness (number of deposited layers), indicating the enhancement of the degree of crystallinity of the formed TiO_2 phase. This result is confirmed by XRD analysis, see Fig. 56.4. XRD patterns of TiO_2 films obtained by grazing incident angle show the appearance of a small and broad peak corresponding to the (101) plane after the deposition of one layer then afterwards, further peaks including (101), (004), (200), (105), (211), and (204) planes attributed to anatase phase appeared. The above peaks become sharper and narrower, indicating the improved crystallinity mainly after the deposition of the fourth layer. The same plots have been observed in 6, 7, 8, and 9.

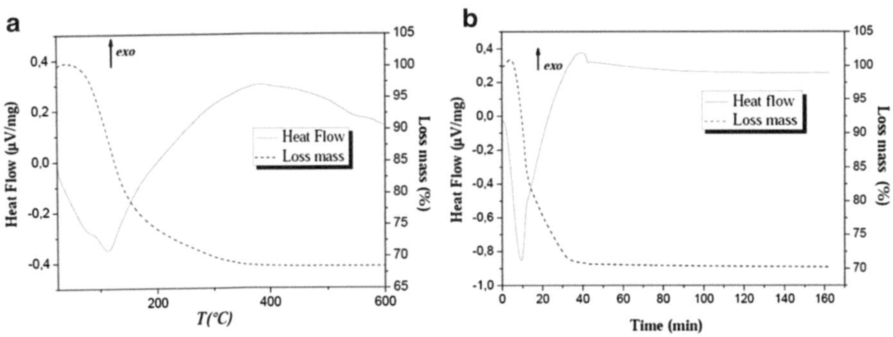

Fig. 56.2 Thermal analysis curves of the xerogel recorded with 10 °C/min in an inert atmosphere, (**a**) DTA and TGA curves and (**b**) Isotherm at 450 °C

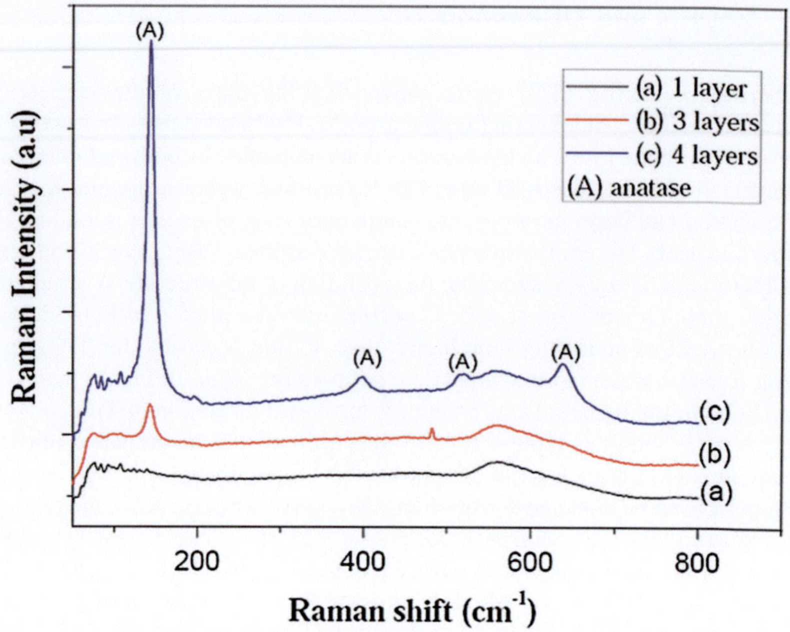

Fig. 56.3 Raman spectra of films with various layers

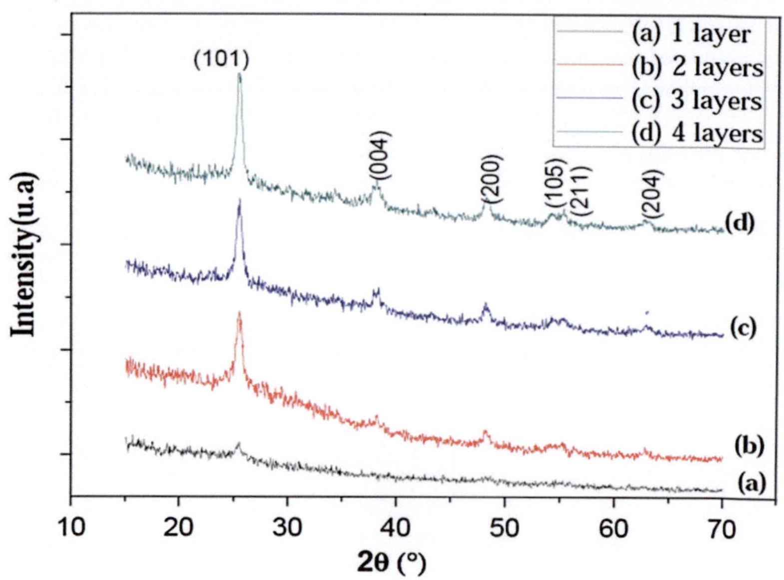

Fig. 56.4 XRD patterns of TiO_2 films deposited on glass substrate with various layers

The crystallite size D was determined from the diffraction peak broadening by employing the Scherrer's equation [20]:

$$D = \frac{0.89 \times \lambda}{\sqrt{\Delta_{hkl}^2 - \Delta_{\mathrm{inst}}^2}} \times \frac{1}{\cos\theta} \quad (56.1)$$

where λ is the wavelength of X-ray beam (Cu-Kα = 1.5406 Å), Δ_{hkl} is the full width at half maximum (FWHM) of (hkl) diffraction peak, Δ_{inst} is the FWHM corresponding the instrumental contribution, and θ is the Bragg angle. It is observed that all the films exhibit grains/particles in the nanoscale range 10–18 nm, in agreement with some results reported in the literature by Mechiakh et al. and Yongjun Chen et al. [31, 32].

In order to study the texture of films at various thickness, the degree of preferred orientation $T_{(hkl)}$ was used. This factor can be calculated using the following relation [23, 24]:

$$T_{(hkl)} = \frac{[I_{\mathrm{m}}(hkl)/I_0(hkl)]}{(1/N)\sum_1^N I_{\mathrm{m}}(hkl)/I_0(hkl)} \quad (56.2)$$

where $T_{(hkl)}$ is the texture coefficient of the (hkl) plane, $I_{\mathrm{m}}(hkl)$ the measured intensity of (hkl) plane, $I_0(hkl)$ the corresponding recorded intensity of a standard TiO_2 using JCPDS Card No. 00-004-0477, and N the preferred direction of growth. From the plots in Fig. 56.5, it is observed that the texture coefficient for all peaks is around the unity ($T_{(hkl)} \sim 1$), which means that there is no preferred orientation for

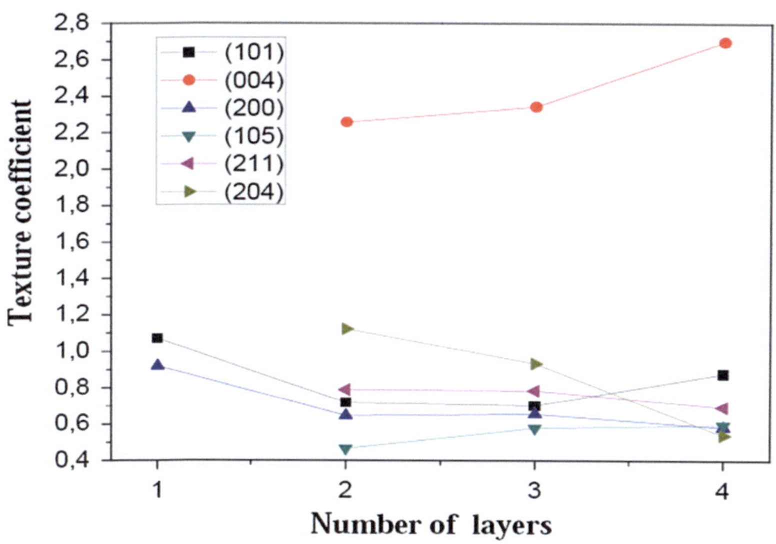

Fig. 56.5 Texture coefficient as a function of number of layers

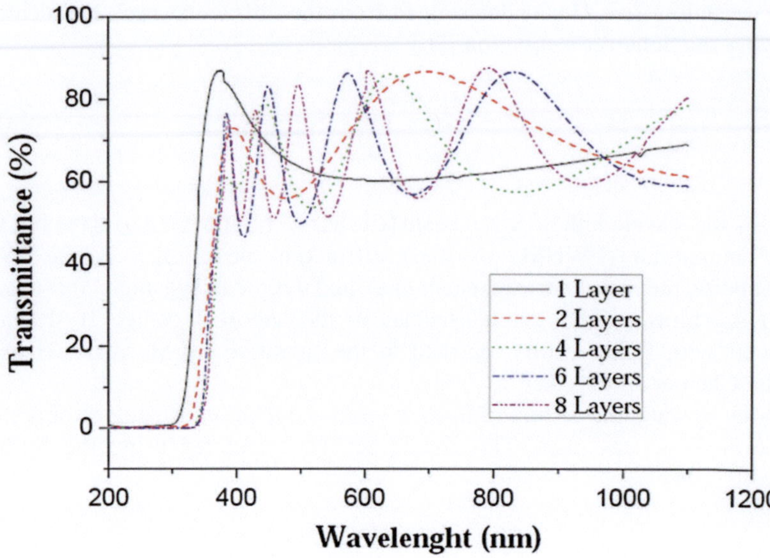

Fig. 56.6 UV–Vis transmittance spectra of TiO_2 thin films for various layers annealed at 450 °C

crystallization except for (004) peak, where $T_{(hkl)} = 2,5$. This indicates that a slight preferred orientation along (004) plane occurs.

XRD Rietveld analysis confirms that all the deposited films crystallize within anatase phase having a tetragonal structure with the space group $I4_1/amd$ (141).

The transmittance $T(\lambda)$ were measured at normal incidence in UV–Vis spectral region of 190–1,100 nm, Fig. 56.6. The spectra reveal very pronounced interference effects in the transparency region 380–1,100 nm with sharp falls at the edge of the band. The appearance of interferences indicates a good homogeneity of the deposited films.

$$\alpha(\lambda) = (1/d)\ln[1/T(\lambda)] \tag{56.3}$$

where d is the film thickness and $T(\lambda)$ the film transmittance. The indirect optical band gap E_g is calculated using the well-known energy-exponential relation [25]:

$$(\alpha h\upsilon) = C\left(h\upsilon - E_g\right)^2 \tag{56.4}$$

where C is a constant. Figure 56.7 presents the variation of $(\alpha h\upsilon)^{0.5}$ as a function $(h\upsilon)$. From the plots, the value of the band gap is determined by extrapolating the linear region (region) to $(\alpha h\upsilon) = 0$. The fit gives an E_g value of 3.30 eV, which corresponds to the band gap anatase in agreement with the values reported in the literature [26, 27].

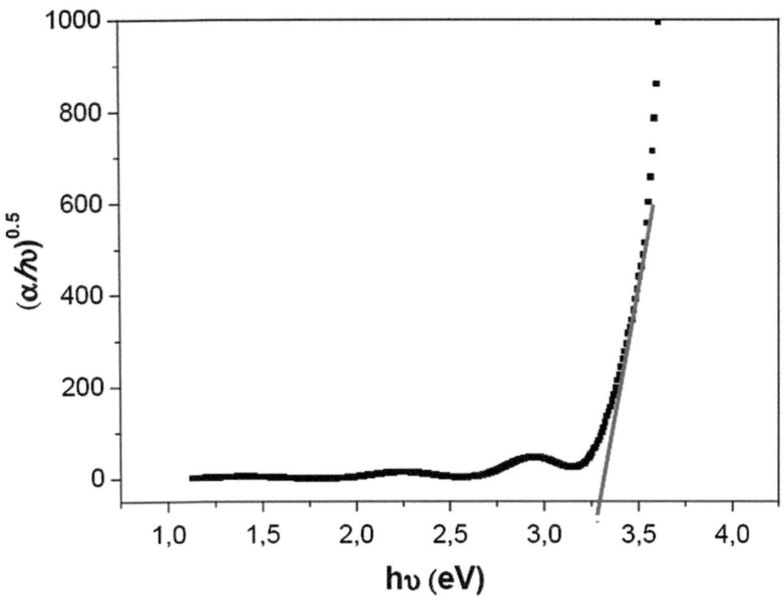

Fig. 56.7 $(\alpha h\upsilon)^{0.5}$ as function of $(h\upsilon)$ plot for TiO_2 film with four layers

The optical refractive index $n(\lambda)$ of the film and the thickness have been determined from the transmittance spectrum following the methods of Manifacier [28] and Swanepoel [29]:

$$n_\lambda = \sqrt{s + \sqrt{(s^2 - n_s^2(\lambda)}}. \tag{56.5}$$

$$S = \frac{1 - n_s^2(\lambda)}{2} + 2n_S \frac{T_{\max}(\lambda) - T_{\min}(\lambda)}{T_{\max}(\lambda) T_{\min}(\lambda)} \tag{56.6}$$

$$d = \frac{(2K + 1)\lambda}{4n_\lambda} \tag{56.7}$$

where n_s is the refractive index of substrate, d is the thickness, T_{max} and T_{min} are the maximum and the minimum of the envelope. Figure 56.8 shows that the refractive index of TiO_2 films increases with increasing the number of dipping and becomes stable at 2.50 at $\lambda = 400$ nm, which represents the refractive index of the anatase phase [30]. It is also observed that the film thickness increases linearly with the number of dipping. 150, 220, and 300 nm for 2, 3, and 4 layer, respectively (Fig. 56.8), which is in good agreement with values reported by some authors [31, 32]. Therefore, from the linear fit of $d(N)$, the layer thickness was estimated, for one layer the thickness value was found to be 80 nm. It is important to note that this thickness is about three times higher than that obtained by Ryu et al. [32] and

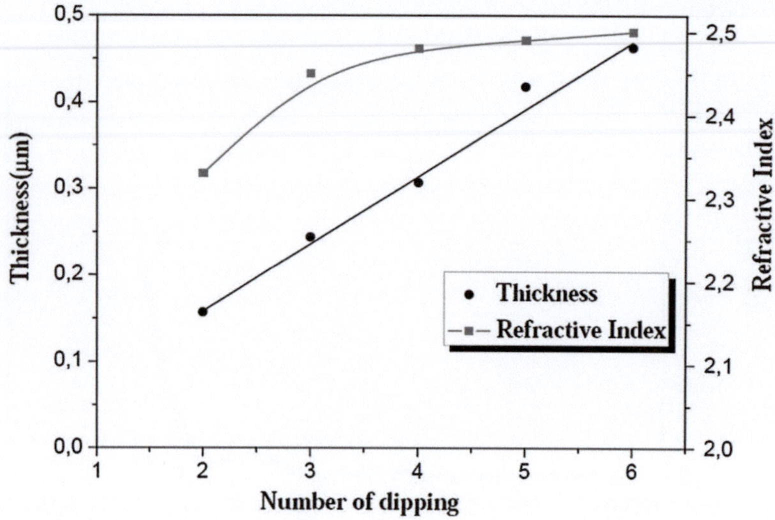

Fig. 56.8 Thickness and refractive index as function as number of dipping

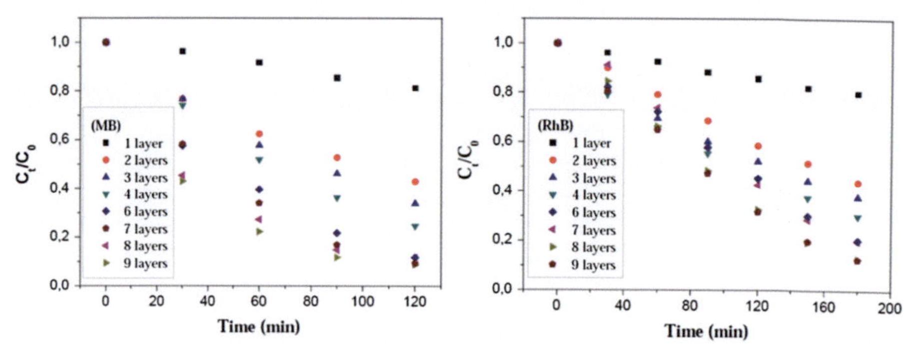

Fig. 56.9 Photocatalytic decomposition of methylene blue (MB) and rhodamine B (RhB) dyes

30 times lower than that obtained by Yongjun Chen et al. [32], but the average calculated crystallite size is in good agreement, i.e., 12–18 nm.

Figures 56.9 and 56.10 illustrate the photocatalytic decomposition of MB and RhB dyes in aqueous solution by pure TiO_2 thin films under UV light irradiation, respectively. As seen in Fig. 56.9, MB dye decomposes continuously during the entire irradiation time for all films. It is important to note that the photocatalytic removal efficiency of TiO_2 film increases with increasing layer number reaching 92 % for 7, 8, and 9 layer number of initial MB concentration. While for RhB, we showed the same degradation kinetics that observed with MB but with a value less than 87 %.

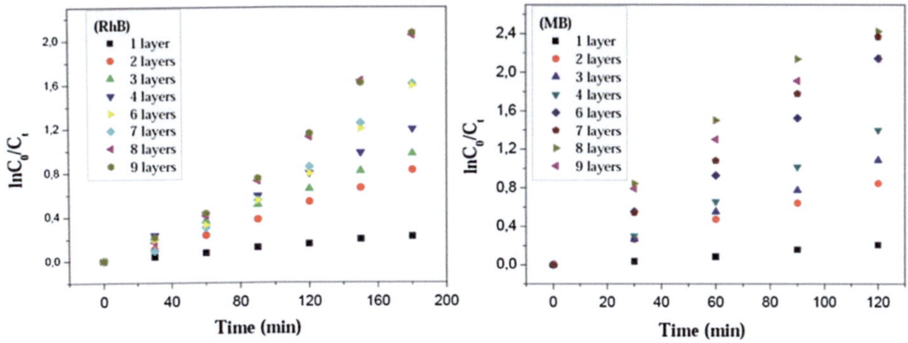

Fig. 56.10 Corresponding plots for reaction rate constant (k_{app}) for methylene blue (MB) and rhodamine B (RhB) dyes under UV–Vis light irradiation

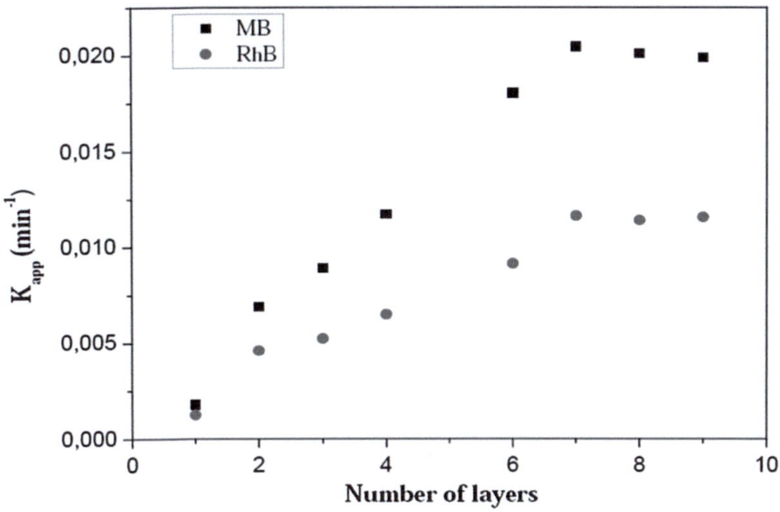

Fig. 56.11 Reaction rate constant (k_{app}) versus the number of layer for methylene blue (MB) and rhodamine B (RhB) dyes

The photocatalytic reaction rate constant (k_{app}) was determined from the plots represented in Fig. 56.10 and using the pseudo-first-order reaction equation [33]:

$$\ln(C_0/C_t) = k_{app}t \tag{8}$$

where C_0 is the initial concentration of dye solution and C_t the concentration of dye at different time interval under UV irradiation. Moreover, in order to show the variation of the constant k_{app} as function of the film thickness, the values of k_{app} versus the layers number are presented in Fig. 56.11. The plots show that k_{app} increases linearly with the film thickness and became stable after seven layers. This behavior is also observed in the case of RhB dye. It is also found that the k_{app} value performs 10 times for layer number greater than 7. The main explanation of the performance of k_{app} with the number of layer can be summarized in three different processes:

1. The absorption efficiency of UV light increases by increasing the layers number which increases the *e*/*h* pair production.
2. The increase of film thickness leads to an increase in the amount of mass matter which increases the adsorption process.
3. It was also observed that the surface roughness increases with layers number leading to the increase of the degradation efficiency of dyes. The TiO_2 films with a high rough surface present a higher efficiency of degradation than those with a smoother surface. The porous surface increases the active surface sites, thus enhancing the catalytic activity of the films, as has been established by several authors [35, 36].

It is important to mention that the rate of degradation (k_{app}) has a limit when the layers number increases. This means that increasing layers number (more than 7 in this study) will not improve the film efficiency in accordance with the work of Wu et al. [34].

More than that, the degradation rate is higher in the case of MB dye than RhB dye. This difference may be attributed to:

1. Chemical reactivity in it even functional groups of target molecules with $HO^{\bullet}$.
2. The adsorption of the dye or not on the surface of photocatalysis.

56.4 Conclusion

Nanocrystalline transparent TiO_2 thin films have been prepared by dip-coating technique. The films thickness has been controlled by the number of dipping. The films structure reveals the stability of single and pure anatase phase. Optical results show that the refractive index increases with the dipping number and reaches a stable value at 2.5. It is also found that the layers number enhances the light absorption, the amount of deposited matter, and the surface roughness which leads to the increase of the dyes degradation efficiency. In this study, the degradation efficiency presented a saturation behavior after seven layers. The rate of degradation k_{app} is higher in the case of MB than RhB.

References

1. Hagfeldt A, Gratzel M (1995) Light-induced redox reactions in nanocrystalline systems. Chem Rev 95:49
2. Sakthivel S, Kisch H (2003) Daylight photocatalysis by carbon-modified titanium dioxide. Angew Chem Int Ed 42:4908

3. Sharma RK, Bhatnagar MC, Sharma GL (1998) Mechanism in Nb doped titania oxygen gas sensor. Sens Actuators B Chem 46:194
4. Luchi K, Ohko Y, Tatsuma T, Fujishima A (2004) Cathode separated TiO_2 photocatalyst applicable to photochromic device responsive to backside illumination. Chem Mater 16:1165
5. Banfi G, Degiorgo V, Ricard D (1998) Non-resonant optical nonlinearities of semiconductor nanocrystals. Adv Phys 47:447
6. Yan Y, Chaudhuri SR, Sarkar A (1996) Synthesis, characterization and optical properties of stacked porous thin films derived from sol-gel process. J Am Ceram Soc 79:1061
7. Honda K, Fujishima A (1972) Electrochemical photolysis of water at a semiconductor electrode. Nature 238:37
8. Kormann C, Bahnemann DW, Hoffmann MR (1988) Preparation and characterization of quantum-size titanium dioxide. J Phys Chem 92:5196
9. Goswami DY (1997) A review of engineering developments of aqueous phase solar photocatalytic detoxification and disinfection processes. J Sol Energy Eng 119:101
10. Kumar KJ, Raju NRC, Subrahmanyam A (2011) Thickness dependent physical and photo catalytic properties of ITO thin films prepared by reactive DC magnetron sputtering. Appl Surf Sci 257:3075–3080
11. Wu C-Y, Lee Y-L, Lo Y-S, Lin C-J, Wu C-H (2013) Thickness-dependent photocatalytic performance of nanocrystalline TiO_2 thin films prepared by sol–gel spin coating. Appl Surf Sci 280:737–744
12. Yang TS, Shiu CB, Wong MS (2004) Structure and hydrophilicity of titanium oxide films prepared by electron beam evaporation. Surf Sci 548:75
13. Marco de Lucas MC, Fabreguette F, Collin S, Bourgeois S (2000) MOCVD growth of TiO_2 thin films on single crystal GaAs substrates. Int J Inorg Mater 2:255
14. Jung S-C, Kim B-H, Kim S-J, Imaishi N, Cho Y-I (2005) Characterization of a TiO_2 photocatalyst film deposited by CVD and its photocatalytic activity. Chem Vapor Deposition 11(3):137–141
15. Damiriv D, Bally AR, Ballif C, Homes P, Schmid PE, Sanjines R, Levy F, Parvulescu VI (2000) Photocatalytic degradation of phenol by TiO_2 thin films prepared by sputtering. Appl Catal B Environ 25:83
16. Hoyer P (1996) Formation of titanium dioxide nanotube array. Langmuir 12:1411
17. Burnside SD, Shklower V, Barbe C, Comte P, Arendse F, Brookes K, Grätzel M (1998) Self-organization of TiO_2 nanoparticles in thin films. Chem Mater 10:2419
18. Smith WS (1990) Principles of materials science and engineering. McGraw-Hill, Singapore, pp 826–830
19. Almeida RM, Christensen EE (1997) Crystallization behavior of SiO_2-TiO_2 sol-gel thin films. J Solgel Sci Technol 8:409–413
20. Cullity BD (1978) Elements of X-ray diffraction, 2nd edn. Addison-Wesley, Reading
21. Danion A, Disdier J, Guillard C, Abdelmalek F, Jaffrezic-Renault N (2004) Characterization and study of a single-TiO_2-coated optical fiber reactor. Appl Catal Environ 52:213–223
22. Ryu CS, Kim M-S, Kim B-W (2003) Photodegradation of alachlor with the TiO_2 film immobilised on the glass tube in aqueous solution. Chemosphere 53:765
23. Lemlikchi S, Abdelli-Messaci S, Lafane S, Kerdja T, Guittoum A, Saad M (2010) Study of structural and optical properties of ZnO films grown by pulsed laser deposition. Appl Surf Sci 256:5650–5655
24. Patil PS, Chigare PS, Sadale SB, Seth T, Amalnerkar DP, Kawar RK (2003) Thickness-dependent properties of sprayed iridium oxide thin films. Mater Chem Phys 80:667–675
25. Hong WQ (1989) Extraction of extinction coefficient of weak absorbing thin films from special absorption. J Phys D Appl Phys 22:1384
26. Sen S, Mahanty S, Roy S, Heintz O, Bourgeois S, Chaumont D (2005) Investigation on sol–gel synthesized Ag-doped TiO_2 cermet thin films. Thin Solid Films 474:245–249
27. Chrysicopoulou P, Davazoglou D, Trapalis C, Kordas G (1998) Optical properties of very thin (<100 nm) sol–gel TiO_2 films. Thin Solid Films 323:188

28. Manifacier JC, Gasiot J, Fillard JP (1976) A simple method for the determination of the optical constants n, k and the thickness of a weakly absorbing thin film. J Phys E Sci Instrum 9:1002
29. Swanepoel R (1983) Determination of the thickness and optical constants of amorphous silicon. J Phys E Sci Instrum 16:1214
30. Kingery WD, Bowen HK, Uhlmann DR (1976) Introduction to ceramics, 2nd edn. Wiley, New York
31. Mechiakh R, Meriche F, Kremer R, Bensaha R, Boudine B, Boudrioua A (2007) TiO_2 thin films prepared by sol–gel method for wave guiding applications: correlation between the structural and optical properties. Opt Mater 30:645–651
32. Chen Y, Dionysiou DD (2006) Correlation of structural properties and film thickness to photocatalytic activity of thick TiO_2 films coated on stainless steel. Appl Catal Environ 69:24–33
33. Fogler HS (2005) Elements of chemical reaction engineering, 4th edn. Upper Saddle River, Prentice Hall
34. Wua C-Y, Leeb Y-L, Lob Y-S, Lina C-J, Wua C-H (2013) Thickness-dependent photocatalytic performance of nanocrystalline TiO_2 thin films prepared by sol–gel spin coating. Appl Surf Sci 280:737–744
35. Aprile C, Corma A, García H (2008) Enhancement of the photocatalytic activity of TiO_2 through spatial structuring and particle size control: from subnanometric to submillimetric length scale. Phys Chem Chem Phys 10:769–783
36. Guillén-Santiago A, Mayén SA, Torres-Delgado G, Castanedo-Pérez R, Maldonado A, Olvera ML (2010) Photocatalytic degradation of methylene blue using undoped and Ag-doped TiO_2 thin films deposited by a sol–gel process: effect of the ageing time of the starting solution and the film thickness. Mater Sci Eng B 174:84–87

Chapter 57
Electrochemical Behavior of Silicon in Light and Dark Environments in 0.5 M Sulfuric Acid

M. Bounoughaz, H. Boukercha, A. Khemmari, M. Allaoua, and E. Salhi

Abstract The aim of this work is to study the effect of lights, dark, and cooling time on the electrochemical properties of the photovoltaic silicon working electrodes in 0.5 M sulfuric acid medium. The silicon samples were heated until 1,200 °C in appropriate oven and cooled with the injection of argon gas at a different flow rates that allow the decrease of the temperature to 650 °C, respectively, after 2, 4, and 8 h. By corrosion potential vs. time, resistance polarization technique and Tafel plot curves, we have studied the behavior of silicon electrodes, and we have noticed that the electrochemical properties of silicon were affected by temperature of the medium, by the cooling time and by the lighting power. The microscopic examination of the electrodes shows a significant change in the morphology of materials after the polarization tests.

Keywords Silicon • Corrosion • Tafel plot • Wafers • Heat treatments

57.1 Introduction

Energy production from fossil origin constitutes a potential source of pollution for the atmosphere, for waters (rivers, seas, and oceans) and for soils [1–4]. Its use has greatly contributed to the proliferation of greenhouse gas emissions. Solar energy is clean and must occupy a significant place in the future in the production of

M. Bounoughaz (✉) • H. Boukercha • A. Khemmari
Département de Chimie, Université M'Hamed Bougara (UMBB), Boumerdes 35000, Algérie

Equipe Electropolymérisation and Corrosion, Laboratoire de Mise en Forme et Traitement des Polymères Fibreux, Université M'Hamed Bougara (UMBB), Boumerdes 35000, Algérie
e-mail: moussa_bounoughaz@yahoo.fr

M. Allaoua
Département de Chimie, Université M'Hamed Bougara (UMBB), Boumerdes 35000, Algérie

SONATRACH, Division Laboratories, TRH, Algiers, Algeria

E. Salhi
Ecole Nationale Polytechnique, Algiers, Algeria

I. Dincer et al. (eds.), *Progress in Clean Energy, Volume 1*,
DOI 10.1007/978-3-319-16709-1_57

electricity. Currently, most solar cells are made of silicon [1, 3, 5]. Silicon dioxide (SiO_2) is available in abundance in the earth's crust (28 %). It is available in the quartz and in the sand (silica). For its application in photovoltaic cells and in semiconductors, the raw material (sand, quartz) is refined at high temperature with the presence of carbon according to reaction 1 and the obtained silicon is metallurgical grade [6–8]

$$SiO_2 + 2C(2,000\,^\circ C) \rightarrow Si + 2CO^{\uparrow} \tag{57.1}$$

Secondly, the powder of silicon reacts with hydrochloric acid and form the trichlorosilane (reaction 2).

$$Si + 3\,HCl(300\,^\circ C) \rightarrow SiHCl_3 + H_2 \tag{57.2}$$

Thereafter, the pure silicon on the solid state is obtained by reacting at high temperature trichlorosilane liquid with hydrogen (Reaction 3):

$$2\,SiHCl_3 + 2\,H_2 \rightarrow 2\,Si + 6\,HCl \tag{57.3}$$

The performance of solar panels on silicon, as reported by researchers [9, 10] was seriously affected by acidic atmospheres. Similarly, the sun light radiation contributes to the degradation of the silicon solar panels. In urban and industrial areas, the residual atmosphere and/or local can be acidic or basic depending on the pollutants emitted into the atmosphere. For example, exhaust gas from vehicles and factories contains a large amount of carbonic dioxide and it participates in the acidification of moisture. The carbon dioxide dissolves in the moisture and forms the carbonic acid (H_2CO_3) which leads to acidification of the medium.

The change in the environment, particularly when it becomes acid, affects the performance of the solar panels. The acidic medium accelerates the degradation of the panel and leads to the degradation of substrate made in silicon.

This study focuses on exploring the effect of the structure of silicon and the effect of light and dark by using electrochemical techniques [11–13]. The silicon samples were heated until 1,200 °C in appropriate oven and cooled with the injection of argon gas at different flow rates that allow the decrease of the temperature to 650 °C, respectively, after 2, 4, and 8 h.

57.2 Experimental Setup

The silicon samples, after preparation, were subjected to a heat treatment of 1,200 ° C in an arc oven. The inert atmosphere was controlled by the flux of argon. The samples were maintained for 2 h at 1,200 °C before the beginning of the cooling process. Three rates of cooling were selected by varying the flow rate of argon. Samples were cooled from 1,200 to 650 °C after 2, 4, and 8 h.

Table 57.1 Electrochemical parameters fixed for tests

Techniques	Parameters
Corrosion potential (E_{corr} mV/s	Time $\geq$ 5,000 s
Linear resistance polarization (LPR)	Scan rate of potential = 0.16 mV/s
	Scanned potential = $\pm$20 mV/E_{corr}
Linear voltametry (Tafel plot)	Scan rate of potential = 0.16 mV/s
	Scanned potential = $\pm$250 mV/E_{corr}

Three working electrodes were made by welding each silicon sample to metallic conductors by the intermediate of conductor glue. One side of the silicone was kept free. The other side was covered by polymeric resin in a cylindrical mold.

The electrochemical measurements were carried out in a three electrodes cell system that consist of silicon as a working electrode, a saturated calomel electrode (SCE) as reference electrode and two rods of carbon glass as a counter electrode. The electrochemical measurements were performed on a set formed by potentiostat/galvanostat 273A EG & G and frequency analyzer, model Schlumberger SI 1260. The tests were completed using the software's M352 (EGG), and Zplot and Zview (Solartron).

Before the electrochemical measures, the working electrode was polished using emery papers of 1200 grade. The working electrode was subsequently polished by alumina slurry on a polishing cloth, and it was rinsed by distilled water and finally cleaned in ultrasonic bath. The electrolyte used in the experiments was sulfuric acid of 0.5 M of concentration. The experimental parameters used in all the electrochemical tests are shown in Table 57.1.

The morphology of the working electrodes made in silicon was examined with an optical microscope OLYMPUS GX 71 equipped with a digital camera.

57.2.1 Environments

Photovoltaic cells are located in different environments such as industrial, rural, urban, desert, and sailor. Different sulfuric acid solutions were used to simulate the acidic atmosphere that characterizes the industrial environments.

57.3 Results and Discussion

57.3.1 Free Corrosion Potential Measures

The measure of the potential of corrosion at open circuit versus the immersion time was realized in a sulfuric acid solution of 0.5 of H_2SO_4MM. The potential of corrosion may be considered stable when the change oscillates in the order of

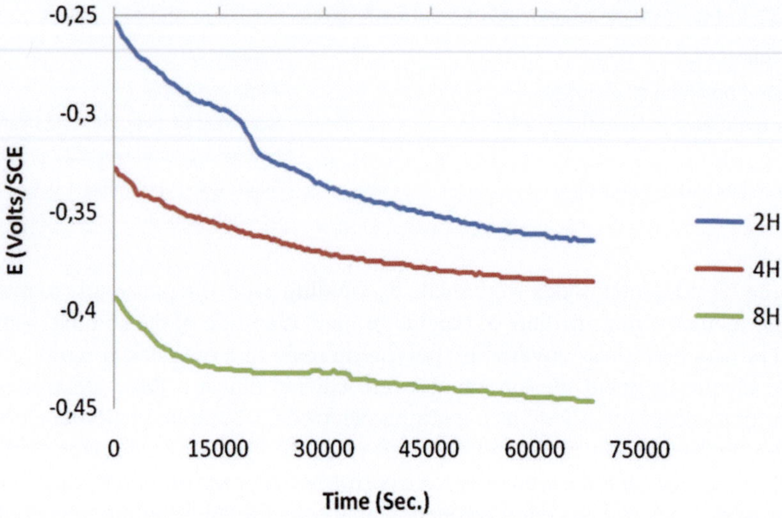

Fig. 57.1 Evolution of the free corrosion potential with immersion time. Working electrode of silicon cooled after 2, 4, and 8 h and under illumination

2 mV/min. Figures 57.1 and 57.2 present the evolution of the free potential of silicon working electrodes designed by 2H, 4H, and 8H. The measures were taken under illumination and under darkness.

Measurements of the potential of corrosion were carried out at a range of temperatures: 21 °C (ambient), 25, 35, and 45 °C. The obtained results are presented in Table 57.2.

The results presented in Table 57.2 show that the potential of corrosion of silicon cooled during 8 h is more electronegative than the others cooled during 2 and 4 h. This observation is valid for all the potentials measured under the used temperatures.

57.3.2 Linear Polarization Resistance Technique

This technique allows measuring the corrosion rate from the resistance of polarization as well as the potential and the currents of corrosion. The electrochemical parameters obtained by this technique are included in Table 57.3. Corrosion rate is expressed in millimeters per year, the resistances of polarization in Ohms cm^2 and corrosion current density is expressed in μA per Sq. cm. The shape of the LPR curves is shown in Fig. 57.3.

As shown in the results of Table 57.3, the polarization resistance of the electrode of silicon 8H in acid medium and under light effect at 21 °C of is the highest. Its value is 11,760 Ω cm^2. This value corresponds to the lower current density

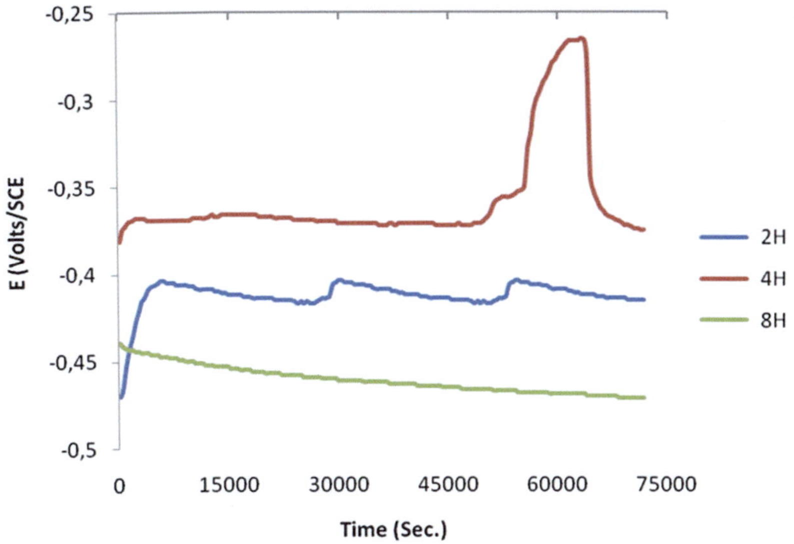

Fig. 57.2 Evolution of the free corrosion potential with immersion time. Working electrode of silicon cooled after 2, 4, and 8 h and under darkness

Table 57.2 Free corrosion potential of silicon at different values of temperature

Silicon working electrodes	Temperature (°C)	E_{corr} (mV/SCE)	
		Under illumination	Under darkness
2H	21	−364.1	−395.0
	25	−365.6	−393.6
	35	−364.7	−398.0
	45	−358.6	−395.3
4H	21	−385.8	−375.3
	25	−384.9	−344.4
	35	−393.9	−381.1
	45	−393.5	−402.6
8H	21	−450.1	−471.1
	25	−453.5	−480.2
	35	−458.0	−483.5
	45	−451.8	−489.6

(1.846 μA/cm^2). By comparing the density of corrosion current obtained by the electrode 2H, we note that it is equal to 13.079 μA/cm^2 which yields a low resistance (1,657 Ω cm^2) at the same temperature.

By increasing the temperature from 25 to 35 °C and to 45 °C, we noticed the reduction of resistance and the increase in the rate of corrosion and the corrosion current density I_{corr} regardless to the nature of the electrode. We can advance that

Table 57.3 Electrochemical parameters form LPR curves

Silicone electrodes	T (°C)	Electrochemical parameters E_{corr} (mV/s)	R_p (Ω cm^2)	I_{corr} (μA/cm^2)	V_{corr} (mm/an)
2H	Blank (21 °C)	−364.1	1,657	13.079	128.4×10^{-3}
	25	−365.6	1,312	16.503	162.1×10^{-3}
	35	−364.7	1,078	20.22	198.7×10^{-3}
	45	−358.6	1,057	20.53	201.7×10^{-3}
4H	Blank (21 °C)	−385.8	9,114	2.383	23.40×10^{-3}
	25	−384.9	8,320	2.60	25.63×10^{-3}
	35	−393.9	5,630	6.448	38.14×10^{-3}
	45	−393.5	3,366	6.448	63.33×10^{-3}
8H	Blank (21 °C)	−450.1	11,760	1.846	18.12×10^{-3}
	25	−453.5	1,099	1.975	19.40×10^{-3}
	35	−458	6,265	3.465	34.03×10^{-3}
	45	−451.8	5,441	3.990	39.19×10^{-3}

Electrodes of silicone cooled during 2, 4, and 8 h. Medium made from 0.5 M H_2SO_4

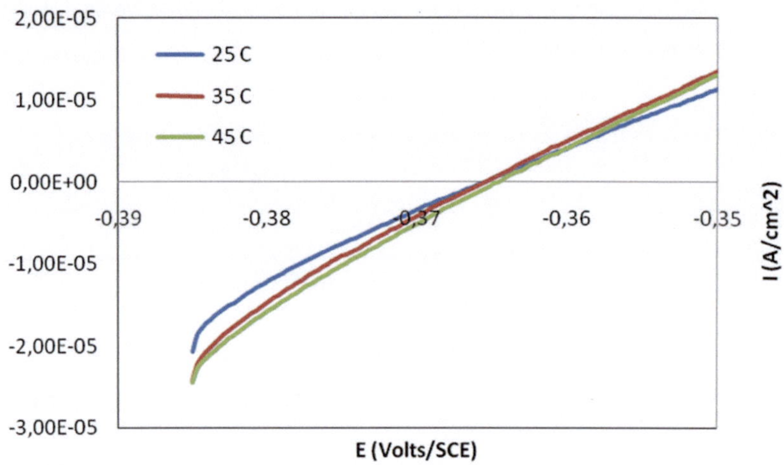

Fig. 57.3 LPR curves of silicon 2H electrodes, under light and in 0.5 M of H_2SO_4

with the increase of temperature, the silicon electrodes become less resistant against corrosion. The electrode 8H under light at 25 °C is more resistant in the acidic medium.

57.3.3 Tafel Plots

This technique involves the scan of cathodic potential toward the anodic potential with scan rate of 0.1 mV/s. The interpretation of the obtained curves by using specialized software that allows to find the corrosion potential and the logarithm of current density of the exchange. The Tafel curves log $(I) = f(E)$ plotted in illuminated conditions are illustrated in Figs. 57.4, 57.5, and 57.6. The obtained results are given in Table 57.2.

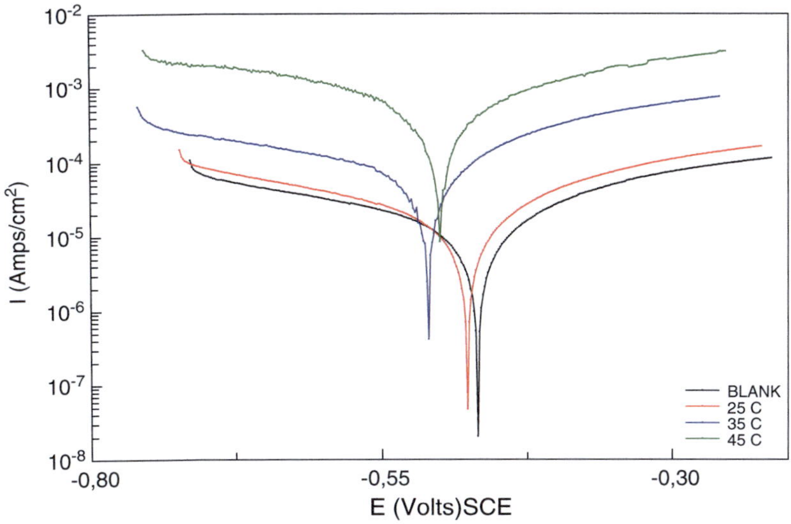

Fig. 57.4 Effect of temperature on the behavior of Tafel curves. Electrode of silicon cooled during 2H and under light conditions. Electrolyte made by 0.5 M H_2SO_4

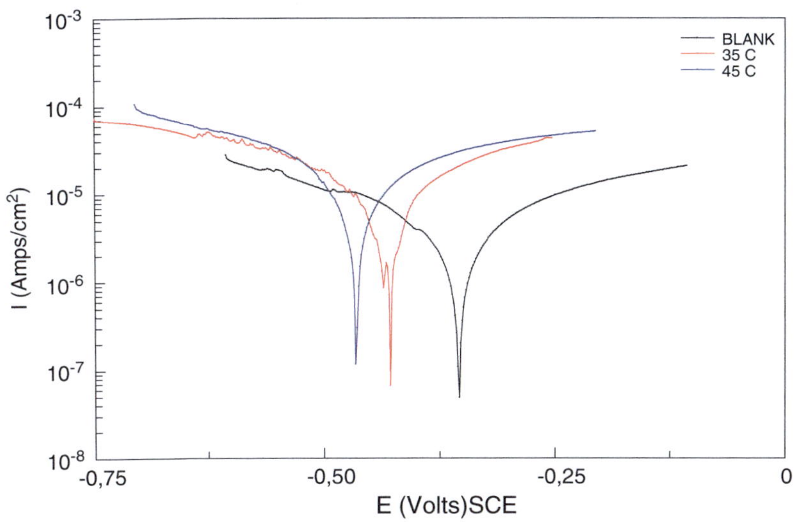

Fig. 57.5 Effect of temperature on the behavior of Tafel curves. Electrode of silicon cooled during 4H and under light conditions. Electrolyte made by 0.5 M H_2SO_4

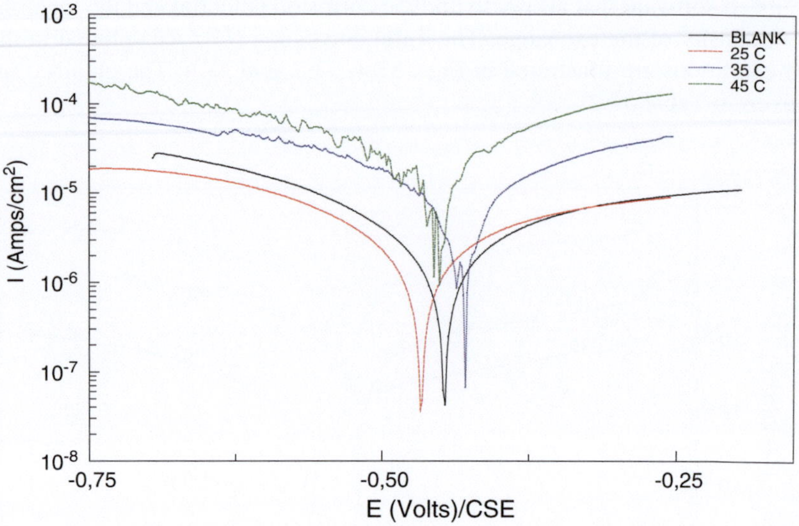

Fig. 57.6 Effect of temperature on the behavior of Tafel curves. Electrode of silicon cooled during 2H and under light conditions. Electrolyte made by 0.5 M H_2SO_4

The working electrode cooled during 2 h indicates that its corrosion potential is weakly affected by temperatures. Its corrosion potential is around −480 mV/CSE (Fig. 57.4). For the 4H working electrode (Fig. 57.5 and Table 57.4), we noticed that the corrosion potential was moved to more negatives values. Similarly, the 8H working electrode shows the temperature has no appreciable effect on the change of the potential electrode. We noted only sensible changes in the current density (see Table 57.4).

57.4 Conclusion

The obtained results indicate that the silicone cooled during 8 h with argon gas is the most resistant against corrosion in the acidic medium of 0.5 M H_2SO_4. Under darken conditions; silicon electrodes are less active than under light conditions. The temperature affects the current of corrosion, and the potential of electrode is weakly affected.

Table 57.4 Electrochemical parameters form Tafel curves

Silicon electrode	T (°C)	Electrochemical parameters				
		V_{corr} (mm/Y)	E_{corr} (mV/ECS)	I_{corr} (μA/cm²)	βa (mV)	βc (mV)
2H	Blank 21 °C	29.456	−485.5	2.9×10^{-3}	671.32	328.46
	25	9.368	−433.05	9.5×10^{-3}	270.41	302.1
	35	50.121	−484.61	5.0×10^{-3}	587.56	550.36
	45	67.6044	−488.53	6.8×10^{-3}	452.7	299.96
4H	Blank 21 °C	0.1596	−285.19	1.622×10^{5}	582.64	256.51
	25	0.164	−299	1.669×10^{-5}	739.66	260.52
	35	0.0029	−332.56	3.044×10^{-7}	385.13	344.1
	45	0.0037	−517.37	1.57×10^{-7}	258.69	158.33
8H	Blank 21 °C	0.082	−491.5	8.33×10^{-6}	543.3	405.59
	25	0.029	−404.6	3.01×10^{-6}	164.3	208.18
	35	25.062	−421.49	2.5×10^{-3}	361.51	309.49
	45	0.286	−469.99	2.912×10^{-5}	335.43	435.76

Electrodes of silicone cooled during 2, 4, and 8 h. Medium made from 0.5 M H_2SO_4

References

1. Schwetz KA (2000) Silicon carbide based hard materials. In: Riedel R (ed) Handbook of ceramic hard materials, vol 2. Weinheim, Wiley, pp 683–748
2. Cook GS, Little JA, King JE (1994) Corrosion of silicon carbide ceramics using conventional and electrochemical methods. Br Corros J 29(3):183–185
3. Herrmann M, Schilm J, Michael G, Adler J (2003) Corrosion behaviour of different technical ceramics in acids, basic solutions and under hydrothermal conditions, in cfi/Ber. DKG 80(4): E27–E34
4. Schwetz A, Hassler J (2002) Stability of high technology ceramics against liquid corrosion, cfi/Berichte. DKG 79(11):D14–D19
5. Divakar R, Seshadri SG, Srinivasan M (1989) Electrochemical techniques for corrosion rate determination in ceramics. J Am Ceram Soc 72(5):780–784
6. Andrews A (2005) Electrochemical corrosion measurement of sintered silicon carbide and liquid phase sintered silicon carbide ceramics. M.Sc. dissertation, University of the Witwatersrand, South Africa
7. Jones DA (1996) Principles and prevention of corrosion, 2nd edn. Prentice-Hall, Upper Saddle River
8. Bard AJ, Parsons R, Jordan J (1985) Standard potentials in aqueous solutions, monographs. In: Bard AJ (ed) Electroanalytical chemistry and electrochemistry. Marcel Dekker, New York, pp 189–206, 566–580, 587–629
9. Pourbaix M, Muylder JV, Besson J, Kunz W (1958) Atlas of electrochemical equilibria in aqueous solution. NACE, Houston, pp 449–463
10. Can A, Herrmann M, McLachlan DS, Sigalas J, Adler J (2006) Densification of liquid phase sintered silicon carbide. J Eur Ceram Soc 26:1707–1713
11. ASTM G5-94 (2000) Making potentiostatic and potentiodynamic anodic polarisation measurements. In: Annual books of ASTM Standards, vol 03.02. ASTM International, Conshohocken. pp 57–67

12. ASTM G3-89 (2000) Conventions applicable to electrochemical measurements in corrosion testing. In: Annual books of ASTM Standards, vol 03.02. ASTM International, Conshohocken. pp 38–42
13. Andrews A, Herrmann M, Sephton M (2006) Electrochemical corrosion testing of solid state sintered silicon carbide in acidic and alkaline environments. J South Afr Inst Min Metall 106:269–276

Chapter 58
Monte-Carlo Modelling and Experimental Study of Radon and Progeny Radiation Detectors for Open Environment

Sofia Kottou, Dimitrios Nikolopoulos, Ermioni Petraki, Debabrata Bhattacharyya, Paul B. Kirby, Tamara M. Berberashvili, Lali A. Chakhvashvili, Paata J. Kervalishvili, and Panayiotis H. Yannakopoulos

Abstract Solid state nuclear track detectors (SSNTDs) have been widely used as sensors of radon and progeny in long-term dosimetry because they exhibit high detection properties while their cost is very low. Alpha-particle energy calculating codes, specific for every incident particle, increase the Monte-Carlo simulation time significantly. The expression of the alpha particle energy as a function of the distance travelled in SSNTD CR-39 was recently introduced as an alternative approximation for the simulation method. This chapter focused on modelling the response of bare CR-39 detectors to alpha-particles emitted by radon and progeny, through Monte-Carlo methods. In order to determine the efficiency of a combined use of bare CR-39 and cup-type detectors in radon measurements, theoretical and experimental CR-39 efficiency factors for alpha-particles were calculated. Modelling rendered calculation of effective volume for CR-39 detector, based on energy and angular distributions of alpha-particles emitted due to decay of radon and

S. Kottou (✉)
Medical Physics Department, Medical School, University of Athens, Mikras Asias 75, Athens 11527, Greece
e-mail: skottou@med.uoa.gr

D. Nikolopoulos • P.H. Yannakopoulos
Department of Electronic Computer Systems Engineering, TEI of Piraeus, Petrou Ralli & Thivon 250, Aigaleo 12244, Greece

E. Petraki
Department of Electronic Computer Systems Engineering, TEI of Piraeus, Petrou Ralli & Thivon 250, Aigaleo 12244, Greece

Kingston Lane, Uxbridge, Middlesex, London UB8 3PH, UK

D. Bhattacharyya • P.B. Kirby
Manufacturing and Materials Department, Cranfield University, Cranfield, Bedfordshire MK43 0AL, UK

T.M. Berberashvili • L.A. Chakhvashvili • P.J. Kervalishvili
Georgian Technical University, 77 Kostava str., Tbilisi 0175, Georgia

I. Dincer et al. (eds.), *Progress in Clean Energy, Volume 1*,
DOI 10.1007/978-3-319-16709-1_58

progeny. The relationship between equilibrium factor F and the recorded track density values ratio (of bare and cup-enclosed SSNTDs, respectively) R was calculated. The sensitivity factor k_B for bare CR-39 was found equal to $k_B = (4.6 \pm 0.6)$ [tracks $\times$ cm^{-2}]/[kBq $\times$ m^{-3} $\times$ h] (assuming the Jacobi's steady-state model), a value not significantly different from the corresponding k_R cup-type value for radon and progeny.

Keywords Radon and progeny radiation • SSNT detectors • Equilibrium factor • Recorded track density

58.1 Introduction

Radon (^{222}Rn) is a naturally occurring radioactive gas generated by the decay of radium (^{226}Ra) which is present in soil, rocks, building materials and waters [1, 2]. Following the decay of radium, a fraction of radon emanates and migrates through diffusion and convection. After migrating, part of radon escapes to the atmosphere and waters and disintegrates to a series of short-lived decay products (progeny) (^{218}Po, ^{214}Bi, ^{214}Pb and ^{214}Po). Radon and its short-lived progeny disintegrate through α- and β-decay. In specific ^{222}Rn undergoes α-decay with $\lambda_0 = 2.093 \times 10^{-6}$ s^{-1}, ^{218}Po α-decay with $\lambda_1 = 3.788 \times 10^{-3}$ s^{-1}, ^{214}Pb β-decay with $\lambda_2 = 4.234 \times 10^{-4}$ s^{-1}, ^{214}Bi β-decay constant with $\lambda_3 = 5.864 \times 10^{-4}$ s^{-1} and ^{214}Po α-decay with $\lambda_4 = 4.234 \times 10^{3}$ s^{-1} [1].

Solid state nuclear track detectors (SSNTDs) have been widely used as sensors of radon and progeny in long-term dosimetry because they exhibit high detection properties while their cost is very low. During the last decade, researchers have developed analytical models or adopted Monte-Carlo techniques for the calculation of the sensitivity of bare SSNTD CR-39 and LR-115 SSNTDs to alpha-particles emitted by radon and progeny [3–9]. Analytical models and Monte-Carlo coding conceived all existing models and experimental data regarding the internal sensitivity of CR-39 and LR-115, as well as the dependence on critical incident angle and etching time. Although published results are promising, up-to-date, a standard methodology still not exists. For example, Sima [10] developed software for the realistic calculation of the sensitivity of various etched track radon monitors. Nikezic et al. [11] developed software which simulated alpha-particle detection of CR-39 detectors. Despite the scientific work in this subject area [12–15], it is still an open issue to determine accurately alpha energy and angle distributions at the surface of open CR-39 SSNTDs due to decay of radon and progeny. Most important is the estimation of a realistic calibration factor of bare CR-39 SSNTDs. One of the major disadvantages of such simulations up-to-date is that all approaches included energy calculating codes specific for every incident particle. This increased the simulation time significantly. To reduce simulation time, the energy of the alpha-particles as a function of the distance travelled in CR-39 was recently introduced as an alternative approximation [16, 17].

Among others, SRIM (stopping and range of ions in matter) is a group of programs which may calculate all interactions of ions with matter and for this reason has attracted attention. The core of SRIM is a program called transport of ions in matter (TRIM). SRIM and TRIM were developed by James F. Ziegler and Jochen P. Biersack about 1983 and they are continuously upgraded. SRIM is based on a Monte-Carlo simulation method. Input parameters are (1) the type of ion and its initial energy in the range of 10 eV to 2 GeV and (2) the target material being considered. SRIM output includes lists and diagrams. The programs were developed so they can be interrupted any time and resume later with a convenient GUI. These characteristics make SRIM very practical. Using SRIM 2008 software, alpha range and stopping power data were easily extracted and the data can be used reliably.

In indoor environments ^{222}Rn is not necessarily in equilibrium with its short-lived progeny and for this reason the equilibrium factor F serves as a fare compromise for identifying the status of equilibrium between parent ^{222}Rn and remaining short-lived progeny [1]. Continuous measurement of F is time-consuming and requires active instruments. Several researchers investigated combined uses of bare and cup-enclosed SSNTDs for long-term estimation of F [6, 9, 18–21].

This chapter focuses on modelling the response of bare CR-39 detectors to alpha-particles emitted by radon and progeny, through Monte-Carlo methods and the use of the latest version SRIM2013. In order to determine the efficiency of bare CR-39 SSNTDs for a combined use of bare CR-39 and cup-type detectors in radon measurements, theoretical and experimental CR-39 efficiency factors were calculated for alpha-particles emitted by radon and progeny in air.

58.2 Materials and Methods

58.2.1 Calculation of Distance Versus Alpha Energy

If an alpha-particle of energy E_1 is emitted within an effective distance, l, from the surface of a CR-39 detector, then it will hit the detector with energy E_2 due to interactions with molecules of the surrounding medium. The distance l must be smaller or equal than the range of alpha-particles in the medium. The distance l is equal to the difference between range r_1 of the alpha-particle of energy E_1 and r_2 of the particle of energy E_2 [22, 23]

$$l = r_1 - r_2 \tag{58.1}$$

If the alpha-particle is generated through a decay scheme and E_1 is its initial energy, then all possible distance values l from CR-39 surface will be

$$l = r_{\max} - r \quad (58.2)$$

where $r_{\max}$ is the corresponding maximum range and r is the range corresponding to a hit-energy E.

To properly define l, all range data of alpha-particles' energies between 0.5 and 10 MeV were extracted from SRIM2013 output tables after properly adjusting for several baffling characters. The range data were then fitted to the nonlinear Eq. (58.3), considering the absence of straggling [16, 17]:

$$r = \alpha \cdot E + b \cdot E^2 + c \cdot E^3 \quad (58.3)$$

By replacing r from Eq. (58.3) in Eq. (58.2), the distance l versus alpha energy can be obtained:

$$l = r_{\max} - \left(\alpha \cdot E + b \cdot E^2 + c \cdot E^3\right) \quad (58.4)$$

Inversely, the energy of alpha-particles at a distance l from the point of emission can be calculated by the reciprocal of Eq. (58.4).

58.2.2 Monte-Carlo simulation

For the Monte-Carlo simulation, the actual dimensions of the CR-39 detectors employed in radon and progeny measurements were inserted, namely of $1 \times 1\ \text{cm}^2$ surface area and 0.5 mm thickness. CR-39's surface was put on the $x - y$ plane. Since alpha-particles originating from radon and progeny may have different initial energies (^{222}Rn 5.49 MeV, ^{218}Po 6.00 MeV, ^{214}Po 7.68 MeV), a surface area was considered vertical to the z-axis, with dimensions $(1+2{\cdot}r) \times (1+2{\cdot}r)$ [10], where r was taken equal to the corresponding alpha-particle range (4.09 cm for alpha-particles of ^{222}Rn, 4.67 cm for alpha-particles of ^{218}Po and 6.78 cm for alpha-particles of ^{214}Po) [8]. The effective volume around CR-39 (Fig. 58.1) was then calculated by Eq. (58.5) [10]:

$$V = r \times (1 + 2 \cdot r) \times (1 + 2 \cdot r) \quad (58.5)$$

Outside the effective volume, alpha-particles do not reach CR-39's surface since the distance travelled is larger than the alpha range. Only alpha-particles generated within volume V are detectable. For the simulation, inside the effective volume, a random point P1 (X1, Y1, Z1) was computationally created. At this point, an alpha-particle from radon's decay was considered to be emitted with random cylindrical angles θ, φ. Depending on the emission angles, a hit-state was calculated, namely being 1 if the alpha could hit CR-39's surface and 0, otherwise. For an alpha-particle of hit-state 1, the hit-angles θ_h, φ_h to the XY entrance surface of CR-39 were calculated from the Cartesian coordinates of point P1, the angles θ, φ and the

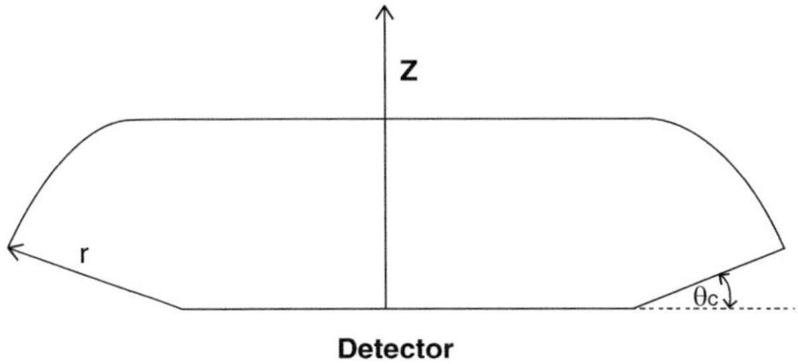

Fig. 58.1 Effective volume for CR-39 for alpha-particles emitted due to decay of radon and progeny. r is the corresponding range, namely 4.09 cm for alpha-particles of ^{222}Rn with initial energy of 5.49 MeV, 4.67 cm for alpha-particles of ^{218}Po with initial energy of 6.00 MeV and 6.78 cm for alpha-particles of ^{214}Po with initial energy of 7.68 MeV

dimensions of CR-39 surface. Since CR-39 does not register alpha-particles with hit-angles above a critical angle, θ_c, alpha-particles were accepted and processed further only if $\theta_h \geq \left(\frac{\pi}{2}\right) - \theta_c$, otherwise it was rejected. An accepted alpha-particle could be registered, and if so, the distance l travelled in CR-39's surrounding medium was calculated again from the Cartesian coordinates of P1, the hit-angles and CR-39's dimensions. From l, θ_h, φ_h data, the arrival point P2 (X2, Y2, Z2) with respect to the detector's plane was calculated thereafter. In parallel, the hit-energy E_h was calculated from the output SRIM2013 for the distance l according to Eq. (58.4). Having calculated hit-data E_h, θ_h, φ_h for an accepted alpha-particle, the distance, d, travelled inside CR-39 was finally calculated from SRIM2013 output. Iterating the above procedure, the energy-angle distribution of various radon's decay alpha-particle showers was also calculated.

Figure 58.2 presents the flow-diagram of the Monte-Carlo simulation.

58.2.3 *Equilibrium Factor,* F

Radon's equilibrium factor, F, is defined as the ratio of the equilibrium equivalent concentration of radon (A_e) over the actual activity concentration of radon in air (A_0), namely [1]:

$$F = A_e / A_0 \tag{58.6}$$

Equilibrium equivalent concentration is determined by the following equation [1]

$$A_e = 0.106 \cdot \left(A_1^a + A_1^u\right) + 0.513 \cdot \left(A_2^a + A_2^u\right) + 0.381 \cdot \left(A_3^a + A_3^u\right) \tag{58.7}$$

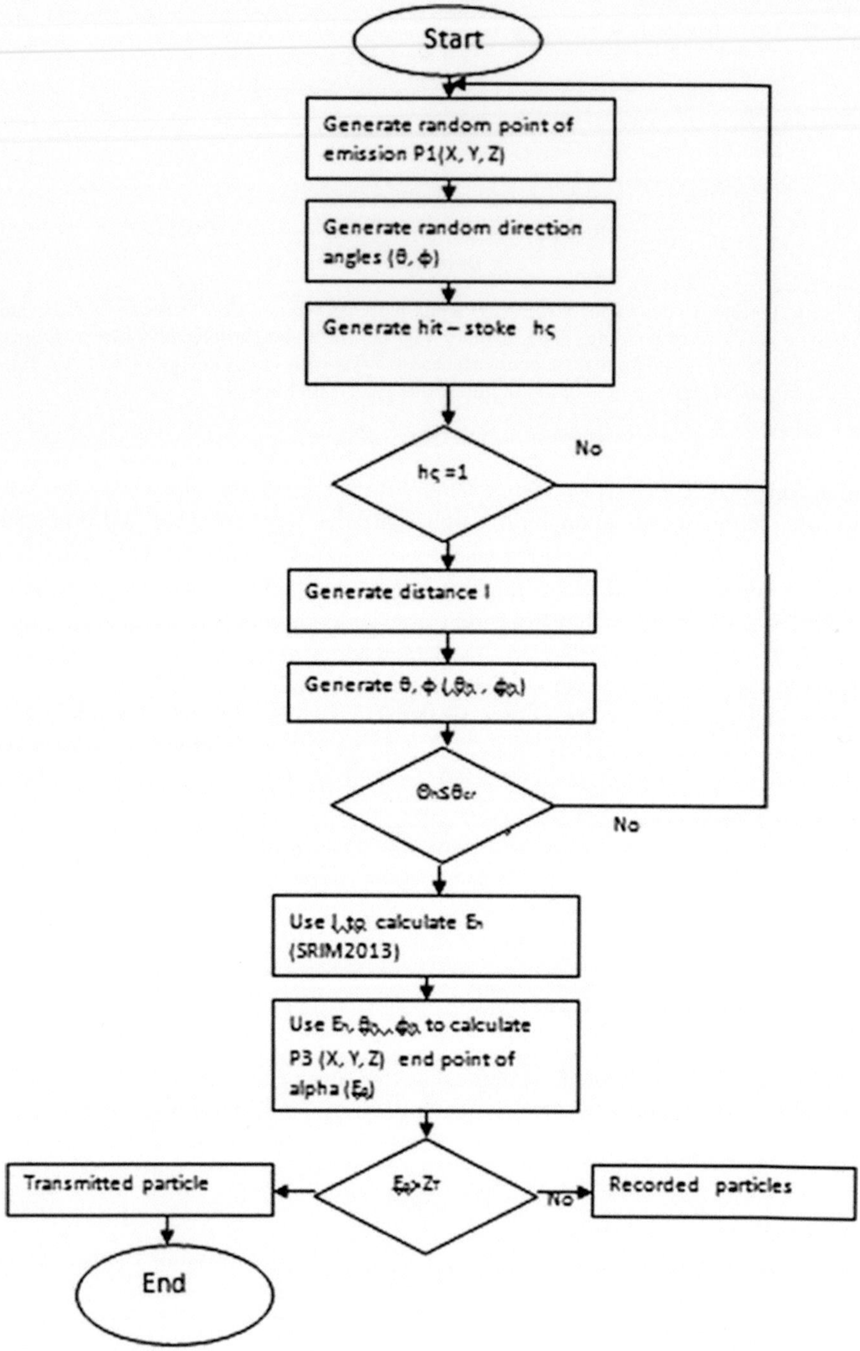

Fig. 58.2 Flowchart of Monte-Carlo simulation of alpha-particles of radon's decay travelling around and within CR-39 detectors

and hence

$$F = \frac{0.106 \cdot (A_1^a + A_1^u) + 0.513 \cdot (A_2^a + A_2^u) + 0.381 \cdot (A_3^a + A_3^u)}{A_0} \quad (58.8)$$

Superscripts a and u distinguish the contribution of each one of the two states of radon progeny (attached, unattached), subscripts 1, 2 and 3 correspond to ^{218}Po, ^{214}Pb, ^{214}Bi and A_0, A_i^x ($x = a$, u and $i = 1, 2, 3$) (Bq/m^3) represent measured concentrations of radon and progeny, respectively (It is well established that ^{214}Bi and ^{214}Po are always in equilibrium due to the very short half life of ^{214}Po (164 ms), so we will simply use A_3 to denote both the air borne concentrations of ^{214}Bi and ^{214}Po).

Assuming radioactive disintegration, ventilation and deposition as the sole processes of removal of radon progeny in ambient air, A_i^x ($x = a$, u and $i = 1, 2, 3$) can be calculated as [9, 24]:

$$A_i^x = d_i \cdot A_{i-1}^x \quad (58.9)$$

Parameter d_i reported by Faj and Planinic [9] can be expressed as:

$$d_i = \frac{\lambda_i}{\lambda_i + \lambda_v + f_i^a \cdot \lambda_i^{d,a} + (1 - f_i^a) \cdot \lambda_i^{d,u}} \quad (58.10)$$

where λ_ν represents the ventilation rate, $\lambda_i^{d,x}$ ($x = a$, u and $i = 1, 2, 3$) is the deposition rate constant of each one of the attached and unattached progeny and

$$f_i^a = \frac{A_i^a}{\sum A_i^a + A_i^u} \quad (58.11)$$

is the attached fraction of progeny i. Neglecting the attachment of ^{218}Po, ^{214}Pb, ^{214}Bi nuclei F may be calculated as [9]:

$$F = 0.105 \cdot d_1 + 0.516 \cdot d_1 \cdot d_2 + 0.380 \cdot d_1 \cdot d_2 \cdot d_3 \quad (58.12)$$

Faj and Planinic [9] calculated d_i as a function of λ_ν employing the Carnado's formula. The solution enabled calculation of F as a function of λ_ν, namely $F = F(\lambda_\nu)$. Similar approach has been followed previously as well [18, 19].

In actual conditions, however, attachment of unattached progeny to aerosol and humidity particles may differ and this affects progeny concentrations A_i^x ($x = a$, u and $i = 1, 2, 3$). According to recent publications [25–27], the deposition and attachment rate constants of attached and unattached progeny differentiate in high-humidity environments due to peaking of water droplets and for this reason, symbolisation $\lambda_i^{d,x}$ ($x = a$, u and $i = 1, 2, 3$) was adopted. Presuming however, only typical low-humidity ambient room environments, under a Jacobian steady-state with complete mixing, $\lambda_i^{d,u}$ and $\lambda_i^{d,a}$ can be considered approximately constant for indoor room conditions [6, 9, 18–21]. In such conditions, attachment and deposition rates are equal between unattached and attached nuclei and hence, symbolisation $\lambda^{d,x}$ ($x = a$, u) could be employed. According to Porstendorfer et al. [28], in typical rooms no differences are usually addressed between ambient electrical charged and neutral progeny clusters in attaching to aerosols and depositing to surfaces. Under this perspective, the deposition rates of attached and unattached progeny to surfaces are equal. Employing symbolisation of Porstendorfer et al. [28], the terms $f_i^a \cdot \lambda_i^{d,a} = f_i^a \cdot \lambda^{d,a}$ and $\left(1 - f_i^a\right) \cdot \lambda_i^{d,u} = \left(1 - f_i^a\right) \cdot \lambda^{d,u}$ in Eq. (58.10) represent the deposition rates of attached and unattached nuclei, namely

$$q^a = f_i^a \cdot \lambda^{d,a} \tag{58.13}$$

where q^a is the symbol for the deposition rate of all attached progeny and

$$q^u = f_i^u \cdot \lambda^{d,u} \tag{58.14}$$

by symbolising q^u the deposition rate of all unattached progeny in the sense of Eq. (58.13). Assuming a steady-state Jacobian model and complete mixing, concentrations of attached and unattached nuclei can be calculated then as [28]:

$$A_i^a = \frac{(1 - R_{i-1})\lambda_i \cdot A_{i-1}^a + X \cdot A_i^u}{\lambda_v + \lambda_i + q^a + X} \tag{58.15}$$

and

$$A_i^u = \frac{\lambda_i \cdot A_{i-1}^u + R_{i-1} \cdot \lambda_i \cdot A_i^a}{\lambda_v + \lambda_i + q^u + X} \tag{58.16}$$

where R_i is the recoil fraction of progeny i, X is the attachment rate to aerosols and $i = 1, 2, 3$. $R_1 = 0.8$ while $R_2 = R_3 = 0$ [1]. Employing Eqs. (58.15–58.16) in Eq. (58.8), F can be calculated as a function of λ_v, $\lambda^{d,u}$, $\lambda^{d,a}$ and X, namely $F = F\left(\lambda_v, \lambda^{d,u}, \lambda^{d,a}, X\right)$, considering $\lambda^{d,u}$ and $\lambda^{d,a}$ equal for all progeny . The latter approximation was employed by Eappen et al. [6] to determine the upper and lower bounds of F, the average values of F and the related uncertainties.

It is very important that both approaches for the calculation of A_i^x ($x = a$, u and $i = 1, 2, 3$), namely Eq. (58.9) for Faj and Planinic [9] and Eqs. (58.15–58.16) for Eappen et al. [6] yield to similar final approximations for the most probable relation of modelled values of F versus measured progeny concentrations A_i^x ($x = a$, u and $i = 1, 2, 3$). This relationship can be employed for the determination of F versus the recording efficiency between bare and cup-type SSNTDs (R). According to Faj and Planinic [9], this relationship follows the exponential law

$$F = \alpha \cdot e^{-bR} \tag{58.17}$$

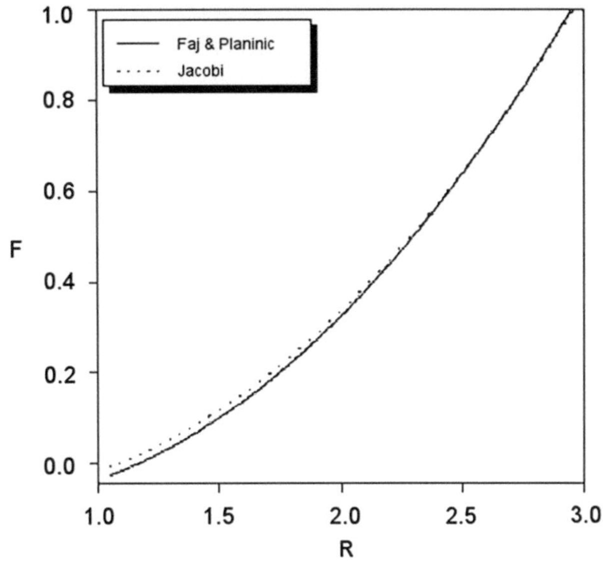

Fig. 58.3 Relationship between equilibrium factor F and the track ratio R

where

$$R = T_B / T_R \tag{58.18}$$

and T_B, T_R are the recorded track density values of bare and cup-enclosed SSNTDs, respectively. Similar were also the results reported by other investigators [18–21]. Figure 58.3 presents the best approximations of F versus R according to the model of Faj and Planinic [9] and according to the model of Eappen et al. [6] (Jacobi's model [29]). Excellent coincidence is observed for all values of R.

58.2.4 Equilibrium Factor, Theoretical Approach

Let's assume a twin CR-39 detector system, namely a bare CR-39 SSNTD and a similar enclosed in a cup. The detector inside the cup records tracks attributable to time integrated ^{222}Rn concentration and the detector outside records tracks due to both ^{222}Rn and its progeny. While radon's concentration is unequivocally estimated, it is not so direct to estimate the progeny's equilibrium factor and potential alpha energy concentration (PAEC) from the track density of bare detectors. When the environment predominantly consists of radon and its progeny, a unique relationship as the one of Eq. (58.17) can be established between equilibrium factor values and the ratio of the cup to bare detector track densities [6, 9, 18–21].

For calibrated CR-39 cup-type detectors, T_R will relate linearly to the concentration A_0 of ^{222}Rn outside the cup. On the other hand, the track density T_B of bare

CR-39 will be proportional to the ambient concentration of all a-emitting nuclei, namely to A_0 of ^{222}Rn, A_1 of ^{218}Po and A_4 of ^{214}Po. If k_R and k_B are the sensitivity factors [tracks $\times$ cm^{-2}]/[kBq $\times$ m^{-3} $\times$ h] of cup-type and bare CR-39, respectively, then

$$T_R = k_R \cdot A_0 \tag{58.19}$$

and

$$T_B = k_B \cdot (A_0 + A_1 + A_3) \tag{58.20}$$

since $A_3 = A_4$. Equation (58.18), according to Eqs. (58.19–58.20) can be written as

$$R = k \cdot (1 + s_1 + s_3) \tag{58.21}$$

where

$$k = k_B / k_R \tag{58.22}$$

is the sensitivity factor ratio, $s_1 = A_1/A_0$ and $s_3 = A_3/A_0$. Importantly, Eq. (58.21) calculates R from the concentration ratios s_1 and s_3.

According to Eqs. (58.8, 58.21–58.22), if the concentrations A_i^x ($x = a$, u and $i = 1, 2, 3$) are known from measurements, equilibrium factor F, as well as s_1 and s_3 can be calculated. If additionally the sensitivity factors k_B and k_R are known, then k can be determined, and hence R. In this manner, the relationship between F and R can be established.

58.2.5 Equilibrium Factor, Experimental Approach

Several active radon and progeny measurements have been conducted in Greek dwellings. Numerous measurements were performed with EQF3023 (EQF) of Sarad Instruments Gbhm. This instrument allows continuous 2 h cycle measurement of radon and progeny nuclei, the latter discriminated for their attached or unattached mode. From the active database, several actual values of A_0 and A_i^x ($x = a$, u, $i = 1$, 2, 3) were employed. Additional value sets were calculated as averages at the 95 % confidence interval, under the constraint of employing only partial values of a certain dwelling measurement-set during each calculation. From these actual A_i^x ($x = a$, u, $i = 1$, 2, 3) measurement sets, equilibrium factor F values were calculated according to Eq. (58.8).

Passive radon measurements are being conducted with a cup-type CR-39 detector which was properly calibrated [30]. Detector has well-established linear response to radon exposure. Detector sensitivity factor has been experimentally defined and found equal to $k_R = (4.62 \pm 0.33)$ [tracks $\times$ cm^{-2}]/[kBq $\times$ m^{-3} $\times$ h].

From k_R and the actual measurements of A_0, T_R was calculated according to Eq. (58.19).

58.3 Results and Discussion

Track density of bare CR-39 detectors was calculated by means of combining the real measurements of EQF with results derived via Monte-Carlo methods. More specifically, A_1 and A_3 were calculated from EQF measurements considering that $A_i = A_i^u + A_i^a$, $i = 1, 3$. From these and the corresponding A_0 values, the concentration ratios were calculated as s_1 and s_3. Since k_B is not easily measurable, Monte-Carlo methods were employed for its determination. The following steps were followed:

1. The distance l travelled by alpha-particles prior to hitting CR-39 was calculated versus alpha energy through SRIM2013 for the whole alpha-particle energy range of radon's decay chain. The relationship

$$l = r_{max} - \left(3.34773 \cdot E + 0.34937 \cdot E^2 + 0.02142 \cdot E^3\right) \tag{58.23}$$

   was employed, where $r_{max} = 4.09$ cm for alpha-particles originating from ^{222}Rn, $r_{max} = 4.67$ cm for alpha-particles originating from ^{218}Po and $r_{max} = 6.78$ cm for alpha-particles originating from ^{214}Po.
2. Random emission points of ^{222}Rn, ^{218}Po and ^{214}Po were generated around CR-39 and their travelling direction vectors were calculated.
3. From the direction vectors of (2), the hit-data (l, θ_h, φ_h) were calculated.
4. For alpha-particles with l inside an effective volume, incident energy E_h was calculated from the reciprocal of Eq. (58.23) under the constraint $\theta_h \geq \left(\frac{\pi}{2}\right) - \theta_c$.
5. From hit-data (E_h, θ_h, φ_h) the range and end points in CR-39 were calculated.
6. Steps (1)–(5) were iterated for N_0 particles of ^{222}Rn, ^{218}Po and ^{214}Po.
7. From steps (1)–(6) the number of recorded particles of ^{222}Rn, N_0^{rec}, ^{218}Po, N_1^{rec} and ^{214}Po N_4^{rec} were calculated.

To estimate realistic values of N_0 for ^{222}Rn, ^{218}Po and ^{214}Po (denoted as $N_{0,i}$), the following equation was employed

$$N_{0,i} = A_i \cdot V_i \cdot t_{exp} \tag{58.24}$$

where $A_i = A_i^u + A_i^a$, V_i is the sensitive volume's dimensions, t_{exp} is an assumed value for the exposure time (30 days) and $i = 0, 1, 4$. From Eq. (58.24) and the Monte-Carlo output the recorded particles N_i^{rec}, $i = 0, 1, 4$ were calculated. From N_i^{rec} the track density of bare CR-39 detectors was calculated as:

Table 58.1 Characteristic value sets of F and R

Equilibrium factor (F)	Ratio (R)
0.3384	1.5852
0.3137	1.1319
0.3379	1.1273
0.2562	1.0043
0.2865	1.0838
0.3148	1.5200
0.2532	1.7734
0.3035	2.2800
0.2678	1.8526
0.2540	1.0117
0.2628	1.1810
0.3249	1.0240
0.3137	1.0051
0.2781	1.0200
0.3345	1.2846
0.3957	1.4425
0.3164	1.1810
0.5709	2.3939
0.6748	2.7557
0.7413	2.7330

$$T_{\mathrm{B}} = \frac{N_0^{\mathrm{rec}} + N_1^{\mathrm{rec}} + N_4^{\mathrm{rec}}}{S} \tag{58.25}$$

where S is the area of the employed CR-39 detectors, namely 1 cm^2.

From Eqs. (58.20 and 58.25) the total sensitivity factor k_{B} of bare CR-39 detectors was calculated as

$$k_{\mathrm{B}} = \frac{T_{\mathrm{B}}}{(A_0 + A_1 + A_3)} = \frac{N_0^{\mathrm{rec}} + N_1^{\mathrm{rec}} + N_4^{\mathrm{rec}}}{S \cdot (A_0 + A_1 + A_3)} \tag{58.26}$$

Table 58.1 presents characteristic value sets of F and R, according to the methodology already described. It may be recalled that the F values were calculated from actual EQF measurements, whereas the R values were calculated applying the appropriate mathematical formulas. The results of Table 58.1 are presented graphically in Fig. 58.4.

The relationship between F and R has similarities to that of Fig. 58.3. For this reason, the data of Table 58.1 were fitted to the exponential model Eq. (58.17), namely to $F = \alpha \cdot \mathrm{e}^{-bR}$. Fitting gave $\alpha = 0.1663$, $b = -0.4820$ with $r^2 = 0.90$. These data are in accordance to the published results of Faj and Planinic [9] and Eappen et al. [6]. It is noted that the latter publication represents a critical review of the subject together with other results. Differences are due to differences in the

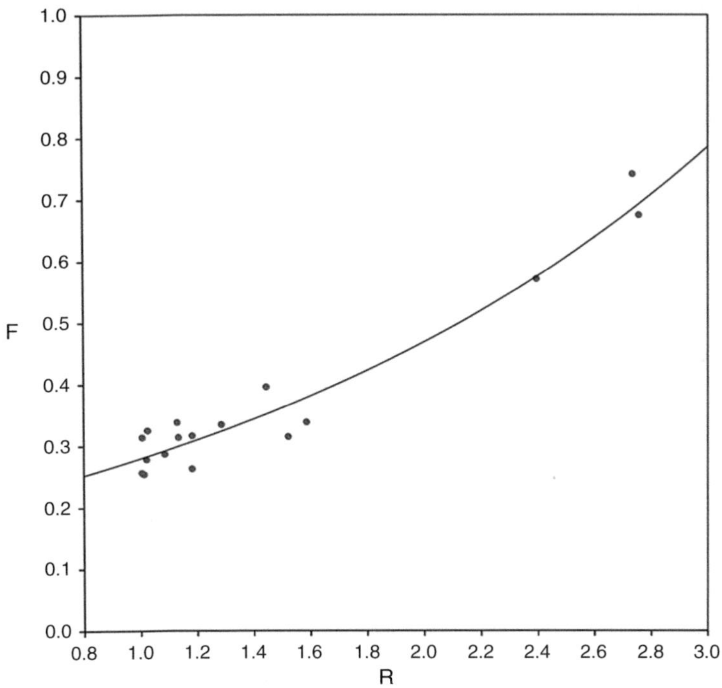

Fig. 58.4 Relationship of *F* and *R* according to Table 58.1

sensitivity factor of the employed cup-type detectors of this study and those of the other studies.

From the data of Table 58.1, sensitivity factors of bare CR-39 SSNTDs were calculated according to Eq. (58.26). Average k_B of this study was found equal to $k_B = (4.6 \pm 0.6)$ [tracks $\times$ cm^{-2}]/[kBq $\times$ m^{-3} $\times$ h]. This value does not differ significantly from the value of k_R. The latter implies from Eq. (58.22) that $k \sim 1$. This finding is very important. Indeed, Faj and Planinic [9] assumed equal values for k_B and k_R. The present study verifies this result. Similar was also the outcomes of Eappen et al. [6]. Related publications gave also comparable results [18–21].

All these findings could be explained by the fact that CR-39 registers alpha-particles from radon and progeny identical either if enclosed in a cup or bare. Observed track density differences are attributable only to the fact that cup type CR-39 detectors register radon alpha-particles only, while bare CR-39 SSNTDs register alpha-particles from all radon alpha-emitters. Future work will employ other expressions of F namely those that take into account the unattached fraction in terms of PAEC.

58.4 Conclusion

This study reported a newly developed Monte-Carlo simulation tool for modelling the CR-39 SSNT detectors efficiency. Simulation combined Monte-Carlo techniques, experimental data and the latest version of SRIM (SRIM2013) software program group. Modelling rendered calculation of effective volume for CR-39 detector, based on energy and angular distributions of alpha-particles emitted due to decay of radon and progeny. The relationship between equilibrium factor F and the recorded track density values ratio (of bare and cup-enclosed SSNTDs, respectively) R was calculated. The sensitivity factor k_B for bare CR-39 was found equal to $k_B = (4.6 \pm 0.6)$ [tracks $\times$ cm^{-2}]/[kBq $\times$ m^{-3} $\times$ h] (assuming the Jacobi's steady-state model), a value not significantly different from the corresponding k_R cup-type value for radon and progeny.

Acknowledgement This work has been co-financed by the Seventh Framework Programme, Grant agreement no: 294299, Acronym SENS-ERA.

References

1. Nazaroff WW, Nero AV (1988) Radon and its decay products in indoor air. Wiley, New York. ISBN 0-471-62810-7
2. UNSCEAR United Nations Scientific Committee on the Effects of Atomic Radiation (2008) Sources and effects of ionizing radiation. United Nations Scientific Committee on the Effects of Atomic Radiation. UNSCEAR 2008 Report to the General Assembly with Scientific Annexes, United Nations, New York
3. Stajic J, Nikezic D (2011) Detection efficiency of a disk shaped detector with a critical detection angle for particles with a finite range emitted by a point-like source. Appl Radiat Isot 70(3):528–532
4. Yu KN, Nikezic D (2011) Long-term measurements of unattached radon progeny concentrations using solid-state nuclear track detectors. Appl Radiat Isot 70(7):1104–1106
5. Nikezic D, Yu KN (2010) Long-term determination of airborne concentrations of unattached and attached radon progeny using stacked LR 115 detector with multi-step etching. Nucl Instrum Methods Phys Res A 613(2):245–250
6. Eappen KP, Mayya YS, Patnaik RL, Kushwaha HS (2006) Estimation of radon progeny equilibrium factors and their uncertainty bounds using solid state nuclear track detectors. Radiat Meas 41:342–348
7. Sima O (2001) Monte Carlo simulation of radon SSNT detectors. Radiat Meas 34 (1–6):181–186. doi:10.1016/S1350-4487(01)00147-0
8. Nikezic D (1994) Determination of detector efficiency for radon and radon daughters with C39 track detector a Monte Carlo study. Nucl Instrum Methods Phys Res A 344:406–414
9. Faj Z, Planinic J (1991) Dosimetry of radon and its daughters by two SSSN detectors. Radiat Prot Dosimetry 35(4):265–268
10. Sima O (1995) Computation of the calibration factor for the cup type SSNTD radon monitor. Radiat Meas 25(1–4):603–606
11. Nikezić D, Kostić D, Krstić D, Savović S (1995) Sensitivity of radon measurements with CR-39 track etch detector—a Monte Carlo study. Radiat Meas 25(1–4):647–648

12. Kappel RJA, Keller G, Nickels RM, Leiner U (1997) Monte Carlo computation of the calibration factor for 222Rn measurements with electrochemically etched polycarbonate nuclear track detectors. Radiat Prot Dosimetry 71(4):261–268
13. Nikezić D, Yu KN (2000) Monte Carlo calculations of LR115 detector response to 222Rn in the presence of 220Rn. Health Phys 78(4):414–419
14. Rehman F-U, Jamil K, Zakaullah M, Abu-Jarad F, Mujahid SA (2003) Experimental and Monte Carlo simulation studies of open cylindrical radon monitoring device using CR-39 detector. J Environ Radioact 65:243–254
15. Rickards J, Golzarri J-I, Espinosa G (2010) A Monte Carlo study of radon detection in cylindrical diffusion chambers. J Environ Radioact 101(5):333–337. doi:10.1016/j.jenvrad.2010.01.003
16. Rezaae MR, Nejad R (2012) Response of CR- 39 detector to radon in water using Monte Carlo simulation. Iran J Med Phys 9(3):193–201
17. Rezaae MR, Sohrabi M, Negarestani A (2013) Studying the response of CR-39 to Radon in non-polar liquids above water by Monte Carlo simulation and measurement. Radiat Meas 50:103–108
18. Planinic J, Faj Z (1989) The equilibrium factor F between radon and its daughters. Nucl Instrum Methods Phys Res A 278:550–552
19. Planinic J, Faj Z (1990) Equilibrium factor and dosimetry of radon by a nuclear track detector. Health Phys 59(3):349–351
20. Amgarou K, Font L, Baixeras C (2003) A novel approach for long-term determination of indoor 222 Rn progeny equilibrium factor using nuclear track detectors. Nucl Instrum Methods Phys Res A 506:186–198
21. Abo-Elmagd M, Mansy M, Eissa HM, El-Fiki MA (2006) Major parameters affecting the calculation of equilibrium factor using SSNTD-measured track densities. Radiat Meas 41:235–240
22. Nikezić D, Yu KN (1999) Relationship between the ^{210}Po activity incorporated in the surface of an object and the potential α-energy concentration. J Environ Radioact 47(1):45–55
23. Paul H (2006) A comparison of recent stopping power tables for light and medium-heavy ions with experimental data, and applications to radiotherapy dosimetry. Nucl Instrum Methods Phys Res B 247(2):166–172
24. Cliff KD, Wrixon AD, Green BMR, Miles JCH (1983) Radon daughter exposures in the U. K. Health Phys 45:323–330
25. Nikolopoulos D, Vogiannis E (2007) Modelling radon progeny concentration variations in thermal spas. Sci Total Environ 373(1):82–93
26. Nikolopoulos D, Vogiannis E, Petraki E, Zisos A, Louizi A (2010) Investigation of the exposure to radon and progeny in the thermal spas of Loutraki (Attica-Greece): results from measurements and modeling. Sci Total Environ 408:495–504
27. Nikolopoulos D, Vogiannis E, Petraki E, Kottou S, Yannakopoulos P, Leontaridou M, Louizi A (2013) Dosimetry modelling of transient radon and progeny concentration peaks: results from in situ measurements in Ikaria spas, Greece. Environ Sci Process Impacts 15:1216–1227
28. Porstendorfer J, Pagelkopf P, Grundel M (2005) Fraction of the positive ^{218}Po and ^{214}Pb clusters in indoor air. Radiat Prot Dosimetry 113(3):342–351
29. Jacobi W (1972) Activity and potential a-energy of ^{222}Rn and ^{222}Rn daughters in different air atmospheres. Health Phys 22:441–450
30. Nikolopoulos D, Louizi A, Petropoulos N, Simopoulos S, Proukakis C (1999) Experimental study of the response of cup-type radon dosemeters. Radiat Prot Dosimetry 83(3):263–266

Chapter 59
Adsorption of Methylene Blue from Aqueous Solution by Natural Clays

Habiba Belbekiri and Meriem Belhachemi

Abstract The objective of this study is the valorization of certain local natural clays by their use in treatment of industrial wastewater. For this purpose, we carried out discoloration tests of effluents by adsorption using two natural clays from two different deposits, one located in Maghnia (MC) and the other in the region of Bechar (BC). The adsorption tests were conducted on a basic dye methylene blue (MB). The effects of temperature, initial dye concentration, contact time, and solution pH on adsorption were studied. The adsorption capacity increased with an increase in adsorbate dosage and a decrease in ionic strength. The equilibrium time was found to be 90 min for full equilibration. Langmuir isotherm model fitted well the equilibrium data for the two sorbents (BC and MC) comparing to the Freundlich isotherm models. The monolayer adsorption capacity of AB and AM for methylene blue (MB) was found to be 223.714 and 510.204 mg/g, respectively. Also, the adsorption processes were endothermic and spontaneous in nature.

Keywords Algerian clays • Adsorption • Dye • Water treatment

59.1 Introduction

Many industries such as the textile, cosmetic, food, leather, pharmaceutical, paper, and printing industries consume large quantities of water to color their products [1]. The effluents from these industries and from manufacturing are highly colored and undesirable, even at very low concentrations in the case of some dyes. Discharge of the dye wastewater into receiving water bodies prevents photosynthetic activity in

H. Belbekiri (✉)
Laboratoire de Fiabilité des matériaux et des structures en région sahariennes, Faculté des Sciences et Technologie, Université de Béchar, B.P. 417, Béchar 08000, Algeria
e-mail: belbekirih@yahoo.fr

M. Belhachemi
Laboratoire de Fiabilité des matériaux et des structures en région sahariennes, Faculté des Sciences et Technologie, Université de Béchar, B.P. 417, Béchar 08000, Algeria

Laboratoire d'étude physico-chimique des matériaux et application à l'environnement, Faculté de Chimie, USTHB, B.P. 32 El Alia, Bab Ezzouar 16111, Algeria

I. Dincer et al. (eds.), *Progress in Clean Energy, Volume 1*,
DOI 10.1007/978-3-319-16709-1_59

aquatic life by reducing sunlight penetration [2]. In addition, some dyes or their metabolites are either toxic or mutagenic and carcinogenic [3]. The conventional methods for removing dyes include adsorption, coagulation and flocculation, oxidation, ion exchange, reverse osmosis, and ultra filtration [4]. Due to economic considerations, adsorption has gained favor in recent years due to proven efficiency in the removal of pollutants from effluents to stable forms for the above conventional treatment methods [5]. This led to a search for cheaper, easily obtainable materials for the adsorption of dye [6]. Clays such as sepiolite, zeolite, montmorillonite, smectite, and bentonite are being considered as alternative low-cost adsorbents. The wide usefulness of clays is a result of their specific surface area, high chemical and mechanical stability, and variety of surface and structural properties. The chemical nature and pore structure usually determine the sorption ability of clays [7]. In relatively recent years, there has been an increasing interest in utilizing natural clay materials for the removal of toxic metals and some organic pollutants from aqueous solutions [8].

This chapter examines the elimination of MB present in an aqueous solution using MC and BC as adsorbent material. In this study, the effects of various parameters including contact time, temperature, and solution pH on the adsorption were investigated. The kinetic and thermodynamic parameters were also calculated to determine rate constants and adsorption mechanism. The experimental data were fitted into Langmuir and Freundlich equations to determine which isotherm gives the best correlation to experimental data.

59.2 Experimental

59.2.1 Materials

The clay samples used in this study were MC and BC. They were obtained from the north and south, respectively, region of Algeria. The sorbent solids were collected and washed. They were grounded and sieved into different particle sizes ranging in diameter range of 0.5–1 mm for BC and MC.

59.2.2 Dye

Methylene blue was chosen as adsorbate in this study as a model molecule for basic dyes. A stock solution of the dye was prepared by dissolving an accurately weighed amount of MB in ultrapure water to give a concentration of 1 g/L. Experimental solutions of different concentrations were obtained by diluting the stock solution with ultrapure water. The characteristics of methylene blue are: $C_{16}H_{18}ClN_3S$ chemical formula, 319.86 g/mol molecular weight and 665 nm maximum wavelength.

59.2.3 *Batch Mode Adsorption Study*

MB solutions were prepared with ultrapure water at natural pH. Batch experiments were carried in a glass beaker by shaking a fixed mass BC or MC (20 mg) with 20 mL diluted solution (20–800 mg/L). After agitation, the solution was centrifuged at 5,000 rpm with a required time at 25 °C. Then, the MB concentration in the supernatant solution was analyzed using a Spectronic 21 MV spectrophotometer by monitoring the absorbance changes at a wavelength of maximum absorbance (665 nm). Each experiment was carried out in duplicate and the average results are presented. Calibration curves were obtained with standard MB solutions using ultrapure water as a blank. Mass capacity of adsorption, q_e, is calculated from the difference between the initial and the final MB concentration as followed in Eq. (59.1).

$$q_e = (C_0 - C_e)V/W \tag{59.1}$$

where C_0 and C_e (mg/L) are the concentrations of MB initially and at time (t), respectively, V (L) is the volume of the solution, and W (g) is the mass of adsorbents used.

59.3 Results and Discussion

59.3.1 *Effect of Contacts Time*

It is essential to evaluate the contact time required to reach equilibrium. In fact, a kinetic study helps to determine the nature of the process and serves as a reference for the assessment of the residence time required for the adsorption process. Hence, experiments were conducted, varying contact time from 10 to 120 min for both samples. Figure 59.1 shows the effect of contact time at initial dye concentration of 200 mg/L. The adsorption capacity increases with an increase in contact time and reaches a plateau at 90 min for BC and MC. Moreover, it is seen that the adsorption of MB is very rapid in the first 10 min and then slowly declines with time until equilibrium. The initial rapid phase may be due to an increase in the number of vacant sites available at the initial stage. Similar trends have been reported for the adsorption of MB by giant duckweed [6].

59.3.2 *Adsorption Kinetics*

In order to investigate the adsorption kinetics of MB, four kinetic models, namely pseudo-first-order, pseudo-second-order, Bangham equation, and intraparticle diffusion models, were used in this study.

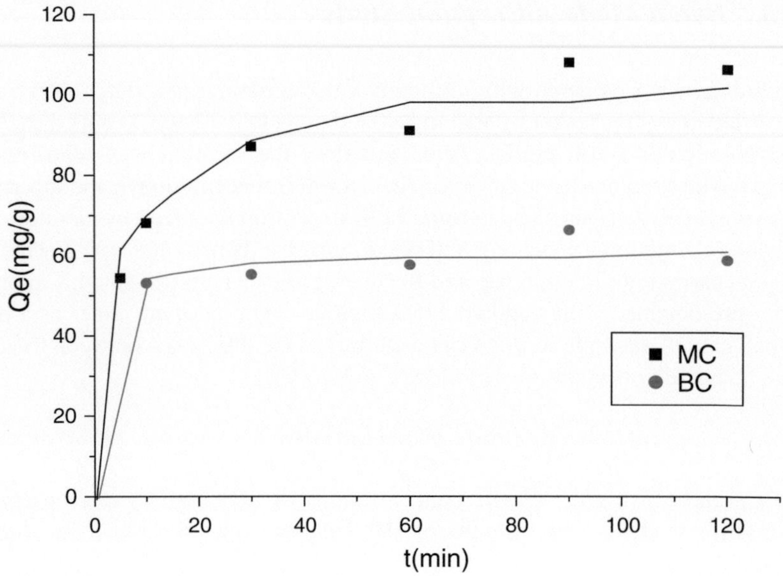

Fig. 59.1 Effect of contact time and initial dye concentration on adsorption of MB by MC and BC (adsorbent dosage 1 g/L, natural pH, 25 °C)

59.3.2.1 Pseudo-First-Order Model

The pseudo-first-order Lagergren equation [9] is given by:

$$\mathrm{d}q_t/\mathrm{d}t = k_1(q_\mathrm{e} - q_t) \tag{59.2}$$

where q_t and q_e (mg/g) are the amounts adsorbed at time t and equilibrium, respectively, and k_1 (min^{-1}) is the pseudo-first-order rate constant for the adsorption process. The integrated linear form of Eq. (59.3) can be expressed as:

$$\log(q_\mathrm{e} - q_t) = \log q_\mathrm{e} - (k_1/2.303)t \tag{59.3}$$

From the plots of log ($q_\mathrm{e} - q_t$) vs. t, it was observed that the pseudo-first-order model was applicable for the initial 10 min (Fig. 59.2a). Thereafter, the experimental data deviated from theory (results not shown).

Ho et al. [10] reported that the sorption data were well represented by the Lagergren first-order model only in the first stage where rapid sorption took place, confirming that it was not appropriate to use the Lagergren kinetic model to predict the adsorption kinetics of MB by adsorbents for the entire adsorption process. The rate constants k_1 and calculated equilibrium adsorption capacities $q_\mathrm{e,cal}$ obtained from the slopes and intercepts of the plots are given in Table 59.1. The correlation coefficient values R^2 were 0.825 and 0.443 for MC and BC, respectively.

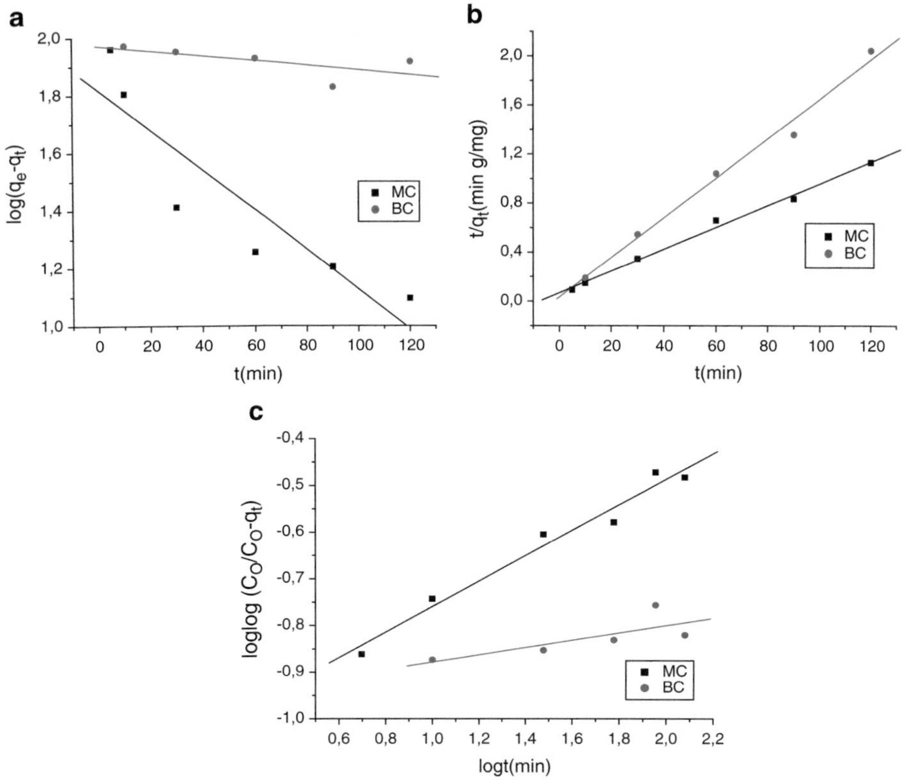

Fig. 59.2 Kinetic models for adsorption of MB by MC and BC: (**a**) pseudo-first-order model; (**b**) pseudo-second-order model; (**c**) Bangham equation model; (adsorbent dosage 1 g/L, natural pH, 25 °C)

The $q_{e,cal}$, calculated values and the experimental values $q_{e,exp}$, are not similar, showing that the adsorption kinetics for the entire process did not follow the pseudo-first-order model.

59.3.2.2 Pseudo-Second-Order Model

In order to characterize the adsorption kinetics, several kinetic models were applied to the experimental data. However, in the last years, the pseudo-second-order kinetic model has been considered to be the most appropriate. The pseudo-second-order equation can be expressed following Eq. (59.4) [10].

$$dq_t/dt = k_1(q_e - q_t)^2 \tag{59.4}$$

Table 59.1 Kinetic parameters for adsorption of MB by MC and BC

	Pseudo-first-order model				Pseudo-second-order model				Bangham equation model		
	$q_{e,exp}$ (mg/g)	k_1 (1/min)	$q_{e,cal}$ (mg/g)	R^2	k_2 (g/mg min)	$q_{e,cal}$ (mg/g)	h (g/mg mn)	R^2	k_0 (mL/g L)	α	R^2
BC	66.33	19.16	93.56	0.443	2.13	62.5	29.41	0.988	0.25	0.078	0.580
MC	108.01	0.016	65.78	0.825	0.0012	112.61	14.74	0.992	0.214	0.27	0.970

where k_2 (g/mg min) is the pseudo-second-order rate constant. After integration, Eq. (59.4) becomes:

$$t/q_t = 1/k_2 q_e^2 + (1/q_e)t \tag{59.5}$$

The initial sorption rate h (mg/g min) is defined as:

$$h = k_2 q^2 \tag{59.6}$$

By plotting t/q_t against t for different initial concentrations, straight lines were obtained (Fig. 59.2b). The rate constants k_2 and calculated equilibrium adsorption capacities $q_{e,cal}$ obtained from the intercepts and slopes of the plots are given in Table 59.2. For all initial concentrations studied, the correlation coefficients were close to unity. In addition, the values of $q_{e,cal}$ showed good agreement with the experimental data. Thus, the sorption could be approximated more appropriately by the pseudo-second-order model, supporting the assumption of chemisorptions as the rate-limiting mechanism through sharing or exchange of electrons between sorbent and sorbate [10]. Similar results have been reported in the literature [11].

59.3.2.3 Bangham Equation

Three sequential steps are involved in the adsorption of sorbate onto sorbent: film diffusion, particle diffusion, and adsorption on the pore surface [12]. Among these steps, the third one is very fast and is not considered as a rate-limiting step. The overall adsorption process may be controlled by the film or particle diffusion step. The Bangham equation is applied in this study to check whether pore diffusion is the only limiting step or not. The Bangham equation is given as

$$\log\left(\log([C_0/(C_0 - q_{tm})])\right) = \log(k_0 m/2.303) + \alpha \log t \tag{59.7}$$

where C_0 (mg/L) is the initial dye concentration, m (g/L) is the adsorbent mass used per liter of solution, V (mL) is the volume of the solution, and α (<1) and k_0 (mL/g L) are Bangham constants. The parameters of Bangham model as well as the correlation coefficient are listed in Table 59.1. The experimental data are not fitted well by the Bangham equation (Fig. 59.2c), the plots have low correlation coefficients, indicating that the diffusion of MB into pores of the adsorbent is not the only rate-controlling step [13].

59.3.3 Adsorption Isotherms

It is important to study the equilibrium adsorption isotherm for the design of the adsorption system. Two adsorption isotherm models, namely Langmuir and Freundlich models were used to fit the adsorption experimental data in this study.

The Langmuir isotherm [14] assumes that sorption occurs at specific homogeneous sites within the adsorbent and that the capacity of the adsorbent is finite. The Langmuir equation is represented as:

$$q_e = K_L q_m C_e / (1 + K_L C_e) \tag{59.8}$$

where q_e (mg/g) is the amount of dye adsorbed at equilibrium, C_e (mg/L) is the equilibrium dye concentration in the solution, q_m (mg/g) is the monolayer adsorption capacity, and K_L (L/mg) is the Langmuir isotherm constant, which is related to the affinity of the binding sites and the energy of adsorption. The essential characteristics of the Langmuir isotherm can be expressed in terms of a dimensionless constant separation factor or equilibrium parameter R_L [15], which is defined as:

$$R_L = 1/(1 + K_L C_0) \tag{59.9}$$

where C_0 (mg/L) is the initial dye concentration and K_L (L/mg) is the Langmuir isotherm constant. The R_L value indicates whether the type of isotherm is unfavorable ($R_L > 1$), linear ($R_L = 1$), favorable ($0 < R_L < 1$), or irreversible ($R_L = 0$).

The linear plots of C_e/q_e versus C_e are used to determine the value of q_m (mg/g) and K_L (L/mg).

The Freundlich isotherm [16] is based on the assumption of a heterogeneous surface with a nonuniform distribution of adsorption heat over the surface. It is an empirical equation expressed as

$$q_e = K_F C^{1/n} \tag{59.10}$$

where K_F (mg $L^{1/n}$ $mg^{1/n}$/g) is the Freundlich constant and $1/n$ is the heterogeneity factor, which is related to the capacity and intensity of the adsorption, respectively.

A plot of log q_e against log C_e gives a straight-line graph with $(1/n)$ as the slope and log K_F as the intercept. Both K_F and n determine the curvature and the steepness of the isotherm.

Relevant isotherm parameters and the values of q_m, K_L, R_L, K_F, $1/n$, and R^2 for all the experiments for removal of MB are presented in Table 59.2.

The linear calculations reveal that the equilibrium data agree well with the Langmuir equation comparing to Freundlich isotherms, for AB and AM adsorbents. This reasonable fit of the mode is indicated by a high correlation coefficient $R^2 > 0.98$. The comparison of correlation coefficients (R^2) of the linearized form of the two isotherm models indicates that the Langmuir model yields a better fit for the experimental equilibrium adsorption data than the Freundlich isotherm models.

According to the values of $R_L < 1$, all the systems show favorable adsorption of MB. The low values of R_L indicate high and favorable adsorption of methylene blue on to MC and BC. The maximum capacity obtained from Langmuir equation are 510.2 and 223.7 mg/g at 25 °C for MC and BC, respectively. Table 59.3 compares between the adsorption capacities of AM, AB, and other adsorbents from the literature. From Table 59.3 the maximum capacity of AM and AB was found to be comparable to those of some other adsorbent.

Table 59.2 Isotherm parameters obtained for adsorption of MB by MC and BC

Isotherm	Parameter	Temperature (°C)	BC values	MC values
Langmuir	q_m (mg/g)	25	223.714	510.204
		45	175.438	196.078
	K_L (L/mg)	25	0.194	0.143
		45	0.792	0.127
	R^2	25	0.794	0.884
		45	0.984	0.993
	R_L	25	0.0251	0.0337
		45	0.0063	0.0378
Freundlich	K_F (mg $L^{1/n}$ $mg^{1/n}$/g)	25	46.581	63.755
		45	39.241	45.897
	$1/n$	25	0.480	0.669
		45	0.309	0.258
	R^2	25	0.664	0.888
		45	0.834	0.912

Table 59.3 Comparison of sorption capacities of MB between MC, BC, and other related sorbents

Adsorbent	Q_m (mg/g)	References
Clay of Maghnia (MC)	510.204	This study
Clay of Bechar (BC)	223.714	This study
Softstem bulrush	53.8	Y. Li et al. (2008)
Rice husk	40.5	V. Vadivelan et al. (2005)
Banana peel	20.8	G. Annadurai et al. (2002)
Orange peel	18.6	G. Annadurai et al. (2002)

59.4 Thermodynamic Studies

The thermodynamic parameters of the adsorption process are obtained from experiments at various temperatures using the following equations:

$$\Delta G = -RT \ln K_e \tag{59.11}$$

where R is the universal gas constant (8.314 J/mol K), T is temperature (K), and K_e is the distribution coefficient for the adsorption calculated from the following equation:

$$K_D = C_a/C_e \tag{59.12}$$

C_a is the amount of dye (mg) adsorbed on the adsorbent per liter of the solution at equilibrium and C_e concentration (mg/L) of the dye in the solution.

Table 59.4 Thermodynamic parameters for adsorption of MB on MC and BC

		Temperature (K)		
Sorbent solid	Parameters	298	308	318
BC	K_e	1.355	1.128	0.950
	ΔG (J/mol)	−754.188	−308.918	134.352
	ΔH (J/mol)	−13,993.434		
MC	ΔS (J/mol K)	−44.427		
	K_e	1.437	1.217	1.042
	ΔG (J/mol)	−898.422	−503.922	−109.422
	ΔH (J/mol)	−12,654.522		
	ΔS (J/mol K)	−39.45		

The enthalpy (ΔH) and entropy (ΔS) parameters are estimated from the following equation:

$$\ln K_e = \frac{-\Delta H}{RT} + \frac{\Delta S}{R} \tag{59.13}$$

ΔH and ΔS were calculated from the slope and intercept of Van't Hoff plots of ln K_e versus $1/T$ (see Fig. 59.3). The results are listed in Table 59.4.

The obtained values for Gibbs free energy change (ΔG) are −898.422, −503.922, and −109.422 J/mol for MB adsorption on MC at 25 °C, 35 °C, and 45 °C, respectively. The negative ΔG values indicate a spontaneous nature of the adsorption. However, ΔG values are, −754.188, −308, and 134.352 J/mol for MB adsorption on BC at 25 °C, 35 °C, and 45 °C, respectively. The obtained ΔG values indicate that increasing temperature leads to decrease in feasibility of adsorption at higher temperatures.

The negative ΔH indicates the exothermic nature of the adsorption and also its magnitude gives information on the type of adsorption, which can be either physical or chemical. Therefore, the ΔH values show that the adsorption processes of MB on adsorbents were taken place via chemisorption. The negative values of ΔS suggest a decrease in randomness at the solid/solution interface [17] (Fig. 59.3).

59.5 The Effect of pH on the Adsorption Process

The effect of solution pH on the adsorption was investigated in the pH range 2.3–11 (adjusted with 0.1 mol/L HCl or NaOH solution) using a pH meter. The solution pH was an important parameter for the adsorption process owing to its impact on the surface binding sites of the clays and the ionization process of the dye molecule [11]. Figure 59.4 shows the dependence of MB dye removal on solution pH.

The amounts of MB adsorbed on each adsorbent in the acidic medium were slightly high as compared to those in the basic medium. In the present study, as the

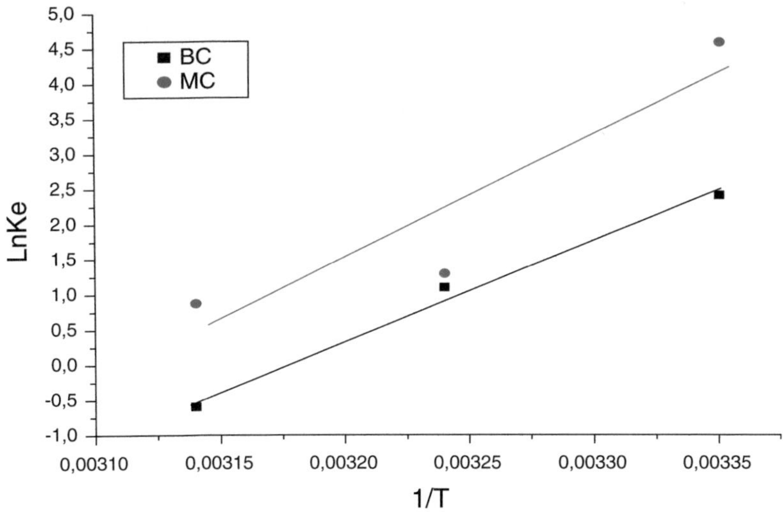

Fig. 59.3 Plot of ln K_e vs. $1/T$ for the estimation of thermodynamic parameters for adsorption of MB on MC and BC

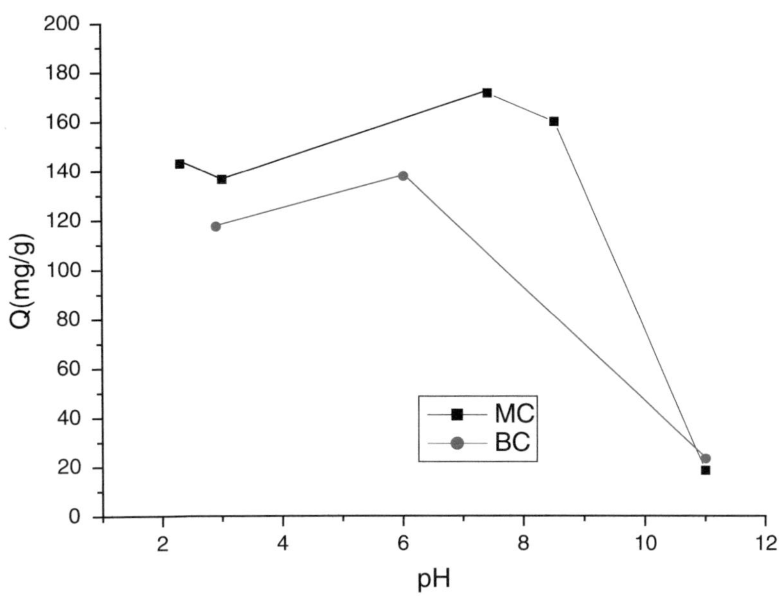

Fig. 59.4 Effect of solution pH on adsorption of MB by BC and MC ($C_0 = 200$ mg/L, adsorbent dosage 1 g/L, 25 °C)

initial solution pH increased of 2.3–7, the number of negatively charged active sites increased. Therefore, the electrostatic attraction between the negatively charged adsorbent and positively charged MB ions increased, resulting in an increase in the adsorption on each adsorbent. In the pH range 7–11, the adsorption of MB decreases on each adsorbent, may be the OH^- ion can competitively exclude the adsorption of MB by exchanging with anion on the surface or in the interlayer region of the clay.

59.6 Conclusion

Adsorption of MB using BC and MC was performed in aqueous solutions. The amount of MB adsorbed on the MC was higher than that on the BC. The adsorption capacity was affected by various parameters including contact time, effect of temperature, and pH solution. Kinetic studies showed that the adsorption process followed the pseudo-second-order model. The experimental data were evaluated by Langmuir and Freundlich isotherms. Equilibrium data were well fitted to Langmuir isotherm model. The maximum adsorption capacities of BC and MC for methylene blue were 223.714 and 510.204 mg/g, respectively, at 25 °C. These values are in a high range for methylene blue adsorption when comparing to other adsorbents. The adsorption was spontaneous owing to the negative value of ΔG. The abilities of all the tested adsorbents to adsorb MB were relatively low in the alkaline medium as compared to those in the acidic medium. BC and MC are easily available and inexpensive, suggesting that it could be employed as an alternative adsorbent for the removal of MB.

References

1. Vijayaraghavan K, Yun YS (2008) Biosorption of C.I. Reactive Black 5 from aqueous solution using acid-treated biomass of brown seaweed Laminaria sp. Dyes Pigm 76:726–732
2. Han MH, Yun YS (2007) Mechanistic understanding and performance enhancement of biosorption of reactive dyestuffs by the waste biomass generated from amino acid fermentation process. Biochem Eng J 36:2–7
3. Chen KC, Wu JY, Huang CC, Liang YM, Hwang SCJ (2003) Decolorization of azo dye using PVA-immobilized microorganisms. J Biotechnol 101:241–252
4. Kiran I, Akar T, Ozcan AS, Ozcan A, Tunali S (2006) Biosorption kinetics and isotherm studies of Acid Red 57 by dried cephalosporium aphidicola cells from aqueous solutions. Biochem Eng J 31:197–203
5. Sun Q, Yang L (2003) The adsorption of basic dyes from aqueous solution on modified peat-resin particle. Water Res 37(7):1535
6. Waranusantigul P, Pokethitiyook P, Kruatrachue M, Upatham ES (2003) Kinetics of basic dye (methylene blue) biosorption by giant duckweed (Spirodela polyrrhiza). Environ Pollut 125:385

7. Juang RS, Lin SH, Tsao KH (2002) Mechanism of sorption of phenols from aqueous solutions onto surfactant-modified montmorillonite. J Colloid Interface Sci 254(2):234
8. Boyd SA, Shaobia S, Lee J, Mortland MM (1988) Pentachlorophenol sorption by organoclays. Clay Miner 35(125):130
9. Lagergren S, Svenska BK (1898) Zur theorie der sogenannten adsorption geloester stoffe. Veterskapsakad Handlingar 24:1–6
10. Ho YS, McKay G (1999) Pseudo-second-order model for sorption processes. Process Biochem 34:451–465
11. Ncibi MC, Mahjoub B, Seffen M (2007) Kinetic and equilibrium studies of methylene blue biosorption by Posidonia oceanic (L.) fibres. J Hazard Mater 139:280–285
12. Dogan M, Ozdemir Y, Alkan M (2007) Adsorption kinetics and mechanism of cationic methyl violet and methylene blue dyes onto sepiolite. Dyes Pigm 75:701
13. Tutem E, Apak R, Unal CF (1998) Adsorptive removal of chlorophenols from water by bituminous shale. Water Res 32:2315
14. Langmuir I (1916) The constitution and fundamental properties of solids and liquids. J Am Chem Soc 38:2221–2295
15. Hall KR, Eagleton LC, Acrivos A, Vermeulen T (1966) Pore-and solid-diffusion kinetics in fixed-bed adsorption under constant-pattern conditions. Ind Eng Chem Fund 5:212–223
16. Freundlich HMF (1906) Uber die adsorption in lasungen. Z Phys Chem 57:385–470
17. Desai M, Dogra A, Vora S, Bahadur P, Ram RN (1997) Adsorption of some acid dyes from aqueous solutions onto neutral alumina. Indian J Chem A 36(11):938

Chapter 60
Application of a Binary Teaching Learning-Based Algorithm to the Optimal Placement of Phasor Measurement Units

Abdelmadjid Recioui, Hamid Bentarzi, and Abderrahmane Ouadi

Abstract In recent years, the placement of phasor measurement units (PMUs) in wide area measurement systems has gained much attention. This work presents a binary teaching learning-based optimization (BTLBO) algorithm for the optimal placement of PMUs. The optimal PMU placement problem is formulated to minimize the number of PMUs installation subject to full network observability at the power system buses. The proposed method is applied to the IEEE14-bus, 30-bus, 57-bus-118 bus systems. The results show that the whole system can be observable with installing PMUs on less than 25 % of system buses. To validate the approach, the results are compared with those reported in literature and has shown that the BTLBO algorithm is efficient for solving the optimal PMU placement problem.

Keywords PMU placement • TLBO • Optimization • Observability

Nomenclature

ABC	Artificial bee colony
ACO	Ant colony optimization
GA	Genetic algorithms
GPS	Global positioning system
OPP	Optimal placement problem
PMU	Phasor measurement unit
PSO	Particle swarm optimization
TLBO	Teaching learning-based optimization

A. Recioui (✉) • H. Bentarzi • A. Ouadi
Laboratory Signals and Systems, Institute of Electrical and Electronic Engineering, University M'hamed Bougara of Boumerdes, Avenue de l'indépedance, Boumerdes, Algeria
e-mail: recmad2006@yahoo.fr

I. Dincer et al. (eds.), *Progress in Clean Energy, Volume 1*,
DOI 10.1007/978-3-319-16709-1_60

60.1 Introduction

Today with the development of the global positioning system (GPS), communication network, and digital signal processing technique, it is possible to monitor better the operation of power systems. Secure operation of any power system is closely related to the measurement and monitoring of the system operating conditions. Phasor measurement units (PMUs) can give real-time phasors of bus voltages and branch currents in a wide-area electric network. The phasors from different nodes, which refer to the same time–space coordination, can improve performance of monitoring and control systems by enhancing power flow calculation, state estimation, transient stability analysis, and frequency stability analysis [1].

PMUs have become the measurement technique of the best choice in electric power systems. They provide positive sequence voltage and current measurements synchronized with accuracy of a microsecond. The deployment of this device can improve performances of power system protection, monitoring, and control systems [2, 3]. PMUs placed at all substations allow direct measurement of the state of the network. However, PMU placement on each bus of a system is difficult to be achieved due to either cost factor or nonexistence of communication facilities in some substations. Moreover, as a consequence of Ohm's Law, when a PMU is placed at a bus, neighboring buses also become observable, this implies that a system can be made observable with a less number of PMUs than the number of buses. So it is neither economical nor necessary to install PMUs at all nodes of a wide-area interconnected network.

In recent years, many investigators have presented different methods for finding the minimum number and optimal placement of PMUs for making a power system completely observable [4]. A bisecting search method is implemented to find the minimum number of PMUs to make the system observable. The simulated annealing method is used to randomly choose the placement sets to test for observability at each step of the bisecting search. Xu and Abur [5] used integer programming to find the minimum number and locations of PMUs. Engineers and mathematicians have developed a variety of optimization algorithms like PSO, GA, ABC, and ACO to solve this problem [6–10].

Recently, the teaching learning-based optimization (TLBO) technique proposed by Rao et al. (2011, 2012), Rao and Savsani (2012) and Rao and Patel (2012) has been used successfully in a number of N-P optimization problems [11–14]. However, TLBO does not have a binary version that is able to optimize binary problems, and since the optimal placement of PMUs is a binary optimization problem, a binary teaching learning-based optimization (BTLBO) based method for finding the minimum required number of PMUs is proposed in this research work.

60.2 The Optimal PMU Placement Problem

PMUs provide two types of measurements: bus voltage phasors and branch current phasors. Depending on the type of PMUs, the number of channels used for measuring voltage and current phasors will vary. Here, it is assumed that each PMU has enough channels to record the bus voltage phasor at its associated bus and current phasors along all branches that are incident to this bus.

The PMU placement optimization (OPP) is to minimize the number of PMUs by optimizing their locations, while keeping all the nodes voltage phasors observable. An example of an optimally set of PMUs placed in a 14-bus system is shown in Fig. 60.1. In this system, there are three PMUs placed at buses 2, 6, and 9, respectively. Bus 7 is the only zero injection bus. The PMU at bus 2 can not only measure the voltage phasor of bus 2, but also the current phasors of branches 2-1, 2-3, 2-4, and 2-5. Using Ohm's law, the voltage phasors at buses 1, 3, 4, and 5 can be obtained from the branch currents and the voltage at bus 2.

By determining voltage phasors at buses 1, 2, 3, 4, and 5, the current phasors of branches 1-5, 3-4, and 4-5 can be calculated. Following the same manner, PMU at bus 6 can provide the voltage phasor at bus 6 and the current phasors of branches 6-5, 6-11, 6-12, and 6-13, thus allowing the calculation of the voltage phasors at buses 5, 11, 12, 13, and the current phasor of branch 12-13. PMU at bus 9 can give the voltage phasor at bus 9 and the current phasors of branches 9-4, 9-7, 9-10, 9-14, and allow the calculation of the voltage phasors at buses 4, 7, 10, 14, and the current phasors of branches 4-7. When voltage phasors of buses 10, 11, 13, and 14 are known, current phasors of branches 10-11 and 13-14 can now also be calculated. Using the known current phasors of branches 4-7 and 9-7, and the zero injection at bus 7, the current phasor of branch 7-8 can be derived using the Kirchhoff's Current Law. The only remaining unknown voltage phasor at bus 8 can now be calculated by using the voltage phasor at bus 7 and the current phasor of branch 7-8. Thus, the entire system becomes observable by placing only three PMUs at buses 2, 6, 9 and by considering the zero injection at bus 7 [15, 17, 19, 20].

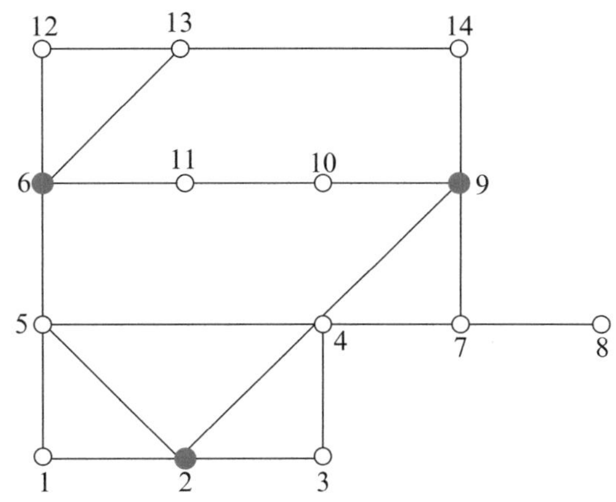

Fig. 60.1 IEEE 14-bus system with three PMUs

60.2.1 Observability Analysis

Observability is a fundamental component of many applications in power system such as real-time state estimation. Observability of electric power system is to study whether there is enough measurement in system in order to measure the state of electric power system including the amplitude value of the voltage and current as well as the phase angle between them.

Power system observability has been defined by graph theory [3]. An N-bus power system is represented as a graph $G = (V, E)$, where V is a set of graph vertices containing all system buses, and E is a set of graph edges containing all system branches [4]. Before developing the rules of observability, a definition of flow and injection measurements are necessary.

Injection is a variable load or source so that an injected current is supplied to the bus. Zero injection buses are the buses from which no current is being injected into the system [16]. Flow measurements are those measurements taken from some power system instruments and allow the calculation of line current magnitude and real/reactive power flowing through the line, where they are installed. So these measurements help for reducing the total number of PMUs to complete observability of the system.

60.2.2 Observability Rules for the Optimal Placement Problem

Observability is generally divided by numerical and topological states. For the present work, we use observability topology analysis method based on PMUs according to the observability rules [16, 18]:

- *Rule 1*: If a PMU is placed at a bus, this bus and its all neighboring buses can be observed.
- *Rule 2*: For a zero injection node which is observed, if all connected nodes are observable except one, then the unobserved node can be observed.
- *Rule 3*: If all the nodes connected to a zero injection node are observable, then the zero injection node can be observed too.

60.2.3 Problem Formulation

A numerical method is presented here to solve the problem [15, 16]. For an N-bus system, the PMU placement problem can be formulated as follows:

Objective function

$$(J) = \mathrm{Min}\sum_{i}^{n} x_i \,\text{such that}\; f(X) \geq \hat{i} \tag{60.1}$$

where X is a binary decision variable vector, whose entries are defined as:

$$x_i = \begin{cases} 1 & \text{if Bus } i \text{ has a PMU} \\ 0 & \text{otherwise} \end{cases} \tag{60.2}$$

$f(X)$ is a vector function, whose entries are non-zero if the corresponding bus voltage is solvable using the given measurement set and zero otherwise. î is a vector whose entries are all ones [5, 15].

60.2.4 Constraints

In order to form the constraint set, the binary connectivity matrix A, whose entries are defined as below, will be formed first:

$$A_{k,m} = \begin{cases} 1 & \text{if} \quad k = m \\ 1 & \text{if} \quad k \text{ and } m \text{ are connected} \\ 0 & \text{if} \quad \text{otherwise} \end{cases} \tag{60.3}$$

Based on the measurement of a PMU placing on one bus, the voltage phasors of this bus and its neighboring buses can be calculated, we can get

$$F(X) = A \times X \tag{60.4}$$

Constraint functions defined by Eqs. (60.1)–(60.4) ensure full network observability while minimizing the total number of the PMUs.

The procedure for building the constraint equations (vector function $F(X)$ will be presented for four possible cases [5]:

Case 1: *A system with no PMUs or conventional measurements*: In this case, the PMU, the flow and the injection measurements are ignored in order to make all bus voltage phasors solvable for this case.

Case 2: *A system with some flow measurements*: This case considers the situation where some flow measurements may be present.

Case 3: *A system with both injection measurements* (*some of which may be zero injection measurements*) *and flow measurements*: In this case, injection measurements, whether they are real-time measurements or zero injections, are treated in the same way.

Case 4: *A system with flow, injection, and PMU measurements*: This case considers the most general situation where both conventional flow or injection measurements and PMU measurements may be present, but not enough to make the entire system observable. To build the constraints for this case is simple. After forming the constraint equation $F(X)$ according to the procedure described above, simply replace all the x_i by 1, where i represents the bus with an already-installed PMU.

60.3 The Teaching Learning-Based Optimization

TLBO is population-based method. In this optimization algorithm, a group of learners is considered as population and different design variables are considered as different subjects offered to the learners and learners result is analogous to the "fitness" value of the optimization problem. In the entire population, the best solution is considered as the teacher. The working of TLBO is divided into two parts: "Teacher phase" and "Learner phase," which are explained below.

60.3.1 Mechanism of TLBO

Most of the metaheuristic methods are inspired from nature, i.e., they mimic the behavior of nature. For example, in Genetic Algorithm inspired from Darwin's theory, the strongest is the one who survive. Particle Swarm is inspired from the movement of a flock of bird, a school of fish, or a swarm of bees that are looking for food. Artificial bee colony simulates the intelligent forging of honey bee swarm. Ant colony shows how ants search for food and how to find an optimal way to it, etc. They prove their effectiveness in solving many engineering optimization problems but each one of them requires its own algorithm-specific control parameters. For example, GA uses mutation rate and crossover rate. Similarly, PSO uses inertia weight, social, and cognitive parameters. The improper tuning of algorithm-specific parameters either increases the computational effort or yields the local optimal solution. Considering this fact, recently Rao et al. [11, 12], Rao and Savsani [13], and Rao and Patel [14] introduced the TLBO algorithm which does not require any algorithm-specific parameters. In this way, TLBO obtain global solutions for continuous nonlinear functions with less computational effort and high consistency [11].

TLBO is a teaching-learning process inspired algorithm based on the effect of influence of a teacher on the output of learners in a class. Teacher and learners are the two vital components of the algorithm and describes two basic modes of the learning, through teacher (known as teacher phase) and interacting with the other learners (known as learner phase). The output in TLBO algorithm is considered in terms of results or grades of the learners which depend on the quality of teacher. So, teacher is usually considered as a highly learned person who trains learners, so that they can have better results in terms of their marks or grades. Moreover, learners also learn from the interaction among themselves which also helps in improving their results [11, 12].

Teacher phase: In this phase, the best student is chosen from the population (the class) according to the fitness function and set as a teacher. Since the teacher is the highest learned person in the class, he puts effort to disseminate knowledge among students, so that he tries to bring the mean level of the class up to his level, the new mean of the class depends on two things:

- The ability of the teacher, i.e., his method in teaching is good or bad and this is represented by a factor $\dot{\tau}_{\mathrm{f}}$ called "teaching factor," it can be 1 or 2 (those values are concluded from experiments).
- The ability of the student to receive and understand concepts from his teacher.

Learner phase: as known, when a student does not understand his teacher or he wants to have more knowledge, he will interact with one of their fellow students. If he finds his friend better than himself, he will learn from him otherwise he will not.

60.3.2 BTLBO for Optimal PMU Placement

Using the BTLBO, an algorithm for the optimal placement of PMUs can be obtained:

1. Build the binary connectivity matrix using the one line diagram and constraint modification when injection, flow measurements or already existing PMUs.
2. Define the optimization parameters; the population size (pop_size), the design variables (N_buses), and the number of generation (N_gen), the elite size (elite_size).
3. Generate random solutions within boundaries of the system.
4. Check that random solutions satisfy the inequality constraint of buses $f(x) = (A \times X) \geq \hat{i}$, modify those not satisfying the constraints.
5. Calculate the fitness of each solution based on the objective function of Eq. (60.1).
6. Set the best solution as teacher of the population.
7. For each student apply teacher phase, replace x by x_{new} if it gives better fitness function (less) otherwise keep the old one.
8. If the duplicate solutions exist, then it is necessary to modify the duplicate solutions in order to avoid trapping in the local optima. In the present work, duplicate solutions are modified by mutation on randomly selected dimensions of the duplicate solutions before executing the next generation without losing the observability.
9. For each student apply learner phase, replace x by x_{new} if it gives better fitness function (less) otherwise keep the old one.
10. Remove duplicate solutions keeping the constrained satisfied (observability).
11. Replace (elite_size) bad solutions by (elite_size) elite solutions.

12. Again, remove duplicate solutions keeping the constrained satisfied (observability), then determine the teacher.
13. Repeat from step (7) for maximum number of iterations.
14. Set the best solution $x_{teacher}$ as the final solutions.

60.4 Results and Discussions

This section contains various simulation examples about the PMU placement algorithm presented earlier, which are carried out using the standards IEEE power systems with 14-bus, 30-bus, 57-bus, and 118-bus shown in Figs. 60.1, 60.2, 60.3, and 60.4, respectively. MATLAB software has been used to solve the optimal placement problem.

60.4.1 Effect of Considering Zero Injections

Two sets of simulations have been carried out on the IEEE power systems, which are initially assumed to have no flow measurements and no existing PMUs. In the first set of simulations, zero injection is simply ignored while in the second set, it is used as existing measurement.

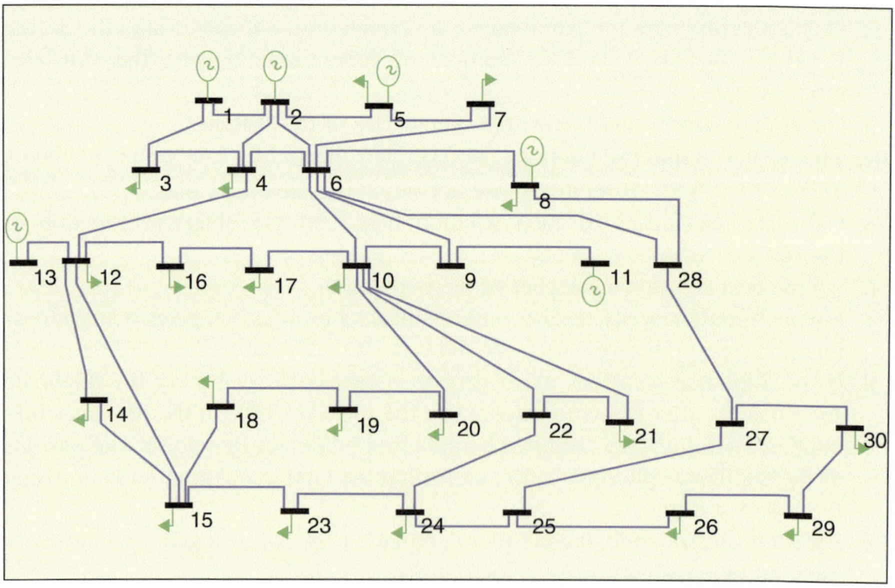

Fig. 60.2 The IEEE 30 bus test system

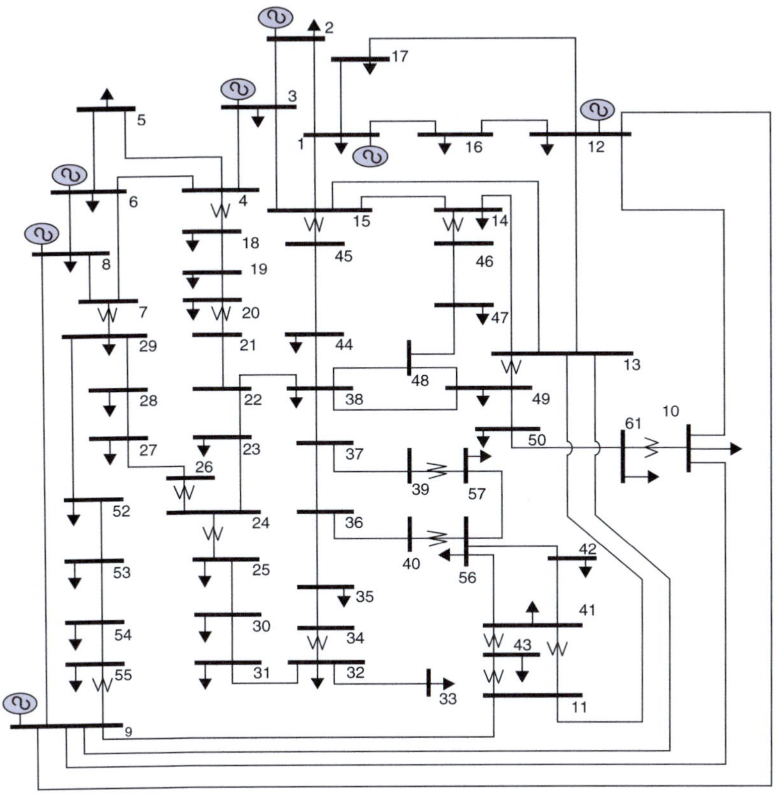

Fig. 60.3 The IEEE 57 bus test system

At first, consider the four systems with and without zero injection buses. The specifications regarding the number and locations of the zero injection buses are given in Table 60.1. The outcomes of the BTLBO algorithm are shown in Table 60.2.

It should be noted that in both situations, the BTLBO converged to the optimal set of the PMUs to place in the system satisfying the constraint of full observability. The results show a reduction in the number of PMUs required when considering zero injections. In some situations, like the case of the IEEE 118-bus power system, a significant reduction in the number of PMUs has been noticed. This can be considered to be a huge cut on the budget allocated for the deployment of the PMUs in the system.

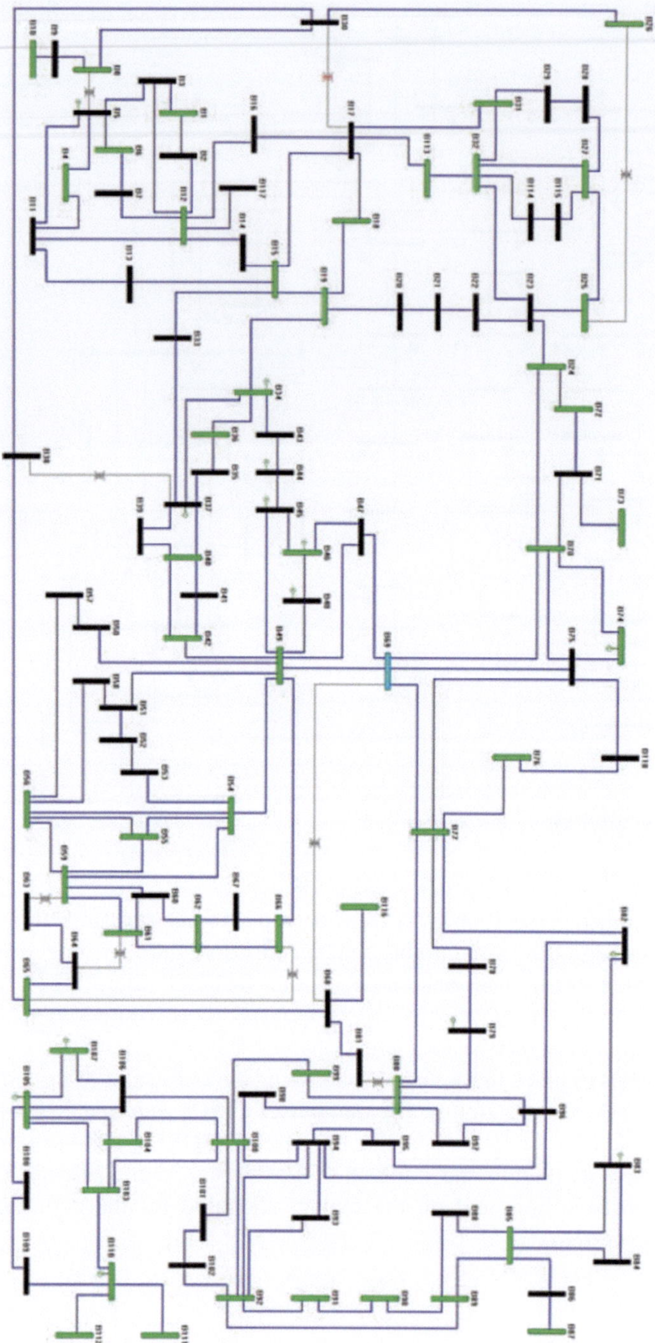

Fig. 60.4 The IEEE 118 bus test system

Table 60.1 Zero injection bus specifications for the different systems

System	Number of branches	Number of zero injections	Zero injection buses
IEEE 14-bus	20	1	7
IEEE 30-bus	41	5	6, 9, 11, 25, 28
IEEE 57-bus	78	15	4, 7, 11, 21, 22, 24, 26, 34, 36, 37,39, 40, 45, 46, 48
IEEE 118-bus	179	10	5, 9, 30, 37, 38, 63, 64, 68, 71, 81

Table 60.2 BTLBO optimal PMU placement for the different systems

System	Number and location of PMUs without injection	Number and location of PMUs with injection
IEEE 14-bus	4: (2,6,8,9)	3: (2,6,9)
IEEE 30-bus	10: (1, 5, 10, 11, 12, 18, 23, 26, 28, 30)	7: (1, 5, 10, 12, 18, 20, 27)
IEEE 57-bus	17: (1, 6, 9, 15, 19, 20, 24, 25, 28, 32, 36, 38, 41, 46, 50, 53, 57)	11: (1, 5, 13, 19, 25, 29, 32, 38, 41, 51, 54)
IEEE 118-bus	32: (1, 5, 9, 12, 13, 17, 21, 25, 29, 34, 37, 42, 45, 49, 52, 56, 62, 64, 70, 71, 76, 77, 80, 85, 87, 91, 94, 101, 105, 110, 114, 116)	28: (3, 8, 11, 12, 17, 21, 27, 31, 32, 34, 39, 42, 45, 49, 53, 56, 62, 72, 75, 77, 80, 85, 87, 90, 94, 102, 105, 110)

60.4.2 Effect of Considering Existing PMUs

Here, simulations have been carried out using IEEE 57-bus system. Three PMUs are added in the system one at a time. The locations of these PMUs in the system are assumed to be placed at buses 1, 9, and 29. Simulation results are shown in Table 60.3.

It should be noted that the required number of PMUs is reduced from 11 to 9. Hence, PMU measurements reduce the number of required PMUs to make the entire system observable.

60.4.3 Comparison of Results of BTLBO with Different Algorithms

For the sake of comparative evaluation, the presented approach is compared with some of the well-known conventional and metaheuristic algorithms used to solve the problem.

Table 60.3 BTLBO optimal PMU placement with existing PMUS at different locations

Existing PMU	Number of required PMUs	Location of PMUs
None	11	1, 4, 10, 17, 19, 22, 26, 28, 29, 36, 39
1 (bus 1)	10	5, 13, 19, 25, 29, 32, 38, 41, 51, 54
2 (bus 1 and 9)	10	5, 13, 19, 28, 30, 33, 38, 41, 51, 53
3 (bus 1, 9, and 29)	9	5,13, 20, 25, 32, 38, 41, 51, 53

Table 60.4 Comparison between the BTLBO optimal PMU placement and some state-of-the-art techniques

Test systems	14-Bus	30-Bus	57-Bus	118-Bus
Binary TLBO algorithm (present work)	*3*	*7*	*11*	*28*
ABC optimization	3	7	13	29
Integer programming	3	8	12	28
Topology-based genetic algorithm	3	7	11	29
Binary PSO algorithm	3	7	13	29

Integer linear programming presented in [16] and [18] has been implemented using the IEEE 14-bus, 30-bus, 57-bus, and 118-bus systems. TOMLAB/MINLP and MILP software package has been used to solve the Integer Linear/Nonlinear Programming problem. The approach is also compared with some other algorithms such as Tabu search [8], branch and bound [6], non-dominated sorting genetic algorithm, graph theoretic procedure, dual search, and artificial bee colony optimization [19]. Table 60.4 shows the comparison study results.

It is observed from Table 60.4 that the proposed method and all the other methods lead to the same number of PMUs for system observability for the IEEE 14-bus system. For IEEE 30-bus system, the performance of the Binary TLBO, GA, and Binary PSO are identical and these algorithms produce better results than integer linear programming. For 57-bus system, the performance of the Binary TLBO and GA are alike, and BTLBO outperforms the Binary PSO algorithm. For 118-bus system, BTLBO and integer programming are similar and BTLBO outperforms GA and binary PSO algorithms.

Overall, the BTLBO converged to the best set of PMUs for all the cases while the other algorithms converged for some cases and failed in the others.

60.5 Conclusions

This work attained two objectives. First, the TLBO method is investigated and a proposed binary version is developed. Second, the developed method is applied for determining optimal locations for PMUs to ensure complete system observability.

TLBO has a different way to improve the level of the population, first it develops students by developing the best student in the class and then it allows students to interact between each other in order to learn from each other randomly. So, TLBO tries first to find the optimal solution in the vicinity of the teacher like ACO after searching between the others like GA and PSO. Besides, BTLBO has been developed in this project starting from the origin TLBO in the real space.

Placement of PMUs can be carried out using different criteria depending on the objective of the investigator. In this chapter, the main focus is to make the entire system observable by optimal placement of PMUs. Various scenarios have been considered when the system is assumed to be observable by placing PMUs only. While this appears impractical today, it may be applied in the future when these devices become standard equipment. Besides, the placement problem is considered for a system with existing measurements using PMUs. Case studies which are carried out on IEEE test systems indicate that strategically PMUs placed at roughly one third of the system buses, the entire system can be observable with only PMUs. Furthermore, zero injections, and/or existing PMU can significantly reduce the required number of PMUs for a given power system.

References

1. Phadke AG, Thorp JS (2008) Synchronized phasor measurements and their applications. Springer, Boston
2. Sharma NK, Tiwari AN, Verma KS, Singh SN (2011) Applications of phasor measurement units (PMUs) in electric power system networks incorporated with facts controllers. Int J Eng Sci Technol 3(3):64–82
3. Baldwin TL, Mili L, Boisen MB, Adapa R (1993) Power system observability with minimal phasor measurement placement. IEEE Trans Power Syst 8(2):707–715
4. Mosavı M-R, Akhyanı A-A, Rahmati A (2012) A PMU placement optimal method in power systems using modified ACO algorithm and GPS timing. Electr Rev 88(8):346–349
5. Xu B, Abur A (2005) Optimal placement of phasor measurement units for state estimation. Final project report, PSERC
6. Ahmadi A, Alinejad-Beromi Y, Moradi M (2010) Optimal PMU placement for power system observability using binary particle swarm optimization and considering measurement redundancy. Electr Power Syst Res 89:1–10
7. Kazem M, Vahedi H (2011) Optimal placement of PMUs in power systems using mixed integer non linear programming and bacterial foraging algorithm. ECTI Trans Electr Eng Electron Commun 9(1):150–156
8. Jiangnon P, Yuanzhang S, Wang HF (2006) Optimal placement for full network observability using Tabu search algorithm. Int J Electr Power Energy Syst 23:1525–1526
9. Melosovic B, Begovic M (2003) Nondominated sorting genetic algorithm for optimal phasor measurement placement. IEEE Trans Power System 18:69–75
10. Marın FJ, Garcıa-Lagos F, Joya G, Sandoval F (2003) Genetic algorithms for optimal placement of phasor measurement units inelectrical networks. Electron Lett 39(19):1403–1405
11. Rao RV, Savsani VJ, Vakharia DP (2011) Teaching-learning-based optimization: a novel method for constrained mechanical design optimization problems. Comput Aided Design 43(3):303–315

12. Rao RV, Savsani VJ, Vakharia DP (2012) Teaching-learning-based optimization: an optimization method for continuous non-linear large scale problems. Inform Sci 183(1):1–15
13. Rao VJ, Savsani JB (2012) Teaching learning based optimization algorithm for constrained and unconstrained real parameter optimization problems. Eng Optim: doi:10.1080/0305215X.2011.652103
14. Rao RV, Patel VK (2012) Multi-objective optimization of combined Brayton and inverse Brayton cycles using advanced optimization algorithms. Eng Optim: doi:10.1080/0305215X.2011.624183
15. Xu B (2006) Optimal monitoring and visualization of steady state power system operation. Ph.D. Dissertation, Texas A&M University, College Station
16. Dambhare S, Dua D, Gajbhiye RK, Soman SA (2008) Optimal zero injection considerations in PMU placement: an ILP approach. Indian Institute of Technology, Karnataka
17. Bei X, Yoon YJ, Abur A (2008) Optimal placement and utilization of phasor measurements for state estimation. Texas A&M University, College Station
18. Mohammad S, Javidi MH (2011) A survey on topological observability of power systems. In: IEEE Power Engineering and Automation Conference, Wuhan, 8–9 Sept 2011
19. Mohammadi-Ivatloo B (2009) Optimal placement of PMUs for power system observability using topology based formulated algorithms. J Appl Sci 9(13):2463–2468
20. Tankasala GR, Sanisetty S, Vala VK (2012) Optimal placement of phasor measurement units for state estimation using artificial intelligence techniques. Int J Sci Eng Res 3(2):1–5

Chapter 61
Innovative Process of Essential Oil Extraction: Steam Distillation Assisted by Microwave

Naima Sahraoui and Chahrazed Boutekedjiret

Abstract This study focuses on the extraction of essential oil of *Thymus pallescens*, using a new extraction process developed in our laboratory: steam distillation assisted by microwaves also called microwave steam distillation (MSD). This process is a combination of traditional techniques, namely, a steam distillation (SD) and an innovative technology, a microwave heating. Indeed, heating by microwaves helped initiate and build the mass and heat transfer inside the plant matrices outward which results in a considerable reduction in the extraction time. Kinetic study of extraction, optimization of operating conditions, and antioxidant activity of essential oil were conducted. The selected operating parameters are the steam flow rate and the microwave heating power.

The yield obtained by microwave steam distillation is comparable to that obtained by the conventional steam distillation, while the extraction time is greatly reduced: 5 min for MSD extraction against 20 min for the SD. The best performance was obtained with a power of 400 W and steam flow rate of 10 g·min^{-1}. Determination of antioxidant activity by DPPH test of *Thymus pallescens* essential oil obtained by both processes showed that the method assisted by microwaves has an influence on the ability of essential oil to inhibit DPPH radical. Indeed, the essential oil extracted by this process is more effective than that obtained by conventional steam distillation. Moreover, the inhibition rate of DPPH radical is more important for essential oil than that of synthetic antioxidant (BHT and BHA) for concentrations higher than 400 mg·L^{-1}. In addition, the values obtained show that the essential oil has a better antioxidant activities in comparison with its main component, namely, the carvacrol.

N. Sahraoui (✉)
Laboratoire des Sciences et Techniques de l'Environnement, École Nationale Polytechnique, BP 182, El Harrach, Alger 16200, Algeria

Département de Génie des Procédés, USTHB, BP 32, El Alia, Bab Ezzouar, Alger 16111, Algeria
e-mail: sahraouinaima65@yahoo.fr

C. Boutekedjiret
Laboratoire des Sciences et Techniques de l'Environnement, École Nationale Polytechnique, BP 182, El Harrach, Alger 16200, Algeria
e-mail: chahrazed.boutekedjiret@g.enp.edu.dz

I. Dincer et al. (eds.), *Progress in Clean Energy, Volume 1*,
DOI 10.1007/978-3-319-16709-1_61

Keywords Microwave extraction • Green process • Essential oil • Antioxidant activity

Nomenclature

AA	The absorption of the tested oil or substance solution at $t = 20$ min
AB	Absorption of the blank sample at $t = 0$ min
BHA	Butylated hydroxyanisole
BHT	Butylated hydroxytoluene
DPPH	1,1-Diphenyl-2-picrylhydrazyl
EOs	Essential oils
I (%)	Inhibition of free radical DPPH
K_E	Extraction rate constant (min^{-1})
MSD	Microwave steam distillation
P	Microwave power (W)
Q	Steam flow rate ($g \cdot min^{-1}$)
RI^a	Retention indices cited in the literature
RI^b	Retention indices relative to C_5–C_{28} *n*-alkanes on HP5MS capillary column
SD	Steam distillation
Y_{EOt}	Essential oil yield at time *t* (gEO/g dry matter)
$Y_{EO\infty}$	Essential oil yield at infinite time (gEO/g dry matter)

61.1 Introduction

The advent of new technical, regulatory constraints of food security and the growing consumer interest in natural products strongly stimulate research of "clean" process allowing the extraction of new ingredients with high added value.

Traditional processes for extracting essential oils, hydrodistillation and steam distillation, are relatively simple, but have several disadvantages: long operating time, significant energy consumption besides the risk of thermal degradation, and hydrolysis of aromatic molecules that can give the essential oil a different smell from that of the raw material. Conventional solvent extraction processes require very long time of treatment and can cause the loss of volatile compounds upon removal of the solvent and the presence of residual solvent in the extracts [1, 2].

Because of the disadvantages of these conventional techniques, food manufacturers have, for several years, focused their research to new processing technologies. Combined with the evolution of technology, this research aims to increase the extraction process, reduce energy consumption and amounts of solvents while ensuring product quality, and meet stringent environmental criteria.

Among the techniques developed in recent years, extraction methods assisted by microwave can be an alternative to conventional methods due to the quality of the extracts recovered, reducing processing times and energy consumption [3].

Steam distillation assisted by microwaves (MSD) is a new extraction process developed in our laboratory [4, 5]. This process is a combination of traditional techniques, namely, a steam distillation (SD) and an innovative technology, a microwave heating. Indeed, heating by microwaves helped initiate and build the mass and heat transfer inside the plant matrices outward which results in a considerable reduction in the extraction time.

The aim of this work was to investigate the efficiency of this method, for the extraction of thyme essential oil in order to increase the quality of the extracts while reducing the costs of exploitation (reduction of processing time, energy consumption). The effectiveness of this innovative method in extraction of thyme essential oil has been evaluated and compared to conventional steam distillation in terms of extraction time and yield. On the other hand, the antioxidant activity of thyme oil extracted by both processes was evaluated and compared for that of synthetic antioxidant, namely, BHA and BHT, and carvacrol witch is the main component of essential oil.

61.2 Extraction Procedure

61.2.1 Plant Material

The plant material used for this study was collected from "Ain Defla" area located 160 km west of Algiers (Algeria) and identified according to the Flora of Algeria [6], at the Agronomic Higher National School of Algiers, as being *Thymus pallescens*. The leaves were air-dried, and the moisture content, determined by Dean-Stark method [7], was 12.5 %.

61.2.2 Apparatus and Procedure

Two processes were used to extract thyme essential oil: microwave steam distillation (MSD) and conventional steam distillation (SD).

61.2.2.1 Microwave Steam Distillation Apparatus and Procedure

MSD apparatus (Fig. 61.1) is a multimode microwave reactor operating at 2.45 GHz with a maximum delivered power of 1,000 W variable in 10 W increments. During experiment time, temperature and power were recorded and controlled. An electrical steam generator and a condenser placed outside the microwave oven are connected to a cartridge containing *Thymus pallescens* via Pyrex connecting tubes. The condenser is connected to a receiving Florentine flask which is preferably a separating funnel to enable the continuous collection of condensate essential oil and water. The cartridge containing 20 g of *Thymus pallescens* is subjected to microwave heating; steam passes through the sample,

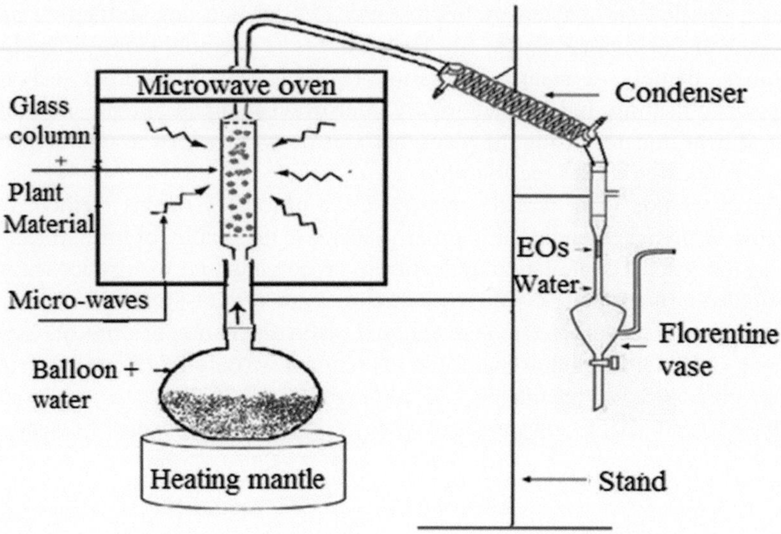

Fig. 61.1 Microwave steam distillation apparatus

evaporates, and carries the essential oil directed toward the condenser and the Florentine flask. The extraction was continued until no more essential oil was obtained. The essential oil is collected, dried with anhydrous sodium sulfate, and stored at 4 °C until used. Extractions were performed at least three times, and the mean values were reported.

61.2.2.2 Steam Distillation Apparatus and Procedure

For rigorous comparison, the same glassware and same operating conditions have been used for conventional steam distillation (the same process but without the use of microwaves). In this system (Fig. 61.2), the vapor produced by the steam generator crosses the plant, charged with the essential oil, and then passes through the condenser to a receiving Florentine flask. The extraction was continued until no more essential oil was obtained. The essential oil is collected, dried with anhydrous sodium sulfate, and stored at 4 °C until used. Extractions were performed at least three times, and the mean values were reported.

61.2.3 Parametric Study

In order to study the role of the factors involved in MSD of thyme essential oil, optimization of operating conditions was carried out using a parametric study. The main factors are steam flow rate Q (g min^{-1}) and microwave power P (W).

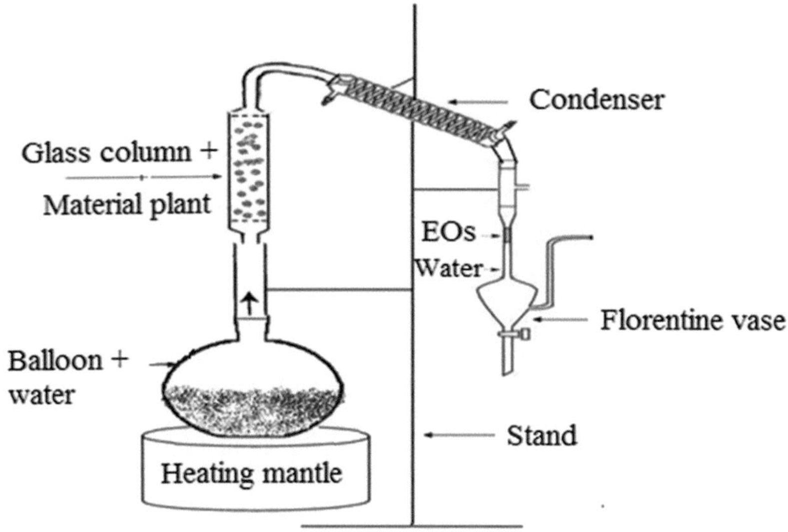

Fig. 61.2 Steam distillation apparatus

61.2.4 Scavenging Effect on DPPH Radical

The antioxidant activity of thyme oils and tested substances (namely, carvacrol, thymol, and eugenol which are the main components of thyme essential oil and synthetic antioxidants BHA and BHT) was measured in terms of hydrogen-donating or radical-scavenging ability, using the stable radical 1,1-diphenyl-2-picrylhydrazyl (DPPH) as a reagent [8, 9]. Fifty microliters of various concentrations of the sample in methanol (essential oil, component of essential oil, and control substance) were added to 2 mL of a 6 μM methanolic solution of DPPH. Absorbance measurements were read at 517 nm after 20 min of incubation time at room temperature. Absorption of a blank sample containing the same amount of methanol and DPPH solution acted as the negative control. All determinations were performed in triplicate. The percentage inhibition of the DPPH radical by the samples was calculated according to the formula Eq. (61.1):

$$\%\,\text{Inhibition} = [(AB - AA)/AB] \times 100 \tag{61.1}$$

where AB is the absorption of the blank sample at $t = 0$ min and AA is the absorption of the tested oil or substance solution at $t = 20$ min [10].

61.3 Results and Discussion

61.3.1 Parametric Study

Several steam flow rates (6, 8, 10, 12, 14, and 16 g·min^{-1}) and power (200, 300, 400, and 500 W) were tested. From the results of Table 61.1, we find that the increase in the flow of steam water has a negative effect on the essential oil yield. Indeed, this later increases to 2.35 ± 0.07 % for a flow rate passing from 4 to 10 g·min^{-1} and then decreases as the steam flow increases beyond 10 g·min^{-1}. A high flow can involve the creation of a bypass, having for consequence a reduced contact between the vapor and the plant and thus decreasing the yield of essential oil. A low flow would be insufficient to extract the totality of essential oil. The optimum was obtained with vapor flow of 10 g·min^{-1}.

Thyme essential oil yield evolution increases up to an optimum value with the power augmentation. Thus, when the power increases from 200 to 400 W, the yield of *Thymus pallescens* essential oil increases from 1.48 ± 0.05 % to 2.45 ± 0.06 %, respectively, and then decreases for the rest of the power values. An appropriate microwave irradiation power is important to ensure that the essential oil is extracted quickly; however, the power should not be too high that would involve the destruction of the vegetable matter. The best performance is obtained for 400 W.

61.3.2 Essential Oil Yield and Kinetic Extraction

The yields obtained by the two processes are similar: 2.45 ± 0.10 % for MSD and 2.35 ± 0.10 % for SD. The difference observed is the duration of extraction. An extraction time of 5 min by MSD provides yields comparable to those obtained after 20 min by SD. Yield is expressed in gram of essential oil per gram of dry matter.

Figure 61.3 shows the variation of the extraction yield according to the extraction time. We can observe that the extraction kinetics of essential oil for both MSD and SD processes have the same extraction profile characterized by two distinct phases. However, extraction speeds are different depending on the process used. The first step which corresponds to the essential oil extraction is faster for MSD

Table 61.1 Results of a parametric study

Q (g·min^{-1})	Yield (%)	P(W)	Yield (%)
4	1.36 ± 0.07	200	1.48 ± 0.05
6	1.50 ± 0.05	300	1.98 ± 0.06
8	1.86 ± 0.06	400	2.45 ± 0.06
10	2.35 ± 0.07	500	2.35 ± 0.05
12	2.00 ± 0.07	700	1.46 ± 0.07
14	1.67 ± 0.08		
16	1.20 ± 0.05		

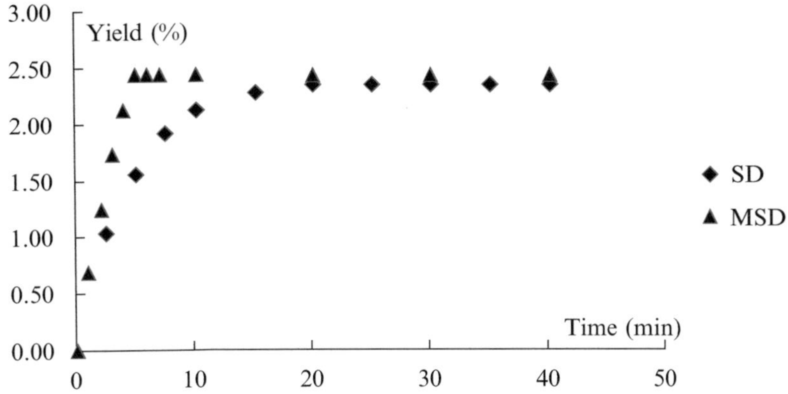

Fig. 61.3 Yield profile of essential oil obtained by MSD and SD as function of time

process with 5 min against 20 min for SD. The rapid increase in the yield during the first step for MSD extraction suggests that the essential oil is easily accessible by the steam. Indeed, the microwave radiations distend the plant cells and lead to the rupture of the glands and cell receptacles. The steam passes easily through the sample, evaporates, and carries the essential oil. In the case of the SD, the cell oils are not broken; the water vapor must diffuse through the walls of the cells to extract oil. The phenomenon of diffusion is slower that slowed down the kinetics of extraction. The second phase which corresponds to a horizontal line marks the end of the extraction.

In order to confirm this result, we determined the extraction rate of the two processes. For that, we assumed that the kinetics of extraction followed a first-order law where the extraction rate is given by an Eq. (61.2):

$$-Ln\left[1-\frac{Y_{EOt}}{Y_{EO\infty}}\right]=K_Et \tag{61.2}$$

where

K_E the extraction rate constant (min^{-1})
Y_{EOt} the essential oil yield at time t (gEO/g dry matter)
$Y_{EO\infty}$ the essential oil yield at infinite time (gEO/g dry matter)

Figure 61.4 representing the variation of $\left[-Ln\left[1-\frac{Y_{EOt}}{Y_{EO\infty}}\right]=K_Et\right]$ according to time highlights a line with a slope K_E for the two processes, showing that the extraction speed evolves according to first-order law. The extraction rate of MSD equal to 0.46 min^{-1} is twice more significant than that of the SD (0.23 min^{-1}); therefore, MSD is clearly faster than conventional SD with an undeniable gain of time of 75 %.

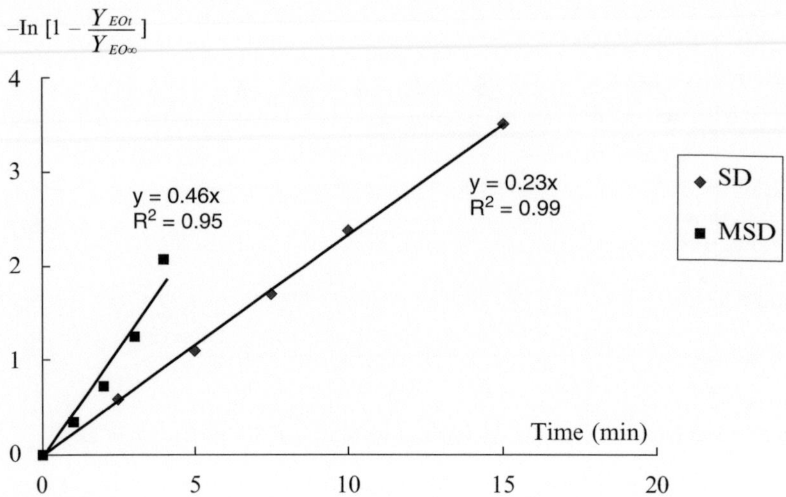

Fig. 61.4 Evolution versus time of $\left[-Ln\left[1-\frac{Y_{EOt}}{Y_{EO\infty}}\right]=K_Et\right]$ for MSD and SD

61.3.3 Essential Oil Composition

Thymus essential oil obtained by MSD and SD was analyzed by GC and GC-MS and compared in terms of chemical composition and relative amount of each component. The results are shown in Table 61.2. It was observed that the chemical compositions of the oils extracted by the two processes are qualitatively comparable. Nevertheless, differences are observed in the contents of the identified components. Carvacrol is the main component in both cases with 77.78 % for MSD and 86.70 % for SD. The contents of ortho-cymene and eugenol are more significant in the oil recovered by MSD 4.47 % against 0.14 % for ortho-cymene and 2.26 % against 0.29 % for eugenol.

61.3.4 Antioxidant Activity

As it is shown in Table 61.3, evolution of antioxidant activity depending on the concentration of thyme essential oil extracted by MSD is important. It increases with the concentration and reaches 95.65 ± 0.52 % for a concentration of 1,000 mg·L^{-1}. This oil is more effective than that obtained by SD. On the other hand, we can see that it is also more effective than its majority compounds (carvacrol, thymol, and eugenol) and synthetic antioxidants (BHT and BHA) for concentrations higher than 400 mg·L^{-1}.

The difference observed in the antioxidant activity of essential oil obtained by MSD and SD could be due to the differences in the chemical composition of these oils.

Table 61.2 Chemical composition of *Thymus pallescens* EOs obtained by MSD and SD

			SD		MSD	
N°	Compound	RI[a]	RI[b]	(%)	RI[b]	(%)
1	Alpha-pinene	926	929	0.02	928	0.18
2	1-Octen-3-ol	978	974	0.19	974	0.19
3	Beta-pinene	980	982	0.08	981	0.09
4	Myrcene	991	992	0.06	992	0.06
5	Ortho-cymene	1,022	1,021	0.14	1,023	4.47
6	Para-cymene	1,026	1,026	0.19	1,026	0.14
7	Limonene	1,031	1,028	0.04	1,028	0.04
8	Gamma-terpinene	1,062	1,064	0.08	1,064	0.08
9	1-Nonen-3-ol	1,072	1,077	0.15	1,077	0.16
10	Terpinolene	1,088	1,095	0.04	1,086	0.06
11	Linalool	1,098	1,100	3.37	1,101	3.32
12	Trans-pinocarveol	1,139	1,135	0.08	1,135	0.11
13	Borneol	1,167	1,162	0.19	1,162	0.15
14	1-Terpinen-4-ol	1,177	1,174	0.34	1,174	0.31
15	Carvacrol methyl ether	1,244	1,239	0.62	1,240	0.88
16	Thymol	1,290	1,287	1.93	1,287	1.5
17	Carvacrol	1,298	1,312	86.70	1,315	77.78
18	Eugenol	1,356	1,354	0.29	1,356	2.26
19	α-Copaene	1,376	1,375	0.04	1,376	0.17
20	β-Bourbonene	1,384	1,384	0.03	1,385	0.05
21	α-Gurjunene	1,409	1,408	0.4	1,410	0.51
22	β-Caryophyllene	1,418	1,418	0.2	1,420	0.55
23	β-Gurjunene	1,432	1,432	0.36	1,434	0.05
24	Aromadendrene	1,439	1,437	0.49	1,438	1.01
25	α-Humulene	1,454	1,455	0.27	1,457	0.25
26	γ-Muurolene	1,477	1,472	0.1	1,474	0.18
27	Viridiflorene	1,493	1,490	0.21	1,492	0.44
28	α-Muurolene	1,499	1,495	0.03	1,496	0.05
29	γ-Cadinene	1,513	1,508	0.13	1,509	0.21
30	δ-Cadinene	1,524	1,517	0.26	1,518	0.27
31	Spathulenol	1,576	1,575	0.45	1,577	0.20
32	T-Cadinol	1,640	1,634	0.05	1,632	0.07
Total				97.53		95.79

Note:
[a]Retention indices cited in the literature [11, 12]
[b]Retention indices relative to C_5–C_{28} *n*-alkanes on HP5MS capillary column. The components are listed in order of elution from the HP5MS column

In addition, the fact that essential oil is more effective than its majority compounds shows that its antioxidant activity would not be due only to its majority compounds but probably to a synergism between these constituents and other components in small amounts.

Table 61.3 Antioxidant activity of *Thymus pallescens* EOs obtained by SD and MSD. Carvacrol, thymol, eugenol, BHT, and BHA

C ($mg{\cdot}L^{-1}$)	Inhibition of free radical DPPH: I (%)						
	SD	MSD	Carvacrol	Thymol	Eugenol	BHT	BHA
100	39.38 ± 0.30	62.40 ± 2.82	49.24 ± 0.28	46.59 ± 0.22	85.52 ± 1.42	91.11 ± 0.30	88.07 ± 1.03
200	55.92 ± 0.41	79.71 ± 0.54	54.77 ± 2.74	55.89 ± 1.23	87.34 ± 1.09	91.53 ± 1.04	88.41 ± 0.51
400	61.45 ± 0.49	92.01 ± 0.78	67.42 ± 0.65	71.74 ± 0.55	87.62 ± 0.98	91.75 ± 0.35	88.76 ± 0.49
600	91.59 ± 0.25	94.83 ± 0.90	72.39 ± 0.35	75.32 ± 0.25	88.53 ± 1.10	92.86 ± 0.36	90.80 ± 1.06
800	93.41 ± 0.29	95.45 ± 0.17	75.82 ± 0.57	79.19 ± 0.49	89.50 ± 0.78	93.04 ± 1.82	91.20 ± 0.89
1,000	93.52 ± 0.92	95.65 ± 0.52	79.23 ± 1.98	86.35 ± 0.85	90.21 ± 0.55	93.13 ± 0.11	91.51 ± 0.02

The data obtained from the evaluation of antioxidant activity of *Thymus pallescens* in this research are highlighting the considerable potential of this plant as an antioxidant food additive.

61.4 Conclusion

The aim of this work was to confirm the efficiency of the microwave in the extraction process of essential oil and to explain how they highly accelerated the extraction and improve the properties and quality of the essential oil. The kinetic study showed that microwave steam distillation is most powerful than conventional steam distillation with a profit of 75 % in terms of duration of extraction and gives an essential oil having a better antioxidant activity highlighting the considerable potential of this oil as a natural antioxidant additive.

Acknowledgment Authors wish to thank sincerely Prof. Farid Chemat, the Director of GREEN Extraction Team from Université d'Avignon et des Pays de Vaucluse, INRA, UMR 408 Avignon, France, to have made us benefit from his competences in the field of the microwave-assisted extraction.

References

1. Chemat F (2011) Eco-Extraction du Végétal: procédés innovants et solvants alternatifs. Dunod, Paris, p 336. ISBN 978-21-005654-3-6
2. Lebovka N, Vorobiev E, Chemat F (2011) Enhancing extraction processes in the food industry. CRC, Cambridge, p 570. ISBN 978-14-398459-3-6
3. Chemat F, Cravotto G (2013) Microwave-assisted extraction for bioactive compound. Springer, NewYork, p 238. ISBN 978-1-4614-4829-7
4. Sahraoui N, Abert Vian M, Bornard I, Boutekedjiret C, Chemat F (2008) Improved microwave steam distillation apparatus for isolation of essential oils. Comparison with conventional steam distillation. J Chromatogr A 1210:229–233
5. Sahraoui N, Abert Vian M, El Maataoui M, Boutekedjiret C, Chemat F (2011) Valorization of *citrus* by-products using microwave steam distillation (MSD). Innov Food Sci Emerg Technol 12(2):163–170
6. Quezel P, Santa S (1963) Nouvelle flores d'Algérie et des régions désertiques méridionales, vol 2. CNRS, Paris
7. Official Method JA 2A-46 (1993) American Oil Chemist' Society. Champaign
8. Burits M, Bucar F (2000) Antioxidant activity of *Nigella sativa* essential oil. Phytother Res 14:323–328
9. Sahin F, Cakmakci R, Kantar F (2004) Sugar beet and barley yields in relation to inoculation with N_2-fixing and phosphate solubilizing bacteria. Plant Soil 265:123–129
10. Mansouri A, Embarek G, Kokkalou E, Kefalas P (2005) Phenolic profile and antioxidant activity of the Algerian ripe date palm fruit (*Phoenix dactylifera*). Food Chem 89:411–420
11. Arctander S (1994) Perfume and flavor chemicals. Allured, Carol Stream
12. Adams RP (2007) Identification of essential oils components by gas chromatography/mass spectrometry, 4th edn. Allured, Carol Stream

Chapter 62
Dynamics Molecular Simulation of the Mechanical and Electronic Properties of Polyethylene/Nanotubes Nanocomposites

K. Kessaissia, A. Ali Benamara, M. Lounis, and R. Mahroug

Abstract Nanocomposites with polymer matrix are reinforced by SWNTs, one identified as excellent candidates for applications: electronics, photovoltaics, and mechanics.

In this direction this work presents a study of the interactions of the composite carbon nanotubes/polymers. In the first part, the effects of functionalization and chirality (the length and the diameter) of the nanotubes on the Young modulus and on interaction energies, one simulated and determined by molecular dynamics, and DFT hile basing itself on physical model R.V.E of the nanocomposites SWNT/polyethylene. Results of interaction energies and of the Young modulus (longitudinal and transverse) validate the tendency of the experimental results reported in the literature. An increase of lengths and a reduction in the diameters of the nanotubes imply an increase in interaction energies and of the Young modulus, which means good mechanical behavior nanocomposites.

Keywords Carbone nanotube • SWNT • RVE • DFT polyethylene • Elastic and mechanical properties • Young modulus

Nomenclature

AFM	Atomic force microscope (microscope à force atomique)
CVD	Chemical vapor deposition (dépôt chimique en phase vapeur)
DFT	Density functional theory (théorie de la fonctionnelle de densité)
LDA	Local density approximation (approximation densité locale)

K. Kessaissia (✉) • M. Lounis • R. Mahroug
Department of Technology, Faculty of Technology, University of Khemis Miliana, Ain Defla, Algeria
e-mail: karimaelctro_10d@yahoo.fr

A.A. Benamara
Department of Physics, Faculty of Sciences, University of Chlef, Chlef, Algeria

I. Dincer et al. (eds.), *Progress in Clean Energy, Volume 1*,
DOI 10.1007/978-3-319-16709-1_62

MEB	Microscope électronique à balayage
MET	Microscope électronique en transmission
MM	Mécanique moléculaire
MD	Dynamique moléculaire
MM3	Molecular mechanics force field 3 (champ de force de mécanique moléculaire
PMMA	Polyméthacrylate de méthyle
PmPV	Poly(*p*-phenylènevinylène-co-2,5-dioctoxy-*m*-phenylènevinylène)
PP	Polypropylène
PS	Polystyrène
PVAc	Polyvinyl acétate
P(s-BuA)	Poly(styrène-co-acrylate de butyle)
SWNT	Single-wall carbon nanotube (nanotube monoparoi)
MWNT	Multiwall carbon nanotube (nanotube multiparoi)
vdW	Van der Waals
VGCF	Vapor-grown carbon fiber (fibre de carbone vapo-déposée)
DMA	Analyse mécanique dynamique)
SEM	Scanning electron microscopy
CNRS	Centre National de la Recherche Scientifique
CRM	Centre de Recherche sur la Matière Divisée
CNT	Carbone nanotube
PEEK	Polyéther éther cétone
PES	Polyéther sulfone
PA	Polyamides
PC	Polycarbonate
ABS	Polyacrylonitrile/butadiène styrène
POM	Polyacétal

62.1 Introduction

The craze for nanomaterials is expected of their development. Special properties resulting from an increase in surface area and decreased length characteristics are considered. In the case of nanocomposite particulates, in addition to a conventional mixing effect, the addition of nano-sized particles of various matrices cause a change of properties, in particular mechanical, electrical, and of the matrix itself, and simultaneously effect greater capacity especially at low particle sizes. The understanding of phenomena that appear at the interface between the inclusions and the matrix, and the analysis of the influence of the characteristic dimensions of the various phases on the mechanical behavior is found to be of particular interest for the study of these materials.

The improved property of materials is a key industrial performance. More efficient materials reduce the amount of material used to perform a function (stiffness, transmission of current, heat, etc.). One solution is to combine materials with different properties to take advantage of each. These composites consist of two, three, or more elementary materials that each have a role and a function specifies a matrix which gives the overall shape and liaison between the various components and reinforcements that provide the technical functions [1–10].

In the past 10 years, new nanoscale objects also having a fiber-type long winding inspired many works on polymer matrix nanocomposites: carbon nanotubes.

Carbon is an almost essential chemical element that enters into the composition of many organic substances as well (amino acids) and also inorganic (carbides, carbonates). Prior to 1985 there were only two crystalline phases of the rare carbon, hard and insulating the more abundant is called the diamond, the other ductile and conductive called graphite.

Carbon nanotubes are expected and will likely be one of the greatest technological revolution sources. Dice 70, micrographs of nanotubes are visible; they are called *hollow cores* because they are the central tube on which are deposited layers of carbon to form VGCF (vapor-grown carbon fibers, fiber carbon spray-marks). We paid little attention to detailed analysis of their structure by Iijima [1, 2] in 1991. This study marks the beginning of intensive research on carbon nanotubes The study of composite nanotube is in itself an area of research.

Regarding micromechanical models for electrical percolation properties, we conducted two studies on polyethylene matrix composites reinforced with carbon nanotube monoparpois (SWNTs) in different lengths.

62.2 Physical Model

Elastic constitutive models are developed for nanotube composites/polyethylene with and without functionalization, to predict the functionalization effect of the mechanical stiffness and elastic properties of nanocomposites [3]. We use the equivalent continuum models (molecular dynamics) for predicting the Young's moduli and shear moduli of the compounds of the amorphous and crystalline polyethylene matrix systems, and various length nanotubes. Figure 62.1 is a model of atomic structure by the functionalization of composite carbon nanotube/ polyethylene.

The purpose of the molecular model is to establish the molecular structure of the balanced system. Simulations of the molecular dynamics (MD) were therefore used to determine the equilibrium structures of the nanotube composite and polymer nanocomposites. Carbon nanotube-polymer considered in this work contains nanotubes of large and small lengths as shown in Figs. 62.2 and 62.3. The dashed boxes in the figure contain a representative volume element that is simulated by molecular dynamics [5, 6]. In each compound, the nanotubes are sufficiently separated to prevent the polymer interactions CNT–CNT. The nanotube composite in Fig. 62.1, for example, contains zigzag nanotubes (10.0) that periodically pass through the folded length of the simulation cell. In our model the nanotube is embedded in an amorphous matrix of polyethylene (PE), represented by units CH2-CH2. Specifically,

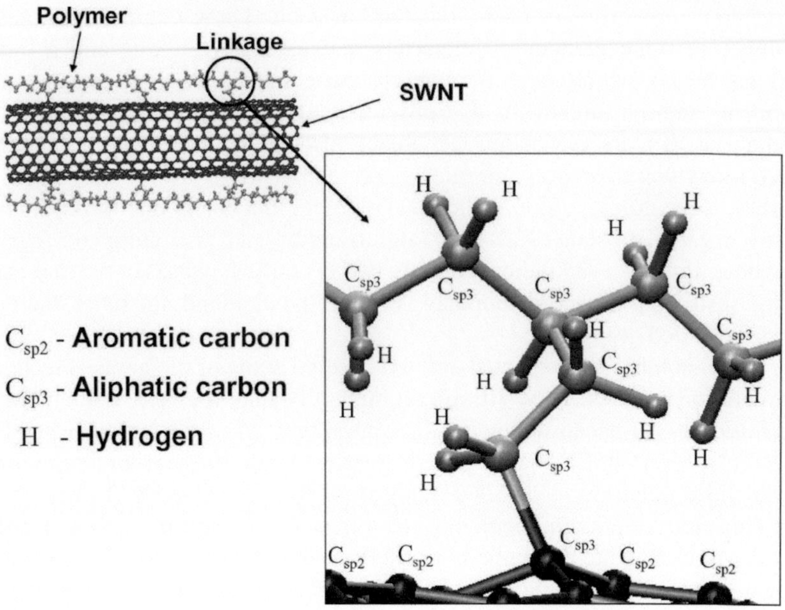

Fig. 62.1 Atomic modeling structure by the functionalization of the polyethylene/nanotube composite [8]

the polyethylene matrix contains six channels of 25-CH2 monomers. For example, Fig. 62.1 shows the application of a stress on the cell simulation for the different directions (x and y). The mechanical response of polymer-nanotube composites submitted to mechanical loading is provided by the stress–strain curves. In the current work, the stress–strain curves of the CNT composite polyethylene obtained by molecular dynamics simulations are presented [2, 13].

62.3 Molecular Mechanics and Dynamics

In a molecular dynamics simulation, the infinite crystal is represented by a cell of the same symmetry as the crystal, subject to periodic boundary conditions; individual movements of particles in the cell are studied by integrating Newton's equations [4].

As part of this work, we consider only the degrees of freedom of the position of Cartesian coordinates of the atoms. In classical simulations, the Hamiltonian molecular system takes the following form [5, 6].

$$H(p,r,m,s) = K(p,m) + v(r,s) \tag{62.1}$$

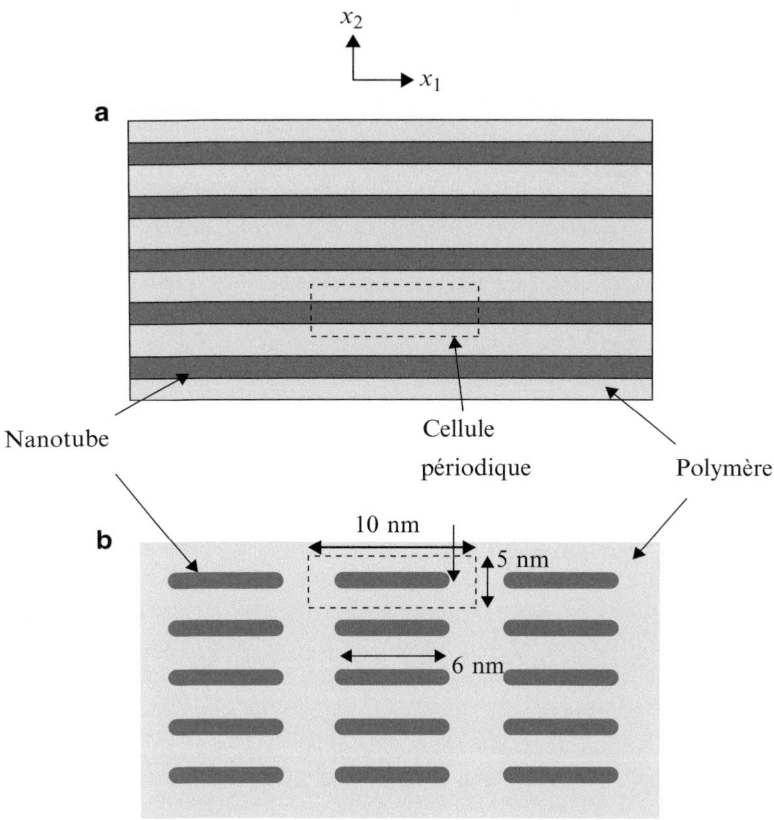

Fig. 62.2 Diagrams of polymer nanocomposite carbon nanotubes: (**a**) big length nanotubes (**b**) small length nanotubes [8]

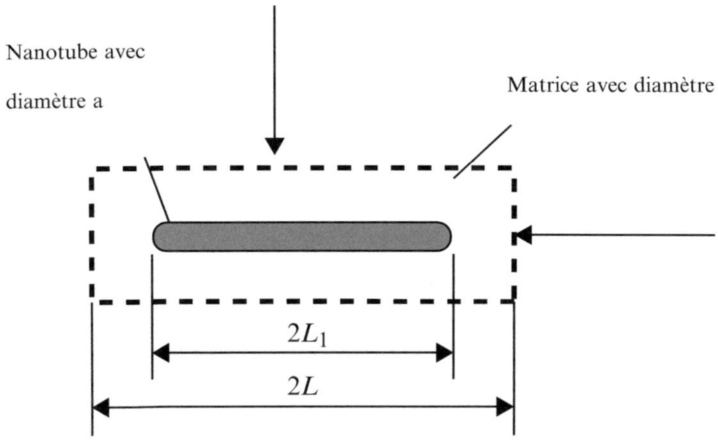

Fig. 62.3 Model (VER cell) nanocomposite [8]

The first term represents the kinetic energy

$$K(p,m) = \sum_{i=1}^{N} \frac{p_i^2}{2m_i} = \sum_{i=1}^{N} \frac{1}{2} m_i v_i^2 \tag{62.2}$$

The potential describing the interaction of the atoms in an organic material is given in many forms. In the system of interactions nanotube/polyethylene carbon involving only carbon and hydrogen, this potential is called the Terssoff–Brenner potential [7, 8] and is widely used for related interactions. It can be expressed by masses and is independent of the coordinates of the particle.

The second term is the potential energy, also called the interaction function (field strength), which describes the interaction energy as a function of particle coordinates r

$$v(r,s) = v(r_1, r_2, \ldots r_N; s_1, s_2, \ldots s_M) \tag{62.3}$$

$$v = \sum_{i} \sum_{j(>i)} \left[v_R(r_{ij}) - B_{ij} v_A(r_{ij}) \right] \tag{62.4}$$

Or r_{ij} is the distance between atoms i and j, V_R and V_A are the attractive and repulsive interactions, and B_{ij} the coupling between the atom i and the atom j for length and angles ENTERED atoms. The force f_i acting on a particle i is given by the following equation,

$$f_i = -\frac{\partial}{\partial r_i} v(r_1, r_2, \ldots r_N) \tag{62.5}$$

We use Newton's equations to describe the evolution of atoms in phase space. Other choices exist such as the Lagrange or Hamilton equations. Newton's equations are used in the form

$$\frac{dr_i(t)}{dt} = v_i(t) \tag{62.6}$$

and

$$\frac{dv_i(t)}{dt} = \frac{f_i(t)}{m_i} \tag{62.7}$$

The atomic velocities are indicated by v_i, the force acting on the atom i and the time t by $f_i(t)$. Newton's equations are only valid for the Cartesian coordinates of a particle r_i of mass m_i. In molecular dynamics simulations, the equations are "integrated" numerically in time.

The integration of these equations is done by dividing the path into a series of discrete states separated by very short intervals of time whose length defines the integration step.

The displacement of an atom during the time interval is thus given by Eq. (62.8).

$$\vec{r}_i(t+\Delta t) = \vec{r}_i(t) + \vec{V}_i\left(t + \tfrac{1}{2}\Delta t\right)\Delta t \tag{62.8}$$

It is then possible to determine the acceleration from Newton's second law and for each atom speed:

$$\vec{V}_i\left(t + {}^{\Delta t}/_{2}\right) = \vec{V}_i\left(t - {}^{\Delta t}/_{2}\right) + \vec{\gamma}_i(t)\Delta t \tag{62.9}$$

The speed determination provides the position of the atom with the equation previously set at the time $(t+\Delta t)$. Repeating this procedure at discrete time intervals depending on the speed of the identification results in the path [9].

62.3.1 Minimization of Energy

Molecular mechanics allows us to minimize the energy in order to obtain low-energy configurations of our molecular system and reduce too large initial forces that lead to an aberrant trajectory.

We used two methods of gradient and Newton's method using the potential energy and the following derivative with r. The disadvantage of these methods is that they follow almost exactly the gradient of potential energy by moving in the configuration space.

$$\Delta r_i \approx -\frac{\partial}{\partial r_i} v(r_1, r_2, \ldots r_N) \tag{62.10}$$

In this way, it moves almost exclusively down on the energy hypersurface, and the minimum usually reaches the local minimum closest to the initial configuration [14].

62.3.2 Molecular Mechanics

Molecular mechanics and molecular structure are likened to a set of harmonic oscillators (system composed of balls and springs); harmonic functions are associated with a series of potential functions [9]. The sum of these functions is then expressed in the form of a force field molecular.

$$E_{tot} = E_{liée} + E_{non-liée} \tag{62.11}$$

62.4 Results and Discussion

The composite studied is obtained by grafting the six channels of PE monomers each consisting of 25 (density 0.71 g/cm^3) in a zigzag; SWNT has different lengths of a temperature 298 K.

62.4.1 Minimization of Composites with Nanotubes of Different Lengths—Results

For minimization, we use three methods of minimization (steepest gradients, conjugate gradient, Newton) and we obtain the potential energy and the van der Waals energy (repulsive) increase with increasing lengths of carbon nanotubes. Figure 62.4 shows the change in potential energy and the van der Waals energy in the function of the lengths of carbon nanotubes.

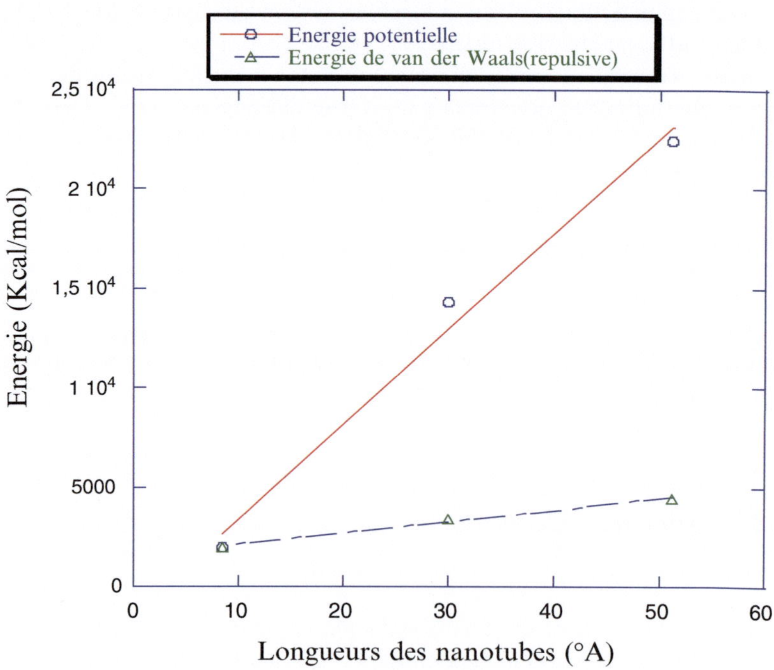

Fig. 62.4 Energy variation according to the lengths of the nanotubes

62.4.2 Molecular Dynamics at Constant Pressure for Composites with Nanotubes of Different Lengths—Results

A molecular dynamics simulation was performed on a model of composite PE/CNT to track changes in the cell over time. The molecular dynamic investigations, subset NPT (number of particles, pressure, and constant temperature), were performed in a simulation box at room temperature 298 K; incorporating Newtonian equations of motion for the simulations was DM done according to the Verlet algorithm [9], with no integration of 1 fs, for a period of 50 ps, and a cutoff number (R_c) equal to 8.50. We used the Parnillo Berendsen method for controlling the temperature and pressure.

Figure 62.5 represents the stress–strain curves of the longitudinal and transverse composites of carbon nanotubes/PE of different lengths.

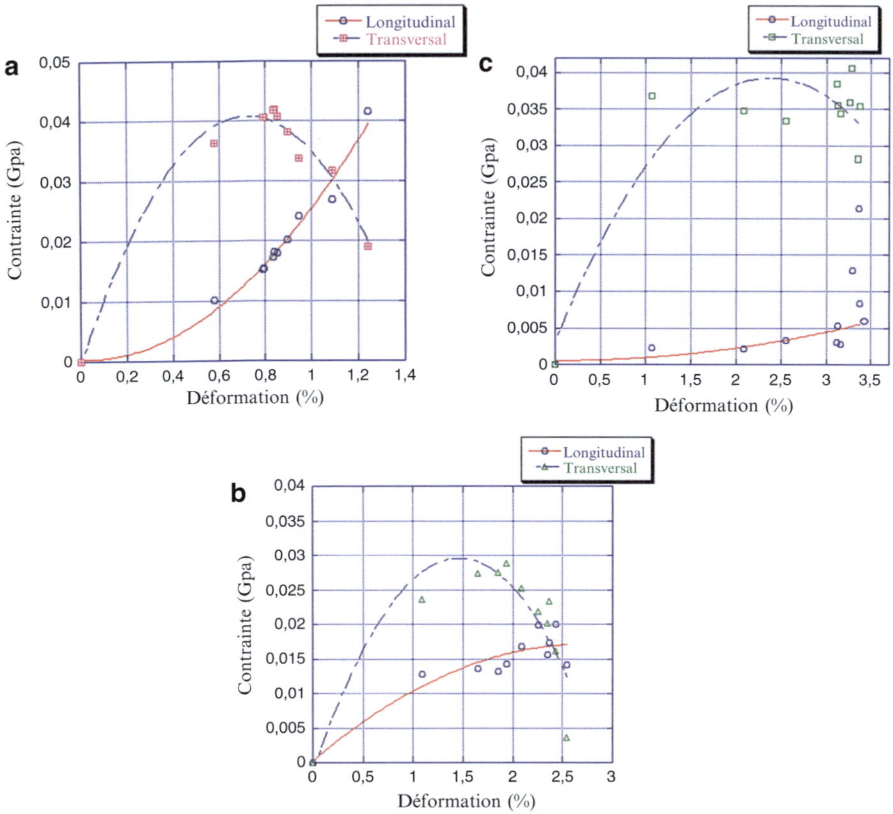

Fig. 62.5 Longitudinal and transverse stress–strain curves in a composite with (**a**) CNT is long (51.12)A; (**b**) CNT medium length (34.08) A; (**c**) CNT short length (8.52)A

The stress in the transverse direction decreases in the three composites because of the crack of the carbon nanotube; that is, the material is more brittle in this direction than in the other direction and requires a long-term deformation. The alignment of CNTs in the polymer matrix can significantly improve the mechanical performance of the composite. The mechanical properties are not linearly proportional to the deformation of CNT, but there is a performance threshold [10].

According to our calculation results and experimental results from [10, 11], we conclude that the main cause of small deformations is the entanglement of carbon nanotubes and poor dispersion CNT in the polymer matrix , whereas the good dispersion of the carbon nanotubes in the longitudinal direction requires very long CNTs. Our results are comparable to results obtained by [8] (e.g., Figs. 62.5 and 62.6).

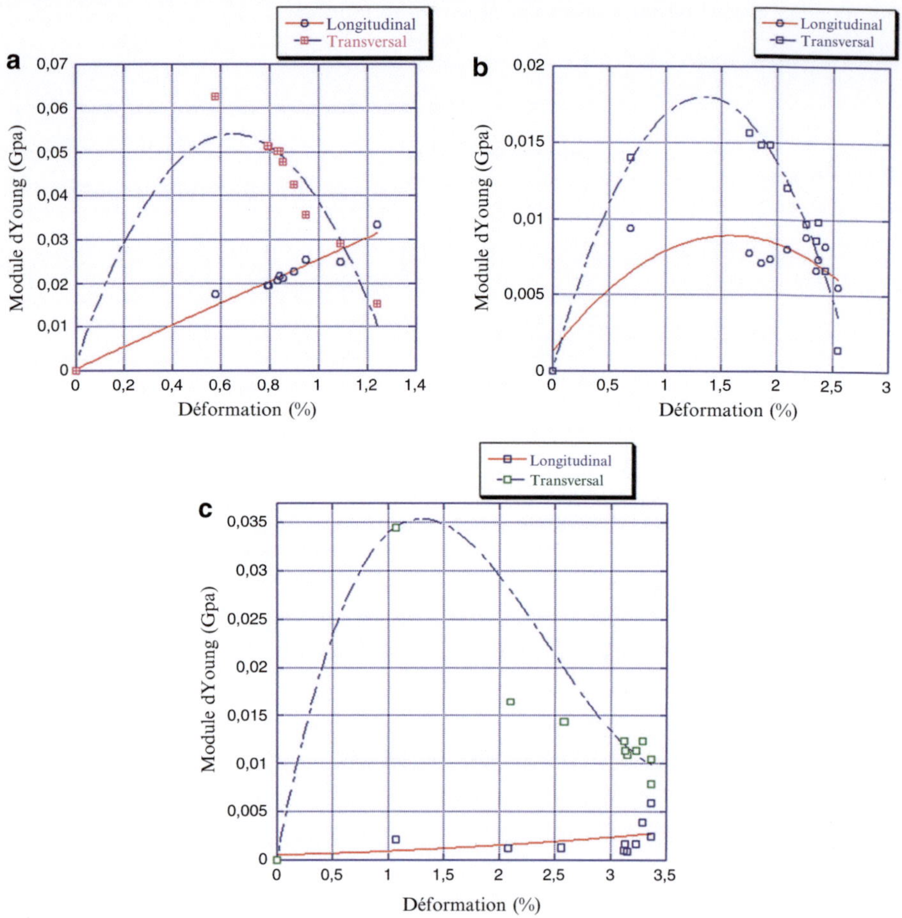

Fig. 62.6 Curve of Young's moduli longitudinal and transverse according to the deformation of the composite: (**a**) carbon nanotubes of high length/PE; (**b**) carbon nanotubes medium length/PE; (**c**) carbon nanotubes small length/PE

To understand the improvement of the mechanical properties of our composite (polyethylene/SWNT zigzag), we ascend the results of Young's modulus for the CNT–PE composite of different lengths which are represented by Fig. 62.6.

62.4.2.1 Young Modulus

Figure 62.6 represents the Young's moduli of the three composites and presents the curves of longitudinal and transverse Young's moduli for composite carbon nanotube/PE different lengths.

In Fig. 62.6a, the longitudinal Young's modulus increases with the deformation until reaching a value of 0.035 GPa and 1.2 % strain. The transverse Young's modulus shows a short-term high value (close to 0.06 GPa) for a deformation of 0.6 %. Then we observe that this module decreases to the value 0.015 GPa; the difference between the two curves shows that the capacity in the longitudinal direction is more resistive with respect to the transverse direction.

In Fig. 62.6b, the longitudinal and transverse Young's modulus increases with strain up to 1.6 % with a module of 0.009 GPa for the longitudinal direction and up to 0.017 GPa and 1.4 % for the management section. Then we observe that these modules reduce to the values 0.006 GPa and 0.002 GPa, respectively.

Figure 62.6c shows the variation of Young's modulus in both directions. For the longitudinal direction, we can say that the Young's modulus increases up to 0.0052 GPa to a deformation of 3.49 % as opposed to the transverse direction where the Young's modulus increases short term up to 0.035 Gpa and 1.1 % strain deformation. Then, we observe that this module decreases to 0.01 Gpa. The composite that contains short lengths of nanotubes is more fragile than the composite with high nanotube lengths.

From the experimental results of [12], the Young's modulus decreases due to transverse cracks of the carbon nanotubes, which verifies our results.

Our results show that the Young's modulus increases with increasing length of the carbon nanotubes; the reinforcement in the longitudinal direction is better than that obtained in the transverse direction. Theoretically, our results are comparable to the experimental results obtained by [12].

62.5 Conclusion

This work has allowed us to present a study of the mechanical behavior of polymer nanocomposites and carbon nanotubes, in order to facilitate their use in electronics knowing that CNTs can be used as reinforcements for drivers or dissipate static electricity in manufacturing equipment hard drives or solid-state, for example. Thus, combining the properties of PMMA and properties of SWNTs, we have an electronically functional reinforced polymer, electrically and thermally conductive applicable to solar cells, LEDs, displays, and so on.

To predict the effect of functionalization on the different properties of nanocomposites, we applied the model to the physical R.V.E. nanocomposite polyethylene (low density amorphous)/SWNT (zigzag).

The simulation of the thermal properties (interaction energy) and lifts (stress–strain and Young's modulus) of the composite PE/CNT by molecular dynamics is a field for discussion in the literature.

Several results are noteworthy in this work. We have observed that increasing the length of CNTs leads to an increase of the interaction energy as well as Young's modulus.

Through the results in the literature, our results are explained by the crack of CNTs and the poor dispersion of nanotubes in the polyethylene matrix. These explanations could be a result of this work, trying to bring in a first step CNTs cracked in our model and to study their impact on the mechanical behavior of the nanocomposite obtained.

The effect of the dispersion of CNTs would constitute a second line of research. We can also study the problem of the orientation of CNTs in the polymer matrix and the presences of defects in CNTs. We conclude that the prospects of this work are many and we cannot list them all.

References

1. Allaoui A (2005) Comportement mécanique et électrique des enchevêtrements de nanotubes de carbone. Thèse de doctorat, Ecole Centrale, Paris, MSS/MAT CNRS UMR N 8579, mai, p 83
2. Gao G (1998) Large scale molecular simulations with application to polymers and nano-scale materials. Thèse de doctorat, California Institute of Technology, March, 1998
3. Odegard GM, Frankland SJV, Gates TS (2005) Effect of nanotube functionalization on the elastic properties of polyethylene nanotube composites. AIAA J 43(8):1828–1835
4. Valavala PK, Odegard GM (2005) Modeling techniques for determination of mechanical properties of polymer nanocomposites. Rev Adv Mater Sci 9:34–44
5. Baaden M (2003) "Dynamique moléculaire in silico", complément de cours. CNRS UPR9080, Avril
6. Gou J, Lau K (2005) Modeling and simulation of carbon nanotube/polymer composites. In: Michael R, Wolfram S (eds) A chapter in the handbook of theoretical and computational nanotechnology, vol 1. American Scientific Publishers, Valencia, pp 1–33. ISBN 1-58883-042
7. Pregler SK, Hu Y, Sinnott SB (2002) Engineering the fiber-matrix interface in carbon nanotube composite. University of Florida, CHE-0200838
8. Frankland SJV, Harik VM, Odegard GM, Brenner DW, Gates TS (2003) The stress–strain behavior of polymer–nanotube composites from molecular dynamics simulation. Compos Sci Technol 63(11):1655–1661
9. Mediani A (2008) Etude des propriétés physiques de polyoxyéthylène et de Poly(Ethylène Glycol) par dynamique moléculaire. Thèse de magistère. U.ST.0. Février
10. Aïssa B (2006) Composite polymère nanotube de carbone. Introduction à l'électronique plastique. INRS, Novembre
11. Dalmas F (2005) Composites à matrice polymère et nanorenforts: flexibles: Propriétés mécaniques et électriques. Thèse de Doctorat, INP de GRENOBLE, novembre, p 140–142

12. Andrews R, et al (2003) Nanotube composite Matetrial. Copyright by held by Centre for Applied Energy Research, University of Kentucky, Lexington
13. Feng XQ, Shi DL, Huang YG, Hwang KC (2007) Micromechanics and multiscale mechanics of carbon nanotubes-reinforced composites. University of Illinois, Urbana
14. Namilae S (2004) Deformation mechanisms at atomic scale: role of defects in thermomechanical behavior of materials. The Florida State University

Chapter 63
Numerical Simulation of Nonlinear Waves in Numerical Wave Tank Using Boundary Element Method

Rezvan Alamian, Rouzbeh Shafaghat, Mustafa Ramzan-nejad, and Seyed Jalal Miri

Abstract Increasing world population and consequent increase in fossil fuel consumption emerge the necessity of looking for new sources of energy, resources that are clean, cheap, and renewable. The extracted energy from the sea is one of the alternatives of fossil fuels. This energy is available either in the form of tidal or wave energy. Wave energy converters are devices which convert the permanent sea waves to electrical energy. In order to achieve this technology and according to the terms of the sea conditions, developing the numerical modeling in a numerical wave tank is very important for modeling the system and analyzing its performance in different circumstances. So the sea conditions can be applied to this system. Simulation of linear and nonlinear waves in numerical wave tank has been one of the most important topics of interest to researchers and scholars working in the areas of the marine industries. Extensive numerical studies have been conducted using different methods in the past years. Among them, boundary element method is one of the most useful methods for modeling the boundary value problems, particularly for problems related to the hydrodynamic of marine structures. It is evident from literature that this method is very efficient and produces very accurate results for problems considered in this field. In this paper, due to capabilities of the boundary element method and using the Euler-Lagrangian method, nonlinear waves generated by a piston wave maker in a two-dimensional wave tank are investigated. For time evolving of wave in the simulation and solving the free-surface boundary condition equations, the conventional Euler-Lagrangian method is used too. For evaluation of results, a sample wave is chosen for simulation, and its result is compared with a result of an analytical method. Finally, a wave with characterization of average Caspian Sea waves is selected for simulation. The simulations have shown that if the wave maker oscillates with the average wave height and period of the Caspian Sea waves, what will be the characteristics of the generated waves, then the extractable energy potential is obtained from it.

R. Alamian (✉) • R. Shafaghat • M. Ramzan-nejad • S.J. Miri
Faculty of Mechanical Engineering, Babol Noshirvani University of Technology, Dr. Shariati Street, Babol, Iran
e-mail: rezvanalamian@gmail.com

I. Dincer et al. (eds.), *Progress in Clean Energy, Volume 1*,
DOI 10.1007/978-3-319-16709-1_63

Keywords Numerical wave tank • Boundary element method • Wave energy • Wave-maker calibration

Nomenclature

A	Amplitude of the motion of the piston wave maker
$C(P)$	Collision angle in point P on the boundary
E	Energy density, Wm^{-2}
g	Acceleration by gravity, ms^{-2}
$G(P, Q)$	Green's function
h	Depth, m
H	Wave height, m
k	Wave number
L	Length, m
n	Normal vector
P	Source point
Q	Field point
s	Displacement of piston wave maker
T	Wave period, s
u	Fluid velocity
x	Horizontal axial position, m
y	Vertical axial position, m

Greek Letters

∇^2	Laplace operator
Ø	Potential function
Γ	Integral boundary
ω	Angular frequency
ρ	Density, kgm^{-3}
λ	Wavelength, m
η	Wave profile at free surface, m

Subscripts

m	Model
p	Prototype
P	Piston wave maker
t	Wave tank

63.1 Introduction

Increasing the need for energy in all industries and the high costs associated with energy production have led the human beings to use the natural energies for different applications, rather than artificial and man-made energies. In the meantime, the sea waves as one of the main sources of energy is a subject of interest to many researchers, and fundamental studies are done on the methods of extracting the energy. Sea waves are caused by the wind streams, which flow over the sea surface. So far, many studies are done by applying the wave motion theories in this field [1]. Using numerical methods, there are examples of these activities [2]. The first results were discussed by Newman in 1976 [3]. Theoretical and applied studies in the field of numerical and analytical methods for wave energy converters were continued, and then, also making the model as the most effective practical work in this field was considered. The structure of these models is based on receiving the sea wave energy; with absorbing a percentage of this energy and transferring it to the generator, the energy needed for various purposes is provided. Many problems in fluid dynamics are investigated using numerical modeling. In order to study the behavior of marine structures, waves encountering them should be simulated. Usually period and amplitude of waves are considered as indicators of water surface waves, and, somehow, in investigating a wave-maker behavior, it is tried to relate these parameters for a wave maker to them for water waves. In recent decades, many studies on the simulation of waves produced by a wave maker are performed both experimentally and numerically. The purpose of many of these studies is investigating the wave production and analysis of wave interactions with floating bodies in the wave tank. Using the boundary element method to investigate the wave profile at the free surface, solving free-surface equations, applying Euler-Lagrangian method, and eliminating numerical instabilities in this method are very important. In this field, Longuet-Higgins and Cokelet were the first researchers who studied the nonlinear surface waves using the boundary element method [4]. They used Euler-Lagrangian method in their work for evolving the free surface for the first time. In 1984, Lin in his PhD thesis examined the waves generated by a wave maker with both analytical and boundary element methods [5]. He compared his results with the results of an experimental data. Using the boundary element method, Ryu et al. investigated the linear and nonlinear waves along with surface potential flow in 2003 [6]. They also presented their results for waves in waters with intermediate depths. In 1998, Celebl et al. simulated linear and nonlinear waves in a three-dimensional wave tank with boundary element and Euler-Lagrangian methods [7]. They eliminate the numerical instabilities in the Lagrangian part of the problem with the help of smoothing and re-meshing techniques. In 2001, Grilli et al. investigated the wave breaking phenomenon numerically, using the boundary element and Euler-Lagrangian methods [8]. Their solution domain was based on a three-dimensional wave and had an arbitrary shape on the bottom. Dommermuth and Yue with simulating nonlinear waves presented a new method to improve the free surface and remove numerical instabilities in 1987 [9].

In this article, with regard to the establishment of the wave energy converter in the Caspian Sea, we focused on the simulation of waves with characteristics of Caspian Sea waves. A sample period and amplitude of wave maker have been chosen with respect to the average period and amplitude of the Caspian Sea waves. In this paper, it is shown how the generated waves react if the mentioned wave maker oscillates with these parameters. Moreover, the extractable energy by wave energy converter from the generated wave is calculated.

63.2 Mathematical Equations

63.2.1 Boundary Element Method

In this method, the governing differential equations become integral identities which are applied on the surface or boundary; the boundary is divided into small segments, and, like the other numerical methods, it will result in a system of linear algebraic equations which have a unique solution. This method is simply applicable to any boundary with complex geometry and can be easily adapted to it. The boundary element method that has been used in this paper is based on the isoparametric quadratic elements. The governing differential equation is the Laplace equation which is expressed as follows [10]:

$$\nabla^2\phi = \frac{\partial^2\phi}{\partial x^2} + \frac{\partial^2\phi}{\partial y^2} = 0 \tag{63.1}$$

where ∇^2 is the Laplace operator, Ø is the potential function, and x and y are the Cartesian coordinate axes. Given the boundary conditions, Ø could be solved by the boundary integral equation:

$$C(P)\phi(P) + \int_{\Gamma} \frac{\partial G(p,Q)}{\partial n}\phi(Q)d\Gamma(Q) = \int_{\Gamma} G(P,Q)\frac{\partial\phi(Q)}{\partial n}d\Gamma(Q) \tag{63.2}$$

In the above equation, Γ is integral boundary, $G(P,Q)$ is Green's function, $C(P)$ is collision angle in point P on the boundary, and P and Q are the source and field points, respectively. The matrix form of Eq. (63.2) is presented as follows:

$$[A]\phi = [B]\frac{\partial\phi}{\partial n} \tag{63.3}$$

When determining every single element of matrixes A and B, applying known boundary values and putting known potential and its derivatives to one side and their unknown values to the other side, the above system of equations can be solved.

63.3 Wave Simulation in Numerical Wave Tank

In this problem, unstable waves generated by the wave maker in a wave tank have been studied numerically. Regular and linear reciprocating motion of a wave maker along the axis causes wave generation.

63.3.1 *Boundary Condition*

The geometry of the problem, including a wave tank with a wave-maker system, is shown in Fig. 63.1. Solution domain is limited by four borders including wave-maker piston in the left, tank floor in the bottom, tank wall in the right, and free surface on top.

Sinusoidal motion of the piston wave maker is expressed by the following equation:

$$x_p(t) = -\frac{A}{2}\cos(\omega t) \tag{63.4}$$

where A is the amplitude of the motion of the piston wave maker and ω is the angular frequency of the sinusoidal motion.

63.3.2 *Boundary Condition on the Wave Maker* $S_L(t)$

Boundary condition on the surface of the wave maker is Neumann boundary condition that is expressed as follows:

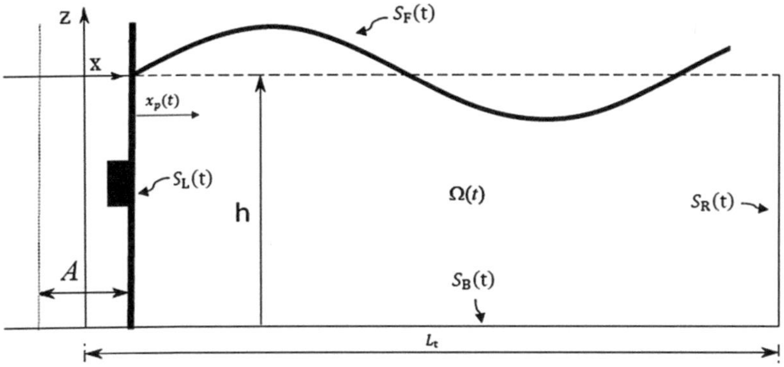

Fig. 63.1 Geometry of the piston wave maker and its boundaries

$$\frac{\partial \phi}{\partial n} = \nabla \phi \cdot n = \dot{x}_p(t) \quad x \in S_L(t) \tag{63.5}$$

In the equation above, n is normal vector that is perpendicular to the surface of the piston wave maker and directed into the fluid domain. Taking the time derivative of Eq. (63.4), sinusoidal velocity relation of piston is determined,

$$\dot{x}_p(t) = \frac{A}{2}\omega \sin(\omega t) \tag{63.6}$$

63.3.3 *Boundary Condition on the Bottom and Right-Wall Side of the Wave Tank*

Neumann boundary condition applies at both borders, and the following kinematic boundary condition is used for both of them:

$$\frac{\partial \phi}{\partial n} = \nabla \phi . n = 0 \quad x \in S_B(t),\ S_R(t) \tag{63.7}$$

63.3.4 *Initial Condition*

In time $t = 0$, initial conditions are as follows:

$$\left.\begin{array}{l} \phi(x,0) = 0 \\ \eta(x,0) = 0 \end{array}\right\} \Rightarrow \quad x \in S_F(t) \tag{63.8}$$

63.3.5 *Kinematic Boundary Condition on the Free Surface*

Kinematic boundary condition is obtained by assuming that fluid cannot pass through the free surface. In the Lagrangian view of the free surface, for a fluid particle, $P(x, y)$, on the free surface, kinematic boundary condition of the free surface can be written as [11]:

$$\frac{Dx}{Dt} = u = \nabla \phi \Rightarrow \left\{ \begin{array}{l} \dfrac{Dx}{Dt} = u = \phi_x \\ \dfrac{Dy}{Dt} = v = \phi_y \end{array} \right\} \quad x \in S_F(t) \tag{63.9}$$

where in the above equation $\frac{D}{Dt} = \frac{\partial}{\partial t} + \nabla Ø \cdot \nabla$ is the material derivative from the Lagrangian view and $u = \nabla Ø = \left(Ø_x, Ø_y\right)$ is fluid velocity on the free surface.

63.3.6 Dynamic Boundary Condition on the Free Surface

The dynamic free-surface boundary condition is obtained by the Bernoulli equation and assuming the continuity of pressure along the free surface. So, the pressure at the free surface must be equal to the atmospheric pressure above it which results in the general form of the dynamic boundary condition on the free surface:

$$\frac{\partial \phi}{\partial t} + \frac{1}{2}|\nabla \phi|^2 + gy + \frac{P_f}{\rho} = 0 \quad x \in S_F(t) \tag{63.10}$$

In the above equation, g is the acceleration by gravity. The usual form of the dynamic free-surface boundary condition is defined by pressure in terms of relative pressure in which the surface pressure is zero $(P_f = 0)$. Finally, the dynamic free-surface boundary condition can be expressed in the form of Lagrangian as below:

$$\frac{D\phi}{Dt} = \frac{1}{2}|\nabla \phi|^2 - gy \quad x \in S_F(t) \tag{63.11}$$

63.3.7 Time Evolving of the Free Surface

To evolve the free surface at time $(t + \Delta t)$, the Lagrangian form of kinematic and dynamic free-surface boundary conditions are integrated with respect to time,

$$\vec{x}^{(k+1)} = \vec{x}^{(k)} + \Delta t\left(\vec{\nabla}\phi\right) \quad \vec{x} = (x, y) \tag{63.12}$$

$$\phi^{(k+1)} = \phi^{(k)} + \Delta t\left(\frac{1}{2}|\nabla \phi^{(k)}|^2 - gy^{(k)}\right) \tag{63.13}$$

To solve the above equations, numerical Runge-Kutta fourth-order method is used. According to the above descriptions, the Lagrangian points and nodes have equal velocity, and it should be noted that our view is totally Lagrangian; thus, the nodes along the free surface are elevated along the x and y directions, both.

63.3.8 Smoothing and Re-meshing of Free Surface

At each time step of solving the problem, we need to re-mesh the free surface for maintaining regular intervals between the elements on the free surface. When evolving the free surface at each step, Lagrangian points are transferred to their new position, thereby causing congestion of these points in areas with high gradient and instability of solution. Accordingly, cubic spline method for interpolation of the Lagrangian points on the free surface and redistributing them is used.

Furthermore, for eliminating the sawtooth instability at each time step, Chebyshev five-point smoothing scheme is used. Emerging this type of instability can have either physical or mathematical reasons. One reason that could be raised about it is the existence of singularity in the corner point of the problem's geometry. This instability first was proposed by [4]. They used the Chebyshev five-point scheme for eliminating the instability.

63.4 Wave Energy and Power

As noted before, the waves as a source of energy is of utmost importance. In this section, the analysis and calculation of energy from sea waves is investigated. In Fig. 63.2, the main characteristics of a wave are shown.

The total energy contained in waves has two types: the potential energy due to the displacement of the water surface and the kinetic energy which is stored by the motion of all particles. The transfer of this energy has a great importance in order to evaluate and extract the available power for the wave energy converters.

Wave energy and power density are calculated according to the following equation in terms of wave characteristics. But it should be noted that almost 50 % of this energy can be extracted:

$$E_{\text{density}} = \frac{\rho g H^2}{8} \tag{63.14}$$

$$P_{\text{density}} = \frac{E_{\text{density}}}{T} = \frac{\rho g H^2}{8T} \tag{63.15}$$

In the above equations, ρ is density of water, H is wave height, and T is wave period. Mostly, the wave energy at each wavelength (λ) is defined per unit width of wave crest which is calculated in Eq. (63.16). The ideal extractable power can be calculated from Eq. (63.17).

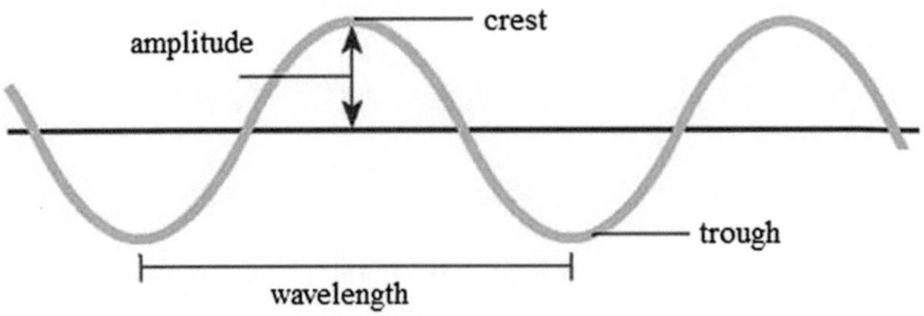

Fig. 63.2 The main characteristics of the wave

$$E_{\text{wavefront}} = \frac{\rho g H^2 \lambda}{8} \tag{63.16}$$

$$P_{\text{ideal}} = \frac{1}{16\pi} \rho g^2 H^2 T(W/m) = 1915 H^2 T \tag{63.17}$$

It is observed from the above equations that the generated power depends on the wave period and amplitude.

63.5 Constraints on Wave Generation

The wave height and period usually are two factors identified as indicators of water surface waves; it is important to know how the generated waves react if a wave maker oscillates with these parameters. Basically, to produce a desired wave by a wave maker, two parameters should be controlled:

1. Piston or paddle velocity
2. Spatial range of the piston or paddle's reciprocating motion

Apparently, there is no limitation in generating waves with different heights and periods in theory, but practically there are two major limitations in the way of achieving this goal:

1. Velocity limitation on wave-maker plate
2. Limitations related to the nature of the resulting wave which leads to instabilities and wave breaking

63.6 Results and Discussion

This section presents the results of the written numerical code. The waves are simulated in a numerical wave tank with length of $L_t = 40m$ and depth of $h = 1m$. To avoid reflection of waves at the end of the tank, the tank is considered large enough. For validation, the result from the solution of the boundary element method is compared with the result of the analytical solution of the piston wave maker [12] (Fig. 63.4). Analytical solution for wave potential and the profile of a piston wave maker are expressed with the following equations:

$$\phi(x, y, t) = \frac{4s \tanh kh \sinh kh}{\omega(2kh + \sinh 2kh)} \cosh k(y + h) \cos(kx - \omega t) \tag{63.18}$$

$$\eta(x, t) = \frac{4s \sinh^2 kh}{2kh + \sinh 2kh} \sin(kx - \omega t) \tag{63.19}$$

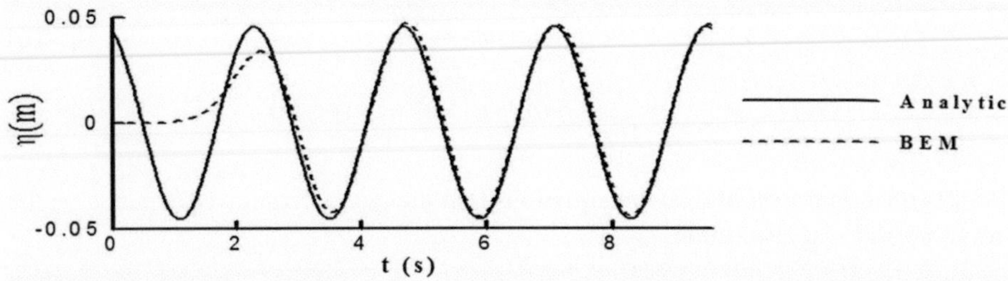

Fig. 63.3 Comparison of analytical solution and numerical solution of the boundary element method at $T = 2.4\,\text{s}$ and $A = 0.1\,\text{m}$

Results for wave profiles from $t = 0$ to $t = 4T$ for both boundary element and analytical methods are compared which is presented in Fig. 63.3. It can be seen that both models are in good agreement and they produce identical results.

The aim of this article is to simulate the Caspian Sea waves in a numerical wave tank. Caspian Sea is the largest lake on the earth with a surface area of 436,340 km^2 which is placed at the north of Iran. This sea has a depth of about 180–1,000 m in the southern basin and 2.6 m in the northern basin. The wave period varies between 2 and 8 s in the southern basin of the Caspian Sea. Also, the wave height changes between 0.5 and 3 m in different seasons [13]. For simulating the Caspian Sea waves in a wave tank, wave height and period must be scaled. First of all, the relationship between the parameters in terms of the scale model and actual samples should be examined. In marine systems, where the free surface is important, to keep constant the important factors which cannot be scaled (such as acceleration by gravity), the Froude number is used. The relationship between the parameters of the model experiment with the prototype based on the Froude similarity law is expressed as Eqs. (63.20) and (63.21) for wave height and period. In these equations, Lm is wave height of wave tank, Lp is wave height of the Caspian Sea, Tm is wave period of wave tank, and Tp is wave period of the Caspian Sea. The scale ratio for simulation of wave tank is chosen to be 1:12 with respect to the model considered to be established in the real wave tank.

$$\frac{L_m}{L_p} = \text{Scale} \tag{63.20}$$

$$\frac{T_m}{T_p} = \sqrt{\text{Scale}} \tag{63.21}$$

Average wave height and period of Caspian Sea are selected for simulation. With regard to the above equations, wave period of 1.5 s and wave height of 0.12 m are chosen for the wave tank modeling. Figures 63.4 and 63.5 show wave amplitude with respect to distance from wave maker and time, respectively.

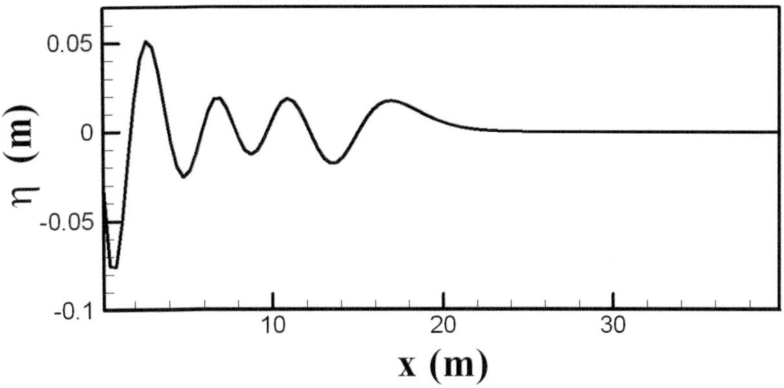

Fig. 63.4 Wave amplitude with respect to distance at $T = 1.5\,\mathrm{s}$ and $A = 0.12\,\mathrm{m}$

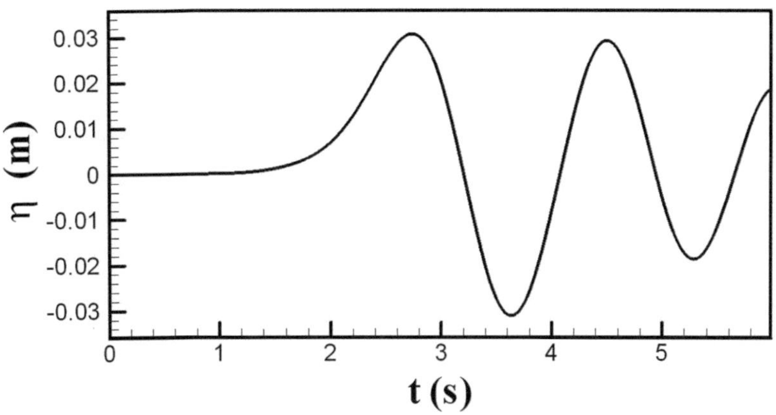

Fig. 63.5 Wave amplitude with respect to time at $T = 1.5\,\mathrm{s}$ and $A = 0.12\,\mathrm{m}$

Analyzing Fig. 63.5, the period and height of the wave are obtained 1.73 s and 0.06 m, respectively. These values are extracted after two wavelengths where the effects of wave maker are negligible. It can be seen that the values are smaller than the wave-maker ones. From Eq. (63.17), the ideal extractable energy from this wave is equal to 12 W/m.

In the future work, same simulation and calculations will be done to the wide range of wave period and heights from minimum to maximum values for the Caspian Sea. It can be very helpful to evaluate the sea, before placing the model in it. Moreover, the domain will be extended to obtain the wave breaking criteria near the wave maker. It will define the boundaries where the wave maker can be applicable.

63.7 Conclusions

In this paper, the piston wave maker and the wave generated by it in the numerical wave tank have been simulated by the boundary element method and mixed Euler-Lagrangian method. Time-dependent kinematic and dynamic free-surface boundary conditions for improvement of the free-surface wave profile and velocity potential are used. In order to eliminate numerical instability in the free surface, re-meshing and smoothing techniques are applied at each time step.

Finally, the results from the simulation of piston wave maker with regard to wave period and height of the Caspian Sea have been presented. It is shown that the wave period and height of the generated waves are smaller than the wave-maker ones.

References

1. Cruz J (2008) Ocean wave energy: current status and future perspectives. Springer, Berlin
2. Alves M, Brito-Melo A, Sarmento A (2002) Numerical modelling of the pendulum ocean wave power converter using a panel method. Proceedings of the Twelfth International Offshore and Polar Engineering Conference, Kitakyushu, Japan, pp 655–661
3. Newman J (1976) The interaction of stationary vessels with regular waves. Proceedings of the 11th symposium on naval hydrodynamics, pp 491–501
4. Longuet-Higgins MS, Cokelet E (1976) The deformation of steep surface waves on water. I. A numerical method of computation. Proc R Soc Lond A Math Phys Sci 350:1–26
5. Lin WM (1984) Nonlinear motion of the free surface near a moving body. Ph.D. Thesis, Department of Ocean Engineering, Massachusetts Institute of Technology
6. Ryu S, Kim M, Lynett PJ (2003) Fully nonlinear wave-current interactions and kinematics by a BEM-based numerical wave tank. Comput Mech 32:336–346
7. Celebl M, Kim M, Beck R (1998) Fully nonlinear 3-D numerical wave tank simulation. J Ship Res 42:33–45
8. Grilli ST, Guyenne P, Dias F (2001) A fully non-linear model for three-dimensional overturning waves over an arbitrary bottom. Int J Numer Methods Fluids 35:829–867
9. Dommermuth DG, Yue DK (1987) Numerical simulations of nonlinear axisymmetric flows with a free surface. J Fluid Mech 178:195–219
10. Becker AA (1992) The boundary element method in engineering: a complete course. McGraw-Hill, London
11. Koo W, Kim M-H (2004) Freely floating-body simulation by a 2D fully nonlinear numerical wave tank. Ocean Eng 31:2011–2046
12. Tanizawa K (2000) The state of the art on numerical wave tank. Proceedings of 4th Osaka colloquium on seakeeping performance of ships, pp 95–114
13. Alamian R, Shafaghat R, Miri SJ, Yazdanshenas N, Shakeri M (2014) Evaluation of technologies for harvesting wave energy in Caspian Sea. Renew Sustain Energy Rev 32:468–476

Chapter 64
Modeling of Flow Around a Solid Body in Free and Restricted Fluidal Media

Munir Suner

Abstract Design has become one of the most essential tasks for human beings throughout the history of mankind. The accurate modeling and analysis of solid bodies moving in free fluid media (such as canals) for velocity, pressure, and forces have become quite significant. In conjunction with this, it is a challenging task to design solid bodies with complex geometries and structures in multidimensional form and it needs the utmost care and attention. Moreover, it is inevitable to develop and design some new systems that will enable us to reduce the energy requirements to minimum levels because nowadays saving energy has become one of the hottest issues for attaining sustainability. In this study, energy-efficient and -effective solid bodies are designed and analyzed by developing a two-dimensional model to study the distributions of flow velocity, pressure, and forces in free and restricted environments. The generalized model solutions are achieved in MATLAB. The results are then discussed for optimum conditions.

Keywords Hydrodynamics design • Potential flow • General solution • Flow around body • Analytical solution

Nomenclature

$F(z)$	Complex potential function (m)
F	Force (N)
j	Number of the intervals used in the series sum of the solid surface (–)
J	Number of channels used to perform rigid channel of the solid surface (–)
m	Number of parts on the side wall of the body between the two end points (–)
n	Cider on the wall between the two endpoints of the number of parts (–)
$\vec{n}$	Unit vector normal of the surface (–)
q	Magnitude of a two-dimensional double (m^2/s)

M. Suner (✉)
Maritime Faculty, Istanbul Technical University, Tuzla, Istanbul, Turkey
e-mail: msuner@itu.edu.tr

I. Dincer et al. (eds.), *Progress in Clean Energy, Volume 1*,
DOI 10.1007/978-3-319-16709-1_64

Q	Source and severity of the well flow rate (m^3/s)
P	Pressure (kg/ms^2)
P_0	Ambient pressure (N/m^2)
u	Components of velocity on the x-axis (m/s)
V	Velocity (m/s)
v	Components of velocity on the y-axis (m/s)
$y(i)$	Position of Doubleon y-axis (m)

Greek Letters

μ	Fluid dynamic (absolute) viscosity (kg/ms)
ρ	Density of the fluid (kg/m^3)
σ	Surface area in the normal direction (m^2)
ϕ	Potential function (m^2/s)
ψ	Stream function (m^2/s)

64.1 Introduction

In order to improve and enhance the welfare of people and enable them to benefit from liquids and gases, usage and utilization of liquids and gases have often become significantly prominent issues among people. Such problems have always kept people busy throughout the history of mankind. Gradually more complicated and sophisticated systems for energy efficiency have been invented much more effectively throughout time and many unsolved problems have been dealt with in various ways in this respect. All liquid and gas systems are utilized to meet the corresponding needs. In this chapter, it has been thought that designing a complex solid body will provide us an energy-efficient system. A very limited number of publications have drawn our attention and some of those studies that are closely related to the subject matter are summarized within the scope of this chapter.

During research, [19] the analysis of the body weight of a pressure-resistant submarine, its structural properties, vibration, tension, analysis of its stationary state, and analysis of its structural fatigue reality were done by means of a computer program. In another survey [15], the direct matrix and replacement technique was utilized in order to solve its potential of speed by using a normal gradient while examining a two-dimensional submerged vessel. Blair and others explained the construction and manufacturing steps of the new sort of submarine bearing a fish-cage system that can turn on the surface so as to eliminate biological contamination and pollution as well as for other purposes [1]. During his doctoral thesis study, Durgun [3] examined and reviewed the effects of interior side walls and the bottom resistance of a free-surfaced rectangular canal. In other words, he reviewed the blockage effect and considered how to calculate ΔV, the rates of velocity increase of those geometrical shapes such as ellipsoid, oval, and Rankine ovoid for the shape of the ship both theoretically and experimentally. In another study [31], the design of the laminar border layer that increases on the Rankine oval by constantly moving fast at a steady speed in unpressured hydraulic fluid or in incompressible fluid was revealed numerically.

In still another study [13], the amount of unpressured potential flow on the closed substances was predicted by adopting an alternative approach. Other research conducted by Szymczak and his coworkers [23] indicated the analysis of the body weight of a submarine, its structural properties, vibration, tension, analysis of its stationary state, and analysis of its structural fatigue reality by means of a computer program. In their study, Sahin and Hyman [21] calculated the flow around the Rankine ovoid that moves at a certain depth under a free water surface by benefiting from the theory of potential current. Research on how to determine the size of an adjustable vehicle was conducted by Huggins and Packwood [6]. Best fit sizes were found for a submarine that submerges into the depths of the Atlantic Ocean autonomously bearing a slow drift coefficient. The survey conducted by Virgin and Erickson [24] was performed in order to reveal and introduce the dynamic movements of those floating substances much more definitely and accurately. In an essay published by Sahin and coworkers [22], the three-dimensional flow around an asymmetrical object submerged into water at a certain depth was calculated by making use of the theory of three-dimensional flow and by means of an analytical and numerical method comprised of the Green function. A study to provide us with the combination of versatile dynamic problems bearing vital importance, such as control of a modern submarine, its static stationary, and its hydrodynamic performance control system, were explained and exposed by Papoulias and others [35]. The losses of stationary state at high and medium speeds were calculated in the research by adopting the nonlinear Hopf-Forking analysis method. Other research [18] examined Rankine and similar geometrical shapes, and the pressure and speed around the Rankine oval by utilizing a general computer-assisted method. Janson [7] calculated the forces of lifting and forces of free surface flows by making use of potential flow panel methods during one of the studies, and thus he developed and introduced a method to optimize ship body shapes automatically. Zungber and Lee [32] studied the random movements of a rigid substance submerged into an incompressible hydraulic fluid in a restricted three-dimensional hydraulic medium of fluid. As a result of this research, the global outlets of the weak solutions were proven in this problem. In another study conducted by Madsen and others [11], the design of a submarine vehicle and management duties of the system were explained. As a result of that research, the model of a mini-submarine was constructed on the objective result pedigree, basic plan, and logical diagram.

In their research, Wu and Chwang [33] worked on the three-dimensional hydrodynamic model of a two-piece underwater towing vehicle. Kouh and Suen [8] studied issues based on three-dimensional potential flow theory by taking a high-level panel method as the basis for their research, and thus they made the dual distribution and Gaussian field measurement suitable for the potential flow problem. In their study, Wu and Chwang [25] suggested a hydrodynamic underwater system attached to a primary wire, a two-piece vehicle capable of doing easy maneuvers, towed actively in the horizontal position, and vertically bearing control surfaces, and they discovered the mass of the vehicle by making use of three-dimensional potential flow theory. In their research, Xu and the others [30] obtained experimentally the drifting resistance generated during the fixed speed horizontal movement by means of the Rankine ovoid and as a function of the Froud and Reynolds count number. Zakharov and others [34] developed and introduced a new method to provide numerical solutions to a two-dimensional surface and potential flows based on a conformal sum and Fourier combination. During a survey conducted by Djerid and his coworkers [2] they examined the turbulence properties of the flow around the cylinder of the Reynolds number 140,000. During research conducted by Niazman and Renksizbulut [14], the viscous flow around a rotating globe-shaped object submerged into water was examined.

In research conducted by Mahfouz and Haddara [12], the impact of total predictions of hydrodynamic parameters and by accelerating and slowing down an underwater robot vehicle floating on the level of the sea surface with random sea waves was examined. Perrault and coworkers [16] tested the sensitivity to changes in the hydrodynamic parameters of typical autonomous underwater vehicle (AUV) response characteristics. In the research conducted by Evans and Nahon [26], they explained the dynamic modeling of an underwater autonomic system and made some evaluations for the attacking angle 360° such as some outer forces affecting the submarine and the control plane for the outer impact forces. In their research, Gimanov and Sotiropoulos [4] developed a hybrid Cartesian control method for the simulation of the flows in three-dimensional complicated geometric-shaped substances and they utilized the Navier–Stokes's equation to solve the problems regarding movable three-dimensional substances in irregular incompressible fluids. In their study regarding similar matters, Li and Lee [9] designed an inspector that is applicable to a nerve network for the submerging behavior of an AUV. Wu and his coworkers [27] suggested a new type of vehicle to be used instead of a controllable underwater system, which can be towed by working on an alternative system experimentally. In another study, Xu and others [29] presented results of their theory on wave drifts for the Rankine ovoid movements in horizontal, steady, fast, and regular layered liquid. During their research studies, by Liu and others [10], they did experiments to control those varying forces on propellers and they put a barrier in the middle in order to evoke those eddy flows so that they could test and measure the characteristics of their movements. During research conducted by Rubin [20], he developed an analytical formula to display the cavitation effect distribution on a long rigid rocket that contained the cavitation impact. In one research study, done by Xu et al. [28], a linear theory on the internal waves generated in the stratified fluid with a pycnocline was presented, and the influence of the internal wave on the flow fields at the free surface was represented by the velocity gradient fields resulting from the internal waves generated by motion of the Rankine ovoid. Hsu et al., in their study, adopted an opened shallow cylindrical shell and deep-diving submersible vehicle, GUPPY, to investigate [5]. Rapoport and Rubin showed that for penetration velocities V below a critical value Vs, the target material maintains full contact with the projectile's surface, the drag force is due solely to the resistance of plastic flow in the target, and it is independent of the velocity V [17].

When the above studies performed in the literature are summarized in general, it can easily be said that the potential flow related to complex solid body designing in the way modeled in this study cannot be found.

The purpose of this chapter is to make a meaningful contribution to solving the hydrodynamic problems regarding complex systems and increase the energy efficiency that can perform all sorts of research and investigation inside those macro- and microengineering systems by implementing analytical methods.

Problem: A mathematical modeling activity has been initiated by making use of the theory of two-dimensional potential flow. Afterwards digital modeling studies have been actualized. The hydrodynamics of the complex solid body in a restricted fluidal medium has been examined by employing potential flow theory. Later, the flow function of the generalized different and complex forms of the solid body was acquired by benefit of the source-well (doubles) and therefore velocity, pressure, and force fields have been obtained analytically. The hydrodynamic design of a solid body, which has complex form, reveals itself predominantly in this study. The purpose of this study is to establish a general analytical modeling of stream function, potential function, pressure and velocity fields, and hydrodynamic design of a solid body shaped in any geometrical function in a open and restricted fluid medium.

64.2 Mathematical Modeling

64.2.1 Definition of the Problem

The geometrical shape of complex solid body supposed to be any solid body to be studied within the scope of this chapter is in the shape of a complex solid body, which combines with functions as $f_1(x), f_2(x), f_3(x), \ldots, f_n(x)$. The movements of two-dimensional permanent potential flow of a solid body moving in an ideal and incompressible hydraulic fluid flowing in a channel is examined by employing analytical methods. The solid body can move in free fluidal media or restricted fluidal media in the same or opposite directions to that of fluid flow.

In the first place uniform flow, source-well, and double flows are utilized in order to examine the movements of the two-dimensional flows that occur around the solid body in an infinite flow environment. The matter is dealt with first in a limitless flow environment and afterwards in a restricted flow medium that occurs in two-dimensional channel.

64.2.2 Hypothesis

Fluid is steady, ideal and incompressible, ($\mu = 0,\ \rho = \rho_0 =$ constant), motion ispatential, velocity field $\left(\vec{\omega} = 0,\ \vec{V} = \mathrm{grad}\phi\right)$, and steady flow $\left(\frac{\partial}{\partial t} \equiv 0\right)$

Here μ stands for dynamic viscosity (kg/ms), ρ mass density (kg/m^3), V velocity (m/s), ω angular velocity (Rad./s), t time, and (s) potential function.

Accordingly, the potential function of the movement is obtained from the continuity equation and from the solution of the Laplace differential equation $\Delta\phi = 0$.

64.2.3 Theory of Two-Dimensional Potential Flow

As known from the theory of potential flow, $z = x + iy = r \cdot e^{i\theta}$ as it is supposed to be the function of the complex potential $F(z)$, potential function ϕ, flow function ψ, field of velocity V, field of pressure P, and field of force F are provided as stated below in Eqs. (64.1) through (64.6). The distributions of velocity and pressure obtained by means of flow functions and potential functions belonging to the flow functions are acquired with the help of the MATLAB program and revealed here.

The function of the complex potential is comprised of the flow function and velocity potential.

$$F(z) = \phi + i\psi \tag{64.1}$$

Velocity is derived from a potential and it is stated as in Eq. (64.2)

$$\vec{V} = u\vec{i} + v\vec{j} = \nabla\phi = \frac{\partial\phi}{\partial x}\vec{i} + \frac{\partial\phi}{\partial y}\vec{j} = \frac{\partial\psi}{\partial y}\vec{i} - \frac{\partial\psi}{\partial x}\vec{j} \tag{64.2}$$

If the resultant of the velocity and Bernoulli theorem is written, it becomes:

$$\begin{aligned} \left|\vec{V}\right|^2 &= u^2 + v^2,\quad p_1 + \frac{\rho}{2}\left|\vec{V}_1\right|^2 = p_2 + \frac{\rho}{2}\left|\vec{V}_2\right|^2 \Rightarrow p_2 \\ &= p_0 + \frac{\rho}{2}\left(\left|\vec{V}_2\right|^2 - \left|\vec{V}_1\right|^2\right) \end{aligned} \tag{64.3}$$

The equation for the field of force can be written as shown in Eq. (64.3)

$$\vec{F} = \iint p\vec{n} \cdot d\sigma = \iint\left(p_0 + \frac{\rho}{2}\left(\left|\vec{V}_2\right|^2 - \left|\vec{V}_1\right|^2\right)\right)\vec{n} \cdot d\sigma \tag{64.4}$$

Here $P_1 = P_\infty$ and $V_1 = U_\infty$ is obtained. If there is more than one flow together, their complex potential functions are added in accordance with the principles of the theory of superposition. Equation (64.5) and flow function, field of velocity, field of pressure, and field of force are arranged according to this newly obtained function.

$$F(z) = F_1(z) + F_2(z) + F_3(z) + \cdots + F_n(z) \tag{64.5}$$

General flow, source, double, general function-double are use to create a solid body, which is combined from different functions and has a complex form, and can be generated with this potential theory. The stream function of flow around a solid body, velocity field, and pressure distribution can be generated analytically for free and in the restricted area, field of velocity, and field of pressure again take place in the content of the following expressions as is known from this theory.

64.3 Modeling Theorem in Free Fluidal Media

64.3.1 Complex Potential Function of Solid Body in Free Fluidal Media

It is assumed that the solid body has a complex shape and certain boundaries as shown in Fig. 64.1. If the theory can be applied on this form with double and general flow, it can generate a complex potential function and solid body modeling.

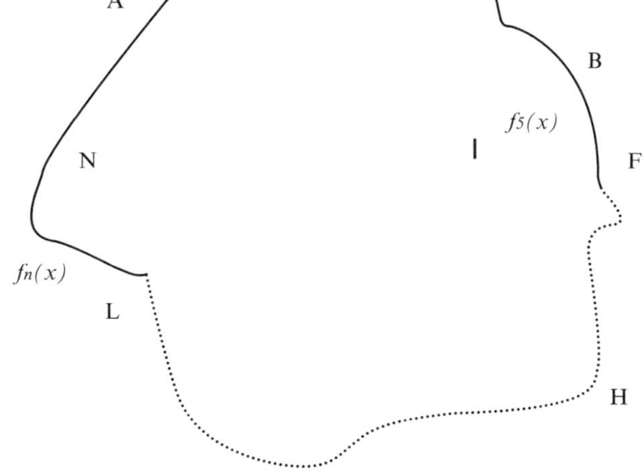

Fig. 64.1 Complex body boundary

A complex potential function of the flow around it in a free medium is as in Eq. (64.6).

$$F_{AN}(z) = F_{AB}(z) + F_{BC}(z) + \cdots + F_{LN}(z)$$
$$= U_{\infty} z + \lim_{n\to\infty} \sum_{i=1}^{n} \frac{q e^{i\alpha(i)}}{z \mp z_{AB}(i)} + \lim_{n\to\infty} \sum_{i=1}^{n} \frac{q e^{i\alpha(i)}}{z \mp z_{BC}(i)} + \cdots + \lim_{n\to\infty} \sum_{i=1}^{n} \frac{q e^{i\alpha(i)}}{z \mp z_{LN}(i)} \tag{64.6}$$

64.3.2 Stream and Potential Function of Solid Body in Free Fluidal Media

From this complex potential function, a streamlined and potential function of potential flow can be derived as in Eqs. (64.7) and (64.8).

$$\begin{aligned}
\psi(x, y) = U_{\infty} \cdot y &- \lim_{n\to\infty} \sum_{i=1}^{n} q \left(\frac{\left(y - y(i)_{AB}\right)}{\left(\left(x - x(i)\right)^2 + \left(y - y(i)_{AB}\right)^2\right)} e^{i\alpha(i)} \right) \\
&- \lim_{n\to\infty} \sum_{i=1}^{n} q \left(\frac{\left(y - y(i)_{BC}\right)}{\left(\left(x - x(i)\right)^2 + \left(y - y(i)_{BC}\right)^2\right)} e^{i\alpha(i)} \right) \\
&- \lim_{n\to\infty} \sum_{i=1}^{n} q \left(\frac{\left(y - y(i)_{CD}\right)}{\left(\left(x - x(i)\right)^2 + \left(y - y(i)_{CD}\right)^2\right)} e^{i\alpha(i)} \right) \\
&- \lim_{n\to\infty} \sum_{i=1}^{n} q \left(\frac{\left(y - y(i)_{DE}\right)}{\left(\left(x - x(i)\right)^2 + \left(y - y(i)_{DE}\right)^2\right)} e^{i\alpha(i)} \right) \\
&- \lim_{n\to\infty} \sum_{i=1}^{n} q \left(\frac{\left(y - y(i)_{EF}\right)}{\left(\left(x - x(i)\right)^2 + \left(y - y(i)_{EF}\right)^2\right)} e^{i\alpha(i)} \right) \\
&- \cdots - \lim_{n\to\infty} \sum_{i=1}^{n} q \left(\frac{\left(y - y(i)_{LN}\right)}{\left(\left(x - x(i)\right)^2 + \left(y - y(i)_{LN}\right)^2\right)} e^{i\alpha(i)} \right)
\end{aligned} \tag{64.7}$$

With this modeling different channel constructions can be designed. After establishing the boundaries of the channel, the channel solid wall is formed by placing doubles on its limits. In Fig. 64.2a, it can be seen that the ellipse and parabolic solid body are combined, and in Fig. 64.2b, it can be seen that the ellipse and parabolic solid body have a perforated oval with it,

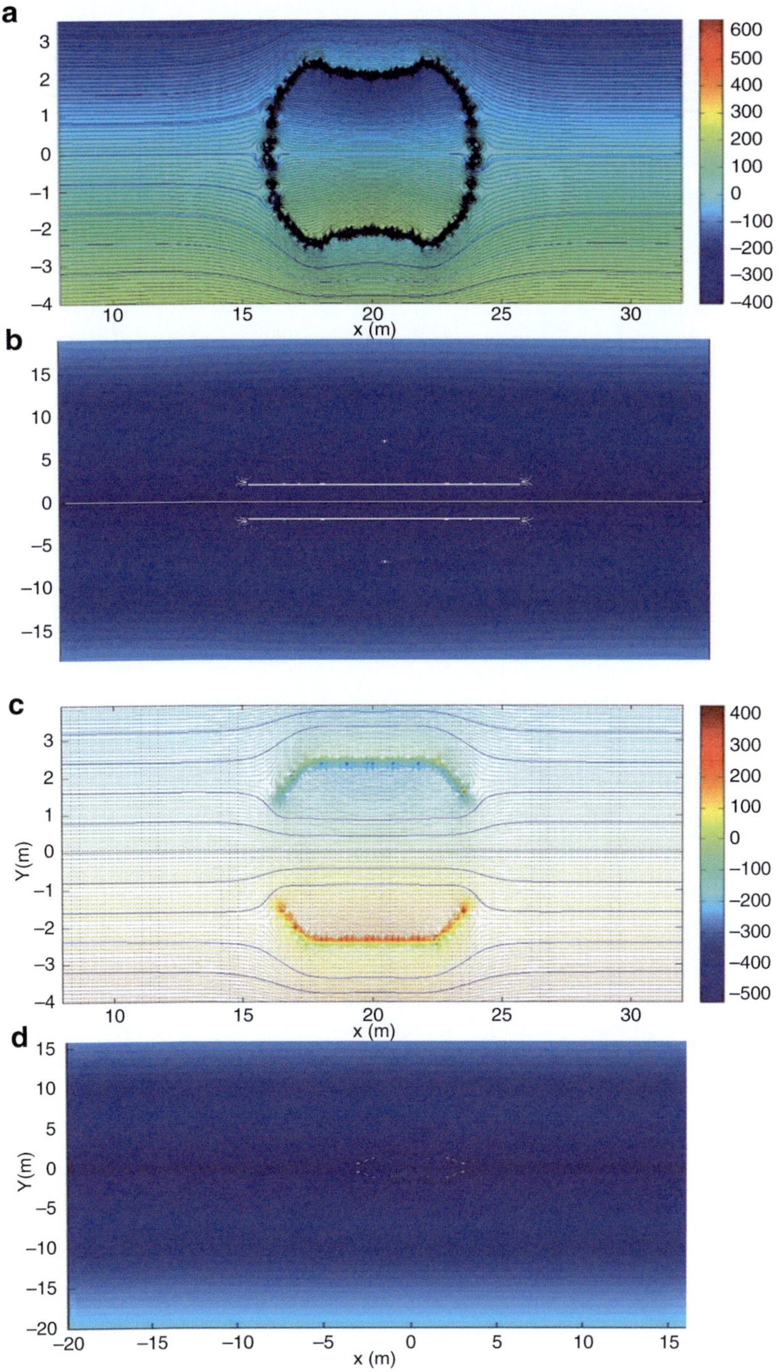

Fig. 64.2 Flow around: (**a**) ellipse and parabolic, (**b**) perforated ellipse, (**c**) smooth and linear function solid body, and (**d**) red cell in free fluidal medium

$$\begin{aligned}\phi(x,y) &= U_\infty \cdot x + \lim_{n\to\infty}\sum_{i=1}^{n} q\left(\frac{(x - x(i)_{AB})}{\left((x - x(i))^2 + (y - y(i)_{AB})^2\right)^2} e^{i\alpha(i)}\right)\\ &+ \lim_{n\to\infty}\sum_{i=1}^{n} q\left(\frac{(x - x(i)_{BC})}{\left((x - x(i))^2 + (y - y(i)_{BC})^2\right)^2} e^{i\alpha(i)}\right)\\ &+ \lim_{n\to\infty}\sum_{i=1}^{n} q\left(\frac{(x - x(i)_{CD})}{\left((x - x(i))^2 + (y - y(i)_{CD})^2\right)^2} e^{i\alpha(i)}\right)\\ &+ \lim_{n\to\infty}\sum_{i=1}^{n} q\left(\frac{(x - x(i)_{DE})}{\left((x - x(i))^2 + (y - y(i)_{DE})^2\right)^2} e^{i\alpha(i)}\right)\\ &+ \lim_{n\to\infty}\sum_{i=1}^{n} q\left(\frac{(x - x(i)_{EF})}{\left((x - x(i))^2 + (y - y(i)_{EF})^2\right)^2} e^{i\alpha(i)}\right)\\ &+ \cdots + \lim_{n\to\infty}\sum_{i=1}^{n} q\left(\frac{(x - x(i)_{LN})}{\left((x - x(i))^2 + (y - y(i)_{LN})^2\right)^2} e^{i\alpha(i)}\right)\end{aligned} \tag{64.8}$$

64.3.3 Velocity Field Around Solid Body in Free Fluidal Media

If it is used in Eq. (64.2), it can be derived in terms of velocity field equations as provided in Eq. (64.9).

$$\begin{aligned}\vec{V} = &\left\{\begin{array}{l} U_\infty + \lim_{n\to\infty}\sum_{i=1}^{n}\left[q\left[\left(\dfrac{(y - y(i))_{AB}{}^2 - (x - x(i))^2}{\left((x - x(i))^2 + (y - y(i))_{AB}{}^2\right)^2}\right) e^{i\alpha(i)}\right]\right] \\ + \cdots + \lim_{n\to\infty}\sum_{i=1}^{n}\left[q\left[\left(\dfrac{(y - y(i))_{LM}{}^2 - (x - x(i))^2}{\left((x - x(i))^2 + (y - y(i))_{LN}{}^2\right)^2}\right) e^{i\alpha(i)}\right]\right] \end{array}\right\}\vec{i} \\ + &\left\{\begin{array}{l} q\lim_{n\to\infty}\sum_{i=1}^{n}\left[\left(\dfrac{2(x - x(i))(y - y(i))_{AB}}{\left((x - x(i))^2 + (y - y(i))_{AB}{}^2\right)^2}\right) e^{i\alpha(i)}\right] \\ + \cdots + q\lim_{n\to\infty}\sum_{i=1}^{n}\left[\left(\dfrac{2(x - x(i))(y - y(i))_{LM}}{\left((x - x(i))^2 + (y - y(i))_{LN}{}^2\right)^2}\right) e^{i\alpha(i)}\right] \end{array}\right\}\vec{j}\end{aligned} \tag{64.9}$$

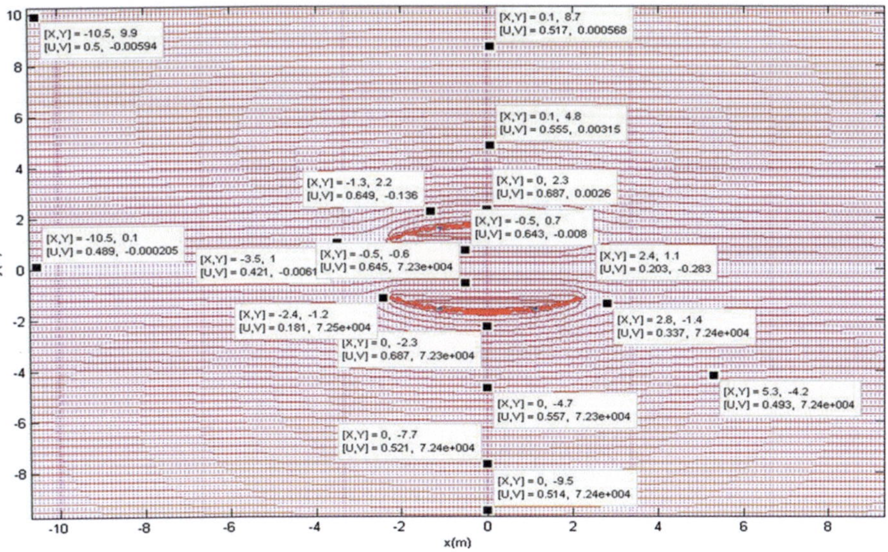

Fig. 64.3 Velocity and pressure field around perforated ellipse oval

The magnitude of velocity can be seen in Eq. (64.10). If it is applied on the perforated ellipse oval, the distribution of velocity will be as shown in Fig. 64.3.

$$
\left|\vec{V}\right|^2 = \left\{ \begin{array}{l} U_\infty + \lim\limits_{n\to\infty} \sum\limits_{i=1}^{n} \left[q \left[\left(\dfrac{(y - y(i))_{AB}{}^2 - (x - x(i))^2}{\left((x - x(i))^2 + (y - y(i))_{AB}{}^2\right)^2} \right) e^{i\alpha(i)} \right] \right] \\ + \cdots + \lim\limits_{n\to\infty} \sum\limits_{i=1}^{n} \left[q \left[\left(\dfrac{(y - y(i))_{LN}{}^2 - (x - x(i))^2}{\left((x - x(i))^2 + (y - y(i))_{LN}{}^2\right)^2} \right) e^{i\alpha(i)} \right] \right] \end{array} \right\}^2 + \left\{ \begin{array}{l} q \lim\limits_{n\to\infty} \sum\limits_{i=1}^{n} \left[\left(\dfrac{2(x - x(i))(y - y(i))_{AB}}{\left((x - x(i))^2 + (y - y(i))_{AB}{}^2\right)^2} \right) e^{i\alpha(i)} \right] \\ + \cdots + q \lim\limits_{n\to\infty} \sum\limits_{i=1}^{n} \left[\left(\dfrac{2(x - x(i))(y - y(i))_{LN}}{\left((x - x(i))^2 + (y - y(i))_{LN}{}^2\right)^2} \right) e^{i\alpha(i)} \right] \end{array} \right\}^2 \tag{64.10}
$$

If Eqs. (64.3) and (64.4) are used for producing pressure distribution and force field around a body it can be seen as Eq. (64.11), and pressure distribution on surface perforated ellipse oval it can be seen in (Fig. 64.4).

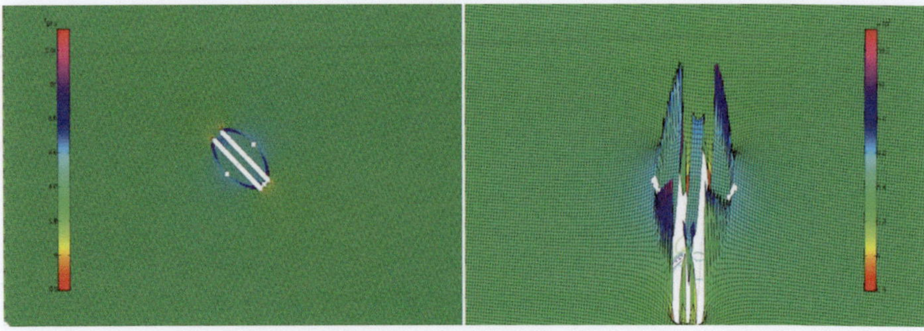

Fig. 64.4 Pressure distribution on surface perforated ellipse oval

$$P = p_\infty + \frac{\rho}{2}\left\{\begin{array}{l} U_\infty + \lim\limits_{n\to\infty}\sum\limits_{i=1}^{n}\left[q\left[\left(\dfrac{(y-y(\dot{I}))_{LM}{}^{2}-(x-x(\dot{I}))^{2}}{\left((x-x(\dot{I}))^{2}+(y-y(\dot{I}))_{LM}{}^{2}\right)^{2}}\right)e^{i\alpha(i)}\right]\right] \\ +\cdots+\lim\limits_{n\to\infty}\sum\limits_{i=1}^{n}\left[q\left[\left(\dfrac{(y-y(\dot{I}))_{AB}{}^{2}-(x-x(\dot{I}))^{2}}{\left((x-x(\dot{I}))^{2}+(y-y(\dot{I}))_{LN}{}^{2}\right)^{2}}\right)e^{i\alpha(i)}\right]\right] \end{array}\right\}^{2}$$

$$+\frac{\rho}{2}\left\{\begin{array}{l} q\lim\limits_{n\to\infty}\sum\limits_{i=1}^{n}\left[\left(\dfrac{2(x-x(\dot{I}))(y-y(\dot{I}))_{AB}}{\left((x-x(\dot{I}))^{2}+(y-y(\dot{I}))_{AB}{}^{2}\right)^{2}}\right)e^{i\alpha(i)}\right] \\ +\cdots+q\lim\limits_{n\to\infty}\sum\limits_{i=1}^{n}\left[\left(\dfrac{2(x-x(\dot{I}))(y-y(\dot{I}))_{LN}}{\left((x-x(\dot{I}))^{2}+(y-y(\dot{I}))_{LN}{}^{2}\right)^{2}}\right)e^{i\alpha(i)}\right] \end{array}\right\}^{2} \tag{64.11}$$

$$F = \iint_A p\vec{n}\,d\sigma$$

$$= \iint_A \left\{\begin{array}{l} p_\infty + \dfrac{\rho}{2}\left\{\begin{array}{l} U_\infty + \lim\limits_{n\to\infty}\sum\limits_{i=1}^{n}\left[q\left[\left(\dfrac{(y-y(\dot{I}))_{LM}{}^{2}-(x-x(\dot{I}))^{2}}{\left((x-x(\dot{I}))^{2}+(y-y(\dot{I}))_{LM}{}^{2}\right)^{2}}\right)e^{i\alpha(i)}\right]\right]+\cdots \\ +\lim\limits_{n\to\infty}\sum\limits_{i=1}^{n}\left[q\left[\left(\dfrac{(y-y(\dot{I}))_{AB}{}^{2}-(x-x(\dot{I}))^{2}}{\left((x-x(\dot{I}))^{2}+(y-y(\dot{I}))_{LN}{}^{2}\right)^{2}}\right)e^{i\alpha(i)}\right]\right] \end{array}\right\}^{2} \\ +\dfrac{\rho}{2}\left\{\begin{array}{l} q\lim\limits_{n\to\infty}\sum\limits_{i=1}^{n}\left[\left(\dfrac{2(x-x(\dot{I}))(y-y(\dot{I}))_{AB}}{\left((x-x(\dot{I}))^{2}+(y-y(\dot{I}))_{AB}{}^{2}\right)^{2}}\right)e^{i\alpha(i)}\right]+\cdots \\ +q\lim\limits_{n\to\infty}\sum\limits_{i=1}^{n}\left[\left(\dfrac{2(x-x(\dot{I}))(y-y(\dot{I}))_{LN}}{\left((x-x(\dot{I}))^{2}+(y-y(\dot{I}))_{LN}{}^{2}\right)^{2}}\right)e^{i\alpha(i)}\right] \end{array}\right\}^{2} \end{array}\right\}\vec{n}\,d\sigma \tag{64.12}$$

64.4 Modeling Theorem in Restricted Fluidal Media

64.4.1 *Complex Potential Function of Solid Body in Restricted Fluidal Media*

It is assumed that the solid body has a complex shape and certain boundaries as shown in Fig. 64.5. If the theory can be applied in restricted fluidal media around the solid body, which has complex shape, the complex potential function can be written as in Eq. (64.13). A solid wall is generated with symmetry, where j is changed from 1 to m.

$$\begin{aligned}
F_{AN}(z) &= F_{AB}(z) + F_{BC}(z) + \cdots + F_{LM}(z) \\
&= U_{\infty} z + \lim_{n\to\infty} \sum_{i=1}^{n} \frac{q e^{i\alpha(i)}}{z \mp z_{AB}(i)} + \lim_{n,m\to\infty} \sum_{i=1}^{n} \frac{q e^{i\alpha(i)}}{z \mp z_{BC}(i)} \\
&\quad + \cdots + \lim_{n,m\to\infty} \sum_{i=1}^{n} \frac{q e^{i\alpha(i)}}{z \mp zN(i)} \\
&\quad + \lim_{n,m\to\infty} \sum_{J=1}^{m} \sum_{i=1}^{n} \frac{q e^{i\alpha(i)}}{z \mp z_{AB}(i) \pm j \cdot bi} + \lim_{n,m\to\infty} \sum_{J=1}^{m} \sum_{i=1}^{n} \frac{q e^{i\alpha(i)}}{z \mp z_{BB}(i) \pm j \cdot bi} \\
&\quad + \cdots + \lim_{n,m\to\infty} \sum_{J=1}^{m} \sum_{i=1}^{n} \frac{q e^{i\alpha(i)}}{z \mp zN(i) \pm j \cdot bi}
\end{aligned} \tag{64.13}$$

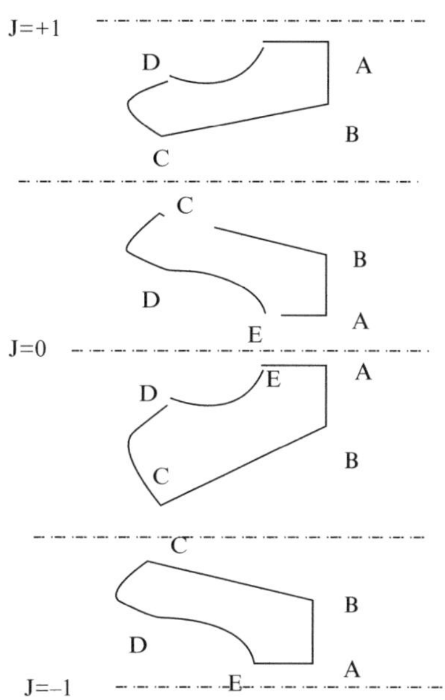

Fig. 64.5 Produced flow around a solid body in restricted fluidal media

64.4.2 *Stream and Potential Function of Solid Body in Restricted Fluidal Media*

An equation based on the equation complex potential function o, flow, and potential function can be derived as in Eqs. (64.14) and (64.15).

$$
\begin{aligned}
\psi(x,y) = {} & U_\infty \cdot y - \lim_{n\to\infty}\sum_{i=1}^{n} q\left(\frac{(y-y(i)_{AB})}{\left((x-x(i))^2+(y-y(i)_{AB})^2\right)}e^{i\alpha(i)}\right) - \lim_{n\to\infty}\sum_{i=1}^{n} q\left(\frac{\left(y-y(i)_{BC}\right)}{\left((x-x(i))^2+\left(y-y(i)_{BC}\right)^2\right)}e^{i\alpha(i)}\right)\\
& - \lim_{n\to\infty}\sum_{i=1}^{n} q\left(\frac{(y+y(i)_{AB})}{\left((x-x(i))^2+(y+y(i)_{AB})^2\right)}e^{i\alpha(i)}\right) - \lim_{n\to\infty}\sum_{i=1}^{n} q\left(\frac{\left(y+y(i)_{BC}\right)}{\left((x-x(i))^2+\left(y+y(i)_{BC}\right)^2\right)}e^{i\alpha(i)}\right)\\
& - \lim_{n\to\infty}\sum_{i=1}^{n} q\left(\frac{\left(y+y(i)_{CD}\right)}{\left((x-x(i))^2+\left(y+y(i)_{CD}\right)^2\right)}e^{i\alpha(i)}\right) - \lim_{n\to\infty}\sum_{i=1}^{n} q\left(\frac{(y+y(i)_{DE})}{\left((x-x(i))^2+(y+y(i)_{DE})^2\right)}e^{i\alpha(i)}\right)\\
& - \lim_{n\to\infty}\sum_{i=1}^{n} q\left(\frac{(y+y(i)_{EF})}{\left((x-x(i))^2+(y+y(i)_{EF})^2\right)}e^{i\alpha(i)}\right) - \cdots - \lim_{n\to\infty}\sum_{i=1}^{n} q\left(\frac{(y+y(i)_{LM})}{\left((x-x(i))^2+(y+y(i)_{LM})^2\right)}e^{i\alpha(i)}\right)\\
& - \lim_{n\to\infty}\sum_{J=1}^{m}\sum_{i=1}^{n} q\left(\frac{(y-y(i)_{AB}-jb)}{\left((x-x(i))^2+(y-y(i)_{AB}-jb)^2\right)}e^{i\alpha(i)}\right) - \lim_{n\to\infty}\sum_{J=1}^{m}\sum_{i=1}^{n} q\left(\frac{\left(y-y(i)_{BC}-jb\right)}{\left((x-x(i))^2+\left(y-y(i)_{BC}-jb\right)^2\right)}e^{i\alpha(i)}\right)\\
& - \lim_{n\to\infty}\sum_{J=1}^{m}\sum_{i=1}^{n} q\left(\frac{\left(y-y(i)_{CD}-jb\right)}{\left((x-x(i))^2+\left(y-y(i)_{CD}-jb\right)^2\right)}e^{i\alpha(i)}\right) - \cdots - \lim_{n\to\infty}\sum_{J=1}^{m}\sum_{i=1}^{n} q\left(\frac{\left(y-y(i)_{LN}-jb\right)}{\left((x-x(i))^2+\left(y-y(i)_{LN}-jb\right)^2\right)}e^{i\alpha(i)}\right)\\
& - \lim_{n\to\infty}\sum_{J=1}^{m}\sum_{i=1}^{n} q\left(\frac{(y-y(i)_{AB}+jb)}{\left((x-x(i))^2+(y-y(i)_{AB}+jb)^2\right)}e^{i\alpha(i)}\right) - \lim_{n\to\infty}\sum_{J=1}^{m}\sum_{i=1}^{n} q\left(\frac{\left(y-y(i)_{BC}+jb\right)}{\left((y-x(i))^2+\left(y-y(i)_{BC}+jb\right)^2\right)}e^{i\alpha(i)}\right)\\
& - \lim_{n\to\infty}\sum_{J=1}^{m}\sum_{i=1}^{n} q\left(\frac{\left(y-y(i)_{CD}+jb\right)}{\left((x-x(i))^2+\left(y-y(i)_{CD}+jb\right)^2\right)}e^{i\alpha(i)}\right) - \cdots - \lim_{n\to\infty}\sum_{J=1}^{m}\sum_{i=1}^{n} q\left(\frac{\left(y-y(i)_{LN}+jb\right)}{\left((x-x(i))^2+\left(y-y(i)_{LN}+jb\right)^2\right)}e^{i\alpha(i)}\right)\\
& - \lim_{n\to\infty}\sum_{J=1}^{m}\sum_{i=1}^{n} q\left(\frac{(y+y(i)_{AB}-jb)}{\left((x-x(i))^2+(y+y(i)_{AB}-jb)^2\right)}e^{i\alpha(i)}\right) - \lim_{n\to\infty}\sum_{J=1}^{m}\sum_{i=1}^{n} q\left(\frac{\left(y+y(i)_{BC}-jb\right)}{\left((x-x(i))^2+\left(y+y(i)_{BC}-jb\right)^2\right)}e^{i\alpha(i)}\right)\\
& - \lim_{n\to\infty}\sum_{J=1}^{m}\sum_{i=1}^{n} q\left(\frac{\left(y+y(i)_{CD}-jb\right)}{\left((x-x(i))^2+\left(y+y(i)_{CD}-jb\right)^2\right)}e^{i\alpha(i)}\right) - \cdots - \lim_{n\to\infty}\sum_{J=1}^{m}\sum_{i=1}^{n} q\left(\frac{\left(y+y(i)_{LN}-jb\right)}{\left((x-x(i))^2+\left(y+y(i)_{LN}-jb\right)^2\right)}e^{i\alpha(i)}\right)\\
& - \lim_{n\to\infty}\sum_{J=1}^{m}\sum_{i=1}^{n} q\left(\frac{(y+y(i)_{AB}+jb)}{\left((x-x(i))^2+(y+y(i)_{AB}+jb)^2\right)}e^{i\alpha(i)}\right) - \lim_{n\to\infty}\sum_{J=1}^{m}\sum_{i=1}^{n} q\left(\frac{\left(y+y(i)_{BC}+jb\right)}{\left((x-x(i))^2+\left(y+y(i)_{BC}+jb\right)^2\right)}e^{i\alpha(i)}\right)\\
& - \lim_{n\to\infty}\sum_{J=1}^{m}\sum_{i=1}^{n} q\left(\frac{\left(y+y(i)_{CD}+jb\right)}{\left((x-x(i))^2+\left(y+y(i)_{CD}+jb\right)^2\right)}e^{i\alpha(i)}\right) - \cdots - \lim_{n\to\infty}\sum_{J=1}^{m}\sum_{i=1}^{n} q\left(\frac{\left(y+y(i)_{LN}+jb\right)}{\left((x-x(i))^2+\left(y+y(i)_{LN}+jb\right)^2\right)}e^{i\alpha(i)}\right)
\end{aligned}
\tag{64.14}
$$

In Fig. 64.6a it can be seen the streamlines of the applied on red cell, 64.6b in perforated red cell, 64.6c perforated rankine oval shaped.

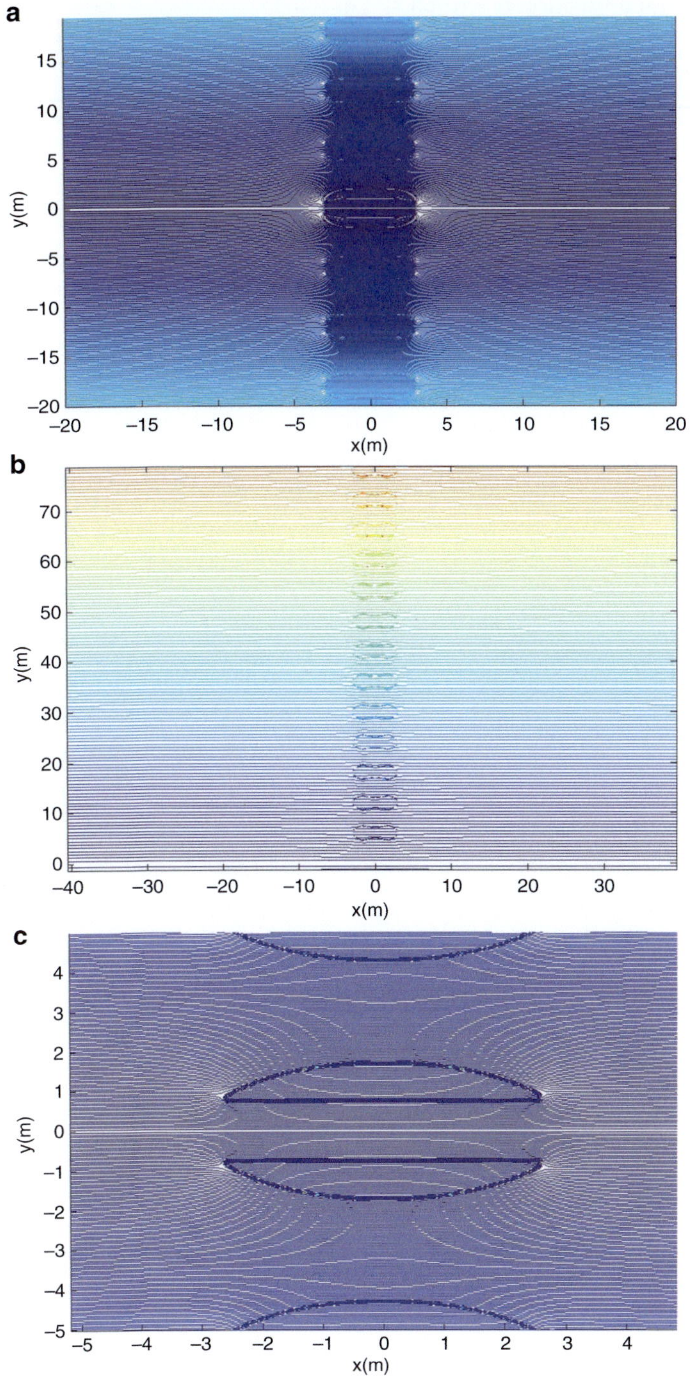

Fig. 64.6 Solid body in restricted fluidal media: (**a**) red cell, (**b**) perforated red cell, and (**c**) perforated ellipse oval

The potential function is written as Eq. (64.15)

$$
\begin{aligned}
\phi(x,y) = U_{\infty}\cdot x &+ \lim_{n\to\infty}\sum_{i=1}^{n} q\left(\frac{(x-x(i))}{\left((x-x(i))^2+(y-y(i)_{AB})^2\right)}e^{i\alpha(i)}\right) + \lim_{n\to\infty}\sum_{i=1}^{n} q\left(\frac{(x-x(i))}{\left((x-x(i))^2+(y-y(i)_{BC})^2\right)}e^{i\alpha(i)}\right)\\
&+ \lim_{n\to\infty}\sum_{i=1}^{n} q\left(\frac{(x-x(i))}{\left((x-x(i))^2+(y-y(i)_{CD})^2\right)}e^{i\alpha(i)}\right) + \lim_{n\to\infty}\sum_{i=1}^{n} q\left(\frac{(x-x(i))}{\left((x-x(i))^2+(y-y(i)_{DE})^2\right)}e^{i\alpha(i)}\right)\\
&+ \lim_{n\to\infty}\sum_{i=1}^{n} q\left(\frac{(x-x(i))}{\left((x-x(i))^2+(y-y(i)_{EF})^2\right)}e^{i\alpha(i)}\right) + \cdots + \lim_{n\to\infty}\sum_{i=1}^{n} q\left(\frac{(x-x(i))}{\left((x-x(i))^2+(y-y(i)_{LN})^2\right)}e^{i\alpha(i)}\right)\\
&+ \lim_{n\to\infty}\sum_{i=1}^{n} q\left(\frac{(x-x(i))}{\left((x-x(i))^2+(y+y(i)_{AB})^2\right)}e^{i\alpha(i)}\right) + \lim_{n\to\infty}\sum_{i=1}^{n} q\left(\frac{(x-x(i))}{\left((x-x(i))^2+(y+y(i)_{BC})^2\right)}e^{i\alpha(i)}\right)\\
&+ \lim_{n\to\infty}\sum_{i=1}^{n} q\left(\frac{(x-x(i))}{\left((x-x(i))^2+(y+y(i)_{CD})^2\right)}e^{i\alpha(i)}\right) + \lim_{n\to\infty}\sum_{i=1}^{n} q\left(\frac{(x-x(i))}{\left((x-x(i))^2+(y+y(i)_{DE})^2\right)}e^{i\alpha(i)}\right)\\
&+ \lim_{n\to\infty}\sum_{i=1}^{n} q\left(\frac{(x-x(i))}{\left((x-x(i))^2+(y+y(i)_{EF})^2\right)}e^{i\alpha(i)}\right) + \cdots + \lim_{n\to\infty}\sum_{i=1}^{n} q\left(\frac{(x-x(i))}{\left((x-x(i))^2+(y+y(i)_{LN})^2\right)}e^{i\alpha(i)}\right)\\
&+ \lim_{n\to\infty}\sum_{J=1}^{m}\sum_{i=1}^{n} q\left(\frac{(x-x(i))}{\left((x-x(i))^2+(y-y(i)_{AB}-jb)^2\right)}e^{i\alpha(i)}\right) + \lim_{n\to\infty}\sum_{J=1}^{m}\sum_{i=1}^{n} q\left(\frac{(x-x(i))}{\left((x-x(i))^2+(y-y(i)_{BC}-jb)^2\right)}e^{i\alpha(i)}\right)\\
&+ \lim_{n\to\infty}\sum_{J=1}^{m}\sum_{i=1}^{n} q\left(\frac{(x-x(i))}{\left((x-x(i))^2+(y-y(i)_{CD}-jb)^2\right)}e^{i\alpha(i)}\right) + \cdots + \lim_{n\to\infty}\sum_{J=1}^{m}\sum_{i=1}^{n} q\left(\frac{(x-x(i))}{\left((x-x(i))^2+(y-y(i)_{LN}-jb)^2\right)}e^{i\alpha(i)}\right)\\
&+ \lim_{n\to\infty}\sum_{J=1}^{m}\sum_{i=1}^{n} q\left(\frac{(x-x(i))}{\left((x-x(i))^2+(y-y(i)_{AB}+jb)^2\right)}e^{i\alpha(i)}\right) + \lim_{n\to\infty}\sum_{J=1}^{m}\sum_{i=1}^{n} q\left(\frac{(x-x(i))}{\left((x-x(i))^2+(y-y(i)_{BC}+jb)^2\right)}e^{i\alpha(i)}\right)\\
&+ \lim_{n\to\infty}\sum_{J=1}^{m}\sum_{i=1}^{n} q\left(\frac{(x-x(i))}{\left((x-x(i))^2+(y-y(i)_{CD}+jb)^2\right)}e^{i\alpha(i)}\right) + \cdots + \lim_{n\to\infty}\sum_{J=1}^{m}\sum_{i=1}^{n} q\left(\frac{(x-x(i))}{\left((x-x(i))^2+(y-y(i)_{LN}+jb)^2\right)}e^{i\alpha(i)}\right)\\
&+ \lim_{n\to\infty}\sum_{J=1}^{m}\sum_{i=1}^{n} q\left(\frac{(x-x(i))}{\left((x-x(i))^2+(y+y(i)_{AB}-jb)^2\right)}e^{i\alpha(i)}\right) + \lim_{n\to\infty}\sum_{J=1}^{m}\sum_{i=1}^{n} q\left(\frac{(x-x(i))}{\left((x-x(i))^2+(y+y(i)_{BC}-jb)^2\right)}e^{i\alpha(i)}\right)\\
&+ \lim_{n\to\infty}\sum_{J=1}^{m}\sum_{i=1}^{n} q\left(\frac{(x-x(i))}{\left((x-x(i))^2+(y+y(i)_{CD}-jb)^2\right)}e^{i\alpha(i)}\right) + \cdots + \lim_{n\to\infty}\sum_{J=1}^{m}\sum_{i=1}^{n} q\left(\frac{(x-x(i))}{\left((x-x(i))^2+(y+y(i)_{LN}-jb)^2\right)}e^{i\alpha(i)}\right)\\
&+ \lim_{n\to\infty}\sum_{J=1}^{m}\sum_{i=1}^{n} q\left(\frac{(x-x(i))}{\left((x-x(i))^2+(y+y(i)_{AB}+jb)^2\right)}e^{i\alpha(i)}\right) + \lim_{n\to\infty}\sum_{J=1}^{m}\sum_{i=1}^{n} q\left(\frac{(x-x(i))}{\left((x-x(i))^2+(y+y(i)_{BC}+jb)^2\right)}e^{i\alpha(i)}\right)\\
&+ \lim_{n\to\infty}\sum_{J=1}^{m}\sum_{i=1}^{n} q\left(\frac{(x-x(i))}{\left((x-x(i))^2+(y+y(i)_{CD}+jb)^2\right)}e^{i\alpha(i)}\right) + \cdots + \lim_{n\to\infty}\sum_{J=1}^{m}\sum_{i=1}^{n} q\left(\frac{(x-x(i))}{\left((x-x(i))^2+(y+y(i)_{LN}+jb)^2\right)}e^{i\alpha(i)}\right)
\end{aligned}
\tag{64.15}
$$

The velocity field and pressure distribution, and force field in restricted fluidal media can be generated analytically as in the free fluidal medium with Eqs. (64.2) and (64.3), when the velocity formula is applied . The velocity field around the body is illustrated in Fig. 64.7.

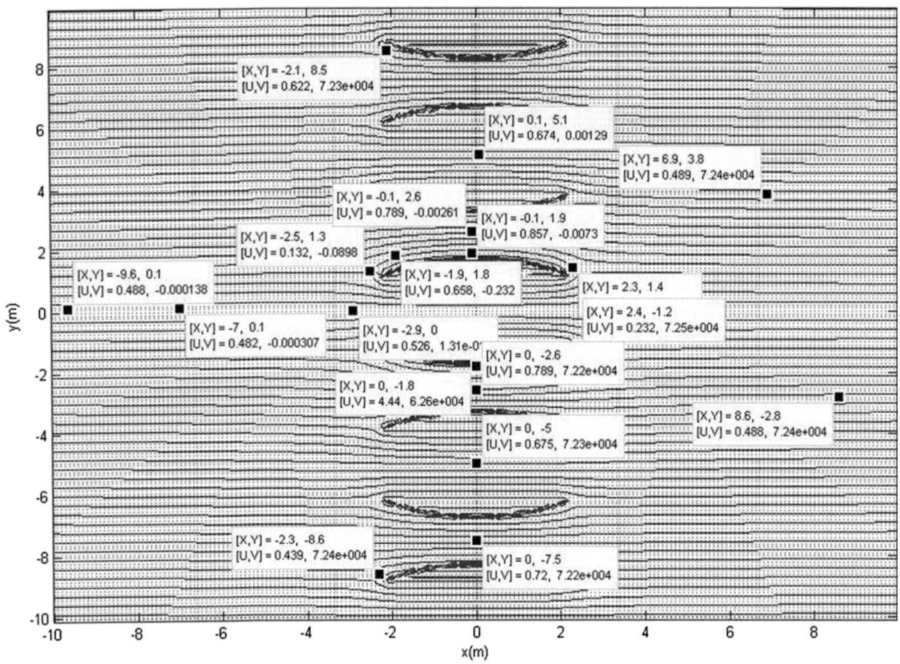

Fig. 64.7 Velocity field around solid body in restricted fluidal media

The velocity field and pressure distribution, and force field in restricted fluidal media can be generated analytically the same as in free media with Eqs. (64.2) and (64.3), and the applied velocity formulas in Eqs. (64.14) and (64.15). The pressure distribution is provided in Figs. 64.8a, b.

With this method any different channel can be constructed. Primarily, the channel boundary is formulated analytically. Then the boundary wall of the channel is made rigid. The channel flow and pressure field of flow in the channel can be seen in Figs. 64.9a, b.

64.5 Conclusions

The model, which has any geometrical solid body, is generated analytically. It has been used in a new theorem. With this theorem, the complex potential function of flows around a body, and stream function, the velocity field is derived. Then pressure and the force field around the solid body for free and restricted fluidal

a

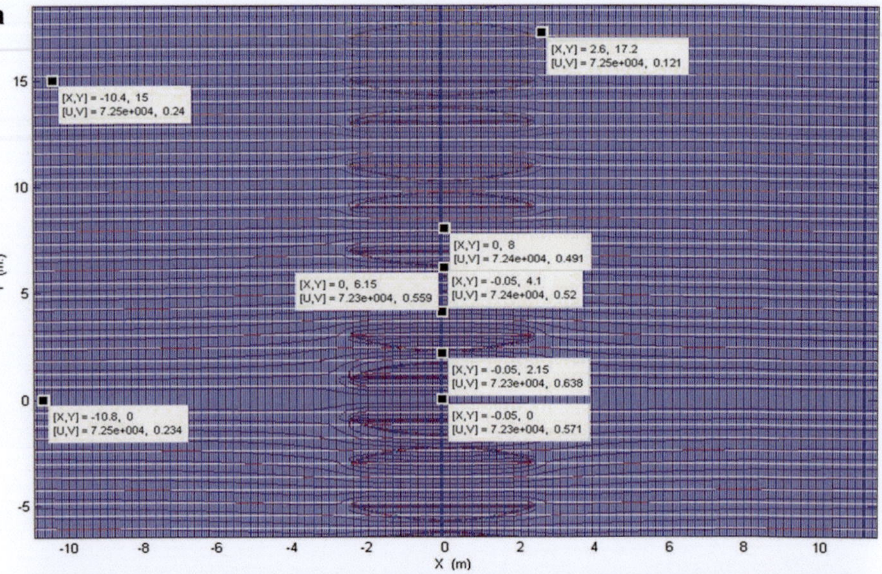

b

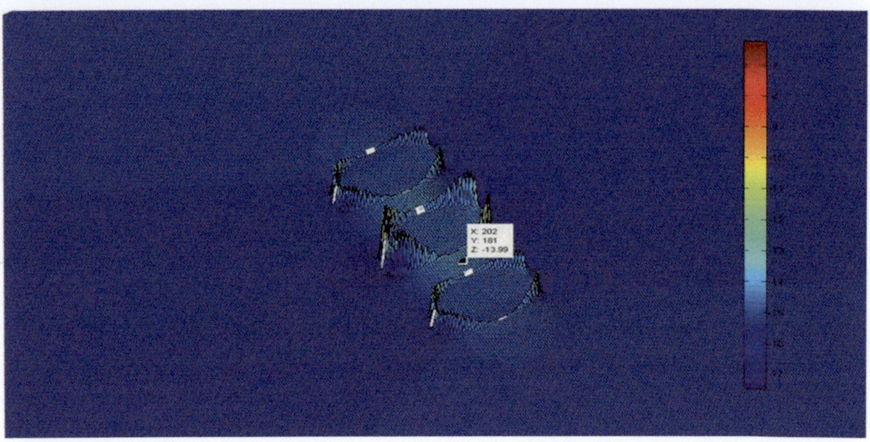

Fig. 64.8 Pressure distribution on solid body in restricted fluidal media: (**a**) with value, and (**b**) pressure surface

media are generated. After that, this function is applied in free fluidal media on bodies with different geometry, and illustrated using MATLAB software. In the finalized media, flow solid medium geometry and flow function with potential function have been obtained and confirmed with the Matlab program in terms of its validity.

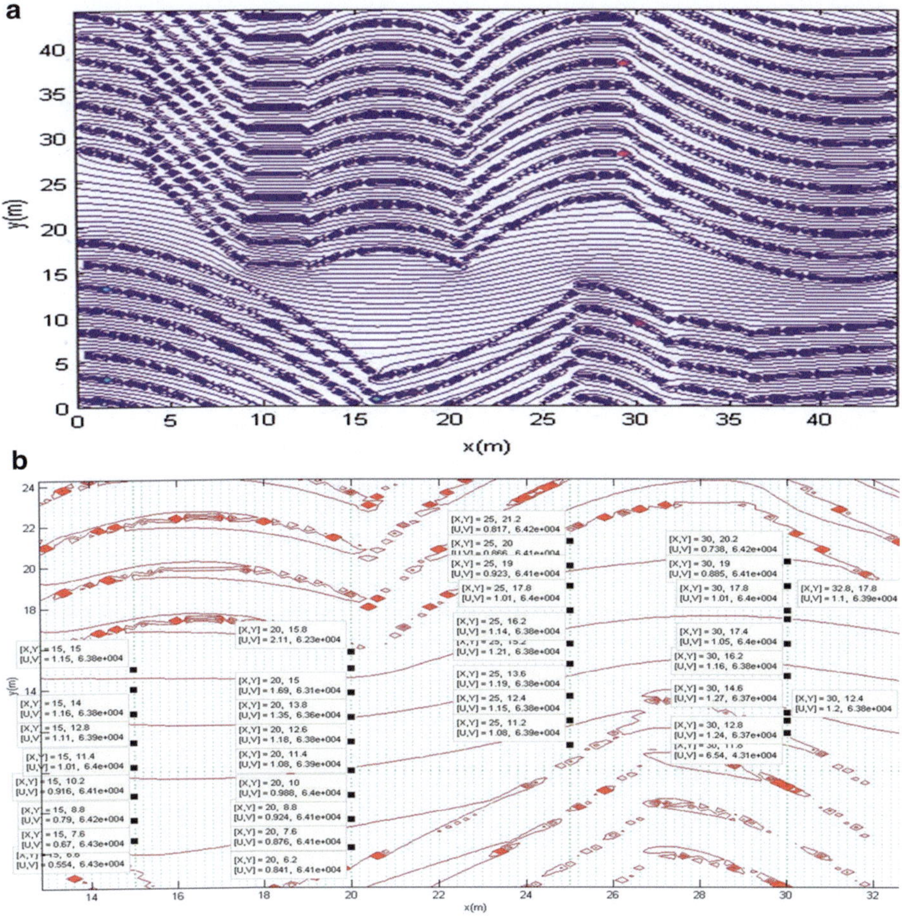

Fig. 64.9 Channel construction: (**a**) channel flow, and (**b**) pressure field of flow in channel

References

1. Blair A, Campbell R, Grant PT (1982) A submersible fish cage that can be rotated on the surface to remove biofouling and for other purposes. Aquaculture 29:177–184
2. Djerid H, Baraza M, Perin R, Harran G (2003) Near wake turbulence properties around a circular cylinder at high Reynold number. Flow Turbul Cumbust 71:19–33
3. Durgun O (1983) Kanallarda blokaj etkisinin belirlenmesi, İTÜ, Fen Bilimleri Enstitüsü, İstanbul, Doktora tezi, İstanbul, 002177
4. Gilmanov A, Sotiropoulos F (2005) A hybrid cartesian immersed body method for simulating flows with 3D, geometrically complex, moving bodies. J Comput Phys 207:457–492
5. Hsu CH, Liang CC, Shiah SW, Jen CY (2005) A study of stress concentration effect around penetrations on curved shell and failure modes for deep-diving submersible vehicle. Ocean Eng 32:1098–1121

6. Huggins A, Packwood AR (1994) The optimum dimensions for a long-range, autonomous, deep-diving, underwater vehicle for oceanographic research. Ocean Eng 21:45–56
7. Janson CE (1997) Potential flow panel method for calculation of free surface flow with lift. Doctor savhandlingar vid chalmers Tekniska Hogskola, 1277 ISBN: 91-7197-469-5
8. Kauh JS, Suen J-B (2001) A 3D potential-based and designgularized high order panel method. Ocean Eng 28:1499–1516
9. Papoulias FA, Bateman CA, Ornek S (1995) Dynamic loss of stability in debth control of submersible vecihles, Applied Ocean Research 17:205–216
10. Liu Z, Xio Y, Tu CX (2012) Method to control unsteady force of submarine propeller based on the control of horseshoe vortex. J Ship Res 56:12–22
11. Madsen HØ, Christensen P, Lauridsen K (2000) Securing the operational reliability of an autonomous mini-submarine. Reliab Eng Syst Saf 68:7–16
12. Mahfouz AB, Haddara MR (2003) Effect of the damping and excitation on the identification of the hydrodynamic parameters for an underwater robotic vehicle. Ocean Eng 30:1005–1025
13. Mraccek CP, Kim MJ, Mook DT (1992) Three-dimensional potential flow by a vorticity-panel method. Comput Fluids 21:31–42
14. Niazmand H, Renksizbulut M (2003) Surface effect on transient three-dimensional flow around rotating at moderate Reynolds numbers. Comput Fluids 32:1405–1433
15. Ohring S, Telste J (1980) The direct matrix imbedding technique for computing three-dimensional potential flow about arbitrarily shaped bodies. Comput Methods Appl Mech Eng 21:315–336
16. Perrault D, Bose N, Young SO, Williams CD (2003) Sensitivity of AUV response to variations in hydrodynamic parameters. Ocean Eng 30:779–811
17. Rapoport L, Rubin MB (2009) Separation and velocity dependence of drag forces applied to a rigid ovoid of Rankine nosed projectile penetrating an elastic-perfectly-plastic-target. Int J Impact Eng 36:1012–1018
18. Goel NS, Gang FA (1995) general and computationally fast method for investigating incrossible flow. Int Commun Heat Transfer Mass Transfer 22:425–434
19. Reynolds T, Lomacky O, Krenzke M (1973) Design and analysis of small submersible pressure hulls. Comput Struct 3:1125–1143
20. Rubin MB (2012) Analytical formulas for penetration of a long rigid projectile including the effect of cavitation. Int J Impact Eng 40–41:1–9
21. Sahin I, Hyman M (1993) Numerical calculation for the flow of submerged bodies under a free surface. J Comput Phys 20:339–347
22. Sahin I, Hyman MC, Nguyen TC (1994) Three-dimensional flow around a submersible body in finite depth water. Appl Math Model 18:611–619
23. Szymzak WG, Rogers JCW, Solomon JM, Berger AE (1993) A numerical algorithm for hydrodynamic free boundary problem. J Comput Phys 106:319–336
24. Virgin LN, Erickson BK (1994) A new approach to the over turning stability of floating structure. Ocean Eng 21:67–80
25. Wu J, Chwang AT (2001) Investigation on a two part underwater maneuverable towed system. Ocean Eng 28:1070–1096
26. Evans J, Nahon M (2004) Dynamics modeling and performance evaluation of an autonomous underwater vehicle. Ocean Eng 31:1835–1858
27. Wu J, Ye J, Yang C, Chen Y, Tian H, Xiong X (2005) Experimental study on a controllable underwater towed system. Ocean Eng 32:1803–1817
28. Xu Z, Chen X, Strurova IV (2006) Surface characters of internal waves generated by Rankine ovoid. Acta Mech Sinica 22:417–423
29. Xu Z, Li O, Gorodtsov VA (2008) Wave drag of rapidly and horizontally moving Rankine ovoid in uniformly stratified fluid. Prog Nat Sci 18:723–727
30. Xu ZT, Zhou XW, Chen X (2002) Drag increment due to internal waves generated by Rankine ovoid. Prog Nat Sci 12:849–853

31. Zubarev VM (1984) Laminar boundary layer on the moving surface of Rankine oval. Fluid Dynamics 19:490–493 589
32. Zungber MD, Lee HC (2000) Global existence of weak solution for viscous incompressible flow around a moving rigid body in three dimensions. J Math Fluid Mech 2:219–266
33. Wu J, Chwang AT (2000) 3-D simulation of a two-part underwater towed system. Ocean Eng 27:455–472
34. Zakharov VE, Dyachenko AI, Vasilyev OA (2002) New method for numerical simulation of a nonstationary potential flow of incompressible fluid with a free surface. Eur J Mech B-Fluids 21:283–291
35. Papoulias FA, Bateman CA, Ornek S (1995) Dynamic loss of stability in debth control of submersible vecihles Applied Ocean Research 17:205–216

Chapter 65
Effect on Wall Shear Rates of Taylor Vortex Instabilities Induced by Progressive Variation of the Inner Cylinder

Emna Berrich, Fethi Aloui, and Jack Legrand

Abstract Taylor–Couette flows are frequently reencountered on engineering processes, such as for cooling the rotors of electric motors by admitting cold air and improving the mixing in the combustion area, for the injection of fuel into the cylinders of gasoline engines, for the injection of just-prepared emulsions on diesel engines, etc. On its simplest configurations, Taylor–Couette systems (TCS) are widely studied when one or both cylinders steadily rotate. Limited studies are available as concerned on the acceleration-deceleration of one or both cylinders. This paper deals with the effect of acceleration-deceleration imposed to the inner cylinder of the CTS on flow instabilities. Imposing progressive periodic transitional motions through the cylindrical CTS walls alters the stability of the flow. The objective of this work is to study the flow dynamics and interactions between wall and vortices while imposing progressive periodic motions. We used the electrodiffusion (ED) method, known also as the polarography technique. The apparition and disappearance of the regime transitions, due to the acceleration-deceleration of the inner cylinder speed, on the frequency response of the ED sensor were illustrated and discussed. The regimes' instabilities and the corresponding Taylor numbers of the first and the second transitions of TC flow were determined. The results show coherence between the temporal wall velocity gradient (determined from the mass transfer rates) and the flow turbulent characteristics, particularly at the transitions between the Couette flow (CF) regime, the wavy vortex flow (WVF), and the modulated wavy vortex flow (MWVF). Instantaneous mass transfer rate and velocity gradient have been compared for the CF, the WVF, and the

E. Berrich
LUNAM Université, Université de Nantes, École des Mines de Nantes, CNRS, GEPEA, UMR6144, 4 rue Alfred KASTLER, BP20722, Nantes Cedex 03 44307, France

F. Aloui (✉)
LAMIH CNRS UMR 8201, Department of Mechanics, University of Valenciennes et du Hainaut-Cambrésis (UVHC), Le Mont Houy, Valenciennes Cedex 9 59313, France
e-mail: Fethi.Aloui@univ-valenciennes.fr

J. Legrand
LUNAM Université, Université de Nantes, CNRS, GEPEA, UMR6144, CRTT, BP 406, Saint-Nazaire Cedex 44602, France

I. Dincer et al. (eds.), *Progress in Clean Energy, Volume 1*,
DOI 10.1007/978-3-319-16709-1_65

MWVF regimes. The apparition of small then large vortices, when the Taylor number increases, i.e., from laminar regime to turbulent one, affects the mass transfer and the wall shear rate. Sensor inertia effect and a slight hysteresis effect, due to sudden periodic flow velocity decrease, were detected. The ED results allowed testing the robustness of the inverse method, based on the inversion of the convection-diffusion equation, on mass transfer to determine the wall shear rate.

Keywords Convection-diffusion • Taylor vortices • Acceleration-deceleration • Hydrodynamic instabilities • Wall shear rate • Mass transfer • Electrodiffusion method • Inverse method

Nomenclature

A	Surface of the probe (m^2)
C	Concentration (mol/m^3)
D	Coefficient of diffusion ($m^2\ s^{-1}$)
e	Gap width $e = R_{out} - R_{int}$
F	Frequency of oscillation (Hz)
I	Current (A)
l	Probe length
N	Velocity in revolution per minute
Pe	Péclet number
R	Radius (m)
Re	Reynolds number $Re = \frac{\Omega R \delta}{\upsilon}$
S	Wall shear stress (s^{-1})
Sh	Sherwood number $Sh = \frac{i\,l}{n F C_0 A D}$
t	Time (s)
Ta	Taylor number $Ta = \frac{\tau_\upsilon}{\tau_{cent}} = \frac{\Omega R_{int} \delta}{\upsilon} \sqrt{\frac{\delta}{R_{int}}} = Re_i \sqrt{\frac{\delta}{R_{int}}}$
u	Velocity ($m\ s^{-1}$)
x, z	Coordinates (m)

Greek Symbols

Ω	Angular velocity of the rotating plate ($rad \cdot s^{-1}$)
υ	Kinetic viscosity ($m^2\ s^{-1}$)
η	Radial ratio
Γ	Aspect ratio

Superscripts

*	Dimensionless

Subscripts

C	Critical
ED	Electrodiffusion
i	Inner
q.l.	Quasilinear
Lev	Lévêque approach
Sob	Sobolik et al. [17] approach

65.1 Introduction

Taylor–Couette system is a simple system composed of two coaxial cylinders. However, it presents a rich variety of flow regimes with hydrodynamic transitions from laminar flow to turbulent one. It occurs in many practical situations and is of great interest in many processes such as engineering processes and medical applications. It is widely used for the depleted uranium enrichment, as a membrane or filter on bioreactors, and for blood detoxification. Taylor vortices have been shown to have excellent mixing characteristics in reactors that processed biological systems such as cell cultures, blood plasmapheresis, and oxygenation [1]. They can be also useful on the injection of fuel into the cylinders of gasoline engines, for cooling the rotors of electric motors by admitting cold air and improved the mixing in the combustion area. Dluska et al. [2] proposed the integration of the Taylor–Couette flow (TCF) contactor with diesel engines for the injection of just-prepared emulsions. Usually, fuel is injected into the engine from storage tanks. Emulsions from tanks include surfactants for stability. During combustion, surfactants generate new undesirable emissions. Simple and multiple water fuel emulsion preparation in a liquid-liquid contactor with Taylor–Couette flow (TCF) contactor is considered as an alternative diesel fuel for improving engine performance and emission characteristics.

Taylor–Couette (rotor-stator) system seems to be an efficient system for the elimination of the "clogging" phenomena in the filtration processes. This is due to the centrifugal forces and to shear induced by the speedy and controlled rotation which generates the vortex apparition. The optimization of such process needs a well knowledge and understanding of the flow structures, in particular in the vicinity of the wall regions which interested us.

A small radius ratio η (i.e., large gap width) can delay the transitions, or even the second instability does not take place ($\eta < 0.65$, [5]). The same analysis is highlighted if the aspect ratio Γ is sufficiently small ($\Gamma < 6$, [3]). The geometrical characteristic impact is more significant on the second instability than on the first one [3–6]. From the second instability apparition, the flow is characterized by a multiplicity of flow states [4]. Several combinations between the axial wave number and the azimuthal one are then possible. An important aspect ratio ($\Gamma > 40$) promotes the multiplicity of the flow states which depend not only on the Taylor number but also on the flow history [7, 8]. Cole [8] studied the effect of cylinder length on the Taylor vortex and secondary wavy vortex instabilities. He observed that the critical Reynolds number for TVF is in good agreement with theoretical predictions for infinite cylinders even for aspect ratios as low as $\Gamma = 8$. However, he found that the transition to WVF is very sensitive to aspect ratio. In particular, one needs $\Gamma > 40$ to obtain quantitative agreement with the theory for infinite cylinders ($\Gamma \to \infty$).

Taylor [9] was the first who studied the stability of a circular Couette flow contained between rotating cylinders. He observed and calculated by a linear stability analysis the critical rotation speed (a critical Taylor number) at which Couette flow becomes unstable. Coles [4] highlighted the multiplicity of the wavy flow in TC flow. He noted that more than 20 flow states were seen and that every flow state is characterized by an axial wavelength and an azimuthal wave number. These flow states strongly depend on the initial conditions and on the manner in which the inner cylinder was accelerated to the final speed. At the end of the 1970s, researches confirmed a theoretical analysis which concluded that a finite number of instabilities, two or three, are sufficient to lead to chaos or weak turbulence [10]. Cognet [7] proposed regime classification for CT system ($\eta = 0.908$) depending on the dimensionless Taylor number defined as

$$Ta^* = \frac{Ta}{Ta_c} \tag{65.1}$$

where

$$Ta = Re_\theta \sqrt{\frac{R_{in}}{e}} \tag{65.2}$$

with

$$Re_\theta = \frac{\rho R_i \Omega}{\upsilon} \tag{65.3}$$

Note the following:

1. Couette flow (CF): $Ta^* < 1$
2. First transition $Ta^* = 1$
3. Wavy vortex flow (WVF): $1 < Ta^* < 1.2$
4. Second transition $Ta^* = 1.2$
5. Modulated wavy Taylor vortex flow (MWVF): $1.2 < Ta^* < 9$

Taking into account our geometrical characteristics, the critical Taylor number characterizing the first instability apparition can be approximated by linear estimation from similar ratio geometrical parameters studied by Roberts [11]. For $\eta = 0.975$, $\mu = 0$, $\alpha = 2\Pi e/\lambda = 3.13$, it is approximately equal to $Ta_c = 41.79$.

65.2 Problem Statement

The few studies realized on the acceleration-deceleration effect on the flow have been limited to the onset of Taylor vortex flow [12] or restricted to only extreme conditions, i.e., sudden start and quasi-steady flow [13, 14]. Interesting visualizations of acceleration-deceleration were realized by Coles [4] and Ahlers et al. [15]. The lack of studies of the acceleration-deceleration effect on Taylor–Couette flow motivated us to carry out the present investigation. We focus on flow dynamics and interactions between wall and vortices while imposing progressive periodic motions. We tried to identify the regime transitions caused by the acceleration-deceleration rotor speed and its effects on mass transfer and wall shear stress at the inner wall of the stator.

65.3 Experimental Apparatus and Method

A schematic view of the Taylor–Couette system constructed and used in GEPEA laboratory is shown on Fig. 65.1. It is composed of two coaxial cylinders with a radius ratio $\eta = R_{int}/R_{out} = 0.97$ and an aspect ratio $\Gamma = L/e = 150$. It is made of plexiglass to assure the visualization of the flow hydrodynamics in the gap using a Kalliroscope electrochemical solution. The inner cylinder rotates while the outer one remains stationary. The inner cylinder speed is imposed by varying the frequency, the amplitude, and the offset of the motor controller to obtain accelerating velocity which decreases suddenly and increases again progressively. This phenomenon is repetitive in order to investigate the hysteresis effect (Fig. 65.2).

Until recently, many studies were done for the experimental determination of the shear rate. One of the well-known methods generally used is the polarography technique [8]. It is a noninstructive method based on a fast oxidoreduction reaction at the electrodiffusion (ED) sensor surface. It is generally recommended for the local wall shear rate measurements. From the current delivered by the electrode (the sensor), the mass transfer rate can be calculated using the Faraday law. The study of the frequency response of the ED sensor allows the determination of the flow transitions at the wall vicinity. Velocity gradients are determined from the mass transfer rate using different equations relating mass transfer and wall shear stress: Leveque method [16], Sobolik et al. method [17], and the inverse method [18]. Circular sensors were used to study the Taylor–Couette flows in the gap [19] and then three-segment sensors [20]. A microphotography of the three-segment sensor used is presented on Fig. 65.3. The multi-segment sensor was constructed according to the method described by Sobolik et al. [21]. The sensor was fixed flush to the inner wall of the outer cylinder of the Taylor–Couette system (Fig. 65.3). An electrolyte solution of equimolar ferri-ferrocyanide with a concentration of 25 mol/m^3, potassium sulfate (K_2SO_4) (130 mol/m^3), 40 % of glycerin, and 2.5 % of Kalliroscope

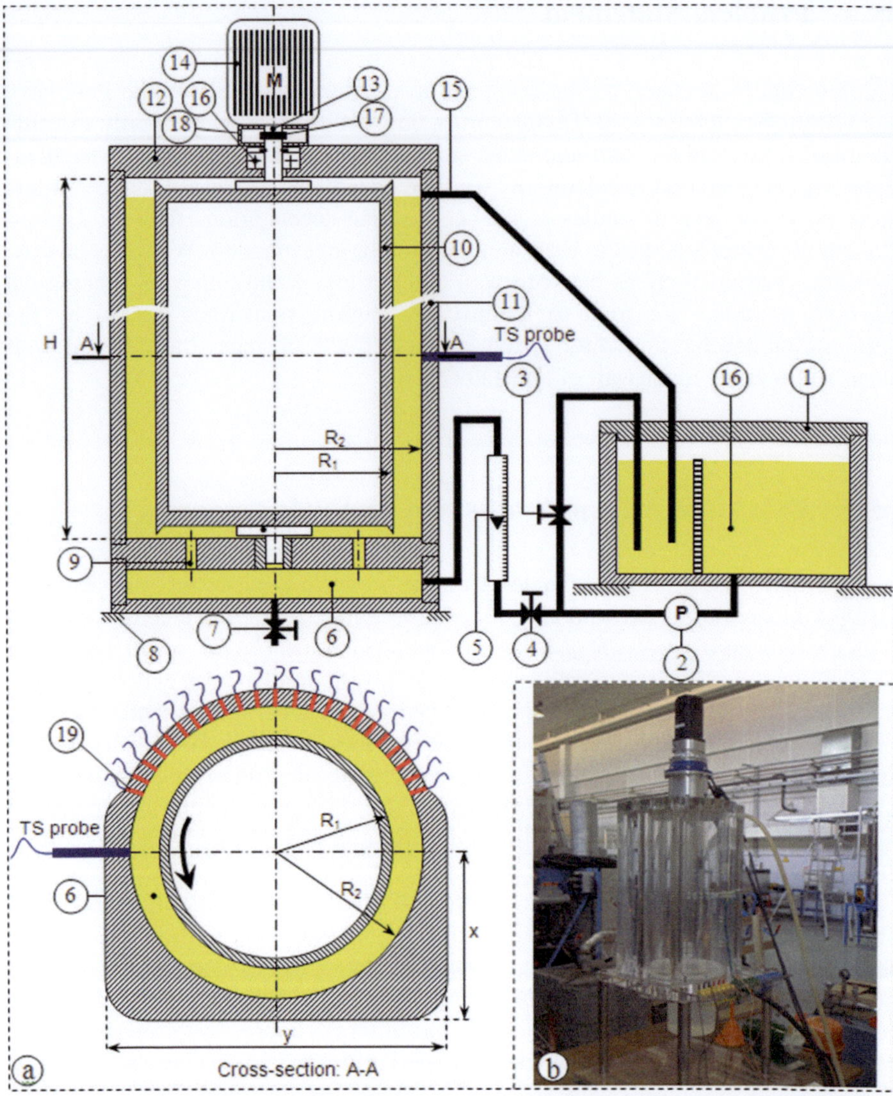

Fig. 65.1 Experimental setup of Taylor–Couette (**a**) Schematic view of the Taylor–Couette with its three-segment and circular probes' location: 1 storage tank; 2 pump; 3 by-pass; 4 flow rate control valve; 5 flow meter; 6 CT feeder tank; 7 drain valve; 8 support; 9 supply holes for the homogeneous filling of the CT gap (8 holes); 10 CT inner cylinder; 11 CT outer cylinder; 12 flabby; 13 motor shaft; 14 electrical motor; 15 overflow; 16 ball bearing; 17 elastic coupling; 18 protection; 19 circular probe. (**b**) Photography of the setup

was used. The supporting electrolyte, i.e., the solution of potassium sulfate (K_2SO_4), allows eliminating the migration effect. The Kalliroscope is reflective flakes, used for flow visualization with a concentration of 2.5 % [22]. The glycerin was used to modulate the viscosity and then the Taylor number.

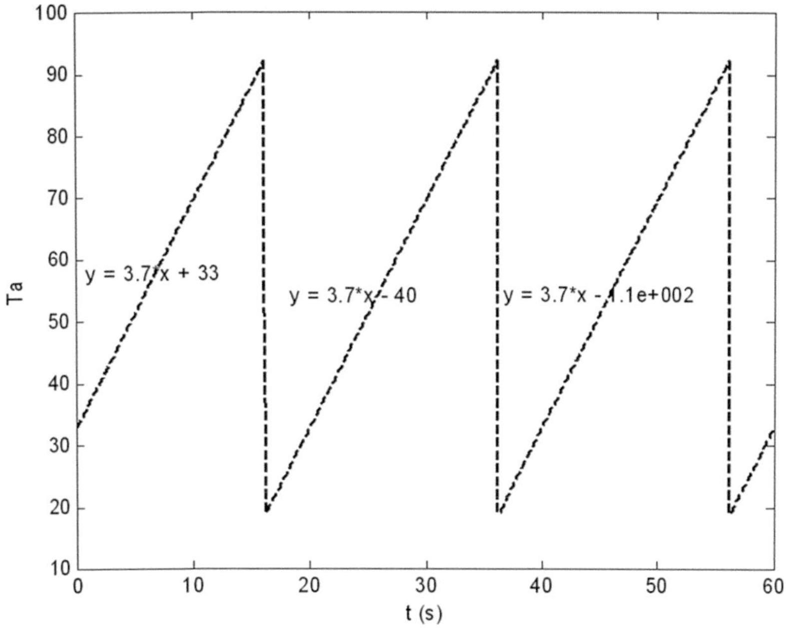

Fig. 65.2 Instantaneous time evolution of the Taylor number

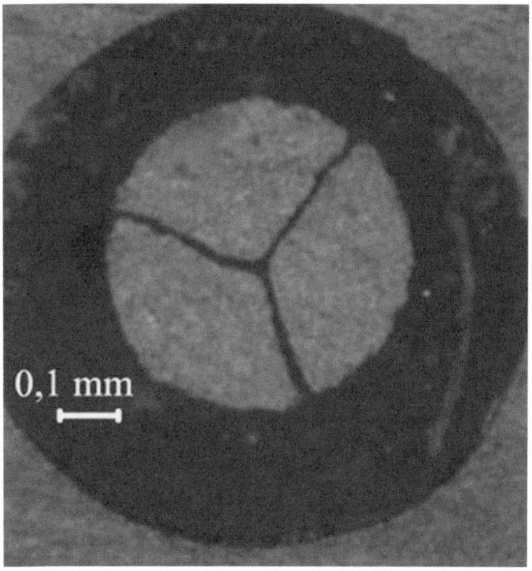

Fig. 65.3 Microphotography of the three-segment probe used

The few studies realized on the acceleration–deceleration effect on the flow have been restricted to only extreme conditions, i.e., sudden start and quasi-steady flow [13, 14]. In our case studies, the initial experimental conditions corresponded to impose a fully developed laminar Couette flow (quasi-steady conditions). Then, a sudden increase of the inner cylinder rotation speed corresponding to a Taylor number characterizing the modulated Taylor wavy vortex flow or any other "superior" (according to the values of Ta) flow regimes was imposed. For example, a laminar Couette flow corresponding (in steady case) to $Ta = 18$ was firstly imposed. An inner cylinder rotation corresponding (in the steady case) to $Ta = 93$ was then imposed.

65.4 Analysis of Instabilities

When the cylinder rotation speed increases, a first flow transition can be identified on the Taylor–Couette flow. It corresponds to the critical Taylor number Ta_{c1}. A series of secondary transitions appear when the flow velocity increases. They are characterized by modulated azimuthal waves, which are periodical to the circumferential direction (modulated wavy vortex flow). These different regimes have been detected, observed, and studied. The corresponded vortex apparition and disappearance impact on the local mass transfer and wall shear rate are presented and discussed.

Our attention is focused on the case of linear acceleration-deceleration for Taylor number Ta range between 18 and 93 (Fig. 65.2). This corresponds to a linear velocity gradient, valid on the laminar Couette flow, range from 41 to 202 s^{-1}. The flow regimes which can be attended strongly depend on the geometrical characteristics of the CT system and on the fluid nature (its viscosity). Indeed, the geometrical CT system characteristics (radius ratio and the aspect ratio) are extremely important because they directly affect the Taylor numbers of transition and the multiplicity of the flow states. Our geometry choice is justified by two aims. The first one is to assure a gap as small as possible in order to compare the experimental wall shear rate to the quasilinear solution:

$$S_{\text{quasi-linear}} = 2\pi N R_{\text{int}} / (60\ (R_{\text{out}} - R_{\text{int}})) \tag{65.4}$$

where N is the velocity in revolution per minute. This allows us to well localize the first regime of laminar Couette flow and to qualify our CT system. The second purpose is that we focus on transition effect on the velocity gradient at the wall vicinity.

The inner cylinder was accelerated linearly from the rest or from a chosen initial speed to a predetermined speed taking into account the mechanical apparatus limitation. This paper presents a study of acceleration from a lower Taylor number characterizing the laminar Couette flow (quasi-steady conditions) to a Taylor number characterizing the modulated Taylor wavy vortex flow according to our visualization.

65.5 Results and Discussion

65.5.1 Instabilities' Effect on Mass Transfer Rates

The time evolution of the Taylor number is illustrated on Fig. 65.2. As we can see, the Taylor number varied from 18 to 93. According to Cognet [7] and Roberts' [11] works, this range corresponds to 3 flow regimes: CF, TVF, and MWVF. The range ($18 < Ta < 41.79$) corresponds to the Couette flow characterized by a linear velocity gradient. This characteristic was used to test if the velocity gradient calculated by an inverse method well follows the quasilinear solution in CF regime.

The first instability is approximately at $Ta = Ta_{c1} = 41.79 \pm 0.2$. A flow characterized by steady toroidal cells appeared. At $Ta^* = Ta/Ta_{c1} = 1.2$, a second instability was observed. This corresponds to the TVF disappearance and the apparition of the modulated wavy vortex flow MWVF (Fig. 65.4). The flow instabilities lead to shear rate and mass transfer coefficient variations. The study of these variations allows the observation of the instabilities and the determination of the corresponding Taylor numbers.

Fig. 65.4 Kalliroscope visualization of modulated wavy vortex flow (MWV flow)

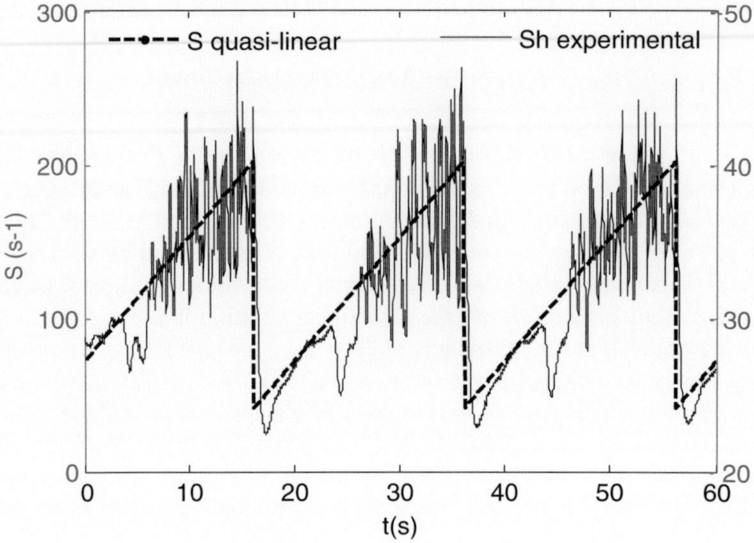

Fig. 65.5 Instantaneous time evolution of mass transfer rates and quasilinear velocity gradient

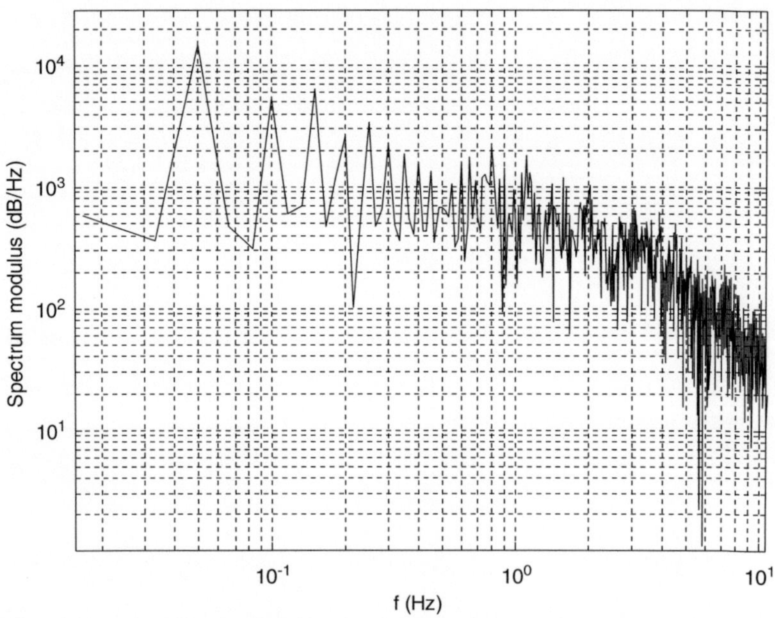

Fig. 65.6 Spectral analysis (FFT absolute) of the instantaneous mass transfer rates

The mass transfer rate variation (Figs. 65.5 and 65.6) illustrates the two main instabilities. Figure 65.5 presents the instantaneous time evolution of the Sherwood number (mass transfer).

The Sherwood number is defined by

$$Sh = \frac{il}{nFC_0AD} \tag{65.5}$$

where i is the sum of the intensities current, l the probe length, n the ions number, F the Faraday number, C_0 is initial concentration, A the probe area, and D the coefficient of diffusion.

The "history effect" of the acceleration-deceleration on the mass transfer rate evolution is thus studied. In fact, the mass transfer evolution follows the acceleration-deceleration flow. However, a delay shift is detected. It is maybe due to the ED probe inertia or to a slight hysteresis effect generated by the flow inertia. For the three periods illustrated (Fig. 65.5), the mass transfer rate values are similar for a fixed Taylor number. The quasi-steady velocity gradient (presented on Fig. 65.5) is valid for laminar Couette flow and no longer available when a flow transition appears.

The spectral analysis (FFT absolute) of the instantaneous mass transfer rates (the Sherwood number) is presented on Fig. 65.6. The fast Fourier transform (FFT) absolute allows the detection of vortices. Indeed, the peaks correspond to the vortex apparition.

The mass transfer in function of the dimensionless Taylor number is presented on Fig. 65.7. The mass transfer function of the dimensionless Taylor number

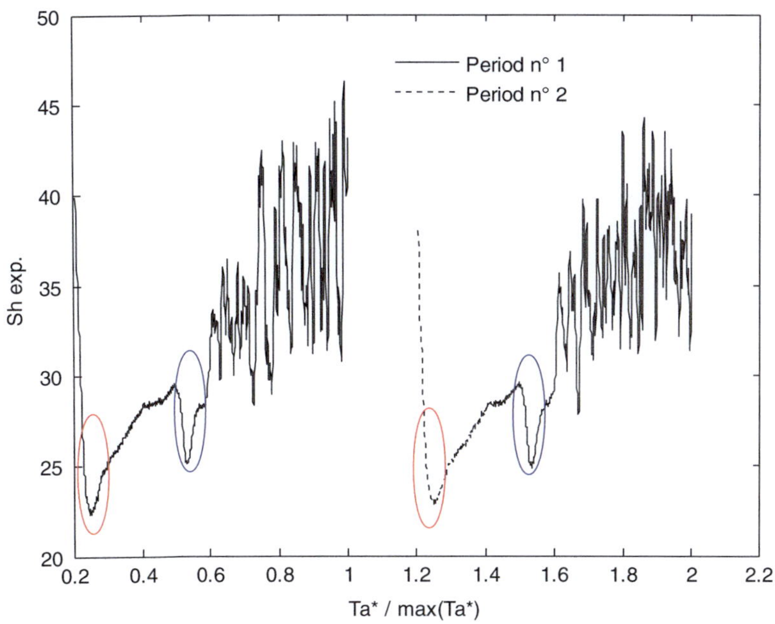

Fig. 65.7 Instantaneous mass transfer rates in function of the dimensionless Taylor number (only two periods are presented for clarity)

($Ta^* = Ta/Ta_c$) demonstrates that the mass transfer is a repetitive phenomenon, i.e., the frequency response of the three-segment sensor follows the hydrodynamic flow acceleration-deceleration imposed. The series of periodic imposed sudden inner cylinder speed decrease generate a mass transfer decrease slightly damped. During the sudden deceleration, a slight shift of the mass transfer rate is detected. This is maybe due to the sensor inertia effect (first ellipse). Indeed, the sensor creates a diffusion layer as a filter (capacitive effect) around the electrode (at the wall vicinity). This accompanied the sudden and fast flow deceleration of the flow from Ta equal to 93 ± 0.2 to 18 ± 0.2. When the Taylor number increases, a second large decrease of the mass transfer rate (second ellipse) is detected. It corresponds to the first instability apparition.

65.5.2 Instabilities' Effect on Instantaneous and Local Wall Shear Rates

Instantaneous time evolution of wall shear rates for three successive periods is illustrated on Fig. 65.8a. The wall shear rate was determined according to the Leveque solution [23] and the Sobolik et al. [17] solution. The wall shear rate evolution is linear for the CT flow regime (Fig. 65.8b). It follows the quasilinear solution defined by Eq. (65.1). However, a decrease is noted, with a sharp slope, accompanied a fast and vertical decrease imposed to the speed of the CT inner cylinder (first ellipse on the Fig. 65.8b). This is maybe due to the sensor inertia effect. Indeed, the sensor presents a filter which delays the frequency response of the sensor for strong and fast flow velocity decreases from Ta equal to 93 to Ta equal to 18 in less than 0.01 s.

This decrease is detected by different methods used for the determination of the velocity gradient at the wall vicinity from mass transfer measurements. It progressively disappears, while the wall shear rate becomes equal to the quasilinear solution defined by Eq. (65.1). When the Taylor number increases, the wall shear rate deviates progressively from the quasilinear solution. This is predicted because the linear approach is only valid for laminar Couette flow (Fig. 65.9). In fact, this characteristic is used to experimentally validate the inverse method. While the flow velocity imposed is suddenly decreased, the Taylor number decreases from 92.2 (at $t = 16.14$ s for the first period, Fig. 65.9) to 18 (at $t = 16.22$ s). At $t = 16.85$ s, the shear rate calculated from the frequency response of the sensor (from the mass transfer rates) using the Sobolik et al. [17] solution begins to follow the linear shear rate (S q.l.).

The sensor response follows the quasilinear shear rate (S q.l.) after 0.63 s. Thus, the flow is rearranged on approximately 0.60 s for an imposed decrease of 80 ms (Fig. 65.8b). When the Taylor number increases, a second large decrease of the wall shear rate (second ellipse on the Fig. 65.8) is detected by the different ED methods tested (Figs. 65.8 and 65.10). This decrease appears at Ta equal to 41.3 according to

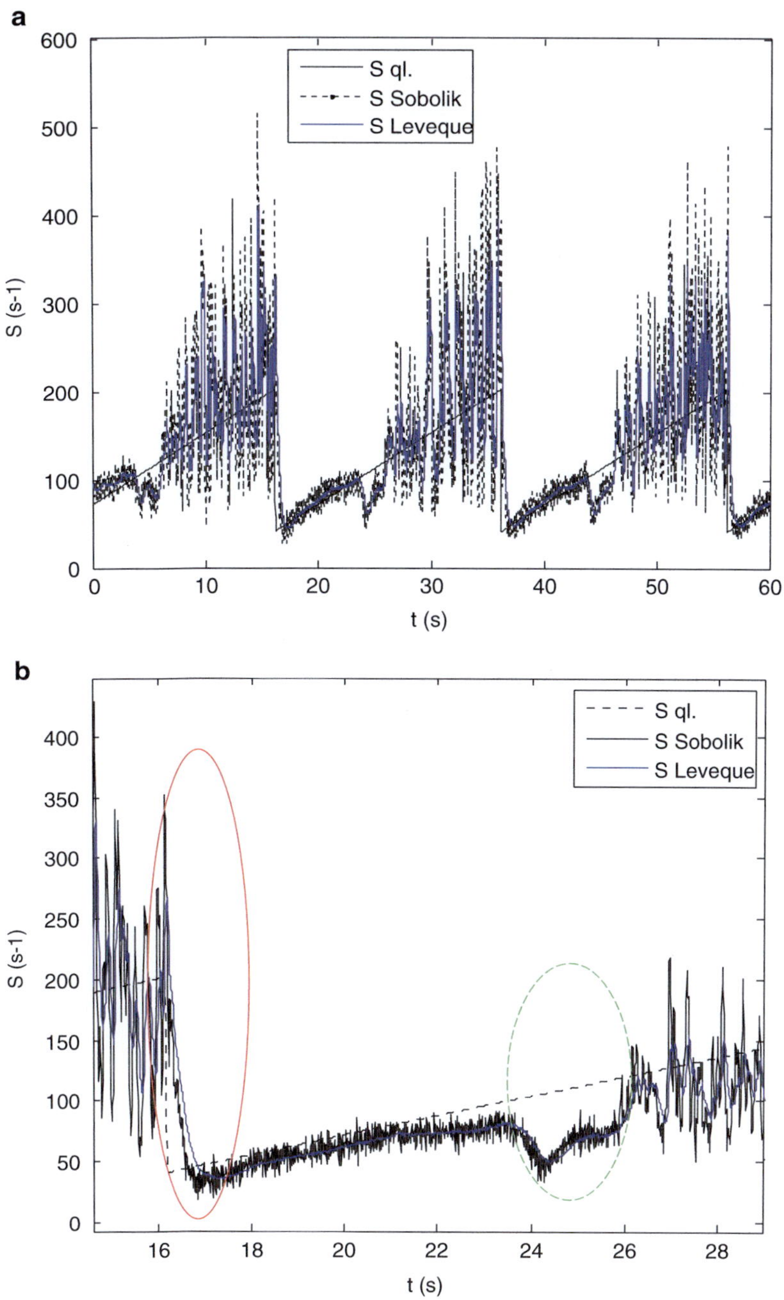

Fig. 65.8 (**a**) Instantaneous time evolution of wall shear rates. (**b**) Zoom on the CF regime succeeded by the TWF: instantaneous time evolution of wall shear rates (LTWF: $t = 22$ s $\rightarrow Ta = 41.4$ until $t = 26$ s $\rightarrow Ta = 56.2$). The wall shear rate was determined according to the inverse method [18] and compared with the quasilinear solution

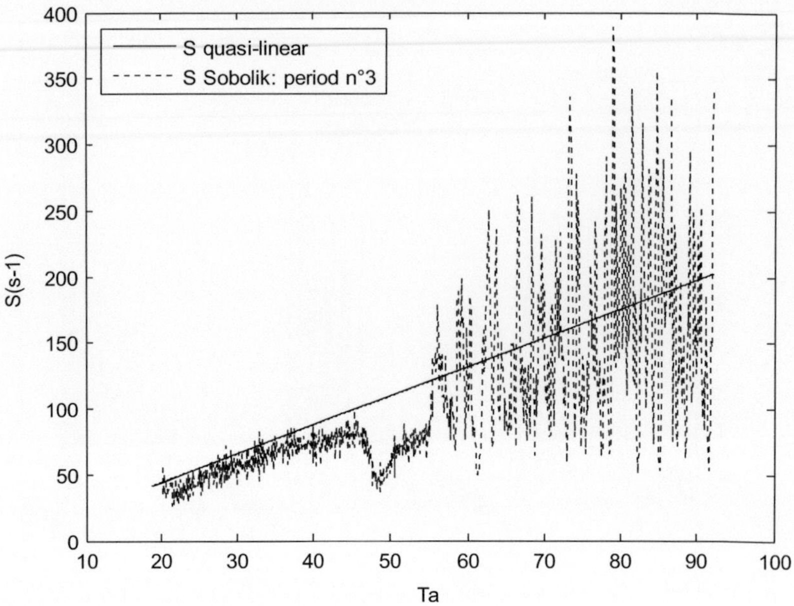

Fig. 65.9 Zoom on the third period of the instantaneous wall shear rate function of the Taylor number

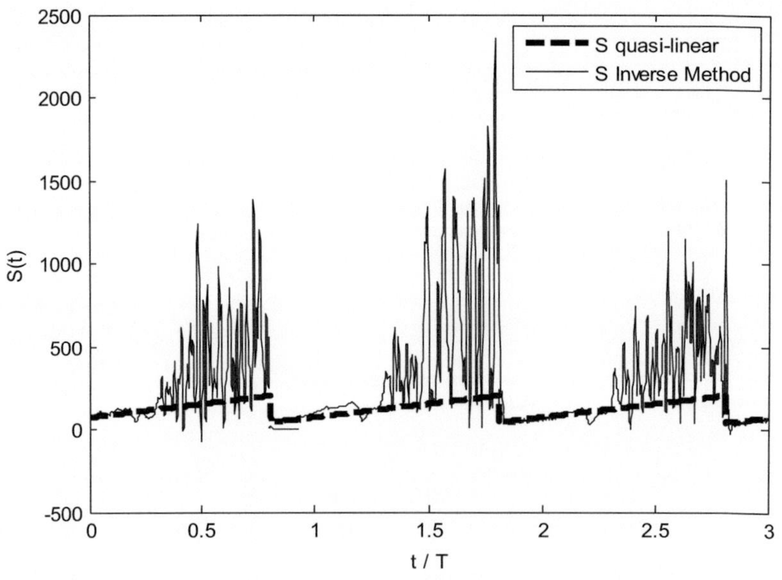

Fig. 65.10 Instantaneous wall shear rate function of the dimensionless time

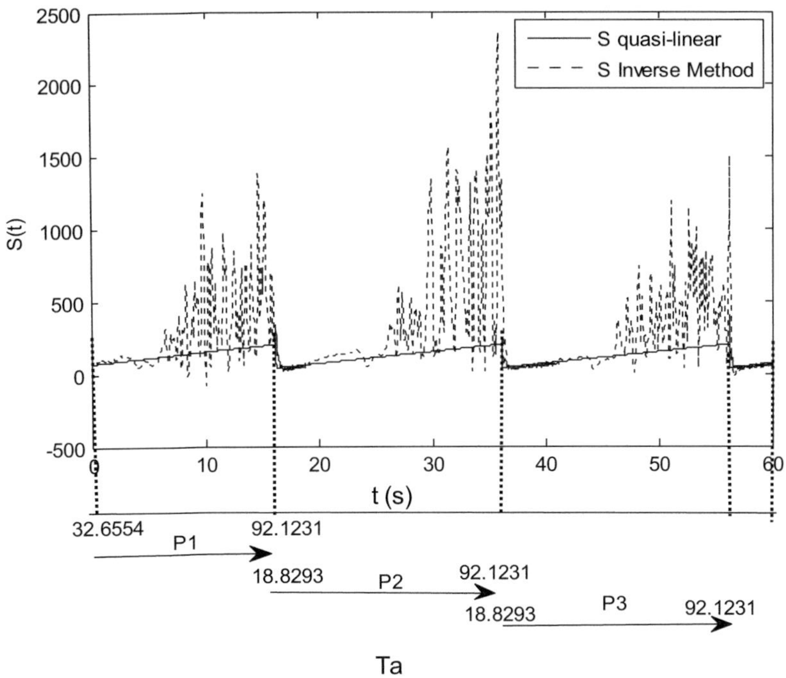

Fig. 65.11 Instantaneous wall shear rate function of the Taylor number

the Sobolik et al. [17] solution (Fig. 65.9). It becomes significant at Ta equal to 46.1 according to the inverse method shear rate evolution (Fig. 65.11).

A local minimum value of the wall shear stress occurs at $Ta = 41.3$. Thus, the decrease characterizes the first instability apparition at Ta approximately equal to $Ta_{c1} = 41.3$ which corresponds to the first critical Taylor number. The decrease continues to develop until $Ta = 48.3$. The first peak corresponds to $Ta = 55$ ($Ta^* = Ta/Ta_{c1} = 1.33$). This characterizes the apparition of the second instability. Then the wall shear rate presents a series of peaks characterizing the modulated wavy Taylor vortices (Figs. 65.9 and 65.11). The flow migrates to the inflow region near the inner cylinder. The quasilinear solution (S q.l.), between $41.3 < Ta < 55$, is bigger than the shear rate determined according to the Sobolik et al. solution [17]. The sensor is localized on the outflow regions of the Taylor cells.

The inverse method [18] is a sequential method based on the inversion of the convection-diffusion equation and on mass transfer to determine the wall shear rate. One of the points of merit of this work is that it is the first time that the inverse method is tested for such type of flow hydrodynamics characterized by fast and strong variations. The wall shear rate was determined according to the inverse method [18] and compared with the quasilinear solution. We have initialized the sequential method using the Sobolik et al. [17] solution. The time of calculation was more than 4 h for the case presented here. The inverse method is in agreement with the wall shear rates (Figs. 65.10 and 65.11) for laminar Couette flow.

When the flow velocity imposed decreases suddenly, the Taylor number decreases from 92.2 (at $t = 56.14$ s for the third period, Fig. 65.2) to 18.8 (at $t = 56.22$ s). In Fig. 65.11, the arrows indicate the velocity gradient evolution during a period of flow velocity imposed to the inner cylinder. A slight hysteresis effect is detected. In fact, the evolution of the velocity gradient does not decrease immediately, i.e., it does not follow immediately the sudden decrease imposed to the flow. Indeed, the wall shear rate increases, due to the hysteresis effect, in the beginning of the imposed flow decrease and then decreases until being equal to the quasilinear solution (the laminar Couette flow regime). The spectral analysis (Absolute Fast Fourier Transform (FFT)) of the instantaneous wall shear rates is determined by Sobolik et al. [17] approach, and the inverse method for the first period (a) and the second period (b) is presented in Fig. 65.12. It illustrates the vortex detection.

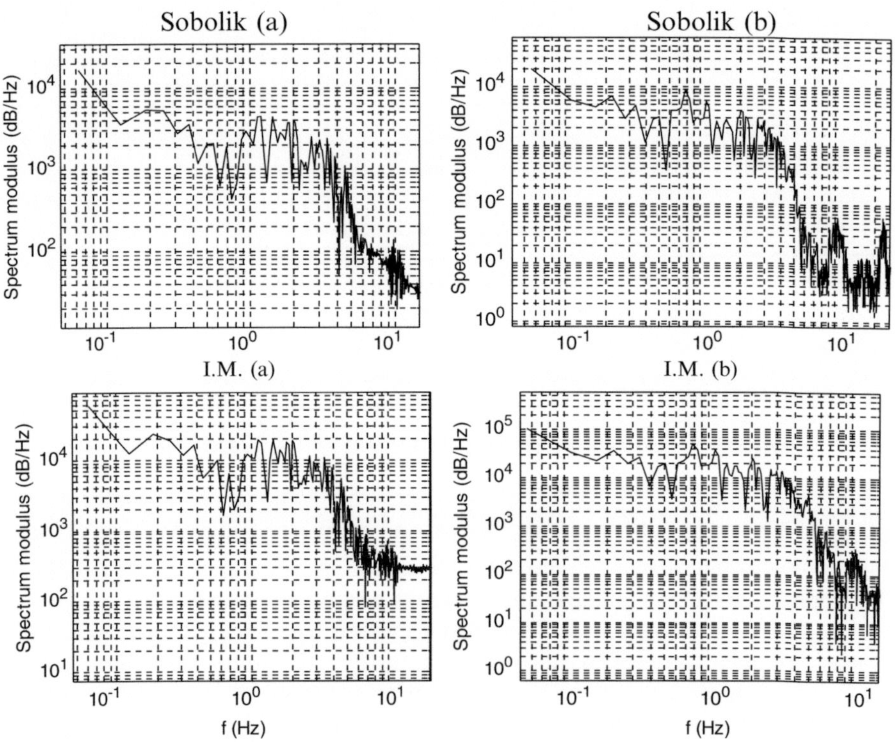

Fig. 65.12 Spectral analysis (FFT absolute) of the instantaneous wall shear rates determined by Sobolik et al. [17], and the inverse method M.I. (*a* first period, *b* second period)

65.6 Conclusions

A deep characteristic of the flow structures, in particular in the vicinity of the wall regions, is illustrated via the analysis of instantaneous and local mass transfer and velocity gradient evolutions with time variation of Taylor numbers. Indeed, the analysis of the instantaneous mass transfer rates and the instantaneous velocity gradient allows the study of the vortex apparition, development, and disappearance. The observation of the apparition and disappearance of the first and the second transitions and thus the observation of three successive flow regimes, i.e., the Couette flow, the LTVF flow, and the MWTVF flow, were realized. The analysis of the three-segment sensor frequency response of the acceleration-deceleration hydrodynamic flow variations allows concluding three main points. Firstly, it illustrates the presence of the sensor inertia effect (a capacitive effect) as a filter which delay the mass transfer and the velocity gradient comparing to the imposed fast and sudden decrease of the shear rate from Ta equal to 93 to Ta equal to 18 in less than 0.01 s. Secondly, the mass transfer rate function of time and/or function of the Taylor numbers vary differently along a period of flow hydrodynamic. This allows the definition of every regime range. For three successive flow periods, the mass transfer presents the same evolution. Thus, the mass transfer rate evolution follows the repetitive phenomenon imposed to the flow hydrodynamics. A slight hysteresis effect was detected during the sudden decrease imposed to the CT flow. Thirdly, the inverse method initialized using the solution of Sobolik et al. [17] was experimentally validated for this type of acceleration-deceleration flow hydrodynamics and for this flow history. The instantaneous wall shear rate then correctly follows the quasilinear linear solution along the laminar Couette flow. The first instability detected at Ta equal to 41.3. It corresponds to a second large drop. The wall shear rate then presents a series of beaks. This illustrates the second instability apparition and the modulated Taylor vortex development.

Acknowledgment This work was supported by the National Agency of Research of France "Agence Nationale de la Recherche (ANR), France," under the Grant n° ANR-08-BLAN-0184-02.

References

1. Beaudoin G, Jaffrin MY (1987) High efficiency plasmapheresis using rotating membrane device. Life Support Syst 5:273–278
2. Dluska E, Hubacz R, Wronski S (2006) Simple and multiple water fuel emulsions preparation in helical flow. Turkish J Eng Env Sci 30:175–182
3. Dominguez-Lerma MA, Ahlers G, Cannel DS (1984) Marginal stability curve and linear growth rate for rotating Couette-Taylor flow and Rayleigh-Bernard convection. Phys Rev 27:856–860
4. Coles D (1965) Transition in circular Couette flow. J Fluid Mech 21:385–425
5. Di Prima RC, Eagles PM, Ng BS (1984) Effect of radius ratio on the stability of Couette flow-Taylor vortex flow. Phys Fluids 27(10):2403–2411

6. Mehel A (2006) Etude expérimentale d'un écoulement diphasique de Couette-Taylor. Thèse de doctorat, Université de Nantes, Ecole Centrale de Nantes
7. Cognet G (1984) Les étapes vers la turbulence dans l'écoulement de Couette-Taylor entre cylindres coaxiaux. Journal de Mécanique Théorique et Appliquée, Numéro spécial, p 7–44
8. Cole JA (1976) Taylor-vortex instability and annulus-length effects. J Fluid Mech 75(part 1):1–15
9. Taylor GI (1923) Stability of a viscous liquid contained between two rotating cylinders, Series A, vol. 223. Philosophical Transaction of the Royal Society, London, p 289–343
10. Ait Aider A (2008) Instabilities and transition to chaos in flows between concentric cylinders. Int J Chem React Eng 6:A98
11. Roberts PH (1965) The solution of the characteristic value problems. Proc R Soc Lond A 283:550–556
12. Park K, Gerard L, Donnelly RJ (1981) Determination of the transition in Couette flow in finite geometries. Phys Rev Lett 47:1448
13. Burkhalter JE, Koschmider EL (1973) Steady supercritical Taylor vortex flow. J Fluid Mech 58:547
14. Burkhalter JE, Koschmider EL (1974) Steady supercritical Taylor vortices after sudden starts. Phys Fluids 17:1929
15. Ahlers G, Cannel DS, Dominguez-Lerma MA (1983) Possible mechanism for transitions in wavy Taylor-vortex flow. Phys Rev A27:1225
16. Hanratty TJ, Campbell JA (1983) Measurement of wall shear stress. In: Goldstein RJ (ed) Fluid mechanics measurements. Hemisphere, Washington
17. Sobolik V, Wein O, Cermak J (1987) Simultaneous measurement of film thickness and wall shear stress in wavy flow of non-Newtonian liquids. Collect Czechoslov Chem Commun 52:913–928
18. Rehimi F, Aloui F, Ben NS, Doubliez L, Legrand J (2006) Inverse method for electrochemical diagnostics of flows. Int J Heat Mass Transf 49:1242–1254
19. Legrand J, Coeuret F, Billon M (1983) Structure dynamique et transfert de matière liquide-paroi dans le cas de l'ecoulementlaminaire tourbillonnaire de Couette–Poiseuille. Int J Heat Mass Transf 26(7):1075–1085
20. Sobolik V, Benabes B, Cognet G (1995) Study of Taylor–Couette flow using a three-segment electrodiffusion probe. J Appl Electrochem 25:441–449
21. Sobolik V, Mitschka P, Menzel T (1986) Method of manufacture of segmented probe with circular cross-section. Czechoslov Pat AO, 262–823
22. Abcha N, Latrache N, Crumeyrolle O, Mutabazi I (2006) Application de la Vélocimétrie par Images de Particules "PIV" à l'étude des vortex de Taylor. Congrès Francophone de Techniques Laser, CFTL 2006, Toulouse, 19–22 septembre
23. Leveque MA (1928) Les lois de transmission de la chaleur par convection. Ann Min 13:381–412

Chapter 66
Void Fraction Influence Over Aqueous Foam Flow: Wall Shear Stress and Core Shear Evolution

Rogelio Chovet and Fethi Aloui

Abstract In this study, the two main transport characterization problems of the foam flow are studied: foam flow stability, through the evolution of the velocity at the core of the foam, and rheology, with the study of the wall shear stress over the lateral walls, for different void fractions. The same velocity profile (block flow, mean velocity 1.75 cm/s) is imposed to the foam flow, at the inlet of the channel, for several void fractions (air/water relation) going from 55 to 85 %. Later on these ones are passed through a singularity (fence) to study the different behaviours induced by the particular properties of each case. The velocity fields, the lateral liquid film thickness and the lateral wall shear stress fields are obtained and compared with one another to comprehend and remark the difference in such a complex flow. The results show that as we move closer to very dry foams the shear at the foam core increases and its velocity becomes higher. However, the wall shear stress at the lateral wall does not present big deviations from one void fraction to the other.

Keywords Foam flow • Wall shear stress • Energy • PIV • Conductimetry method • Void fraction • Efficiency

Nomenclature

e	Film thickness
$\vec{e}_x$	Unit vector in the x-coordinate direction
Q_g	Gas flow rate
Q_l	Liquid flow rate
Q_t	Total foam flow rate
u	Axial velocity component

R. Chovet • F. Aloui (✉)
LAMIH CNRS UMR 8201, Department of Mechanics, University of Valenciennes et du Hainaut-Cambrésis (UVHC), Campus Le Mont Houy, Valenciennes Cedex 9 59313, France
e-mail: Rogelio.Chovet@gmail.com; Fethi.Aloui@univ-valenciennes.fr

I. Dincer et al. (eds.), *Progress in Clean Energy, Volume 1*,
DOI 10.1007/978-3-319-16709-1_66

$\vec{u}$ Axial velocity vector component
v Vertical velocity component
x Axial axe direction
y Vertical axe direction
z Depth axe direction

Greek Letters

β Quality (foam void fraction)
τ Wall shear stress
μ_l Dynamic viscosity of the liquid phase

Abbreviation

PIV Particle image velocimetry

66.1 Introduction

Since the beginning of the twentieth century we have been experiencing a phenomenon that has changed the way we look at energy. Exponential human population opened the mind of the scientific community and created a movement whose only purpose is to innovate topics related to all fields of energy (from creation to consumption). One of many original ideas is the use of "not-regular" fluids over "regular" applications to improve the efficiency of these ones. Foam studies have been around since 1887, when Lord Kelvin exposed the packing and structure of the ideal bubble to bubble configuration. Nowadays, due to the energy challenge we are facing, people start to take into consideration this kind of fluids to improve industrial processes. Its unusual rheology properties, low density and important interfacial surface, gives foam flow many interesting uses: assisted oil extraction and heat exchange, among others.

Aqueous foams are complex fluids made out of a dispersion of gas bubbles in a soapy liquid. These bubbles present a wide distribution of sizes, randomly mixed and arranged. There are at least four length scales at which we can consider the properties of a foam:

- The observer's scale, of the order of metres; the foam has the appearance of a soft, opaque solid.
- The millimeter scale, the bubbles can be distinguished; there are a small number of local geometry rules, which describe how the bubbles pack together to form the foam organization [1].
- The micron scale, which shows how liquid is distributed between the bubbles.

- The nanometer scale, at which the molecular structure of the interfaces appears; the presence of particular molecules, surfactants, which positions themselves at the liquid layer between bubbles, lowers the liquid films energy and makes the foam formation easy.

When treated as a macroscopic system, foam elements undergo no significant thermal fluctuation, by means of which it might explore alternatives to the local minimum of energy in which it finds itself [2]. It is not considered to be in a metastable state, since it continually evolves according to the coarsening process, air exchange between neighbour bubbles [3]. However, this is a very slow process. The foam therefore stays very close to a true equilibrium except where a local topological change may take place.

When the topological structure of the foam is altered, it may be brought to a configuration in which there is a violation of the Plateau's rule by the introduction of a *forbidden* vertex. This configuration dissociates rapidly and a new structure is formed. For a 2D flow, the possibilities are rather simple: The so-called T1 process eliminates the fourfold vertex and forms two threefold ones, while the T2 process is associated with the disappearance of a bubble, also known as coalescence. Both processes are sudden events in which the surface energy of the foam drops abruptly and the energy loss is dissipated as heat (Hutzler et al., 2008) [4]. The bubble rearrangement leads to local shear flow of the liquid inside the foam films, resulting in the dissipation of energy and shear rate-dependent contribution to the macroscopic stress.

Drainage is described as a phenomenon by which liquid flows out of a foam [5]. It may look to be a relatively straightforward fluid problem dealing with the flow between bubbles, but is instead a very complex one, with length scales ranging from nanometres for surfactant molecules to micrometres for films, to millimetres for bubbles, to centimetres for bulk foams. Two main flow stages can be distinguished: the drainage of the films between the bubbles and then the liquid flows to the centre of the channels towards the Plateau junctions. The flow inside the liquid film is due to several phenomena: the gravity force, the capillarity force and the pressure force between the Plateau regions and the film between the bubbles.

A foam may be classified as a dry or wet foam according to the liquid content, which may be represented by the liquid volume fraction. This ranges from much less than 1 % to about 30 %. At each extreme (the dry and the wet limits) the bubbles come together to form a structure which resembles one of the classic idealized paradigms of nature's morphology: the division of cells in the dry limit and the close packing of spheres in the wet limit.

Rheological properties of foams, such as elasticity, plasticity and viscosity play a major role in foam production, transportation and application. If foam is subjected to a small shear stress, it deforms like a soft solid. This response can be characterized with a visco-elastic modulus. For applied low yield stress the visco-plastic flow sets in. In this regime, foams behave like a shear-thinning fluid. This means that their effective viscosity is a decreasing function of the shear rate [6]. Additional rheological phenomena arise at the contact between the bubbles and the confining solid walls of a channel [7]. If the surface of the solid wall is smooth enough, the foam tends to slip on the wall over a very thin liquid film. In this case, the velocity of the first layer of bubbles in contact with the wall does not match to what is observed with simple liquids [8]. The rheological properties of foams are complex not only because both elastic and viscous responses are non-linear functions of the applied stress but also because shear localization may occur under certain conditions.

66.2 System Description

The system, comprising a stabilized liquid circuit (water with surfactant) and a gas circuit (pressurized air), is injected into the liquid through a porous media into the horizontal test section. The main straight duct is made in a 3.2 m long transparent Plexiglas with a square section of 21 mm × 21 mm. The measurement section is located at the middle of the duct at 1.3 m from the conducts entry. Figure 66.1 shows the schematic view of the fence located at this part of the duct. This one allows us to study the behaviour of the foam flow when faced against a pattern change and its reorganization once through the singularity [9].

66.3 Measurement Techniques

Foam regimes in straight ducts can be qualified in three groups [10]:

- *One-dimensional flow*: for this regime the flow behaves as a whole, it moves like a block or a piston. The velocity vectors have one uniform axial component in the flow direction:

$$\vec{u} = u \cdot \vec{e}_x = cte \cdot \vec{e}_x \tag{66.1}$$

- *Two-dimensional flow*: it is obtained when the established flow has an axial velocity component that only depends on the *y-coordinate*:

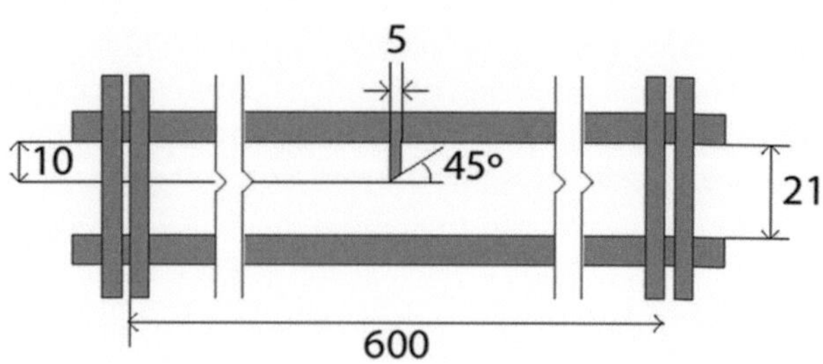

Fig. 66.1 Lateral view of the fence [dimensions in mm]

$$\vec{u} = u(y) \cdot \vec{e}_x \tag{66.2}$$

- *Three-dimensional flow*: this one is obtained when the established flow velocity vector presents an axial component that depends on the *z*- and *y-coordinates*:

$$\vec{u} = u(y, z) \cdot \vec{e}_x \tag{66.3}$$

For a foam flow made out of a known gas flow "Q_g" and a known liquid flow "Q_l", we can define the void fraction expression as the measure of the empty space inside the foam:

$$\beta = \frac{Q_g}{Q_t} \tag{66.4}$$

where "Q_t" corresponds to the foam flow such as $Q_t = Q_l + Q_g$.

For our experiments, measurements were undertaken in the first regime (one-dimensional flow regime) using a mean flow velocity of 1.75 cm/s and with a void fraction of 55, 65, 75, and 85 %. The liquid and gas flow rates were measured with a group of rotameters (Brooks).

The particle imaging velocimetry technique was used to obtain the foam flow behaviour, velocity fields and profiles. It is a non-intrusive optic method capable of obtaining the displacement of particles in a plane. This displacement is determined by the comparison of two instantaneous position fields of particles inside. The gas/liquid interface is darker than the rest of the flow. Therefore, the bubbles contour can be used to obtain the movement of the foam flow as a whole. The system used is a TSI set. It uses a software Insight 4G to treat the images. Due to the opacity of the foam flow, the measurements were done for the bubbles flowing over the lateral channel wall. The laser is produced by a double impulsion laser (ND-YAG) with a wavelength of 532 nm (green) and a frequency of 7.25 Hz. Figure 66.2 shows the schematic representation of the setting arrangement [11].

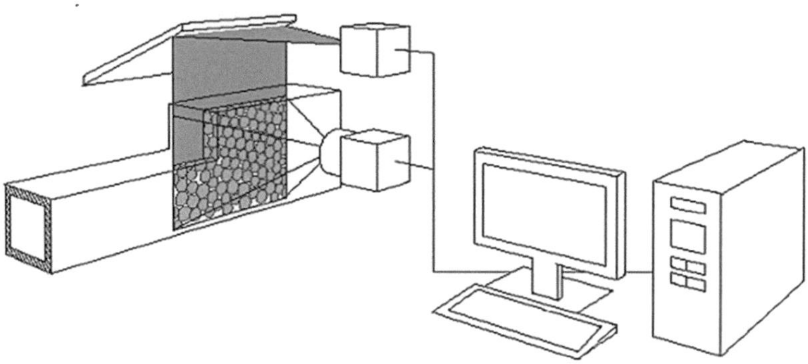

Fig. 66.2 PIV setting arrangement

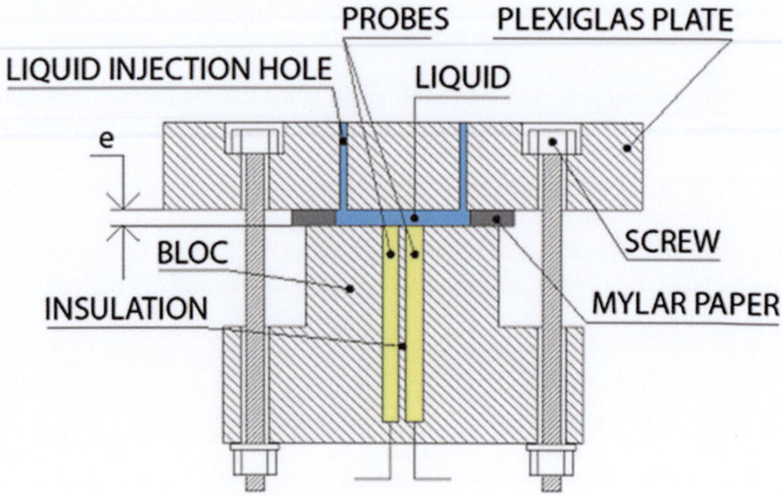

Fig. 66.3 Calibrate system for the conductimetry probe

The stability experiments conducted were not aimed to accurately locate and determine the T1 and T2 rearrangement. Contrariwise, they were interested in obtaining the velocity changes that these ones may generate in the foam flow and therefore induce a shear over its core. The PIV is a method than can accurately give us these results by analyzing the velocity fields of our foam flow at the wall.

A conductimetry block located at the lateral wall was used to obtain the liquid film thickness evolution. A frequency generator is used to emit a sinusoidal alternating voltage between the conductimetry electrodes (between 50 and 100 kHz) to neglect the electrodes' polarization phenomenon. The block used to measure the liquid film thickness is rigid and cannot be moved. Therefore, to calibrate it, an auxiliary system was created. This circuit is represented in Fig. 66.3. The thickness e between the plate and the block can be adjusted. Then, the liquid fills the space between the plate and the block. The experimental relation gives the output voltage as a function of the thickness e and establishes it for every pair of probes [12].

66.4 Results and Discussion

The two main transport characterization problems (stability and rheology) are undertaken by presenting the results in terms of streamlines, velocity fields and profiles, liquid film thickness evolution over the lateral wall and finally the wall shear stress over the lateral wall for a foam flow with a block velocity of 2 cm/s and a void fraction of 55, 65, 75 and 85 %.

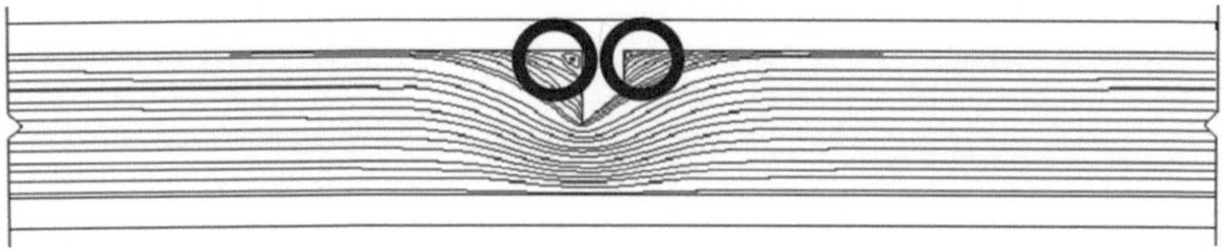

Fig. 66.4 Streamlines of the mean velocity fields

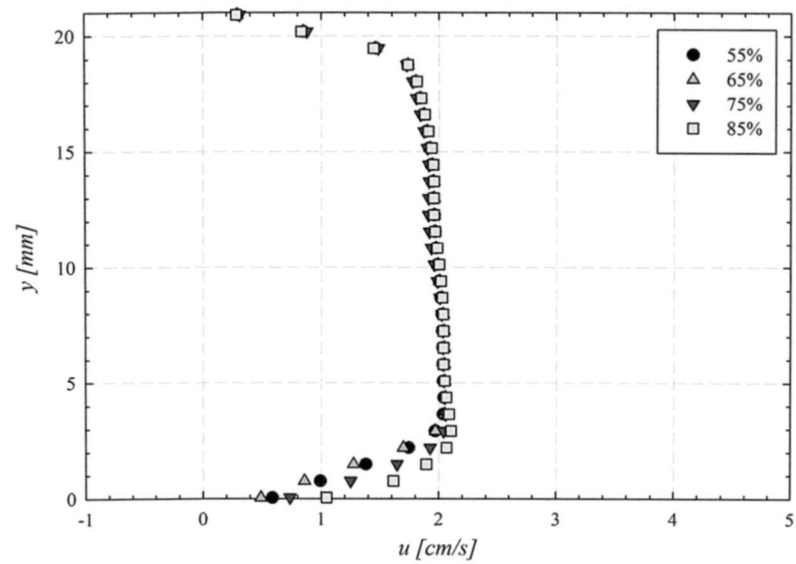

Fig. 66.5 Profile of the axial velocity component upstream of the singularity for all cases

Figure 66.4 shows the streamlines of the mean velocity fields for the foam flowing through the singularity. The presence of the fence causes the formation of dead zones in the immediate vicinity. In these regions the flow becomes slower and it can even get stagnated. In this first approach the foam can be divided in two regions: the principal flow and the dead zones where the bubbles can be completely motionless.

One of the most important aspects of this study was to assure that all foam flows had the exact initial conditions except for the change in the void fraction. This means same bubble size and same velocity profile at the entrance of the conduct. For the bubbles, the same porous medium was used in all cases. It has a porosity of 40 μm and creates bubbles with a diameter of 0.5 mm approximately. Figure 66.5 shows the velocity profile of all cases upstream of the singularity, far away enough to not be influenced by it. It can be seen that they all present the one-dimensional regime with a mean velocity of 1.75 cm/s.

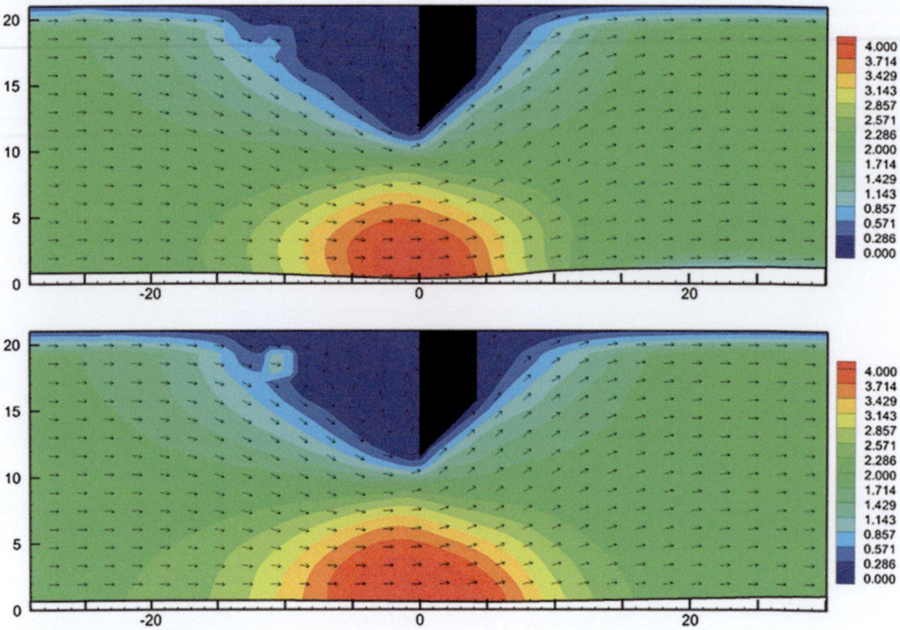

Fig. 66.6 "u" [cm/s] velocity fields for 55 % (*top*) and 85 % (*bottom*)

The "*u*" velocity fields for the void fractions of 55 and 85 % are represented in Fig. 66.6. These two specific cases were selected because they accurately represent the two limits of foam flow, dry and wet limits respectively. Also, in these two images the shear difference it is easily noted. Despite having the same velocity profiles upstream and downstream of the singularity, the velocity fields, while the foam passes through the singularity, are quite different. The dryer foam is more influenced by the singularity than the wet one; it dissipates more energy, in the axial direction, via viscous friction and stores less via its elastic properties. Therefore, it is easier to shear the dryer foam than the wet one. The maximum velocity is achieved for both cases right under the singularity with a value of around 4 cm/s. As the dry foam is more influenced by the velocity it also takes more time to achieve an equilibrium state, around 16 mm for the 11 mm of the wet foam.

The *v* velocity fields for the same void fractions are shown in Fig. 66.7. For both cases, 55% and 85%, the behavior downstream of the singularity is similar. They return to its initial state at about the same distance from the fence. Upstream of the singularity everything changes. Dry foam resists more to change than the wet one. Therefore, dry foam tends to shear less than the wet one. This is due to the gravity effects and the densities differences. Wet foams are denser than dry foams. Therefore, it will be more influenced by the gravitational force. In both cases a maximum value of 1.2 cm/s is obtained at the exit of the singularity and a minimum value of −0.8 at the enter of this one.

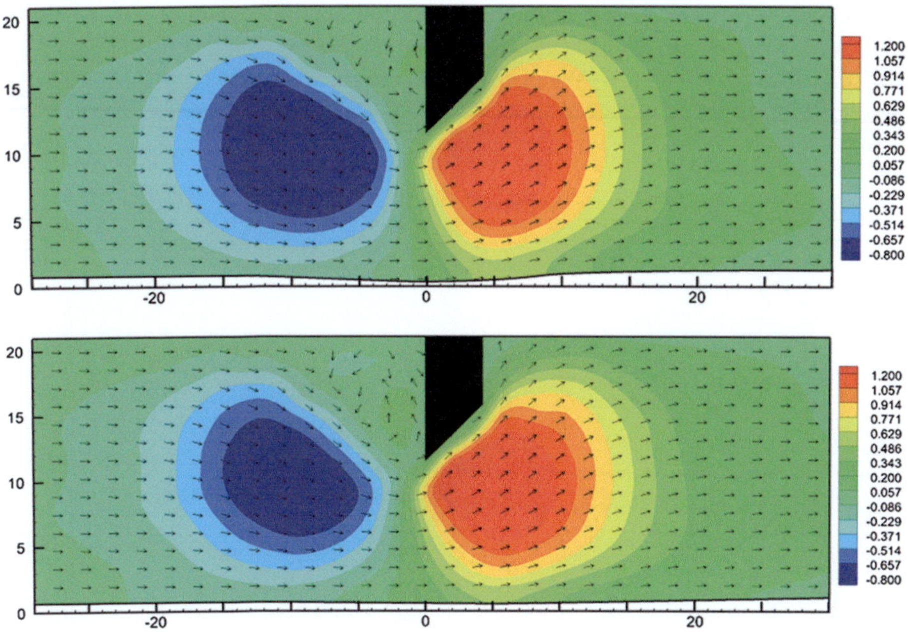

Fig. 66.7 "v" [cm/s] velocity fields for 55 % (*top*) and 85 % (*bottom*)

Figure 66.8 represents the comparison of axial velocity profiles for all cases (55, 65, 75 and 85 %) as they pass though the singularity starting at −25 mm and finishing at 25 mm from it. An acceleration can be remarked as we move closer to the fence ($x = 0$), reaching a maximum velocity of 4.3 cm/s. After this sudden acceleration, the flow returns to equilibrium state with a velocity close to the one it had before entering the singularity. The presence of death zones is clearly noticed near the singularity where the axial velocity tends to 0. The absence of negative velocities indicate that there are no recirculation zones and the foam flow stagnates near the fence. Except for the driest foam (85 %), all other curves present the same appearance putting into evidence once more the facility for the dry foam to shear more than the wet one. This difference is about 15 % in its most deviated points.

Over the same plane the vertical velocity profiles were obtained and they are shown in Fig. 66.9. As in a nozzle, the reduction of the section creates a vertical acceleration towards the passage section below the singularity. As the foam leaves the fence it fully occupies the rest of the duct. This phenomenon is a consequence of its low density and high active surface. At the same time, it gradually decelerates reaching the one-dimensional regime, in which the vertical velocity tends to zero. As for the shear difference, it can be deducted from the velocity profiles that upstream of the singularities wet foam tends to shear more in the vertical direction, due to gravitational forces, than dry foam, and once the obstacle is passed the phenomenon inverses and the dry foam becomes faster and shears more than the wet foam. Regardless of the different void fractions, the evolution of the mean value of the liquid film over the lateral wall remains approximately the same. As an example, Figure 66.10 shows the instant behaviour of the liquid film thickness for a void fraction of 75 % as bubbles pass over the conductimetry probe. From these signals, the maximum and averaged values are not significant. They are a result of

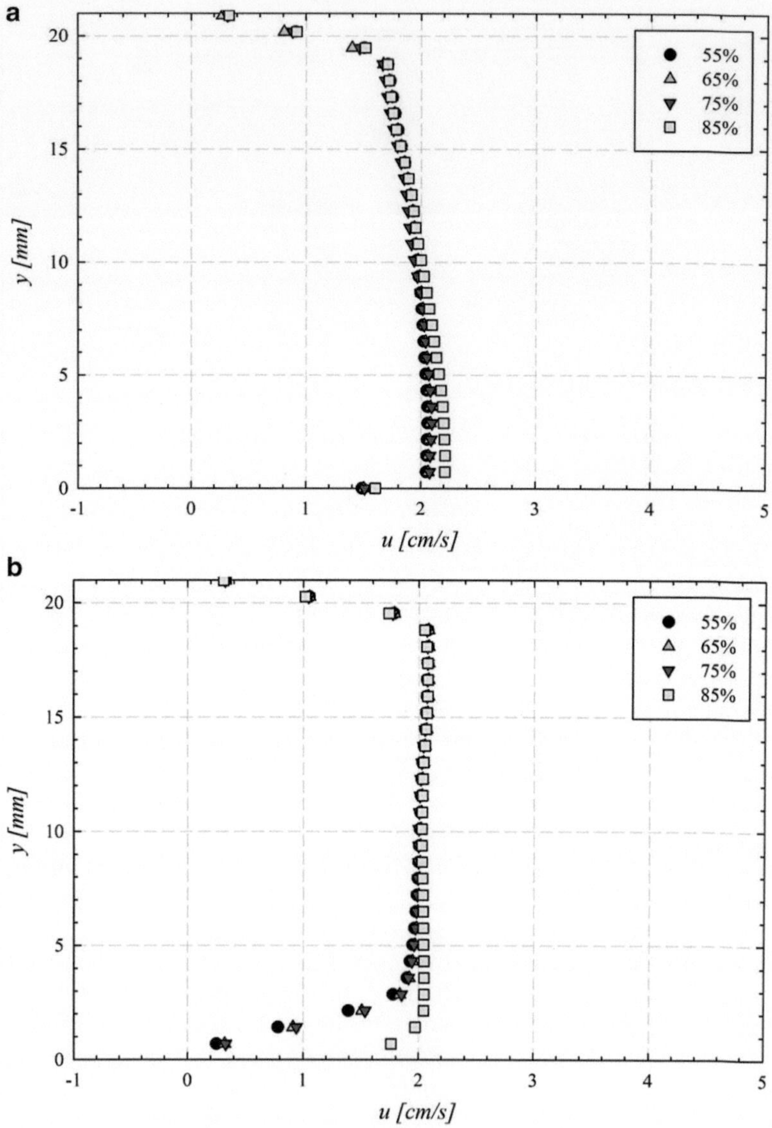

Fig. 66.8 "u" [cm/s] velocity profiles at (**a**) $x = -25$, (**b**) $x = 25$, (**c**) $x = -15$, (**d**) $x = 15$, (**e**) $x = -5$, (**f**) $x = 5$ and (**g**) $x = 0$

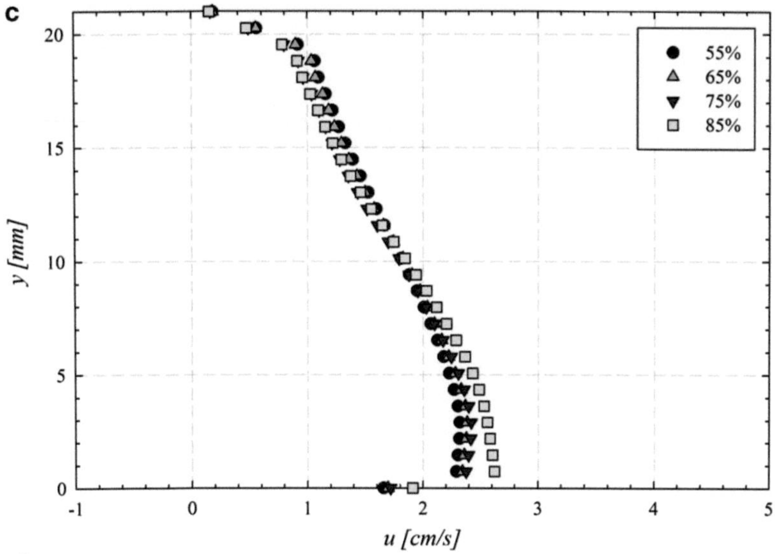

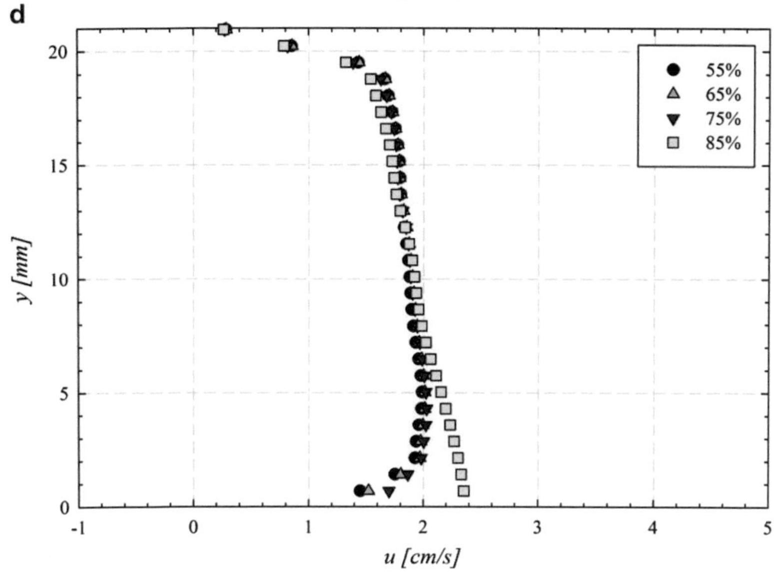

Fig. 66.8 (continued)

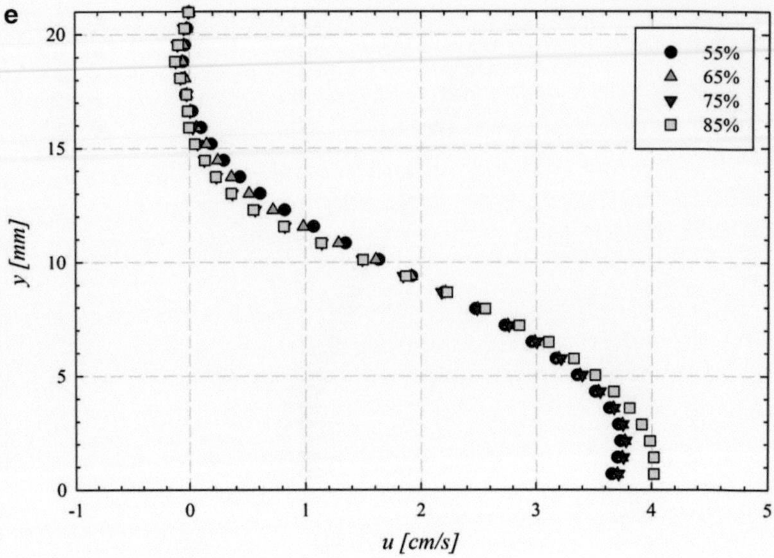

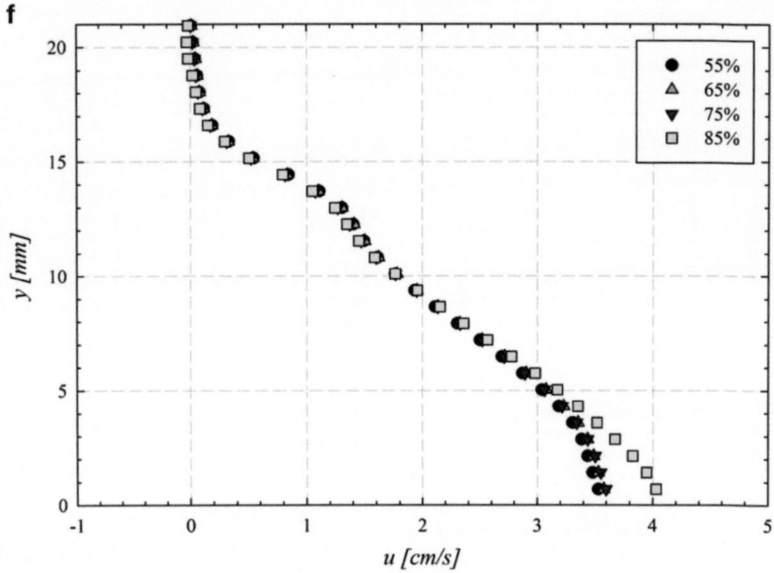

Fig. 66.8 (continued)

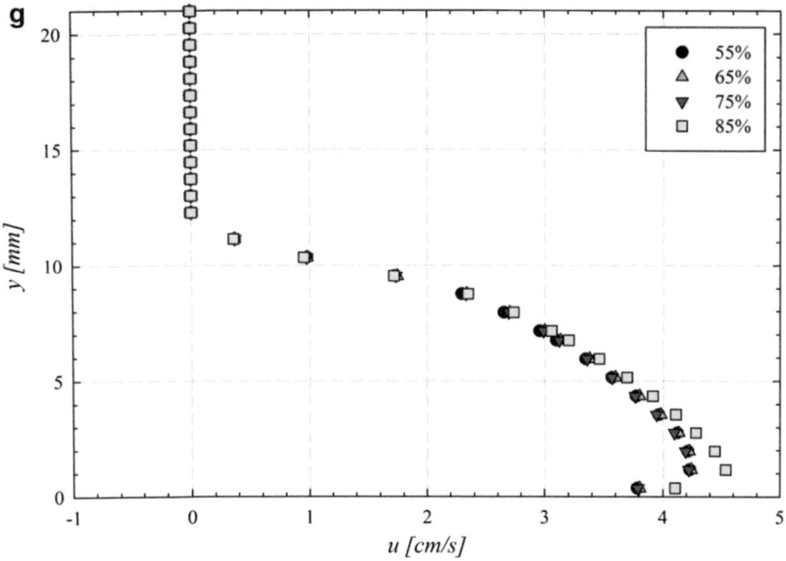

Fig. 66.8 (continued)

the passage of a bubble and not the actual conduction of the liquid film. Therefore, only the measurements of the minimal thickness were analyzed.

The evolution of the average liquid film thickness along the lateral wall is represented in Fig. 66.11. It can be noted that, due to the drainage forces (gravity, capillarity and disjoining pressure), the liquid film closest to the bottom of the channel is thicker (around 60 μm) than the one at the top (around 5 μm). Its evolution can be adjusted to a curve with the form $y = 82.33 \times e^{-0.904}$. The thickness of the liquid film at the bottom of the channel depends on the void fraction but does not make part of this study.

Some assumptions were made to calculate the wall shear stress over the lateral wall along the test section: The liquid film thickness remains constant along the axial length, it is equal to the theoretical curve adjusted to the experimental results and the wall velocity gradient is equal to the axial bubble velocity over the wall, shown in Fig. 66.5. The value of the wall shear stress is obtained with the expression

$$\tau(y) = \mu_l \frac{\delta u}{\delta y} \tag{66.5}$$

where $\mu_l = 1.002 \times 10^{-3}$ is the liquid dynamic viscosity (same as the water), δu is the wall velocity gradient and $\delta y = e$ is the thickness of the liquid film.

Figure 66.12 shows the wall shear stress field over the lateral wall for both limits, wet (55 %) and dry (85 %). Though the velocity fields and profiles of the axial component showed us a difference in terms of shear and behaviour of these two void fractions, the wall shear stress difference is not appreciated. Maximum values are achieved away from singularity at the top of the channel (4 Pa). The axial velocity takes a minor part in the wall shear stress and it is the liquid film thickness that influences the behaviour of this one the most. Therefore, in those places where the liquid film tends to a minimum value (5 μm) the wall shear stress tilts to its maximum.

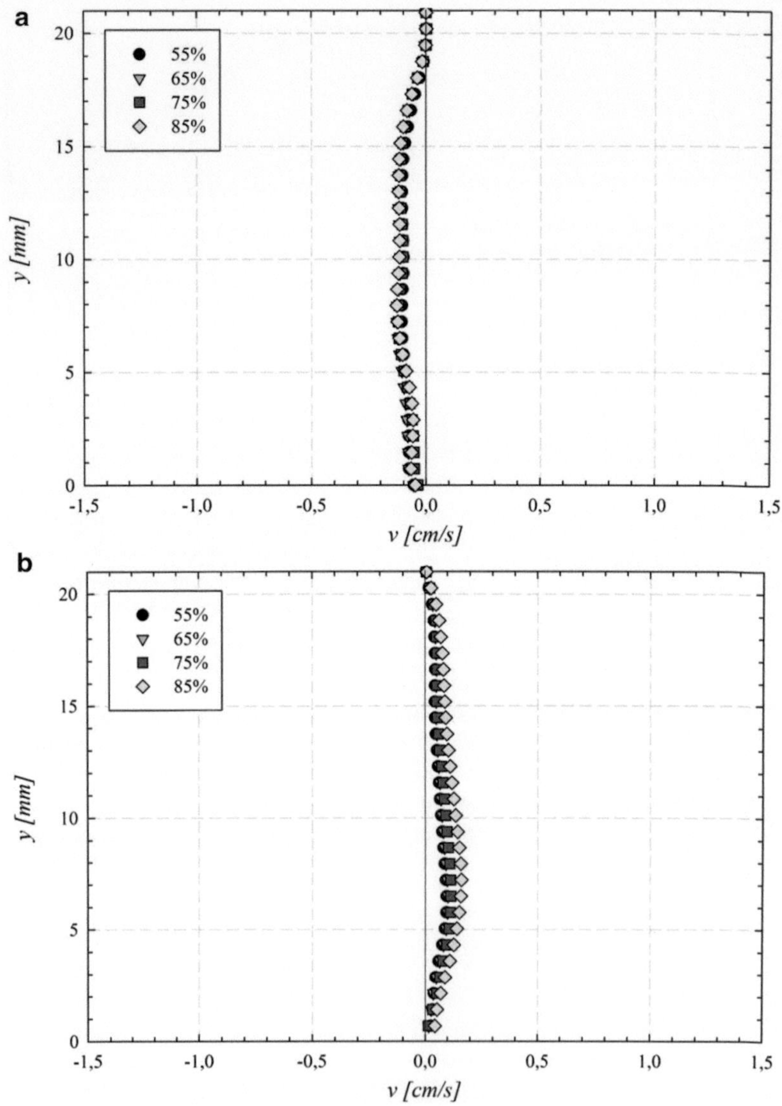

Fig. 66.9 "v" [cm/s] velocity profiles at (**a**) $x=-25$, (**b**) $x=25$, (**c**) $x=-15$, (**d**) $x=15$, (**e**) $x=-5$, (**f**) $x=5$ and (**g**) $x=0$

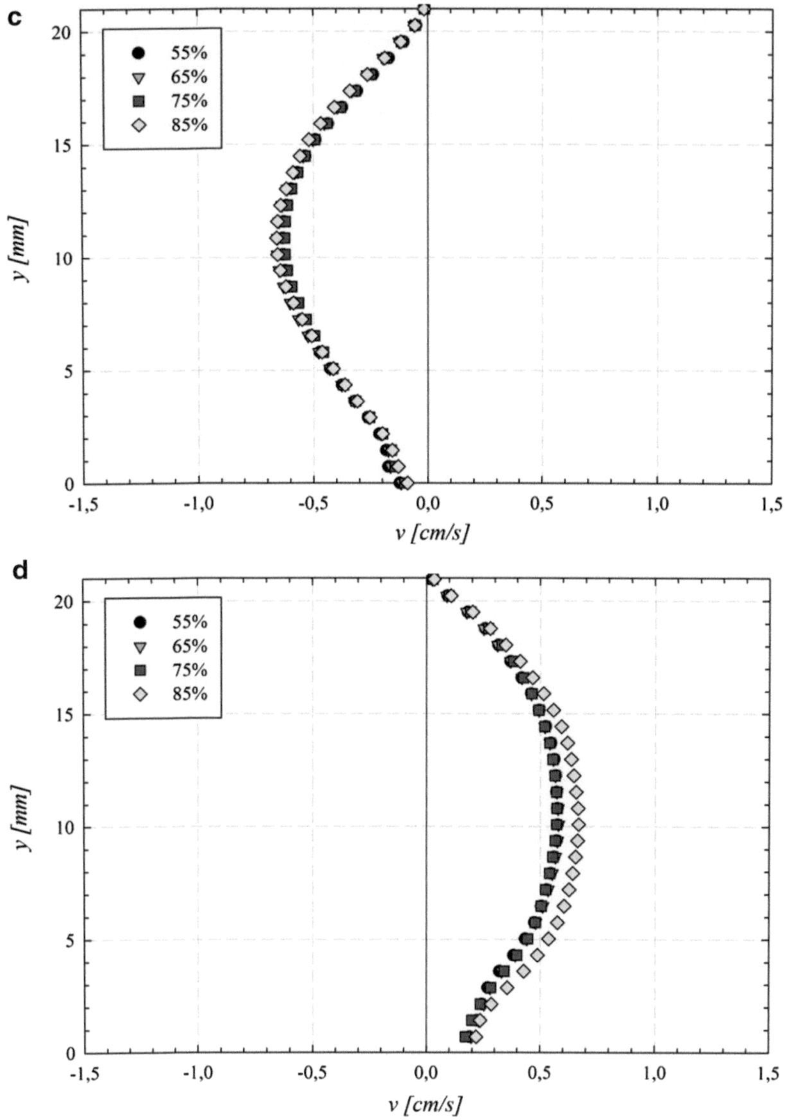

Fig. 66.9 (continued)

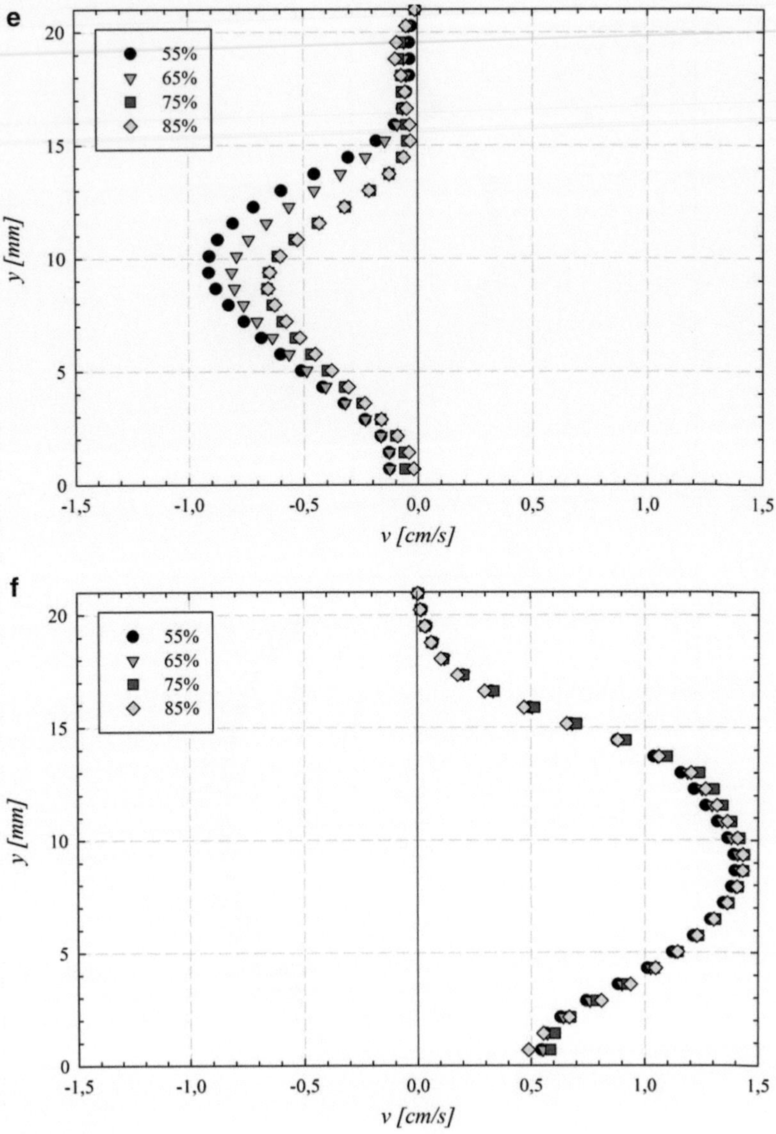

Fig. 66.9 (continued)

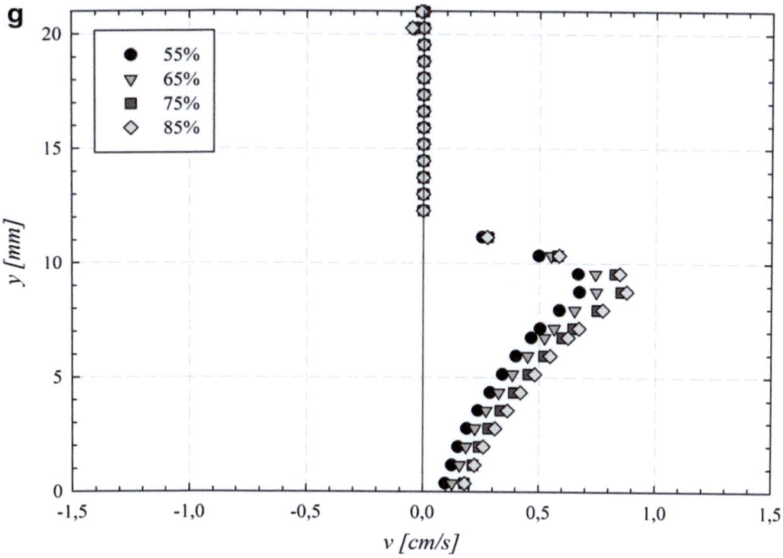

Fig. 66.9 (continued)

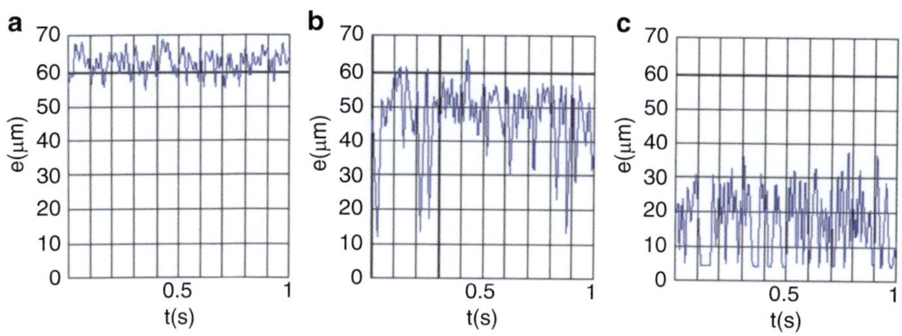

Fig. 66.10 Instant evolution of the liquid film thickness over the channel's lateral wall at the (**a**) liquid/foam interface; (**b**) middle of the lateral wall and (**c**) top side of the lateral wall

From the previous fields, the profiles of the wall shear stress over the later wall along the length of the test channel were extracted and they are displayed in Fig. 66.13. The maximum value is obtained at both upstream and downstream of the singularity where the flow is in a one-dimensional regime with a mean velocity of 1.75 cm/ and at the top of the channel where the liquid film presents its minimal value (5 μm). By moving closer to the singularity, the velocity affects the wall shear stress, and despite having smaller thickness at the top of the wall the maximum stress is created near the maximum values of velocities (4 cm/s) towards the middle, between the fence and the

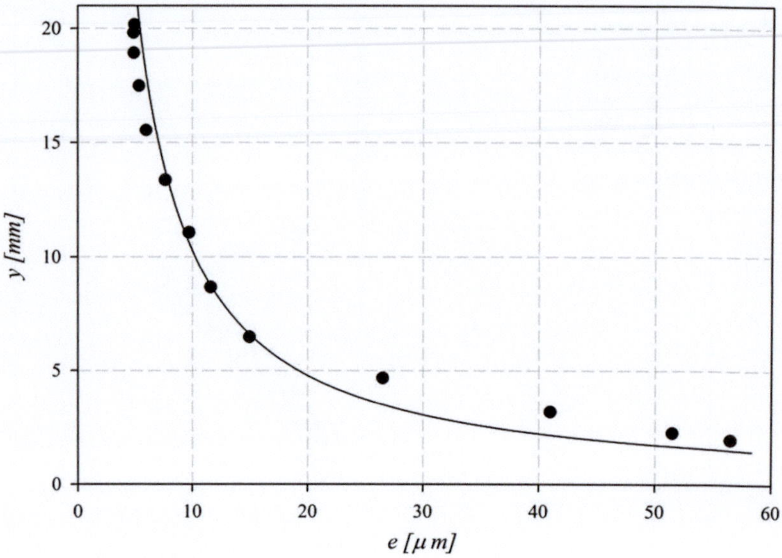

Fig. 66.11 Evolution of the average liquid film thickness along the lateral wall of the channel

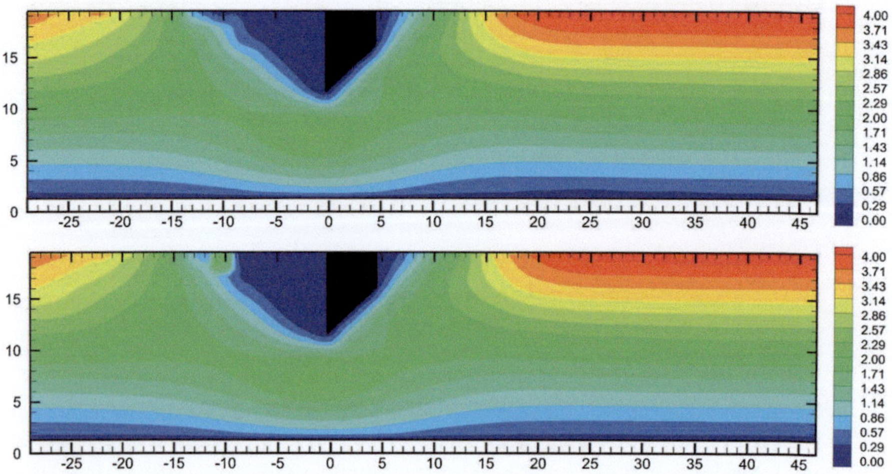

Fig. 66.12 Wall shear stress [Pa] fields for $\beta = 55$ % (*top*) and $\beta = 85$ % (*bottom*)

bottom of the channel. In this region a deviation of around 5 % is noted for the void fraction difference.

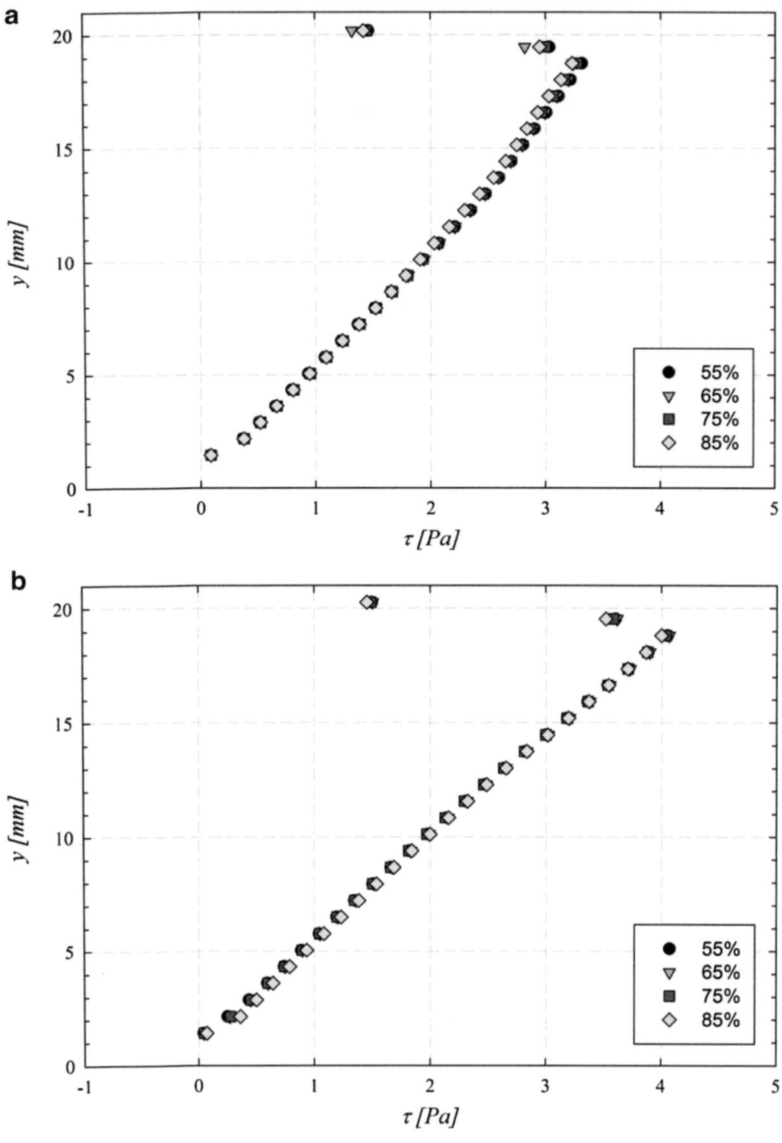

Fig. 66.13 "τ" [Pa] profiles at (**a**) $x = -25$, (**b**) $x = 25$, (**c**) $x = -15$, (**d**) $x = 15$, (**e**) $x = -5$, (**f**) $x = 5$ and (**g**) $x = 0$

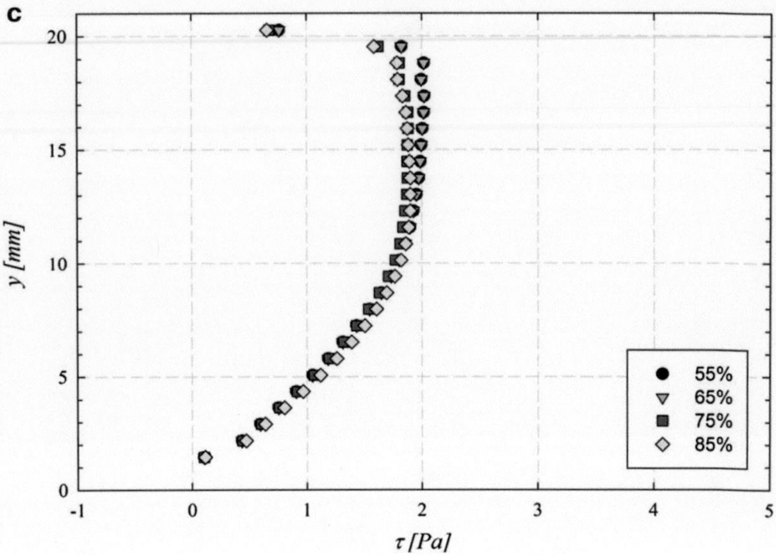

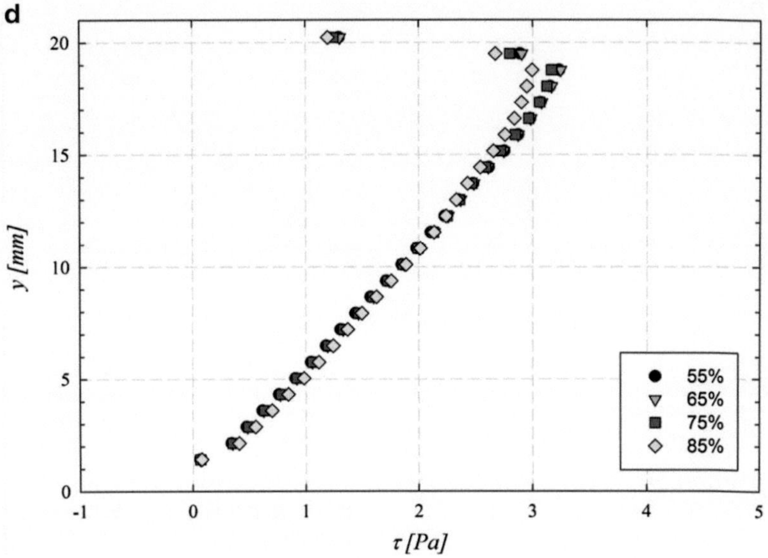

Fig. 66.13 (continued)

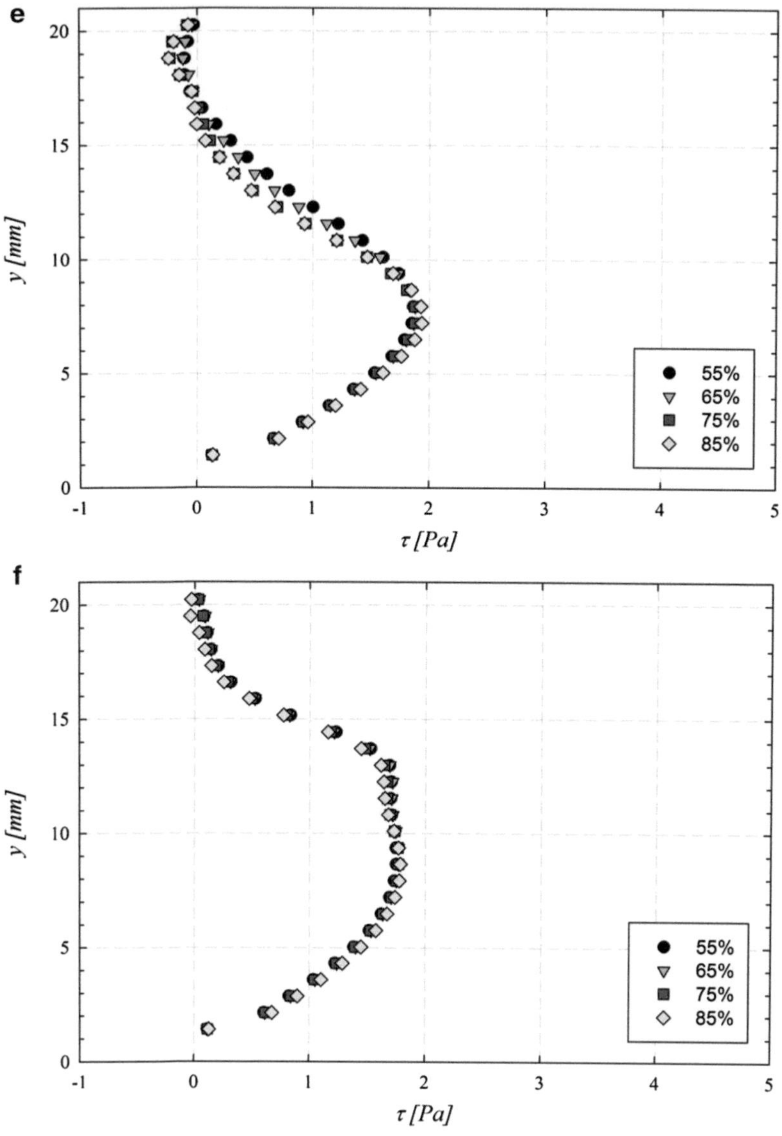

Fig. 66.13 (continued)

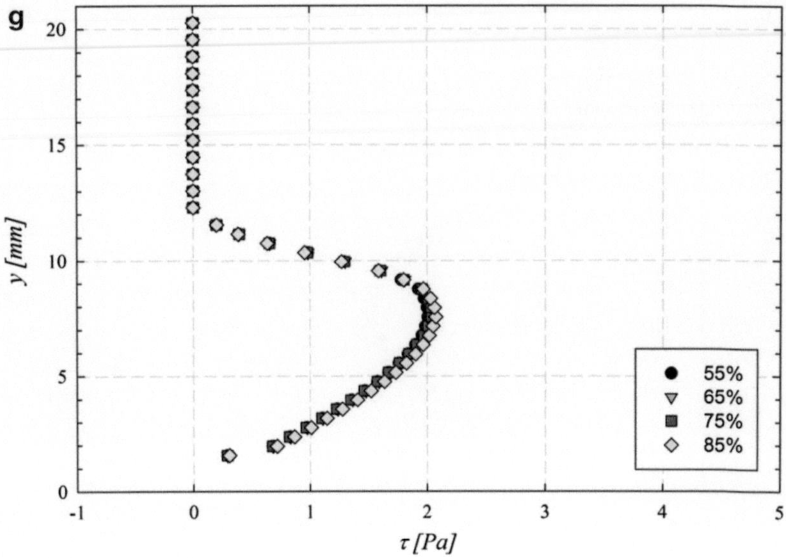

Fig. 66.13 (continued)

66.5 Conclusions

In this study, we characterize and analyze the velocity fields, liquid film thickness and wall shear stress of an aqueous foam flow with a mean velocity of 1.75 cm/s and void fractions of 55, 65, 75, and 85 %. The mean goal was to attack the two main problems of any foam flow inside a canalization, its rheological and stability properties. The measurements were made over an aqueous foam flowing through a fence with the PIV and conductimetry methods. The results put into evidence the complexity of this flow. The rheological and stability properties are extremely sensitive to the condition of the flow and its surroundings.

The singularity considerably changes the foam flow structure reorganization in its vicinity with an axial acceleration when approaching the fence and deceleration when exiting this one. The dry foam tends to move faster, therefore more shear, in the axial direction than the wet foam. For the vertical velocity component, the gravity and the foam density play a major role in the behavior. When the negative, or downward, movement sets in the wet foam (denser) tends to accelerate more than the dry foam. In the opposite case (upward or positive movement) is the dryer one that moves faster towards the top of the channel.

The obtaining of the liquid film at the lateral wall, with the conductimetry method, allowed the estimation of the wall shear stress. Despite the difference of void fraction, both the liquid film thickness evolution and the wall shear stress do not suffer any important change in their behaviour. Due to drainage forces, the maximum values for the liquid film thickness over the lateral wall were found at the bottom of the channel and the minimum at the top. The axial velocity does not have a big impact over the maximum value of the wall shear stress. This one is inversely proportionally to the liquid film thickness. The bigger stress is obtained away from the fence, where the foam flow presents one uniform axial component, near the top of the channel, for the smaller thickness. Close to the singularity the axial velocity influences the wall shear stress and a small difference between the dry and wet foam can be noted.

If foam flow can teach something is that problems need to be seen from both the macroscopic and the microscopic points of view. In a foam, the distribution of molecular particles inside the liquid films can change the rheological properties of the whole. The same approach needs to be taken in all subjects concerning energy and its efficiency. Even small changes in a system can improve its performance. Foam flows present complex properties but once they are understood the possibilities are infinite. This study is one step closer in understanding those properties and improving the efficiency of foams used in some industrial processes.

References

1. Plateau JAF (1861) Mem Acad Roy Sci Belg vol. 33. 5th and 6th Series
2. Weaire D, Hutzler S (1999) The physics of foams. Clarendon, Oxford
3. Durian DJ, Weitz DA, Pine DJ (1990) Dynamics and coarsening in three-dimensional foams. J Phys Condens Matter 2:433–436
4. Hutzler S, Saadatfar M, van der Net A, Weaire D, Cox S (2008) The dynamics of a topological change in a system of soap films. Colloids Surf A Physicochem Eng Asp 323:123–131
5. Bikerman JJ (1973) Foams. Springer, New York
6. Weaire D (2008) The rheology of foam. Curr Opin Colloid Interface Sci 13(3):171–176
7. Tisné P, Aloui F, Doubliez L (2003) Analysis of wall shear stress in wet foam flows using the electrochemical method. Int J Multiphase Flow 29(5):841–854
8. Marze S, Langevin D, Saint-Jalmes A (2008) Aqueous foam slip and shear regimes determined by rheometry and multiple light scattering. J Rheol 52:1091
9. Aloui F, Madani S (2007) Wet foam flow under a fence located in the middle of a horizontal duct of square section. Colloids Surf A Physicochem Eng Asp 309:71–86
10. Blondin E, Doubliez L (2002) Particle imaging velocimetry of a wet aqueous foam with an underlying liquid film. Exp Fluids 32(3):294–301
11. Aloui F, Madani S (2008) Experimental investigation of a wet foam flow through a horizontal sudden expansion. Exp Thermal Fluid Sci 32:905–926
12. Tisné P, Doubliez L, Aloui F (2004) Determination of the slip layer thickness for a wet foam flow. Colloids Surf A Physicochem Eng Asp 246:21–29

Chapter 67
Signal-Processing Technology for Rotating Machinery Fault Signal Diagnosis

Byung Hyun Ahn, Yong Hwi Kim, Jong Myeong Lee, Jeong Min Ha, and Byeong Keun Choi

Abstract The acoustic emission (AE) technique is widely applied to develop early fault detection systems, on which the problem of a signal-processing method for an AE signal is mainly focused. In the signal-processing method, envelope analysis is a useful method to evaluate the bearing problems and the wavelet transform is a powerful method to detect faults occurring on rotating machinery. However, an exact method for the AE signal has not been developed yet. Therefore, in this chapter two methods are given: Hilbert transform and discrete wavelet transform (IEA), and DET for feature extraction. In addition, we evaluate the classification performance with varying the parameter from 2 to 15 for feature selection DET and 0.01–1.0 for the RBF kernel function of SVR; the proposed algorithm achieved 94 % classification accuracy with the parameter of the RBF 0.08, 12 feature selection.

Keywords Acoustic emission • Signal processing • Hilbert transform • Fault classification • Feature selection

67.1 Introduction

Eshleman proposed a hardware-based signal envelope technique in which signals are passed through a capacitor to produce a demodulated time waveform. The Fast-Fourier-transform (FFT)-based Hilbert transform is the traditional method for deriving the signal envelope and has been widely used in roller bearing diagnostics. More recently, wavelet transform has been used for signal demodulation and optimal bandpass filter (BPF) design. In summary, the Hilbert and wavelet transform methods are promising methods for signal demodulation. These methods, however, cannot enhance a weak signature from a noisy signal, and cannot detect early stage defects.

The problem of signal denoising has a strong connection to machine component prognostics. Denoising and the extraction of a weak signature are crucial to fault

B.H. Ahn (✉) • Y.H. Kim • J.M. Lee • J.M. Ha • B.K. Choi
Department of Energy & Mechanical Engineering, College of Marine Science, Gyeongsang National University, Cheondaegukchi-Gil 38, Tongyeong, Gyeongnam 650-160, Republic of Korea
e-mail: bgchoi@gnu.ac.kr

I. Dincer et al. (eds.), *Progress in Clean Energy, Volume 1*,
DOI 10.1007/978-3-319-16709-1_67

prognostics, where the features are often very weak and masked by noise. Prognosis is achieved by detecting the defect at its initial stage, and alerting maintenance personnel to such before it develops into a catastrophic failure. The standard approach for extracting signals from a noisy background is to design an appropriate filter that removes the noise components and at the same time allows the desired signal to go through unchanged. Based on the noise type and application, different filters can be designed to conduct denoising [1, 2].

Acoustic emission (AE) is the phenomenon of transient-elastic–wave generation due to the rapid release of strain energy caused by a structural alteration in a solid material under mechanical or thermal stress. The generation and propagation of cracks are among the primary sources of AE in materials. AE transducers are designed to detect the very-high–frequency (>50 kHz) stress waves that are generated when cracks extend under load. The application of the AE technique in the condition monitoring of rotating machinery has been increasing of late. This is particularly true for bearing defect diagnosis and seal rubbing. The technique, however, has a major drawback: the attenuation of the signal and, as such, the AE sensor, has to be close to its source. It is often practical, however, to place the AE sensor on the nonrotating member of the machine, such as the bearing or gear casing. Therefore, the AE signal originating from the defective component will suffer severe attenuation before reaching the sensor [3–8].

To remove background noise in the AE signal and to find the relationship between the AE signal and the fault mechanism, envelope analysis and wavelet transform were applied for the detection of defects in the rolling element bearing, gear, and cracked shaft. In the original power spectrum of the AE signal, the defect mechanism cannot be identified because the frequency range is over a decade kHz. It can indicate the defect frequency in the power spectrum, however, through envelope analysis with the Hilbert transform [9], after which the defect can be evaluated. Envelope analysis has a disadvantage, though: it has difficulty predicting or specifying which resonant modes of the neighboring structures will be excited. It will be costly and unrealistic in practice to find the resonant modes by performing experiments on rotating machinery that may also change under different operational conditions. In addition, it is difficult to estimate how these resonant modes are affected in the assembly of a complete bearing and mounting in a specific housing, even if theoretically, the resonant frequencies of the individual bearing elements can be tested or calculated. As such, most researchers choose the bandpass range as an option. To make up for the disadvantage of this option, wavelet analysis was included in the process of envelope analysis in this chapter because the discrete wavelet transform (DWT) has a denoising function and a filter effect focused on the impact signal. To make up for the weakness of BPF in envelope analysis, DWT was substituted with BPF. The signal by DWT was separated into different bandwidths by decomposition level, and was adapted to the signal with impact. In addition, to increase the effect of DWT, a novel mother function was developed through a fatigue crack growth test because the ability of DWT depends on the pattern of the mother function. In the wavelet toolbox of MATLAB, most of the patterns of the mother function are not corrected with AE signals.

67.2 Algorithm

See Fig. 67.1

67.3 Intensified Envelope Analysis (IEA)

Envelope analysis is an important signal-processing technique that helps in the identification of bearing defects by extracting the characteristic defect frequencies (which may not be present in the direct spectrum) from the vibration signal of the defective bearing. Every time a defect strikes its mating element, a short-duration pulse is generated, which excites the resonances periodically at the characteristic frequency related to the defect location. The resonances are thus amplitude-modulated at the characteristic defect frequency. By demodulating these resonances, a signal indicative of the bearing condition can be recovered. In practice,

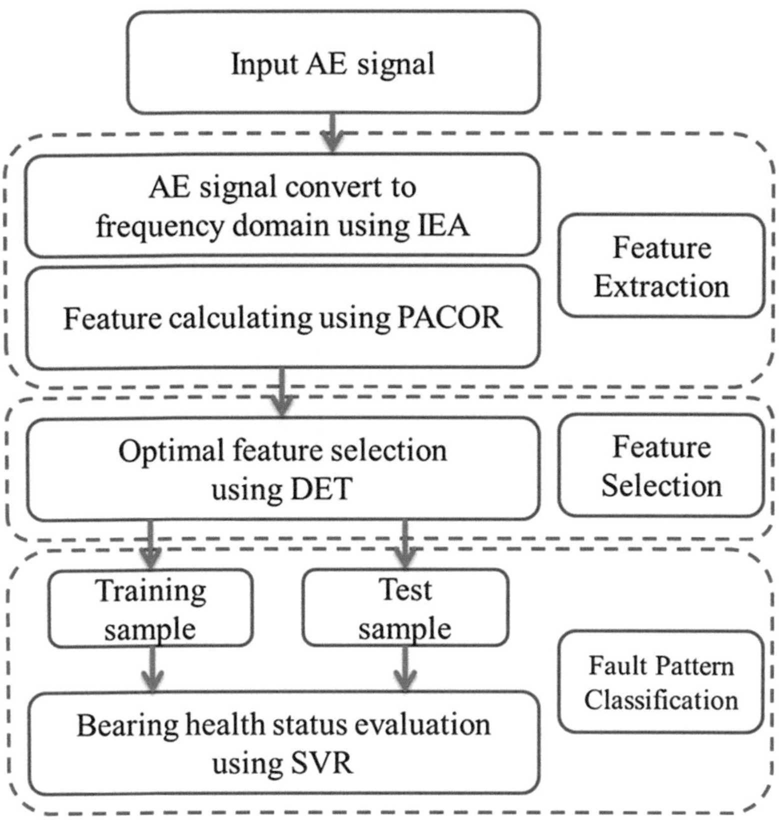

Fig. 67.1 Proposed bearing fault diagnosis scheme using acoustic emission signal

the signal is bandpass filtered around one of the resonant frequencies, thus eliminating most of the unwanted vibration signals from other sources. This bandpass-filtered signal is then demodulated by an envelope detector, in which the signal is rectified and smoothed by lowpass filtering to eliminate the carrier or bandpass-filtered resonant frequency. The spectrum of the envelope signal in the low-frequency range is then obtained to determine the characteristic defect frequency of the bearing.

Envelope analysis typically refers to the following procedure sequence: (1) bandpass filtering, (2) wave rectification, (3) Hilbert transform or lowpass filtering (LPF), and (4) power spectrum. The purpose of BPF is to reject the low-frequency, high-amplitude signals associated with the *i*th mechanical vibration components, and to eliminate the random noise outside the passband. Theoretically, the best bandpass range includes the resonance of the bearing components. Theoretically, in HFRT (high-frequency resonance technique) analysis, the best bandpass range includes the resonance of the bearing components. This frequency can be found through impact tests or theoretical calculations involving the dimensions and material properties of the bearing. It is very difficult, however, to predict or specify which resonant modes of the neighboring structures will be excited. It will be costly and unrealistic in practice to find the resonant modes by performing experiments on rotating machinery, which may also change under the different operational conditions. In addition, it is difficult to estimate how these resonant modes are affected in the assembly of a complete bearing and mounting in a specific housing, even if theoretically, the resonant frequencies of the individual bearing elements can be tested or calculated. As such, most researchers choose the bandpass range as an option. To make up for the disadvantage of envelope analysis, wavelet analysis was substituted with BPF in envelope analysis, as shown in Fig. 67.1

The wavelet theory was introduced as a tool for the analysis of transient, nonstationary, or time-varying phenomena. Wavelet analysis is also called wavelet transform. There are two kinds of wavelet transform: continuous wavelet transforms (CWT) and discrete wavelet transforms (DWT). CWT is defined as the sum of all the times of the signal multiplied by the scaled shifted versions of the wavelet function. To use CWT, one signal can be decomposed into a series of "small" waves belonging to a wavelet family. The wavelet family is composed of scaling functions deduced by the father wavelet and of wavelet functions deduced by the mother wavelet.

The decomposition coefficients can therefore be determined through convolution, and can be implemented using a filter (low- or highpass filter). The decomposition process can be iterated, with the successive approximations being decomposed in turn, so that one signal is broken down into many lower-resolution components. Furthermore, DWT has a denoising function and a filter effect focused on the impact signal. To make up for the weakness of BPF in envelope analysis, DWT was intercalated into the typical envelope analysis, particularly between BPF and wave rectification. The signal by DWT was separated into different bandwidths by decomposition level, and was adapted to the signal with impact (Fig. 67.2).

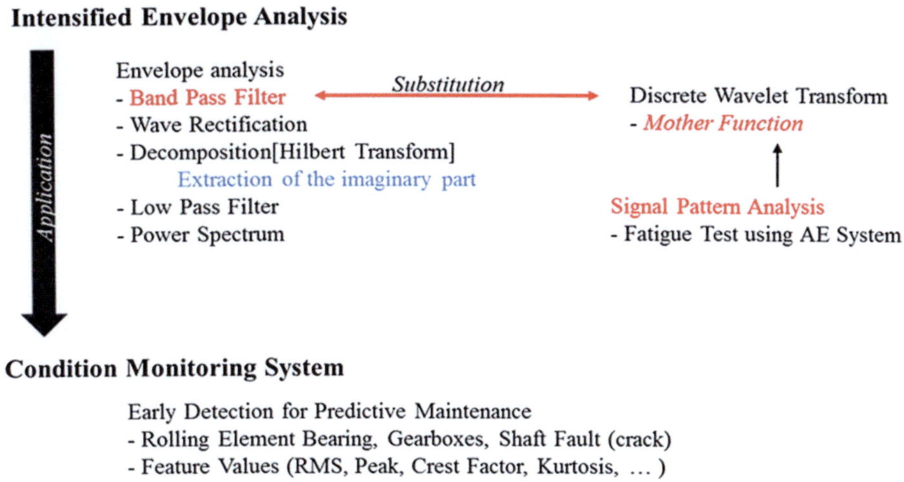

Fig. 67.2 Concept of IEA

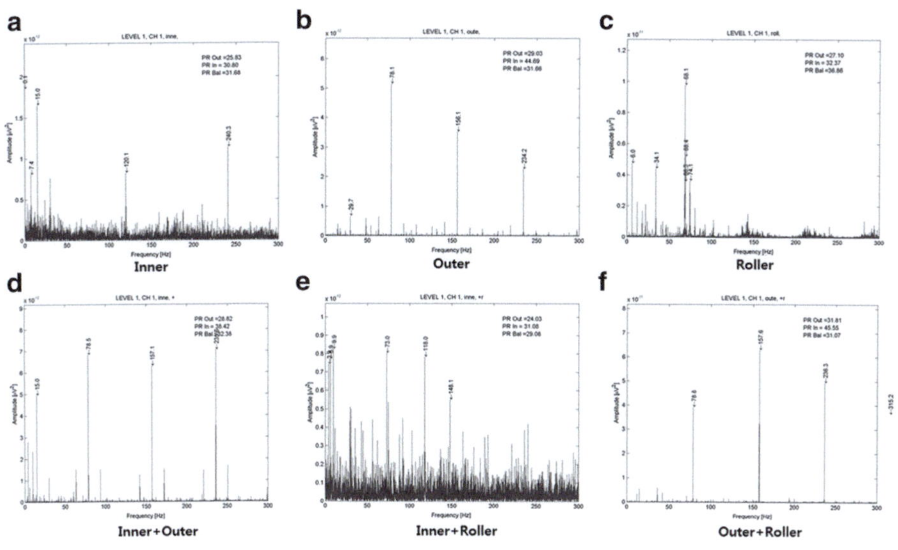

Fig. 67.3 Each defective bearing spectrum of fault signal

67.4 Feature Extraction

Figure 67.3 shows the correlation between the spectra of the type of fault. As shown in Fig. 67.4 are independently of each type of bond type between the spectra appears it can be seen that no duplication. In this paper, it extracts the peak signal appears in the spectrum to a particular vector to diagnose the condition of bearings.

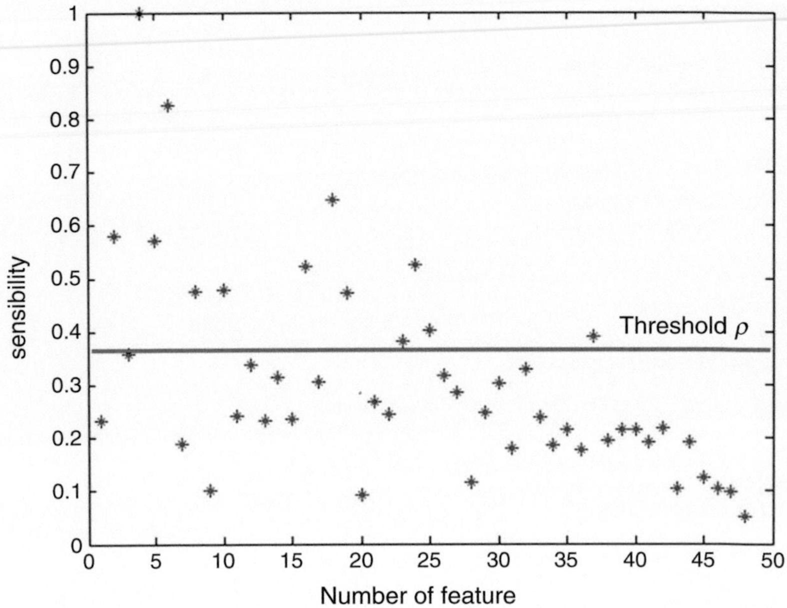

Fig. 67.4 Number of features

Past and current specimen samples on the basis of the input correlation between the sample models a representation of the LPC (linear predictive coding). However, LPC coefficients contained in the output signal by the change of noise have severe drawbacks. In this chapter, the signal may result in the acquisition of these phenomena taking into account the noise is stronger than the partial correlation coefficients (PARCOR). LPC is used as a feature vector [10].

67.5 Feature Selection

Supposed C pattern to $\omega_1, \omega_2, \ldots, \omega_c$,

$$\left\{ p^{(i,k)} i = 1, 2, \ \ldots, c; \ k = 1, 2, \ \ldots, N_i \right\} \tag{67.1}$$

At $p^{(i,k)}$, i is the kth feature, N_i is the number of i.

The average deviation of all features of i is as below,

$$S_i = \frac{1}{2} \frac{1}{N_i} \sum_{j=1}^{N_i} \frac{1}{N_i - 1} \sum_{k=1}^{N_i} \left| p^{(i,j)} - p^{(i,j)} \right| \tag{67.2}$$

The average deviation of S_i is

$$S_w = \frac{1}{c}\sum_{i=1}^{c} S_i \tag{67.3}$$

Equation (67.2) changes to Eq. (67.3)

$$S_w = \frac{1}{c}\sum_{j=1}^{N_i} \frac{1}{N_i - 1}\sum_{k=1}^{N_i} \left| p^{(i,j)} - p^{(i)} \right| \tag{67.4}$$

Here

$$\mu^{(i)} = \frac{1}{N}\sum_{i=1}^{N_i} p^{(i,k)} \tag{67.5}$$

i is average value

With each different condition pattern cf average deviation $\omega_1, \omega_2, \ldots, \omega_c$, is

$$S_b = \frac{1}{c}\sum_{i=1}^{c} \left| \mu^{(i)} - \mu \right| \tag{67.6}$$

Here

$$\mu = \frac{1}{c}\sum_{i=1}^{c} \frac{1}{N_i}\sum_{k=1}^{N_i} p^{(i,k)} \tag{67.7}$$

All of the c condition pattern is the feature average. Inside the condition pattern is less than average deviation S_w; the between-average deviation pattern condition is more; the average indicates the optimal feature very well. Next is regarding distance evaluation.

$$J_A = \frac{S_b}{S_w} \tag{67.8}$$

67.6 Support Vector Regression

Support vector regression (SVR) theory is developed based on the principle of SVM and is suitable for time series prediction. It aims to obtain a function $f(x)$ that can predict the output within the error limit of ε. In addition, the estimation function $f(x)$ can be as flat as possible to ensure a good generalization property and variance [11].

This function is presented as
Maximize

$$-\frac{1}{2}\sum_{i=1}^{n}\sum_{j=1}^{n}\left(a_i - a_i^*\right)\left(a_j - a_i^*\right)K\left(x_i x_j\right) \tag{67.9}$$

$$\in \sum_{i=1}^{n}\left(a_i + a_i^*\right) - \sum_{i=1}^{n}\left(a_i - a_i^*\right) \tag{67.10}$$

Subject to

$$\sum_{i=1}^{n}\left(a_i - a_i^*\right) = 0,\ a_i,\ a_i^* \in [0, C] \tag{67.11}$$

$$\overline{w} = \sum_{SV_s}\left(\overline{a}_i - \overline{a}_i^*\right)z(x_i) \tag{67.12}$$

$$\overline{b} = -\frac{1}{2}\sum\nolimits_{SV_s}\left(\overline{a}_i - \overline{a}_i^*\right)\left[K\left(x_y x_i\right) + \left(x_s x_j\right)\right] \tag{67.13}$$

$$f(x) = \sum\nolimits_{SV_s}\left(\overline{a}_i\overline{a}_i^*\right)K\left(x_i x_j\right) + \overline{b} \tag{67.14}$$

$K\left(x_i x_j\right) = \varphi(x_i)\cdot\varphi(x)$ is a symmetric positive defined kernel function given by Mercer's theorem, and the kernel function can map a low-dimensional vector to high feature space through some nonlinear function. In this chapter, the popular radial basis function (RBF) is adopted and its mathematical formula is given as [12]:

$$K\left(x_i x_j\right) = \exp\left(-\frac{\left|\left|x_i - x_j\right|\right|^2}{2\sigma^2}\right) \tag{67.15}$$

67.7 AE Signal Acquisition

Table 67.1 shows properties of the AE transducer (PAC), amplifier, and data acquisition board to obtain the AE signal.

Table 67.1 Properties of AE system

2 Channel AE system on PCI-board	18 Bit A/D conversion
	5M samples/s rate
AE sensor (Wideband type)	Peak sensitivity: −62 dB
	Operating freq. range: 100 ~ 1,000 kHz
	Resonant frequency: 650 kHz
	Directionality: +/−1.5 dB
Preamplifier gain	Wide dynamic range <90 dB
	20/40/60 dB selectable gain

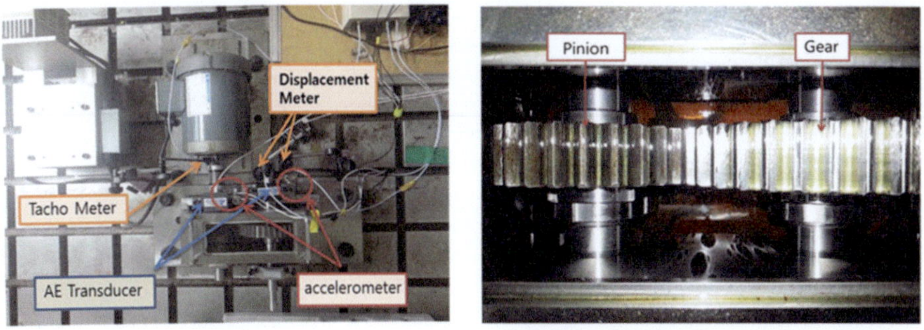

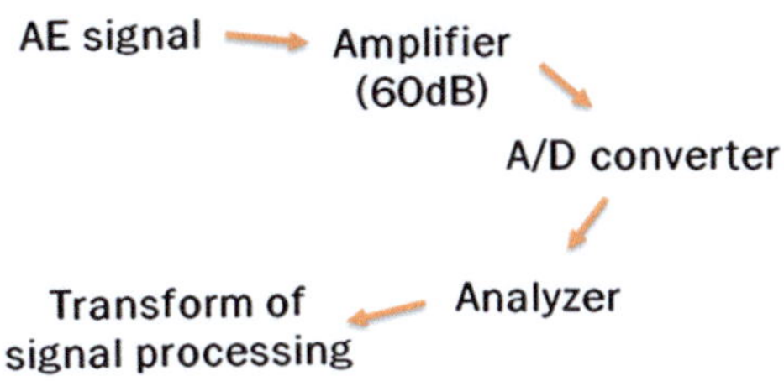

	Gear	Pinion
No. of teeth	38	25
Speed of shaft	15 Hz	

Fig. 67.5 Acoustic emission signal acquisition procedure

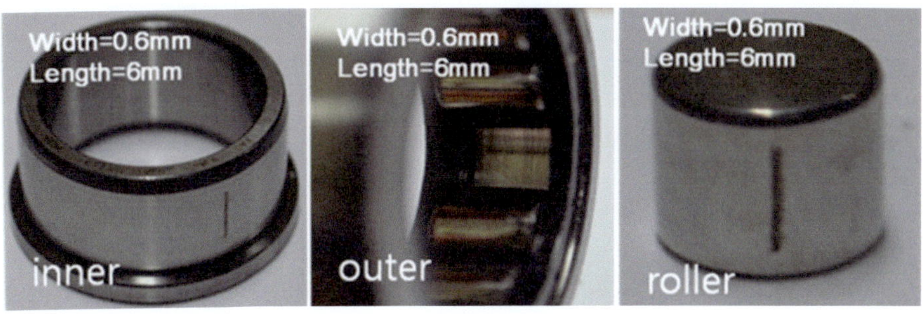

Fig. 67.6 Type of defective bearing

Figure 67.5 shows the acoustic emission signal system procedure. AE sensors are attached to the housing of the drive shaft and driven shaft. The test is conducted on the no-load 15 Hz.

Figure 67.6 indicates three type of bearing fault on each component of roller bearing (FAG NJ206E-TVP2). Local single fault type is made by using electric discharge machining for AE data acquisition.

67.8 Result of Signal Processing

Figure 67.7 and Table 67.2 show optimal classification performance on the AE signal using the proposed algorithm in this chapter. The training sample is 25 and the test sample is 36 of each fault type. Then the performance average result is

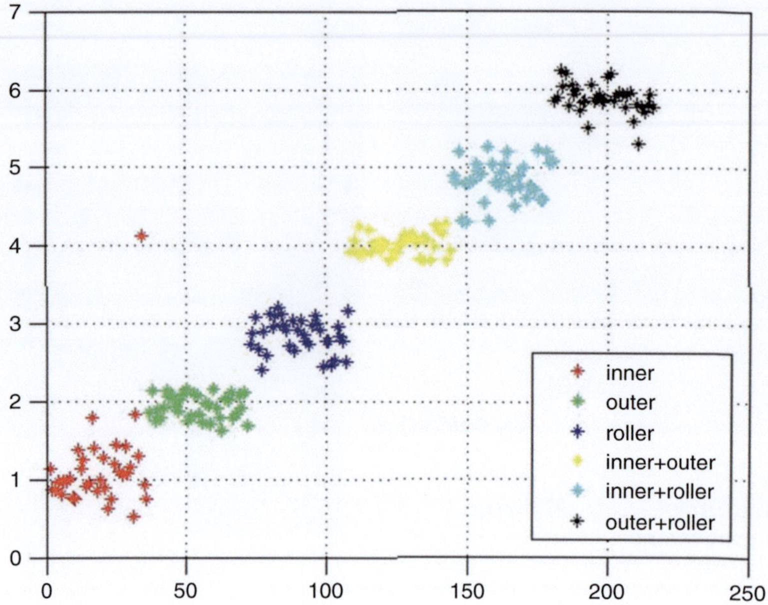

Fig. 67.7 Fault classification performance

Table 67.2 Fault classification performance

	Inner	Outer	Roller	Inner +Outer	Inner +Roller	Outer +Roller	Performance (%)
Inner	33	2	0	1	0	0	91.66
Outer	0	36	0	0	0	0	100
Roller	0	4	32	0	0	0	88.89
Inner +outer	0	0	0	36	0	0	100
Inner +roller	0	0	0	3	33	0	91.67
Outer +roller	0	0	0	0	1	35	97.23

94.9 %. And the optimal parameter of the RBF kernel is 0.01–0.1; the number of features 2–15 was conducted. The experiment was conducted without noise and it is required for optimal classification; the parameter number is 0.08 and the number of features is 12.

By signal processing, classification performance is improved and the result shows the case of bearing faults in the graph, and the result has reliability of the fault classification.

Also fault classification is conducted under reduced noise and the performance is evaluated so this processing can be applied to the AE signal system. Therefore, it can be analyzed for rotating machinery fault signal diagnosis using the acoustic emission signal.

References

1. Mobley RK (2002) An introduction to predictive maintenance, 2nd edn. Elsevier, New York
2. Sullivan GP, Pugh R, Melendez AP, Hunt WD (2010) Operations & maintenance—best practices: a guide to achieving operational efficiency (Rel. 3.0). U.S. Department of Energy, Richland
3. Robert B, Antoni J (2011) Rolling element bearing diagnostics. Mech Syst Signal Process 25 (2):485–520
4. Baydar N, Ball A (2003) Detection of gear failures via vibration and acoustic signals using wavelet transform. Mech Syst Signal Process 17(4):787–804
5. Jeong HE, Gu DS, Kim HJ, Tan A, Kim YH, Choi BK (2007) The application of AE transducer for the bearing condition monitoring of low-speed machine. KSNVE07S-09-03
6. Kim HJ, Gu DS, Jeong HE, Tan A, Kim E, Choi BK (2007) The comparison of AE and acceleration transducer for the early detection on the low-speed bearing. KSNVE07S-09-04
7. Loutas TH, Sotiriades G, Kalaitzoglou I, Kostopouls V (2009) Condition monitoring of a single-stage gearbox with artificially induced gear cracks utilizing on-line vibration and acoustic emission measurements. Appl Acoust 70:1148–1159
8. Entezami M, Hillmansen S, Roberts C (2010) Wind turbine condition monitoring system. University of Birmingham, Birmingham
9. Wang D, Miao Q, Kang R (2009) Robust health evaluation of gearbox subject to tooth failure with wavelet decomposition. J Sound Vib 324:1141–1157
10. Lei Y, He Z, Zi Y, Qiao H (2007) Fault diagnosis of rotating machinery based on multiple ANFIS combination with GAs. Mech Syst Signal Process 21:2280–2294
11. Smola AJ, Scholkopf B (2004) A tutorial on support vector regression. Stat Comput 14:199–222
12. Deriche M (2005) Bearing fault diagnosis using wavelet analysis, computers, communications, & signal processing with special track on biomedical engineering. CCSP 2005. 1st International Conference on, pp 197–201

Index

I. Dincer et al. (eds.), *Progress in Clean Energy, Volume 1*,
DOI 10.1007/978-3-319-16709-1

G

I

S

MIX
Papier aus verantwortungsvollen Quellen
Paper from responsible sources
FSC® C105338

FSC
www.fsc.org

If you have any concerns about our products,
you can contact us on
ProductSafety@springernature.com

In case Publisher is established outside the EU,
the EU authorized representative is:
Springer Nature Customer Service Center GmbH
Europaplatz 3, 69115 Heidelberg, Germany

Printed by Libri Plureos GmbH
in Hamburg, Germany